Oct '84

PETROLEUM GEOLOGY

of the

UNITED STATES

PETROLEUM GEOLOGY

of the

WILEY-INTERSCIENCE

A DIVISION OF JOHN WILEY & SONS

New York • Chichester • Brisbane • Toronto • Singapore

UNITED STATES

KENNETH K. LANDES

Professor of Geology
University of Michigan

10 9 8

Library of Congress Catalogue Card Number: 77-101975

SBN 471 51335 0

Printed in the United States of America

PREFACE

The preface to the second edition of my "Petroleum Geology," published by Wiley in 1959, contained the following paragraph.

"A considerable part of the first edition, which covered the distribution of oil and gas fields throughout the world and future oil supplies, has been eliminated from the second edition.

This material will be expanded and published in a separate volume, "Petroleum Geology of World Oil Fields," which is now in preparation."

On starting this global task ten years ago I began with the United States. Apparently I fell in love with my own words, since the manuscript grew like Topsy. Concurrently, time began to run out on me. Therefore my worldwide project terminates here as an exclusively domestic operation.

The objectives of this geographic approach to petroleum geology have been to determine in what places and under what conditions hydrocarbons are known to have accumulated. This information can assist in the search for new deposits. Geologic history, stratigraphy, and structure are all germane to the occurrence of petroleum.

A possible by-product in reading the pages that follow is an increased awareness of the future petroleum possibilities in the United States. Almost every year since Colonel Drake drilled U. S. No. 1, the annual production of total petroleum (crude oil, natural gas, and natural gas liquids) has risen, and 1968 was no exception. Concurrently the capital in the bank, in the form of known recoverable reserves, also has risen.

This book does not cover the fundamental principles and techniques of petroleum geology. It is intended to supplement and not to supplant standard texts in this field.

Ann Arbor, Michigan
February 1969

Kenneth K. Landes

ACKNOWLEDGMENTS

Many man-hours were spent by fellow geologists in reading critically and writing constructive comments on chapters of the manuscript that they kindly agreed to cover. The cooperation was wonderful, and the result is a greatly improved book.

To these fine people, my sincere thanks: W. Lynn Kreidler, New York Geological Survey (Chapter 1), Arthur A. Socolow and staff, Pennsylvania Geological Survey (Chapter 2), Kenneth N. Weaver, Maryland Geological Survey (Chapter 3), Paul H. Price, West Virginia Geological Survey (Chapter 4), James L. Calver, Virginia Geological Survey (Chapter 5), H. C. Milhous, Tennessee Division of Geology (Chapter 6), E. N. Wilson, Kentucky Geological Survey (Chapter 7), Warren L. Calvert, Phillips Petroleum Company (Chapter 8), John B. Patton and T. A. Dawson, Indiana Geological Survey (Chapter 9), D. C. Bond and staff, Illinois Geological Survey (Chapter 10), Gar Ells, Michigan Geological Survey (Chapter 11), K. O. Emery, Woods Hole Oceanographic Institution (Introduction to Division I, and Chapter 12), Wilson M. Laird and Sidney Anderson, North Dakota Geological Survey (Chapter 13), Duncan J. McGregor, South Dakota Geological Survey (Chapter 14), Eugene C. Reed, Nebraska Geological Survey (Chapter 15), Edwin D. Goebel, Kansas Geological Survey (Chapter 16), William C. Hayes, Missouri Geological Survey (Chapter 17), Robert H. Dott, American Association of Petroleum Geologists, and John A. Taylor, Hiawatha Oil Company (Chapter 18), H. Garland Hershey, Iowa Geological Survey (Chapter 19), Clarence V. Babcock, Florida Geological Survey (Introduction to Division III, and Chapter 20), Donald B. Moore, Alabama Geological Survey (Chapter 21), John T. Rouse and Robey H. Clark, Mobil Oil Corporation (Chapters 22, 25, 26, and 27), Norman F. Williams, Arkansas Geological Commission (Chapter 23), James P. Spillers, Louisiana Mineral Board (Chapter 24), Peter T. Flawn, Texas Bureau of Economic Geology (Introduction to Division IV), Louis Franklin, Independent (Chapter 28), Herbert H. Bradfield, Consulting (Chapter 29), Paavo Puumala and Suzanne Takken, Mobil Oil Corporation (Chapter 30), Theodore S. Jones, Union Oil of California (Chapter 31), William H. Curry, III, Consulting (Introduction to Division V), Sherman A. Wengerd, University of New Mexico (Chapter 32), N. Wood Bass, U. S. Geological Survey (Chapter 33), William H. Curry, Jr., Consulting (Chapter 34), George R. Wulf, Consulting (Chapter 35), Armand J. Eardley, University of Utah (Chapter 36), George C. Hardin, Jr., and geologists in Amarillo office, Kerr-McGee Corporation (Chapter 37), D. Keith Murray, Sinclair Oil and Gas Company (Chapter 38), E. C. Doell, Standard Oil of California (Introduction to Division VI, and Chapter 42), Gordon B. Oakeshott, California Division Mines and Geology (Chapter 40), and John M. Sweet and staff, Atlantic Richfield Company (Chapter 41).

I am also indebted to the many individuals and organizations who courteously granted permission to use copies of published illustrations. At the very beginning of the project the Bulletin of the American Association of Petroleum Geologists, the Oil and Gas Journal, and Oil World gave me blanket permission to use illustrations appearing in their publications. The regional and local geological associations and societies throughout the United States also were generous. Individual authors were completely cooperative. All of the publications and authors involved are individually acknowledged in the illustration legends.

I conclude this listing of gracious people with thanks and appreciation to my wife, Susan Beach Landes. She not only typed and corrected the manuscript but also put up with me during my ten years of gestation and *accouchement!*

K. K. Landes

CONTENTS

PETROLEUM GEOLOGY

of the

UNITED STATES

Conterminous United States, showing areas covered in Divisions I to VI (except Alaska).

INTRODUCTION

Subdivisions. The United States has been arbitrarily separated into six divisions numbered I to VI (Frontispiece). These divisions encompass groups of states (except for Texas) and are designated as follows: I—Eastern; II—Central; III—Eastern and Central Gulf; IV—Texas; V—Rocky Mountains; and VI—Pacific. It will be observed that the principal basis for this division is geographic; geologic factors are also involved, but to a lesser degree. Each petroleum-producing state is discussed as a separate chapter, with the exception of Texas, to which seven chapters are devoted. The chapters are numbered consecutively from 1 to 42. The use of state boundaries instead of geologic provinces is due to the availability of all statistical and much geologic information by political rather than lithologic or physiographic units.

Oil and gas are being produced in 32 out of the 50 states. The 18 currently unproductive states are divided among four of the six divisions as follows: Eastern, 12; Central, 2; Rocky Mountain, 1; and Pacific, 3. Within a division only the producing states are covered in detail, the nonproductive states being discussed briefly in a concluding chapter.

Among the six divisions, Texas produced 35 per cent of the 1967 output; Eastern and Central Gulf was second with 27 per cent; Pacific, Central and Rocky Mountain were third, fourth, and fifth with 12 to 11 per cent each; and Eastern was sixth with 4 per cent.

General References. The following annual publications were consulted throughout the preparation of the manuscript: Forecast Review issue, published in late January or early February by the Oil and Gas Journal; Forecast–Review issue, published February 15 by World Oil; North American Development issue, published in June in the Bulletin of the American Association of

Petroleum Geologists; "The Oil Producing Industry in Your State," published by the Independent Petroleum Association of America; "Gas Facts," by the American Gas Association; and the "Minerals Yearbook," Vol. 2 (Fuels) and Vol. 3 (Area Reports) by the U. S. Bureau of Mines.

Maps frequently used include Geologic Map of the United States, and Oil and Gas Map of the United States, by the U. S. Geological Survey; Tectonic Map of the United States, by the U. S. Geological Survey and the American Association of Petroleum Geologists; and Basement Rock Map of North America, by the American Association of Petroleum Geologists and the U. S. Geological Survey.

Division I

EASTERN

The Eastern Division includes the states that lie between the Atlantic Ocean and the Mississippi River, north of the Gulf Coastal Plain (Figure I.1). Here, in western Pennsylvania, the petroleum industry of the United States was born with the completion of the Drake well in 1859. Between 1859 and the early 1900s the Eastern Division produced by far the greater part of the domestic oil output. For example, in 1900 it accounted for 91 per cent of the oil produced. Two-thirds of a century later, in 1967, the percentage had dropped to less than 4 per cent. However, the actual production in barrels of petroleum in 1967 was double that in 1900.

Within the division Illinois stands at the top, with about one half of the 120 million barrel output in 1967. Other important states, with production to the nearest million barrels in parentheses, are Kentucky (15), Michigan (14), Indiana (11), Ohio (10), Pennsylvania (4), and West Virginia (3).

The dominating geological feature of the Eastern Division is the Appalachian geosyncline, which cuts across the area from northeast to southwest west of the Great Smoky–Blue Ridge sector of the Appalachian Highlands (Figure I.2). The western rim of the Appalachian geosyncline is the Cincinnati anticline, which forks at the northern end to produce the Michigan basin. West of the Cincinnati arch is the Eastern Interior coal basin. The Atlantic Costal Plain overlaps the eastern part of the area east of the Appalachian Highlands from New Jersey southwest into Georgia, where it merges with the Gulf Coastal Plain.

All of the oil and gas production so far obtained in the Eastern Division has been from Paleozoic rocks, ranging in age from Pennsylvanian to Cam-

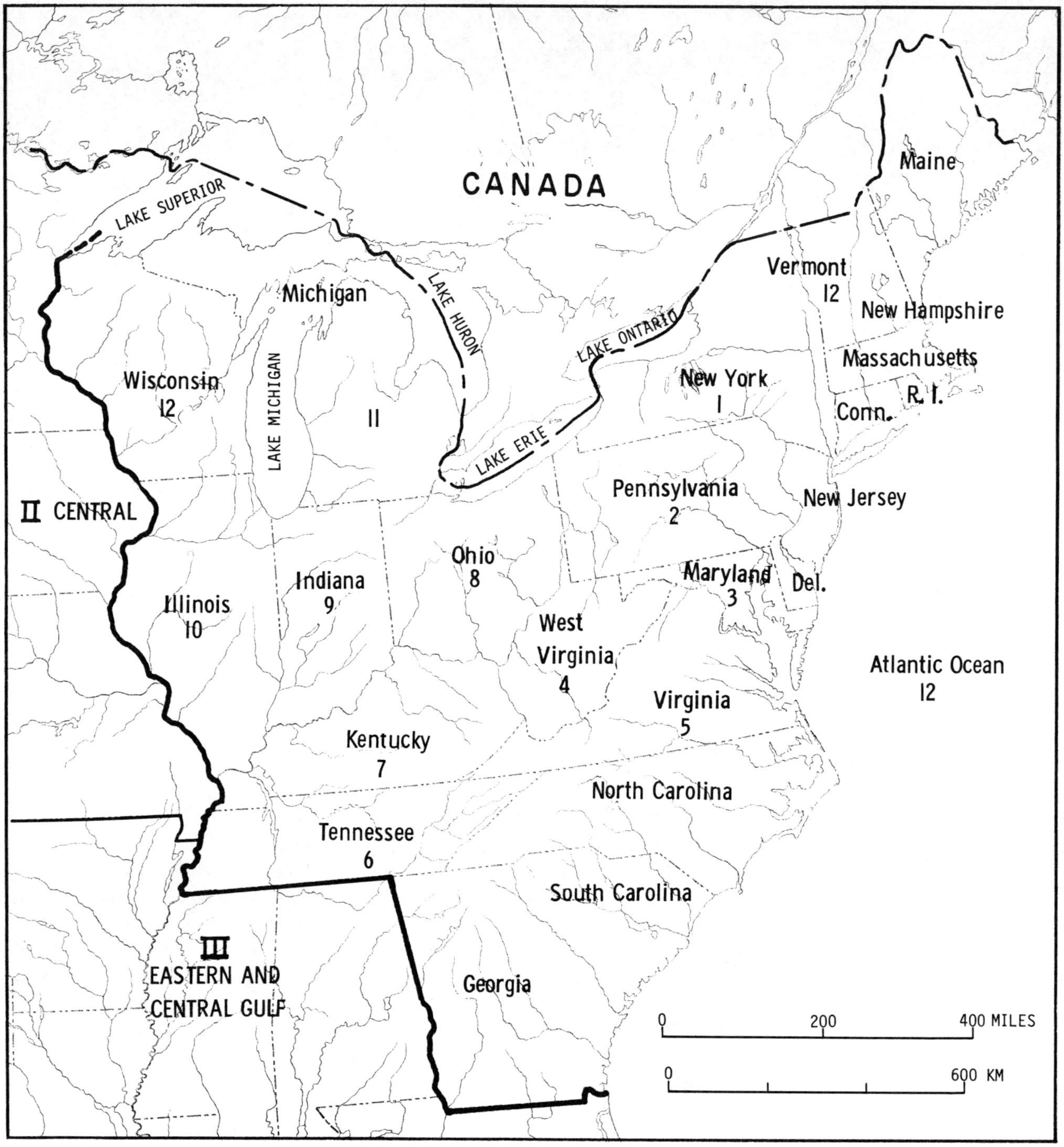

FIGURE I.1 Eastern Division. Index map. Arabic numerals are chapter numbers.

brian. Except for the coastal plain most of the bedrock is either Precambrian or Paleozoic. The youngest Paleozoics are Permian rocks that crop out along the axis of the Appalachian geosyncline in southwestern Pennsylvania and western West Virginia. A relatively minor volume of Triassic sediment and traprock occurs along the eastern border of the crystalline belt. The coastal plain sediments were deposited during the Cretaceous and Tertiary. The crystalline rocks exposed to the east and northeast of the Appalachian geosyncline in the Appalachian Highlands are both Precambrian and Paleozoic in age; to the northwest in Wisconsin and the adjacent part of Michigan they are Precambrian.

Because the Eastern Division is based to a greater

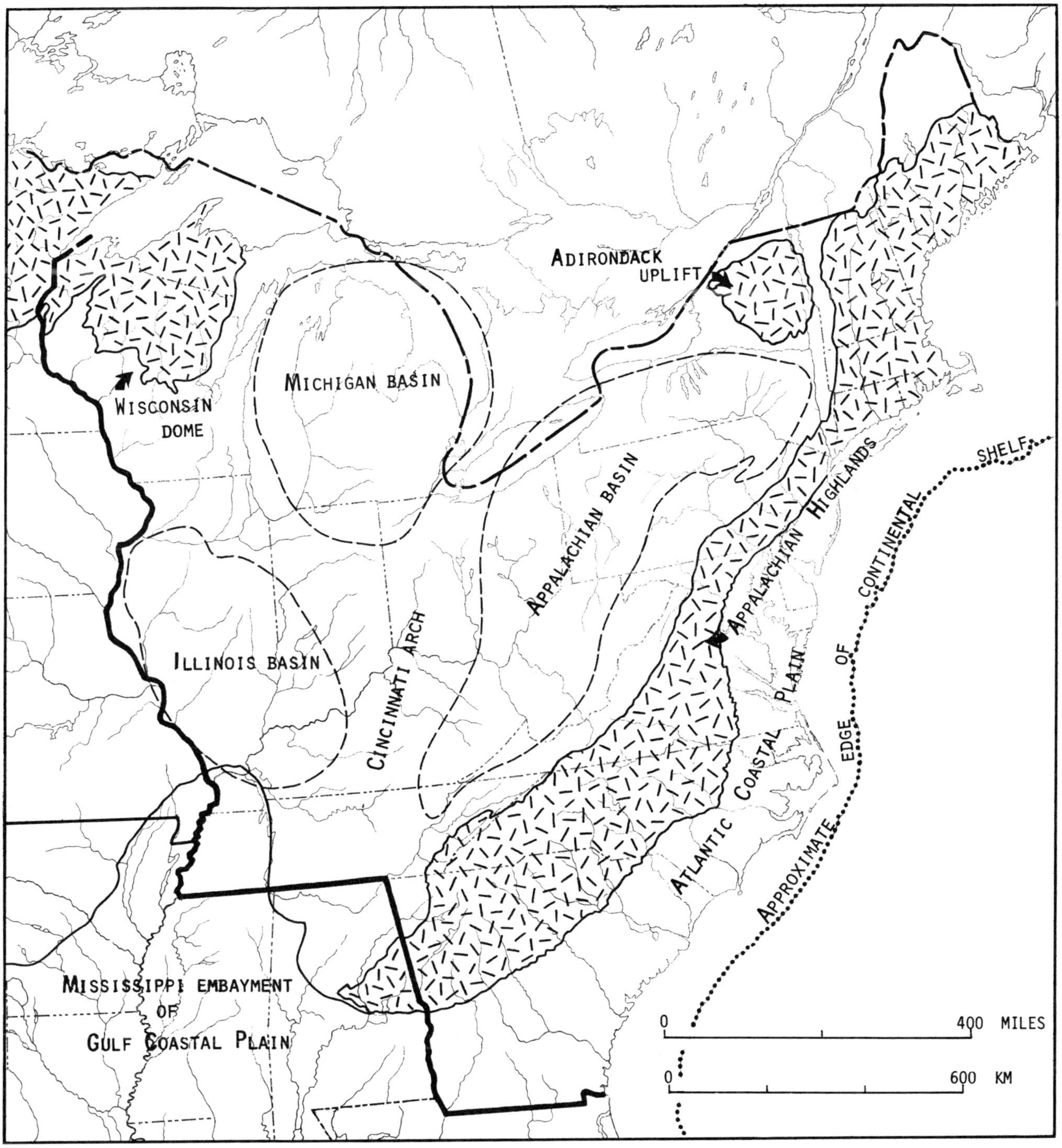

FIGURE I.2 Eastern Division. Major tectonic features. Sedimentary basins are outlined by dashed lines. Pattern designates areas of predominantly crystalline rock, including Precambrian basement plus (in eastern part of Division) Paleozoic igneous and metamorphic rock. Appalachian Highlands area also includes narrow basins of Paleozoic sedimentary rock, parallel to regional strike, which are not shown on map.

extent on the geographical location of state boundaries than on types of bedrock, there is overlap geologically with neighboring divisions. The Eastern Interior coal basin crosses the Mississippi River into eastern Iowa and Missouri of the Central Division. An oil field in Missouri is in this basin. Westernmost Tennessee is in the Gulf Coast geologic province, but these rocks are as yet unproductive. Relatively insignificant production, mainly gas, is obtained in the Gulf Coast states of Mississippi and Alabama by drilling through the overlapping Mesozoic sediments into Paleozoic rocks.

The general order of discussion of the states in this division will be from north to south along the Atlantic Coast, then northward and westward inland. The possibilities for discoveries in the currently nonproductive states are considered in the concluding chapter.

Chapter

1

NEW YORK

The oil production of New York State is relatively insignificant, amounting to less than 0.1 per cent of the total United States production each year. However, the fields are remarkably long-lived, and New York has been a consistent producer of oil for many years. Furthermore, the state has considerable gas production, and the demand is so great that exploration continues to be active and has met with considerable success.

Even before Colonial times the oil springs of Allegany County (Figure 1.1) were exploited by the Indians and, subsequently, by the white settlers (Herrick, 1949). In 1821 shallow gas was struck near Fredonia and piped into the local hotel. This is the first record of the transportation of natural gas in the United States. The first successful oil well was completed at Limestone, New York, in 1865 (Kreidler, 1953). The first major discovery was the Allegany County (Richburg) field in 1879. This field eventually was extended into Pennsylvania, and the Bradford field of Pennsylvania grew northward into Cattaraugus County, New York. In 1882 oil production reached a peak of 6,685,000 barrels in New York and then declined steadily until 1919, when secondary recovery by waterflooding became legal and production climbed to a second peak. The yield in 1967 was nearly 2 million barrels, an increase of over 200,000 barrels from the year before. In addition three new gas fields were discovered during the year (Kreidler et al., 1968). Gas output in recent years has been increased by fracturing the pay zone.

STRUCTURE. The Appalachian geosyncline extends into western New York State. The axis of the geosyncline rises to the north, so that it can be said that western and central New York contains the "prow" of this great structural trough. The regional dip in western New York is southward at a rate of between 30 and 50 feet per mile. The youngest Paleozoic formations are in western New York along the Pennsylvania line; Precambrian rocks are exposed in the Adirondack Mountains in northeastern New

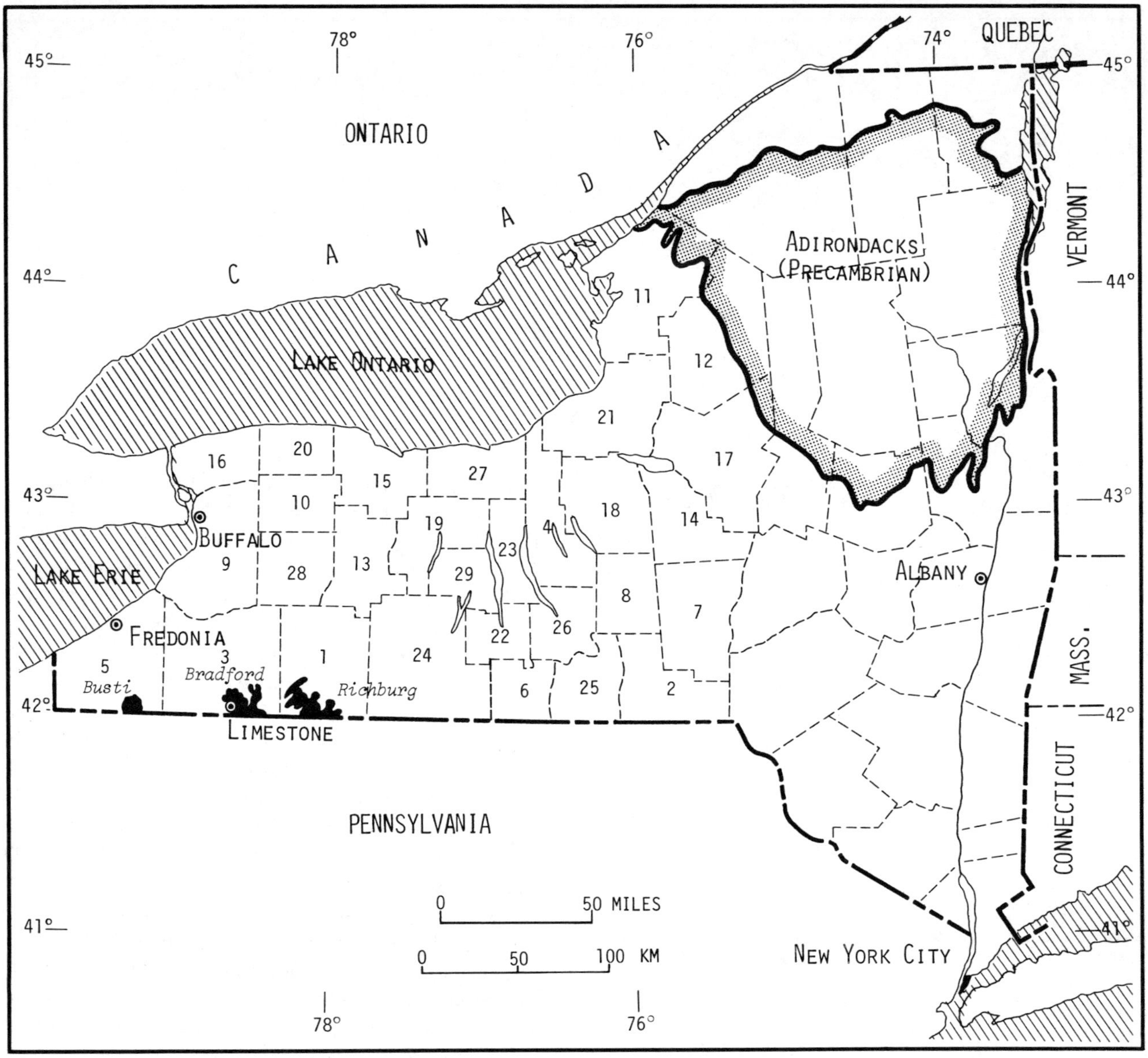

Figure 1.1 New York. Index map. County Key:

1 Allegany	6 Chemung	10 Genesee	14 Madison	18 Onondaga	22 Schuyler	26 Thompkins
2 Broome	7 Chenango	11 Jefferson	15 Monroe	19 Ontario	23 Seneca	27 Wayne
3 Cattaraugus	8 Cortland	12 Lewis	16 Niagara	20 Orleans	24 Steuben	28 Wyoming
4 Cayuga	9 Erie	13 Livingston	17 Oneida	21 Oswego	25 Tioga	29 Yates
5 Chautauqua						

York (Figure 1.1). A narrow belt of lower Paleozoic sediments lies north of the Adirondacks in the St. Lawrence River valley. In southwestern New York are found the alternating anticlines and synclines with axes trending northeast for which western Pennsylvania is famous. The subsurface section in this area contains three important hydrocarbon reservoir rocks. The oil reservoirs are the Bradford and other sands of Late Devonian age. Gas occurs in considerable volume in two older formations, the Oriskany Sandstone, of Early Devonian age, and the Medina Sandstone, which is Silurian (Figure 1.3).

The oil fields of New York State (Figure 1.2) are confined largely to Cattaraugus and Allegany Counties, which lie just north of the Pennsylvania line in western New York. Most of the pools lie on the Bradford anticline, which crosses the line from Pennsylvania. Nodes along the axis of this anticline are the Knapp Creek dome just across the line in Pennsylvania and the Alfred and

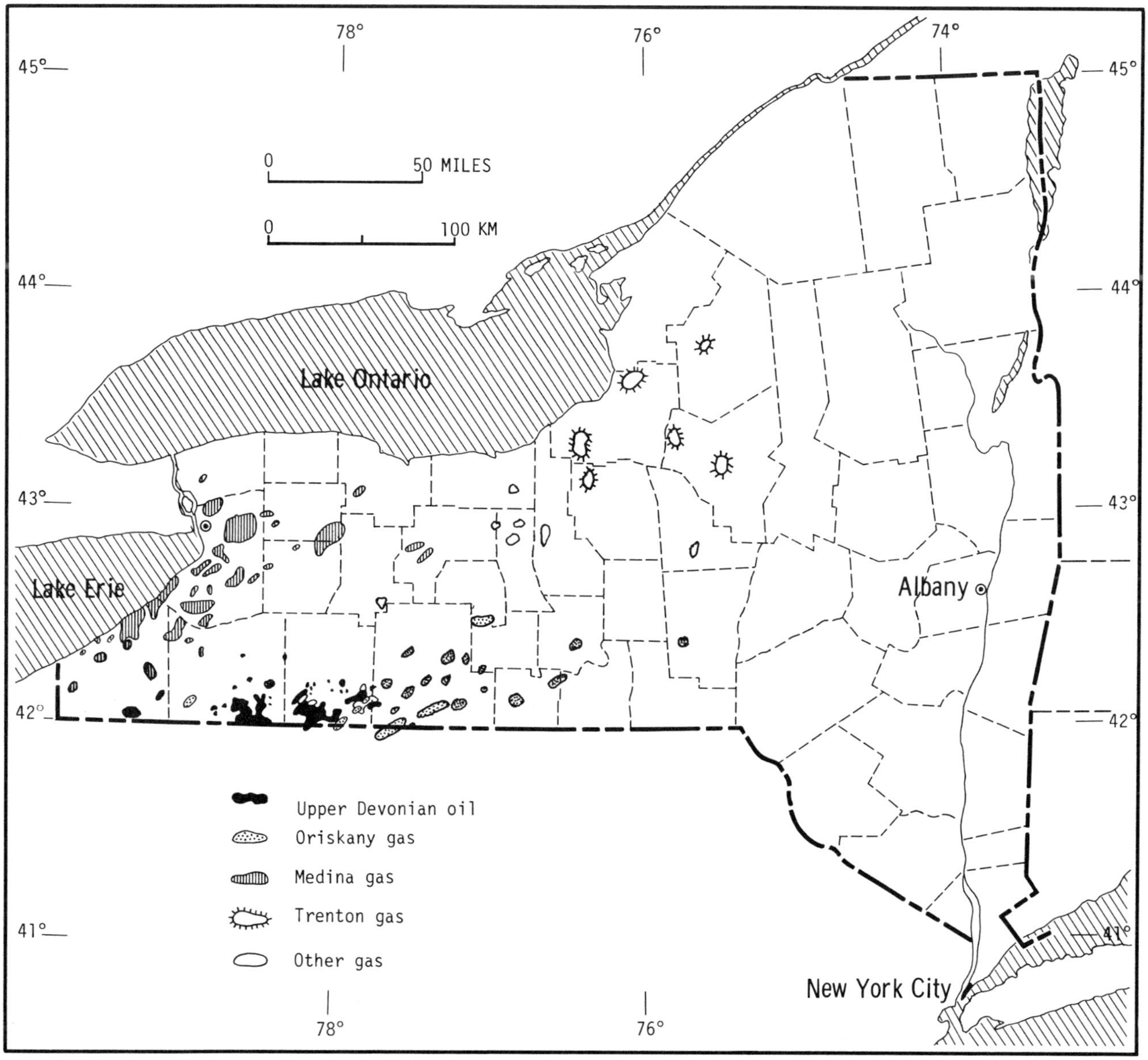

FIGURE 1.2 New York. Oil and gas fields. Abandoned, inactive, and storage fields not shown. *After William L. Kreidler (1953).*

Sharon anticline in Allegany County. The Bradford and Richburg pools are the largest oil fields in New York State.

RESERVOIRS. The reservoirs, more than 16 in number, are medium to fine-grained sandstones and graywackes lying in the "Chemung" section of the Upper Devonian. The names of the most important reservoir "sands" are given in Figure 1.3. Most of New York's oil production has come from the Bradford third sand (Cattaraugus County) and the Richburg pay (Allegany County). Next in yield is the Chipmunk sand (Cattaraugus County). A newer oil pool, the Busti in Chautauqua County, produces from the Upper Devonian Glade sand.

Although the oil pools are structurally high regionally, the local accumulations are not noticeably structure controlled. The reservoirs are extremely lenticular, and little or no water is present beneath the oil. Updip the reservoirs become either gas bearing or impermeable because of increased shale content.

The source rocks may be the dark shales that underlie the Chemung section. The seals are the shales within the Chemung which surround the reservoir sandstone lenses.

Most of the gas fields fall into three categories, depending upon the age of the reservoir rock. The largest are the Medina (Silurian) Sandstone fields of western New York (Figure 1.2). The principal reservoir is the

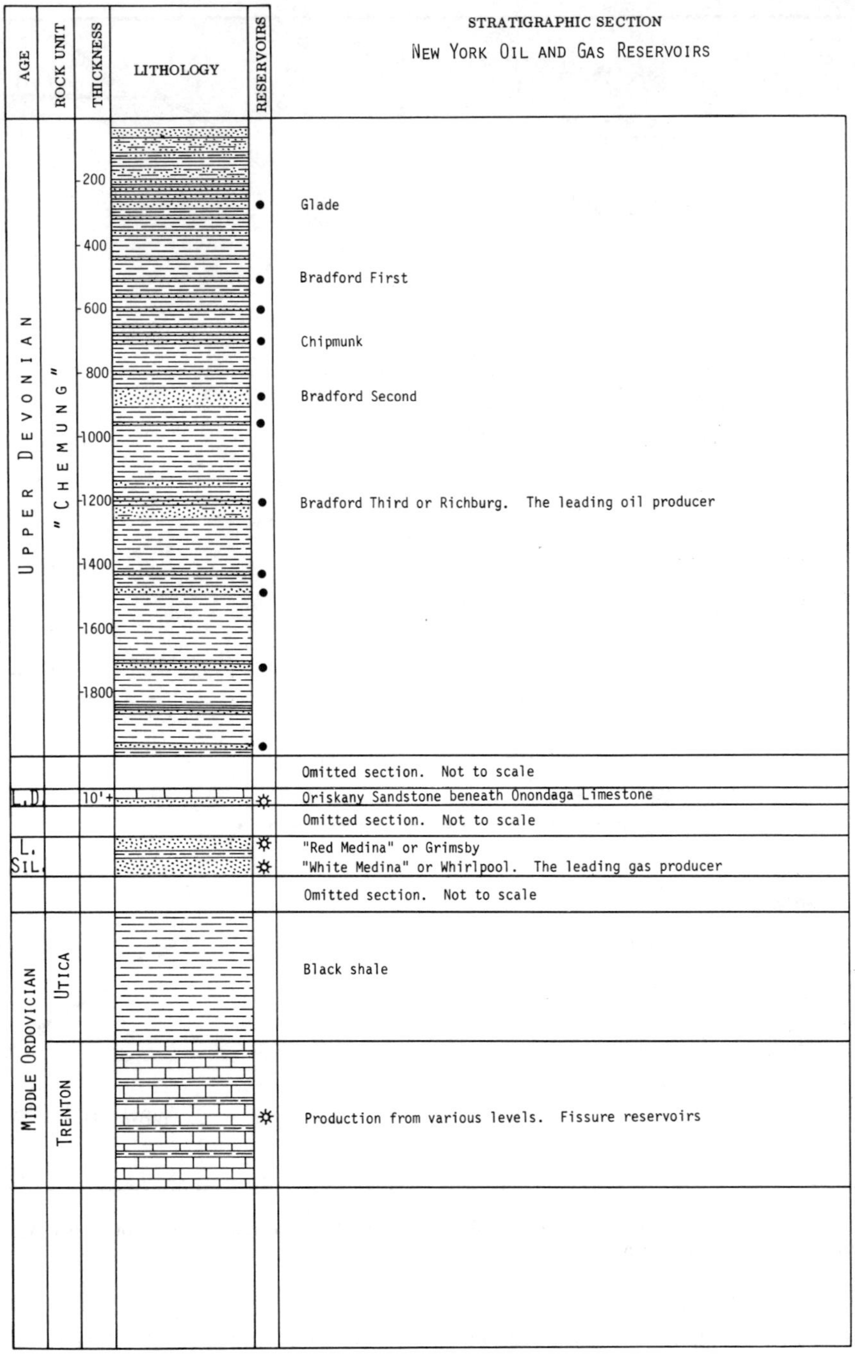

Figure 1.3 **Stratigraphic section for oil- and gas-producing area of New York, with oil and gas reservoirs designated.** ***Upper Devonian section after William L. Kreidler (1953).***

"White" Medina, which lies below the "Red" Medina. These sandstones contain the largest known gas reserves in the state. Accumulation is in lenticular bodies of porous sand surrounded by impermeable rock.

Second in gas production and known reserve is the Oriskany (Lower Devonian) Sandstone, which is present in Pennsylvania and southwest New York (Figure 1.4). Trapping is by anticline and, along the northern ("pinchout") edge of the Oriskany Sandstone, by updip termination of effective porosity (Harding, 1966). The overlying richly organic calcareous ooze that lithified into the Onondaga Limestone is a likely source rock for the Oriskany gas.

The third gas source is the Trenton (Ordovician) For-

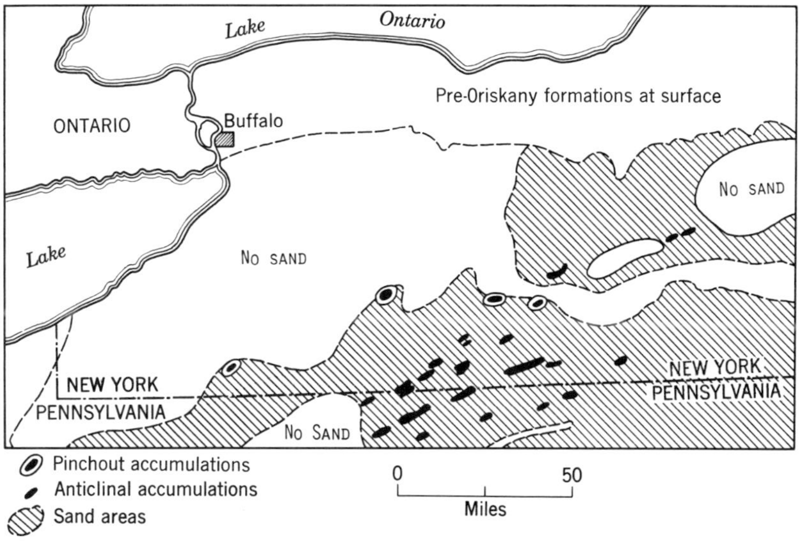

FIGURE 1.4 Oriskany Sandstone gas fields in New York, with type of trap designated. ***After F. H. Finn (1949).***

mation, which has been productive in central New York, south and east of Lake Ontario. The Trenton reservoirs are fractured limestone layers lying between thin shale beds.

A relatively small amount of gas is obtained from other Devonian and Silurian reservoirs. Many depleted fields have been converted into gas storage pools.

Gas was found in an Onondaga reef in southwestern Steuben County in 1967. Both the reef and the underlying Oriskany Sandstone contain commercial gas. "This is the first time a subsurface reef has been reported in the upper Onondaga" (Kreidler et al., 1968, p. 941).

Virtually unexplored are the Oriskany and Trenton oil and gas possibilities in southeastern New York (Kreidler, 1957; Harding, 1959). Here the drilling depths will be greater than in the present producing areas, but they are not prohibitive. Still older Cambro-Ordovician formations continue to interest the prospector (Weaver, 1965). Even the Cambrian Potsdam Sandstone is considered to have reservoir potential (Weaver, 1964; Roth, 1968, p. 1709).

Lake Erie offshore from New York was in the plans to be opened up for exploration in early 1968 (Oil and Gas Jour., Jan. 8, 1968, p. 34–35). This underwater acreage lies between onshore gas fields in New York and offshore gas fields in Ontario.

REFERENCES CONSULTED IN PREPARING CHAPTER 1

Harding, Richard W., 1966.

Herrick, John P., 1949.

Kreidler, William L., 1953, 1957, 1968.

Kreidler, W. Lynn, Van Tyne, Arthur M., Richards, Horace G., and Grow, George C., Jr., 1968.

Roth, Ernest E., 1968.

Weaver, O. D., 1964, 1965.

Chapter

2

PENNSYLVANIA

Pennsylvania has two major economic advantages as a producer of oil and gas: (1) its oil is exceptionally high grade; and (2) both its oil and its gas lie within the highly industrialized northeast part of the country, where transportation costs are minimum and marketing opportunities are maximum. Although now accounting for less than 0.14 per cent of the domestic oil production, Pennsylvania has had a long and glorious history as a petroleum state and even today is producing more than 4 million barrels of premium grade crude oil each year. Furthermore, Pennsylvania has an annual output of about 90 billion cubic feet of valuable natural gas; this is about 0.5 per cent of the United States total.

The occurrence of seeps in Pennsylvania, especially along Oil Creek, central Venanago County (Figure 2.1), and the drilling of the first oil well in the United States in that vicinity in 1859 are described elsewhere (Dickey, 1959; Giddens, 1941). From 1859 to 1895 Pennsylvania led the states in petroleum production; its peak year was 1891, with an output of 31,424,000 barrels. Since 1895 Pennsylvania has been surpassed by more western states. In the final decade of its first century as an oil and gas producer, exploration in Pennsylvania was stimulated by a series of highly profitable gas discoveries in the relatively deep Oriskany Sandstone and overlying Onondaga Chert. The first of these was the completion of a wildcat well in Clinton County in 1950 that had an initial potential production of 145 million cubic feet, the largest gas well ever drilled in the eastern United States. This was the discovery well of the Leidy field (Figure 2.1). During the next year the Driftwood pool in Cameron County was discovered; by 1959 this had grown into a field (the Punxsutawney–Driftwood) nearly 50 miles long. A further impetus to gas exploration and production was the adoption of hydraulic fracturing of the Onondaga Chert as standard procedure in the mid-fifties. Exploration is still active in Pennsylvania; wildcatting in 1967 resulted in three oil and two gas discoveries (Lytle et al., 1968).

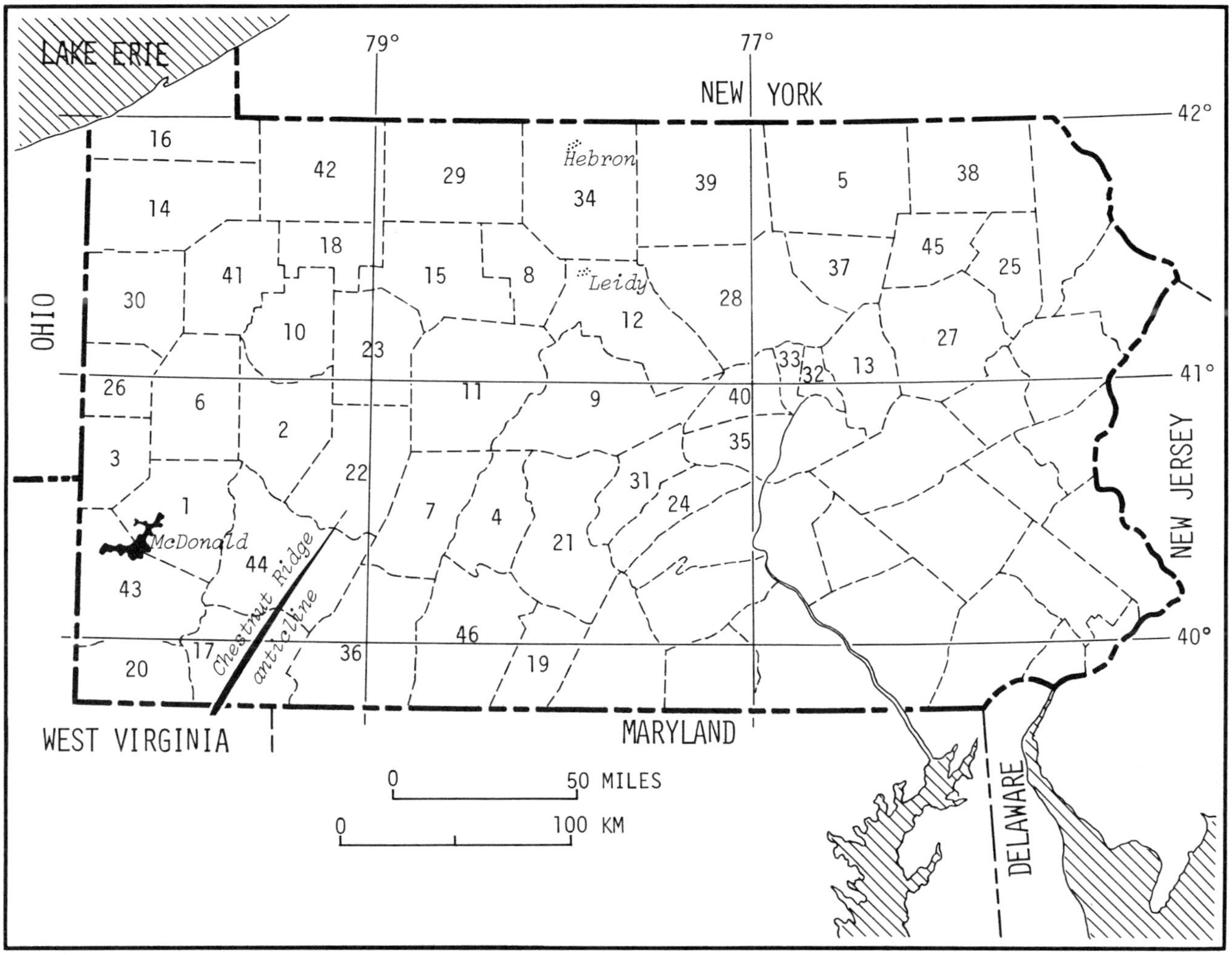

FIGURE 2.1 Pennsylvania. Index map. County Key:

1	Allegheny	7	Cambria	14	Crawford	21	Huntingdon	28	Lycoming	34	Potter	40	Union
2	Armstrong	8	Cameron	15	Elk	22	Indiana	29	McKean	35	Snyder	41	Venango
3	Beaver	9	Centre	16	Erie	23	Jefferson	30	Mercer	36	Somerset	42	Warren
46	Bedford	10	Clarion	17	Fayette	24	Juniata	31	Mifflin	37	Sullivan	43	Washington
4	Blair	11	Clearfield	18	Forest	25	Lackawanna	32	Montour	38	Susquehanna	44	Westmoreland
5	Bradford	12	Clinton	19	Fulton	26	Lawrence	33	Northumberland	39	Tioga	45	Wyoming
6	Butler	13	Columbia	20	Greene	27	Luzerne						

The storage of gas in depleted fields is big business in Pennsylvania. In fact, this state is the leading one in volume stored (Lytle, 1963).

The oil and gas fields of Pennsylvania (Figure 2.2) occupy a belt nearly 200 miles long and from 25 to 125 miles wide, crossing the western side of the state. The total number of named fields exceeds 300, of which over half are gas fields. The greatest concentration of gas fields is on the east side of the petroliferous zone, but many others are scattered across the zone to the westward. The eastern production limit used to be approximately the 70 per cent isocarb for gas and the 65 per cent isocarb for oil, but recent gas discoveries in Bedford and Lackawanna Counties have been far to the east of the 70 per cent isocarb.

GEOLOGY. The oil and gas deposits of Pennsylvania occur in the Appalachian geosyncline (Figure 2.2) which is a great spoon-shaped trough with a northeast-southwest trend, plunging to the southwest from the prow in western New York State to the deepest point, close to the southwestern corner of Pennsylvania. Most of the oil and gas fields lie along the axis and the west flank of the Appalachian geosyncline. The trough is furrowed by a great series of parallel anticlines and synclines that are also parallel to the geosynclinal axis.

The principal reservoirs (Figure 2.3) in Pennsylvania

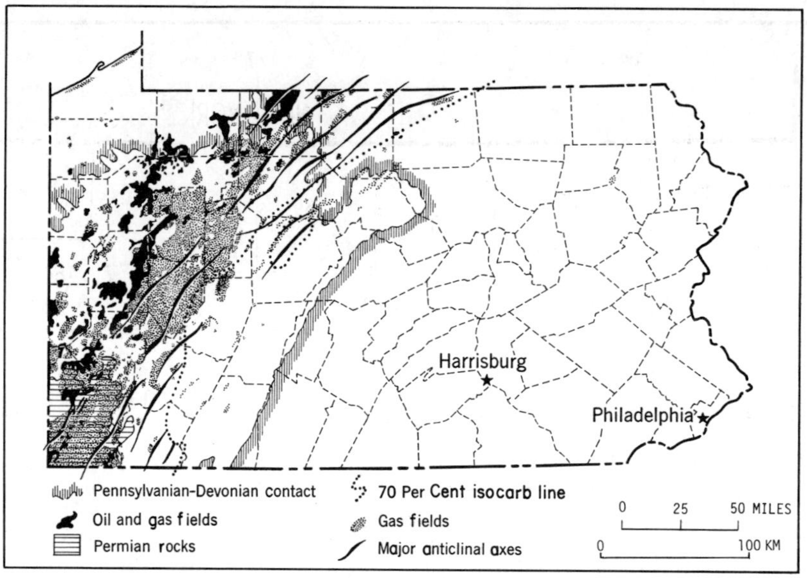

Figure 2.2 Pennsylvania. Oil and gas fields and major structural features.

belong in either the Mississippian or the Devonian System, although some oil has been obtained from Pennsylvanian rocks in the southwestern part of the state. The most important of these reservoirs are the Venango and Warren–Bradford sequences of the Upper Devonian, and the Oriskany and Medina pay zones, which produce in the deep zone. Altogether more than 70 distinct oil and gas reservoirs have been recognized and named in Pennsylvania.

The reservoirs consist of friable, porous sandstones as well as some brittle fractured rocks. In some instances where fractures constitute the major porosity present, hydraulic fracturing has proved to be an effective production stimulant and has given the region a significant boost in development.

Traps and Fields. Some oil and gas accumulations in this state are anticlinal. Most of the sands in the Appalachian geosyncline are lenticular, however, so that entrapment is due in part at least to the termination of porosity or permeability over and around the periphery of these lenses. The sand deposits tend to be elongate; some appear to be beach-bar deposits parallel to old shorelines, but others run normal to the shorelines and probably represent stream channel fillings, especially in the distributaries of deltas. In most fields the lenticularity is due to the deposition of distinct sand bodies, but in a few a facies change from sand to shale has been responsible for the trapping. Hence stratigraphic traps involving pinchouts and facies changes are common. All the production in the *Venango* district (Sherrill et al., 1941) in northwestern Pennsylvania comes from lenticular sands that are generally less than 1000 feet below the surface (Figure 2.4). The Venango Group of sands lies within the uppermost series of the Devonian System. It is about 350 feet in thickness and consists of alternating sands and shales. No single sand stratum can be traced throughout the district, but four persistent sand zones separated by sections of shale can be mapped. These sands are designated, in descending order, the First, Red Valley, Second, and Third. Although the sands within each of these members are lenticular, appearing and disappearing across the district, the intervening shales are continuous. The trapping in the Venango district is entirely stratigraphic. The lenticularity of the sand bodies is due to original deposition, and the four sand members differ not only in stratigraphic position but also in the environment during sedimentation. The small, irregular sand bodies of the First sand trend roughly parallel to the old shoreline, modified by deltaic deposits. The Red Valley consists of discontinuous beach-bar sands. The producing sands in the Second sand formation are also beach-bar deposits parallel to the shore. These sands may have been distributed by longshore currents. The Third type consists of long, narrow, and sometimes pebbly sand bars, some parallel to the shoreline and others not; these appear to be deltaic fluviatile deposits. Figure 2.4, upper drawing, shows the oil pools producing from the Third Stray sand and the structure as contoured on top of that sand. Locally the rocks dip to the southwest, and the pools occur along this slope without any apparent regard for the rock structure. A similar situation is obvious in the lower drawing, which shows the structure and distribution of oil fields in the Third sand.

Pennsylvania's most notable field, and one of the great

oil fields of the United States, is the *Bradford* (Fettke, 1938), which lies mostly in McKean County, but extends a short distance north into western New York. This field was discovered in 1871 and since then has produced more than half a billion barrels of oil. It still accounts for the greater part of Pennsylvania's annual oil output, even after nearly 90 years of continuous production. This pool reached its peak in 1881 with a production of nearly 23 million barrels, which was 83 per cent of the country's entire output for that year. By 1906 Bradford production had declined to 2 million barrels. However, in 1890 the operators had discovered that water breaking through rusted casing in abandoned wells was reaching the oil sand and increasing the yield of adjacent oil wells. This led to the development of artificial water drive, which, although illegal at the time,

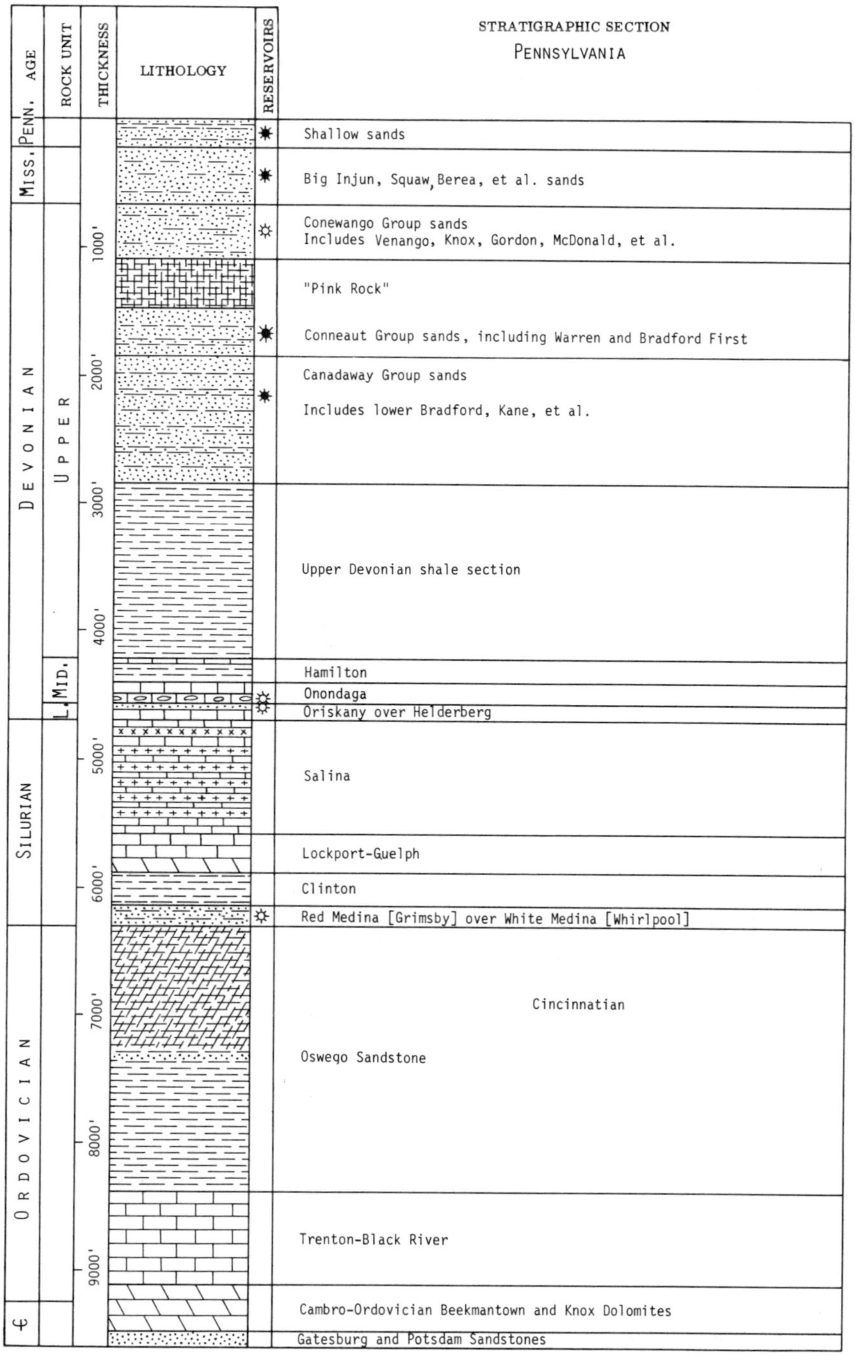

FIGURE 2.3 **Pennsylvania column, and major oil and gas reservoirs.** ***Courtesy Pennsylvania Topographic and Geologic Survey.***

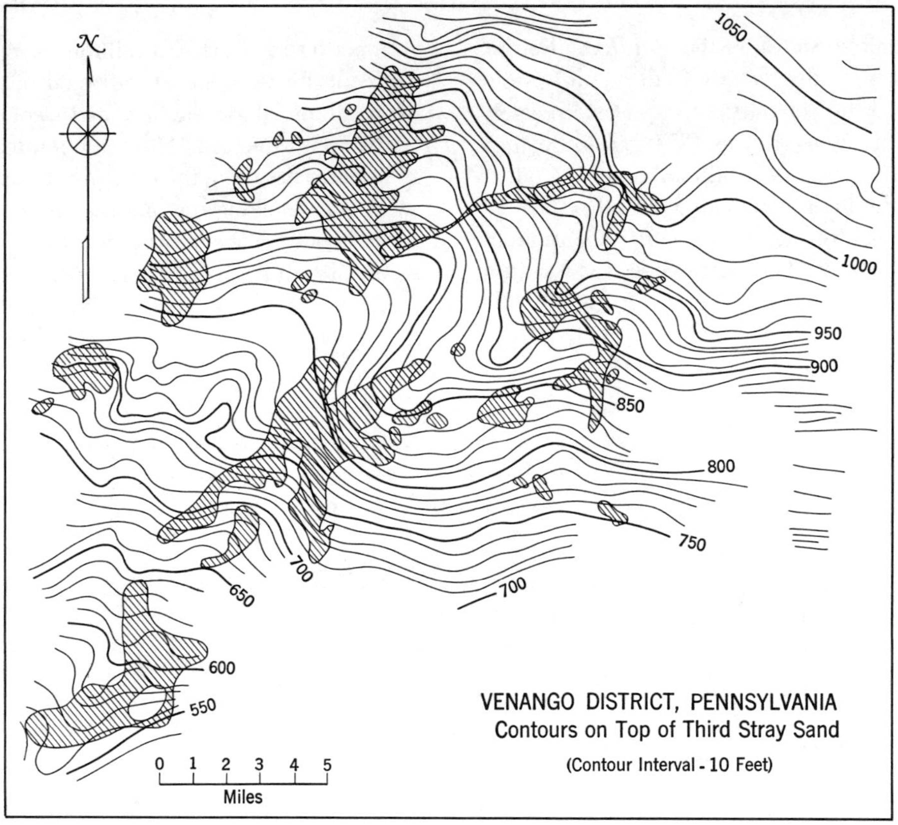

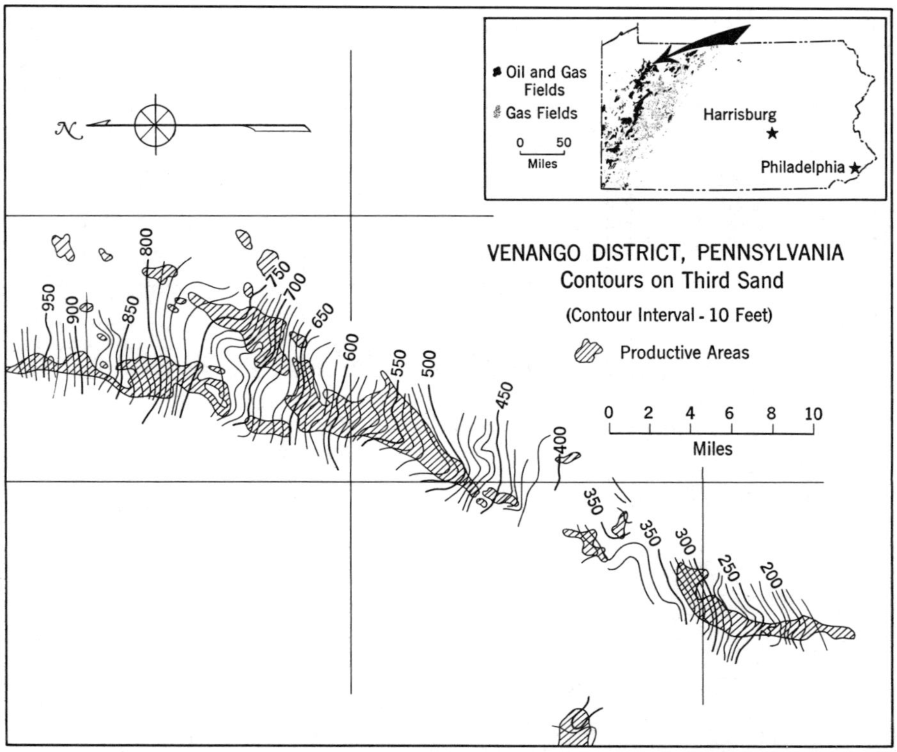

FIGURE 2.4 Oil trapping in the Venango district, northwestern Pennsylvania. Oil pools designated by diagonal lines. *Courtesy American Association of Petroleum Geologists. After R. E. Sherrill et al. (1941).*

led to a gradual increase in Bradford production after 1906. In 1921 waterflooding was legalized by the state legislature, and by 1937 production had risen to nearly 17 million barrels for a second peak in the production curve for this field.

The Bradford pool is relatively shallow, with the reservoirs lying at depths ranging from 1000 to 1700 feet. The Bradford sand is more widespread than most Pennsylvania reservoir rocks, and is up to 90 feet in thickness. The sand body, actually sandstone and siltstone lenses of a delta complex, carries across the large Bradford anticline (Figure 2.5), which explains the effectiveness of the water drive. However, approximately 50 per cent of the original oil in place remains there.

The *Music Mountain* pool (Fettke, 1941) in McKean County was opened in 1937. Although this pool lies on

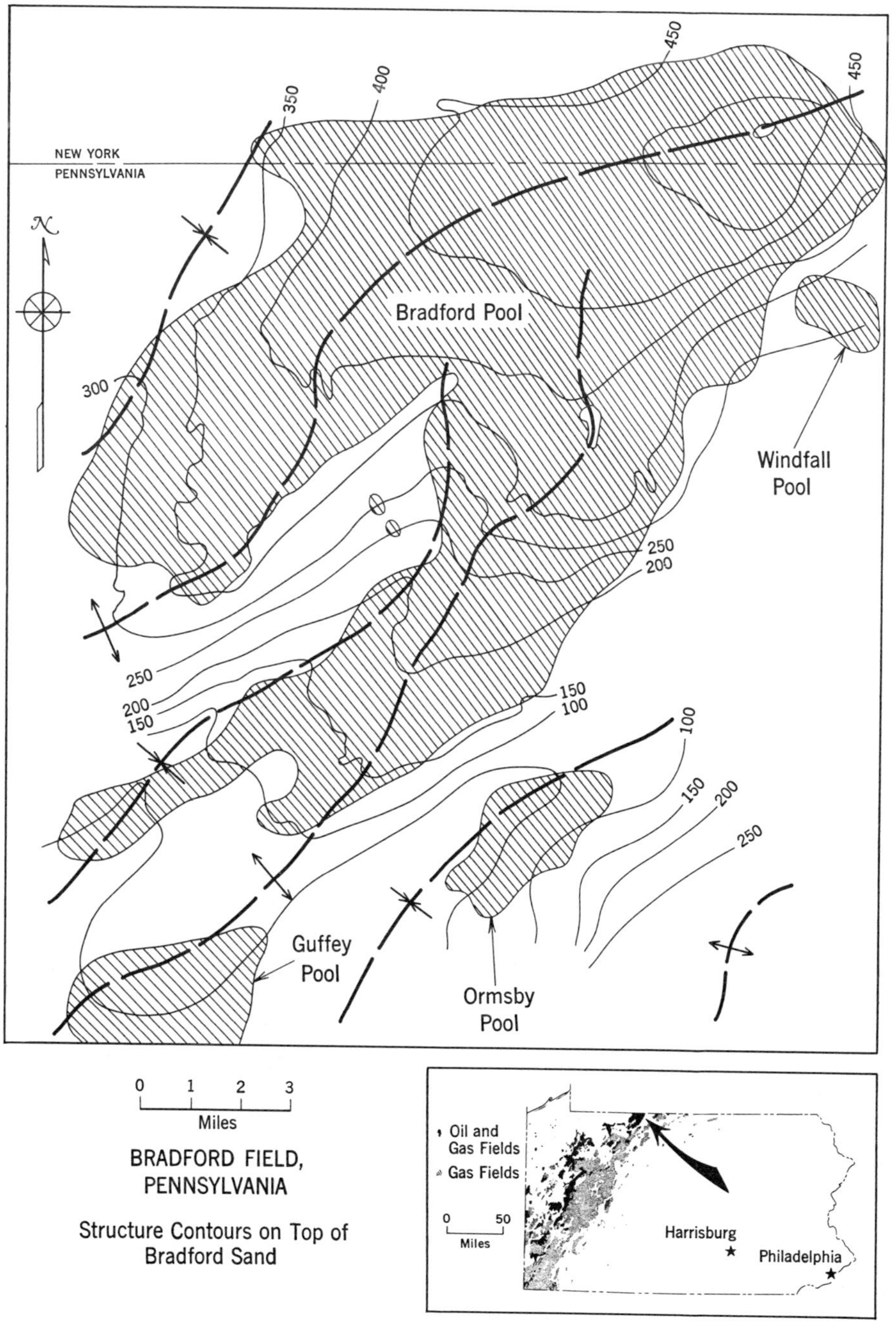

FIGURE 2.5 Bradford field, Pennsylvania-New York. Producing areas (diagonal lines) superimposed upon structure contours. Contour interval, 50 feet. ***Courtesy Pennsylvania Topographic and Geologic Survey. After Charles R. Fettke (1934).***

the southwest flank of the Bradford anticline, accumulation is due entirely to varying permeability. The reservoir rock is an Upper Devonian sandstone that occurs in a shoestring lens, "probably the remains of an ancient offshore bar," about 4 miles long and 800 to 2000 feet wide, with an average thickness of 35 feet.

Second only to Bradford in production is the *McDonald* field of Allegheny and Washington Counties (Figure 2.1), which, with three adjacent fields, forms a producing area 15 miles long by 1 to 4 miles wide (Ingham, 1949). The field, discovered in 1891, contains two wells that had initial productions estimated to be close to 15,000 barrels per day. The principal reservoirs are sands of Late Devonian age, elongate lenticular bodies trending northeast-southwest, which were probably originally deposited as offshore bars and irregular sheetlike sands. Trapping is due primarily to the lenticularity of the sands.

The gas reservoirs of western Pennsylvania are divided into two classes, the shallow and the deep sand fields. The terms "shallow" and "deep" refer more to stratigraphic level than to actual depth. The shallow sand fields produce from the Upper Devonian and younger reservoir rocks, and stratigraphic conditions, especially lenticularity, control gas trapping in the same way as they control oil accumulation in the Venango district previously described (Ingham et. al., 1956). Although most of the shallow sand gas fields lie in or immediately to the east of the western Pennsylvania oil district, one small field discovered in 1956, the *Harvey Lake,* is near Scranton in Lackawanna County in far eastern Pennsylvania (Grow, 1964). It lies just 4 miles west of the western edge of Pennsylvania's anthracite coal-producing district. The discovery of Harvey Lake proved that gas can exist in rocks where the carbon ratios run as high as 90. It was discovered by a wildcat that first explored the Oriskany horizon (the Oriskany was absent) at 8200 feet and then was plugged back to 3600 feet, where it was completed as a gas well in sands in the Chemung Group of Late Devonian age. The trap is caused by reservoir lenticularity along the axis of a northeastward plunging anticline (Grow, 1964).

The heating value of the gas diminishes from west to east in Pennsylvania. The gas along the northern and eastern fringes of the Appalachian province is almost entirely methane (Roth, 1968, p. 1713).

Shallow sand fracturing has been very successful in the gas field of western Pennsylvania. In Armstrong County this practice resulted in a fortyfold increase in gas production. (Lytle, 1958).

The deep sand pools of Pennsylvania produce from the Oriskany Sandstone of Early Devonian age and the overlying mid-Devonian Onondaga cherty Limestone (Finn, 1949).

Deep sand gas production began in 1930 after Oriskany discoveries in Tioga and Potter Counties (Ebright et al., 1949) to the east of the Bradford field. The search for deep gas was further accelerated by the discovery of the prolific Leidy field in Clinton County, north-central Pennsylvania, in 1950. An intensive drilling campaign led to the development by the end of 1959 of eight fields containing a total of 37 gas pools producing from Oriskany Sandstone and Onondaga Chert.

Most gas wells in these reservoirs tend to have high initial yields and short lives, although Medina wells are long lived. It has been stated that "the average life of a gas field here is only $2\frac{1}{2}$ to 5 years. After depletion, they are usually converted to storage reservoirs" (Gardner, 1960, p. 57). The occurrence of gas in fractures in the Oriskany Sandstone and especially in the Onondaga Chert is probably the reason for the rapid depletion of both individual wells and gas pools. For the same reason hydraulic fracturing has opened production in seemingly tight sections of reservoirs.

Although the Oriskany Sandstone is not continuous regionally it has enough local continuity so that accumulation of the gas is commonly anticline controlled. Therefore the wildcatters have found good hunting along the axes and flanks of such famous Pennsylvania plateau anticlines as the Chestnut Ridge and the Laurel Hill.

Typical of the Oriskany accumulations in Pennsylvania is the *Hebron* gas field (Reeves, 1936), which lies 35 miles southeast of Bradford (Figure 2.1). Trapping is anticlinal, with the gas field occupying a local swelling on the Hebron anticline, which extends across northern Pennsylvania and southwestern New York for 75 miles or more. The Hebron dome has a closure of 250 feet and a structural relief of 1500 feet. The dome is an elongate ellipse, and the gas field is 12 miles long and $1\frac{1}{2}$ miles wide.

The *Leidy* field (Ingham, 1954; Ebright and Ingham, 1951), the first of the large Oriskany gas accumulations to be discovered, occupies the Leidy dome, a node on the Wellsboro anticline (Figure 2.1). Some wells had exceptionally large initial flows, probably because of the intersection of fractures. The pressures were also abnormally high, up to 4200 pounds per square inch from a reservoir depth of 6000 feet.

Future. Pennsylvania is far from depleted as an area for further exploration. Her most promising frontier is in depth, the third dimension. Drilling to greater depths in the plateau area will reach the hitherto largely unexplored Cambro-Ordovician section. Of the few wells that have tested these older rocks, most have found encouraging oil or gas shows. The Ordovician contains the Oswego Sandstone in the upper part. The Middle Ordovician has the Black River–Trenton Formation, which has been so productive to the west on the Cincinnati arch, and in the Lower Ordovician are the Beekmantown Dolomite and the Cambro-Ordovician Knox Dolomite. The Upper Cambrian contains the Gatesburg and Potsdam Sandstones, both of which have yielded good shows of oil and gas. The plateau area is replete with anticlinal structures; in addition some thousands of feet of section wedges out to the westward, no doubt creating innumerable stratigraphic traps (Ingham, 1960).

There are, of course, more Oriskany Sandstone and Onondaga Chert discoveries to be made in the plateau area. Onondaga reservoirs beyond the limits of the Oriskany Sandstone should not be ignored in this search.

Exploration can also be continued to the eastward. Only a few of the many anticlines in the Appalachian Valley and Ridge province have been tested through the Ordovician, with its potential gas reservoirs (Hull, 1965). Gas discoveries in Bedford and Lackawanna Counties show that this is not completely barren territory. The deepest part of the Appalachian basin is on the east side (Roth, 1968, p. 1711; Woodward, 1964).

The same general area in northwestern Pennsylvania where Colonel Drake discovered oil at 69½ feet still has shallow oil possibilities (Weaver, 1966). Lower Ordovician and Cambrian rocks are inadequately explored in western Pennsylvania as well as in the central and east-central parts of the state (Wagner, 1966). "The introduction of hydraulic fracturing in the Warren area (in northwestern Pennsylvania) may make economical the drilling and production of the fine-grained, low permeability sands" (Lytle, 1965, p. 43). This could also mean a second look at other parts of western Pennsylvania in which exploration has been discouraged by tight sands.

Pennsylvania, with less shoreline than any other Great Lakes state, was the first to explore offshore. A potential gas producer was drilled just off the northwest corner of Erie County (and of Pennsylvania) in 1958. An initial rate of 200,000 cubic feet of gas per day was reported after fracturing the Red Medina (Silurian) Sandstone at a depth of 2500 feet. The offshore area has been inactive since then, but in early 1968 Pennsylvania carried out a big lease-sale that should lead to intensive exploration of the eastern Lake Erie floor out to the Canadian boundary. Canada has been producing from her side of the lake for years.

REFERENCES CONSULTED IN PREPARING CHAPTER 2

Ashley, G. H., 1938.
Dickey, Parke A., 1959.
Ebright, John R., Fettke, Charles R., and Ingham, Albert I., 1949.
Ebright, John R., and Ingham, Albert I., 1951.
Fettke, Charles R., 1938, 1941.
Finn, Fenton H., 1949.
Gardner, Frank J., 1960.
Giddens, Paul, 1941.
Grow, George C., 1964.
Hull, Joseph P. D., Jr., 1965.
Ingham, Albert I., 1949, 1954, 1960.
Ingham, A. I., Lytle, W. S., Matteson, L. S., and Sherrill, R. E., 1956.
Lytle, William S., 1958, 1963, 1965.
Lytle, William S., and Heeren, Lillian A., 1964.
Lytle, William S., et al, 1968.
Reeves, John R., 1936.
Roth, Ernest E., 1968.
Sherrill, R. E., Dickey, P. A., and Matteson, L. S., 1941.
Wagner, Walter R., 1966.
Weaver, O. D., 1966.
Woodward, Herbert P., 1958, 1964.

Chapter

3

MARYLAND

The axis of the Appalachian geosyncline passes through West Virginia to the west of Maryland, but the eastern flank of this trough crosses western Maryland. Maryland became a gas-producing state in 1949, when a successful well was completed about 5 miles southeast of Oakland in Garrett County (Figure 3.1) near the western edge of the "panhandle" (Amsden, 1954; Singewald, 1954). This was the discovery well of the *Mountain Lake Park* field, which developed along the axis of the Deer Park anticline, a major fold crossing Garrett County from northeast to southwest. Production is obtained from depths in the neighborhood of 5000 feet; the reservoir rocks are the Middle Devonian Huntersville Chert and the underlying Ridgely (Oriskany) Sandstone of the Lower Devonian (Figure 3.2). As in Pennsylvania the Huntersville gas is stored in fractures cutting through the chert formation.

Subsequently, in 1953, a second gas field, the *Accident*, was found to the north of Oakland in northwestern Garrett County. The field was named after the Accident anticline, on which it lies, rather than for the conditions attending discovery. The reservoir rocks here are the same as those in the Mountain Lake Park field, but the depths are greater, between 7000 and 8000 feet.

In 1965, a new area of production was discovered on the east flank of the Accident anticline. Three wells, all producers, have been drilled in this field, *Negro Mountain*. One of these wells was drilled to a depth of 8700 feet, making it the deepest producing well in Maryland (Weaver, 1966 et seq., 1968).

A fourth producing area was discovered in 1966 in eastern Allegany County in the Green Ridge State Forest. As in the other fields, the production is from the Oriskany-Huntersville interval. This field, the *Green Ridge*, is the southwesterly extension of the Artemas field in Bedford County, Pennsylvania.

Up to this time, the deep pre-Devonian rocks of western Maryland have not been adequately tested. The

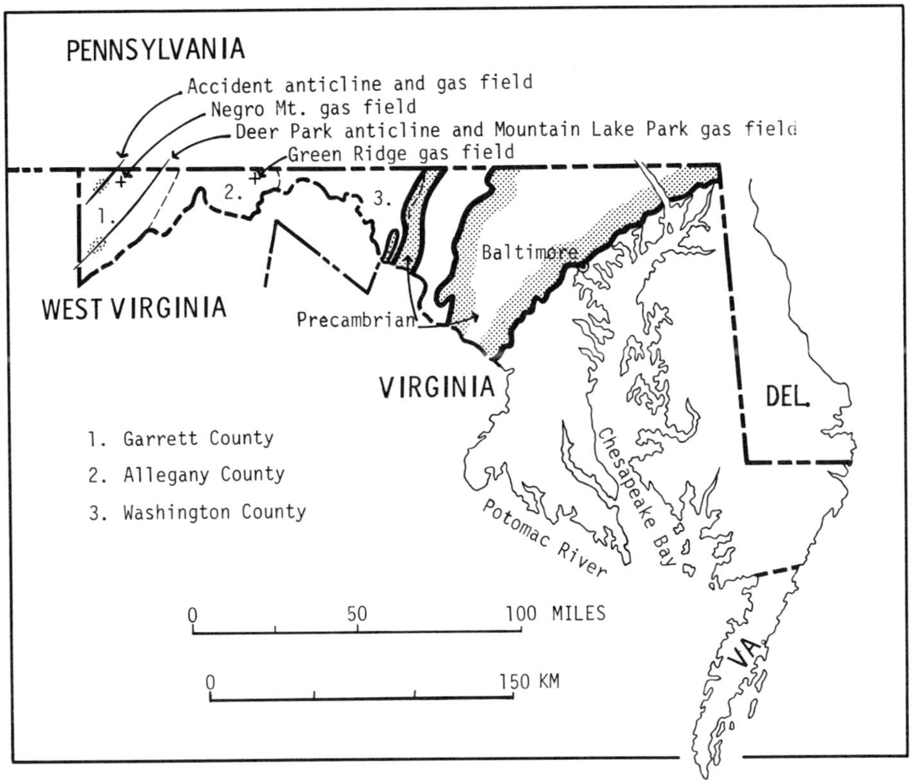

FIGURE 3.1 Maryland. Index map.

oldest formation penetrated thus far is the uppermost Tuscarora (Amsden, 1954, plate 3).

REFERENCES CONSULTED IN PREPARING CHAPTER 3

Amsden, Thomas W., 1954.
Singewald, Joseph T., Jr., 1954.
Weaver, Kenneth N., 1966 et seq., 1968.

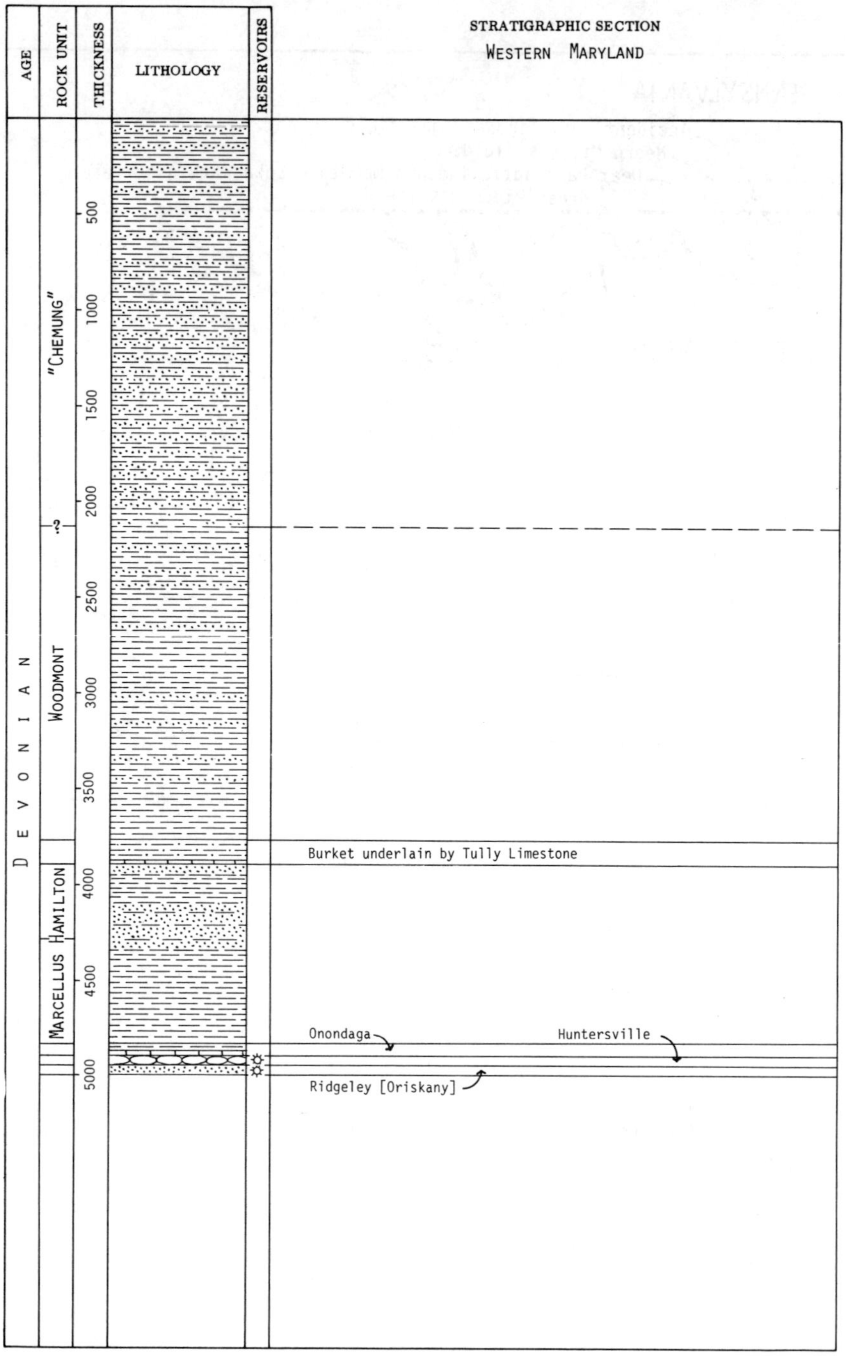

FIGURE 3.2 Maryland. Stratigraphic column, with gas reservoirs. ***After Thomas W. Amsden (1954).***

Chapter

4

WEST VIRGINIA

West Virginia is an important gas state, producing more than 1 per cent of the country's annual total, but its oil yield amounts to only about 0.1 per cent. This state is almost as old as Pennsylvania as a producer of petroleum. Both gas and oil were struck during the drilling of salt wells in West Virginia in the latter part of the eighteenth century, and one enterprising salt producer began using gas to fire his furnaces *circa* 1807. However, gas seeps were known much earlier; George Washington owned a "burning spring," which he willed to the public (Roth, 1968, p. 1702). The first well drilled for the purpose of obtaining oil was started shortly after the completion of the Drake well and became a producer in early 1860 at a depth of 303 feet. In 1899 West Virginia surpassed Pennsylvania in oil output but did not remain ahead for long. It reached its production peak in 1900 with an output of more than 16 million barrels. The 1967 figure was about $3\frac{1}{2}$ million barrels.

In recent decades gas discoveries have dominated the exploration picture. The large Elk-Poca gas field (Figure 4.1) in Jackson, Putnam, and Kanawha Counties was discovered in the early 1930s. It produces from the Oriskany Sandstone. In 1944 the first of the Oriskany gas accumulations in the major folds to the east of the older fields was found in West Virginia, after similar discoveries in Pennsylvania and New York. Now the emphasis is on Oriskany and Newburg exploration; nonetheless, many new discoveries each year continue to be made in the younger "traditional" gas sands (Headlee, 1960, p. 105).

Geology. The Appalachian geosyncline petroliferous province (Figure 4.2) crosses from southwestern Pennsylvania into West Virginia and continues in a southwesterly direction, covering a path from 40 to 70 miles wide, across the western end of the state into Kentucky. Along the northwest edge the oil and gas field belt overlaps the southeastern corner of Ohio. The deepest part of the Appalachian trough is in northern West Virginia, adjacent to the southwestern corner of Pennsylvania. In

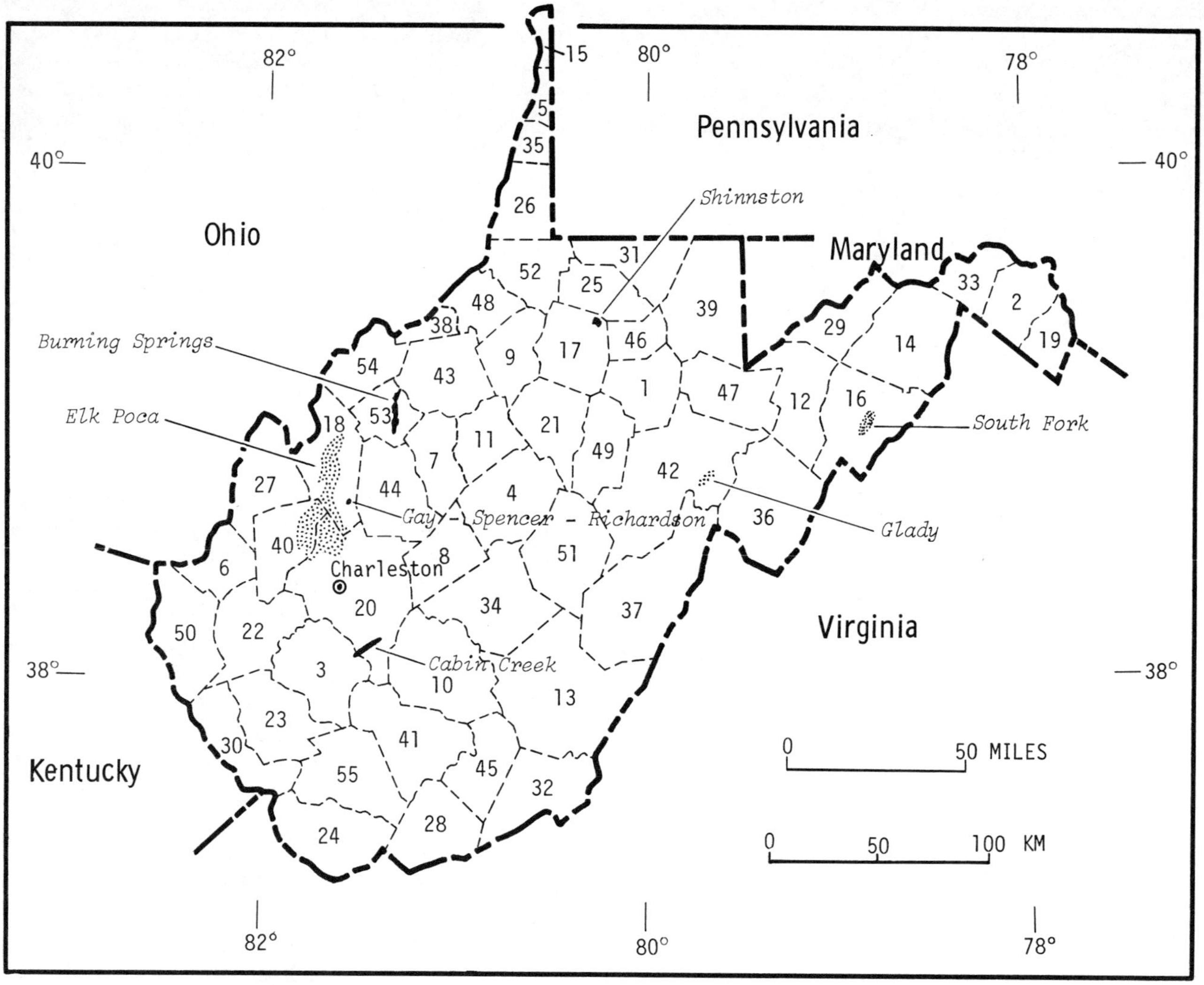

Figure 4.1 West Virginia. Index map, including oil and gas fields named in text. County Key:

1	Barbour	9	Doddridge	17	Harrison	25	Marion	33	Morgan	41	Raleigh	49	Upshur
2	Berkeley	10	Fayette	18	Jackson	26	Marshall	34	Nichols	42	Randolph	50	Wayne
3	Boone	11	Gilmer	19	Jefferson	27	Mason	35	Ohio	43	Ritchie	51	Webster
4	Braxton	12	Grant	20	Kanawha	28	Mercer	36	Pendleton	44	Roane	52	Wetzel
5	Brooke	13	Greenbrier	21	Lewis	29	Mineral	37	Pocahontas	45	Summers	53	Wirt
6	Cabell	14	Hampshire	22	Lincoln	30	Mingo	38	Pleasants	46	Taylor	54	Wood
7	Calhoun	15	Hancock	23	Logan	31	Monongalia	39	Preston	47	Tucker	55	Wyoming
8	Clay	16	Hardy	24	McDowell	32	Monroe	40	Putnam	48	Tyler		

this area occur Permian formations that are the youngest rocks to be found in the Appalachian province. The subsurface section contains reservoir rocks of Pennsylvanian, Mississippian, Devonian, and Silurian age (Figure 4.3); the most important here are sands in the Mississippian and Upper Devonian. Mississippian reservoirs of note include, proceeding downward, the Greenbrier Limestone (Youse, 1964) and the Big Injun, Weir, and Berea Sandstones. In central West Virginia the Greenbrier is the most important pay zone. It is thick (120 to 260 feet) and present everywhere. "Spectacular" increases in productivity have been obtained by hydraulic fracturing (Haught, 1964). Below the Mississippian formations are various discontinuous sands in the Catskill and Chemung of the Upper Devonian, including the Benson (Figure 4.3). The Middle Devonian (Onondagan) Huntersville Chert and the immediately underlying Lower Devonian Oriskany Sandstone are also productive of gas, especially in the more strongly folded sediments in the eastern half of West Virginia (Patchen, 1968a). The

Newburg sand (Silurian) is an important gas pay in the west-central part of the state (Patchen, 1968b).

Traps and Selected Fields. The trapping of oil and gas in West Virginia has been due in some instances to the presence of pronounced anticlines and widespread porosity in the reservoir rocks, and in others to varying permeability caused by the lenticularity of the oil and gas sands. The anticlinal accumulation of oil is best illustrated by the *Burning Springs* field (so named because of a gas seep on fire) on the Volcano–Burning Springs anticline, which occupies part of four counties in northwestern West Virginia. This fold is 25 miles long and asymmetrical, with dips on the flanks ranging from 20 to 60°. Unlike most flexures in the Appalachian geosyncline, this one does not parallel the geosynclinal axis but runs due north and south, cutting across the "grain" of the larger feature. Although much of the area covered by the Volcano–Burning Springs anticline is unproductive because of the tight nature of the sands, a large number of successful oil wells have been drilled close to the top of the anticline, most of them immediately to the west of the axis. The discovery well for West Virginia was drilled here. The principal reservoir in the Burning Springs field is the Salt Sand section in the Pottsville (Lower Pennsylvania).

The *Cabin Creek* field (Wasson and Wasson, 1927), 20 miles southeast of Charleston, illustrates the trapping of oil on a monocline because of the wedgout of the reservoir sand in all directions. The reservoir rock, a Berea Sandstone lens, is a typical shoestring, 12 miles long by $\frac{1}{2}$ to 2 miles wide. It lies parallel to a nearby synclinal axis. No water is present; oil occupies the downdip edge of the sand body, and gas the upper edge. Another Berea shoestring is the *Spencer–Richardson–Bee Run–Yellow Creek* oil and gas trend (Heck, 1941), a discontinuous sand body extending for 60 miles in a northeast-southwest direction in northwest West Virginia. The *Shinnston* oil pool (Reger, 1941) to the northeast produces from a much less elongate lens of Upper Devonian sand. However, this sand is likewise water-free, and it lies on a monoclinal slope with gas concentrated along the updip edge. The so-called synclinal pools of West Virginia are described elsewhere (Landes, 1959; Davis and Stephenson, 1929). In some of these the oil has accumulated in the vicinity of a plunging synclinal axis with gas above extending across neighboring anticlines, and in others the sandy lenses with adequate porosity happened to be structurally low.

The trapping of gas in the Oriskany Sandstone is largely due to varying permeability in western West

Figure 4.2 West Virginia. Oil and gas fields and major tectonic features.

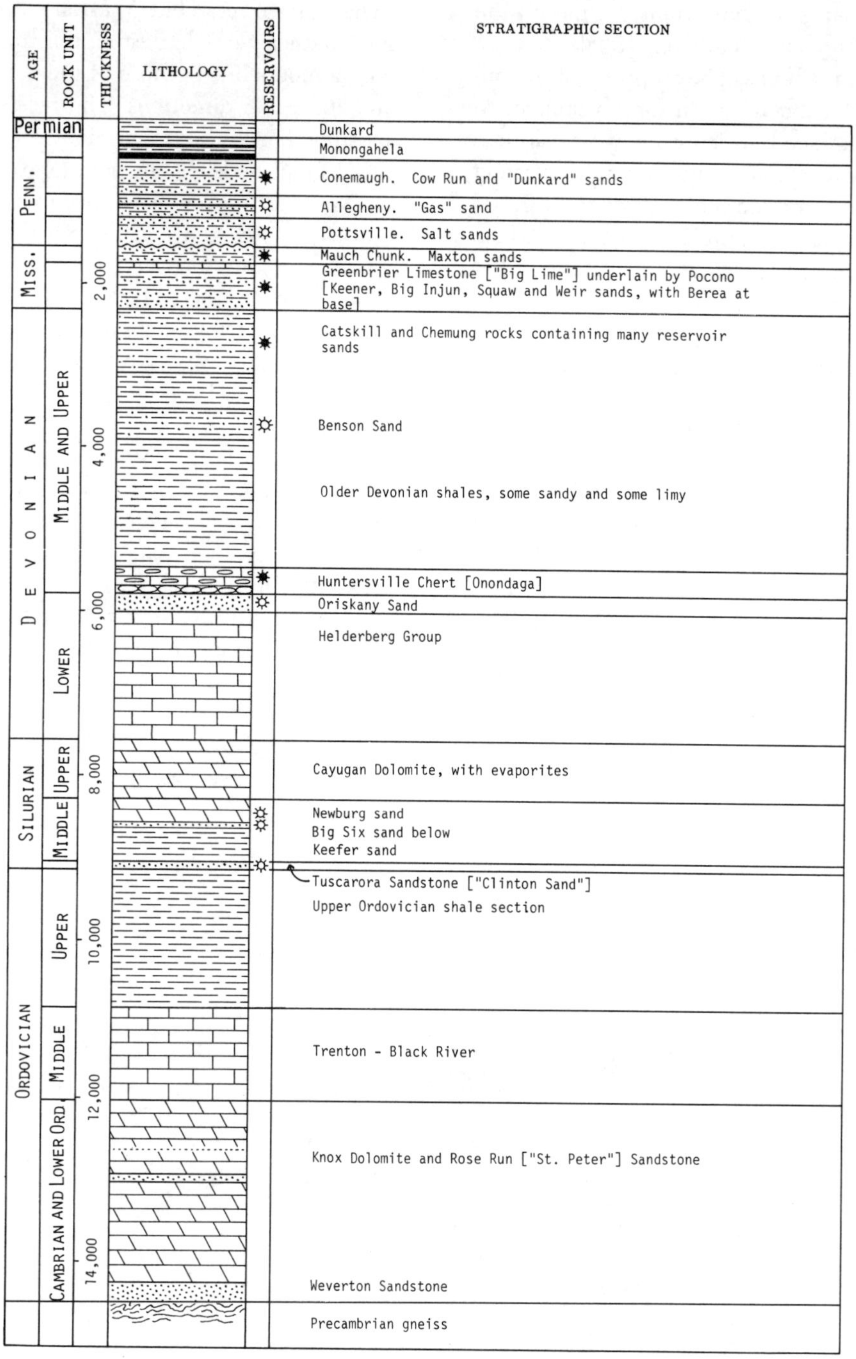

Figure 4.3 West Virginia. Stratigraphic column with some oil and gas reservoirs noted. *After Robert E. Bayles (1956) and William K. Overby (1962).*

Virginia, where the strata are nearly flat lying. Farther to the east, however, in the belt of major folds where the rocks were once believed to be too metamorphosed for oil or gas occurrence, all pools so far discovered are in anticlinal traps. The Oriskany Sandstone in West Virginia is a sheet sandstone from 50 to 200 feet in thickness that underlies the Huntersville Chert and overlies the Helderberg Limestone.

In western West Virginia the *Elk-Poca* field has been an extremely important producer of Oriskany gas (Price et al., 1957, p. 345). The eastern boundary of this field is the water line below the gas, and the western and southern boundaries are formed by the sand either pinching out or changing in facies to a rock of inadequate permeability. The Oriskany gas in western West Virginia is wet, with a little oil also present in the reservoir.

The drilling of major anticlines in eastern West Virginia has resulted in the discovery of numerous Huntersville Chert–Oriskany Sandstone dry gas fields. The largest of these is the *Glady* in Pocahontas and Randolph Counties. This field stretches along the axis of the Blackwater anticline for a distance of 16 miles; the maximum width is 2 miles. Production is from both the Huntersville Chert and the Oriskany Sandstone at depths of 6100 to 6200 feet. As a general rule these reservoirs are hydraulically fractured to increase the initial flow. The *South Burns Chapel* field, which straddles the Monongalia–Preston County line in northern West Virginia, was discovered in these same reservoirs in 1964 (an earlier test with a flow of 2 million cubic feet was completed in 1962; it turned out to be on the southeastern edge of South Burns Chapel). Subsequently more than 30 gas wells have been drilled, to an average depth of 7800 feet. This field occupies the crest and higher flanks of the Chestnut Ridge anticline (Patchen, 1968a).

Future. The younger, relatively shallow reservoirs will continue indefinitely to be important to the oil and gas industry in West Virginia. However, the deeper, more prolific reservoirs will increase in importance with new discoveries across the years. The brightest star is the Oriskany Sand; its importance has increased with the realization that production where the sand is tight can be stimulated by fracturing and acidizing. Numerous parts of West Virginia have not been tested adequately at the Oriskany level. For example, in a two-county (Barbour and Upshur) area in north-central West Virginia only one well, drilled in 1956, has penetrated the Oriskany. It found the sand tight; however, good gas production is obtained from this reservoir northeast across the Barbour County line (Haught, 1965, p. 15).

The Upper Silurian Newburg sand became the principal target in "deep" gas well drilling in west-central West Virginia in 1966 (Fenstermaker, 1966; Tarver, 1966). Activity has centered on western Kanawha County, where the Newburg lies at an average depth of about 1 mile, but extends northward into Jackson County and southward into Boone County.

> Early deep wells to this sandstone did not find gas in commercially attractive quantities, but modern completion techniques of acidizing and fracturing have greatly increased production . . . A recent wildcat in Jackson County discovered oil and gas in the Newburg, the State's first Silurian oil well (Patchen, 1968b, p. 1).

Accumulation in the Newburg has taken place principally where there is updip loss of porosity on anticlinal noses, and on anticlines.

Gas has also been found in the deeper Tuscarora Sandstone at 11,000 feet in eastern Monongalia County (Latimer, 1968). Potential reservoirs are present in the pre-Silurian section as well (Reeves and Price, 1950; Haught, 1953).

Northeastern West Virginia lies in the Appalachian Valley and Ridge province, which was once considered to be poor hunting because of greater metamorphism in this belt of strongly folded rocks (Roth, 1968, p. 1712). The *South Fork* gas field along the top of the Whip Cove anticline in eastern Hardy County was discovered in 1964. The field is at least $4\frac{1}{2}$ miles long. The pay zone is the Oriskany, here over 200 feet in thickness. "The discovery of this field provides one more, though now hardly necessary, proof that any structurally favorable area in eastern West Virginia is worth testing for gas production" (Price and Haught, 1964). By mid-1968 a total of 16 gas fields producing from Huntersville Chert and/or underlying Oriskany Sandstone had been found in northeastern West Virginia (Patchen, 1968a).

South of the main Huntersville-Oriskany gas area the geology becames really complicated. A well drilled in southwestern Pendleton County in 1960 to a depth of 13,000 feet went through a major thrust fault at 10,035 feet and ended at total depth in the same Middle Ordovician formation that it had previously passed through at a depth of 1690 feet (Perry, 1964).

Seven waterfloods were operating in the state during 1967 (Lytle, 1968, p. 957). Additional secondary recovery projects are possible (Wesson, 1967).

REFERENCES CONSULTED IN PREPARING CHAPTER 4

Bayles, Robert E., 1956.
Davis, Ralph E., and Stephenson, Eugene A., 1929.
Fenstermaker, Charles D., 1966.
Haney, Gene, 1968.
Haught, Oscar L., 1953, 1964, 1965.
Headlee A. J. W., 1960.
Heck, E. T., 1941.
Landes, Kenneth K., 1959.
Latimer, I. S., 1968.
Lytle, William S., et al., 1968.
Overby, William K., Jr., 1962.
Patchen, Douglas G., 1968a, 1968b.
Perry, William J., Jr., 1964.
Price, Paul H., and Haught, Oscar L., 1964.
Price, P. H., Tucker, R. C., and Haught, Oscar L., 1957.
Reeves, Frank, and Price, Paul H., 1950.
Reger, David B., 1941.
Roth, Ernest E., 1968.
Tarver, Earl A., 1966.
Wasson, Theron, and Wasson, Isabel B., 1927.
Wesson, James A., 1967.
Youse, Arthur C., 1964.

Chapter

5

VIRGINIA

Virginia is bisected from north to south by the Blue Ridge sector of the Appalachian Highlands, a wide band of igneous and metamorphic rocks with some sedimentary rocks (Figure 5.1). These are Triassic, Paleozoic, and Precambrian in age. To the east are the young (Cretaceous to Recent), nearly flat-lying sediments of the Atlantic Coastal Plain; to the west are the highly folded Paleozoic rocks of the Appalachian Valley and Ridge province. Still farther west, including all or parts of eight counties in southwesternmost Virginia, is the eastern edge of the Appalachian Plateau province. So far the oil and gas discoveries have been entirely in the Paleozoic rocks of western and southwestern Virginia.

The sedimentary rocks in western Virginia range from Middle Pennsylvanian down through the Paleozoic section to the Precambrian (Figure 5.2). The Appalachian Valley rocks are mainly lower Paleozoic, in some places repeated by overthrusting. Along the western edge of the state, northeast of Tazewell County, Devonian, Mississippian, and Pennsylvanian rocks lie at the surface.

The first discovery in Virginia was the small Early Grove gas field (Averitt, 1941) near Bristol along the Scott–Washington County line, and about 4 miles north of the Tennessee line, which began producing in 1931. Additional gas discoveries were made subsequently in some of the counties contiguous to the Kentucky and West Virginia boundaries as far northeast as the northwest corner of Rockingham County. One small oil field, the Rose Hill in western Lee County, was discovered in 1942, and a second, the Ben Hur (also in Lee County), was found in 1963.

The gas reservoirs are principally sandstones and limestones in the Mississippian section and fractured shale in the Upper Devonian, but the gas field in Rockingham County produces from the Lower Devonian Oriskany Sandstone (Figure 5.1). The oil reservoirs tapped by the Rose Hill field are Trenton and Moccasin limestones of Middle Ordovician age.

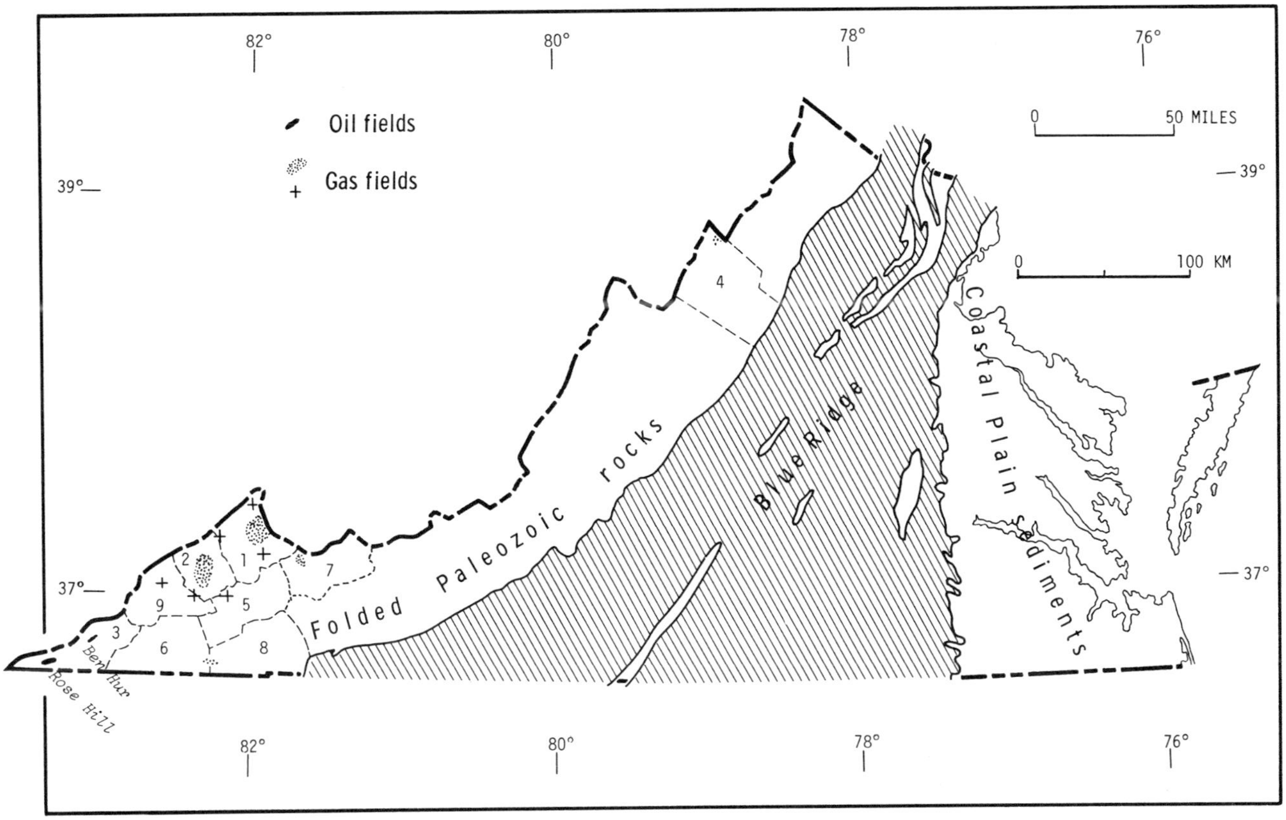

Figure 5.1 Virginia. Index map with oil and gas fields. County Key:

1 Buchanan
2 Dickenson
3 Lee
4 Rockingham
5 Russell
6 Scott
7 Tazewell
8 Washington
9 Wise

The small *Rose Hill* oil field (Young, 1958; Miller, 1948, 1954; Miller and Fuller, 1954) is in the highly folded and faulted Appalachian Valley belt. Substantial oil seepages that had been known for many years led to the original drilling in this area. The discovery well was completed in May, 1942, with a first-day production of 90 barrels but a settled production of approximately 8 barrels per day. About 30 additional producers were subsequently drilled and an equal number of dry holes. This field reached a peak in 1947 with a yield for that year of 61,000 barrels. By 1967 the output was down to 1870 barrels from 4 wells (Young, 1968).

An unusual feature about the Rose Hill accumulation is that the Trenton Limestone pay zones, the principal reservoir, occurring at depths of between 900 and 1110 feet, lie some distance below an overthrust sheet containing earlier Paleozoic rocks. Most of the successful oil wells, however, were drilled directly into the stationary block in fensters ("windows") created by erosion of the overriding block. Since the limestone does not contain any visible porosity, the accumulation must be in fractures. The seal that keeps the oil in the fractures does not appear to be either overlying rock or fault gouge. "Cementation by carbonate minerals from circulating ground waters is postulated to account for the impervious nature of the fractured limestone and dolomite updip from the oil pool. Migration of the oil to its present location not only followed the folding and overthrusting of the Appalachian orogeny but followed precipitation from water circulating along the most open channels and the resultant cementation of the fractured and brecciated rocks" (Miller, 1956). Minor oil production has been obtained from the upper member of the Mocassin Limestone at a depth of 1608 feet.

The *Ben Hur* field, about 16 miles northeast of Rose Hill, is similar in its geology to the Rose Hill occurrence. Here the Trenton pays in the discovery well produce from depths of 1595 to 1611 and 1632 to 1642 feet. Ben Hur produced 1621 barrels of oil in 1967, from two wells (Young, 1968).

In addition to Early Grove, the discovery gas field of Virginia, southwest Virginia contains several small gas fields and a number of additional gas wells scattered over Wise, Dickenson, Russell, Tazewell, and Buchanan Coun-

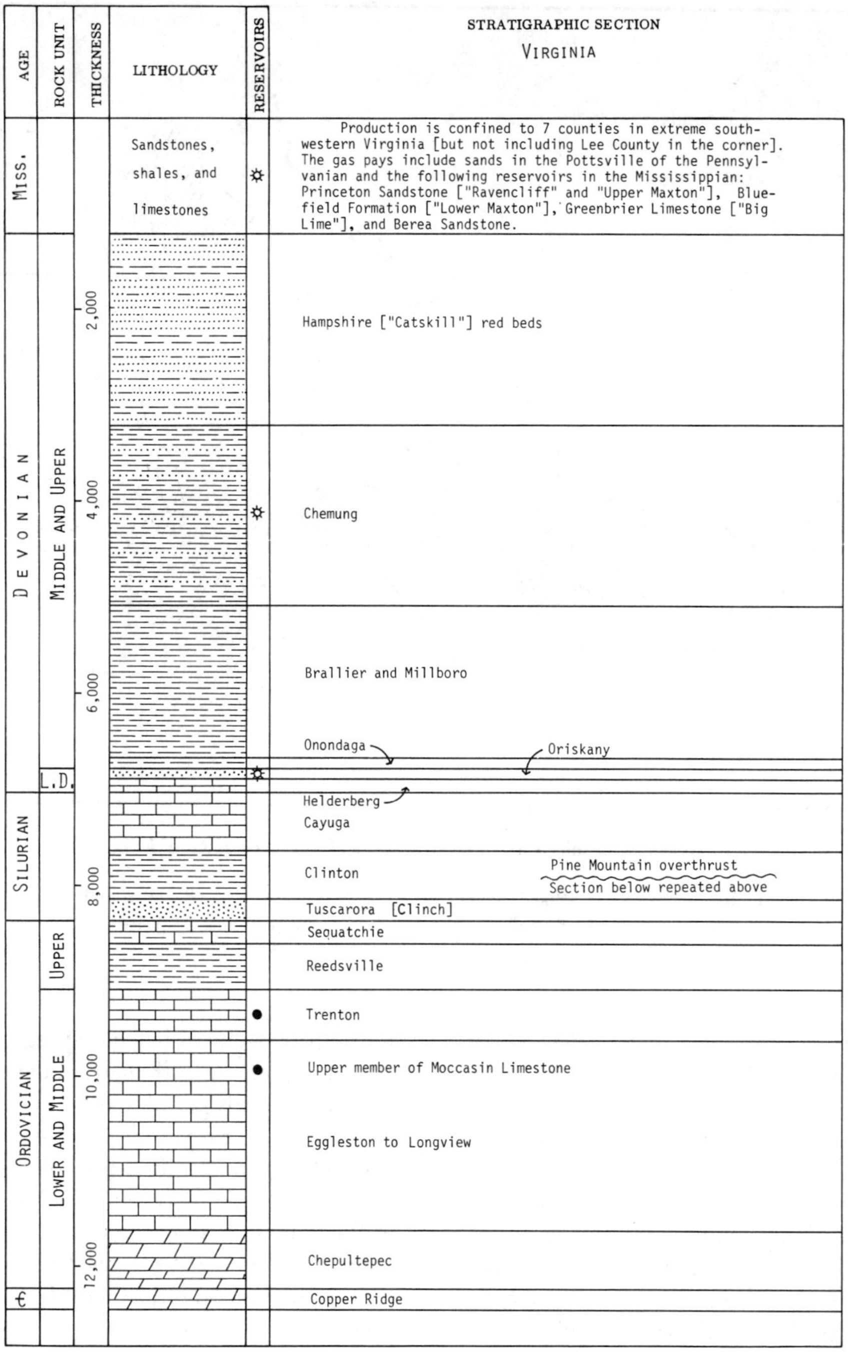

FIGURE 5.2 **Stratigraphic column for Virginia oil- and gas-producing area, with known pay zones designated.**

ties (Figure 5.1). The reservoirs, all Mississippian, are the Ravencliff in the Princeton Sandstone, the Maxton sands in both the Princeton and the Bluefield Formations (Huddle et al., 1956), the "Big Lime" (Greenbrier), and the Berea Sandstone. Some production has also been obtained from fractures in Upper Devonian shale.

The *Bergton* gas field (Young and Harnsberger, 1955; Price, 1942, p. 275), near the northwestern corner of Rockingham County, is a continuation of Oriskany gas production from New York across Pennsylvania, Maryland, and West Virginia into the western edge of Virginia. The trap is a node on the Crab Run anticline, which crosses the county from north to south. The depth to the reservoir ranges from 2600 to 4000 feet in the Bergton field. The effective porosity in the Oriskany Sandstone appears to be mostly in fractures rather than in intergranular openings.

The annual gas production in Virginia is in the neighborhood of 4 billion cubic feet. Of this total, about 60 per cent comes from Tazewell County and virtually all of the rest from Buchanan and Dickenson Counties (Young, 1968).

Prospects for future discoveries in Virginia include Oriskany accumulations in anticlines along the western edge of the state, where this formation is present and is under adequate cover; additional Mississippian pools, also in western Virginia (Wilpolt and Marden, 1960); and oil deposits of the Rose Hill type in southwestern Virginia (Miller and Brosgé, 1954).

REFERENCES CONSULTED IN PREPARING CHAPTER 5

Averitt, Paul, 1941.
Harnsberger, W. T., 1955.
Huddle, J. W., Jacobsen, Eloise T., and Williamson, A. D., 1956.
Le Van, D. C., 1962.
Miller, Ralph L., 1948, 1954, 1956.
Miller, Ralph L., and Brosgé, William P., 1954.
Miller, Ralph L., and Fuller, J. Osborn, 1954.
Price, Paul H., 1942.
Wilpolt, Ralph H., and Marden, Douglas W., 1960.
Young, David M., 1958, 1968.
Young, Robert S., and Harnsberger, W. T., 1955.

Chapter

6

TENNESSEE

The production of oil and gas in Tennessee is insignificant. This state had some oil output from 1866 to 1907, and it resumed production in 1916 in relatively minor amounts. The peak year was 1926, but the total production then was only 67,140 barrels; in 1967 the yield was down to less than 8000 barrels. About 60 million cubic feet of gas was produced during the latter year, mostly from Morgan County (Van Den Berg et al., 1968, p. 967).

The geology of Tennessee is dominated by the Nashville dome (Figure 6.1). This feature has been referred to by Wilson and Stearns (1963) as a "swelling" near the southwestward end of the exposed Cincinnati arch. It has bulged upward about 850 feet since Early Ordovician time. To the north the Cincinnati arch axis plunges downward to form the Cumberland saddle of southern Kentucky. On the east edge of the state is the Tennessee Valley, consisting of highly folded and faulted Ordovician and older rocks. The west side of the state is veneered by Cretaceous and Tertiary sediments of the Gulf embayment, an area that has been described as having some petroleum possibilities (Wilson, 1955; Moritz, 1948).

The oil and gas fields of Tennessee, both current and abandoned, have been found in ten counties (Figure 6.1) in the northern part of the state close to the Kentucky line. This area is a continuation of the Cincinnati arch–Cumberland saddle district of south-central Kentucky. These fields produce from Mississippian, Silurian, and Ordovician (Figure 6.2) limestones (Milhous, 1960).

The commercial gas fields so far discovered are in four counties on the eastern side of the district. Accumulation in Tennessee appears to be due to varying permeability within the reservoir rocks and also to the presence of small folds.

Although 47 test wells were drilled in 1967, there "were no significant discoveries of either oil or gas" (Van Den Berg et al., 1968, p. 967). So far, at least, attempts to expand the producing area west of the Cincinnati arch

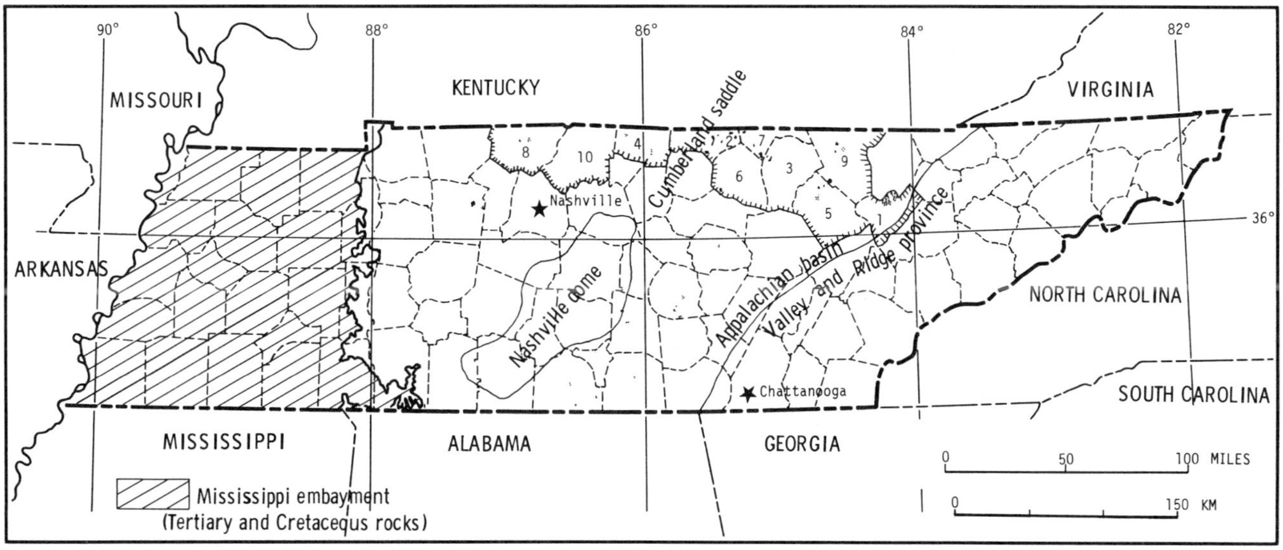

Figure 6.1 Tennessee. Index map, also showing major tectonic divisions and principal producing counties, past and present. County Key:

1	Anderson	3	Fentress	5	Morgan	7	Pickett	9	Scott
2	Clay	4	Macon	6	Overton	8	Robertson	10	Sumner

axis in southern Kentucky into Tennessee have not met with any great success. Large anticlines in the thrust belt of eastern Tennessee, however, are considered to be potentially productive (Weaver, 1964.)

REFERENCES CONSULTED IN PREPARING CHAPTER 6

Milhous, H. C., 1960.

Moritz, Carl A., 1948.

Van Den Berg, Jacob, Swann, D. H., Carpenter, G. L. Milhous, H. C., and Nosow, Edmund, 1968.

Weaver, O. D., 1964.

Wilson, Charles W., Jr., 1955.

Wilson, Charles W., Jr., and Stearns, Richard G., 1963.

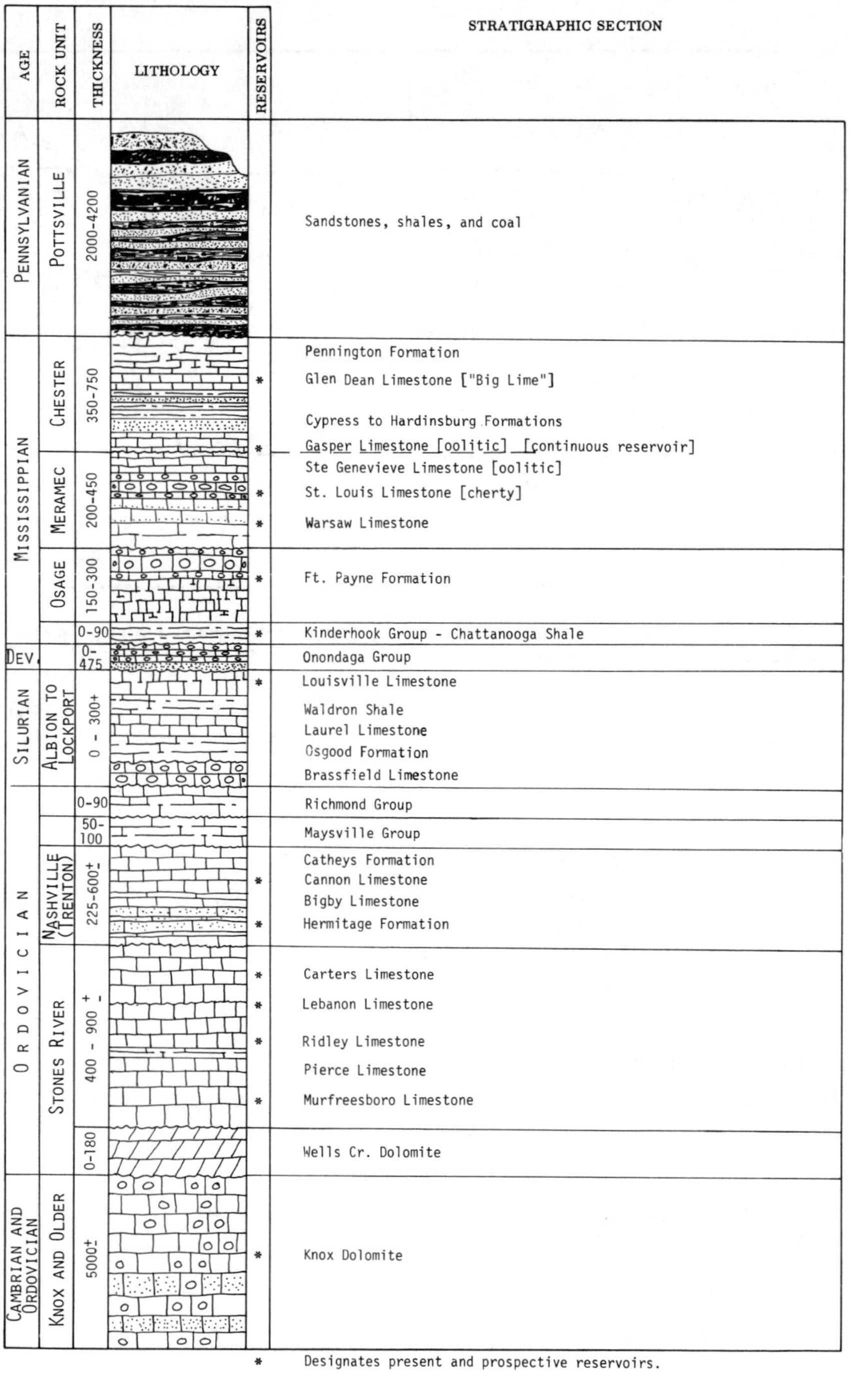

Figure 6.2 Tennessee. Stratigraphic column, showing present and prospective pay zones. *After H. C. Milhous (1960, p. 93).*

Chapter

7

KENTUCKY

Currently Kentucky is producing about 0.5 per cent of the annual domestic oil output and 0.4 per cent of the gas. Oil was first discovered by design in this state in 1860 and natural gas in 1889; however, oil from a brine well in McCreary County (Figure 7.1) was reported in 1819 (E. N. Wilson, letter dated 19 October, 1968), and oil from a similar well in Cumberland County was bottled and sold for medicinal purposes as early as 1829. Oil seepages are quite common, especially in the southern and eastern parts of the state. Continuous production started in 1883. The first production peak was reached in 1921. Thereafter the output declined until the Illinois basin development spread into western Kentucky; then the Kentucky production climbed again to a new and higher peak in 1946. A transient "boom" in Green County in south-central Kentucky in the 1950s made 1959 a record year, with a production of over 27 million barrels of oil, but by 1967 the annual yield was back to slightly more than 15 million barrels.

Structure. Kentucky is divided geologically by the axis of the Cincinnati arch, which crosses the state from north to south at its widest part (Figure 7.2). In northern Kentucky the axis of the arch rises over the large Jessamine dome, and at the south edge of the state the arch rises again toward the summit of the Nashville dome of central Tennessee. The intervening sag in south-central Kentucky is known as the Cumberland saddle. To the east of the Cincinnati arch is the continuation of the Appalachian geosyncline, which enters northeastern Kentucky from southwestern West Virginia. This area is known as the eastern Kentucky coal basin. Although the stratigraphy is similar to that in the geosyncline to the northeast, the sediments are finer grained, and more carbonate rocks are present. The section thins toward the Cincinnati arch. The parallel elongate folds of the Appalachian geosyncline to the northeast become less distinct in the eastern Kentucky coal basin. The folds are flatter and have been altered by the structural effects

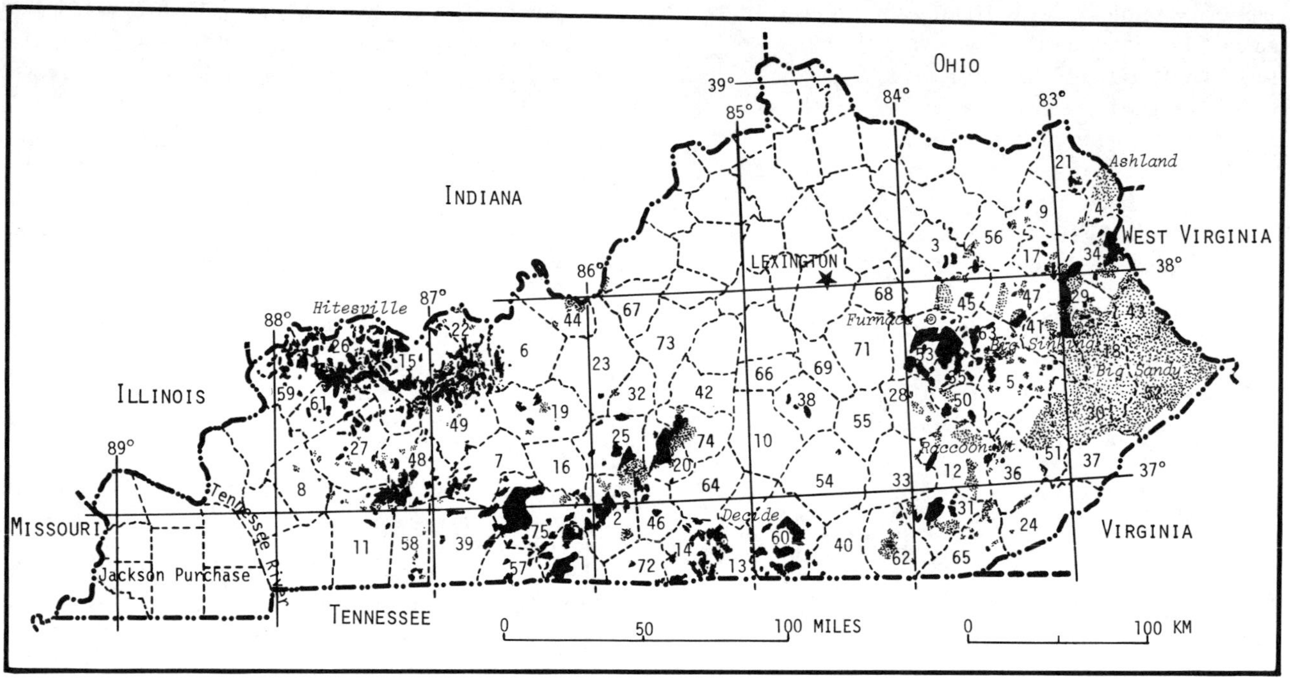

FIGURE 7.1 Kentucky. Index map with oil and gas fields. County Key:

64	Adair	8	Caldwell	18	Floyd	28	Jackson	38	Lincoln	47	Morgan	57	Simpson
1	Allen	9	Carter	69	Garrard	70	Jefferson	39	Logan	48	Muhlenberg	74	Taylor
2	Barren	10	Casey	19	Grayson	29	Johnson	40	McCreary	73	Nelson	58	Todd
3	Bath	11	Christian	20	Green	30	Knott	71	Madison	49	Ohio	59	Union
65	Bell	68	Clark	21	Greenup	31	Knox	41	Magoffin	50	Owsley	60	Wayne
4	Boyd	12	Clay	22	Hancock	32	Larue	42	Marion	51	Perry	75	Warren
66	Boyle	13	Clinton	23	Hardin	33	Laurel	43	Martin	52	Pike	61	Webster
5	Breathitt	14	Cumberland	24	Harlan	34	Lawrence	44	Meade	53	Powell	62	Whitley
6	Breckenridge	15	Davies	25	Hart	35	Lee	45	Menifee	54	Pulaski	63	Wolfe
7	Butler	16	Edmonson	26	Henderson	36	Leslie	46	Metcalfe	55	Rockcastle		
67	Bullitt	17	Elliott	27	Hopkins	37	Letcher	72	Monroe	56	Rowan		

of movement along the Irvine–Paint Creek and Pine Mountain fault belts (Figure 7.2).

West of the Cincinnati arch is the western Kentucky coal basin, a part of the Illinois basin which is illustrated in Figure 10.2. Western Kentucky contains the south rim of this regional basin. It is also crossed by an east-west fault zone known on this side of the arch as the Shawneetown–Rough Creek. The oldest exposed Paleozoic rocks in Kentucky, which are Black River–Chazy in age, crop out in the center of the Jessamine dome, and the youngest rocks (Late Pennsylvanian) occur in the centers of the two coal basins.

More than half of the counties of Kentucky produce oil or gas. Most of the counties as yet nonproductive are either (1) on or surrounding the Jessamine dome (Figure 7.2), where Ordovician rocks lie at the surface, or (2) in the southwestern tip, where the Paleozoic rocks are blanketed by Cretaceous and Eocene sediments of the Mississippi embayment. This area, lying between the Tennessee and the Mississippi Rivers, is known as the "Jackson Purchase" region.

The oil and gas fields of Kentucky fall naturally into three districts, each one belonging to a different tectonic province. The fields of eastern Kentucky belong to the Appalachian province and are a continuation of similar accumulations occurring in Ohio, New York, Pennsylvania, West Virginia, and Virginia. The oil fields in the south-central part of the state lie in two distinct groups, one on the northwest limb of the Cincinnati arch, opposite the Cumberland saddle, and the other in a similar position on the southeastern flank of the arch. The western Kentucky fields belong geologically with those of the Illinois basin.

Western Kentucky is the leading oil-producing district with three-fourths of the total state output in 1967 (Van Den Berg et al., 1968, p. 973). Eastern Kentucky is second with 19 per cent, and the central district produced 4 per cent. Five of the six leading oil counties are in western Kentucky, with Henderson County foremost. Lee County in eastern Kentucky is first in its district and second in the state. Eastern Kentucky produces three-fourths of the state's gas, with virtually all the remainder coming from western Kentucky

All of the reservoirs found to date are in the Paleozoic, and most are in the Mississippian System. The Chesterian and the Meramecian Series dominate in the west, but pay zones occur throughout the Mississippian System in the east.

Eastern Kentucky. The oil and gas fields of eastern Kentucky produce from sandstone, carbonate rock, and shale reservoirs ranging in age from Early Pennsylvanian to Cambrian (Figure 7.3). Mississippian pays include Greenbrier Limestone (or "Big Lime") and the deeper Big Injun, Weir, and Berea sandstones. Gas is produced in Pike County from oölitic and dolomitic zones in the Greenbrier (Youse, 1964). Other important reservoirs are porous dolomites of Devonian and Silurian age (Freeman, 1951) and sandstones of Niagaran (Silurian) age. Ordovician and Cambrian pays include (downward) the St. Peter Sandstone, the Ordovician part of the Knox Dolomite, and the Antietem Sandstone below the Tomstown Dolomite. The Cambro-Ordovician Knox Dolomite has various types of porosity, including vugs, oölitic zones, and fractures (McGuire, 1965a). The Antietem Sandstone of the Lower Cambrian became productive of apparently indigenous oil in 1967, when a well drilled in Boyd County began pumping 32 barrels of oil a day from slightly dolomitic sandstone zones at a depth of 7600 feet. This is the deepest producer in Kentucky (Van Den Berg et al., 1968, p. 973). The Antietem is probably the basal Cambrian sandstone in Boyd County, where it may be as much as 1500 feet thick.

As in other parts of the Appalachian geosyncline, some accumulations are due to anticlinal structure and others to varying permeability. One example of an eastern Kentucky oil field is the *Big Sinking* field (Freeman, 1941) in Lee County. It lies about 5 miles south of the Irvine–Paint Creek fault zone, but apparently there is little relationship between the faulting and the occurrence of oil. The reservoir rock at Big Sinking is Niagaran (Silurian) dolomite, which has several zones of high porosity because of exposure at the surface in pre-Hamilton (Devonian) time. Trapping has been due to the tilting and truncation of the Niagaran dolomites with subsequent overlap and sealing by Devonian shale.

According to Billingsley (1934, p. 509), accumulation in the Estill and Lee Counties oil fields has been made possible by the sealing of sandstone and dolomitic reservoirs by updip wedging. Much of the recent oil production in eastern Kentucky can be credited to successful waterflood operations in the older fields.

The eastern Kentucky gas fields occur both inter-

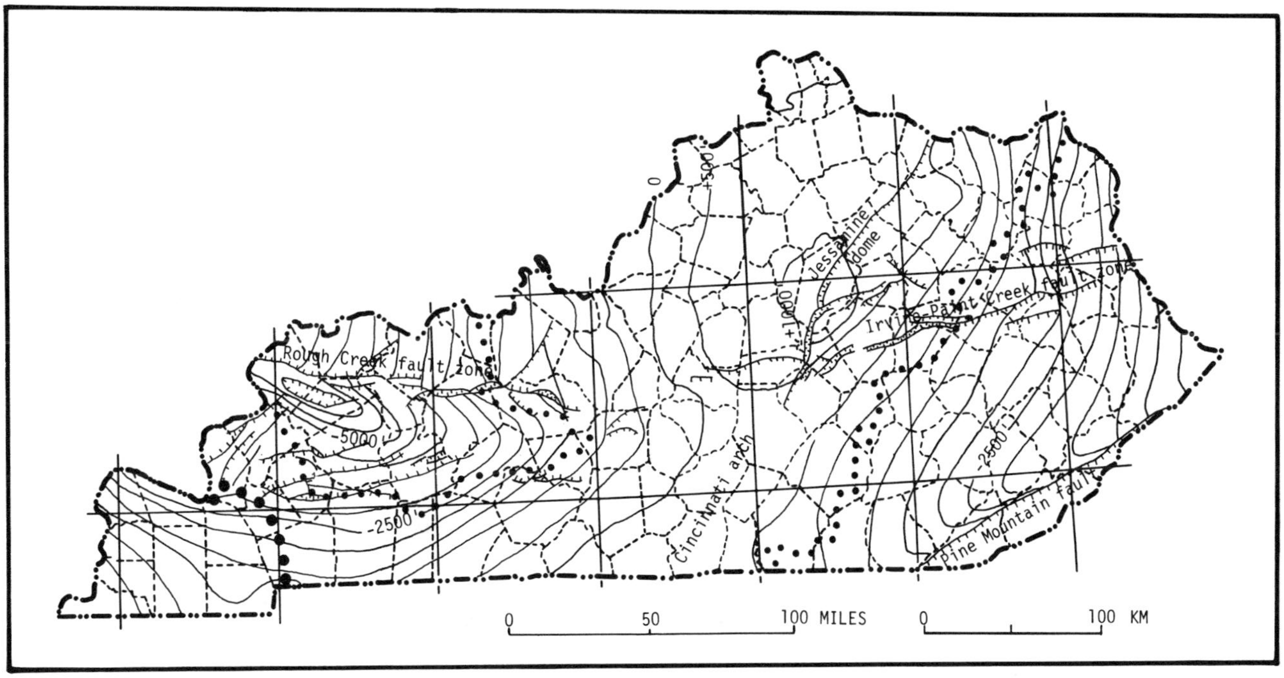

Figure 7.2 Kentucky. Tectonic map. Large dots in far western Kentucky mark northern edge of overlapping Cretaceous rocks in Mississippi Embayment. Dotted lines to north and east show southern and western, respectively, limits of Pennsylvanian rock outcrops. Structure contours are based on top of Trenton; interval, 500 feet. Only a part of the mapped faults in western Kentucky is shown. *After George V. Cohee et al. (1962).*

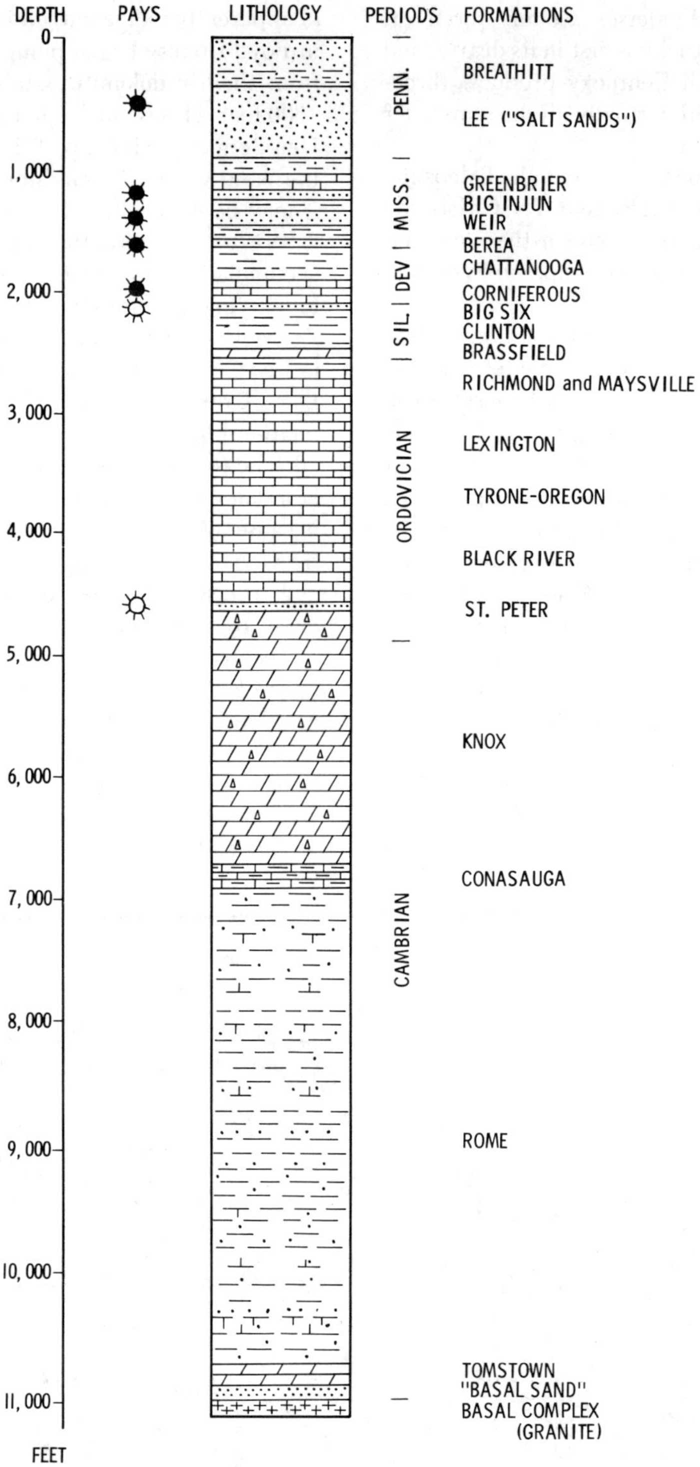

FIGURE 7.3 **Eastern Kentucky stratigraphy. Pay zones in Breathitt County and vicinity.** ***Courtesy George R. Thomas (1960, p. 19), E. N. Wilson, and the Kentucky Geological Survey.***

spersed among the oil fields and in isolated accumulations to the east of the oil-bearing rocks. The gas is trapped in the structurally high areas where the reservoir rocks contain adequate porosity. Pike County is the leading producer of gas, mainly from the Big Lime, but also from the deeper Berea and the Devonian shale (Thomas and Nosow, 1960; Greenfield, 1958). Devonian shale gas, which is apparently not structurally controlled, is tapped by more than 4200 wells in the *Big Sandy* gas field of Pike, Floyd, Martin, and Knott Counties and in the *Ashland* field of Boyd County (Hunter, 1962). The *Furnace* field of Powell County has some large gas wells that produce from the St. Peter Sandstone. This gas runs 42 per cent carbon dioxide. The *Trapp* gas field in southern Clark County was the first commercial St. Peter Sand field. The *Raccoon Mountain Consolidated* field in Clay and Laurel Counties produces from Trenton, Black River, and Knox reservoirs on a well defined anticline (Wilson, 1968). Eastern Kentucky contains a number of anticlines as yet untested (Bergin, 1956; Welch, 1955). Other possible traps, suggested specifically for accumulations in porous Knox Dolomite, include erosional remnants beneath a regional unconformity, truncated porosity zones, fault seals, and lenticular permeability. (McGuire, 1965a).

The older rocks are inadequately explored in eastern Kentucky. In addition to St. Peter and uppermost Knox, the Rose Run sand (Thomas, 1960; McGuire, 1968), Copper Ridge Dolomite, Rome Formation, and now the Tomstown Dolomite and Antietem Sandstone merit testing. The basal Cambrian sand has excellent porosity, adequate thickness, and wide lateral extent (Nosow, 1961).

Central Kentucky. The central Kentucky district is productive only from Lincoln County southward. It straddles the axis of the Cincinnati Arch, with the greater part of the production to date on the west side. Reservoirs range in age from Meramecian (Mississippian) to Cambrian, with Silurian carbonate rocks the most prolific (Figure 7.4). The principal traps in the Silurian reservoirs occur where porosity zones have been truncated by erosion and sealed by the deposition of the Devonian New Albany Shale (Chattanooga) immediately above the unconformity (Nosow, 1959; Hagan, 1960).

The beveled reservoirs beneath this unconformity include Middle Devonian limestones and sands ("Corniferous"), as well as three Silurian pay zones, the Lego (Louisville), Laurel, and Brassfield. Still deeper reservoirs are the Ordovician Trenton–Black River–Chazy and the Cambro–Ordovician. Some of the Ordovician production is from fracture porosity. The *Decide* pool in Clinton County east of the arch produces from an Upper Ordovician reef limestone encased in younger shale (Perkins, 1955).

The phenomenal oil boom in Green County west of the arch axis that accounted for more than 10 million barrels in 1959 from depths of 350 to 500 feet started in early 1958 (Barnes, 1959). The reservoir is the Laurel Dolomite of Silurian age beneath the New Albany Shale. Porosity averages 12 per cent, and the thickness of the pay zone averages about 12 feet with a maximum of 25 feet. Spectacular as it was, the Green County boom was short lived; by 1965 the annual yield for this county had dropped to a low of 136,000 barrels. Nonetheless, the success of the Green County play led to many other shallow discoveries in south-central Kentucky (Nosow, 1960, p. 966), especially in Metcalfe, Adair, Cumberland, and Monroe Counties.

The barren middle and northern parts of central Kentucky do not have to remain unproductive just because a part of the stratigraphic column has been stripped off by erosion across the Jessamine (or Lexington) dome. A number of shallow gas zones have been found in shaly limestones of Middle Ordovician age; some of these are exploited locally. Potential reservoirs in the Cambrian, Cambro-Ordovician, Ordovician, Silurian, and even Devonian sections have adequate cover for oil accumulation outward from the center of this dome. Garrand County, south of Lexington, probably became productive in late 1968, when a wildcat flowed 7 million cubic feet of gas daily from a pay zone in the Rome (Cambrian) Sandstone at a depth of 4600 feet (Oil and Gas Jour., Sept. 9, 1968, Newsletter). Nosow (1961, p. 74) calls attention to shallow Devonian possibilities around the rim of the Lexington dome. Schmidt and Warner (1964) discuss Knox (Cambro-Ordovician) possibilities in northern Kentucky south of Cincinnati.

Western Kentucky. The oil fields in the western Kentucky coal basin resemble those to the northwest in southwestern Indiana and southeastern Illinois. This area is one of overthrust faulting and relatively sharp folding along the Rough Creek fault system, and igneous intrusions in the mineralized area (Rehn, 1956, 1968; Mullins, 1968). Away from the Rough Creek fault system, the Kentucky sector of the Illinois basin has gentle folds and normal faults similar to those of adjacent Indiana and Illinois. The reservoir rocks (Figure 7.5) are Pennsylvanian and Mississippian sandstones and porous oölitic Mississippian limestones (Fugate, 1956). The Pennsylvanian pay zones are chiefly channel sandstones in the basal section that were deposited on an erosion surface at the top of the Mississippian rocks (Clark, 1961, Figure 2). More than a dozen oil and gas pays occur in the upper, middle, and lower Chesterian Series in the Mississippian. One of these is the Bethel channel sand, which extends across western Kentucky for more than 100 miles. At least seven fields occur within the distributary network of this channel system (Reynolds and Vincent, 1967). The slightly older Ste Genevieve, also in the Upper Mississippian, contains at least five oölitic and sandy pay zones.

Accumulation is largely anticline controlled, but the distribution across the structure may depend on local porosity, especially in the sandstones (Bruce, 1949; Ingham, 1948). Some of the anticlines have faults on the flanks (Wood, 1955) or are bounded by faults (Avila

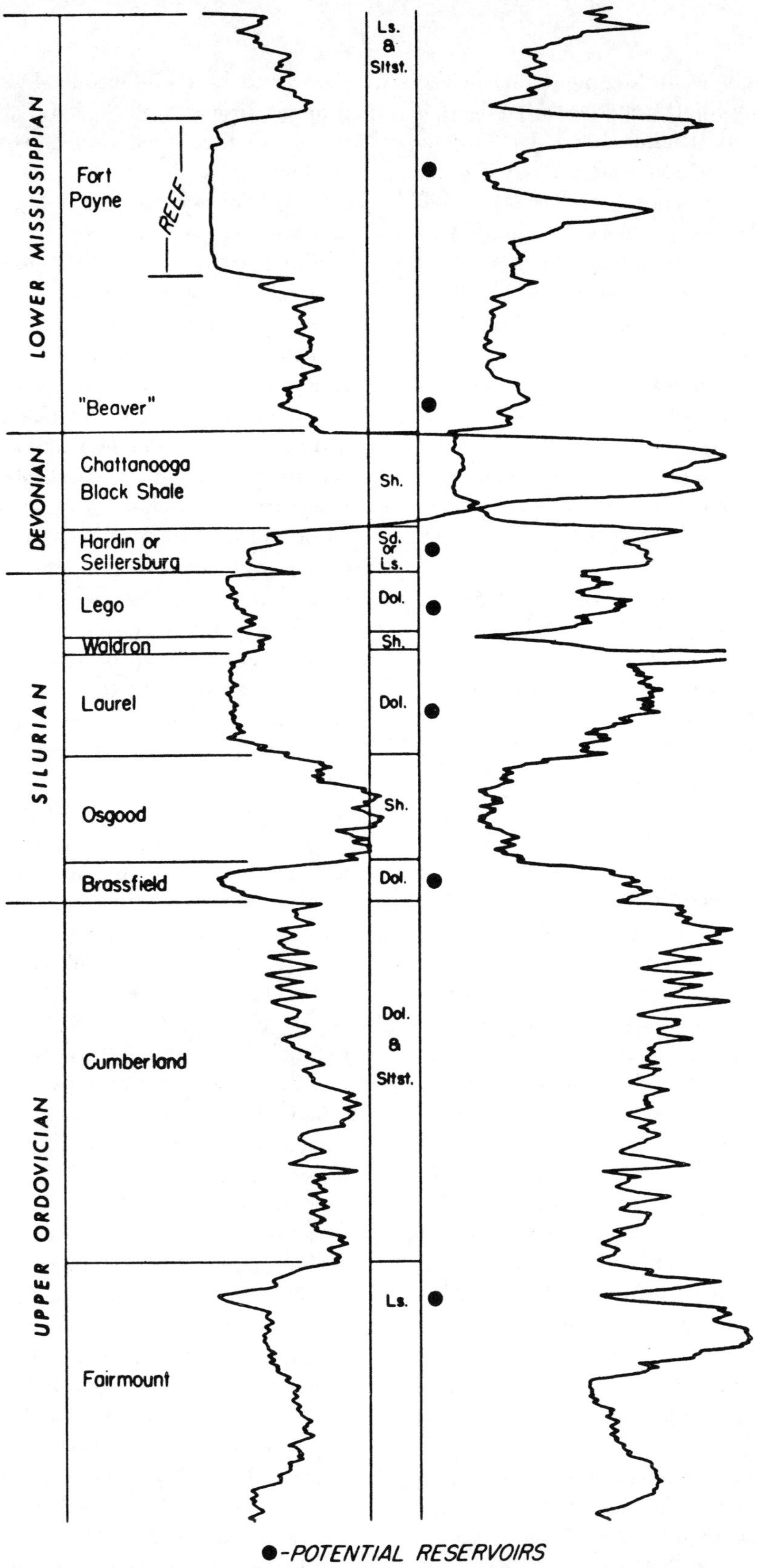

FIGURE 7.4 South-central Kentucky stratigraphy. Composite log for northern Monroe County, showing potential reservoirs. *Courtesy Hunt Perkins (1968, Figure 2, p. 48) and the Kentucky Geological Survey.*

et al., 1963). McGuire (1965b) has noted that the broad areas of production in western Kentucky are generally associated with major gravity highs that probably reflect "deep faulting or other basement tectonics."

The largest known productive anticline in western Kentucky is the *Hitesville* (Bybee, 1948) dome in Union County (Figure 7.6). This structure is irregular in plan and has a closure of less than 100 feet. The anticline measures 5 by 6 miles and the oil field about 2 by 5 miles. It was discovered in 1943 by the third well drilled to test a structure that had been explored previously by both core drill and seismograph. The structure map shown is contoured on the McCloskey D zone, which is the most important oil-producing unit in the section. Seven separate porous zones occur in the McCloskey Member of the Ste Genevieve (Mississippian). In addition the Chesterian Series of the Upper Mississippian contains five oil reservoirs, and a little production has been obtained from a Pennsylvanian sandstone. The reservoirs at Hitesville are either large in area, with the oil partly water driven, or are discontinuous lenses with dissolved gas drive. Accumulations in the latter type of reservoir are erratically distributed across the Hitesville structure.

During the period from 1963 to 1965 Hopkins and surrounding counties south of the Rough Creek fault zone in western Kentucky experienced a drilling boom that had considerable success (Barnes, 1963). During 1963 alone nearly 300 exploratory tests were drilled, resulting in the discovery of 17 oil and gas pools and 52 new reservoirs or extensions (Rose, 1965a). For many years the Rough Creek fault zone had been considered to be the southern limit of Illinois basin production (Rose, 1965b); hence this activity demolished one more mental barrier to discovery and development. Perhaps the next barrier to go will be the one limiting the stratigraphic depth of the potential pay zones to the Upper Mississippian (Stanonis, 1968).

As in eastern Kentucky, a considerable portion of the current production in western Kentucky can be attributed to waterflood operations in the older fields.

JACKSON PURCHASE. The westernmost tip of Kentucky, veneered by Tertiary and Cretaceous clastic sediments of the Mississippi embayment, is the second area in size (after north-central Kentucky) without oil or gas production. One well drilled through this veneer in Calloway County found a few hundred feet of Devonian and Silurian sedimentary rocks overlying a thick section of Ordovician, including Trenton or Kimmswick (Sargent, 1964) and Black River, and over 1400 feet of Cambro-Ordovician Knox and older rocks (Nosow, 1961, p. 72–74). Only a start has been made in exploring the Paleozoic section in the Jackson Purchase region, for so far only about three dozen test wells have been drilled in this eight-county area.

Period	Series	Group	Formation
PENN	Upper Middle Lower		Henshaw Lisman Carbondale Tradewater Caseyville
MISSISSIPPIAN	Chesterian Meramecian Osagian Kinderhookian		Grove Church Kinkaid Degonia Clore Palestine Menard Waltersburg Vienna Tar Springs Glen Dean Hardinsburg Golconda Cypress Paint Creek Bethel Renault Aux Vases Ste Genevieve St. Louis Salem Warsaw Fort Payne New Providence Rockford
DEVONIAN			New Albany Beechwood Sellersburg Grand Tower Clear Creek Backbone Bailey
SILURIAN	Cayugan-Niagaran Alexandrian		Decatur Brownsport Louisville Waldron Laurel Osgood Brassfield
ORDOVICIAN	Cincinnatian Champlainian Canadian	Knox	Maquoketa Kimmswick Decorah Plattin Pecatonica Joachim Dutchtown St. Peter Everton Black Rock Smithville Powell Cotter Jefferson City Roubidoux Gasconade
CAMB	St. Croixan		Eminence Potosi Elvins Bonneterre Lamotte

FIGURE 7.5 Western Kentucky stratigraphy. Nomenclature used in Kentucky sector of the Illinois basin. ***Courtesy Illinois and Indiana-Kentucky Geological Societies (1968, p. 22).***

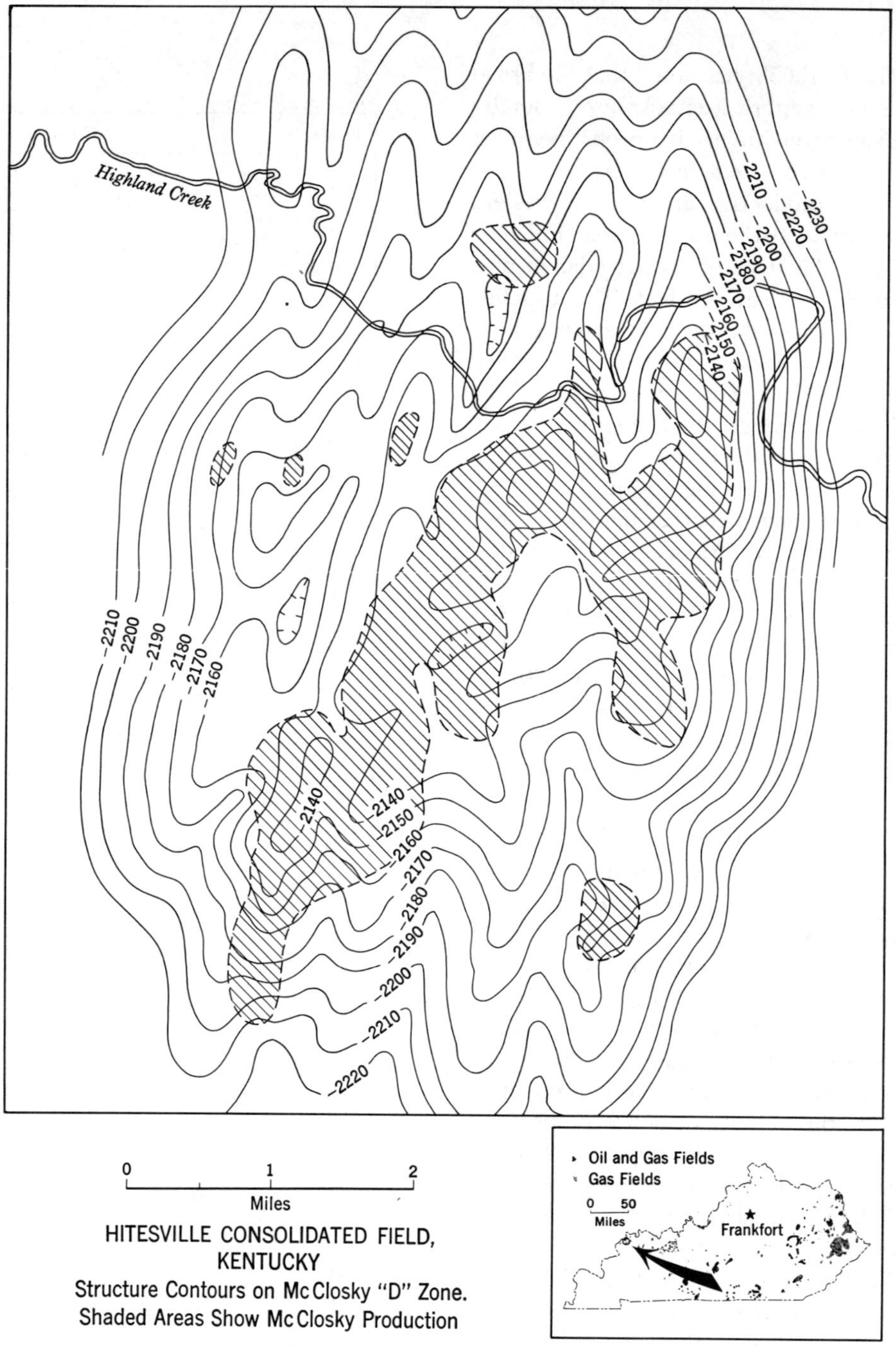

FIGURE 7.6 Hitesville field, Union County. Structure map. Contour interval, 10 feet. ***After H. H. Bybee (1948). Courtesy American Association of Petroleum Geologists.***

REFERENCES CONSULTED IN PREPARING CHAPTER 7

Avila, John, Barnes, John M., and Nuttall, Brandon (panelists), 1963.

Barnes, John M., 1959, 1963.

Bergin, M. J., 1956.

Billingsley, J. E., 1934.

Bruce, Clemont H., 1949.

Bybee, H. H., 1948.

Clark, R. E., 1961.

Freeman, Louise Barton, 1941, 1951.

Fugate, George W., Jr., 1956.

Greenfield, R. E., 1958.

Hagan, Wallace W., 1960.

Hunter, Coleman D., 1962.

Ingham, W. I., 1948.

McGuire, William H., 1965a, 1965b, 1968.

Mullins, Allan T., 1968.

Nosow, Edmund, 1959, 1960, 1961.

Perkins, Jerome Hunt, 1955, 1968.

Rehn, E. E., 1956, 1968.

Reynolds, Douglas W., and Vincent, James K., 1967.

Rose, William D., 1965a, 1965b, 1968.

Sargent, Don C., 1964.

Schmidt, R. G., and Warner, R. A., 1964.

Stanonis, Frank L., 1968.

Thomas, George R., 1960.

Thomas, George R., and Nosow, Edmund, 1960.

Van Den Berg, Jacob, Swann, D. H., Carpenter, G. L., Milhous, H. C., and Nosow, Edmund, 1968.

Welch, Stewart W., 1955.

Wilson, Edward N., 1968.

Wood, E. Boyne, 1955.

Youse, Arthur C., 1964.

Chapter

8

OHIO

Production in Ohio, which at one time was the foremost oil-producing state in the Union, has dwindled both actually and relatively to such an extent that in recent years this state has produced only 0.2 to 0.6 per cent of the country's total. Gas production in Ohio in 1967 was 0.2 per cent of the U. S. output. Oil was first produced commercially in 1860 in Washington County (Figure 8.1) in the southeastern corner of the state. Continuous production started in 1876. Discovery of gas in what soon became the great Lima-Indiana oil and gas field in northwestern Ohio occurred in 1884. According to Orton (1888), oil seeps into wells and excavations in the Findlay area antedated discovery by 50 years. The Ohio part of the Lima-Indiana field reached maximum production in 1896, when the Trenton (Ordovician) reservoir rock yielded more than 20 million barrels of oil and the total for the state peaked at approximately 24 million barrels. The next major discovery in Ohio was that of the Clinton gas field in 1887. In recent years the most active exploration has been in central Ohio, where reservoir objectives have ranged in age from Mississippian to Cambrian. The most sensational result of this search has been the discovery of commercial oil in pre-Trenton rocks. The central Ohio boom, which started in 1961, raised the state production figure to a second peak of nearly 16 million barrels in 1964. The yield in 1967 was just under 10 million barrels.

Tectonics. The greater part of Ohio lies on the west limb of the Appalachian geosyncline. The structurally deepest rocks are in the southeastern corner of the state, which is closest to the axis of the Appalachian trough (Figure 8.2). From this area the strata rise to the north and especially to the west toward the axis of the Findlay arch, the righthand fork of the Cincinnati arch, which crosses western Ohio from south to north. In northern Ohio the Findlay arch turns northeast and forms the backbone of the southwestern Ontario peninsula. The Cincinnati arch complex began its major upwarping in the Silurian period (Scotford, 1964). The crest is wide,

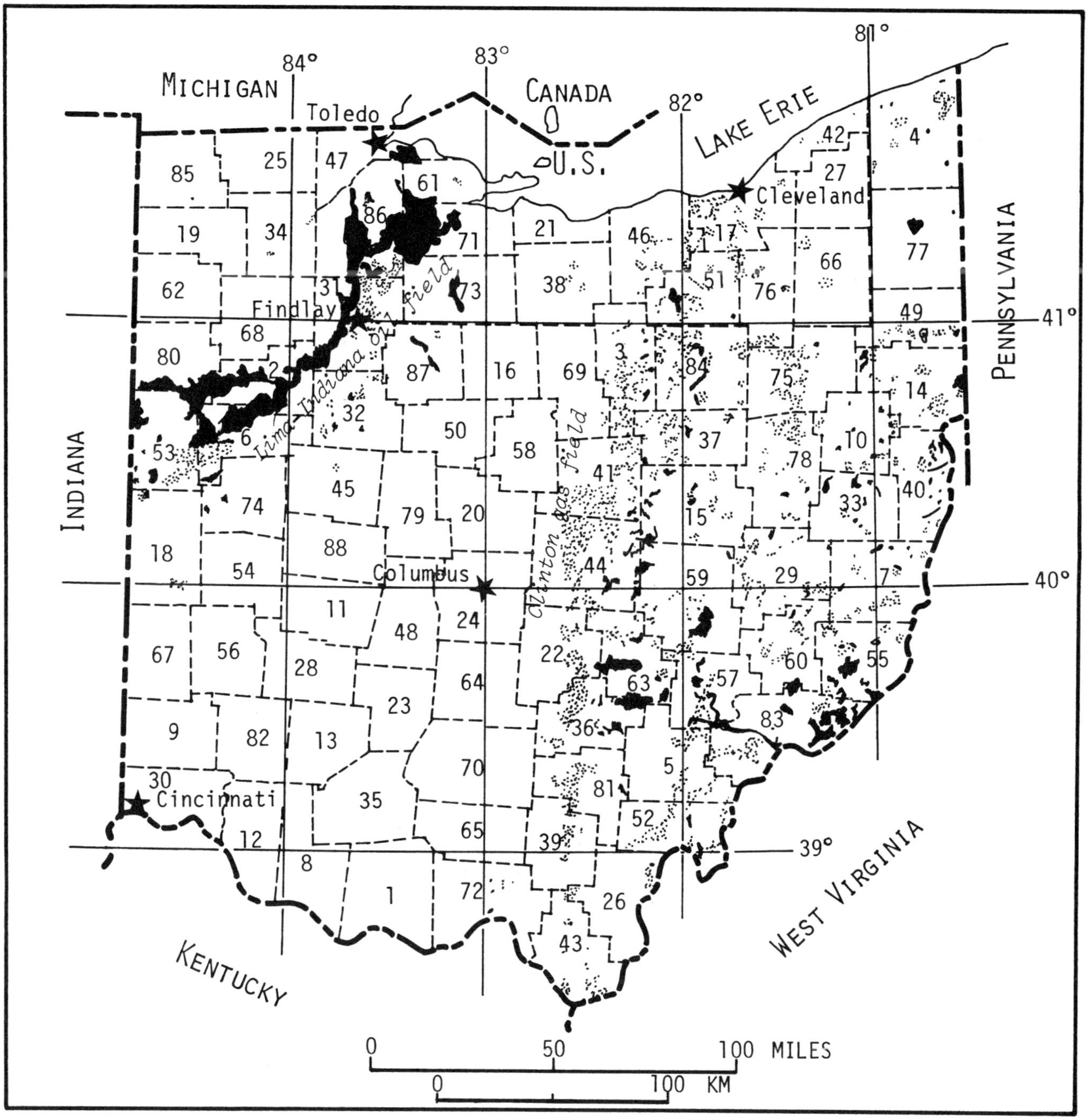

FIGURE 8.1 Ohio. Index map. County Key:

1	Adams	13	Clinton	26	Gallia	39	Jackson	52	Meigs	64	Pickaway	76	Summit
2	Allen	14	Columbiana	27	Geauga	40	Jefferson	53	Mercer	65	Pike	77	Trumbull
3	Ashland	15	Coshocton	28	Greene	41	Knox	54	Miami	66	Portage	78	Tuscarawas
4	Ashtabula	16	Crawford	29	Guernsey	42	Lake	55	Monroe	67	Preble	79	Union
5	Athens	17	Cuyahoga	30	Hamilton	43	Lawrence	56	Montgomery	68	Putnam	80	Van Wert
6	Auglaize	18	Drake	31	Hancock	44	Licking	57	Morgan	69	Richland	81	Vinton
7	Belmont	19	Defiance	32	Hardin	45	Logan	58	Morrow	70	Ross	82	Warren
8	Brown	20	Delaware	33	Harrison	46	Lorain	59	Muskingum	71	Sandusky	83	Washington
9	Butler	21	Erie	34	Henry	47	Lucas	60	Noble	72	Scioto	84	Wayne
10	Carroll	22	Fairfield	35	Highland	48	Madison	61	Ottawa	73	Seneca	85	Williams
88	Champaign	23	Fayette	36	Hocking	49	Mahoning	62	Paulding	74	Shelby	86	Wood
11	Clark	24	Franklin	37	Holmes	50	Marion	63	Perry	75	Stark	87	Wyandot
12	Clermont	25	Fulton	38	Huron	51	Medina						

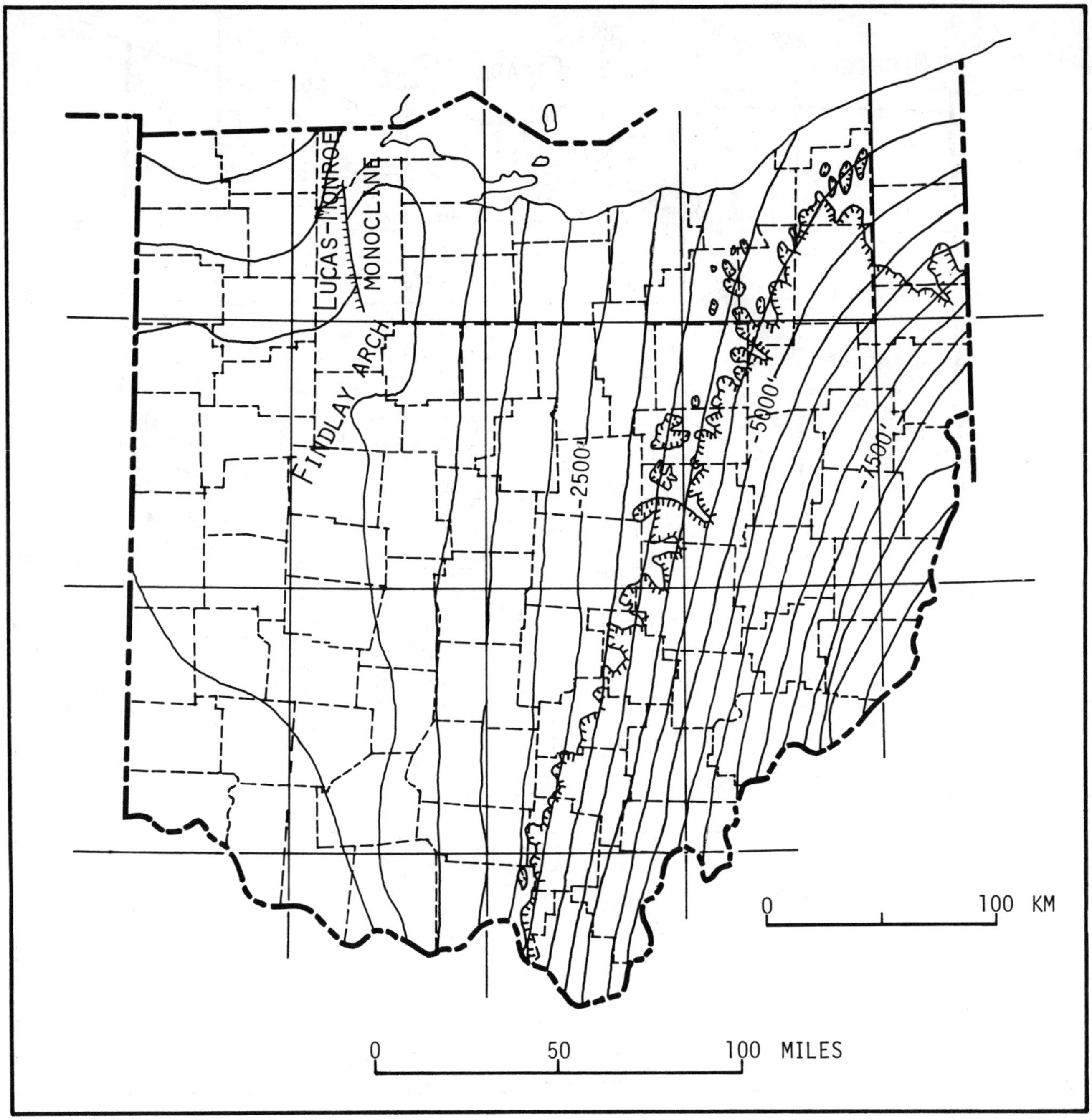

FIGURE 8.2 Structure map of Ohio contoured on top of Trenton. Interval, 500 feet. Dotted line is eastern edge of Pennsylvania sedimentary rocks at the bedrock surface. Hachured lines are faults. *After George V. Cohee et al. (1962).*

and dips are exceptionally gentle. There is a growing tendency to refer to this feature as the Indiana-Ohio platform (Owens, 1967); it was so named by Green (1957).

The oil fields of Ohio can be divided geographically into two groups. In the central and eastern part of the state the accumulations are associated with the Appalachian geosyncline, and the fields belong geologically with those of Pennsylvania and West Virginia. However, there are differences between the local structures on the west limb of this geosyncline and those to the east in the axial zone. In the latter area the strata are folded into prominent parallel anticlines and synclines, whereas in eastern Ohio (outside of Columbiana County) anticlines are few and are mostly of the plunging type with a southeast trend. Most of the accumulations of oil and gas in the other group of fields, in the western third of Ohio, occur along the highest parts and down the west flank of the Findlay arch.

Reservoir Rocks. The reservoir rocks in Ohio (Figure 8.3) are both sandstones and limestones (dolomites) ranging in age from Pennsylvanian to Cambrian. The youngest are Carboniferous sandstones, usually grouped together in statistical tables as "shallow sands." They occur only in southeastern Ohio and are similar to the reservoirs of the same age in adjoining West Virginia.

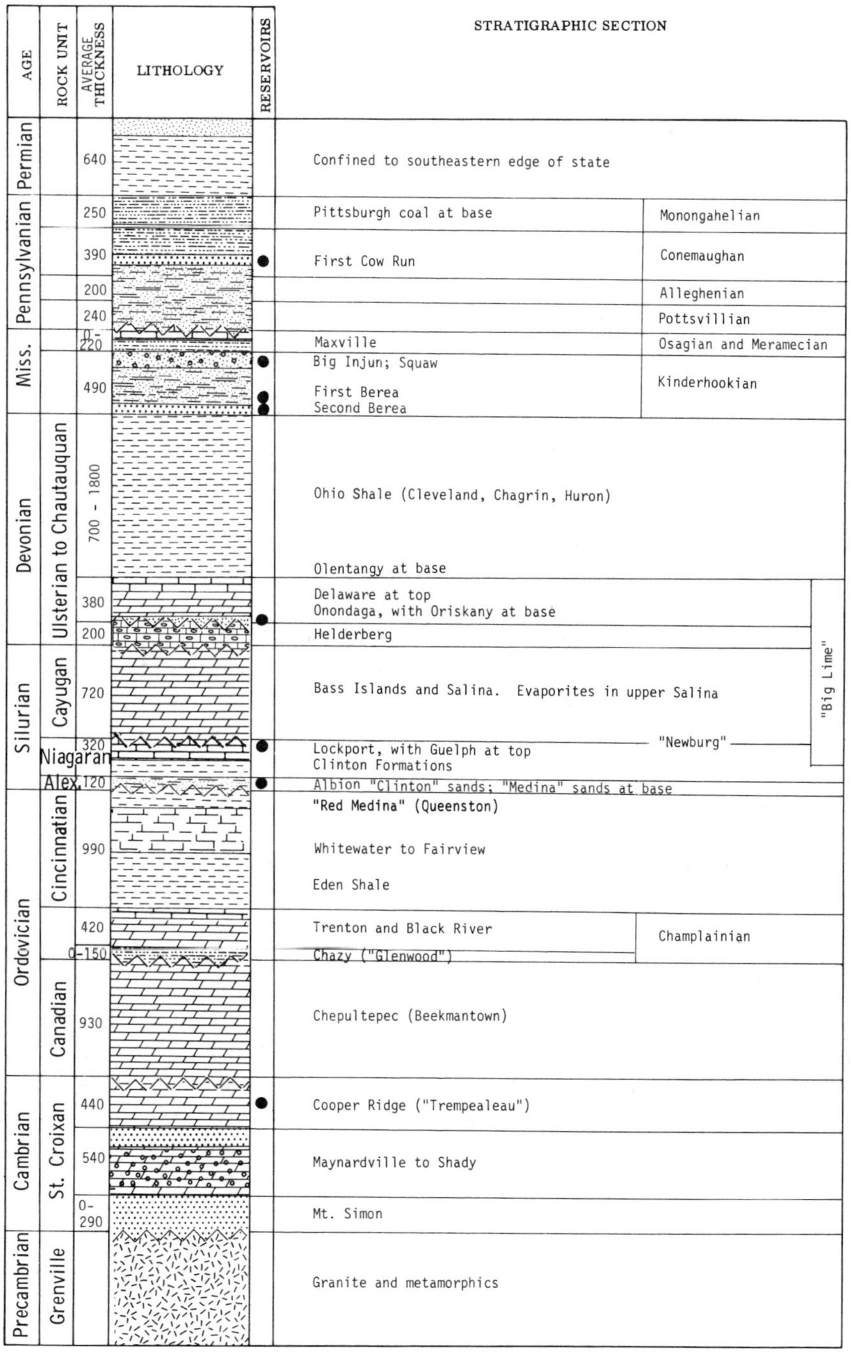

Note: the same reservoir symbol is used for both oil and gas.

Figure 8.3 Stratigraphic column in Ohio, showing oil and gas reservoirs, by Warren L. Calvert, 1960. Alex. = Alexandrian. *Courtesy Ohio Geological Survey.*

The Lower Mississippian Berea sandstones (the "Second Berea" is a sand at the base of the Bedford Shale) and older rocks rise to within reasonable drilling depths in northern and central Ohio. Much of the older production in northern Ohio came from the Berea sands (Pepper et al., 1954; de Witt, 1949); today most Berea production is due to secondary recovery by waterflooding.

The erratically distributed Oriskany Sandstone (Hall, 1952) of the Lower Devonian produces both oil and gas in a number of fields in eastern Ohio. Older and deeper is the "Newburg sand," which is in reality porous dolomite, lying within or immediately above the Niagaran Group of Silurian age. This reservoir contains oil and gas in commercial amounts in north-central Ohio. In the Lower Silurian is the "Clinton" sandstone reservoir (Pepper et al., 1955; Alkire, 1952; Swain, 1950), which actually occurs in the Albion (Medina) Group underlying the true Clinton. Long famous as a gas reservoir, this formation is still a major objective of wells drilled in Ohio. Not only gas but also "Pennsylvania grade" crude oil are the potential rewards of this exploration. Hocking, Portage, Stark, and Ashtabula Counties have been areas of profitable "Clinton" discovery and development in recent years.

The western Ohio pools on the Findlay arch produce chiefly from the Trenton Formation of Ordovician age. Rooney (1966) states that the top of the Trenton is an erosional unconformity across the Lima-Indiana field and to the north and west; to the south and east of the field there appears to have been continuous deposition. The exploration of pre-Trenton rocks (Sitler, 1963; Calvert, 1962; Summerson, 1962; Shearrow, 1959; Wasson, 1932) in western Ohio resulted in the discovery of oil in relatively small amounts as far back as 1909 (Sitler and Wehmeyer, 1962, p. 111), but it was not until the period from 1959 to 1961 that enough oil was found in these rocks in Ohio and southwestern Ontario (Burgess, 1963) to spur an intensive and successful search that will be described subsequently.

Central and Eastern Ohio. Because of the thicker section resulting from the regional dip to the southeast into the Appalachian geosyncline, southeastern Ohio has many more known and possible reservoirs than has central and western Ohio. Thus Morgan County has 16 known oil and gas sands (Magbee and Alkire, 1954, p. 8).

Trapping in eastern Ohio is due largely to the updip wedgout of reservoir sands on the flank of the eastward dipping monocline (O'Rourke, 1941). Most of the Oriskany gas fields, for example, which are confined by the distribution of the Oriskany to eastern and northeastern Ohio, occur along the west boundary of the sandstone where it wedges out updip (Finn, 1949). In a few Oriskany gas fields the trapping has been anticlinal (Lockett, 1948). Accumulation is also anticlinal in the *Mayfield* pool (Cuyahoga County), which produces gas from the porous "Newburg" dolomite. This large gas field, which was discovered in 1938 and which lies only 16 miles to the east of the center of Cleveland, occupies a broad structural dome (Rothrock, 1949).

The elongate *"Clinton" gas field* (Figure 8.1), which extends from Lawrence County in southern Ohio northward to Lake Erie in Cuyahoga County, owes its existence to the updip shaling of the Medina reservoir rock. The *Moreland* oil field illustrates two different types of variable permeability trapping in two separate reservoirs (Multer, 1963). This field is in central southern Wayne County, crossing into northern Holmes County. The lower reservoir, the "Clinton," produces oil and gas from a pool 12 miles long (north-south) and $\frac{1}{2}$ mile (average) wide. The pays in the Clinton are elongate bodies, probably originally offshore bars, of fine sandstone interbedded with shale. Both intergranular and fracture porosity are present. The younger reservoir, the "Newburg," is vuggy, fossiliferous, and fractured dolomite occurring in zones in the lower "Big Lime," another Ohio drillers' term. The Newburg pays lie in the lower Cayugan–Upper Niagaran section of the Upper Silurian. The Newburg pool at Moreland overlaps the east side of the northern end of the Clinton pool. It produces gas exclusively. Trapping is due to the lenticularity of the porous dolomite zones.

It was not until 1961 that Ohio began to produce significant quantities of oil from pre-Trenton rocks. The year before, reconnaissance gravity, magnetic, and seismic surveys were made in Morrow County (Stanton, 1966). Morrow County is in central Ohio, between the Lima-Indiana field to the west and the productive belts to the east, in hitherto nonproductive country. The Morrow County play was triggered in June, 1961, by a wildcat, the No. 1 Myers, drilled in the central-western part of the county, which came in for 200 barrels daily under choke of 39° API oil from a depth of about 3000 feet. The reservoir, Cambrian dolomite, had an oil-saturated pay zone 123 feet in thickness (Calvert, 1964a). The field thus discovered, the *Denmark*, was soon surrounded by other fields, especially to the south, east, and northeast (Figure 8.4). By March, 1964, the daily production from this area had risen to more than 30,000 barrels (Adams, 1964).

The dolomite reservoir in Morrow and adjacent counties is the Copper Ridge (or Trempealeau) at the top of the Cambrian (Calvert, 1964b). It has both vuggy and fracture porosity (Sutton, 1965). Accumulation has taken place in leached, weathered, and buried hills, erosion remants of Upper Cambrian dolomite that were submerged beneath the Early Ordovician sea and covered by Glenwood Shale and younger formations

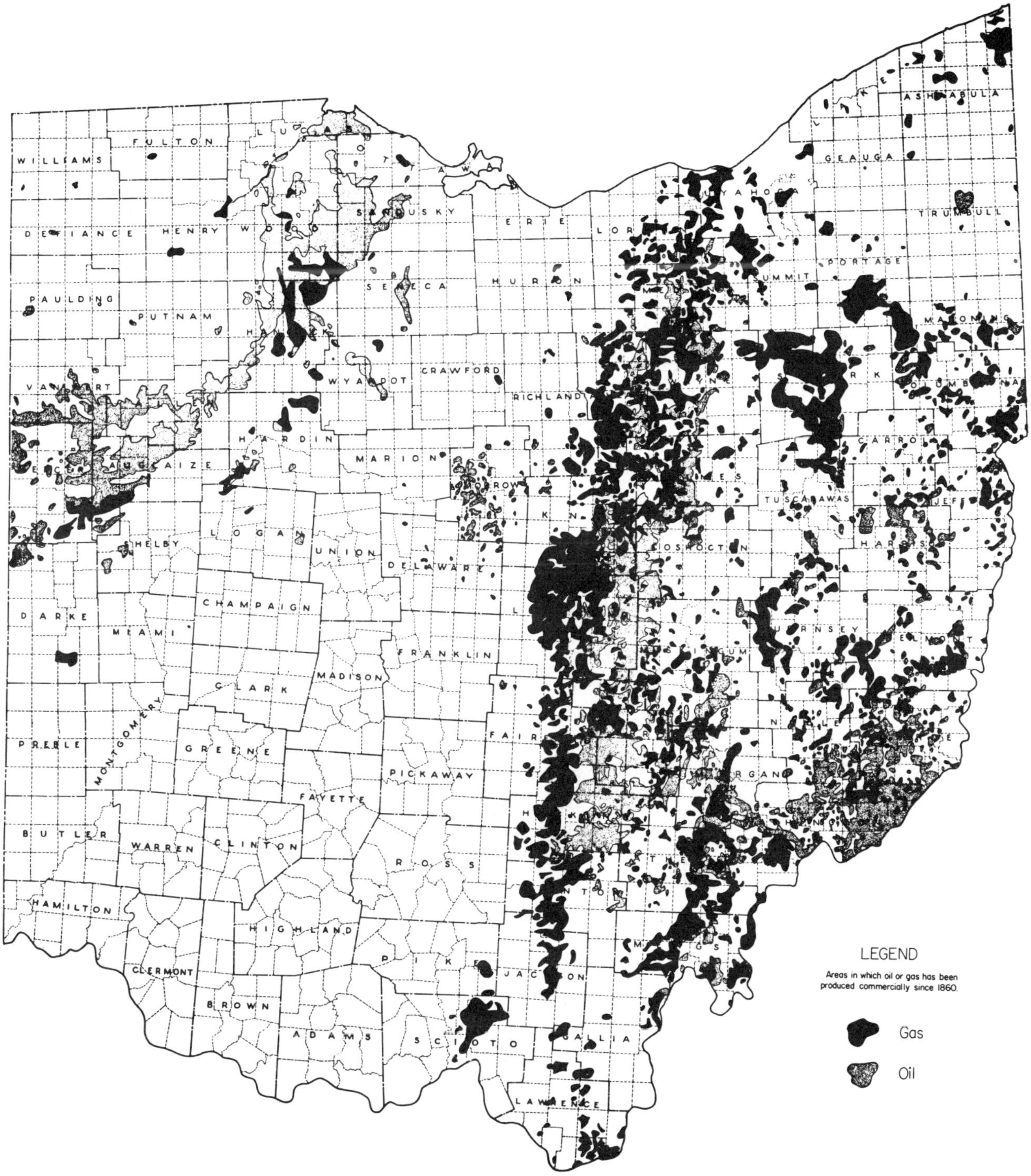

COURTESY
STATE OF OHIO
DEPARTMENT OF NATURAL RESOURCES
DIVISION OF GEOLOGICAL SURVEY, 1966

OIL & GAS FIELDS MAP OF OHIO

10 0 10 20 30 miles
Scale

FIGURE 8.4

which supplied the essential seal. The relief of the top of the reservoir across these accumulations may be more than 100 feet (Burgess, 1963, p. 128).

Western Ohio. The Trenton production in the Lima-Indiana district is due to irregularly distributed dolomitization porosity enlarged by weathering before the Cincinnatian submergence. The map (Figure 8.5) of the Lima-Indiana oil and gas district shows that the hydrocarbons have accumulated along a broad belt extending from the top of the Kankakee arch in eastern Indiana northeastward into northwestern Ohio, crossing the axis of the Findlay arch and continuing some distance down the east flank (Carman and Stout, 1934, p. 522). Carman and Stout (1934, p. 528) have noted a very close correlation between magnesium content and oil production. No oil is found in the Trenton in this area unless the magnesium carbonate content is at least 20 per cent, and the best production occurs where the rock is a true dolomite. Most of the oil and gas in the Lima-Indiana district is found in the uppermost 20 to 30 feet of the dolomitized Trenton Limestone. Northwestern Ohio has produced about three-fourths of the total oil output of this district, and eastern Indiana the remainder.

Future. The continuing discovery of new oil and gas fields in Ohio testifies to the probability that the state's resources of these mineral fuels are far from exhausted. The recent pre-Trenton discoveries mean that virtually the entire western half of the state, where most of the surface rock is older than the reservoirs in eastern Ohio, is fair game for intensive hunting. The equivalents of these older formations are also present in eastern Ohio, but at considerably greater depth (Ohio Geol. Survey, 1959).

The Cambrian discoveries in central Ohio encouraged drilling into these reservoirs both to the north and to the south. By the end of 1968 the belt of Cambrian fields had been extended north to Erie County and across Lake Erie into Ontario. Much infilling remains to be done

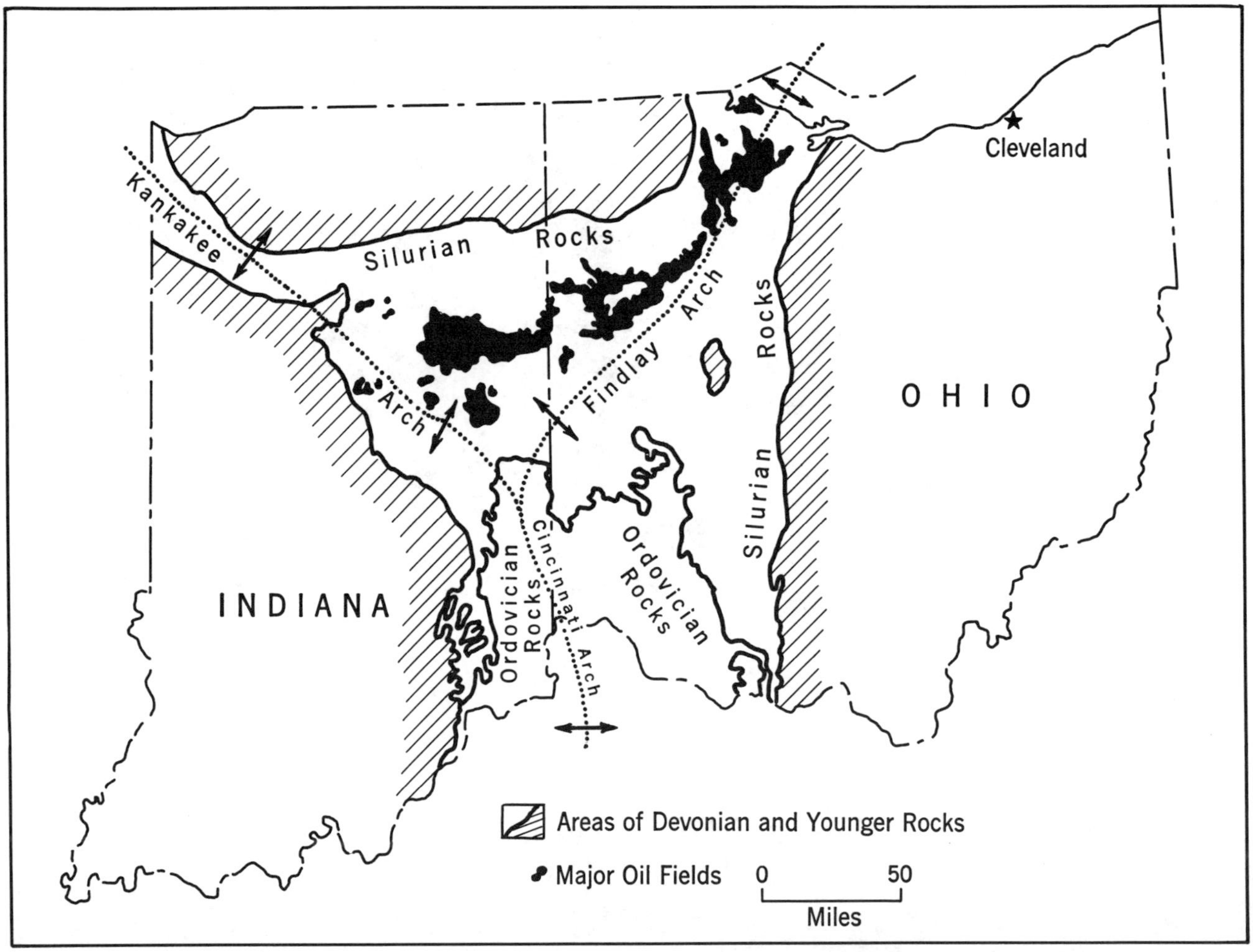

FIGURE 8.5 Lima-Indiana district, Indiana and Ohio, showing location of major oil fields in respect to regional structure and areal geology.

between Morrow County and Lake Erie, and to the south, from Morrow County to the Ohio River.

Owens (1967, p. 3) calls attention to the hinge belt that separates the Appalachian basin from the Indiana-Ohio platform. He suggests a possible concentration of structural, stratigraphic, and reef traps along this belt. The hinge was in existence in the Precambrian; the line is drawn from Sandusky Bay to Clermont County on the basis of differences in lithologies in the basement rock (McCormick, 1961, Figure 1, p. 2).

Subsurface studies by Calvert (1964a) show that the sub-Trenton rocks (Knox and older Cambrian formations) that crop out in Virginia and Tennessee wedge out and are truncated to the north to a thin section of basal Paleozoic arkose in southwestern Ontario. This means that successively older potential Cambrian reservoirs subcrop beneath the Knox unconformity from south to north across central Ohio. Hydrocarbon accumulations are possible beneath the unconformity where anticlinal nosing or laterally lenticular permeability occurs.

The erratic Rose Run sand in the Lower Ordovician disappears completely updip on the west flank of the Appalachian basin in southeastern Ohio, according to Shearrow and Preston (1965). No doubt most of the stratigraphic traps in this area are yet to be found.

For many decades householders and small industries in Ohio have burned shale gas obtained at relatively shallow depths from holes bored into the Upper Devonian Ohio Shale (Figure 8.3). This gas is stored in fractures, and the wells, although of limited daily output, are remarkably long lived. A systematic program of hydraulic fracturing of the more promising zones in this shale section of more than 1300 feet in eastern Ohio might expand materially the profitable production of shale gas in the state (Shearrow, 1968).

Offshore exploration in shallow Lake Erie, already successful on the Canadian side of the lake, should expand the Ohio reserves. Especially inviting is the possible sublake extension of the Clinton gas field (Donnan, 1957).

During 1965 there were 31 waterflood and 18 gas injection recovery projects in operation (Oil and Gas Jour., Nov. 8, 1965, p. 148–149). Principal pay zones being swept were the Berea and the Clinton. No secondary recovery projects are underway in the Trenton reservoir, in spite of the probability that more oil remains in the Lima–Indiana field than has been removed.

REFERENCES CONSULTED IN PREPARING CHAPTER 8

Adams, John, 1964.
Alkire, Robert L., 1952.
Burgess, R. J., 1963.
Calvert, Warren L., 1962, 1964a, 1964b, 1964c.
Carman, J. Ernest, and Stout, Wilber, 1934.
de Witt, Wallace, Jr., 1949.
Donnan, Bryson C., 1957.
Finn, Fenton H., 1949.
Green, Darsie A., 1957.
Hall, John F., 1952.
Lockett, J. R., 1948, 1968.
Lytle, William S. et al., 1968.
Magbee, Byron D., and Alkire, Robert L., 1954.
McCormick, George R., 1961.
Multer, H. Gray, 1963.
Norling, Donald L., Bendler, E. Perry, Buschman, William J., Jr., Johnson, Allan W., 1968.
Ohio Geological Survey, 1959.
O'Rourke, E. V., 1941.
Orton, Edward, 1888.
Owens, Gordon L., 1967.
Pepper, James F., de Witt, Wallace, Jr., and Demarest, David F., 1954.
Pepper, James F., de Witt, Wallace, Jr., and Everhart, Gail M., 1955.
Rooney, Lawrence F., 1966.
Roth, Ernest E., 1968.
Rothrock, Howard E., 1949.
Scotford, David M., 1964.
Shearrow, George G., 1959, 1968.
Shearrow, George G., and Preston, Arthur F., 1965.
Sitler, Guy, 1963.
Sitler, Guy F., Jr., and Wehmeyer, Karl L., Jr., 1962.
Stanton, E. C., Jr., 1966.
Summerson, Charles H., 1962.
Sutton, Emmett, 1965.
Swain, James F., 1950.
Wasson, Isabel B., 1932.

Chapter

9

INDIANA

Recent annual oil production in Indiana has been in the neighborhood of 0.3 per cent of the country's total. Oil was discovered on the Indiana side of the Ohio–Indiana boundary in Trenton Dolomite in 1889, 3 years after the initial discovery of the Lima–Indiana district in northwestern Ohio. The state reached its first production peak in 1904, after which the Trenton pools began to decline. However, the east-central Indiana Trenton district still accounts for more than one-fourth of the state's cumulative oil production. The southwestern Indiana oil district, which in prior years had never yielded any significant volume of oil, became productive in 1937, when the Illinois basin boom got under way. Since then many new oil pools have been discovered in that area (Figure 9.1), with the result that a new record was established in 1953, when nearly 13 million barrels of oil was produced.

There is still some successful probing of the Trenton reservoir in northeastern Indiana, and a few small producers (especially of gas) are drilled each year in that area.

Geology. The regional geology of Indiana is dominated by the Kankakee arch (Figure 9.2), the northwest prong of the Cincinnati arch. The strata dip to the southwest from the axis of the Kankakee arch into the Illinois basin, which reaches its lowest point in southern Illinois not far west of the southwestern tip of Indiana. The structural environment of southwestern Indiana is best seen on two illustrations, Figures 10.2 and 10.3, in the next chapter. These give, by structure contour map and cross section, respectively, the regional picture of the Illinois basin. The Eastern shelf of the basin is in southwestern Indiana (Figure 9.2).

The Illinois basin is interrupted in eastern Illinois by the La Salle anticline, and the southwestern Indiana oil fields are largely concentrated in the "pocket" between Illinois and Kentucky where the axis of the anticline would cross the corner of Indiana if it continued this far south. However, at its southern end the La Salle

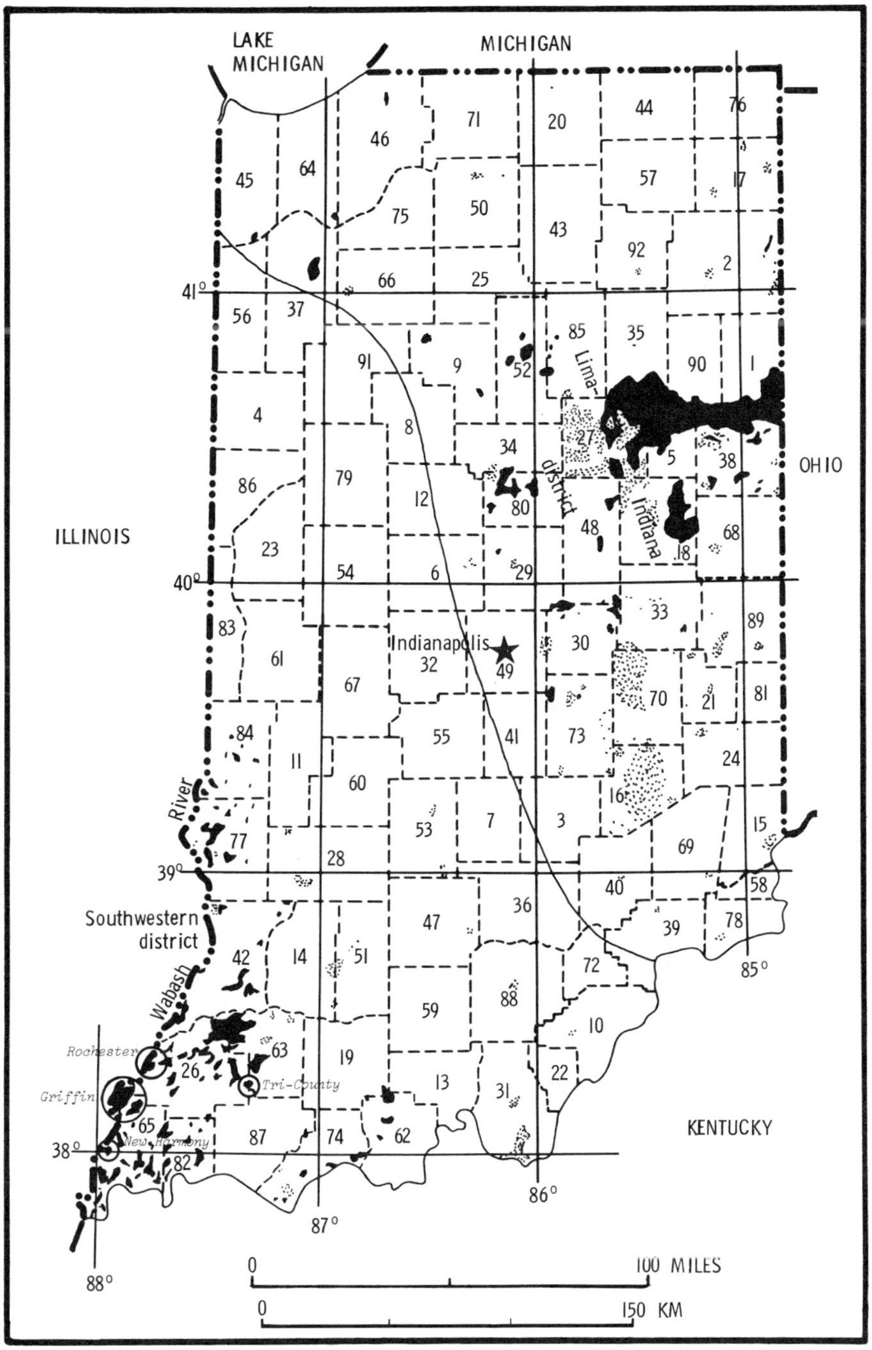

FIGURE 9.1 Indiana. Index map with oil and gas fields. Line separates Lima-Indiana and Southwestern districts. County Key:

1 Adams
2 Allen
3 Bartholomew
4 Benton
5 Blackford
6 Boone
7 Brown
8 Carroll
9 Cass
10 Clark
11 Clay
12 Clinton
13 Crawford
14 Daviess
15 Dearborn
16 Decatur
17 DeKalb
18 Delaware
19 Dubois
20 Elkhart
21 Fayette
22 Floyd
23 Fountain
24 Franklin
25 Fulton
26 Gibson
27 Grant
28 Greene
29 Hamilton
30 Hancock
31 Harrison
32 Hendricks
33 Henry
34 Howard
35 Huntington
36 Jackson
37 Jasper
38 Jay
39 Jefferson
40 Jennings
41 Johnson
42 Knox
43 Kosciusko
44 Lagrange
45 Lake
46 La Porte
47 Lawrence
48 Madison
49 Marion
50 Marshall
51 Martin
52 Miami
53 Monroe
54 Montgomery
55 Morgan
56 Newton
57 Noble
58 Ohio
59 Orange
60 Owen
61 Parke
62 Perry
63 Pike
64 Porter
65 Posey
66 Pulaski
67 Putnam
68 Randolph
69 Ripley
70 Rush
71 St. Joseph
72 Scott
73 Shelby
74 Spencer
75 Starke
76 Steuben
77 Sullivan
78 Switzerland
79 Tippecanoe
80 Tipton
81 Union
82 Vanderburgh
83 Vermillion
84 Vigo
85 Wabash
86 Warren
87 Warrick
88 Washington
89 Wayne
90 Wells
91 White
92 Whitley

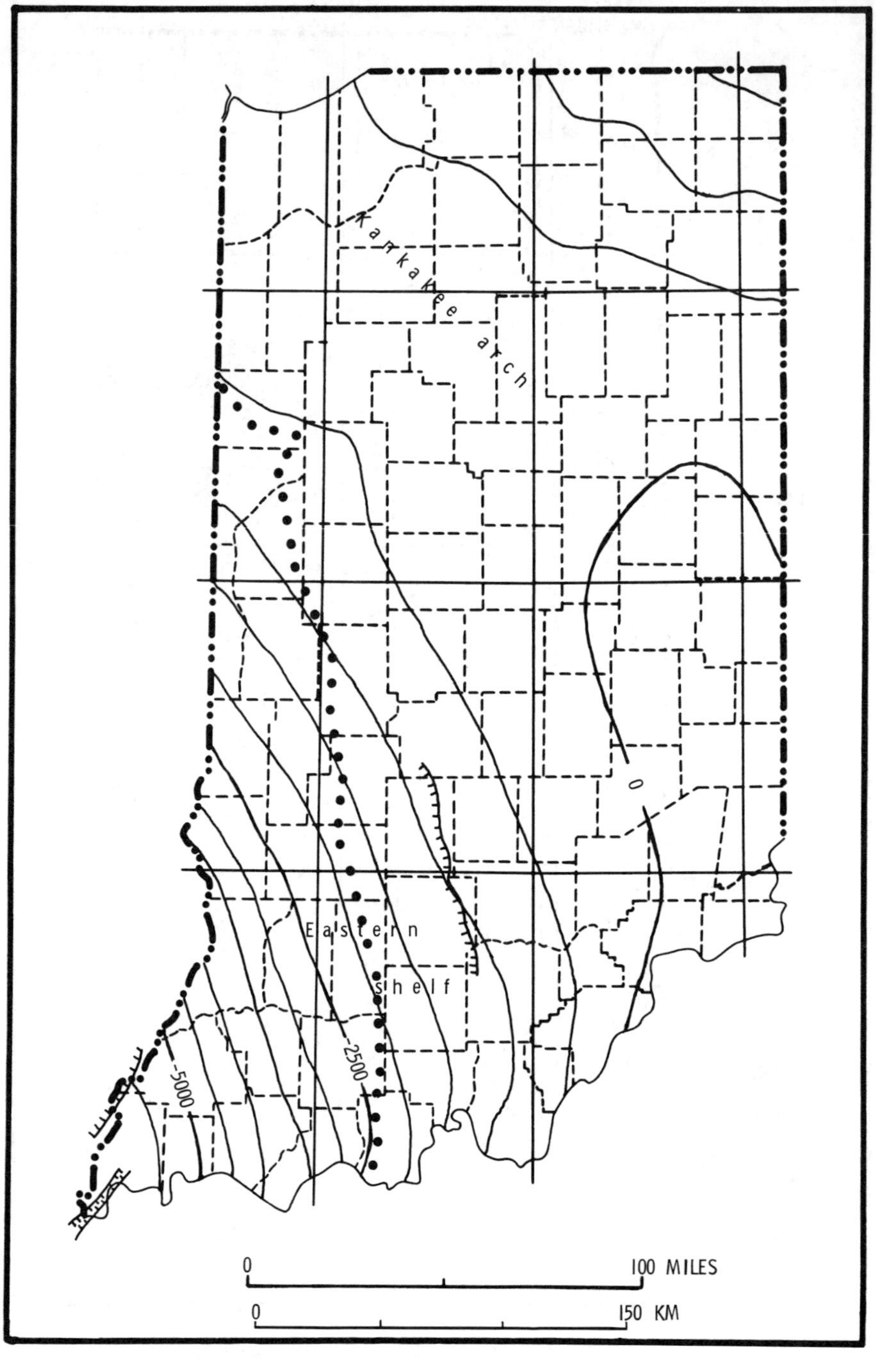

FIGURE 9.2 Structure map of Indiana contoured on top of Trenton. Interval, 500 feet. Dotted line is eastern edge of Pennsylvanian sedimentary rocks at the bedrock surface. Hachured lines are faults. ***After George V. Cohee et al. (1962).***

anticline noses downward rapidly and for the most part has lost its identity by the time it reaches the Wabash River.

The leading field in southwestern Indiana is *Griffin*, immediately east of the Wabash River in Gibson and Posey Counties. Discovered in 1938, the field had produced nearly 66 million barrels of oil by the end of 1967 and was still accounting for 23 per cent of the state's annual output.

Table 9.1 lists the known oil reservoir rocks in Indiana.

Table 9.1 Named Stratigraphic Units in Indiana Showing Oil-Producing Zones and Waterflooded Zones

System	Series	Formation or group	Lithology	Producing Zones	Waterflooded Zones
Pennsylvanian	Conemaugh	Undifferentiated	sh., ss., ls., coal		
	Allegheny	Undifferentiated	sh., ss., ls., coal		
	Pottsville	Undifferentiated	sh., ss., thin coal	x	x
Mississippian	Chester	Kinkaid Ls.	ls., sh.		
		Degonia Ss.	ss., sh.	x	x
		Clore Ls.	ls., ss., sh.	x	x
		Palestine Ss.	ss., sh.	x	x
		Menard Ls.	ls., sh.		
		Waltersburg Ss.	ss., sh.	x	x
		Vienna Ls.	ls., sh.		
		Tar Springs Fm.	ss., sh.	x	x
		Glen Dean Ls.	ls., sh.		
		Hardinsburg Fm.		x	x
		Golconda Ls.	ls., sh.		
		Big Clifty (Jackson) Fm.	ss., sh.	x	x
		Beech Creek (Barlow) Ls.	ls.		
		Cypress Fm.	ss., sh.	x	x
		Paint Creek Fm.	ls., ss., sh.	x	
		Bethel Fm.	ss., sh.	x	x
		Renault Fm.	ls., ss., sh.	x	x
		Aux Vases Fm.	ls., sh., ss.	x	x
	Meramec	Ste Genevieve Ls.	ls., sh., ss.	x	x
		St. Louis Ls.	ls., sh.	x	
		Salem Ls.	ls.	x	x
		Harrodsburg Ls.	ls., sh.	x	
	Osage	Borden Group	slt., sh., ls., ss.	x	
	Kinderhook	Rockford Ls.	ls.		
		New Albany Sh.	sh.		
Devonian	Undifferentiated	Undifferentiated	ls., dol.	x	
Silurian	Undifferentiated	Undifferentiated	ls., dol., sh.		
Ordovician	Cincinnatian	Undifferentiated	sh., ls.		
		Cynthiana Fm.			
	Mohawkian	Trenton Ls.	ls., dol.	x	
		Black River Ls.	ls., dol.	x	
	Chazyan	Joachim Dol.	dol.		
		St. Peter Ss.	ss.		
	Canadian	Knox Dol.	dol.		
Cambrian	St. Croixan	Eau Claire Fm.	slt., sh., ss., dol.		
		Mt. Simon Ss.	ss.		
Precambrian					

Carpenter and Smith, 1962, p. 5.

Nineteen of the 21 in the tabulation produce in southwestern Indiana. The reservoirs range in age from Pennsylvanian to Devonian, with most of the production coming from the 17 Mississippian reservoirs listed. A single field such as Griffin may have as many as eight superimposed pay zones, all within 3000 feet of the surface. The natural gas reservoirs are in the same stratigraphic units and in addition are in Pennsylvanian rocks above the youngest oil reservoir. The Pennsylvanian reservoirs and most of those in the Upper Mississippian are sandstones, whereas the older reservoirs are carbonate rocks. Included among the Mississippian limestone reservoirs that are productive in this area are the McClosky (Ste Genevieve) oölite zones, which are prolific producers of oil in the Illinois basin. Devonian production is obtained in Sullivan and Vigo Counties to the north of the "pocket" district.

Traps. East-central and northeastern Indiana has produced more than 100 million barrels of oil from the Ordovician Trenton and Black River Limestones. The trapping of oil in dolomitized zones in these limestones is described in the discussion of the Lima–Indiana field in Chapter 8 on Ohio.

In southwestern Indiana trapping in some pools is controlled or influenced by local structure, but in most pools accumulation has been brought about by the lenticularity of the sandstone reservoir rock. Gibson and adjacent counties in Indiana are blessed with oil-containing sand lenses encased in shale. Some of the sands are elongate, as much as 5 miles long, 2000 feet wide, and 50 feet in maximum thickness. These were probably offshore bars (Fritz, 1968). Stratigraphically the sands are classified as Big Clifty (Jackson Formation) in the Chester Series of the Mississippian. The Devonian fields occupy domes, two of which were found by utilizing coal drill core data, and others by seismograph surveys (Lundy, 1958). The anticlinal structure is the result of the "draping" of the Devonian reservoirs over Silurian reefs by differential compaction. In the *Tri-County* field and the *New Harmony* field, which overlaps the Illinois boundary, differential cementation of the sandstone reservoirs has been effective in limiting the distribution of oil, but accumulation within the producing fields has been due primarily to structure (Wanenmacher and Gealy, 1930; Cohee, 1942, p. 1606). The Waltersburg (Upper Mississippian) sand pool of the New Harmony field, however, is a sand bar, with trapping caused mainly by pinch out of the sand deposit (Skolfield, 1968).

Most significant pools in southwestern Indiana have been rejuvenated by waterflooding operations. Secondary recovery in this state rose above 50 per cent of the total annual production in 1963, and the percentage has increased subsequently (Carpenter, 1965). The reservoirs undergoing flooding are noted in Table 9.1.

Future. Five new fields were discovered in Indiana in 1967 (Van Den Berg et al., 1968). Two of these produce from Mississippian and three from Ordovician reservoirs. This situation can be expected to continue indefinitely; new fields with Mississippian pay zones will be found in the Indiana sector of the Illinois basin, and virtually the entire state is open to discoveries in Ordovician, Cambro-Ordovician, and Cambrian reservoirs. These older reservoirs are especially inviting in the two barren belts that cross Indiana: (1) from east to west at the north end of the state, and (2) from northwest to southeast between the Lima–Indiana district and the southwest (Illinois basin) district (Figure 9.1). They are also prime targets at greater depth in producing areas.

There is no reason why the Trenton Limestone should not be locally dolomitized, porous and permeable, and oil-and gas-bearing in traps in parts of Indiana not within the old Lima–Indiana field. Rooney (1966) has presented convincing evidence of a widespread erosional unconformity at the top of the Trenton over all of Indiana except the southeastern part. One result of the subaerial exposure to weathering was the development of porosity at and near the top of the Trenton section, ranging from vuggy to cavernous. The northern counties are especially inviting, for here the Trenton is largely dolomite and much more porous than to the south, where limestone predominates (Patton and Dawson, 1968).

Northeastern Indiana is crossed by the Kankakee arch. The part of the state lying northeast of the arch axis is on the southwestern flank of the Michigan basin. Here, in central Wabash County, the *Urbana* oil field was discovered in mid-1966. The pay zone is the weathered top of the Trenton at a depth of a little more than 1000 feet. Porosity is due to both vugs and fractures. The trap is a node on a long anticline that creases the north flank of the Kankakee arch. The crest of the arch is 30 miles south; the northerly dip into the Michigan basin is 40 to 50 feet per mile (Avila, 1967).

The pre-Trenton section in Indiana includes the St. Peter Sandstone, Knox Dolomite, Eau Claire Formation, and Mount Simon Sandstone (see Table 9.1). All of these are possible reservoirs for oil and gas. "The St. Peter sandstone has been considered a good prospect for oil or gas, but remarkably few shows of either have been reported . . . " (Gutstadt, 1958, p. 9). Below the Ordovician rocks is the pre-Chazy unconformity. In eastern Indiana, at least, the porous Knox (Cambro-Ordovician) Dolomite beds are overlain by Ordovician Glenwood Shale. The geologic picture is very similar to that in Morrow County in central Ohio (see p. 48). Furthermore, it has been discovered belatedly that a deep "Black River" pay zone in an old (1904) field in Jay County in eastern Indiana is actually "an erosional remnant of Knox dolomite that juts upward through the Glenwood shale and into the overlying impervious Black River limestone" (Dawson, 1967, p. 130). In the near future there should be a lot of searching in Indiana for other weathered Knox Dolomite hills sealed in by Glenwood Shale.

The still deeper Cambrian formations are virtually untested, although the Eau Claire "exhibits rapid local changes in porosity which might serve to localize accumulations of oil and gas" and the "Mt. Simon has ideal reservoir characteristics of the blanket sand type" (Gutstadt, 1958, p. 9).

Although the current production of virgin gas is not significant in Indiana, the underground storage of natural gas brought in from states to the southwest is a major undertaking (Dawson and Carpenter, 1963). The search for new storage sites is a continuing one.

REFERENCES CONSULTED IN PREPARING CHAPTER 9

Avila, John, 1967.
Carpenter, G. L., 1965.
Cohee, George V., 1942.
Dawson, T. A., 1967.
Dawson, T. A., and Carpenter, G. L., 1963.
Fritz, Arthur J., 1968.
Gutstadt, Alan M., 1958.
Lockett, Robert, 1968.
Lundy, Curtis L., 1958.
Patton, J. B., and Dawson, T. A., 1968.
Rooney, Lawrence F., 1966.
Skolfield, John L., 1968.
Van Den Berg, Jacob, Swann, D. H., Carpenter, G. L., and Nosow, Edmund, 1968.
Wanenmacher, Joseph N., and Gealy, Wendell B., 1930.

Chapter

10

ILLINOIS

Illinois is the westernmost state in Division I. It is bordered on the west by states assigned to Division II (Figure 10.1). This state accounted for a little less than 2 per cent of the United States oil production in 1968, ranking tenth. However, its 60 million barrel oil yield in 1967 was slightly more than the output for that year of all the other states in Division I combined.

In early years the search for oil and gas in Illinois was aided by exploration for coal. Not only were seeps observed in core drill holes, but also the enormous amount of stratigraphic elevation data obtained in both coal boreholes and mines made possible detailed structure maps throughout the area covered by the Eastern Interior coal basin (Illinois basin). Illinois has produced gas commercially since 1885 and oil continuously since 1904. The first oil peak was reached in 1908 with a yield of nearly 34 million barrels. Subsequently, production declined until 1936, when new discoveries in the Illinois basin led to a spectacular increase and a second, much higher peak in 1940. During that year, production reached nearly 148 million barrels of oil, and Illinois was in fourth place among oil-producing states. Since then production has tapered off again, down to 60 million barrels in 1967.

Natural gas has never been an important marketed commodity in Illinois. It has been mostly solution gas produced with oil and either flared (in earlier days) or used locally in field operations (Bond et al., 1968). Gas in gas caps, and in separate nonassociated deposits, has been marketed, but not in any volume since 1965 (Illinois State Geol. Survey staff, 1968, p. 35). The biggest activity in natural gas in Illinois involves storing it for winter withdrawals. Gas from other states is pumped into underground reservoirs in 24 places in Illinois. Most of this gas is stored in sandstone reservoirs of Cambrian and Ordovician age.

Tectonics. The Illinois basin (Figure 10.2) is a regional downwarp, covering not only all of Illinois (except the

1	Adams	52	Lee
2	Alexander	53	Livingston
3	Bond	54	Logan
4	Boone	55	McDonough
5	Brown	56	McHenry
6	Bureau	57	McLean
7	Calhoun	58	Macon
8	Carroll	59	Macoupin
9	Cass	60	Madison
10	Champaign	61	Marion
11	Christian	62	Marshall
12	Clark	63	Mason
13	Clay	64	Massac
14	Clinton	65	Menard
15	Coles	66	Mercer
16	Cook	67	Monroe
17	Crawford	68	Montgomery
18	Cumberland	69	Morgan
19	DeKalb	70	Moultrie
20	De Witt	71	Ogle
21	Douglas	72	Peoria
22	Du Page	73	Perry
23	Edgar	74	Piatt
24	Edwards	75	Pike
25	Effingham	76	Pope
26	Fayette	77	Pulaski
27	Ford	78	Putnam
28	Franklin	79	Randolph
29	Fulton	80	Richland
30	Gallatin	81	Rock Island
31	Greene	82	St. Clair
32	Grundy	83	Saline
33	Hamilton	84	Sangamon
34	Hancock	85	Schuyler
35	Hardin	86	Scott
36	Henderson	87	Shelby
37	Henry	88	Stark
38	Iroquois	89	Stephenson
39	Jackson	90	Tazewell
40	Jasper	91	Union
41	Jefferson	92	Vermilion
42	Jersey	93	Wabash
43	Jo Daviess	94	Warren
44	Johnson	95	Washington
45	Kankakee	96	Wayne
46	Kane	97	White
47	Kendall	98	Whiteside
48	Knox	99	Will
49	Lake	100	Williamson
50	La Salle	101	Winnebago
51	Lawrence	102	Woodford

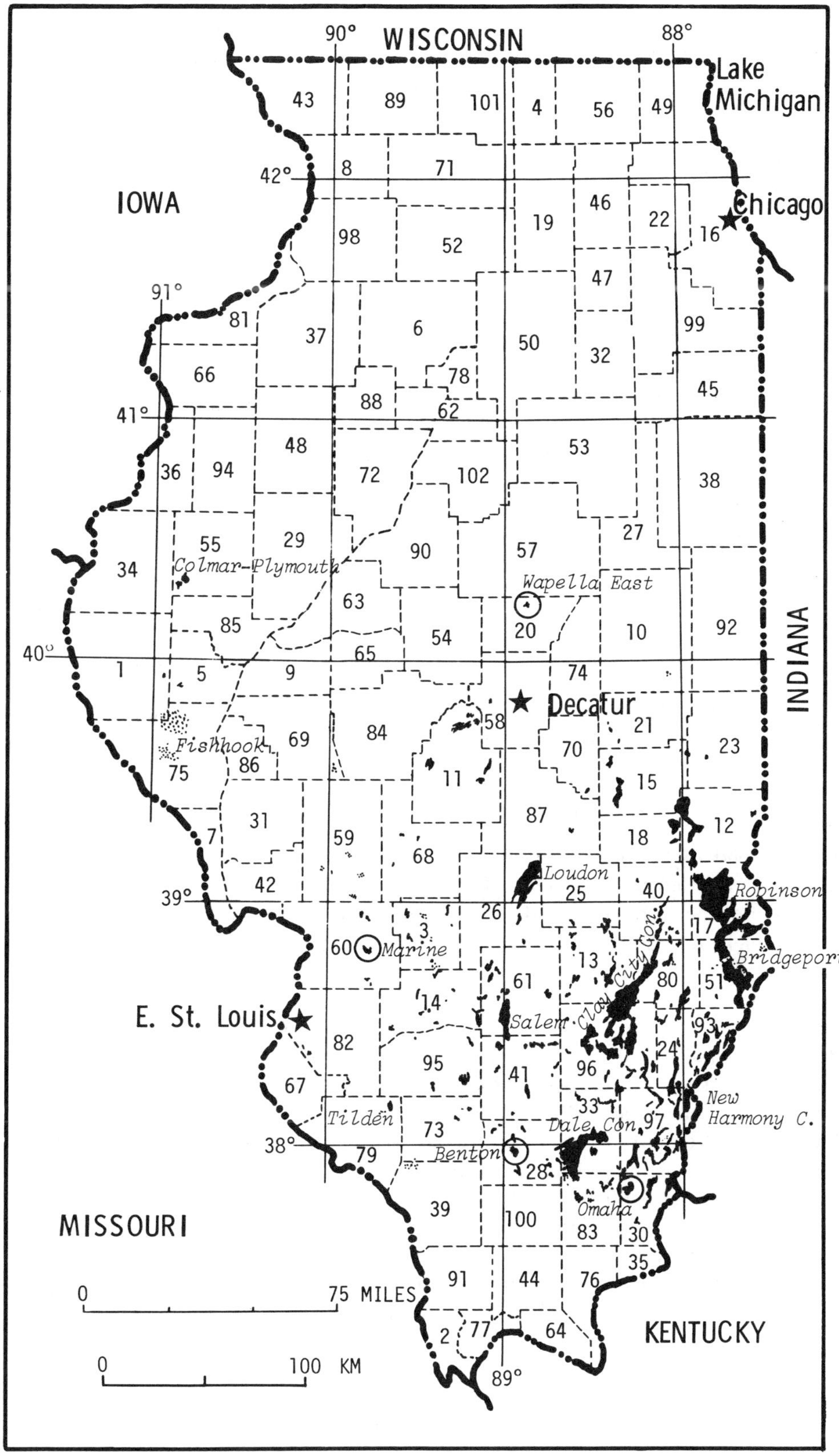

Figure 10.1 Illinois. Index map with oil and gas fields. County Key on left.

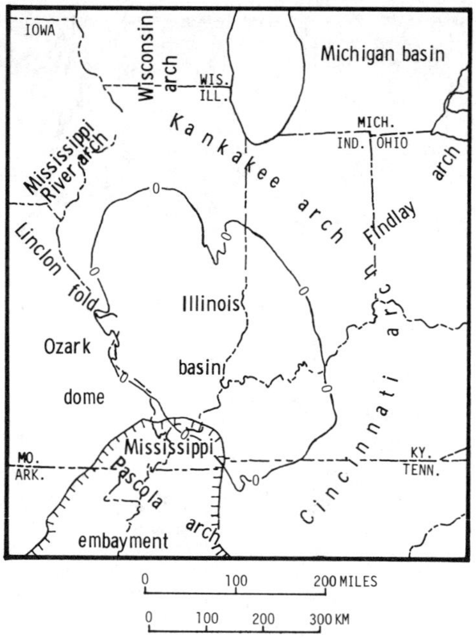

FIGURE 10.2 Structural environment of Illinois basin. Zero line is structural contour on top of Devonian limestone (Figure 10.3). Hachured line outlining Mississippi embayment is outer edge of Cretaceous-Cenozoic sediments. ***After David H. Swann and Alfred H. Bell (1958, Figure 1, p. 448). Courtesy American Association of Petroleum Geologists.***

northern end), but also southwestern Indiana and a part of western Kentucky. It is the locale for all of the Illinois oil, much of the Indiana oil, and a considerable part of the Kentucky oil. The tectonic map of the Illinois basin shows a spoon-shaped basin with the long axis of the depression extending in a direction a little west of north and rising to the north. The Precambrian basement rock floor is covered by 2000 to more than 13,000 feet of sedimentary strata. The deepest part is close to the juncture of the Wabash and Ohio Rivers, where Illinois, Indiana, and Kentucky meet. Wells drilled to the basement out in the basin are few; so far the greatest local relief found has been 800 feet in 8 miles. Scattered buried monadnocks are a probability (Bradbury and Atherton, 1965). From the deepest point in the basin the strata ascend gently toward the Ozark uplift on the west, the Wisconsin shield to the north, the Cincinnati arch to the east, and the Pascola arch to the south. The southwestern rim of the basin was penetrated for a short distance by the Mississippi embayment, which resulted in the deposition of much younger sediments unconformably over Paleozoic rocks.

The eastern flank of the Illinois basin is interrupted northeast of the deepest part by the La Salle anticline, which generally parallels the basin axis both in direction and in plunge. Two much more local, but nonetheless prominent folds, the DuQuoin monocline and the Woodlawn block made up of the Salem–Louden anticline, lie on the west flank of the basin. The deepest part of the Illinois basin is called the Fairfield basin. It is creased from north to south by several anticlinal folds, the most prominent of which is the Clay City anticline (Figure 10.4). The southern flank of the basin is cut by many faults, only a few of which are shown on the structure contour map. This fault belt continues across southernmost Illinois into Kentucky on the east and Missouri on the west. Most of the faulting took place toward the end of the Paleozoic era, as did the intrusion of peridotite dikes and the deposition of fluorite and other minerals in the same area.

Figure 10.3 is a west-east cross section of the basin from the Mississippi River to the top of the Cincinnati arch in southeastern Indiana. The vertical scale is highly exaggerated as can be seen by noting the outlines of the true scale section beneath. Dips are measured in feet per mile, rather than degrees. The cross section brings out the basinward thickening of most of the sedimentary units. It also shows some of the major folds that crease the flanks of the basin; however, the major anticlinal interruption to the regional dip basinward, the La Salle anticline, does not extend as far south as the line of the section.

Figure 10.4 is a structure contour map of the Illinois sector of the basin based on the top of the Trenton (Champlainian) Limestone. It shows several of the structural features that appear on the cross section, plus the La Salle anticlinal belt. Much greater structural detail can be seen in large-scale structure maps (too large to show here) contoured on widespread, readily recognized stratigraphic units such as the base of the uppermost Devonian and Kinderhookian (Mississippian) New Albany Shale (Schwalb, 1968) or the base of the Chesterian Beech Creek (or Barlow) Limestone (Bristol, 1968). Since the latter structural datum lies in the middle of the numerous Mississippian reservoirs in the basin, it has been penetrated by many boreholes, making possible an unusually accurate and detailed structure map.

The Illinois basin is filled with nearly 3 miles of Paleozoic sediment. Most of it is marine, shallow water deposit, laid down over a quarter of a billion years as the basin sank at an average rate of "only two-thirds of an inch per thousand years" (Swann, 1968, p. 4). The contributing land areas were the arches on the west, north, and east; the seas came in from the southwest until the end of the Paleozoic, when the development of the Pascola arch completed the closing of the basin. During the Paleozoic era "the entire basin was emergent at least 50 times, and the marginal areas were dry land about 100 times" (Swann, 1968, p. 3).

The La Salle anticline and related structures began

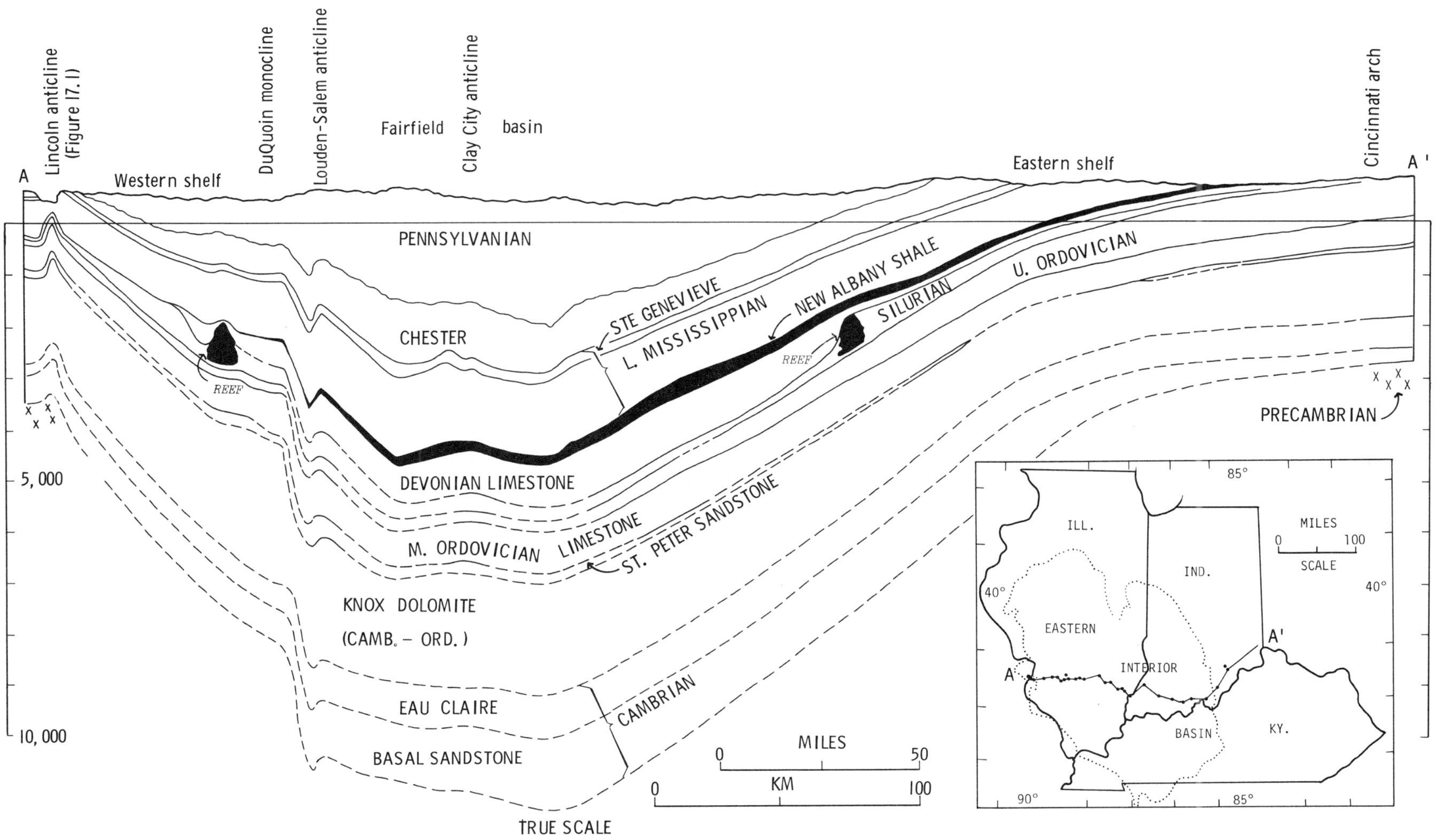

FIGURE 10.3 Illinois basin, west to east cross section. The eastern flank of the basin here is south of main La-Salle anticline. *After Illinois Geological Society and cooperating organizations (1951, Figure 147, p. 488). Courtesy American Association of Petroleum Geologists.*

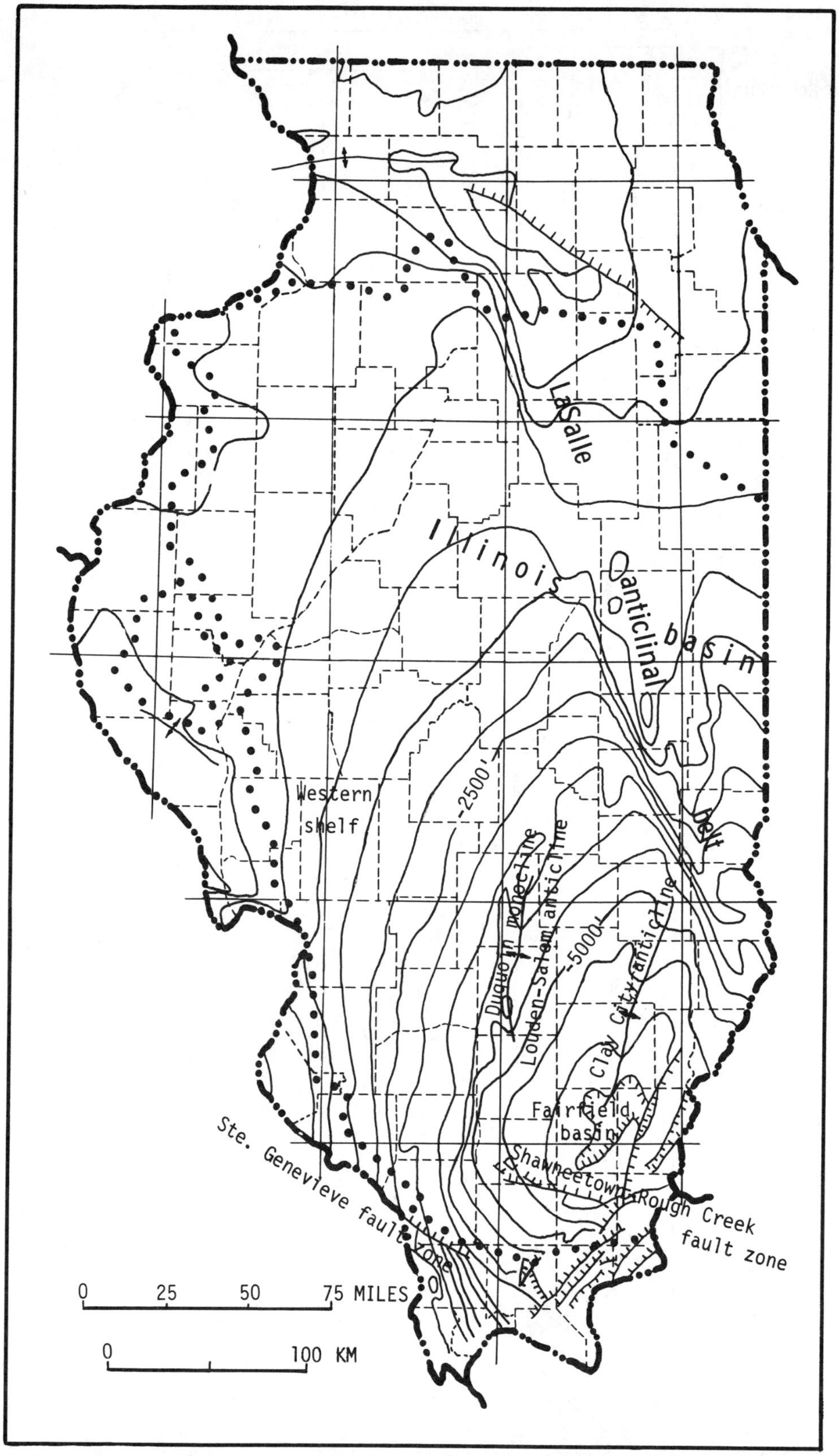

Figure 10.4 Illinois structure map contoured on top of Trenton. Interval, 500 feet. Hachured lines are faults; only a part of the mapped faults in southern Illinois is shown. Dotted line is outer edge of Pennsylvanian age rocks. *After George V. Cohee et al. (1962).*

to form before Pennsylvanian deposition. The entire area was uplifted and eroded after Mississippian sedimentation. Folding continued during and after the deposition of the Pennsylvanian strata, so that coal beds crossing the anticlinal complex show an uplift of about 1000 feet, whereas the Ordovician rocks have a structural relief of 2500 feet (Clegg, 1965).

Extensive preglacial erosion planed off the surface in the Illinois area, exposing successively older rocks outward from the center of the basin, especially to the north toward the Wisconsin boundary. Subsequently, the great ice sheets deposited an average thickness of 100 feet of drift over practically the entire state of Illinois. As a consequence, the two most widely used and most successful prospecting methods for petroleum and natural gas are subsurface geology and the reflection seismograph.

Stratigraphy. The stratigraphic column, showing the important pay zones, for southern Illinois (the only part of the state that is petroleum productive) is given in Figure 10.5 (Swann, 1967). The dominant bedrock subcropping beneath the glacial drift in Illinois is Pennsylvanian in age. The outer boundary of Pennsylvanian rocks is shown by a dotted line in Figure 10.4. Many use this contact line to delineate the basin, even though the basin flanks can be considered to extend to the tops of the surrounding arches. To the north of the Pennsylvanian rocks in Illinois are the older Paleozoics down to Champlainian in age. Near the southern end of the state Mississippian rocks fringe the Pennsylvanian at the outcrop.

Figure 10.5 also shows the many unconformities resulting from the 50 or more emergences of the entire basin. One partial emergence occurred during Chesterian deposition; the result was that Chesterian sediments were not deposited over most of the northern half of Illinois and hence do not appear there either on the sub-Pennsylvanian geologic map or at the subcrop fringing the Pennsylvanian sediments.

Lithologically, the 3 vertical miles of subsurface Paleozoic rock consists chiefly of dolomite "followed by limestone, shale, sandstone, chert, anhydrite, and coal, in that order" (Swann, 1968, p. 3). The mining of Pennsylvanian coal in Illinois is big business; the coal exceeds oil in dollar value of annual production by more than 30 per cent.

The oil pay zones range in age from Pennsylvanian to Champlainian. Not shown in Figure 10.5 are the deposits of Pleistocene age, the glacial drift that veneers most of Illinois. Two zones in the drift yield about 60 million cubic feet of gas daily in northeastern Illinois. Most of this gas appears to have originated in buried peat and organic silt, but some of the deeper gas may have come from underlying Pennsylvanian coal beds (Meents, 1968).

So far, about 52 per cent of the Illinois oil has come from the Upper Mississippian (Chesterian) sandstones (Swann and Bell, 1958, p. 447). Within the Chesterian, the Waltersburg and Tar Springs in the upper part are important reservoirs, but most of the Chesterian production has come from the lower part of the sandstone section, such as the Cypress and Yankeetown Sandstone (Benoist) reservoirs. Another 26 per cent has come from limestones and sands in the Valmeyer Series of the Mississippian, especially the McClosky pay zone in the Ste. Genevieve Formation. These two parts of the Mississippian account for nearly 79 per cent of the oil so far produced in Illinois.

Pennsylvanian reservoirs have contributed another 16 per cent of the Illinois oil, mostly from along the La Salle anticline, where the state reached its first production peak in the early years of the century. Most of the remaining petroleum yield has come from reservoirs lying in the Middle Silurian–Middle Devonian section, including reef limestones, dolomites, and some sandstones. Pay zones in the Grand Tower Limestone (Middle Devonian) have supplied the bulk of the Devonian oil produced in Illinois. This is essentially a carbonate rock unit; a basal sandstone member, the Dutch Creek, is discontinuous but widespread. Dolomitic members of the Grand Tower, including especially the Geneva, have furnished most of the oil, but the Dutch Creek is oil bearing in the deepest part of the basin (Meents and Swann, 1965). The only Ordovician formation that has produced so far is the Galena Group (Trenton), which is the great reservoir in the Lima-Indiana district to the east.

Traps and Selected Fields. Illinois has six fields classified as giant (Oil and Gas Jour., Feb. 5, 1968, p. 162). At the top of the list with 643 million barrels is the field called either Southeastern Illinois (Illinois Geol. Survey) or Old Illinois (Oil and Gas Jour.). This is actually an aggregation of fields overlying the La Salle anticline, mainly in Crawford and Lawrence Counties, including especially the Robinson and Bridgeport fields, which are giants in their own right. The other five fields, with cumulative production to the nearest million barrels in parentheses, are Salem (334), Louden (328), Clay City (260), New Harmony (135), and Dale (88).

In terms of current production, the six leading fields in decreasing order of output during 1967 are Southeastern Illinois, Louden, Clay City, Dale, Salem, and New Harmony. These fields accounted for 62 per cent of the total Illinois yield for the year.

All of these fields are anticlinal traps, with permeability controlling distribution in some pay zones. A part of the New Harmony oil comes from sand bars.

Swann and Bell (1958) note that much Illinois oil has come from two north-trending belts or hinge lines, the La Salle and the DuQuoin. The fields along the axis of the La Salle anticlinal belt, toward its southern end, are

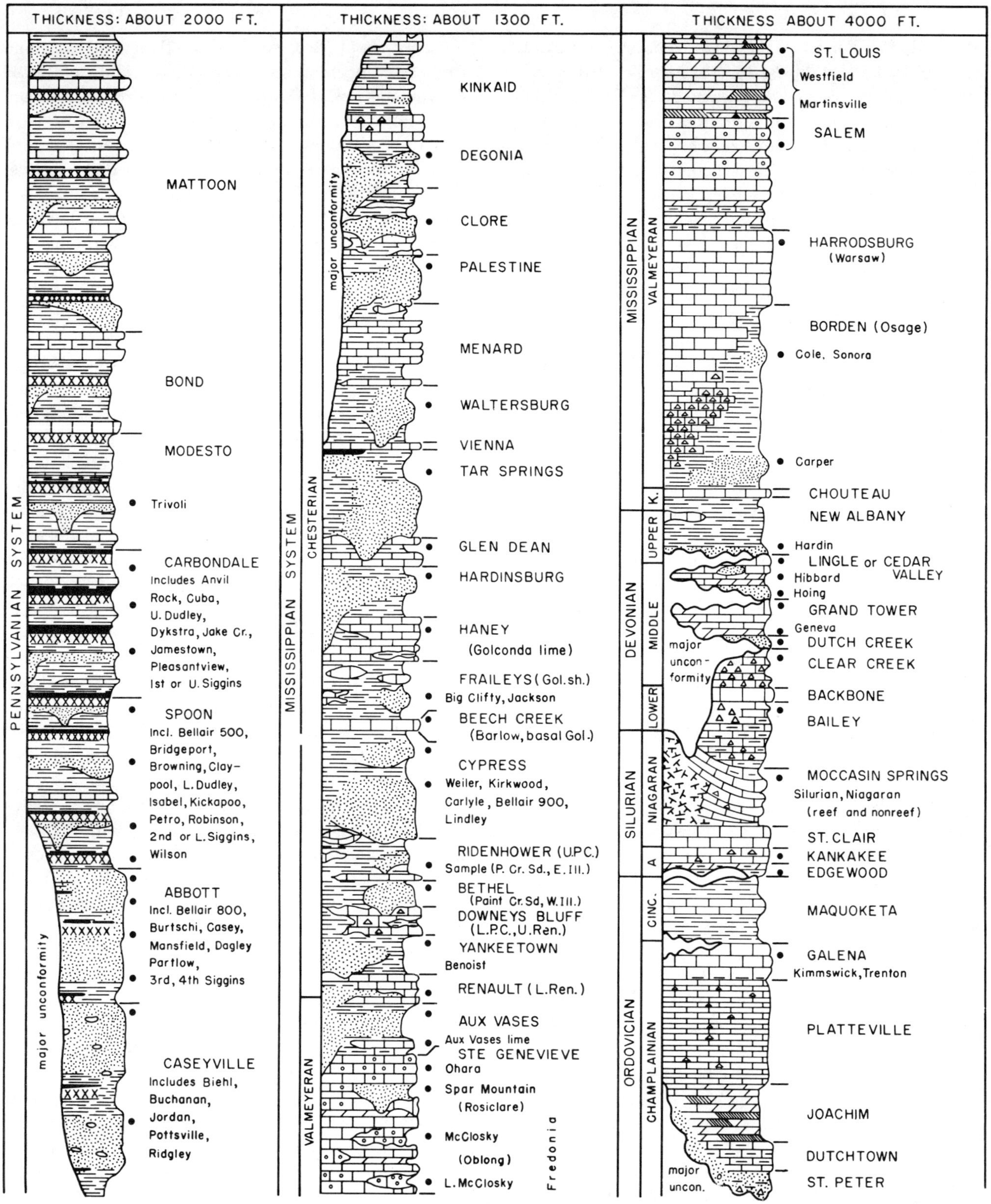

FIGURE 10.5 Generalized geologic column of southern Illinois. Black dots indicate oil (and gas) pay zones. Formation names are in capitals; other pay zones are in lower case. Not shown are about 4000 feet of Lower Ordovician strata and all Cambrian rocks below the St. Peter. Abbreviations: K.-Kinderhookian; A.-Alexandrian; Cinc.-Cincinnatian; Gol.-Golconda; P. C.-Paint Creek; and Ren.-Renault. Vertical scale variable. *By David H. Swann, in Jacob Van Den Berg (1968, Figure 2, p. 7). Courtesy Illinois State Geological Survey.*

the oldest. The largest of these, Bridgeport and Robinson (Figure 10.1), began producing in 1906. Both Pennsylvanian and Mississippian reservoirs are present. The largest of the DuQuoin hinge belt fields are the Louden, discovered in 1937, and the Salem (1938). These occur along the axis of the Salem–Louden anticline and produce principally from Mississippian pay zones.

Between the two hinge belts are the "basin" fields, which also have produced much oil. Many of the newer fields, including Clay City, are in this deeper part of the basin. A much smaller number of fields lie on the western shelf. Only minor production has been obtained on the southern flank of the basin.

The *Southeastern Illinois* producing belt overlies the top and flanks of the La Salle anticline, extending about 70 miles from the vicinity of the common corner of Coles, Clark, and Cumberland Counties south-southeast across Crawford and Lawrence Counties. The southward plunging anticlinal axis crosses the Wabash River into Indiana in southeastern Lawrence–northeastern Wabash Counties. Southeastern Illinois has produced, since the first discovery in 1904, nearly one-fourth of the total output of Illinois and is still yielding about one-sixth of the annual production.

The pay zones, 20 or more in number, range in depth from about 300 to 2000 feet. They are Pennsylvanian sandstones and Mississippian sandstones and limestones. It has been estimated that about 88 per cent of the Pennsylvanian production in Illinois has come from the La Salle anticlinal belt (Wanless, 1962, p. 52). The carbonate rock reservoirs are beneath the sand producing section. Pay zone thicknesses range from 2 to 40 feet (Van Den Berg, 1968, p. 39). Trapping is anticlinal.

Major development of the La Salle anticlinal belt began in 1906 with the discoveries of the Robinson field in Crawford County and the Bridgeport field in Lawrence County (Figure 10.1). A structure map of a part of the *Bridgeport* field is shown in Figure 10.6 (Blatchley, 1913). At the time that this map was prepared, the field had six oil and gas pay zones, ranging from the Bridgeport sand of Pennsylvanian age to the McClosky oölitic limestone of Mississippian age. The field is on top of the La Salle anticline; production is limited on both the steeper west flank and the east flank by salt water. Some of the sands are lenticular and therefore do not produce across the entire anticline. To the north and south production ends where the fold axis plunges into a saddle. The major fields occupy elongate nodes along the La Salle axis; toward the north the axis splits locally into two offshoots that rejoin farther north (Clegg, 1965).

In the deep basin, to the southwest of the southeastern Illinois district, are numerous fields, including two that ranked third and fourth in oil production for 1967. These are Clay City Consolidated, with more than 11 per cent of the state yield, and Dale Consolidated, with over 7 per cent. The *Clay City Consolidated* field extends into parts of four counties: Clay, Wayne, Richland, and Jasper. It overlies the broad Clay City anticline and associated structures (Lee, 1939). Discovered in 1937, Clay City is now producing from 11 pay zones, ranging in depth from 2175 to 3000 feet. All of these reservoirs are in the Mississippian, ranging from Waltersburg in the Chesterian to Warsaw in the Valmeyeran. In addition oil has been found in one well in Devonian limestone at 4350 feet. The principal pays are sands, down to 3000 feet, and limestones, including oölitic limestones below. The most prolific pays are McClosky oölitic limestone at 3050 feet and Aux Vases Sandstone at 2940 feet. The average pay zone thickness is 11 feet, the maximum 17 feet. Trapping is anticlinal, but distribution is limited in the sands by lenticularity and in the oölitic limestones by variable permeability.

The *Dale Consolidated* field is in southwestern Hamilton County, with some overflow into Franklin and Saline Counties. The discovery well was located on an anticline long known through coal development. It was completed in 1940, producing 115 barrels a day from Cypress, Benoist, and Aux Vases sands at 3215 feet depth. Subsequently two more sand pay zones were found uphole and three limestone reservoirs downhole, all in the Mississippian (Figure 10.7). The Aux Vases Sandstone is most prolific. It is a fine-grained, silty sandstone 30 to 35 feet in thickness with an average porosity of 20 per cent and 140 md permeability (Webb, 1968). The trap is a wide, irregular northeasterly trending anticline with dips on the flanks averaging about 50 feet per mile. Distribution of oil across the fold is not uniform because of lenticularity of the pay sands. Dale Consolidated is the scene of many waterflood projects, most of which are sweeping the Aux Vases Sandstone.

The *Benton* field to the west of Dale Consolidated in west-central Franklin County (Figure 10.1) is also in the "basin" district. It occupies a much smaller anticline (Howell, 1948). The principal pay zone at Benton is the Tar Springs Sandstone of Chesterian age. The trapping is anticlinal, but at the south end of the anticline there is abrupt gradation from sand to shale. Since the Benton field lies in an active coal-mining district, it was necessary to drill many of the wells through either operating or abandoned coal mines. For the operating mines, accurate maps were available, and the wells were located so as to pass through the centers of pillars at least 40 feet across instead of through mine openings. Where the mine had been abandoned and no accurate map showing the rooms and pillars was available, considerable trouble occurred until a technique was developed for meeting this situation.

The western hinge belt, consisting of the Salem-Louden anticline and the DuQuoin monocline (Figure 10.4), contains two giant fields after which the anticline

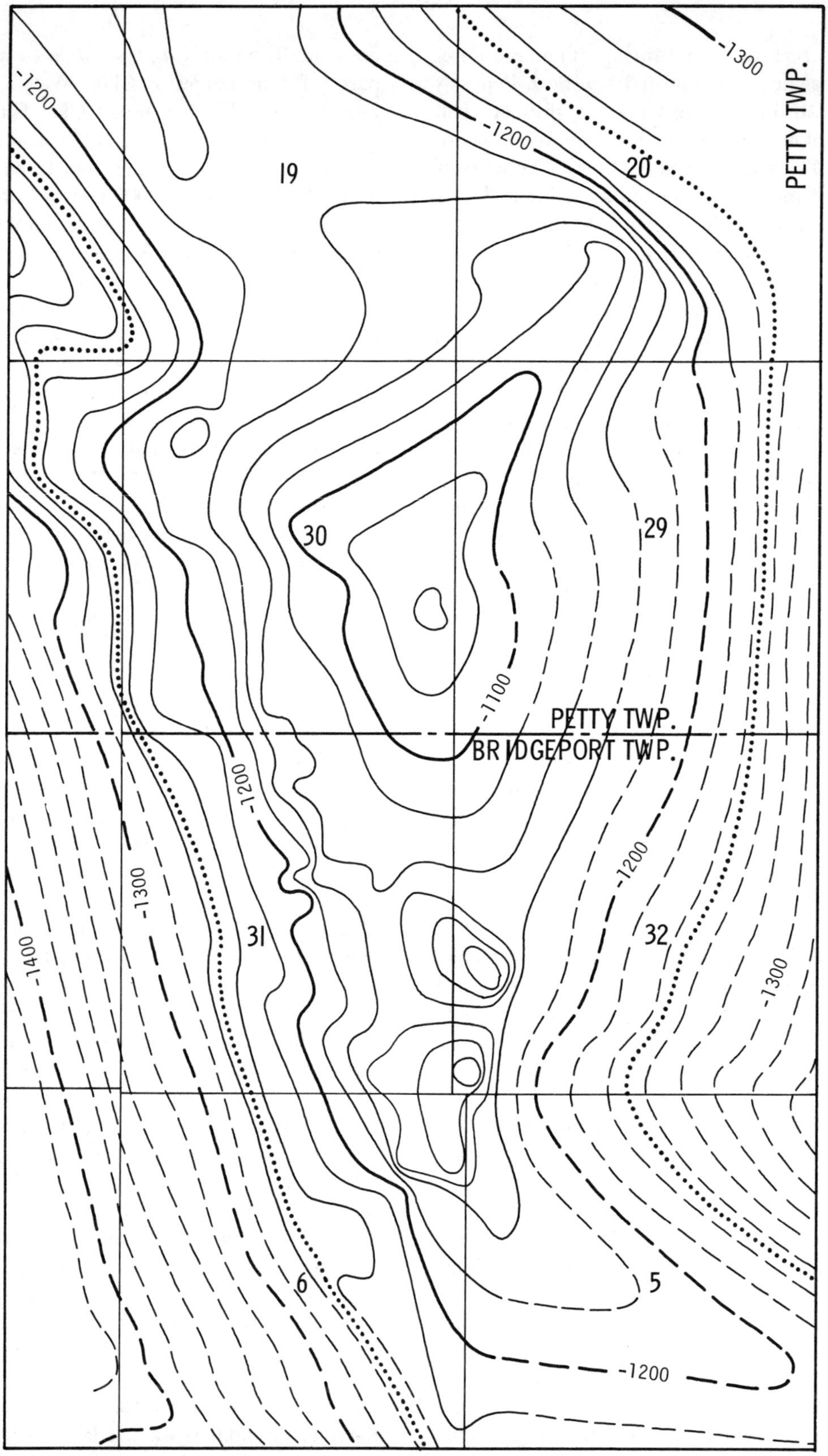

FIGURE 10.6 Structure contour map, based on top of McClosky sand, of a part of the Bridgeport field in west-central Lawrence County, southeastern Illinois. Interval, 20 feet. Sections are 1 mile (1.61 km) square. Original oil-water interface was at about −1250 feet (dotted line). A gas cap was present on the higher nodes and also on the terrace in the northeast quarter of section 31. *After Raymond S. Blatchley (1913, Plate 11).*

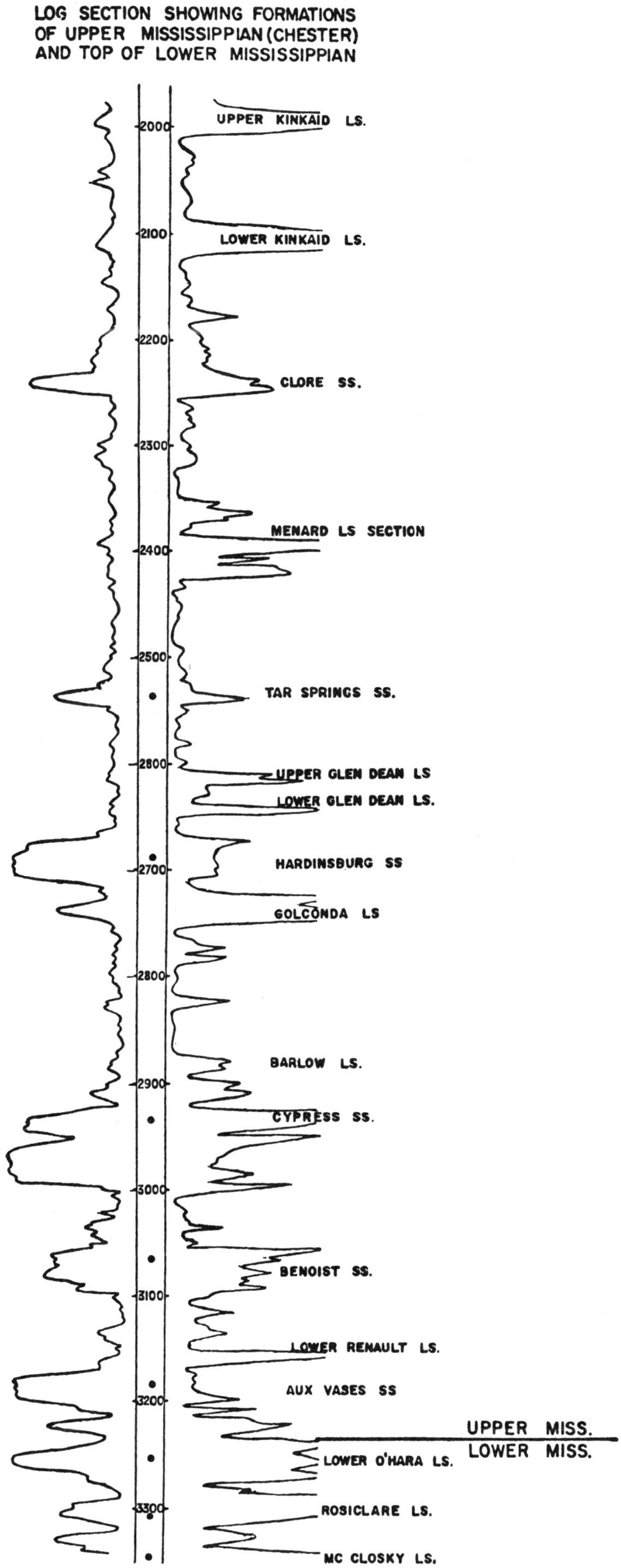

FIGURE 10.7 Dale Consolidated field, Hamilton and Saline Counties, Illinois. Electric log, with pay zones indicated by solid black circles in depth column. *Courtesy Wilton H. Webb (1968) and the Illinois and Indiana-Kentucky Geological Societies.*

is named. The *Salem* field, mainly in southwestern Jefferson County, was discovered on July 1, 1938 (Arnold, 1939); 6 months later it ranked seventh in the United States on the basis of daily production, and 2 years later it was second. The field covers an area of about 14 square miles, and the Salem anticline (Figure 10.8), on which it lies, has closure at reservoir level of more than 200 feet. Although permeability controls to some extent the distribution of oil in the reservoirs, the trapping is entirely anticlinal. Two sandstones, the Benoist at 1780 feet and the Aux Vases at 1825 feet of Chesterian (Late Mississippian) age, separated by only a few feet of shale and possessing a common oil-water contact, form the highest and most important group of reservoirs. These pay zones are interrupted by tight silty or shaly streaks, and only about half of the total thickness

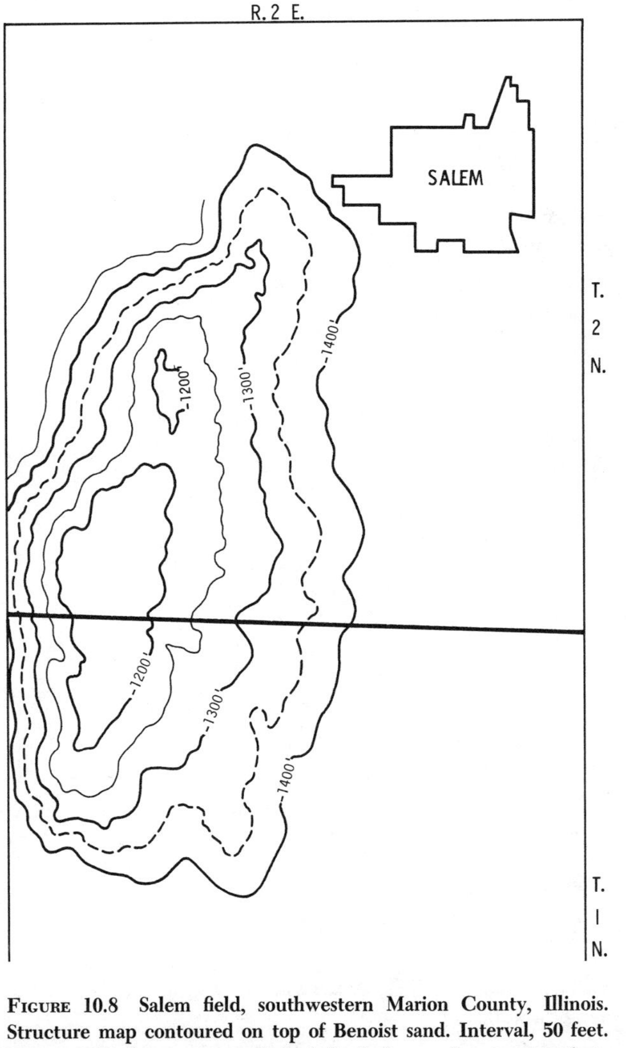

FIGURE 10.8 Salem field, southwestern Marion County, Illinois. Structure map contoured on top of Benoist sand. Interval, 50 feet. Minus 1350 foot contour (dashed line) shows oil-water interface. Township is 6 miles (9.66 km) square. ***Courtesy H. H. Arnold, Jr. (1939, Figure 5, p. 1,362) and the American Association of Petroleum Geologists.***

of 120 feet is effective reservoir rock. The next lower reservoir, at 2050 feet, is the McClosky oölitic limestone, also of Mississippian age. This is the most productive reservoir in Illinois but, in the Salem field, is second in importance to the Chesterian sandstones. Lenses of porous oölitic limestone average about 15 feet in thickness in the Salem field. A sandy dolomite of Devonian age at 3440 feet is productive beneath about half of the field; this pay zone contributed greatly to the high yield during the first 2 years, but the output decreased rapidly after reaching this peak. Trenton Limestone produces beneath a quarter of the field from a depth of 4500 feet.

The *Louden* field in northeastern Fayette County was discovered in 1937 after a seismograph survey was made "over a large block of Illinois acreage acquired as a result of limited surface geology. The seismic exploration delineated a large anticline of low relief" (Lyons, 1948, p. 461). The pay zones include Mississippian Cypress, Bethel, Benoist, and Aux Vases sands at depths of 1500 to 1600 feet, the McClosky Limestone at 1785 feet, a basal Mississippian sandstone (Carper) at 2830 feet, plus Devonian Geneva Dolomite at 3000 feet and Ordovician Trenton Limestone at 3905 feet. The uppermost three sands, especially the Cypress, have been the most productive. To date, the Geneva pay zone is the only one of the deeper discoveries to become an important reservoir.

The original pre-discovery seismograph structure map, based on the top of the Trenton, is shown in Figure 10.9. It was found to be "in substantial agreement" with subsequent structure maps contoured on various pay zones. The Louden anticline occupies a major node on the regional anticline about 36 miles north (center to center) of the Salem node. Between these giant fields are six smaller fields strung along the anticlinal axis. The producing area at Louden is about 15 miles long (north-northeast), and the maximum width is nearly 5 miles. The fold has more than 100 feet of closure. Trapping is anticlinal, but the oölitic McClosky limestone pay zone has erratic permeability.

Westward and northward from the DuQuoin hinge line is the western shelf of the Illinois basin. Oil and gas fields are present on this shelf, but the density is much less than to the eastward in southern Illinois. No doubt one reason is that the prolific Chesterian rocks thin out and disappear on the western and northern flanks of the basin.

The *Wapella East* field in northern DeWitt County lies close to both the center of the state and the northward ascending Illinois basin axis (Howard, 1963, 1964). The discovery well was completed in December, 1962, at a depth of 1117 feet; in its first 4 years of production the yield exceeded 1 million barrels. Attention was first attracted to this relatively northern site as a result of

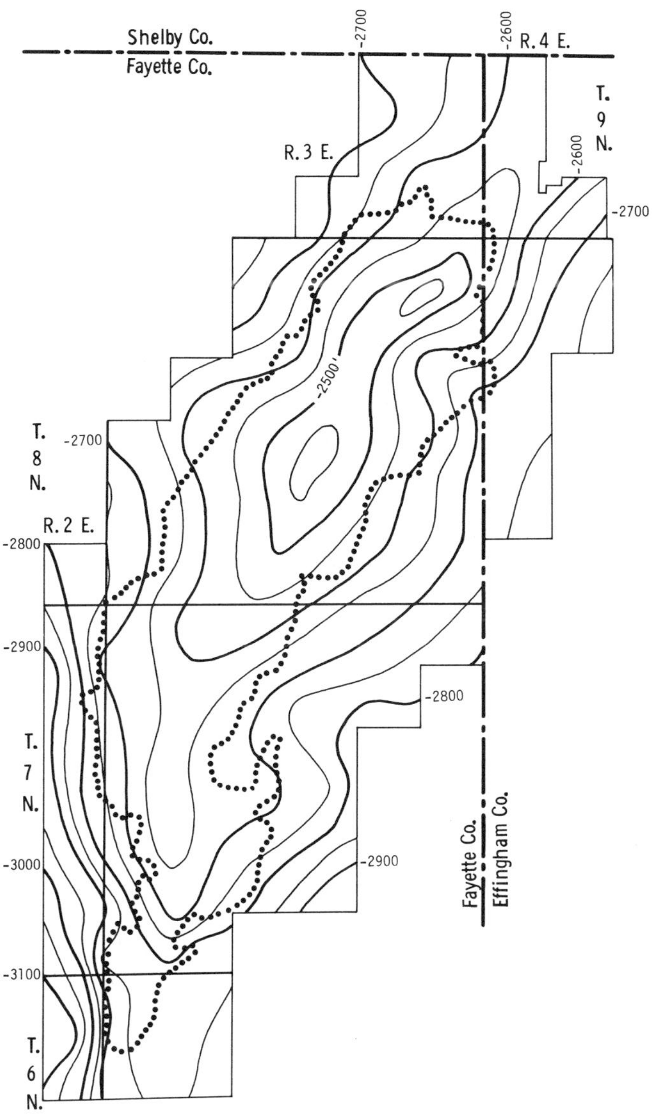

Figure 10.9 Louden field, northeastern Fayette County, Illinois. Structure map of top of Trenton, based on seismograph survey before drilling. Contour interval, 50 feet. Dotted line encloses producing area as of 1947. Township squares are 6 miles (9.66 km) on a side. *Courtesy Paul L. Lyons (1948) and the Society of Exploration Geophysicists.*

structure tests drilled by a gas company in search of possible storage reservoirs. The discovery well was drilled midway between the two structurally highest test holes (Harris, 1968). The pay zone is crystalline Silurian dolomite having numerous vugs and vertical fractures and averaging 22 feet in thickness. Insignificant production has also been obtained from immediately overlying Devonian limestone. The trap is an almost circular anticline (Figure 10.10), with a closure of about 100 feet, which is superimposed upon a southward plunging structural nose. The highest well drilled the top of the pay at less than 302 feet; the oil-water interface is at approximately −340 feet. Many consider this reservoir to be a Niagaran reef. It "appears to be a flat lying type (biostrome) and may be more or less continuous over several square miles" (Harris, 1968, p. 244).

Four counties to the west of Wapella East, and high on the western shelf, is the *Colmar-Plymouth* field which has produced over $4\frac{1}{2}$ million barrels since its discovery in 1914. This hyphenated field is largely in southwestern McDonough County. Production is obtained from Devonian Hoing Sandstone at a depth of 450 feet. Trapping is brought about by a combination of sand lenticularity and anticlinal structure (Van Den Berg, 1968, p. 26).

During the late 1950's shallow oil and gas accumulations were found, likewise high on the western shelf, in Pike, Adams, and Brown Counties from 30 to 50 miles south of Colmar–Plymouth. The pays are dolomitized zones in Silurian limestone. In the *Fishhook* gas pool (Meents, 1958), which lies mainly in northern Pike County, a probable 10 billion cubic foot gas reserve has been developed at an average depth of less than 500 feet. Seventy successful gas wells were drilled over an elliptical area $5\frac{1}{2}$ miles long and $2\frac{1}{2}$ miles wide. The trap is an anticline, with the same shape as the gas-producing area, and a closure of about 100 feet. Subsequently, two

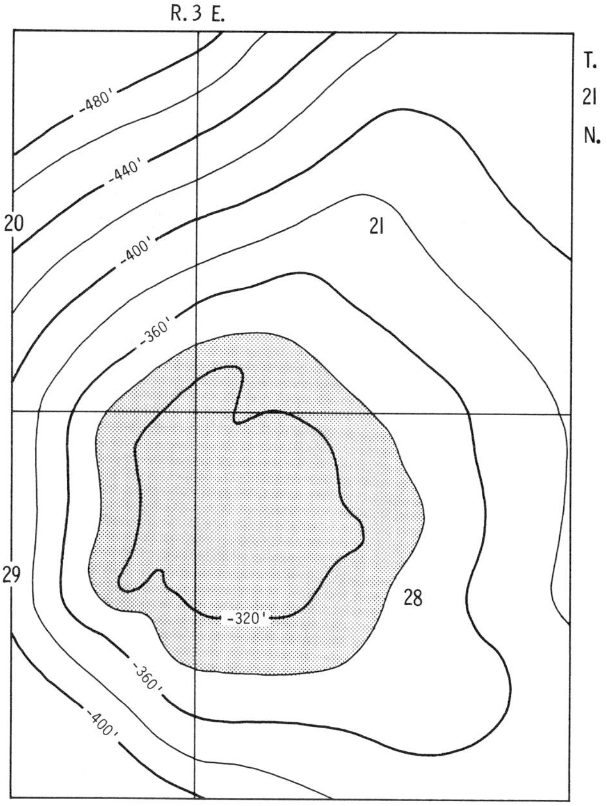

Figure 10.10 Wapella East oil field, northern DeWitt County, Illinois. Structure map contoured on top of Silurian reservoir. Interval, 20 feet. Sections are 1 mile (1.61 km) on a side. Stippled pattern shows producing area. *Courtesy Lloyd A. Harris (1968, p. 245) and the Illinois and Indiana-Kentucky Geological Societies.*

small oil fields were discovered on opposite sides of the Adams-Brown county line (Howard, 1961).

Niagaran (Silurian) reefs (Morgan, 1952; Bell, 1951; Lowenstam, 1950) first became productive in 1943 with the discovery of the *Marine* pool. This field has subsequently (through 1967) produced over 11 million barrels of oil. Marine is in east-central Madison County, about 25 miles northeast of St. Louis, Missouri. It lies near the western border of the Illinois basin. The discovery was due in part to geophysical exploration. Figure 10.11 shows the structure of the Marine pool as contoured on top of the Devonian (Lowenstam, 1948). The underlying reef is horseshoe shaped, and the top has 120 feet of closure. The deeper rocks, belonging to the Ordovician System, conform to the regional dip, which is to the east. The younger rocks reflect the reef structurally, but to a diminishing extent upward. Figure 10.12 is a hypothetical cross section of the reef. The oil has accumulated in a coquina-like detrital limestone, which mantles the reef. As many as four porous zones are present, and in addition some oil is obviously stored in a network of interconnecting fissures that extend not only through the reef limestone but also into the overlying Devonian limestone. The latter has a domal structure caused by draping over the reef.

Near the southern end of the western shelf, in northeastern Randolph County, is another Silurian reef field, the *Tilden*. Between discovery in 1952 and the end of 1967 this reef yielded over $3\frac{1}{2}$ million barrels of oil.

On the faulted southern flank of the Illinois basin is the unusual *Omaha* pool (English and Grogan, 1948). This oil pool is in northwestern Gallatin County near the southeastern corner of Illinois and about 20 miles southeast of the giant Dale Consolidated field described earlier. Omaha, discovered in 1940, has produced from seven pay sands, of which three are Pennsylvanian and four are Mississippian. Depths range from 385 to 2402 feet. Trapping is in an extremely symmetrical dome

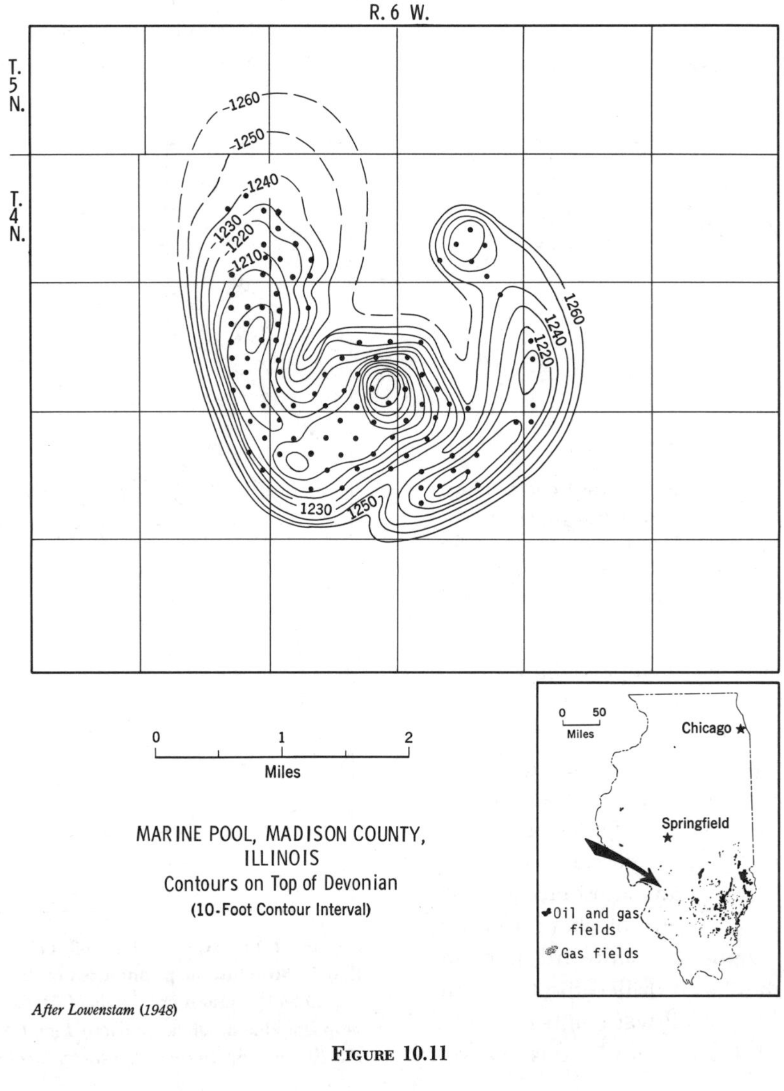

MARINE POOL, MADISON COUNTY, ILLINOIS
Contours on Top of Devonian
(10-Foot Contour Interval)

After Lowenstam (1948)

FIGURE 10.11

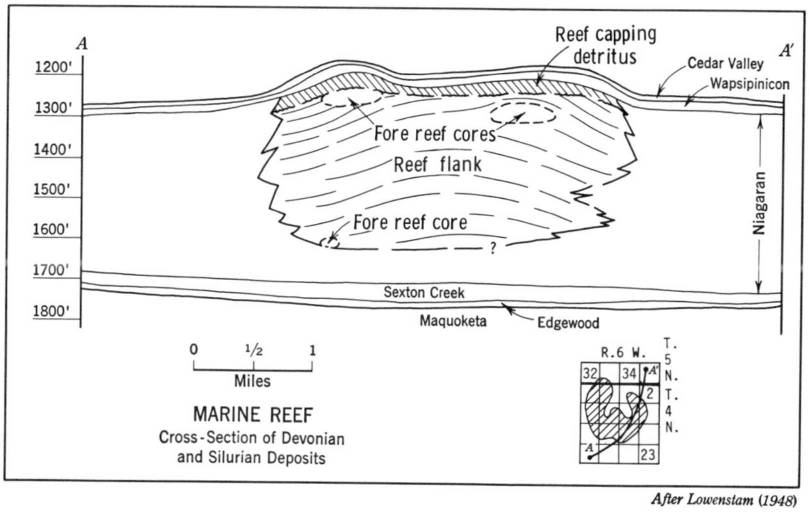

FIGURE 10.12

(Figure 10.13) that has been intruded by igneous sills and dikes and is, in all probability, the result of upward pushing on the part of a laccolith or other type of subjacent intrusion.

The most prolific type of stratigraphic trap so far developed in the Illinois basin is created by the lenticularity of the Chesterian Waltersburg Sandstone (Swann, 1951). The Waltersburg accumulations have been found in southeastern Illinois and southwestern Indiana (the "Lower Wabash area"). The elongate sand-bar-shaped lenses of the Waltersburg tend to line up in belts with a northeast to north trend. One such trend starts at the foot of the Omaha dome just described (Figure 10.13) and proceeds northeasterly to and across the Wabash River into Indiana. A parallel trend lies to the southeast, and farther north in Illinois are two trends, one of which parallels the Wabash and the other, to the west, extends almost straight north (Howard, 1967, Figure 6, p. 14).

Figure 10.13 shows the southwestern end of the trend that takes off from the Omaha dome. This is on the faulted south flank of the Illinois basin. The Waltersburg sand pools occur "in diverse structural situations, not only on and near the crests of anticlines and the upthrown sides of faults, but also on unfaulted regional monoclines, low on the flanks of anticlines, and on the downthrown sides of faults" (Swann, 1951, p. 2561). They do appear to avoid the structurally lowest areas.

The sand lenses occur in an interval between the Menard and Vienna Limestones that is occupied elsewhere by gray carbonaceous shale. Some of the sand bodies are completely surrounded by this shale (source rock?). The sand is well sorted, and Swann postulates the deposition of the barlike bodies "in a near-shore or shore-line environment by wave current action" (p. 2561).

The *New Harmony Consolidated* field of White County, Illinois, and Posey County, Indiana, is a giant field covering 24,650 acres with 5 Pennsylvanian sandstone reservoirs and 17 Mississippian pays of which the lowest 6 are carbonate rocks. The trapping here is anticlinal, but a Waltersburg sand trend crosses the anticline and is productive from a half-dozen lenses. The largest of these is a bar-shaped sand body that extends in a northwest direction, almost at right angles to the trend. This pool has been described by Skolfield (1968). The discovery well was completed in 1939 at a depth of about 2200 feet. The narrow sand bar has been completely delineated by wells drilled to deeper pay zones. The producing area is 725 acres, but other Waltersburg pools in this vicinity bring the aggregate area of Waltersburg production up to 1170 acres. The sand has 19.2 per cent porosity and 475 md permeability. Its maximum thickness is 60 feet. The pool is now in an advanced stage of waterflooding, which has been unusually successful.

FUTURE. The known reserves of Illinois are locked up largely in the Mississippian formations and in nearby older and younger rocks within the "basin" and "west flank" districts. Some, at least, of the deposits yet to be discovered lie at greater stratigraphic depths.

Five small fields were discovered in Illinois in 1967. One is producing from a Silurian pay zone, none from Ordovician or Cambrian reservoirs. The Ordovician and Cambrian section in the deeper part of the Illinois basin probably is more than 6000 feet in thickness. In total volume these rocks exceed the younger Paleozoics. The latter have produced nearly 3 billion barrels of oil; production to date from the older Paleozoics has been relatively insignificant and has come from near the top of the Ordovician. Within the Ordovician and Cambrian section are marine dolomite, sandstone, shale, and lime-

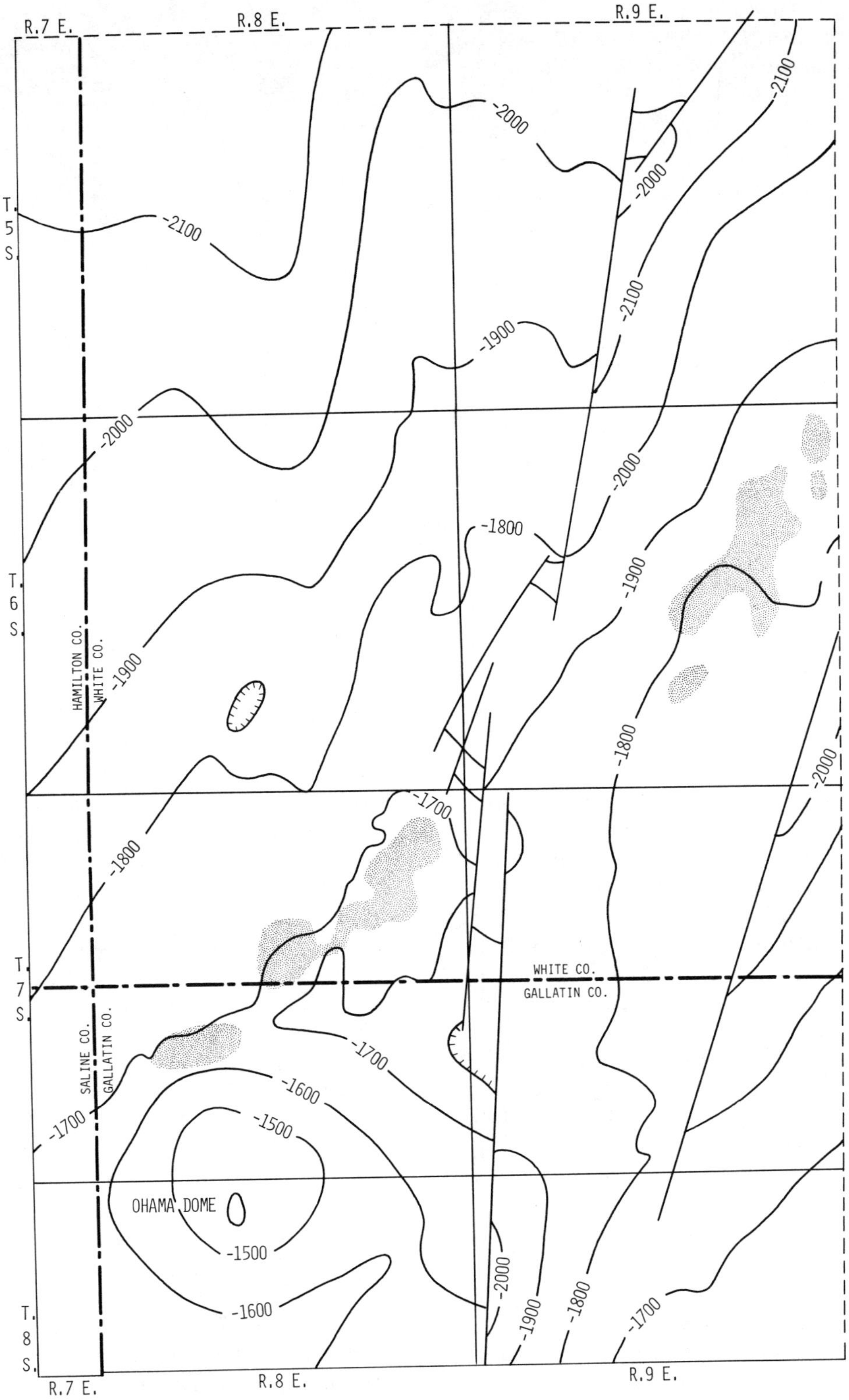

Figure 10.13 Structure contour map of a part of southeastern Illinois, based on Mississippian Menard Limestone, which overlies erratic Waltersburg Sandstone. Interval, 100 feet. Waltersburg sand oil pools are shown by stippled pattern. Six north to northeast striking faults terminate the contours in eastern half of map area. Townships are 6 miles (9.66 km) on a side. *After David H. Swann (1951, left half Figure 4, p. 2566). Courtesy Illinois State Geological Survey.*

stone beds, capable of furnishing both source and reservoir rocks. The maximum depth to probe clear to the basement is about 14,000 feet (Swann et al., 1966).

In "Deep Oil Possibilities of the Illinois Basin," Bell et al (1964) discuss the more likely Ordovician and Cambrian reservoirs as follows:

> More production may be found in the Middle Ordovician Galena Limestone (Trenton), thus extending the present productive regions. In addition, new production may be found in narrow, dolomitized fracture zones in the tight limestone facies on the north flank of the basin.
>
> At the base of the Middle Ordovician, the St. Peter Sandstone and the underlying Everton Group have much porous rock, but the best possibilities are in the eastern part of the basin, where the St. Peter breaks up into isolated sand bodies.
>
> The Lower Ordovician and Cambrian Knox Dolomite Megagroup (Arbuckle, Ellenburger) is a prolific producer in regions where the production is in part related to the sub-Tippecanoe unconformity, which truncates the Knox . . . . Zones of porous dolomite are present . . . . [Later: a well (Union Oil Co. Cisne Community No. 1) drilled in central Wayne County in 1967 recovered oil in a drill stem test in Knox Dolomite at about 7800 feet. Although the oil found was not commercial, "the test shows that oil can be present in the Knox in the central part of the basin" (D. C. Bond, letter dated 11 November, 1968).]
>
> The Eau Claire Formation, beneath the Knox, is essentially sandstone north of the deep part of the basin, dolomite with some oolitic limestone to the west, and shale to the east. The area in the deep part of the basin where these three rock types interfinger is the most favorable area for deep oil production. Although the Mt. Simon Sandstone at the base of the column is the thickest fluid-yielding unit in the region, it appears to offer fewer production possibilities than other units.

There still remain, as possibilities for the future, accumulations high on the basin flanks. In recent years much has already been done in this regard. The Omaha pool, bizarre as it may be in terms of trap genesis, and other fields nearby constitute good evidence that hydrocarbons have been available for entrapment on the south flank. On the west flank the Marine pool moved the production periphery westward; it jumped still farther with the serendipitous discovery of an oil field in St. Louis County, Missouri (see p. 119). The northwest rim has been approached by the previously described fields in McDonough, Brown, Adams, and Pike Counties (Howard, 1961). So far Wapella East is the farthest point north in the central part of the basin, but this situation could change (Garland and Hoovler, 1965). There is still room for discovery between the present periphery and the rim, but the greatest area for further testing is the largely barren belt between these outposts and the field-rich inner basin. Even in the basin interior, however, further discoveries will be made. There are pay zones within the prolific Mississippian section not yet adequately explored laterally. For example, the St. Louis Limestone (Figure 10.5) "is a potential target in much of the Illinois basin" (Bristol and Howard, 1966).

Hydraulic fracturing has been quite successful in many instances in Illinois. It not only has increased the output of producing wells but also has converted dry holes into producers (Whiting, 1956).

About 72 per cent of the current Illinois production is due to waterflooding (Lawry and Mast, 1968). Illinois, during its boom years, did not join other major oil-producing states in instituting production controls. In consequence much reservoir energy was wasted, and primary recoveries were minimal. Waterflooding operations are therefore unusually successful here because of the large quantity of oil left behind and available for salvage.

Because of the strategic location of northern Illinois in respect to the winter market for natural gas, the search for possible underground storage sites in this state will no doubt continue. In 1967 Buschbach and Bond stated: "The total volume of gas in underground storage in the state will probably be increased by 50 per cent or so within the next few years" (1967, p. 18). In Illinois, unlike many other areas where semidepleted gas fields can be and are converted into storage fields, it has been necessary to use aquifers for over 90 per cent of the gas storage (Buschbach and Bond, 1967, p. 8). A trap for the gas (anticline, fault, facies change, etc.) is still essential.

REFERENCES CONSULTED IN PREPARING CHAPTER 10

Arnold, H. H., Jr., 1939.
Bell, Alfred H., 1951.
Bell, Alfred H., Atherton, Elwood, Buschbach, T. C., Swann, David H. 1964.
Blatchley, Raymond S., 1913.
Bond, D. C., Bell, A. H., and Meents, W. F., 1968.
Bradbury, James C., and Atherton, Elwood, 1965.
Bristol, Hubert M., 1968.
Bristol, H. M., and Howard, R. H., 1966.
Buschbach, T. C., and Bond, D. C., 1967.
Clegg, Kenneth E., 1965.
English, R. M., and Grogan, R. M., 1948.
Garland, Robert F., and Hoovler, Paul B., 1965.
Harris, Lloyd A., 1968.
Howard, Richard H., 1961, 1963, 1964, 1967.
Howell, J. V., 1948.
Illinois and Indiana–Kentucky Geological Societies, 1968.
Illinois State Geological Survey Staff, 1968.
Lawry, T. F., and Mast, Richard F., 1968.
Lee, Lynn K., 1939.

Lowenstam, H. A., 1948, 1950.
Lyons, Paul L., 1948.
Meents, Wayne F., 1958, 1968.
Meents, Wayne F., and Swann, David H., 1965.
Morgan, Donald A., 1952.
Schwalb, Howard R., 1968.
Skolfield, John L., 1968.
Swann, David H., 1951, 1967, 1968.
Swann, David H., and Bell, Alfred H., 1958.
Swann, David H., Buschbach, T. C., Atherton, Elwood, and Frye, J.C., 1966.
Van Den Berg, Jacob, 1968.
Van Den Berg, Jacob, Swann, D. H., Carpenter, G. L., Milhous, H. C., and Nosow, Edmund, 1968.
Wanless, Harold R., 1962.
Webb, Wilton H., 1968.
Whiting, Lester L., 1956.
Willman, H. B., and others, 1967.

Chapter

11

MICHIGAN

In recent years, Michigan has contributed about 0.5 per cent of the annual petroleum production in the United States. Oil was first discovered in this state at Port Huron (Figure 11.1) in 1886, but it was not until 1926 that any substantial production, stemming from the discovery the year before of oil at Saginaw, was obtained. The Muskegon field was found in 1927, and Mt. Pleasant, the first of the important basin fields, was discovered in 1928. The year of maximum oil production so far obtained was 1939, when the yield was $23\frac{1}{2}$ million barrels. A second, but lower, peak of 19 million barrels was reached in 1961 after the discovery of the rich Trenton–Black River (Ordovician) pools on the southern flank of the basin in 1957. Michigan produced a little less than 14 million barrels in 1967.

Geology. The regional structure of the Southern Peninsula of Michigan, which contains all the oil and gas fields so far discovered (Cohee and Landes, 1958), is very simple. This area is an almost perfectly circular structural basin (Figure 11.2). The crystalline basement rocks, Precambrian in age, crop out around the north rim of the basin in Wisconsin, the Northern Peninsula of Michigan, and Ontario (Cohee, 1965). In the center of the basin, which is in the Southern Peninsula, the top of the Precambrian rock lies at a probable depth of 14,000 feet. Erosion has not uncovered the basement rocks around the south rim of the Michigan basin, but formations as old as Ordovician are exposed there (in Ohio and Indiana). The southern end of the Michigan basin lies inside the fork formed by the Kankakee and Findlay branches of the Cincinnati arch (Newcombe, 1934); the top of the arch forms the south rim. The southeastern flank of the Michigan basin contains the Howell anticline and fault (Figure 11.2); less prominent anticlines and terraces occur on the other flanks.

Toward the center of the Michigan basin the youngest widespread bedrock formations are Jurassic and Pennsylvanian sediments covered by up to 800 feet of glacial

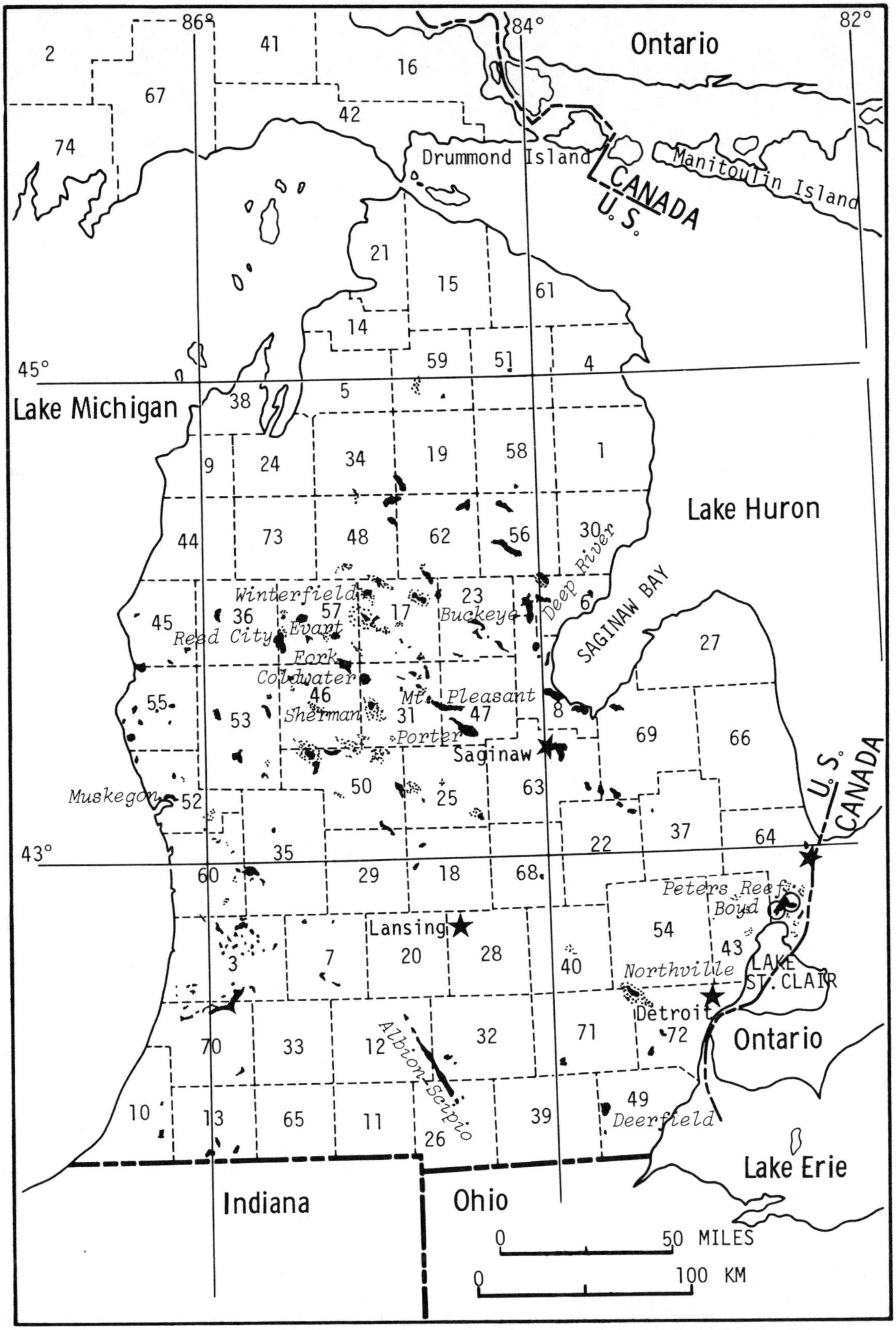

FIGURE 11.1 Southern Peninsula of Michigan, and a part of the eastern end of Northern Peninsula. Index map with oil and gas fields. County Key:

1	Alcona	12	Calhoun	22	Genesee	33	Kalamazoo	44	Manistee	54	Oakland	64	St. Clair
2	Alger	13	Cass	23	Gladwin	34	Kalkaska	45	Mason	55	Oceana	65	St. Joseph
3	Allegan	14	Charlevoix	24	Grand Traverse	35	Kent	46	Mecosta	56	Ogemaw	66	Sanilac
4	Alpena	15	Cheboygan	25	Gratiot	36	Lake	47	Midland	57	Osceola	67	Schoolcraft
5	Antrim	16	Chippewa	26	Hillsdale	37	Lapeer	48	Missaukee	58	Oscoda	68	Shiawassee
6	Arenac	17	Clare	27	Huron	38	Leelanau	49	Monroe	59	Otsego	69	Tuscola
7	Barry	18	Clinton	28	Ingham	39	Lenawee	50	Montcalm	60	Ottawa	70	Van Buren
8	Bay	19	Crawford	29	Ionia	40	Livingston	51	Montmorency	61	Presque Isle	71	Washtenaw
9	Benzie	74	Delta	30	Iosco	41	Luce	52	Muskegon	62	Roscommon	72	Wayne
10	Berrien	20	Eaton	31	Isabella	42	Mackinac	53	Newaygo	63	Saginaw	73	Wexford
11	Branch	21	Emmet	32	Jackson	43	Macomb						

drift. The areal geologic map shows a succession of older rocks in circular patterns surrounding the Pennsylvanian outcrop. The oldest rocks to reach the bedrock surface in the Southern Peninsula are Silurian, but in the Northern Peninsula the exposed Paleozoic section extends down to the Precambrian. Within the basin several rock groups, notably the evaporite-containing Salina (Silurian) and Detroit River (Devonian), thicken greatly between outcrop zone and basin center. During these times, and to a somewhat lesser extent during several other sedimentation periods, the Michigan basin was sinking concurrently with the deposition of clastic, organic, and chemical sediment.

Figure 11.3 is a graphic presentation of the stratigraphic column for the Southern Peninsula of Michigan, with the principal oil and gas reservoir rocks designated by appropriate symbols (Michigan Geol. Survey, 1964). A few ephemeral wells have been completed in glacial drift and Pennsylvanian sediments (Champion, 1968). The youngest producing reservoir of any importance is the Michigan "stray" sandstone. This formation has been found to be productive only in the west-central part of the Southern Peninsula. It has yielded 42 per cent of the state's total gas production, reaching an annual peak of nearly 20 billion cubic feet in 1947. Subsequently the decline has been rapid, and since 1957 the annual yield has been less than 1 billion cubic feet. However, the Michigan "stray" is still of great importance to the gas industry as a reservoir for gas storage (Vary et al., 1968). The "stray" has produced a little oil.

Another sandstone reservoir, the Berea, also of Mississippian age, has yielded relatively minor amounts of both gas and oil. This thin sandstone, and its western Michigan facies, thin sandy dolomites in the upper part of the Ellsworth Shale, have been found to be productive in the Saginaw Bay area and in central-western Michigan adjacent to the Lake Michigan shore. The Berea and the Michigan "stray" are virtually the only sandstone reservoirs productive in the state of Michigan; currently more than 99 per cent of the state's petroleum is coming from carbonate rock reservoirs.

By the end of 1967 about 84 per cent of Michigan's cumulative oil production had come from three Devonian reservoirs, which are, in order of increasing age, the Traverse, Rogers City–Dundee, and Detroit River. Over the same period Ordovician rocks yielded 14 per cent of the total. By 1967, however, the Devonian formations were producing only 35 per cent of the state's annual petroleum output, and the Ordovician output had risen to 59 per cent (Michigan Geol. Survey, 1968). These rocks also accounted for 34 per cent of the gas yield.

Traverse production has been found over the Southern Peninsula except in the northern part and in southeastern Michigan. The Traverse Group consists of a series of interbedded limestone and shale formations with the greater part of the production coming from limestones in the upper part of the group. The Dundee (which, for statistical purposes, is combined with the overlying Rogers City Limestone) has accounted for more than two-thirds of the state's oil and is now producing about 25 per cent of the annual yield. The Dundee–Rogers City producing belt is from three to five counties wide across central Michigan. These limestones, where productive, have porosity due either to secondary leaching or to local dolomitization. In some fields, as at Coldwater, although the wells produce from the Rogers City Limestone the porosity continues downward into the underlying Dundee so that the two formations constitute a single reservoir (Criss and McCormick, 1954).

The underlying Detroit River Group contains three pay zones: the Reed City near the top, the "Sour" in the middle of the Lucas evaporite section, and the Richfield near the base of the Lucas. The Reed City is productive in central and western Michigan; the Sour zone (so named because it yields high-sulfur gas and oil) and the Richfield produce in central and eastern Michigan (Hautau, 1954). The *Reed City* field (Upp, 1968) in Osceola and Lake Counties has produced from all three Detroit River pays, and the higher Dundee and Traverse as well. It is currently a combined secondary recovery and gas storage operation.

Silurian carbonate rocks have become the state's leading source of gas (57 per cent in 1967). Both gas and oil occur in the Salina salt measures, especially in the

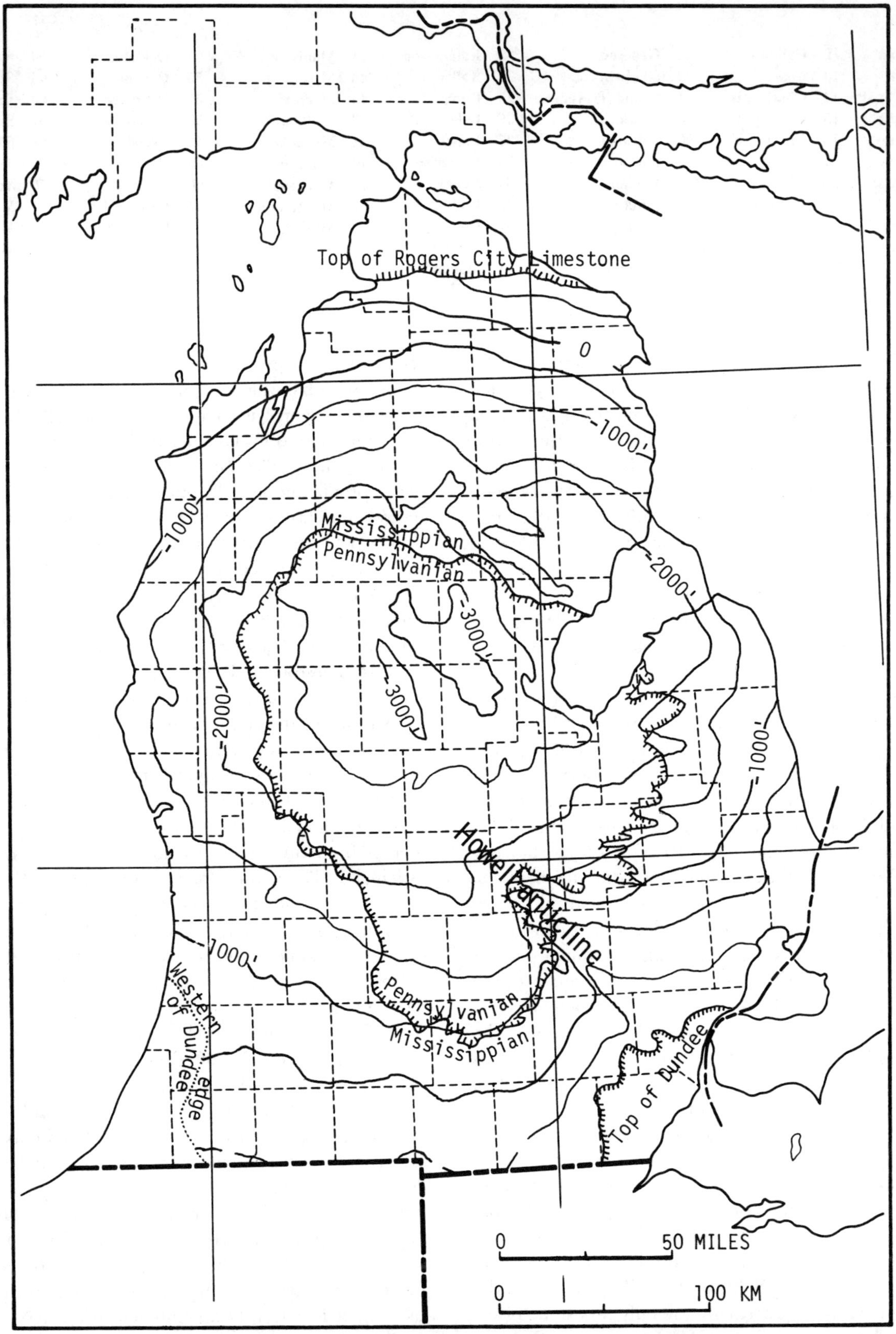

Figure 11.2 Michigan. Southern Peninsula structure map contoured on top of Rogers City-Dundee (Devonian) Limestone. Interval, 500 feet. The Pennsylvanian-Mississippian contact circling the basin center is also shown. *After George V. Cohee (1948).*

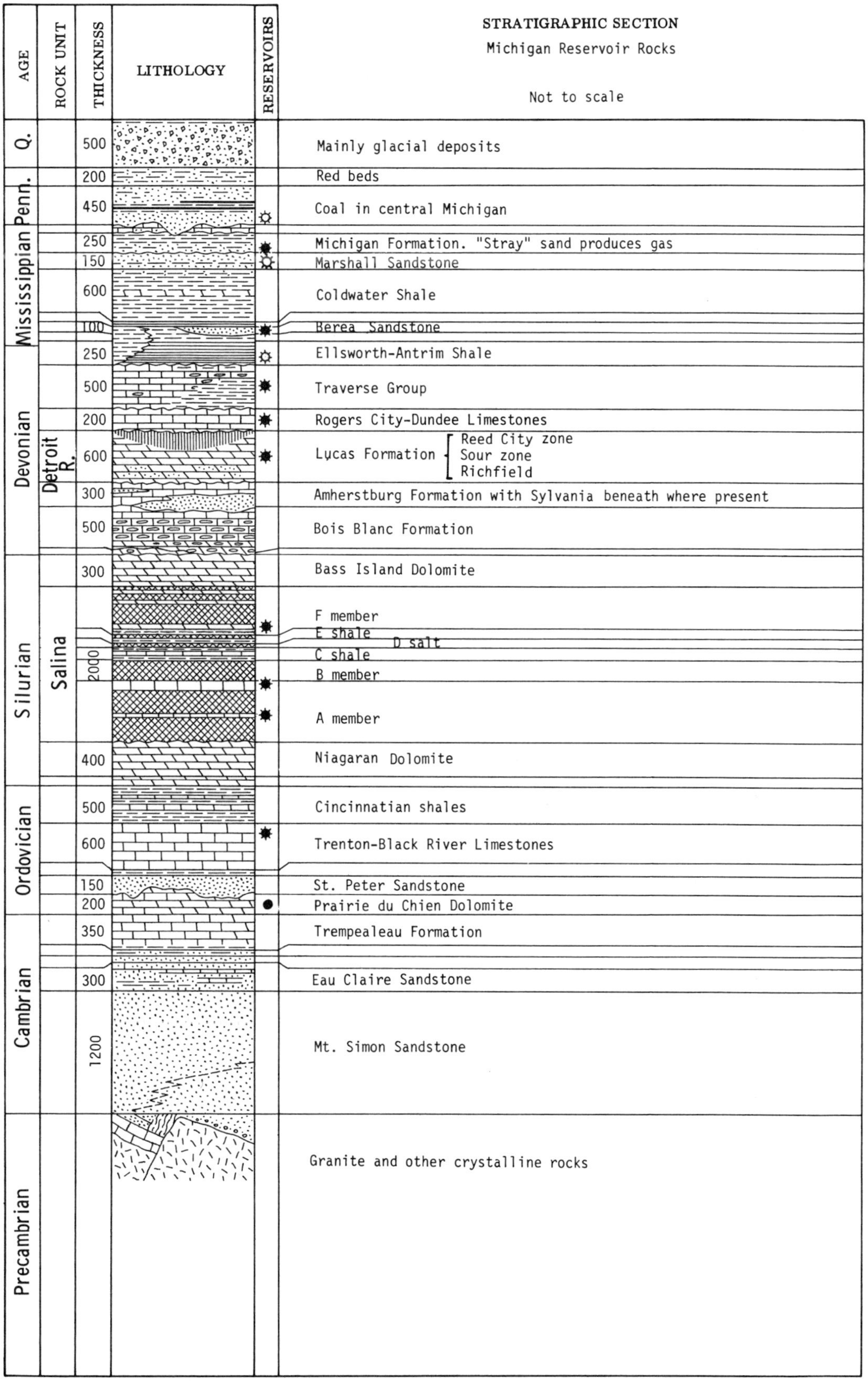

Figure 11.3 Stratigraphic column, with oil and gas reservoirs. *After Michigan Geological Survey (1964) and others.*

limestones in the lower part of the section, and in the underlying late Niagaran reef-type structures. So far, production from these reservoirs has been largely confined geographically to restricted areas in the southeastern and southwestern parts of the Southern Peninsula.

Relatively prolific production has recently been developed in the Trenton and Black River Formations of Ordovician age in central and eastern-southern Michigan. As in the Trenton reservoirs to the south in the once famous Lima–Indiana field (Figure 11.5), porosity is confined to local zones of dolomitization, and therefore the oil is similarly restricted.

Traps. The trapping of hydrocarbons in Michigan is both anticlinal and due to varying permeability wedgouts. The central and southeastern parts of the Michigan basin are crossed by a series of northwest-southeast anticlinal trends. Six or more of these have been mapped, and much of the Michigan basin oil so far discovered has accumulated in the structurally higher places along the anticlinal axes. In the northern part of the Southern Peninsula, the trends appear to change direction to northeast-southwest. The folds in southwestern Michigan have no apparent dominant structural trend. As a general rule, the larger producing Michigan anticlines are from 4 to 6 miles in length and from 1 to $1\frac{1}{2}$ miles wide. They tend to be asymmetrical, with the deeper flank toward the basin, probably because of continued basin sagging after the anticline was formed. The dips, even on the steeper side, rarely exceed 200 feet to the mile, and most of the closures fall between 50 and 100 feet.

The producing anticlines in Michigan were discovered by subsurface methods. Not only does glacial drift cover the bedrock over most of the basin area, but also in places it is too thick for satisfactory use of the reflection seismograph in structure mapping.

Typical anticlinal accumulations in Michigan are the Muskegon, Buckeye, and Porter oil fields. The *Muskegon* oil field (Newcombe, 1932) was discovered in 1927 and

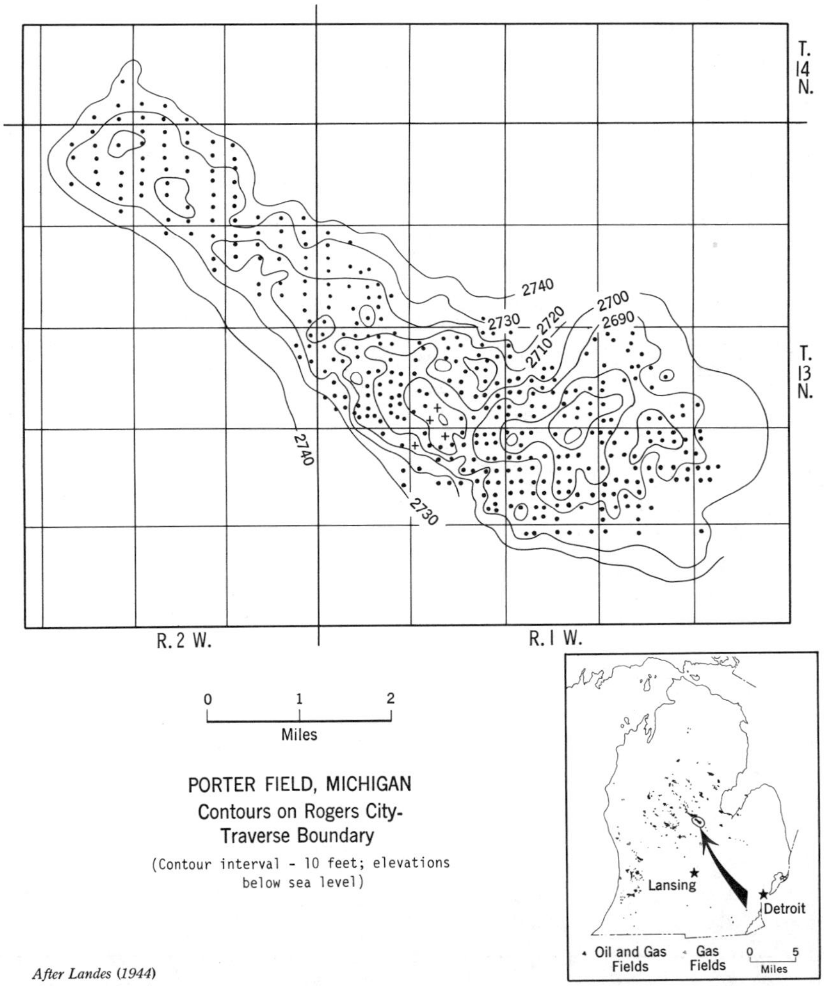

After Landes (1944)

Figure 11.4

was the first field of any appreciable size to be found in the state. The anticlinal trap is an irregular dome with some 60 to 70 feet of closure. The reservoir rocks are porous limestones of Devonian age. *Buckeye* (Addison, 1940) consists of two oil fields, North and South Buckeye, occupying separate domes. Although the two domes are comparable in size and closure, the production of the north dome has been nearly four times as great as that of the south dome. This difference is attributed to changes in porosity and permeability over the short distance separating them. The *Porter* field (Landes, 1944; Helmboldt, 1968), discovered in 1931, was the leading Michigan field in cumulative production until the development of Albion–Scipio; it now ranks second. It lies in southwestern Midland County close to the center of both the Southern Peninsula and the Michigan sedimentary basin. The northwest-southeast trend of anticlines on which the Porter field lies was discovered through the study of brine well records. The structure map for Porter (Figure 11.4) is contoured on the top of the Rogers City Limestone of Devonian age. The maximum closure of the Porter anticline is only about 50 feet. The greater part of the production from this field has come from the Dundee Limestone (Devonian), which underlies the Rogers City Limestone, a dark brown organic rock that may have been the source rock for the oil in the underlying Dundee. The upper part of the Dundee, where the oil occurs, is porous because of pre-Rogers City emergence and leaching. Trapping is due to the anticlinal structure, but distribution on the anticline has been controlled by local permeability.

The *Northville* field (Amoruso, 1957; Miller, 1954; Checkley, 1968), which spreads over the common corner of Wayne, Washtenaw, and Oakland Counties, is of interest for two reasons: (1) the oil discovery well, completed in early 1954, was at that time much the largest Trenton reservoir producer ever drilled in the state, with an initial yield of 625 barrels daily flowing through a restricted choke; and (2) the field has produced from reservoirs belonging to three geological periods—the Ordovician, Silurian, and Devonian—which is unusual in Michigan. The short-lived Dundee (Devonian) and the Salina–Niagaran (Silurian) pay zones produced gas, although two or three Niagaran wells have yielded oil in quantity. The Trenton and Black River rocks, of Ordovician age, produce some oil but mostly gas. As elsewhere in the north-central United States, the Trenton–Black River is productive only where it has been dolomitized; beyond the narrow zone of dolomitization the limestone is nonporous and nonproductive. The trap at Northville is an anticline faulted on the northeast side. Ordrovician production is obtained from fractured and dolomitized rock on the east flank. The Northville anticline plunges gently to the northwest and is considered to be a southeast extension of the much larger Howell anticline, which is developed best in Livingston County (Figure 11.1).

Much of the natural gas in Michigan has come from sands of Mississippian age (Ball et al., 1941). A new type of trap for this state, reef accumulation, was discovered in St. Clair County in 1952. Subsequently more than thirty reef pools have been discovered, mostly in St. Clair County, but also in Calhoun County to the west in central-southern Michigan and in Allegan County in the southwestern part of the state (Ells, 1967). These reefs occur at the top of the Niagaran (Silurian) section and jut upward into the lower evaporites of the Salina group. They have considerable vertical relief and relatively small area. Oil and gas occur in both the unstratified biohermal reef rock and in the overlying lower Salina Dolomite. Gravity meters have had considerable success in locating these reefs (Ferris, 1961), but because the reef is small and the gravity anomalies may not be centered on it, several probes may be necessary to hit the reef above the oil-water interface. The ratio of Niagaran reefs that are oil or gas bearing is unusually high, perhaps as great as 90 per cent. It is interesting to note that reefs had been productive across the St. Clair River in southwestern Ontario for many years before the first discovery was made on the Michigan side of the international boundary.

Buried Niagaran reefs in southeastern Michigan have been described by Ells (1967). These reefs are algal-stromatoporoid biohermal mounds. The Boyd field reef in St. Clair County has about 300 feet of relief on the Niagaran surface (Ells, 1967, p. 22). Lower Salina carbonates and evaporites lap upon, and are draped over, the bioherm. The oil and gas are stored in porous zones in the upper part of the reef and in the overlapping basal Salina dolomite. Evaporite beds above the reef dolomite supply the seal. Sharma (1966), in a detailed study of the *Peters* field reef, also in St. Clair County to the northeast of the Boyd field, concluded that the source rocks are deeper Silurian shales. The compaction of these shales forced the hydrocarbons upward into the lowermost Salina dolomite. Presumably these fluids then migrated updip beneath the evaporite seal to traps formed by the biohermal bulges, where accumulation took place both in the basal Salina dolomite and in the topographically high reef rock. Peters "is the largest reef associated oil field" in Michigan (Kiddoo, 1968).

In the most widespread type of varying permeability trap in Michigan the porosity is lenticular because of local dolomitization of limestone (Landes, 1946). This type of trapping has been mentioned before, in connection with the accumulation of oil in the Lima–Indiana field of Ohio and Indiana and the Northville field in Michigan. The *Deep River* field, an earlier example in

Michigan, has been described elsewhere (Landes, 1948; Lundy, 1968). According to Addison (letter, 1945), the Sherman, Coldwater, Fork, Winterfield, and Evart fields of Michigan are other illustrations of the accumulation of oil in openings produced during dolomitization.

The biggest event in Michigan oil history since the discovery of the first large oil field in the central basin area was the completion of the discovery well of the *Albion–Scipio* oil field trend in 1957 (Ells, 1962, 1967; Ells and Ives, 1959; Ives, 1958, 1960; Buehner and Davis, 1968). Within 5 years the annual production of oil along this trend exceeded 50 per cent of the state's output. The reservoir is dolomitized Trenton and Black River Limestones, and the surprising feature is not so much the discovery of this prolific series of pools as the fact that it was not found until 70 years after the development of the great Lima–Indiana Trenton field a few miles to the south. In the intervening years the only Trenton discoveries of importance were the Deerfield pool (Lindberg, 1948), a relatively small field in Monroe County near the southeastern corner of Michigan, and the Northville field previously described.

The Albion–Scipio trend is a narrow belt of oil and gas production extending, with some gaps, for about 35 miles across parts of three counties, Calhoun, Jackson, and Hillsdale. The oil and gas pays occur in both the Trenton and the underlying Black River, but only where these limestones have been dolomitized. Structure contours drawn on the Trenton top and overlying rocks show that the producing areas are in shallow synclines where the limestone has been fractured as well as dolomitized. The N30°W direction of the production zone is assumed to be due to the presence in the underlying basement rock of a major fracture (fault?) slicing through the southern flank of the Michigan basin. It is postulated that waters containing magnesium ions in solution ascended along this crack and altered to dolomite the Trenton–Black River Limestone neighboring the walls of the channelway. The presence of barren stretches along the trend is probably due to the fact that in those areas the crack was tight, and hence the dolomitizing solutions were diverted to more permeable sections to the northwest and southeast. In other words, the pools are in greatly elongated porosity cupola traps.

Future. Michigan has enough unexplored reservoirs and inadequately tested but promising territory to be in the oil and gas production business for a long time to come. The present and potential pay zones will be considered from youngest to oldest.

The Antrim Shale of earliest Mississippian age has not been developed to any extent. This richly organic shale, up to 600+ feet in thickness, contains fissure gas, as do similar shales in Ohio, Kentucky, Kansas, and elsewhere. A field in Otsego County is reported to have produced 340 million cubic feet of gas between 1940 and 1950. The gas was used by a local industry (Oil and Gas Jour., Nov. 8, 1965, p. 212). A second Antrim gas field in Otsego County was discovered in 1965. The Antrim Shale could also be exploited as an oil shale; uranium is a possible by-product (Swanson, 1960). Some thought has been given to oil extraction from the shale by *in situ* combustion.

Of the seven new field discoveries in 1967, four were in the Devonian (one Traverse, three Dundee), two were reefs in the Silurian, and one was the Berea Sandstone of Mississippian age (Ells and Layton, 1968, p. 979). The preponderance of Devonian discoveries leads to the conclusion that much oil remains to be found in and surrounding the cental basin in the "traditional" pay zones. The Berea discovery, although a small one, may encourage further exploration of this reservoir. In addition to the inner basin Devonian prospects, there could be oil- or gas-occupied traps in Devonian and older reservoirs where they thin down and wedge out to the south as they approach the axes of the Kankakee and Findlay arches and also upflank toward the northern rim of the basin.

Lindberg (1963?) calls attention to the wide, barren gap between the Devonian fields in central lower Michigan and the fields producing from Silurian and Ordovician reservoirs in southeastern and southern Michigan. This belt has been explored, but the dry hole density is not adequate to condemn it. Possible reservoirs in this belt include the as yet unproductive Sylvania Sandstone and carbonate rocks of Bois Blanc, Bass Island, and Late Salina age, as well as younger Devonian and older Siluro-Ordovician pay zones.

The recent discoveries of commercially important accumulations of oil and gas in Silurian and Ordovician reservoirs point to the probable development of many new fields producing from these rocks. According to Sharma (1966), the Silurian reefs in the Michigan basin "occur in the tectonic hinge belt" of the basin. This belt also contains salt anticlines and perhaps facies changes as well. Presumably the hinge belt completely encircles the inner basin. So far, reef fields have been found only in southeastern Michigan, Calhoun County in south-central Michigan, and Allegan and Muskegon Counties in southwestern Michigan. Allegan County also contains Silurian salt anticline accumulations (Ells, 1959, 1967). There is considerable room for additional hinge-inspired hydrocarbon accumulations not only between these known occurrences but also along the much greater arc to the north.

The Ordovician Trenton–Black River, now supplying the greater part of Michigan's hydrocarbon output, is certainly the most glamorous local prospect. Cohee (1945) has noted that all of the Trenton fields so far

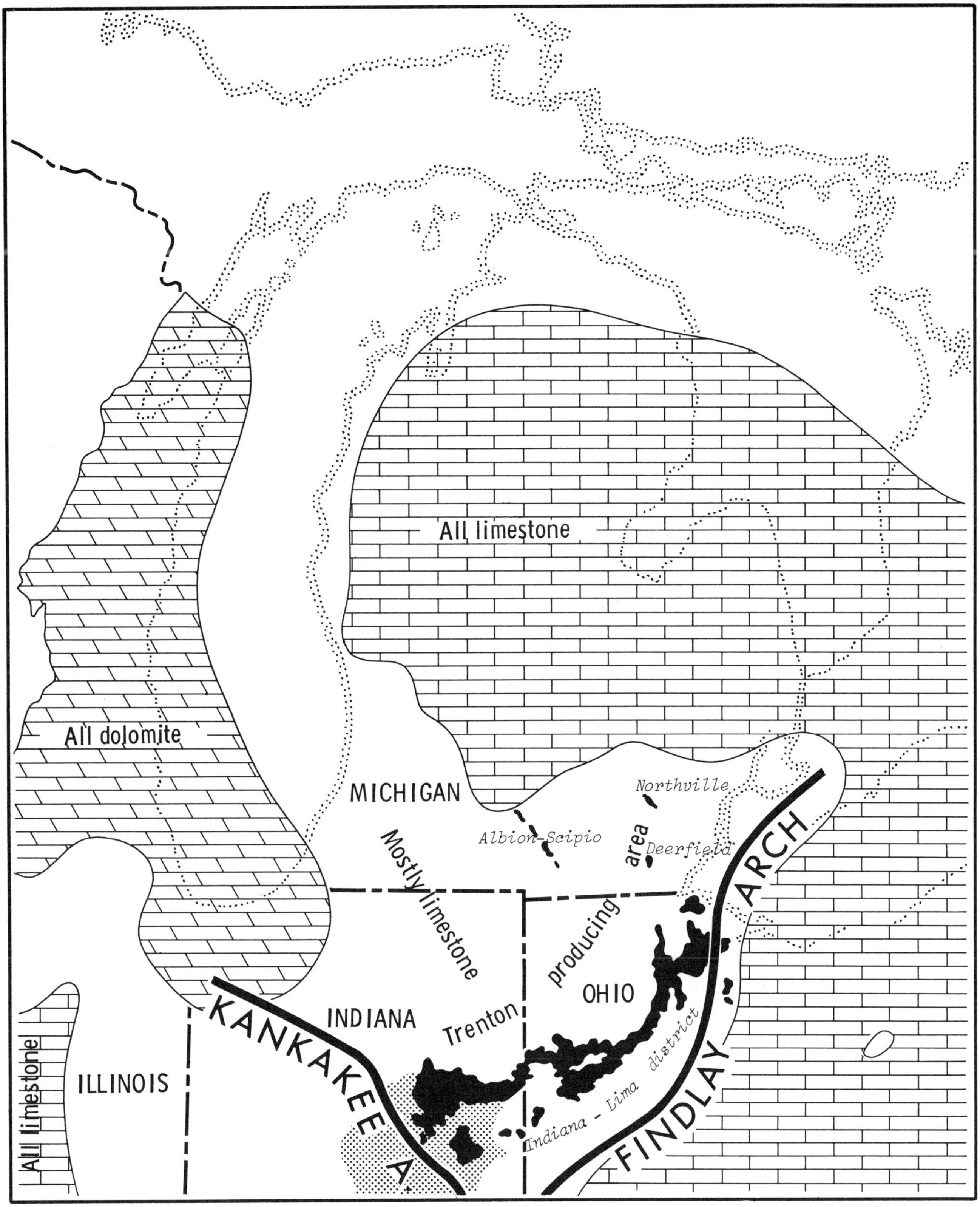

FIGURE 11.5 Relationship between Trenton (Ordovician) fields in Indiana-Ohio-Michigan and lithology of Trenton-Black River section. *After George V. Cohee (1945).*

discovered in the north-central part of the United States occur only where limestone has been locally dolomitized. The area of the Trenton–Black River Limestone, locally dolomitized, extends northward from the Lima–Indiana field into southern Michigan (Figure 11.5) and similar lithologies occur high on the north flank of the basin. These areas are therefore considered to be favorable for Trenton prospecting. Probing for trends parallel to the Albion-Scipio field so far has not been successful, but efforts continue.

In recent years there has been much interest in exploring the large barren area to the north of the basin fields. Here is not only the basin hinge line, preferred locale for traps in Silurian reservoirs, but also, in the northern part of the Southern Peninsula, the proper lithology within the Trenton–Black River section for porosity. That oil occurs in the Trenton on the north flank of the Michigan basin is attested to by the presence of small oil wells on Manitoulin Island, Ontario, and the logging of a good show of oil in a well drilled on Drummond Island in Michigan's Northern Peninsula. Although the Trenton is relatively shallow along the southern coast of the Northern Peninsula, small oil fields along this belt are by no means impossible. More promising, however, is the northern part of the Southern Peninsula, where an exploration program is now underway.

Exploration beneath producing pay zones has been inadequate throughout the state. For example, only one well has been drilled into Ordovician rocks at Porter, Michigan's second largest field and an anticlinal accumulation. One dry hole does not authenticate a drought. The possibility of fracture porosity traps here has been suggested by Helmboldt (1968, p. 140).

Below the Trenton and Black River the section consists principally of sandstones and sandy dolomites with an average thickness of about 2000 feet. Although possible reservoirs are abundant, organic sediments that could have functioned as source rocks are scarce. However, a one-well field in southeastern Hillsdale County produced nearly 8000 barrels of oil from a Prairie du Chien (Early Ordovician) Dolomite pay zone between its discovery in 1960 and its abandonment in 1965, indicating that pre-Black River oil is a possibility. The Glenwood Shale, which caps the Cambrian pay zones in central Ohio, is both thin and erratic in distribution in Michigan (Cohee, 1945).

REFERENCES CONSULTED IN PREPARING CHAPTER 11

Addison, Carl C., 1940
Amoruso, John J., 1957.
Ball, Max W., Weaver, T. J., Crider, H. D., and Ball, Douglas S., 1941.
Buehner, J. H., and Davis, S. H., Jr., 1968.
Champion, Beverly L., 1968.
Checkley, W. G., 1968.
Cohee, George V., 1945, 1948, 1965.
Cohee, George V., and Landes, Kenneth K., 1958.
Criss, C. R., and McCormick, R. L., 1954.
Ells, Garland D., 1959, 1962, 1967.
Ells, Garland D., and Ives, Robert E., 1959.
Ells, G. D., and Layton, F., 1968.
Ferris, Craig, 1961.
Hautau, Gordon H., 1954.
Helmboldt, D., 1968.
Ives, Robert E., 1958, 1960.
Kiddoo, Gordon, 1968.
Landes, Kenneth K., 1944, 1946, 1948.
Lindberg, George D., 1948, 1963?
Lundy, C. L., 1968.
Michigan Basin Geological Society, 1968.
Michigan Geological Survey, 1964, 1968.
Miller, Lee S., 1954.
Newcombe, R. B., 1932, 1934.
Sharma, G. D., 1966.
Swanson, Vernon E., 1960.
Upp, J. E., Jr., 1968.
Vary, J. A., Elenbaas, J. R., and Johnson, M. A., 1968.

Chapter

12

NONPRODUCTIVE STATES

The states so far nonproductive in the Eastern District are Wisconsin on the northwestern corner, all of the New England states, and New Jersey, Delaware, the two Carolinas, and Georgia. (Figure I.1). All of these, except the southern and eastern edges of Wisconsin, are floored with either Precambrian crystalline rocks (Figure I.2) or surrounding Upper Cambrian sandstones derived from these crystalline rocks. Southern Wisconsin contains outcropping Ordovician sediments, and on the east side, inland from Lake Michigan, is a belt about 50 miles wide of Silurian and Ordovician rocks. All of these Paleozoic sections thin and become more clastic toward the Wisconsin arch, where the Precambrian rocks are exposed. For these reasons commercial deposits of oil or gas are improbable in Wisconsin.

With one minor exception all of the New England states consist of igneous and metamorphic rocks ranging in age from Carboniferous to Precambrian (Figure I.2). Even the Pennsylvanian coals in the minuscule Narragansett basin in Rhode Island are so high in carbon ratio as to be in part graphitic. The one exception is the northwestern corner of Vermont, which is penetrated by the narrow southern end of the Champlain–St. Lawrence sedimentary basin (Taylor, 1964). Here are mostly marine sedimentary rocks ranging in age from Middle Ordovician to Early Cambrian. The glacial drift is thin, rarely exceeding 50 feet. So far, wildcatting has resulted only in shows of oil or gas. Commercial gas has been obtained in this sedimentary basin about 100 miles to the north in the Quebec Lowlands. However, the successful exploitation to date has been of accumulations in the glacial drift and not from the underlying Paleozoic bedrock (Maughan, 1964). The latter is in all probability the source of the gas, and many tests have encountered shows in the bedrock section.

The greatest potential area for future development is the submerged coastal plain. The continental shelf (Emery, 1966b), has much greater potential than the emergent plain. The area is larger, the section thicker, and (probably) the stratigraphic column is more diversified in terms of possible source and seal rocks, as well as reservoirs (Emery, 1966a). The entire Eastern Division, northeast of New Jersey as well as southwest, is fringed by a continental shelf ranging in width from a minimum of 25 miles off Cape Hatteras to 250 miles off Cape Cod. South of Cape Hatteras the width of the shelf averages about 70 miles. Submarine canyons off the Florida coast have yielded samples of reefal limestones (Uchupi and Emery, 1967, 1968) of Early Cretaceous age at depths in excess of 16,000 feet. Downwarps of the continental shelf have produced at least two areas of thicker sedimentation, which may be basins. One of these is the Baltimore Canyon area off New Jersey; the other is Georges Bank off Cape Cod. Both were under consideration for geophysical surveys in 1968 (Petroleum Engineer, September, p. 110). A wildcat on bleak Sable Island northeast of Georges Bank and 100 miles off Nova Scotia obtained a gas show at a depth of 14,300 feet. The volume of Mesozoic and Cenozoic rocks beneath the continental shelf appears to be some hundreds of thousands of cubic miles. Maher (1967) suggests that Lower Cretaceous and Upper Jurassic strata are the most likely stratigraphic locale for oil and gas deposits.

The emergent Atlantic Coastal Plain (Figure I.2) starts at Lower New York Bay and continues southwestward, ever widening to a maximum of about 200 miles in southwestern Georgia, where it merges with the Gulf Coastal Plain. Between these end points the plain covers all of Delaware and the eastern parts of Maryland, Virginia, and North and South Carolina. This belt is nonproductive; the oil and gas fields of Maryland and Virginia lie in the Appalachian Valley across the Blue Ridge from the Atlantic Coastal Plain. The onshore sedimentary section is thin between New Jersey and South Carolina, but farther southwest it becomes adequate for the containment of commercial oil and gas accumulations. A major anticline, the Chattachoochie, crosses southwestern Georgia; oil seeps have been reported along its trend (Sever, 1964–1965). The density of dry holes drilled in this area is too low to condemn it.

Another anticline, the Cape Fear arch, parallels the North Carolina coast line southwest of Cape Hatteras. Here, however, the granite basement rises to within 700 feet of the surface (Oil and Gas Jour., May 2, 1966, p. 171).

REFERENCES CONSULTED IN PREPARING CHAPTER 12

Emery, K. O., 1966a, 1966b, 1967.
Heezen, Bruce C., 1968.
Maher, John C., 1967.
Maughan, J. B., 1964.
Sever, Charles W., 1964–1965.
Taylor, Earle F., 1964.
Uchupi, Elazar, and Emery, K. O., 1967, 1968.

Division II

CENTRAL

The Central Division of the United States (Figure II.1) is largely a geographic one, lying between the Mississippi River on the east and the eastern edges of the Rocky Mountain states on the west. To the north is the international boundary with Canada, and the southern limits are the Oklahoma-Texas and Missouri-Arkansas boundaries. In the early days of the domestic petroleum industry the only producing states in this area were Kansas and Oklahoma, and it was referred to as the "Midcontinent" district.

Tectonically the greater part of the Central Division has been stable throughout the post-Proterozoic history (Ham and Wilson, 1967), so that plains dominate the topography. However, the Precambrian shield, which has been eroded into the Laurentian Upland (Figure II.2), penetrates the northeastern sector of the Central Division. The southern corners of South Dakota contain in the east the Sioux uplift and in the west the Black Hills, an outlier of the Rocky Mountains. The Ozark uplift in southern Missouri and northern Arkansas exposes Precambrian rock (Rudman et al., 1965), which creates a submountainous topography. The greatest tectonic activity has been in southern Oklahoma, where the Wichita, Arbuckle, and Ouachita orogenies took place during the Paleozoic era.

The plains inside this perimeter of ancient rock outcrops are underlain by broad sedimentary basins separated by structural divides or arches. The largest of the basins are the Williston, Denver, Salina, Forest City, Anadarko, and Arkoma (Figure II.2). Much of the plains area is either actual or prospective oil country, regardless of whether it is tectonically basin or arch. Except locally, as well as around the fringes of the plains, the sedimentary rock thick-

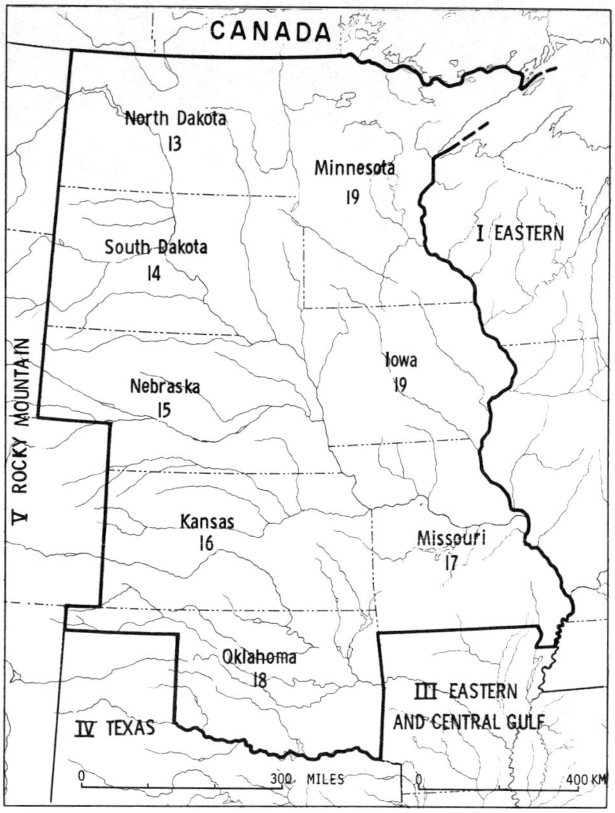

FIGURE II.1 Central United States. Index map of Division II. Numbers refer to chapters.

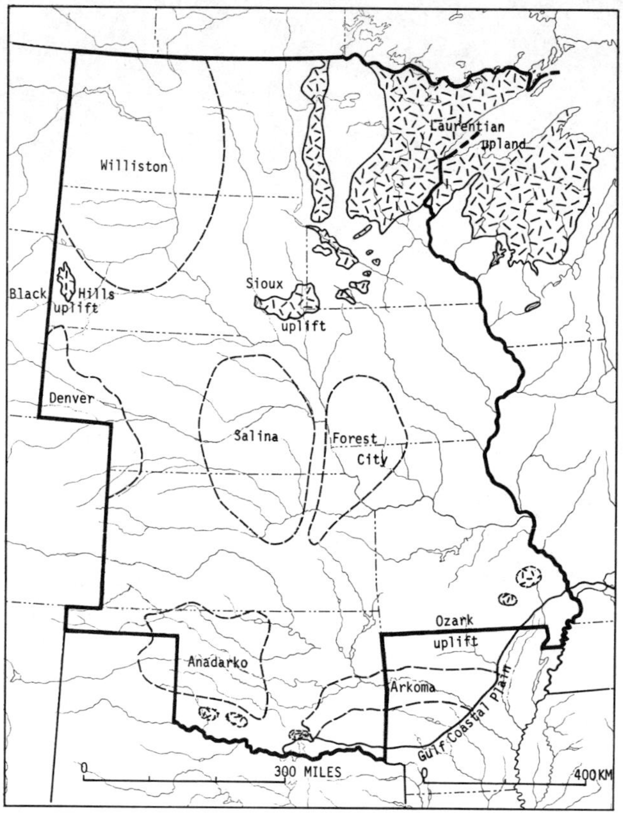

FIGURE II.2 Central Division. Major tectonic features. Sedimentary basins are outlined by dashed lines. Pattern shows outcrops of basement rock. Laurentide upland is southern extension of Canadian shield.

ness is adequate for the occurrence of oil and gas deposits.

The surface rocks in the eastern and southern parts of the Central Division are largely Paleozoic; Cretaceous rocks prevail in the western and northern areas. The surface along the western edge of the province is veneered by late Tertiary continental deposits. Cretaceous sediments of the Gulf Coastal Plain overlap the Paleozoic rocks in southeastern Oklahoma, and early Tertiary sediments of the Mississippi embayment (a northward extension of the Gulf Coastal Plain) cover the surface in the southeastern corner of Missouri.

Geologically, Division II continues northward into Manitoba and Saskatchewan, Canada. Along the western boundary, central geology crosses the North Dakota line into northeastern Montana, and Rocky Mountain geology impinges upon Division II in western South Dakota. Southern Oklahoma geology extends westward into the Texas Panhandle and eastward into Arkansas; also there are sedimentary basins along the Red River that are productive in both southern Oklahoma and north Texas.

REFERENCES CONSULTED IN PREPARING INTRODUCTION TO DIVISION II

Ham, William E., and Wilson, James Lee, 1967.
Rudman, A. J., Summerson, C. H., and Hintze, W. J., 1965.

Chapter

13

NORTH DAKOTA

North Dakota's first oil production was in 1951, with 25,000 barrels. Fifteen years later, in 1966, production reached a peak of 27,168,000 barrels. In 1967 the figure had fallen to 25,185,000 barrels, which was less than 1 per cent of the total U.S. production for that year. The oil came from more than 90 fields in western North Dakota (Figure 13.1). There is some production along the Canadian border from both sides of the international boundary.

North Dakota is a relatively minor producer of natural gas (Anderson and Eastwood, 1968). An unusual occurrence is the presence of appreciable quantities of nitrogen gas in five oil fields on the Nesson anticline (Marchant, 1966).

Tectonics. The major structural feature of North Dakota is the Williston basin (Figure 13.2), which it shares with northwestern South Dakota, a large part of northeastern Montana extending southward almost to the Wyoming boundary, southern Saskatchewan, and the southwestern corner of Manitoba. This basin is a large, roughly elliptical downwarp in the earth's crust, covering somewhere between 134,000 (Laird, 1956) and 240,000 square miles (Barnes, 1953), depending upon what is used as the boundary. The larger area is bounded on the northeast by the Canadian shield, on the southeast by the Sioux uplift, on the southwest by the Black Hills uplift, and on the west by the Big Snowy uplift.

The Williston basin is a typical sedimentary basin, which sagged intermittently to permit the accumulation of unusual thicknesses of sedimentary rocks, including evaporites. In the central part the surface is floored with Tertiary beds, which dip gently toward the middle of the basin, indicating that the latest sagging took place subsequently to the deposition of the youngest rocks in the basin. The pre-Tertiary stratigraphic section includes about 6000 feet of Mesozoic rocks, mostly clastics, and another 5000 to 7000 feet of Paleozoic sediments, which are chiefly carbonate rocks and evaporites.

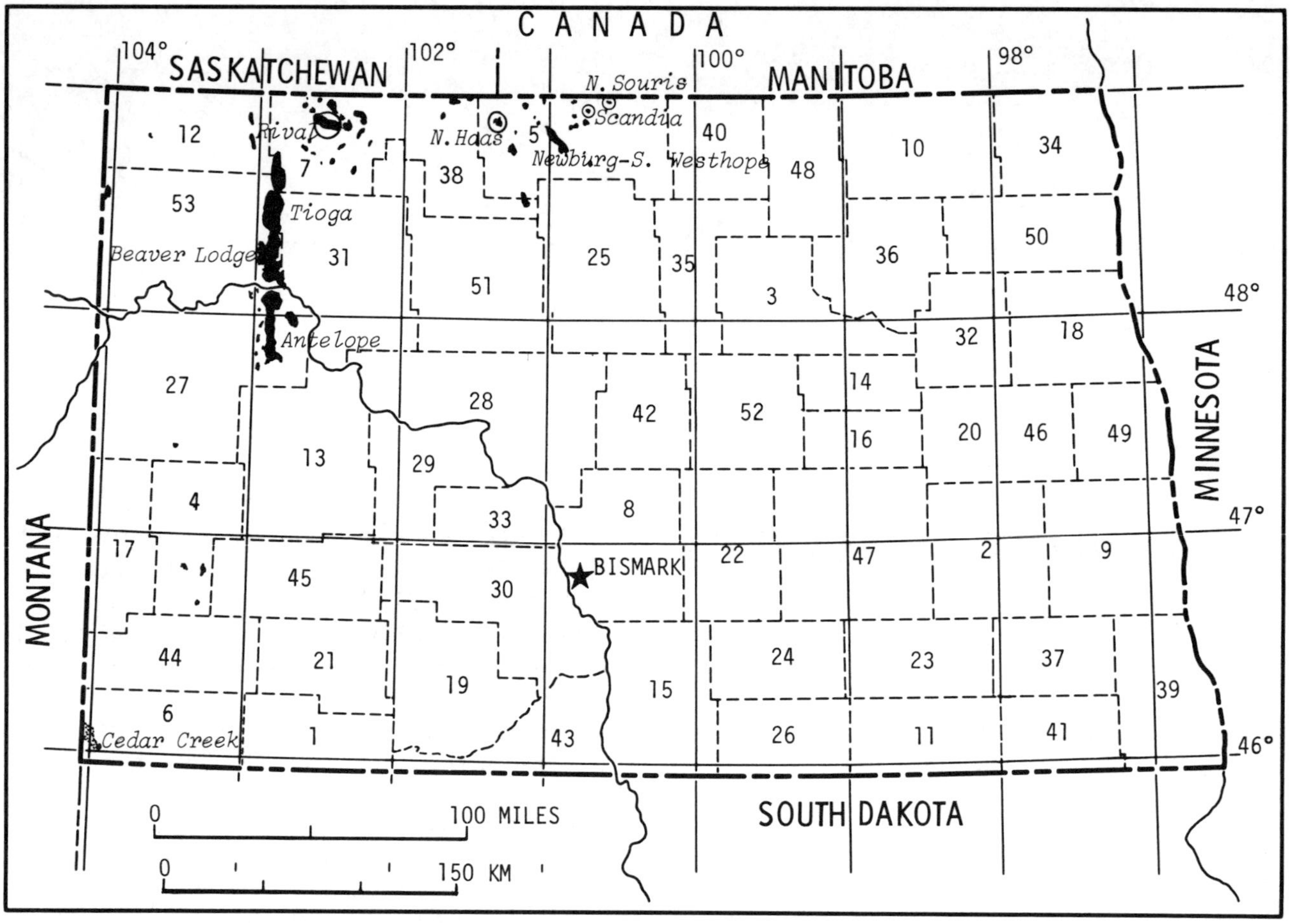

Figure 13.1 North Dakota. Index map with oil and gas fields. County Key:

1	Adams	9	Cass	17	Golden Valley	25	McHenry	33	Oliver	40	Rolette	47	Stutsman
2	Barnes	10	Cavalier	18	Grand Forks	26	McIntosh	34	Pembina	41	Sargent	48	Towner
3	Benson	11	Dickey	19	Grant	27	McKenzie	35	Pierce	42	Sheridan	49	Traill
4	Billings	12	Divide	20	Griggs	28	McLean	36	Ramsey	43	Sioux	50	Walsh
5	Bottineau	13	Dunn	21	Hettinger	29	Mercer	37	Ransom	44	Slope	51	Ward
6	Bowman	14	Eddy	22	Kidder	30	Morton	38	Renville	45	Stark	52	Wells
7	Burke	15	Emmons	23	LaMoure	31	Mountrail	39	Richland	46	Steele	53	Williams
8	Burleigh	16	Foster	24	Logan	32	Nelson						

The Williston basin is productive in each of the two Canadian provinces and three states that it overlaps, although so far the South Dakota production is relatively insignificant. Gas discovery and development started in 1913 along the Cedar Creek anticline in eastern Montana. The first oil discovery in the Williston basin was the successful wildcat that opened up the Virden (Daly) field in southwestern Manitoba in 1951.

The area covered by the basin in North Dakota is 51,600 square miles. The maximum known thickness of the sedimentary rock column is 15,128 feet in McKenzie County in the western part of the state (Carlson and Anderson, 1965). The discovery well for North Dakota was completed on the Nesson anticline (Figure 13.2) in April, 1951, in what became known as the Beaver Lodge field (Harrison, 1954). It reached a total depth of 11,744 feet, obtaining its first production from the Interlake (Silurian) Formation at a depth of 11,630 to 11,660 feet. The initial production rate was 300 barrels daily of 50° gravity oil through a $\frac{3}{4}$ inch choke. Three months later this well was recompleted at a depth of 10,490 to 10,530 feet in the Duperow (Devonian) Formation for a yield of 200 barrels daily of 40° oil. In November, 1952 this exploratory test was converted to a Madison (Mississippian) producer. By the end of 1967, the Beaver Lodge field had produced more than 55 million barrels, the largest cumulative production in the state. It was also the largest field in annual yield, with over 4.2 million barrels.

During 1952 five additional fields were found in North Dakota, four of which lie upon the Nesson anticline. Subsequently discoveries have been made every year, both on the Nesson anticlinal complex (North Dakota Geol. Soc., 1959) and in other parts of western North Dakota (Figure 13.1).

Except for the Cedar Creek anticline, which cuts across the southwestern corner of the state, the Nesson anticline is the only prominent upfold in the North Dakota part of the Williston basin. About one-fourth of the fields so far discovered in North Dakota lie over the top or on the flanks of this structural feature, and these fields account for the greater part of the production to date.

The deepest part of the Williston basin, as mapped on the Piper Formation of Middle Jurassic age (Sandberg, 1959), is about 50 miles west of the southern end of the Nesson anticline. This fold therefore lies low on the eastern flank of the basin. It has a north-south trend and is plunging gently to the south. The anticline creates a steepening of the basinward dip on the west side and a parallel plunging syncline (White Earth syncline) on the east side. It is more than 100 miles long and is breached topographically by the valley of the Missouri River, which exposes the gentle arching of the sedimentary layers.

The Nesson anticline is actually a large but low upfold complex containing many nodes and intervening saddles. The largest node on the trend is the Beaver Lodge, which has a closure of more than 100 feet and a closed area covering about 14 square miles. The crest of the Nesson anticline plunges to the south at about 30 feet per mile, so that the northernmost producing field produces from the Madison (Mississippian) reservoir at an elevation that is about 2000 feet higher than the elevation of this reservoir in the southernmost field.

Stratigraphy. More than 15,000 feet of sedimentary

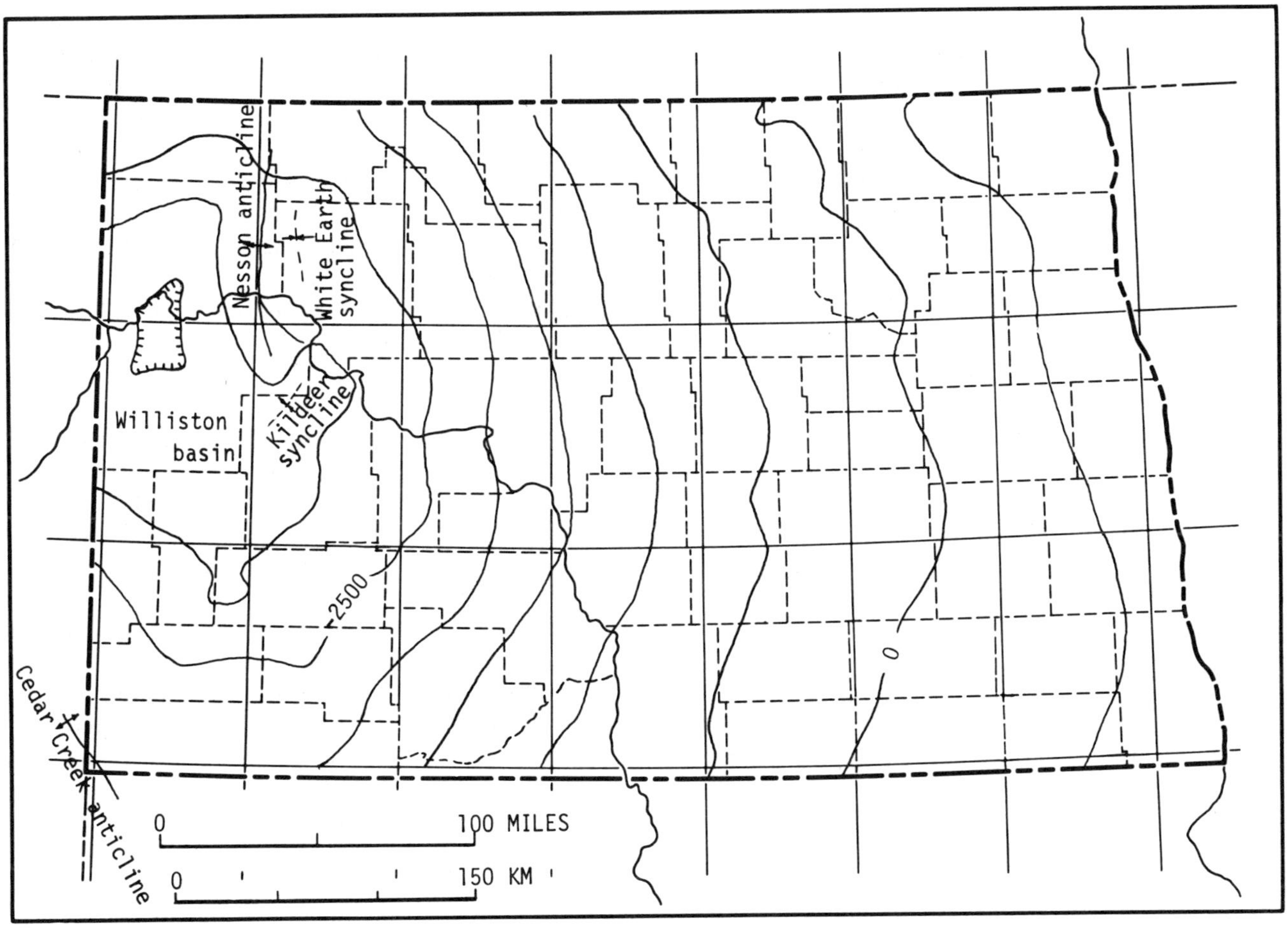

Figure 13.2 North Dakota. Tectonic map. Contoured on base of Colorado Shale. Interval, 500 feet. *After George V. Cohee et al. 1962.*

rock (Figure 13.3) overlies the Precambrian basement in the Williston basin in western North Dakota. Of this total thickness approximately half is Cenozoic and Mesozoic rock and the other half Paleozoic. There is a major unconformity between the Mesozoic formations and the Paleozoic rocks; the Mesozoic sediments overlie the truncated Paleozoic formations over most of North Dakota. Over half of the Cenozoic-Mesozoic section is Late Cretaceous in age, and the Mississippian dominates the Paleozoic section, constituting about 40 per cent. So far, with two minor exceptions, all of the reservoirs found to be productive in the North Dakota part of the Williston basin are in the Paleozoic section, ranging from a Pennsylvanian sandstone to the Late Cambrian—Early Ordovician Deadwood Sandstone.

The youngest reservoir so far discovered is a sand near the base of the Pierre (Upper Cretaceous) Shale that produces gas in the Cedar Creek gas field of Montana–North Dakota. Another gas sand in the Upper Cretaceous is the Eagle. The only other Mesozoic pay as yet is a sandstone in the basal Triassic section that produces oil in Bottineau County.

The highest Paleozoic reservoir is the basal Pennsylvanian Tyler Sandstone. This is productive in Billings and Stark Counties in southwestern North Dakota (Foster, 1961).

By far the most prolific source of oil in North Dakota is the Madison Group of the Lower Mississippian (Smith et al., 1958). It was estimated by Carlson et al. (1960) that 98 per cent of all of the oil produced in North Dakota through 1958 came from carbonate rock reservoirs lying within this group. The Madison, from the top down, consists of a thick evaporite section, a massive bedded carbonate rock section, and a thin bedded, cherty, argillaceous limestone. The evaporites and the carbonate rocks interfinger, however, because of simultaneous deposition in different parts of the basin. The oil and gas pays are carbonate rocks lying both within and below the evaporite section.

The Madison reservoir rock in the Beaver Lodge field has been described as

> . . . a fine to medium grain fragmental limestone and dolomitic limestone. Grains include fossil fragments, crystal fragments, and fecal pellets. Originally, porosity was intergranular, but later it was increased by solution and dolomitization, and decreased by recrystallization and cementation (Towse, 1957).

As many as three superimposed porous zones, each from 30 to 50 feet thick, may be present. However, pressure studies show that the zones are interconnected, probably by means of fractures occurring in the denser, finer grained carbonate rocks between the porous intervals. These fractures also contribute to the reservoir capacity of the Madison. To the north and east of the major Nesson anticline fields, oil occurs in the evaporite section of the upper Madison, and here the anhydrite beds act as a seal for the hydrocarbons in the porous interbedded carbonate layers. This stratigraphic section is referred to as the Mission Canyon Formation (Harris et al., 1966).

The Three Forks Formation of Late Devonian age yields oil from fractures in a part of the Antelope field on the east side of the Nesson anticline. This occurrence is described on p. 94. Limestone and dolomitic limestone of the Duperow Formation, also Late Devonian in age but older than the Three Forks, are productive in several fields along the Nesson anticline. It is assumed to be a fractured carbonate rock type of porosity.

The first reservoir to produce at Beaver Lodge was the Interlake Dolomite of Silurian age. It occurs there at a depth of more than 11,400 feet. Because of an abnormally high gas–oil ratio (more than 7000 to 1) this reservoir has been developed as a condensate pool. The Interlake produces oil in the Antelope field.

A fossiliferous limestone in the Red River Formation of Middle Ordovician age, which is an important oil reservoir in the Montana part of the Williston basin and along the Cedar Creek anticline in North Dakota, was being developed in the Beaver Lodge and Tioga fields in 1968. At Beaver Lodge the Ordovician production consists largely of gas and condensate. These fluids are produced from the underlying Winnipeg and Deadwood Sandstones as well. Shales within the Deadwood may be the source rocks (Lochman–Balk and Wilson, 1967).

Traps and Selected Fields (North Dakota Geol. Soc., 1962). The greatest concentration of oil fields in North Dakota is along the axis and on the flanks of the Nesson anticline (Laird and Folsom, 1956). In addition, two fields overlie the axis of the Cedar Creek anticline, which cuts across the extreme southwestern corner of the state. Forty-four per cent of the traps described in the symposium volume, "Oil and Gas Fields of North Dakota," occur on these two major upfolds. Twenty-eight (80 per cent) of the traps on the Nesson anticline are the result of anticlinal closure. However, in seven of these traps, the distribution of oil across the anticline has been affected by local changes in permeability. Permeability changes are also important in the accumulation in the two fields on the Cedar Creek anticline. The remaining 20 per cent of the Nesson anticline traps are noses on the flanks of the major anticline where the updip sealing was brought about by decreasing permeability.

A structure contour map of the *Tioga* field (Figure 13.4) is included as an example of anticlinal accumulation on the Nesson upfold. This field lies along the axis of the anticline immediately to the north of the Beaver Lodge field, where the discovery well for the North

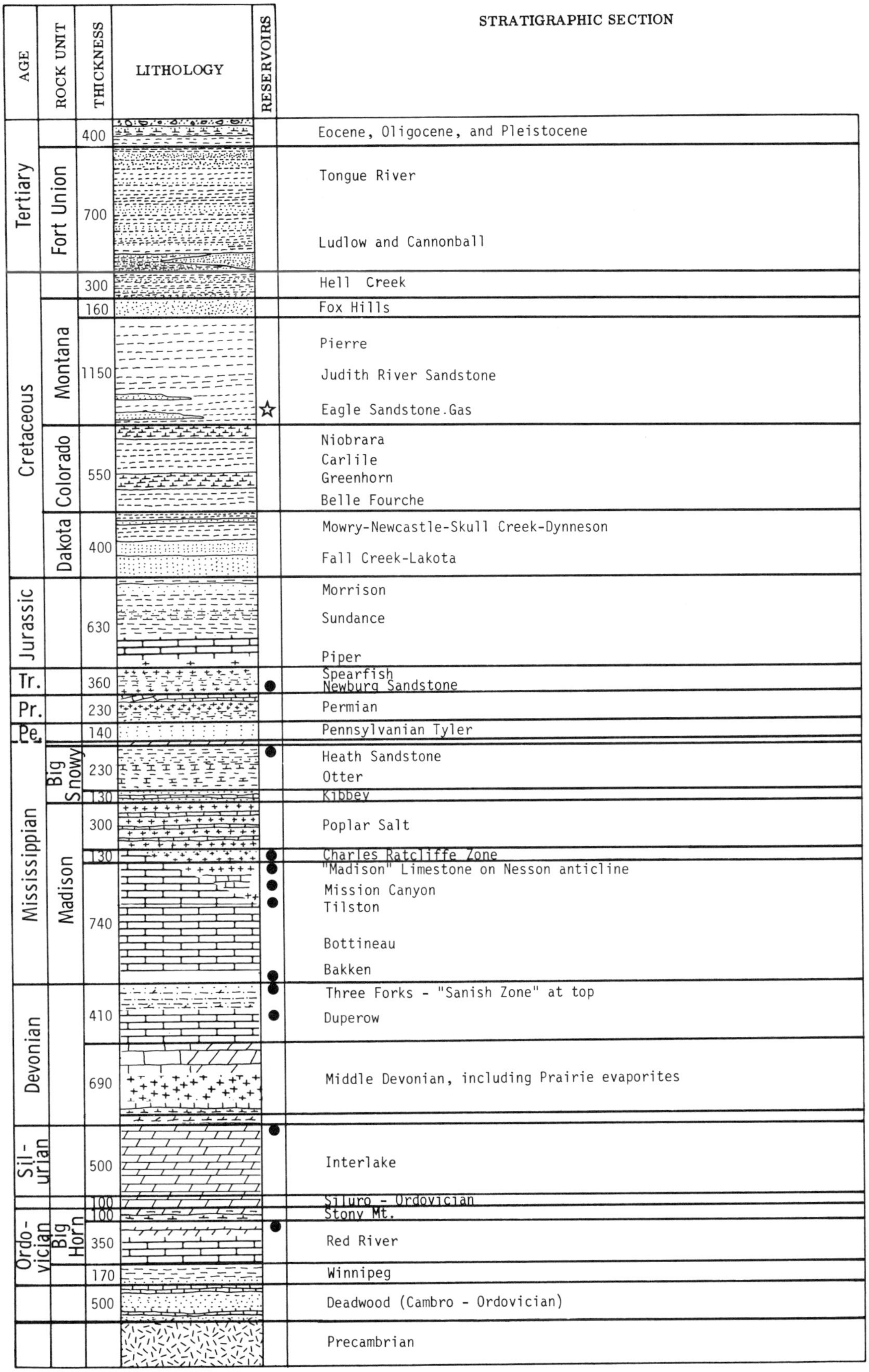

FIGURE 13.3 **North Dakota. Stratigraphic column, showing principal reservoirs.** ***Courtesy North Dakota Geological Survey.***

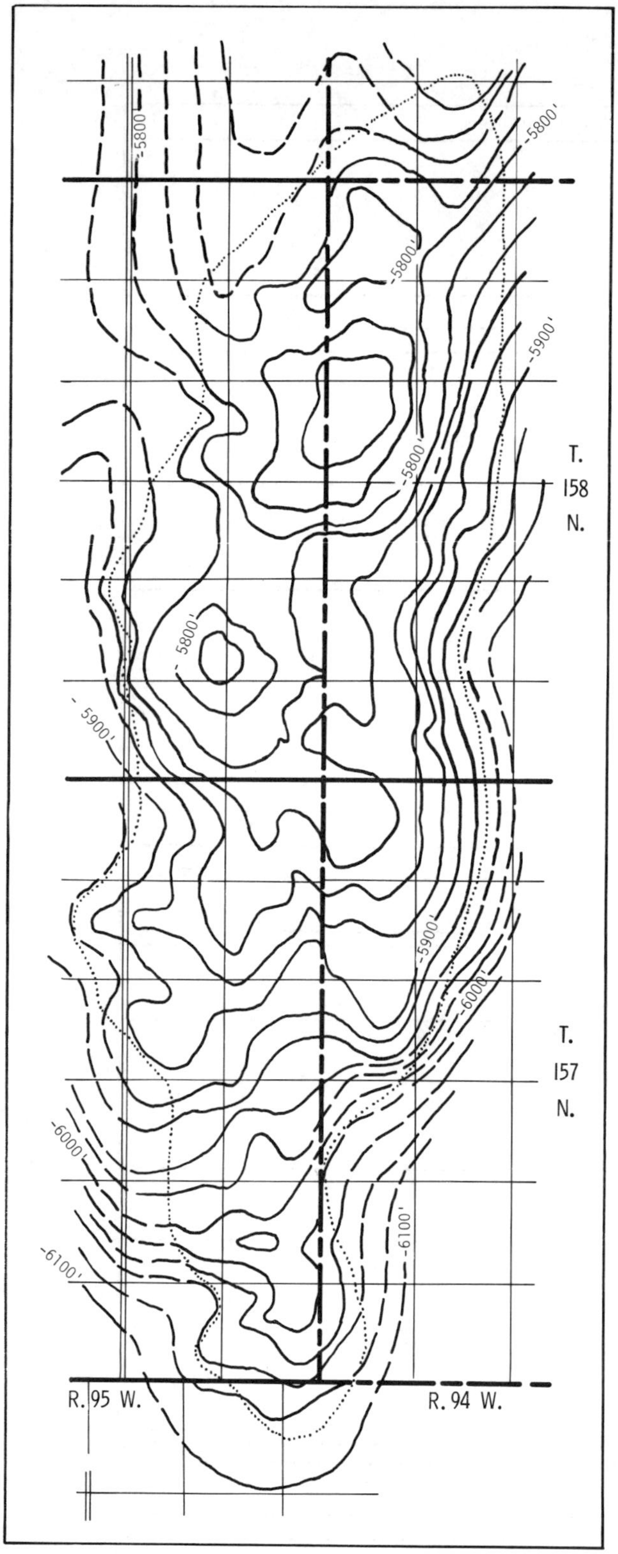

Figure 13.4 **Tioga field, Williams County, North Dakota. Structure contours are based on top of pay zone (Rival subinterval in the Mississippian). Contour interval, 20 feet. Dotted line encloses productive area. Squares are sections, 1 mile (1.61 km) on a side.** ***Courtesy North Dakota Geological Society (1962, p. 201 et seq.).***

Dakota sector of the Williston basin was drilled. It extends from Williams County eastward into Mountrail County and northward into Burke County (North Dakota Geol. Survey, 1960). Tioga field has a proved area in excess of 20,000 acres, and during its first decade (1952–1961) produced nearly 45½ million barrels of oil. The Nesson anticline is plunging to the south; the difference in elevation between the lowest producing well at the southern end of the field and the highest producing well at the northern end is about 370 feet. The principal reservoir rock is limestone, occurring near the top of the Madison (Figure 13.3) of Early Mississippian age. It is fine grained with pinpoint porosity; the porosity percentage is 7.8. Production was established in the Devonian Duperow and the Ordovician Red River by deeper drilling in 1967 (Olson et al., 1968).

The *Antelope* field (Figure 13.1) in northeastern McKenzie County is southeast of Beaver Lodge. It occupies an elongate node on a spur of the Nesson anticline. Five different pools have been developed, in Mississippian, Devonian (three), and Silurian reservoirs. Second only to the Madison in productivity is the uppermost Devonian reservoir, referred to as the Sanish zone, at the top of the Three Forks (Murray, 1968). This zone consists of a thin, discontinuous dolomitic sandstone at the top; downward and laterally the Sanish beds are dense dolomites with intercalated waxy shales.

The trapping in the Antelope field is anticlinal for all pays, but in the Sanish zone the distribution of the oil on the anticline is controlled by porosity and permeability. At first it was assumed that the Sanish oil was stored in the erratic sandstone at the top of the zone, but this idea was abandoned when it was found that some wells in the sand were poor producers whereas excellent wells may have penetrated no sand whatsoever (Murray, 1968, p. 58). Furthermore, it was discovered that the greatest yield from the Sanish pay came from wells drilled on the steepest (northeast) flank of the Antelope anticline. Murray (1968) concludes that the porosity and permeability in this reservoir are due to tension fracturing of the Sanish zone rocks on the steep limb of the fold. It is further concluded "that the overlying Bakken Shale (lowermost Mississippian) is the immediate, as well as the ultimate, source of this oil."

The 50 traps described (56 per cent) that are not related to either the Nesson or the Cedar Creek anticline are mainly in Burke, Renville, and Bottineau Counties, along the Canadian border in western North Dakota. Among these, 7 (14 per cent) are anticlinal accumulations, with permeability controlling localization in the closed area in 4 instances. Twenty-one pools (42 per cent) are anticlinal noses with impounding taking place below an updip permeability barrier. The largest number of pools, 22 (44 per cent), are stratigraphic traps. Most

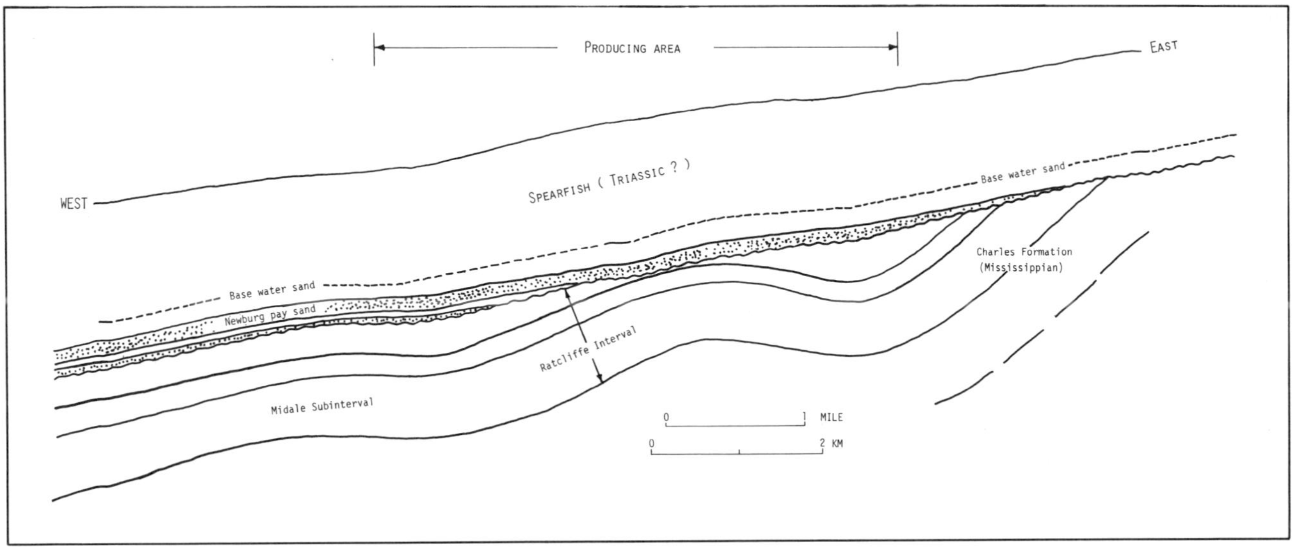

FIGURE 13.5 Newburg field, Bottineau County, North Dakota. West to east cross section, showing wedgout of Newburg pay sand. ***Courtesy North Dakota Geological Society (1962, insert following p. 142).***

of these underlie an unconformity. The *Newburg* field, Bottineau County, is an outstanding example of this type of trapping. The cross section (Figure 13.5) shows that the principal reservoir is a basal Triassic sandstone that wedges out updip on the east flank of the Williston basin. Production is also obtained from a fragmental limestone occurring in the Ratcliff interval of the Madison Group (Mississippian) along its subcrop beneath the pre-Triassic erosional and angular unconformity. In two other Bottineau County fields, the *North Souris* and the *Scandia,* oil stored in Mississippian limestones is sealed updip by anhydrite infilling immediately beneath the pre-Triassic unconformity and overlying Triassic shale. In the *Rival* pool of Burke County, the updip dam is a facies change from porous limestone to tight anhydrite without benefit of overlying unconformity. The reservoir in the *North Haas* field in Bottineau County is a lens of sandstone in the Madison that is completely surrounded by anhydrite.

FUTURE. Only a relatively small segment of the part of the Williston basin lying in North Dakota has been tested with anything approaching an adequate degree of wildcat density. Even the much drilled areas, such as the Nesson and Cedar Creek anticlines, have not been tested thoroughly for pre-Mississippian reservoirs, even though oil has been found in Ordovician, Silurian, and Devonian formations on the Nesson anticline. Likewise, there has not yet been an adequate search for accumulations in Jurassic and Cretaceous rocks. Discontinuous sands in the Cretaceous Dakota Group, especially the Newcastle, which are important oil reservoirs in eastern Wyoming, have been traced into the Williston basin of North Dakota (Reishus, 1968; Wulf, 1964b). The Muddy Sandstone, which is so productive in the Bell Creek field in southeastern Montana, may be present in North Dakota as well (Biggs and McGregor, 1968, p. 53). This erratic sand occurs in the Dakota Group at the base of the Mowry Shale. It is similar to the Newcastle Sandstone, but slightly younger.

Facies change from sandstone to shale has been noted in a six-county area near the southwestern corner of the state by Anderson (1967). Depths to the Newcastle here are in the 5000 foot range. This same corner is considered to have good possibilities in other reservoirs as well, including the Upper Cretaceous Eagle Sandstone, the Tyler Sandstone (basal Pennsylvanian), and the Mississippian Madison and Ordovician Red River Formations. In addition a now abandoned field produced from the Bakken–Three Forks section of Mississippian–Devonian age, and shows of oil have been found in the Interlake (Silurian).

What may well offer the best hunting is the wide band on both the east and the west flanks of the Williston basin, wherein Devonian, Mississippian, and Triassic reservoirs wedge out beneath unconformities. Already large deposits of oil have been found on the northeast flank, especially in Bottineau County; in all probability these same conditions continue southward on the east flank (Folsom and Anderson, 1955; Pye, 1955). Likewise, the older reservoirs wedge out against the northeast flank of the Cedar Creek anticline on the west flank of the Williston basin. Included in this updip disappearance are Devonian reservoirs that are productive along the Nesson anticline (Anderson, 1963).

Reef trends in the Ratcliffe zone of the Madison

mapped in Williams County, northwestern North Dakota, may have additional oil fields (Hansen, 1966).

Permeability pinchouts are also attractive, even though more difficult to find. Sand lenses in the Tyler Formation of basal Pennsylvanian age have been described by Foster (1961). Considerable oil has already been found in North Dakota in porous limestone downdip below anhydrite seals. Additional accumulations may exist in Bottineau and nearby counties because of updip facies change from carbonates to evaporites in the Mississippian Charles and Mission Canyon Formations (Wakefield, 1965; Eastwood, 1962). A widespread porosity zone in the upper Mission Canyon is truncated, sealed by overlap, and productive here and there along a southeast-trending belt extending from southeastern Saskatchewan across the southwestern corner of Manitoba into eastern Bottineau County, North Dakota (Kinard, 1964), where it is least explored.

REFERENCES CONSULTED IN PREPARING CHAPTER 13

Anderson, Sidney B., 1963, 1966, 1967.

Anderson, Sidney B., and Eastwood, William P., 1968.

Barnes, T. R., 1953.

Biggs, Charles A., and McGregor, Alexander A., 1968.

Billings Geological Society, North Dakota Geological Society, and Saskatchewan Geological Society, 1964.

Carlson, C. G., and Anderson, S. B., 1965.

Carlson, Clarence G., Bakken, Wallace E., and Kume, Jack, 1960.

Eastwood, William P., 1962.

Folsom, Clarence B., and Anderson, Sidney B., 1955.

Foster, Frank W., 1961.

Hamke, J. R., Marchant, L. C., and Cupps, C. Q., 1966.

Hansen, Alan R., 1966.

Harris, Steven H., Land, Cooper B., Jr., and McKeever, John H., 1966.

Harrison, Ray L., Jr., 1954.

Kinard, John C., 1964.

Laird, Wilson M., 1956.

Laird, Wilson M., and Folsom, Clarence B., Jr., 1956.

Lochman-Balk, Christina, and Wilson, James Lee, 1967.

Marchant, L. C., 1966.

Murray, George H., Jr., 1968.

North Dakota Geological Society, 1959, 1962.

North Dakota Geological Society and Saskatchewan Geological Society, 1956, 1958.

North Dakota Geological Survey, 1960.

Olson, Charles F., Curry, William, H., III, Norwood, E. Earl, and Slocum, R. C., 1968.

Pye, Willard D., 1955, 1958.

Reishus, Mark, 1968.

Sandberg, Dorothy T., 1959.

Smith, G. Wendell, Summers, George E., Jr., Wallington, Dale, and Lee, Jean L., 1958.

Towse, Donald, 1957.

Wakefield, W., W., 1965.

Wulf, George R., 1964b.

Chapter

14

SOUTH DAKOTA

South Dakota became a producing state in 1954 (Agnew and Gries, 1960) with the discovery of the Buffalo field (Figure 14.1) close to the northwest corner of the state. Twenty-eight producers have been completed in this field. Another small field, the Barker Dome, was discovered in 1955 on the southwestern edge of the Black Hills. The state's first production peak was reached in 1960 with a total output of 281,000 barrels; in 1966 the yield was 220,000 barrels.

Three major structural features dominate the geology of South Dakota. Two of these are uplifts, the Black Hills uplift in the southwestern part of the state, and the Sioux uplift near the southeastern corner. Erosion has uncovered Precambrian rock in both of these areas, as well as just off the eastern edge of northeastern South Dakota. The third major structural feature is the Williston basin, which extends southward into western South Dakota from North Dakota. The Cedar Creek anticline, a sharp elongate upfold in western Montana, continues southeastward across the southwest corner of North Dakota into the northwest corner of South Dakota. Minor basins, so far as extent in South Dakota is concerned, are the Denver in the southwestern corner of the state and the Kennedy to the east, across the Chadron arch (Figure 14.1). It is probable that the Kennedy basin is an extension southward of the Williston basin (Duncan J. McGregor, letter dated 18 October, 1968).

Large areas in eastern and south-central South Dakota are floored with sedimentary formations of Cretaceous age. Tertiary sedimentary rocks underlie Cretaceous rocks in large areas in south-central and northwestern South Dakota and veneer the surfaces of higher relief in western South Dakota except in the areas of the Black Hills uplift, where they are for the most part absent. The pre-Tertiary subsurface section extends from the Upper Cretaceous to the Precambrian. In northwestern South Dakota, where the basement lies at greatest depth, the sedimentary section is approximately 10,000 feet thick;

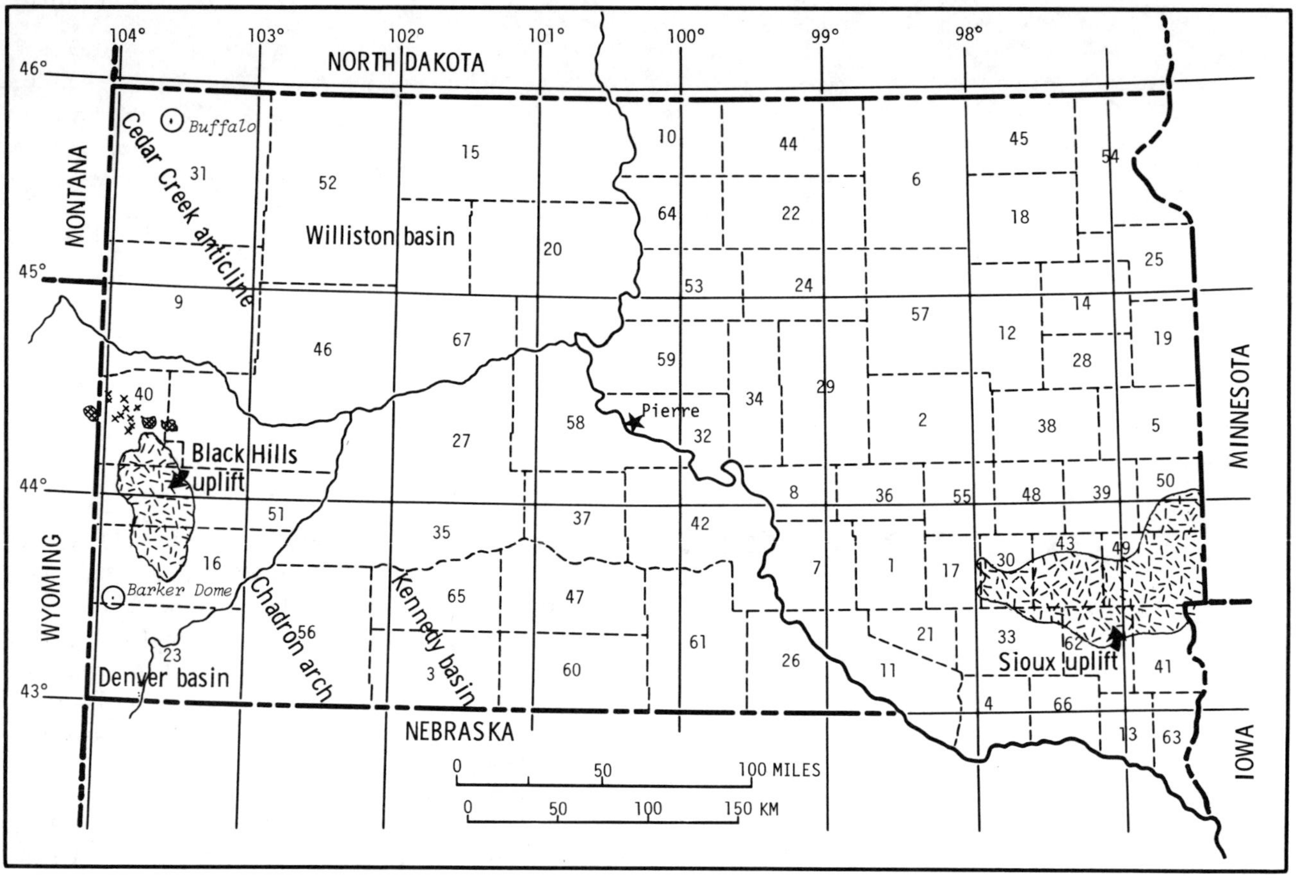

FIGURE 14.1 South Dakota. Index map with oil and gas fields. Two large pattern areas are outcrops of Precambrian basement rock in cores of uplifts. Small areas north of the Black Hills Precambrian designated by pattern and crosses are exposures of Tertiary intrusives. *After George V. Cohee et al. (1962).* County Key:

1	Aurora	11	Charles Mix	21	Douglas	31	Harding	41	Lincoln	50	Moody	59	Sully
2	Beadle	12	Clark	22	Edmunds	32	Hughes	42	Lyman	51	Pennington	60	Todd
3	Bennett	13	Clay	23	Fall River	33	Hutchinson	43	McCook	52	Perkins	61	Tripp
4	Bon Homme	14	Codington	24	Faulk	34	Hyde	44	McPherson	53	Potter	62	Turner
5	Brookings	15	Corson	25	Grant	35	Jackson	45	Marshall	54	Roberts	63	Union
6	Brown	16	Custer	26	Gregory	36	Jerauld	46	Meade	55	Sanborn	64	Walworth
7	Brule	17	Davison	27	Haakon	37	Jones	47	Mellette	56	Shannon	65	Washabaugh
8	Buffalo	18	Day	28	Hamlin	38	Kingsbury	48	Miner	57	Spink	66	Yankton
9	Butte	19	Deuel	29	Hand	39	Lake	49	Minnehaha	58	Stanley	67	Ziebach
10	Campbell	20	Dewey	30	Hanson	40	Lawrence						

available well information indicates that Paleozoic rocks here are about 4100 feet in thickness and are covered by approximately 6000 feet of Mesozoic and Cenozoic sediments. Because of the fact that sedimentation throughout much of the Paleozoic and Mesozoic was influenced by nearby emergent areas, facies changes are the rule rather than the exception.

The *Buffalo* oil field in northwest-central Harding County consists of one large and three small producing pools (Sandberg and Prichard, 1964). They occupy nodes on the Cedar Creek anticline near its southern terminus. Two pay zones occur about 60 feet apart at depths of 8600 to 8700 feet. The reservoir is Red River (Ordovician) carbonate rock.

The *Barker Dome* field lies on the southwestern flank of the Black Hills (Figure 14.1). It was discovered in 1955, by the eighth well drilled on this surface-mappable anticline. It is now a four-well field. The pay zone is a sand in the Minnelusa Formation (Pennsylvanian) at a depth of about 1400 feet (Gries, 1964).

The greatest expectations for further discovery in South Dakota center on the western half of the state, where the sedimentary basins have an adequate stratigraphic thickness. The greatest of these basins is the

Williston; the South Dakota sector has an area of 25,000 square miles (Sandberg and Prichard, 1964, p. 157), nearly one-third of the state. So far the only discovery in this vast area has been the Buffalo field, but in spite of many exploratory wells the dry hole density is still very low. A near discovery took place in southwestern Dewey County in 1964, when noncommercial production was obtained from Red River Dolomite. This formation is believed to have the greatest potential in the Williston basin of South Dakota. Other good possibilities include the Winnipeg Formation (Ordovician), which underlies the Red River, Silurian and Devonian rocks, and the Madison Group of Mississippian age (Sandberg and Prichard, 1964, p. 158). Broad wedgout belts of these Paleozoic rocks, which lie high on the flanks of the Williston basin in western and northern South Dakota, should contain varying permeability traps (Wulf and Gries, 1963). Structural traps are no doubt present also, but the localization of oil in such traps is likely to be controlled by varying permeability due to facies changes.

The rest of western South Dakota is occupied by the Black Hills uplift, which is the eastern rim of the Powder River basin of Wyoming, the northernmost tip of the Denver basin, the north-plunging axis of the Chadron arch, and the northern sector of the Kennedy basin, which is mostly in Nebraska. So far the four-well Barker Dome field has been the only discovery in this considerable area. Bolyard and McGregor (1966) point out that Lower Cretaceous sands contain oil in quantity trapped below updip permeability barriers in the Powder River basin immediately to the west in Wyoming, and that similar traps are present in the Lower Cretaceous sands of western South Dakota. The Minnelusa (Pennsylvanian and Permian) Formation, a leading source of oil in eastern Wyoming, is looked upon with favor in the belt surrounding the Black Hills (Conley, 1962). Wulf (1964a) recommends exploration of the Chadron arch to the southeast. Across the arch the southern extension of the Williston basin (Kennedy basin) is inadequately tested.

REFERENCES CONSULTED IN PREPARING CHAPTER 14

Agnew, Allen F., and Gries, John Paul, 1960.

Bolyard, Dudley W., and McGregor, Alexander A., 1966.

Conley, Jack N., 1962.

Gries, John Paul., 1964.

Hamke, J. R., Marchant, L. C., and Cupps, C. Q., 1966.

Olson, Charles F., Curry, William, H. III, Norwood, E. Earl, and Slocum, R. C., 1968.

Sandberg, C. A., and Prichard, G. E., 1964.

Schoon, Robert A., 1965.

Wulf, George R., 1964a.

Wulf, George R., and Gries, J. P., 1963.

Chapter

15

NEBRASKA

Nebraska had its first recorded production in 1939, and by 1962 the annual output figure had risen to a peak of nearly 25 million barrels. The yield for 1967 was 13½ million barrels, less than 0.5 per cent of the national total. The oil and gas fields are scattered over 15 counties in western Nebraska and 1 county (Richardson) in the extreme southeastern corner of the state (Figure 15.1).

The discovery oil well for Nebraska was completed in 1939 in what became known as the Falls City pool in Richardson County. This was the first production in the Forest City basin. The first western Nebraska discovery was made in 1949, when a wildcat well in Cheyenne County found the Gurley oil pool. This discovery extended the productive part of the Denver basin into the southwestern corner of the Nebraska "panhandle." In June, 1950, the first discovery was made on the east side of the Cambridge arch in Harlan County in south-central Nebraska. The Cambridge arch is the northwestern extension of the Central Kansas uplift, which had been productive for many years in north-central Kansas. Most of the Nebraska oil and gas production comes from fields in the Denver basin.

Regional Tectonics. Western Nebraska is crossed from northwest to southeast by the Chadron-Cambridge arch (Figure 15.2). This broad uplift area connects with the Black Hills uplift to the northwest in western South Dakota, and in the other direction merges with the Central Kansas uplift. An even broader spur, the Siouxana arch, takes off at right angles from the Chadron-Cambridge arch in central western Nebraska and connects with the Sioux uplift in southeastern South Dakota. To the north between the diverging axes of these two uplifts is the Kennedy basin, which also extends into South Dakota.

From the top of the Chadron arch in northwestern Nebraska, the Precambrian floor dips southwestward into the Denver basin (Martin, 1965) at an average inclination of about 50 feet per mile. However, in southwestern

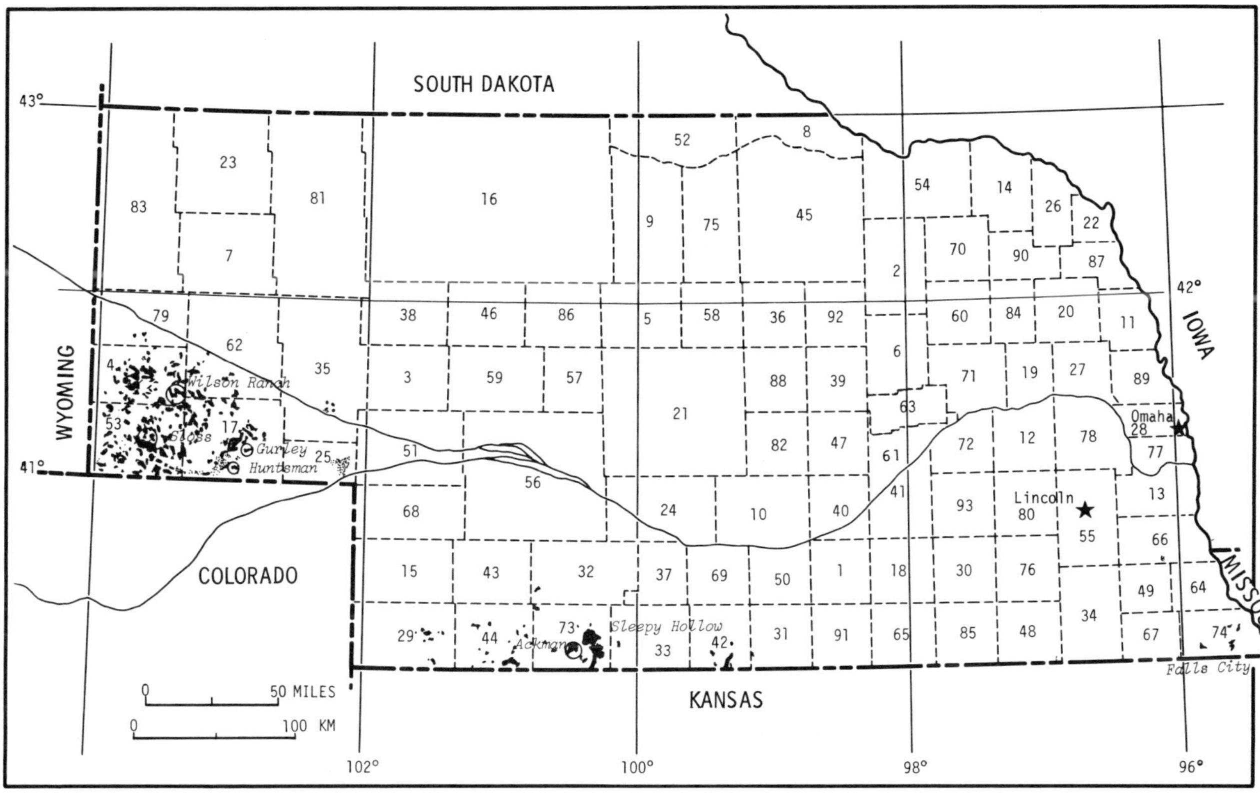

FIGURE 15.1 Nebraska. Index map with oil and gas fields. County Key:

1	Adams	15	Chase	29	Dundy	42	Harlan	55	Lancaster	68	Perkins	81	Sheridan
2	Antelope	16	Cherry	30	Fillmore	43	Hayes	56	Lincoln	69	Phelps	82	Sherman
3	Arthur	17	Cheyenne	31	Franklin	44	Hitchcock	57	Logan	70	Pierce	83	Sioux
4	Banner	18	Clay	32	Frontier	45	Holt	58	Loup	71	Platte	84	Stanton
5	Blaine	19	Colfax	33	Furnas	46	Hooker	59	McPherson	72	Polk	85	Thayer
6	Boone	20	Cuming	34	Gage	47	Howard	60	Madison	73	Red Willow	86	Thomas
7	Box Butte	21	Custer	35	Garden	48	Jefferson	61	Merrick	74	Richardson	87	Thurston
8	Boyd	22	Dakota	36	Garfield	49	Johnson	62	Morrill	75	Rock	88	Valley
9	Brown	23	Dawes	37	Gosper	50	Kearney	63	Nance	76	Saline	89	Washington
10	Buffalo	24	Dawson	38	Grant	51	Keith	64	Nemaha	77	Sarpy	90	Wayne
11	Burt	25	Deuel	39	Greeley	52	Keyapaha	65	Nuckolls	78	Saunders	91	Webster
12	Butler	26	Dixon	40	Hall	53	Kimball	66	Otoe	79	Scotts Bluff	92	Wheeler
13	Cass	27	Dodge	41	Hamilton	54	Knox	67	Pawnee	80	Seward	93	York
14	Cedar	28	Douglas										

Nebraska and northwestern Kansas the westward dip flattens out, and even reverses locally, across the northeastern extension of the Las Animas arch of southeastern Colorado. This sag between the two arches has been termed the Southwest Nebraska–Northwest Kansas basin (Monahan and Rutledge, 1959). It is actually the northern tip of the Hugoton embayment of western Kansas.

To the east of the Cambridge arch and southeast of the Siouxana arch is the Central Nebraska basin, which is the north end of the Salina basin. This basin is bounded on the east by the Nemaha uplift; across the Nemaha uplift in the southeastern corner of the state is the Forest City basin.

The depth to the Precambrian basement floor ranges from less than 500 feet on the Nemaha uplift to below 9000 feet in the southwestern corner of the Nebraska panhandle.

STRATIGRAPHY. Quaternary deposits and Tertiary rocks cover all of central and western Nebraska, except in the northwest corner, where Cretaceous rocks crop out on the south flank of the Black Hills uplift, and in the major river valleys in north-central and southern Nebraska, where erosion has uncovered underlying Cretaceous formations. Cretaceous sediments also cover the surface in eastern Nebraska, except the southeastern corner, where Pennsylvanian rocks crop out.

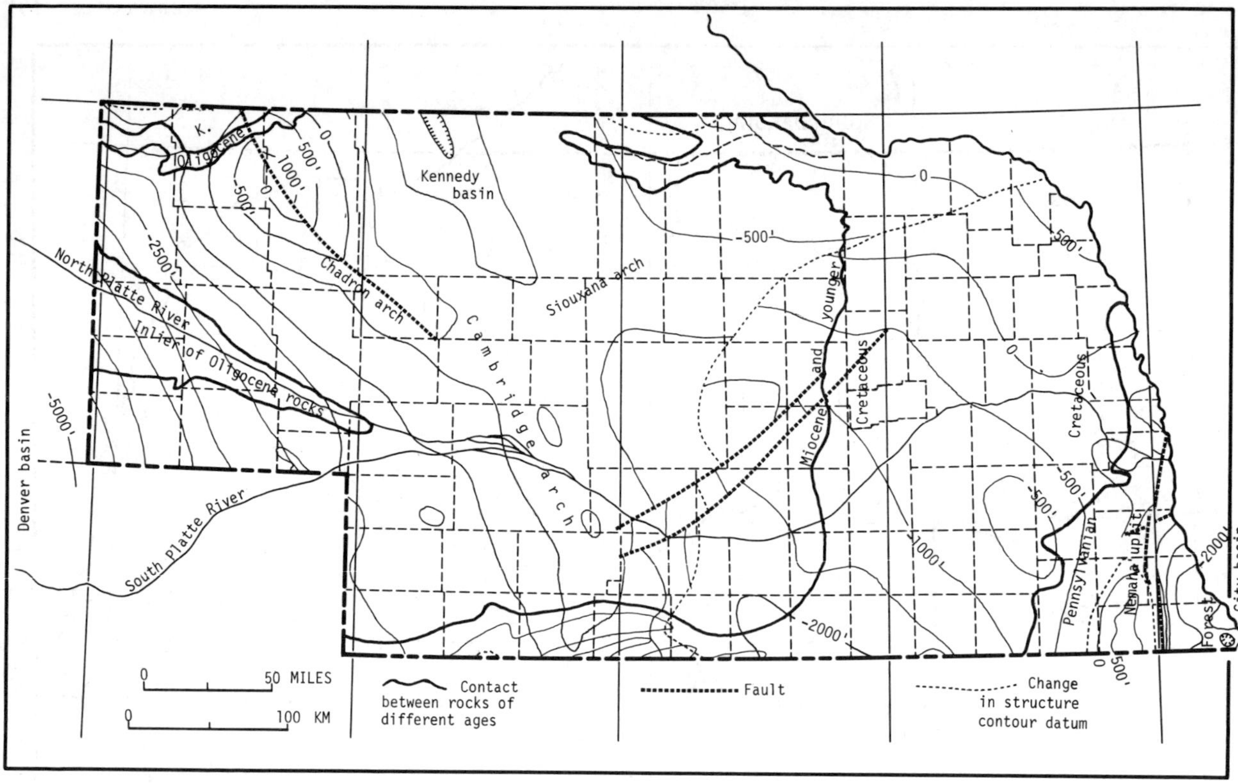

Figure 15.2 Nebraska. Tectonic and simplified geologic map. Structure contours based on Precambrian floor in western Nebraska, and top of Fernvale ("Trenton"-Ordovician) to the east. Interval, 500 feet. *After George V. Cohee et al. (1962).*

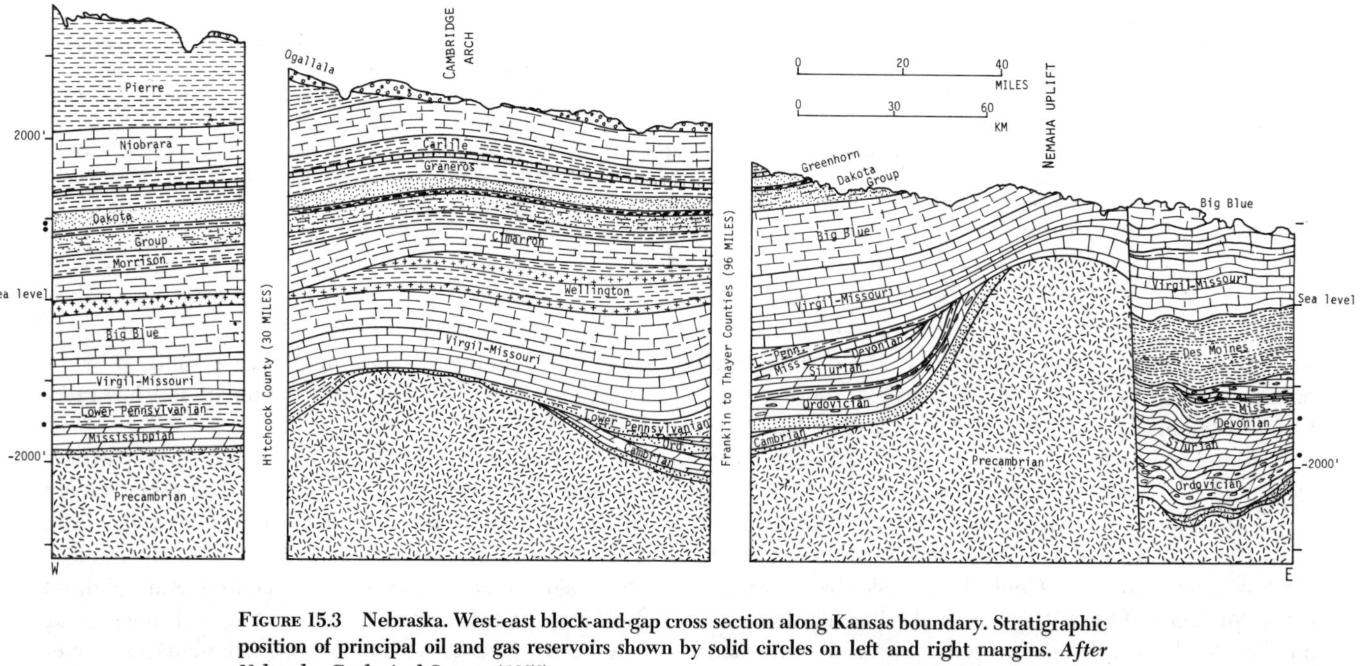

Figure 15.3 Nebraska. West-east block-and-gap cross section along Kansas boundary. Stratigraphic position of principal oil and gas reservoirs shown by solid circles on left and right margins. *After Nebraska Geological Survey (1950).*

Although the composite stratigraphic section (Figure 15.3) shows that almost all of the geologic periods since the Precambrian are represented in the Nebraska subsurface, a pre-Pennsylvanian section is limited to the Forest City, Salina, and Southwest Nebraska–Northwest Kansas basins. In the very northwestern corner of the panhandle area Mississippian rocks occur beneath the Pennsylvanian. However, elsewhere in the panhandle and western Nebraska the Pennsylvanian rocks rest upon the Precambrian basement. This condition exists also in Pawnee County in southeastern Nebraska along the axis of the Nemaha uplift.

The reservoir rocks so far discovered in Nebraska range in age from Cretaceous to Early Ordovician. Sandstones in the Dakota Group of Cretaceous age are the reservoir rocks in the Nebraska panhandle, which lies on the northeast flank of the Denver basin. The principal sands are the D and J sands, the same as in northeastern Colorado. To the eastward, fields on the Cambridge arch produce from Pennsylvanian sediments. The most prolific reservoirs are limestones in the Lansing–Kansas City Group and a basal Pennsylvanian sandstone that immediately overlies the Precambrian and is in part granite wash (Svoboda, 1962). A basal sandstone produces oil on the north flank of the Sleepy Hollow field in east-central Red Willow County, and to the northwest near the corner of the county (Haynes, 1967, p. 1126). Most of the production in the Forest City basin of southeastern Nebraska has come from Siluro-Devonian carbonate reservoirs, especially the Hunton. Some production has also been obtained from the Ordovician, including the Viola and the St. Peter ("Wilcox") Sandstone.

Traps and Selected Fields. Accumulation in Nebraska is due to both stratigraphic and structural trapping. In the Nebraska panhandle the emphasis is on varying permeability, with accumulation taking place in lenticular Cretaceous sands. However, anticlinal traps are also present. The *Huntsman* field in Cheyenne County, the largest in the state in cumulative gas production, occupies an anticline with closure in excess of 100 feet.

The *Sloss* field in Kimball County in the panhandle illustrates stratigraphic trapping in the Nebraska section of the Denver basin. Here the basinward dip off the Chadron arch is westerly at about 35 feet per mile. Two lenticular sands, the J1 (upper) and J2 (lower), occurring within the Dakota (Cretaceous) group are oil bearing. Figure 15.4 shows the relative positions of these sands in an east-northeast cross section through the field. The upper sand is a single lens underlying most of the productive area. The lower reservoir is split into separate lenses. The average pay thickness is only about 9 feet, but the porosity averages nearly 18 per cent. Until the discovery and development of the Sleepy Hollow field in Red Willow County the Sloss was the largest oil field in Nebraska. Secondary recovery by waterflooding commenced in mid-1958.

A short distance west of the valley fill fields is an area producing from marine sand bars (Exum and Harms, 1968). This area covers two townships in adjacent parts

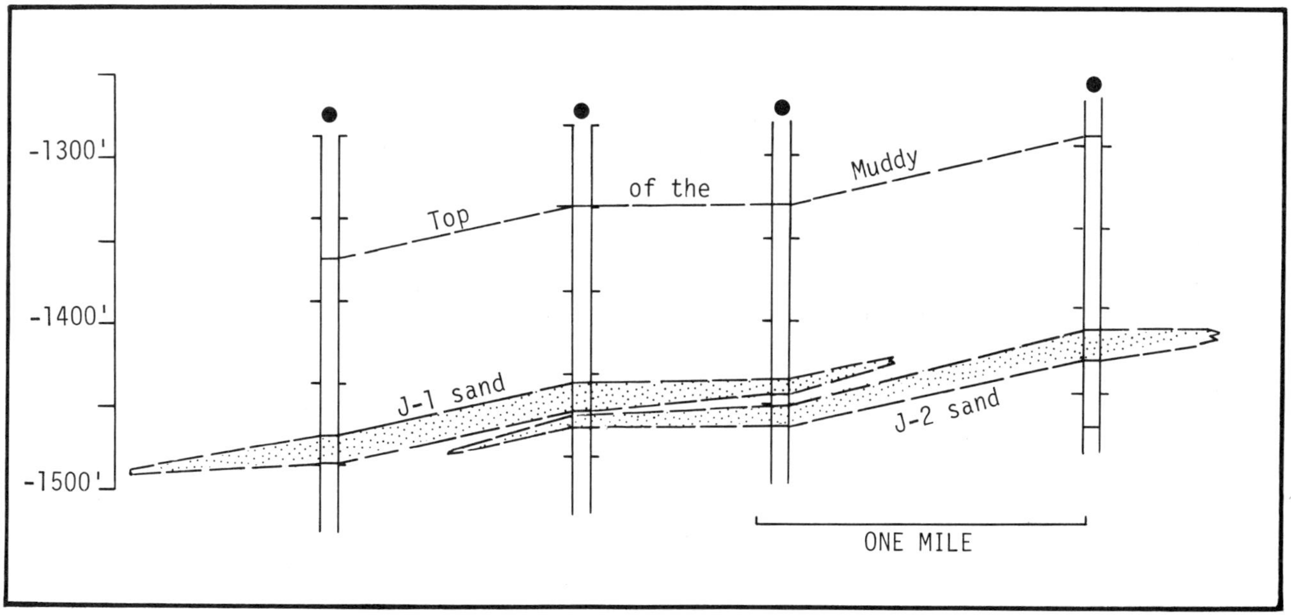

Figure 15.4 Sloss field, Kimball County in western Nebraska, cross section illustrating lenticular sandstone traps. *Courtesy Rocky Mountain Association of Geologists (1962, p. 329) and the Pan-American Petroleum Corporation.*

of Banner, Morrill, and Cheyenne Counties. The pay zones are the upper and lower J1 sands, which are separated by only a few feet of shale. The upper J1 sand bars trend northwest, whereas the lower ones trend northeast. In three fields, such as the Wilson Ranch oil field in southeastern Banner County, production is obtained from both sands in the vicinity of the point where they cross at right angles. Entrapment in the marine bars, unlike the valley fill oil fields, is independent of structural closure, and most of the sand bodies are entirely oil filled.

Seven fields have been discovered along a north-south trend in Cheyenne and Morrill counties, in which accumulation has taken place in valley fill sandstone bodies in the J interval (Harms, 1966). This ancient valley fill is chiefly porous and permeable sandstone; it is about 1500 feet wide and 50 feet thick. The "shoestring" sand is continuous for at least 20 miles, but trapping of oil has taken place only where the sand body crosses a southwesterly trending anticlinal nose.

So far, most of the accumulations of oil and gas discovered elsewhere in Nebraska have been on anticlines. The fields on the flanks of the Cambridge arch in southwestern Nebraska are in the main low-relief anticlinal traps. The Ackman (Larson, 1962) and Sleepy Hollow fields in Red Willow County are typical of this district. The *Ackman* field was discovered after subsurface and seismic work. It lies about 10 miles to the southwest of the axis of the Cambridge arch (Figure 15.5). The maximum closure of the Ackman dome is only about 35 feet.

To the northeast of the Ackman field and closer to the axis of the Cambridge arch (but still on the west flank) is the more recently discovered and considerably larger *Sleepy Hollow* oil field. This field had produced nearly 20 million barrels of oil by the end of 1967. Waterflooding commenced in 1966. Both the Ackman and Sleepy Hollow anticlines have relatively steep east flanks that are local reversals of the regional western dip. These east dips may be reflections of faulting in the basement rock.

As can be seen in Figure 15.5, several other, and so far smaller, fields have been discovered in this vicinity. The principal oil-producing reservoirs are limestone zones lying within the Lansing–Kansas City Group of Pennsylvanian age. The second and third (downward) zones are the most important reservoirs in the Ackman field. Fractures contribute importantly to the reservoir volume of these limestones. Furthermore, the fissure porosity appears to be maximum in the anticlinal areas, indicating that the same diastrophic movements that created the upfolds on the flanks of the Cambridge arch also contributed to the cracking of the brittle rocks in the stratigraphic section.

The Sleepy Hollow field has as an additional reservoir a basal Pennsylvanian sand. According to Larson (1962, p. 2088), this probably consists of reworked older sand (including granite wash) that was "washed in and deposited as a sand bar in a localized basinal area during early Pennsylvanian time."

Future. Except for the southwestern panhandle area, southwestern Nebraska proper, and the southeastern corner of the state, Nebraska has had relatively little exploration. Therefore this state offers opportunities for new discoveries laterally as well as deeper in the basin areas. Almost every year the productive periphery of the Nebraska sector of the Denver basin moves northward through new discoveries. The same thing is happening to the west in the Wyoming sector of the basin; a coalescing of the two productive areas along the state line can be anticipated.

In the northwestern corner of the state, the Chadron arch, with a structual relief of 3000 feet (Wulf, 1964a) and an area of more than 1800 square miles, has a deep-drill-hole density of only one well per 150 square miles. Possible reservoirs range in age from Cretaceous to Permo–Pennsylvanian. No doubt both anticlinal node and flank wedgout traps are present.

Likewise, across the Cambridge arch in southwestern Nebraska, pre-Pennsylvanian erosion has truncated potential reservoirs of Mississippian and Cambrian age (Svoboda, 1958). Barr and Bieber (1961) have postulated the possible presence of Permian reef reservoirs on top of the Chadron arch in northwestern Nebraska. By far the largest virtually unexplored area is the Central Nebraska basin, where there is a thick section beveled by pre-Pennsylvanian erosion on the basin flanks (Reed, 1954; Gilmore, 1952).

The oldest producing area in Nebraska, the Forest City basin (Anderson and Wells, 1968), has been referred to as one of the nation's "least-tested petroleum provinces" (Svoboda, 1960). Beneath the outcropping Pennsylvanian sediments are both known and potential reservoirs in the Silurian (Hunton) and Ordovician (Viola and St. Peter).

REFERENCES CONSULTED IN PREPARING CHAPTER 15

Anderson, K. H., and Wells, J. S., 1968.
Barr, C. R., and Bieber, A. M., 1961.
Beebe, B. W., 1968a.
Exum, F. A., and Harms, J. C., 1968.
Gilmore, R. L., 1952.
Harms, J. C., 1966.
Haynes, Edward H., 1967.
Larson, Wilbert S., 1962.
Martin, Charles A., 1965.
Monahan, Rex, and Rutledge, Robert D., 1959.
Parker, John M., ed., 1962.

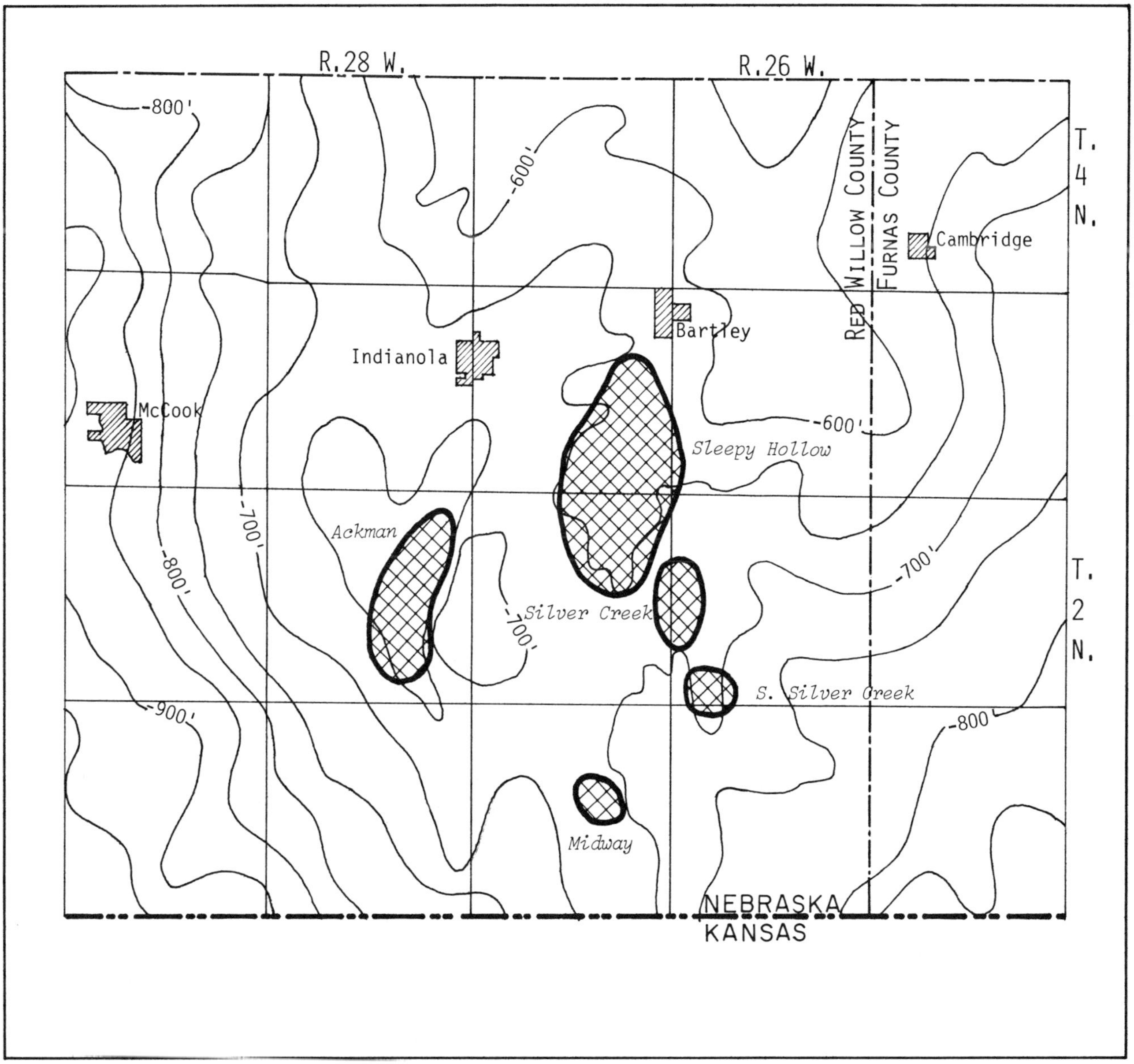

Figure 15.5 Regional structure and oil fields, Red Willow County, Nebraska. Oil fields designated by crosshatched pattern. Structure map based on top of Lansing-Kansas City Groups (Missouri Series). Interval, 50 feet. Township squares are 6 miles (9.66 km) on a side. *After Wilbert S. Larson (1962, Figure 3, p. 2081). Courtesy Wilbert S. Larson and the American Association of Petroleum Geologists.*

Reed, E. C., 1954.
Reed, E. C., Svoboda, R. F., Prichard, G. E., and Fox, Jeannette, 1958.
Rocky Mountain Association of Geologists, 1955, 1962.
Svoboda, Richard F., 1958, 1960, 1962.
Volk, Richard W., 1968.
Wulf, George R., 1964a

Chapter

16

KANSAS

Oil was first struck in Kansas at a depth of 275 feet in Miami County in the 1860s (Anderson and Wells, 1968). The first recorded oil production in the state was a yield of 1000 barrels in 1889. The highest peak so far reached was in 1956 with a production of 124,204,000 barrels. During 1968 the state produced about 3 per cent of the total domestic output, ranking seventh in the United States. Recently about one-fifth of the annual Kansas petroleum yield has been due to secondary recovery (Beene and Oros, 1967).

The commercial production of gas began in 1893 in Allen County in southeastern Kansas (Figure 16.1). Subsequently over 11 trillion cubic feet has been produced, the greater part of it from the Hugoton field in the southwestern part of the state (Merriam and Goebel, 1968). In 1967 Kansas was in fifth place in the production of gas, with a yield which was 5 per cent of the total for the entire country. So far, 92 counties out of 105 in the state either are now producing oil or gas or have produced oil or gas in the past (Figure 16.2). Although Kansas has been an important producer of oil and gas for nearly three-quarters of a century, many new fields are found in this state each year. Seven wildcat wells discovered significant reserves in 1967 (Porter, 1968).

The earliest discoveries of both oil and gas were made close to the eastern edge of the state, and for the first 33 years after 1889 the oil production came largely from eastern Kansas. During that time the great El Dorado and Augusta pools of Butler County (Figure 16.1) and the prolific shoestring sand fields of Greenwood County were discovered and developed. In 1923 oil was found in the Fairport pool of Russell County in western Kansas more than 100 miles to the northwest of the nearest oil field. About this time the development of the great Hugoton gas field in the southwestern corner of the state also got under way. Subsequently, discoveries have been made throughout the western two-thirds of the state, which for some years now has produced more oil than eastern Kansas.

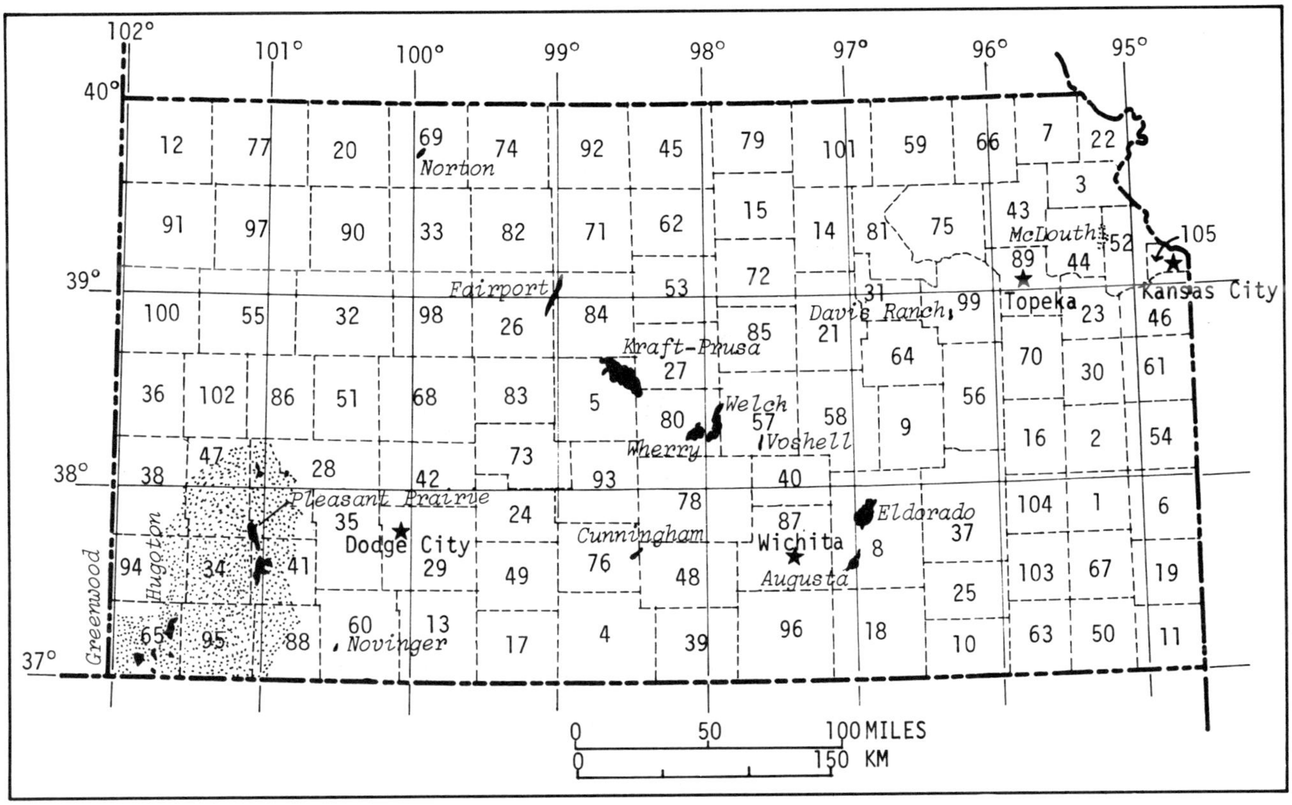

FIGURE 16.1 Kansas. Index map, also showing oil fields named in text. County Key:

1	Allen	16	Coffee	31	Geary	46	Johnson	61	Miami	76	Pratt	91	Sherman
2	Anderson	17	Comanche	32	Gove	47	Kearny	62	Mitchell	77	Rawlins	92	Smith
3	Atchison	18	Cowley	33	Graham	48	Kingman	63	Montgomery	78	Reno	93	Stafford
4	Barber	19	Crawford	34	Grant	49	Kiowa	64	Morris	79	Republic	94	Stanton
5	Barton	20	Decatur	35	Gray	50	Labette	65	Morton	80	Rice	95	Stevens
6	Bourbon	21	Dickinson	36	Greeley	51	Lane	66	Nemaha	81	Riley	96	Sumner
7	Brown	22	Doniphan	37	Greenwood	52	Leavenworth	67	Neosho	82	Rooks	97	Thomas
8	Butler	23	Douglas	38	Hamilton	53	Lincoln	68	Ness	83	Rush	98	Trego
9	Chase	24	Edwards	39	Harper	54	Linn	69	Norton	84	Russell	99	Wabaunsee
10	Chautauqua	25	Elk	40	Harvey	55	Logan	70	Osage	85	Saline	100	Wallace
11	Cherokee	26	Ellis	41	Haskell	56	Lyon	71	Osborne	86	Scott	101	Washington
12	Cheyenne	27	Ellsworth	42	Hodgeman	57	McPherson	72	Ottawa	87	Sedgwick	102	Wichita
13	Clark	28	Finney	43	Jackson	58	Marion	73	Pawnee	88	Seward	103	Wilson
14	Clay	29	Ford	44	Jefferson	59	Marshall	74	Phillips	89	Shawnee	104	Woodson
15	Cloud	30	Franklin	45	Jewell	60	Meade	75	Pottawatomie	90	Sheridan	105	Wyandotte

The bedrock is widely exposed in eastern Kansas, and most of the earlier finds of oil and gas in this area were the direct result of surface structure mapping. Modern and ancient floodplains, dunes and loess deposits, and Tertiary rocks of continental origin mask the bedrock over much of the western two-thirds of the state, so that most discoveries there have been due to core-drill surveys, subsurface geology, and geophysical exploration.

Geographically, the Kansas oil fields are included in the northern Midcontinent province; geologically, Kansas lies within the Western Interior coal basin. This coal basin is bounded on the north by the Wisconsin shield, on the east by the Ozark uplift, and on the south by the Ouachita and Arbuckle-Amarillo uplifts. The western boundary is indefinite. The Carboniferous coals do not extend very far west, but the structural basin continues westward until it merges with features belonging to the Rocky Mountain geosyncline.

The oil and gas field map of Kansas (Figure 16.2) is convincing evidence that this state is bountifully supplied with (1) source beds to furnish oil and gas (2) reservoir rocks in which these fluids could accumulate, (3) seal rocks to retain the hydrocarbons within the reservoir rocks, and (4) traps to capture the oil and gas in deposits of commercial value.

Structure. Although the areal geologic map of Kan-

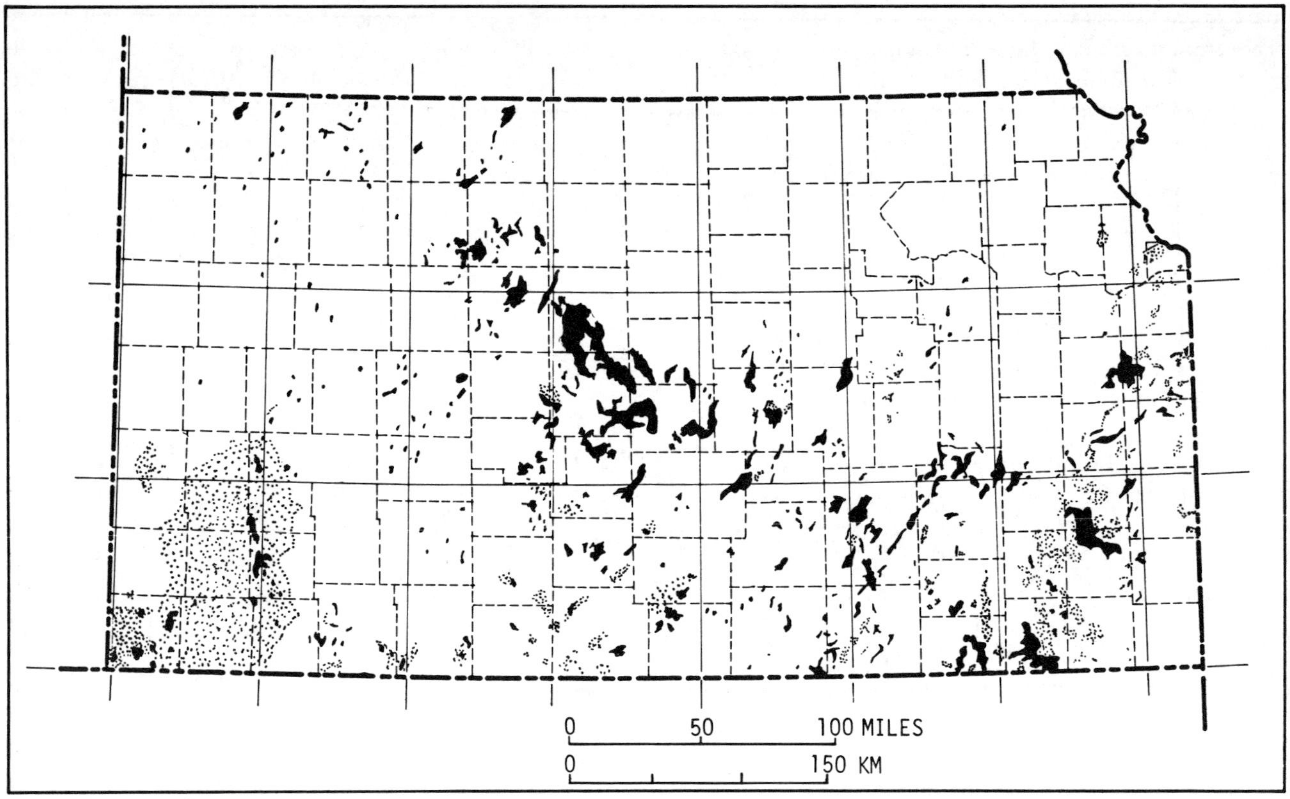

FIGURE 16.2 Kansas. Oil and gas fields. Oil fields solid black; gas fields stippled.

sas shows the rock groups outcropping in bands that cross the state in a general north-south direction, with the oldest in the east and the youngest in the west, the truncation of westward dipping strata in eastern Kansas and the general increase in topographic elevation across western Kansas make the three-dimensional picture of Kansas geology by no means simple. The subsurface section contains some great unconformities, including especially the ones that preceded and followed the deposition of the Mississippian sediments.

The most striking tectonic features in Kansas geology today are the Nemaha and the Central Kansas uplifts (Figure 16.3). The Nemaha uplift crosses the state from north to south, with the axis plunging to the south. It was pushed up in post-Mississippian time, and extensive erosion in the Early Pennsylvanian stripped off not only older Pennsylvanian sediments, but also the Mississippian and Ordovician, so that along the axis in the northern and higher parts of the uplift the later Pennsylvanian sediments rest directly upon Precambrian crystalline rock. Farther down the flanks, or farther south on the axis, the same sediments overlie tilted and truncated Ordovician and Mississippian formations. Because the structure of the Nemaha uplift is reflected in the overlying sediments, this great anticlinal ridge can be mapped by surface geologic methods.

The Central Kansas uplift is a much broader arch, but it has had a similar structural history. It also has a crystalline rock core and is bordered by arched and beveled Ordovician and Mississippian sediments. It is crossed from northwest to southeast by three parallel anticlinal axes. To the south the Central Kansas uplift narrows into the Pratt anticline.

Five basins containing a much more complete section of sedimentary rock surround the two uplifts. East of the Nemaha uplift, the Forest City basin occupies northeastern Kansas and extends into northwestern Missouri, southeastern Nebraska, and southwestern Iowa (Anderson and Wells, 1968). It is primarily a Pennsylvanian feature. The Bourbon arch separates this basin from the Cherokee basin of southeastern Kansas and northeastern Oklahoma. Between the Nemaha uplift and the Central Kansas uplift are the Salina basin in the northern half of the state and the Sedgwick basin in the southern half. South and west of the Central Kansas uplift is the Hugoton embayment or Dodge City basin, which is the northern extension of the Anadarko basin of western Oklahoma.

The Precambrian basement rock lies within a few hundred feet of the surface along the axis of the Nemaha ridge where it crosses from Kansas into Nebraska; it has a maximum depth of more than 8000 feet in the lowest part of the Dodge City basin in southwestern Kansas. Therefore the base of the sedimentary column lies within reasonable drilling depth throughout the state.

The most important tectonic province in terms of oil production today is the Central Kansas uplift. The Hugoton embayment is the leading area for gas production.

Many different types of traps have been found in Kansas. The most abundant of those so far discovered is the anticline. In many instances, however, the eccentricities and permeability of the reservoir rock or rocks have played a role in the distribution of the gas or oil across the upfold.

Another widespread type of trap in Kansas is due to truncation and overlap at unconformities (Jewett, 1960). Especially important in this regard is the angular unconformity beneath the Pennsylvanian rocks, which assists in the trapping of oil and gas on both the Nemaha and the Central Kansas uplifts.

Reservoirs (Merriam and Goebel, 1959; Hilpman, 1958; Moore et al., 1951). The geological formations in Kansas range in age from the Miocene–Pliocene (Tertiary) Ogallala Formation, a continental deposit that blankets most of western Kansas, through the Cretaceous chalky limestones and sandstones and a fairly complete Paleozoic section, to the Precambrian. Oil reservoirs (Figure 16.4) occur in every Paleozoic period, and even in the Precambrian itself. The leading reservoir in cumulative production is the Cambro–Ordovician Arbuckle Dolomite and the underlying basal sand known as the Reagan. These reservoirs have produced about 45 per cent of the state's oil to date. Second in importance, with about a third of the Kansas oil production to its credit, is a series of limestones, plus a basal sandstone, lying within the Pennsylvanian System of rocks. Mississippian reservoirs, the Viola (Ordovician) Limestone, and Ordovician sandstones belonging in the Simpson Group rank third, fourth, and fifth, respectively, in cumulative production. Other producing rocks are the Permian limestones (which are outstanding in gas production), the "Hunton" Limestone of Siluro–Devonian age, and the weathered and fissured granitic rocks of Precambrian age immediately underlying the Paleozoic section.

So far as gas is concerned, "Permian rocks have yielded about 74 per cent of the cumulative production, Pennsylvanian rocks 15 per cent, and Mississippian rocks 9 per cent; the other 2 per cent is from lower Paleozoic rocks" (Merriam and Goebel, 1968, p. 1548). More detailed descriptions of oil and gas reservoirs, from youngest to oldest, follow.

Because of the regional structure and topography of the state, the formations cropping out in eastern and central Kansas are buried at sufficient depth to the west

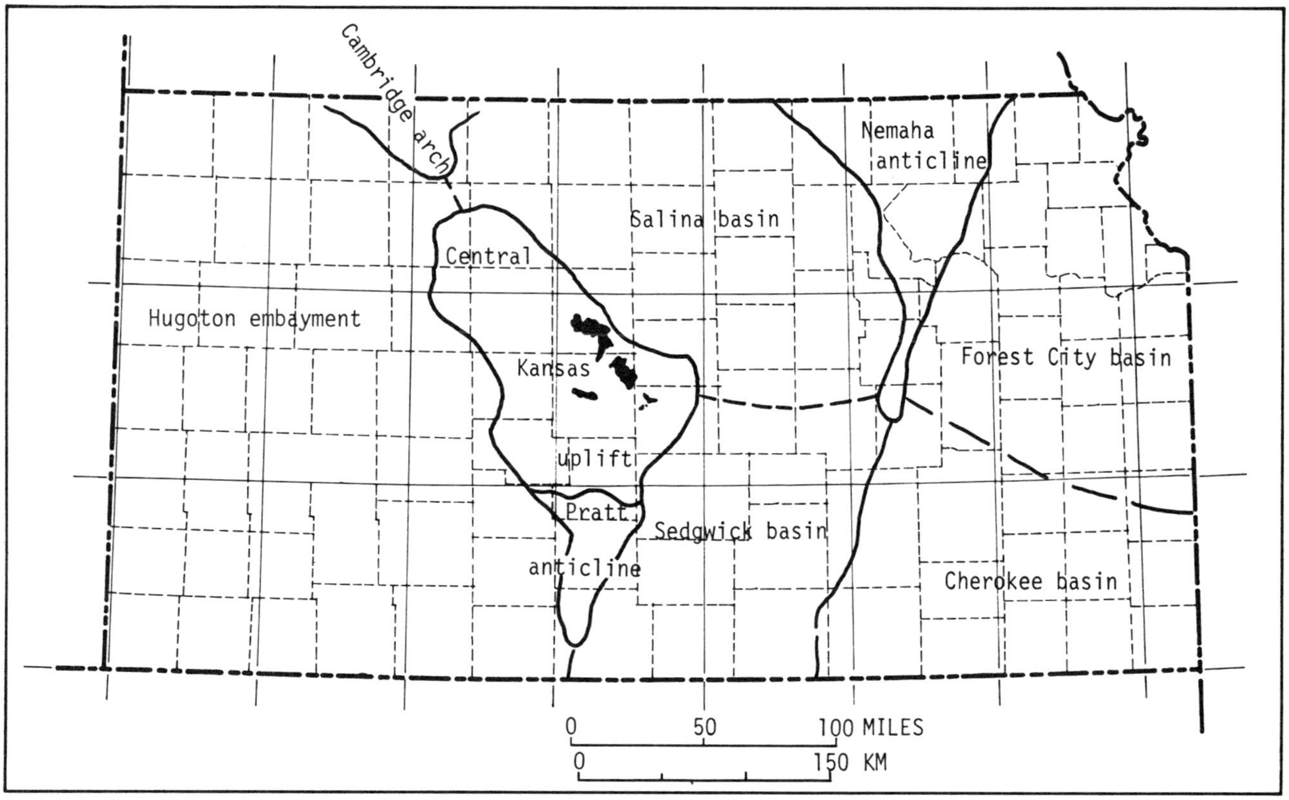

Figure 16.3 Kansas. Tectonic map, also showing fields producing in part from Precambrian rock. *Tectonic features from Paul Hilpman (1958).*

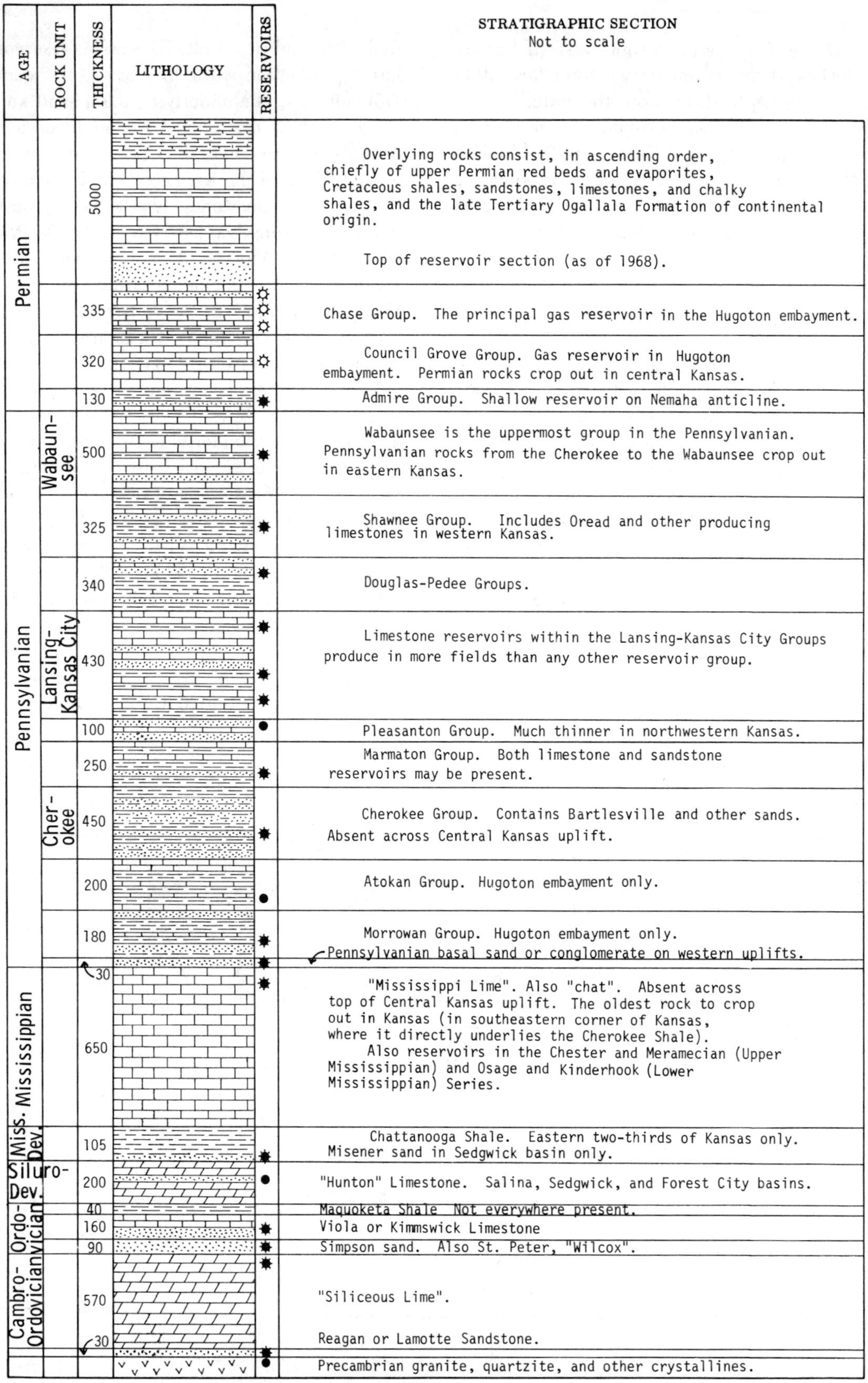

FIGURE 16.4 Kansas reservoir rocks. *After Raymond C. Moore et al. (1951), Paul L. Hilpman (1958), and Daniel F. Merriam and Edwin D. Goebel (1968).*

of the outcrop area to be reservoir rocks for oil or gas wherever they have adequate porosity and permeability. Rocks older than the oldest beds to outcrop in Kansas occur at the surface to the south in the Arbuckle Mountains of southern Oklahoma, and to the east in the Ozarks of Missouri and Arkansas. By the same token, the Cretaceous sediments that crop out in western Kansas contain productive reservoir rocks in Colorado and Wyoming.

The youngest commercial reservoirs so far developed in Kansas are the prolific gas-producing dolomites of the Chase and Council Grove Groups of the Lower Permian in the Hugoton embayment in the southwestern part of the state. Two sandstones in the Admire Group of lowest Permian have produced some oil in the El Dorado field of Butler County and elsewhere along the Nemaha uplift.

The Pennsylvanian System is very well developed in Kansas, and every group within it contains reservoir rocks. Most of these reservoirs, except in the lowermost Pennsylvanian, are carbonate rock. The youngest of the Pennsylvanian groups, the Wabaunsee, crops out in the Flint Hills in eastern Kansas and produces oil on the Central Kansas uplift and its southern extension, the Pratt anticline. In Morton County in the extreme southwestern corner of the state Wabaunsee reservoirs contain gas. Limestone reservoirs in the next lower group, the Shawnee, have similar distribution. The Douglas-Pedee Groups contain limestones that produce oil locally on the Central Kansas uplift and on the Pratt anticline.

The Lansing–Kansas City Groups have carbonate rock reservoirs that are productive in more fields and in more counties than any other reservoir group. The easternmost production from these formations is on the west flank of the Cherokee basin; they yield oil from there to the northwestward across the Nemaha anticline, the Sedgwick basin, and the Pratt anticline–Central Kansas uplift–Cambridge arch province into western Nebraska. Gas has been found in this part of the section in Morton County.

Lenticular sands in the Pleasanton Group have produced oil and gas in eastern Kansas. Both limestones and sandstones in the Marmaton Group are reservoir rocks in the Cherokee and Forest City basins, and in scattered counties in south-central and western Kansas.

Lenticular sandstones in the Cherokee Shale Group have produced large quantities of oil throughout the Cherokee basin and on the southern flank of the Forest City basin. The most important of these sandstones is the Bartlesville, which is not only lenticular but also in many instances extremely elongate, resulting in the so-called shoestring sand fields. The Cherokee Group contains several stray sands, including the Squirrel, so named by drillers because it "jumps around so" in the section. At the base of the Cherokee Shale Group (and the Pennsylvanian) in contiguous corners of Sedgwick and Harvey Counties is a local oil-productive conglomeratic sandstone, the Burgess (Morgan and Torline, 1964).

In southeastern Kansas the Cherokee Shale lies directly upon Mississippian limestone. However, in the Hugoton embayment of southwestern Kansas, two older Pennsylvanian series, the Atokan and the Morrowan, underlie the Cherokee Shale and overlie the Mississippian formations. The Morrowan has been found recently to be productive of both oil and gas in a number of counties in the Hugoton embayment. The Atokan produces a little oil in Meade and Morton Counties.

At the base of the Pennsylvanian section on the Central Kansas uplift is a sandstone or conglomerate that is an important reservoir rock for oil and gas. It does not occur, as would appear from the chart (Figure 16.4), between the Morrowan sediments and the Mississippian rocks, but on the contrary lies at the base of whatever post-Cherokee rocks may be present.

For many years the weathered surface of the Mississippian limestone section has been an important oil and gas reservoir on the west flank of the Cherokee basin and in the Sedgwick basin. This reservoir is ordinarily referred to as the Mississippi lime, but it may also be called simply the Chat, the nickname applied to chert in the Tri-State (Joplin) zinc district. Farther west, in west-central and southwestern Kansas, oil and gas are obtained from porous carbonate pay zones within the Mississippian section, including Chesterian (Kornfeld, 1968) and Meramecian (Upper Mississippian) and Osage (Lower Mississippian) rocks (Euwer, 1964).

All knowledge of pre-Mississippian stratigraphy in Kansas has been obtained from the subsurface. The Chattanooga Shale of partly Early Mississippian but mostly Late Devonian age is present in the eastern two-thirds of Kansas only. The Misener Sandstone may or may not be present at the base of this unit. The latter is a reservoir rock for both oil and gas in the Sedgwick basin.

Three of the eastern Kansas basins, the Salina, Sedgwick, and Forest City, contain a limestone unconformably underlying the Chattanooga that, although it may be referred to as "Hunton," is not the equivalent of the Hunton Limestone of southern Oklahoma. This limestone sequence has been referred to as "Devonian carbonates" (Amsden et al., 1968, Figure 6, p. 925). Much more widespread in its extent is the Viola Limestone of Ordovician age. This limestone has its greatest development in the Salina and Sedgwick basins and on the flanks of the Pratt anticline and the Nemaha uplift. Sandstones in the Middle Ordovician Simpson Formation (also known as the St. Peter and the "Wilcox") are oil and gas pays in the Sedgwick basin, on the Nemaha and Pratt anticlines, and on the southern and southwestern flanks of the Central Kansas uplift.

Omnipresent in Kansas, except where eroded off the

higher parts of the Nemaha and Central Kansas uplifts, is the Arbuckle Group of carbonate rocks, often referred to as the "Siliceous Lime." Porous zones at and near the top of this cherty limestone and dolomite group are highly productive reservoirs over most of Kansas. The Arbuckle, high on the west flank on the Central Kansas uplift and the Cambridge arch, is underlain by a basal Paleozoic sandstone (Miller, 1968), known variously as the "Basal sandstone," the Reagan (?), or "granite wash" because the sand grains are derived from the underlying granitic basement rock.

The discovery of oil in commercial quantities in the Precambrian of the Central Kansas uplift in 1932 really confounded the textbook writers. The discovery well produced at a rate of 1800 barrels daily from fractured Precambrian quartzite. Subsequently more than 50 producing wells have been completed in the basement rock in this area (Walters, 1953; Cole et al., 1961; Farquhar, 1957). In every instance the fractured rock producing oil is topographically high, and in all probability oil has migrated into these cupolas from topographically lower flanking sediments.

The distance of migration of the oil and gas caught in Kansas traps probably can be measured in feet in the case of the shoestring sand fields of eastern Kansas. Where sheet porosities exist, as in the uppermost Arbuckle, such distances are measurable in scores of miles.

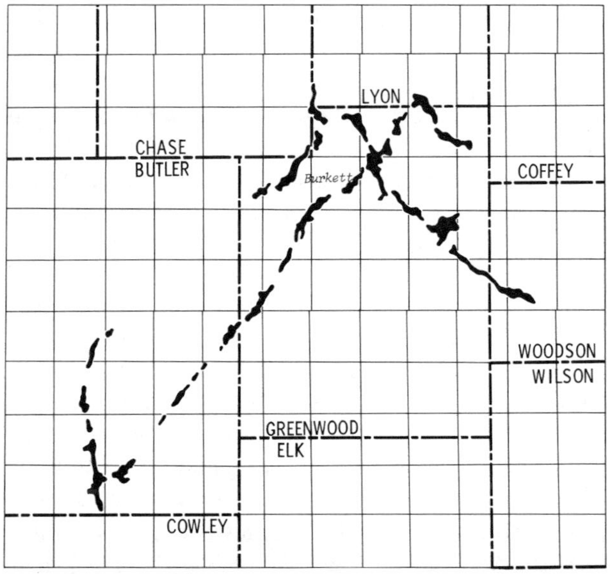

FIGURE 16.5 Shoestring sand pools of Greenwood and adjacent counties. Shows location of Burkett oil field. Small squares are townships, 6 miles (9.66 km) on a side. *After Paul L. Hilpman (1958).*

DISTRICTS, TRAPS, AND SELECTED FIELDS

CHEROKEE BASIN. Eastern Kansas is the type locality for shoestring sand pools (Bass, 1936; Cheyney, 1929). The largest, best known, and most prolific of these occur in Greenwood and surrounding counties in the Cherokee basin of southeastern Kansas. The elongate lenticular sand bodies occur in the lower part of the Cherokee Shale of Pennsylvanian age. At least two different ages of sand deposition are involved; where the map (Figure 16.5) shows what appears to be intersecting sand trends, in reality a slightly younger shoestring sand with a totally different trend crosses above an older sandstone body. Even on the same trend, however, the elongate sand bodies are completely surrounded by shale, which separates one from the other. The enclosing Cherokee Shale is a highly organic rock containing, somewhat higher in the section, the major coal beds of Kansas and southwestern Missouri.

Geologists have long suspected that the oil in the sand lenses was squeezed out of the surrounding organic shale. This theory has recently been supported by chemical evidence that nonreservoir facies of the Cherokee rocks were the source of the petroleum within the sandstone lenses (Baker, 1962).

Shown on Figure 16.6 is a block diagram of the *Burkett* shoestring sand body (location given on Figure 16.5). Each oil well in the field is shown as the intersection of two sandstone "fences." The stippled pattern indicates the reservoir sand. The best explanation so far advanced for these features is that they are remnants of offshore sand bars (Bass, 1936). The importance of shoestring sand bodies as a source of oil in Kansas is shown by the fact that Greenwood County, where most of the oil has been obtained from them, ranks fifth among Kansan counties in cumulative production, with a total approaching a quarter of a billion barrels.

FOREST CITY BASIN (Anderson and Wells, 1968; Brown, 1968). To the northeast in Anderson County in the Forest City basin are similar but smaller shoestring sands occurring both in the Cherokee Shale and in younger shales of the Pennsylvanian System (Charles, 1941; Kansas Geol. Survey, 1927; Rich, 1923, 1926). Charles believes that some at least of these shoestring sand bodies are more likely to be stream channel fillings than offshore bars.

A typical Forest City basin anticlinal trap is illustrated by *Davis Ranch* oil pool in Wabaunsee County (Smith and Anders, 1951). The structure at the surface shows a closure of only 20 feet. However, contours drawn on the productive Viola Dolomite show a north-south anticline 2 miles long and $^3/_4$ mile wide with a closure of approximately 80 feet. The fold is decidedly asym-

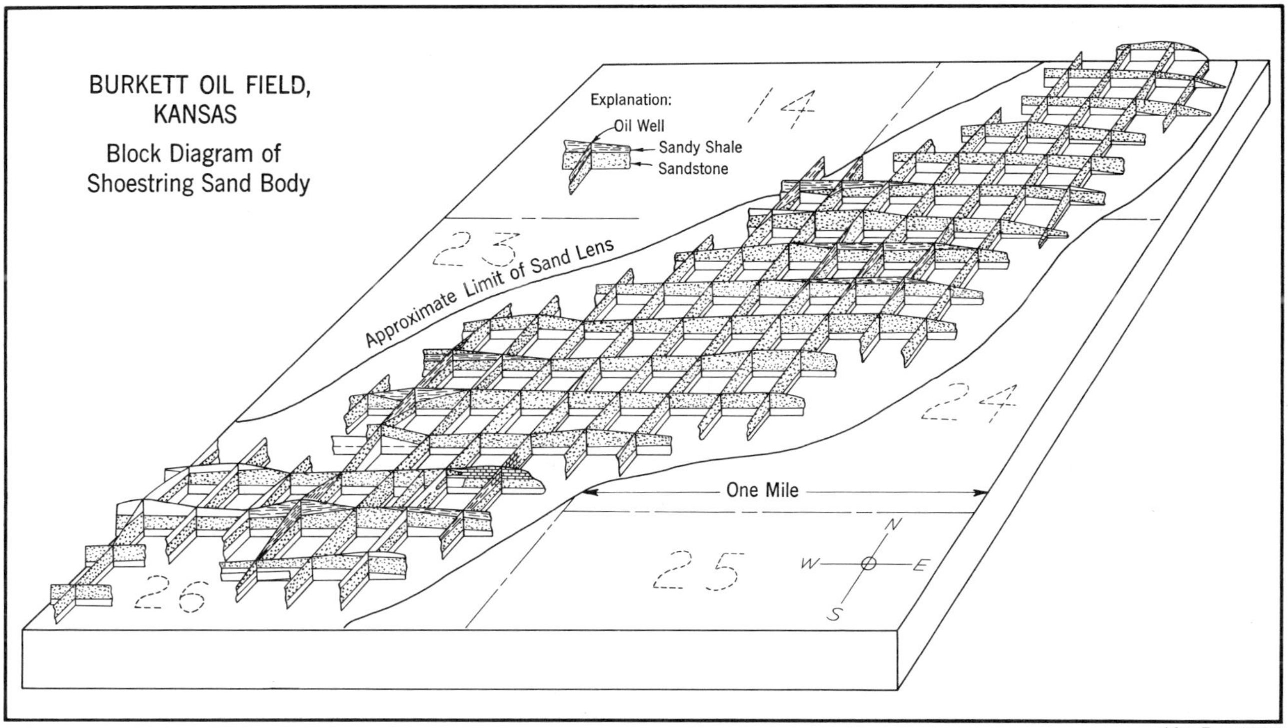

After Bass (1936)

FIGURE 16.6

metrical with a westerly dip of 120 feet to the mile, and an easterly dip of 350 feet per mile into a reverse fault.

In another part of the Forest City basin, the *McLouth* gas field (Lee and Payne, 1944) shows anticlinal trapping with an unusual subsequent history. The reservoir is a Pennsylvanian sandstone lying at the base of the Cherokee Shale. Apparently the gas, with minor amounts of oil, was first trapped in an anticline, and then the oil beneath the gas dried out so that it made an effective seal between the underlying water and the overlying gas. Subsequently regional tilting took place, but because of the locked-in nature of the hydrocarbons in the anticline they have been unable to adjust to the new structural setting.

Although the Forest City basin is one of the older oil- and gas-producing districts, new discoveries continue to be made there, such as on the Alma anticline, which lies a few miles east and parallel to the axis of the Nemaha anticline (Merriam, 1960). Two new counties, Jackson and Antchison, near the northeastern corner of the state, became modest hydrocarbon producers in 1963–1964.

NEMAHA UPLIFT. Nearly every oil pool along the axis of the Nemaha uplift occupies a structural dome. Some of these are typical bald-headed structures with trapping due to a combination of anticlinal folding and overlap by impervious cap rock.

The *El Dorado* field (Fath, 1921; Reeves, 1929) of Butler County, which was discovered in 1914 and flooded the domestic market in 1916, has been an important source of oil in Kansas ever since, standing first in cumulative production (over 270 million barrels) and returning briefly in 1962 to first place in annual production through a resurgence in output brought about by secondary recovery. This field directly overlies the Nemaha "granite ridge" axis. The structure of the Pre-cambrian surface is reflected upward, producing a prominent upfold in the sedimentary rocks. The El Dorado anticline was first mapped by means of surface geology. The closure in the outcropping Lower Permian and Upper Pennsylvanian formations is about 150 feet, but below the pre-Pennsylvanian angular unconformity the truncated Ordovician strata have much steeper dips, and the total structural relief of these older rocks is in the neighborhood of 1400 feet. The principal reservoir, the so-called Stapleton zone, consists of eroded truncated edges of Viola, Simpson, and Arbuckle rocks immediately beneath the Pennsylvanian nonconformity.

SEDGWICK BASIN. On the northwest side of the Sedgwick basin there is considerable trapping of the stratigraphic type, where the Mississippian and Ordovician reservoirs, dipping off the Central Kansas uplift, have been truncated and overlapped by Pennsylvanian sediments. Examples include the Zenith (Imbt, 1941), Wherry, Welch, and other fields (Clark et al., 1948; McNeil, 1941). The *Cunningham* oil and gas field of Pratt and Kingman Counties occupies an anticline with a northeast-southwest trend nearly 7 miles in length and about 2 miles wide. This fold lies high on the west flank of the Sedgwick basin, a short distance to the east of the Pratt anticline. The Cunningham anticline plunges to the southwest and is steep sided, with about 200 feet of structural relief. It contains four reservoirs, of which the upper two produce oil and the lower two gas. The greatest production has come from the Lansing reservoir of Pennsylvanian age; the second reservoir is the Viola of Ordovician age. The gas reservoirs are the deeper Ordovician Simpson Sandstone and Arbuckle Dolomite.

Also in the Sedgwick basin are the Voshell and other anticlinal accumulations. The *Voshell* field (Hiestand, 1933) occupies an elongate north-south anticline with about 150 feet of closure. The fold appears to be as full of oil as the closure will permit. A fault along the west flank limits the production in that direction but has not been a cause of accumulation. The reservoir rocks are three limestones and sandstones ranging in age from Mississippian to Cambro-Ordovician.

Salina Basin. This basin lies between the Nemaha and Central Kansas uplifts in northern Kansas and southern Nebraska. It is crossed on the eastern side by the Abilene arch, which parallels the Nemaha axis. Today the least explored part of Kansas is the Salina basin. The reason for this relatively low density of exploratory wells is that none of the fields discovered in that area to date is a major oil or gas producer. However, there does not appear to be any significant differences in structure and stratigraphy between this basin and other, more prolific basins in Kansas. Because of its size and the depth of the floor, an unusually complete stratigraphic section is present, and a series of pre-Pennsylvanian reservoir rocks are known to be truncated and overlapped by younger rocks around the rim, especially on the east flank of the Central Kansas uplift. A number of oil and gas seeps have been observed in this area, especially in Smith County, during the past century, many of which antedate the storage of petroleum fuels underground by man. These occurrences at the surface, directly above the wedgout edges of deeper potential reservoirs, may be significant (Landes and Jewett, 1939).

Central Kansas Uplift and Extensions. The Central Kansas uplift merges to the northwest with the Cambridge arch of Nebraska. A spur to the south is called the Pratt anticline. The Salina and Sedgwick basins lie to the east, and the Hugoton embayment borders the uplift on the west. The Central Kansas uplift has been an area of search and discovery since 1924. Important finds have been made in recent years on the Cambridge arch in western Nebraska.

Anticlinal nodes on the Central Kansas uplift are important traps for oil. The discovery well for this area, which opened the *Fairport* field (Allan and Valerius, 1929) in 1923, was drilled on a sharp elongate anticline with a north-south trend that was recognized by surface mapping of the Cretaceous rocks. The structure is asymmetric, dipping much more steeply to the west than to the east, and increasing several-fold in magnitude with depth (Figure 16.7). Oil occurs in four porous zones in Pennsylvanian limestone. To the northwest of the Fairport pool, on the southeastern end of the Cambridge arch, is the *Norton* field, another anticlinal accumulation. This field, which is about 3 miles long and 1 mile wide, produces from both the Arbuckle Dolomite and the underlying Reagan Sandstone (Merriam and Goebel, 1954). The anticline has a northeast-southwest trend. On the northwest side, pre-Pennsylvanian erosion beveled the Reagan Sandstone and uncovered the underlying Precambrian rock. The Arbuckle itself is eroded across the top; trapping is due to a combination of upfold and impervious overlap.

The *Kraft-Prusa* field of Barton County (Walters and Price, 1948), one of the larger fields on the Central Kansas uplift with more than 100 million barrels of cumulative production, illustrates several different types of traps, of which the most important is the angular unconformity occurring beneath the Pennsylvanian section. The youngest reservoirs are Pennsylvanian limestones which overlie "hills" on the Precambrian surface and in which the trapping is anticlinal. The next older reservoir rock is the unconformity sand; here the trapping is entirely due to varying permeability, the porous and permeable sand on the flanks of the uplift being overlain by relatively impervious Pennsylvanian sediment. The reservoir that has produced the most oil, however, is the arched and truncated Arbuckle Dolomite lying immediately beneath the angular unconformity.

Many new fields have been discovered on the Pratt anticline in south-central Kansas in recent years. In these fields the trapping is due to porosity wedgouts, mainly through truncation and overlap, but in some instances the termination of permeability is the result of original deposition (Kansas Geol. Soc., 1956).

Walters (1958) discusses the distribution of oil and gas pools in the Arbuckle in Central Kansas. He concludes that these hydrocarbons traveled northward for distances of many miles within the porous upper Arbuckle and that their separate entrapment is due to upward spilling from traps passed through en route to their present resting places.

Hugoton Embayment. The northern spur of the Anadarko basin of western Oklahoma and the Texas Panhandle is referred to as the Hugoton embayment or Dodge City basin. It is bounded on the east by the Cambridge arch–Central Kansas uplift–Pratt anticline and on the west by the Las Animas arch of eastern Colorado. The Hugoton embayment is the largest basin

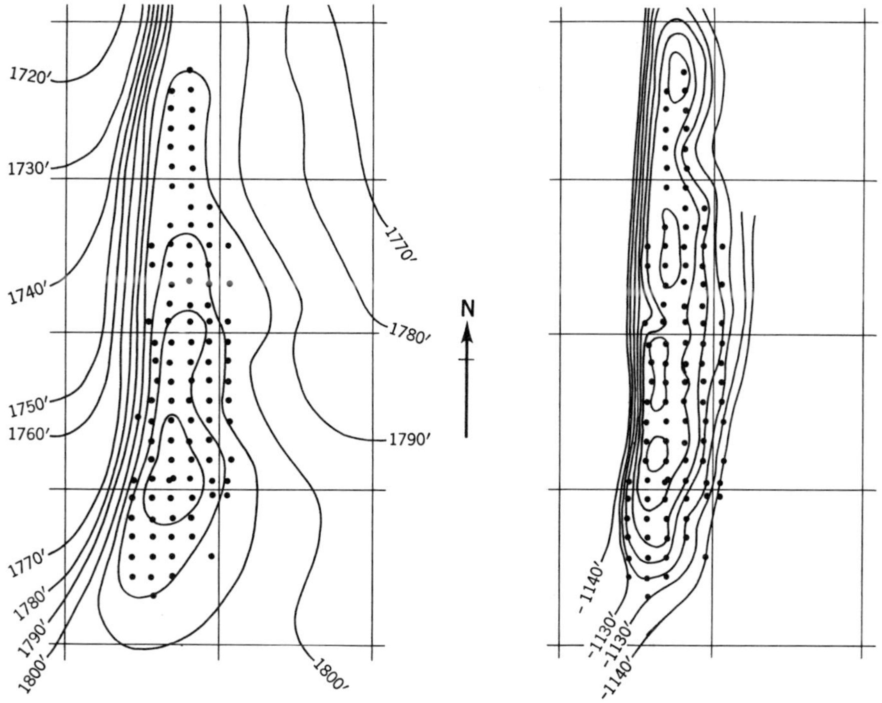

FIGURE 16.7 Fairport oil field, northwestern Russell County, Kansas. Structure contours drawn on outcropping Cretaceous rocks (left) and Pennsylvanian pay zone (right). Interval, 10 feet. Squares are 1 mile (1.61 km) on a side. *After Thomas H. Allan and M. M. Valerius (1929). Courtesy American Association of Petroleum Geologists.*

in Kansas, covering all of the western end of the state.

The gas fields of the southwestern corner of Kansas illustrate stratigraphic trapping on a regional scale. The largest of these fields and one of the great gas accumulations in the world is the *Hugoton* (Furbush, 1959; Garlough and Taylor, 1941), discovered in 1922. This field covers more than 4000 square miles in Kansas alone and contains nearly 4000 producing gas wells. The regional dip is to the eastward toward the axis of the Hugoton embayment. The reservoirs are four limestone and dolomite formations in the Chase Group of the Lower Permian (Figures 16.4 and 16.8). A fifth dolomite, the Herington Limestone, lying at the very top of the Chase Formation, is an additional reservoir locally. Trapping is brought about by facies changes that result in an updip reduction of permeability. The gas is both underlain and overlain by water (Mason, 1968). The porous limestone and dolomites become increasingly less permeable to the westward because of gradual increase in shale and siltstone content. This is illustrated in Figure 16.9, which shows by means of electric logs the updip diminution of porosity to the westward. In all probability the enormous volume of gas that has accumulated in this decreasing permeability trap is due to the great size of the downdip gathering area to the east and southeast.

Immediately to the west of the western edge of the Hugoton field is the *Greenwood* gas field of Morton County, Kansas, Baca County, Colorado, and Texas County, Oklahoma, which is second in size among Kansas gas fields. The trapping here is very similar to that of Hugoton, except that the pay zones of Greenwood are nine separate porous limestone beds ranging in thickness from 1 foot to 30 feet and lying within the Virgil Series of the Upper Pennsylvanian (Wingerter, 1959, 1968). Obviously, the same general environmental conditions that supplied fine clastics to the sea floor in this general area existed during both Late Pennsylvanian and Early Permian time, so that the carbonate rocks of the Anadarko basin lens out westward into impermeable limestone and clastic strata.

It was not until 30 years after the discovery of the Hugoton field that the first oil was found within the confines of the gas field, in southern Finney County, Kansas. Two years later the first oil discoveries were made in Haskell and Morton Counties. Subsequently, additional oil fields have been found in southwestern Kansas both within and beyond the boundaries of the Hugoton and Greenwood gas accumulations. So far, the oil reservoirs have been mainly in Lower Pennsylvanian rocks, especially the Marmaton and Morrow Groups, and in the Meramecian Series of the Upper Mississippian. The latter is very important in the *Pleasant Prairie* oil

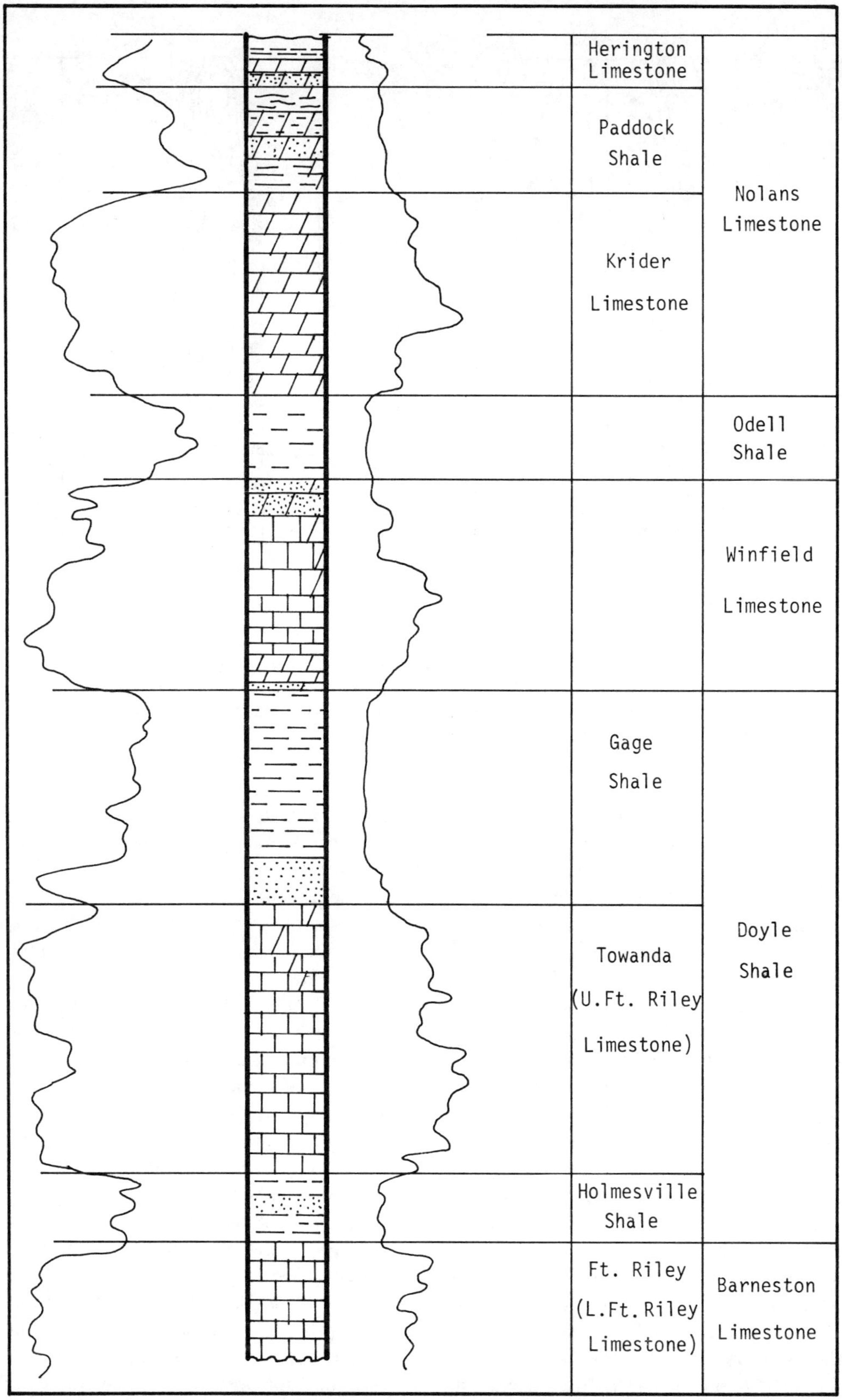

FIGURE 16.8 Hugoton gas field, southwestern Kansas. Typical log through gas reservoirs. The thickness of the section illustrated is approximately 200 feet. *Courtesy Malcolm A. Furbush (1959) and the Kansas Geological Society.*

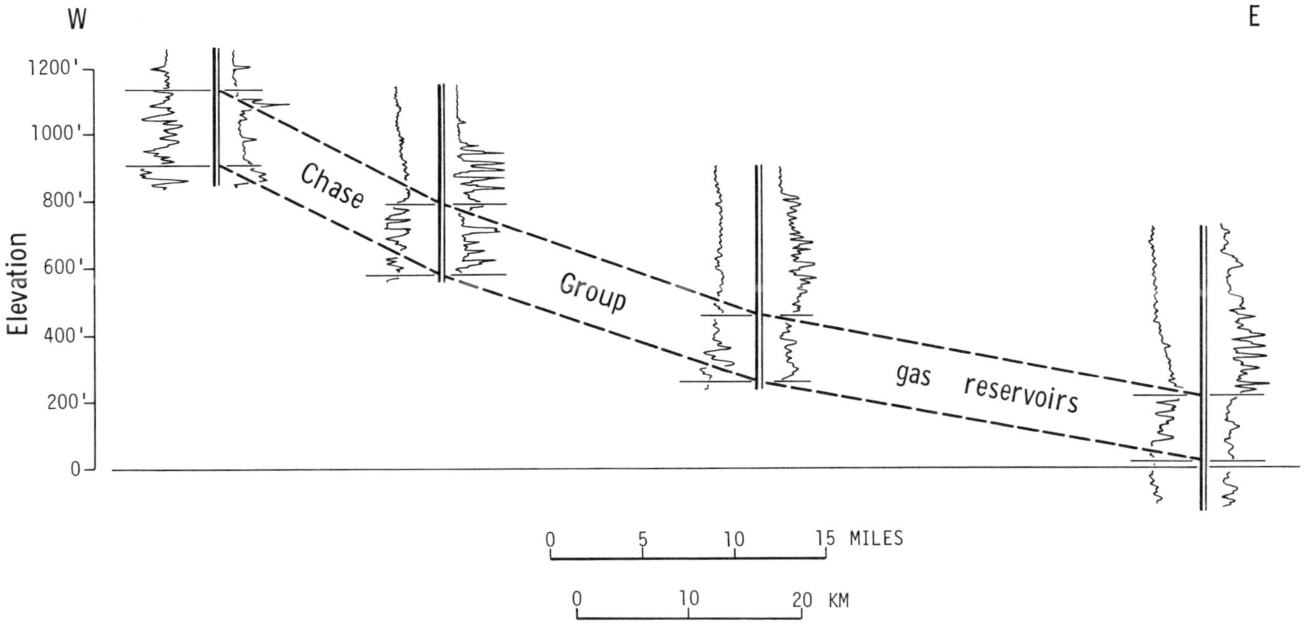

FIGURE 16.9 Hugoton gas field, southwestern Kansas. Cross section showing updip diminution in porosity of reservoirs that traps the gas. ***Courtesy Malcolm A. Furbush (1959) and the Kansas Geological Society.***

field (Figure 16.1) in Finney and Haskell Counties, where the pay is an oölitic limestone in the St. Louis Formation.

Concurrently, both oil and gas have been found elsewhere in the Hugoton embayment, including eastward to Meade County and northward to the Nebraska line. Discoveries in these areas range from Upper Pennsylvanian reservoirs downward through Cherokee, Atoka, and Morrow sediments of the Lower Pennsylvanian, and into the "Mississippi lime" and the Arbuckle.

In spite of the high percentage of carbonate rock reservoirs in Kansas, limestone reefs and reef accumulations are scarce. The first reefoid reservoir in the state was discovered in the Hugoton embayment in western Meade County in 1951. The field, the *Novinger* (Renfroe, 1954), is about 3 miles long and 1 mile wide, with a north-south trend. The reservoir is a body of porous, oölitic, and fragmental limestone occurring in the Marmaton Group in the Pennsylvanian. The thickest production section, 37 feet, occurs along the axis of the field.

FUTURE (Merriam and Goebel, 1956, 1960; Levorsen, 1958). Although Kansas is one of the older oil states, the volume of its known hydrocarbon reserves continues to mount, in spite of withdrawals, practically every year. None of the seven major tectonic divisions (Figure 16.3) of Kansas can be considered to be completely explored, either stratigraphically or geographically.

The possible reservoirs range in age from Cretaceous to Precambrian. There is as much as 3000 feet of Cretaceous rocks in northwestern Kansas, including potential reservoirs such as the Dakota and Cheyenne Sandstones (Merriam, 1958), which are an important source of oil to the northwest in northeastern Colorado and the Nebraska panhandle.

Proceeding down the time column, there are, in addition to the well known Upper Pennsylvanian pay zones, reservoir rocks in the Cherokee, Atoka, and Morrow Series in the Lower Pennsylvanian, which are largely unexplored in western Kansas.

Mississippian rocks have yet to be as thoroughly explored in western Kansas as in the eastern part of the state. Permeability wedgout traps in Mississippian reservoirs are present on the flanks of the Central Kansas uplift (Euwer, 1964; Kornfeld, 1968). The Mississippian is still capable of surprises in eastern Kansas: "Recently, several significant gas discoveries have been made along the trend of the Mississippian thickening of the Forest City basin, particularly in Leavenworth County, Kansas" (Anderson and Wells, 1968, p. 279-280).

The Devonian, Silurian, and Upper Ordovician rocks, which are confined chiefly to the basin areas, have not yet been tested adequately in the basins because of greater relative depth and difficulty in identifying them. The most ubiquitous of all reservoirs, the Arbuckle Formation, is apparently present everywhere throughout Kansas except high on the major uplifts. It has not yet been explored in southwestern Kansas, where it lies at greatest depth. Even where productive, there may be deeper Arbuckle oil beneath impermeable shale stringers (Mullins and Ireland, 1968).

Although Kansas was a leader in commercial production from the Precambrian, there has been little or no serious attempt to explore these rocks where they are topographically high and flanked by producing basins (like the Nemaha "ridge"). A listing of about 2500 boreholes that reached the Precambrian in Kansas indicates that most of them penetrated these rocks only a few feet (Cole et al., 1965). A feldspathic sandstone of probable Precambrian age has been described (Scott, 1966); if it actually is Precambrian, porosity in these ancient rocks is no longer completely dependent upon fractures, weathering, and wash.

Trapwise, porosity pinchouts on anticlinal noses hold considerable promise. Three out of the seven substantial fields discovered in 1967 had this type of accumulation (Porter, 1968).

The area of greatest promise is the Hugoton embayment. The density of wells testing the pre-Permian is low in southwestern Kansas. To the north the density of boreholes reaching any depth is also low. This embayment has both a large area and a stratigraphic section that includes potential reservoirs ranging in age from the Cretaceous to and into the basement.

Recent interest in thermal recovery projects in the shallow low-gravity oil deposits of easternmost Kansas may lead to an increase in both annual production and reserves in this state (Goebel, 1966).

REFERENCES CONSULTED IN PREPARING CHAPTER 16

Allan, Thomas H., and Valerius, M. M., 1929.

Amsden, T. W., Caplan, W. M., Hilpman, P. L., McGlasson, E. H., Rowland, T. L., and Wise, O. A., Jr., 1968.

Anderson, K. H., and Wells, J. S., 1968.

Baker, Donald R., 1962.

Bass, N. Wood, 1936.

Beene, Douglas L., and Oros, Margaret L., 1967.

Brown, Donald M., 1968.

Charles, Homer H., 1941.

Cheyney, A. E., 1929.

Clark, Stuart K., Arnett, C. L., and Royds, James S., 1948.

Cole, Virgil B., Merriam, Daniel F., Franks, Paul C., Hambleton, William W., and Hilpman, Paul., 1961.

Cole, Virgil B., Merriam, Daniel F., and Hambleton, William W., 1965.

Euwer, Robert M., 1964.

Farquhar, O. C., 1957.

Fath, A. E., 1921.

Furbush, Malcom A., 1959.

Garlough, John L., and Taylor, Garvin L., 1941.

Goebel, Edwin D., 1966.

Goebel, E. D., Hilpman, P. L., Beene, D. L., and Noever, R. J., 1961.

Hiestand, T. C., 1933.

Hilpman, Paul Lorenz, 1958.

Imbt, W. C., 1941.

Jewett, J. M., 1960.

Kansas Geological Society, 1956, 1959, 1960, 1965.

Kansas Geological Survey, 1927.

Kansas Geological Survey, Annual publication.

Kornfeld, Joseph A., 1968.

Landes, Kenneth K., and Jewett, John M., 1939.

Lee, Wallace, and Payne, Thomas G., 1944.

Levorsen, A. I., 1958.

McNeil, Harold E., 1941.

Mason, John W., 1968.

Merriam, Daniel F., 1958, 1960, 1963.

Merriam, Daniel F., and Goebel, Edwin D., 1954, 1956, 1959, 1960, 1968.

Miller, Robert R., 1968.

Moore, Raymond C., Frye, John C., Jewett, J. M., Lee, Wallace, and O'Connor, Howard G., 1951.

Morgan, F. W., and Torline, M. E., 1964.

Mullins, Claude A., and Ireland, W. C., 1968.

Porter, James L., 1968.

Reeves, John R., 1929.

Renfroe, Charles A., 1954.

Rich, J. L., 1923, 1926.

Scott, Robert W., 1966.

Smith, R. Kenneth, and Anders, Ellis L., Jr., 1951.

Walters, Robert F., 1953, 1958.

Walters, Robert F., and Price, Arthur S., 1948.

Wingerter, Hugh R., 1959, 1968.

Chapter

17

MISSOURI

Petroleum and natural gas have been produced intermittently in Missouri since 1888, but the annual yield has never been large. Oil has been produced continuously since 1936, when the Clarke-Miller oil field (Figure 17.1) came in. The Missouri oil production reached a new peak in 1966 with 97,000 barrels for the year. Oil and gas are produced in six counties in western Missouri and in one county at the eastern end of the state. Two of the counties in western Missouri, Caldwell and Clinton, yield gas only.

Southern and central Missouri are dominated by the Ozark uplift, which brings rocks ranging in age from Devonian to Precambrian to the surface. To the northwest and northeast are sedimentary basins floored mainly with Pennsylvanian rocks. In the southeastern corner, including the "Bootheel," is the northern end of the Mississippi embayment, containing Tertiary and Cretaceous sedimentary formations overlapping folded and faulted Paleozoics.

The basin to the northwest of the Ozark uplift is the Forest City, which covers contiguous parts of Missouri, Kansas, Nebraska, and Iowa (Anderson and Wells, 1968). It is here that all but one of the Missouri fields are located. The *Clarke–Miller* field in southwestern Jackson County, south of Kansas City, produces from the "Squirrel" sand of the upper Cherokee (Lower Pennsylvanian). Trapping is anticlinal. The *Tarkio* field in Atchison County in the northwestern corner of Missouri produces from a sand in the lower Cherokee. A series of sands in the Lower and Middle Pennsylvanian carry the gas in Clinton County (Anderson and Wells, 1964).

To the northeast of the Ozark uplift is the Illinois basin with its western rim in eastern Missouri. Here the only oil field in the state east of northwestern Missouri was discovered in 1952 by a gas company searching for a possible underground gas reservoir (McCracken, 1955). So far this field, the *Florissant,* is the only one in the Illinois basin west of the Mississippi. It is north of St.

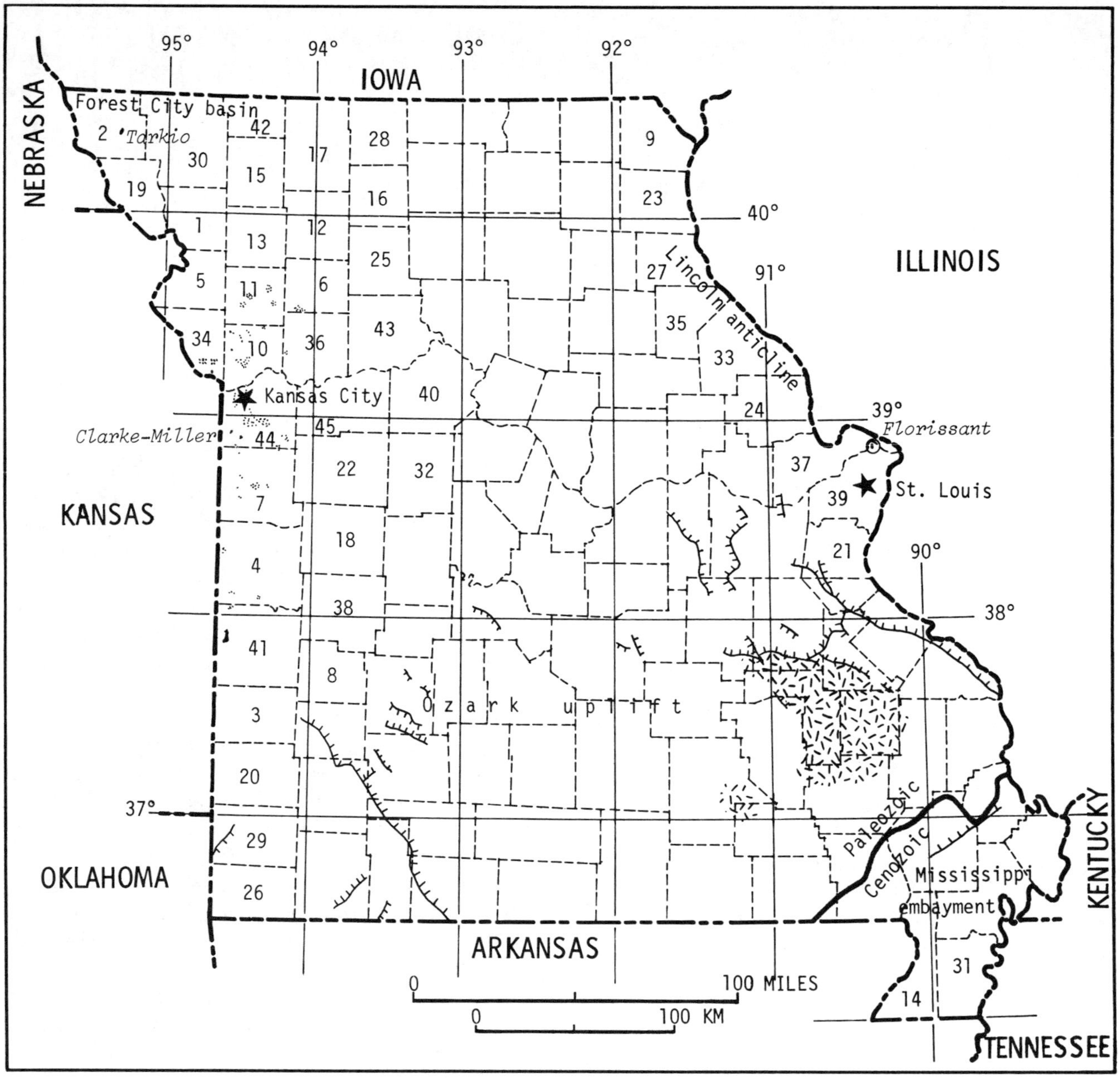

FIGURE 17.1 Missouri. Index map with oil and gas fields. Pattern shows Precambrian rock at surface. County Key:

1	Andrew	7	Cass	14	Dunklin	20	Jasper	25	Livingston	31	Pemiscot	37	St. Charles
2	Atchison	8	Cedar	15	Gentry	21	Jefferson	26	McDonald	32	Pettis	38	St. Clair
3	Barton	9	Clark	16	Grundy	22	Johnson	27	Marion	33	Pike	39	St. Louis
4	Bates	10	Clay	17	Harrison	45	Lafayette	28	Mercer	34	Platte	40	Saline
5	Buchanan	11	Clinton	18	Henry	23	Lewis	29	Newton	35	Ralls	41	Vernon
6	Caldwell	12	Davies	19	Holt	24	Lincoln	30	Nodaway	36	Ray	42	Worth
43	Carroll	13	DeKalb	44	Jackson								

Louis, in northeastern St. Louis County. The pay zone lies at the top of Kimmswick (Ordovician) Limestone at a depth of about 1300 feet. The trap is a dome on the downthrown side of a fault. About 45 wells have been drilled in the field, and their production has made St. Louis County the leading county in the state in annual oil yield. Incidentally, the gas company, although diverted by the oil strike, has also been able to meet its original objective; gas storage wells continue to be completed in this area (Porter, 1968). The storage reservoir is the St. Peter Sandstone, 500 feet below the Kimmswick oil.

Northwest of the Florissant field is the Lincoln anticline (Figure 17.1), a prominent fold with a northwest trend. It is more than 100 miles in length. So far all attempts to find oil or gas on this structure have been unsuccessful.

The Bootheel has also lured the wildcatter. Seeps have been noted in southeastern Missouri. Beneath the Cenozoic and Mesozoic embayment sediments are folded and faulted Paleozoics. So far, very few wells have explored the Paleozoic rocks.

REFERENCES CONSULTED IN PREPARING CHAPTER 17

Anderson, K. H., and Wells, J. S., 1964, 1968.

McCracken, Earl, 1955.

Porter, James L., 1968.

Chapter

18

OKLAHOMA

In recent years Oklahoma has produced between 7 and 8 per cent of the annual domestic output of both petroleum and natural gas. It was fourth in oil production and third in gas in 1968. The discovery and development of oil fields in southern Kansas focused attention on the possibilities of Indian Territory (now northern Oklahoma) as early as 1884 (Rister, 1949), but exploration was seriously impeded by the fact that most of the land belonged to various Indian tribes and hence leasing was both involved and difficult. Except for a salt borehole in Mayes County (Figure 18.1) that flowed an average of 10 barrels of oil a day for more than a year from a depth of 70 feet (John A. Taylor, letter dated 5 August, 1968), the first recorded production was in 1889, in northeastern Rogers County (Woodruff and Cooper, 1930). The discovery well was drilled at an oil spring and produced $1\frac{1}{2}$ barrels daily from a depth of 36 feet. However, the first important strike occurred near Bartlesville in Washington County in northeastern Oklahoma in 1897, when the discovery well for the Bartlesville field was completed. This well flowed 150 barrels of oil per day from a depth of 1300 feet, but litigation with the U. S. Department of the Interior kept the well capped until 1903. After that date development was rapid, many discoveries were made, and Oklahoma soon became a major producer. Its record year so far in crude oil yield was 1927, when 277,775,000 barrels of oil was produced; the production in 1967 was about 215,000,000 barrels. A total of 72 counties (out of 77) has produced oil or gas or both in this state (Figure 18.2).

Regional Tectonics. Oklahoma contains examples of practically all types of tectonic features. It has major uplifts, basins of deposition, geosynclines that have been tightly folded and thrust faulted, some profound unconformities, and convergence on a grand scale.

The major positive features in Oklahoma (Figure 18.3) are the Ozark uplift, which reaches its highest point in southern Missouri with its western flank extending into

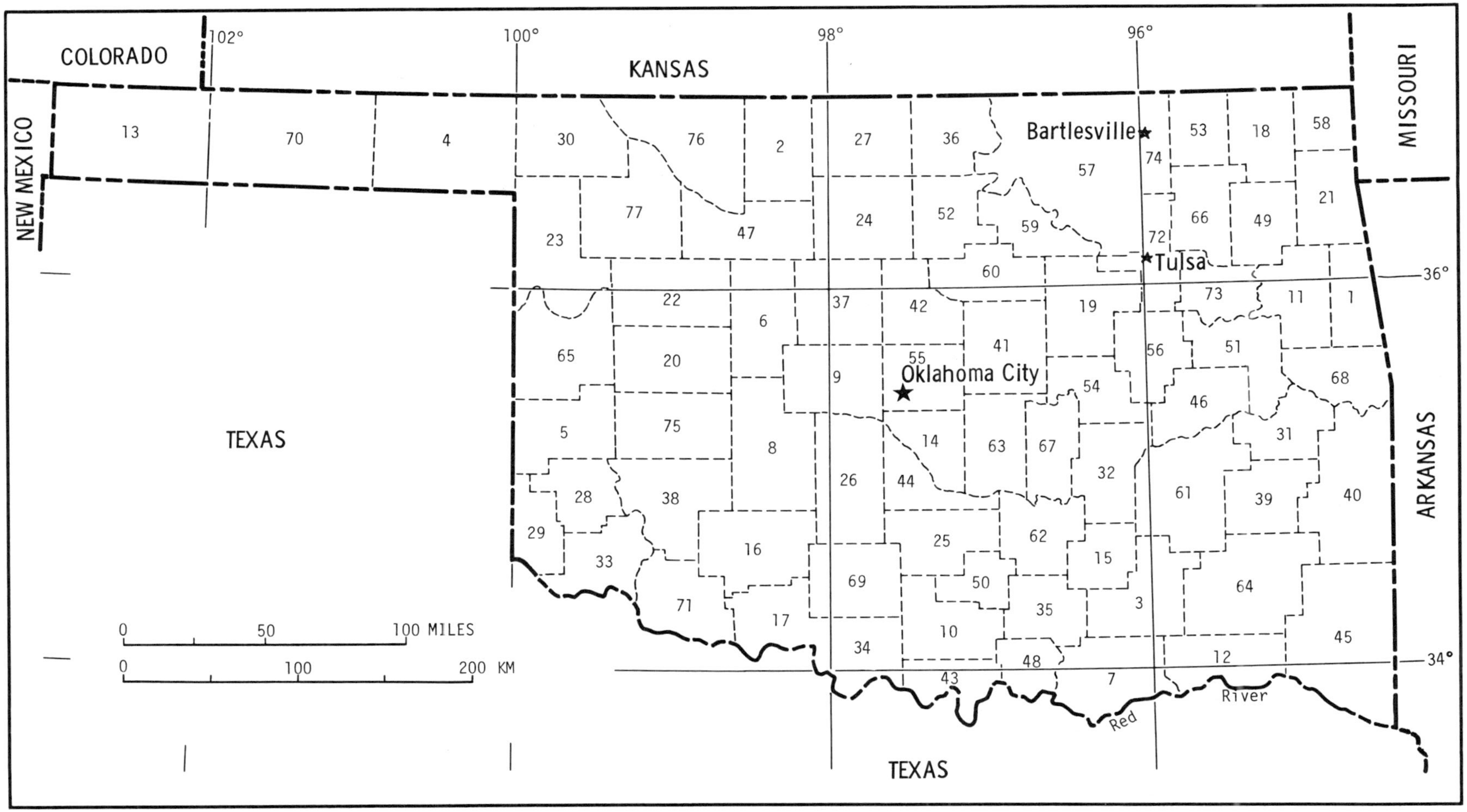

Figure 18.1 Oklahoma. Index map. County Key:

1 Adair
2 Alfalfa
3 Atoka
4 Beaver
5 Beckham
6 Blaine
7 Bryan
8 Caddo
9 Canadian
10 Carter
11 Cherokee
12 Choctaw
13 Cimarron
14 Cleveland
15 Coal
16 Comanche
17 Cotton
18 Craig
19 Creek
20 Custer
21 Delaware
22 Dewey
23 Ellis
24 Garfield
25 Garvin
26 Grady
27 Grant
28 Greer
29 Harmon
30 Harper
31 Haskell
32 Hughes
33 Jackson
34 Jefferson
35 Johnston
36 Kay
37 Kingfisher
38 Kiowa
39 Latimer
40 LeFlore
41 Lincoln
42 Logan
43 Love
44 McClain
45 McCurtain
46 McIntosh
47 Major
48 Marshall
49 Mayes
50 Murray
51 Muskogee
52 Noble
53 Nowata
54 Okfuskee
55 Oklahoma
56 Okmulgee
57 Osage
58 Ottawa
59 Pawnee
60 Payne
61 Pittsburg
62 Pontotoc
63 Pottawatomie
64 Pushmataha
65 Roger Mills
66 Rogers
67 Seminole
68 Sequoyah
69 Stephens
70 Texas
71 Tillman
72 Tulsa
73 Wagoner
74 Washington
75 Washita
76 Woods
77 Woodward

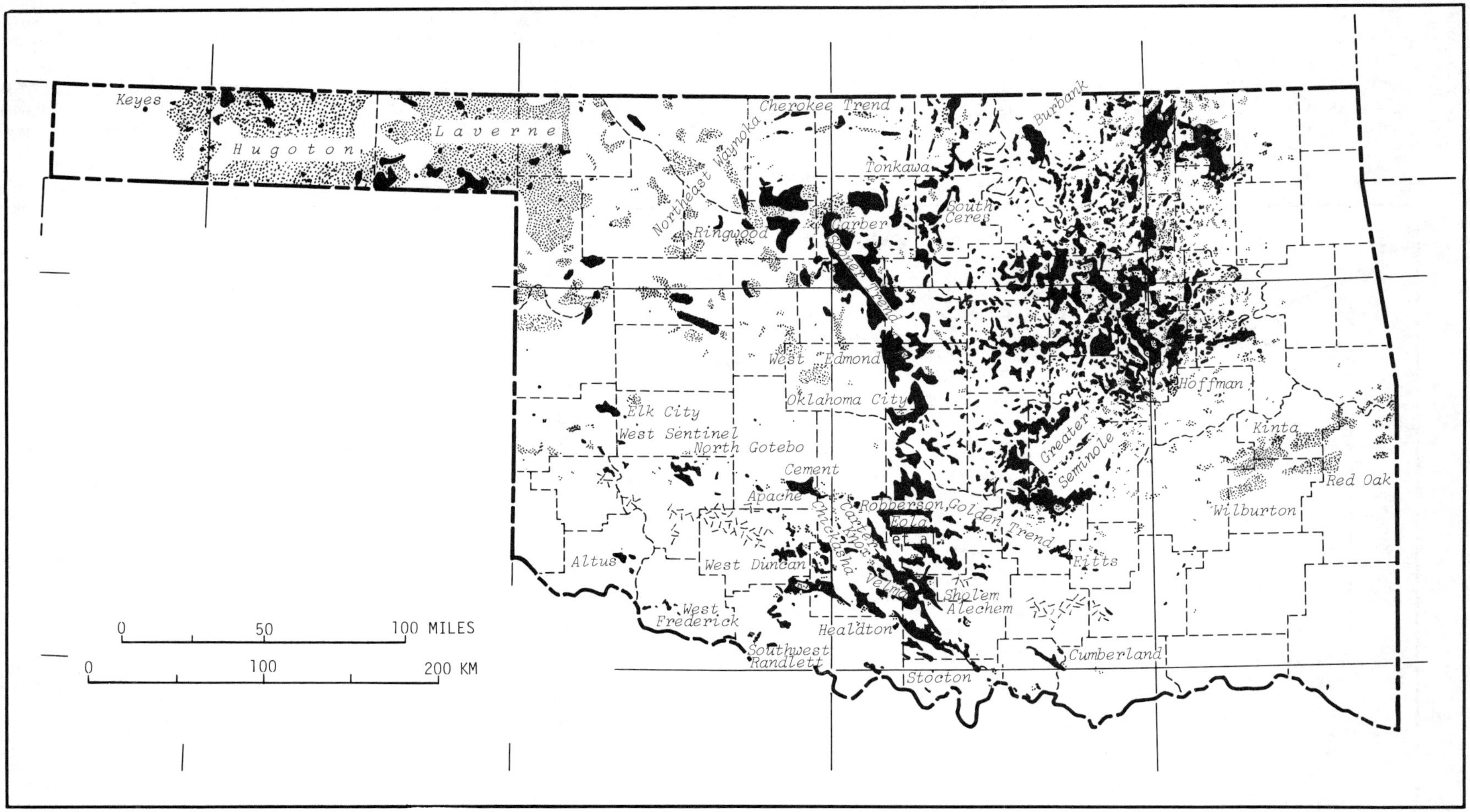

FIGURE 18.2 Oklahoma. Oil and gas fields, also showing larger basement rock outcrops. Names of fields discussed in text that are in congested areas not shown. For these fields see location within county in text, and position of county in Figure 18.1.

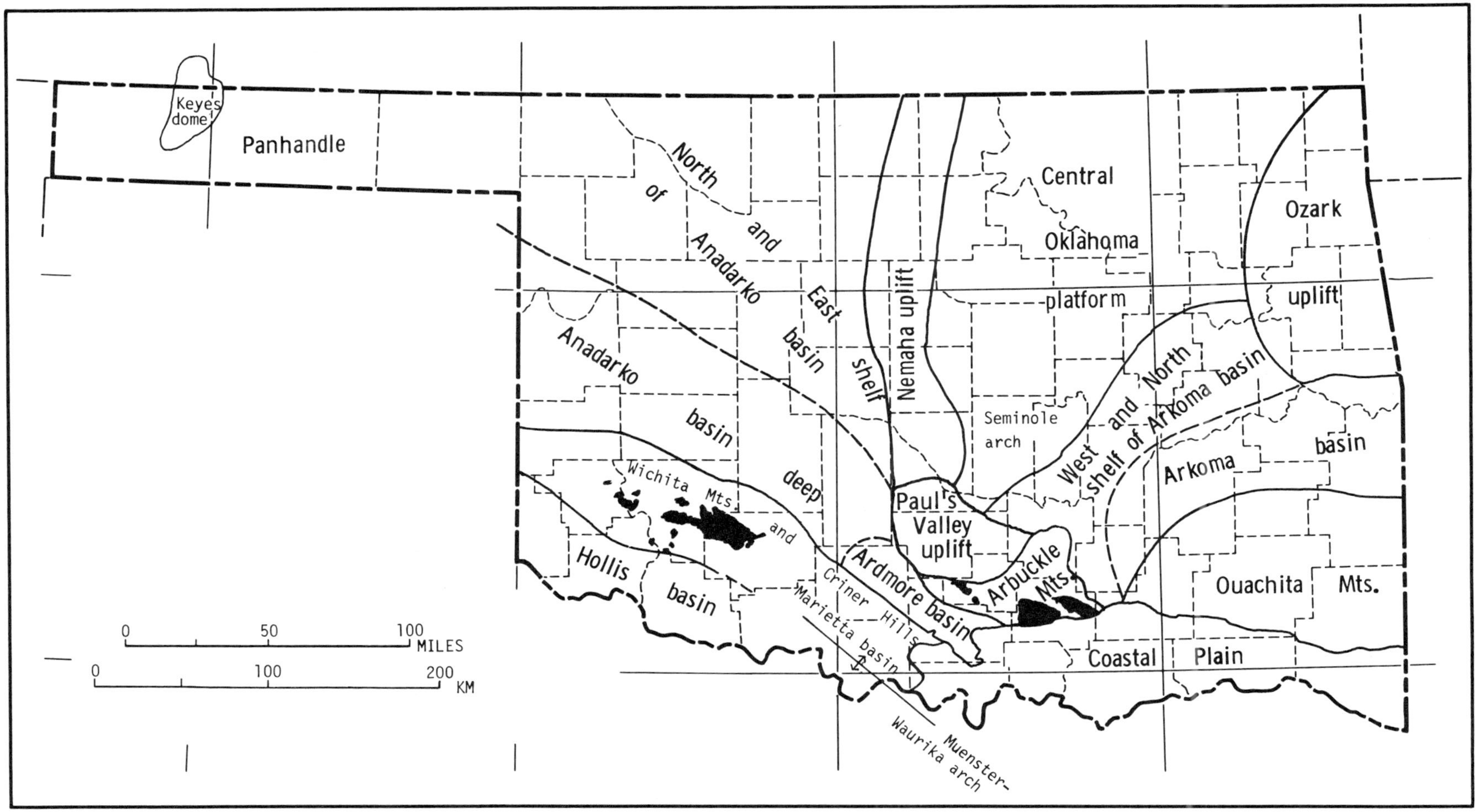

FIGURE 18.3 Oklahoma. Tectonic map. The larger basement rock areas are shown in solid black. *After F. L. Mackey and Jack P. F. Bowles, Jr. (1963) and George V. Cohee et al. (1962).*

northeastern Oklahoma, and three mountainous ranges that cross southern Oklahoma. The Ouachita geosyncline and uplift crosses Arkansas and extends westward into the eastern part of southern Oklahoma. It is succeeded to the west by the Arbuckle Mountain geosyncline and uplift, and the Wichita Mountains lie still farther to the west (Ham et al., 1965). Major negative features include the Arkoma basin, the westward extension of the Arkansas basin of Arkansas, which lies between the Ozark uplift and the Ouachita Mountains in east-central Oklahoma, and the great Anadarko basin, which covers northwestern Oklahoma, the Oklahoma panhandle, a part of the Texas Panhandle, and the western end of Kansas, where it is known as the Hugoton embayment. Three smaller basins, the Ardmore, Marietta, and Hollis, lie to the south of the Arbuckle and the Wichita Mountains in southern Oklahoma. The westernmost of these basins, the Hollis, is the eastward extension of the Palo Duro basin, which lies south of the buried Amarillo Mountains in the Texas Panhandle.

Three intermediate tectonic features are of utmost importance in respect to oil production. These are the Nemaha uplift, the Paul's Valley uplift, and the Central Oklahoma platform. The Nemaha uplift, which locally elevates the basement crystalline rocks some hundreds of feet above their normal level, is a continuation of the Nemaha "ridge" in Kansas. Although buried to a greater depth in Oklahoma, this uplift is very similar tectonically, even to the presence of a series of faults along the crest with the down-dropped side on the east. Steep dips on the west side of the Nemaha uplift lead directly into the Anadarko basin. The Paul's Valley uplift connects the Nemaha uplift with the Arbuckle tectonic province. The Central Oklahoma platform covers a large area in central and northeastern Oklahoma. It is bordered on the west by the Nemaha uplift, on the southeast by the rim of the Arkoma basin, and on the east by the Ozark uplift. The Seminole arch is in the southern end of the Central Oklahoma platform.

A major downwarp of the Gulf of Mexico embayment extended north far enough so that the eastern part of southernmost Oklahoma was invaded by Cretaceous seas during the early history of the embayment. As a result the Precambrian and Paleozoic rocks are overlapped by Cretaceous sediments in that area (Figure 18.3).

Stratigraphy. The geologic history in Oklahoma has been marked by breaks in the depositional record, accompanied by diastrophism, uplift, and erosion, which have resulted in several major unconformities. The greatest of these is the one beneath the Mississippian rocks; the pre-Chattanooga formations are truncated over a wide area. The Ouachita geosyncline was an active trough of deposition in Pennsylvanian time, so that, while sediments 2700 feet thick were being deposited in northern Oklahoma, 23,000 feet of rock was being laid down close to the axis of the Ouachita trough. The Pennsylvanian sediments also change in lithology from north to south; in northern Oklahoma they are mainly shales and limestones, and in the geosyncline they are shales and sandstones.

Rocks cropping out in Oklahoma range in age from Quaternary river deposits and dune sands to Precambrian crystallines. However, the greater part of the state is floored by rocks of either Pennsylvanian (eastern half of state) or Permian (western half) age. The pre-Pennsylvanian rocks are confined to the cores of the uplift areas previously described, and the post-Permian rocks are either continental deposits of late Tertiary age, found mainly in the panhandle of Oklahoma, or marine Cretaceous rocks of the Gulf Coastal Plain.

The reservoir rocks of Oklahoma range in age from Early Cretaceous to Cambrian. Both sandstones and carbonate rocks yield oil and gas in this state. Figure 18.4 shows the position of the better known reservoirs in a generalized stratigraphic section compiled from charts published by Branson and others, with additional help from Jordan's lexicon of subsurface stratigraphic names (Jordan, 1957).

One major reason for the prolificacy of many of the Oklahoma oil pools is the large number of reservoirs that can be found in a single field. For example, the Cushing, East Seminole, and St. Louis fields produce from at least 6 separate formations. Garber has produced from as many as 22 different reservoirs. Many of the Oklahoma reservoir rocks do not crop out at the surface.

The youngest reservoir so far developed is the Bilbo sand in the Paluxy Formation of Early Cretaceous age, which has produced a little oil in southeastern Oklahoma. Local sands in the Leonardian Group (Permian) are productive in the southeastern part of the Anadarko basin. Upper Wolfcampian (Lower Permian) carbonate rocks are the gas reservoirs in the Hugoton field of southwestern Kansas and the Oklahoma panhandle. Two sands in the Wolfcampian are productive in Garfield County. Sandstones in this part of the stratigraphic column are also productive in south-central Oklahoma.

The Pennsylvanian sediments in Oklahoma have a maximum thickness of many thousands of feet of predominantly sandstone and shale. Formations range in age from the top to the base of the Pennsylvanian. The sands are lenticular, and many have very limited lateral extent. Numerically the Pennsylvanian reservoirs exceed the total of all other reservoirs in Oklahoma; their cumulative production is also considerable.

The youngest "granite wash" reservoirs lie close to the top of the Pennsylvanian deposits. Also in the upper part of the Pennsylvanian, extending from the Virgilian

Generalized Stratigraphic Section

SYSTEM	SERIES	GROUP		FORMATION OR RESERVOIR ROCK NAME (Asterisk designates rock unit producing oil or gas)
T.				Ogallala. Late Tertiary continental deposits in western Oklahoma
CRET.	Gulfian	Colorado		Occurs both in Cimarron Co. in panhandle and in southeast Oklahoma.
CRET.	Comanchean			Dakota Sandstone in Cimarron Co., Paluxy Sandstone in southeast; Bilbo sand*
J.				Cimarron Co.
TR.		Dockum		Panhandle only
PERMIAN	Ochoan			Outcropping red beds in western Oklahoma
PERMIAN	Guadalupean	Whitehorse El Reno		
PERMIAN	Leonardian			Local sands* in lower part produce in southeast Anadarko basin
PERMIAN	Wolfcampian	Chase	Pontotoc*	Gas reservoirs* in Hugoton field
PERMIAN	Wolfcampian	Council Grove	Pontotoc*	
PERMIAN	Wolfcampian	Admire	Pontotoc*	
PENNSYLVANIAN	Virgilian	Wabaunsee to Douglas	Hoxbar*	"Granite wash" * Hoover sand * Endicott sand * Tonkawa (Stalnaker) sand*
PENNSYLVANIAN	Missourian	Lansing	Hoxbar*	Perry sand *
PENNSYLVANIAN	Missourian	Kansas City	Hoxbar*	Cottage Grove sand Hogshooter limestone * Layton sand * County Line lime *
PENNSYLVANIAN	Missourian	Pleasanton	Hoxbar*	Checkerboard lime * Cleveland sand *
PENNSYLVANIAN	Desmoinesian	Marmaton	Deese*	Wayside sand* Peru sand* Oswego Limestone* (Lone Grove sands*)
PENNSYLVANIAN	Desmoinesian	"Cherokee"	Deese*	Prue sand *; Squirrel sand * Skinner sand * Red Fork sand *; Burbank sand *; Earlsboro sand *; Dora sand* Bartlesville sand *; Glenn sand* Booch sand * Burgess sand*
PENNSYLVANIAN	Atokan	U. Dornick Hills		Dutcher Sand*; Gilcrease Sand*
PENNSYLVANIAN	Morrowan	L. Dornick Hills		Wapanucka Limestone Union Valley Formation; Cromwell sand *; Primrose sand*
PENNSYLVANIAN	Springeran			Springer sand *
SIL.-DEV.-MISS.	Chesterian Maramecian			Goddard Shale Caney Shale; "Chester Lime" *; Manning Limestone Sycamore Limestone
SIL.-DEV.-MISS.	Osagean Kinderhookian			Boone Chert; "Mississippi chat" *; "Mississippi lime" *; Arkansas Novaculite Woodford Shale; Misener sand*
SIL.-DEV.-MISS.	Oriskanian to Albion	Hunton*		Bois d' Arc Limestone *
ORDOVICIAN	Cincinnatian Trentonian			Sylvan Shale Viola Limestone
ORDOVICIAN	Black Riverian Chazyian	Simpson*		Bromide sand *; "Wilcox sand" * McLish sand * Oil Creek sand *
ORDOVICIAN	Canadian	Arbuckle*		"Siliceous lime" *; Turkey Mt. sand*
CAMBRIAN	Croixian	Arbuckle*		
CAMBRIAN	Croixian	Timbered Hills		Reagan sand *
	Precambrian crystallines			Granites, porphyries, basic igneous rocks, metasediments

FIGURE 18.4 **Stratigraphic section, Oklahoma.** *Data derived from various sources, including especially Carl C. Branson (1956) and Louise Jordan (1957).*

Series to the Marmaton Group in the upper Desmoinesian, are several important sandstone reservoirs and four productive limestones, including the Oswego (Richardson, 1965). Included in this part of the column is the Hoxbar group, which produces in the southeastern part of the Anadarko basin. Below the Marmaton in northeastern Oklahoma is a series of prolific sands of lower Desmoinesian or "Cherokee" age. The best known of these are the Burbank, Bartlesville, and Glenn. The Cherokee sands are highly lenticular (Figure 18.5); the general relationship among four such sands has been depicted by Branson (1957). Also of Desmoinesian age are the Deese Sands, which are large producers of oil in both the Paul's Valley uplift area and the Ardmore basin. The greater part of the Pennsylvanian production, as well as an important share of the Oklahoma output of oil up to the present time, has come from Cherokee and Deese Sands.

The Atokan (Six, 1962), Morrowan, and Springeran (Jacobsen, 1959) Series at the base of the Pennsylvanian section contain reservoir sands in the Central Oklahoma platform district. In addition, sands in the Morrowan Series have been found in recent years to be productive of both oil and gas in northwest Oklahoma and in the panhandle.

The "Mississippi lime" has produced large quantities of oil annually in eastern Oklahoma for more than half a century (Hayes, 1956). The principal reservoir here is a cherty section belonging in the Osagean Series, which lies immediately below the Pennsylvanian-Mississippian unconformity. Another important Mississippian reservoir locally is the Misener sand at the base of the system. From the early 1960s much new oil and gas has been found in the Mississippi lime (Meramec and Osage) on the eastern flank of the Anadarko basin in a four-county area (Garfield, Logan, Kingfisher, and Major) covering some 800 square miles. This many-field complex, the *Sooner Trend,* is shown on Figure 18.2. Farther west, in the Oklahoma panhandle, Upper Mississippian pay zones, including the Manning Limestone and the "Chester lime," have been found to be productive.

The Bois d'Arc Limestone (Oxley, 1959; Culp and Barrett, 1957) belonging in the Hunton Group of Siluro-Devonian age is an important source of oil in the West Edmond field on the Nemaha uplift in central Oklahoma. In the latter 1960s Hunton limestone pay zones became prime drilling targets in the Anadarko basin in western Oklahoma (Harvey, 1968). Wells exceeding 100 million cubic feet of gas daily have been completed, and fields with large gas reserves have been developed.

The Viola Limestone of Trenton (Ordovician) age is an important carbonate rock reservoir in the midcontinent (Skarda, 1956), especially where fractured. Slightly lower in the section is the Simpson Group, which

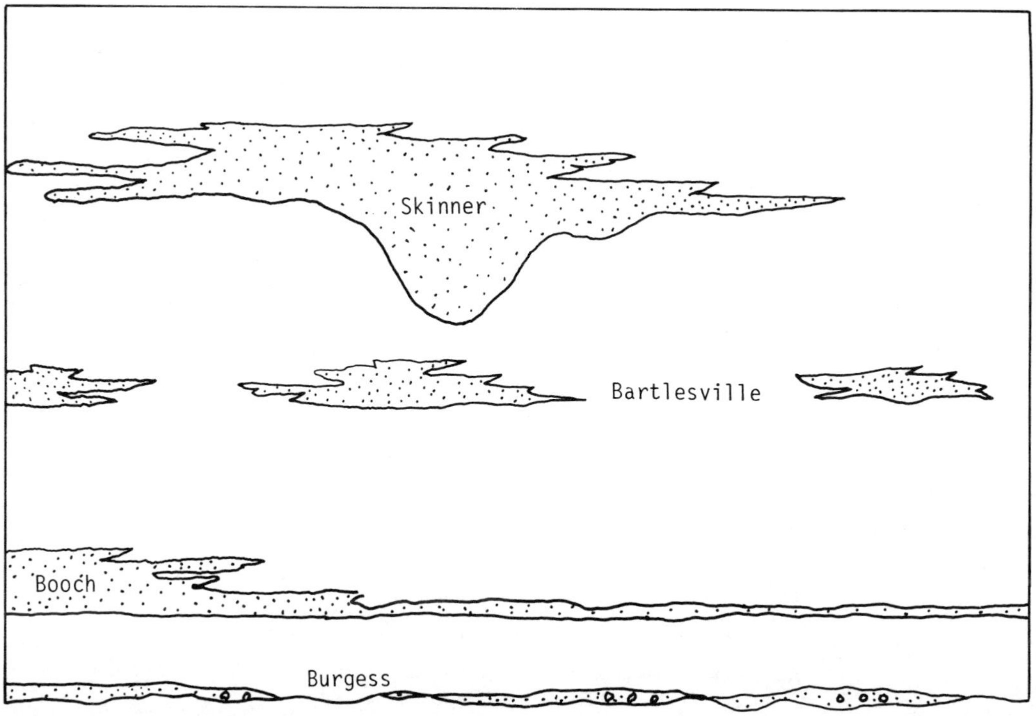

FIGURE 18.5 Cross section of Pennsylvanian sandstone bodies on northeastern Oklahoma platform. ***Courtesy Carl C. Branson (1957) and World Oil.***

contains sandstone reservoirs of phenomenal productivity in Oklahoma. Local names applied to these sandstones include Bromide, "Wilcox," and McLish. The cumulative production from Simpson sands in eastern Oklahoma exceeds a hundred million barrels.

Oklahoma is not only the type locality for the Arbuckle Limestone of Cambro-Ordovician age but also is a leading state in oil production from that reservoir. This thick group of carbonate rocks dips away in all directions from its outcrop in the Arbuckle Mountains, disappearing beneath younger formations. It is present throughout the state except in uplift areas where it has been eroded off locally. A frequently used drillers' term for the Arbuckle is "Siliceous Lime."

The Reagan Sandstone is a basal Cambrian (and Paleozoic) deposit that overlies the Precambrian floor and is productive in a few areas.

PETROLEUM-PRODUCING DISTRICTS

There is no universally accepted division of the oil and gas fields of Oklahoma by districts. In the regional survey that follows, the fields are grouped into districts based upon the tectonic map (Figure 18.3), which in turn is a synthesis of maps published by Cohee (1961) and others. The districts thus derived from the tectonic map are (1) Central Oklahoma platform, (2) Arkoma basin, (3) Nemaha uplift, (4) Paul's Valley uplift, (5) Southern Oklahoma, and (6) Anadarko basin.

Central Oklahoma Platform (Fitts, 1951). The Central Oklahoma platform has also been referred to as the Northeast Oklahoma platform and the North-central Oklahoma shelf. It is the discovery district for Oklahoma. To the north the Central Oklahoma platform merges with the Cherokee basin of southeastern Kansas, in which the first commercial oil discoveries for the midcontinent region were made. In the northern part of this district, as in the Cherokee basin, the emphasis is on Cherokee (Pennsylvanian) sand reservoirs (Clayton, 1965). Here the strata are dipping westward so that the depths to these reservoirs become greater from east to west. In the southern part of the Central Oklahoma platform district the greater part of the production comes from pre-Pennsylvanian reservoirs.

Among the famous producing sands within the Cherokee Shale are, going down the section, the Prue, Skinner, Red Fork, Burbank, Earlsboro, and the well-known Bartlesville. The Red Fork sand, which is the approximate equivalent of the Burbank, occurs in elongate oil-containing lenses from the Osage City field in southwestern Osage County south-southeastward across contiguous parts of Pawnee, Creek, and Tulsa Counties (Wright, R., 1941). The largest lens is more than 10 miles long and varies in width from one well location to half a mile. It differs from many other shoestring sand bodies in that it contains some water; in fact, where the elongate sand lens crosses a syncline, it is filled with water. There is no water drive, however, and the water present appears to be completely sealed in along with the oil. The most logical conclusion as to the source of the oil is that it entered the porous and permeable sandstone during the compaction of the surrounding organic Cherokee Shale. The reason for the water may be that insufficient oil was injected into the sand to displace completely all the water present. After the Red Fork discovery many other shallow oil and gas fields were found in this part of Oklahoma (Dille, 1952).

Farther to the west in northern Oklahoma is Osage County (Clinton, 1957; Riggs et al., 1954) with its many oil fields, including the famous *Burbank* (Bass et al., 1942; Sands, 1927; Markham and Lamar, 1937; Beckwith, 1928). The Burbank and South Burbank pools produce from the Burbank sand, which lies at a depth of about 2800 feet and has an average thickness of 57 feet. This sand occurs in the lower part of the Cherokee Shale unit of Pennsylvanian age. In these pools the sandstone reservoir rock, although lenticular in shape, is much greater in fluid capacity than the oil accumulation and hence contains water below the oil. The regional dip is westward, and the oil has been impounded by the updip lensing of the Burbank sand to the east. In the main Burbank pool, above the level of the oil-water interface, there does not appear to be any relationship between the occurrence of oil and the local structure, but in the South Burbank pool a gas cap has accumulated in the structurally highest parts of the hydrocarbon-filled reservoir rock. In recent years the Burbank field and Osage County in general have enjoyed a second "boom" because of a big jump in production figures brought about by secondary recovery. Currently about 76 per cent of the Burbank output is credited to waterflooding. The cumulative production for the field is in excess of $\frac{1}{2}$ billion barrels (Largent, 1968).

The great *Glenn* pool lies to the south of Burbank in eastern Creek County (Wilson, 1929). This field, discovered in December, 1965, was the first major oil pool in Oklahoma. The reservoir rock is the Glenn (Bartlesville) Sandstone, which also lies within the Pennsylvanian Cherokee shale. The depth of the reservoir is about 1500 feet. Local structure is not important in controlling accumulation in this reservoir rock, but the regional structure and the lenticularity of the sand are of utmost importance. As in the Burbank pool, the regional dip is to the westward and the oil has been trapped by the updip lensing out of the sandstone. However, there is some production from the deeper Mounds ("Wilcox") Sandstone of Ordovician age, in which accumulation is entirely anticlinal.

The Prue sand lens lying at the top of the Cherokee shale has trapped and is producing oil in the *Davenport* (White, 1941) and other fields in Lincoln County in the western part of the northern end of the Central Oklahoma platform. Slightly lower in the Cherokee section is the Skinner sand, which is also an important reservoir in this area.

The *Cushing* field (Riggs et al., 1958; Weirich, 1929) in western Creek County is another field of large cumulative production. The structure is a pre-Pennsylvanian bald-headed anticline 20 miles in length overlapped by Pennsylvanian sediments. The Pennsylvanian formations are likewise anticlinal in structure because of recurrent movement. The reservoirs include seven Pennsylvanian sandstones and, beneath the pre-Pennsylvanian unconformity, Ordovician sands and the Cambro-Ordovician Arbuckle Dolomite. The Bartlesville Sandstone, which lies within the Cherokee Shale (Pennsylvanian), has been oil bearing over the entire anticlinal crest, covering an area of 28 square miles. The Ordovician sands produced only from the tops of the domes superimposed upon the anticlinal axis, and the Arbuckle carried oil on one dome only where the overlying Ordovician sandstones had been removed during the pre-Pennsylvanian erosion interval. By far the greater part of the trapping at Cushing is anticlinal, but in the older reservoirs sealing of the truncated domes by overlap has contributed.

The *Ramsey* pool (Umpleby, 1956; Frost, 1940) is in west central Payne County about 30 miles east of the axis of the Nemaha uplift. Here the trap is an almost circular dome with a maximum structural relief of about 200 feet. The dome has been truncated; about 130 feet of Mississippian limestone is absent across the top of the fold as the result of pre-Pennsylvanian erosion. The reservoirs are, in descending order, the Misener-Hunton, First "Wilcox," and Second "Wilcox." The east side of the dome is cut by a north-south fault on which 700 feet of horizontal displacement and 120 feet of vertical displacement have taken place. Both the domal folding and the faulting occurred in several steps. Accumulation apparently began with the development of the first closure and continued until after the trap had been developed to its present size. The fault permitted migration of oil from the Second Wilcox upward into the First Wilcox; it also permitted complete cross-fault migration in the Misener-Hunton reservoir, so that the level of the oil-water interfaces is the same on both sides.

To the southwest of the Ramsey pool and close to the Nemaha uplift in southeastern Logan and northeastern Oklahoma Counties is the *Arcadia–Coon Creek* oil pool (Carver, 1948). Figure 18.6 consists of two maps of this pool at different levels. The upper shows the structure contoured on top of the Oswego Limestone of Pennsylvanian age, which lies at a depth of about 5000 feet. There is no suggestion of closure at this stratigraphic level. The lower drawing shows the structure as contoured on top of the "Second Wilcox" Sandstone, which is Ordovician in age and lies at a depth of approximately 6000 feet. The Second Wilcox is the producing formation in the Arcadia–Coon Creek pool. It is a white sandstone 150 feet thick, consisting of rounded and frosted quartz grains. At the producing level the structure is an anticline about 3 miles long by 1 to $1\frac{1}{2}$ miles wide. The anticline has two distinct domes, but the oil-water interface lies approximately 4850 feet below sea level, which is low enough to enclose both domes so that production is continuous across the saddle. The effective oil sand thickness from the water level to the top of the anticline is 58 feet, out of a total closure of 70 feet.

The dying-out upwards of the anticline at Arcadia–Coon Creek is due to the thinning of the post-Wilcox formations across the crest of the anticline. The anticline disappears entirely above the unconformity at the base of the Pennsylvanian. Mississippian sediments are missing completely across the crest of the anticline, as a result of pre-Pennsylvanian erosion and truncation.

Trapping in lenticular sandstones is not confined to the northern end of the Central Oklahoma platform. It is also found to the south, in the central and southern parts of the district. Examples include the Dora and Olympic pools. The *Dora* pool (Ingham, 1941) is in the so-called Greater Seminole district but is not a structural accumulation as are most of the others in this area. The producing sand, locally called the "Dora," is in the Thurman Formation of Early Pennsylvanian age. It occurs at depths ranging from 2750 to 3000 feet and has a maximum thickness of about 100 feet. The Dora sand deposit is an irregularly shaped lens lying on a northwest-dipping monocline. The reservoir rock is completely surrounded by shale, and it is believed that these shales are the source rock for the oil.

The *Olympic* pool (Dillard, 1941; Tillotson, 1938) in southwestern Okfuskee and northwestern Hughes Counties produces from a sand in the Senora Formation of Pennsylvanian age that is slightly younger than the Thurman. Accumulation is very similar to that at Dora. The reservoir is a sand bar measuring about $6\frac{1}{2}$ miles long with a maximum width of $1\frac{1}{4}$ miles. The sandstone interfingers with shale at its lateral margins.

Both stratigraphic and structural trapping are illustrated in the *East Tuskegee* pool (Borden and Brant, 1941), which lies north of the Olympic pool in southeastern Creek County. Oil in the Ordovician "Wilcox" Sandstone is confined to two structural domes, but the higher Misener (basal Mississippian) sand is tight on the apices of the domes so that production from that reservoir is confined to the flanks.

One of the greatest concentrations of oil in the country

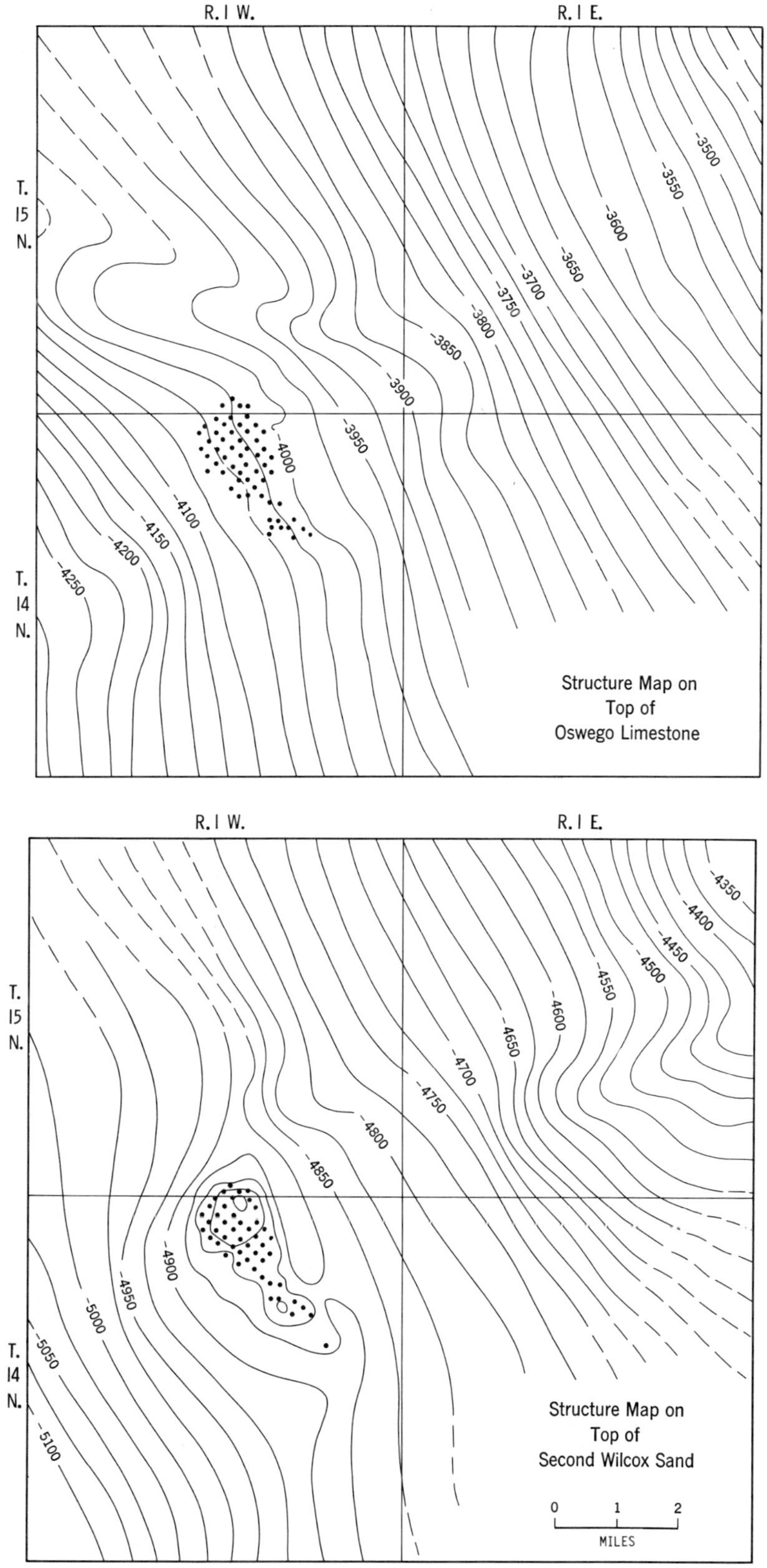

Figure 18.6 Arcadia-Coon Creek oil field, Oklahoma and Logan Counties, Oklahoma. Structure contour maps based on Oswego Limestone (upper drawing) and Second Wilcox Sand (the reservoir). Contour interval, 25 feet. *After G. E. Carver (1948).*

has been in the *Greater Seminole* district of Seminole and Pottawatamie Counties (Tanner, 1956; Levorsen, 1929; Rau and Ackley, 1939). Among the more important pools in the Greater Seminole district are the *Seminole City, Searight, Earlsboro, Bowlegs, Little River, Mission,* and *St. Louis*. The discovery and development of these pools in 1926 and the years immediately thereafter led to serious overproduction in the midcontinent. At the height of production the Greater Seminole district yielded 500,000 barrels of oil daily. Oil accumulation in this area is largely in Devonian, Silurian, and Ordovician reservoirs. The trapping is anticlinal, with the domes at the Ordovician level having closures of several hundreds of feet (Figure 18.7, right). However, the outcropping Pennsylvanian rocks show little or no closure (Figure 18.7, left), a fact that explains why these fields were not discovered earlier.

The *Hoffman* (Hamric, 1961) gas accumulations occur along the hinge line between the Central Oklahoma platform and the Arkoma basin. Although tectonically this area lies on the southeastern edge of the Central Oklahoma platform, the gas captured probably came out of the Arkoma basin. The structure consists of a series of broad anticlines with faults on the east, or basin rim, side. The displacement produced by the faulting is downward on the east. Although some oil and gas have been obtained from Simpson sands in the Ordovician, by far the greatest production in this area has been gas from both the Salt (Bartlesville) sand in the Cherokee of Pennsylvanian age and the Cromwell sand of the Morrowan Series in the lower Pennsylvanian. Further gas production has come from sands in the pre-Bartlesville Cherokee and in the Atokan Group lying between the Cherokee and the Morrowan sediments.

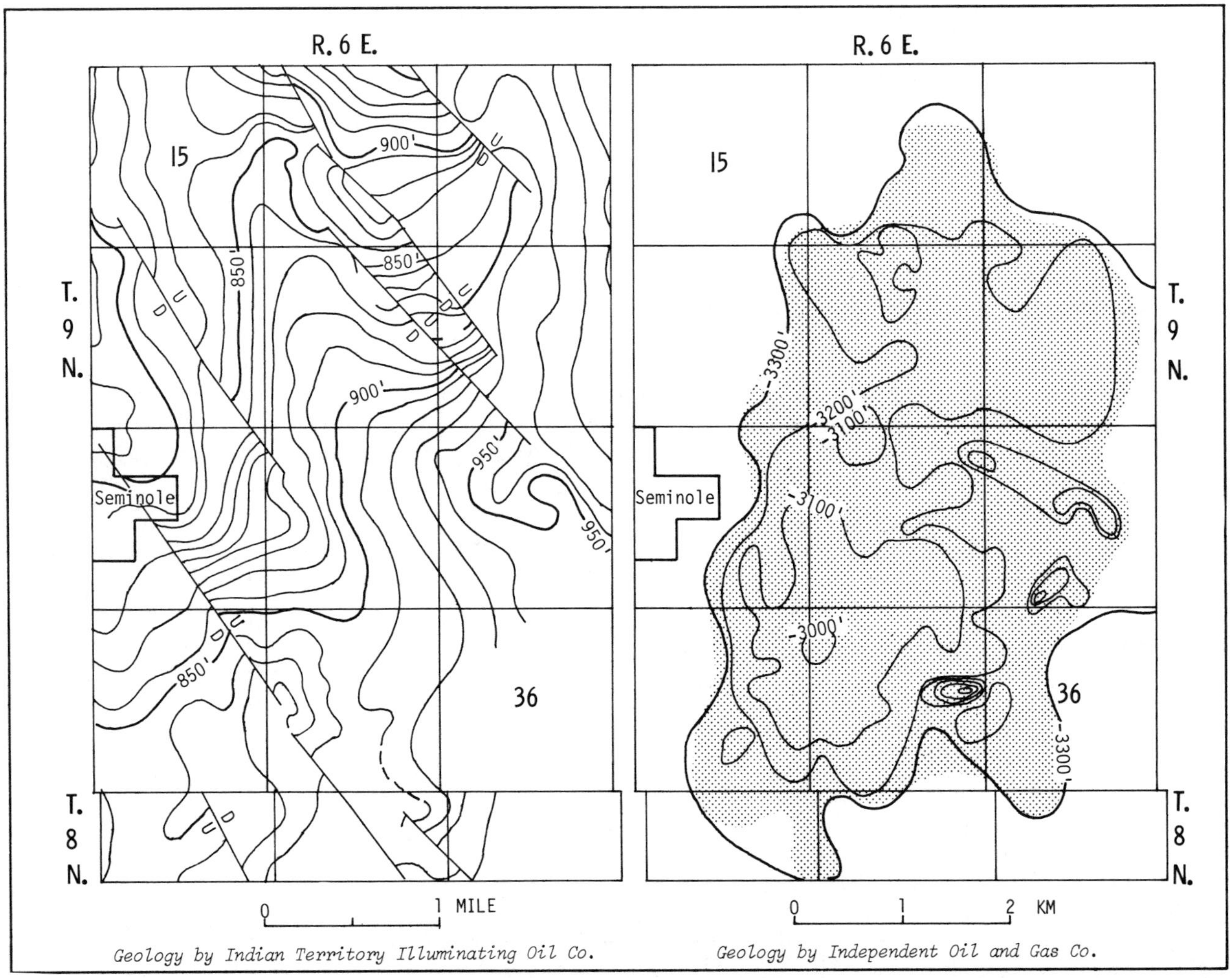

FIGURE 18.7 **Seminole pool, Seminole County, Oklahoma. Left: Structure map based on outcrops of the Pawhuska Limestone. Contour interval, 10 feet. Right: Subsurface structure map of same area based on the Viola Limestone. Contour interval, 100 feet. Productive area is stippled.** ***From A. I. Levorsen (1929, Figure 11, p. 34 J) and the American Association of Petroleum Geologists.***

Trapping is anticlinal, with gas concentrating in the nodes lying along the axes of the large anticlines.

Arkoma Basin (Oklahoma City Geol. Soc., 1963; Bartlett, 1966; Woncik, 1962; Thralls et al., 1962; Planalp, 1961; Wellman, 1951; Colton, 1935). The name of this basin is derived from the first three letters in Arkansas and the last three letters in Oklahoma. It is a fair arrangement, for each state owns approximately half of the basin. In Arkansas it has been known as the Arkansas basin; and before the word Arkoma was coined, the Oklahoma end of the basin was referred to as the western part of the Arkansas basin. Another commonly used term was the McAlester coal basin, and this name is still applied to the western half of the Oklahoma part of the Arkoma basin (Koinm and Dickey, 1967).

The Arkoma basin is almost entirely a gas province. Gas was first discovered commercially in the Arkansas side of the basin in 1902; the first important discovery on the Oklahoma side was made in 1910. A revival of interest took place in this area in the late 1950s and led to many new discoveries. During the years 1961 and 1962, 20 new fields were discovered each year in the Oklahoma part of the district.

In Arkansas and extreme eastern Oklahoma the Arkoma basin is sandwiched between the Ozark uplift and the Ouachita Mountains. To the westward the basin is limited by the Arbuckle Mountains and Paul's Valley uplift, and to the northwest by the Central Oklahoma platform and the Seminole arch. Between these higher features and the deeper part of the basin are the west and north shelves of the Arkoma basin. The boundary between the Arkoma tectonic province and the Ouachita Mountain area in southeastern Oklahoma is generally accepted to be the Choctaw fault. Hydrocarbon occurrences on the southeastern side of this fault are discussed subsequently. In progressing southward in Arkansas and southeastward in Oklahoma across the Arkoma basin, the folding of the sediments increases from gentle through moderate to intense before reaching the thrust-faulted Ouachita province. The carbon ratios in the zones of moderate and intense folding are high for oil, a fact that probably explains the predominance of gas in this area. The Arkoma basin is therefore comparable to the wide gas belt that intervenes in western Pennsylvania between the gently folded oil-bearing Appalachian rocks and the intensely folded barren rocks to the east.

The reservoirs in the Oklahoma sector of the Arkoma basin currently productive are nine in number, of which eight are Pennsylvanian and one is Simpson (Ordovician). Most of the gas comes from lenticular, fine-grained sandstones of Atokan (Early Pennsylvanian) age (Branan, 1968).

Most of the traps so far discovered are anticlines with moderate to steep dips. However, the lenticularity of the sand bodies controls the distribution of the gas across the anticline. Faulting is common in the Arkoma basin, especially toward the Ouachita Mountain area.

Many of the Arkoma basin anticlines can be mapped adequately at the surface by strike and dip readings, since the closures even at the surface may be measured in hundreds and in some places in thousands of feet. Photogeology is also an effective technique for surface-structure mapping. However, the axes of the anticlines may shift through a considerable distance with depth. Also in the deeper part of the basin imbricate thrust faults further confuse the structural picture in the subsurface.

Two of the largest fields in the Arkoma basin are the Red Oak–Norris in Latimer and LeFlore Counties and the Kinta in Haskell County. The *Red Oak–Norris* is in the deeper part of the basin, where the faulting is extremely complex. Accumulation is in the Brazil anticline, which is expressed at the surface as a valley. The area proved productive is 20 miles long (east-west) and 6 miles wide. The principal reservoir, an Atokan sandstone, lies at a depth of 11,000 to 13,000 feet (Six, 1968).

The *Kinta* gas field, as now constituted, is the result of the merging of six gas fields, one of which was the original Kinta. Tectonically the field is on the northern shelf of the Arkoma basin. It has an estimated reserve of over a trillion cubic feet of gas. The pay zones are lenticular sands of Early Pennsylvanian age. The structure is anticlinal, with normal growth faults responsible for thicker sections of Atokan shale on the down blocks (Woncik, 1968a).

A third field, the *Wilburton* in western Latimer County, is of interest because a part of the southern edge extends across the surface trace of the Choctaw fault (Branan, 1968, Figure 3, p. 1620). Gas wells in this area were drilled through the fault into the underthrown Arkoma basin section beneath.

At least two minor oil discoveries have been made in the Oklahoma part of the Arkoma basin. One wildcat drilled in Coal County struck oil in a sand in the Morrow Series, and another well in the same county found oil in a Simpson (Ordovician) sand.

Nemaha Uplift (Bale and Williams, 1951). The Nemaha uplift extends into Oklahoma from Kansas and occupies a relatively narrow belt southward from Kay County across parts of Grand, Garfield, Noble, Logan, Oklahoma, and Cleveland Counties. It abuts at the southern end against the Paul's Valley uplift. The Nemaha uplift is bounded on the east by the Central Oklahoma platform and on the west by the Anadarko basin.

The first discovery in the Oklahoma part of the Nemaha uplift was the Garber field in 1916. Subse-

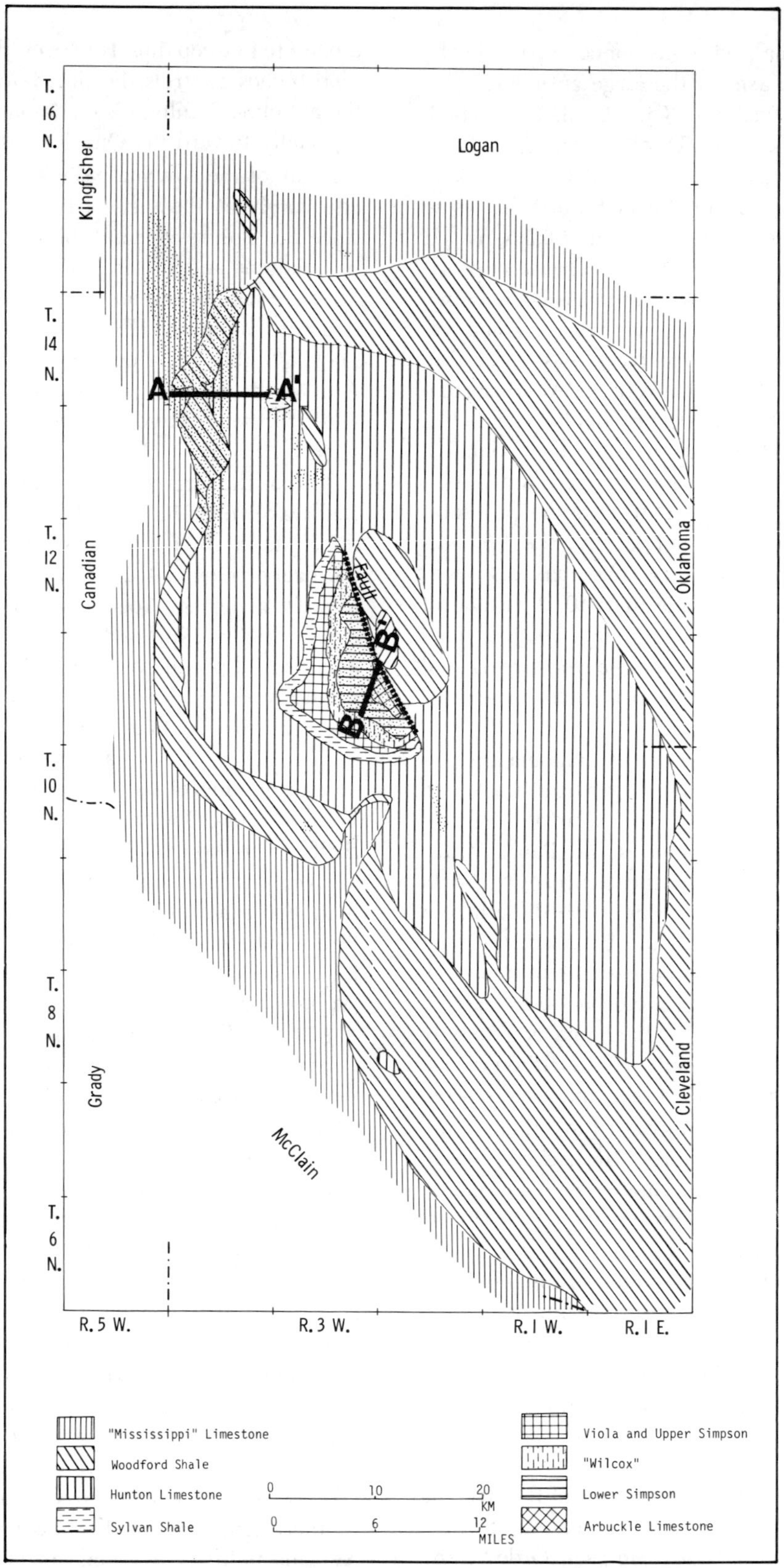

Figure 18.8 Pre-Pennsylvanian areal geologic map of Oklahoma and Cleveland counties and parts of adjacent counties to the west and north. Shows location of sections through West Edmond (*A-A′*) and Oklahoma City (*B-B′*), which appear on Figure 18.9. *After D. A. McGee and H. D. Jenkins (1946). Courtesy the authors (per McGee) and the American Association of Petroleum Geologists.*

quently many smaller fields and such big fields as Tonkawa, Oklahoma City, and West Edmond have been discovered. In spite of its narrow width, the density of exploration wells, and the often expressed opinion that this province had been completely exploited, in 1962 the Nemaha uplift was the scene of 22 new pool discoveries, which constituted 19 per cent of the state total for that year.

In Oklahoma, as in Kansas, the Nemaha uplift is a linear arrangement of buried hills with granite cores, with the surface strata reflecting the presence of the deeper nodes by gentle dome-shaped anticlines. The geologic history of this belt in Oklahoma has been summarized by Bale and Williams (1951) as follows:

> Quiet deposition continued from Cambrian through the Devonian; then emergence and development of sharp structures along an arcuate line of release; tilting and excessive peneplanation; Mississippian deposition over the area except for some unburied hills; re-emergence with structural wrinkling and renewal of pre-Mississippian faulting; then shallow-sea deposition of Pennsylvanian and Permian beds in an almost uninterrupted sequence; and lastly, rejuvenation of earlier folding.

One result of this history of deposition, folding, erosion, and further deposition is a profound unconformity at the base of the Pennsylvanian (Figure 18.8).

Various types of trapping occur in the Nemaha uplift area. Some traps are simple anticlines with the hydrocarbons occurring both on the flanks and across the top of the fold within a given reservoir rock. In many instances, however, the upfold has been truncated to varying depths so that additional traps are produced by truncation and overlap. Because of different periods of movement and the presence of unconformities in the section, trapping in the youngest and the oldest reservoirs may be anticlinal, with wedgout of reservoirs creating the traps at intermediate stratigraphic levels.

The greatest oil field in terms of cumulative production in Oklahoma, and one of the ten largest oil fields in the United States, is the *Oklahoma City* field (McGee and Clawson, 1932), which lies along the axis of the Nemaha uplift in Oklahoma County. The structure is a so-called bald-headed anticline with a core of truncated Arbuckle Limestone of Cambro-Ordovician age (Figure 18.9, lower). The Arbuckle strata are upfolded here in an elliptical anticline that is underlain by a node of Precambrian crystalline rock. Surrounding the sub-Pennsylvanian crop of the Arbuckle in the axial area of the anticline are successively younger Ordovician strata to the west, but to the east a major fault has dropped Mississippian limestone down to the Arbuckle level. Pennsylvanian strata unconformably overlie the truncated Ordovician rocks and also reflect in a subdued manner the anticlinal structure of the older rocks. The discovery reservoir was the Arbuckle Limestone with the oil trapped at the top of the dome beneath the unconformity; stratigraphically above the Arbuckle are at least 20 pay zones, ranging in age from Ordovician to Late Pennsylvanian. The most prolific have been the basal Simpson (Ordovician) sands, which are truncated beneath the Pennsylvanian unconformity to the west of the "bald-head." More than ½ billion barrels of oil has come from this reservoir, and the Simpson oil still present is estimated at over 1 billion barrels (Gatewood, 1968).

In the *West Edmond* field (McGee and Jenkins, 1946; Swesnik, 1948) the reservoir rock is the Bois d'Arc Limestone of Hunton (Siluro-Devonian) age, which has been tilted to the west toward the Anadarko basin, truncated, and sealed by overlapping Lower Pennsylvanian shale (Figure 18.9, upper).

The *Garber* field (Cary, 1955; Gish and Carr, 1929), which lies to the north of the West Edmond field, is also prolific. It is an elliptical dome with a north-south trend bounded by a fault on the east side. The Garber field not only is the first to be discovered along the Oklahoma sector of the Nemaha uplift but also is notable for the fact that it has 22 superimposed reservoir rocks. These range in age from a sandstone in the Chase Group of the Lower Permian to the Arbuckle of Cambro-Ordovician age.

Likewise typical of the Nemaha uplift is the *Tonkawa* field (Clark, G. C., 1926). This field also is a truncated and faulted anticline in the pre-Pennsylvanian section that has been overlapped by Pennsylvanian sediments. The Pennsylvanian and Permian rocks are arched across the field and contain a number of producing sandstones (both oil and gas) in which the trapping is entirely anticlinal. However, the principal reservoirs are oil-containing sandstones in the Simpson Group in which trapping is due to a combination of structure and overlap seal. More fields of the same type overlie the Nemaha granite ridge to the north in Kay County (Clark and Daniels, 1929).

The Nemaha district also has sand lens traps. The ultimate in shoestring-shaped lenticular sands occurs in the *South Ceres* field (Neal, 1951), where the Red Fork sand forms a northeast-southwest bar in the Cherokee Shale 10½ miles long with an average width of 1000 feet.

Paul's Valley Uplift. Paul's Valley uplift lies between the Nemaha uplift and the Arbuckle Mountains (Figure 18.3). It is bordered on the west by the Anadarko basin and on the east by the Central Oklahoma platform and the west shelf of the Arkoma basin. The uplift covers Garvin County (except the southwest corner) and parts of McClain, Murray, and Pontotoc Counties.

The Paul's Valley petroleum province, first developed in 1942, led the state in annual production in 1949 and

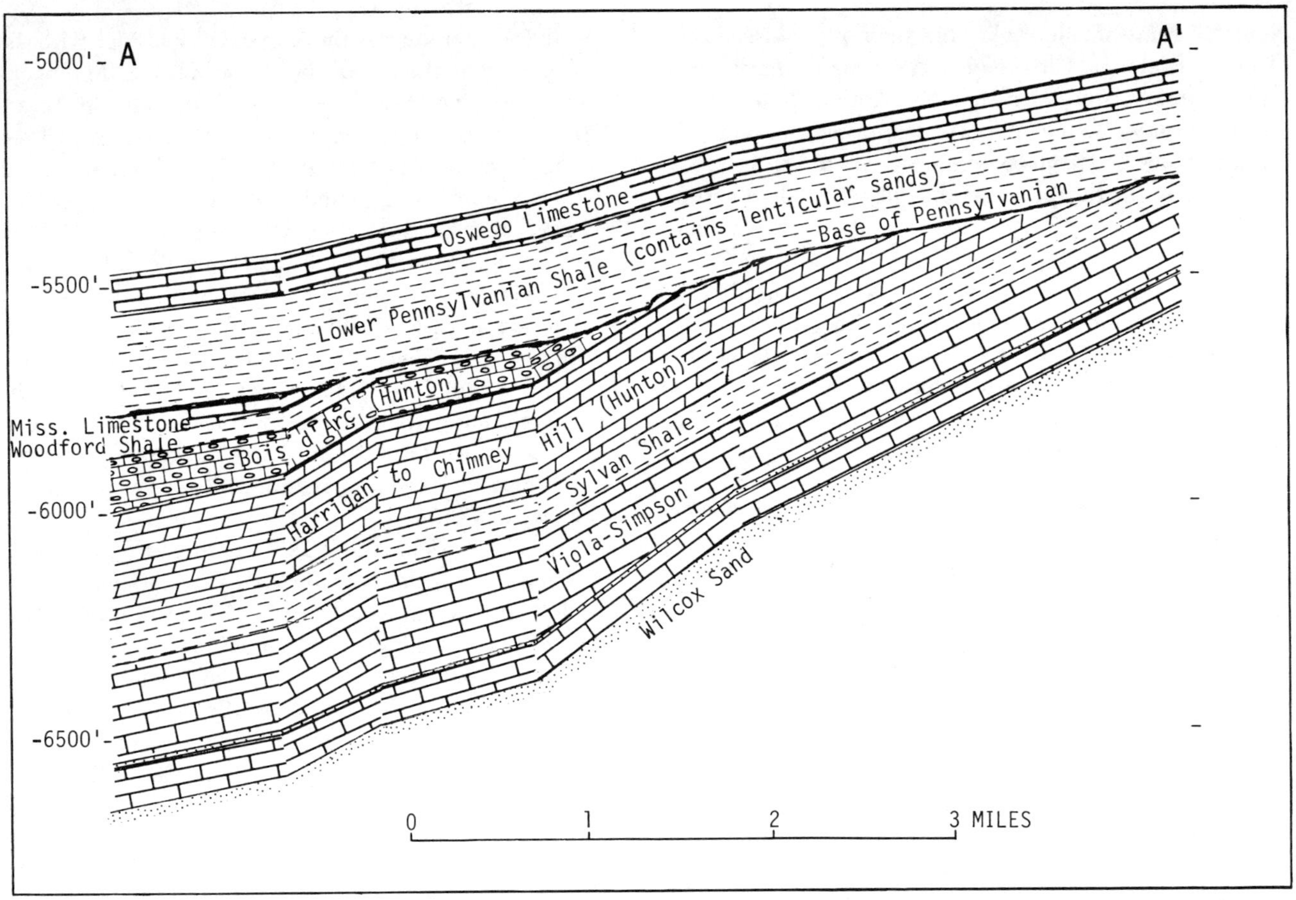

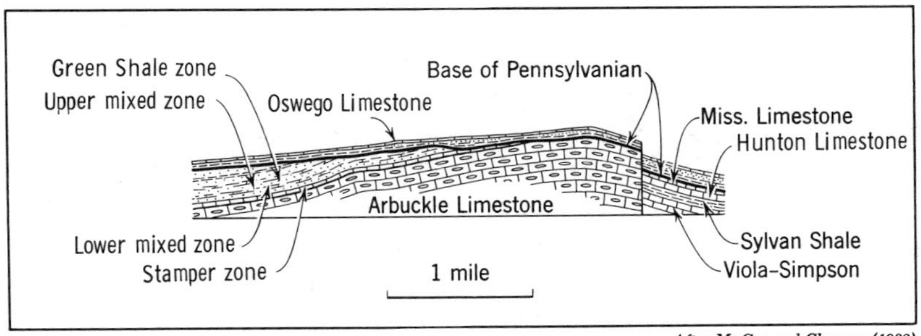

After Mc Gee and Clawson (1932)

FIGURE 18.9 Upper: Cross section, West Edmond field along line *A-A′* on Figure 18.8. *Courtesy D. A. McGee and H. D. Jenkins (1946) and the American Association of Petroleum Geologists.* Lower: Cross section, Oklahoma City field along line *B-B′*, on Figure 18.8. *Courtesy D. A. McGee and W. W. Clawson (1932) and the American Association of Petroleum Geologists.*

was the scene of eight new discoveries in 1962. The greater part of the production has come from the so-called Golden Trend, a linear arrangement of prolific fields the first of which was discovered in 1944.

The uplift of the Paul's Valley area started during Devonian but reached its climax in Early Pennsylvanian (post-Morrow, pre-Des Moines) time, when the sediments were both folded and faulted and the area elevated during Wichita orogeny. Subsequently, the Early Pennsylvanian and older formations were truncated by erosion and submerged beneath the Desmoinesian sea, in which several thousands of feet of shale beds and at least four persistent sandstones belonging to the Deese Formation were deposited. The Deese Formation laps upon the truncated older formations, thinning to extinction high on the uplift. Younger Pennsylvanian and Permian sediments were deposited above the Deese and over the uplift. Westward tilting and minor folding have taken place subsequently.

The *Golden Trend* (Swesnik, 1950; Wheeler, 1950;

Selk, 1951) lies on the flank between the uplift and the Anadarko basin. Oil is produced in this belt from rocks ranging in age from Permian to Ordovician. Some fields produce from Simpson (Ordovician) sands in anticlinal traps, but by far the greater part of the yield has come from highly porous and permeable sandstones lying within the Deese (Pennsylvanian) group. The trap is formed by the abutment of the westward-dipping reservoir sands against the steeper surface of uplifted impervious older rocks (Figure 18.10). This is an excellent example of an unconformity seal produced by offlap.

Southern Oklahoma (Am. Assoc. Petroleum Geologists, 1956, 1959; Ham, 1958; Selk, 1951a; Tomlinson, 1935; Gouin, 1956). This district extends across southern Oklahoma from the Arkansas line to the Texas Panhandle. It includes three mountain areas (Ouachita, Arbuckle, and Wichita-Criner), three basins (Ardmore, Marietta, and Hollis), and the northern edge of the Gulf Coast embayment. Except in the middle of the district, where the Paul's Valley uplift lies immediately north of the Arbuckle Mountains, the southern Oklahoma petroleum province is bounded on the north by two large basins, the Arkoma north of the Ouachitas and the Anadarko north of the Wichitas. Most of the oil production in southern Oklahoma is obtained in the basin areas.

The **Ouachita Mountains** (Chenoweth, 1959, 1960; Wellman, 1951) area has two minuscule oil fields (one of which has been abandoned) and at least one gas well. This subprovince is separated from the Arkoma province on the north by the Choctaw fault. To the south, Cretaceous beds of the Gulf Coastal Plain overlap the Ouachita Mountain rocks. The province extends eastward across southern Arkansas to the Mississippi embayment.

The Ouachita terrane consists of a series of thrust sheets cutting Paleozoic siliceous sediments, including shale, sandstone, novaculite, and chert.

A gas well with a calculated potential of nearly 2 million cubic feet a day was drilled close to the southern border of Latimer County about 14 miles south of the Choctaw fault in 1959 (Branan and Jordan, 1960). The well started in the Stanley Shale of Mississippian age, passed through several thrust fault planes, and struck the gas in the Bigfork Chert (equivalent to the Viola) of Late Ordovician age.

The **Arbuckle Mountain subdistrict** lies in central southern Oklahoma. It is bordered on the northwest by Paul's Valley uplift, on the northeast by the west shelf of the Arkoma basin, on the south by coastal plain sediments that overlap the Arbuckle Mountain rocks and structures, and on the southwest by the Ardmore basin. Although its oil fields are few, the Arbuckle Mountain area contains accessible outcrops of a thick Paleozoic section and is the type locality for many formations that are important reservoir rocks in the subsurface over a wide surrounding area. Precambrian crystalline rocks crop out in the core of the Arbuckles.

The discovery of oil in the *Fitts* pool (Hyatt, 1936) in 1933 opened this structurally complicated area for exploration. The Fitts pool is notable for both its complex structure and its unusual number of reservoirs. Accumulation has taken place in a faulted anticline that

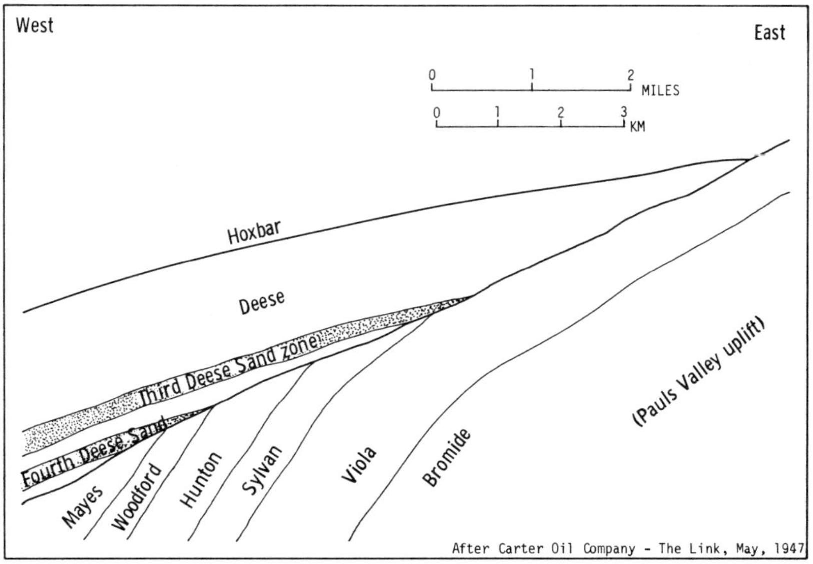

Figure 18.10 **Paul's Valley uplift, south-central Oklahoma. Cross section showing trapping by the termination of sheet sands against the Paul's Valley uplift.** ***Courtesy Humble Oil and Refining Company.***

lies within a graben on the north flank of the Arbuckle Mountain complex. Production is limited on the west and south by faults. The reservoir rocks lie within the Atoka (Pennsylvanian), Hunton (Siluro-Devonian), and the Viola, Bromide, and McLish (all Ordovician) Formations. The Atoka reservoir, a thick sandstone, has produced large quantities of gas as well as oil. The Hunton contains two limestone reservoirs, one at the top and one at the bottom. The Viola Limestone and various sandstones in the Bromide and upper McLish Formations constitute a section nearly 800 feet thick, which is classified as a single reservoir. The fifth or basal McLish sand is 200 feet thick and has an average productive thickness of 40 feet.

The nearby *Jesse* pool (Boyd, 1938) in southeastern Pontotoc and southwestern Coal Counties is very similar structurally and stratigraphically to the Fitts pool. The trapping is due to the presence of a large anticline that is faulted on the south (Arbuckle Mountain) side. The faulting consists of a series of parallel step faults with the down side on the south flank of the anticline. Because of this fault zone, production is confined to the crest and north sides of the fold.

The Fitts and Jesse fields lie on the northeastern fringe of the Arbuckle area on the western edge of the Arkoma basin. To the south of the southeastern corner of the exposed Arbuckles are the *Cumberland* (Cram, 1948; Gahring, 1959; Womack, 1956) and other fields that produce from folded Paleozoics where they are covered by Cretaceous sediments of the Gulf Coastal Plain. The trap in the Cumberland field is a faulted anticline lying in a down-faulted block. The reservoirs are various sandstones belonging in the Simpson Group of Ordovician age.

Similarly, just north of the western end of the Arbuckle Mountain exposed area, the *Eola* (Swesnik and Green, 1950) and *Robberson* fields produce from Simpson sands in tightly folded and faulted anticlines hidden from surface view by Permian red beds.

The **Ardmore Basin** (Tomlinson and McBee, 1959; Shaw, 1955; Tomlinson, 1952–1953) district lies to the south of the Arbuckle Mountains and extends northwest between the Arbuckle and Wichita uplifts, where it merges into the southeastern end of the Anadarko Basin. Like the north side of the Arbuckles, the Ardmore district is one of intense orogeny, including faulting with displacements measured in the thousands of feet.

The largest fields in the Ardmore basin are in the center and toward the westward end in the narrow belt between the *en échelon* Arbuckle and Wichita uplifts. The south rim of the basin is the Wichita Mountain–Criner Hills axis. To the southeast, the basin disappears beneath overlapping Cretaceous sediments of the Gulf Coastal Plain.

A deep drilling program during and immediately after World War II in the western Carter County–Stephens County part of the Ardmore basin resulted in the discovery of valuable reservoirs; some new fields were found and some of the older ones were rejuvenated. An example of a spectacularly revived field is the *Velma* (Rutledge, 1956; Selk, 1951b, p. 603; Mallory, 1949) in Stephens County, originally developed in 1917, which, as the result of new deep discoveries, became the most productive field in Oklahoma in 1949 with an output of nearly 10 million barrels of oil. The Velma structure appears at the surface as a prominent northwest-southeast anticlinal ridge formed by red Permian sandstone. The older production came from arched Permian rocks, but the newer and more copious yield is being obtained from Pennsylvanian, Mississippian, and Ordovician reservoirs that carry oil on the east flank of the anticline. A profound angular and erosional unconformity occurs beneath the veneer of Permian and Upper Pennsylvanian formations. The older rocks are complexly folded and thrust faulted.

The *Sholom Alechem* field (Billingsley, 1956; Selk, 1951b) nearby was discovered in 1923 but did not become an important oil source in Oklahoma until the discovery and development of deep Springer (lowest Pennsylvanian) sands in 1948. As a result the cumulative production for this field doubled within 5 years to make Sholom Alechem a major oil field in the state. The Sholom Alechem structure is a closely compressed anticline with a closure of about 1500 feet at the level of the deeper reservoirs. It is crossed by several faults, but they do not limit production. Accumulation is anticlinal in the Deese Sands above the unconformity at the top of the Springer, and in the deeper (and most prolific) Springer sands below the unconformity (Figure 18.11). However, distribution across the anticline is controlled by sand lenticularity. Trapping in the upper Springer sands is due to truncation at the unconformity.

The **Wichita Mountains-Criner Hills axis** (Harlton, 1963; Chase, 1956) lies to the south of the Anadarko and Ardmore basins and to the north of the Hollis and Marietta basins. To the west this uplift zone disappears as a surface feature beneath Permian red beds, but continues as the buried Amarillo Mountains across the Texas Panhandle. To the east it is overlapped by Cretaceous sediments of the Gulf Coastal Plain. The Wichita Mountains in southwestern Oklahoma expose basement rock. The Criner Hills to the southeast are sculptured from Paleozoic sediments. These two topographic features are connected structurally by the axis of the anticlinorium.

Toward the western end of this subprovince, and far down the south flank of the Wichita Mountain uplift, is the *Altus* field (Ryniker et al., 1959) in Jackson County. Here the reservoir rocks are four granite wash units

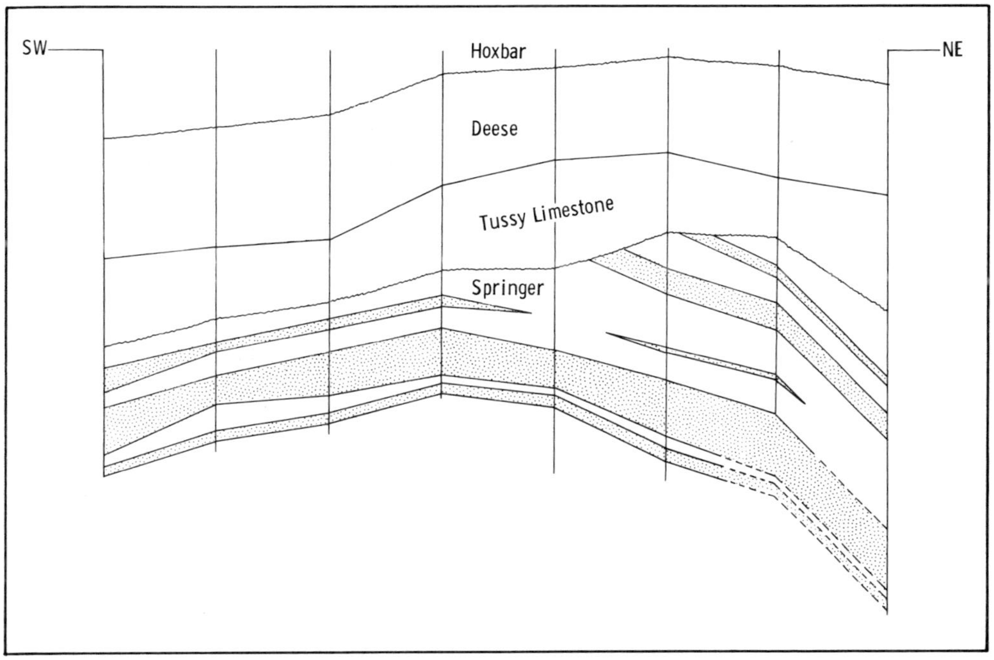

FIGURE 18.11 Cross section through Sholom Alechem field in Stephens and Carter Counties, showing truncated Springer Sandstone reservoirs beneath unconformity. Length of section, 5 miles. ***Courtesy H. R. Billingsley (1956) and the American Association of Petroleum Geologists.***

occurring in Wolfcampian sediments of Early Permian age. Accumulation is anticlinal, but production is limited down the northeast flank of the fold by a fault.

The buried north flank of the Wichita Mountain area also has been the scene of oil and gas accumulations. The *North Gotebo* gas field in southern Washita and northern Kiowa Counties is an example. The producing trend is about 9 miles long (east-west) and ½ mile wide. The Gotebo thrust fault, which controls the shape of the field and most of the accumulation, may be considered the boundary between the Wichita Mountains and the Anadarko basin (Takken, 1968). Hunton and younger pre-Pennsylvanian rocks are thrust over the Springer Formation (Upper Mississippian), which contains the pay sands. In most wells these sands are stratigraphically upside down because of overturning beneath the thrust. Trapping is due principally to faulting, but unconformities also serve as updip barriers.

About 30 miles to the east-southeast, in the southwestern sector of Caddo County, is the *Apache* oil field (Selk, 1951b). This deposit also occurs in a highly deformed belt on the northeast flank of the uplift. Ac cumulation has taken place in the Simpson Sandstone, which here has an average thickness of 20 feet of pay. The trap is an anticline with 700 feet of productive closure. The fold is overturned with dips of 25 to 30° to the south on the south flank and locally exceeding 90° on the north flank.

Farther to the southeast the anticlinorium axis plunges enough so that reservoir rocks can overlie the crest of the uplift at adequate depth to contain commercial accumulations of oil and gas. A row of oil fields more than 40 miles in length lies along the axial zone of the anticlinorium in this area. Near the western end of this line is the *West Duncan* field (Putman, 1956) on a buried Wichita Mountain spur that trends to the northeast or approximately at right angles to the anticlinorium axis. The principal reservoir is a sandstone in the Upper Pennsylvanian. The trap is an asymmetrical anticline with a steep west dip and a structural relief on the west of at least 600 feet.

Farther to the southeast is the *Healdton* field, which is referred to as the discovery that initiated large-scale oil development in southern Oklahoma (Gouin, 1956). The Healdton accumulation directly overlies the Wichita Mountain–Criner Hills axis. The trap is an elongated elliptical anticline that is also a topographic feature of considerable prominence. Erosion has exposed a core of Lower Permian rock circled by Upper Permian sandstones. The deep core of the anticline is a Precambrian granite hill. The major reservoirs are (1) Pennsylvanian sands, of which the upper ones are lenticular but the lowest sand (the "Fourth") is considerably more extensive; and (2) the lowest of three dolomitic zones in the Arbuckle section. These Ordovician strata are much more tightly folded and faulted than the Pennsylvanian (Latham, 1968).

The producing belt continues in the Criner Hills area. One example is the *Southwest Ardmore* field (Hale, 1959; Jordan, 1963), which produces from sandstones in southeastern Carter County and limestones in the Simpson Group of the Middle Ordovician. The trap is an elongate anticline with a fault limiting production on the southwest side. Erosion of the surface has exposed younger Ordovician formations along the axis of the anticline.

The **Marietta Syncline** (Reeves, 1960, 1961; Reeves and Mount, 1960) is a structural sag lying between and parallel to the Criner Hills axis on the northeast and the Muenster-Waurika arch on the southwest. It appears to prow out against the Wichita Mountain uplift to the northwest, and it submerges beneath the Cretaceous coastal plain sediments to the southeast. So far only a few small pools have been found in the Marietta syncline area. The *Stockton* field in northwestern Love County produces from sandstones of Pennsylvanian age. Farther to the northwest, commercial oil has been discovered in the Hunton (Devonian) Limestone beneath the Pennsylvanian unconformity.

The **Muenster-Waurika arch** (McBee and Vaughan, 1956), which crosses the Red River from north Texas, trends northwesterly until it disappears against the Wichita Mountain uplift. It is a low structural ridge separating the Marietta basin from the eastern end of the Hollis basin. The arch is cored with basement rock of probable Early and Middle Cambrian age (Ham et al., 1964) and Arbuckle Limestone, which were eroded into hills and buried beneath the sea in Pennsylvanian time so that Middle Pennsylvanian sediments overlie the much older rocks. Permian formations underlie the surface. A nest of small pools has been discovered on both sides of the Red River in Montague County, Texas, and Jefferson County, Oklahoma. The oil and gas accumulations are concentrated in nodes on the arch axis. Hydrocarbons moving up the flanks have become trapped (1) in Pennsylvanian sands, (2) in bioclastic limestones deposited around the fringe of the Arbuckle islands, and (3) within the upper part of the Arbuckle subcrop.

Other fields overlie the Muenster-Waurika arch to the northwest in central and southeastern Cotton County.

The Hollis basin (Laing, 1963; Sears, 1951, 1952) is in reality the eastern end of the Palo Duro basin of northwest and Panhandle Texas south of the buried Amarillo Mountains. It covers all or parts of four counties along the Red River in western Oklahoma and two counties across the Red River in Texas (where it is referred to as the eastern Palo Duro basin). The Hollis basin is bounded on the north by the Wichita Mountain uplift and on the south by the Red River uplift.

A few oil fields have been found on both sides of the Red River boundary line, with the greatest concentration to date in Tillman County, Oklahoma. The *West Frederick* field (Markley, 1959) in this county produces from the Canyon Limestone of Pennsylvanian age and the Ordovician Arbuckle Dolomite. The trap is an upfold that appears to be a node along an anticlinal trend running westward from central Frederick County, crossing the Red River and northwestern Willbarger and northern Hardeman Counties, Texas. Similar nodes along this trend are occupied by other oil accumulations.

To the east, in Cotton County, is the *Southwest Randlett* field (Cipriani, 1956) which is also an anticlinal accumulation. Here the principal production comes from an Upper Pennsylvanian sandstone at a depth of 1600 feet.

The Oklahoma part of the **Gulf Coastal Plain** (Wellman, 1951) has been probed in the search for traps in the overlapped Paleozoic rocks that lie on the south flank of the Ouachita Mountain subprovince; fields thus found were described earlier (p. 137). Cretaceous sediments of the coastal plain also mask Paleozoic rocks in the eastern parts of the Arbuckle basin, the Criner Hills arch, and the Marietta basin; some success has attended drilling through the Cretaceous rocks in these areas as well.

The coastal plain sediments are principally clastic and loosely consolidated materials of Early Cretaceous (Comanchean) age. A few small, shallow oil pools have been developed in the coastal plain sediments in Marshall County.

Anadarko Basin (Oklahoma City Geol. Soc., 1963; Adams, 1965; Wheeler, 1951). The Anadarko basin petroleum district is the largest of the Oklahoma districts. It is the greatest both in area and in depth. The Anadarko basin occupies the V-shaped part of Oklahoma lying north of the Wichita Mountain uplift and west of the Paul's Valley and Nemaha uplifts. It continues to the northwest far beyond the state borders, covering the Texas Panhandle north of the buried Amarillo Mountains, the panhandle of Oklahoma, and southwestern and far western Kansas (Hugoton embayment). However, the deepest part of the Anadarko basin is in southwestern Oklahoma; it rises to the southeast into the Ardmore basin area, and to the west into the Texas Panhandle. The basin is asymmetrical, with the steeper flank on the south plastered onto the Wichita uplift. This brings the axis not far north of the north edge of the Wichita Mountain subprovince.

The Oklahoma part of the Anadarko basin is covered at the surface mainly by Upper Permian red beds. Also present are late Tertiary and Quaternary continental deposits. In Oklahoma the Anadarko basin can be divided for descriptive purposes into three subprovinces: (1) the deep basin (Anadarko Deep), (2) the east and north shelf (North Shelf of the Anadarko basin), and (3) the Oklahoma panhandle. The greatest concentrations of oil fields occur (1) at the eastern end of the deep basin where it merges with the Ardmore basin, and (2) in the Sooner Trend high on the northeast shelf. Gas fields dominate the shelf to the west, the panhandle area, and the deep basin itself.

The **Anadarko Deep** (Adkison, 1960; McNeal, 1953; Lang, 1951) includes all of the Anadarko basin in Oklahoma south of the synclinal axis and the inner, steeper part of the north flank. The line of demarcation between the inner basin and the east and north shelf is the hinge area where the southward dip moderates. The south flank of the basin laps upon the Wichita Mountains. To the east the deep basin narrows between the Wichita Mountains–Criner Hills axis and the Paul's Valley uplift and becomes the Ardmore basin.

The maximum depth of the Anadarko basin can be estimated from a deep well drilled in the Elk City pool area in 1959. This location is in the deeper part of the basin, a short distance south of the synclinal axis. It was and still is (in early 1969) the second deepest well to be drilled in the United States, reaching a total depth of slightly more than 24,000 feet. The oldest rocks penetrated were either Lower Pennsylvanian Morrowan (Oil and Gas Jour., Jan. 28, 1963, p. 163) or Upper Mississippian Chester (Roberts, 1960, p. 763). The interval between the top of the Mississippian and the top of the Arbuckle farther to the southeast in the basin is about 4000 feet, which, by simple extrapolation, would place the Arbuckle surface at a depth of at least 35,000 feet in the deepest parts of the Anadarko basin. The thickness of the Arbuckle and underlying Reagan Sandstone is not known this far west, but it appears safe to assume that the maximum thickness of the sedimentary cover in the Anadarko basin is in the 35,000 to 40,000 foot range.

The greater part of the trapping in the oil and gas fields so far discovered in the deep Anadarko basin is anticlinal. One example is the *Carter-Knox* field (Reedy, 1958, 1968; Reedy and Sykes, 1962; Pate, J. H., 1953), which lies close to the northwestern extremity of the Ardmore basin. This field has had three birthdays so far. The first was the discovery of oil in a relatively shallow Permian sandstone in 1922. Twenty-four years later gas, and subsequently oil, were found in volume in a sandstone within the Springeran Series of the lowermost Pennsylvanian. This led to the development of an oil field nearly 10 miles long and about 1 mile in width. The third birthday was in January, 1956, when an exploratory well drilled to below 15,000 feet struck an estimated open flow potential of 31,000,000 cubic feet of gas containing over 5000 barrels of condensate per day. This deep reservoir is a series of sandstones in the Bromide Formation of the Simpson Group (Ordovician). The Permian and Pennsylvanian sandstone accumulations are in an elongate anticline with a northwest-southeast trend that terminates on the northeast side in a thrust fault with a throw of several thousands of feet (Figure 18.12). The Ordovician accumulation, on the other hand, has taken place in a relatively simple anticline, the axis of which lies about a mile to the southwest of the anticlinal axis in the younger formations.

The *Chickasha* gas field (Wallace, 1955) is about 10 miles to the northwest of Carter-Knox. This field also is elongate, with a length of approximately 10 miles, but the trend is nearly north-south. It was discovered in 1922 and since then has produced gas from sands in both the Permian and the Pennsylvanian. Oil was discovered in 1954 in shallow Permian sands that had been ignored in earlier drilling. Gas trapping in the Pennsylvanian reservoirs is in an anticline that was discovered by surface mapping. However, the Permian sandstones are lenticular, and consequently the distribution of hydrocarbons in these sands is erratic.

One of the oldest oil fields in the inner Anadarko basin is the *Cement* (Harlton, 1960), discovered in 1916. Since then a continuing series of new discoveries in deeper reservoirs has maintained production across the years. This field is in southeastern Caddo County, a short distance northwest of the Chickasha gas field. The oil field is about 16 miles long, and trapping is anticlinal. Production is limited on the north by a fault of considerable magnitude.

The *West Sentinel* oil field (Gelphman, 1960) lies in far southwestern Oklahoma about midway between the Amarillo-Wichita axis and the Anadarko basin syncline axis. The south flank of the Anadarko basin is notable for the thick section of granite wash that was washed into the basin from the emergent basement rock of the uplift during Pennsylvanian time. The granite wash in the West Sentinel field is a mixture of pebbles and sand with some shale. The pebbles are derived from both igneous rock and cherty carbonate rocks. The sand is arkosic material loosely cemented with calcium carbonate. The porous and permeable granite wash sediments constitute the principal reservoir at West Sentinel. The trap appears to be an offshore bar on a horst block.

Nearby *Elk City* field (Beans, 1952; Wilgus, 1950) is very close to the deepest part of the Anadarko basin with the synclinal axis lying just a short distance to the north. Oil has accumulated here in an elongate anticline with a closure of about 500 feet. The axis of the Elk City anticline is parallel to that of the Anadarko syncline. The field is about 12 miles long and 5 to 6 miles wide; average dips on the flanks are 4 to 5°. The principal reservoirs are granite wash strata that are interbedded with shales and limestones identified as Missourian (Upper Pennsylvanian).

In 1962 the north flank of the deep basin became an important hunting ground with the discovery of oil and

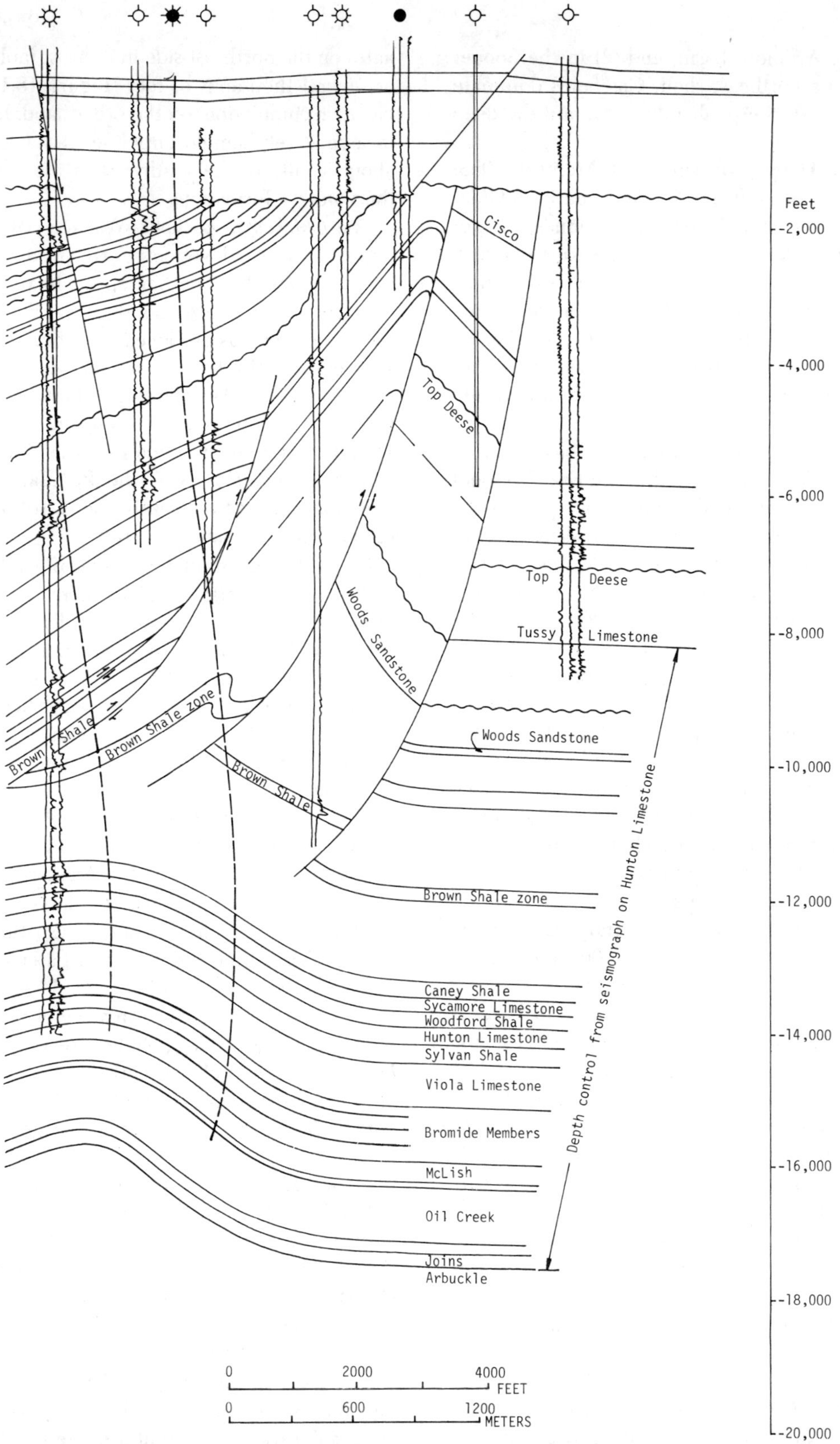

FIGURE 18.12 Carter-Knox oil field, Grady and Stephens Counties, Oklahoma. Southwest-northeast cross section. *After Harold J. Reedy and Howard A. Sykes (1962), p. 103. Courtesy of authors (per Harold J. Reedy) and the Tulsa Geological Society Digest.*

gas in the Oswego Limestone of Middle Pennsylvanian age in Custer and Dewey Counties (Biddick, 1963). By early 1963 a northwest-southeast trend 36 miles in length had been developed across these two counties. Gas, gas condensate, and oil are being produced from depths mostly in the 9000 to 10,000 foot range.

The hinge line that separates the steeper regional dips of the inner basin from the more gentle dips of the **east and north shelf** extends northwestward from the Paul's Valley uplift to and beyond the northeastern corner of the Texas Panhandle. The shelf area is one of both anticlinal and stratigraphic traps (Pate, J. D., 1959).

The regional dip on the shelves is about 50 feet to the mile southwesterly. On the east side of the shelf, west of the Nemaha uplift, there are at least 18 distinct reservoirs consisting of both sandstone and carbonate rock and ranging in age from Pennsylvanian to Ordovician (Jordan, 1962; Woncik, 1961). Discovered to date are 10 Pennsylvanian pay zones, 5 Mississippian reservoirs, the Hunton and Misener of Devonian age, and the Simpson in the Ordovician. Varying permeability traps predominate in this area. They are mainly of two types, lenticular sands in the Pennsylvanian, and fracture porosity traps in the "Mississippi lime" reservoir rocks. The *Ringwood* field (Lillibridge, 1950) in Major County to the north is a stratigraphic trap in which the reservoir, the Manning Limestone of Mississippian age, pinches out completely updip from a maximum thickness of about 150 feet.

The so-called *Cherokee Trend* (Morrisey, 1956) in Alfalfa and Grant Counties immediately south of the Kansas line is a series of pools in which the reservoir rock is a lenticular sandstone in the Cherokee Group of Pennsylvanian age. This sandstone was deposited in narrow bars with an east-west trend; it is productive of both oil and gas (Pate, J. D., 1959, p. 48). The *Northeast Waynoka* field (Barby, 1962) in the next county west (Woods) produces gas from the Cottage Grove Sandstone of Late Pennsylvanian age. This sand wedges out updip to the north (Figure 18.13), but occurs over a large area to the south. The gas at Northeast Waynoka has migrated up the regional dip and has become trapped in structural noses below the wedgout.

The northwest corner of Oklahoma (Jordan et al., 1959) immediately east of the panhandle has been referred to as the "gas corner," but scattered oil discoveries in 1963 gave promise that this area could also become an important source of oil. The largest producing district is the *Laverne* (Pate, J. D., 1962, 1968), which spreads from Harper County over into Beaver County in the Oklahoma panhandle. There are four reservoirs in this district, three of them sandstones belonging to the Pennsylvanian System, and the fourth and deepest a limestone of Chester (Mississippian) age. The most prolific reservoir is the youngest, which lies in the upper Pennsylvanian; so far it has been found to be productive of gas only. This sandstone occurs in bars completely surrounded by shale and has a maximum thickness of 210 feet. The next older sandstone, the Tonkawa (also of the Upper Pennsylvanian), produces gas and some oil from four sand bars. The deepest sand so far productive is in the Morrow Series of the Lower Pennsylvanian; again accumulation is in sand bars. A little oil and a considerable quantity of condensate are produced as well as gas. Chester limestones produce in this district where the reservoir has been truncated and overlapped by lowermost Pennsylvanian rocks. The unusual situation in the Laverne district is that a single well may penetrate as many as all four of these stratigraphic traps superimposed one upon the other.

The Oklahoma **Panhandle** (Barby, 1961; Panhandle Geol. Soc., 1961; Moore, 1957; Beebe, 1957) is a geographic rather than a geologic subprovince. It consists of three counties, which are, from west to east, Cimarron, Texas, and Beaver. The great *Hugoton* gas field of southwestern Kansas crosses the Oklahoma panhandle through Texas County and continues into the Texas Panhandle, where it abuts against the Amarillo gas field. In the Oklahoma panhandle the gas accumulation is in the same Permian formations as in the Kansas sector of the Hugoton field, and the trapping is due to the same cause, namely, the updip wedgout of porosity to the westward (Mason, 1968).

The Anadarko basin continues westward into the Texas and Oklahoma panhandles north of the buried Amarillo Mountains at least as far as the Cimarron arch, which crosses the eastern side of Cimarron County, culminating in the northeast corner of the county in the Keyes dome (Figure 18.3). The hinge line that separates the deep basin from the north shelf cuts across Beaver County from southeast to northwest. Consequently, northeastern Beaver County has the same geological conditions as Harper County immediately to the east, as demonstrated by the Laverne district just described. Therefore, in the eastern part of the panhandle gas and oil are obtained from pre-Permian rocks, especially Early Pennsylvanian (Morrow) sandstones (Totten and Horn, 1968, p. 1526), instead of from the Permian reservoirs that are so productive of gas in Texas County.

A large gas field, by other than Hugoton standards, is the *Keyes*, which occupies the Keyes dome in northeastern Cimarron County. This field, like the Greenwood field to the north in the southwestern corner of Kansas, produces mainly from a Morrow sand in the Lower Pennsylvanian. The reservoir sand is a sheet deposit, and trapping is anticlinal. Some oil and gas have been obtained from younger, lenticular, Pennsylvanian sands in the Keyes field.

The successful search for pre-Permian reservoirs in the

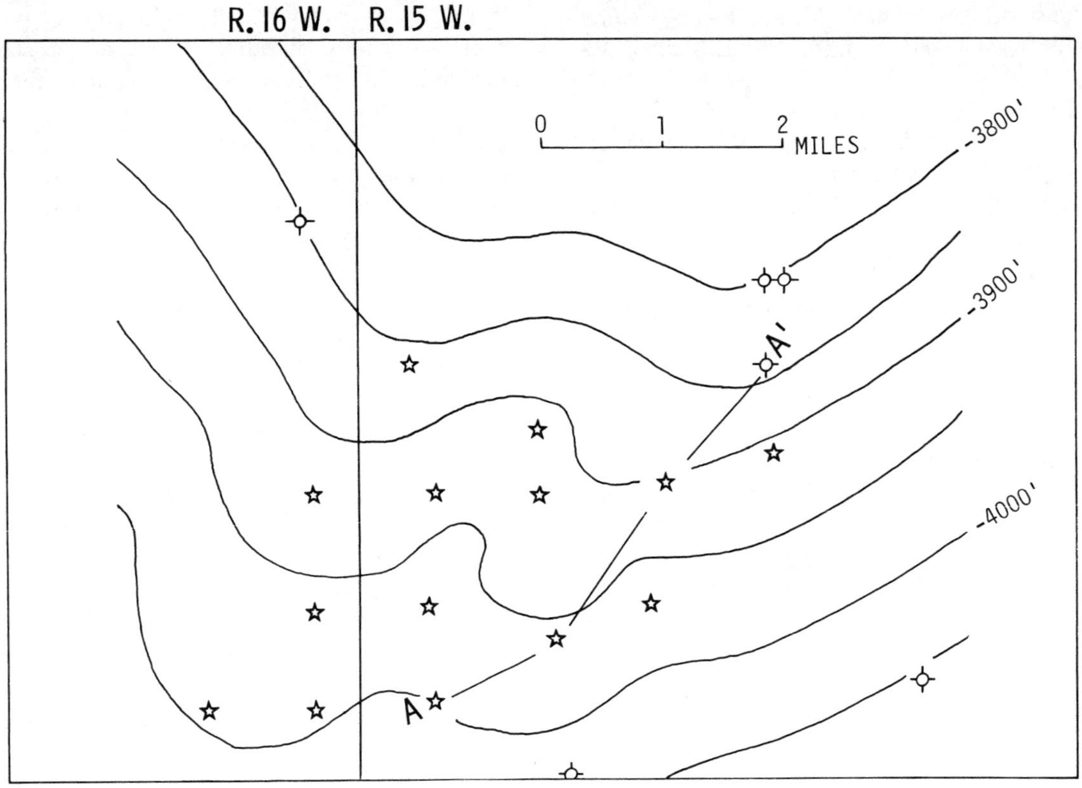

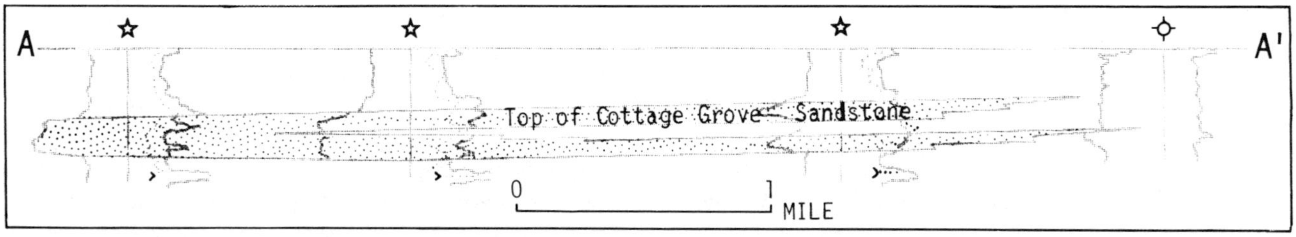

FIGURE 18.13 Northeast Waynoka field, Woods County. *Courtesy B. G. Barby (1962) and the Tulsa Geological Society.* Upper: structure contour map, based on top of Cottage Grove Sandstone. Interval, 50 feet. Lower: cross section along line *A-A'* shown in upper drawing.

greater Hugoton district made the Oklahoma panhandle one of the most active areas of exploration in the state during the early 1960s. Discoveries have been made beneath the Permian gas reservoirs in Texas County, as well as in Cimarron County to the west and Beaver County to the east.

FUTURE. All of the petroleum provinces and subprovinces in Oklahoma are areas of continuing interest and discovery. However, the greatest room for expansion in all three dimensions is in the Anadarko basin, where exploration in reality has just begun (Wroblewski, 1967). The enormous area and a stratigraphic column containing many reservoirs through a maximum vertical distance measurable in miles produce a cubic potential that can be described only as enormous. As of mid-1968, "Deep exploratory drilling now dominates the hinge line and the deeper portion of the Anadarko basin, principally for deep Hunton gas production" (John A. Taylor, letter dated 5 August, 1968). Especially inviting is the relatively shallow Hunton carbonate rock wedge-edge on the north shelf (Logsdon and Brown, 1967). In the southeastern part of the Anadarko deep basin the Arbuckle has not been tested adequately in such high-relief anticlines as the Carter–Knox (Reedy, 1968, p. 1490). Much exploring remains to be done in the truncated and overlapped reservoirs on the east shelf (Klinger and Ash, 1967).

Southern Oklahoma still has many inadequately explored areas. Largely unknown is the possible potential of the Ouachita uplift of both Oklahoma and Arkansas (Pitt, 1960; Goldstein and Flawn, 1958). The other major uplifts in southern Oklahoma, the Arbuckle (Caswell,

1957) and the Wichita, have proved to have oil hidden in the complex structures down flank from the outcropping basement rock cores. The southern Oklahoma basins, especially the Marietta (Reeves and Mount, 1960) and Hollis, have not been tested thoroughly as yet.

Even northern and eastern Oklahoma, the oldest producing parts of the state, still have a promising future, including the possibility of locating buried reefs (Cassidy, 1968, Harbaugh, 1957). Included among the recent new finds were widely scattered discoveries of Arbuckle production. The buried Precambrian hills have not been adequately tested for possible oil accumulations in fractures.

The shallow reservoirs in northeastern Oklahoma also are considered possibilities for further discoveries (Wright, L. M., 1952). On the other hand, in the Ozark uplift in the northeastern corner of the state the stratigraphic column starts with Lower Mississippian rocks at the grass roots and the number and thickness of potential reservoirs are minimal. Therefore this area is not considered to have much oil and gas potential (Huffman, 1951).

Secondary recovery is adding greatly to the life expectancy of many of the older fields. A waterflood program projected for the Pennsylvanian reservoirs in the giant Healdton field will add 80 to 100 million barrels to the ultimate recovery.

REFERENCES CONSULTED IN PREPARING CHAPTER 18

Adams, William L., 1965.
Adkison, W. L., 1960.
American Association of Petroleum Geologists, 1956, 1959.
Bale, Hubert E., and Williams, George O., 1951.
Barby, B. G., 1961, 1962.
Bartlett, Charles S., 1966.
Bass, N. W., Goodrich, H. B., and Dillard, W. R., 1942.
Beans, Robert J., 1952.
Beckwith, H. T., 1928.
Beebe, B. W., 1957.
Biddick, James R., 1963.
Billingsley, H. R., 1956.
Borden, Joseph L., and Brant, Ralph A., 1941.
Boyd, W. Baxter, 1938.
Branan, C. B., Jr., 1968.
Branan, C. B., Jr., and Jordan, Louise, 1960.
Branson, Carl C., 1956, 1957.
Campbell, Ralph L., 1968.
Carver, G. E., Jr., 1948.
Cary, L. W., 1955.
Cassidy, Martin M., 1968.
Caswell, C. A., 1957.
Chase, G. W., Frederickson, E. A., and Ham, W. E., 1956.
Chenoweth, Phillip A., 1959, 1960.
Cipriani, D., Jr., 1956.
Clark, Glenn C., 1926.
Clark, Stuart K., and Daniels, James I., 1929.
Clayton, John Mason, 1965.
Clinton, R. P., 1957.
Cohee, George V., et al., 1961.
Colton, Earl G., 1935.
Cram, Ira H., 1948.
Culp, Eugene, and Barrett, Ed., 1957.
Dillard, W. Reese, 1941.
Dille, Glen S., 1952.
Fitts, L. E., Jr., 1951.
Frost, V. L., 1940.
Gahring, Ross R., 1959.
Gatewood, Lloyd E., 1968.
Gelphman, Norman Ray, 1960.
Gish, Wesley G., and Carr, Raymond M., 1929.
Goldstein, August, Jr., and Flawn, Peter T., 1958.
Gouin, Frank, 1956.
Hale, G. Carl, 1959.
Ham, William E., 1958.
Ham, William E., Denison, Rodger E., and Merritt, Clifford A., 1964, 1965.
Hamric, Burt E., 1961.
Harbaugh, John W., 1957.
Harlton, Bruce H., 1960, 1963, 1966.
Harvey, Ralph L., 1968
Hayes, J. A., Jr., 1956.
Huffman, George Garrett, 1951.
Hyatt, Don L., 1936.
Ingham, W. I., 1941.
Jacobsen, Lynn, 1959.
Jones, Fred B., Jr., 1968.
Jordan, Louise, 1957, 1962, 1963.
Jordan, Louise, Pate, J. Durwood, and Williamson, Sidney R., 1959.
Klinger, Robert, and Ash, Richard, 1967.
Koinm, David N., and Dickey, Parke A., 1967.
Laing, William E., 1963.
Lang, Robert C., III, 1951.
Largent, B. C., 1968.
Latham, Jack W., 1968.
Levorsen, A. I., 1929.
Lillibridge, Harry E., 1950.
Logsdon, Truman, and Brown, Alton R., 1967.
McBee, William, Jr., and Vaughan, Leslie G., 1956.
McGee, D. A., and Clawson, W. W., Jr., 1932.
McGee, D. A., and Jenkins, H. D., 1946.
McNeal, Robert P., 1953.
Mallory, William W., 1949.
Markham, E. O., and Lamar, L. C., 1937.
Markley, L. C., 1959.
Mason, John W., 1968.

Moore, Carl A., 1957.
Morrisey, Norman S., 1956.
Neal, Eugene P., 1951.
Oklahoma City Geological Society, 1963.
Oklahoma Geological Survey, 1928.
Oxley, Marvin L., 1959.
Panhandle Geological Society, 1961.
Pate, J. Durwood, 1959, 1962, 1968.
Pate, Joe H., 1953.
Pitt, William D., 1960.
Planalp, Roger N., 1961, 1968.
Putman, D. M., 1956.
Rau, H. L., and Ackley, K. A., 1939.
Reedy, Harold J., 1958, 1968.
Reedy, Harold J., and Sykes, Howard A., 1962.
Reeves, C. C., Jr., 1960, 1961.
Reeves, C. C., Jr., and Mount, J. Russell, 1960.
Richardson, W. E., 1965.
Riggs, C. H., Eakin, J. L., Jr., Larman, J. Heath, and Johnston, K. H., 1958.
Riggs, C. H., Wey, E., and Maude, J. V., 1954.
Rister, Carl Coke, 1949.
Roberts, M. C., 1960.
Rutledge, Richard B., 1956.
Ryniker, Charles, Shortridge, C. G., and Maxwell, R. W., 1959.
Sands, J. Merville, 1927.
Sears, Joe M., 1951, 1952.
Selk, Erwin L., 1951.
Shaw, Richard F., Jr., 1955.
Six, David A., 1962, 1968.
Skarda, Everett, 1956.
Skelton, Alan G., and Skelton, Martha B., 1942.
Summers, Fred C., Jr., 1968.
Swesnik, Robert M., 1948, 1950.
Swesnik, Robert M., and Green, Thomas H., 1950.
Takken, Suzanne, 1968.
Tanner, William F., 1956.
Thralls, Hugh M., Robb, George L., and Schisler, A. J., 1962.
Tillotson, Allen W., 1938.
Tomlinson, C. W., 1935, 1952–1953.
Tomlinson, C. W., and McBee, William, Jr., 1959.
Totten, Robert B., and Horn, Paul H., 1968.
Tulsa Geological Society Digest, 1962.
Umpleby, S. S., 1956.
Wallace, P. A., 1955.
Weirich, T. E., 1929.
Wellman, Dean C., 1951.
Wheeler, Robert R., 1950, 1951.
White, Stanley B., 1941.
Wilgus, W. L., 1950.
Wilson, W. B., 1929.
Womack, Jerry L., 1956.
Woncik, John, 1961, 1962, 1968a.
Woodruff, E. G., and Cooper, C. L., 1930.
Wright, Leo M., 1952.
Wright, Randall, 1941.
Wroblewski, E. F., 1967.

Chapter

19

NONPRODUCTIVE STATES

All of the states in Division II are productive except Minesota and Iowa.

Minnesota. South of the Precambrian exposures in Minnesota, and locally in patches along the western boundary, are sedimentary rocks of Paleozoic and Cretaceous age. These include sandstones, shales, and carbonate rocks. The sandstones have excellent porosity and permeability and would make ideal reservoir rocks. However, they are flooded with fresh water and hence make excellent aquifers instead. "Minnesota is justly listed among the states of no importance from the standpoint of possible oil and gas production" (Stauffer, 1926).

Iowa. Although a ten-barrel oil well drilled in 1963 (Kornfeld, 1963) on the Keota dome in Washington County in southeastern Iowa (Figure 19.1) did not turn out to be the discovery well of a commercial oil pool, it focused attention on this state as a possible oil producer. However, the drilling flurry that followed failed to find oil or gas. The oil zone on the Keota dome is in the Pecatonica Limestone Member of the Platteville (Ordovician) Formation. It was struck at a depth of 960 feet. The original objective of the boring was the St. Peter Sandstone, which was to be explored, not for oil or gas, but for gas storage possibilities. The Keota dome had been mapped by the geologists of the Iowa Geological Survey primarily from water well data. It has an area of approximately 3000 acres, with a structural closure of about 120 feet (Hershey, 1963). The only recent activity in Iowa has been the drilling of gas storage wells on the Keota dome (Porter, 1968), meeting the original objective.

The tectonic map (Figure 19.1) shows that eastern and northern Iowa are structurally high, and only in southwestern Iowa, which is on the northeastern flank of the Forest City basin, does the basement rock lie at any great depth. The Mississippi River arch in eastern Iowa is a structural ridge that connects the Ozark uplift with the Wisconsin arch and separates the Illinois basin from the

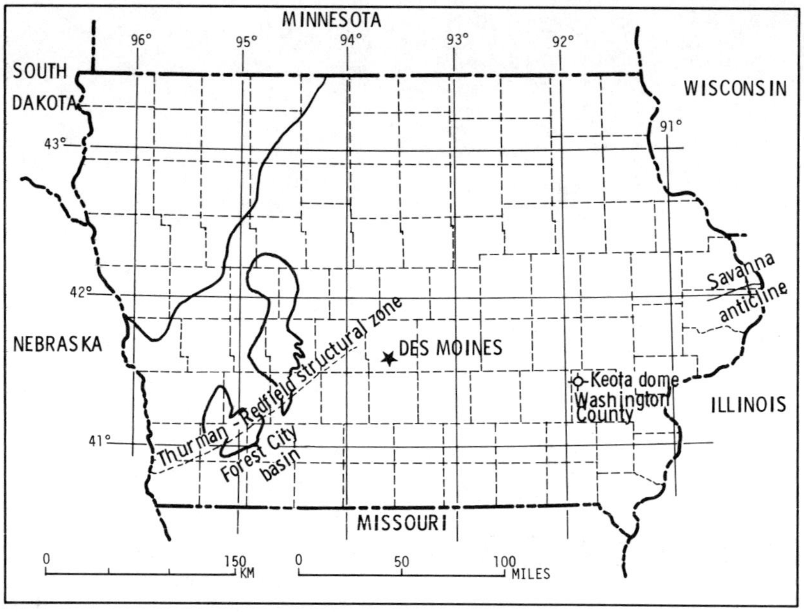

FIGURE 19.1 Iowa. Index map. Heavy lines in western Iowa show eastern edge and outliers of Cretaceous rocks. The regional dip is toward the Forest City basin in the southwestern corner of the state.

Forest City basin. Across northern Iowa the strata are rising toward the Wisconsin shield and the Sioux uplift.

The Forest City basin produces oil and gas to the south of southwestern Iowa in southeastern Nebraska, northeastern Kansas, and northwestern Missouri. However, the closest production to the Keota dome well is in Illinois about 100 miles to the southeast across the Mississippi River arch and high on the western flank of the Illinois basin. Here the Colmar-Plymouth oil field has produced, since its discovery in 1914, more than 4½ million barrels of oil from Devonian sandstone at a depth of 450 feet.

REFERENCES CONSULTED IN PREPARING CHAPTER 19

Anderson, K. H., and Wells, J. S., 1968.
Hershey, H. Garland, 1963.
Kornfeld, Joseph A., 1963.
Porter, James L., 1968.
Stauffer, Clinton R., 1926.

Division III

EASTERN AND CENTRAL GULF

The Eastern and Central Gulf Division (Murray, 1961) consists of four states, Florida, Alabama, Mississippi and Louisiana, that lie along the Gulf of Mexico between the Atlantic Coast and Texas, and a fifth state, Arkansas, that is inland immediately north of northern Louisiana (Figure III.1). The principal petroleum producer in this block is Louisiana.

Geologically, the five states lie either exclusively or partly within the Gulf embayment. To the east this embayment merges with the Atlantic Coastal Plain. The belts of Cenozoic and Mesozoic sediments appearing at the surface are delineated in Figure III.1. Louisiana and Florida are entirely within the Cenozoic belt. The Mesozoic belt appears as a wedge from the west in Arkansas; it crosses the northeastern corner of Mississippi and curves across northwestern central Alabama. Western and northern Arkansas are floored with Paleozoic sediments. The greater part of northern Alabama is in the southern Appalachian Highlands.

The Cenozoic and Mesozoic bands of the Eastern and Central Gulf Division continue southwestward across Texas. This state is treated as a separate unit because of its large area and productive capacity.

Tectonically, the Eastern and Central Gulf Division is dominated by the Gulf Coast geosyncline. Gulfward dipping Cenozoic and Mesozoic sediments lap upon the continental shelf like shingles on a roof. The degree of dip tends to increase with depth because of gulfward increase in stratigraphic thicknesses brought about in large part by growth faults described in Chapter 24. Thus the sedimentary formations in the subsurface along the coast line have many times the thickness measured at the inland outcrop. Furthermore, the

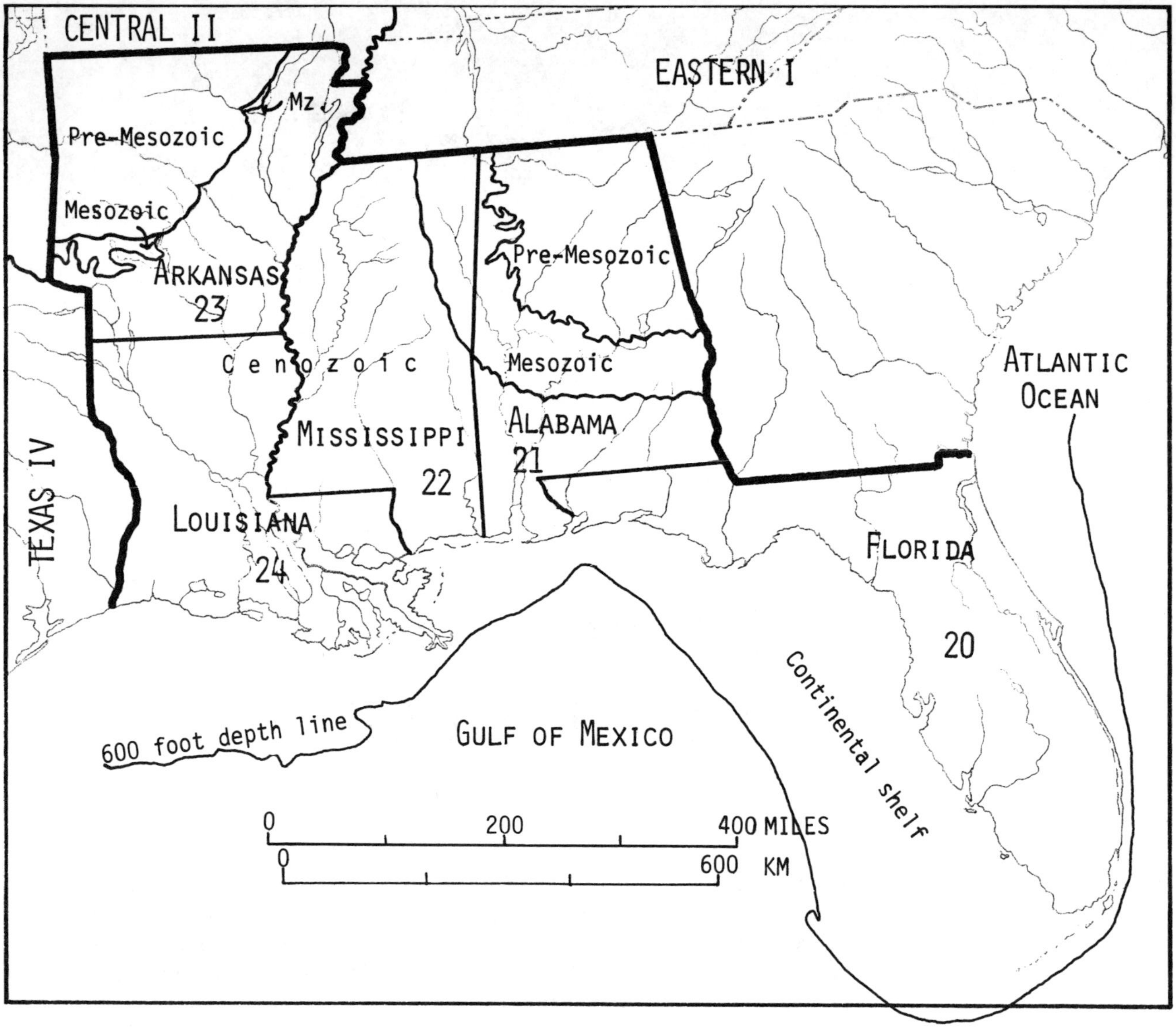

FIGURE III.1 Eastern and Central Gulf Coast of the United States. Index map of Division III. Also shows major geological boundaries (Mz = Mesozoic). Numbers refer to chapters.

subsurface section contains formations that never reach the surface because of overlap by younger sediments.

In addition to the gulfward regional dip, the Gulf Coastal Plain sediments are also faulted. The major faults are strike faults, with the down block mostly, but not exclusively, on the Gulf side. Salt flowage structures are also present, especially west of Florida (Andrews, 1960).

The gulfward dip is expressed in Mesozoic and Cenozoic sediments and in the deep Paleozoic erosion surface. However, the productivity of the Paleozoic sediments near the areas of outcrop is insignificant, except for gas in northwestern Arkansas, and there have been almost no discoveries beneath a Mesozoic cover. Various reasons have been suggested for this near barreness; perhaps the most plausible is the degree of metamorphism in the Paleozoic section. The Mesozoic section contains excellent productive reservoirs that have been successfully developed both within the outcrop belt and gulfward beneath overlapping Cenozoic sediments. By the same token the older Cenozoic sediments produce downdip where covered by younger deposits. The most prolific reservoirs of all are Miocene sands where buried beneath younger Cenozoic sediments in onshore and offshore Louisiana. A considerable volume of the younger sediments was brought to the sea by the ancient and the present Mississippi River (Moody, 1967).

Much of the total area of the Eastern and Central Gulf Division is offshore on the continental shelf (Figure III.1). This sector has been successfully exploited in Louisiana (Atwater, 1959). The volume of the known reserves of oil and gas offshore is great; the amount yet to be found (Meyerhoff, 1967) may be astronomic.

Geophysics first became an important exploration tool for possible hydrocarbon traps in the Gulf Coast of Louisiana and Texas in the early 1920s. Geophysical techniques are still of paramount importance in searching for oil and gas deposits in the Eastern and Central Gulf Coast. Subsurface geology, especially the recognition and identification of sedimentary units through microscopic examination of fossils and minerals in cuttings and cores, has attained a high level of development and use in this area also. Surface geology has been successful in discovering many salt domes and in structure mapping elsewhere, especially in the older rock belts.

REFERENCES CONSULTED IN PREPARING INTRODUCTION TO DIVISION III

Andrews, Donald I., 1960.

Atwater, Gordon I., 1959.

Hardin, Frank R., and Hardin, George C., 1961.

Meyerhoff, A. A., 1967.

Moody, Clarence L., 1967.

Murray, Grover E., 1961.

Uchupi, Elazar, and Emery, K. O., 1968.

Chapter

20

FLORIDA

Florida has produced oil since 1943, when the Sunniland field was discovered. A second field, the Forty-Mile Bend, was discovered in 1954 and abandoned in 1955. The third and largest field to date, Sunoco-Felda, was discovered in 1964. By 1967 this field was producing at the rate of more than $1\frac{1}{4}$ million barrels a year.

Rocks at the surface of the Florida peninsula consist of Quaternary and Tertiary sediments of dominantly marine origin. In the western part of the state, Quaternary sands and shales of deltaic affinity predominate at the surface, and the section is similar to that of Alabama and Mississippi.

The subsurface section of Florida contains many thousands of feet of Cretaceous and Tertiary rocks. A predominantly carbonate facies characterizes most of this sequence of rocks in the peninsula, whereas a clastic facies prevails in the panhandle. The Suwannee Channel of northwestern Florida controlled the boundary between these facies during Early Cretaceous time (Applin and Applin, 1944, p. 1679–1680), as well as during the deposition of Tertiary sediments (Chen, 1965, Figures 24, 28, 33 and 37). At succcessively higher stratigraphic markers this feature migrated to the northwest; it was probably partly structural and partly topographic in origin (Applin, 1952, p. 1162). The carbonate-clastic facies boundary, as generalized after the various Tertiary maps by Chen (1965) and Upper Cretaceous maps by Applin and Applin (1947, map No. 1) and Hull (1962, Figure 2, p. 120), is depicted in Figure 20.1.

In peninsular Florida the limestone-anhydrite-shale sequence of subsurface rocks is abundantly supplied with porous reservoir rocks and cap rocks, but unfortunately potential source beds are considered to be less plentiful. Though clastics predominate in the stratigraphically equivalent sequence of most of the mainland of northwestern Florida, a more marine section is present in the far western part of the state and also undoubtedly occurs in offshore areas.

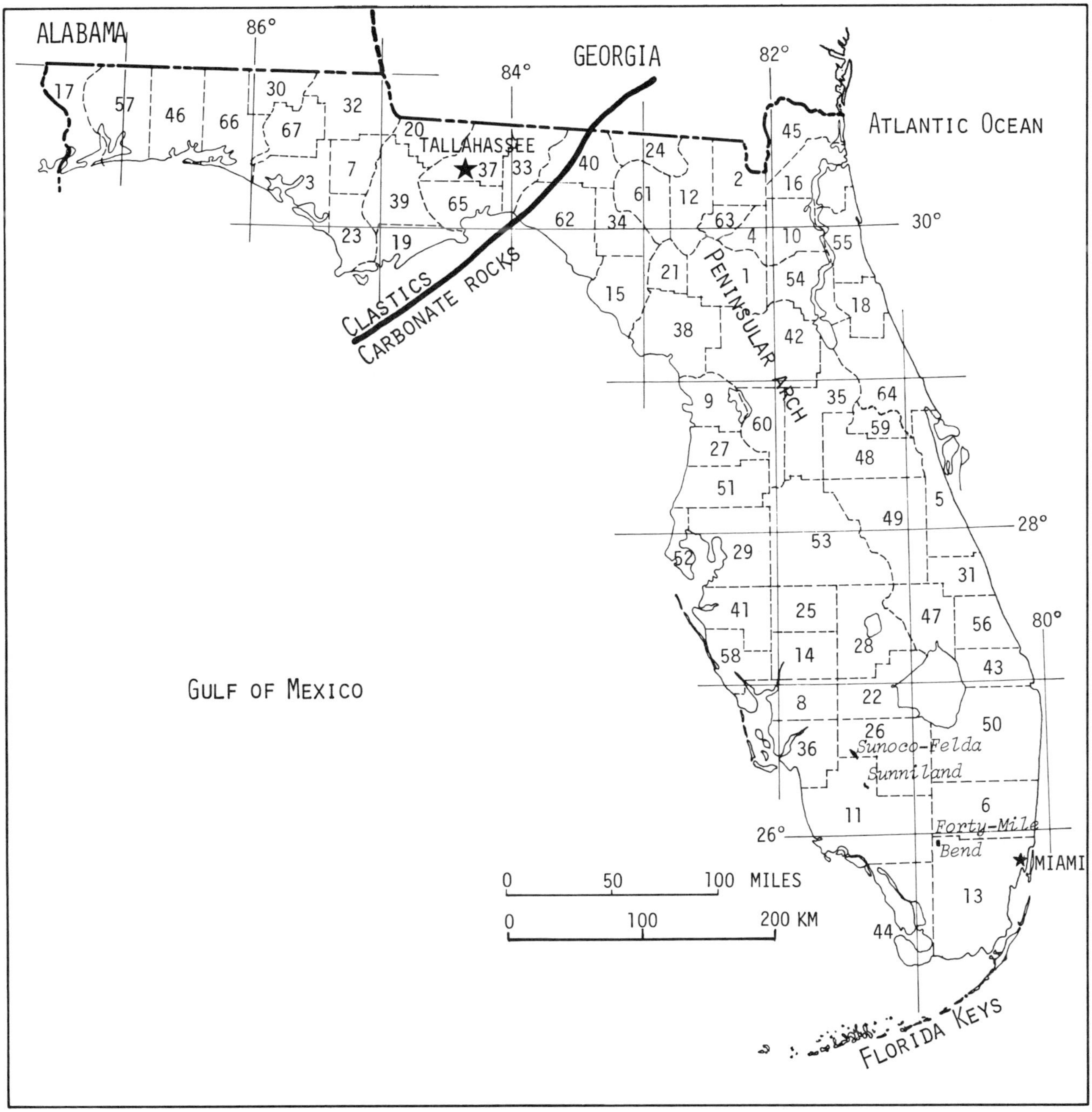

FIGURE 20.1 Florida. Index map with oil fields. The facies boundary separating clastic sediments from carbonate rocks is by Clarence V. Babcock (letter dated 25 September, 1968) *after Chih Shan Chen (1965, Figures 24, 28, 33, and 37), Paul L. Applin and Esther R. Applin (1947, Map no. 1), and Joseph P. D. Hull, Jr. (1962, Figure 2)*. County Key:

1	Alachua	11	Collier	21	Gilchrist	31	Indian River	41	Manatee	50	Palm Beach	59	Seminole
2	Baker	12	Columbia	22	Glades	32	Jackson	42	Marion	51	Pasco	60	Sumter
3	Bay	13	Dade	23	Gulf	33	Jefferson	43	Martin	52	Pinellas	61	Suwanee
4	Bradford	14	DeSoto	24	Hamilton	34	Lafayette	44	Monroe	53	Polk	62	Taylor
5	Brevard	15	Dixie	25	Hardee	35	Lake	45	Nassau	54	Putman	63	Union
6	Broward	16	Duval	26	Hendry	36	Lee	46	Okaloosa	55	St. Johns	64	Volusia
7	Calhoun	17	Escambia	27	Hernando	37	Leon	47	Okeechobee	56	St. Lucie	65	Wakulla
8	Charlotte	18	Flagler	28	Highlands	38	Levy	48	Orange	57	Santa Rosa	66	Walton
9	Citrus	19	Franklin	29	Hillsborough	39	Liberty	49	Osceola	58	Sarasota	67	Washington
10	Clay	20	Gadsden	30	Holmes	40	Madison						

The most prominent structural feature of Florida is the Peninsular arch (Figure 20.1), which forms the backbone of the northern and central parts of the peninsula. The greater part of the peninsula, the Keys, the submerged shelf, and the islands to the east constitute the Florida-Bahama platform (Owens, 1960). The maximum known thickness of the sedimentary veneer in this area exceeds 19,000 feet. The Tertiary and Cretaceous section is as great as 15,000+ feet with Paleozoic rocks (mainly Ordovician clastic) underlying the Lower Cretaceous section.

The *Sunniland* field lies in Collier County on the west side of Florida near the southern tip. Permeability in the Sunniland Limestone, which is the reservoir rock, is

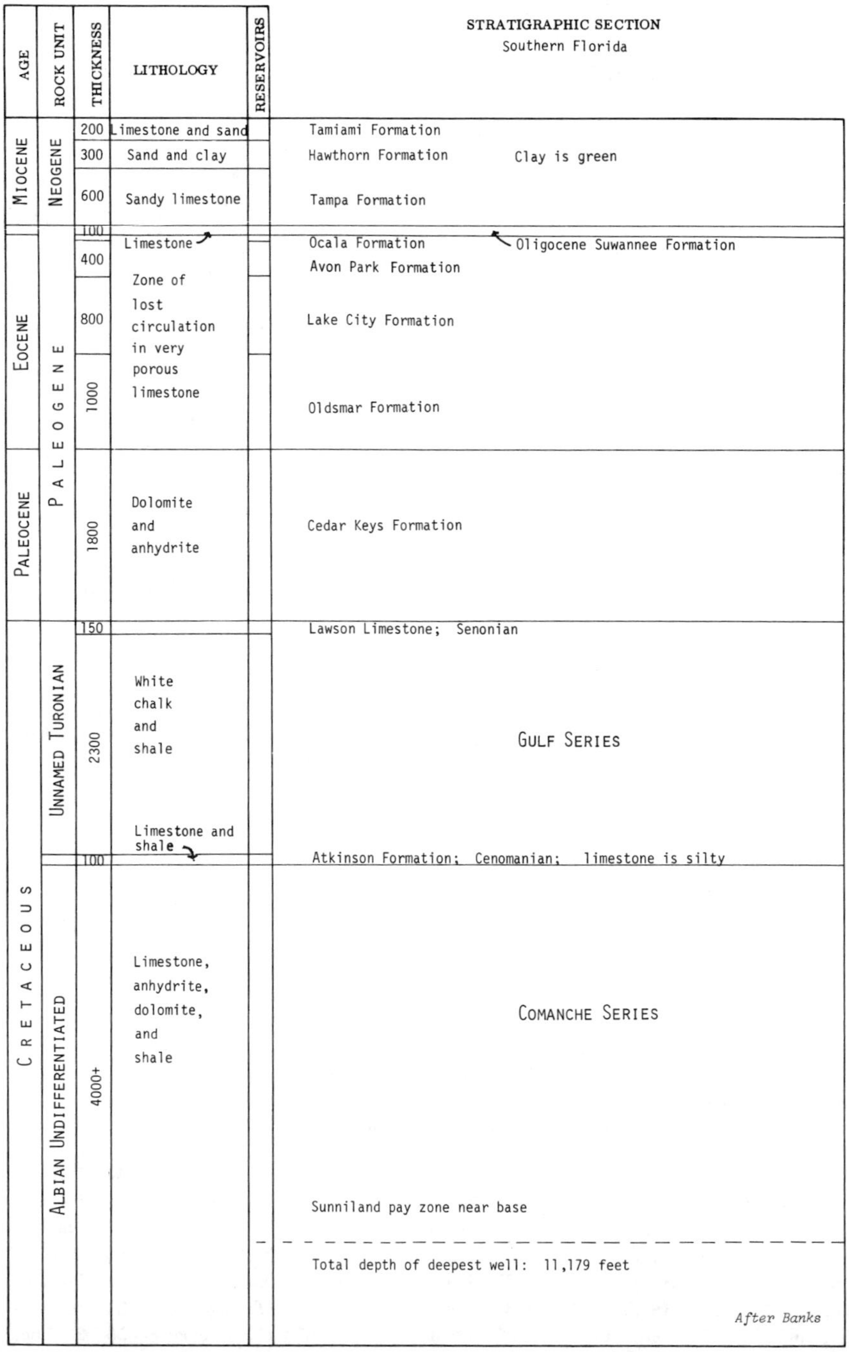

FIGURE 20.2 Stratigraphic section, southern Florida. *Courtesy Joseph E. Banks (1960) and the American Association of Petroleum Geologists.*

mostly due to disoriented rudistids and rudistid fragments. This rock is both overlain and underlain by evaporite sediments. The pay zone has a net thickness of 40 feet; it lies near the base of the known Cretaceous section, where the rocks are Comanchean (Early Cretaceous) in age (Figure 20.2). The discovery well was completed at a total depth of 11,626 feet. The trap is an elliptical dome with a northwest-southeast trend. Although the dome has a relief of 150 feet, the proved closure in the reservoir is only 36 feet (Puri and Banks, 1959). The producing area is $3\frac{1}{2}$ miles long and averages about 1 mile in width (Murray, 1961, p. 461). On December 31, 1967, the cumulative production from this field since its discovery 24 years ago was 11,130,134 barrels of oil.

The short-lived *Forty-Mile Bend* field is in Dade County about 40 miles west of Miami and about 40 miles southeast of the Sunniland field. The reservoir rock here is also the permeable Comanchean Sunniland Limestone, with a part of the reservoir consisting of dolomitized reef rock (Banks, 1960). In the nineteen months that this field was operated before its abandonment in 1955, it produced 32,888 barrels of oil.

The *Sunoco–Felda* field is in western Hendry County and northernmost Collier County (Figure 20.1), about 18 miles north of Sunniland. The discovery well flowed 111 barrels daily from a 20 foot pay zone at 11,485 feet. Location was made on subsurface geology that became available by prior drilling in the area. The reservoir is a medium-grained, brown, porous limestone that has been named the Roberts zone of the Sunniland Limestone. It appears to correlate with the lowest productive interval in the Sunniland field (Babcock, 1966, p. 16). The Roberts zone is overlain by a fine-grained limestone referred to as the "vertical permeability barrier." Updip trapping is due to a permeability pinchout (Babcock, 1965) on a nose mapped by gravity survey (Oglesby, 1967). As of December 31, 1967, the cumulative production from this field during its slightly more than 2 years of operation was 2,685,931 barrels of oil.

About 400 wildcat wells have been drilled in various parts of Florida with only two exploratory holes, the discovery wells for the Sunniland and Sunoco-Felda fields, so far resulting in lasting success. A major handicap in oil hunting in southern and eastern Florida is the presence of high-velocity and cavernous limestones that impede seismic exploration (Owens, 1960). However, well control indicates the presence in southern Florida of a northwest-trending belt of porosity in the Sunniland Limestone 30 to 50 miles wide and 200 miles long, passing through Sarasota, Charlotte, Lee, Hendry, Collier, Broward, and Dade Counties (Roberts and Vernon, 1961, p. 4). This circumstance, in combination with the presence of shows in most of the wells drilled in the southern part of the state and the occurrence of oil in commercial quantities in two fields, with a third marginal field in line to the south, strongly suggests that there may be additional beads on the string.

A second area of promise, as yet unfulfilled, is the Florida panhandle, which stretches westward south of central and eastern Alabama. It is most unlikely that the clastic sediments in this area, which are so productive along the Gulf Coast farther west, are completely barren here. Also it is probable that the Louann salt, with its satellitic salt intrusions and other flowage structures, which is present in the subsurface from Texas to Alabama, continues eastward into the western panhandle (Marsh, 1967). In late 1968 Jurassic production in Alabama (see p. 160) jumped 80 miles from Choctaw County to Escambia County (Figure 21.1). Another hop, this one of less than 5 miles, would bring Jurassic production into another Escambia County, the one at the western end of the Florida panhandle.

The submerged continental shelf, which has maximum widths of 75 miles in the Atlantic off northern Florida and 150 miles in the Gulf of Mexico off the peninsula (Figure III.1), has as much area as emergent Florida or even more. Seismic surveys have shown the possibility of reefs in Upper Cretaceous rocks in the northeastern Gulf of Mexico (Antoine and Harding, 1965). Although numerous dry wildcats have been drilled off Florida's west coast, the area yet to be explored is large.

REFERENCES CONSULTED IN PREPARING CHAPTER 20

Antoine, John W., and Harding, James L., 1956.

Applin, Paul L., 1952.

Applin, Paul L., and Applin, Esther R., 1944, 1947.

Babcock, Clarence, 1965, 1966.

Banks, Joseph E., 1960.

Chen, Chih Shan, 1965.

Hull, Joseph P. D., Jr., 1962.

Marsh, Owen T., 1967.

Murray, Grover E., 1961.

Oglesby, Woodson R., 1967.

Owens, Harold, 1960.

Puri, Harbans S., and Banks, Joseph E., 1959.

Roberts, Wendell L., and Vernon, Robert O., 1961.

Sargent, Don C., 1964.

Chapter

21

ALABAMA

The discovery of oil in Mississippi in 1939 stimulated prospecting in southern Alabama, and in 1944 the discovery well for the Gilbertown field was completed. Subsequent discoveries of importance are the Pollard field in 1952, the Citronelle field in 1955, and Smackover production in two fields in Choctaw County in 1967 (Browning, 1968, p. 1029). Alabama's oil production reached a peak in 1966 with a petroleum output of more than 8 million barrels. The discovery of gas in northwestern Alabama antedated the first oil by over 30 years. The Fayette gas field (Figure 21.1) was discovered there in 1909 and was abandoned in 1917. Other small gas fields found during this period in northwestern Alabama were likewise short lived. However, gas production was resumed in the early 1950's after new discoveries in Marion County.

Geology. Two major physical divisions of the United States, the Appalachian Highlands and the Gulf Coastal Plain, extend into Alabama. The Appalachian rocks occupy northeastern and northern Alabama (Figure 21.1); to the south and southwest they are overlapped by Upper Cretaceous rocks of the Gulf Coastal Plain. On the southeastern side of the Appalachian province in Alabama the rocks are Precambrian crystallines, in the central part of the province they are tightly folded and faulted Paleozoic sediments, and in northwestern Alabama the exposed rocks are gently folded Paleozoics, as in western Pennsylvania.

Northwestern Alabama and northeastern Mississippi are occupied by the Black Warrior basin (Pike, 1968; Everett, 1954; Mellen, 1947). This basin is bounded on the east by the highly folded Paleozoic sediments, on the south and southwest (beneath the embayment overlap) by the southeasterly extension of the Ouachita Mountain system, and on the north by the Nashville dome on the axis of the Cincinnati arch. The rocks exposed at the surface in the basin are Pennsylvanian (Metzger, 1965) and Mississippian in age.

1 Autauga
2 Baldwin
3 Barbour
4 Bibb
5 Blount
6 Bullock
7 Butler
8 Calhoun
9 Cherokee
10 Chilton
11 Choctaw
12 Clarke
13 Coffee
14 Colbert
15 Conecuh
16 Covington
17 Crenshaw
18 Cullman
19 Dale
20 Dallas
21 DeKalb
22 Escambia
23 Etowah
24 Fayette
25 Franklin
26 Geneva
27 Greene
28 Hale
29 Henry
30 Houston
31 Jackson
32 Jefferson
33 Lamar
34 Lauderdale
35 Lawrence
36 Limestone
37 Lowndes
38 Macon
39 Madison
40 Marengo
41 Marion
42 Marshall
43 Mobile
44 Monroe
45 Montgomery
46 Morgan
47 Perry
48 Pickens
49 Pike
50 Russell
51 St. Clair
52 Shelby
53 Sumter
54 Talladega
55 Tuscaloosa
56 Walker
57 Washington
58 Wilcox
59 Winston

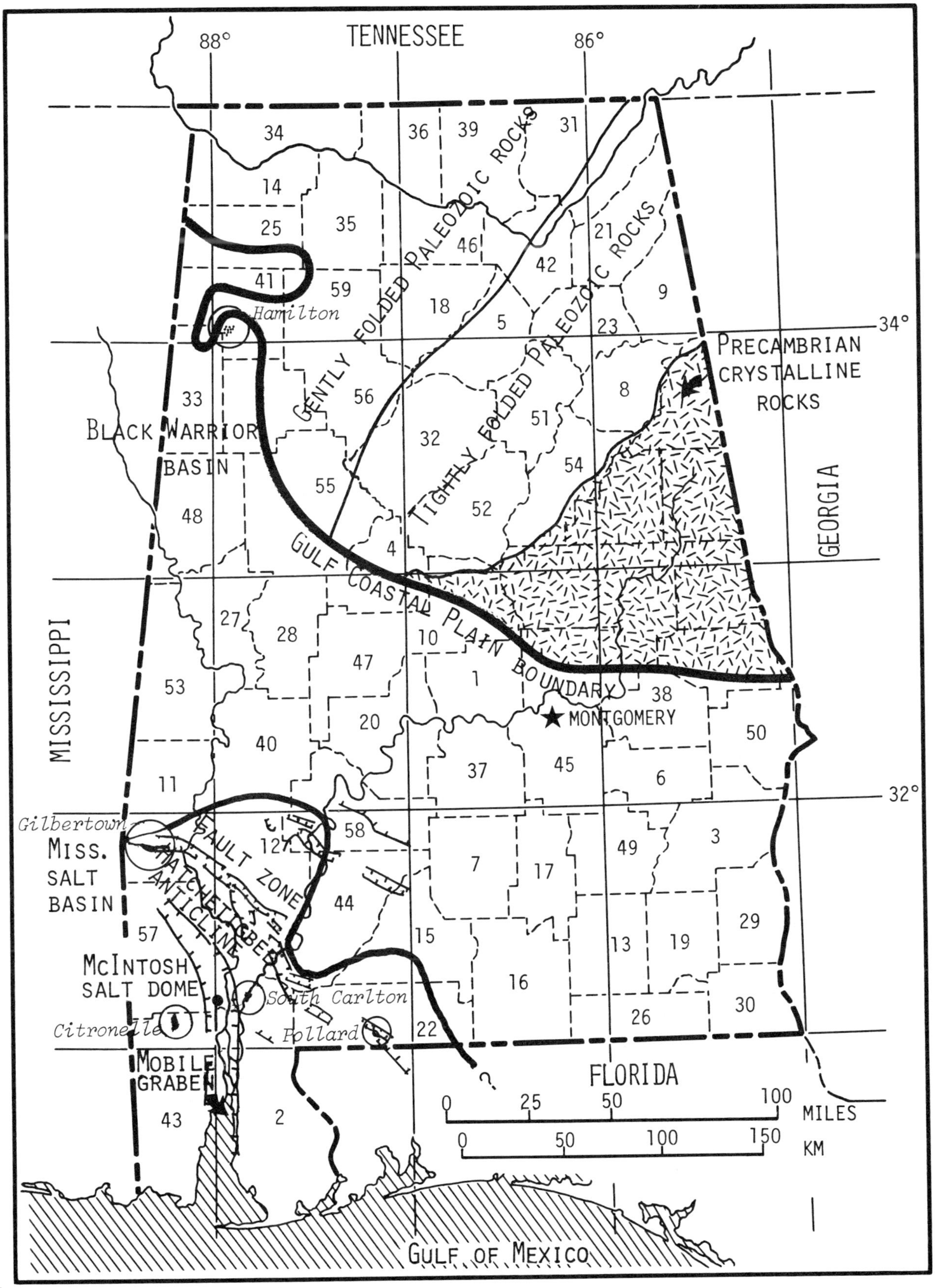

FIGURE 21.1 Alabama. Index, tectonic, and oil and gas field map. County Key on left.

The coastal plain sediments curve around the southwestward plunging Precambrian and Paleozoic rocks in central Alabama. The Upper Cretaceous outcrops form an inner band averaging about 60 miles in width; to the south and west are the Tertiary sediments, ranging in age from Eocene to Pleistocene, in similar parallel bands; and Quaternary sand and gravel deposits occupy the shoreline belt.

The Lower Cretaceous sediments, which have been explored by drilling in the southern part of the Alabama coastal plain, wedge out to the north and northeast. Hence they fail to crop out between the Upper Cretaceous sediments and the overlapped Paleozoic and Precambrian rocks.

The Mississippi salt basin crosses the state line and extends into southwestern Alabama for a maximum distance of about 100 miles. Within this salt basin are the Hatchetigbee anticline (Figure 21.1) and the Gilbertown-Pollard fault zone, which trends southeasterly across the southwestern corner of the state and into the Florida panhandle. This is a continuation of the Pickens-Gilbertown fault zone of Mississippi, which in turn connects westward with the south Arkansas faults and the Mexia-Talco fault zones of northeastern and southern Texas. So far, one piercement salt dome has been found in the Alabama part of the Mississippi salt basin. This is the McIntosh dome in southern Washington County, where the top of the salt lies at a depth of 410 feet. At least two nonpiercement salt domes have been found. One underlies the South Carlton field; the other was discovered by an oil test well drilled to a depth of 15,350 feet about 20 miles to the northeast of the McIntosh dome and a somewhat lesser distance north of the South Carlton oil field.

The reservoir rocks (Figures 22.3 and 21.3) so far productive in Alabama range in age from Late Cretaceous to Mississippian. The Paleozoic production is confined to the Warrior basin in northwestern Alabama, where sandstones of Chester (Late Mississippian) age contain gas (Welch, 1958). The older oil fields of southern Alabama (Gilbertown, South Carlton, and Pollard) were discovered by drilling through 4000 to 5000 feet of Tertiary sediments into Upper Cretaceous reservoir rocks. The Lower Cretaceous became productive in 1955, when the Citronelle field in northernmost Mobile County was discovered. The pay zones are two sand intervals in the Rodessa Formation at depths in the 10,000 to 11,000 foot range. An oil well was completed in the Paluxy Formation of Early Cretaceous age (Figure 24.4) in Baldwin County in late 1965, although production from this well is marginal. The first Jurassic oil to be found in Alabama was in Smackover carbonate rocks in the Toxey field in Choctaw County in May, 1967. Three months later "excellent Smackover production" was discovered at Choctaw Ridge (Figure 21.1), also in Choctaw County. In late 1968 Jurassic oil was found far below the Upper Cretaceous reservoirs at Pollard.

Traps and Selected Fields. The discovery oil field of Alabama, *Gilbertown* (Current, 1948), extends for about 13 miles across south-central Choctaw County. It is in the western portion of the fault trend shown in Figure 21.1, and three major faults in this area are important factors in trapping the oil at Gilbertown. The primary reservoir is the Eutaw Formation of Late Cretaceous age. Sands in this formation produce oil where permeability has been cut off updip (northward) by faulting across structural noses. These faults are normal, with down-dropped younger Selma Chalk providing the seal. In some parts of the field, however, Eutaw oil has leaked into fractures in the Selma Chalk, so that this formation is also a reservoir. Selma production comes principally from the upper part of the chalk, where fractures are larger and most numerous, immediately above the fault plane (Moore, 1967).

The stratigraphically deeper Upper Cretaceous Tuscaloosa formation is the reservoir rock in the *South Carlton* field. The oil occurs in two sand intervals in the lower Tuscaloosa at depths of 4200 to 4500 feet. The trap is an elliptical dome with a closure of about 80 feet. The salt in the deeper nonpiercement core is Buckner (Jurassic) or older. The salt was penetrated by one well between 14,640 and 15,730 feet TD.

The *Pollard* field (Winter, 1954) is another example of trapping against a fault zone (Figure 21.2). The Pollard fault makes the south side of a graben about $2\frac{1}{2}$ miles wide. It cuts an elliptical anticline close to the axis; the greater part of the production in the Pollard field is obtained from high on the anticline on the southern (upthrown) side of the fault. The reservoir rocks are three productive sand zones in the Upper Cretaceous Tuscaloosa Formation (Figure 21.3).

The youngest of these sands, the Miller at the base of the upper Tuscaloosa, is a lenticular sand confined to the graben side of the fault. The trap here is anticlinal closure and drag against the fault plane. The two lower Tuscaloosa sand reservoirs, the prolific Moye and the Massive, contain oil in the upthrown block, where it has been impounded against the fault plane. It has been suggested that the oil trapped in the Miller came from the nearby Moye by leakage through the fault seal.

The trap in the *Citronelle* field (Jones and McGlamery, 1957) is a nearly circular anticline having a diameter of about 4 miles with a closure of over 100 feet. Unfortunately, the reservoir sands are quite variable in permeability with a tendency toward tightness. Nonetheless, Citronelle is Alabama's foremost field, with 70 million barrels of cumulative production at the end of 1967 and an estimated 50 million barrels in reserve.

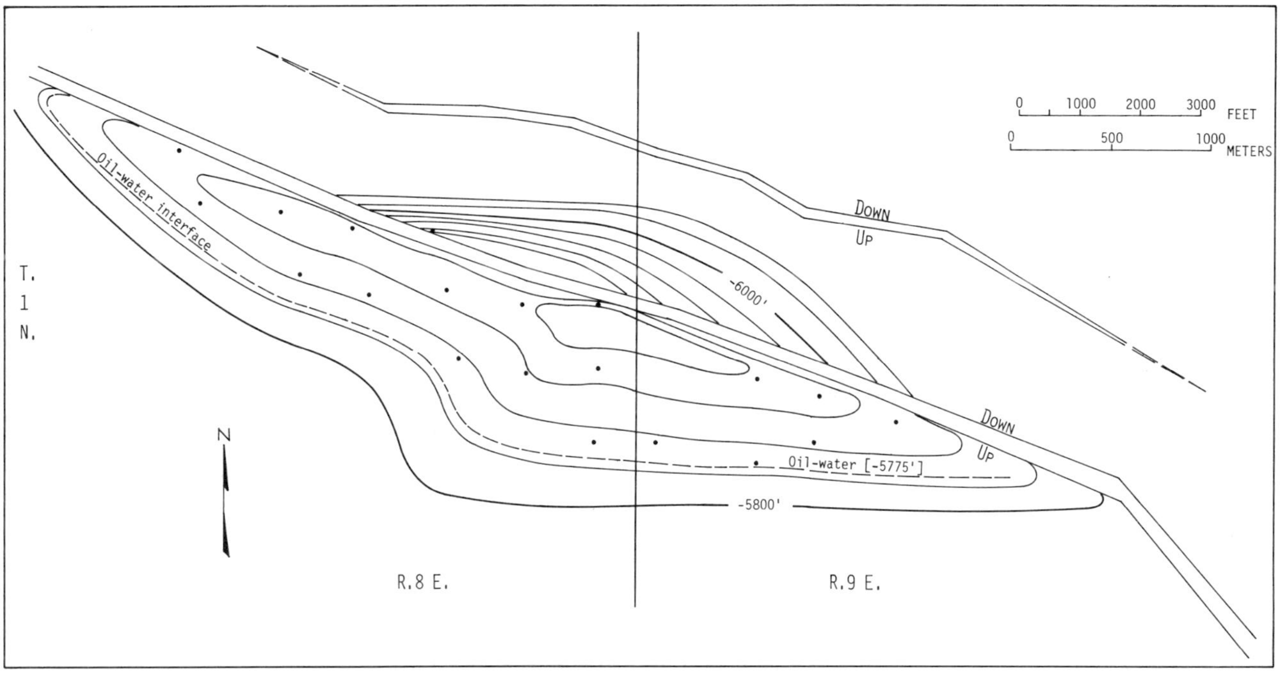

Figure 21.2 Pollard field, Escambia County, Alabama. Structure contour map based on Moye (Upper Cretaceous) Sand. Contour interval, 20 feet. Shows Moye producing wells only. *Courtesy C. Victor Winter (1954, Figure 9, p. 134) and the Gulf Coast Association of Geological Societies.*

Future. Warrior basin in northwestern Alabama and northeastern Mississippi is an area of recurrent interest (Jones, 1960; Everett, 1954). In 1968 the Alabama sector of this basin had one producing field, the Hamilton gas field in Marion County, and another field in the same county, the White House, that was intermittently productive. The reservoir is a sandstone in the Upper Mississippian. At least six other small gas fields in northwestern Alabama have been discovered, developed, and abandoned, all since 1911. This part of the state presents the best opportunity for developing production in Paleozoic rocks, both where these formations lie at the surface and where they are hidden by a not-too-thick cover of overlapping Cretaceous sediments. The Pennsylvanian sediments in the Warrior basin in Alabama thicken rapidly southwestward, reaching over 6000 feet in southern Pickens County. Shows of oil and gas have been obtained in drilling through these sediments throughout the basin. A few thousand barrels of oil has been produced from Ordovician Knox Dolomite immediately across the state line in Monroe County, Mississippi. "The Ordovician and Cambrian rocks of the basin are relatively untested and are considered an important objective for future exploration" (Pike, 1968, p. 1701). Unfortunately, the belt of Paleozoic reservoir rocks that extends southwestward from western Pennsylvania diminishes considerably in productivity south of central Kentucky, perhaps because of either increase in the degree of metamorphism or decrease in the quality of source rocks.

The opportunities for additional discoveries appear to be much greater in Gulf Coastal Alabama than in the Paleozoic area. In the Gulf Coast region both further fault accumulations and salt dome fields are distinct possibilities. The commercial occurrence of hydrocarbons in Alabama in stratigraphic traps is a probability, and Eocene reservoirs may occur in Alabama, as in Mississippi. Younger Cenozoic sediments also may be present in adequate thickness on the Gulf Coast, both onshore and offshore (Figure III.1). It is difficult to believe that all of the oil stored in Upper and Lower Cretaceous reservoirs has already been found in Alabama's few oil fields. Moore (1967, p. 14) calls attention to the potential of fractured chalk.

Today the most promising objectives in Alabama are pre-Cretaceous, especially Jurassic, pay zones. A deep exploratory test well drilled at the southwestern edge of the Pollard field in Escambia County was reported in late 1968 to have flowed 30 million cubic feet of gas plus 1560 barrels of condensate from a Jurassic pay zone (possibly Norphlet) at 15,320 to 15,360 feet (Oil and Gas Jour., Nov. 25, 1968, p. 165). The stratigraphic position of the Norphlet is shown in Figure 22.3. This discovery opens up new vistas for both southern Alabama and panhandle Florida.

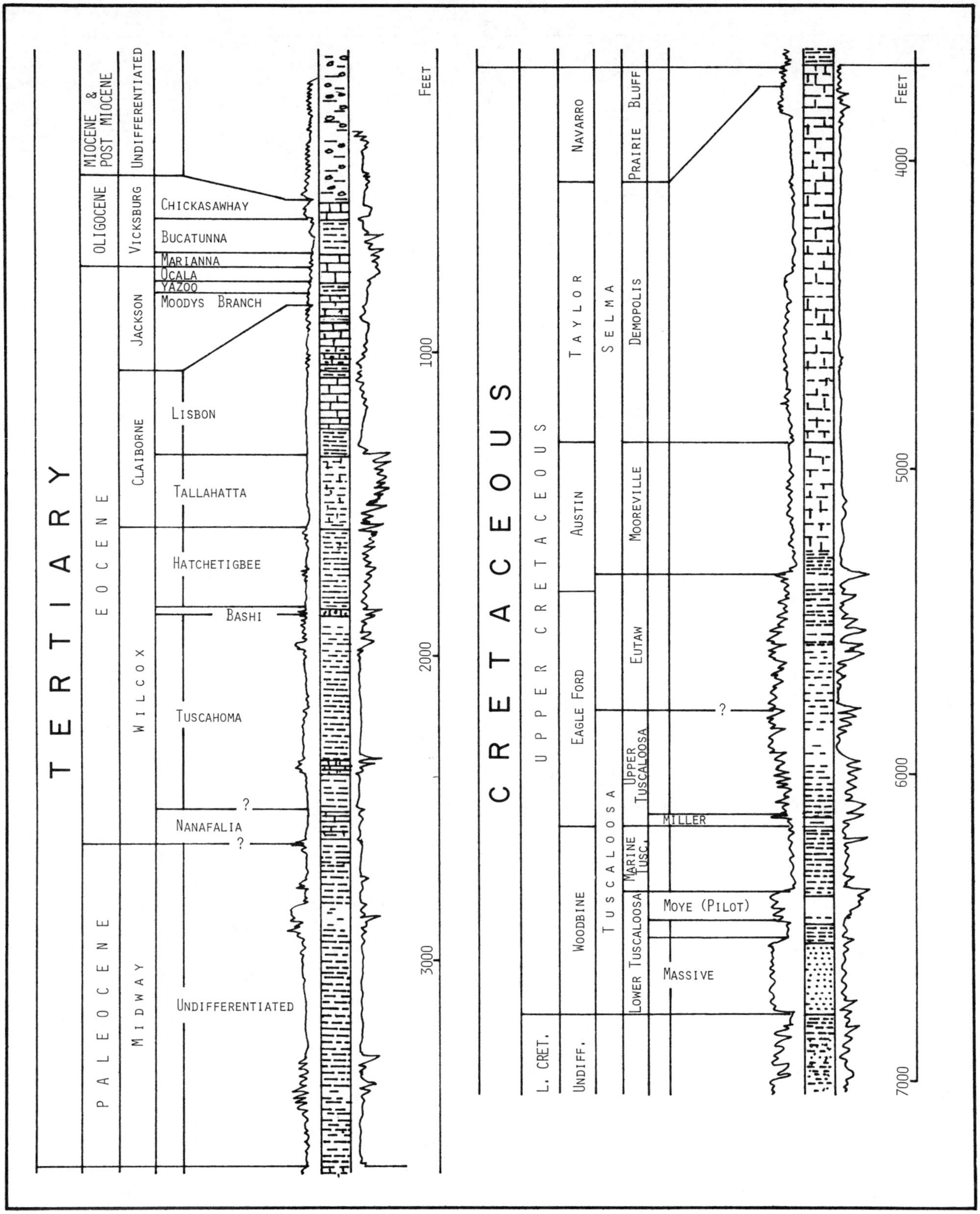

Figure 21.3 Composite electric log and columnar section, Pollard oil field, Escambia County, Alabama. *Courtesy C. Victor Winter (1954, Figure 3, p. 127) and the Gulf Coast Association of Geological Societies.*

REFERENCES CONSULTED IN PREPARING CHAPTER 21

Browning, W. Fleming, 1968.
Current, A. M., 1948.
Everett, Rizer, 1954.
Frascogna, Xavier M. (ed.), 1957.
Jones, Walter B., 1952, 1960.
Jones, Walter B., and McGlamery, Winnie, 1957.
Mellen, F. F., 1947.
Metzger, William J., 1965.
Moore, Donald B., 1967.
Pike, Stuart J., 1968.
Welch, Stewart W., 1958.
Winter, C. Victor, 1954.

Chapter

22

MISSISSIPPI

In recent years the state of Mississippi has produced a little less than 2 per cent of the nation's petroleum and 0.8 per cent of its natural gas. It ranked ninth in oil production in the United States in 1968. The first commercial discovery of hydrocarbons in Mississippi took place in 1926, when the small Amory gas field in Monroe County (Figure 22.1) in northeastern Mississippi was found. The much larger Jackson gas field was discovered in 1930, and the first oil field, the Tinsley (now the largest of Mississippi's three giant fields), became productive late in 1939. This discovery brought about an extensive exploration program that is still continuing. As a result, 38 of the 82 counties in Mississippi now produce hydrocarbons. So far the peak year was 1963, when nearly 59 million barrels of oil was produced. Five wildcats in 1967 discovered new fields (Browning, 1968, p. 1020).

Regional Structure. The geology of Mississippi is dominated by the Mississippi embayment sector of the Gulf Coast embayment. As a result, the oldest extensive surface formations are Cretaceous sediments in the northeastern part of the state. Successively younger Tertiary formations crop out to the south and west, and Quaternary deposits overlie the Tertiary in the Mississippi Valley and along the Gulf Coast shore zone. Figure 22.2 shows the outcrop boundary between the oldest Tertiary sediments and the underlying Cretaceous. The strike of the formations generally parallels this contact line; the dip in the northern half of the state is to the west, but it veers to the southwest in southern Mississippi. The dips range from 25 to 100 feet per mile, increasing with depth (Beebe, 1968, p. 1181). Some of the stratigraphic sections also increase considerably in thickness gulfward into the Gulf Coast geosyncline. Major thickening is present in the Cenozoic sediments above the Eocene and in the Lower Cretaceous and Jurassic sections (Braunstein, 1958, p. 515).

In the very northeastern corner of the state, Paleozoic formations of Mississippian age crop out beneath the

1 Adams
2 Alcorn
3 Amite
4 Attala
5 Benton
6 Bolivar
7 Calhoun
8 Carroll
9 Chicasaw
10 Choctaw
11 Claiborne
12 Clarke
13 Clay
14 Coahoma
15 Copiah
16 Covington
17 DeSoto
18 Forrest
19 Franklin
20 George
21 Greene
22 Grenada
23 Hancock
24 Harrison
25 Hinds
26 Holmes
27 Humphreys
28 Issaquena
29 Itawamba
30 Jackson
31 Jasper
32 Jefferson
33 Jefferson Davis
34 Jones
35 Kemper
36 Lafayette
37 Lamar
38 Lauderdale
39 Lawrence
40 Leake
41 Lee
42 Leflore
43 Lincoln
44 Lowndes
45 Madison
46 Marion
47 Marshall
48 Monroe
49 Montgomery
50 Neshoba
51 Newton
52 Noxubee
53 Oktibbeha
54 Panola
55 Pearl River
56 Perry
57 Pike
58 Pontotoc
59 Prentiss
60 Quitman
61 Rankin
62 Scott
63 Sharkey
64 Simpson
65 Smith
66 Stone
67 Sunflower
68 Tallahatchie
69 Tate
70 Tippah
71 Tishomingo
72 Tunica
73 Union
74 Walthall
75 Warren
76 Washington
77 Wayne
78 Webster
79 Wilkinson
80 Winston
81 Yalobusha
82 Yazoo

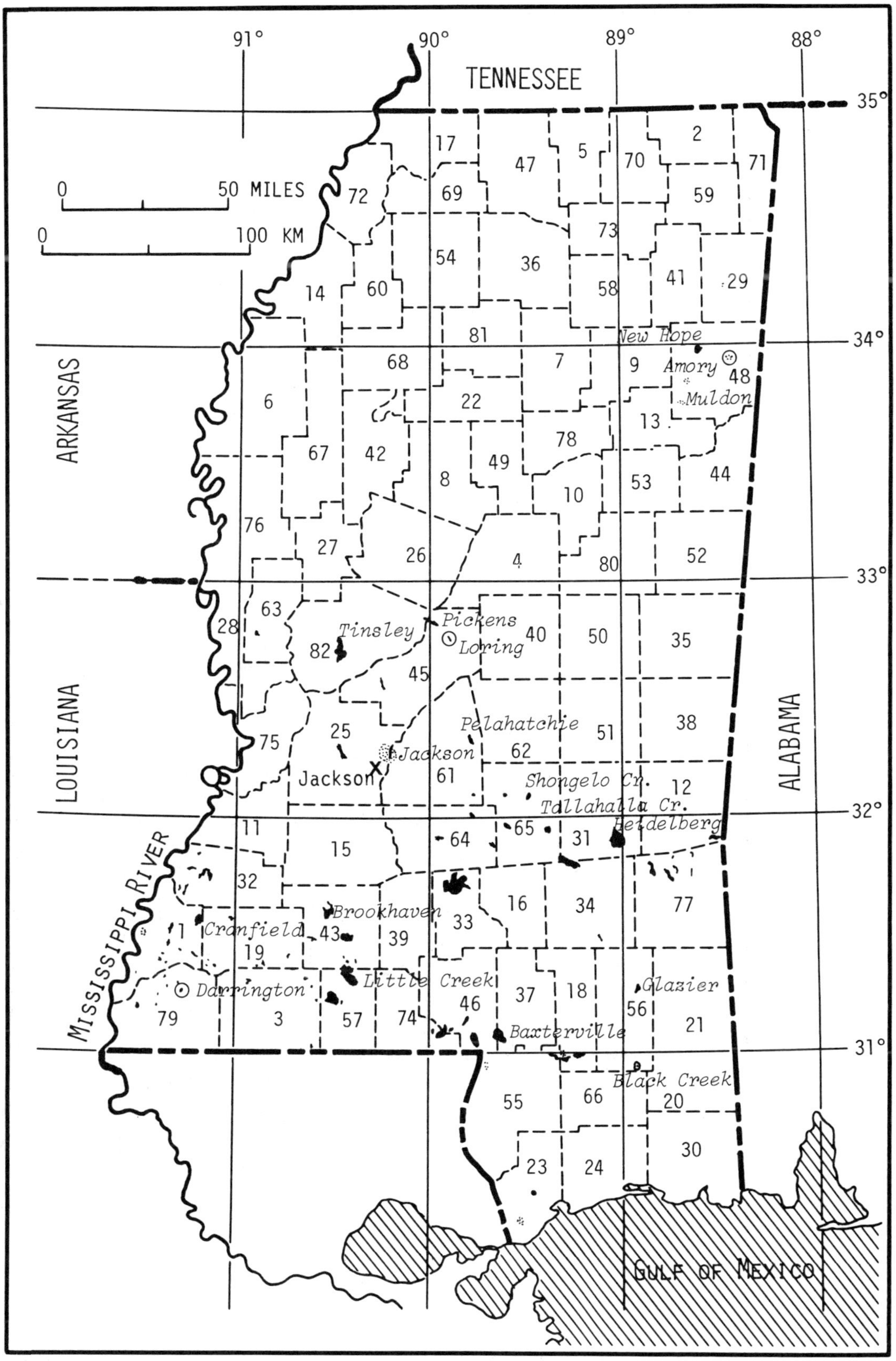

FIGURE 22.1 Mississippi. Index map with oil and gas fields. County Key on left.

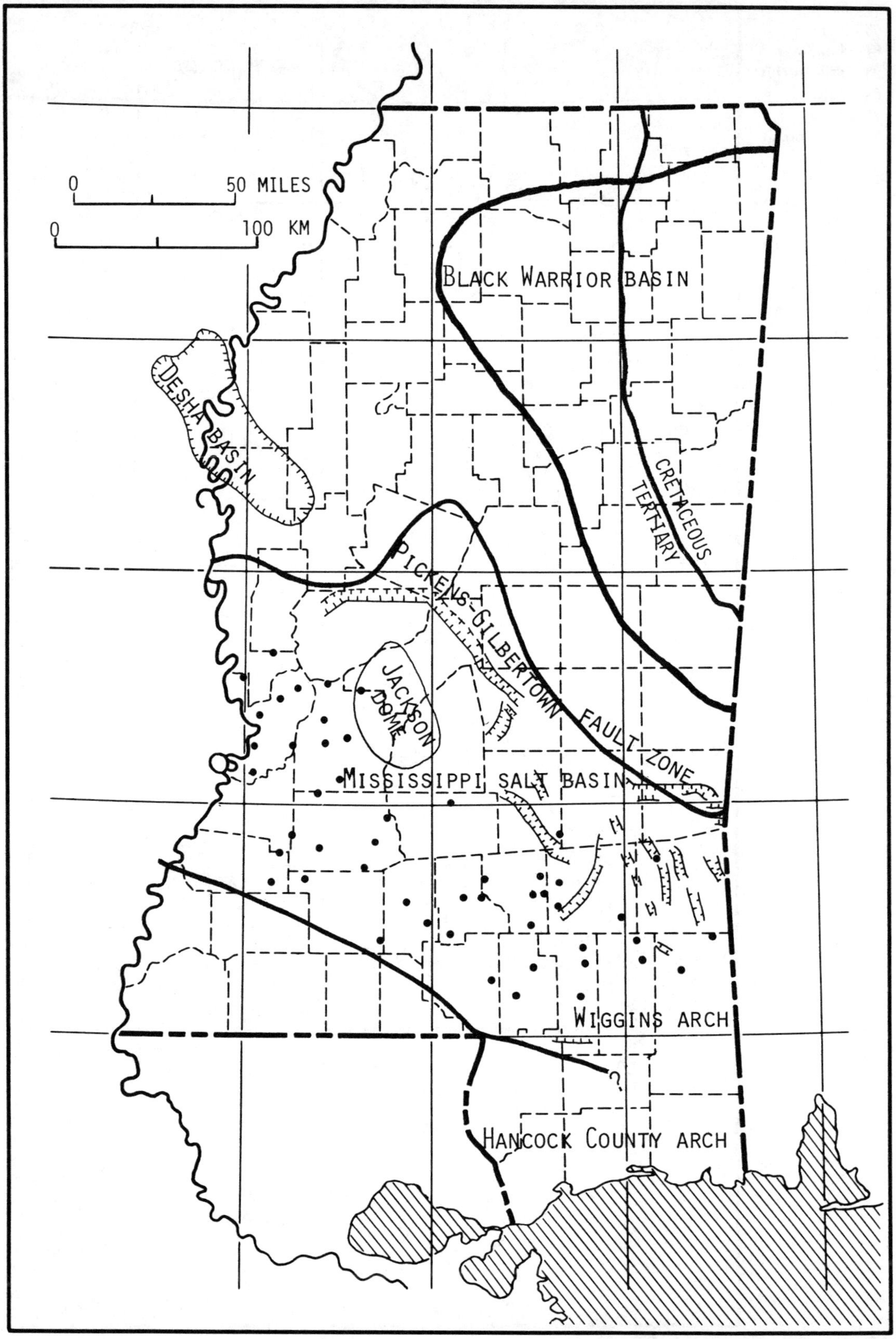

FIGURE 22.2 Mississippi. Tectonic map. Solid circles are known piercement salt domes.

feather edge of Cretaceous sediments. Throughout a considerable area in northern Mississippi, Paleozoic formations lie within less than 6000 feet of the surface. This is the western end of the Paleozoic Black Warrior basin (Mellen, 1947).

The basins within the embayment area are the Desha (Figure 22.2), which extends from southeastern Arkansas into west-central Mississippi, and the Mississippi salt basin. "From the standpoint of the oil industry, by far the most important basin within the eastern Gulf Coast is the Mississippi salt basin, which extends into western Alabama on the east and into central Louisiana on the west" (Braunstein, 1958). This tectonic feature includes, in addition to over 50 known salt domes (Beikman and Drakoulis, 1958), the prominent igneous rock-cored Jackson dome, which covers many square miles and has a closure of more than 1500 feet. To the north and east of the Jackson dome is a zone of faults, the Pickens-Gilbertown-Pollard. This appears to be a segment of a linear fault complex extending north and east from Luling-Mexia-Talco in south and northeast Texas across southern Arkansas ("South Arkansas fault zone") and thence southeasterly across Mississippi into southern Alabama, home of the Gilbertown and Pollard fields. Within the Mississippi salt basin depths to salt range from a few hundred feet to more than 20,000 feet. It is generally assumed that the oil fields producing from structural domes (and this category constitutes a large proportion of the fields in Mississippi) overlie deep-seated salt cores, the intrusion of which was responsible for the doming (Shreveport Geol. Soc., 1946, 1947). Deep exploration of the salt basin by the drill, plus improved seismic techniques, have made possible a classification of salt-formed structures (Dudley J. Hughes, paper read before Gulf Coast Association of Geological Societies as reported by the Oil and Gas Jour., Dec. 9, 1968, p. 150). There are four main groups, which are centripetally situated. Around the rim of the basin are the "periphery salt ridges." Farther downdip occur the "buried salt ridges." Basinward from these, where the mother salt is much thicker, are the "intermediate salt structures," which have had greater uplift. Closer to the basin center the Louann salt reaches maximum thickness, and the "high-relief salt structures" attain great size, both vertically and horizontally.

The Mississippi salt basin barely enters the southeastern extension of Mississippi, which provides a corridor to the Gulf Coast. Just beyond the southern terminus of the basin in this area is the west-east–trending Wiggins arch (Figure 22.2). In the southwestern corner of the extension is the Hancock County arch, which extends over into the eastern Lake Ponchartrain area of neighboring Louisiana.

The coastal salt dome province of Louisiana comes to within 50 miles of the southwestern corner of Mississippi and trends southeasterly, apparently passing to the south of the Mississippi shoreline on the Gulf of Mexico.

Stratigraphy. Figure 22.3 gives the stratigraphic sections explored by drilling in both northeastern Mississippi and western and southern Mississippi. The northeastern area is far enough inland on the Gulf embayment so that only the oldest Tertiary (Eocene) is present at the surface, and the underlying Cretaceous sediments lap upon rocks of Paleozoic age. The greater part of the relatively minor production in this area to date has been gas from Mississippian sandstones in the Black Warrior basin. The oldest reservoir so far discovered is of Cambro-Ordovician age; it produced oil briefly from a now abandoned one-well pool in Monroe County.

The lefthand columnar section in Figure 22.3 applies to southern and western Mississippi, where most of the state's oil and gas production has taken place. The section so far explored ranges from marine Pleistocene down to the salt that underlies the Jurassic Smackover Formation. Most of the cumulative production in Mississippi has come from Cretaceous reservoirs on the north and south flanks of the salt basin that crosses the state from west to east.

Much the most important Tertiary reservoir known in Mississippi is the Wilcox Sandstone of the lower Eocene. The greatest development of this reservoir has been in four counties, Adams, Franklin, Jefferson, and Wilkinson, in the southwestern corner of the state. In this area 21 new Wilcox oil fields were discovered in 1966–1967 (Guy, 1967; Gulmon, 1967; Browning, 1968). The Wilcox lithology is that of rapidly deposited clastic sediments with the sands lensing laterally into shales in short distances. Braunstein refers to these accumulations as "structurally modified stratigraphic traps" (1958, p. 520).

Over the Hancock arch in southwestern Mississippi a large section of Tertiary, including some of the lower Miocene, all of the Oligocene, and the Eocene down to the Wilcox, is missing both in Hancock County and in adjacent Louisiana to the southwest.

Rocks of Late Cretaceous age are the most prolific reservoirs. Included, from youngest to oldest (with Texas equivalents in parentheses), are Selma (Navarro and Taylor), Eutaw (Austin and Eagle Ford), and Tuscaloosa (Woodbine).

Mississippi's most unusual reservoir is the "Gas Rock" of youngest Cretaceous age (Selma and younger). It is a limestone (in some fields a sandstone) ranging in thickness, where present at all, from 50 to over 1500 feet. It is confined largely to uplifts that were relatively high in Late Cretaceous time, and it unconformably overlies not only truncated older Cretaceous rocks but Upper

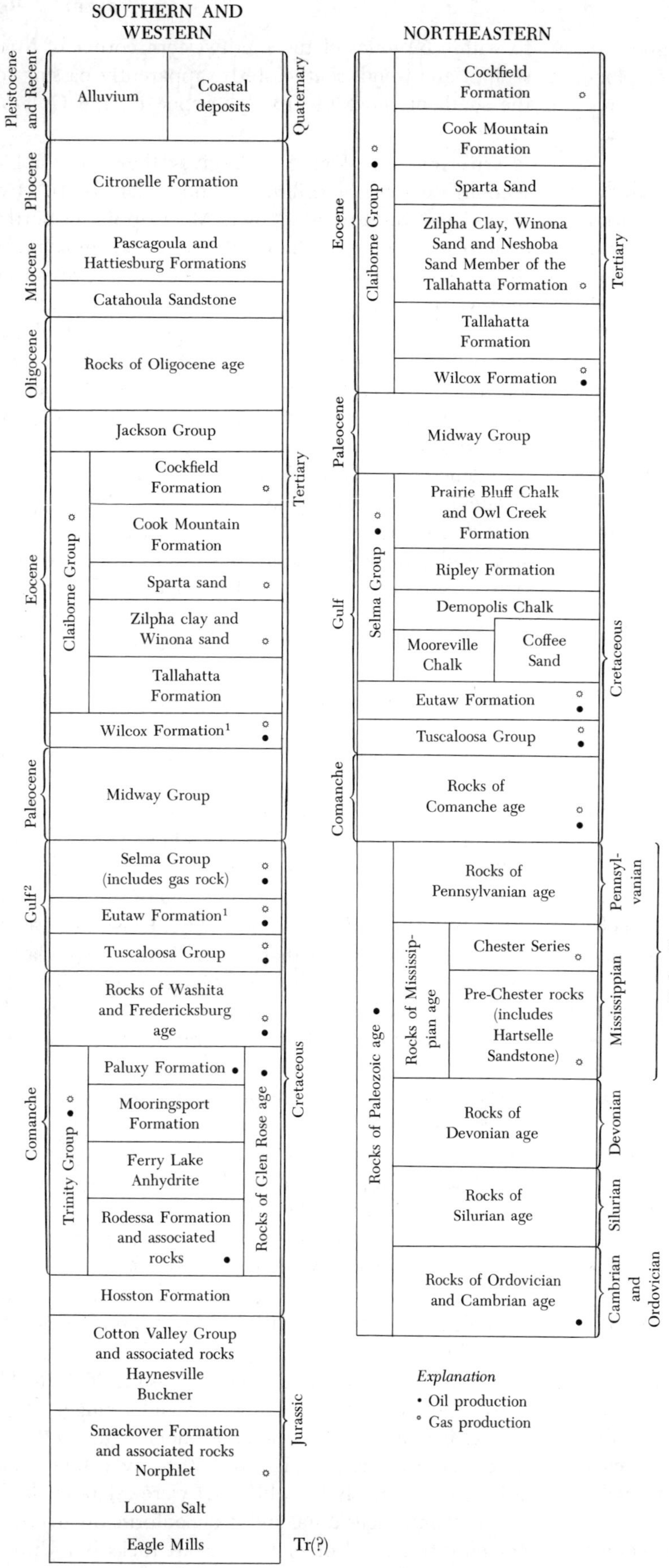

Explanation

• Oil production

° Gas production

FIGURE 22.3 Diagrammatic composite stratigraphic sections of rocks penetrated by drilling in Mississippi. From Helen M. Beikman and Sophie Drakoulis (1958). Courtesy U. S. Geol. Survey.

Footnotes

[1] *Names of some of the producing zones within these formations are listed below. These names follow the usage of the Mississippi State Oil and Gas Board.*

Wilcox Formation

2nd	Baker	McShane
4400	Ball	Netterville
4600	Barksdale	Ogden
5200	Benbrook	Parker
5500	Blake	Pearline
5800	Harmon	Ratcliff
Armstrong	Kendall	Robinson
Artman	McKittrick	Walker
		Wilson

Eutaw Formation

Christmas *Stanley*

[2] Texas names used by some geologists in Mississippi for rocks of Gulf age, from oldest to youngest, are the Woodbine Formation, Eagle Ford Shale, Austin Chalk, Taylor Marl, and Navarro Group.

Jurassic rocks as well. This erratic reservoir has produced more than one-fourth of Mississippi's gas.

The Eutaw Formation contains the leading reservoirs on the north flank of the basin, where faulting has contributed to the trapping. The Eutaw at the outcrop is sandstone, but down the regional dip to the south the sand is succeeded by a chalk facies.

The lower Tuscaloosa sands are most important, especially on the south flank of the salt basin (Eisenstatt, 1960, p. 206). They "tend to be lenticular and erratic . . . and the traps may be considered as stratigraphically modified structural traps" (Braunstein, 1958, p. 518).

Between the mid-1940s and the early 1960s almost 40 fields were discovered in which the reservoir rocks are in part or exclusively of Early Cretaceous age. Sands in the Paluxy and Glen Rose have been the principal producers. These fields also lie within the salt basin (Philpott, 1960, p. 1603).

Before 1963 the only Jurassic production in Mississippi was gas and condensate from the Smackover Formation in two wells in the Loring pool of Madison County. The discovery well was completed in May, 1954, testing 4,600,000 cubic feet of noncombustible gas (7 percent H_2S) and 223 barrels of 49.4° distillate per day at 12,100 to 12,500 feet (Mississippi Geol. Soc., 1957, p. 70). Trapping is anticlinal. Originally the gas was flared, but later it was used for the extraction of sulfur (Mellen, 1966, p. 96).

Commercial oil was discovered in Smackover Sandstone at 14,150 feet in a wildcat well drilled in northeastern Smith County in September, 1963 (Gardner, 1963b). This field became known as Bienville Forest. "The discovery of low gas-oil ratio sweet oil in dolomitic limestone" at Bienville Forest "initiated an intensification of leasing, exploration, and development in the southeastern states" (Mellen, 1966, p. 97). This exploration is continuing throughout the Gulf Coast. The most active Smackover field in Mississippi at the close of 1967 was the Tallahalla Creek in southeastern Smith County, where more than 1000 barrels per day of oil was reported coming from two pay zones in the Smackover.

The Smackover Formation averages over 1000 feet in thickness in Mississippi. It is variable in character, both laterally and vertically. The requisite porosity for storing hydrocarbons may be in (1) oölitic limestone, especially at the top of the formation, where it may be highly porous; (2) intergranular dolomitic limestone; (3) vuggy and fractured limestones; and (4) sandstones within the carbonate rock section (Mellen, 1966, p. 97).

Another chapter in Mississippi oil history was written in 1967, when an exploratory well in eastern Rankin County yielded an initial production of 450 barrels a day of 51° oil at a depth of 17,160 feet in the Norphlet. This is not only the deepest oil production in Mississippi but also the first discovery in the Norphlet, which underlies the Smackover and overlies the Louann Salt (Bishop, 1967).

By the end of 1967 at least 18 additional Jurassic fields had been discovered in Smith and nearby counties. The Jurassic oil production rose from about 500,000 barrels in 1964 to an estimated 8 million barrels in 1967 (Oil and Gas Jour., Dec. 18, 1967, p. 117). The producing formations for these fields are, from the top of the Jurassic down, the Cotton Valley (9 fields), Smackover (12), and Norphlet (1). Three fields produce from both the Cotton Valley and the Smackover (Mellen, 1966; Browning, 1968). The largest Jurassic field is Bay Springs in central western Jasper County. This field produces both oil and gas from the Cotton Valley at depths around 14,550 feet; the oil output in 1967 was nearly 4 million barrels, placing Bay Springs (discovered in early 1965) in second place among Mississippi's oil fields for that year.

Traps and Selected Fields. (Mississippi Geol. Soc., 1952, 1957; Thomas, 1950). A compilation of the Mississippi oil and gas fields (except those of Tertiary age) published in 1957 shows that over 90 per cent are structural traps of one type or another. Just the opposite conditions exists, however, in the Tertiary fields, where the trapping is due in large part to lenticularity of Wilcox sand deposits.

In the Mississippi salt basin anticlines are the predominant trap in Cretaceous pay zones, and deep-seated salt flowage is generally considered to be responsible for the upward arching. One field, Glazier, overlies a known piercement dome. The Jackson dome has an igneous rock core.

Anticlines are also the principal type of trap in the Jurassic accumulations so far discovered. Barber Creek, a Smackover field in southern Scott County, is an example (Beebe, 1968c, p. 1219).

Many of the anticlines in the Mississippi salt basin are faulted; in addition, the Pickens-Gilbertown fault zone, which crosses southern Mississippi from west to east, is the cause of the trapping in Pickens and four other fields (Figure 22.2). The Muldon field in the Black Warrior basin is on a horst block (Figure 22.9).

Varying permeability brought about by lenticularity of reservoir sands is also present in the Cretaceous producing belt. In at least four of the producing anticlines the distribution of the hydrocarbons across the anticline is controlled by the lenticularity of the reservoir rocks (Figure 22.5). In Darrington and other fields there is no anticlinal closure; the trapping is the result of lenticularity on a structural nose.

It was only natural that the prominent *Jackson* dome (Figure 22.2) should be one of the first areas to be explored in Mississippi. The first gas well on this structure was completed in February, 1930, and the first oil well in March, 1932. Depths to the pays range from 2350 to 2500 feet. However, only three additional wells proved to be productive of oil, so that petroleum never became of economic importance here. By 1934, 113 wells were producing gas, and the field reached its peak in 1939, when more than 15 billion cubic feet of gas was produced. It declined rapidly thereafter, and the production of virgin gas was discontinued in 1955, after an overall yield of over 117 billion cubic feet (Beebe, 1968c), and a part of the reservoir converted into a gas storage facility. The gas accumulated in a group of nodes at the very top of the large dome, with water underlying the gas. The reservoir is the previously described "Gas Rock," a porous zone at the top of the Upper Cretaceous Selma Chalk. The intrusive igneous rock that forms the core of Jackson dome is syenitic in composition.

The *Tinsley* field of Yazoo County in west-central Mississippi not only is the first oil field to be discovered in this state (1939), but also leads all other fields in Mississippi in cumulative production, — more than 178 million barrels by the end of 1967. Tinsley is in southwest central Yazoo County in the loess hills, which lie east of the alluvial valley of the Mississippi–Yazoo Rivers (Mississippi Geol. Soc., 1957, p. 124). The field is on the northern flank of the Mississippi salt basin. The anticlinal structure was discovered originally by surface mapping done by the Mississippi Geological Survey during a locally sponsored county mineral survey, but before testing the structure was checked by a reflection seismograph survey. The discovery well was completed in September, 1939, at 4560 feet in an Upper Cretaceous sand ("Woodruff") of probable Navarro (Selma) age. It flowed 235 barrels a day of 35° oil. Subsequently six deeper Upper Cretaceous sands, five in the Eutaw and one in the upper Tuscaloosa, were discovered to be oil productive at depths down to slightly below 6100 feet. In 1951 a test was drilled to 14,427 feet, 400 feet in the Smackover. It discovered sour gas with an estimated open flow rate of 2,500,000 cubic feet per day in a sandstone–limestone interval between 14,019 and 14,042 feet. The well was shut in until 1954, when it was temporarily abandoned.

The structure of the Tinsley anticline, as contoured on the top of the "Woodruff" zone, is shown in Figure 22.4. Accumulation has taken place in a complexly faulted elongate anticline that probably overlies a deep-seated salt dome. The oil is structurally trapped except along a part of the west flank of the anticline where the sand lenses out. It also becomes less permeable at the north end of the field. The Woodruff sand has a maximum thickness of 100 feet.

The *Baxterville* field was first in Mississippi in 1967 in production for the year, and was second only to Tinsley in cumulative production, which at that time exceeded 117 million barrels. The greater part of this field is in the southwestern corner of Lamar County, but the western edge extends into Marion County. It lies on the southern flank of the salt basin, slightly north of the Wiggins arch (Mississippi Geol. Soc., 1957, p. 16). The location of the discovery well (drilled in 1944) was made after a seismic survey of an earlier known prominent minimum-gravity anomaly. The trap at Baxterville is similar to Tinsley in that it is a faulted dome believed to be formed by deep-seated salt movement. In area it is about 50 per cent larger than Tinsley, covering nearly 13,000 acres. The Baxterville field produces gas from a sandstone section in the Eutaw–upper Tuscaloosa at 7600 to 8000 feet, and oil from reservoir sands in the lower Tuscaloosa (8600 to 9000 feet). Commercial gas was found in Wilcox sand (Eocene) at depths of 4800 to 5200 feet in 1951. This gas is confined to the higher parts of the anticline.

The *Brookhaven* field was also found by seismic exploration after a gravity survey had located a gravity minimum. This field is in northwestern Lincoln County on the southwestern edge of the salt basin (Mississippi Geol. Soc., 1957, p. 26). This field produces from basal sands of the Tuscaloosa Formation of Late Cretaceous age (Womack, 1950). The pay zones lie at depths in the 10,000 to 10,300 foot range. The trap is an elliptical north-south anticline 6 miles long and 3½ miles wide with a closure at the top of the basal Tuscaloosa of 400 feet. Three major faults cross the dome, producing a central graben area, but the faulting has had little or no effect on oil accumulation. The interesting feature of the Brookhaven field is that the producing sands not only are lenticular but also are confined entirely to the closed part of the dome. The stippled pattern in Figure 22.5 shows the area underlain by the most widespread sand zone in the reservoir rock sequence. The outer edge of this stippled area is the zero isopach line, and the aggregate sand thickness increases inward to between 60 and 70 feet. The dry area near the center and on the northwest flank of the dome, where there is no effective sand thickness in any of the reservoir zones, covers approximately 1200 acres.

Cranfield, like Baxterville, has relatively minor Wilcox gas production in addition to deep Cretaceous oil and gas. It lies in northeastern Adams County and northwestern Franklin County, on the southwestern flank of the Mississippi salt basin. The discovery well, drilled in 1943, flowed 106 barrels daily of 56° distillate from a sand in the lower Tuscaloosa Formation at a depth of 10,275 feet. A second discovery was completed 3 years later in a Paluxy (Lower Cretaceous) sand at 11,790 feet, producing gas with some distillate. The first discovery well also found gas in the "5800 foot" sand of the Eocene Wilcox Formation. Subsequently other Wilcox gas sands

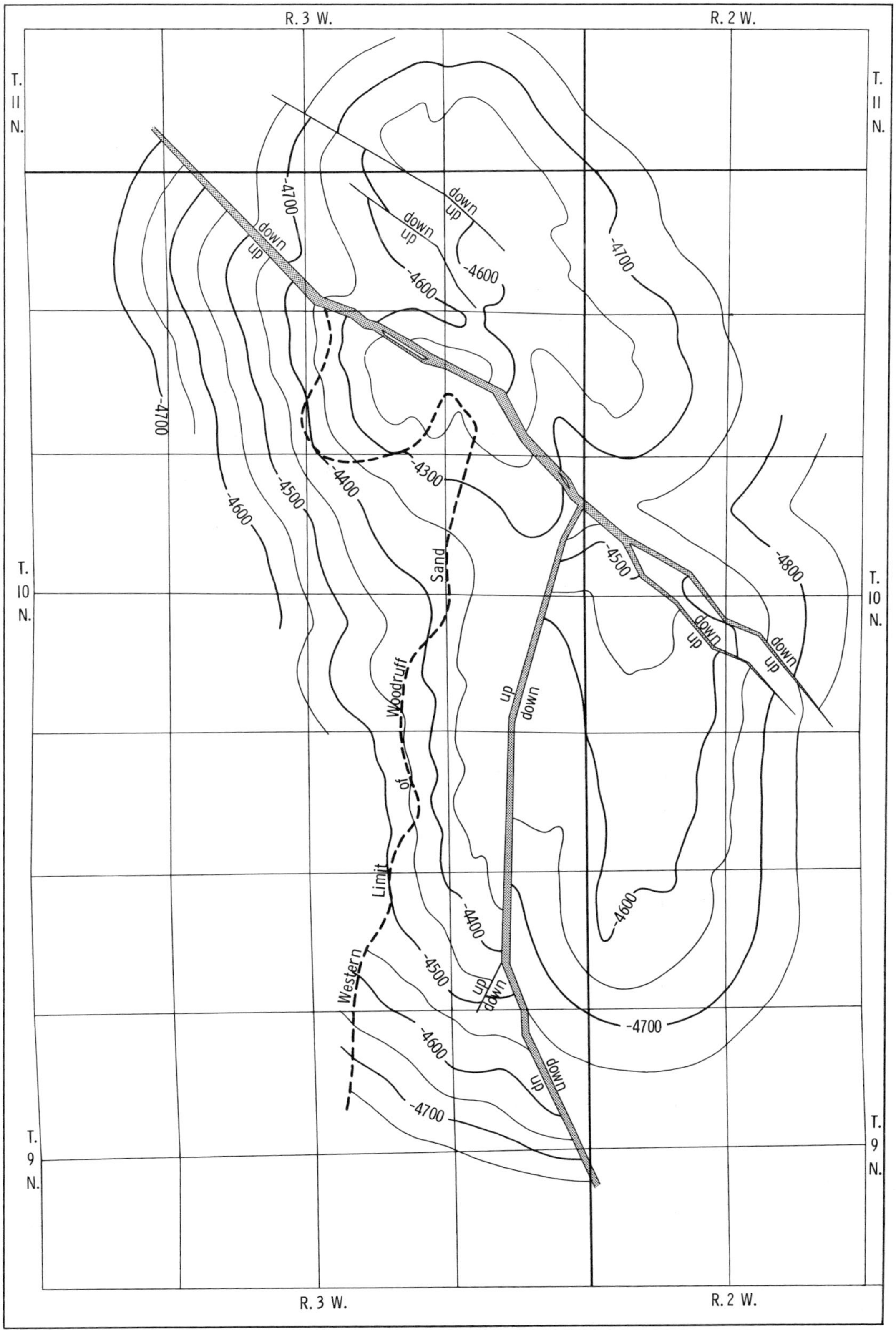

Figure 22.4 Tinsley field, Yazoo County, Mississippi. Structure map contoured on top of Woodruff zone (Upper Cretaceous, probably Navarro). Interval, 50 feet. Sections are 1 mile (1.61 km) on a side. Original oil-water contact was at approximate elevation −4700 feet. *Courtesy Mississippi Geological Society (1957, p. 124).*

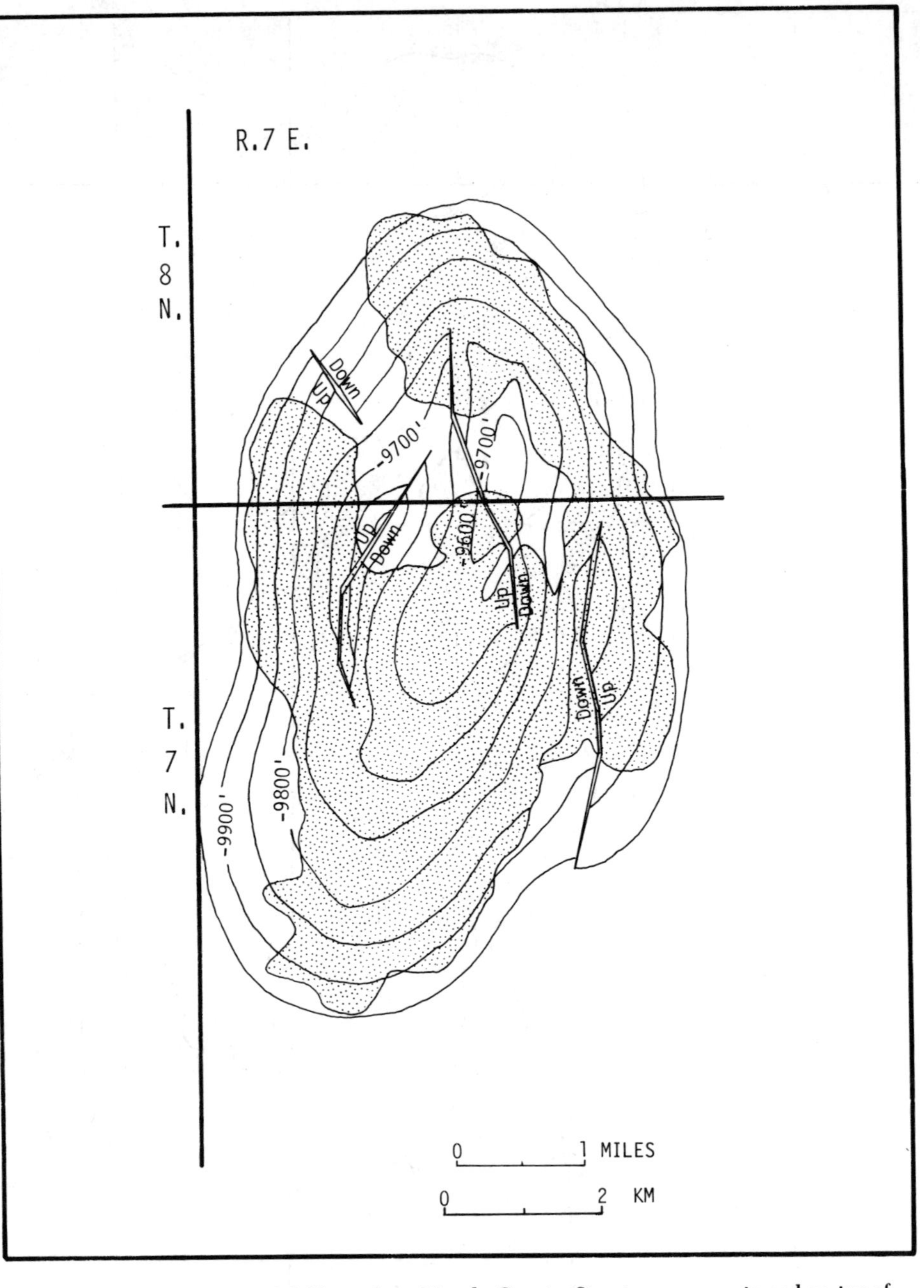

FIGURE 22.5 Brookhaven oil field, northern Lincoln County. Structure map contoured on top of basal Tuscaloosa section. Interval, 50 feet. Stippled area shows maximum extent of reservoir sands. ***Courtesy Robert Womack, Jr. (1950) and the American Association of Petroleum Geologists.***

were discovered at 4400, 5200, and 5500 feet and given names according to those depths.

The trap at Cranfield is a typical, almost circular dome overlying a deep-seated salt dome. The closure at the lower Tuscaloosa level is over 250 feet (Mississippi Geol. Soc., 1957, p. 42–43). It is slightly faulted on the northeast flank. The closure on the "5800 foot" Wilcox sand, the most productive of the Eocene reservoirs, is about 200 feet. Faulting is not evident at that level (Mississippi Geol. Soc., 1952, p. 36–39).

The *Little Creek* field is in Lincoln and Pike counties. The accumulations here are also in lower Tuscaloosa sands, but the bodies are elongate bars rather than nearly circular lenses. The sand thickness averages 40 feet, with a maximum of about 80 feet, and lies at a depth of approximately 10,300 feet. Trapping takes place where the sand bars cross gentle southward plunging noses (St. John and Lusk, 1959, p. 1341; Eisenstatt, 1960).

The *Glazier* field in Perry County is a small, highly faulted dome overlying the top of a piercement salt body. The discovery well was drilled in 1951 on a strong minimum-gravity anomaly. The reservoir rocks are sands in the upper Tuscaloosa Formation of Late Cretaceous age, lying at depths of 7050 to 7200 feet. The top of the intrusive salt is at 10,070 feet. The Glazier field reached its maximum production of slightly less than 50,000 barrels a year 2 years after discovery, when three wells were producing. Three years later it was back to a single producing well. Across the state to the west from Perry County in Claiborne County are three piercement salt domes, one of which is reported to be producing natural gas (Bicker, 1966).

The *Heidelberg* field in the southeastern corner of Jasper County is one of the three oil fields in Mississippi classified as giant. It ranks after Tinsley and Baxterville, with a cumulative production of nearly 85 million barrels by the end of 1967. In addition, it had estimated recoverable reserves of more than 25 million barrels. Heidelberg lies on the northeastern flank of the Mississippi salt basin. It is another example of successful search after detailed seismic surveys of a prominent minimum-gravity anomaly (Mississippi Geol. Soc., 1957, p. 62-63). The discovery well was completed in early 1944 with production from the Upper Cretaceous Eutaw Formation at a depth of 4970 feet. Subsequently three additional discovery wells were drilled into reservoirs in the underlying Tuscaloosa, but the two lower pay zones have been abandoned. Both the Eutaw and the upper Tuscaloosa reservoirs are highly porous and permeable sands lying between shales.

The trap at Heidelberg (Figure 22.6) is a complexly faulted anticline with a structural relief of at least 600 feet. It has been described as a "classic example of complex graben pattern with subsidiary compound–offset fault system" (Murray, 1966, p. 467). Accumulation has taken place in two areas separated by a nonproductive graben that cuts north-northwest across the highest part of the dome (Currie, 1956, p. 2). In both parts of the Heidelberg field oil has been impounded updip against graben-bordering faults. Salt has been penetrated at a depth of 9538 feet, or about 4570 feet below the Eutaw pay zone.

The Pickens-Gilbertown fault zone on the north flank of the Mississippi salt basin has created traps in both Mississippi and Alabama. The general rule in these fault fields is that accumulation occurs in the upthrown block, which is on the Gulf side of the fault plane. The oil has moved updip from the south until trapped by the fault barrier.

The *Pickens* field in northern Madison and northeastern Yazoo Counties is typical of this type of trap (Figure 22.7). Pickens lies about 25 miles south of the north edge of the Mississippi salt basin. The discovery well was completed in 1940, after core drill and seismic exploration. It pumped 240 barrels per day of 40° oil from 4890 feet; the pay zone is a sand in the Eutaw. Ten years later a one-well pool that produced for 3 years was discovered in a sand zone in the Glen Rose (Lower Cretaceous) at a depth of 7730 feet (Mississippi Geol. Soc., 1957, p. 92). The Eutaw pay is a fine- to medium-grained, very porous and permeable sandstone. The Glen Rose reservoir is a tight, silty sandstone. The Eutaw reservoir is closed laterally where the fault plane crosses southward plunging synclines; in other words the oil has accumulated where anticlinal noses abut against the fault plane.

Typical of the stratigraphic traps in the Eocene reservoir fields of southwestern Mississippi is the *Darrington* oil field of northern Wilkinson County (Gulmon et al., 1957). As can be seen in Figure 22.8, the Eocene sands shale out along a concave belt that crosses a southward plunging structural nose to create a perfect trap for oil moving northward up the dip of the plunging anticline. Production depths are around 7100 feet.

The Paleozoic reservoirs in the Black Warrior basin in northeastern Mississippi are principally sands in the Chester Series of Mississippian age. Although this province was opened up in 1926 with the finding of the *Amory* gas field (which is now being used for storage purposes) at a depth of 2400 feet, most of the discoveries since then have been small gas fields that were either shut in or soon exhausted. In most instances the trapping has been anticlinal, but in the *Muldon* field in Monroe County the gas accumulated against updip faults on a tilted horst block (Figure 22.9). The production limit downdip to the south in the gas-water contact. The displacement on the faults is several hundred feet (Mississippi Geol. Soc., 1957). The faults are normal, dipping away from the horst block at about 65°. The principal reservoir is the Sanders Sandstone of the Chester (Mississippian) Series (Knight, 1954). It is approximately 100 feet thick and is productive wherever drilled in the horst above the gas-water contact. In the first 5 years after discovery the Muldon field produced more than 5½ billion cubic feet of gas and 58,000 barrels of condensate from eight wells.

Currently, oil production in the entire Warrior basin is confined to the one-well *New Hope* field in northwest-central Monroe County, Mississippi, which has produced 7800 barrels of oil (Pike, 1968, p. 1701). The pay zone is Ordovician Knox Dolomite.

Future. Mississippi has much oil and gas yet to be discovered. No doubt many more hidden permeability traps, such as channel sands (Craft, 1966), remain to be found in the Eocene section of southwestern Mississippi. Exploration in the coastal counties and offshore has been inadequate. The water depth over a considerable area of the continental shelf is less than 600 feet (Figure III.1).

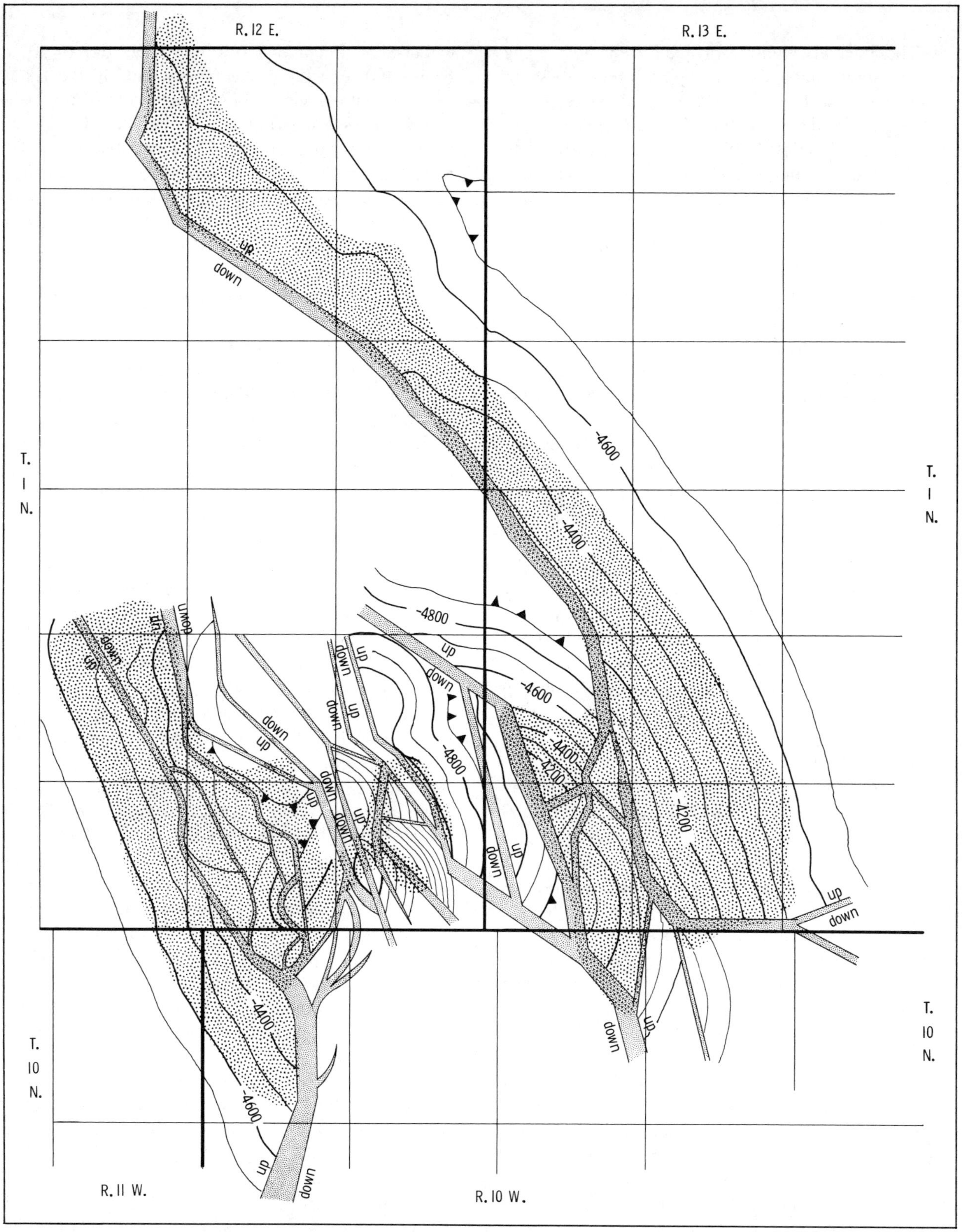

Figure 22.6 Heidelberg field, Jasper County, Mississippi. Structure map contoured on top of Eutaw Formation. Interval, 100 feet. Sections are 1 mile (1.61 km) on a side. Pattern designates two producing areas that constitute the field. *Courtesy Mississippi Geological Society (1957, p. 62).*

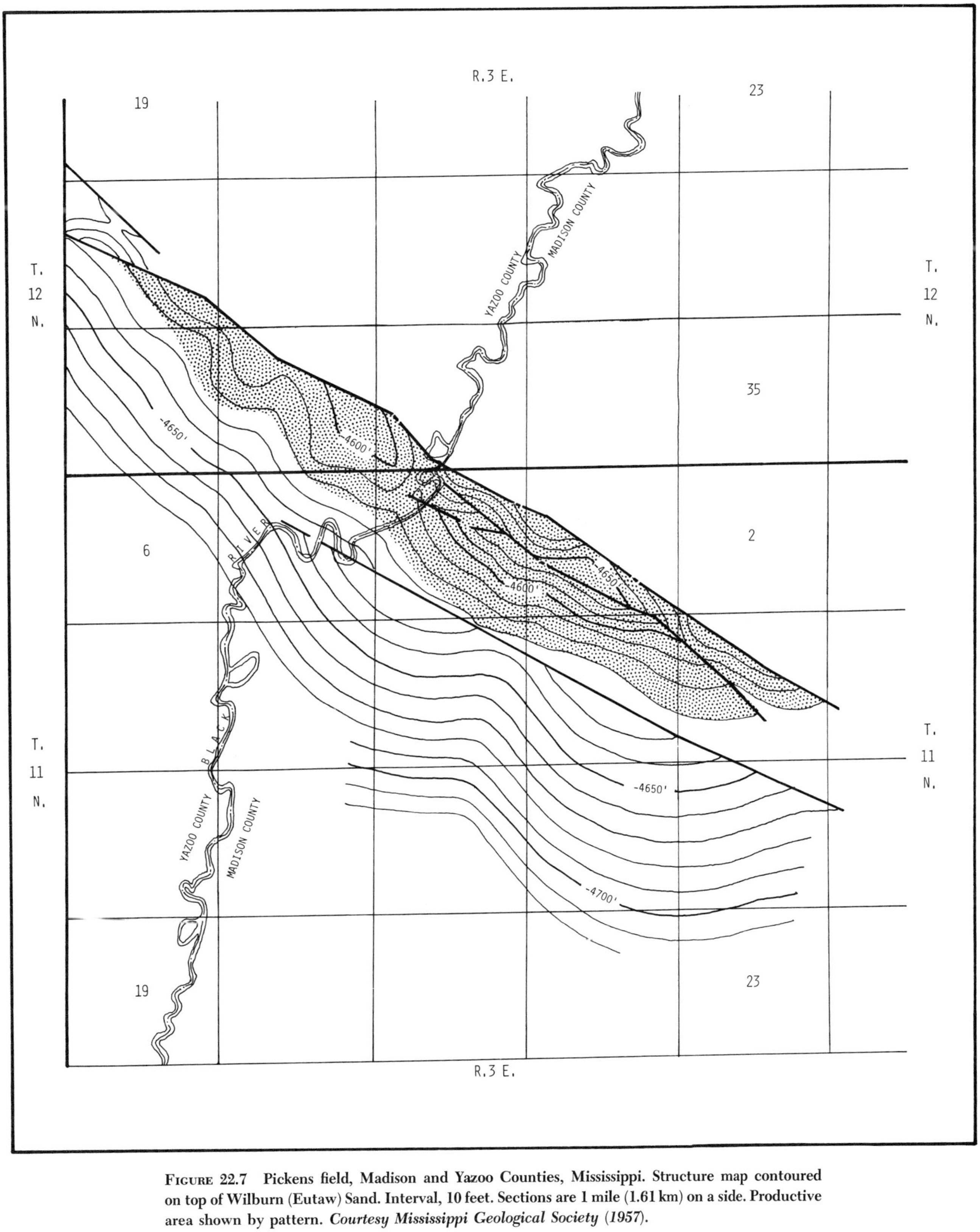

FIGURE 22.7 Pickens field, Madison and Yazoo Counties, Mississippi. Structure map contoured on top of Wilburn (Eutaw) Sand. Interval, 10 feet. Sections are 1 mile (1.61 km) on a side. Productive area shown by pattern. *Courtesy Mississippi Geological Society (1957).*

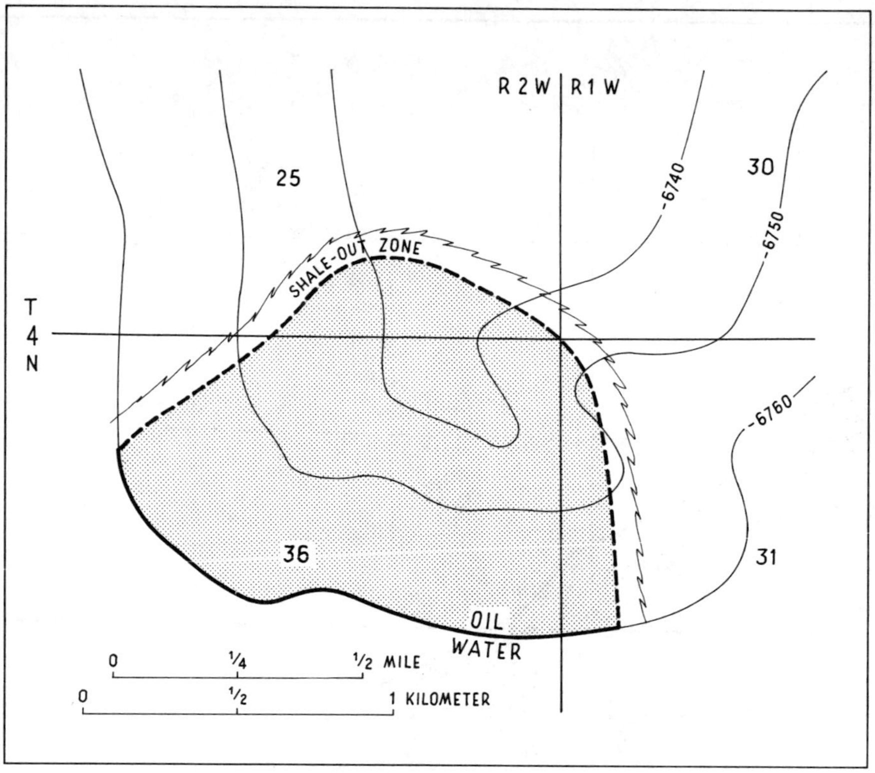

Figure 22.8 Darrington field, Wilkinson County, Mississippi. Structure contour and reservoir lithologic map. Contour interval, 10 feet. Datum: top of shale overlying reservoir sand. Productive area is stippled. *After G. W. Gulmon et al. (1957, Figure 2, p. 73). Courtesy the authors and the Oil and Gas Journal.*

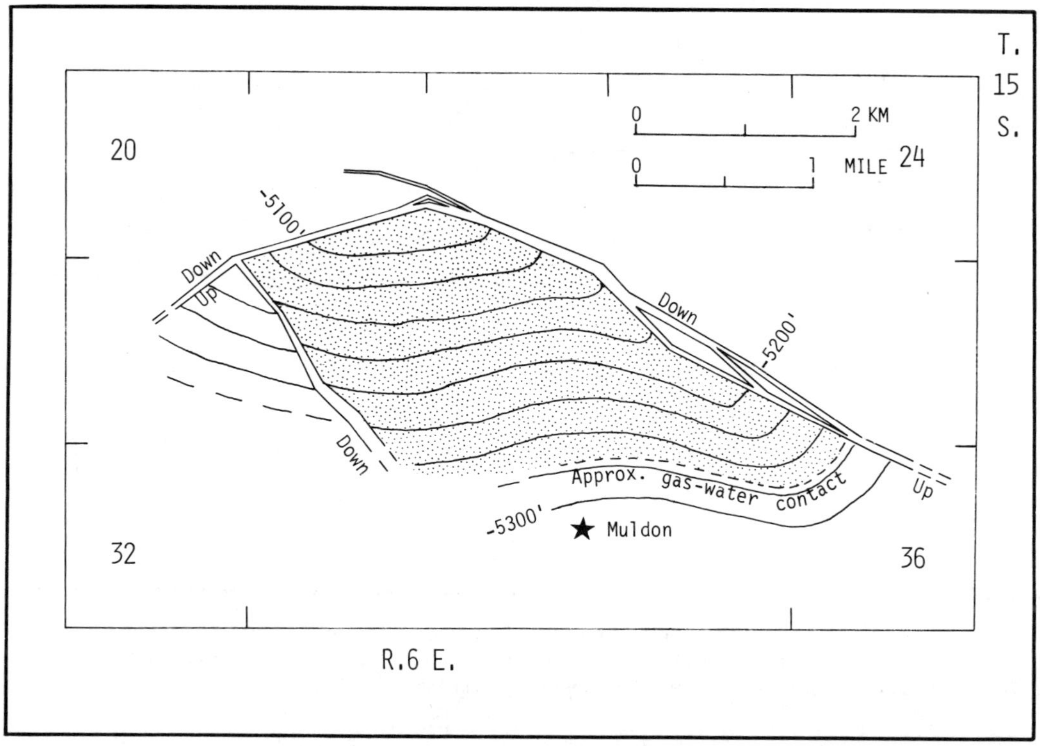

Figure 22.9 Muldon gas field, southwestern Monroe County in northeastern Mississippi. Gas-producing area stippled. Structure contours are based on the Sanders Sand. Interval, 25 feet. *Courtesy Mississippi Geological Society (1957, v. 1).*

Lower Tuscaloosa reservoirs still afford good hunting in the salt basin, as in Claiborne County (Dinkins, 1966). About 40 piercement salt domes in Mississippi have not yet been adequately explored for flank accumulations (Bike, 1960).

Possibilities for Jurassic production below accumulations in Cretaceous reservoirs were verified by the discovery of Cotton Valley oil at greater depth in the older Quitman field. The Smackover below the Cotton Valley has become a major objective. The Tallahalla Creek field in southeast-central Smith County was discovered by a successful test drilled into a Cotton Valley reservoir at 15,620 feet in 1966. The next year a 1 mile southeast extension was completed in the Smackover for 424 barrels a day of 45° oil at 16,470 feet. Similar strikes, this time only 3 months apart, took place in nearby Shongelo Creek field. An exploratory well drilled in 1965 on the Wiggins arch in southern Perry County reached a total depth of 20,130 feet in Louann Salt. It was completed for 1.4 million cubic feet daily of gas containing 76.8 per cent H_2S from the Smackover Formation at a depth of 20,000 feet (Guy, 1967, p. 1105-1106). This one-well field, the Black Creek, has a sulfur capability of about 45 long tons per day. It was shut in "while construction of a sulfur extraction plant was being planned" (U. S. Bur. Mines, Minerals Yearbook, v. 3, 1966, p. 448).

The presence of high-quality oil in the hitherto rarely penetrated Norphlet Formation presents an exciting prospect. This strike was made in the Pelahatchie field in eastern Rankin County, close to the center of the Mississippi salt basin. Pelahatchie is an older field that produces a modest amount of oil annually from four Lower Cretaceous reservoirs, Paluxy, Rodessa, Slizo, and Hosston sands. Trapping is anticlinal. The Pelahatchie deep discovery was completed in mid-1967 for 1440 barrels per day of 51° gravity oil from 17,152 to 17,160 feet. A north offset was also completed as a good producer from the same zone (Karges, 1968). The pay is a gray, fine- to medium-grained sand with good porosity and permeability. The oil column in the producing sand is 75 to 80 feet. Karges (1968, p. 101) suggests that this reservoir may be a basal Smackover sand. Regardless of whether it is Norphlet or overlying basal Smackover, an intensive search for hydrocarbons in this pay zone in the Gulf Coast region can be anticipated.

The abundant anticlines of the Warrior basin invite further exploration of the Paleozoic section present, especially the Ordovician and Cambrian rocks (Pike, 1968, p. 1701). Stratigraphic traps are also a possibility (Jones, 1960). So far, exploration in the Mississippi sector of this basin has resulted mostly in small gas fields and three minor oil fields. An exception is the Muldon field previously discussed, which has produced considerable gas along with some oil. It may be that, because of eometamorphism (early or incipient metamorphism) in the geologic past, many of the reservoirs are close to the petroleum "floor" or even below it.

REFERENCES CONSULTED IN PREPARING CHAPTER 22

Beebe, B. W., 1968c.
Beikman, Helen M., and Drakoulis, Sophie, 1958.
Bicker, Alvin R., Jr., 1966.
Bike, Peter B., 1960.
Bishop, William F., 1967.
Braunstein, Jules, 1958.
Browning, W. Fleming, 1968.
Craft, W. E., 1966.
Currie, John B., 1956.
Dinkins, Theo H., Jr., 1966.
Eisenstatt, P., 1960.
Gardner, Frank J., 1963b.
Gulmon, G. W., 1967.
Gulmon, G. W., Hansen, H. E., and Simms, B. L., 1957.
Guy, Samuel C., 1967.
Jones, Walter B., 1960.
Karges, H. E., 1968.
Knight, Wilbur H., 1954, 1955.
Mellen, F. F., 1947, 1966.
Mississippi Geological Society, 1952, 1957.
Murray, Grover E., 1966.
Philpott, T. H., 1960.
Pike, Stuart J., 1968.
St. John, F. B., Jr., and Lusk, Tracy W., 1959.
Shreveport Geological Society, 1946, 1947.
Thomas, E. P., 1950.
Womack, Robert Jr., 1950.

Chapter

23

ARKANSAS

In 1967 Arkansas contributed less than 0.7% of the nation's total oil production. The yield that year was 21.1 million barrels, an output that ranked this state in fifteenth place in the country. Arkansas also ranked eleventh in gas production in 1967.

Although the northwestern part of Arkansas began producing natural gas in 1902, it was not until 1921, when the El Dorado field (Figure 23.1) in the southern part of the state was discovered, that Arkansas joined the oil-producing states. A year later the prolific Smackover pool, in which some wells were credited with initial productions of as much as 70,000 barrels per day, was found. The resulting flood of oil pushed Arkansas into fourth place among oil states in 1925, the peak year to date, when the production exceeded 77 million barrels. Five new fields were discovered in 1967, of which two are gas fields in the once neglected Arkoma basin in the northwestern part of the state, and the other three are oil fields in the major producing belt of southern Arkansas (Morrow et al., 1968).

Geology and Tectonic Features. Geologically Arkansas can be divided by a diagonal line extending from northeast to southwest across the state (Figure 23.2). The northwest segment (covering about one-half of the state) is known as the Interior Highland region, or the Paleozoic outcrop area of the state. Southern, eastern, and northeastern Arkansas is the area of outcrop of Mesozoic and Cenozoic sediments deposited in the Mississippi embayment of the Gulf Coastal Plain. These younger rocks lap upon the Paleozoic rocks and unconformably overlie them. Cenozoic (Tertiary and Quaternary) sediments are much more extensive at the surface than the Mesozoic rocks, which crop out in a wedge-shaped area in southwestern Arkansas.

The Paleozoic province in northwestern Arkansas has three major tectonic divisions. These are, from north to south, the Ozark uplift, the Arkoma basin, and the Ouachita Mountains. The center of the Ozark uplift is

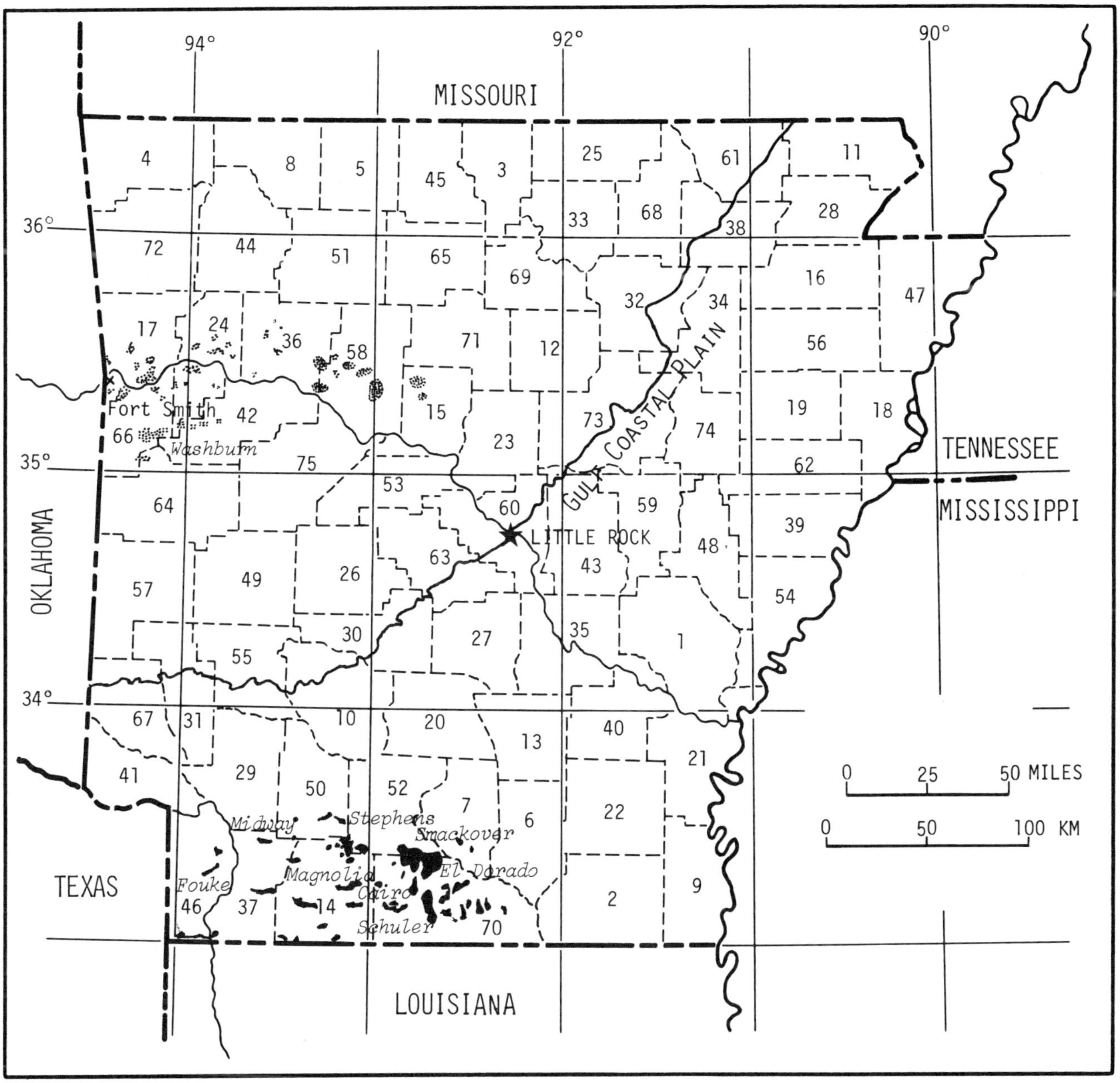

Figure 23.1 Arkansas. Index map, with oil and gas fields. County Key:

1 Arkansas
2 Ashley
3 Baxter
4 Benton
5 Boone
6 Bradley
7 Calhoun
8 Carroll
9 Chicot
10 Clark
11 Clay
12 Cleburne
13 Cleveland
14 Columbia
15 Conway
16 Craighead
17 Crawford
18 Crittenden
19 Cross
20 Dallas
21 Desha
22 Drew
23 Faulkner
24 Franklin
25 Fulton
26 Garland
27 Grant
28 Greene
29 Hempstead
30 Hot Spring
31 Howard
32 Independence
33 Izard
34 Jackson
35 Jefferson
36 Johnson
37 Lafayette
38 Lawrence
39 Lee
40 Lincoln
41 Little River
42 Logan
43 Lonoke
44 Madison
45 Marion
46 Miller
47 Mississippi
48 Monroe
49 Montgomery
50 Nevada
51 Newton
52 Ouachita
53 Perry
54 Phillips
55 Pike
56 Poinsett
57 Polk
58 Pope
59 Prairie
60 Pulaski
61 Randolph
62 St. Francis
63 Saline
64 Scott
65 Searcy
66 Sebastian
67 Sevier
68 Sharp
69 Stone
70 Union
71 Van Buren
72 Washington
73 White
74 Woodruff
75 Yell

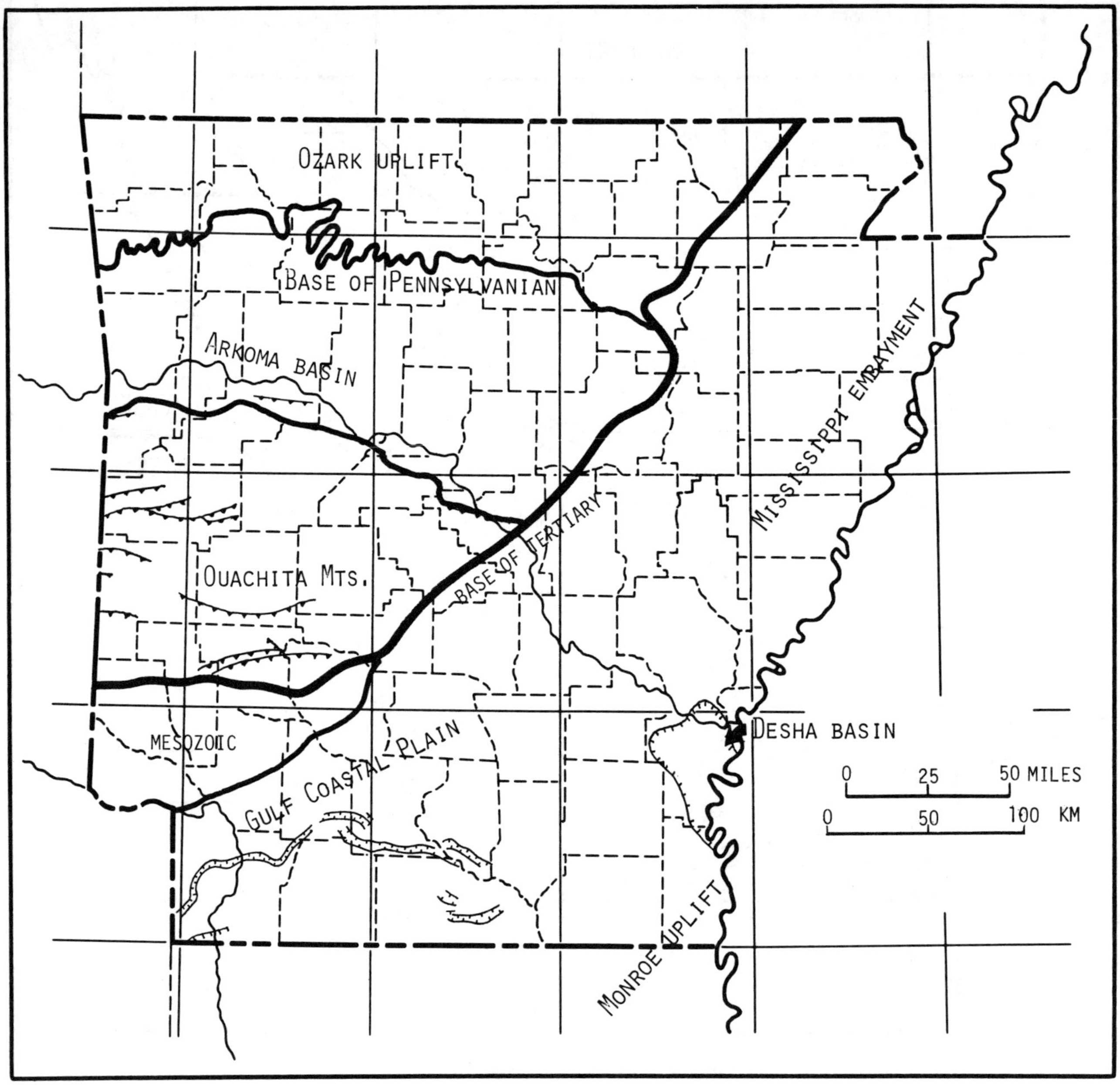

Figure 23.2 Arkansas. Major tectonic features. Shows principal fault zones only.

in southern Missouri. The pre-Pennsylvanian rocks that crop out in mountainous northern Arkansas constitute the south flank of this uplift.

Extending from west to east across the middle of the Paleozoic outcrop area of Arkansas is the Arkoma basin, a wide geosynclinal belt of Pennsylvanian rocks containing commercial coal beds. The Arkoma basin is both a structural downwarp and a topographically low area that is the valley of the eastward flowing Arkansas River. The structure within the basin is similar to that in the folded Appalachians to the east. The folding becomes more intense and the thrust faults grow more numerous from north to south across the basin (Thralls et al., 1962). Eastward the folded Pennsylvanian sediments of the Arkoma basin disappear beneath the Tertiary coastal plain sediments to the northeast of Little Rock, and westerly the basin continues across the state boundary into eastern Oklahoma.

The Ouachita Mountain area consists of intensely folded and thrust-faulted Carboniferous sediments with a core of earlier Paleozoic rocks ranging in age from Devonian to Cambro-Ordovician. The Ouachita belt also is overlapped to the east by the Tertiary sediments of the coastal plain; to the south in Arkansas and eastern

Oklahoma the Ouachita structural complex disappears beneath coastal plain sediments of Mesozoic age.

The regional structure of the coastal plain part of Arkansas, approximately one-half of the state, is less complicated than that of the Paleozoic area, but by no means monotonous. In the Mesozoic area the regional dip is to the south, whereas the Cenozoic rocks generally dip southeasterly toward the axis of the Mississippi embayment. This axis does not vary far in position from the path of the Mississippi River, which is also the eastern boundary of Arkansas. In southeastern Arkansas, however, the dips are northeasterly into the Desha basin (Figure 23.2), which extends northwestward into Arkansas from Mississippi. Southeastern Arkansas is also on the north flank of the Monroe uplift, the highest point of which is in northeast Louisiana. The coastal plain sediments in the southwestern corner of the state are synclinal, with the axis plunging westerly into the East Texas embayment. The south flank of this syncline is also the north flank of the Sabine uplift, which peaks in the northwestern corner of Louisiana. The South Arkansas fault zone lies on the north flank of the syncline.

ARKOMA BASIN

Arkansas shares the Arkoma basin with Oklahoma (see p. 133). Although this area had been the scene of sporadic discoveries of dry natural gas since the turn of the century, it was not until 1958, when five fields were discovered in Arkansas, followed by six fields in 1959, that its gas potential became widely recognized. By 1962 the total number of fields in the Arkansas part of the basin exceeded 60. So far, the Arkoma basin is exclusively a gas-producing province, and the high carbon ratios (82 to 87) of the coals in western Arkansas (Croneis, 1935) are regarded by some geologists as a deterrent to the occurrence of oil in commercial quantities.

The gas fields in the Arkansas sector of the basin lie mainly in an east-west belt 15 to 20 miles wide that crosses parts of seven counties in northwestern Arkansas (Figure 23.1). This belt is bordered on the north by the rim of the basin and the Ozark foothills, and on the south by the basin trough (Branan, 1968, p. 1620–1621). The fields lie entirely within the Arkansas River valley on both sides of the river. The belt extends westward across the Oklahoma line at Forth Smith.

Stratigraphy. The subsurface section so far explored in the Arkansas part of the Arkoma basin ranges mainly from the Hartshorne Sandstone of Pennsylvanian (Desmoinesian) age, down through the Mississippian, Devonian, and Silurian, to the Cotter Dolomite of Ordovician (Canadian) age (Figure 23.3). This section has been described in part by Ballard (1957) and in part by Bartlett (1962). So far, most of the gas reservoirs are sandstones in the Atokan Series, where they appear to be most abundant in the middle and upper parts of the series. Since the Atoka increases in thickness to the south, its maximum may be as great as 19,000 feet or more near the southern boundary of the Arkoma basin. Various local names have been given to the different sandstone reservoirs within the Atoka.

Stratigraphically deeper gas reservoirs occur in limy and sandy sections of the Bloyd and Hale Formations of lowest Pennsylvanian (Morrowan) age. Considerable interest was generated in the latter part of 1962 by the discovery of gas in commercial amounts in several exploratory wells in Siluro-Devonian rocks tentatively correlated with the Hunton Limestone of Oklahoma.

Traps. The gas in northwestern Arkansas has been produced from stratigraphic traps as well as structural traps. Most of the latter are anticlines, some of them of considerable size, ranging from 20 to 30 miles long and up to 12 miles wide (Planalp, 1961). Even larger is the Washburn anticline, which is close to 40 miles in length (Figure 23.4). Three gas fields have been found so far along the axis of this anticline, and it has been suggested that these will eventually be connected into one large gas field (Bartlett, 1962, p. 141; 1968, p. 1658). A cross-section of the anticline (Figure 23.5) shows that the many reservoir sands of the Atoka are cut by thrust faults as they arch over the top of the structure. However, accumulation appears to be principally the result of the anticlinal folding. The closure on the Washburn anticline is 2500 to 3000 feet. It lies in the southern part of the basin, where the folding is more intense than to the north.

In spite of the intensive exploration program in the Arkoma basin during the last few years, there are still opportunities for discovery. At the present time, activity is confined largely to the drilling of hitherto untested or inadequately tested anticlines and the finding of new reservoirs in a section that is loaded with permeable sands in the Pennsylvanian rocks. For the future there are almost infinite opportunities to explore the pre-Pennsylvanian section (Bartlett, 1968). A gas field was discovered near the Missouri line in Benton County at a depth of less than 400 feet in the Everton Formation of Ordovician age in 1967 (Morrow et al., 1968, p. 1007). It is also important that the search for traps other than anticlines be intensified. Lenticular sands and faults both offer additional possibilities in this regard.

ARKANSAS COASTAL PLAIN

The coastal plain half of Arkansas is a part of the widespread Gulf Coast province of Texas, Louisiana, and

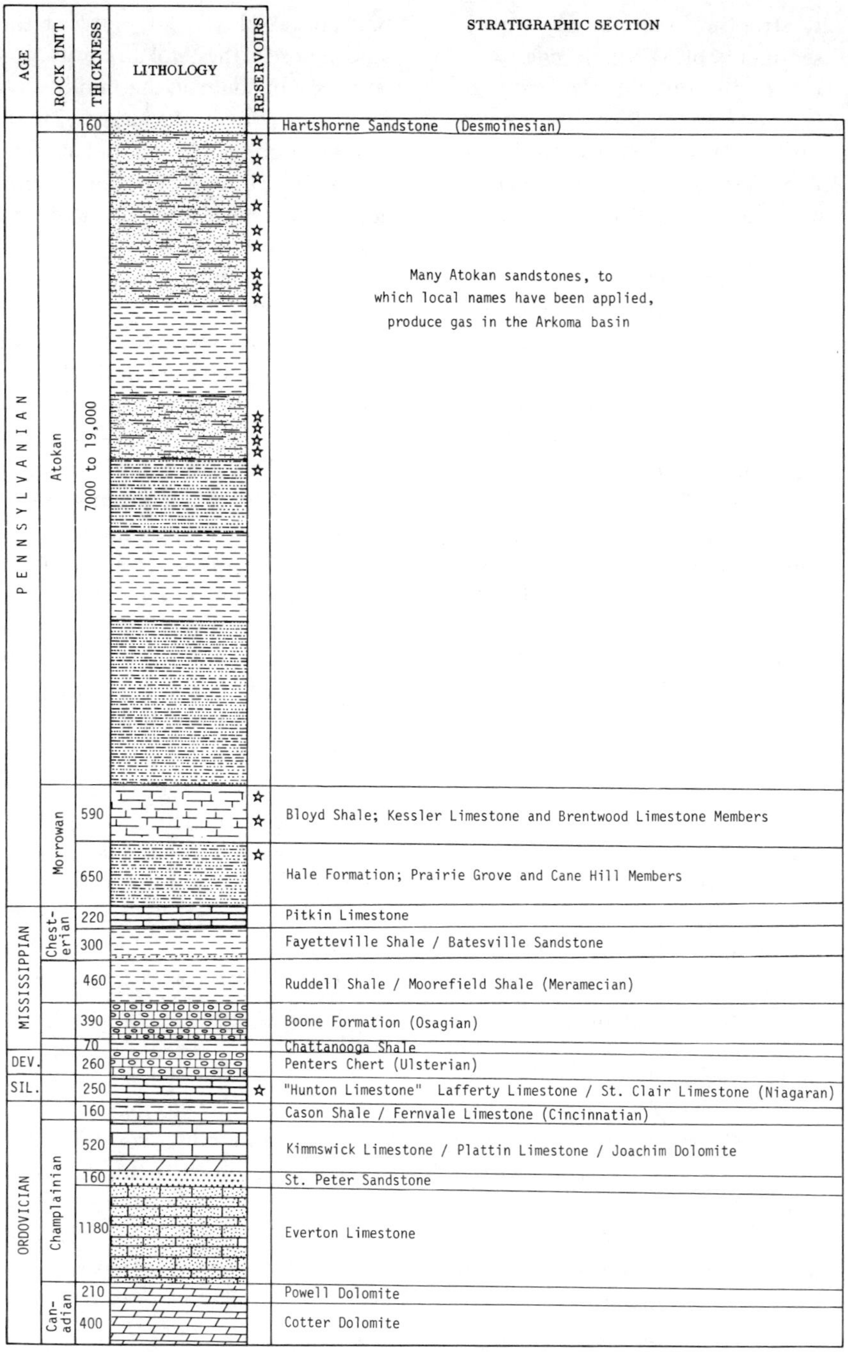

FIGURE 23.3 Stratigraphic section. Arkoma basin (Arkansas-Oklahoma), showing gas reservoirs. *After William Wayne Ballard (1957) and Charles S. Bartlett (1962).*

states to the eastward. Tectonically and stratigraphically, southern Arkansas belongs with northern Louisiana and northeastern Texas. This area contains prolific Mesozoic reservoirs.

Most of the oil fields of Arkansas (Figure 23.1) lie in the southern tier of counties, extending from Miller on the Texas line across Lafayette, Columbia, and Union Counties. The district extends far enough north to include Hempstead, southern Nevada, Ouachita, Calhoun, and Bradley Counties in the second tier north of the Louisiana border.

STRATIGRAPHY (Hundley and Shute, 1963, p. 1101; Bank of the Southwest, 1956; Shreveport Geol. Soc., 1953). The accompanying table gives the stratigraphic column so far penetrated by the drill in seeking oil and gas deposits on the coastal plain of Arkansas. A lithologic

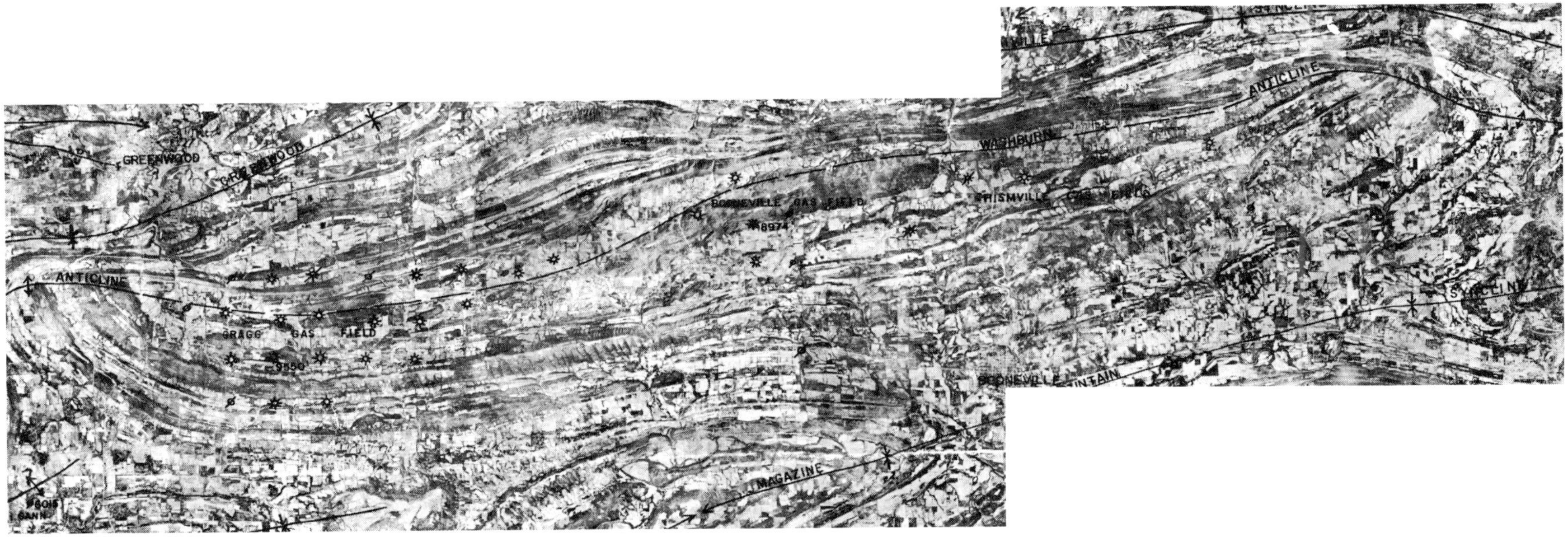

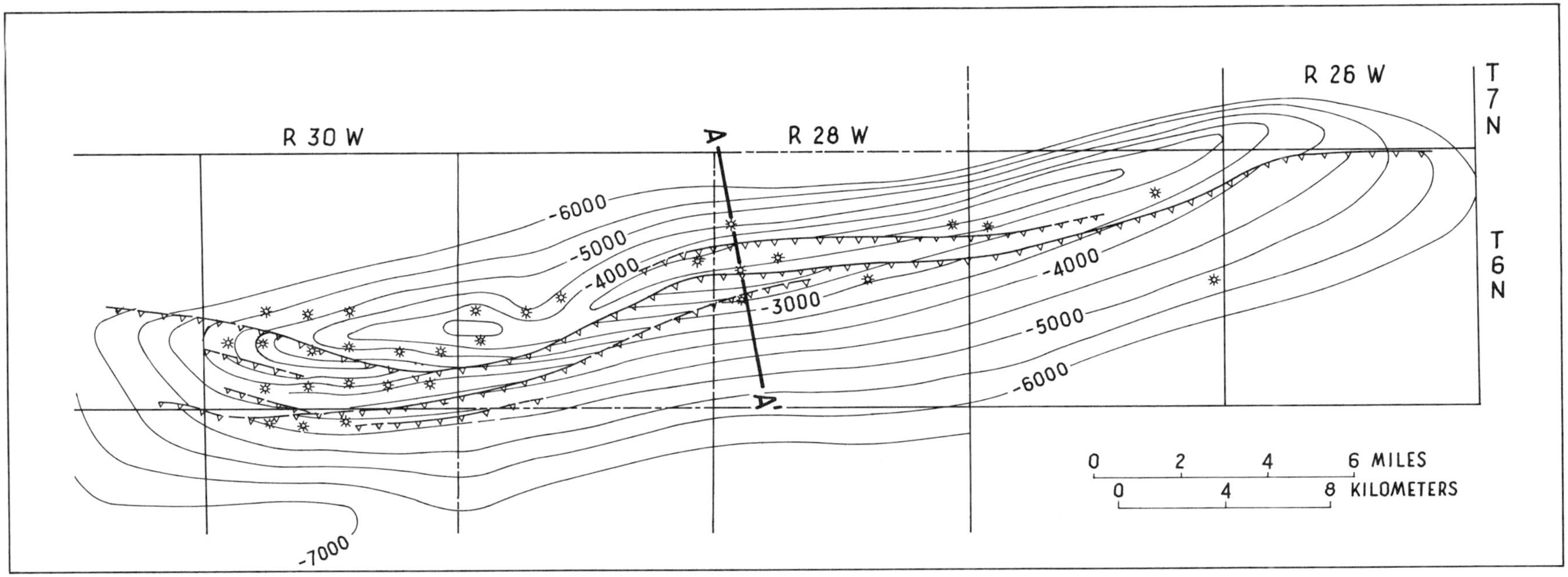

FIGURE 23.4 Upper: Washburn anticline, Sebastian and Logan Counties, Arkansas. Air photo by Edgar Tobin Aerial Surveys. *Courtesy Charles S. Bartlett, Jr., and the Oil and Gas Journal.* Lower: same area and scale as air photograph above. Subsurface structure map, based on Upper Borum zone (see Figure 23.5). Contour interval, 500 feet. Shows major overthrust faults. *Courtesy Charles S. Bartlett, Jr., and the Oil and Gas Journal.*

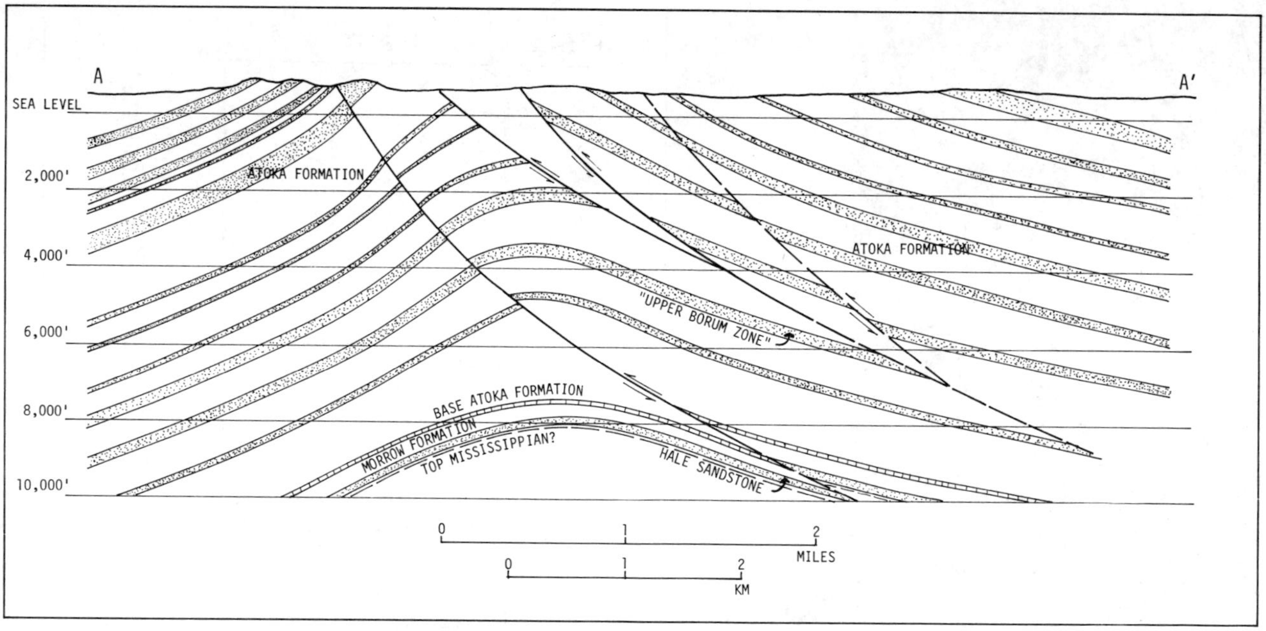

Figure 23.5 Cross section along line A—A′ on Figure 23.4 (lower). Shows gas sands (stippled) of the Atoka (Pennsylvanian). Intervening strata are shales. *Courtesy Charles S. Bartlett, Jr.*

log for the south Arkansas–north Louisiana district is given in Chapter 24 on Louisiana (Figure 24.4).

Generalized Geologic Column

South Arkansas

(From Hundley and Shute, 1963; pre-Smackover section from Scott, Hayes, and Fietz, 1961, and Bishop, 1967)

- Tertiary
  - Eocene
    - Claiborne
    - Wilcox
    - Midway
- Upper Cretaceous
  - Navarro
    - Arkadelphia
    - Nacatoch
  - Taylor
    - Saratoga
    - Marlbrook
    - Annona
    - Ozan (Meakin and Graves)
  - Austin
    - Brownstown
    - Tokio (Blossom)
  - Eagle Ford
  - Tuscaloosa
- Lower Cretaceous (Comanchean)
  - Washita–Fredericksburg (Goodland)
  - Trinity
    - Paluxy
    - Mooringsport
    - Ferry Lake Anhydrite
    - Rodessa (Kilpatrick)
    - James
    - Pine Island
  - Nueva Leon
    - Sligo (Pettet)
    - Hosston
- Jurassic
  - Cotton Valley
    - Schuler
    - Bossier (Jones)
  - Haynesville
    - Buckner
  - Smackover (Reynolds)
  - Norphlet
  - Louann Salt
  - Werner Anhydrite
- Triassic
  - Eagle Mills
- Pennsylvanian
  - Morehouse (North Louisiana)
  - Ouachita facies

The bedrock section involved extends downward from the Eocene through the Cretaceous and Jurassic and into the Pennsylvanian. So far, the reservoir rocks found to be productive in southern Arkansas are confined to the Cretaceous and Jurassic periods. The most prolific in terms of cumulative production have been sands in the Ozan Formation of Late Cretaceous age. Among these reservoirs the Graves sand has been outstanding. The Reynolds Oölite Member of the Jurassic Smackover Formation is second in yield. The Upper Cretaceous Nacatoch Formation ranks third. Other important reservoirs are the Cotton Valley (including the Jones), Tokio (especially the Blossom sand), James, and Rodessa. The

Cotton Valley is Jurassic; the Hosston, James, and Rodessa are Early Cretaceous, and the Tokio is Late Cretaceous in age. Production has also been obtained in southern Arkansas from the Upper Cretaceous Tuscaloosa Formation and from the Lower Cretaceous Goodland, Paluxy, and Sligo reservoirs. All of these are sandstones except the Smackover, which is limestone, and the Sligo, James, Rodessa, and Goodland, which are sandy limestones.

The prolific Smackover has at least six types of porosity as described by Vestal (1950, p. 9). These are (1) inter-oölite porosity in the Reynolds zone, (2) vuggy porosity where the oölites have been dissolved, (3) solution channel porosity, (4) porosity in cavernous coral reef material, (5) interstitial porosity in granular limestone, and (6) fracture porosity. The Reynolds oölitic zone has produced most of the Smackover oil and gas. Unfortunately its porosity lenses out to the north and east of the southwestern part of the state.

Traps and Selected Fields. Although the discovery field for Arkansas, the El Dorado, is a stratigraphic trap on a structural terrace, most of the subsequent discoveries have been anticlinal accumulations, but with distribution of the hydrocarbons within the trap controlled by varying permeability in some fields. Second to the anticlinal trap in importance is the fault trap.

The Smackover field stands at the top of the list in cumulative production; by the end of 1967 it had produced nearly ½ billion barrels of oil. In second, third, and fourth places in cumulative production were the state's remaining giants, Magnolia, Schuler, and El Dorado. During 1967 Magnolia was in first place and Smackover second.

The *Smackover* field (Gibbon, 1952) was discovered on April 22, 1922, when a well drilled "without the benefit of geology" blew-in with volcanic violence, creating a 400-foot-deep crater still visited by the curious. The Smackover field produces from a number of sands in the Upper Cretaceous, the uppermost of which is the Nacatoch, which lies at an average depth of 2200 feet. The oil-bearing sands occur as lenses in a section 200 feet thick. In 1939 this field had a second boom after the discovery of oil in the much deeper Jurassic limestone that became known as the "Smackover lime." This reservoir subsequently was found to be present and productive in other fields in southern Arkansas and northern Louisiana. The major structural features at Smackover are the Norphlet dome on the east, separated by a syncline from the Louann terrace on the west, and a downthrown block to the north. The trapping in the Upper Cretaceous sands at the east end of the field is due to the closure on the dome, but at the western edge trapping in the Meakin reservoir appears to result from the shaling of the sand. Even in the structurally closed part of the field, the distribution of producing wells is controlled to a considerable extent by the porosity and permeability of the reservoirs.

The *Magnolia* field (Shreveport Geol. Soc., 1946), was discovered in 1938 by a well located after surface geology and seismograph work had indicated the presence

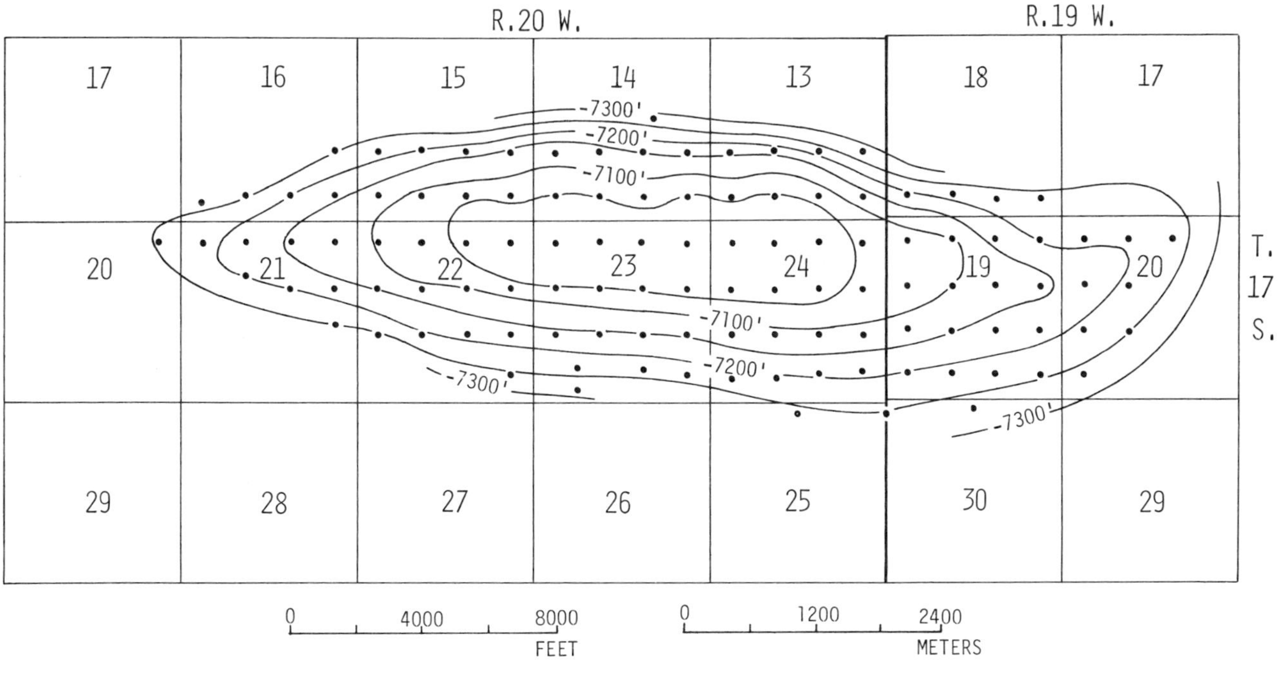

Figure 23.6 **Magnolia field, Columbia County, Arkansas. Structure contour map based on top of the Smackover Limestone. Contour interval, 50 feet.** ***Courtesy Shreveport Geological Society (1946, p. 38).***

of a large anticline. The field lies a short distance east of Magnolia, the county seat of Columbia County. It is 6 miles long and 1½ miles wide. The long axis trends almost due east. The trap is an elliptical, fairly symmetrical anticline (Figure 23.6) with nearly 300 feet of closure above the oil-water interface. The principal reservoir is the Reynolds Oölite Member of the Jurassic Smackover Formation (Figure 23.7), topped about 7350 feet beneath the surface.

A more recent discovery in southern Arkansas is the *Schuler* field (Shreveport Geol. Soc., 1946; Weeks and Alexander, 1942) in Union County, about 18 miles southwest of El Dorado, Arkansas. The principal reservoir is the Jones sand at the base of the Cotton Valley Formation (Jurassic), which produces at depths between 7500 and 7650 feet. The discovery well, which was drilled in 1937, struck a heavy flow of oil in the Jones. Trapping is anticlinal, but lenticularity of the sand is also a factor.

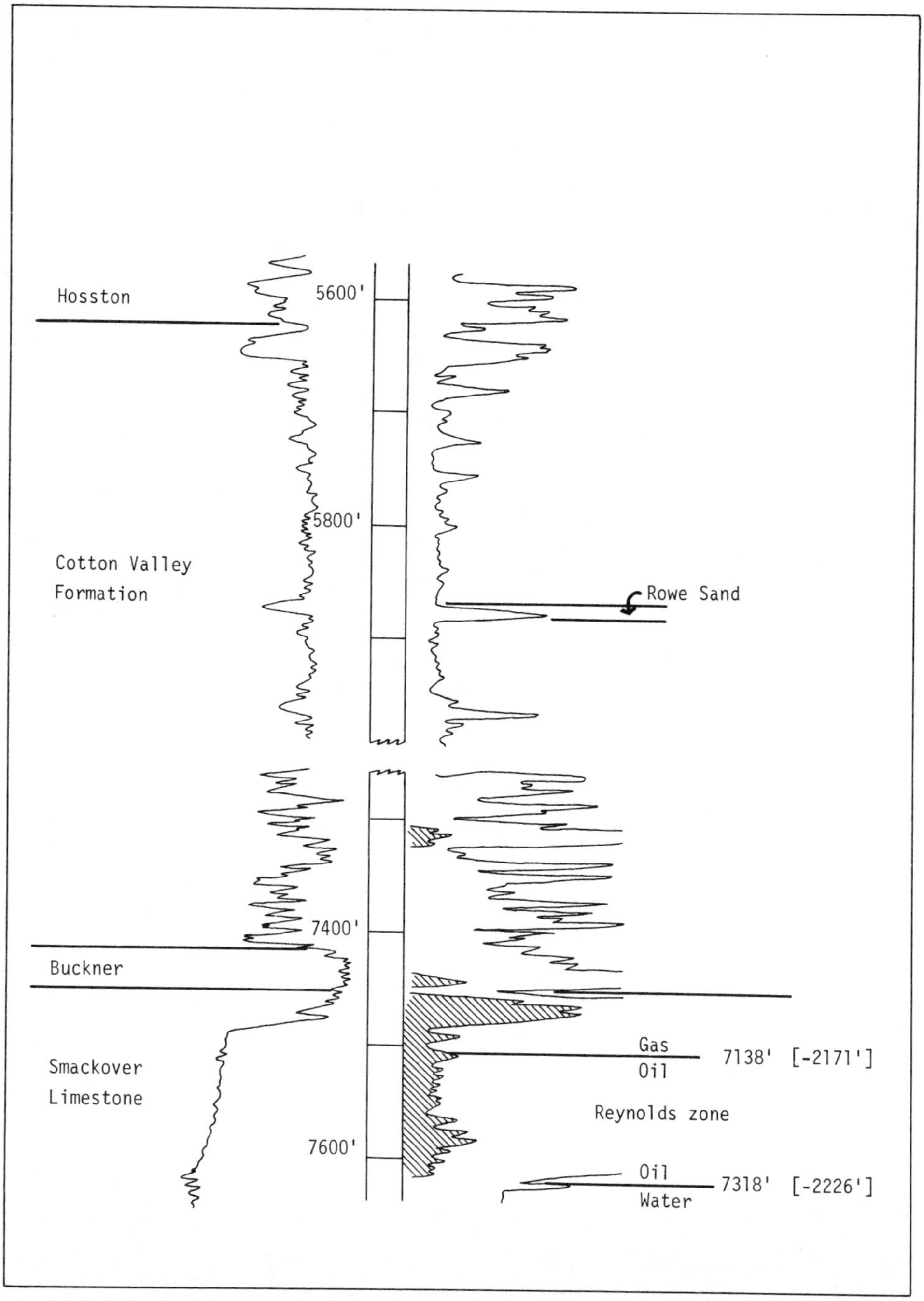

FIGURE 23.7 **Magnolia field, Columbia County, Arkansas. Type electric log showing gas-oil-water distribution in Reynolds reservoir.** ***Courtesy Shreveport Geological Society (1946).***

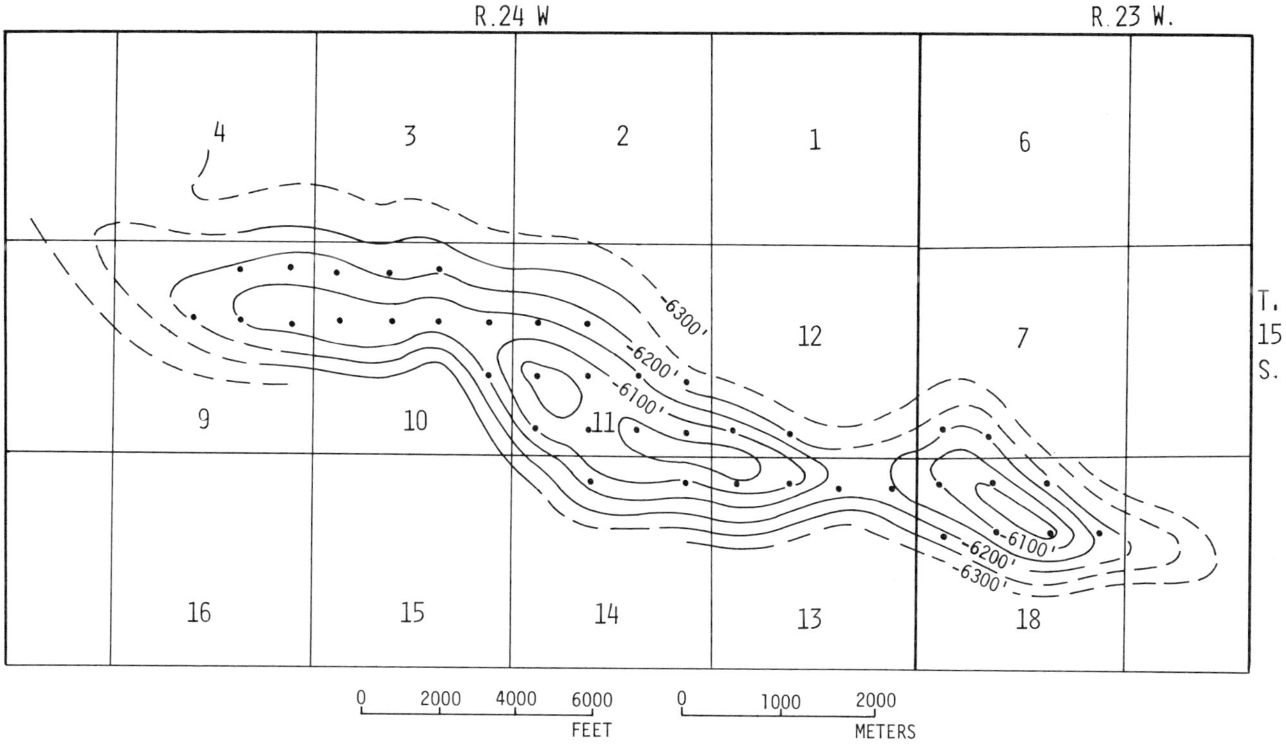

Figure 23.8 **Midway field, Lafayette County, Arkansas. Structure contour map based on top of the Smackover Limestone, Contour interval, 50 feet.** ***Courtesy Shreveport Geological Society (1947, p. 346).***

The subsurface contours show the presence of two domes on the Schuler anticline; oil fills not only both domes but the intervening saddle as well. Repressuring of the Jones reservoir by both gas and water has proved highly successful.

The *Midway* field (Shreveport Geol. Soc., 1947), another anticlinal trap, is in northern Lafayette County in southwestern Arkansas. The trap is an elongate anticline with an east-west trend (Figure 23.8). This structural ridge contains several nodes, but the oil-water interface lies below the lowest of the intervening saddles. The presence of the Midway anticline was suspected from subsurface mapping and confirmed by seismograph surveys. The producing formation is the Reynolds Oölite Member of the Smackover Limestone. Both the oil and the gas produced in this field are sweet; sour gas and oil are the general rule in the other fields in this area. The Midway field was discovered on January 1, 1942, and water injection for pressure maintenance began less than 5 months later.

The relatively small *Cairo* field lies 2 miles northeast of the Schuler field. The oil here is confined to the top and the southwest flank of a faulted elliptical dome. The long axis of the dome has a southwest trend that is at right angles to the axis of the Schuler anticline. The reason for the eccentric distribution of oil on the Cairo anticline is the relatively steep southwest dip of the oil-water interface. Goebel (1950) suggests that this slope toward the neighboring field is due to the heavy withdrawals of gas-distillate, oil, and salt water from the Smackover Formation in the 11-year-older Schuler field.

The *Stephens* field (Shreveport Geol. Soc., 1947) overlaps corners of three counties in the north-central part of the south Arkansas oil-producing district (Figure 23.1), but most of the productive area is in northeastern Columbia County. This field lies immediately to the south of, and is bordered by, the south Arkansas fault zone (Figure 23.2). The structural contours on the top of the James Limestone of Early Cretaceous age, which is one of the three reservoirs in this field, are shown in Figure 23.9. The trap here is a dome closed on the northeast flank by a downfaulted block. The uppermost reservoir is the basal sand of the Upper Cretaceous, which is productive over approximately 700 acres. This sand is unusual in that it is coarse and contains a volcanic matrix. The average depth is 2500 feet. The Lower Cretaceous reservoirs, which lie at depths of 2600 to 2900 feet, have a productive area of 1100 acres. These are in the lower Glen Rose Formation and include the James Limestone and a series of lenticular sands.

The *Fouke* field (Carter Oil Co., 1957) lies athwart the south Arkansas fault zone in Miller County in the south-

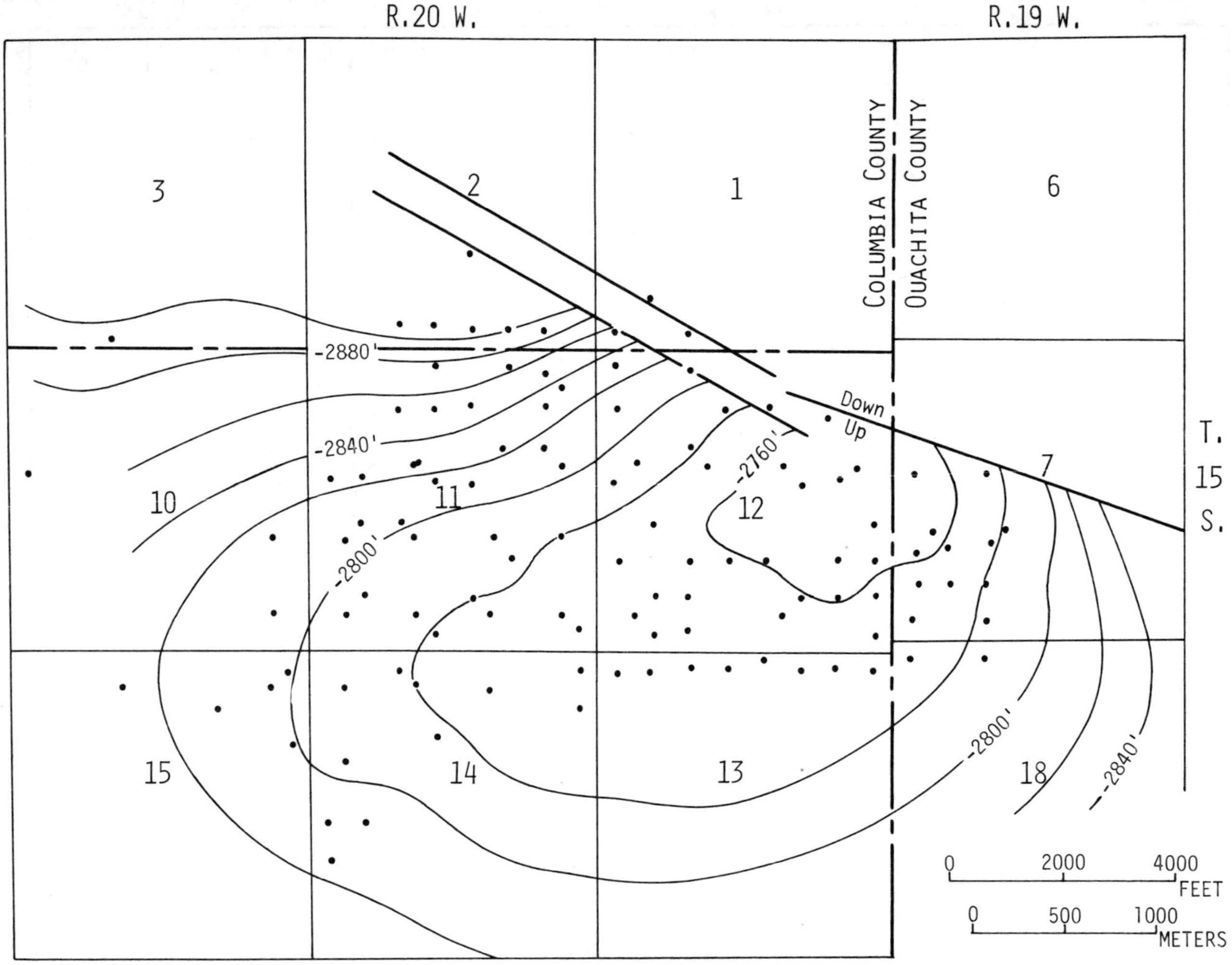

FIGURE 23.9 Stephens field, Columbia and Ouachita Counties, Arkansas. Structure contour map based on top of James Limestone. Contour interval, 20 feet. *Courtesy Shreveport Geological Society (1947, p. 358).*

western corner of the state. As can be seen from the cross section (Figure 23.10), this field has multiple reservoirs and two types of traps. The greater part of the oil produced has been impounded against a northeast-trending fault zone that cuts across a southeastward plunging anticlinal nose. Minor production is obtained in an anticlinal trap in the downfaulted block northwest of the fault plane. The Fouke field is 5 miles long and 1 mile wide. It has had a series of birthdays, starting with the finding of oil in the Paluxy Formation in 1940. The deep Smackover pool was discovered in 1955. This reservoir is at a depth of about 9000 feet and contains gas and condensate.

FUTURE. "The best Smackover hit in years" (Oil and Gas Jour., Oct. 21, 1968, p. 123) was an oil, gas, and condensate discovery well drilled near the southwestern corner of Arkansas in 1968. It gave further encouragement to an already intense Jurassic exploration program extending from Mississippi to Texas.

"Future major new oil and gas production in the south Arkansas–north Louisiana region has to come either from areas not now producing, or from formations deeper than those now normally drilled" (Chapman, 1963, p. 1992). The most likely deeper possibility in southern Arkansas, according to Chapman, is a marine sequence of Pennsylvanian age discovered by a wildcat drilled in Morehouse Parish, Louisiana, and as yet unexplored (Figure 24.1). The well penetrated 1180 feet of gently dipping dark gray marine shales ("Morehouse" Formation). "If the buried Arkansas–Louisiana high was the site of erosion during the time of deposition of the Morehouse beds, nearshore and beach deposits should exist south of it, equivalent in age to the Morehouse sediments" (Chapman, 1963, p. 1993).

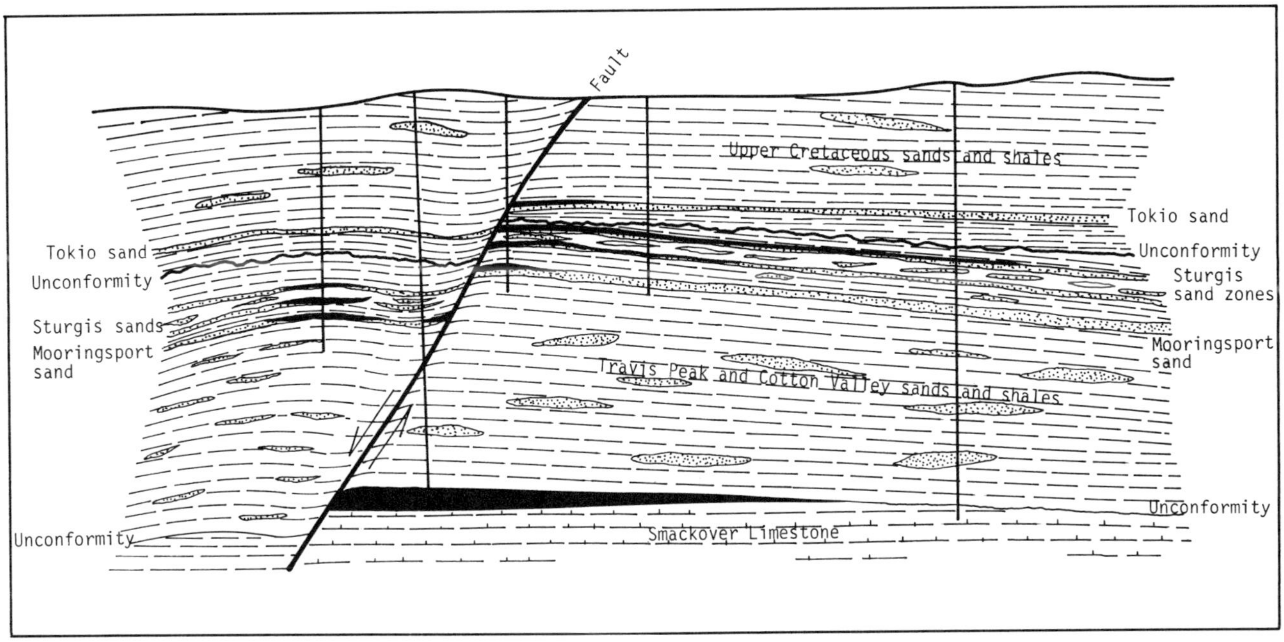

FIGURE 23.10 Fouke field, Miller County, Arkansas. Cross section, showing trapped oil (in solid black) on both sides of fault plane. Width of section, about 2 miles. Deepest well (on right) has total depth of 9204 feet. *Courtesy Humble Oil and Refining Company (Carter Oil Company, 1957, p. 16).*

Eastern Arkansas to the east and northeast of the southwestern Arkansas producing district covers a large area as yet unproductive (Caplan, 1954a, 1954b, 1964; Scull, 1954). This area lies within the Mississippi basin sector of the Gulf Coastal Plain.

Since the Mississippi embayment did not develop until after Lower Cretaceous sedimentation, Jurassic and Comanchean rocks are confined to southernmost Arkansas (Murray, 1961). However, Upper Cretaceous deposits, consisting of normal marine sands and shales, and overlying Tertiary sediments are present in eastern Arkansas. The Cenozoic Desha basin contains a thick Tertiary section, including the Wilcox (Eocene) sand, which is productive to the south in both Louisiana and Mississippi. Unfortunately exploration along the axis of the Mississippi embayment is handicapped by the masking effect of thick Quaternary alluvium deposited by Old Man River.

To the north of the Desha basin folded Paleozoic sediments, which are extensions of the Arkoma basin and Ouachita Mountain rocks on the west and the Warrior basin on the east, underlie the younger sediments in the embayment. Potential reservoir rocks deposited during this era lie within reasonable drilling distance in east-central and northeastern Arkansas. Because the correlative Paleozoic rocks outside the embayment are essentially dry gas producers in Arkansas, it is expected that gas is more likely to be discovered within Paleozoic rocks in the embayment than oil. However, the possibility of the presence of the latter in commercial amounts cannot be discounted.

REFERENCES CONSULTED IN PREPARING CHAPTER 23

Ballard, William Wayne, 1957.
Bank of the Southwest, 1956.
Bartlett, Charles S., Jr., 1962, 1968.
Bishop, William F., 1967.
Branan, C. B., Jr., 1968.
Caplan, William M., 1954a, 1954b, 1964.
Carter Oil Company, 1957.
Chapman, John J.., 1963.
Croneis, Carey, 1935.
Gibbon, Anthony, 1952.
Goebel, Lawrence A., 1950.
Hartman, F. H., 1968.
Hundley, Harland B., and Shute, George C., 1963.
Morrow, Ernest H., Skinner, John H., and Hinyard, Paul, 1968.
Murray, Grover, 1961.
Planalp, Roger N., 1961, 1968.
Scott, K. R., Hayes, W. E., and Fietz, R. P., 1961.
Scull, B. J., 1954.
Shreveport Geological Society, 1945, 1946–1947, 1953.
Thralls, Hugh M., Robb, George L., and Schisler, A. J., 1962.
Vestal, Jack H., 1950.
Weeks, Warren B., and Alexander, Clyde W., 1942.

Chapter

24

LOUISIANA

In 1968 the state of Louisiana (Figure 24.1), was credited with more than 830 million barrels of oil, which was $24\frac{1}{2}$ per cent of domestic production. Louisiana moved ahead of California in 1958 to become second (to Texas) in annual oil production in the United States and has been increasing its lead subsequently. Louisiana was also in second place (again to Texas) in 1968 in the production of natural gas. Its share of domestic production during that year was 33 per cent.

It is customary to consider Louisiana as two oil-producing districts, north Louisiana and the Louisiana Gulf Coast (or south Louisiana). The dividing line is shown on the accompanying map (Figure 24.2). Geologically, the north Louisiana district is a part of a greater district that includes southern Arkansas and northeastern Texas. Similarly, the Louisiana Gulf Coast district is an arbitrary segment of a tectonic division that includes the Texas Gulf Coast to the southwest and the coastal areas of the states to the east.

Ninety-four per cent of the Louisiana oil production of 1968 came from the Gulf Coast area and only 6 per cent from north Louisiana. Coastal Louisiana has 35 giant oil fields (capable of producing over 100 million barrels) to 9 in north Louisiana. However, north Louisiana contains several major gas fields that render this area of considerable importance as a source of natural gas.

Louisiana had no important production of either oil or gas until the discovery of the Jennings salt dome field (Figure 24.2) in 1902. This followed the discovery of Spindletop across the line in Texas in 1901. No important finds were made in north Louisiana until 1904, when gas was discovered in the Caddo–Pine Island field, in which oil was also found the following year. The Caddo field, although now far down the list in annual production, is the second largest field in Louisiana in cumulative output, having produced more than 300 million barrels of oil since discovery.

The first oil field (Creole) to be found in the open

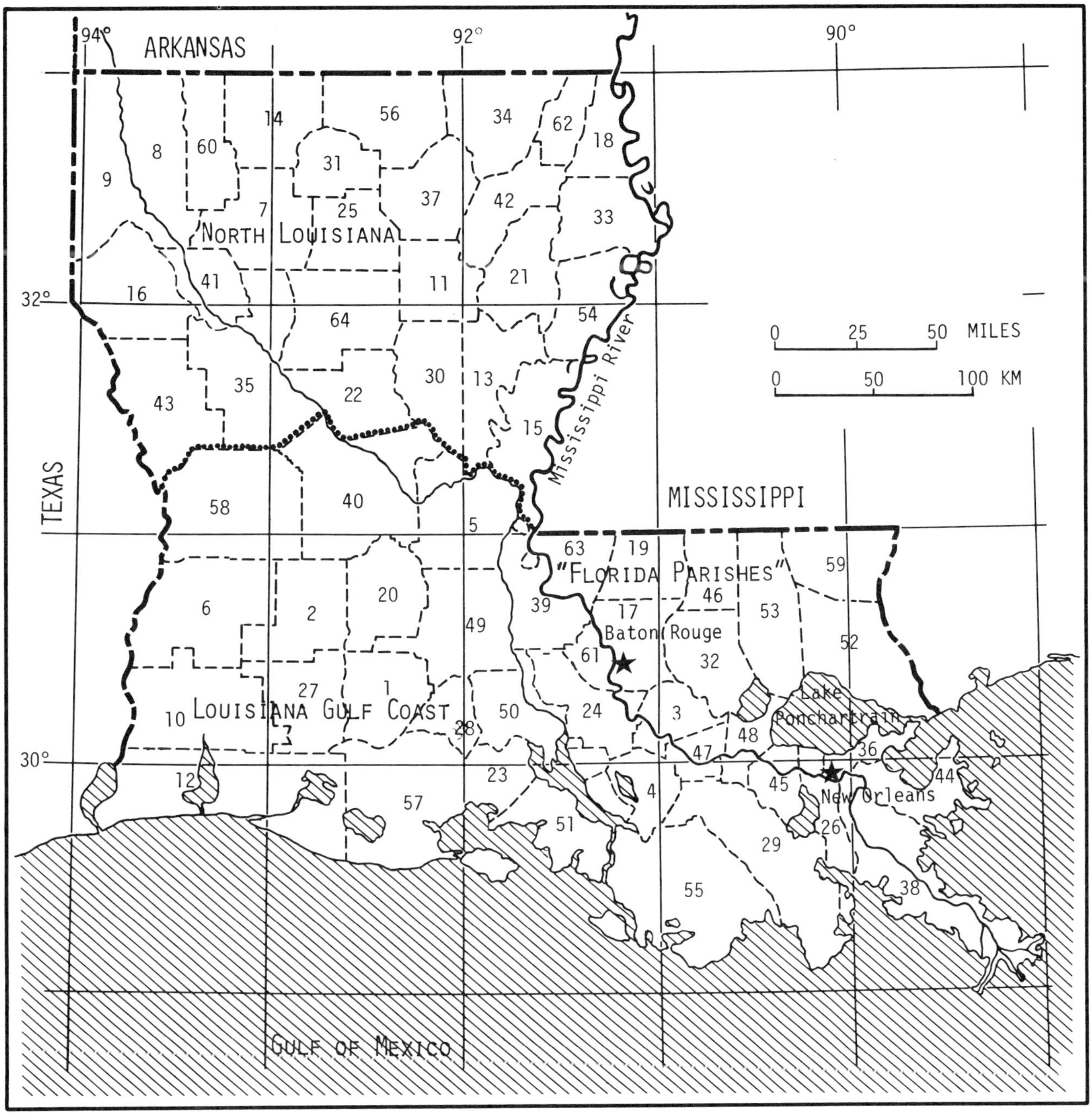

FIGURE 24.1 Louisiana. Index map. Parish Key:

1 Acadia
2 Allen
3 Ascension
4 Assumption
5 Avoyelles
6 Beauregard
7 Bienville
8 Bossier
9 Caddo
10 Calcasieu
11 Caldwell
12 Cameron
13 Catahoula
14 Claiborne
15 Concordia
16 De Soto
17 East Baton Rouge
18 East Carroll
19 East Feliciana
20 Evangeline
21 Franklin
22 Grant
23 Iberia
24 Iberville
25 Jackson
26 Jefferson
27 Jefferson Davis
28 Lafayette
29 Lafourche
30 La Salle
31 Lincoln
32 Livingston
33 Madison
34 Morehouse
35 Natchitoches
36 Orleans
37 Ouachita
38 Plaquemines
39 Pointe Coupee
40 Rapides
41 Red River
42 Richland
43 Sabine
44 St. Bernard
45 St. Charles
46 St. Helena
47 St. James
48 St. John the Baptist
49 St. Landry
50 St. Martin
51 St. Mary
52 St. Tammany
53 Tangipahoa
54 Tensas
55 Terrebonne
56 Union
57 Vermilion
58 Vernon
59 Washington
60 Webster
61 West Baton Rouge
62 West Carroll
63 West Feliciana
64 Winn

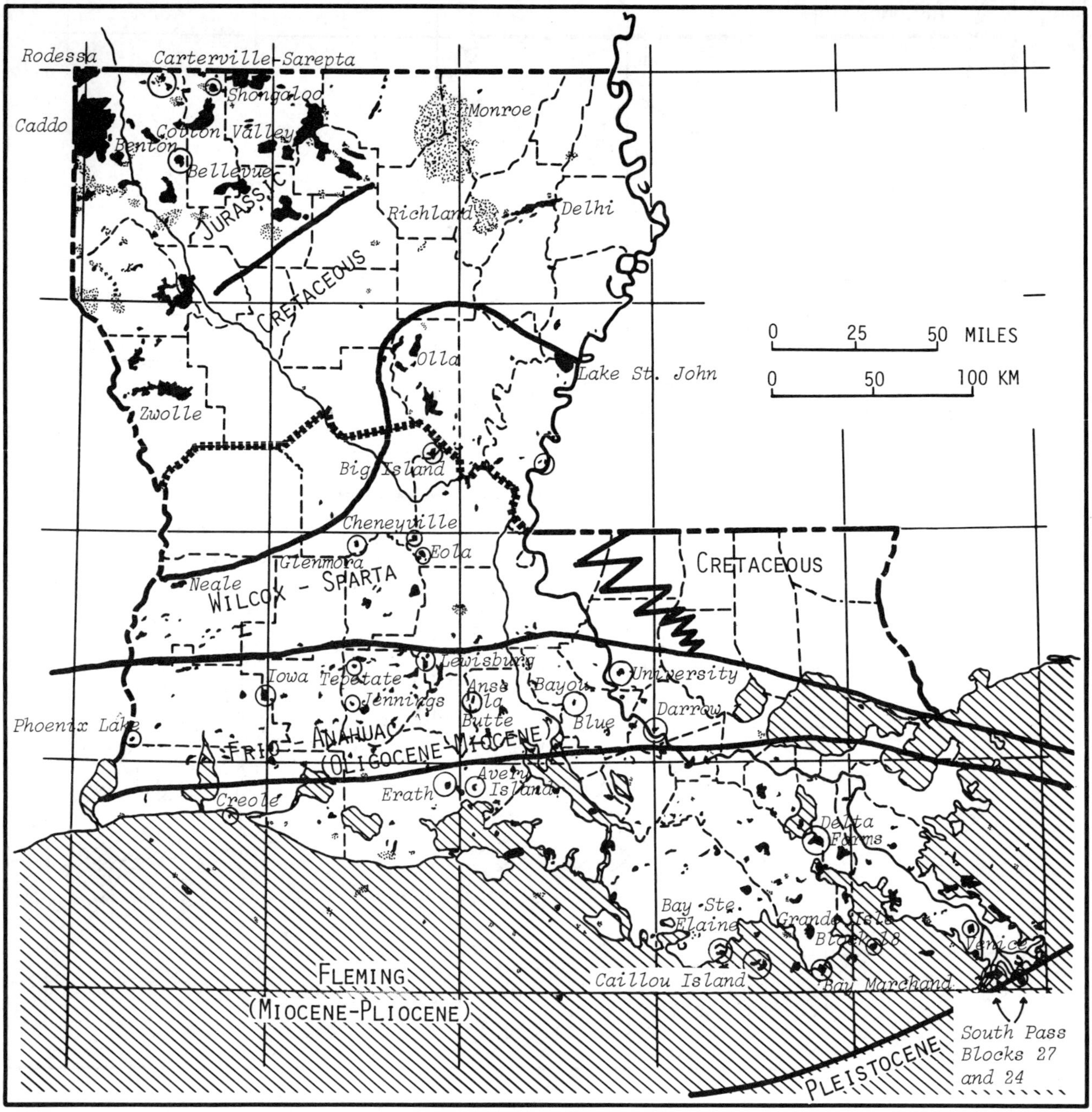

FIGURE 24.2 **Louisiana. Trends and oil and gas fields. The heavy dashed line crossing the midsection is the arbitrary division between North and South Louisiana (or Louisiana Gulf Coast).** ***Trend lines and names courtesy of James P. Spillers (letter dated 21 October, 1968).***

waters of the Gulf of Mexico was discovered in 1938. Since then, many new offshore discoveries have been made, with the result that annually during recent years Louisiana has established a new production record. Sixty-one out of the 64 parishes in the state are productive.

In June, 1967, the oil industry paid $510 million in bonuses for Federal offshore leases in Louisiana (U. S. Dept. Interior, 1968, p. 59).

REGIONAL STRUCTURE. All Louisiana lies within the Gulf Coast geosyncline province. No rocks older than Tertiary are exposed, except around some of the interior salt domes, where Cretaceous rocks have been pushed to the surface. The Tertiary rocks dip southward toward the Gulf; the oldest rocks (Eocene) cover northern Louisiana, and to the south successive bands of younger Tertiary formations cross the state from west to east. The Tertiary bedrock is obscured in southern Louisiana by a wide belt of deltaic sediments, and these deposits continue northward up the Mississippi River valley across eastern Louisiana.

The gulfward dipping geosynclinal sediments are interrupted by a major tectonic feature in northwestern Louisiana, the Sabine uplift (Figure 24.3). This uplift is a dominating influence on accumulation not only in northwestern Louisiana but also in southern Arkansas and northeastern Texas. To the east of the Sabine uplift is a synclinal or graben zone, the North Louisiana salt basin, and then the strata rise once more to cross the Monroe uplift of northeastern Louisiana. The Monroe uplift is less prominent in both relief and area than the Sabine.

As elsewhere in the Gulf Coast, the Miocene sediments thicken greatly southward. Therefore it is possible to

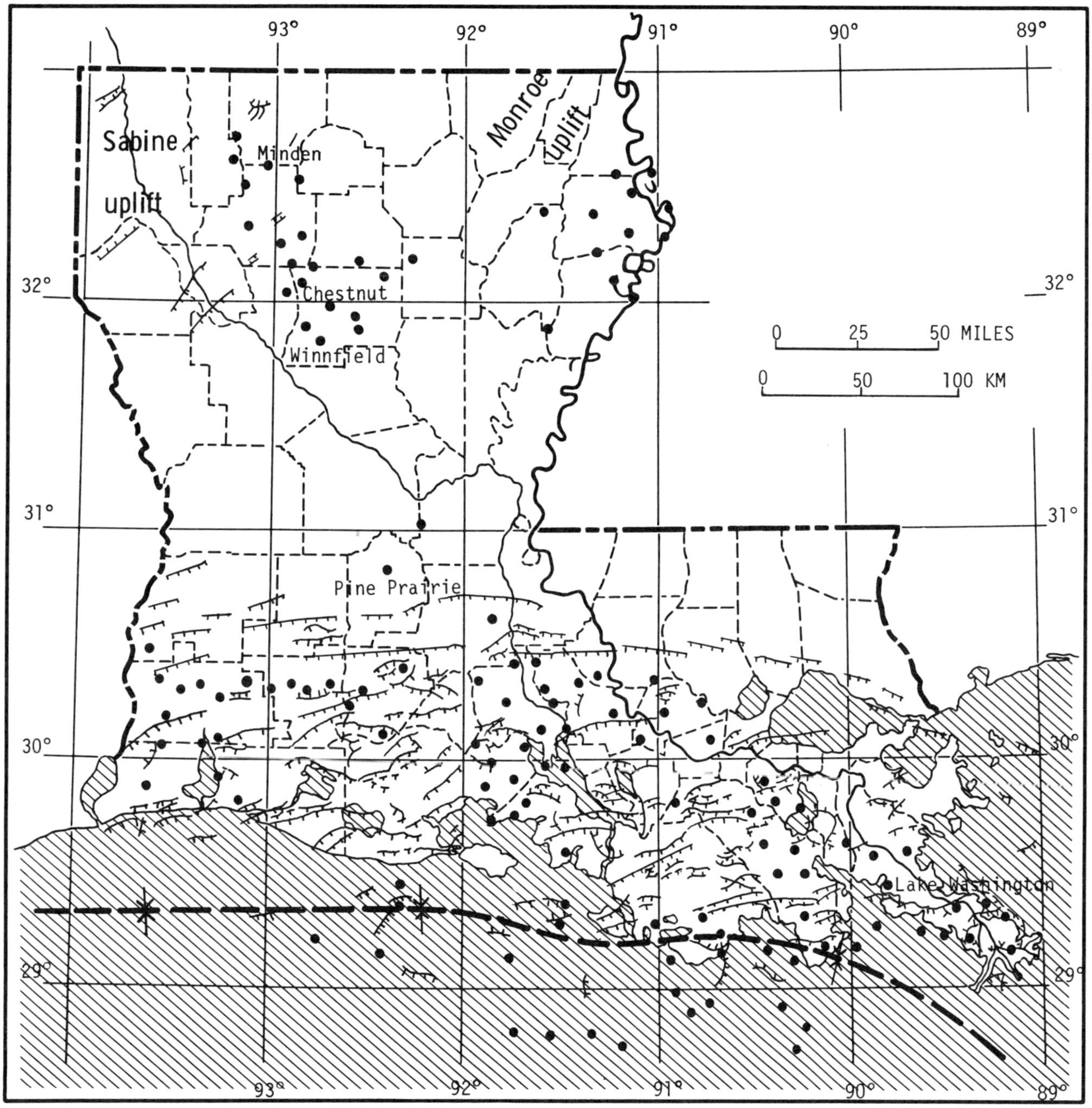

FIGURE 24.3 Louisiana. Tectonic map showing major uplifts, faults, and salt domes (solid circles). Heavy dashed syncline line in gulf is approximate axis of Gulf Coast geosyncline. *In part after Lafayette Geological Society (Lafayette and New Orleans Geological Societies, 1968, Figure 11, p. 390–391).*

drill wells to depths exceeding 20,000 feet in the coastal area without getting below the Fleming (Figure 24.9).

A myriad of major strike (east-west) faults occur in the Anahuac trend of the Louisiana Gulf Coast (Figure 24.3). Large faults are also abundant in the coastal (Fleming) belt and are no doubt as numerous offshore as onshore (James P. Spillers, letter dated 21 October, 1968). Wallace (1962) has remarked aptly that "the number of faults known in south Louisiana is proportional to the number of wells drilled." The Sabine uplift in north Louisiana is crossed by a number of northeast-southwest faults.

Salt domes (Figure 24.3) are abundant in two belts, one along the Gulf Coast and the other in northern Louisiana. The number of known salt domes in Louisiana (as of January 1, 1965) was 179, of which 148 are in the Gulf Coast sector (Louisiana Geol. Survey, 1965). Some of the salt intrusions approach so near the surface, especially in the "Five Islands" locality in Iberia and St. Mary Parishes, that they have been mined for salt for many years. Cap rock sulfur is obtained by the Frasch process at Lake Washington and Garden Island Bay in Plaquemines Parish, at Lake Pelto in Terrebonne Parish, and by an offshore operation at Grand Isle in Jefferson Parish. Other cap rock sulfur deposits have been exploited, depleted, and abandoned.

The tectonically negative area between the Sabine and Monroe uplifts is an interior salt dome district containing 19 known salt intrusions. Some lie close enough to the surface so that solution effects are visible in the topography, and in two places, Winnfield dome and Pine Prairie, the cap rock is exposed, making possible "hard rock" quarries. The gypsum and "limestone" at these two localities are now depleted. At least 2000 feet of collapse took place above the Chestnut salt dome in Natchitoches Parish because of solution of the top of the salt spine. Another interior salt dome district lies on both sides of the Louisiana–Mississippi boundary in the structural valley between the Monroe and Jackson uplifts. Twelve of the known salt domes in this district are in northeastern Louisiana.

The seismograph is the principal exploring implement used in both north Louisiana and in the Gulf Coast district, including the open waters of the Gulf. The gravity meter serves extensively in south Louisiana as a preliminary ("lead-in") tool. The core drill is also used to some extent in northern Louisiana, and subsurface geology is employed in both areas.

NORTH LOUISIANA

Geographically north Louisiana is the part of the state lying north of the 31st parallel, which, east of the Mississippi, separates Louisiana and Mississippi. However, west of the Mississippi this parallel passes through three parishes, Avoyelles, Rapides, and Vernon. For convenience the north boundaries of these parishes are used here to separate north Louisiana from the Louisiana Gulf Coast (Figure 24.2). The resulting division roughly coincides with the contact line between Eocene and post-Eocene sediments.

The western two-thirds of north Louisiana is floored almost entirely with Eocene sediments. Pleistocene deposits overlie the Tertiary rocks in the eastern third of the district.

North Louisiana is notable for its great gas fields (Shreveport Geol. Soc., 1968), which extend in a wide belt from high on the Monroe uplift in northeastern Louisiana westerly to the Texas line (Figure 24.2). The Sabine uplift (Figure 24.3) in northwestern Louisiana contains, across the top and down the east and south flanks, many large oil fields as well as gas fields. The eastern side of north Louisiana, where the southeasterly dip into the Mississippi embayment carries the Eocene sediments beneath a thick section of Pleistocene, is an area in which Eocene oil production predominates.

Caddo (Figure 24.2) is the largest field in north Louisiana in cumulative production and in 1967 was second to Black Lake in annual production. High in cumulative production but no longer a large field in terms of annual yield is the Rodessa, which overlaps northeast Texas.

Geologically north Louisiana and southern Arkansas are continuous, and these areas are frequently combined for statistical purposes also. During recent years the Louisiana part of this oil and gas district has produced more than twice the oil and many times the gas yielded by the Arkansas portion.

Statigraphy (Meade, 1964; Bank of the Southwest, 1956; Martin et al., 1954; Shreveport Geol. Soc., 1953). The sedimentary column so far explored in north Louisiana extends from Recent to Permian in age. The reservoir rocks are many, ranging from upper Eocene to Lower Jurassic (Figure 24.4). It should be noted in using this stratigraphic section that the lower part of the Coahuila (lowest Lower Cretaceous) section, so far unproductive, has been omitted.

Tertiary sandstone reservoirs have produced more oil than those of any other period. The youngest reservoirs are sands in the Claiborne Formation of middle Eocene age. However, the most prolific production has come from Wilcox (lower Eocene) reservoirs (Durham and Smith, 1958); at least 126 fields in central Louisiana obtain oil from sands in this formation. An electric log study (Shreveport Geol. Soc., 1961) of the oil and gas reservoirs of Wilcox age in this area showed the presence of 40 stratigraphically distinct sands, of which 27 lie in

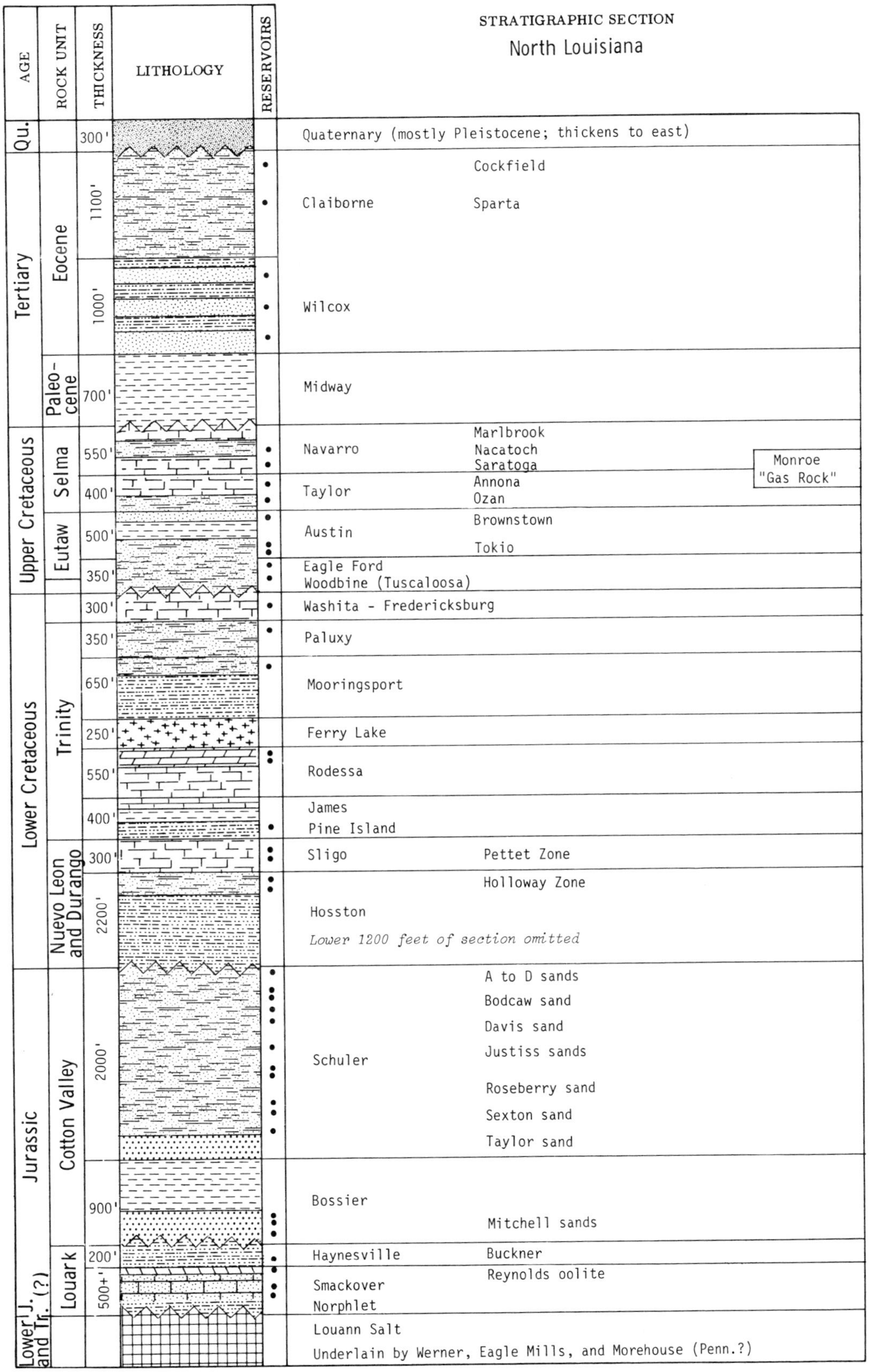

FIGURE 24.4 Stratigraphic section, North Louisiana. A composite. Taken mostly from *James L. Martin et al.* (*1954, Plates 3 and 13*), and partly from *Lafayette and New Orleans Geological Societies* (*1968, Figure 2, p. 380*).

a highly productive zone originally designated as Nebo. Some fields have as many as 12 superimposed Wilcox reservoir sands. Production is also obtained from sands in the overlying Claiborne Formation.

The Navarro and Taylor Groups at the top of the Upper Cretaceous sediments contain a number of reservoirs, including especially the Monroe "gas rock," which is the principal reservoir in the great Monroe gas field. This section also contains the dense fractured chalks that constitute the leading reservoir in Sabine Parish (Woods, 1963). Tuscaloosa reservoirs toward the base of the Upper Cretaceous section are important, especially in the eastern part of north Louisiana. Between the Upper and Lower Cretaceous is a major unconformity; over 7000 feet of uppermost Lower Cretaceous sediments was removed in the uplift areas before the deposition of the first Upper Cretaceous sediments (Philpott, 1952).

The many Lower Cretaceous reservoirs include especially the Paluxy in the upper part, the Rodessa in the middle, and sands in the Sligo and Upper Hosston in the lower part of the section that are the most important gas reservoirs in the synclinal area between the Monroe and Sabine uplifts.

The most numerous reservoirs are those in the Schuler Formation in the upper part of the Jurassic formation. Included in this group is the prolific Bodcaw sand (Sloane, 1958). The Upper Jurassic contains the Smackover reservoir, which is the oil source in the deeper pools on both sides of the Louisiana–Arkansas boundary (Rogers, 1956). Pre-Smackover formations recognized in north Louisiana include the Norphlet, Louann, and Werner (all Jurassic) and the Pennsylvanian Morehouse (Bishop, 1967).

Traps and Selected Fields. Anticlines, faults, truncation with overlap, and one salt dome are types of traps present in northern Louisiana. Up to the present, at least, structural trapping has been responsible for most of the oil accumulations, but many of the large gas fields are the result of truncation and overlap. Furthermore, in many of the anticlinal fields the distribution of hydrocarbons across the structure is controlled in part by lenticularity of the reservoirs.

The *Cotton Valley* field (Martin et al., 1954, Shreveport Geol. Soc., 1947), one of north Louisiana's giants, is a good example of anticlinal trapping. The anticline has two nodes (Figure 24.5). The southern dome is much the higher, with a structural relief of over 800 feet and with average dips of about 400 feet to the mile on the Bodcaw sand (Jurassic), the principal reservoir in the field. The other dome, which lies about 5 miles to the north-northeast, is approximately 400 feet lower. The intervening saddle separating the two domes is only about 100 feet lower than the top of the north dome. The north flank of the north dome is cut by an east-west fault which has elevated that flank nearly 100 feet.

The existence of the Cotton Valley anticline was suspected because of surface exposures in the area and verified by the drilling of three unsuccessful wildcat wells. The discovery well for the field was completed in August, 1922, at a depth of 2560 feet in an Upper Cretaceous sand tentatively identified as the Blossom sand of the upper Austin. This reservoir produced nearly 15,000,000 barrels of oil and considerable gas before exhaustion in 1940. From 1926 on, other discovery wells found oil or gas in various older and deeper reservoirs extending from the Lower Cretaceous Rodessa Formation to the Jurassic Smackover. The most prolific reservoirs were found in the Cotton Valley Group (Jurassic); especially in the Schuler Formation, including, in addition to the Bodcaw sand, the A to D sands above and the Davis sand below. Although many of these reservoir sands vary considerably in permeability and thickness, the trapping is structurally controlled. Interbedded shales contribute the necessary sealing strata.

The Cotton Valley field produces oil, distillate, and gas. A pressure maintenance plan was formulated in 1940, and a cycling unit constructed to repressure the Davis, Bodcaw, and D sands.

The *Olla* field (Shreveport Geol. Soc., 1946) occupies an elongate anticline. The axis of this anticline trends about 30° east of north, and the productive area has a maximum width of 2½ miles and a length of 7 miles. The central and highest node has a structural closure of about 120 feet. The field was discovered in 1940 after subsurface mapping and seismic surveys. It lies in LaSalle Parish in the belt of fields producing from Eocene reservoirs. At least 10 sandstone members, within a stratigraphic range of about 300 feet, in the upper part of the Wilcox have been found to be productive in various parts of the field.

Another elongate anticline is the locale for the *Benton* field in Bossier Parish. The long axis of this fold is 7 miles; the average width is 2½ miles. It was discovered by a regional gravity meter survey followed by seismograph work (Valerius and Von Croy, 1951). The reservoirs in the Benton field are the D and Bodcaw sands in the Cotton Valley Group.

The Eocene producing belt continues eastward from LaSalle Parish across eastern-northern Louisiana and into Mississippi. In this area are many stratigraphic traps caused by sand pinchouts updip and compaction associated with channel sand deposition.

The *Zwolle* field (location shown in Figure 24.2) crosses central Sabine Parish in the southwestern corner of the north Louisiana district. Structurally, it is a much faulted, deep-seated anticlinal complex containing a considerable number of irregularly shaped nodes (Andersen, 1960). The discovery well for the Zwolle district was completed on Thanksgiving day in Novem-

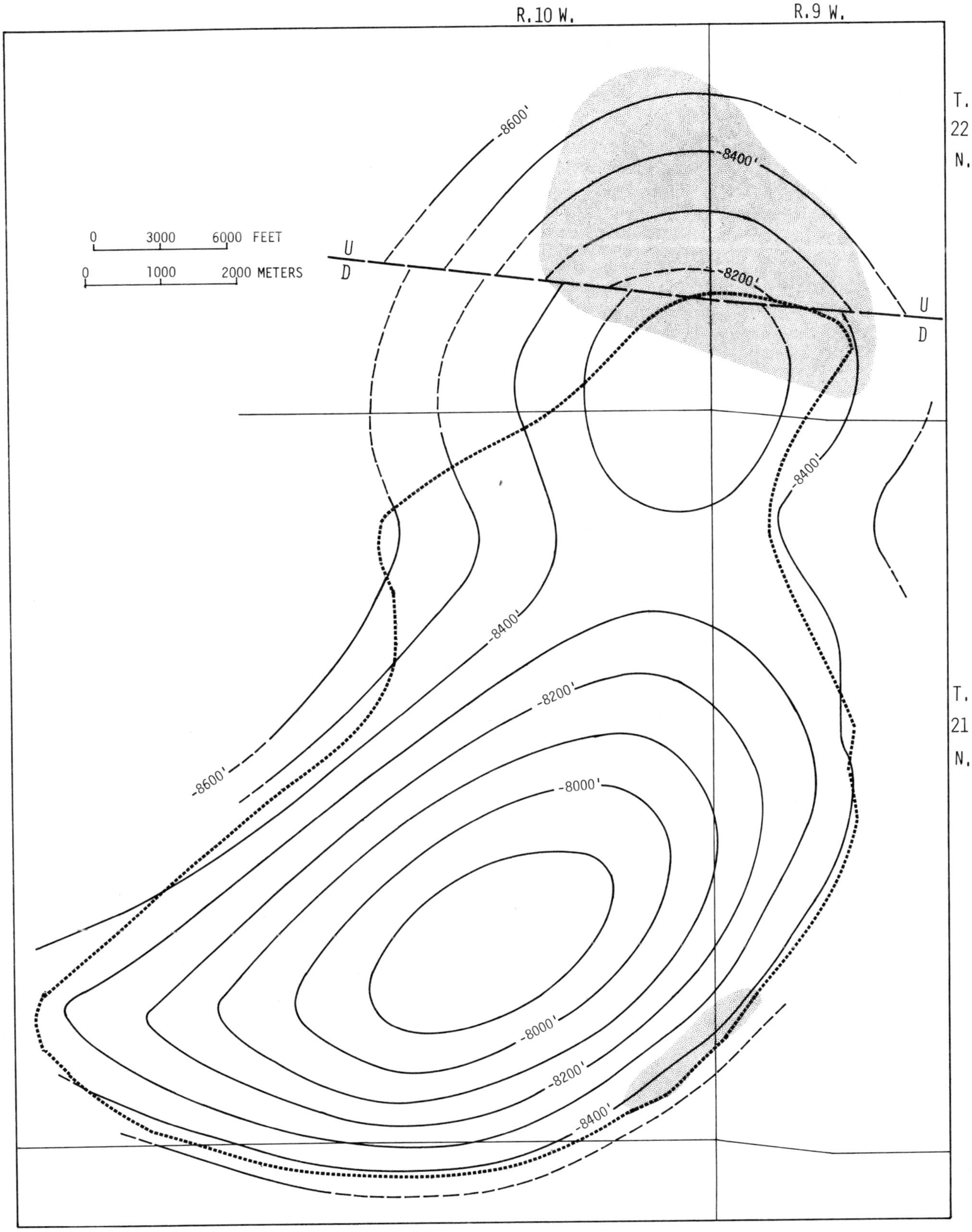

Figure 24.5 Cotton Valley field, Webster Parish, Louisiana. Structure contour map based on the top of the Bodcaw Sand. Contour interval, 100 feet. Stippled pattern shows oil-producing areas in Bodcaw reservoir; dashed line outlines Bodcaw gas production. *Courtesy Louisiana Geological Survey (1954, Plate 17).*

ber, 1928, in the Saratoga Chalk of Late Cretaceous age at a depth of 2400 feet. Subsequently, a second Late Cretaceous chalk reservoir was discovered in the Annona Formation, which lies about 200 feet below the Saratoga. These chalks are productive only where fractured. The erratic production across the years, which reached a maximum of nearly 3,000,000 barrels in 1933, has been due to a considerable part to the difficulties involved in finding one or the other of these reservoirs adequately fractured for commercial production. Although accumulation is structurally controlled, the distribution of oil on the anticlinal complex is dependent upon local fracturing; the productive area may extend as much as 200 feet lower on the flanks of the nodes where the fracturing permits.

Northern Louisiana contains 28 known piercement salt domes, of which 18 are in the north Louisiana salt basin, between the Sabine and Monroe uplifts, and 10 are in the eastern Louisiana part of the Mississippi salt basin (Clark, 1961). Only one of these domes so far has been found to be productive. This is the *Minden* salt dome (Figure 24.3), which is the northernmost one so far found in the north Louisiana salt basin. It was discovered by a well completed in 1957 in the Youngblood sand of the Paluxy Formation (Lower Cretaceous) with an initial potential production of 4,280,000 cubic feet of gas and 55 barrels of condensate per day. Subsequent wells found production in the Rodessa, James, and Hosston Formations (likewise Lower Cretaceous), in addition to the Paluxy. These wells are on the east and west flanks of the salt plug (Figure 24.6). Trapping is due to wedgout of the reservoir sands (Richardson, 1961).

The *Lake St. John* field, one of the more important oil fields in northern Louisiana, overlaps Concordia and Tensas Parishes. It contains seven distinct reservoirs distributed in age from middle Eocene to Early Comanchean. The hydrocarbons obtained range from nearly dry gas through distillate to crude oil (Shreveport Geol. Soc., 1947). The trap is a faulted dome; geophysical surveys have shown the presence of a gravity minimum at the center of the dome, suggesting the presence of a deep-seated salt core.

The *Bellevue* oil field (Crider, 1938) in Bossier Parish also illustrates anticlinal trapping. The principal reservoir rock is the Nacatoch sand, which lies high in the Upper Cretaceous and occurs at Bellevue at the unusually shallow depth of 300 to 400 feet. The structure at the Nacatoch level is a faulted dome with a fairly flat top; its presence had been suspected after surface geological observation, but it was first mapped in detail by a core-drill survey. It has been suggested that the flowage of salt at depths below those so far tested (6147 feet) created the Bellevue dome by forcing the overlying formations up as much as 2000 feet above their normal

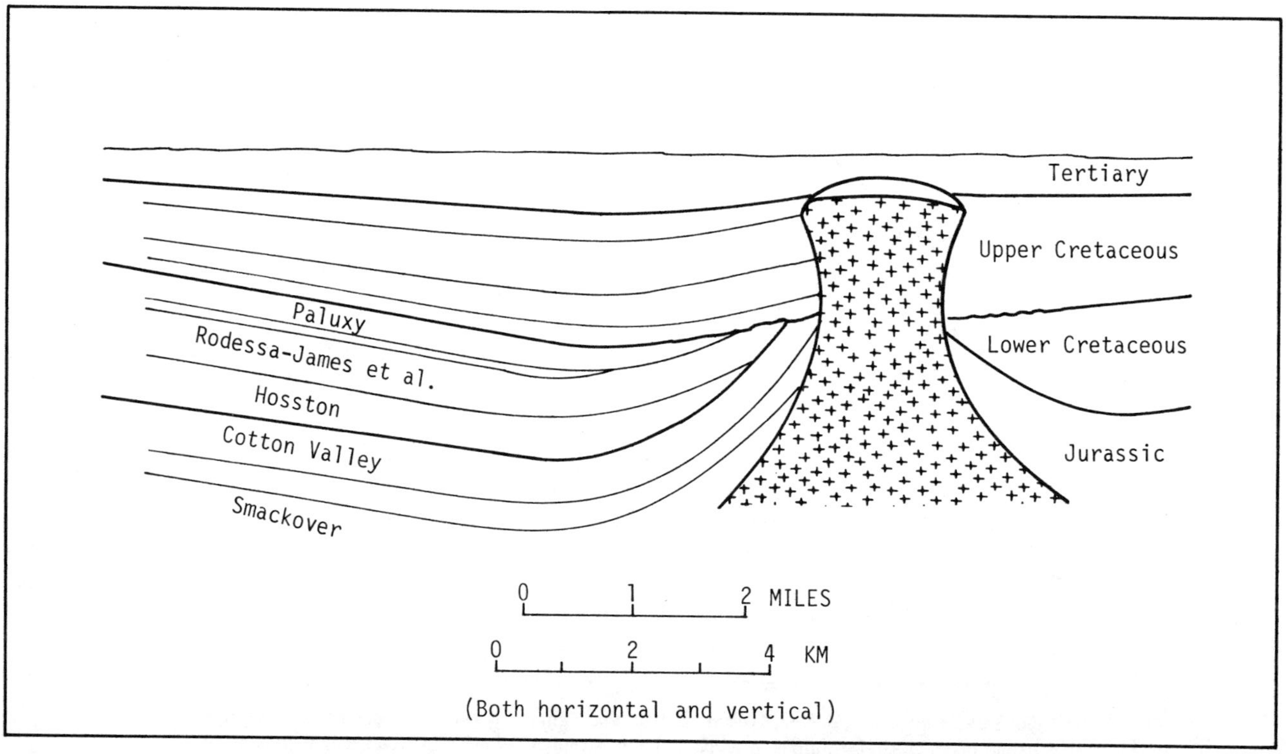

FIGURE 24.6 Minden salt dome, Webster Parish, Louisiana. Cross section showing formations pierced by salt intrusion. ***Courtesy G. C. Clark (1961) and Shreveport Geological Society.***

elevations. An average yield of 11,000 barrels per acre from depths of less than 400 feet makes the Bellevue field a phenomenal one in terms of investment return. The *Carterville-Sarepta* and *Shangaloo* fields (Thomas, 1938) also illustrate anticlinal accumulation, but with distribution of the oil governed to a considerable extent by the degree of permeability of the reservoir sandstones. These fields lie south of the Arkansas line in Bossier and Webster Parishes. Their structure is also credited to salt flowage at depth.

A once great field is the *Rodessa*, which lies on the northwest flank of the Sabine uplift in the very northwesternmost corner of Louisiana (Clark, 1939). The field is unusually elongate, with a northeast-southwest trend that carries it into Arkansas in one direction and Texas in the other. The total length is 33 miles, and the maximum width 4½ miles. Trapping is due to the presence of a major fault having a displacement of about 500 feet at the reservoir level. The southeast side of the fault, the updip side in terms of the Sabine uplift, is the down side; accumulation is in the uplifted block immediately to the northwest of the fault. Oil moving updip from structurally lower parts of northeastern Texas became impounded by the presence of impervious rock across the fault plane. Many minor faults occur in the upfaulted block parallel to the major fault. Oil and gas are produced from several Comanchean reservoirs, of which the principal one is the Rodessa sand of the Trinity Group. Rodessa was discovered in 1930 and was explored and exploited both vigorously and wastefully until the Louisiana legislature passed conservation laws, which became effective on January 1, 1937.

The *Delhi* field and the *West Delhi* field constitute a continuous belt of oil production 20 miles long, extending westerly from northwestern Madison Parish to central Richland Parish and cutting a corner of Franklin Parish en route. The Delhi field was discovered by a regional study of the subsurface geology followed by seismic work in the area. The first of a series of discovery wells was completed in December, 1944, in the Holt sand in the Paluxy (Lower Cretaceous) Formation. Subsequently the field was extended by further borehole exploration to the east, west, and south, and a half-dozen more oil and gas reservoirs were discovered, most of them sands in the Tuscaloosa of Late Cretaceous age. Delhi is another giant, with cumulative production in 1967 of more than 140 million barrels.

As can be seen from the east-west cross section (Figure 24.7), the regional dip is easterly off the Monroe uplift to the west. The trapping is due in part to lenticularity of some of the Tuscaloosa sands, but mainly to truncation of both the Lower Cretaceous Paluxy sands and the Upper Cretaceous Tuscaloosa Formation. The basal Tuscaloosa sand, locally called the Barrier sand, is a sheet sand that overlies the beveled and truncated Paluxy sediments. Where the Paluxy sands come in contact with the Barrier sand, they become a part of the same reservoir (Shreveport Geol. Soc., 1946, p. 426). The Tuscaloosa sands above the Barrier are separate reservoirs. Some production of both oil and gas has been obtained from stratigraphically deeper Lower Cretaceous formations down to and including a sandstone member of the Hosston. The seals are shales interbedded with the sandstone reservoirs.

The large *Monroe* gas field (Wang, 1952; Fergus, 1935) (Figure 24.8) and the somewhat smaller *Richland* (Gordon, 1935) field to the southeast are both examples of truncation, with sealing by overlap. The Lower and Upper Cretaceous sedimentary section, which contains the gas reservoirs, was tilted westward, beveled by erosion, and then buried beneath unconformable Tertiary sediments that trapped the gas in the truncated permeable strata beneath. Later both areas were arched upward so that the structure of the Tertiary rocks shows two large domes. The Monroe field, discovered in 1916 and constituting a major source of natural gas ever since, enjoyed a considerable rejuvenation in the early 1960s, when 300 new wells were completed in 20 months (Wilson, 1962). The principal reasons for this renewed activity were (1) the realization that considerably greater quantities of gas could be obtained by hydraulic fracturing of the Monroe "Gas Rock" (the principal reservoir), (2) the discovery on the north and west sides of the field of a new reservoir sand about 150 feet deeper than the "Gas Rock," and (3) increased prices for the gas as "depression" purchase contracts expired and were renegotiated.

A large quantity of natural gas condensate has been trapped in sand wedgouts along a line extending for 125 miles from the south side of the Monroe uplift westerly across the north flank of the north Louisiana salt basin to the southeastern side of the Sabine uplift near the Texas line (Pate, 1963). This line approximates the shoreline of three producing sands in the Schuler Formation of the Upper Jurassic Cotton Valley Group. Twenty-four stratigraphic trap fields lie along this zone. Closure updip is due to the disappearance of sand. Laterally, closure has developed in two ways: (1) edge water on the flank of a structural nose, and (2) sand lensing into shale because of the sinuous nature of the original shoreline.

The *Colgrade* field (Nelson, 1963) in northeastern Winn Parish occupies most of the southern flank and top of a broad ameboid anticline. The reservoir is a sandstone at the top of the lower Eocene Wilcox Group. Sand lenses into shale on the north and southwestern flanks, limiting production in those directions.

A giant in the making is *Black Lake* field (White, 1966; Wilson, 1967) in northern Natchitoches Parish.

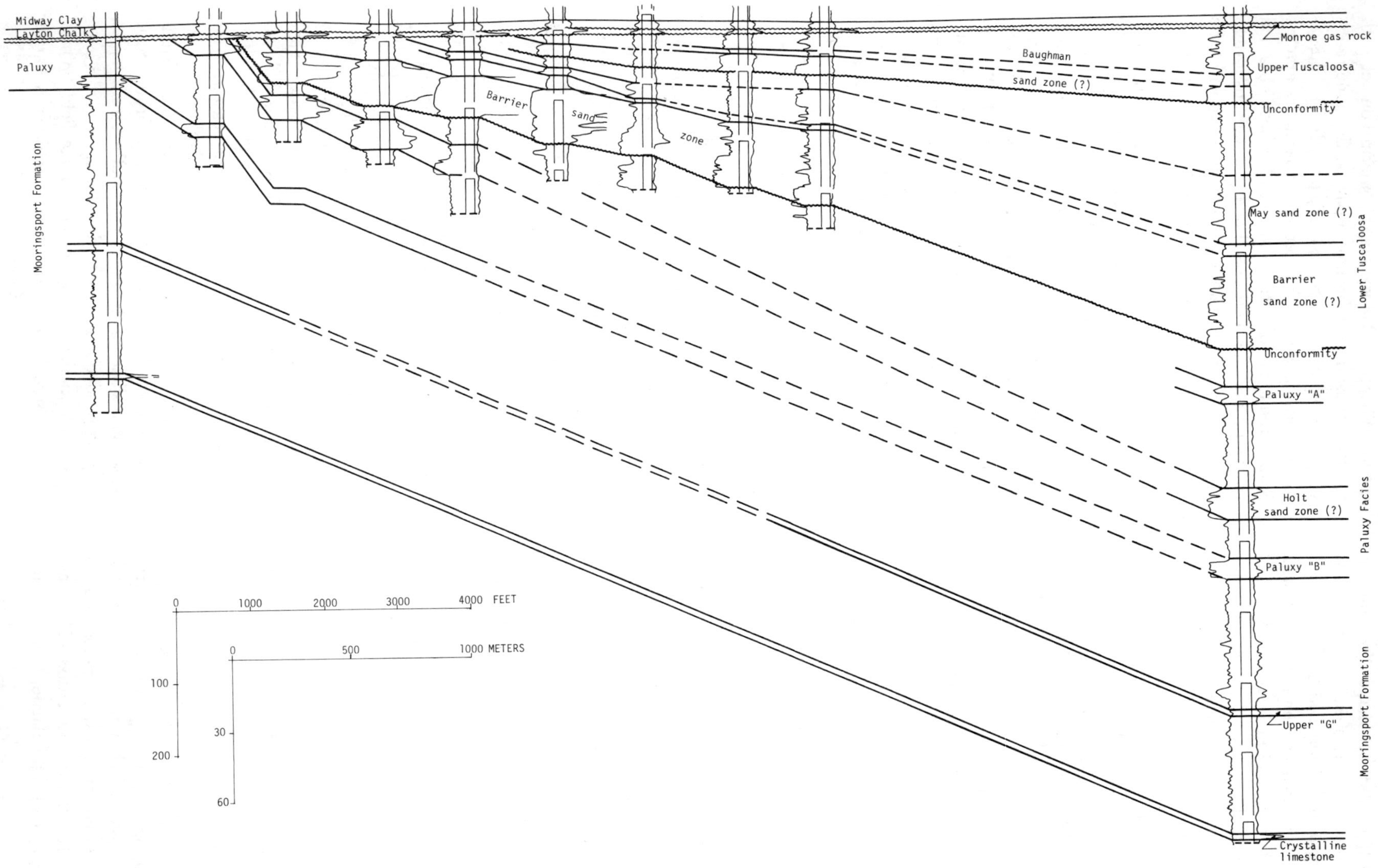

Figure 24.7 North-south cross section, Delhi field, Richland Parish, Louisiana. *Courtesy Shreveport Geological Society (1946, p. 426).*

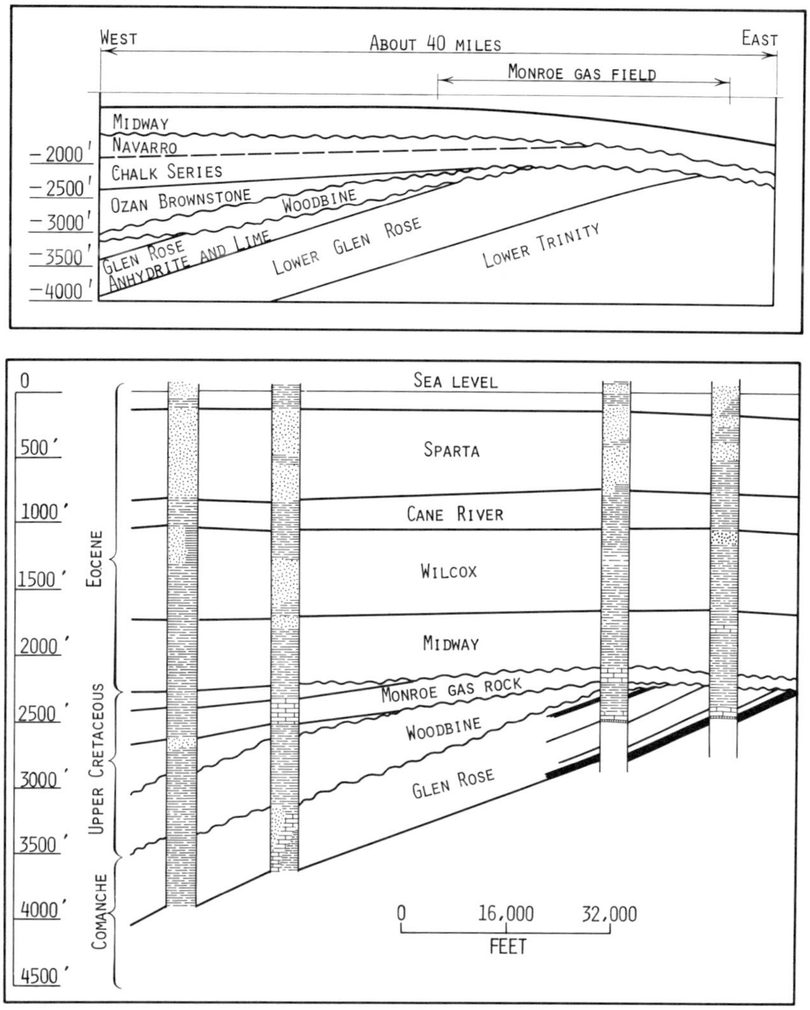

Figure 24.8 Monroe gas field, northern Louisiana. Upper: regional cross section through north-central part of field. Lower: close up, showing reservoir rock. ***From Preston Fergus (1935). Courtesy American Association of Petroleum Geologists.***

Discovered in mid-1964, this field was in first place in northern Louisiana by the end of 1967, with a production of more than $5\frac{1}{4}$ million barrels of oil for that year. It is also a large producer of natural gas and condensate. The reservoir is a reef in the Pettet Formation of Early Cretaceous age. Black Lake field is about 8 miles long (northeast-southwest) and 5 miles wide. Trapping on the north and west is due to updip termination of the reefoid porosity; water underlies the oil in the reef downdip to the south and east. The oil column has a maximum thickness of 35 feet, above which the gas cap ranges up to 125 feet in thickness.

The *Caddo* field is a bald-headed structure with most of the oil and gas coming from Upper Cretaceous reservoirs arched over the buried truncated anticline; the rest of the oil comes from truncated Comanchean formations on the flanks of the anticline beneath the unconformity (Fletcher, 1929). A major reservoir, especially in the northern (Pine Island) sector of the Caddo field, is the Annona Chalk, which is high in porosity but low in visible permeability. The recoverable oil is stored in fractures. Caddo has had two production surges since discovery in 1906. The first of these was due to acidization in the late 1930s, and the second was the result of hydraulic fracturing, which was introduced in 1953 (Billingsley, 1964). Caddo lies close to the top of the Sabine uplift in northwestern Louisiana. It overlaps slightly into northeast Texas.

Future. As long as 15 new fields continue to be discovered each year (Morrow et al., 1968) in northern Louisiana, the future will take care of itself. Most of these new discoveries appear to be made in areas of intermediate field density. There still remain considerable areas in the west-central and northeast parts of north Louisiana without production, but with adequate geology for the occurrence of hydrocarbons in commercial accumulations (Philpott, 1952). The search for elusive permeability wedgouts was successful at Black Lake and will be elsewhere. In both the older and the new areas, deeper potential reservoirs (including pre-Smackover) remain largely untested (Gardner, 1964; Chapman, 1963).

The 27 out of 28 known salt domes as yet nonproductive in north Louisiana present a continuing challenge to the wildcatter. Furthermore, there are without a doubt additional salt domes yet to be discovered in this province. As Clark (1939, p. 16) has pointed out, the dry hole risk is exceptionally high in probing for production on the flanks of piercement domes.

LOUISIANA GULF COAST

The Louisiana Gulf Coast is an arbitrary geographic division comprising all of Louisiana east of the Mississippi River and the part of Louisiana to the west of the river that is not included in north Louisiana (Figure 24.2). It includes 38 parishes and 16 continental shelf areas offshore. Quaternary sediments cover all of the Gulf Coast sector except the northern parishes east of the river and parts of six inland parishes west of the river. Tertiary sediments crop out at the surface in these areas. They are all post-Eocene in age and consist largely of Miocene deltaic sediments overlain by Pleistocene alluvial deposits.

Geographically, the Louisiana Gulf Coast can be separated into the onshore and the offshore districts. Forty-three per cent of the southern Louisiana oil output in 1968 was from offshore deposits. A belt of high oil field density about 70 miles in width extends from the shoreline inland, and from the Texas boundary (see Texas Upper Gulf Coast, p. 245) eastward to the Mississippi River delta.

Offshore the geology is the same, but water depth introduces an economic limit to exploration and development. The water depth limit is not fixed; it was at first a few tens of feet but later was extended to about 200 feet (Atwater, 1959), and subsequently the entire continental shelf, regardless of water depth, has become possible prospective territory. The edge of the shelf is less than 20 miles off the tip of the Mississippi delta but reaches a maximum distance offshore of about 130 miles near the Texas line. Although the field density offshore is nowhere near as great as it is onshore, it is increasing steadily.

New field discoveries in 1967 in the Louisiana Gulf Coast totaled 17, of which 7 were onshore and 10 offshore (Riddle et al., 1968). The coastal area is deep-drilling country with most test and field wells reaching depths of 10,000 to 22,000 feet. A considerable percentage of the deeper fields produce gas and gas condensate.

Among the largest fields in coastal Louisiana are Caillou Island, Bay Marchand, South Pass Block 24, and Timbalier Bay. These lie astride the shoreline and hence produce both onshore and offshore. The Louisiana Gulf Coast has a total of 35 fields classified as giant, of which 22 are onshore.

Much the greater part of the Louisiana Gulf Coast oil and gas accumulation so far discovered has taken place in traps created by the upward intrusion of salt. It is an area where geophysical prospecting has been most successful (Melchior, 1953).

Stratigraphy. The Louisiana Gulf Coast sector is divisible into five zones (Figure 24.2), depending on the age of the subsurface oil and gas reservoirs. Because of the southerly dip, the oldest reservoir rocks are to the north. The Cretaceous belt of north Louisiana crosses the northwestern corner of south Louisiana. Next, moving coastward, is the Wilcox–Sparta trend, which is a continuation southwestward of the Eocene Wilcox and Claiborne production of the east portion of south-central Louisiana and southwestern Mississippi. The field density in the southward extension is greatest in Beauregard Parish (Holland et al., 1952).

The stratigraphic section in the Wilcox–Sparta trend is similar to the upper part of the section for north Louisiana (Figure 24.4). According to Bornhauser (1950), an outstanding change in facies occurs in the Tertiary section from north to south across the trend. The predominantly sand section grades downdip gulfward into shale.

The middle belt (Frio–Anahuac trend) averages only about one parish in width but has a higher field density than does the trend to the north. Acadia Parish with its many oil and gas fields lies in the middle of the productive part of this belt. The stratigraphic positions of the principal reservoirs in this trend are shown in Figure 24.9. The lithology consists of alternating beds of sandstone and shale. The reservoirs range in age from Pliocene (?) to Eocene (Cockfield). Most of the production, however, comes from the Anahuac and Frio Formations. The leading reservoirs in the Anahuac are the Discorbis zone sands in the upper part of the formation; the most important reservoirs in the Frio are in the middle part and include especially the Klump and Homeseeker sands. Multiple pay fields are the rule.

Both the Anahuac and the Frio Formations thicken several-fold in the relatively short distance from north to south across this zone (Paine, 1962, plate 4). Much of this thickening is due to greater deposition on the down blocks of growth faults. This subject is discussed later (p. 210). The thickening is especially notable in the middle and lower Frio sections along the Acadia–St. Landry Parish line. This belt has been termed the

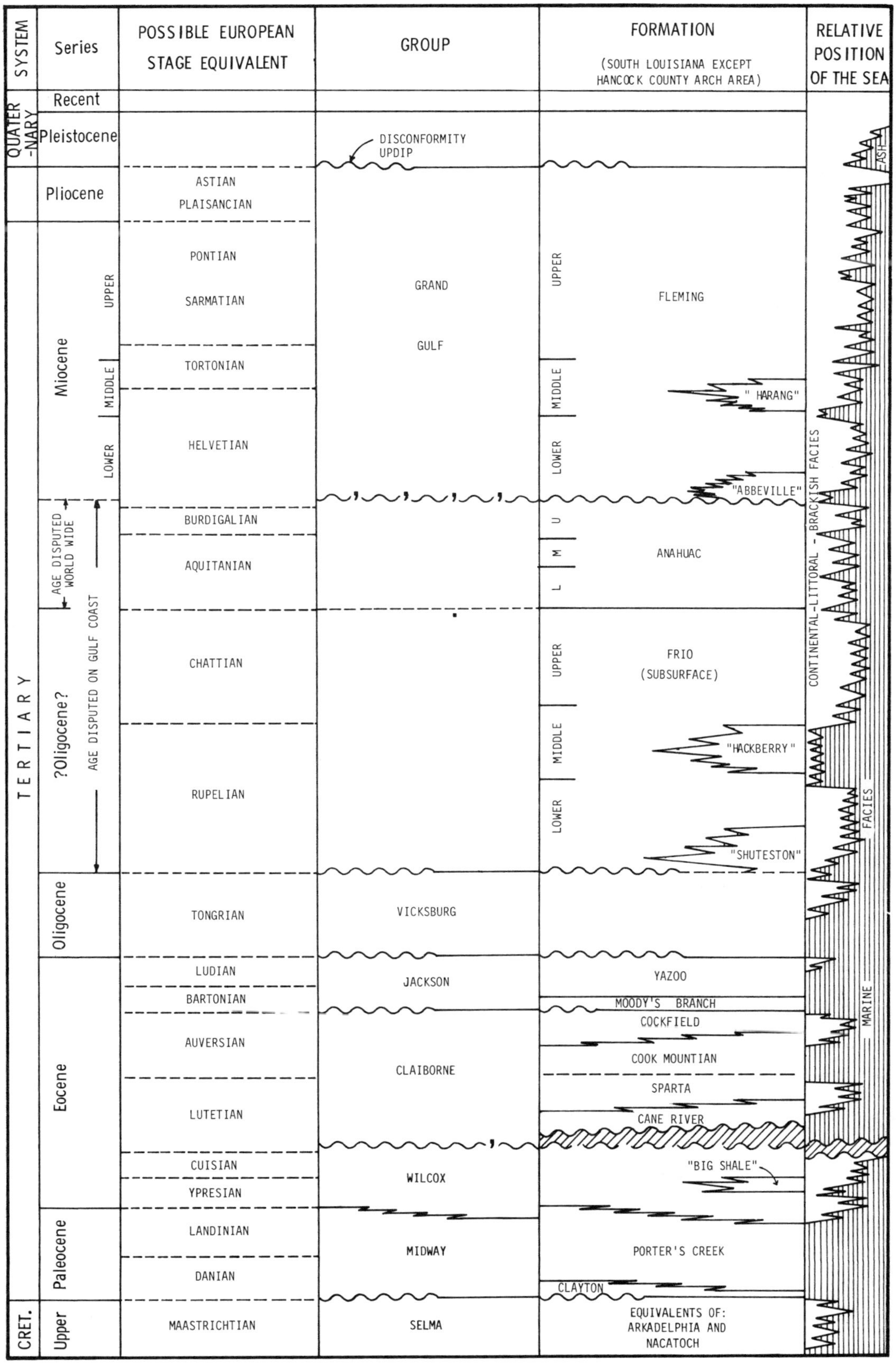

FIGURE 24.9 Generalized stratigraphic chart for Cenozoic of South Louisiana. *Courtesy Lafayette and New Orleans Geological Societies (1968, Figure 3, p. 381) and the American Association of Petroleum Geologists.* A list of diagnostic fossils or fossil assemblages is included in the original columnar section.

"Nodosaria embayment" (Grigg, 1957) after pay zones in the lower Frio.

The eastern ends of the parishes immediately north and south of Lake Pontchartrain are on the western end of the Hancock County arch of southernmost Mississippi. In this area the greater part of the pre-Fleming Tertiary section is missing.

The coastal belt (Fleming trend) is mostly offshore at the Texas boundary, but the northern edge trends slightly north of east, a circumstance that, combined with the southeasterly direction of the shoreline, makes the width on land considerably greater to the east. This trend contains the most productive part of Louisiana (Limes and Stipe, 1959). The Fleming deposits are very thick "in southern Louisiana and adjacent offshore. The wedge of regressive, mainly deltaic, sand, silt, and shale and transgressive marine shale is highly productive of oil and gas." (Rainwater, 1964).

The lithology of the subsurface section has been described as follows:

> Oil test wells in coastal southwestern Louisiana, some of which have been drilled to depths greater than 20,000 feet, log a monotonous sequence of medium- to fine-grained sands interbedded with organic clays or silts, with infrequent zones of marine shells and marl. Few dense limestone beds are recorded, and coarse-grained deposits—coarse sand and gravel—generally do not occur below depths of about 1,000 feet. The character of most of the deposits indicates a deltaic or near-shore marine origin (Jones et al., 1954).

The composite log of the South Pass Block 27 field (Figure 24.10) is typical of the shoreline zone (Smith, 1961). This field lies in offshore Plaquemines Parish between two major distributaries of the Mississippi delta. The log shows more than 13,000 feet of section ranging in age from Recent to upper Miocene. The spontaneous potential curve (left side) of the electric log shows by "kicks" to the left the many sandstones in the section, whereas the resistivity curve on the right shows, by kicks in the opposite direction, the presence of hydrocarbons in the sands. A total of 16 reservoirs was producing at the time this composite log was prepared. All are in the Pliocene–Miocene section. The Venice field in onshore Plaquemines Parish is reported to have "approximately" 37 Miocene reservoir sands (Oil and Gas Jour., July 11, 1955, p. 122).

Offshore Louisiana, especially out from the Mississippi delta, contains a thick marine Pleistocene section (the Pleistocene trend on Figure 24.2) that yields oil and gas. It is believed that the oil occurring in the older Pleistocene sands, at least, is indigenous (Andrews, 1961).

The Gulf Coast district is truly electric log country, because of the predominance of shale and sandstone in the section.

Traps and Selected Fields. Spillers (1965) has made a detailed analysis of south Louisiana traps in terms of estimated ultimate recoveries. A total of 690 fields, both onshore and offshore, was included in this study. Traps are classified into six types:

1. Salt domes, in which intrusive salt has been penetrated by the drill.
2. Circular to elongate domes; underlying flowage salt suspected but not yet drilled.
3. Anticlinal closures associated with growth faults.
4. Fault closures.
5. Regional nose closures.
6. Sandstone pinchouts on the flanks of regional noses.

The first two types of traps, produced by known or probable salt flowage, account for over four-fifths of the estimated ultimate recoveries of oil and two-thirds of the gas and condensate. Nearly half of the known gas and condensate is trapped in the nonpiercement domes, whereas the intrusive domes contain 62 per cent of the oil.

The first oil field in Louisiana, the *Jennings* (Roach, 1943), is in Acadia Parish in the middle (Marginulina) belt. The discovery well for this field was located, because of gas seeps, over what later proved to be a piercement salt dome. It found noncommercial gas in a shallow supercap sand. Shortly afterward, production was obtained from cap rock, and this reservoir was the sole source of oil between 1902 and 1929. During the latter year the first flank production was discovered at a depth of more than 7500 feet. However, extensive exploitation of the dragged-up, complexly faulted, flank sands did not begin until 1936.

The hiatus between the discovery of shallow supercap reservoirs and that of flank reservoirs was even greater at *Anse la Butte* dome (Bates and Wharton, 1943). This salt-cored structure was discovered in 1901, but it was not until 1940 that flank production was found. At Anse la Butte the salt rises to within 160 feet of the surface. The reservoirs range in age from Pleistocene to Miocene and in depth from a few hundred feet to more than 9000 feet.

The first salt dome found to be oil bearing east of the Mississippi River was the *Darrow* dome (Cook, 1938), which lies on the left bank of the Mississippi River, between Baton Rouge and New Orleans. The Darrow dome produces from both supercap and flank Miocene sands.

Although virtually all subsequent discoveries in the open waters of the Gulf of Mexico have been on piercement domes, the first offshore field, the *Creole*, illustrates fault trapping and is described later (p. 212).

All of the various types of salt dome traps (supercap, cap rock, and flank) are illustrated in the Louisiana Gulf

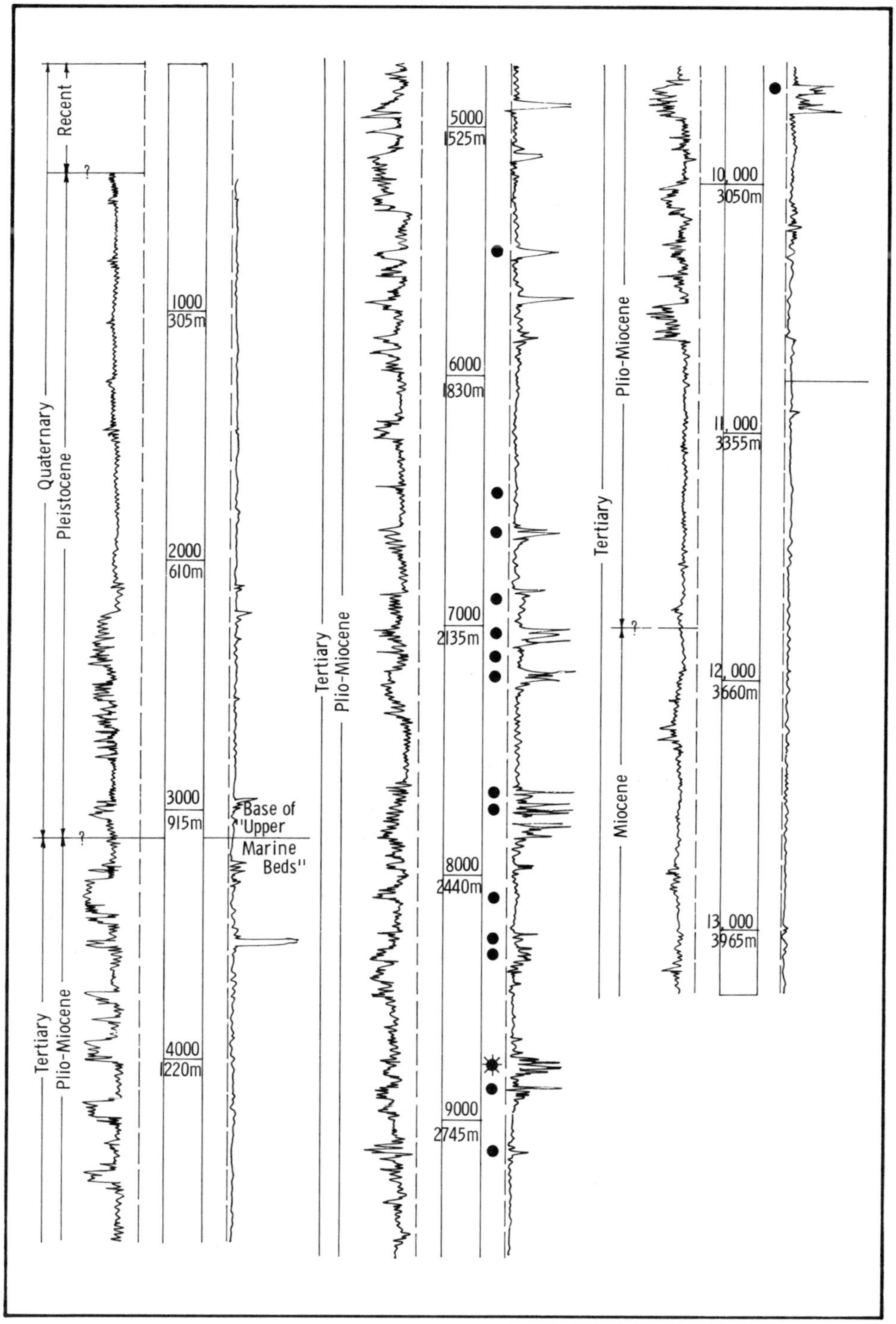

Figure 24.10 Composite type log, South Pass Block 27 field, offshore Louisiana. ***After Darrell A. Smith (1961). Courtesy American Association of Petroleum Geologists.***

Coast. The flank traps include sealing of dragged-up reservoirs by:

1. Pinchout of the sandstones due to either pre-existing lenticularity or drag.
2. Wedgout against dragged-up pre-existing planes of unconformity.
3. Faults.
4. Abutment against the "intrusion" itself. This may be truncation of the reservoir against the overhanging

salt mass, or it may be abutment against the flank of the intrusion. In all cases of sealing by truncation and abutment the actual sealing agent is rarely the salt itself but is usually either the anyhydrite that developed around the periphery of the salt intrusion or diapiric shale (Atwater and Forman, 1959) that was "squirted" upward by flowage in a similar manner to the salt.

The fact that it may take many exploratory dry holes to discover flank production is due to (1) the small area of the different pools, caused by the steep dip of the dragged-up reservoir rocks, and (2) the circumstance that (so far at least) salt dome flank accumulations of hydrocarbons appear to be confined to one or at most two quadrants about the periphery of the intrusion. "There appears to be no consistent preferential geographic location of these prolific segments, and their apparent random positions on the individual structures contribute greatly to the exploration problems of piercement salt domes" (Atwater and Forman, 1959, p. 2592).

Also variable is the depth level at which sealing and accumulation may take place. One deposit of gas and condensate was discovered after drilling through a salt overhang almost 13,000 feet thick. This was at *Cote Blanche Island*, St. Mary Parish. The discovery is credited to "subsurface and geophysical data combined with experience in deep drilling in the nearby Weeks Island salt" (Stern, 1955).

The volume of salt contained within these flowage masses is astronomic.

> The Lake Washington dome containes 48 cubic miles of salt down to a depth of 16,000 feet. If the salt core is projected downward as a cylinder from that depth to a depth of 50,000 feet, which appears to be a reasonable estimate of the depth to the mother salt, it would contain 340 cubic miles of salt. Other salt cores in the area are known to be even larger (Atwater and Forman, 1959, p. 2600).

Exploration along the Gulf Coast shoreline area to depths to and beyond 20,000 feet has shown that some salt domes are cupolas or nodes projecting upward from very much larger salt flowage bodies (massifs).

> The *Marchand–Timbalier–Caillou Island* salt massif, at a depth of 20,000 feet, is more than 27 miles long and slightly more than 13 miles wide. It has an east-west orientation. The massif has an area of 200 square miles, as mapped at a depth of 20,000 feet, and contains 265 cubic miles of salt down to that depth. When the massif is projected downward with vertical sides from that depth to 50,000 feet, it contains 1400 cubic miles of salt (Atwater and Forman, 1959, p. 2601–2602).

A well known salt dome on the Louisiana Gulf Coast is *Avery Island* in Iberia Parish (Bates et al., 1959). This is one of the famous "Five Islands," where five separate salt intrusions have produced low hills that rise above the surrounding sea level marshes. These islands have been areas of salt exploitation, first by brine springs and quarries and subsequently by underground mines, for many decades. Figure 24.11 shows a generalized block diagram of the Avery Island salt dome. The reservoirs are truncated, flanking upper and middle Miocene sands that range in depth from 4500 to 15,600 feet. As can be seen in the diagram, sealing has been brought about by both radial faults and updip termination against the intrusive mass.

Radial faults are of common occurrence surrounding the piercement salt domes. For example, the *Bay Sainte Elaine* oil field in Terrebonne Parish (Schneider, 1959) is surrounded by four major radial faults that are parts of a regional fault. The largest of these, which is on the north flank, is about ½ mile in width at the periphery of the salt mass, but 2 miles to the north the fault zone is 2 to 3 miles wide. These faults have contributed to oil accumulation. Other types of traps are pinchouts against the salt intrusion, and facies changes that result in an updip decrease in permeability.

Peripheral faulting is also present in Gulf Coast salt domes. Figure 24.12 shows faulting of this type at *Bayou Blue* salt dome in Iberville Parish (Mais, 1957). Oil was originally discovered in the radially faulted Pliocene and upper Miocene sands above the apex of the salt intrusion, but subsequently flank production was found from sands of the lower Miocene. The peripheral faults are downthrown away from the salt plug for vertical distances as great as 4600 feet. These faults are not straight but are arcuate and even sinuous. They may join the salt face along the strike or at depth. The largest accumulation so far discovered is on the southwest flank, where the sandstone reservoirs in the downthrown block abut against the steep (faulted) salt face.

Eocene (Cockfield) sand reservoirs produce oil in the *Cheneyville* field in Rapides Parish. Oil accumulation "is localized by a complex mosaic of normal faults surrounding the salt plug" (Fisk, 1940, p. 199). The mosaic is created by a combination of major faults radiating out from the salt core and secondary faults that tend to parallel the periphery of the salt intrusion.

Still another type of piercement salt dome structure is characterized by a central graben. The *Iowa* oil field (Paine, 1962, p. 172-175), which laps over from Jefferson Davis Parish into Calcasieu Parish, is an example. Most of the early production from this field came from shallow sands in an elongate anticline creasing the central graben block. The two faults bounding the graben come together at a depth of 8000 feet, or about 900 feet above the top of the salt plug. Subsequently flank traps that have yielded considerably more oil were discovered.

The simultaneous production of salt and oil from piercement domes has been fairly commonplace in the

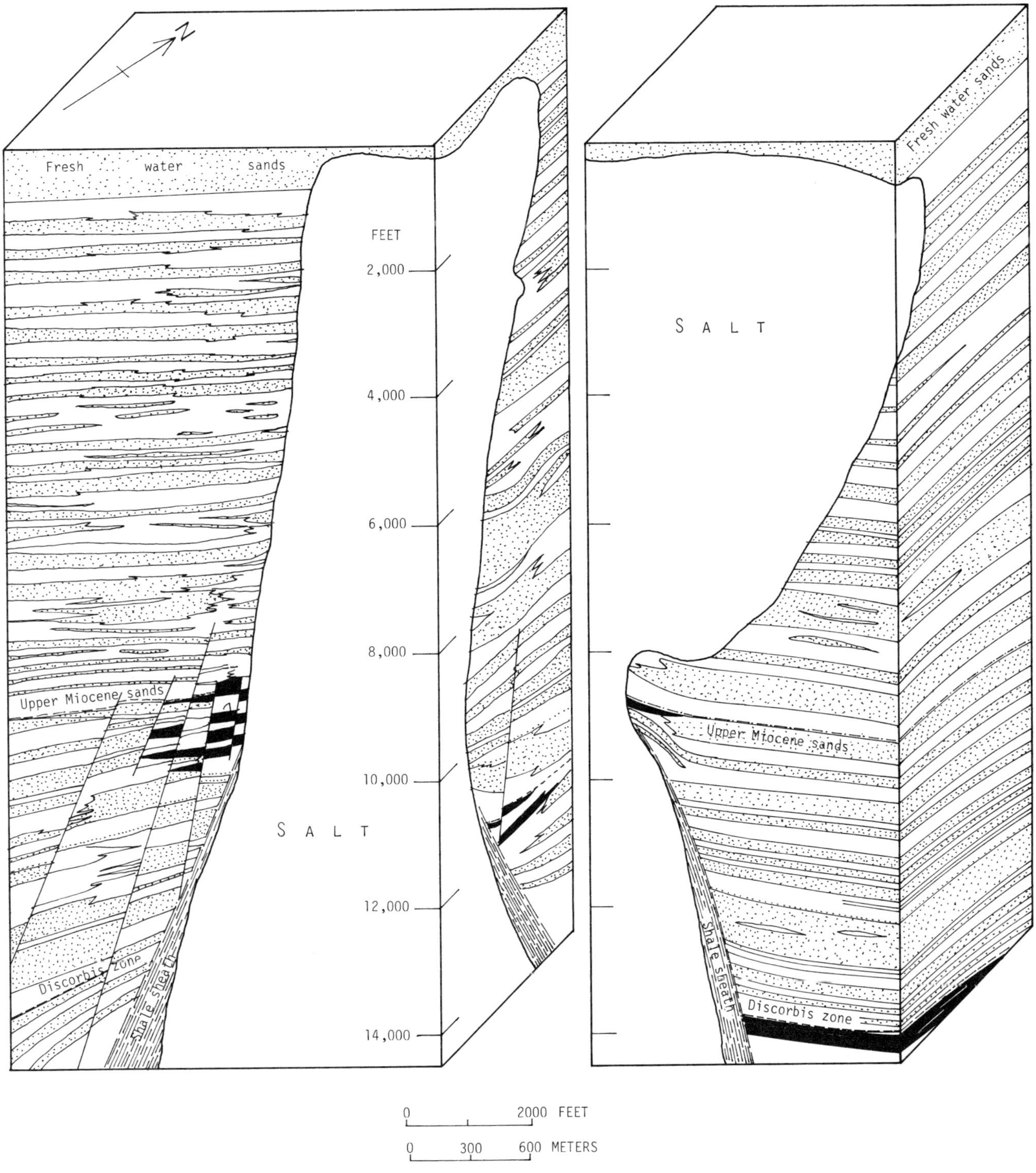

Figure 24.11 Avery Island salt dome and oil field, Iberia Parish, Louisiana. Generalized block diagram showing oil (solid black) trapped in faulted sediments flanking the salt intrusion. *Courtesy F. W. Bates et al. (1959, p. 950) and the American Association of Petroleum Geologists.*

Gulf Coast, but the concurrent production of sulfur and oil is more unusual. Figure 24.13 shows a cross section of the top of the salt intrusion in the *Block 18* field in the Grand Isle area. This field lies about 7 miles offshore in water depths of 45 feet (Atwater, 1959, p. 428). The cross section shows the cap rock overlying the salt and the position of the sulfur-bearing limestone within the cap. The insert map shows the location of the sulfur-

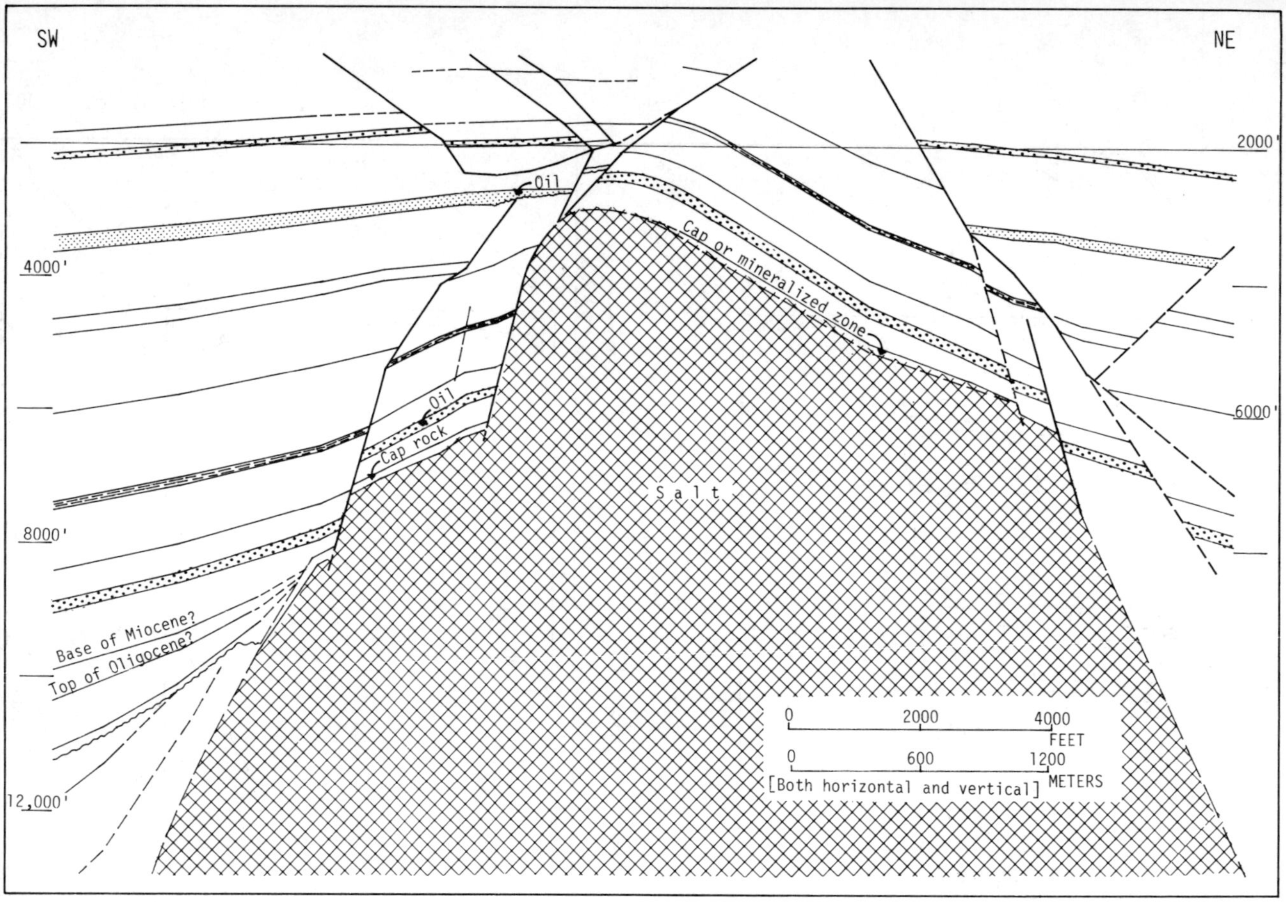

Figure 24.12 Bayou Blue salt dome, Iberville Parish, Louisiana. Cross section showing peripheral faulting. *Courtesy William R. Mais (1957, Figure 14, p. 1934–1935) and the American Association of Petroleum Geologists.*

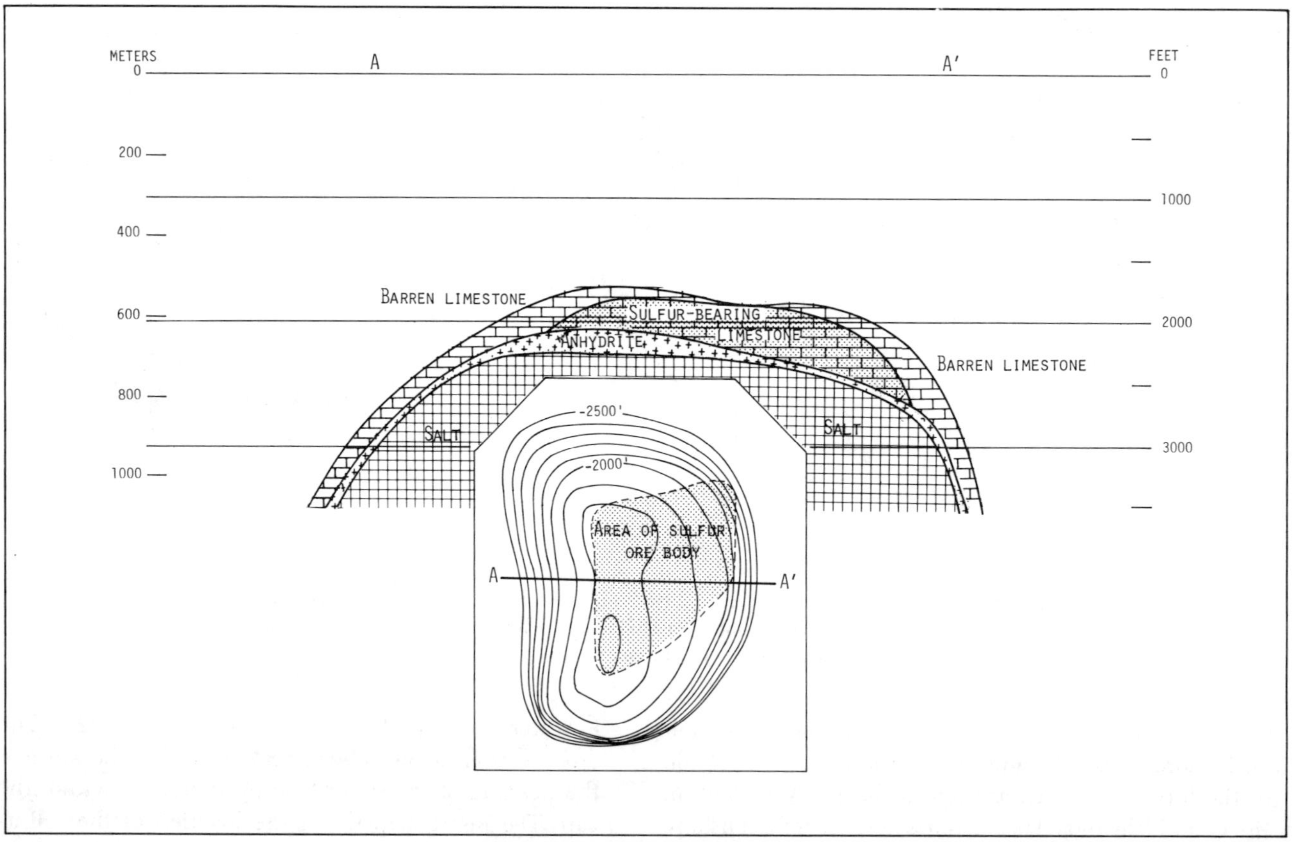

Figure 24.13 Cross section of upper part of Block 18 (Grand Isle) salt dome, showing location of sulfur-bearing limestone. Areal distribution shown in insert, which is a structure contour map based on top of cap rock. Contour interval, 100 feet. Distance AA′, 500 feet. *Courtesy Gordon I. Atwater (1959, p. 431) and the Fifth World Petroleum Congress.*

limestone lens in terms of the structure contours drawn on top of the cap rock. The sulfur is obtained by wells, using the Frasch process; the entire operation is carried on from platforms. Commercial oil production is obtained on the flanks of the dome.

In addition to the many known piercement domes in the Louisiana Gulf Coast, there are also productive anticlines in which the presence of a deep salt dome is suspected (Type 2). In some instances this suspicion may be supported strongly by the presence of a negative-gravity anomaly; in others the nearly circular shape, radial faults, and abrupt character of the anticline imply salt intrusion at depth. An example of the latter type is *Erath* (Dobie, 1964; Steig et al., 1951) in Vermilion Parish. This field is a major source of gas condensate. Figure 24.14 is a structure contour map based on the 11,300 foot reservoir sand. In all there are about 20 sandstone reservoirs, all Miocene in age, lying at depths of 8000 to 12,000 feet. The aggregate sand thickness is over 1000 feet. The trap, a nearly circular anticline, is cut by a dendritic fault that divides it into three principal segments. The first pool was developed around the top of the dome, south of the main fault, after discovery in 1940. Fourteen years later a second accumulation was discovered in the northeast fault block, and the next year the northwest fault block became productive. These two extensions doubled the size of the Erath field.

A similar deep-seated salt dome presumably underlies the *Phoenix Lake* field, which is mostly in Calcasieu Parish but extends westward across the Sabine River into Orange County, Texas (Wilson, 1952). This dome is also faulted with the down side to the southeast, but in this case the productive area lies in the up block to the northwest. The presence of a gravity minimum tends to confirm the conclusion that the domal structure is due to upward flowage of salt at depth.

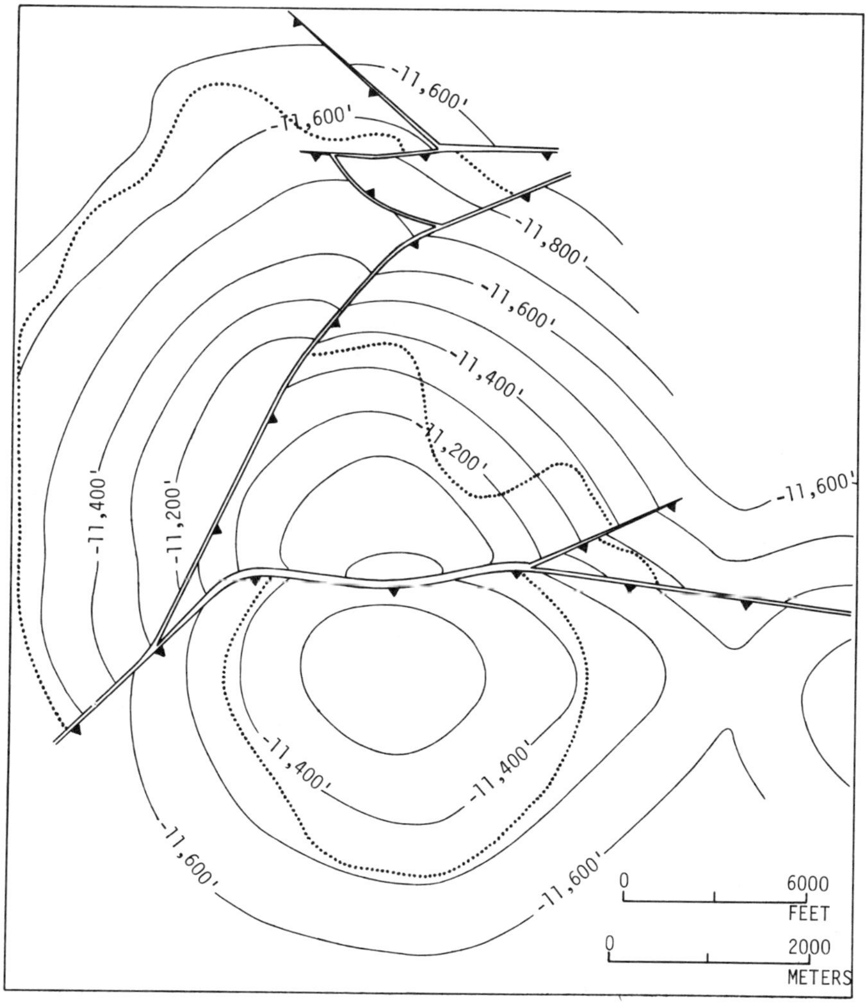

FIGURE 24.14 Erath field, Vermilion Parish, Louisiana. Structure contour map based on 11,300 foot A sand. Contour interval, 100 feet. Dotted lines show pool boundaries. Gas condensate field. *Courtesy C. Walter Dobie (1964) and the Lafayette Geological Society.*

The *Delta Farms* field (Morgan, 1953), as contoured on the Miocene (?) 9350 foot sand, shows a series of gentle domes cut through the middle by a branching fault (Figure 24.15). A gravity low immediately to the northwest suggests deep-seated salt flowage as the cause of this complicated anticlinal structure.

The *Eola* field (Bates, 1941; Fisk, 1940, p. 206-219) is an example of a salt dome satellitic trap. Eola is in Avoyelles Parish in the northern part of the Louisiana Gulf Coast sector. This is in the belt of Eocene production, and the principal reservoir is the Wilcox sand. Figure 24.16 shows the structure contour map drawn on

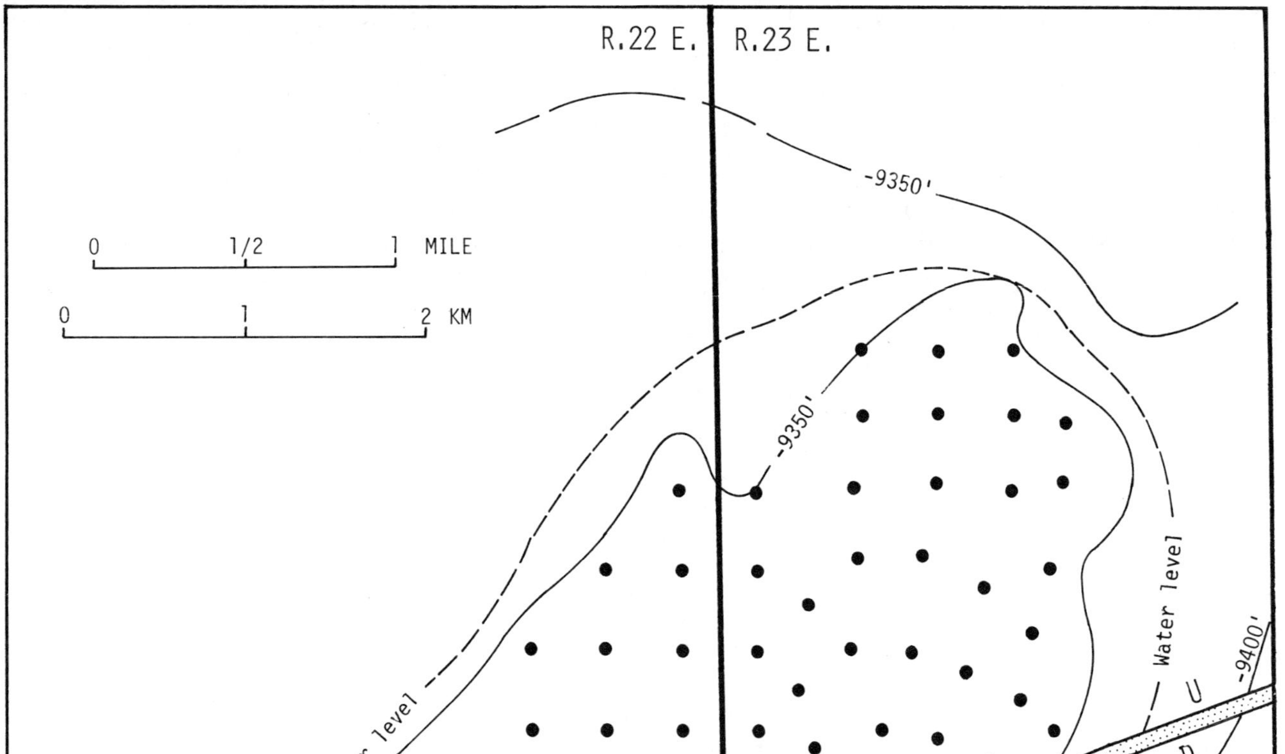

FIGURE 24.15 Delta Farms field, Lafourche Parish, Louisiana. Structure contour map based on 9350 foot sand (Miocene?). Contour interval, 50 feet. *Courtesy A. L. Morgan (1953, p. 2668) and the American Association of Petroleum Geologists.*

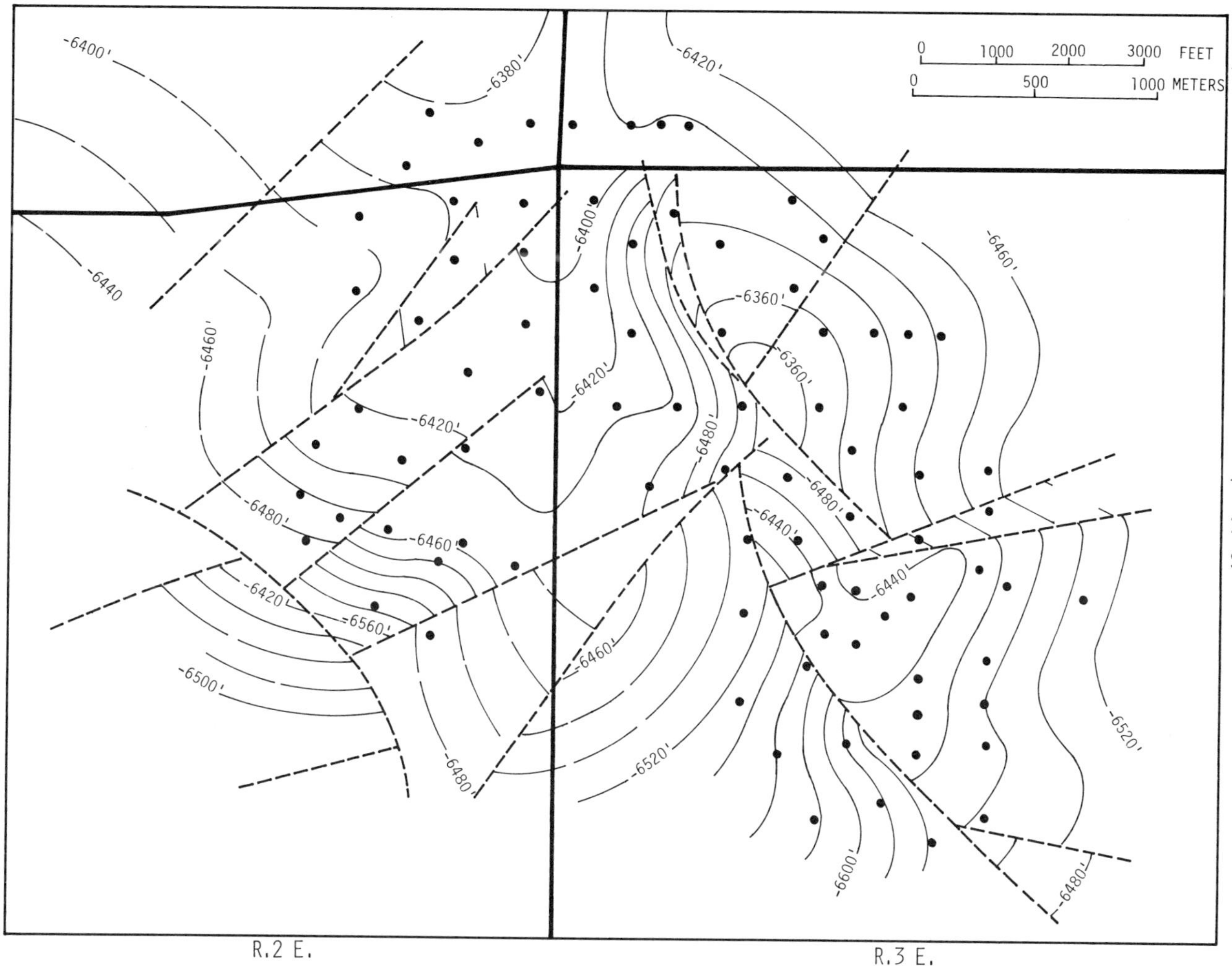

FIGURE 24.16 Eola field, Avoyelles Parish, Louisiana. Structure contour map drawn on top of the Wilcox. All wells shown were completed in the Wilcox. Contour interval, 20 feet. *Courtesy Fred W. Bates (1941, p. 1382) and the American Association of Petroleum Geologists.*

top of the Wilcox. Accumulation has taken place in an oval anticline that is part of a large nose extending to the southeast from the previously described Cheneyville salt dome. The trap owes its closure on the northwest flank to a complex of normal faults that trend northeast-southwest with the down side to the southeast. The tilting of the block to the northwest during faulting increased the structural relief of the Wilcox sand to a maximum of about 400 feet.

Probably deep flowage of salt also has been responsible for trapping at the *Tepetate* field (Lafayette Geol. Soc., 1964, p. 32–32e) in northwestern Acadia Parish. This field was discovered in 1935, after torsion balance and reflection seismograph surveys. The trap is an elongate east-west anticline bordered on the north by a normal down-to-the-coast strike fault. The discovery led to an extensive search elsewhere for "Tepetate type" traps.

The Tepetate field produces from depths ranging from 7500 to 10,000 feet. Eleven Fleming sandstones contain oil and gas across the top of the anticline (Figure 24.17), but the Ortego A is by far the most productive reservoir. The fault on the north flank lies downdip from these accumulations and therefore does not impound any hydrocarbons. Updip lensing out of the Homeseekers A sand on the south flank is responsible for stratigraphic trapping downdip from the main field.

Faulting at Tepetate was accompanied by some concurrent deposition, as shown by the thickening of the reservoir section northward. The field lies in the Oligocene domal trend (Figure 24.19), north of Ocamb's area of growth faulting. Ocamb (1961, p. 154) points out that at Tepetate and elsewhere in the Oligocene belt growth faulting may be present but is relatively minor in extent. Perhaps Tepetate can be considered as a connecting link between Spillers' Type 2 and Type 3 traps; discussion of the latter follows.

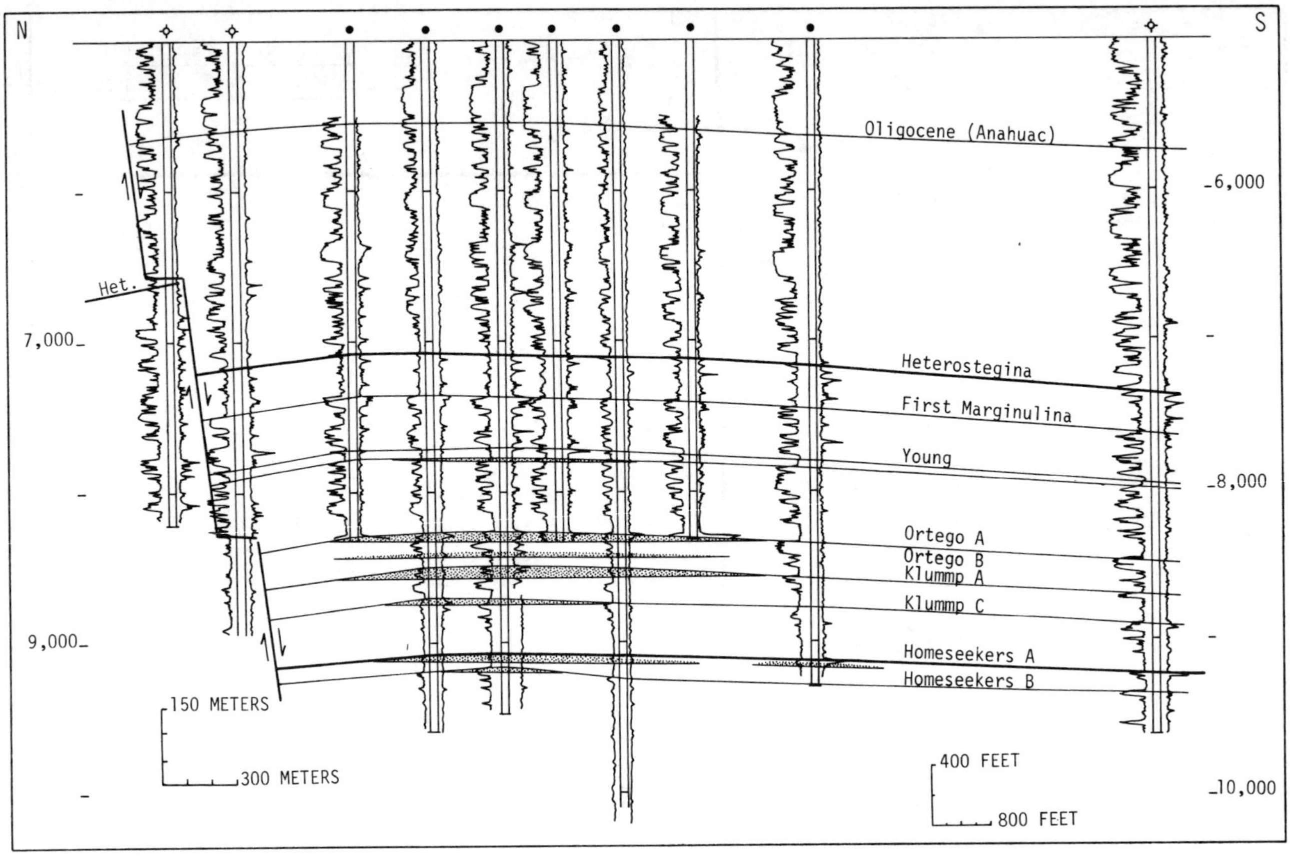

FIGURE 24.17 Tepetate field, Acadia Parish, Louisiana. North-south cross section, showing anticlinal trapping of oil (stippled zones) in down-to-the-gulf fault block. *Courtesy Lafayette Geological Society (1964, p. 32d).*

The next most productive trap in south Louisiana is Type 3 (Spillers, 1965), associated with growth faults. "Growth faults are defined as those faults which have a substantial increase in throw with depth and across which, from the upthrown to the downthrown block, there is a great thickening of correlative section" (Ocamb, 1961, p. 139). The increase in the displacement is due to repetitive faulting; the increase in thickness of section, to contemporaneous sedimentation. Growth faults are also known as contemporaneous faults (Hardin and Hardin, 1961), syndepositional faults, and Gulf Coast type faults, as well as by other names. The concurrent deposition is much greater on the down block (Figure 24.18); a 3000 foot section may be deposited there while only 300 feet of sediment is accumulating on the up block. The Gulf Coast is terraced structurally by normal faults that approximately parallel the shoreline and are mostly down-to-the-coast. The thickening of stratigraphic units in the Gulf Coast geosyncline toward the gulf is well known; growth faults play a major role in this increase in thickness (Lafayette and New Orleans Geol. Societies, 1968, p. 386–387).

Not all faulting in this belt is down-to-the-coast, but most of it is (Ocamb, 1961, p. 143). Where up-to-the-coast growth faulting has taken place, there is a local interruption in the general gulfward thickening.

Although growth faults are a "distinctive feature of south Louisiana geology" (Thorsen, 1963, p. 103), they are not confined to this area. They have been described elsewhere in the Gulf Coast, especially in South Texas (Chapter 27) and to a lesser extent in the Texas Upper Gulf Coast (Chapter 26). Contemporaneous normal faults in Late Mississippian sandstone cropping out in the southernmost Appalachians in central Alabama have been described (Thomas, 1968).

Growth or contemporaneous faults in the Gulf Coast "are largely confined to a belt paralleling the coast and located downdip form the Vicksburg (Oligocene) flexure. . . . In general, the younger the sediment the more numerous are contemporary faults" (Hardin and Hardin, 1961, p. 238). The area of major growth faulting in south Louisiana is shown in Figure 24.19 (Ocamb, 1961, Figure 3, p. 145). It covers only the southern part of south Louisiana. Whether or not a normal fault becomes a

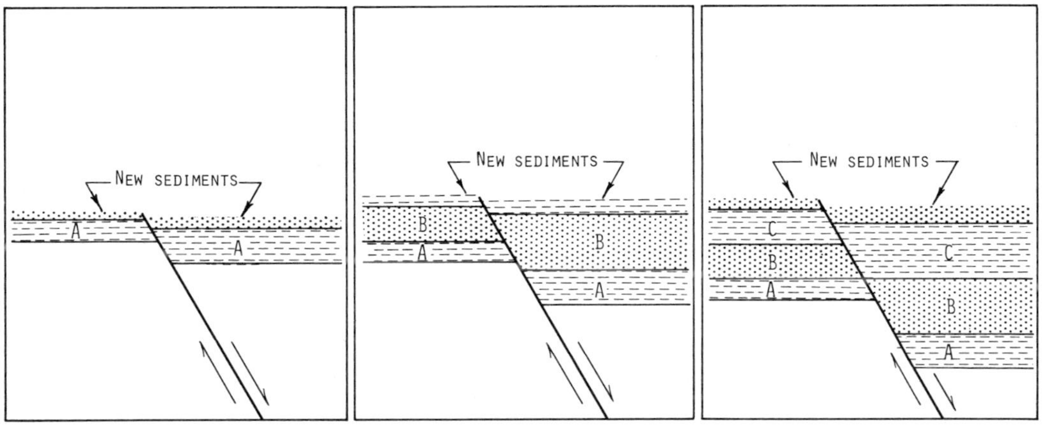

Figure 24.18 Schematic cross sections, illustrating contemporaneous faulting and deposition. ***Courtesy Frank R. Hardin and George C. Hardin, Jr. (1961, Figure 1, p. 240) and the American Association of Petroleum Geologists.***

growth fault is an accident of paleogeography; the happening must take place on a partially submerged coastal plain.

Growth faults are commonly arcuate, with the concave side toward the coast. However, where the fault is up-to-the-coast, the concave side may be toward the interior. Of considerable economic importance is the fact that "anticlines commonly form on the downthrown sides of growth faults and many are important hydrocarbon traps" (Lafayette and New Orleans Geol. Societies, 1968, p. 389).

Authorities on growth faults have noted a conincidence between these faults, and their associated anticlinal trends, with gravity minima. It is quite possible that deep-seated salt flowage is also responsible for the Type 3 trap architecture, in which case all but a small percentage of South Louisiana oil, gas, and condensate production to date has come from traps largely salt-motivated.

A not uncommon feature of normal faulting is "reverse drag" or "rollover" (Figure 24.20). In this case the rock in the down block adjacent to the fault bends downward toward the plane of the fault, instead of upward, because of frictional lag or drag. The result is a reversal of the regional dip, into instead of away from the fault plane.

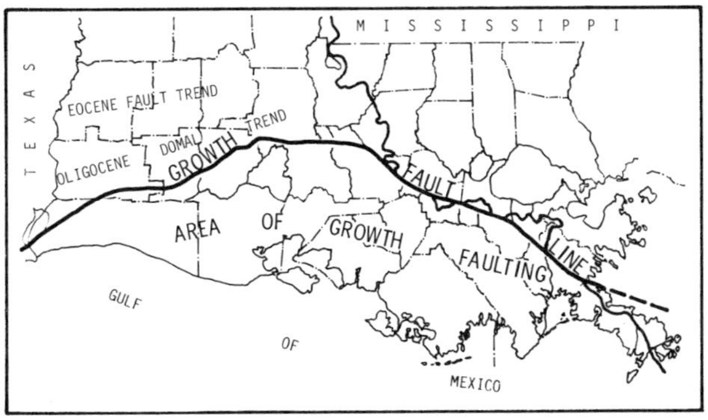

Figure 24.19 South Louisiana, showing area of growth faulting. ***Courtesy Rayburn D. Ocamb (1961, Figure 3, p. 145) and the Gulf Coast Association of Geological Societies.***

Reverse drag is not peculiar to growth faults; Hamblin (1965) describes examples of it visible along fault outcrops in the western Colorado Plateau in Arizona and Utah. A published cross section of the Luling field in South Texas (Brucks, 1929, Figure 4, p. 263) shows rollover at the fault face without any sign of concurrent deposition. Likewise there can be growth faults without rollover (Perkins, 1961). The requisites for trapping can be present without either growth faulting or reverse drag.

Numerous explanations have been offered to explain rollover or reverse drag (Hamblin, 1965). Perhaps the most widely accepted is that there is a rotational movement of the down block as it slides down a curved fault plane. This is readily demonstrated; in fact, it is impossible to move a block down a curved surface *without* creating a tilt of the block into that surface. If the block is flexible, the maximum rollover will be next to the curved surface. In south Louisiana the dip of major growth faults "generally decreases with depth, but may steepen with depth" (Lafayette and New Orleans Geol. Societies, 1968, p. 389). Where the dip lessens with depth, as in Figure 24.20, rollover can be anticipated with any normal fault, growth or nongrowth.

Figure 24.20 (Lafayette and New Orleans Geol. Societies, 1968, Figure 9, p. 387) is a schematic drawing illustrating both growth faulting and possible locales for hydrocarbon traps. The presence of a sealing layer above each accumulation must be assumed. The part played by the rollover in trapping deposits of oil and/or gas is emphasized in this illustration.

The *Lewisburg* field in Acadia and St. Landry Parishes,

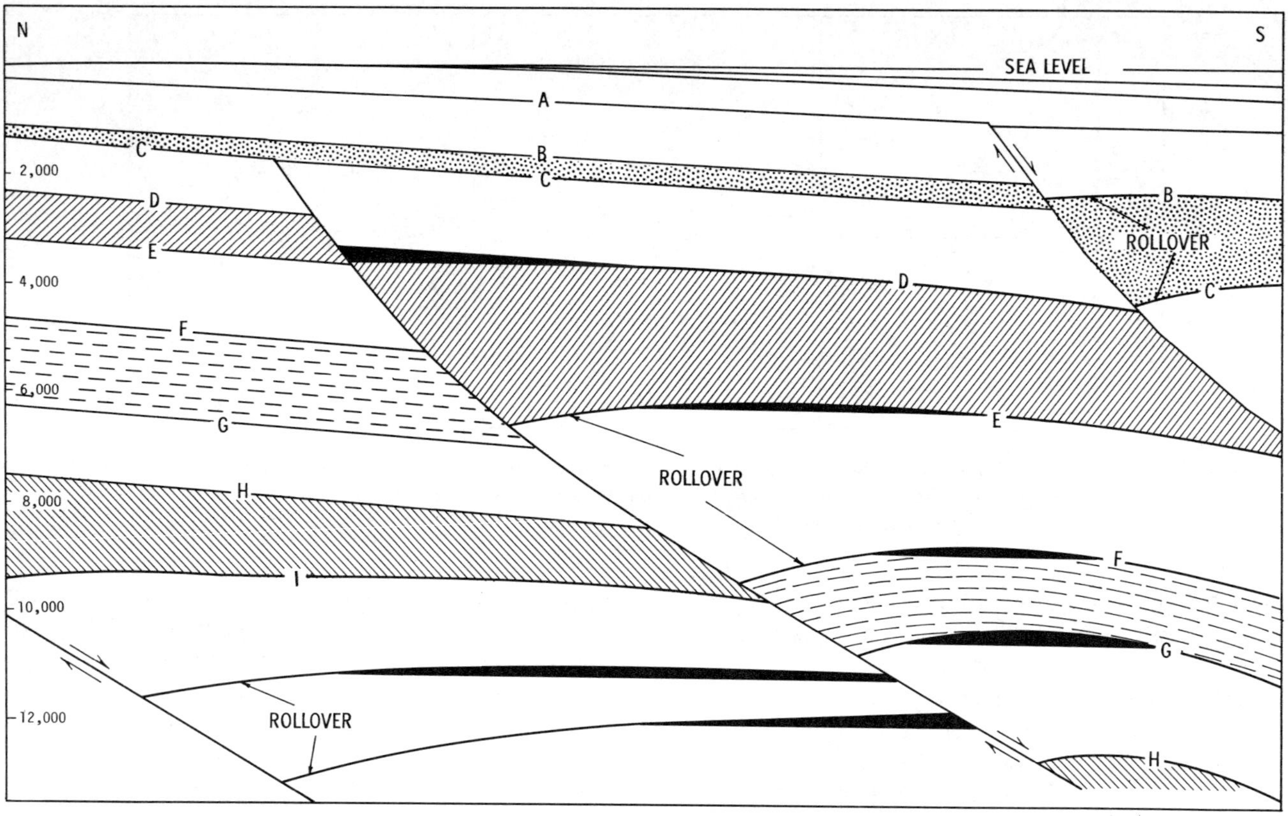

FIGURE 24.20 Schematic cross section from north to south (toward the gulf), showing hydrocarbon accumulations (in black) in traps induced by contemporaneous faulting and deposition. Note (1) development of rollovers; (2) increased thickness of sedimentary units gulfward across the faults; (3) increase in amount of rollover with depth; and (4) general decrease in age of latest fault movement gulfward. ***Courtesy Lafayette Geological Society (Lafayette and New Orleans Geological Societies, 1968, Figure 9, p. 387) and the American Association of Petroleum Geologists.***

which lies close to the northern edge of the growth fault zone (Figure 24.19), provides an excellent example of contemporaneous faulting and deposition. This field has been described by Ocamb and Grigg (1954), Varvaro (1957), and Paine (1962). It was discovered in 1941 after torsion balance, reconnaisance seismograph, and detailed seismograph surveys. Lewisburg is primarily a gas and condensate field, producing from multiple Frio sands in the 10,000 foot range.

The structure on top of the Frio (Figure 24.21), as mapped by Ocamb and Grigg (1954, Figure 4, p. 190), shows five and possibly six down-to-the-coast faults with parallel elliptical anticlines on each down block. The principal reservoirs are the Nodosario sands within the Frio. The isopachous map (Figure 24.22) for the interval between the top of the first Nodosaria sand and the top of the Frio shows a thickening from 1750 to 1900 feet toward the fault in the southernmost anticline, whereas the thickness of the same interval immediately across the fault is only 1300 feet. Obviously deposition took place concurrently with faulting. Furthermore the anticline must be due to rollover, as illustrated in the schematic cross section (Figure 24.19).

Figures 24.21 and 24.22 also show the importance of both structure contour and isopach maps in growth fault trap hunting. Interpretative subsurface geology based on both geophysics and borehole records is essential.

The *Creole* field (Wasson, 1948) in offshore Cameron Parish in southwestern Louisiana is a somewhat spectacular example of fault trapping (Figure 24.23). The productive portion occupies but a small part of a large domed area (probably produced by deep-seated salt flowage) that was first outlined by a seismograph survey. Oil occurs in three sands of middle and early Miocene age ranging in depth from 5100 to 6600 feet. Up-to-the-coast faulting has resulted in the sealing of the sand reservoirs against shale bodies.

Combination trapping is shown in the *University* oil field, which lies east of the Mississippi in East Baton Rouge Parish (Halbouty, 1941). Oil and gas are obtained from four Miocene sandstones ranging in depth from 4300 to 7100 feet. Three of the producing sandstones,

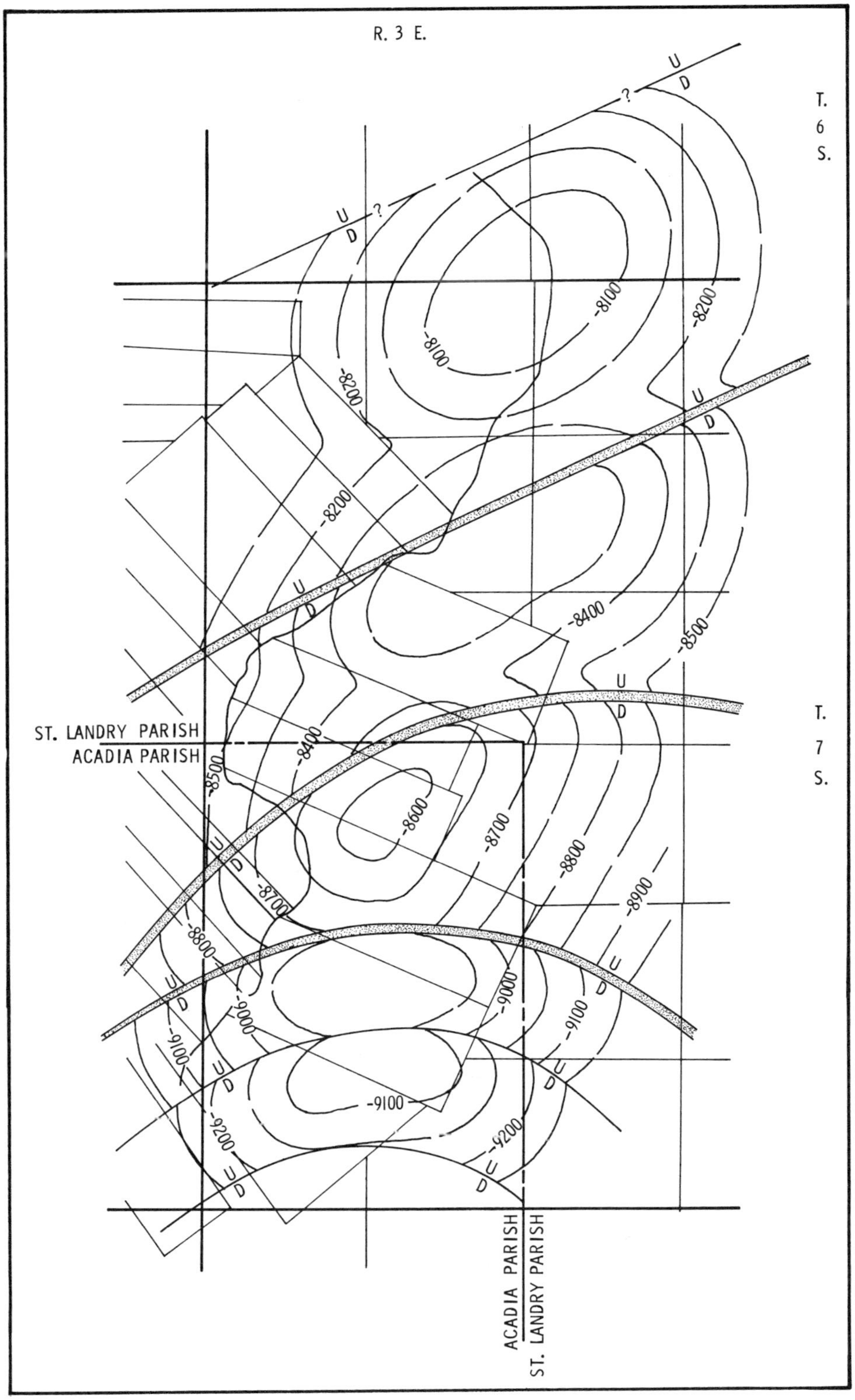

FIGURE 24.21 Lewisburg field, Acadia and St. Landry Parishes, Louisiana. Structure map contoured on approximate top of Frio. Interval, 50 feet. West side of T. 7 S., R. 3 E. is 6 miles (9.66 km).

Higher parts of all five fault block anticlines shown are productive of gas condensate. *Courtesy Ray Ocamb and Robert P. Grigg, Jr. (1954, Figure 4, p. 190), and the Gulf Coast Association of Geological Societies.*

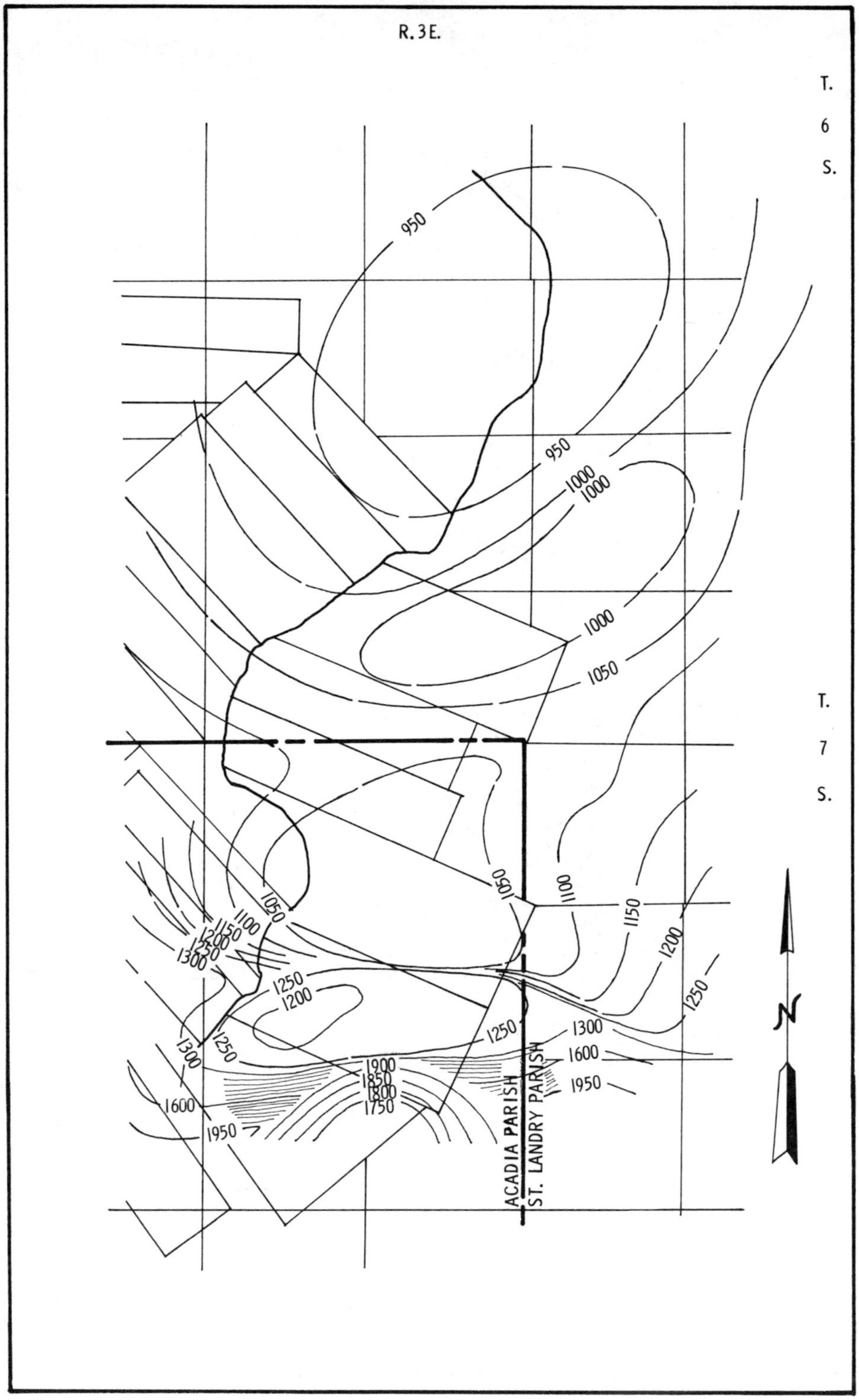

Figure 24.22 Same area as in Figure 24.21. Isopachous map based on thickness of section from the top of the Frio down to the First Nodosaria Sand with all faulting restored. Interval, 50 feet. *Courtesy Ray Ocamb and Robert P. Grigg, Jr. (1954, Figure 8, p. 195), and the Gulf Coast Association of Geological Societies.*

including the main oil-producing sand ("6400 foot"), show structural trapping in a relatively steep-sided dome with a gently rounded top. The "4300 foot" gas sand, however, pinches out to the northwest (Figure 24.24), so that only the southeast half of the dome is productive at this level. The University structure also has probably

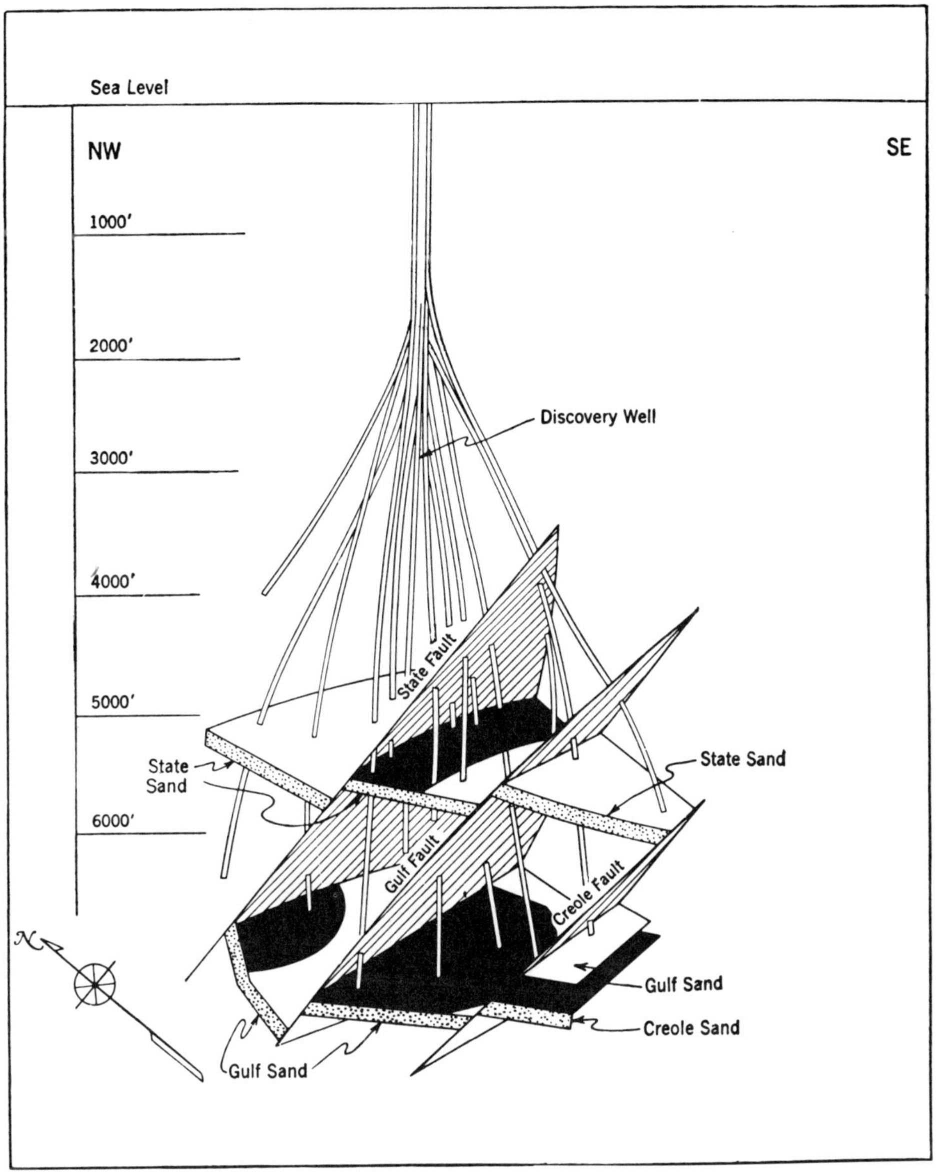

FIGURE 24.23 Model of Creole field, offshore Cameron Parish, Louisiana. Illustrates faulting by up-to-the-coast faults. After Theron Wasson (1948). *Courtesy American Association of Petroleum Geologists.*

been the result of the intrusion of a deep-seated salt mass. Selma (Upper Cretaceous) sediments have been penetrated by the drill at a depth of about 13,000 feet on the west flank of the dome.

True stratigraphic trapping is unusual among the fields so far discovered in the Louisiana Gulf Coast. However, the reservoir sands in the Eocene belt are erratic in thickness; and, where they pinch out updip, trapping can and does take place. The *Glenmora* field (Bates and Copeland, 1952) in Rapides Parish produces from a series of Wilcox sands lying more than 2000 feet below the top of the formation. These sands have erratic permeability, but contrary to the general rule in the Gulf Coast they thin (locally at least) instead of thicken toward the Gulf. Structure contours drawn on formation contacts above the Wilcox show a monocline to the southward with slight nosing. Deeper, at the level of the upper of the two principal Wilcox sand reservoirs, the nose has become a distinct terrace and a closure of about 10 feet appears in the lower major reservoir sand. Production, however, extends beyond the closed area, and it is obvious that erratic permeability is the controlling factor in accumulation.

FUTURE. It would be difficult to find an area with more untested acreage of unusual promise than offshore Louisiana (Rogers and Wilson, 1964). The approximate position of the 600 foot depth contour on the continental shelf is shown in Figure III.1. The fact that more new

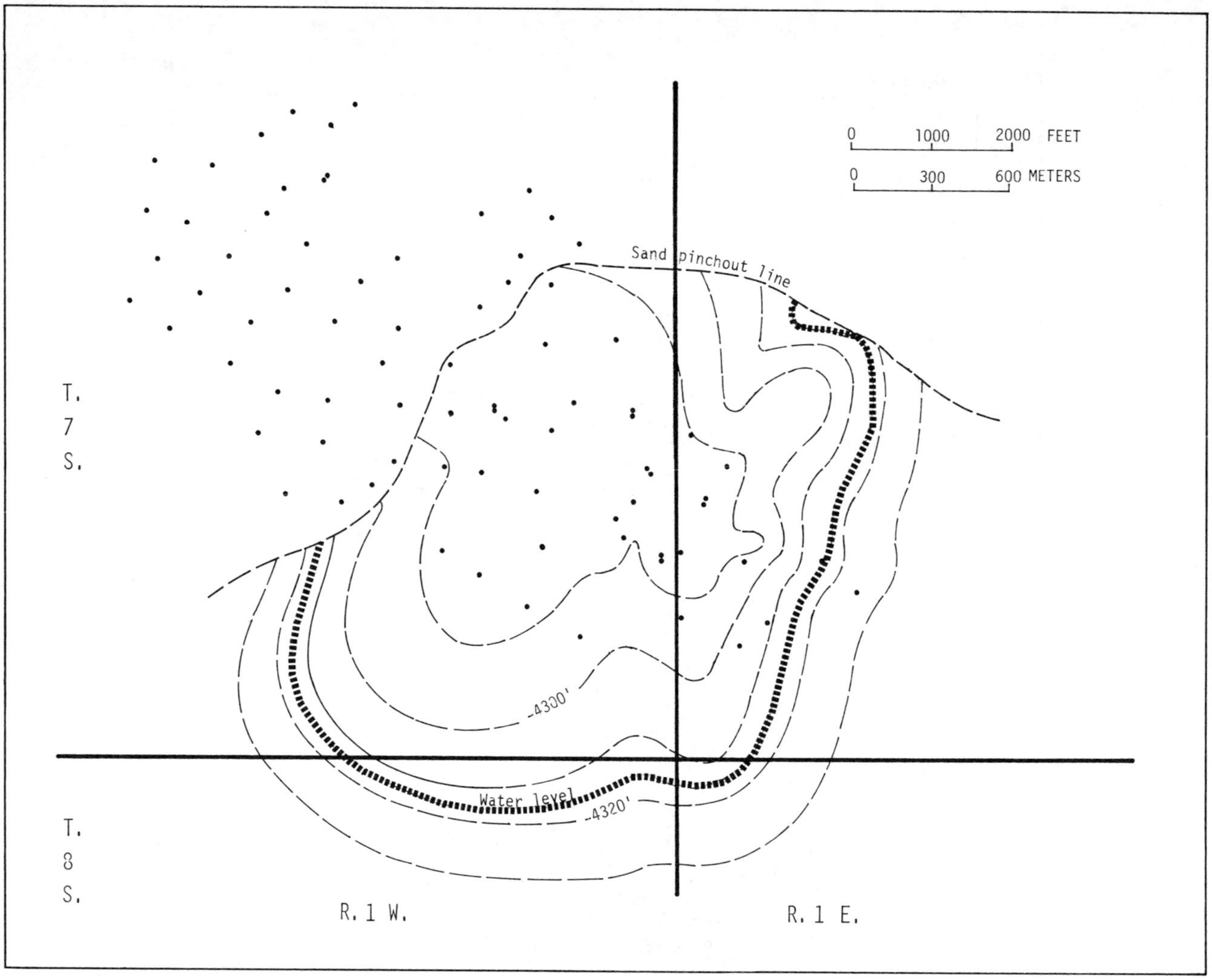

Figure 24.24 University field, East Baton Rouge Parish, Louisiana. Structure contour map based on the −4,300 foot sand, with production in that sand limited by lenticularity. Contour interval, 10 feet. *Courtesy Michel T. Halbouty (1941, Figure 5) and the American Association of Petroleum Geologists.*

fields in Gulf Coast Louisiana were found offshore than onshore in recent years testifies to the fecundity of this vast acreage. Furthermore, whereas *all* of the 86 known salt domes onshore yield hydrocarbons, 21 (as of January 1, 1965) of the 62 known offshore salt intrusions are not yet productive (Louisiana Geol. Survey, 1965). It is not unreasonable to assume that the salt domes will eventually be 100 per cent productive offshore as well as onshore.

Another great but incalculable reserve consists of the pools yet to be found on the flank of already discovered piercement salt domes, both onshore and offshore.

The potential Cretaceous (both Upper and Lower) reservoirs in the Wilcox–Yegua trend are largely untested as yet. The Big Island field in Avoyelles Parish produces from both Wilcox (Eocene) and Tuscaloosa (Upper Cretaceous) sands (Williams, 1962). The presence of Cretaceous reservoirs makes gas and gas distillate production probable.

Actually, Louisiana's already great gas production can be expected to increase throughout the foreseeable future. "Sedimentary rocks in the south Louisiana part of the Gulf Coast geosyncline contain at least 111 trillion cubic feet of gas which is one-quarter of the proved ultimate natural gas reserve of the United States" (Lafayette and New Orleans Geol. Societies, 1968, p. 376). Much of this gas is produced at depths below the probable oil "floor" (Landes, 1967); therefore, as the drill goes deeper, the percentage of hydrocarbons produced as gas and condensate increases. The gas "floor" is an economic level where increased costs and decreased production due to lower porosity (Atwater and Miller,

1965) make the operation nonprofitable.

The small percentage of stratigraphic trap fields in coastal Louisiana previously noted (Lafayette Geol. Soc., 1963; New Orleans Geol. Soc., 1963) is due to the prevalence of depositional facies changes updip from shale to sand, which makes the locating of such traps both difficult and expensive. Understandably, the availability of so many salt domes, and of large anticlines probably formed by salt flowage, all with a good chance of prolific productivity, has motivated most exploration in coastal Louisiana. However, as salt structure discoveries diminish, more attention will be paid to finding the elusive varying permeability traps (Seglund, 1967).

The least productive part of Louisiana to date is the block of eight parishes lying east of the Mississippi River, north of Lake Pontchartrain and both west and south of the state of Mississippi. These are called the Florida parishes (Figure 24.1) because at one time they were part of a political unit known as West Florida (Gardner, 1963a). The area involved is "strategically located between two major producing provinces, the Cretaceous and Wilcox production of south Mississippi to the north and the prolific Miocene production of south Louisiana to the south" (Howe, 1962, p. 121). Except for a single oil field and a one-well gas field in Tangipahoa Parish the only productive areas in the Florida parishes occur where a belt of Mississippi fields cuts across the northeastern corner and where the Fleming trend fields cross the Mississippi River from the west (Howe, 1962). No geological reason why these parishes should be largely unproductive is known. To be sure there are no known piercement salt domes (except in the extreme southwest corner); immediately to the north, however, southwestern Mississippi contains numerous oil fields where accumulation has taken place without benefit of salt intrusion. The leading reservoirs in this area are sands in the Wilcox (Eocene), Tuscaloosa (Upper Cretaceous), and various Lower Cretaceous formations. These sands are present also in the subsurface of the Florida parishes of Louisiana.

REFERENCES CONSULTED IN PREPARING CHAPTER 24

Andersen, Harold V., 1960.

Andrews, Donald I., 1961.

Atwater, Gordon I., 1959.

Atwater, Gordon I., and Forman, McLain J., 1959.

Atwater, G. I., and Miller, E. E., 1965.

Bank of the Southwest, 1956.

Bates, Fred W., 1941.

Bates, Fred W., and Copeland, Robert R., Jr., 1952.

Bates, Fred W., Copeland, Robert R., Jr., and Dixon, Kenneth P., 1959.

Bates, Fred W., and Wharton, Jay B., Jr., 1943.

Billingsley, David L., 1964.

Bishop, William F., 1967.

Bornhauser, Max, 1950.

Brucks, Ernest W., 1929.

Chapman, John J., 1963.

Clark, C. C., 1939.

Clark, G. C., 1961.

Cook, Carroll E., 1938.

Crider, A. F., 1938.

Dinnean, Robert F., 1958.

Dobie, C. Walter, 1964.

Durham, Clarence O., Jr., and Smith, Chester R., 1958.

Fergus, Preston, 1935.

Fisk, H. N., 1940.

Fletcher, Corbin D., 1929.

Gardner, Frank J., 1963a, 1964.

Gordon, Dugald, 1935.

Grigg, Robert P., Jr., 1957.

Halbouty, Michel T., 1941.

Hamblin, W. K., 1965.

Hardin, Frank R., 1968.

Hardin, Frank R., and Hardin, George C., 1961.

Holland, Wilbur C., Hough, Leo W., and Murray, Grover E., 1952.

Howe, Herbert J., 1962.

Jones, Paul H., Turcan, A. N., Jr., and Skibitzke, Herbert E., 1954.

Lafayette Geological Society, 1963, 1964.

Lafayette and New Orleans Geological Societies, 1968.

Landes, Kenneth K., 1967.

Limes, Leonard L., and Stipe, Jack C., 1959.

Louisiana Geological Survey, 1965.

Mais, William R., 1957.

Martin, James L., Hough, Leo W., Raggio, David L. and Sandberg, Adolph, 1954.

Meade, Wade, 1964.

Melchior, Louis F., 1953.

Morgan, A. L., 1953.

Morrow, Ernest H., Skinner, John H., and Hinyard, Paul, 1968.

Nelson, Robert E., 1963.

New Orleans Geological Society, 1963, 1965, 1967.

Ocamb, Rayburn D., 1961.

Ocamb, Ray, and Grigg, Robert P., Jr., 1954.

Paine, William R., 1962.

Pate, Barney F., 1963.

Perkins, Hunt, 1961.

Philpott, T. H., 1952.

Rainwater, E. H., 1964.

Richardson, P. H., 1961.

Riddle, Herschel L., Slingerland, Robert E., and Groenewold, Bernard C., 1968.

Roach, C. B., 1943.

Rogers, Leslie C., and Wilson, Howard M., 1964.

Rogers, R. Lee, 1956.

Schneider, Stephen J., 1959.

Seglund, James A., 1967.

Shreveport Geol. Soc., 1946, 1947, 1953, 1961, 1968.
Sloane, B. J., Jr., 1958.
Smith, Derrell A., 1961.
Spillers, James P., 1965.
Steig, M. H., Nichols, I. K., Gray, Shapleigh G., and Denham, R. L., 1951.
Stern, Allen R., 1955.
Thomas, G. D., 1938.
Thomas, William A., 1968.
Thorsen, C. E., 1963.
U. S. Department of the Interior, 1968.
Valerius, C. N. and Von Croy, Stefan, 1951.
Varvaro, Gasper G., 1957.
Wallace, W. E., ed., 1962.
Wang, Kia Kang, 1952.
Wasson, Theron, 1948.
White, B. Ross, 1966.
Williams, Neil, 1962.
Wilson, Howard, 1962, 1967.
Wilson, R. M., 1952.
Woods, Dalton J., 1963.

Division IV

TEXAS

Texas is not only the leading oil-producing state in this country but also one of the great sources of world oil. During 1967 this state provided 35 per cent of domestic oil production and 9 per cent of the world output. If Texas were a separate country, the rest of the United States would be in a virtual tie (as of 1967) with U.S.S.R. for first place, with Texas fourth, after Venezuela and ahead of Saudi Arabia.

Texas first took the lead within the United States in 1928, and no other state has come close since. The peak year (so far) was 1968, when the yield was 1,140,000,000 barrels. The cumulative production for this state since the Drake well was discovered exceeds 30 billion barrels or more than 36 per cent of the domestic output. A total of 215 Texas counties (out of 254) produce oil or gas, and the number continues to grow.

Texas is also outstanding in gas production, with an output of 40 per cent of the United States total in 1968. About three-fourths of this total came from exclusively gas wells, and the balance from oil and gas wells. The known reserves of both gas and oil in this state are also exceptional. For oil the figure is in the neighborhood of 14 billion barrels, or 45 per cent of the total known U. S. reserves; for gas the comparable figures are 125 trillion cubic feet, and 43 per cent.

Districts. In the separate chapters that follow, Texas has been divided into seven districts (Figure IV.1): Northeast, Upper Gulf Coast, South, West Central, North, Panhandle, and West. The basis for the division is largely geologic, but partly geographic for greater expediency. The divisions are those originally followed by the Bulletin of the American Association of Petroleum

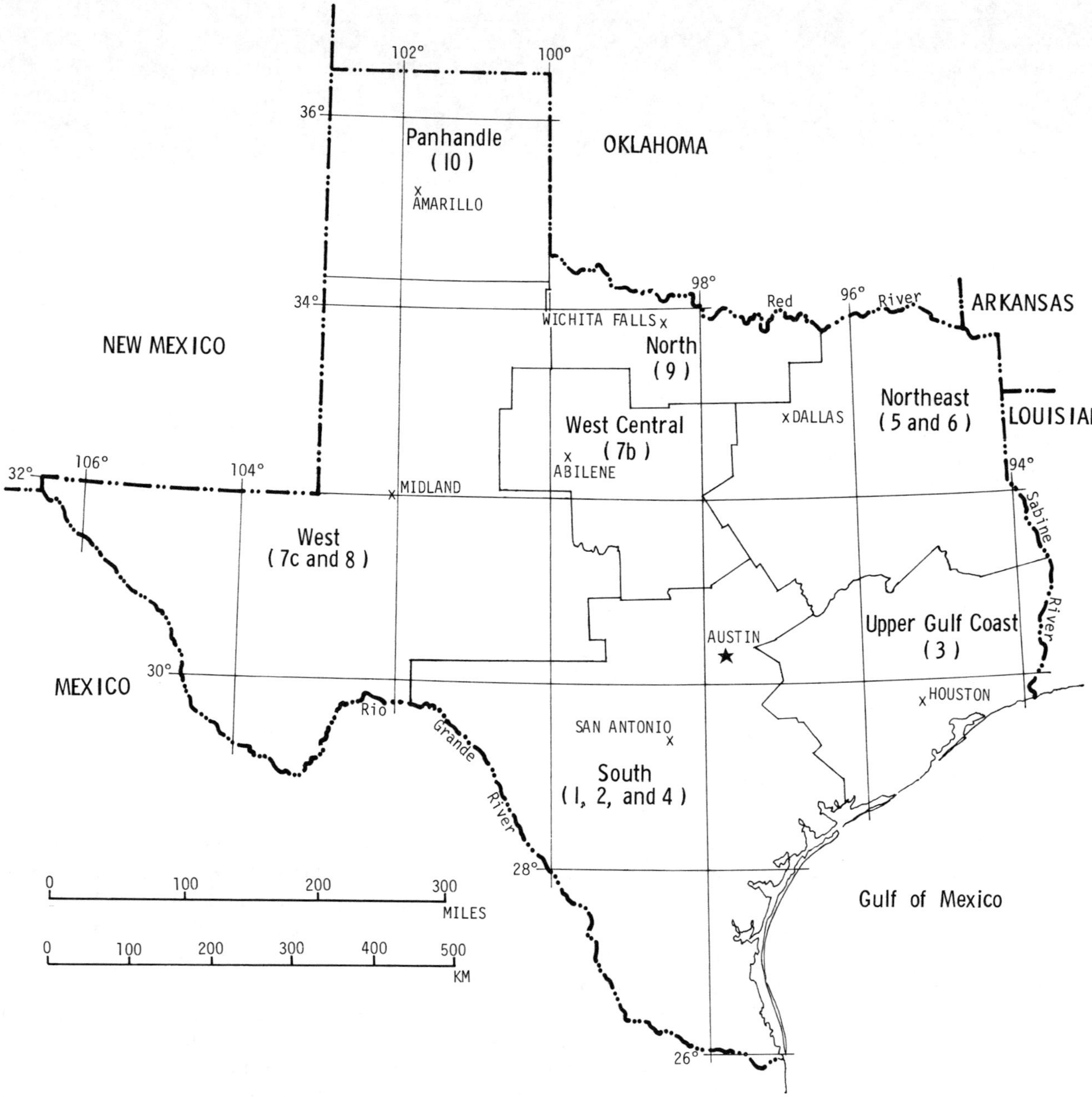

FIGURE IV.1 Texas (Division IV). Index map, showing areas discussed separately in following chapters. Numbers in parentheses are Texas Railroad Commission district numbers.

Geologists in its annual (June) review number devoted to exploration and development. However, beginning with the 1965 review number the Bulletin combined West Central and North Texas into a single district, North Central Texas. The East Texas district of the AAPG Bulletin is here called the Northeast Texas district to avoid confusion with the East Texas oil field. This is the nomenclature used in University of Texas Publication 5116, "Occurrence of Oil and Gas in Northeast Texas" (Herald, 1951). The Texas Railroad Commission (the state oil and gas regulatory body) divides the state into 11 districts (including two subdivisions of number 7), and the scheme followed here does not cross Railroad Commission district boundaries. The Railroad Commission district numbers are shown in parentheses in Figure IV.1.

The three eastern districts (Northeast, Upper Gulf Coast, and South) are a continuation of the southern Arkansas, northern Louisiana, and Louisiana Gulf Coast districts to the east. The four districts to the west are a southwesterly continuation of the Paleozoic producing areas of Oklahoma and Kansas.

The Texas districts vary widely in importance in terms of oil production. In recent years the West Texas district has produced more than 45 per cent of the state's oil.

History (Warner, 1939). Although some oil had been produced previously (the earliest recorded production was in 1889), the first real development of the petroleum industry in Texas took place in 1896 with the discovery of commercial oil in the Corsicana fault trap pool (Rister, 1949) in Navarro County in what is now referred to as the Northeast Texas district (Figure 25.2). However, it was the discovery of the prolific Spindletop field (Figure 26.6) in 1901 that really focused attention upon the potentiality of Texas as an oil-producing state. The years immediately following were exciting ones in the Gulf Coast district, as other large salt dome accumulations were found in rapid succession. The next major event in Texas petroleum history was the discovery of Burkburnett (Figure 28.1) and other large fields nearby between 1911 and 1921. Concurrently, beginning in 1914, salt dome flank sands were found to carry oil in some of the Gulf Coast pools, and exploration took on new impetus in that area. Northeast Texas, the scene of the initial Corsicana discovery, became an important source of oil beginning in 1922. In 1925 West Texas entered the production column in a big way, and a year later oil was discovered in the Panhandle district. The East Texas field, the greatest accumulation yet to be found in the United States, was discovered in 1929. The Conroe field (Figure 26.6), which ushered in the era of deep supercap salt-dome discoveries in the Gulf Coast district, became productive in 1931.

Regional Tectonics (Flawn, 1956; Gibson, 1965). The most striking structural features (Figure IV.2) of Texas are the Gulf Coastal Plain and two uplifts, the Llano and the Marathon. The coastal plain is probably the north flank of a great salt basin now largely covered by the Gulf of Mexico. In any event the sea floor subsided as the Cretaceous and Tertiary sediments were deposited, so that the stratigraphic section along the present shoreline, when completely explored, probably will be found to measure somewhere between 45,000 and 75,000 feet in thickness.

Inland from the coastal plain are the two uplifts, the Marathon in the southern part of western Texas and the Llano in central Texas. These uplifts are similar in that erosion has exposed very old rocks toward the center, and differences in resistance to erosion have caused the development of a fairly rugged topography. Within the coastal plain are lesser downwarps and upwarps and fractures. The westward extension of the coastal plain is known as the Rio Grande embayment (Figure 27.2). The northward sector is the East Texas embayment containing the Tyler basin (Figure 25.3), a nearly circular downwarp that interrupts locally the normal gulfward dip. From the bottom of the Tyler basin the strata rise gently but consistently eastward to the top of the Sabine uplift, a large dome with its highest point in northwestern Louisiana. Toward the inner edge of the coastal plain is a fault zone of great length (Figures IV.2, 25.3 and 27.2). South of the Llano uplift the Balcones fault zone, a major element, trends easterly. Farther east it curves around the Llano uplift, striking first northeast and then north. The Balcones zone appears to lose identity northeast of the Llano uplift, but another and similar fault zone known as the Mexia–Talco, which is offset slightly to the east, takes its place and continues northward. Northwest of the Tyler basin, it curves eastward through a 90° arc, proceeds east across northeastern Texas, and crosses the state line into southwestern Arkansas.

North of the Llano uplift is the Bend flexure or arch, which is in reality a northward plunging spur of the uplift. This warping took place during the Pennsylvanian, and younger Pennsylvanian and Permian sediments were deposited unconformably across the arch so that it is not discernible on the areal geologic map. To the north is the east-west Red River uplift overlapping a part of southern Oklahoma.

West of the Llano region and north of the Marathon uplift is the great Permian basin of western Texas and southeastern New Mexico. This is in reality two major basins, the Midland to the northeast and the Delaware to the southwest, separated by the Central basin platform, a northwest-southeast–trending residual feature that has resulted from the downwarping of the basins on both sides. To the west of the Delaware basin is the Diablo platform; the Midland basin is bordered on the north by the Matador arch.

The Panhandle of Texas is crossed from east to west by the Amarillo uplift, which has a core of Precambrian hills completely buried beneath much younger sediments. This uplift is separated from the Matador arch on the south by the Palo Duro basin. Northeast of the Amarillo uplift is the southwest flank of the Anadarko basin of western Oklahoma.

Stratigraphy. From the geologic map (Figure IV.3) it is possible to divide Texas into two parts, (1) Mesozoic and Cenozoic Texas, and (2) Paleozoic Texas. The former includes all southern and eastern Texas; the outcropping rocks are either Comanchean, Upper Cretaceous, or Tertiary formations that were deposited in an ancient basin and that crop out in bands that roughly parallel the present Gulf Coast shoreline. The rocks increase in age inland. Paleozoic Texas extends over the rest of the state, covering northern and western Texas, including the

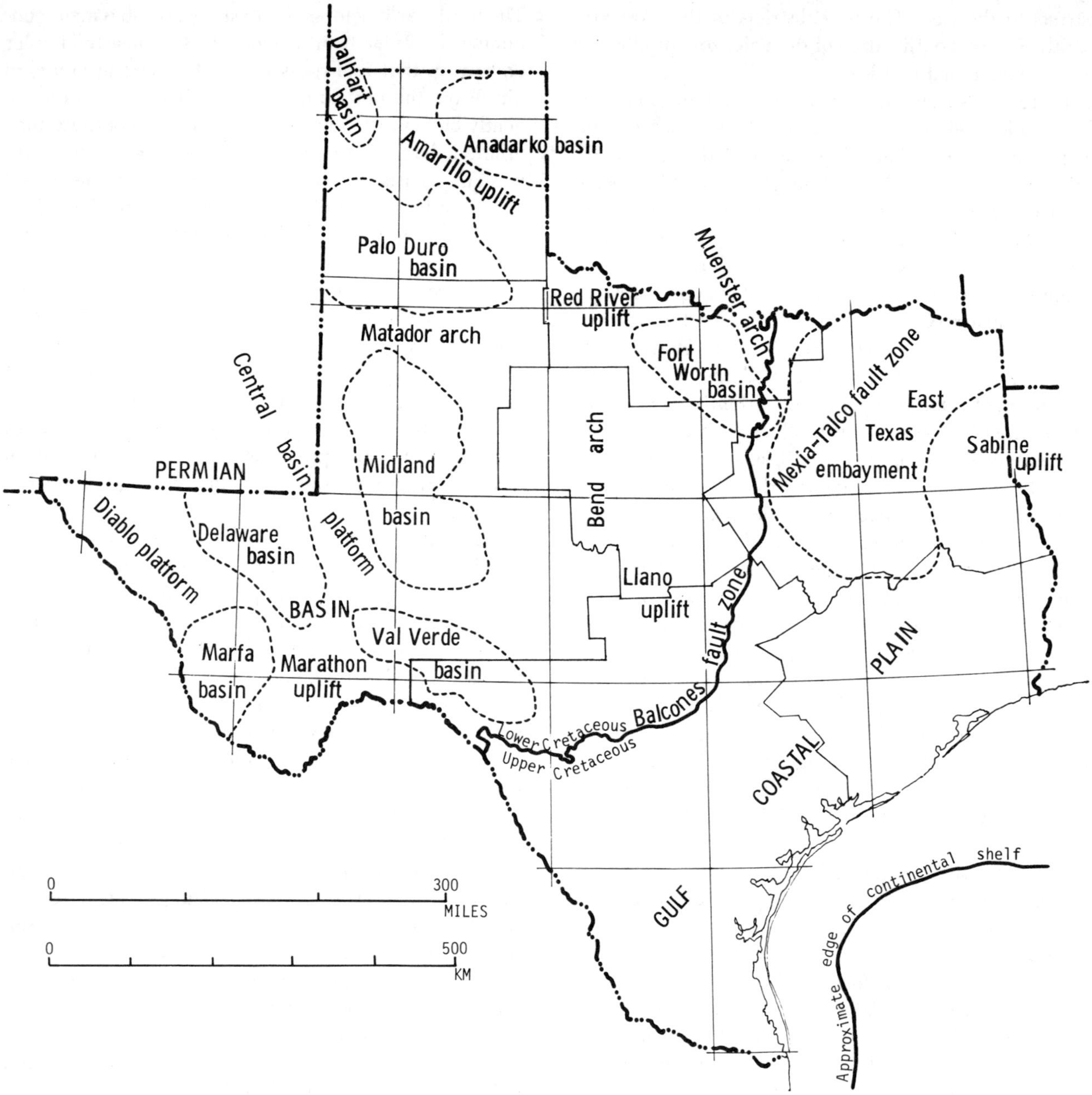

Figure IV.2 Texas. Major tectonic features. Heavy black line crossing state from south to north is contact between Lower and Upper Cretaceous; also the inner edge of Gulf Coastal Plain.

Panhandle. Here Paleozoic rocks, and some Precambrian as in the Llano uplift, lie at the surface except where mantled by a veneer of continental Tertiary or younger fluviatile or eolian sediments. Also covering the Paleozoic rocks in a part of far western Texas are the Tertiary Davis Mountain volcanics.

The stratigraphic section in Paleozoic Texas has Permian rocks at the top in most places. These are largely limestones, dolomites, shales, and evaporites (including salt and even potash), and they reach great thicknesses in the West Texas basins. The Pennsylvanian is especially well represented in the basins flanking the Bend arch with as much as 6000 feet of limestones and clastics. Rocks of Mississippian, Devonian, and Silurian age occur in various parts of this great area but are less abundant. The Ordovician is widely present and is of great importance as a source of petroleum. A monograph (Flawn, 1956) describes the lithology of the Precambrian.

The stratigraphic section in the Gulf Coastal Plain has been explored to below 20,000 feet, but there is without doubt an enormous thickness of rock yet to be penetrated by the bit, especially close to and out beyond the present shoreline. The Quaternary is represented by Recent and Pleistocene muds and sands, and all the Tertiary epochs were times of abundant clastic deposition in this area. Upper and Lower Cretaceous sediments have not been reached by the drill along the shoreline, but have been thoroughly explored in the interior of the coastal plain. The Upper Cretaceous rocks include both clastic and carbonate deposits. The Comanchean (Lower Cretaceous) section contains not only clastics and carbonates but also anhydrite. Beneath the Comanchean is a thick Jurassic section that has been explored by deep drilling in recent years. This section was somewhat unexpected, for the Jurassic deposits in outcrop are relatively insignificant. Below the Smackover Limestone of undoubted Jurassic age is a thick salt and red bed section that may be Jurassic or may be older. Regardless of its geologic age this probably is the formation that supplied the salt for not only the interior but also the coastal salt domes.

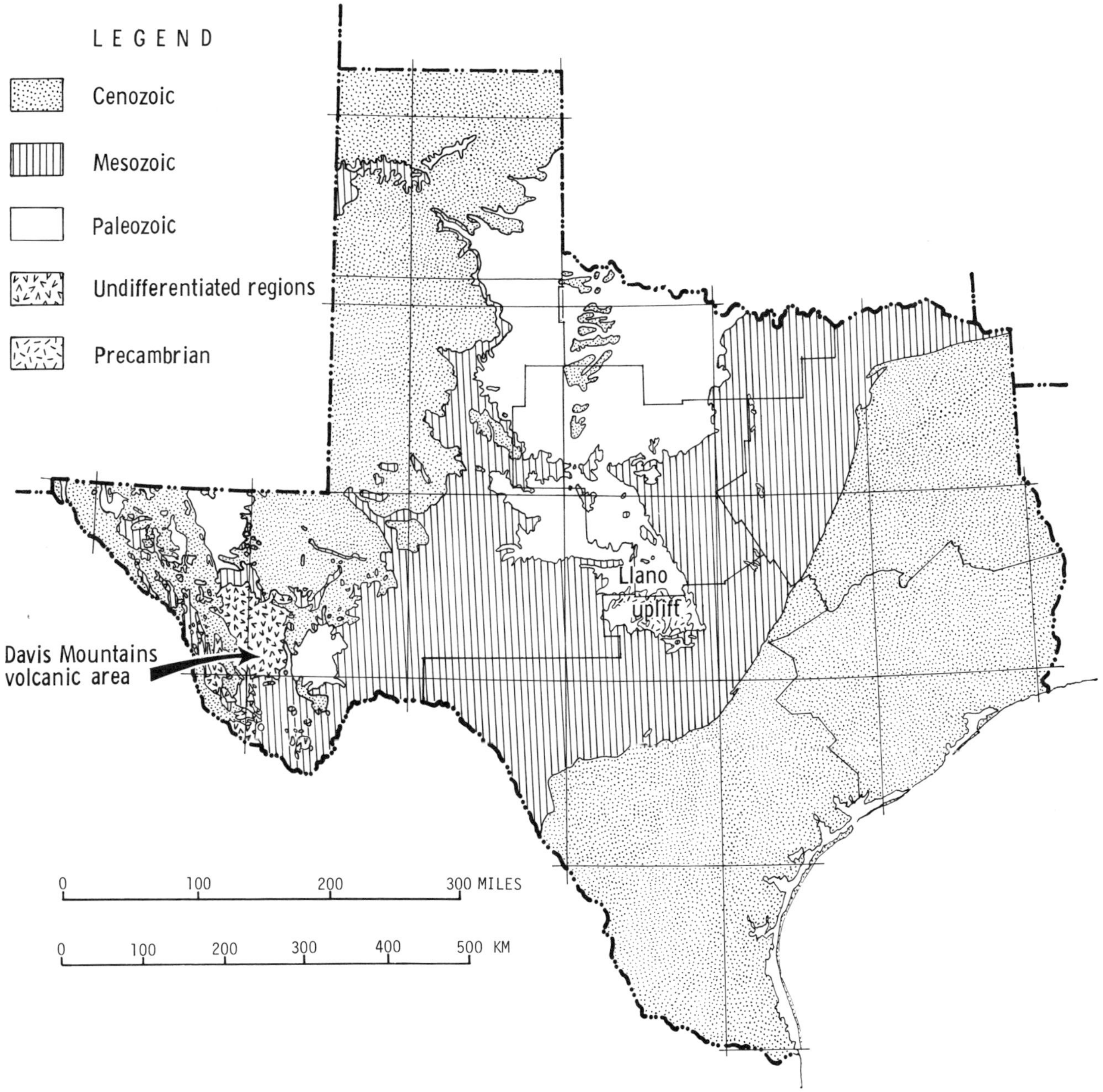

FIGURE IV.3 Texas. Geologic map, showing distribution of rocks in major age divisions and igneous rock areas. Adapted from Geological Highway Map of Texas. *Courtesy Dallas Geological Society (1963).*

The oil and gas reservoirs in Texas range in age from Pliocene to Precambrian. The Tertiary reservoirs are confined to the Gulf Coast belt. Among the well-known producing formations or zones are the Catahoula and younger reservoirs of the Miocene, the Anahuac, Frio, and Vicksburg of the Oligocene, and the Jackson, Claiborne, Wilcox, and Midway of the Eocene. Upper Cretaceous reservoirs include the Navarro, Taylor, and Austin; outstanding is the Woodbine Sand, which carries the oil of the East Texas field and most of the fault pools. The Jurassic Cotton Valley Smackover produces both gas and oil in Northeast Texas. Some production is obtained from the Washita, Fredericksburg, and Trinity of the Lower Cretaceous, especially in Northeast Texas. Permian carbonate rocks, notably the Guadalupian Series, are the principal reservoirs in West Texas and the Panhandle; sandstone reservoirs in the Wolfcamp are important in West Central Texas. The Strawn, Bend, and other Pennsylvanian formations carry oil in North and West Central Texas and farther west. Devonian, Silurian, and Ordovician (especially the Ellenburger limestone) formations are the deep reservoirs of West Texas. Fractured Precambrian volcanic rocks contain oil and gas in the Texas Panhandle.

Exploration. Seeps and other surface signs were responsible for the earlier discoveries in Texas. Areal geologic mapping and photogeology also contributed. For many years geophysical prospecting has been used extensively in the Texas coastal plain. The seismograph was first used here as an exploration tool, and it is still the leading geophysical instrument. The gravity meter is also used. Outside of the coastal plain, subsurface geology is the most common discovery technique, but the seismograph, the gravity meter, and the core drill also are used.

Texas oil and gas fields are shown on the district maps in the chapters that follow. The principal authority for these maps is Vlissides (1964). The only areas of any magnitude without oil and gas fields in Texas are those in which igneous and metamorphic rocks lie at the surface (Figure IV.3) or under a thin sedimentory cover.

REFERENCES CONSULTED IN PREPARING INTRODUCTION TO DIVISION IV

Andrews, Donald I., 1960
Flawn, Peter T., 1956.
Gibson, George R., 1965.
Herald, Frank A., ed., 1951.
Meyerhoff, A. A., 1967.
Moody, Clarence L., 1967.
Murray, Grover E., 1961.
Rister, Carl Coke, 1949.
Southwestern Federation of Geological Societies, 1960.
Vlissides, Sophie D., 1964.
Warner, C. A., 1939.

Chapter

25

NORTHEAST TEXAS

Northeast Texas (Figure 25.1) comprises Districts 5 and 6 of the Texas Railroad Commission (Figure IV.1). It is bounded on the north by Oklahoma, on the northeast by a corner of Arkansas, on the east by Louisiana, on the south by the Upper Gulf Coast, on the southwest by a corner of the South Texas district, on the west by West Central Texas, and on the northwest by North Texas. The Northeast Texas district covers 47 counties, of which 37 are productive. The counties so far dry lie along the western and northwestern fringe of the district (Figure 25.2).

The birthplace of the Texas oil industry was Navarro County, where the first of the great fault line pools, Corsicana, began producing in 1896. After the remaining fault fields had been found, additional finds were slow in coming, and it was not until the discovery of the East Texas field in 1930 that the district once more became prominent. The accelerated exploration brought about by East Texas has resulted in many new oil fields (13 in 1963), and the search is continuing. By the end of 1964, the Northeast Texas district inclued 10 "giant" fields (those which are destined to produce 100 million barrels or more). In 1964 this district accounted for approximately 10 per cent of the overall Texas oil production, with the East Texas field alone supplying two-fifths of that total. The runner-up was the Hawkins field with less than one-fourth of the East Texas field output.

Tectonics. The greater part of Northeast Texas lies within the East Texas embayment, a northward tongue of the Gulf Coastal Plain (Figure IV.2). Northeast Texas tectonics is dominated by the Tyler basin (Figure 25.3). This is a major depression in the embayment, elliptical in shape with its major axis trending slightly east of north. Forming the rim of the basin are the Sabine uplift to the east, in the northwestern corner of Louisiana (see Chapter 24), with a structural relief of some 10,000 feet, the Ouachita Mountains of southeastern Oklahoma to the north, and the Bend arch to the west in the North Texas

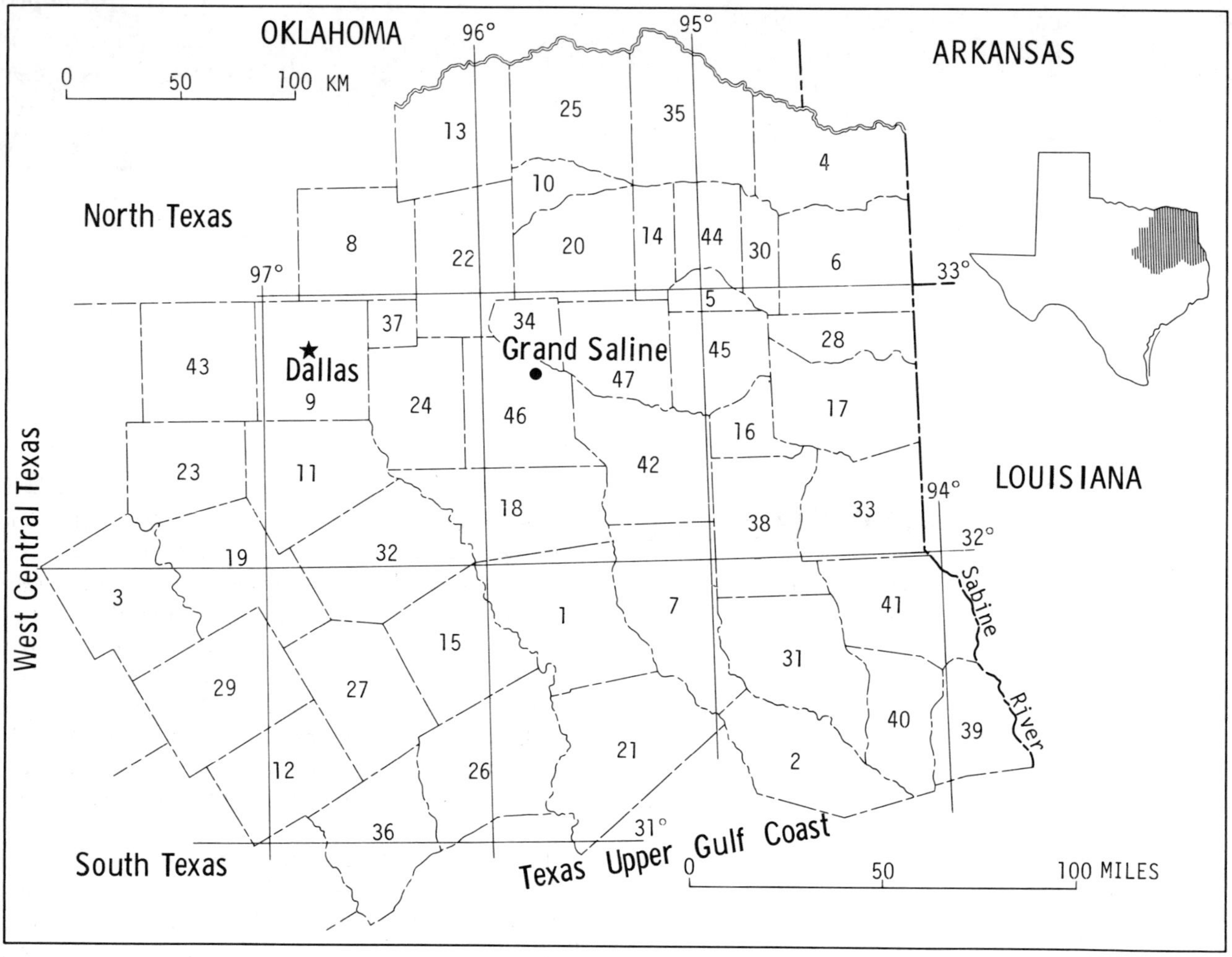

FIGURE 25.1 Northeast Texas. Index map. County Key:

1	Anderson	8	Collin	15	Freestone	22	Hunt	29	McLennon	36	Robertson	43	Tarrant
2	Angelina	9	Dallas	16	Gregg	23	Johnson	30	Morris	37	Rockwell	44	Titus
3	Bosque	10	Delta	17	Harrison	24	Kaufman	31	Nacadoches	38	Rusk	45	Upshur
4	Bowie	11	Ellis	18	Henderson	25	Lamar	32	Navarro	39	Sabine	46	Van Zandt
5	Camp	12	Falls	19	Hill	26	Leon	33	Panola	40	San Augustine	47	Wood
6	Cass	13	Fannin	20	Hopkins	27	Limestone	34	Rains	41	Shelby		
7	Cherokee	14	Franklin	21	Houston	28	Marion	35	Red River	42	Smith		

district. The southern rim is low; it is the structural divide where the normal gulfward dip of the coastal plain sediments is resumed.

The north and west flanks of the Tyler basin are scored by the Mexia–Talco fault zone, which on the west side trends northeast, but in Hunt County veers 60° toward the east and continues into southern Arkansas (see p. 179). This fault zone has an average width of about 5 miles. In simplest terms it consists of a front fault on the gulfward side, a complementing back fault, and an intervening graben. Actually the individual faults are discontinuous, but they overlap each other *en échelon* so that the zone is continuous. Furthermore, multiple parallel faults may divide the graben area into a series of slices, producing structural stair steps down one side and up the other. Fault movements have been recurrent, the first taking place at least as far back as the Early Cretaceous, with the climax occurring during Tertiary time (Hager and Burnett, 1960). An extension northward from the South Texas district of the Balcones fault system, which parallels the Mexia–Talco zone to the west, cuts across central McLennan County. Other faults lie on the west flank of the Sabine uplift (Figure 25.3).

The lowest part of the Tyler basin contains at least 17 piercement salt domes (Eaton, 1956). One salt intrusion, the Grand Saline in Van Zandt County, is being mined for salt, in the course of which enormous rooms have been developed. These rooms are justly famous for the vertical banding to be seen on walls and ceilings (Balk, 1949). The white salt and dark anhydritic layers, originally deposited horizontally, are now vertical because of upward flowage during the salt intrusion (Figure 25.4).

Stratigraphy. The surface rocks in Northeast Texas are exclusively Tertiary and Late Cretaceous in age, except in the westernmost part of the district, where Lower Cretaceous rocks crop out. The general pattern of surface distribution is given by the base-of-Tertiary contact line shown in Figure 25.3. To the east and south of this line successively younger Eocene formations appear; higher Tertiary and Quaternary sediments occur farther to the south, in the Upper Gulf Coast district.

The Upper Cretaceous outcrop bands, successively older to the north and west, parallel the base Tertiary contact line. In this Upper Cretaceous belt some of the most prolific reservoirs, including the Woodbine of the Tyler basin oil fields, can be examined at the outcrop (East Texas Geol., Soc., 1951).

The strike faults of the Balcones system cut Upper Cretaceous rocks in the southwestern corner of the district. To the north and east the Mexia–Talco system strike faults cut both youngest Cretaceous and oldest Eocene formations.

The Lower Cretaceous Washita and Fredericksburg (including the plateau-making Edwards Limestone) sedi-

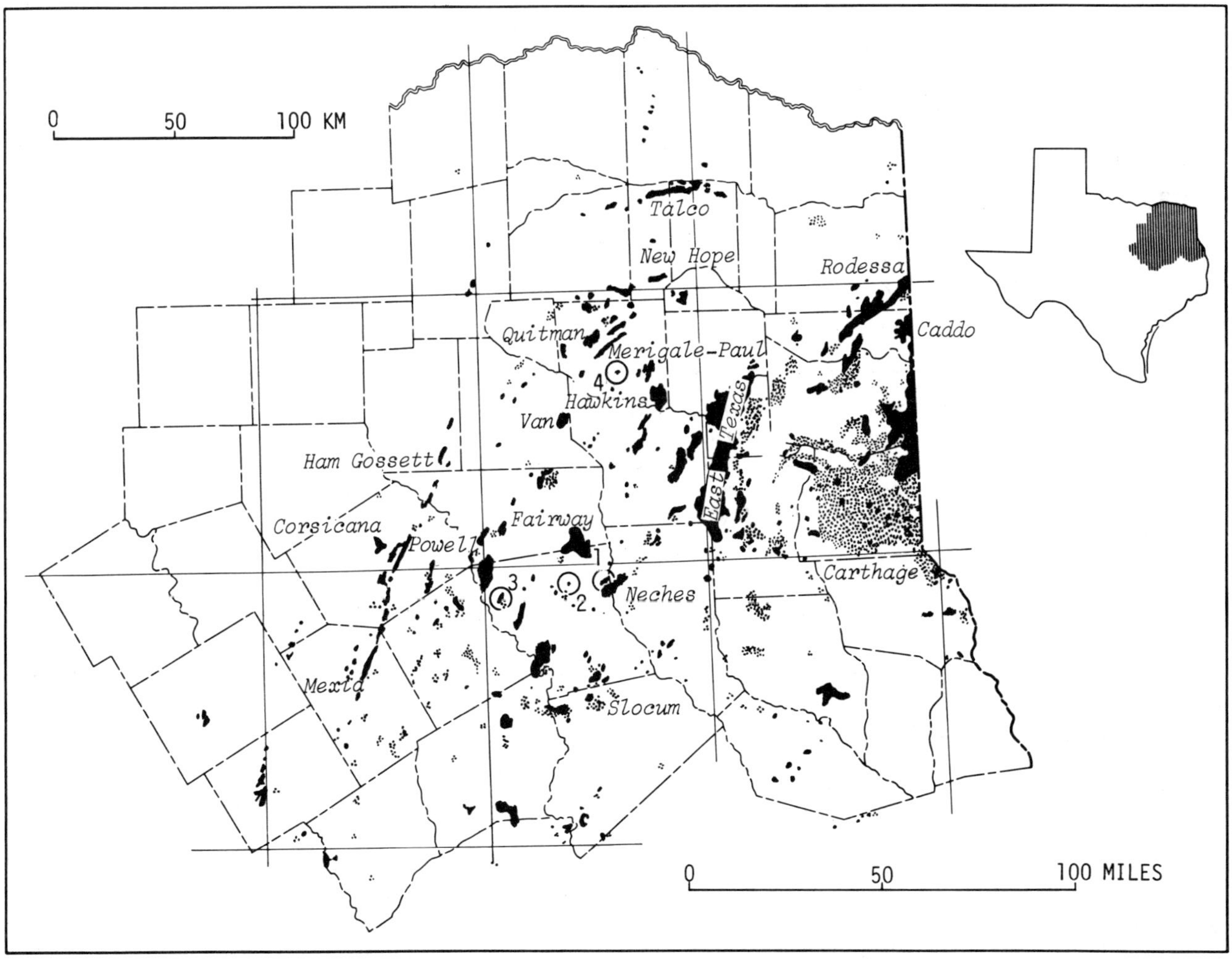

Figure 25.2 Northeast Texas oil and gas fields. Locations of fields described in text are indicated. Numbered circles enclose piercement salt dome fields described: 1, Boggy Creek; 2, Concord; 3, Bethel; 4, Hainesville.

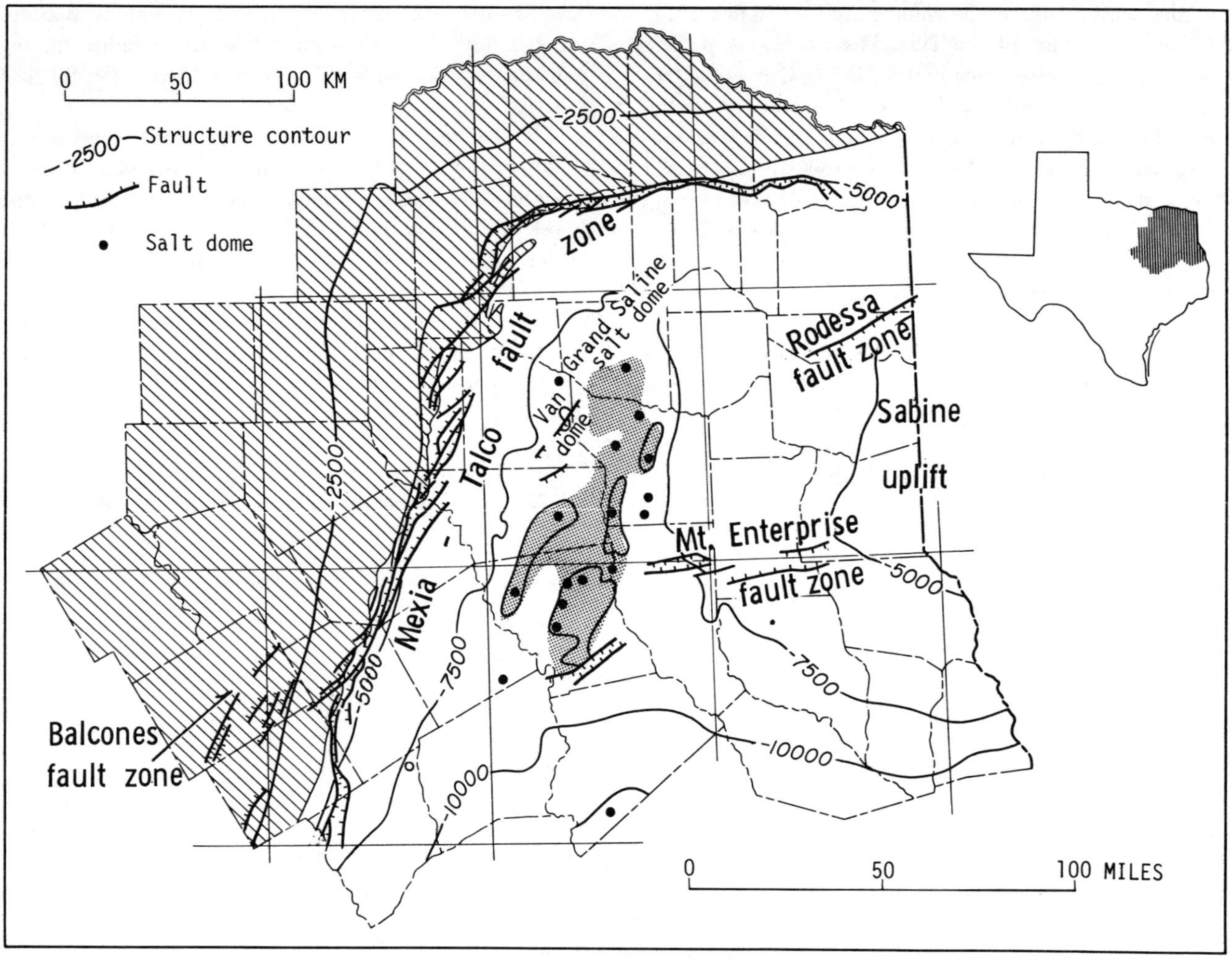

FIGURE 25.3 Northeast Texas. Tectonic and geologic map. Structure contour datum: top of Lower Cretaceous. Contour interval, 2500 feet. Ruled area on west and north is Lower Cretaceous (confined to western edge) and Upper Cretaceous; remaining area is Eocene. Deepest part of Tyler basin, approximately 10,000 feet below datum, is shaded. Entire map area lies within the East Texas embayment sector of the Gulf Coast geosyncline.

ments are confined to the five westernmost counties in the district.

Oil and gas deposits occur in the downdip extension of these outcropping formations, as well as in still older rocks. A generalized chart of the subsurface stratigraphy of Northeast Texas is give in Figure 25.5. This chart is a synthesis of several borehole records published in the Northeast Texas oil and gas monograph edited by Herald (1951). Because of the presence of many unconformities in the section, formations occurring in one part of the district may be absent in another part. Furthermore, a general gulfward thickening of individual units in the stratigraphic section takes place, so that the maximum subsurface section far exceeds the aggregate thickness of the outcropping rocks.

Although in the Tyler basin and elsewhere in the southern part of the district it may be necessary to drill through 4000 feet or more of Eocene sediments, only minor oil production in scattered sands in the Claiborne has been found as yet. The Upper Cretaceous section has been the most prolific source of oil, mainly because of the presence of the Woodbine Sand at the base (Alexander, 1951). The Woodbine is the reservoir rock not only at super-giant East Texas but also at Van, Hawkins, Powell, and Mexia (giants all). Other important reservoirs (Figure 25.5) are the Nacatoch, Taylor, Blossom, and Sub-Clarksville.

The upper part of the Lower Cretaceous section in Northeast Texas includes Washita, Fredericksburg, and especially Trinity reservoirs. The Buda Limestone in the upper Washita produces oil from fractures in Houston County, and in next-door Madison County in the Texas Upper Gulf province (Brundage, 1967). The Paluxy is near the top of the Trinity Group; the upper, middle, and lower Glen Rose and the Travis Peak Formation complete the Trinity (and the Lower Cretaceous) section. In the Glen Rose are the highly productive Bacon, Rodessa, James, and Pettet reservoirs. The Hosston sand is the principal hydrocarbon source in the Travis Peak Formation.

The deep Jurassic rocks include, going down the section, the Cotton Valley, Buckner, and Smackover Formations, highly important producers to the northeast in southern Arkansas and northern Louisiana. So far, of these three, the Cotton Valley and Smackover have been found to be the most productive in northeastern Texas.

Below the Smackover is a shale and salt series, shown in Figure 25.5 as the Eagle Mills Formation of Permian(?) age. The Eagle Mills has been divided into (in descending order) the Norphlet, Louann Salt, and Werner tongues or members. Recently geologists working in the subsurface in northeastern Texas, northern Louisiana, and

FIGURE 25.4 Salt mine in Grand Saline dome, Van Zandt County, Texas. Note nearly vertical bedding of salt on face, due to upward flowage. ***Courtesy Morton Salt Company.***

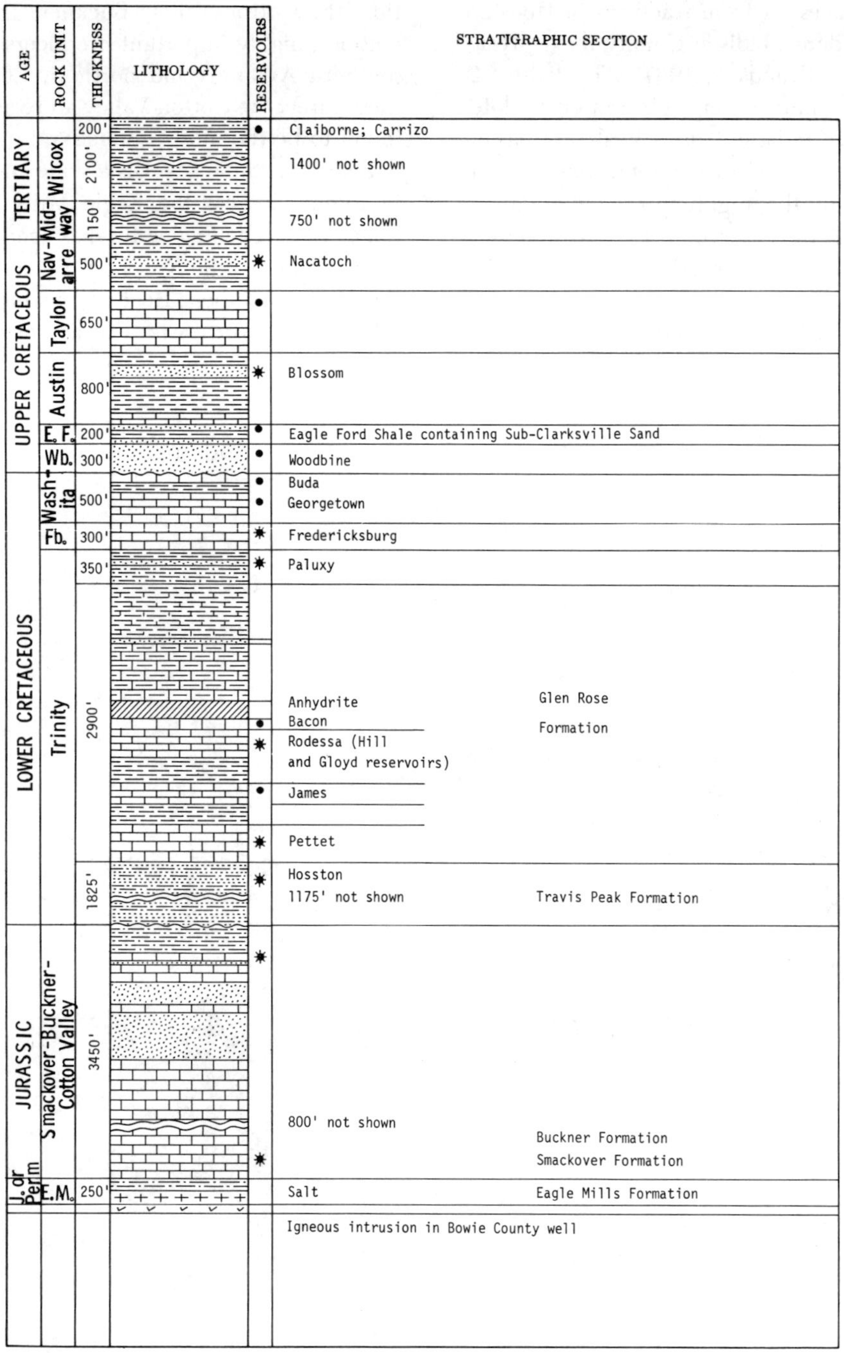

FIGURE 25.5 Northeast Texas. Composite stratigraphic section, showing principal reservoirs.

southern Arkansas have tended to consider these members as Upper Jurassic formations underlying the Smackover Limestone and overlying the Eagle Mills (restricted) red beds of unknown age (Nichols et al., 1968). This group of rocks marks the base of exploration so far in northeastern Texas. In Bowie County, close to the Arkansas line, a wildcat drilled into intrusive igneous rocks under sediments identified as Eagle Mills in age.

TRAPS AND FIELDS (Herald, 1951). In the paragraphs that follow, the relatively unimportant (so far) piercement salt dome fields are discussed first, followed by fields overlying deep-seated salt flowage anticlines. Then come fault traps, reservoir wedgout by truncation as illustrated by the East Texas field, anticlinal traps, and facies change traps.

The East Texas embayment strata have been locally

disturbed by salt domes, both piercement and nonpiercement. The seventeen known piercement domes are confined to the outcrop belt of the Eocene Wilcox and Claiborne Formations. Unfortunately they have not yet achieved the production records of the salt intrusions into much younger Tertiary strata in the coastal belt.

The first and for many years the only oil field in Northeast Texas with a piercement salt trap is the *Boggy Creek* (Eaton, 1951; McLellan et al., 1932). This field, discovered in 1927, has produced about 5½ million barrels of oil and some gas. It is close to the axis of the Tyler basin. The salt lies at a depth of about 2500 feet and has a depressed area across the top (Figure 25.7). This saucer shape extends to the surface and results in exposures of youngest rock in the center of the structure instead of older rock as is the usual rule. Oil and gas are produced from the Woodbine Formation (Cretaceous) at an average depth of 3700 feet in an elongate accumulation on the southeast flank of the dome (Figure 25.6). The reservoir sandstone has been dragged up and pierced by the salt core. At the extreme southwestern end of the field the oil accumulated in attenuated Woodbine sandstone immediately above the salt in a faulted anticline (Eaton, 1951, cross section B–B′ and C–C′, p. 33).

In spite of the accelerated search for interior piercement dome oil brought about by the Boggy Creek discovery in 1927, it was not until 1942 that the next strike was made—on the Concord dome, also in Anderson County. However, this discovery turned out to be a one-well pool that was abandoned a year later. Interest again lagged until 1956, when two successful wildcats were drilled on the Hainesville (Wood County) and Bethel (Anderson County) domes. Subsequently other domes have been found to be oil bearing, so that, by the end of 1964, 9 out of the 17 salt piercement structures either were producing or were potential oil producers. Included in this list was the Concord dome, on which oil was rediscovered in 1960.

The *Bethel* salt dome field in northwestern Anderson County was discovered in 1956 (oil) and 1957 (gas). Figure 25.8 shows the regional structure and a section across the overlying salt plug (Krusekopf, 1959). The structure map shows that the salt intrusion has taken place low on the west flank of a regional syncline. Although the top of the salt lies about 1500 feet below the surface, there is no indication of its presence at the surface. The mushroom cap is more than 2 miles in diameter, and the overhang to the west juts out about 4000 feet. The reservoirs are sands in the Woodbine (Upper Cretaceous), which produce oil, and the Rodessa (Lower Cretaceous) Limestone, which produces gas and condensate. Reserves of gas are estimated to be 95 billion cubic feet (Krusekopf, 1968). All of the oil production comes from the west side of the salt dome. The strata beneath the overhang are nearly flat except close to the stock, where they have been dragged up steeply.

Although exploration of the piercement salt domes has so far resulted in the discovery of only relatively minor oil fields, deep-seated movement of salt on a grand scale has produced some major anticlines in the overlying strata. Three of the giant oil fields of Northeast Texas produce from such upwarps. The presence of salt cores is assumed by the gravity records, which show pronounced minima across the anticlinal crests.

The *Hawkins* field in Wood County is an outstanding example of accumulation above a probable deep-seated salt dome. This field was second in the district only to East Texas in annual production in 1967, having an output of over 17 million barrels. It also ranked second that year after East Texas in cumulative production with a total of 340 million barrels. The Hawkins dome (Wentlandt et al., 1946) was first discovered in 1930 by surface geologic mapping. A seismograph survey carried out in 1935 was inconclusive regarding the presence of a structural trap. A core-drilling program in 1937 confirmed the presence of a dome, and the discovery well was drilled in 1940. Figure 25.9 is a structure map contoured on the oil reservoir rock, the Woodbine Sand. Below the map is a northwest-southeast cross section of the dome. The Woodbine Sand has been uplifted approximately 1200 feet; at the same time tensional strains across the crest of the dome produced the graben faulting shown in the cross section. The average depth of the Woodbine pay is 4850 feet.

The Woodbine Formation at Hawkins is a mixture of intercommunicating sands and interbedded clays and shales. The upper part of the formation is about 30 per cent and the lower part 70 per cent sand. The Woodbine is only 450 feet thick at the crest of the Hawkins anticline on account of thinning across the axis; elsewhere it is 700 feet thick.

The aggregate thickness of the gas and oil column in the Hawkins dome is about 1000 feet. This is more than the thickness of the reservoir rock, and so on the crest of the anticline the entire Woodbine section lies above the base of the gas cap and no oil is found beneath about 400 acres in the structurally highest area. All the faults save one are of insufficient magnitude to interfere with the free movement of oil through the reservoir; the exception is the northeast-trending fault of the east flank of the anticline, which has a maximum displacement of about 375 feet and divides the reservoir into the East Hawkins pool and the West Hawkins pool. The West Hawkins contains about 85 per cent of the producing area.

Another giant field is the *Van* (Betts, 1951; Liddle, 1936) in Van Zandt County. It is 12 miles east of Canton,

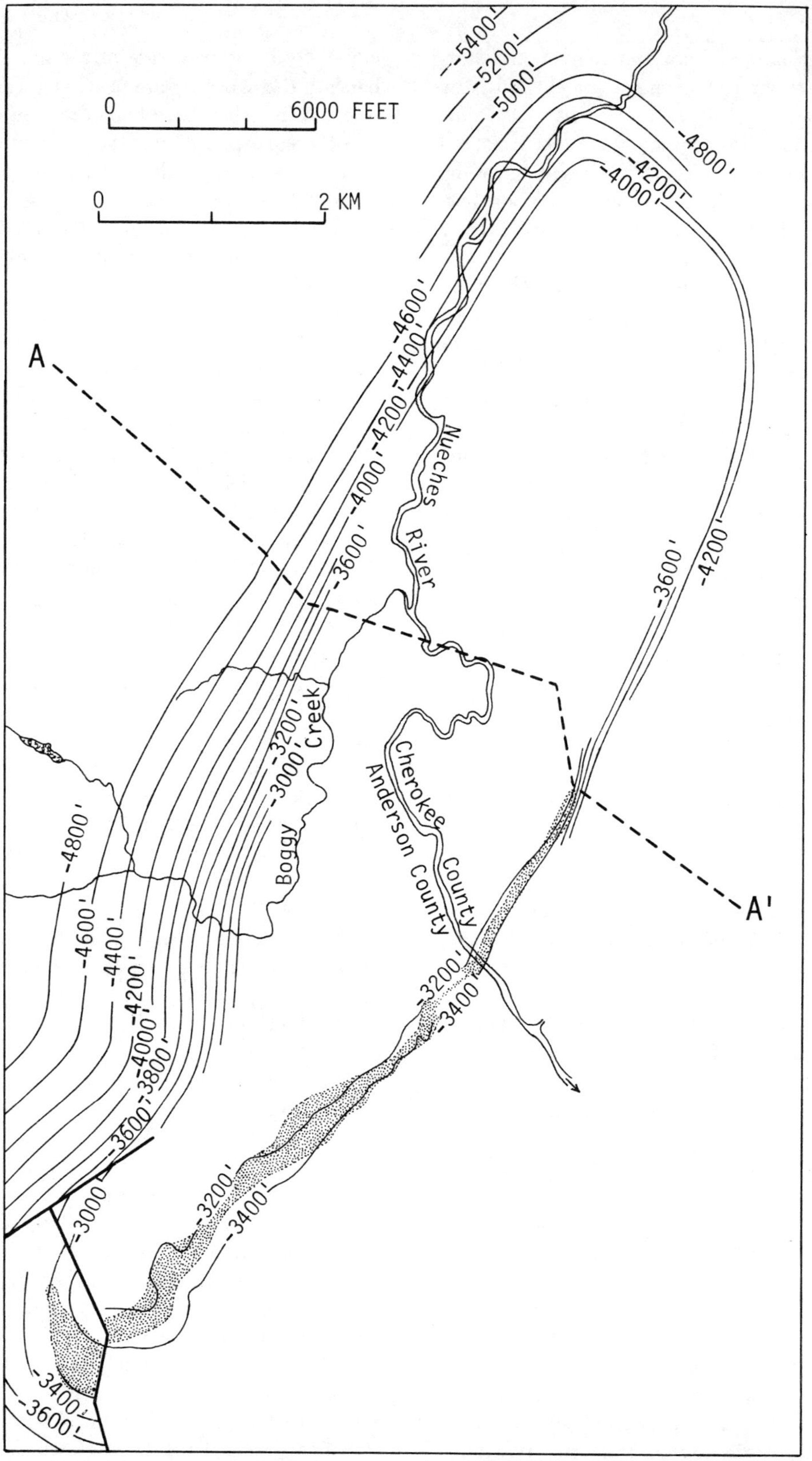

Figure 25.6 Boggy Creek field, Anderson and Cherokee Counties, Texas. Structure contour map based on top of Woodbine (Cretaceous) Sandstone. Contour interval, 200 feet. Blank area in center has no Woodbine because of piercement by salt intrusion (Figure 25.7). Productive area shown by stipple pattern. *After R. W. Eaton (1951). Courtesy author and the Texas Bureau of Economic Geology.*

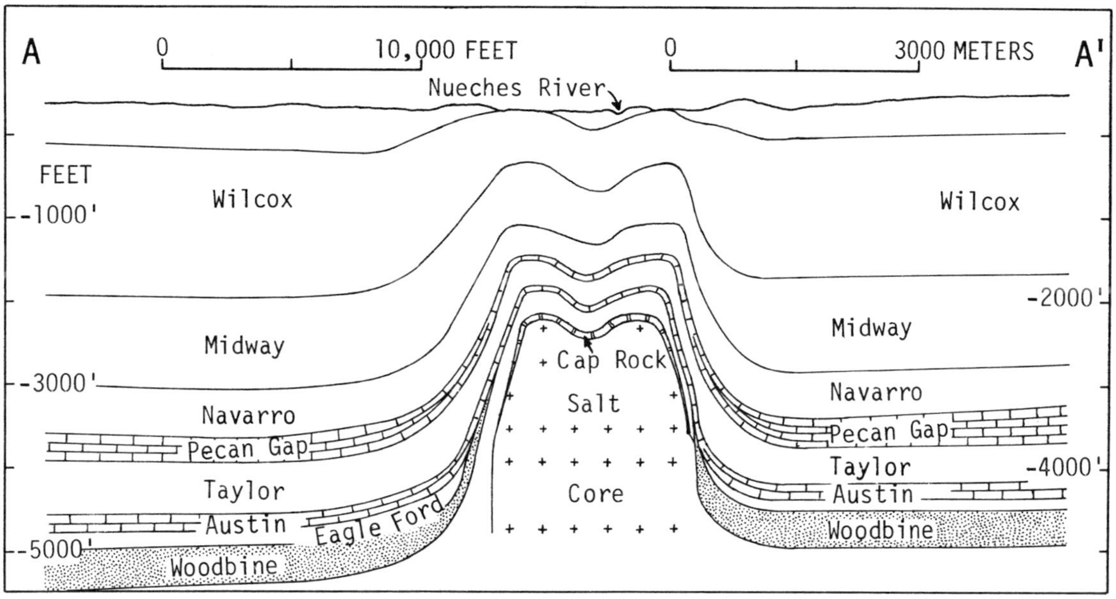

Figure 25.7 **Boggy Creek salt dome. Cross section along line A–A′ in Figure 25.6. Reservoir rock is truncated Woodbine Sandstone on southeast flank of salt dome (right side of cross section).** ***After R. W. Eaton (1951). Courtesy author and the Texas Bureau of Economic Geology.***

the county seat. The Van dome is of such magnitude that it appears on the tectonic map of the United States (Figure 25.3). It was recognized by surface geology, verified by seismograph, torsion balance, and magnetic survey, and detailed by core drilling, all before the discovery well was drilled in 1929. By the end of 1967 Van had produced more than 330 million barrels of oil, placing it third in the Northeast Texas district. The same year Van was fourth in annual yield with over 8 million barrels.

The structure contour map of the Van field (Figure 25.10), based on the top of the principal producing formation, shows the trap to be a complexly faulted dome with the original oil-water interface at an elevation of less than 2500 feet. Most of the Van oil has come from sandstones in the Woodbine Formation of Cretaceous age. The upper sands in this reservoir tend to be lenticular, but in the lower Woodbine they have field-wide continuity. Depths range from 2300 to 3700 feet. The shallow Van reservoir is the Nacatoch sand in the Navarro Formation. The productive area of Van Shallow is about 200 acres compared with over 4500 acres for the Woodbine Van. A deep test drilled in 1952 penetrated 20 feet of salt at a total depth of 11,560 to 11,580 feet.

A third deep-seated salt dome field is the *Quitman*, which in its first 20 years produced more than 55 per cent of the eventual 100 million barrels required for "giant" classification. The Quitman field (Smith, 1951; Scott, 1948) is also on a faulted anticline, but is considerably more elongate than the Van or Hawkins structures. This upfold trends northeastward across west-central Wood County for a distance of about 7 miles and has closure in the neighborhood of 300 feet. A narrow graben lies along the axis of the anticline. The domal structure was discovered by refraction seismograph and verified by a detailed reflection survey. The first test quit at 5040 feet after penetrating 600 feet of Woodbine without finding oil or gas. Seven years later, in 1942, after discoveries of commercial oil in older reservoirs had been made elsewhere, oil was found at Quitman in sands in the Paluxy (Lower Cretaceous) Formation at depths in the 6750 to 7000 foot range. During the next 6 years three additional discoveries were made uphole in Upper Cretaceous reservoirs. The leading producer is still the Paluxy with two sand zones that are field wide in extent, although the individual sandstone strata are discontinuous. Two thousand feet higher, at the base of the Upper Cretaceous, is the Woodbine Sand, which produces in one well from a 9 foot pay zone. Of greater importance are the still younger sandstones in the Eagle Ford Group, the lower "Eagle Ford sand" (at 4600 feet) and the higher "Sub-Clarksville sand" (4300 feet). The Eagle Ford sand is continuous throughout the field, but the Sub-Clarksville sands are irregular and lenticular. Trapping at Quitman is anticlinal, but distribution is controlled by the thickness and permeability of the sands and (in the Woodbine) by faulting.

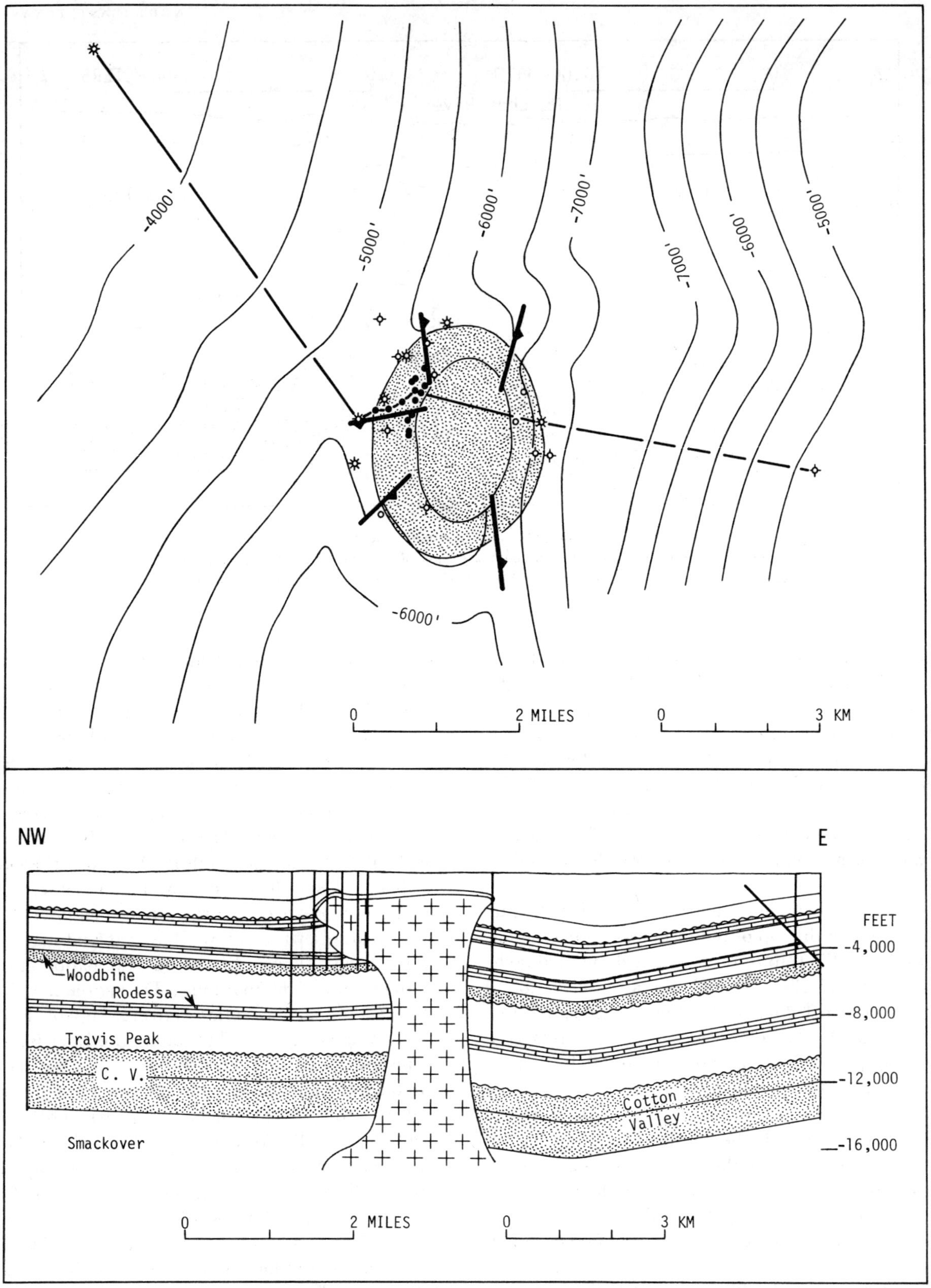

FIGURE 25.8 Bethel salt dome, Anderson County, Texas. Upper: regional structure map based on top of Woodbine. Contour interval, 500 feet. Shaded part is producing area. Line from northwest of field to east of field shows location of cross section below. Lower: cross section along line on map above. Vertical scale designated by depth footages on right. Salt stock shown by plus symbols. *After H. H. Krusekopf (1959). Courtesy author and the Oil and Gas Journal.*

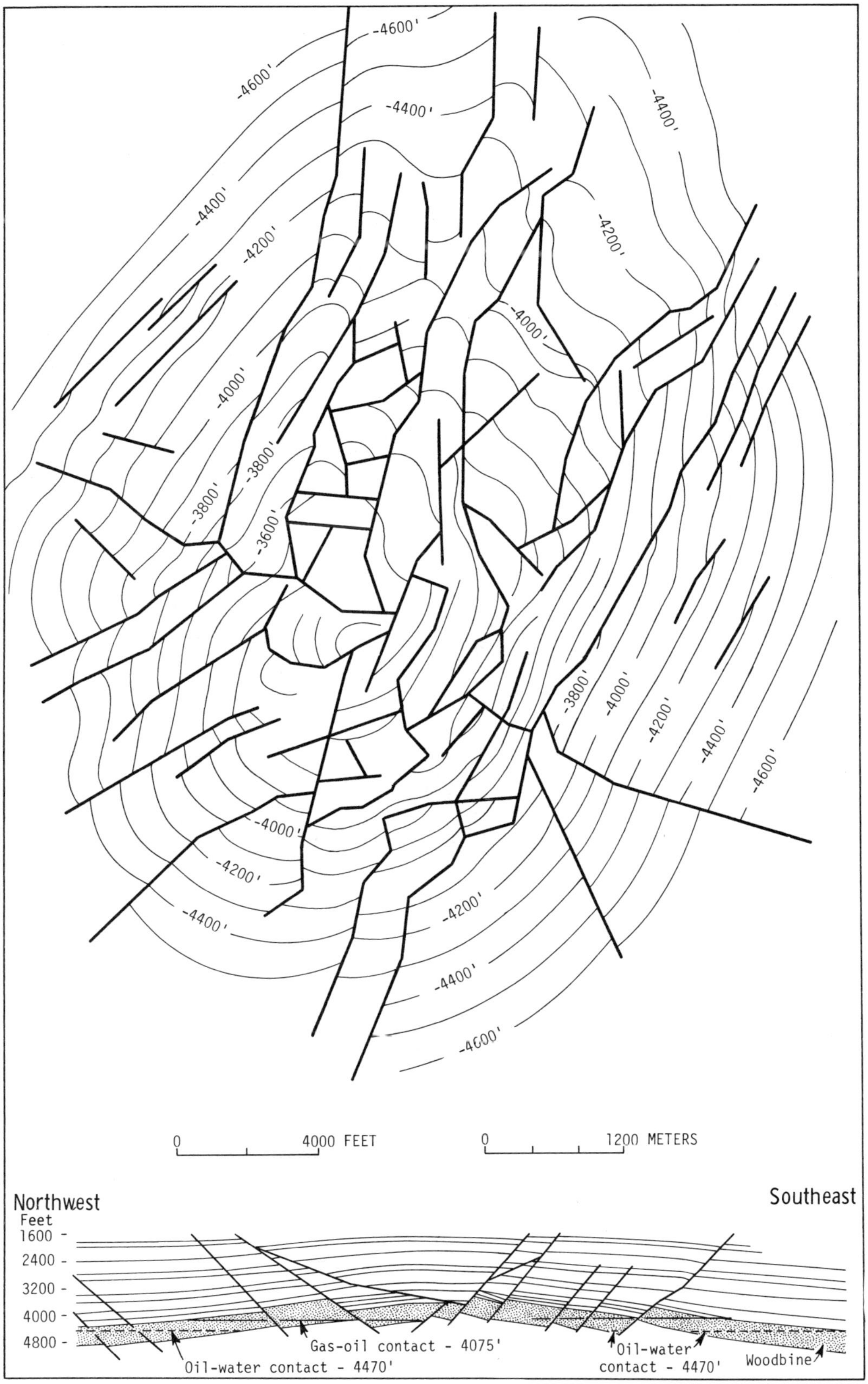

FIGURE 25.9 **Hawkins field, Wood County, Texas. Structure contour map based on top of Woodbine Formation. Contour interval, 100 feet. Below is a cross section showing gas cap and oil-water interface.** ***After E. A. Wendlandt et al. (1946). Courtesy American Association of Petroleum Geologists.***

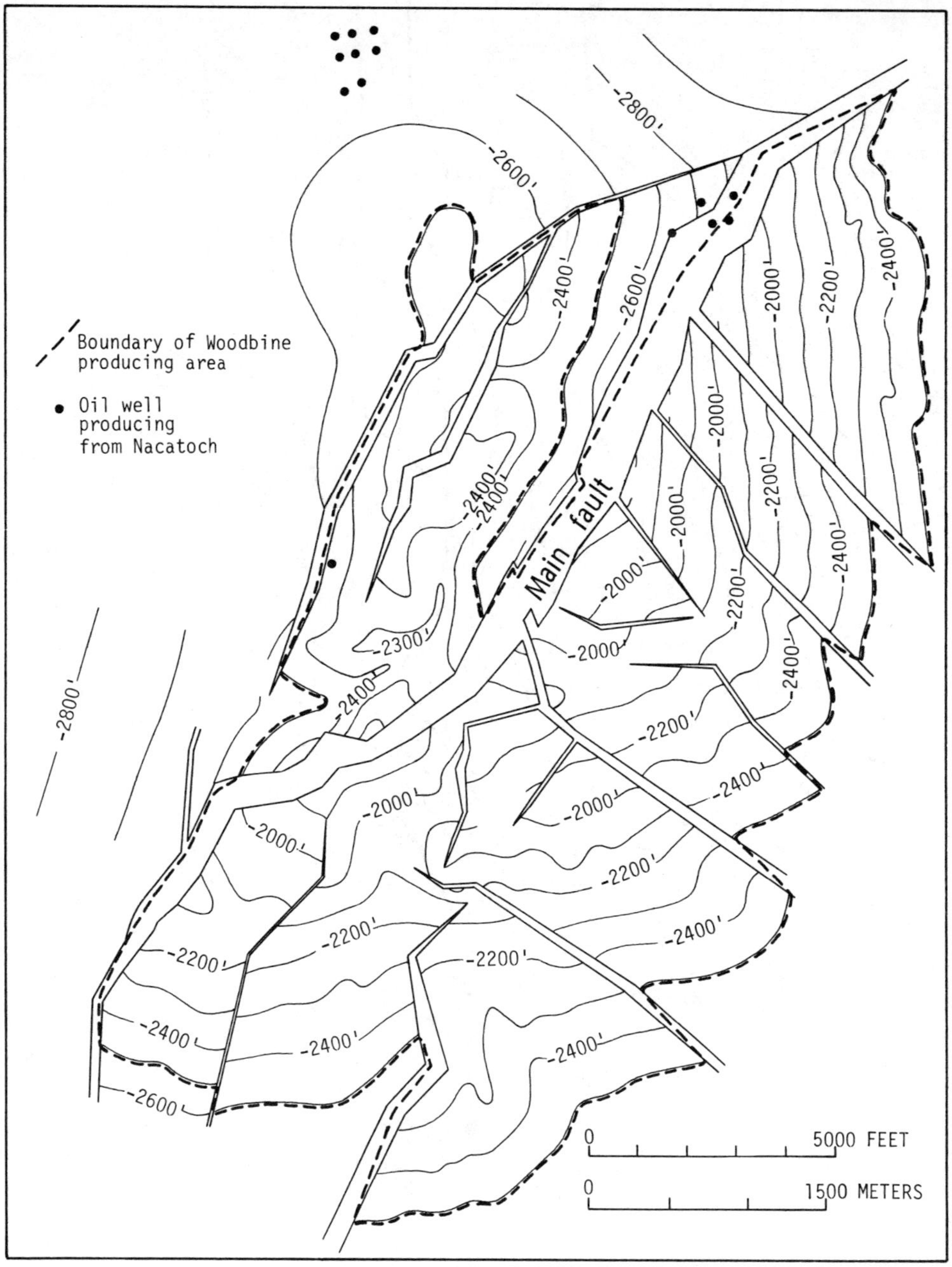

FIGURE 25.10 Van oil field, Van Zandt County, Texas. The field completely surrounds the town of Van. Structure contour map based on top of Woodbine (Cretaceous) Sand. Contour interval, 100 feet. Dashed line encloses Woodbine productive area (Van Deep). Scattered wells shown produce from Nacatoch Sand (Van Shallow). *After Ralph A. Liddle (1936) and Frank A. Herald (1951). Courtesy Texas Bureau of Economic Geology.*

There are other, but smaller, fields in the Northeast Texas district in which the trap has been the result of deep-seated salt intrusion (Read, 1959).

The *Merigale-Paul* field (Moore, 1951; Wright, 1945) is a fault accumulation lying some distance basinward from the Mexia–Talco fault zone and in no way related to it. Merigale–Paul is in Wood County, about 1½ miles southeast of the town of Quitman. The principal reservoir is the Sub-Clarksville sand of latest Cretaceous age at a depth of about 4850 feet. It carries oil only on the upthrown (southeast) side of the fault. The regional dip is southeasterly toward the axis of the Tyler basin; oil moving updip through the reservoir rock was impounded below the fault plane because of the impervious nature of that barrier. Lateral sealing to the southwest is by decreased porosity and permeability due to increase in the volcanic ash content of the reservoir rock in that direction. In the other direction the field is closed by a curving of the fault plane, which causes it to slice across the oil-water interface level.

The Mexia–Talco fault zone (Figure 25.3) dominates the west and north flanks of the East Texas embayment (Lahee, 1929). Strung out along this zone are about 40 fault trap oil fields. Three of them, Mexia, Powell, and Talco, are giant fields. Very little oil or gas has been discovered as yet upflank from this line of faults; Corsicana is an outstanding exception.

The *Mexia* field in Limestone County is the oldest of the three giants; its first birthdate was in 1912, when gas was discovered in the shallow (about 700 feet minimum depth) Nacatoch (Upper Cretaceous) sand (Hudnall, 1951b). The deeper Cretaceous Woodbine Sandstone was found to be oil bearing at depths of 3000 to 3200 feet in 1920, and in the next 47 years this reservoir produced more than 106 million barrels of oil. The cross-section (Figure 25.11) shows the nature of the trap. The primary accumulation has been due to the presence of an elongate anticline on the basinward (southeastern) side of the fault in the upthrown block. The fault sliced off the two Woodbine Sandstone reservoirs high on the northwest flank of the anticline and dropped them 600 feet to below the normal level of the oil-water interface.

The *Powell* field to the north in Navarro County, discovered in 1923, is larger than Mexia and has produced close to 130 million barrels of oil. The reservoirs at Powell are also sands in the Woodbine at depths of about 3000 feet, and the trapping is very similar to that at Mexia.

To the north of the Powell field in Kaufman County is a much newer fault trap field, the *Ham Gossett* (Wiggins, 1954). This field was discovered in 1948 after surface mapping, core drilling, a seismic survey, and three dry holes. The initial reservoir was the Bacon Limestone Member of the Glen Rose Formation of Comanchean age, penetrated at a depth slightly below 6100 feet. Further exploration during the next 5 years discovered ten additional reservoirs, of which five are in the Upper Cretaceous Woodbine, two are in the Glen Rose above the Bacon, and three are in the lower Glen Rose, the oldest being the Rodessa Limestone Member.

The Ham Gossett structure map (Figure 25.12) shows the combination anticlinal and fault trap in the upthrown block that is characteristic of the Mexia–Talco fault zone. However, the exploration of this field led to the first discovery of a commercial accumulation of oil across the fault in the downthrown block. Here the trapping medium is a low-relief elongate anticline paralleling the fault. This discovery opened up possibilities elsewhere.

The largest of the fault trap giants, *Talco,* is also the newest. This field lies on the north flank of the Tyler basin, where the strike of both the rock strata and the fault zone is easterly. The Talco fault was mapped on the surface in 1924, but exploration by drilling was impeded by the failure to discover oil in the Woodbine Sandstone north of the Powell field. However, the discovery of prolific oil in the Woodbine in the East Texas

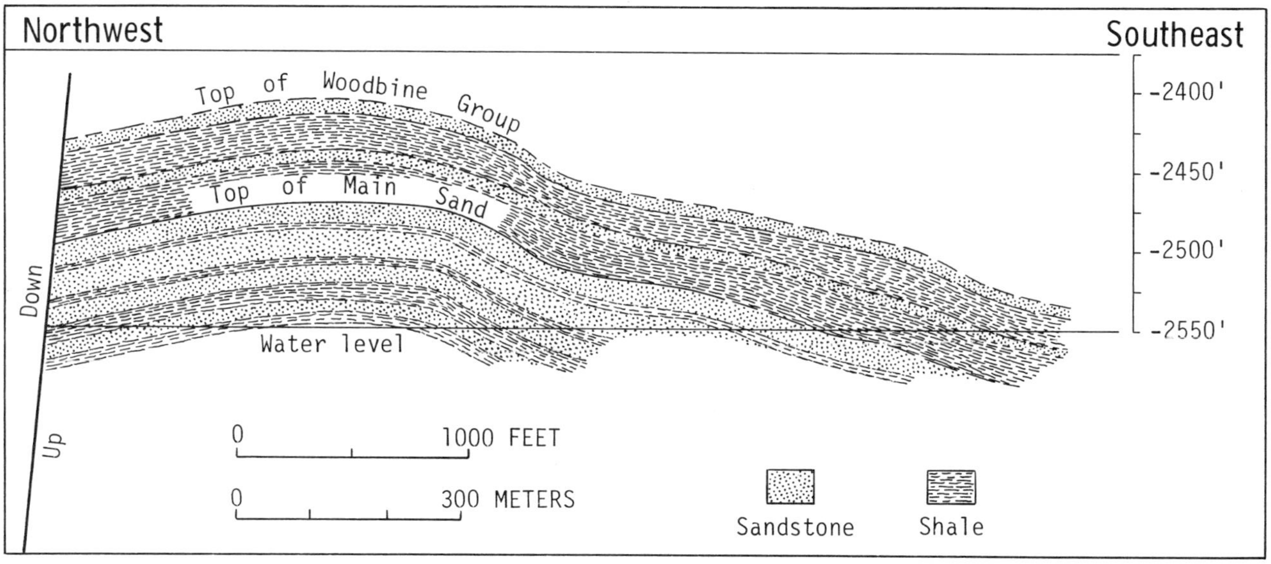

FIGURE 25.11 Mexia field, Limestone County, Texas. Northwest-southeast cross section cutting middle of field, showing faulted anticline. Top of Woodbine Group across fault to northwest lies at elevation −3,035 feet, approximately 600 feet below level on fault footwall. Producing wells have been drilled into the Main Sand above water level. ***Courtesy James S. Hudnall (1951) and the Texas Bureau of Economic Geology.***

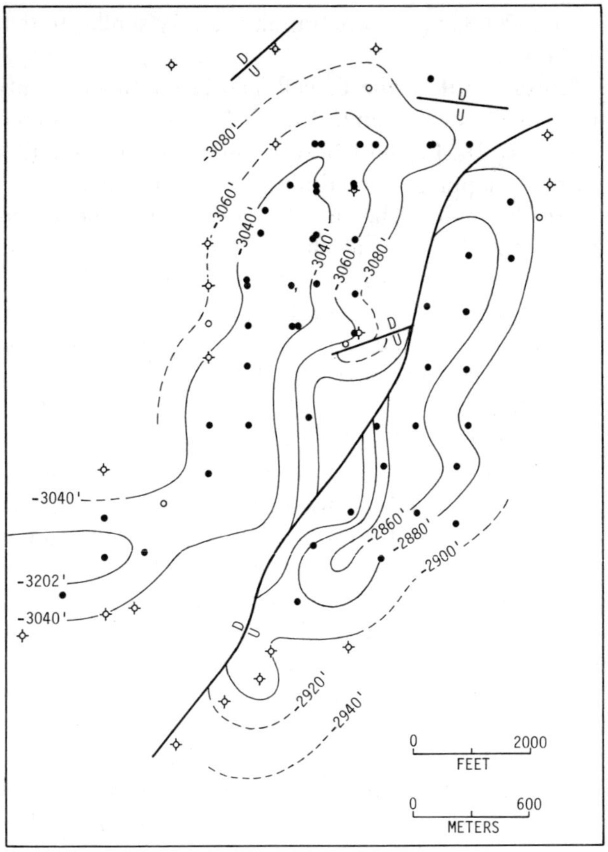

FIGURE 25.12 Ham Gossett field, Kaufman County, Texas. Contour structure map based on top of Woodbine. Interval, 20 feet. ***Courtesy Peter N. Wiggins III (1954) and the American Association of Petroleum Geologists.***

field about 55 miles to the south in 1930 changed the picture. Three dry holes in the years 1933 and 1934 in what became the Talco field failed to find oil in the Woodbine but did prove the existence of a trap. The discovery well was drilled in 1936, finding oil in the Comanchean Paluxy sands at a depth of 4300 feet, about 800 feet below the Woodbine (Wendlandt and Shelby, 1948). The Paluxy reservoir consists of a series of porous and permeable sands that are individually lenticular, but with some degree of intercommunication, so that as a general rule these variable and erratic sands function as a single reservoir. The trap at Talco is the now familiar elongate anticline cut, just over the crest, by a strike fault that drops the formations on the north side over 1000 feet (Figure 25.13). The resulting structural closure is 450 feet. The oil migrating up the regional dip from the south and into the anticlinal flank is stopped just beyond the anticlinal axis by fault gouge backed up by impervious Washita shales across the fault plane.

The graben between the Talco fault and the Sulphur River fault, about 7000 feet to the north, became a subject of interest and exploration after the discovery in 1950 of production on the downthrown side at Ham Gossett to the southwest. As a result, the *Triz-Liz* field was discovered in an anticline in the graben; other discoveries in this area include a small field where the trap is a nose truncated updip (to the north) by the Sulphur River fault.

For more than a third of a century the *East Texas* field was the number one giant of North America, but in 1966 it yielded first place in the United States in annual production to Wilmington in California. The East Texas output in 1967 was still a respectable 48+ million barrels. Its cumulative production at the end of 1967 of 3.76 billion barrels was exceeded, in the free world at least, only by the Burghan field of Kuwait (more than double), Ghawar in Saudi Arabia (3.94 billion), and Kirkuk in Iran (3.77 billion); Agha Jari in Iran was not far behind with 3.4 billion barrels.

The East Texas field (Hudnall, 1951a; Minor and Hanna, 1941) is about 50 miles to the southwest of the Arkansas–Louisiana–Texas corner. It is 43 miles long and has an average width of 5 miles. The field lies along the wedgout zone of the westward dipping Woodbine Sand of Late Cretaceous age. The regional dip is into the Tyler basin. Updip to the east is the Sabine uplift with its highest point in northwestern Louisiana. The East Texas field was discovered on the third try at a depth of 3600 feet by wildcatter C. M. Joiner during depression year 1930. After two additional discoveries had been made some distance to the north, it dawned on the industry that there could be a continuous field connecting the three successful wildcats. Oil men immediately set out to prove this theory, and they did so, to the complete demoralization of the crude oil market. The price of oil fell to below 10 cents a barrel. Out of this experience evolved the proration and conservation laws of today.

The trap in the East Texas field is illustrated in Figures 25.14 and 25.15. Figure 25.15, a cross section, shows the actual wedgout of the sandstone reservoir. Oil moving updip, from an enormous gathering area to the west, through the Woodbine Sandstone, has been impounded along the truncated edge of the Woodbine by impervious Austin Chalk above the unconformity. Figure 25.14, a water encroachment map, shows how the trap is closed in the other two (horizontal) dimensions. The eastern side of the oil field is the wedgout line of the Woodbine. To the west of this line (and its continuation north and south) the Woodbine is present and is thick enough to contain oil in economically producible quantities. The edge of the field to the west is the position of the original oil-water interface. It is, therefore, a structural contour, based on the Woodbine Sand reservoir, and shows the presence of an anticlinal nose, with a slight furrow in the middle, which plunges westward down the flank of

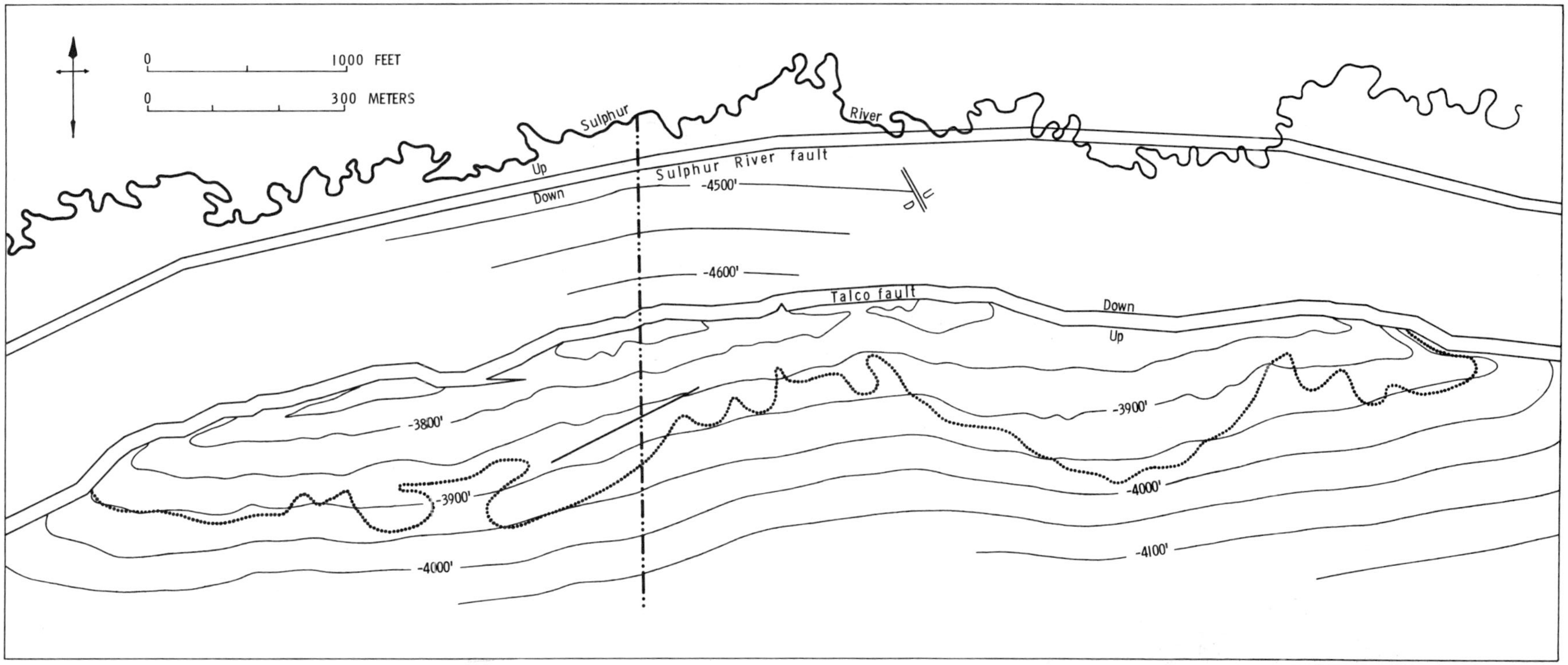

FIGURE 25.13 Structure contour map of Talco field, Franklin and Titus Counties, Texas. Datum: base of Goodland Limestone (cap rock of oil reservoir). Contour interval, 500 feet. Talco fault and dotted line delineate productive area of the Paluxy. *Courtesy T. H. Shelby, Jr., (1951) and the Texas Bureau of Economic Geology. Map modified by Shelby after E. A. Wentlandt and T. H. Shelby (1948).*

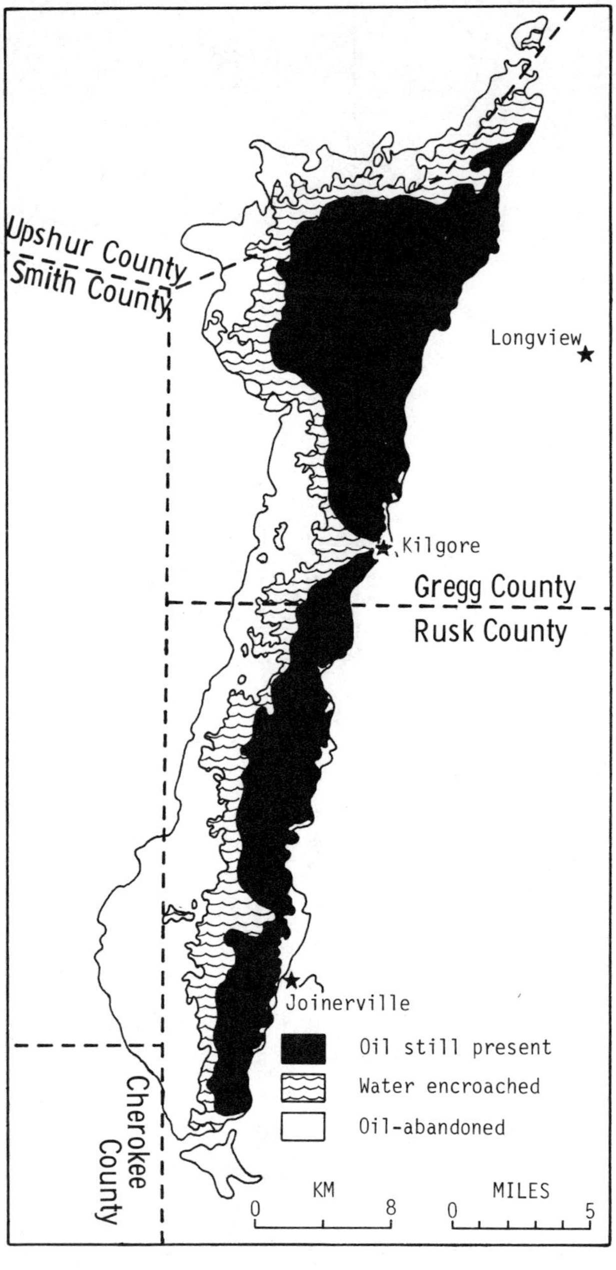

FIGURE 25.14 Water encroachment map, East Texas oil field. The line farthest west marks the position of the oil-water interface when the field was discovered in 1930. Thirty-five years later water encroachment had virtually separated the field into three parts. ***Based on drawings that appeared in the Pan American Petroleum Corporation magazine, Horizons. Courtesy Pan-American Petroleum Corporation.***

the Tyler basin. Below this line to the west, and to the north and south where it intersects the wedgout line, the Woodbine contains water. Above the line oil was originally present in copious quantities, but the edge water has moved updip with production so that the whole area has been completely swept clean of its original oil content. In the intermediate (encroachment) zone oil in the interconnecting pores of the reservoir sandstone is being replaced by water.

The solid black shows what is left of the original field. It will continue to shrink each year until the water drive has pushed the last of the oil into the highest well on the anticlinal nose below the wedgout line.

The Woodbine Formation in this area consists of a series of lenticular sands interstratified with shales and silts. The individual sandstone reservoirs have an average thickness of about 30 feet; the maximum aggregate thickness observed is in the neighborhood of 75 feet. These lenses appear to be interconnecting in the northern part of the field, where most wells are flowing and the sands behave as a common reservoir. In the southern part of the field, however, the sands are much tighter and are not everywhere interconnecting. Here most wells are pumpers, and fracturing of the tight sands improves their productivity. Furthermore, it has become the practice in the southern sector, after a well goes to water, to seal off the water and drill through an underlying shale section from 60 to 150 feet thick into a lower, obviously separate pay sand.

The *New Hope* field (Clayton, 1951) is another Northeast Texas field expected to yield more than 100 million barrels of oil before abandonment. It is in southeastern Franklin County on the north flank of the Tyler basin south of the Talco–Sulphur Bluff fault zone. The trap is a simple elliptical anticline, with a probable closure of 150 to 200 feet. It was discovered by a gravity survey and confirmed by a reflection seismograph survey. The first discovery was in the Bacon (middle Glen Rose of the Lower Cretaceous) Limestone at a depth of 7200 to 7300 feet in 1943; within the year three additional discovery wells were drilled into three deeper reservoirs, with a maximum depth of about 8200 feet.

The north-south section (Figure 25.16) has been drawn across the elliptical upfold along its short axis. The highest reservoir, the Bacon Lime, is also the most extensive, covering 4600 acres. It is an oölitic and fossiliferous porous limestone. The three deeper reservoirs are sandstones with lenticular porosity; their productive acreages range from 30 to 70 per cent of the total field area. They are lower Glen Rose in age.

New Hope had a second birthday in January, 1954, when a Smackover (Jurassic) discovery well was completed at 12,080 to 12,320 feet. By the end of 1965 this deeper pool had produced more than 100 billion cubic feet of gas and 12 million barrels of condensate. Sulfur is also produced as a by-product; a published analysis of the gas shows a hydrogen sulfide content of 13.71 per cent (Chisholm, 1968, p. 1072). The productive zone in the Smackover reservoir has a maximum thickness at the crest of the anticline of 600 feet; the average thickness is 350 feet. The gas is stored in intercrystalline openings

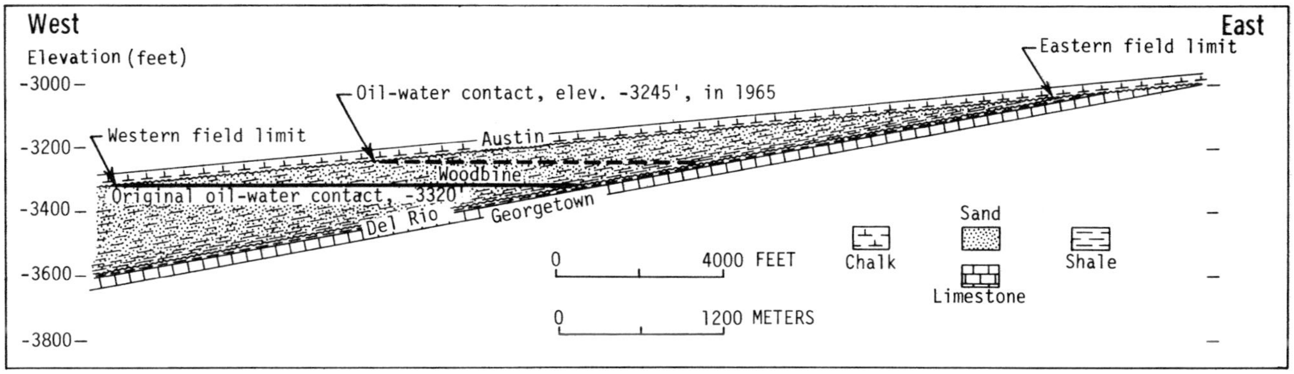

Figure 25.15 East Texas field. Cross section, showing truncation by erosion of Woodbine Sand and overlap by Austin Chalk. The oil is trapped updip in the Woodbine by this wedgout. The closure to the north and south is due to the presence of a westward plunging anticlinal nose, which is outlined by the original oil-water interface in Figure 24.14. *After James S. Hudnall (1951). Courtesy author and the Texas Bureau of Economic Geology.*

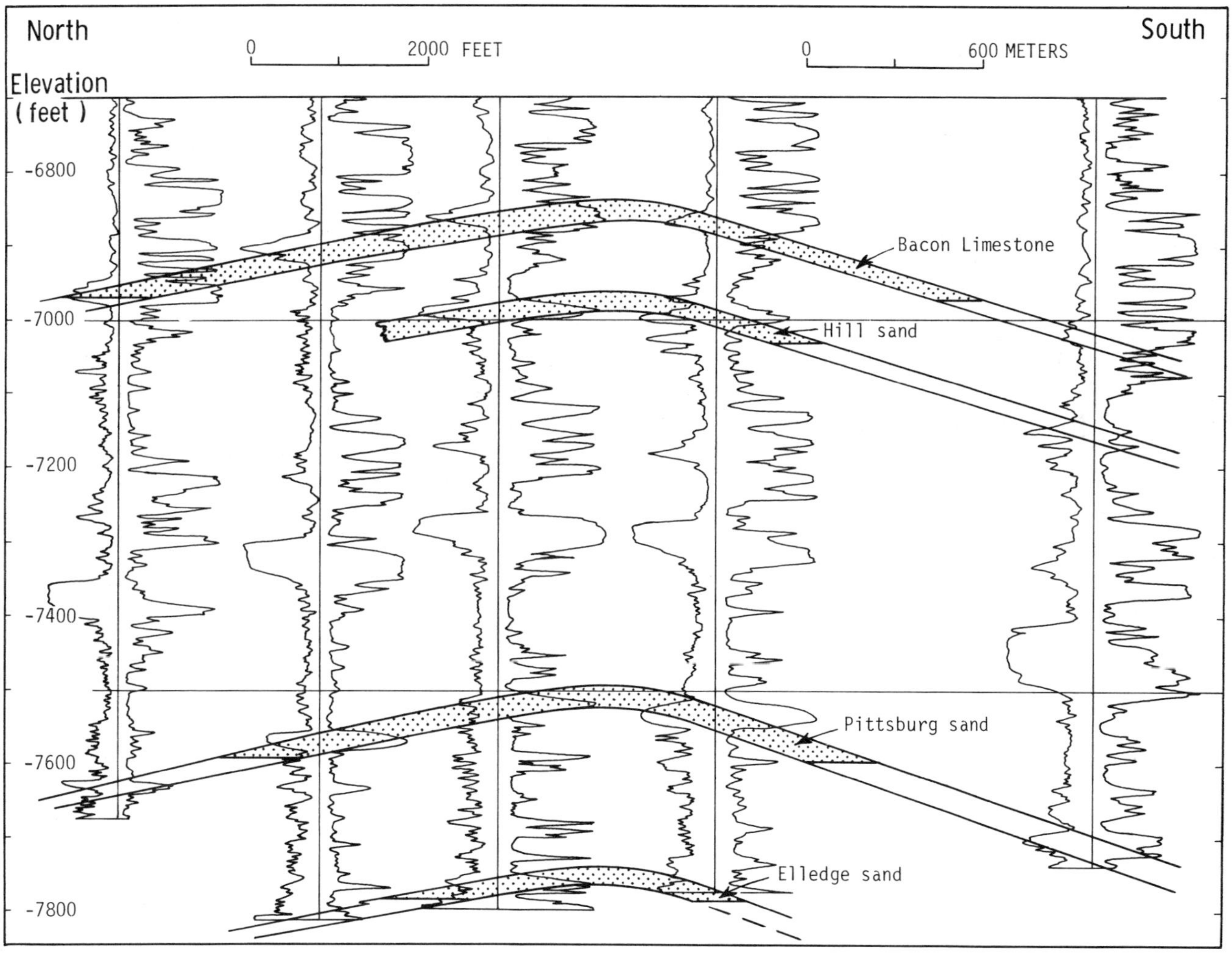

Figure 25.16 New Hope field, Franklin County, Texas. North-south cross section along short axis of elliptical anticline. Stippling designates oil. Reservoirs shown are all Glen Rose (Early Cretaceous) in age. Subsequently gas and condensate were discovered in Smackover (Jurassic) Limestone about 4000 feet below the Elledge Sand. After John M. Clayton (1951). *Courtesy author and the Texas Bureau of Economic Geology.*

supplemented by vugs and fractures in a limestone and dolomite section. The porosity average is about 7 per cent; Chisholm (1968, p. 1072) notes that newer Smackover fields have a reservoir porosity of around 18 per cent. Most subsequent Smackover discoveries have been in a trend along the Talco–Sulphur Bluff–Mexia fault zone, especially in Hopkins County.

Two other fields destined to become giants are the *Neches,* discovered in 1953, and the *Fairway* (1960). The Neches field is bisected by the Neches River, which is the boundary between Anderson and Cherokee Counties in the southern part of the district. It produces mainly from the Woodbine Sand of Late Cretaceous age; the younger Sub-Clarksville (also Upper Cretaceous) is a lesser source. The trap is an elliptical anticline with a north-northeast trend and a closure at the Woodbine level of about 150 feet. The presence of this anticline was indicated by both surface and subsurface observations. The discovery well was drilled 1 mile off the southeast flank of the productive Boggy Creek piercement salt dome (Figure 25.6), and the Neches anticline parallels the elongate salt intrusion (Hunt and O'Connor, 1954).

During 1967, 7 years after discovery, the Fairway field in southeastern Henderson and northeastern Anderson Counties produced nearly 9 million barrels of oil and was third in Northeast Texas in annual output. At that time its reserves were estimated to exceed 167 million barrels. The principal reservoir is a reef in the Lower Cretaceous James Formation lying at a depth of 10,000 feet. The reef is roughly triangular and is oil filled over 23,000 acres. It lies close to the deepest part of the Tyler basin.

Mapping by seismograph and shallow subsurface control before discovery had shown the presence of a "large, sprawling structure" (Perkins, 1964), but the record of the James Limestone as a reservoir elsewhere had not encouraged a thorough exploration of this structure down to the James level. Nonetheless the giant Fairway field was eventually discovered, and the James Limestone has become recognized as a worthwhile exploration objective.

Trapping within the reef is due to faulting on the northwest, porosity pinchout on the northeast, and plunging of the roof of the sealed reef beneath water level elsewhere. Within 5 years of discovery the owners of James Limestone production at Fairway had succeeded in unitizing this pool (Latimer and Oliver, 1964; Latimer, 1965).

So far, three other pays have been discovered at Fairway but are relatively unimportant. These are also in the Lower Cretaceous; two are younger and one is older than the James Limestone. The shallowest pay is the "massive anhydrite," which contains several porous limestone lentils in which the oil is stored. Next downhole is the Rodessa, with oil occurring in a porous limestone at the top of this formation. Beneath the James is the Pettet reservoir, in which oil occurs in an oölitic limestone bed near the top of a predominantly shale section.

On the east side of the Northeast Texas district are two major fields that are partly in Texas and partly in northwestern Louisiana. One of these is the *Caddo,* which is an anticlinal accumulation lying close to the top of the Sabine uplift. Only a relatively small part of the Caddo production has been obtained from the Texas side of the line. To the north is the elongate northeast-southwest *Rodessa* field, which has a greater acreage in Texas than in Louisiana but has produced more oil on the Louisiana side of the line. Rodessa occupies the extreme northwest corner of Louisiana and extends a short distance into Arkansas as well as into Texas. Both Caddo and Rodessa are described in Chapter 24.

The *Carthage* gas field, which covers the greater part of Panola County, is a giant in both area and production. Discovered in 1936, by the end of 1959 it had produced about 4 trillion cubic feet of gas plus 7 million barrels of oil and 40 million barrels of condensate. The reservoirs are four oölitic limestones in the Lower Cretaceous Rodessa at 5400 to 5500 feet and Pettet (92 per cent of the production) at 6200 to 6500 feet. The fifth reservoir is Travis Peak Sandstone, also Lower Cretaceous, at 6500 to 6800 feet. The trap is combination stratigraphic and structural. The field occupies a broad structural terrace on a southwest-plunging nose off the west flank of the Sabine uplift. Distribution of hydrocarbons on the terrace is controlled by the permeability of both the oölitic limestones and the deeper sandstones. The permeable zones tend to occur as "fingers" with a general northwesterly alignment (Rogers, 1968).

The *Good Omen* field (O'Connor, 1955) in southeastern Smith County is a facies change trap of a type similar to the one at Cottonwood Creek, Wyoming (Chapter 34). The reservoir is the uppermost sand in the Woodbine, but this sand becomes shale updip. As can be seen in Figure 25.17, trapping at Good Omen occurs where an updip re-entrant in the sand-shale facies change line is superimposed on a structured nose plunging down the regional dip.

Future. Northeast Texas still has a multitude of prospects, extending in three directions from currently producing areas and zones. The discovery of giant Fairway in 1960 is ample evidence that this district still affords good hunting. It proved the presence of oil-bearing reefs in the Lower Cretaceous James Limestone (Fox, 1963). In all probability other reefs and other varying permeability traps remain to be discovered in intervening areas.

The western fringe cannot be condemned for its barrenness up to this time. Although most of the Upper Cretaceous and some of the Lower Cretaceous reservoirs

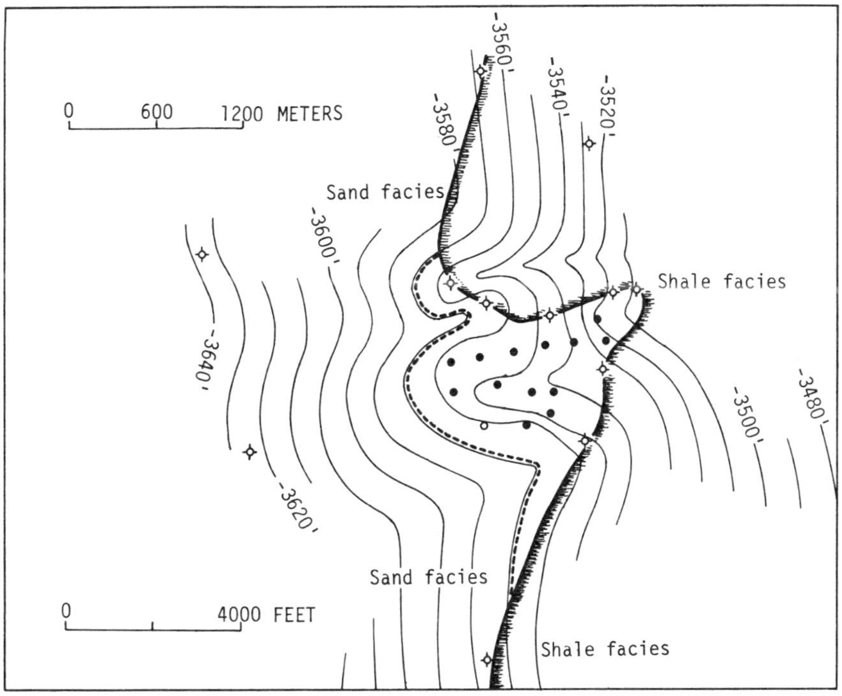

Figure 25.17 Good Omen field, Smith County, Texas. Structural and lithofacies map. Structure contours based on Woodbine Sand. Interval, 10 feet. Dashed line shows original oil-water interface. Whiskered line shows lens-out of sand facies into shale facies. *After J. P. O'Connor, Jr. (1955, p. 47). Courtesy author and the Gulf Coast Association of Geological Societies.*

have been removed by erosion in this area, deeper Lower Cretaceous reservoirs remain, even though this section has been shortened by wedgout against the underlying Paleozoic Ouachita folded belt. There is even an updip wedgout of the Jurassic Cotton Valley Formation along the eastern edge of the western fringe. The post-Paleozoic section has a maximum thickness of 5000 feet or more in the barren belt. Therefore the largely untested overlapped Paleozoic rocks (Morgan, 1965; Flawn, 1958) lie within drillable distances. Three wells in Dallas County penetrated normal (nonmetamorphosed) Pennsylvanian and Ordovician rocks beneath the Cretaceous cover.

Nichols (1964) calls attention to Mesozoic reservoir possibilities, largely untested as yet, along the Angelina–Caldwell trend at the southeastern edge of the district.

The most promising future, however, probably lies in more intensive exploration of deeper reservoirs within the area of major production (Coon, 1957). A surprising number of anticlinal traps have not been tested adequately below the Woodbine. The 40 miles across the anticlinal nose that seals in the Woodbine wedgout at East Texas has not been thoroughly probed for deeper permeability pinchouts or reversals in the regional dip. One wonders whether the same density of Pettet–Travis Peak exploration wells in this area that prevails in Smith County to the west, and in the rest of Rusk County to the east, would not yield a comparable number of new pools beneath the East Texas field itself.

Furthermore, the Travis Peak at the base of the Cretaceous does not mark the bottom of hydrocarbon production. In recent years an intensive exploration program seeking gas, condensate, and oil production in the Jurassic Cotton Valley and (especially) Smackover reservoirs has been quite successful. During 1967, out of 25 new fields discovered in Northeast Texas, 6 were in the Cotton Valley and 14 in the Smackover (Morrow et al., 1968). The happy hunting ground so far has been a broad curved belt inside the Mexia–Talco fault zone extending from the Arkansas line to Freestone County. The search continues; in late 1968 the most active exploration was taking place to the south in a belt crossing Henderson, Navarro, Freestone, and Limestone Counties, where three Smackover gas fields had been discovered earlier (Oil and Gas Jour., Oct. 7, 1968, p. 86–87).

Uphole possibilities, especially in the older fields, should not be ignored. An example is the Van Shallow field (Figure 25.10), which produces from the Nacatoch above the Woodbine in two small areas, one of them within the boundaries of the Woodbine pool. The other is to the north, across the bounding fault.

Heavy oil, lying at a depth of about 500 feet in the

Slocum field of Anderson County, is being exploited by steam flooding. The reservoir is a 25 foot sandstone in the lower Claiborne (Tertiary).

Lastly, when other searches fail (or before then, for that matter), oil seekers can always return to the piercement salt domes and probe some more around the flanks (Kornfeld, 1958).

REFERENCES CONSULTED IN PREPARING CHAPTER 25

Alexander, C. I., 1951.
Balk, Robert, 1949.
Betts, A. J., 1951.
Brundage, Harrison T., 1967.
Chisholm, L. R., 1968.
Clayton, John M., 1951.
Coon, Lester A., 1957.
Dallas Geological Society, 1965.
East Texas Geological Society, 1951.
Eaton R. W., 1951, 1956.
Flawn, Peter T., 1958.
Fox, Bruce W., 1963.
Hager, Dilworth S., and Burnett Claude M., 1960.
Herald, Frank A., ed., 1951.
Hudnall, James A., 1951a, 1951b.
Hunt, V. G., and O'Connor, J. P., Jr., 1954.
Kornfeld, Joseph A., 1958.
Krusekopf, H. H., Jr., 1959, 1968.
Lahee, Frederic H., 1929.
Latimer, J. R., Jr., 1965.
Latimer, J. R., Jr., and Oliver, Fred L., 1964.
Liddle, Ralph A., 1936.
McLellan, H. H., Wendlandt, E. A. and Murchison, E. A., 1932.
Minor, H. E. and Hanna, Marcus A., 1941.
Moore, Hastings, 1951.
Morgan, Henry J., Jr., 1965.
Morrow, Ernest H., Skinner, John H., and Hinyard, Paul, 1968.
Nichols, Paul H., 1964.
Nichols, P. H., Peterson, G. E., and Wuestner, C. E., 1968.
O'Connor, J. P., Jr., 1955.
Perkins, Stanley L., 1964.
Read, John L., Jr., 1959.
Rogers, Robert E., 1968.
Scott, E. R., 1948.
Smith, W. K., 1951.
Wendlandt, E. A., and Shelby, T. H., Jr., 1948.
Wendlandt, E. A., Shelby, T. H., Jr., and Bell, John S., 1946.
Wiggins, Peter N., III, 1954.
Wright, A. C., 1945.

Chapter

26

TEXAS UPPER GULF COAST

The Upper Gulf Coast of Texas is also referred to as the southeastern Texas Gulf Coast and the Texas salt dome province. It is continuous with the South Louisiana district to the east and with the South Texas district to the southwest. It is bounded on the north by the Northeast Texas district.

The Upper Gulf Coast district coincides with the Texas Railroad Commission District 3. It includes twenty-nine counties and three offshore areas (Figure 26.1). Although currently the third district in the state in annual production, with 14 per cent of the total, the Upper Gulf Coast dominated Texas production during the early decades of the present century.

The most productive part of the district is about three counties wide, adjoining and paralleling the shoreline of the Gulf of Mexico. For centuries before the drilling of the discovery well this area was notable for its seepages of oil and gas.

> The locations of seepages of oil and gas were well known to the early Indians, to some of the Spanish explorers, and many of the first white settlers. Mention of some of these seepages, particularly those at Sour Lake [Figure 26.6], was made in publications prior to the Civil War, and many contracts granting the right to explore for oil and gas were made in several of the Gulf Coastal counties in the years immediately after that war (Warner, 1939, p. 185).

It was not until 1901, however, that Captain Lucas drilled the first (in North America) of many 75,000 barrel per day gushers in the Spindletop field (Figure 26.6). The reservoir was salt dome cap rock, and several more prolific fields of this type were discovered during the next 5 years. In addition commercial sulfur deposits have been found in the cap rocks of at least nine salt domes in this area.

The second birthdate in the history of this salt dome district was marked by the discovery in 1915 of flank production from the upturned and truncated deeper strata around the periphery of the Sour Lake salt dome.

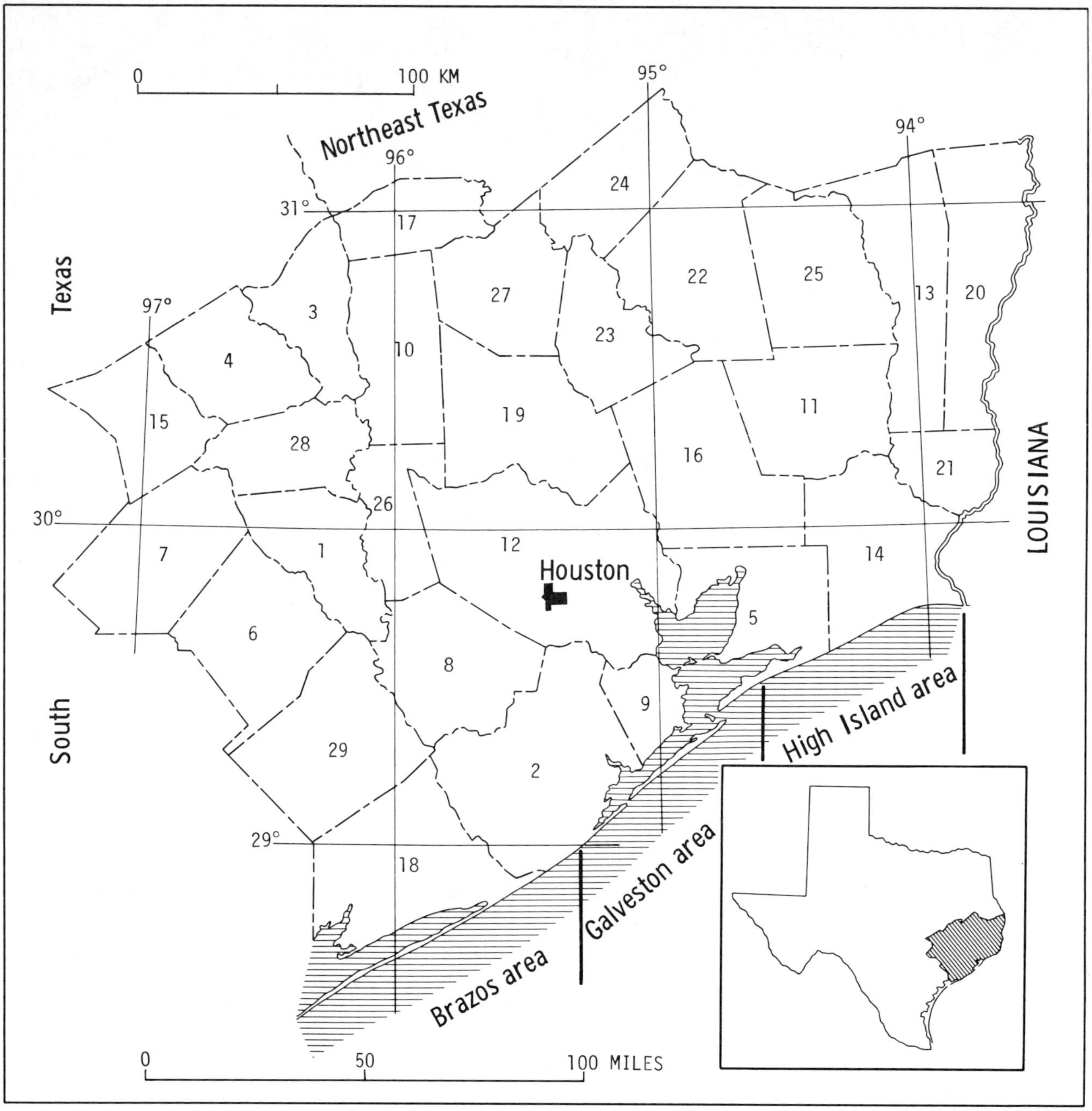

Figure 26.1 Texas Upper Gulf Coast. Index map. County Key:

1 Austin
2 Brazoria
3 Brazos
4 Burleson
5 Chambers
6 Colorado
7 Fayette
8 Fort Bend
9 Galveston
10 Grimes
11 Hardin
12 Harris
13 Jasper
14 Jefferson
15 Lee
16 Liberty
17 Madison
18 Matagorda
19 Montgomery
20 Newton
21 Orange
22 Polk
23 San Jacinto
24 Trinity
25 Tyler
26 Waller
27 Walker
28 Washington
29 Wharton

Ever since, the so-called piercement domes have been explored vigorously for flank reservoirs. The discovery of the Goose Creek field in 1917 initiated production from the arched supercap strata above an assumed salt core at an unknown depth beneath. However, it was not until the development of the large Conroe field in 1931 that the search for this type of accumulation became intensive. Subsequently many fields of the deep-seated salt core type have been found both in the Upper Gulf Coast of Texas and in southern Louisiana.

Through 1965, each of 21 fields in the Upper Gulf Coast of Texas had produced more than 100 million barrels of petroleum. Included in this list is Spindletop, the discovery field. However, the leading field in cumulative production is Conroe with over 400 million barrels of produced oil. Hastings was the leading field in 1965 with nearly 13 million barrels.

Tectonics. Structure contours (Figure 26.2) show that the Tertiary and Quaternary sediments are dipping gently gulfward (Figure 26.3, lower) into the Gulf Coast geosyncline, with the strike closely paralleling the present shoreline. At least one flexure, the Hartburger in Newton County (Bornhauser, 1960), is superimposed upon this monocline. Also, a few strike faults hasten the gulfward descent of these strata, but the strike faults in this district are less abundant than in southern Louisiana

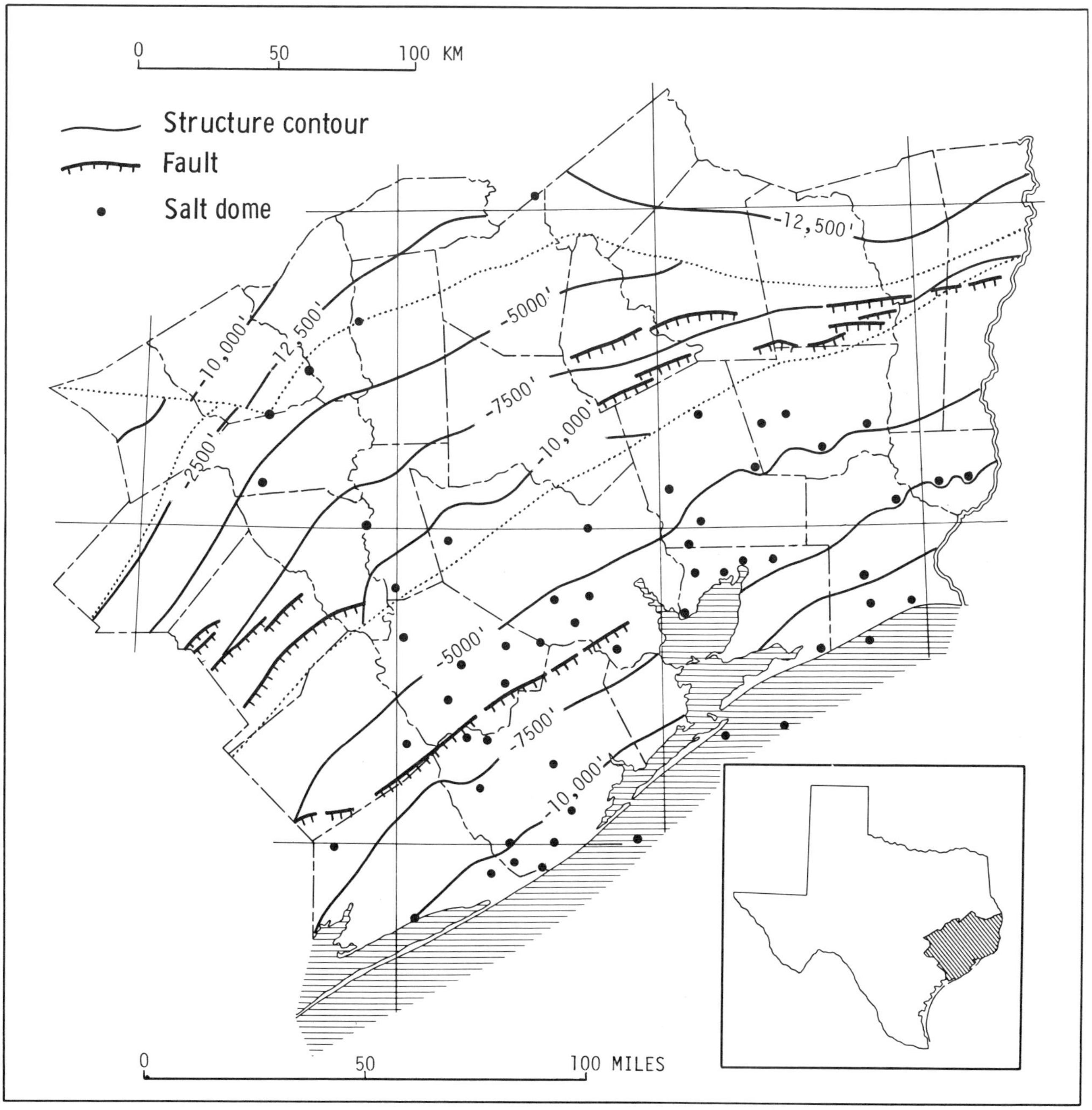

Figure 26.2 Upper Gulf Coast. Tectonic map. Structure contours show regional dip gulfward. Contour interval, 2500 feet. Datum: (1) marker bed near top of Pliocene in coastal belt, (2) top of Lower Cretaceous in intermediate belt, and (3) basement rocks along inland edge.

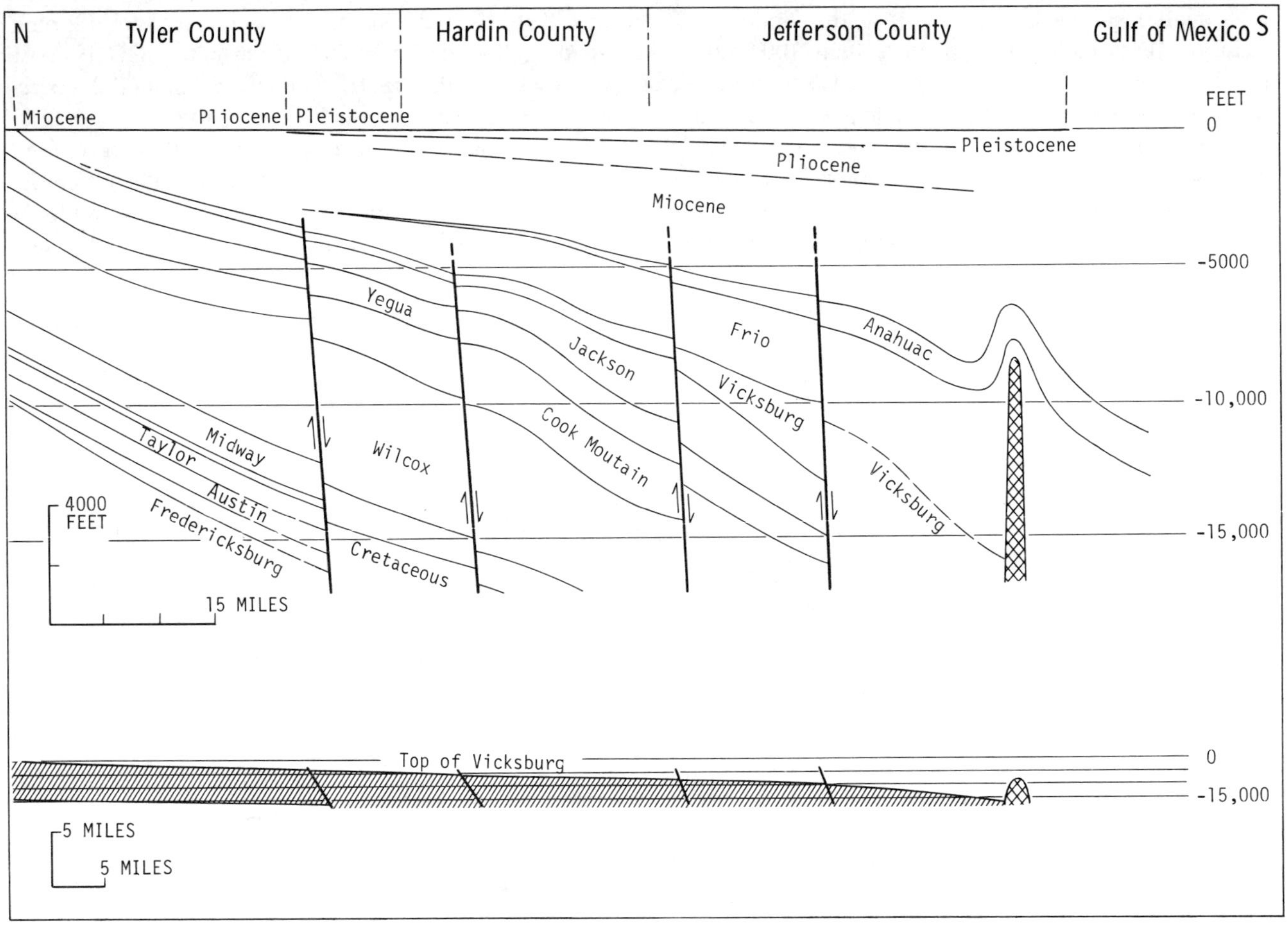

FIGURE 26.3 North to south cross section, southeastern corner of Texas Upper Gulf Coast. Upper section has vertical exaggeration of 13×; lower section is without exaggeration. Patterned prism is a salt dome. *Courtesy Houston Geological Society and the Oil and Gas Journal.*

to the east and on the coastal plain of South Texas to the southwest.

The known salt domes that pierce the northwestern flank of the Gulf Coast geosyncline in the Texas Upper Gulf district are shown in Figure 26.2. No doubt there are many more, both onshore and offshore, yet to be discovered. No one knows for certain the age of the mother salt, which must be present in great volume at depths considerably in excess of 4 miles. The best estimate is that the salt is a downdip continuation of the Eagle Mills Formation, which is salt bearing in northeast Texas, northern Louisiana, and southern Arkansas. The Eagle Mills Formation is unfossiliferous; it underlies the Jurassic Smackover and is probably Jurassic also, but some consider it to be Permian.

So far as is known, the same tectonic picture of monoclinal southerly dips, strike faults, and salt structures continues off-shore in the Texas Upper Gulf district, at least as far as the edge of the continental shelf.

STRATIGRAPHY. The entire Upper Gulf Coast district lies in the belt of Tertiary and Quaternary overlapping sediments. The half of the district lying adjacent to the coast is floored with Quaternary sediments, followed inland by a broad belt consisting of Pliocene, Miocene, and Oligocene deposits. The oldest sediments exposed in the Upper Gulf district are Eocene in age. The westernmost corner of the district in Lee County almost, but not quite, reaches the eastern edge of outcropping Cretaceous rocks.

The Cenozoic formations grow progressively more marine toward the Gulf with the sand series often grading seaward into marine shale. . . . During all of the Cenozoic, the source of clastic sediment is believed to have been the interior continent to the north and northwest much as it is today. . . . Deposition in the Gulf is at present continuing in all the environments from continental to the abyssal deep. Since the beginning of Cenozoic these various zones of deposition have migrated gulfward as the landward side of the gulf basin has been filled. . . . All formations are believed to have been laid down in lens-shaped bodies thinning both up and downdip (Sheets and Cockrell, 1962, p. xi).

The subsurface stratigraphic section, as assembled from thousands of drill records, is shown in Figure 26.4. This chart shows the principal oil and gas reservoirs in the Texas Upper Gulf Coast district. In the subsurface the areally important Recent, Pleistocene, and Pliocene sediments are not differentiated. These poorly consolidated sediments, consisting of sand and shale, are not productive as yet in the Upper Gulf district in spite of production to the east and a known maximum thickness of 7000 feet.

By far the greater part of the production in this district comes from Tertiary reservoirs, although Cretaceous rocks are within drillable distances along the inner edge. Figure 26.5 shows by heavy lines the zones or trends of different reservoir objectives (Dollison, 1966, p. 1229). These trend lines have been superimposed upon a map of the oil and gas fields.

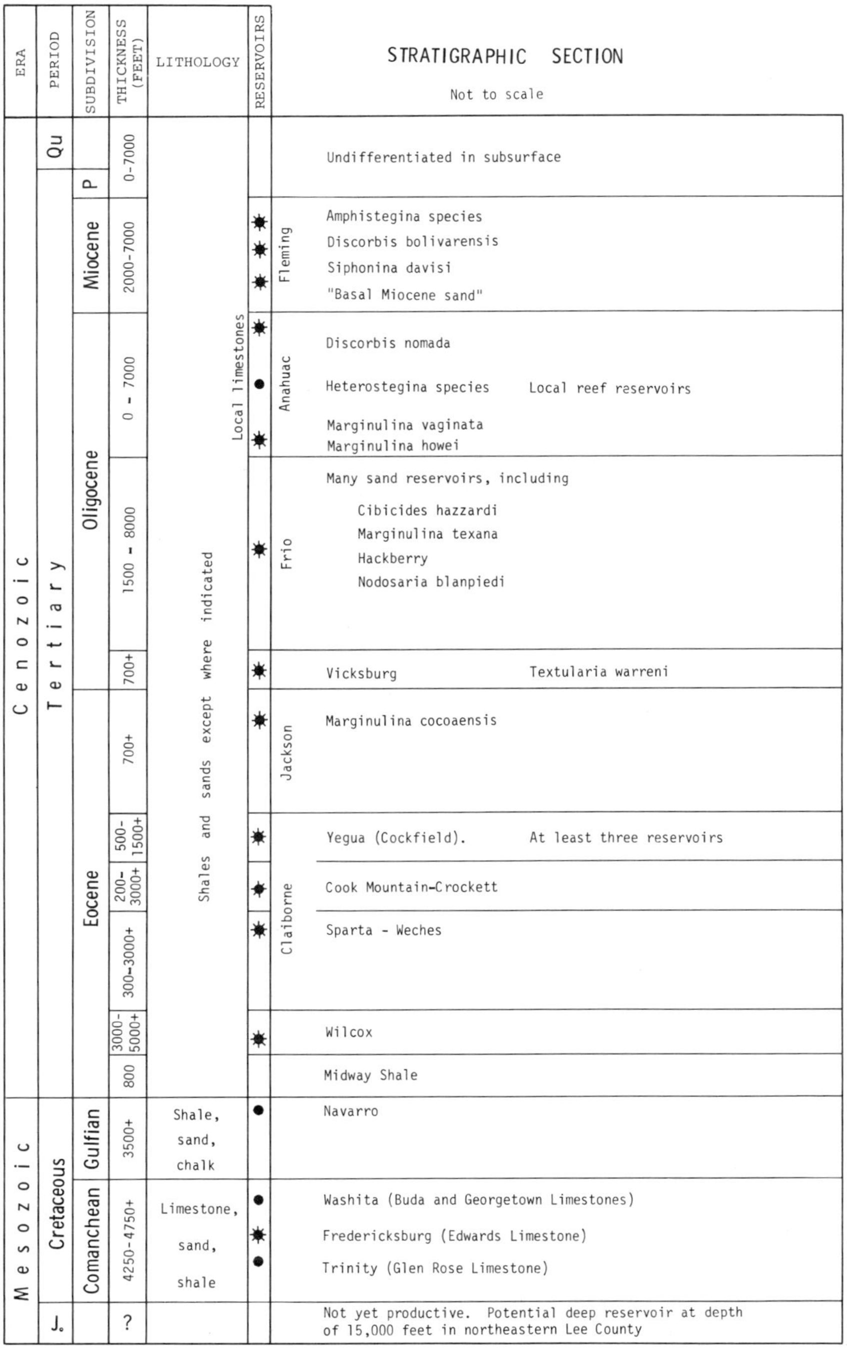

Figure 26.4 Texas Upper Gulf. Stratigraphic section. ***After M. M. Sheets and Ernest Cockrell, Jr. (1962). Courtesy M. M. Sheets and the Houston Geological Society.***

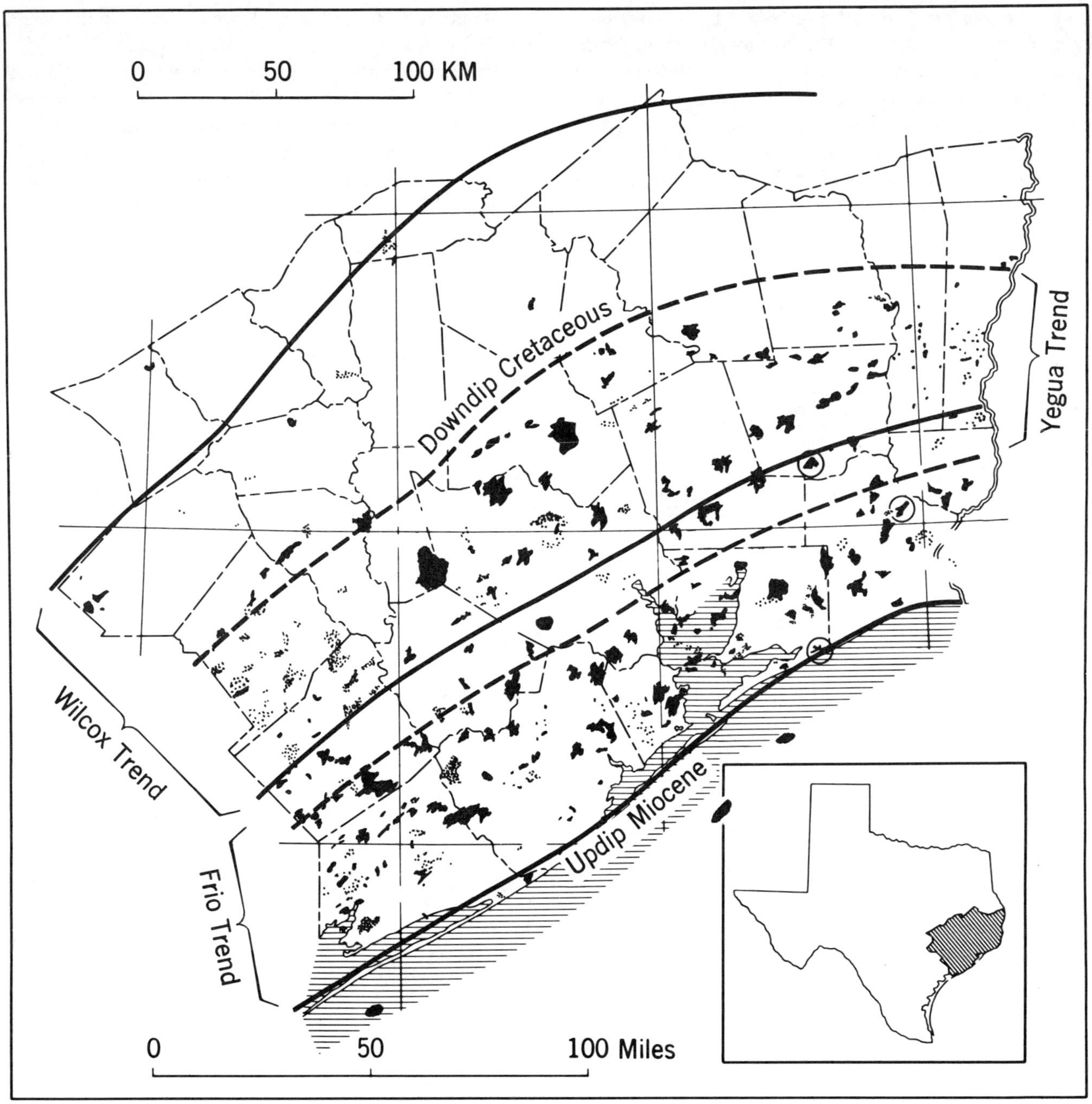

FIGURE 26.5 Texas Upper Gulf. Trend map. *After Robert S. Dollison (1966).*

There are five of these zones, with some overlap where more than one reservoir group may be sought.

The outermost trend, which covers the offshore and a narrow belt inland, is the Miocene. Three named sand zones in the Fleming Formation are the principal reservoirs. The Miocene ranks fourth in production in the Upper Gulf district. Rainwater (1964) describes the regional stratigraphy of the Gulf Coast Miocene as follows:

> Along the Upper Texas coast and in the adjacent offshore region the Miocene is also quite thick and is composed entirely of terrigenous clastic sediments. However, this was interdeltaic area during Miocene time; the thickest section is composed of fine grained neritic deposits much of which was brought there by long-shore currents from the large deltas farther east; there is rapid transition from all-continental to all-marine deposits; and there are relatively few important fluctuations of the shoreline.

The first trend inland, the Frio (Oligocene), is a segment of a producing belt 40 to 60 miles wide that extends nearly 700 miles from Mexico to the state of Mississippi

(Burke, 1958). In the Texas Upper Gulf the Frio zone is about two counties wide. It is the leading reservoir in the district in total production. The greater part of the Frio oil is produced from salt dome traps, and a few giant fields have been responsible for most of this oil (Tolbert, 1958). In recent years the Frio, in the coastal counties where it lies at greatest depth, has become a prolific source of gas and condensate (Williams, 1961). The Frio is a complex of many sand lenses alternating with shale beds. These sands shale out both updip and downdip; the downdip termination lies close inside of, and parallel to, the shoreline (Houston Geol. Soc. Study Group, 1962a, p. 39).

The Frio trend also includes both younger and older reservoirs. Above the Frio is the upper Oligocene (or Miocene-Oligocene) Anahuac Formation with several sand and local limestone reef reservoirs. A Heterostegina reef as much as 377 feet thick is productive around Damon Mound, Nash, and West Columbia piercement salt domes in Brazoria County (Cantrell et al., 1959). In

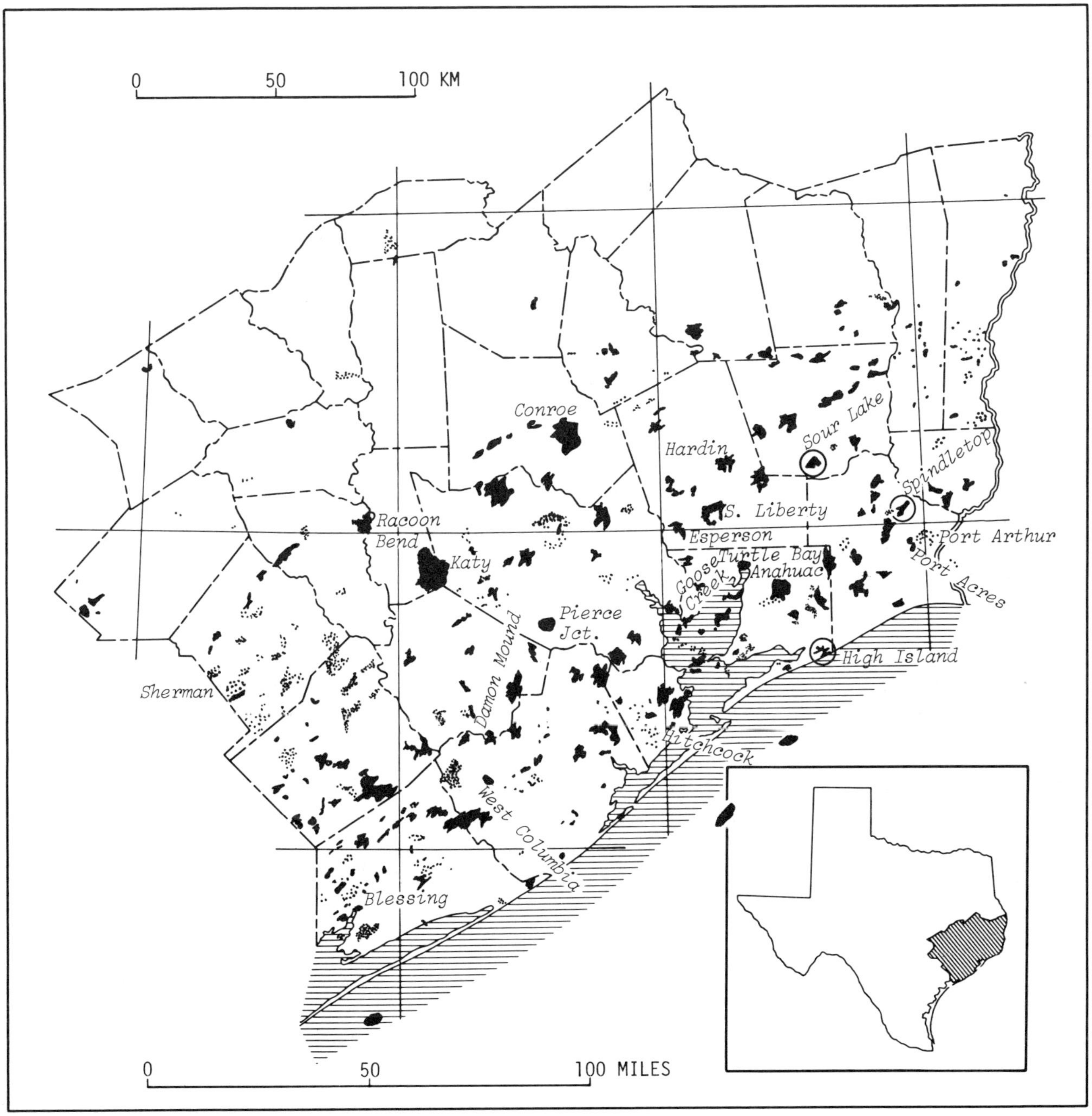

Figure 26.6 Texas Upper Gulf. Oil and gas field map. Fields referred to in text are named.

the lower Oligocene are the Vicksburg sands, which rank fifth in the district in cumulative production.

The Eocene Yegua trend also parallels the shoreline and occupies a two-county-wide belt across the middle of the Upper Gulf Coast district (Figure 26.5). It overlaps the Frio trend on the gulfward side, and about half of the Wilcox belt inland. The Yegua contains at least three distinct sand reservoirs. It ranks second in the district in cumulative production, partly because of the fact that super-giant Conroe field produces from Cockfield sands which are now considered to be Yegua in age. The Yegua and the Wilcox yield the greater part of the Eocene gas in the Upper Gulf district (Hardin and Clark, 1968).

Also exploited in the Yegua trend are younger Eocene sands in the Jackson formation. These sands lie closer stratigraphically to the lower Oligocene Vicksburg than to the Yegua. Below the Yegua are other Claiborne sands (Figure 26.4) that are productive reservoirs in the area of Yeagua-Wilcox trend overlap.

The innermost and widest trend in the Texas Upper Gulf Coast is the Wilcox. This lower Eocene sandstone reservoir has been actively explored in recent years and now ranks third in the district in total production. It consists of up to 5000 feet of alternating sands and shales. Updip the sediments are largely continental and source rock poor. Downdip marine wedges appear, and the hunting improves (Houston Geol. Soc. Study Group, 1962b, p. 35).

The innermost trend, which underlaps the northern half of the Wilcox belt, is the Cretaceous. It occupies the northwestern and northern edges of the district in a belt from one to two counties wide. The reservoirs currently productive are the Navarro in the Upper Cretaceous, and the Buda, Georgetown, Edwards, and Glen Rose in the Lower Cretaceous. So far the most productive has been the Edwards Limestone. The Buda produces from fractures in Madison County (Brundage, 1967).

The Jurassic trend, which lies along the inner edge of the district, is not shown in Figure 26.5. It is not productive as yet; in fact it is virtually unexplored.

Traps And Selected Fields. So far most of the oil and gas fields found in the Upper Gulf Coast of Texas (Figure 26.6) have been onshore. A classification of traps in the district follows:

- Piercement salt dome
  - Cap
  - Flank
- Supercap
  - Piercement
  - Deep-seated
- Anticline
  - Faulted on north flank (Tepetate type)
  - Other types
- Fault
  - Growth
  - Nongrowth
- Lenticular sands

In this district, as in southern Louisiana, the flowage of salt has been the dominating factor in creating traps. In a few of the older fields, cavernous calcitic rock capping shallow salt dome intrusions was a spectacular ("gusher") source of oil. Many salt domes produce trapped flank oil. Arched overlying reservoirs have been prolific producers in some fields, including both those overlying piercement domes and others where the salt is deep but assumed to be present. This assumption may be based on gravity surveys, structure contour patterns, or simply the fact that salt movement seems to be the most logical explanation. It is also probable that deep-seated salt movement created the elongate anticlinal traps that parallel the Gulf shore and are cut on the north flank by a normal gulfward dipping fault.

Lenticular sands, other than those pinched out against salt dome flanks, have not as yet played a major role in this district in trapping hydrocarbons. They are of importance, however, in limiting the distribution of gas and oil across some anticlinal traps.

The shallow piercement salt domes are not notable for their gas production, compared with oil yields. Earlier Teas (1935, p. 683) noted that in the entire Gulf Coast salt dome province, commercial gas had been produced from only seven salt dome fields, "although most of the fields have sufficient gas for local fuel needs." He ascribes the paucity of gas to its escape into the atmosphere through cracks and faults during successive periods of salt uplift.

The offshore search for oil and gas so far has not been anywhere near as rewarding in Texas as in next-door Louisiana. It can be noted that the present distribution pattern of offshore fields in Louisiana shows a general pinchout westward, so that a line drawn along the outer edge of discovered deposits trends northwesterly and approaches the shoreline off the southwest corner of Louisiana. The westward decrease in the percentage of oil fields as against gas fields in this belt also should be noted. One suggested reason for lessened offshore oil productivity westward is a diminution in the size of sand grains away from the mouth of the Mississippi, both present and ancestral.

The highest borehole temperature so far recorded in nonmagmatic areas occurred in a well drilled near the gulf shore in eastern Matagorda County at the southern end of the Upper Gulf Coast district. This well, the Falcon Seaboard Bear Ranch 2A, recorded 525°F at 19,222 feet (Hough and Couvillon, 1966). It was subsequently drilled to 22,515 feet TD. If the postulated gradient of 2°F per hundred feet continues downward, the bottom hole temperature is nearly 600°. In any case, the temperature is considerably above the presumed tolerance for oil (Landes, 1967).

Various oil and gas fields in the Texas Upper Gulf Coast are illustrated and discussed in the following pages as examples of the more important types of traps in this district.

The discovery field of the entire Gulf Coast was the already mentioned *Spindletop*, drilled near Beaumont in Jefferson County in 1901. This fabulous field has been described and discussed in countless articles, symposia, and books. Its discovery, which was both seep and mound inspired, has been described by Clark and Halbouty (1952, p. 4–5):

> Spindletop Springs, south of town, on the banks of the Neches River, was an ideal spot for outings. One afternoon Higgins, who now taught in the Baptist Sunday School, took his class there on a picnic. When the picnic was over, he and the children drove to the round mound a mile or so west. Up on the hill, which rose some fifteen feet above the surrounding prairie, he punched a cane into the ground and then lit a flame as the gas escaped. It was fun for the children and it intrigued Higgins. He made a mental note to come back and find out more about it.

The mound was recognized as a salt-associated feature by Captain Lucas, who in 1896 drilled wells on the Belle Isle salt dome on the Louisiana Gulf Coast in St. Mary Parish and noted the distinctive topography (James P. Spillers, letter dated 21 October, 1968).

The discovery well was drilled by Lucas on top of the mound that had intrigued Higgins. It was a gusher, estimated to flow 75,000 barrels daily from a depth of 1020 feet. During the following 9 months 64 gushers of equal capacity were completed. The oil came from highly cavernous dolomite cap rock (Figure 26.7). This type of reservoir exhausts rapidly, and Spindletop, after yielding a record 17½ million barrels of oil in 1902, dropped to less than half the total of each preceding year for 3 years in a row. Production then diminished more gradually to an average annual yield of about 400,000 barrels from 1914 until 1925. Late that year, probing of the Fleming sands of Miocene age on the salt dome flanks resulted in a second major oil discovery on the south side of the salt dome at a depth of 2515 feet. This successful wildcat was drilled adjacent to, and a mere 1500 feet deeper than, the shallow cap rock field discovered 24 years earlier. Rapid development of the flank pool extended Miocene production downward to a depth of 4800 feet. In 1927 a new production record for Spindletop of over 21 million barrels for the year was reached. Nor did the history of Spindletop end with this new discovery. In 1951, gas condensate was found on the northwest flank of Spindletop dome inside the city limits of Beaumont. Further exploration of the western flank in the mid 1960s resulted in additional discoveries in the Oligocene Frio at depths in the 9000 foot range.

Figure 26.8 is a structure contour map and cross section of the *Damon Mound* oil field in Brazoria County. Damon Mound itself is a conspicuous topographic feature on the Texas coastal plain, rising 83 feet above the surrounding countryside and readily visible from points 5 and 6 miles away. Other criteria of a salt-cored structure were noted by the early settlers. Many of the wells struck highly mineralized water, and a few produced a small amount of gas. The northern part of the hill contained isolated crystals of sulfur, the western flank a 1-acre deposit of sour dirt, and on the eastern slope the limestone cap rock cropped out and was quarried in the early days. First exploration of the Damon Mound cap rock after the Spindletop discovery failed to find any oil. In 1915, however, the search for flank production led to the drilling of a well on the west side of the hill which came in wild, producing an estimated 5 million cubic feet of gas and 100 barrels of oil daily before it sanded up and was abandoned. Two years later the first commercial producer was drilled.

There are two separate oil deposits on the flanks of the Damon salt dome. The larger is on the southwestern side; the other, on the northeastern flank. About 15 per cent of the total acreage of Damon Mound is classed as productive. The structure contour map is based on a limestone marker in the Oligocene. The most productive sands, Heterostegina and Marginulina, are also Oligocene. Upward movement of the salt core dragged up these sands and either pinched them off or faulted them. Dips within the producing areas are in the neighborhood of 40° (Bevier, 1925).

Another salt dome flank trap is *High Island* (Halbouty, 1936), which is about a mile inland from the shore of the Gulf of Mexico in Galveston County. Because of the presence of a hill similar to that at Spindletop, and abundant seeps of sulfurous gas, the testing of this prospect started in 1901, but it was not until 1922 that High Island became a commercial producer. The first reservoir was porous cap rock, but in 1931 deep and prolific flank reservoirs were discovered. The most productive of the flank sands are in the lower Miocene and middle Eocene. Oil has been trapped beneath overhanging cap rock and salt. The beds have been dragged up so as to be both vertical and broken by shattering. Giant *West Columbia* in Brazoria County, on the other hand, has produced most of its oil from pinched sands some distance out from the salt core. The salt flank itself dips outward at a four-to-one slope (Carlton, 1929, p. 455).

One of the most amazing piercement dome salt flank fields is *South Liberty* in Liberty County, which has had at least 14 discovery wells drilled, according to Halbouty (1962). The reason for this number is that the dome is

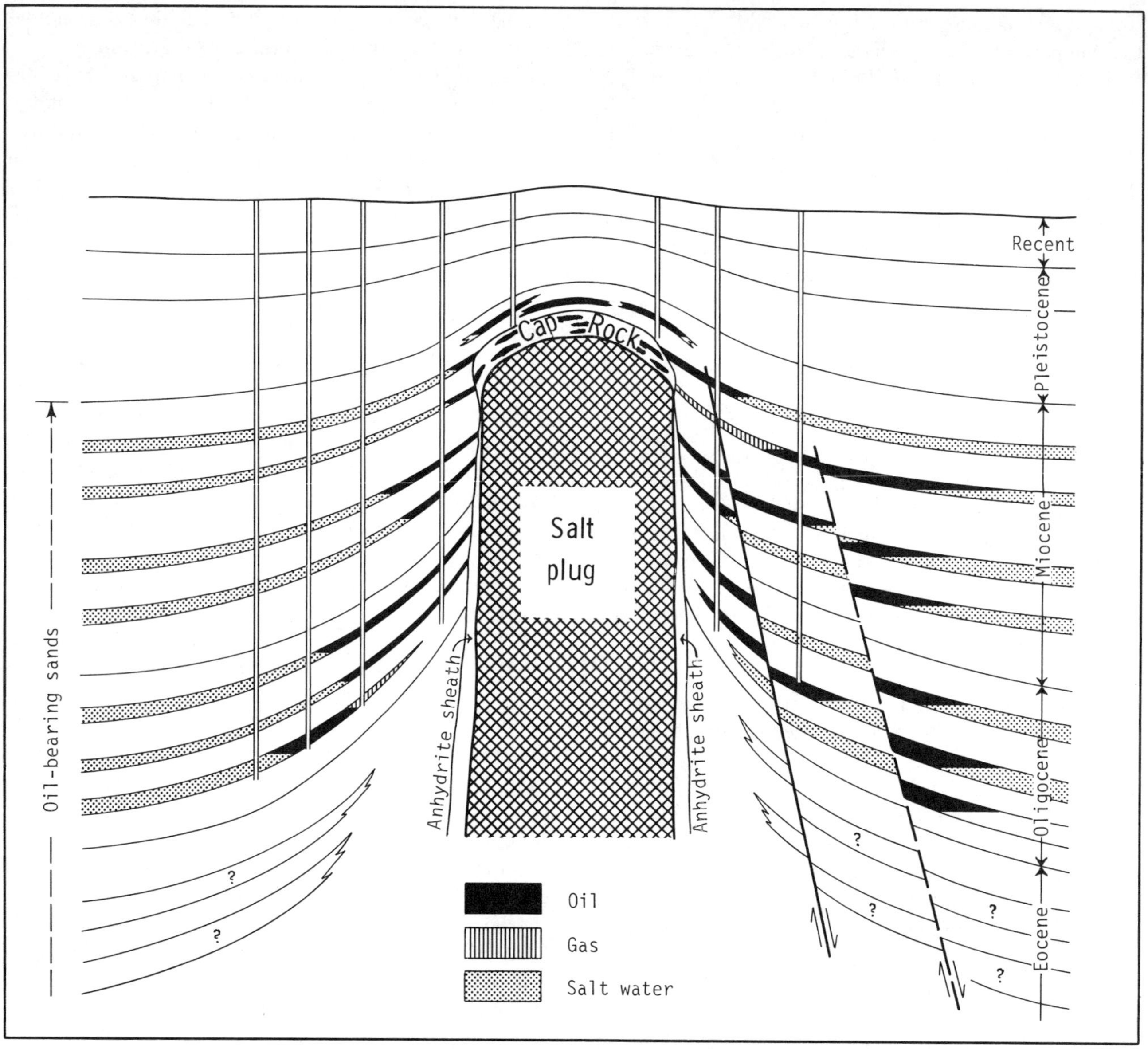

FIGURE 26.7 Salt dome traps (schematic). Shows supercap reservoirs, cavernous cap rock oil, and flank reservoirs (abutting, pinchout, and fault sealed). Sulfur, if present, is in cap rock. ***Courtesy James A. Clark and Michel T. Halbouty (1952), and Random House.***

surrounded by a series of fault blocks each one of which constitutes a separate trap. As a result the salt is surrounded by producing wells, but not continuously; there are unproductive gaps between the fault blocks. The dome is elliptical, with the top of the cap rock coming within 275 feet of the surface. The entire complex of cap rock, salt, and flank sediments is cut by both radial and tangential faults. The reservoir sands range in age from Oligocene Frio down to the Yegua and Cook Mountain in the Eocene. There are at least 25 separate reservoirs in that section. Trapping is due to pinchouts and faulting.

A good illustration of the fact that old salt dome fields never die is *Pierce Junction* (Teas, 1935, p. 700–702) on the southwest outskirts of Houston in Harris County. As far back as 1901, a salt plug was known to underlie this area because of the slight topographic dome and the presence of gas seeps and paraffin dirt. However, it was not until 1921 that oil was discovered at shallow depth on the periphery of the salt intrusion. Subsequent drilling produced a typical halo of discontinuous pools completely surrounding the core, producing from mid-Oligocene sands. Production had declined markedly when in 1954 a well drilled farther out from the core and to the east of the earlier field discovered oil in the deeper Vicksburg (lower Oligocene) sand. During the next 2 years many more successful wells, both discovery and development, were drilled, and a discontinuous outer ring of pools took shape. The greater part of the newly discovered oil is obtained by drilling through salt over-

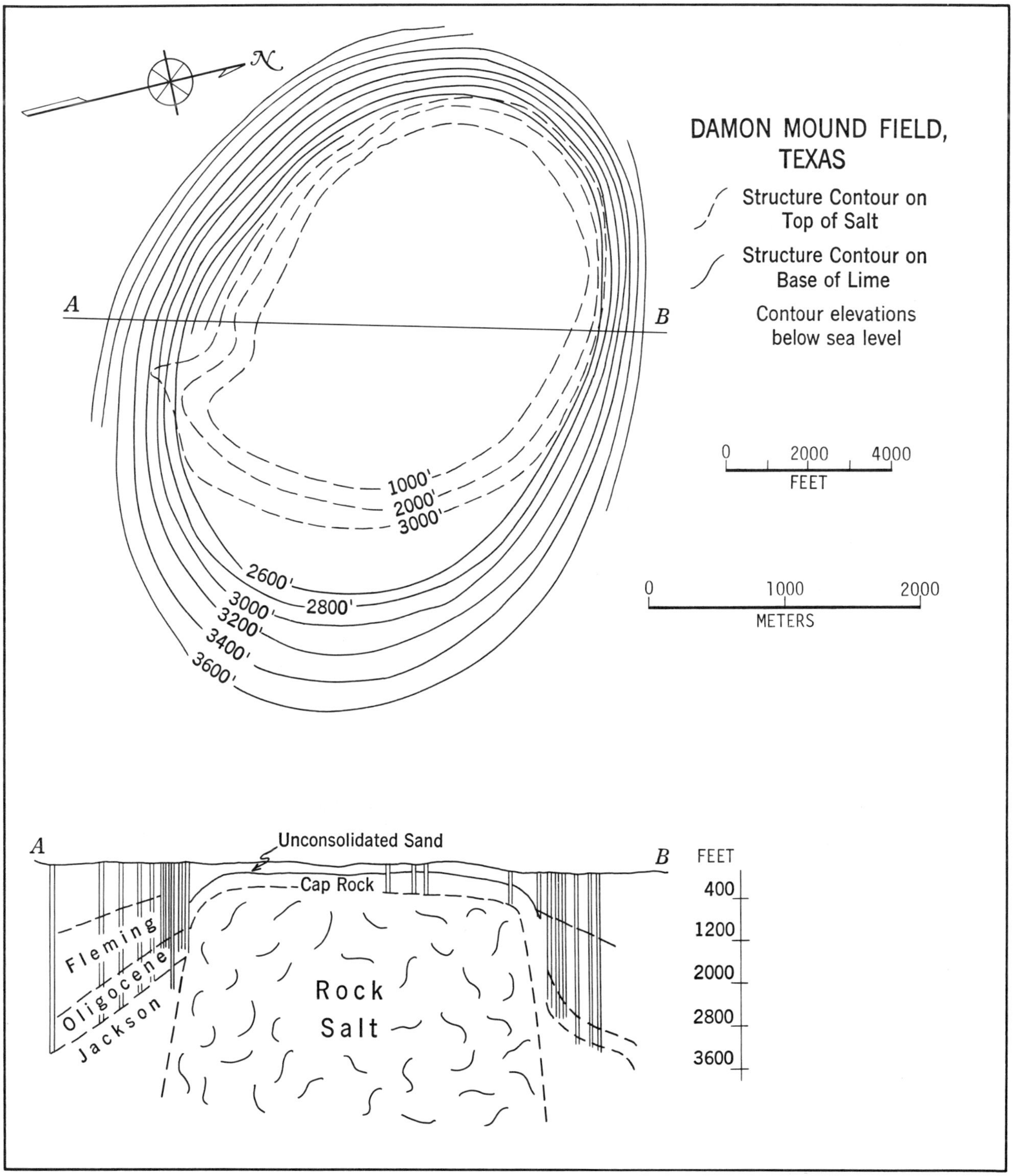

FIGURE 26.8 Damon Mound field, Brazoria County, Texas. Salt dome flank production. *After George M. Bevier (1925). Courtesy American Association of Petroleum Geologists.*

hang where the dragged up sands either pinch out or abut against the nearly vertical, anhydrite-sheathed salt body. The rock structure in the outer ring is a fault block mosaic consisting of more than 40 separate fault blocks. Today there are at least eight producing reservoir sands ranging in age from Miocene to the Yegua (Eocene) and in depth from 3000 to nearly 10,000 feet.

Because spacing regulations do not apply to salt dome structures, some wells have been drilled on leases of town lot size. Diagonal drilling is often followed in order to penetrate the trap below the salt overhang.

The *Esperson Dome* field in southwestern Liberty County is a piercement dome with both supercap and flank production. It was outlined by a torsion balance survey in 1928, followed by the drilling of the discovery well the next year. Because of the many radial faults the structure map of the field, based on one of the supercap reservoirs, looks like an overturned soup plate that has been dropped on a cement floor and then put together again with household cement. Figure 26.9 is a cross section in a north-south direction across the circular dome, showing the fragmented condition of the rock strata overlying and flanking the salt intrusion. The principal reservoirs are sands in the Miocene, the Frio and Vicksburg of the Oligocene, and the Yegua and Crockett in the Eocene. According to Riddell (1962, p. 47),

> Miocene, Frio, and Vicksburg traps are the result of normal faulting and upward flexures associated with a salt dome intrusion. Yegua and Crockett traps were created by truncation of the strata against the salt mass and in some instances by pinchouts of beds over the salt mass. Radial faulting on the flanks has in many cases prevented effective communication between adjacent fault blocks, thus allowing each productive fault block to have its own characteristic oil-water and gas-oil contact.

Because of lack of communication between the oil accumulations on different fault blocks, seven different names have been applied to the separate fault block fields.

The first of the supercap fields to be discovered, *Goose Creek* (Minor, 1925), is in Harris County about 25 miles southeast of Houston. This field was discovered by a well

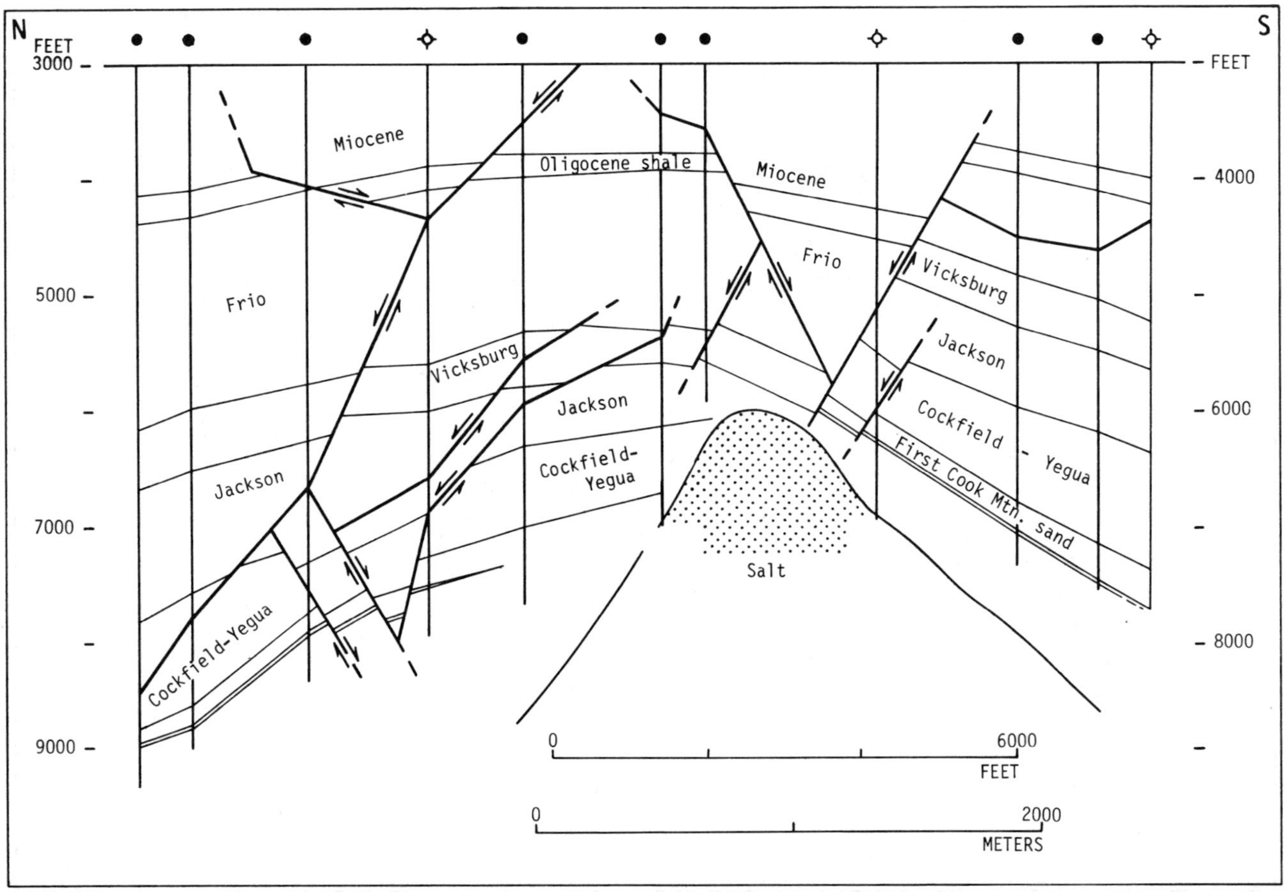

Figure 26.9 Esperson dome, Liberty County, Texas. North-south cross section through salt core, showing fault fragmentation of overlying sediments. *Courtesy John J. Riddell (1962) and the Houston Geological Society.*

drilled in 1908 after a party of oil men, including a geologist, had noticed gas seeps in the area while on a fishing trip. The discovery well was completed at a depth of 1600 feet for an initial production of 800 barrels. Further exploration was discouraging until 1916, when a 10,000 barrel well was brought in, and subsequently some wells were rated as high as 35,000 barrels per day. Although production has been obtained from a number of sandstone reservoirs lying at various depths, no well has yet penetrated the assumed salt core. A notable feature at Goose Creek has been the marked subsidence of the surface overlying the center of the field. The land there originally stood $\frac{1}{2}$ foot to 2 feet above high tide, but after a couple of years of active exploitation the surface subsided until at high tide 2 to 3 feet of water inundated the central part of the oil field. This subsidence is thought to be due to the pumping of sand and water (which causes dehydration of the clays) along with oil from the reservoir (Pratt and Johnson, 1926; Snider, 1927).

The *Conroe* field, discovered in 1931, is the king of the supercap producers. Furthermore, it is the largest field in the entire Upper Gulf Coast in terms of cumulative production, which by the end of 1965 had reached a total of nearly 425 million barrels. Conroe ranked fourth in the district in that year, in spite of its quarter-century age, with a production of $8\frac{3}{4}$ million barrels. This field lies 40 miles north of Houston in Montgomery County. It was the first of the large domal accumulations that presumably overlie a deep-seated salt core. The producing area at Conroe covers 27 square miles. Gas seeps and springs of hydrogen sulfide-bearing water marked the field at the surface. Structure contours based on the top of the main producing sand show an oval shape with a closure between 350 and 400 feet (Figure 26.10). Three normal faults cross the ellipse parallel to the minor axis. The fault pair to the south, which crosses the middle of the dome, creates a keystone graben $\frac{3}{4}$ mile wide. Three interconnecting members of the Cockfield sand (Yegua-Eocene) constitute a single reservoir with a gas cap above and bottom water below. The gas cap is 170 feet in thickness, and the oil occupies an interval of 130 feet (Michaux and Buck, 1936). The gas-oil contact lies at a depth of about 5000 feet.

Another giant field in the Texas Upper Gulf Coast that owes its trapping to a large domed uplift is *Anahuac* in Chambers County. A deep salt core is assumed, but the only positive evidence of its presence is a weak torsion balance minimum (Sealy, 1962, p. 1). The almost circular oil-producing dome is crisscrossed by tensional faults, also assumed to be due to salt flowage. The latest movement in this fault complex took place not long ago. The reservoir sands, at least ten in number, range from upper Miocene to mid-Oligocene. The most prolific are the Marginulina in the lower Miocene and sands in the Frio beneath. The Marginulina reservoir has a closure in the neighborhood of 600 feet.

The *Raccoon Bend* oil field (Teas and Miller, 1936) in Austin County is also supercap accumulation. Oil was discovered in 1928, and this structure likewise was thought to have a salt core. This assumption was verified in 1946 when a well drilled to a depth of 11,447 feet entered salt (McLellan, 1946).

The *Katy* field (Allison et al., 1946), mostly in southeastern Waller County, covers about 30,000 acres and is one of the major gas-condensate accumulations in the United States. The trap is a broad, unfaulted dome. Gas condensate and oil are produced from six or more sandstone reservoirs lying in the Yegua (Eocene) formation between depths of 6250 and 7450 feet. The Katy field operators marketed gas as well as oil and condensate for the first 8 years after discovery in 1935. Then cycling of the gas back into the reservoir after extraction of the condensate was initiated and continued for 24 years. In 1967 the operators of Katy began to market gas again, drawing upon the 6 trillion cubic feet of reserve that had been used as a revolving fund to extract the maximum feasible volume of condensate (Wilson, 1967).

The Upper Gulf Coast, like next-door southern Louisiana, has numerous Tepetate-type anticlinal traps faulted on the inland flank. One example is the *Blessing* field in northwestern Matagorda County. Except for the major fault on the north flank, the elongate anticline (Figure 26.11) is cut by only one fault, which probably joins the major fault at depth. As is the general rule with this type of structure, the major axis of the anticline is parallel with the fault on the inland flank, and accumulation takes place in the anticline, which is in the down block on the gulf side of the big fault. The Blessing field produces from at least seventeen different sands, all of them in the upper and middle Frio of the Miocene-Oligocene (Smith and Goodwyn, 1962, p. 18). A few producing wells have been drilled into the upthrown block north of the fault in the northward plunging nose produced by the truncation of the anticlinal flank.

The *Sheridan* field (Lofton, 1962, p. 192–197) in Colorado County is yet another trap of the Tepetate type, but without any production on the north, upthrown, side of the fault. The productive part of the anticline is an ellipse measuring approximately $6\frac{1}{2}$ by $3\frac{1}{2}$ miles. The youngest reservoir is in the Yegua (Claiborne–Eocene); it is a sand lens underlying about 480 acres out of a total productive area of 14,900 acres. The main reservoirs are thick sands in the Wilcox (lower Eocene). About a dozen different sands in this formation produce oil; the aggregate reservoir thickness ranges from 1200 feet on the southwest flank of the field to 1700 feet on the northeast flank.

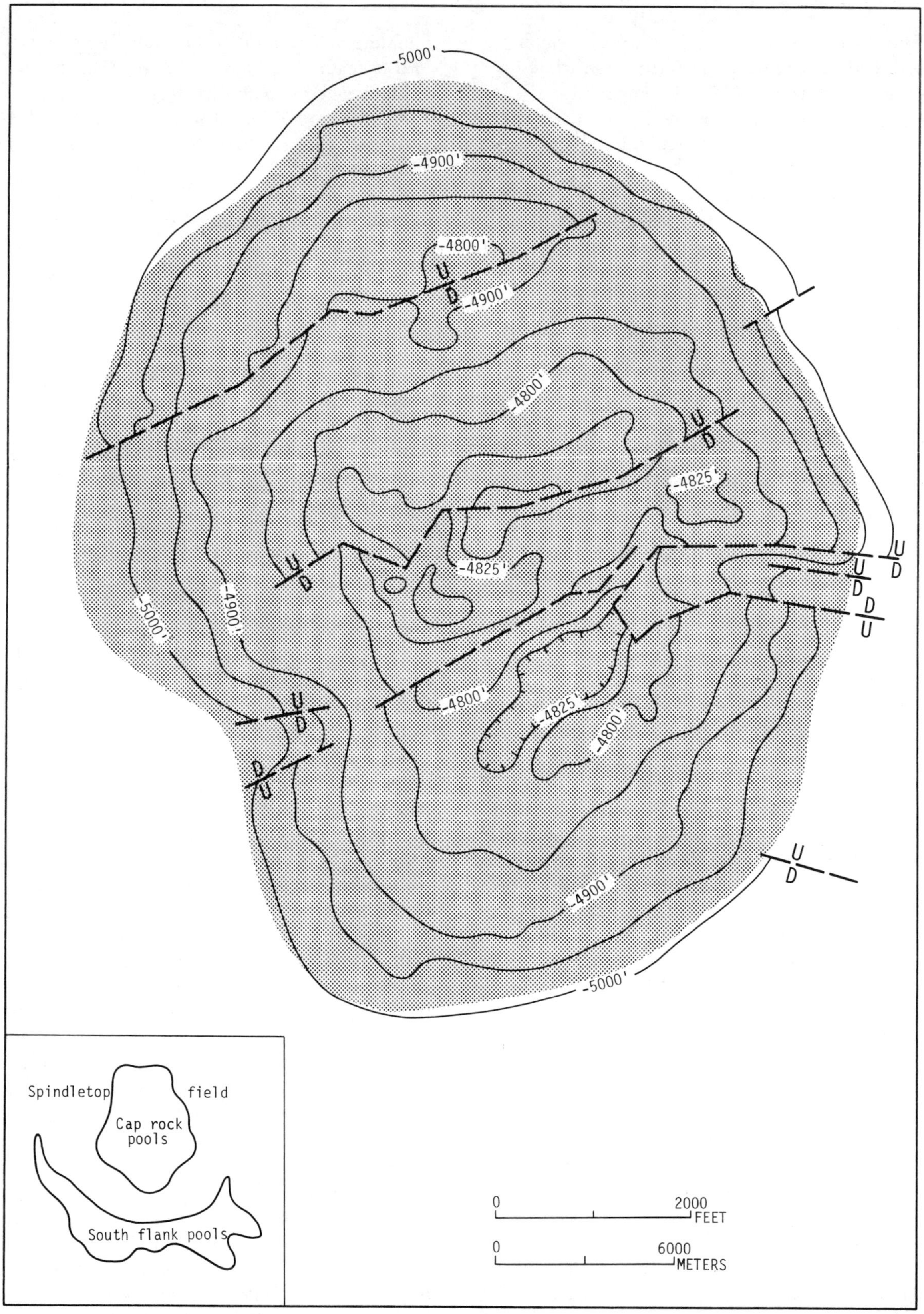

Figure 26.10 Structure contour map, Conroe oil field, Montgomery County. Contour datum is main Conroe sand. Contour interval; 50 feet. Productive area of this reservoir is shaded. Giant supercap trap. Insert at lower left shows map of Spindletop field at same scale. *Courtesy Frank W. Michaux and E. O. Buck (1936, p. 800) and the American Association of Petroleum Geologists.*

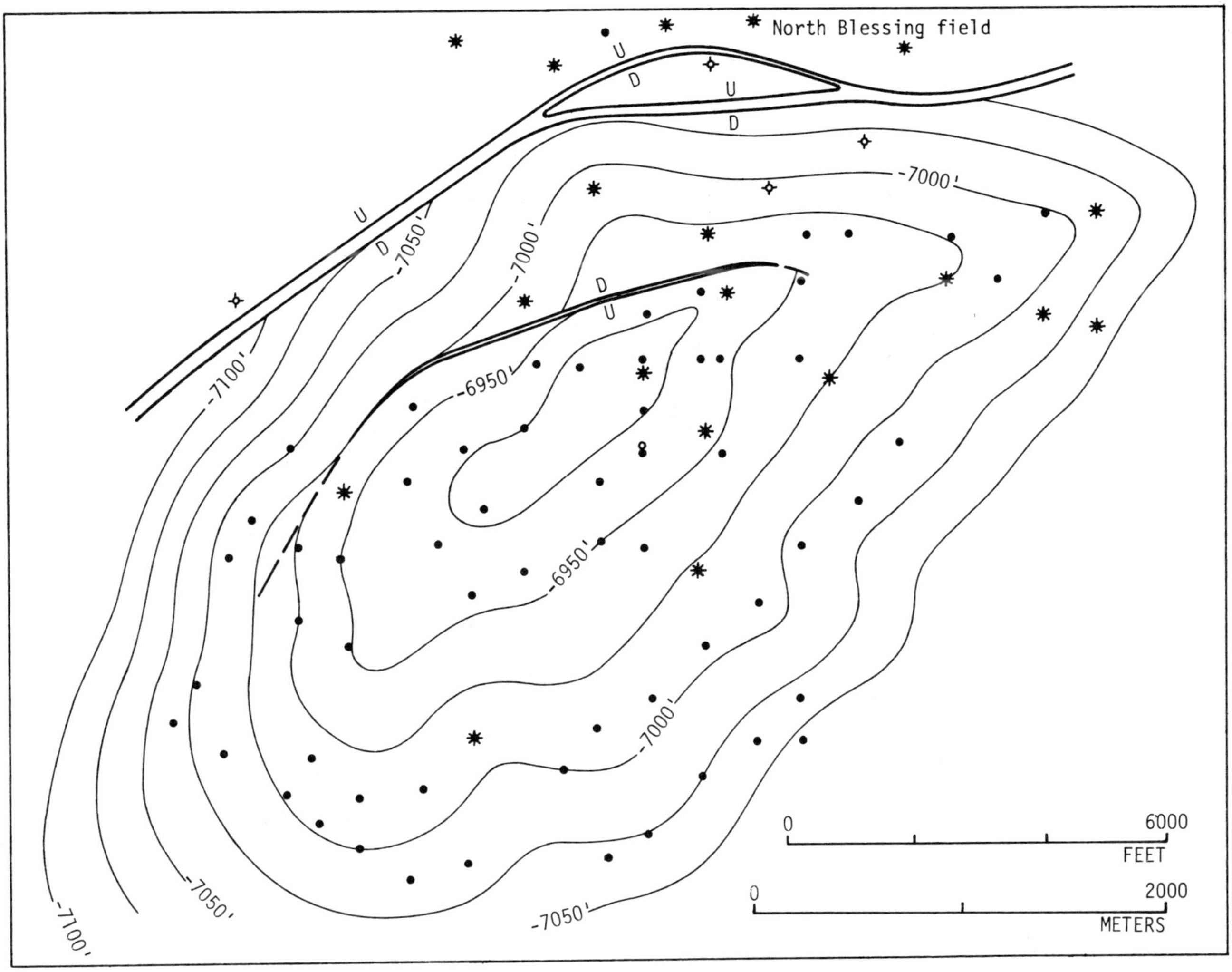

FIGURE 26.11 Blessing field, Matagorda County, Texas. Contours based on F-3 reservoir sand. Interval, 25 feet. *After Fred L. Smith and J. T. Goodwyn (1962, p. 20). Courtesy authors and the Houston Geological Society.*

An unusual type of anticlinal trap is responsible for oil accumulation in the *Turtle Bay* field in Chambers County (Akkerman, 1959). This anticline is not an actual upfold, but is a residual feature produced by local reversal of regional dip into a rim syncline bordering a salt dome.

Varying permeability ("stratigraphic") traps are also present in the Upper Gulf district. Two adjacent fields, Port Acres and Port Arthur in Jefferson County, illustrate three varieties of trap (Halbouty and Barber, 1962, p. 169–173). *Port Acres* is an excellent example of lenticular sand trapping. The cross section (Figure 26.12) shows gas and condensate accumulation in three sandstone lenses, two in the lower Hackberry (Oligocene) and the third in the Vicksburg (also Oligocene). In addition a third Hackberry sand lens produces hydrocarbons in a limited area to the southwest of the cross section illustrated. An interesting feature of the Vicksburg sand production is that producing wells were drilled in the Port Arthur field (or at the edge thereof), but were drilled through the fault to the Port Acres side in order to reach the Vicksburg sand.

Although the major trapping in the *Port Arthur* field is anticlinal, it can be noted that the deepest reservoir (lower Frio) is combination anticlinal and fault trapped, and that the youngest reservoir, seen at the extreme right of the cross section, fails to reach the top of the anticline because of lenticularity. The fault that contributes to the trapping in the Port Arthur field is an excellent example of growth (or contemporaneous) faulting, which is very common in southern Louisiana (Chapter 24) and is also present in South Texas (Chapter 27). The growth fault at Port Arthur, like most of the Louisiana faults of this type, is down-to-the-coast. The sediments in the down

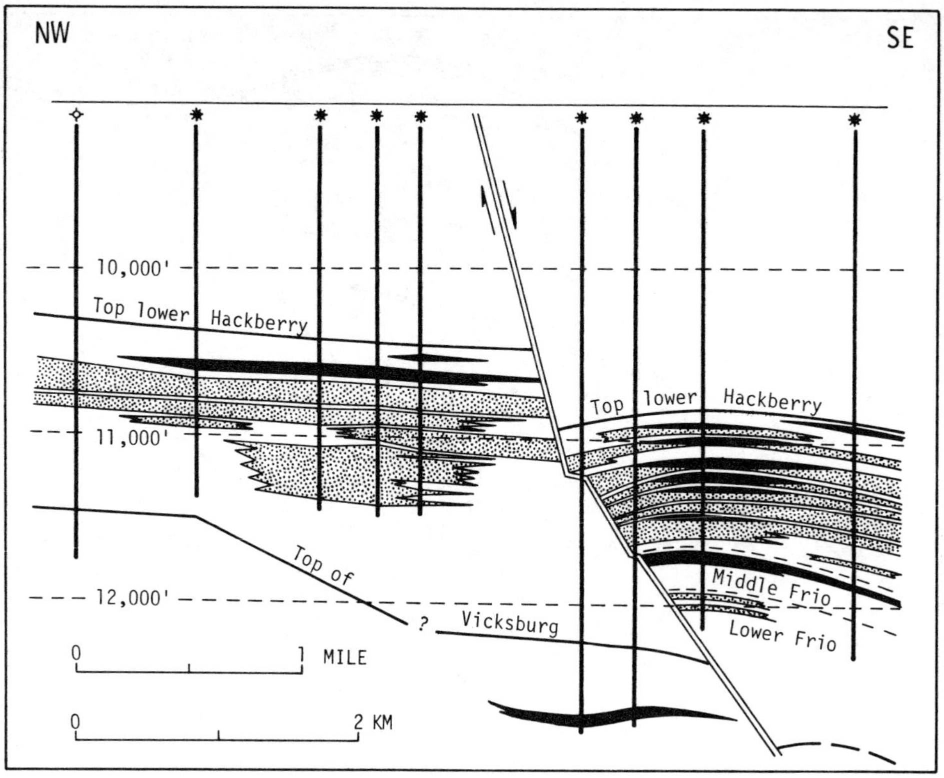

Figure 26.12 Port Acres-Port Arthur fields, Jefferson County, Texas. Cross section, with Port Acres field on left (NW). Gas and distillate accumulations shown in solid black. *After Michel T. Halbouty and Thomas D. Barber (1962, p. 171). Courtesy authors and the Houston Geological Society.*

block increase in thickness toward the fault, and the anticline (or pseudo-anticline) responsible for the accumulation in the Hackberry reservoirs is due to the reverse dip at the fault plane, resulting from contemporaneous deposition and faulting.

The reservoirs so far developed in these fields are all Frio (Oligocene) in age. In terms of annual production, the Port Acres field is larger, producing nearly 34 billion cubic fee of gas and approximately 1,700,000 barrels of distillate in 1961.

The *Hitchcock* field (Reiter, 1959) in Galveston County is an anticlinal trap in which the principal reservoir is absent across the crest because of pinchout but is present on the west, south, and east flanks. The folding, and accompanying strike faulting, are ascribed to elongate ("ridge") salt intrusions at depth. The oil reservoirs are Miocene sands at just below 5000 feet; some gas is produced from Pliocene sediments. Discovery of this field in 1937 was the result of torsion balance and reflection seismograph surveys. No evidence of the presence of a trap occurs at the surface, which is floored with Pleistocene clay. The land is marshy, with a mean elevation above sea level of 1 foot (Halbouty and Simmons, 1941).

The *Hardin* field, in Liberty County, Texas, is a nearly circular anticline probably overlying a deep-seated salt dome. It is faulted across the middle; oil occurs in nine sands, all of Yegua (late Eocene) age, on the downthrown (gulfward) side of the fault. In eight of these sands accumulation has been due to the presence of the anticline, although individual sands may not be present over the entire upfold. The ninth sand, the Davis, which lies in the upper middle of the productive section, was deposited as a barrier beach or offshore bar and is completely surrounded by shale (Casey and Cantrell, 1941). This bar lies along the southeastern flank of the dome; it has a total length of about 11,000 feet and an average width of 1250 feet. Figure 26.13 is a section across the bar near its northeastern end. It is nearly 40 feet thick except at the northwest and southeast ends, where it lenses out between drill holes 200 to 400 feet apart. It can be noted also that the presence of the bar has produced nearly parallel arching in the overlying sediments. Similar overlying arched structures elsewhere have been attributed to differential compaction, the sand compacting less than the surrounding mud (shale).

Future. In addition to continued success by lateral

exploration, the Upper Gulf Coast district can anticipate much more activity, with resultant discoveries, in further exploration of the already known salt domes, the deeper reservoirs, and the offshore area. There are a number of known piercement domes with no production at all as yet. The *Ferguson Crossing* dome on the western edge of Grimes County was discovered in 1934 by a well that penetrated cap rock into salt. Twenty-six years later a 3 million cubic foot discovery gas well was drilled on this dome at a depth of 1700 feet. No doubt other hitherto barren domes will be found productive.

Perhaps of even greater potential is the discovery of new flank accumulations around the sides of piercement domes already producing. The previously described Pierce Junction salt dome is a good example of this possibility. It takes much more than one dry hole to a section to condemn flank production. The map of the *Big Hill* field in Jefferson County (Dollison, 1965, p. 27) shows production of oil and gas on the northwest flank as far as 3 miles from the top of the salt dome, but the dry hole density in the same zone in the other three quadrants is less than one to the square mile.

Only a start has been made in plumbing deeper reservoirs in the already productive parts of the district (McQueen, 1960). The Fort Trinidad field, which overlaps the Madison-Houston county line (thereby crossing the boundary between the Upper Gulf Coast and the Northeast Texas districts), is an example of downward progression of pay zones. Discovered in 1953, with production from an Upper Cretaceous sand and a limestone near the top of the Lower Cretaceous, Fort Trinidad became a multipay field in 1961, when eight deeper Lower Cretaceous reservoirs were discovered. Three of these are in the Edwards; the other five, in the Glen Rose section below (Pampe, 1963).

There is no production as yet from the still deeper Jurassic. This section is still virtually unexplored, however, in spite of the fact that there is a lot of acreage along the northwest side of the district, where the most promising potential reservoir, the Smackover, lies between 15,000 and 20,000 feet in depth.

The Houston Geological Society Study Group (1962a) calls attention to the large possible potential in following the Eocene Yegua and Wilcox reservoirs farther downdip:

> The Yegua and Wilcox formations have great undeveloped potential. In the case of the Yegua downdip sedimentation and stratigraphy must be considered from both regional and local aspects. As more control is afforded, the areas of maximum downdip sand accumulation may be delineated. . . . Wilcox potential suggests the necessity for local study of facies changes rather than extended downdip penetration. . . . Complete penetration of this formation in the updip areas should lead to the discovery of producing trends of good reservoir rocks in communication with marine source beds (p. 35).

Even though the Frio shales out along the present shoreline, the downdip edge of this prolific reservoir has not been completely probed as yet.

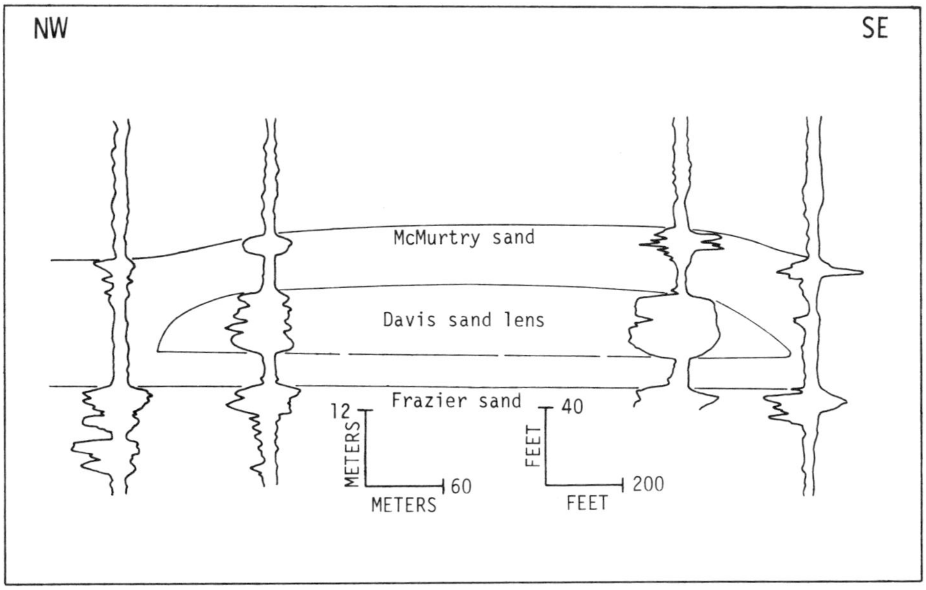

FIGURE 26.13 Hardin field, Liberty County, Texas. Cross section showing Davis sand lens of Yegua (late Eocene) age surrounded by shale. All three sands shown are reservoirs; the Frazier is the leading producer in the field. ***After S. Russel Casey and Ralph B. Cantrell (1941, p. 576). Courtesy authors and the American Association of Petroleum Geologists.***

In spite of the fact that offshore exploration in the Upper Gulf Coast district has not, to date, been as successful as anticipated, two "significant" discoveries were made in 1967 (McGrew, 1968) to add to the score or so of offshore fields discovered previously. The faith of the oil industry in offshore Texas (both Upper Gulf Coast and South districts) was demonstrated by lease bonus payments for Federal acreage of $594 million in May, 1968 (U. S. Dept. Interior, 1968, p. 59).

REFERENCES CONSULTED IN PREPARING CHAPTER 26

Akkerman, R. P., 1959.
Allison, A. P., et al., 1946.
Bevier, George M., 1925.
Bornhauser, Max, 1960.
Brundage, Harrison T., 1967.
Burke, Ray A., 1958.
Cantrell, R. B., Montgomery, J. C., Woodard, A. E., 1959.
Carlton, D. P., 1929.
Casey, S. Russel, and Cantrell, Ralph B., 1941.
Clark, James A., and Halbouty, Michel T., 1952.
Denham, R. L., ed., 1962.
Dollison, Robert S., 1965, 1966.
Halbouty, Michel T., 1936, 1962.
Halbouty, Michel T., and Barber, Thomas D., 1962.
Halbouty, Michel T., and Simmons, Benjamin T., 1941.
Hardin, George C., Jr., and Clark, W. Kenley, 1968.
Hough, Howard, and Couvillon, J. A., 1966.
Houston Geological Society, 1962.
Houston Geological Society Field Trip Committee, 1953.
Houston Geological Society Study Group, 1962a, 1962b.
Landes, Kenneth K., 1967.
Lofton, Coleman L., 1962.
McGrew, Alan R., 1968.
McLellan, H. J., 1946.
McQueen, Henry S., 1960.
Michaux, Frank W., Jr., and Buck, E. O., 1936.
Minor, H. E., 1925.
Pampe, Carl F., 1963.
Pratt, W. E., and Johnson, D. W., 1926.
Rainwater, E. H., 1964.
Reiter, J. O., 1959.
Riddell, John T., 1962.
Sealy, George, 1962.
Sheets, M. M., and Cockrell, Ernest, Jr., 1962.
Smith, Fred L., Jr., and Goodwyn, J. T., Jr., 1962.
Snider, L. C., 1927.
Teas, L. P., 1935.
Teas, L. P., and Miller, Charis R., 1936.
Tolbert, A. M., 1958.
U. S. Department of the Interior, 1968.
Warner, C. A., 1939.
Williams, Neil, 1961.
Wilson, Howard M., 1962.

Chapter

27

SOUTH TEXAS

The South Texas district, sometimes referred to as Southwestern Texas, contains 58 counties (Figure 27.1) and includes Texas Railroad Commission Districts 1, 2, and 4 (Figure IV.I). It is bounded on the southeast by the Gulf of Mexico, on the south and southwest by the Republic of Mexico, on the northwest by the West Texas district, on the north by the West Central district, and on the north and east by the Northeast and the Upper Gulf Coast districts. South Texas ranks second in annual yield to West Texas, with 16 per cent of the state production in 1965.

The Thrall field (Figure 27.8) a serpentine producer in Williamson County, was discovered in 1915. The Luling field, a fault accumulation also in the interior, was opened up in 1922. The first discoveries in the coastal part of the South Texas district were made in Refugio County in 1929. Most of the oil fields in this district occur in the eastern part in a zone extending inland five or six counties from the Gulf of Mexico. The oil fields within this zone tend to lie in trends parallel to the shoreline and major fault zones. First in terms of cumulative production—more than 300 million barrels by the end of 1965—is the Seeligson oil field complex in Jim Wells and Kleberg Counties. This was also the leading producer in 1965. Second place, both in cumulative and in recent production, goes to the Tom O'Conner field in Refugio County, which has reached nearly 300 million barrels in total production.

Tectonics. The northwestern side of the South Texas district is dominated by (1) the Llano uplift, which exposes Precambrian crystalline rocks in the core of the Central Mineral Region; and (2) the plateau formed by the Edwards (Lower Cretaceous) Limestone to the southwest (Figure 27.2). A southeast-trending spur of the Llano uplift, the San Marcos arch, is covered by coastal plain sediments. The regional dip in the coastal plain is gulfward, into the Gulf Coast geosyncline. The subsidence of this great geosyncline, accompanied by deep-

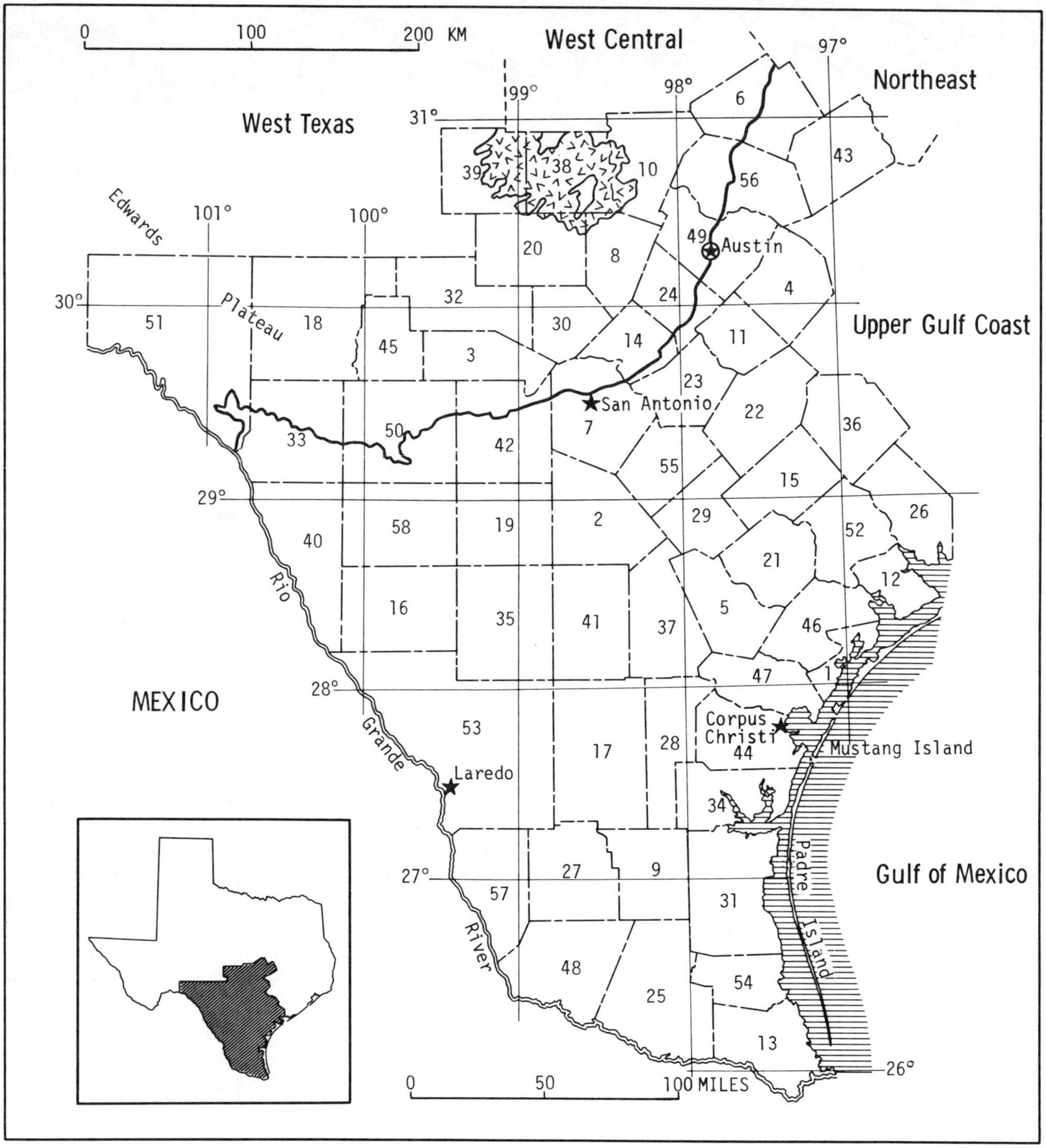

FIGURE 27.1 South Texas. Index map. Bold line crossing district from west to north marks inner edge of Upper Cretaceous sediments. Pattern in Llano and adjacent counties marks outcrop of Precambrian basement rock. County Key:

1 Aransas
2 Atascosa
3 Bandera
4 Bastrop
5 Bee
6 Bell
7 Bexar
8 Blanco
9 Brooks
10 Burnet
11 Caldwell
12 Calhoun
13 Cameron
14 Comal
15 DeWitt
16 Dimmit
17 Duval
18 Edwards
19 Frio
20 Gillespie
21 Goliad
22 Gonzales
23 Guadalupe
24 Hays
25 Hidalgo
26 Jackson
27 Jim Hogg
28 Jim Wells
29 Karnes
30 Kendall
31 Kenedy
32 Kerr
33 Kinney
34 Kleberg
35 LaSalle
36 Lavaca
37 Live Oak
38 Llano
39 Mason
40 Maverick
41 McMullen
42 Medina
43 Milam
44 Nueces
45 Real
46 Refugio
47 San Patricio
48 Starr
49 Travis
50 Uvalde
51 Val Verde
52 Victoria
53 Webb
54 Willacy
55 Wilson
56 Williamson
57 Zapata
58 Zavalla

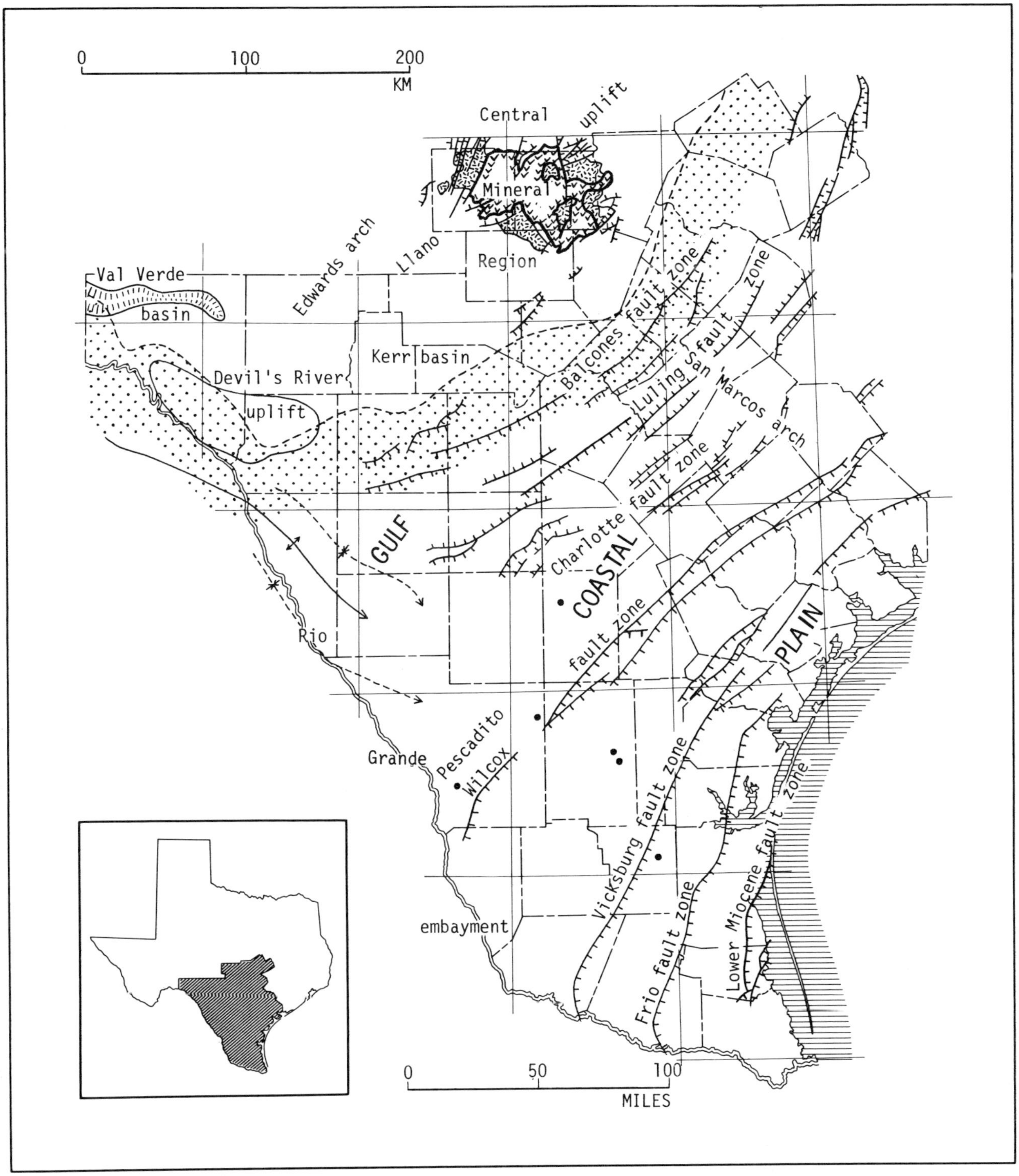

FIGURE 27.2 South Texas. Tectonic map. Shows faults, major anticlines and synclines, uplifts, arches, and basins. Large black dots are salt domes; small dot pattern is subcrop of Paleozoic Ouachita facies. Central Mineral Region shows outcrop of metamorphics (checked area) and intrusive granite (short random dashes). Val Verde basin closed contour is −15,000 feet and Devils River Uplift contour is −5000 feet, both on top of Lower Cretaceous. *After George V. Cohee et al. (1962).*

seated salt movements, resulted in the formation of most oil and gas traps in South Texas (Johnson and Mathy, 1957, p. 212). The gulfward dip is interrupted by faults and flexures and punctured locally by piercement salt domes inland from the coast in the southern part of the district.

The salt domes along the coast, so prominent in the Texas Upper Gulf district, appear to peter out at the

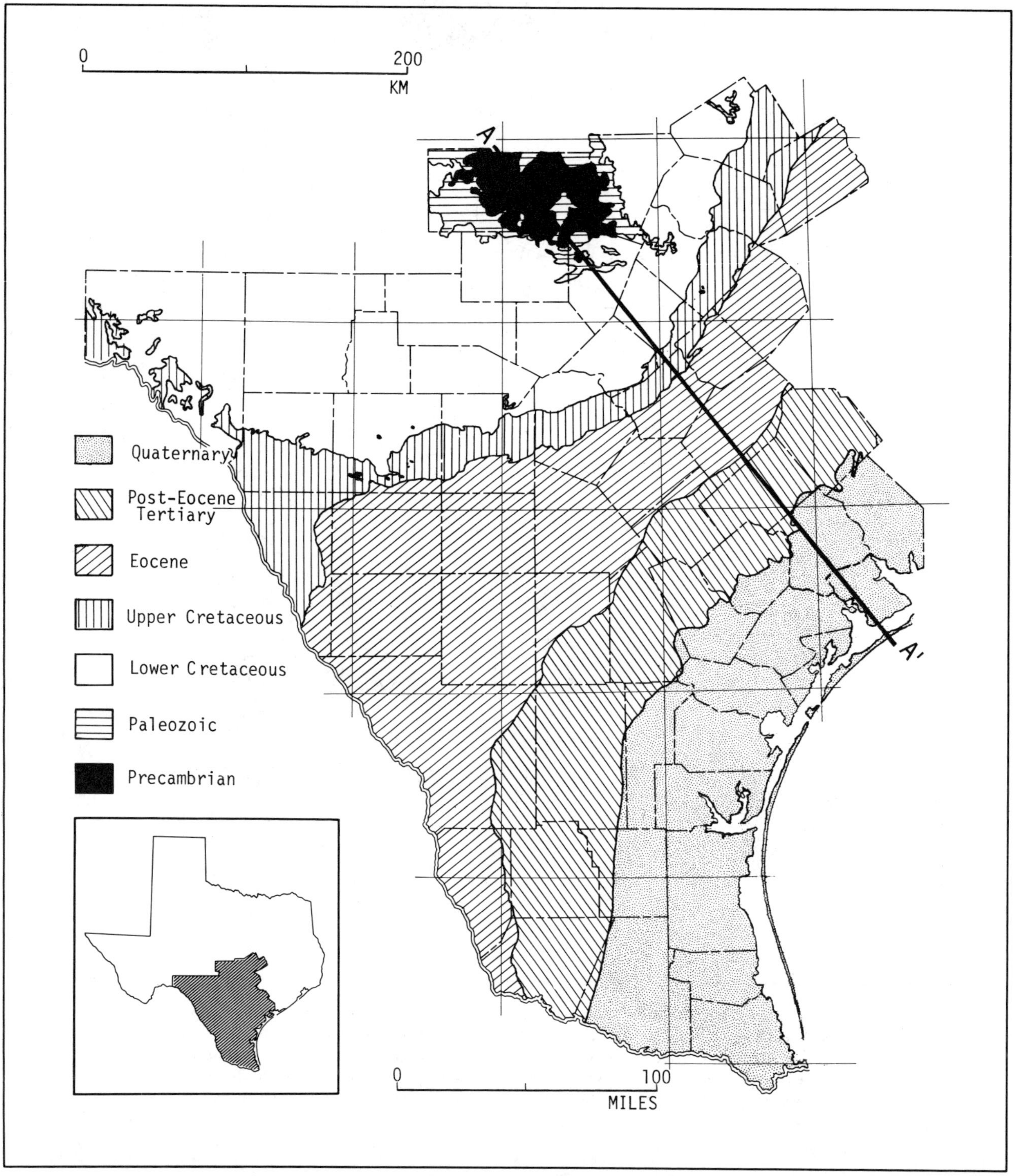

Figure 27.3 South Texas. Geologic map. Also shows location of cross section *A–A'* (Figure 27.4).

border between the two provinces (Figure 26.2). However, shale diapirs continue across the line and contribute to the entrapment of petroleum. "The shale diapirs appear to be a less attractive exploration target than the associated deep-seated structures, suspected of being nonpiercement shale domes, which underlie profitable Lower Frio oil and gas fields such as Una West, Portilla, and Copano" (Brooner, 1966).

South Texas has other structural upwarps and basins in addition to the major tectonic features mentioned above. These are shown on the tectonic map (Figure 27.2). Ups and downs confined to the Edwards plateau include the Val Verde basin, Devil's River uplift, Edwards arch, and Kerr basin. The seaward dip of the sediments is modified by a broad gulfward plunging syncline, the Rio Grande embayment, and a gulfward plunging anticline, the San Marcos arch. The axis of the Rio Grande embayment is close to the river of the same name. The line of cross section *A–A′* (Figure 27.4), shown on the geologic map (Figure 27.3), coincides with the axis of San Marcos arch.

The most prominent surface fault feature is the Balcones zone, which lies along the periphery of the Edwards Plateau, curving around the Llano uplift southeast and east of the Central Mineral Region. The Balcones faults are large, discontinuous, normal faults dipping coastward. Gulfward from the Balcones belt is the Luling segment of the Luling-Mexia-Talco fault zone. The faults in this zone are likewise normal but are down on the inland side, creating a low-relief graben trough between the two belts. Other faults cut the sediments on the monocline, as can be seen on the tectonic map. Especially noteworthy are the Wilcox, Vicksburg, Frio, and lower Miocene fault zones. The Wilcox and Frio zones are most important in terms of oil and gas production. Anticlines and synclines also furrow the monoclinal sediments. In the Rio Grande embayment they tend to strike parallel to the embayment axis.

The tectonic map shows also the subcrop of the Paleozoic Ouachita tectonic belt. These folded Paleozoic rocks are confined to the subsurface, buried below the unconformity at the base of the Mesozoic sediments.

The northern and southern edges of the South Texas district have local abnormally high temperature gradients. The Amerada R. F. Tally No. 1 in east-central Victoria County established what was then a new temperature record of 520° at 23,884 feet TD in 1964 (Bleakley, 1965). This gives a gradient of more than 1.8°F per hundred feet as against a country-wide average of about 1.3°. Subsequently an even hotter well was drilled in nearby Matagorda County at the southern end of the Upper Gulf Coast district. Earlier (in 1957), the discovery well of the huge Pescadito salt dome in southwestern Webb County, about 10 miles east of the Rio Grande River, recorded 450°F at a depth of 14,522 feet.

Stratigraphy. There is more variety in both lithology and geologic age in the surface rocks of South Texas than in any other district in the state. The age ranges from Precambrian in the Llano Mountains at the northern edge of the district to Recent along the Gulf Coast shoreline (geologic map, Figure 27.3). The structural relationship of these rocks can be seen in the cross section (Figure 27.4). The line of this section is shown in Figure 27.3.

The Precambrian metamorphic rocks of the Llano area consist of schist, much of which is graphitic, and gneiss. Intruded into these ancient rocks are great granitic batholiths of younger Precambrian age. The metamorphic rocks are not separated from the igneous in the geologic map, but this distinction has been made on the tectonic map (Figure 27.2). Because the Precambrian rocks are much more resistant to erosion than the surrounding sedimentary formations, they have lagged behind in the base-leveling process during the current cycle of erosion and hence this region is submountainous.

Flanking the arched and eroded Precambrian core of the Llano uplift are Paleozoic beds, including a small amount of Mississippian and a somewhat greater exposure of Ellenburger (Cambro-Ordovician) carbonate rock. The latter contains remnants of a Silurian limestone in collapse structures cropping out in Burnet County on the Llano uplift (Barnes et al., 1966). Still farther out, covering the surface in northwestern South Texas, are Comanchean (Lower Cretaceous) rocks. Included here is the Edwards Limestone, which, because its resistance to erosion is greater than that of the other post-Paleozoic rocks, caps the uplands to form a prominent topographic feature, the Edwards Plateau.

Proceeding gulfward, the next sequence of rock formations is the Upper Cretaceous. The surface contact between the Lower and the Upper Cretaceous in many places is a fault line along which the more resistant Lower Cretaceous limestones have been differentially eroded to produce the Balcones escarpment, named after the fault zone of the same name. The relatively narrow belt of Upper Cretaceous rocks is much more subdued topographically than is the country to the northwest. However, the Austin Chalk outcrop is noticeable because of the sculpturing of a cuesta. The Upper Cretaceous belt also contains a few scattered outcrops of intrusive rock, especially to the east and south of the Central Mineral Region. These intrusives are serpentinized basalt masses. Extrusive serpentinized lavas and pyroclastics may also be present. Pilot Knob in Travis County is an exhumed volcano of Cretaceous age, composed mainly of serpentinized pyroclastics of extrusive origin (Romberg and Barnes, 1954).

The two Tertiary belts shown on the map are the

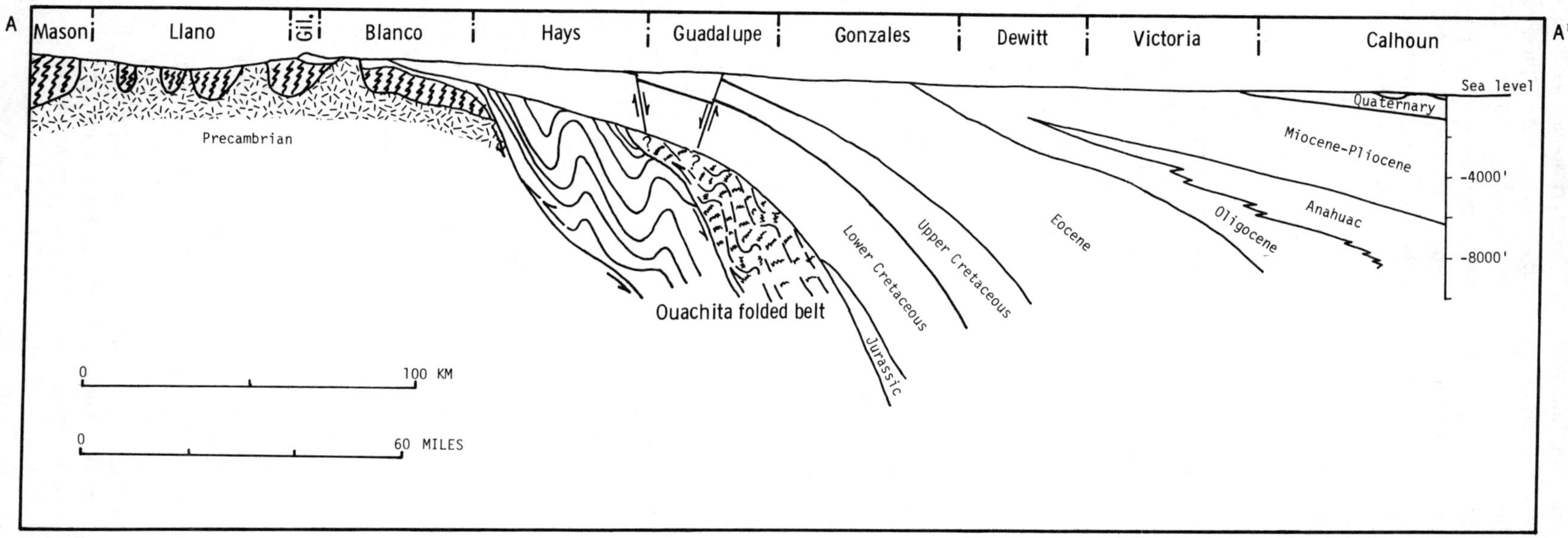

Figure 27.4 Cross section, Llano uplift to the Gulf of Mexico, along line A–A′ in Figure 27.3. Vertical exaggeration: 21×. *Courtesy Dallas Geological Society (1963).*

Eocene and the Oligocene-Miocene-Pliocene. The Eocene belt contains low escarpments here and there, but the younger Tertiary merges with overlapping Quaternary sediments in a low, wide, coastal belt of plains and prairies.

The broad details of the subsurface stratigraphy can be seen in the cross-section (Figure 27.4). The stratigraphic chart (Figure 27.5) shows the principal reservoirs, as well as potential ones. The pre-Lower Cretaceous rocks are largely unexplored in the subsurface. As yet, the Edwards Limestone of Early Cretaceous age is the oldest formation with significant production in the South Texas district (Winter, 1961). This limestone is famous for its fault trap fields, but recently a reef facies development (Figure 27.7) has aroused considerable interest. Hydrocarbons are produced from Upper Cretaceous reservoirs, including the Austin Chalk, serpentinized igneous rock, and sands in the Taylor (San Miguel) and the Navarro (Olmos). The Austin Chalk oil is stored in fractures (Doyle, 1955), whereas the porosity in the serpentine is due largely to alteration, although there may have been some vesicularity in the original igneous rock.

Eocene sands are second only to those in the Oligocene in productivity. One reservoir sand, the Poth, lies within the lower Eocene Midway Shale. The next younger reservoir series is the highly productive Carrizo-Wilcox sandstone section. This stratigraphic unit is a complex

- CENOZOIC
  - Quaternary
    - Recent
    - Pleistocene
  - Tertiary
    - Pliocene
    - Miocene
      - Catahoula
    - Miocene-Oligocene
      - Anahuac
    - Oligocene
      - Frio: leading reservoir in district
      - Vicksburg
    - Eocene
      - Jackson
      - Claiborne: includes Yegua
      - Wilcox: includes Carrizo
      - Midway: includes Poth
- MESOZOIC
  - Cretaceous
    - Upper (Gulfian)
      - Navarro: includes Olmos
      - Taylor: includes San Miguel
      - Austin
      - Eagle Ford
      - Woodbine
    - Lower (Comanchean)
      - Washita
      - Fredericksburg
        - Edwards
        - Paluxy
      - Trinity
        - Glen Rose
  - Jurassic
      - Cotton Valley
      - Smackover
      - Louann Salt
      - Werner
      - Eagle Mills
- PALEOZOIC
  - Carboniferous
  - Cambro-Ordovician
      - Ellenburger
- PRECAMBRIAN
      - Granite, schist, and gneiss

Figure 27.5 South Texas stratigraphic chart.

of interfingering sands and shales, notable for the rapidity of facies change in every direction. The upper formation in the section, the Carrizo,

> is entirely a nonmarine, continental deposit composed chiefly of medium to coarse grained quartz sand with local beds and sections of shale or sandy shale occurring within the formation. The sand is typically friable, subangular, relatively clean, well sorted, very porous and permeable and is characteristically cross-bedded and massive. (Hargis, 1962, p. 131).

The lower Wilcox reservoirs are also sand bodies, but these sediments were deposited, either in deltas or parallel to the shore, as beaches, barrier bars, or shelf deposits (Fisher and McGowen, 1967). The Wilcox lens-out line is shown in Figure 27.6. The Claiborne and the Jackson also are notable for their reservoir sands.

The Oligocene Frio is a record holder in South Texas in both multiplicity of sands (up to 58 separate reservoirs in a single field) and volume of production (more than 1 billion barrels). Boyd and Dyer (1964) studied the available subsurface data over a four-county area (Aransas, Nueces, Refugio, and San Patricio) and concluded (p. 309) that an offshore barrier bar system was established in Frio time.

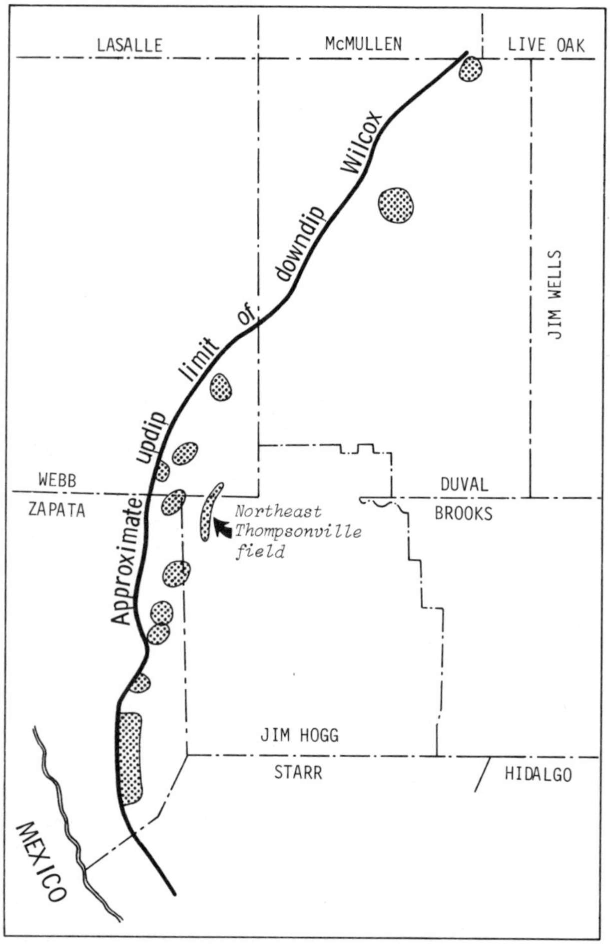

FIGURE 27.6 **Map of southern South Texas, showing distribution of downdip Wilcox gas fields (patterned areas) relative to updip limit of isolated sand facies. Geology of Northeast Thompsonville is described in Figure 27.13.** ***Courtesy Leighton F. Young, Jr. (1966, p. 516) and the American Association of Petroleum Geologists.***

> With the termination of the major Vicksburg transgression, the seas began their slow withdrawal and deposition of Frio clastics commenced. Within this regressive framework of the Frio, a massive sand bar developed which is analogous to the present day Padre–Mustang–St. Joseph–Matagorda Island complex. These Frio bar sands were transported by longshore currents from an area to the southwest where extensive deltas were being built by the ancestral Rio Grande River. Within the Frio bar system continental shelf, bar, and lagoonal environments are recognizable.

Nanz (1954) identified an Oligocene sandstone reservoir in the Seeligson field as being a fresh or brackish water deltaic plain deposit.

As could be expected from the different environments of deposition, Frio sands vary considerably in porosity and permeability. In the predominantly marine sand section in the central and northern parts of the district, porosities of 33 per cent, with permeabilities of 1500 md, have been recorded (Johnson and Mathy, 1957, p. 212). To the south, approaching the Rio Grande embayment, the minimum porosity and permeability figures noted are 22 per cent and 135 md. The recoverable oil in these pay zones ranges from 700 barrels per acre foot in the more porous sand to 300 barrels in the least porous.

Miocene sands contain gas in southeastern Cameron County at the southern tip of the district, as well as offshore.

TRAPS AND FIELDS. The distribution of oil and gas fields in South Texas is shown in Figure 27.8. The producing areas here, as in previously discussed Gulf Coast districts, tend to be aligned in belts that parallel the shoreline and the strike of the subsurface formations. These zones or trends are named after the age of the principal reservoir rocks tapped by the producing wells. There is, of course, some overlap, for the principal reservoirs in one zone may be underlain by downdip extensions of the reservoirs that are most productive in the next trend inland.

The trends in South Texas are shown in Figure 27.7, printed above the oil and gas field map. The farthest inland is the Cretaceous belt, in which the fields produce from reservoirs lying within both Upper Cretaceous and Lower Cretaceous rocks. Next is the Cretaceous-Eocene trend, where most of the shallower fields produce from Wilcox reservoirs and the deeper fields from the Edwards Limestone. Cutting across this trend from the southern corner of the Cretaceous belt, and continuing along the

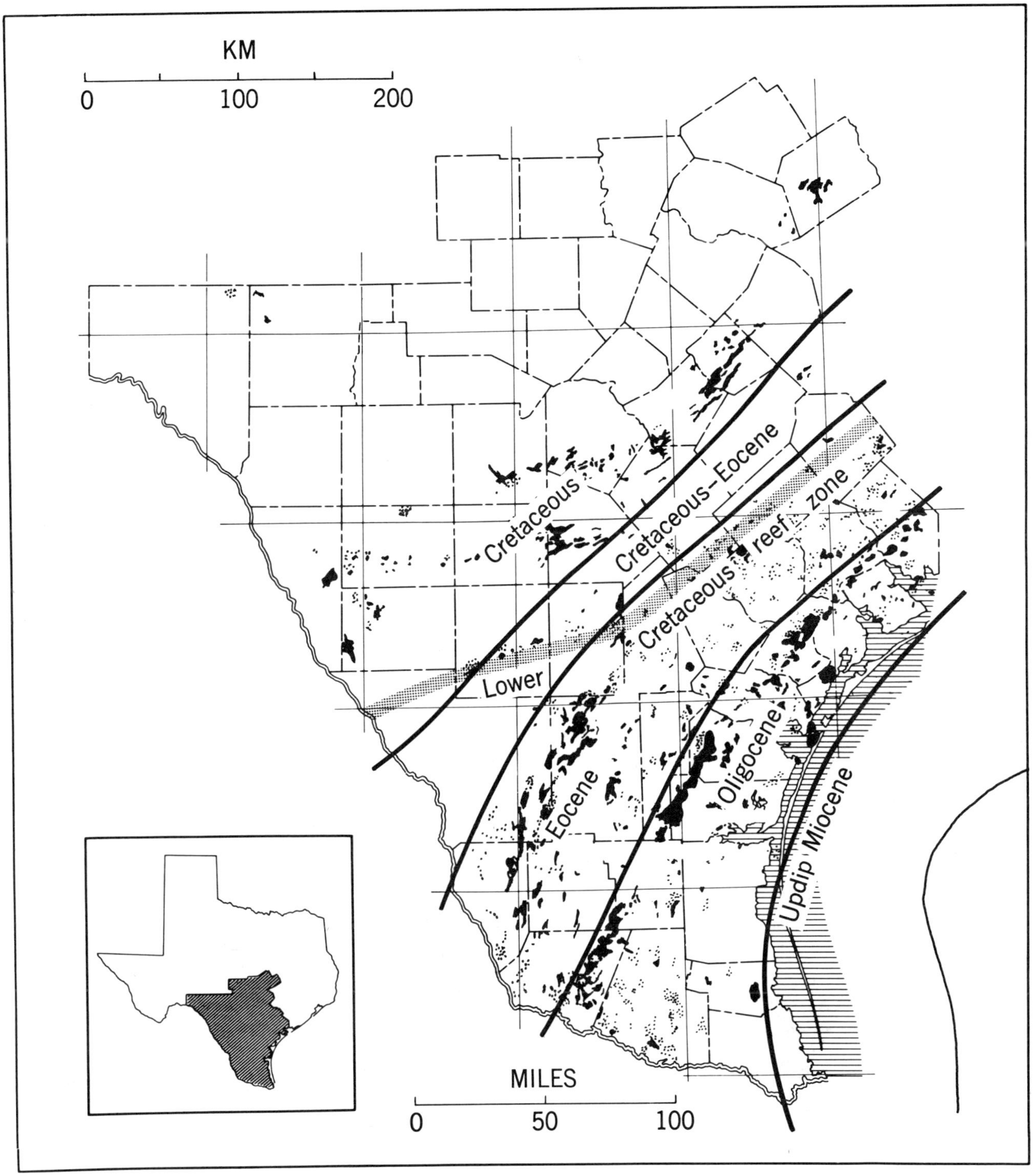

FIGURE 27.7 South Texas. "Trend" map, showing geologic age of principal reservoirs beneath each zone. *From various authors.*

inland edge of the next belt, is the deep Lower Cretaceous reef trend. Next gulfward is the Eocene, or "Wilcox-Jackson," trend, which is densely populated with oil and gas fields. Bordering the Gulf Coast is the Oligocene belt, popularly called the Frio Trend. This zone is the most productive; not only does it contain a large number of fields, but also many of them are of large area and giant stature. Lying offshore, except for the southeastern tip of the district, is the Miocene trend.

Most Cretaceous reservoirs are limestone, whereas the

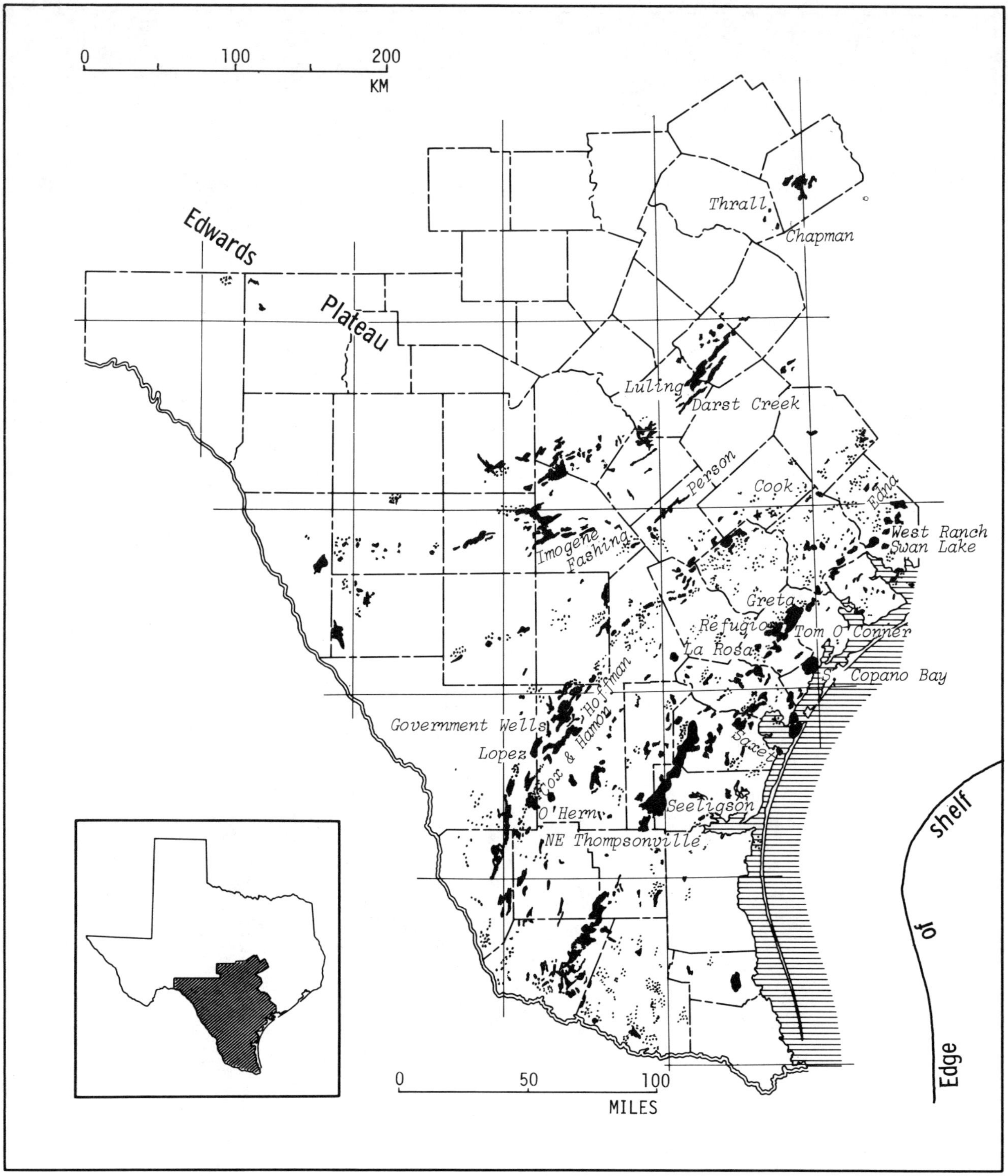

FIGURE 27.8 South Texas. Oil and gas fields map. Fields referred to in text are named.

reservoirs in the Tertiary are sandstone. The traps are mainly anticlinal, fault, or sand pinchout. Piercement salt domes are relatively unimportant (as yet, at least) as traps in South Texas, unlike districts to the northeast. However, it is probable that much of the local structure has been due to deep-seated salt movements. South Texas is second only to Louisiana in the number and economic importance of growth (contemporaneous) faults. This type of faulting is described and discussed in Chapter 24 on Louisiana.

The descriptions of fields and traps that follow are arranged by trends, starting inland and moving gulfward. Many traps in South Texas, especially in the coastal zone, have been discovered by combining geophysical data, especially seismic and gravity, with subsurface information obtained from boreholes.

The oldest production in the Cretaceous trend comes from the Edwards Limestone in the shallow Balcones fault fields, which extend from Bexar to Bastrop Counties. Perhaps the greatest accumulation of fault-impounded oil in the United States has taken place here, in Caldwell and Guadalupe Counties near the crest of the San Marcos arch (Hendy, 1957). Two fields alone in this fault trap province, *Luling* and *Darst Creek*, have together produced 250 million barrels of oil since the discovery of Luling in 1922. Most of this oil has come from depths of less than 2000 feet. Closure is against the hanging wall of up-to-the-coast faults. A geologist first discovered an exposure of the Luling fault on the banks of San Marcos River, and subsequent surface mapping showed the presence of an inlier of lower Wilcox rocks abutting against the fault on the southeast side. A photograph of the surface exposures of this fault is included in the article by Brucks (1929, p. 257).

Darst Creek field (McCallum, 1933) lies a few miles to the south of Luling. The impounding of oil is almost exactly the same in these two fields, although they are on separate faults. The oil occurs at the top of the Edwards (Comanchean) Limestone in a plunging anticline on the upfaulted block. The limestone has been dropped down on the west side of the fault, bringing the younger and relatively impervious Taylor Marl against the broken edge of the eastward dipping Edwards reservoir.

After the discovery of the Luling and Darst Creek fields along what became known as the Luling fault zone, additional discoveries were made both to the northeast and to the southwest. A second line of parallel faults to the south of the Luling zone, the Charlotte fault belt (Figure 27.2), was discovered in 1935, also by surface geology. These faults are likewise up to the southeast. Exploration in 1940 resulted in the discovery of the Imogene field in central Atascosa County (Bolinger, 1953). Additional fault line fields in the Edwards Limestone were found subsequently along this belt (Knebel and Jones, 1957).

The most unusual trapping in South Texas is in altered igneous rock reservoirs, also in the Cretaceous trend. Fourteen fields have produced oil from serpentine. They lie in two groups, one to the northeast of Austin in northeastern Travis County and southern Williamson County, and the other to the southeast of Austin on each side of the Bastrop-Powell line. Both intrusion and extrusion of igneous rock occurred in this area during Late Cretaceous time. The igneous rocks that subsequently became oil reservoirs were probably submarine volcanoes which erupted during Austin time (Simmons, 1967). They were overlain and surrounded by clays and marls belonging to the Taylor (Upper Cretaceous) Formation. The volcanic rocks were altered to serpentine with the

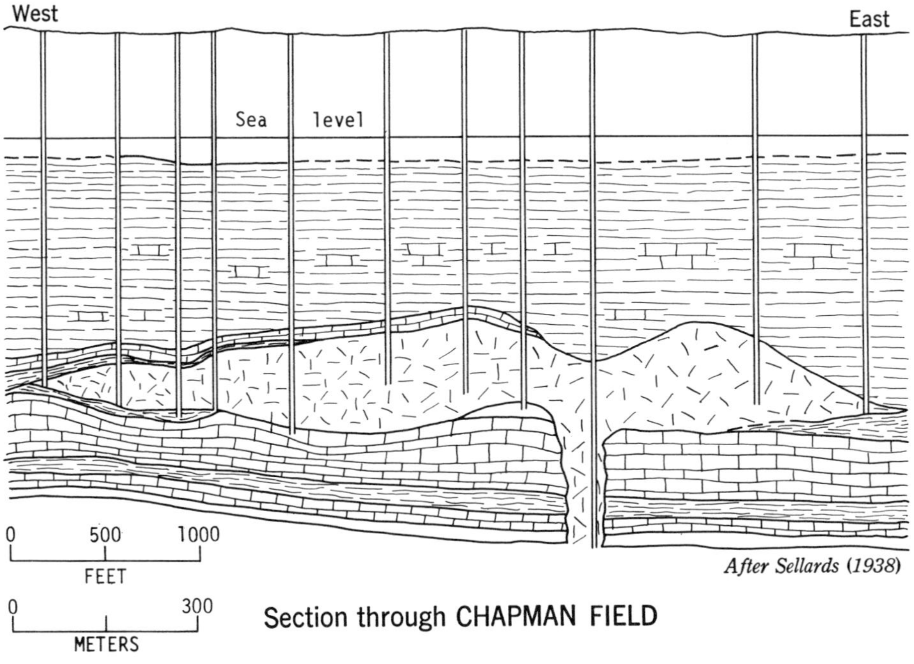

FIGURE 27.9 Chapman field, Williamson County, Texas. Hypothetical cross section through the serpentinized igneous rock that is the reservoir.

development of additional porosity during pre-Taylor exposure. Oil has accumulated in both serpentine and reworked igneous material deposited on the flanks of the intrusions.

Figure 27.9 is a hypothetical cross section of the *Chapman* field (Sellards, 1938) in the southeastern corner of Williamson County. Presumably oil was squeezed out of the compacting, younger, overlying Taylor clay and obtained sanctuary in the underlying porous serpentine. However, some oil found lodgement in the porous capping limestone, as can be seen on the western flank of the igneous mass.

Other, less exotic Upper Cretaceous reservoirs yield oil and gas in the southwestern part of the Cretaceous belt. The "Olmos trend" in Dimmit, Zavala, and Frio Counties produces from the Olmos (Navarro) sand; trapping is due to a combination of reservoir lenticularity and faulting (Glover, 1955).

The Cretaceous-Eocene belt contains shallow Wilcox fields as well as deep Edwards production, both fault trap and reef. The "deep Edwards fault trend" extends from southeastern Atascosa County to the northwest across Karnes County and southeastern Gonzales County (Knapp, 1962). This fault line was found by the discovery in 1956 of the *Fashing* gas field, which illustrates maximum fault trapping with minimum anticlinal assistance. The Fashing fault is an excellent example of growth faults, wherein faulting was contemporaneous with deposition. It is a more unusual type in the Gulf Coast in that the fault movement was up-to-the-coast. The maximum known vertical displacement is 700 feet at the Edwards (Lower Cretaceous) level and 320 feet at the Carrizo (Eocene) level (Keahey, 1968). The Fashing field, which measures about 10 miles by 2 miles, extends from the southeast corner of Atascosa County to the northwest corner of Karnes County (Keahey, 1962). Fashing produces gas and distillate; it also yields sulfur, which is removed from the gas at a rate of $\frac{3}{4}$ ton per million cubic feet of gas processed. Figure 27.10 shows by structure contours the nature of the trap. The northeast-trending fault limits production to the northwest; production comes from pay zones in the upblock. Closure is due in part to a slight nosing of the contours, but mainly to a northwest bulge in the fault line. The reservoirs are two porous zones in the Edwards (Lower Cretaceous) Limestone.

The discovery of Edwards production (mostly gas or gas and condensate) in this area led to a successful hunt for other fields, which is still continuing (Beebe, 1968b). The second discovery was the large Person oil field to the northeast of Fashing in 1959 (Reagan and Foust, 1960; L. L. Jones, 1962).

The "Lower Cretaceous (or Comanchean, or Edwards) reef trend," which crosses parts of three belts, is shown

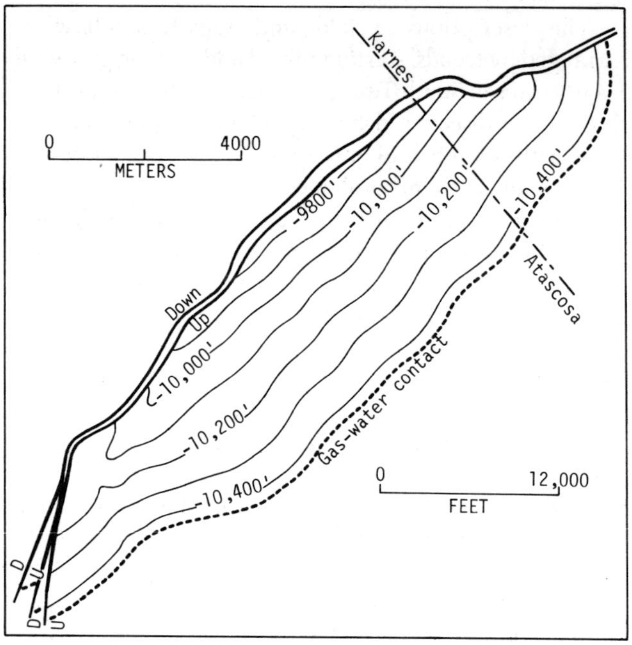

Figure 27.10 Flashing gas field, Atascosa and Karnes Counties, Texas. Structure contour map based on top of the Edwards Limestone. Interval, 100 feet. The gas-water interface, as established by the Texas Railroad Commission at −10,440 feet, is shown by dashed line. Entrapped gas lies between this contact and longitudinal fault at top of anticline. *Courtesy Robert Keahey (1968) and the Gulf Coast Association of Geological Societies.*

on Figure 27.7 as a narrow light gray band. It was discovered in 1954, and during the next 9 years seventeen small fields, mostly gas, were found in this reservoir from Webb to Lavaca Counties (Holden, 1963). Subsequently exploration, however, has been discouraging (Roe et al., 1966). Sandidge (1961) refers to the reservoir as a barrier reef. It appears to have formed during Glen Rose and Edwards time.

The Eocene trend extends southwestward from the end of the Yegua belt in the Texas Upper Gulf district to Zapata County on the Mexican border. Reservoirs in the Midway (Poth sand), Wilcox (Carrizo), Claiborne (Yegua), and Jackson are productive in this belt. Of these, the Wilcox is the greatest source of hydrocarbons.

Sand facies in the Wilcox disappear downdip but reappear in a probably narrow belt extending across Duval County, southeastern Webb County, northwestern Jim Hogg County, and eastern Zapata County (Figure 27.6). By 1966 eleven fields (mainly gas) had been discovered in this downdip Wilcox trend (Young, 1966).

Traps range in variety from textbook anticlines to straight stratigraphic traps. An example of the former type is the *Cook* field, which lies in the recently developed Lower Wilcox trend (F. C. Jones, 1964). The trap (Figure 27.11) is an elliptical anticline superimposed on the step lying between two major down-to-the-coast

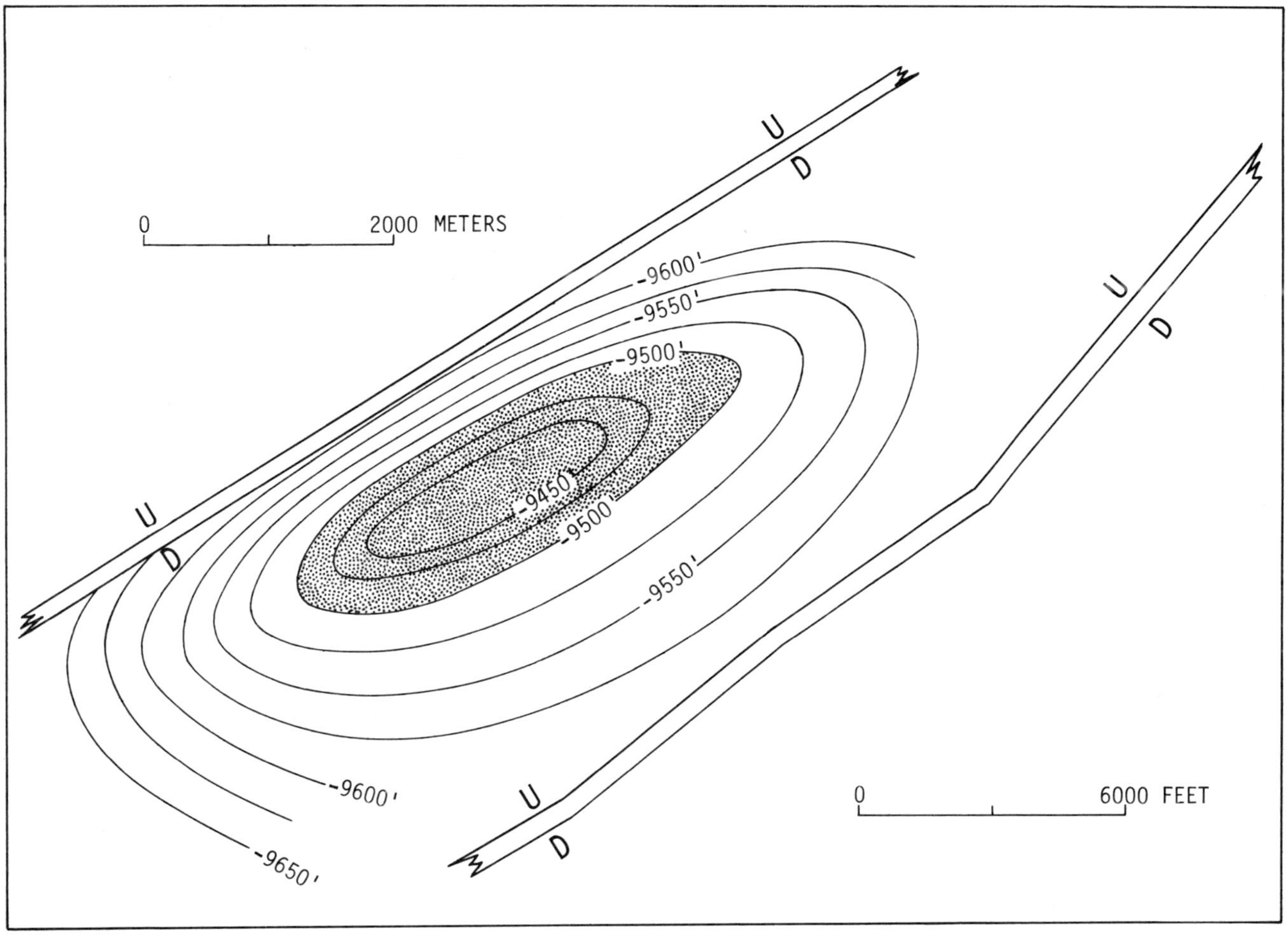

FIGURE 27.11 **Cook gas field, Dewitt County, Texas. Structure contour map based on top of Wilcox Sand. Interval, 25 feet. Producing area stippled.** ***Courtesy R. L. Scheig (1961) and the Gulf Coast Association of Geological Societies.***

normal faults (Scheig, 1961). Although bordered by faults, the anticline itself does not appear to be fractured by faulting, but more drilling may change this picture. Cook is a gas field and affords an excellent example of gas accumulation at the top of an anticline. The shaded area in Figure 27.11 shows the gas-containing part of the trap; the original contact between the gas and the underlying water varied by only 5 feet, from −9497 to −9502 feet elevation. Cook is in central Dewitt County about 5 miles southwest of Cuero. Five sandstones in the Wilcox (Eocene), ranging in depth from 9320 to 10,102 feet, are the reservoirs so far discovered.

The *Cox and Hamon* field (Storm et al., 1961) in southwestern Duval County is an excellent example of a "pure" stratigraphic trap, one in which there has been no contribution by structure, other than regional dip, to the entrapment of the hydrocarbons. Figure 27.12, a structure contour map of the field based on a Claiborne reservoir sand above the massive Pettus, shows that the sand wedgout line, which generally parallels the strike, has enough of an updip curve to seal in the oil throughout the arc of the curve without any help from local folding or faulting. The Cox and Hamon field is about 19,000 feet long and 2200 feet wide. The original oil-water contact was the −2975 foot contour, the lowest one shown in Figure 27.12.

A major gas field, the *Northeast Thompsonville* (Young, 1966), produces from the downdip Wilcox sand facies previously mentioned. The elongate anticlinal trap is in the upfaulted block with the normal fault on the gulfward side (Figure 27.13). Northeast Thompsonville is in the northwestern corner of Jim Hogg County and extends northward into the southeastern corner of Webb County. The field is approximately 8 miles long and 2 miles wide and has an estimated reserve of 500 to 700 billion cubic feet of recoverable gas; with the gas is an average of 3 barrels of condensate per million cubic feet. Although the area is known to be downdip from what was con-

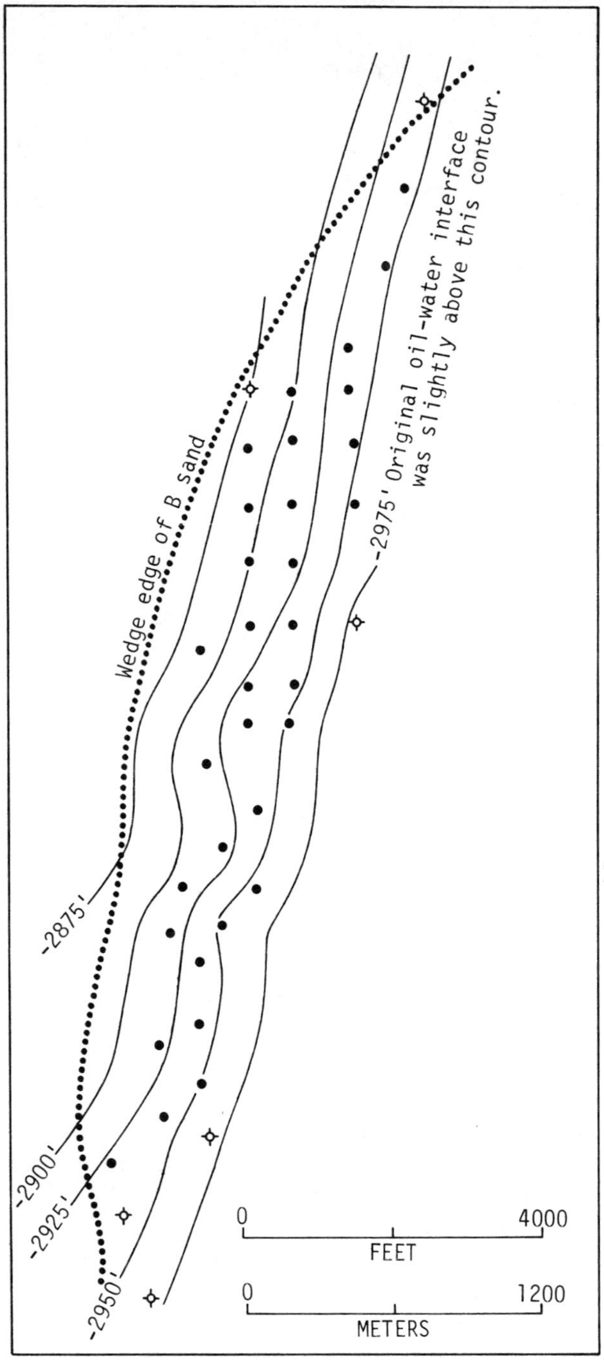

FIGURE 27.12 Cox and Hamon field, Duval County, Texas. Subsurface structure map based on top of B reservoir sand. Contour interval, 25 feet. Shows trapping up monoclinal dip against concave sand wedgout edge (dotted line). *Courtesy Willis Storm et al. (1961) and the Gulf Coast Association of Geological Societies.*

sidered to be the eastern edge of sand facies in the Wilcox, seismic data revealed a large anticline and the decision was made to explore it by drill, gambling on finding adequate permeable sand in the Wilcox. The gamble was successful.

Figure 27.13 shows on the left the structure map of the Northeast Thompsonville field, based on the top of the reservoir sand, which is also considered to be the top of the Wilcox. The anticline is closed on the east by the fault plane which dips gulfward at about 27°. The displacement on this fault ranges from 400 feet at the southern end to a maximum of 1000 feet to the north. North of the Webb County line the main fault bifurcates, and a cross fault, down to the north, confines accumulation in that block to near the top of the anticline. The minimum structural closure is 550 feet, and the gas column is 500 feet. The right half of Figure 27.13 shows the thickness of the reservoir sand. It diminishes from a maximum of over 100 feet near the fault plane on the east to 60 feet down the west flank of the anticline. This sand facies disappears updip to the westward (Figure 27.6), and it has been suggested that the sand was deposited as an offshore bar. Recently gas has been found in the downthrown block.

The happy hunting ground in South Texas for stratigraphic traps is along the Jackson trend (West, 1963). The predominant stratigraphic traps are the result of updip pinchout of about 50 separate sand members through an 1800 foot stratigraphic interval in the Jackson and Yegua sections.

Several thousand exploratory wells drilled during the past forty-two years have resulted in the discovery of some 250 oil and gas fields ranging in depth from 160 to 4000 feet. Fields range in size from Government Wells with over 86 million barrels of oil recovery to date to short lived one well fields (West, 1963, p. 67).

Figure 27.14 shows part of the Jackson trend and the location and structural position of a few fields that have resulted from the updip pinchout of lenticular sands. The pinchout line is also a sedimentation strand line. Two such strand lines are shown in the figure. In every case the updip closure is due to the disappearance of sand in that direction. The lateral closures are brought about in three ways: (1) an updip re-entrant in the pinchout line, which seals in the hydrocarbons even though the regional dip is purely monoclinal; (2) structural nosing down the regional dip, which will cause the oil-water interface contour line to curve into the zero sand thickness line at both ends of the field; and (3) cutoff by a cross or diagonal fault. Actually many of these strand line traps are the result of a combination of two or even three of these trap-contributing situations.

The previously mentioned *Government Wells* field (Trenchard and Whisenant, 1936) in northwestern Duval County is due to a combination of tilting, folding, faulting, and sandstone lenticularity. The trap is closed by regional dip in one direction, by faulting in a second, by folding in a third, and by loss of permeability in a fourth.

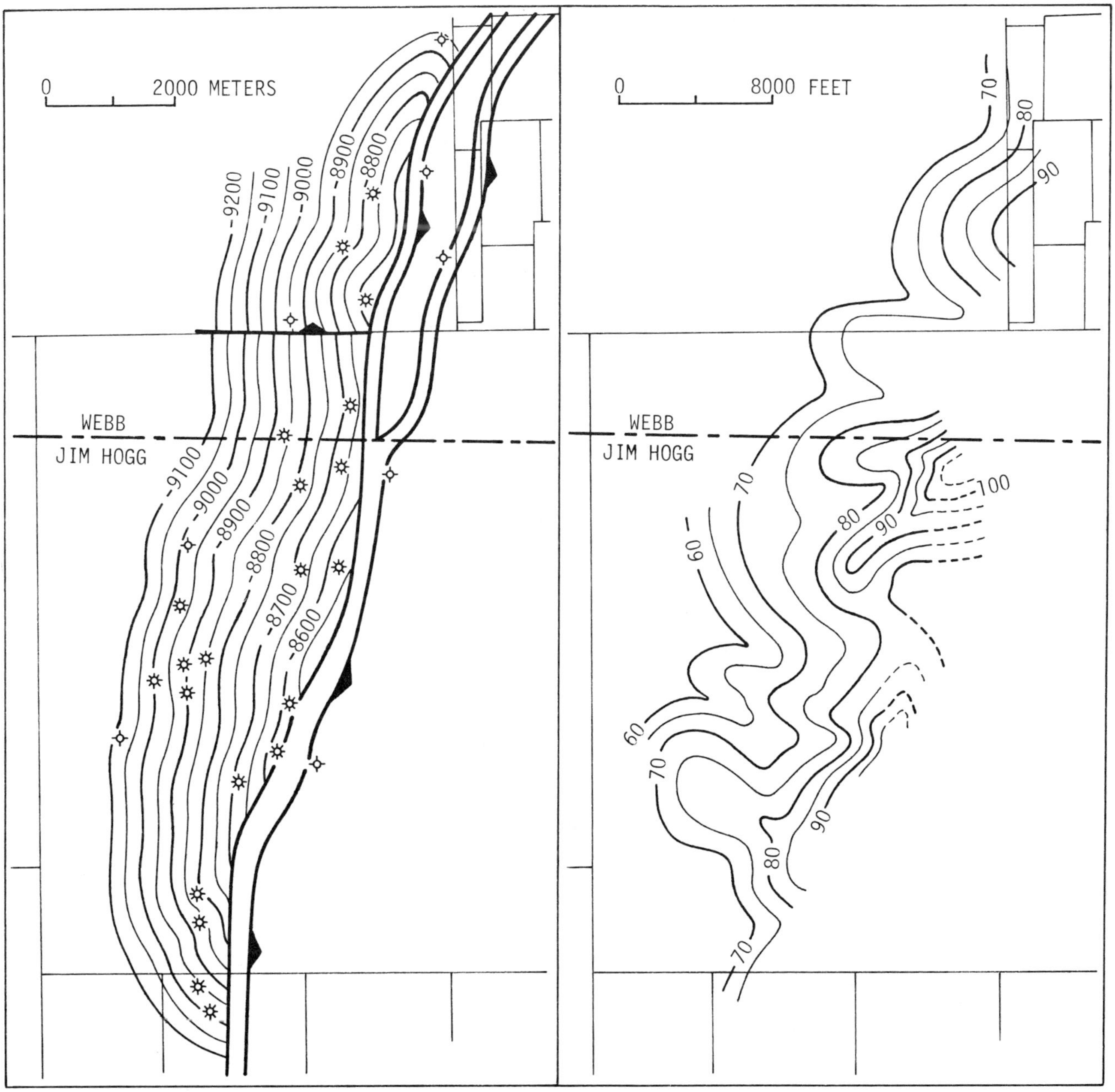

Figure 27.13 Northeast Thompsonville gas field, Jim Hogg and Webb Counties, Texas. Location shown on Figure 27.5. On left is structure map based on top of Wilcox (Hinnant) Sand. Contour interval, 50 feet. On right is isopach map of gross Hinnant Sand. Thickness interval, 5 feet. *Courtesy Leighton F. Young, Jr. (1966, p. 509 and 512), and American Association of Petroleum Geologists.*

The *Lopez* oil field (Best, 1941) and the *O'Hern* field (Barnett, 1941) lie along the boundary between Duval and Webb Counties, and both owe their trapping to updip pinchout along a concave edge of a sand reservoir. The *Hoffman* field (Whitaker, 1940), yet another Duval County field, differs from most of its neighbors in being a structural accumulation. The trap at Hoffman is a broad, low anticline with only about 40 feet of closure in the upper Eocene sandstone reservoirs. At the southern end of the Eocene (Jackson) trend, in what has been termed the Laredo district (McFarland, 1929), both faulting and sand lenticularity (Freeman, 1949) are factors in trapping oil and gas.

The Oligocene, or Frio, trend covers the first row of counties inland, gradually broadening southward so as to cover all or part of three counties at the Mexican

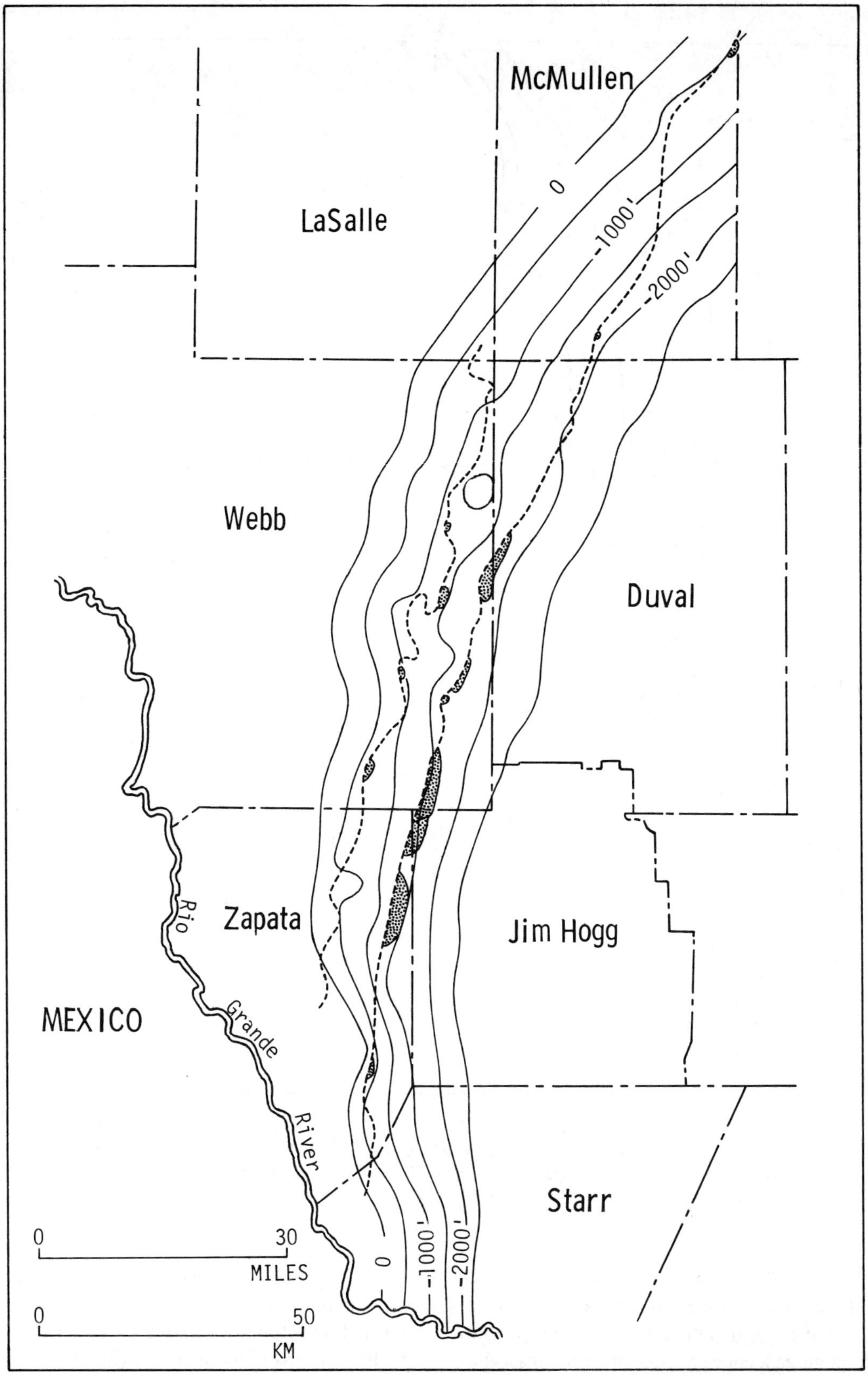

FIGURE 27.14 Jackson trend, South Texas. Regional structure contour map on Pettus Sand (P-2). Interval, 500 feet. Upper Rosenburg (west) and Lopez (east) strand lines shown by dashed lines. Dotted pattern designates oil and gas fields trapped against strand lines. *Courtesy Thomas S. West, Sr. (1963) and the Gulf Coast Association of Geological Societies.*

border. It is the leading source of oil and gas in the South Texas district (Corpus Christi Geol. Soc., 1968, p. 241). Traps include "accumulations along the downthrown segments of down-to-the-coast faults, up-to-the-coast fault closures, closures on the upthrown side of down-to-the-coast faults and salt generated structures" (Johnson and Mathy, 1957, p. 207).

The star of the Oligocene belt is the amazing *Seeligson* field (Deegan, 1945) in Jim Wells and Kleberg Counties. Everything about this field, which was discovered in 1937, is big. It is the largest in the district in both cumulative and annual production, it is one of the great fields of the country in area, and it has produced from as many as 58 separate Frio sand reservoirs. Trapping is anticlinal, but the reservoirs themselves are lenticular. The Seeligson anticline is on the Vicksburg flexure trend, along which many other fields have been discovered. It lies on the downthrown side of the Vicksburg fault zone, where contemporaneous faulting has taken place on a regional scale. The Seeligson anticline has two nodes, the north and the south domes. Most of the producing reservoirs lie in a 2000 foot stratigraphic section between depths of 4300 and 6300 feet. Secondary recovery, by gas injection and waterflooding, started in 1958, sweeping 18 of the larger sand reservoirs.

Other anticlinal traps include the Refugio (Martyn, 1938) and the Greta (Stamey et al., 1936) fields in Refugio County, and the West Ranch (Bauernschmitt, 1944) and the Edna (Kornfeld and Steinberger, 1941) in Jackson County. These fields produce from broad, fairly flat dome-shaped anticlines that may or may not be faulted and are thought to have been formed by deep-seated salt intrusions. Gravity surveys tend to confirm this theory. *Refugio,* one of the major fields in the district, shows indications of several periods of folding. After each period of folding or tilting, the hydrocarbons adjusted to the new structural environment by migration. The *Edna* gas field trap was discovered by surface geologic mapping. The center of the dome is marked by a wooded sand inlier about 6000 feet in diameter.

The *La Rosa* field (Fisher, 1941) in the central part of Refugio County is a typical contemporaneous fault trap complete with reverse drag. Accumulation is in an elongate anticline bordered on the inland side by a down-to-the-coast normal fault (Figure 27.15). In addition to the 5900 foot sand, the discovery reservoir rock, there are five other oil sands and four gas sands, all in the Frio Formation, at La Rosa. Trapping is anticlinal; however, in the deepest oil reservoir, the 6300 foot, the sand grades into shale a short distance west of the anticlinal axis. As is usual in this type of trap, the fault cuts the anticline below the oil-water interface on the inland flank and hence does not itself close the producing part of the trap. The top of the Frio lies at a depth of about 4600 feet; apparently both the folding and the faulting ceased with the end of Frio time, for the younger sediments show no faulting and no comparable folding.

The *Swan Lake* field (Bowers and Denman, 1953) in southwestern Jackson County is another domal trap in a down block. In this case, however, the anticline is bounded on both the west and the north by upfaulted blocks.

The *South Copano Bay* field (Bryant, 1965) is an anticlinal complex fractured by a series of down-to-the-coast faults. The field lies on both sides of the Aransas–Refugio county line out in Copano Bay. It is in the Frio trend, and all of the reservoirs at South Copano are sands within this formation. The leading producer is the Melbourne or K-2 sand, which lies at the top of the middle Frio section. Figure 27.16, a structure contour map with contours based on the top of the Melbourne, shows the distribution of the oil accumulations in South Copano Bay and also illustrates the method of trapping. The oil has accumulated in anticlinal nodes and noses, with single faults, branching faults, and crossing faults closing one or two sides of each trap.

Cutting across the Eocene and Oligocene trends in South Texas is a group of piercement salt domes. Six domes, all in the Rio Grande embayment in the southern part of the district (Figure 27.2), have been proved to be salt cored by the drill (Corpus Christi Geol. Soc., 1957). Four of these domes are producing oil from flank sands, ranging in age from Yegua to Frio. Cumulative productions for these fields range up to more than 3 million barrels. The other two domes have produced oil or gas from supercap reservoirs. In addition, there are many anticlinal traps, here and elsewhere in South Texas, producing from what are probably supercap reservoirs, but salt has not yet been penetrated.

Although most of the Miocene trend is offshore, gas occurs in this belt beneath the onshore southern tip in Cameron County. Furthermore, shallow Miocene reservoirs overlap Oligocene sands along the coastal belt to the north. An example is the *Saxet* field (Poole, 1940), a major accumulation in Nueces County immediately to the west of Corpus Christi. Saxet produces from the Frio and younger Oligocene sands and from Miocene sand. Accumulation has taken place in a large dome that is crossed, high on the northwest flank, by a major fault. The downfaulted block is on the southeast, toward the gulf; production is obtained across the fault on the northwest side. Several minor compensating faults are also present. It is believed that the major fault supplied a channelway for migrating gas, which filled the shallower sands under such high pressure that at least a score of "blowouts" took place during the development of the field.

Future. Every 2.7 days, on the average, a successful

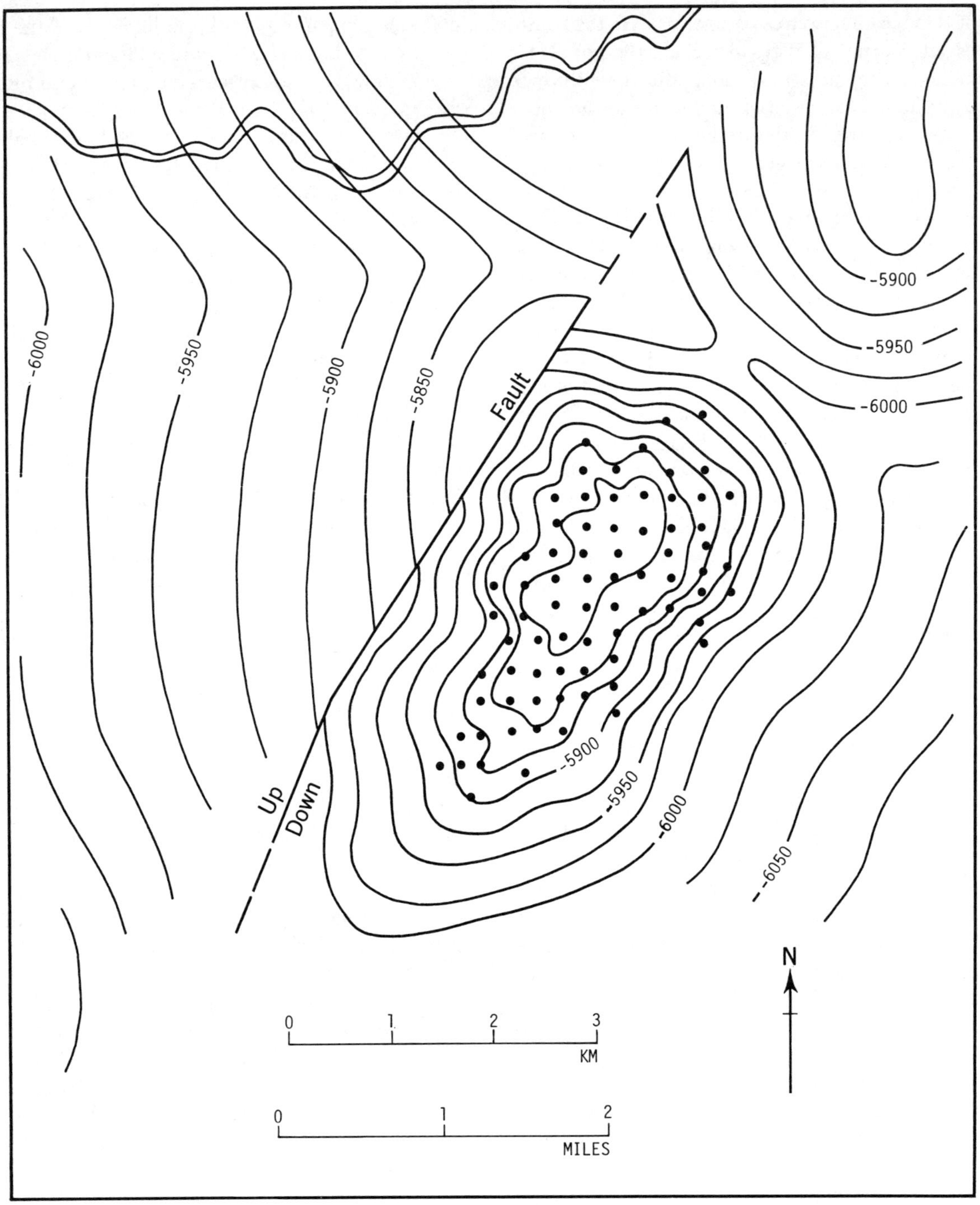

Figure 27.15 La Rosa field, Refugio County, Texas. Structure map based on 5900 foot sand. Contour interval, 25 feet. Entrapment in dome on down-faulted block. *After Barney Fisher (1941). Courtesy American Association of Petroleum Geologists.*

exploratory test was drilled in South Texas during 1967 (Jennings et al., 1968). Comparable discoveries can be anticipated across future years.

So far, little success has resulted from exploration of Paleozoic and Jurassic sections, but the dry hole density is still very low. In 1967 a well flowing from a Mississippian reservoir in Kerr County was reported (Oil and Gas Jour., Aug. 21, 1967, p. 119); this county had not been in the producing column previously. The buried Ouachita facies (position of subcrop shown in Figures 27.2 and 27.4) has been, and will continue to be, a Paleozoic objective. There may even be a possibility that less metamorphosed petroleum-bearing Paleozoic rocks underlie thrust-faulted Ouachita facies rocks (Barnes, 1948, p. 12). The Jurassic Smackover has been probed recently in the Rio Grande embayment (Roe et al., 1966, p. 1209). The Jurassic section is deep here but contains both clastics and carbonate rocks. It is considered to have possibilities over a belt extending southwest from the Upper Gulf district boundary to the Rio Grande (South Texas Geol. Soc., 1951). This belt has an area of 12,000 square miles; beneath it lies an estimated 3000 cubic miles of Jurassic sediments. The Jurassic does not crop out updip in South Texas but wedges out beneath Lower Cretaceous rocks. The opportunities for unconformity, porosity wedgout, anticlinal, and fault traps beneath these 12,000 square miles must be considerable.

Both the Jurassic and the Lower Cretaceous are considered to be potential reservoirs across the San Marcos arch (Halbouty, 1966; Rives, 1962). Further successful downdip exploration of Lower Cretaceous and Upper Cretaceous limestones is also expected.

Most of the new discoveries in the future, as in the past, will be made by continued infilling within the Cretaceous, Eocene, and Oligocene trends. However, some of the salt domes merit further exploration of the flanks. For example, the giant Pescadito salt dome in Webb County (Figure 27.2) has been estimated to be 9 to 10 miles long and 4 to 5 miles wide. The discovery well for salt, drilled in 1957, became productive at a depth of 15,070 feet, beneath nearly 700 feet of anhydrite cap rock. Gas has been found in the Wilcox in the

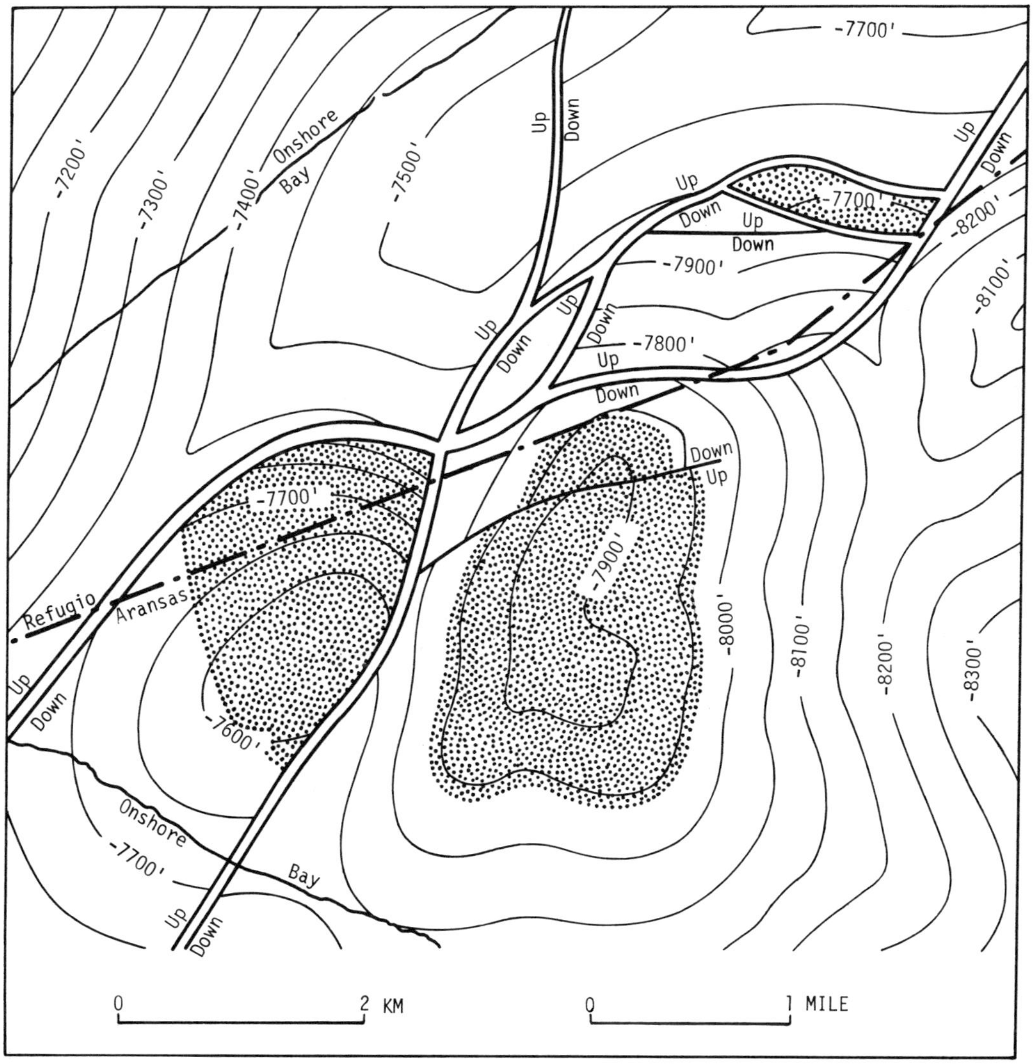

FIGURE 27.16 South Copano Bay gas field, Refugio County, Texas. Structure contour map based on top of Melbourne Sand (K-2 reservoir). Interval, 50 feet. Stippled areas show the larger separately trapped gas accumulations in this offshore field. *Courtesy Leonard C. Bryant (1961, p. 19) and the Gulf Coast Association of Geological Societies.*

supercap section. No doubt it will be a long time before the deep flank possibilities around the 30 mile perimeter of this salt intrusion have been prospected adequately.

The largely offshore Miocene trend has proven gas reserves both onshore in Cameron County and offshore between the mainland and Padre Island. Thick Miocene sands have been discovered offshore; there is also the possibility of finding limestone reef facies in the offshore Miocene section.

The deeper Frio is productive in Corpus Christi and other bays along the South Texas coast, on Mustang Island, and in deep water immediately off Mustang Island (especially of gas). Certainly deep Frio reservoirs will continue to be probed along the shoreline and in the adjacent offshore region. The lower Eocene Wilcox has not been explored adequately off the beaten path in South Texas (Jindra, 1967).

The South Texas district is notable for its large reserves of shallow heavy oil. Recent developments in thermal recovery have made possible economic recovery from many such deposits. Nine thermal recovery projects were either planned or operating in South Texas in the fall of 1967 (Oil and Gas Jour., Oct. 2, 1967, p. 60–61). Parts of the Refugio and Government Wells fields were among the operating projects. More activity in developing dormant heavy oil deposits can be anticipated.

REFERENCES CONSULTED IN PREPARING CHAPTER 27

Barnes, Virgil E., 1948.

Barnes, V. E., Boucot, Arthur J., Cloud, Preston E. Jr., Miller, Richard H., and Palmer, A. R., 1966.

Barnett, Donald C., 1941.

Bauernschmitt, A. J., 1944.

Beebe, B. W., 1968b.

Best, J. Boyd, 1941.

Bleakley, W. B., 1965.

Bolinger, John W., 1953.

Bowers, Emil F., and Denman, R. H., 1953.

Boyd, Don R., and Dyer, Byron F., 1964.

Brooner, Frank I., 1966.

Brucks, Ernest W., 1929.

Bryant, Leonard C., 1965.

Corpus Christi Geological Society, 1957, 1967, 1968.

Dallas Geological Society, 1963.

Deegan, Charles J., 1945.

Doyle, William M., Jr., 1955.

Fisher, Barney, 1941.

Fisher, W. L., and McGowan, J. H., 1967.

Fowler, Phillip, 1956.

Freeman, James C., 1949.

Glover, J. E., 1955.

Halbouty, Michel T., 1966.

Hargis, Richard N., 1962.

Hendy, W. J., 1957.

Holden, Richard N., 1963.

Jennings, A. Ray, Payne, W. R., and Wharton, Ralph, 1968.

Jindra, Roy, 1967.

Johnson, Ray B., and Mathy, Harold Edward, 1957.

Jones, Franklin C., 1964.

Jones, Larry L., 1962.

Keahey, Robert, 1962, 1968.

Knapp, R. W., 1962.

Knebel, Robert M., and Jones, Franklin C., 1957.

Kornfeld, M. M., and Steinberger, C. R., 1941.

Martyn, Phillip F., 1938.

McCallum, H. D., 1933.

McFarland, P. W., 1929.

Nanz, Robert H., Jr., 1954.

Poole, J. C., 1940.

Reagan, M. A., Jr., and Foust, R. T., Jr., 1960.

Rives, John S., 1962.

Roe, N. Charles, Shambaugh, John S., and Wood, John W., 1966.

Romberg, Frederick, and Barnes, Virgil E., 1954.

Sandidge, John R., 1961.

Scheig, Richard L., 1961.

Sebring, Louie, Jr., 1948.

Sellards, E. H., 1938.

Sheldon, I. R., 1936.

Simmons, Kenneth A., 1967.

South Texas Geological Society, 1951, 1962.

Stamey R. A., Montgomery, J. C., and Easton, H. D., Jr., 1936.

Storm, Willis, Hensley, J. H., and Reiss, Bernard W., 1961.

Trenchard, John, and Whisenant, J. Barney, 1936.

West, Thomas S., Sr., 1963.

Whitaker, Harvey, 1940.

Winter, Jan A., 1961.

Young, Leighton F., Jr., 1966.

Chapter

28

WEST CENTRAL TEXAS

West Central Texas includes the twenty-four counties to the north of the Central Mineral Region that are included in Railroad Commission District 7B. It is the only one of the seven districts described in Chapters 25 to 31 that is completely surrounded by other districts. West Central Texas is bounded on the north, east, south, and west by North, Northeast, South, and West Texas, respectively (Figure 28.1).

West Central Texas ranks sixth out of the seven Texas subdivisions in volume of petroleum production with about 4 per cent of the state's total. The production comes from a considerable number of relatively small fields. The district does not have any "giant" fields (those that have produced or are destined to produce more than 100 million barrels). Five discoveries were made here in 1967 (Makins, 1968).

West Central Texas first became prominent in 1917, when the Ranger field in Eastland County was discovered at a depth of 3400 feet by a well that "roared in at an estimated daily flow of 1700 barrels of oil" (Rister, 1949, p. 146). During the same year nearby Desdemona (originally known as Hog Town!) in southern Eastland County also came in with gusto. The next year, 1918, Breckenridge in Stephens County immediately to the north of Eastland County blew in with a gasser that caught fire and burned for 3 days before the blaze was extinguished. Later the discovery well was completed as a 200 barrel oil producer at a depth of 2960 feet (Rister, 1949, p. 161). Although these three fields were spectacular and started a boom that was both notable and notorious, only Breckenridge has approached giant stature in cumulative production.

Many other, but much smaller, discoveries followed the original trio in West Central Texas. The low exploration costs in this district, due to the relatively shallow depth, has encouraged exploration and development by small independent companies and individual entrepreneurs.

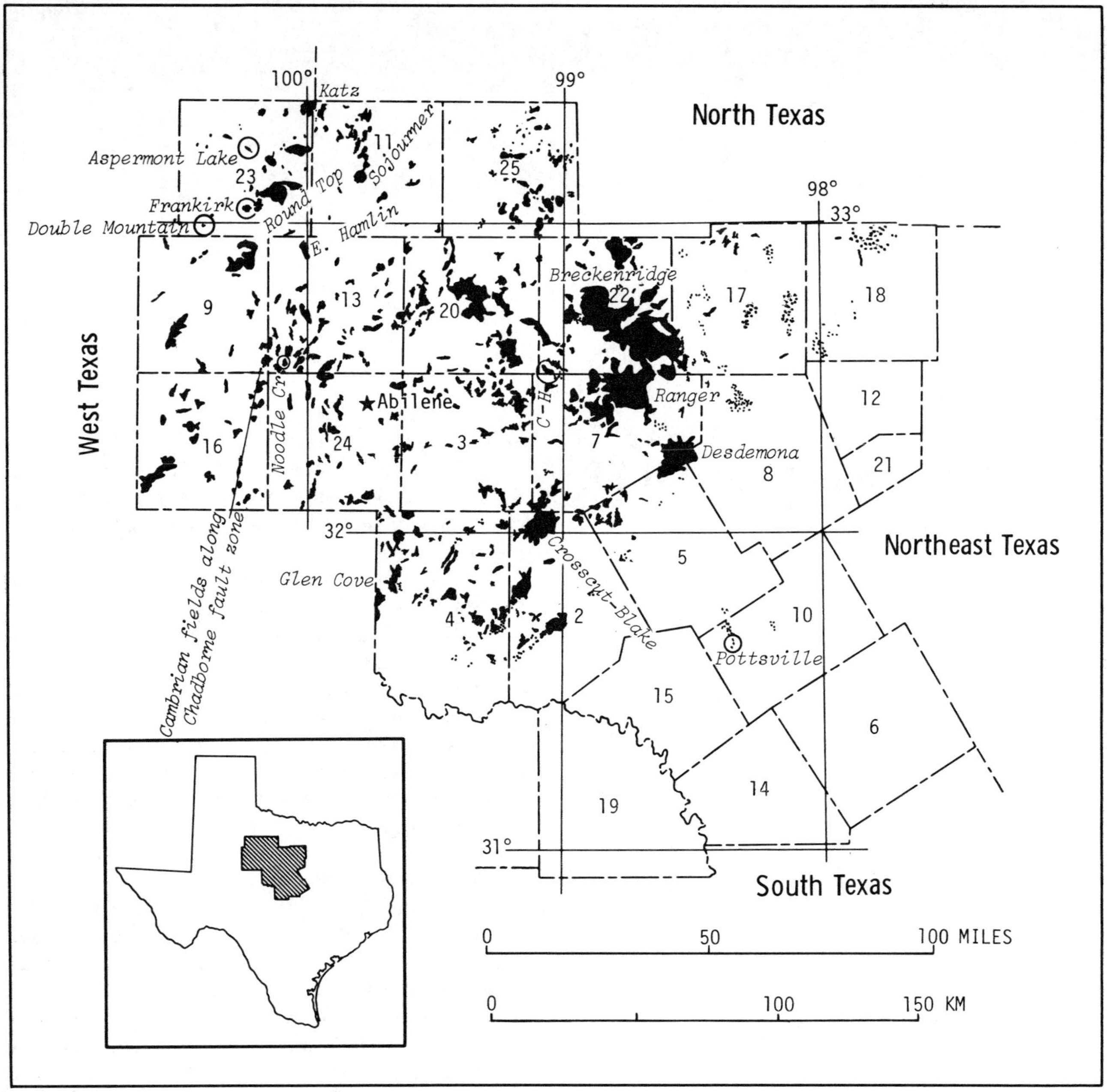

Figure 28.1 West Central Texas. Index and oil and gas field map. Fields referred to in text are named. C-H indicates Calloway-Henry field. County Key:

2	Brown	6	Coryell	10	Hamilton	14	Lampasas	18	Parker	22	Stephens
3	Callahan	7	Eastland	11	Haskell	15	Mills	19	San Saba	23	Stonewall
4	Coleman	8	Erath	12	Hood	16	Nolan	20	Shackelford	24	Taylor
5	Comanche	9	Fisher	13	Jones	17	Palo Pinto	21	Somervell	25	Throckmorton

Tectonics. The regional structure of West Central Texas (Figure 28.2; also 29.3) is dominated by the Bend arch, a flexure running northward from the Llano uplift (Cheney and Goss, 1952). This is a residual feature, left there by downwarps to the west (Midland basin), north (Fort Worth basin), and east (Gulf of Mexico geosyncline). Most of the district, except along the eastern side where the gulfward dip takes over, is on the wide, flattish arch summit. This is shown by the structure contours in Figure 28.2. A northeastward nose off the Llano uplift, the Lampasas arch, influences the Paleozoic rock structure in the southeastern corner of the district. On the west side of the Bend arch is the Concho arch, which was a prominent uplift before Mississippian time (Galley, 1958). Where covered by Lower Cretaceous sediments, the Bend arch and associated structural features are masked by the unconformable veneer.

Stratigraphy. West Central Texas lies west of the coastal plain. Here Paleozoic rocks either are at the surface or underlie a veneer of Lower Cretaceous or Quaternary sediments. Figure 28.2 shows the areal distribution of the major divisions. At the southern tip of the district, on the north flank of the Llano uplift, Cambrian and Ordovician sediments lap upon the Precambrian crystallines. These rocks are succeeded to the northward by sedimentary formations of Mississippian and Pennsylvanian age. The western half of the district (west of the dashed line in Figure 28.2) is floored with Permian sediments.

The eastern part of West Central Texas is covered

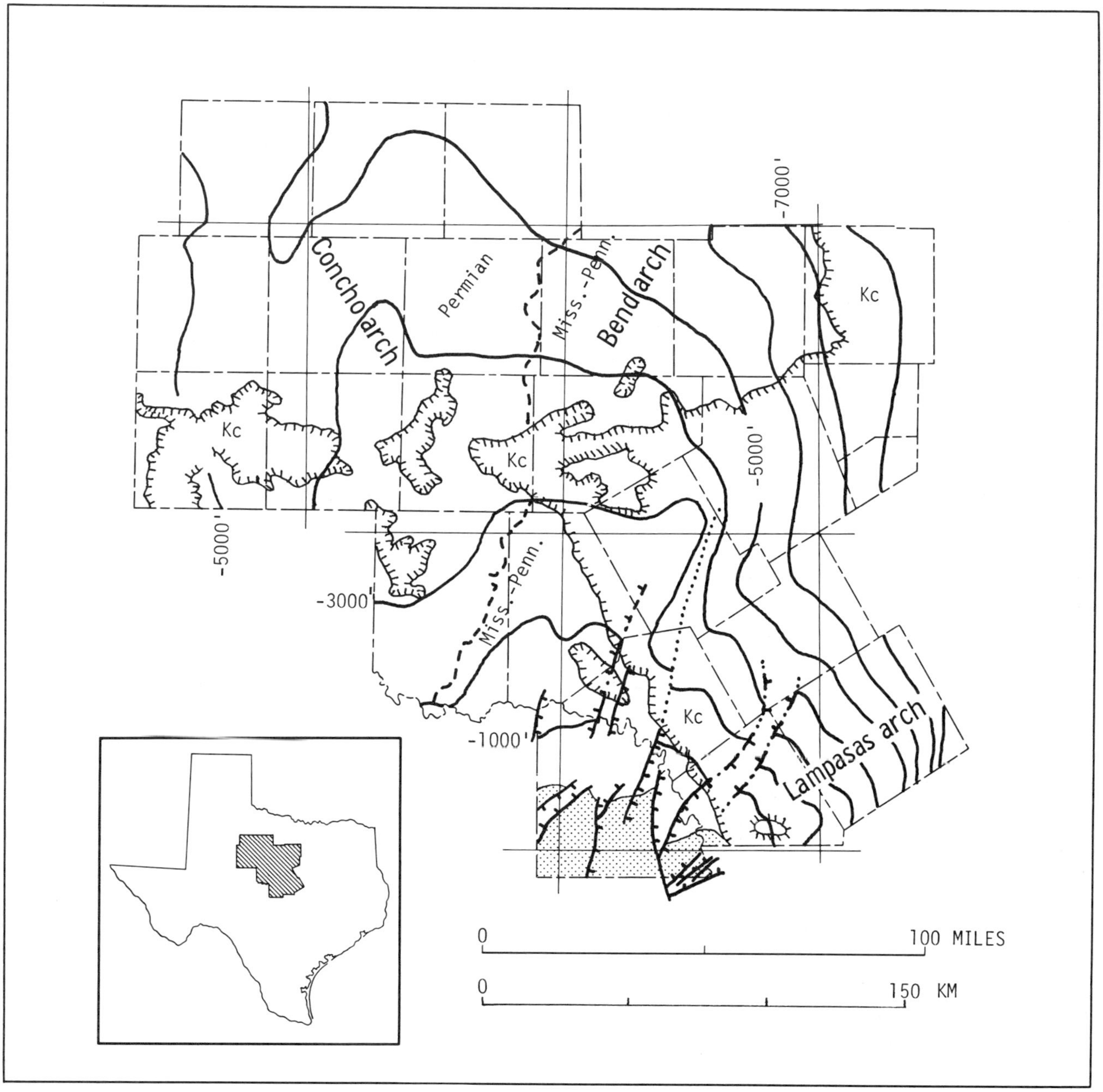

Figure 28.2 West Central Texas. Geologic and tectonic map. Shows boundary between Pennsylvanian and Permian rocks, unconformably overlain by eroded Comanchean (Kc) sediments. Stippled area is lower Paleozoic outcrop on north flank of Central Mineral Region. Structure contours are based on Precambrian basement. Interval, 1000 feet.

by Lower Cretaceous rocks that unconformably overlie the Mississippian and Pennsylvanian formations. The western edge of the Cretaceous belt is retreating eastward by erosion, a circumstance that explains its frayed appearance. Residual outliers of Lower Cretaceous rocks extend westward across the district, overlying Permian as well as Mississippian and Pennsylvanian sediments. To the north of these outliers are similar remnants of Quaternary deposits (not shown on Figure 28.2) veneering the Permian erosional surface.

The subsurface stratigraphy of West Central Texas is shown in Figure 28.3 (Abilene Geol. Soc., Stratigraphic and Study Group, 1960). The uppermost 2000 feet of this section consists of Permian rocks that crop out in the Abilene area. They range in age from Guadeloupe to Clear Fork (Leonard) and consist dominantly of red beds, mostly shales, dolomites, and anhydrites. Below the outcrop section is the lower Leonard Series, which is dominantly dolomitic.

The youngest reservoir rocks are in the Wolfcamp Series in the Lower Permian. The most prolific reservoirs are in the Pennsylvanian, but reservoirs are also present in Mississippian and Cambro-Ordovician formations (Abilene Geol. Soc. Stratigraphic and Study Group, 1960). The Lower Permian and most of the Pennsylvanian reservoirs are lenticular sandstones. It is probable that some Pennsylvanian sands referred to as "Canyon" are actually Cisco or younger in age (Jackson, 1964). The sands in the Bend Series of the lowermost Pennsylvanian tend to have greater lateral continuity than the younger sands. Limestone reefs are locally present in the Pennsylvanian (especially in the Canyon) and in the Mississippian. In addition, a few nonreef limestones are productive, such as the "Caddo" in the Strawn Series in the Lower Pennsylvanian. The oldest formation so far productive consists of the sands and dolomites of Cambrian age.

The subsurface section in West Central Texas is notable for the many unconformities present. These occur both between and even within rock series, especially in the Pennsylvanian. A major unconformity separates Mississippian from Ordovician deposits. The Ellenburger unconformably overlies upper Cambrian dolomite. The basal Paleozoic sandstone, the Hickory, was deposited on Precambrian rocks consisting mainly of schist into which granite has been intruded.

Traps and Fields. The traps of West Central Texas are either anticlinal or due to varying permeability of the reservoir rock. Combinations of the two types are common; the hydrocarbons have accumulated in an upfold, but the distribution across the anticline has been controlled by reservoir rock thickness and permeability. Here, unlike most other parts of Texas, there is no salt in the subsurface section; hence trapping conditions brought about by salt flowage are absent. Although faults are not abundant, they are present, and they play an important role in Cambrian oil accumulations on the west side of the district, along the Chadborne fault zone (Figure 28.1).

The greater part of the oil so far discovered in West Central Texas has been caught in anticlinal traps. One reason for the high percentage is that over much of the district, outcrops are adequate for surface structure mapping. Even though there have been repeated interludes of diastrophism during the geologic history of this area, the larger anticlines, at least, continue upward through successive unconformities to the outcropping Pennsylvanian and Permian rocks. Perhaps the best examples of anticlinal accumulation are to be found in Stephens County, high on the axis of the Bend arch (Esgen, 1929). Near-giant Breckenridge field occupies a large area in the middle of this county, and Ranger overlaps a part of the southeastern corner. These anticlines are relatively low relief features, with closures measurable in tens, rather than in hundreds, of feet, as a general rule. However, some have been found to be oil filled over an area of many square miles.

The *Ranger* field, mostly in Eastland County to the south of Stephens, produced 22 million barrels of oil in its peak year, 1919 (Ellison, 1960, p. 20). The oil accumulated here in an anticlinal complex (Figure 28.4). The four anticlines shown in the illustration are mapped structurally on a black limestone in the Marble Falls Formation, which lies 200 feet above the principal reservoir, the "McClesky sand." The largest of these local anticlines, the New Hope, has a closure of about 90 feet and dips away from the crest at the rate of 250 feet to the mile. The structure map based on an outcropping limestone shows a westward plunging nose, with no closure, and maximum dips of 100 feet to the mile. However, the anticlinal axes of the deeper closed anticline and of the structural nose conform rather closely in position and direction (Reeves, 1923, p. 130).

Reeves (1923, p. 134) lists eight pay sands in the Ranger field, of which three are in the Strawn, one in the Smithwick Shale, and four (of which the lowest is the McClesky) in the Marble Falls Limestone. The Strawn reservoirs lie at depths of 1300 to 2000 feet, and the Marble Falls reservoirs occur mostly in the 3300 to 3500 foot depth range. An estimated 90 per cent of the oil has come from the deeper reservoirs, chiefly the McClesky sand; next in importance is a sandy limestone reservoir lying at the top of the "black lime." Initial productions in the McClesky sand were as high as 11,000 barrels a day.

Figure 28.4 shows not only the subsurface structure but also the thickness of the McClesky sand. The entire area covered by structure contours has been productive from this reservoir, except low on the flanks of the syn-

	SYSTEM	SERIES	MARKER BEDS	PRODUCING ZONES
Paleozoic	Permian	Wolfcamp	Coleman Junction	
			Sedwick	
			Dothan	Dothan
				Moutray
			Camp Colorado (Noodle Creek)	Camp Colorado
				Frye of Shackelford and Jones Counties
				Tannehill
			Saddle Creek	Saddle Creek of Taylor County
				Bluff Creek
				Flippen
				Cook
			?	?
	Pennsylvanian	Cisco	Crystal Falls	Hope of Callahan, Jones, and Shackelford Counties
			Breckenridge	Breckenridge
				King
				Swastika—Serratt—Avis—Ivy of Shackelford County
			Gunsight	Gunsight
				McMillan of Runnels County
				Bunger
		Canyon	Home Creek	Home Creek
				Hardy of Jones County
				Overall of Coleman County
			Ranger	
			Winchell	Winchell of Coleman County
			Adams Branch	Adams Branch
				Cross Plains of Callahan County
			Palo Pinto	Palo Pinto
				Crosscut
				Moran
				Dog Bend of Runnels County
				Morris
			?	?
		Strawn	Upper Capps	Upper Capps—Breneke
				Upper Fry
			Lower Capps	Lower Capps
				Lower Fry
			Goen (Stephens)	Goen
				Harris of Haskell County
				Jennings—? Burson of Haskell and Throckmorton Counties
				Gardner—? Elbert of Haskell and Throckmorton Counties
				Overby, Manning of Haskell and Throckmorton Counties
				Gray—? Witaker of Haskell and Throckmorton Counties
				Sojourner of Haskell and Throckmorton Counties
			Odom	Odom
			Caddo	Caddo
		"Bend"		Lake
			Big Saline	Big Saline
			Marble Falls	Marble Falls
	Mississippian		Barnett	
			Mississippian	Mississippian
	Cambro-Ordovician		Ellenburger Group	Ellenburger
			Wilberns	Wilberns
			Riley	
	Precambrian			

Figure 28.3 **West Central Texas. Composite stratigraphic chart.** ***Courtesy Abilene Geological Society (1960).***

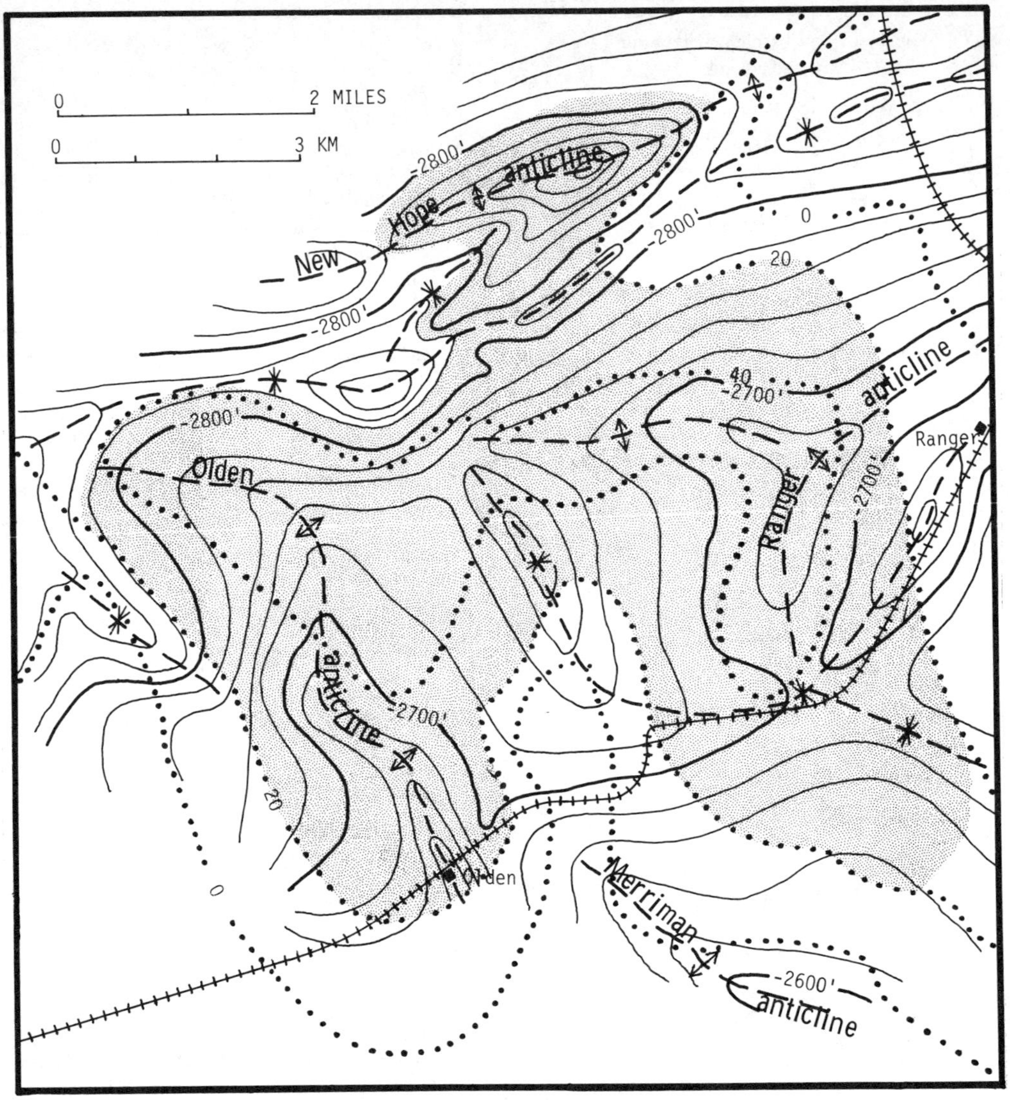

Figure 28.4 Ranger oil field, Eastland County. Structure and McClesky sand thickness map. Structure contours based on black limestone 200 feet above McClesky reservoir. Interval, 20 feet. Dotted lines show McClesky sand thickness. Interval, 20 feet. Productive area for this reservoir is stippled. *After Frank Reeves (1923), Plate XIX.*

clines and where the sand is less than 20 feet thick. Younger reservoirs that are structurally high may of course be productive in areas beyond the McClesky 20 foot isopach. However, these reservoirs are much more erratic in porosity than the McClesky and therefore have nowhere near the total productive area.

The *Katz* field (Maxwell, 1952) in the northeast corner of Stonewall County, where the West Central district has a common corner with the North Texas and the Panhandle districts, is another example of anticlinal trapping. It was discovered in 1951 after a magnetometer survey. Production is obtained from three sands in the Strawn lying at depths between 4800 and 5100 feet. The trap as mapped on top of the Strawn is a north-south asymmetrical anticline, with the steeper dip to the east and a closure of more than 50 feet. Sixteen months after discovery the productive area included nearly 2200 acres. By the end of 1966 the Katz field had produced almost 35 million barrels of oil, making it second among current fields in the district in cumulative production.

Another large, low-relief anticline in the southern part of Stonewall County is occupied by the *Frankirk* field (Vickers and Ratliffe, 1952). This field was discovered by subsurface mapping and a reflection seismograph survey. The reservoir is Ellenburger (Ordovician) Limestone lying at a depth of about 6000 feet. The anticline is triangular in plan, with two nodes.

Across the Bend arch in the southeastern part of the

West Central Texas district in northwestern Hamilton County are the *Pottsville* and *South Pottsville* gas fields. This region is in the zone of northeasterly trending faults radiating outward from the Central Mineral Region (Munn and Riddle, 1957). The Pottsville fields are bordered on the northeast by a normal fault that has dropped the eastern side about 700 feet (Figure 28.5). Accumulation is anticlinal; all that the fault has done has been to limit production on the east flank of the two domes. However, erratic porosity within the upper Marble Falls (Bend) Limestone reservoir rock has resulted in some dry holes high on the nodes. The structural closure of 400 feet in the Pottsville field is unusual for the district.

West Central Texas also has varying permeability traps ranging in type from sand lenses to reef limestones. In the *Crosscut–Blake* district in the northwestern corner of Brown County, the Crosscut sand in the lower part of the Canyon Series occurs in lenticular bodies of various sizes and shapes. Structurally the area is one of small anticlines and anticlinal noses. The Crosscut sand may occupy an entire closed anticline, or it may be confined to only a part of an upfold. In any event, where oil is present in the sand, water is present also and the oil has accumulated at the top of the trap, wherever it may be structurally. However, there is a tendency for the sand to be thicker higher on the structure; this has been ascribed to either differential deposition or stream erosion in the topographically low areas shortly after sand deposition (Klinger, 1941).

Similar lenticular sand trapping occurs farther to the south in west-central Brown County (Storm, 1929). The *East Hamlin* pool in northwestern Jones County produces most of its oil from the lower Swastika of Cisco age. Accumulation is in an anticlinal ridge about 2 miles long that rises more than 40 feet above the oil-water

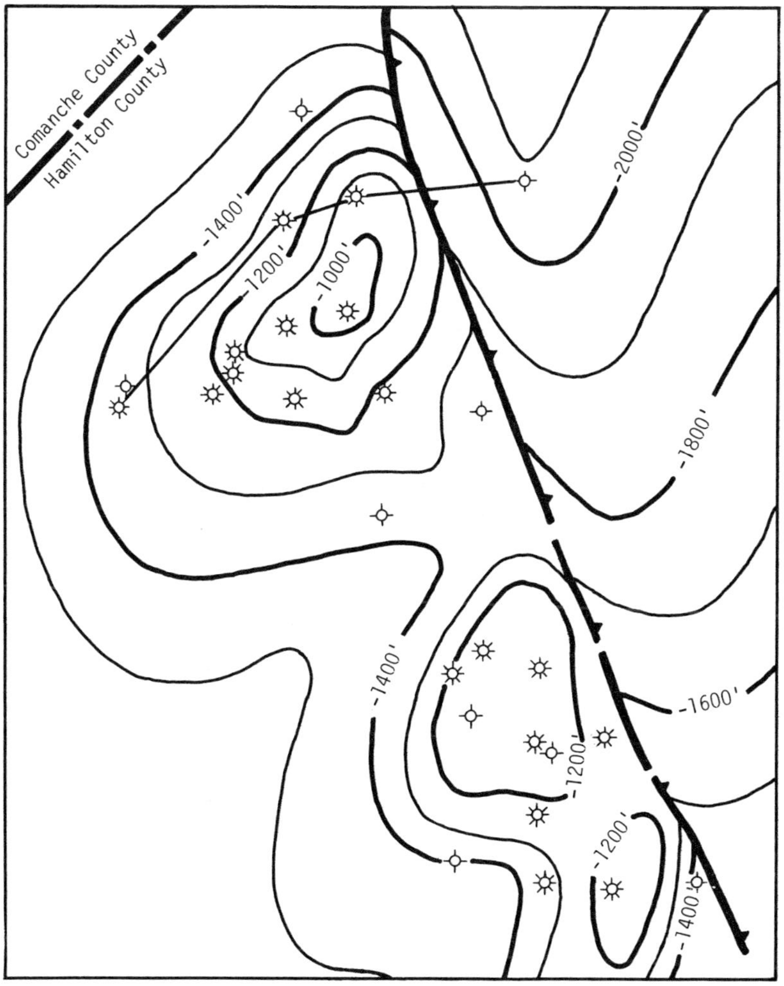

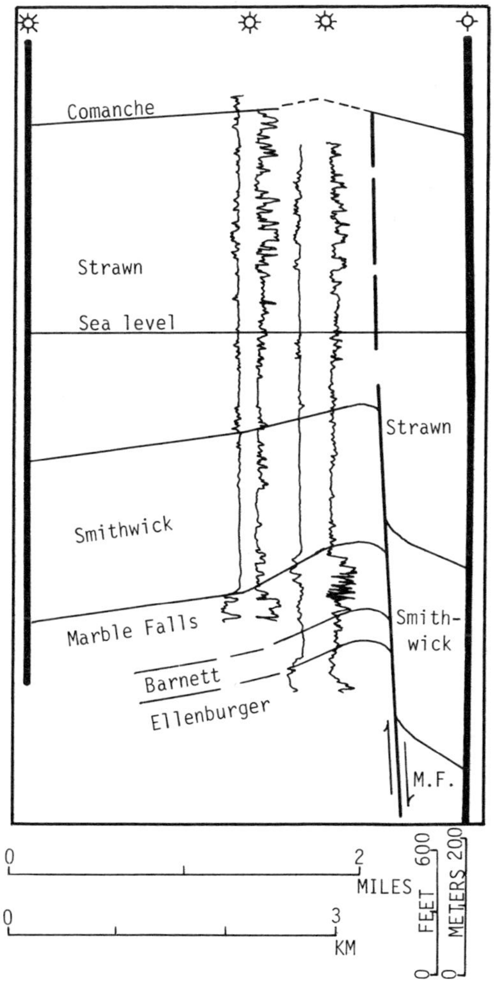

FIGURE 28.5 North and South Pottsville gas fields, Hamilton County, Texas. Left: structure contour map based on top of the Marble Falls (reservoir rock). Interval, 100 feet. West to northeast line is line of cross section shown on right. ***Courtesy J. K. Munn and B. D. Riddle (1957, Plates I and II), and the Abilene and Fort Worth Geological Societies.***

interface. Oil occurs along the west flank of this north-south structure, over the top, and about halfway down the east flank where the reservoir sand pinches out (Wilkinson, 1952). The *Glen Cove* field in west-central Coleman County overlies the gentle west flank of the Bend arch. It produces oil from the Jennings sand of Strawn age. The field is elongate, trending slightly east of north, parallel to the regional strike. The structure map of the Glen Cove field shows that it occupies a narrow terrace; the isopach map of the reservoir sand indicates that the trap is an elongate sand lens suggestive of an offshore bar (Frazell, 1952).

The highly effective combination trap brought about by an anticlinal nose closed updip by a sand pinchout line curving in an opposite direction from the structure contours is illustrated beautifully by the *Sojourner* oil field in central Haskell County (Figure 28.6). The regional dip here is westward off the Bend arch, and the nose is likewise plunging westward down the flank of the arch. As can be noted in Figure 28.6 (left), the sand pinchout line cuts off accumulation on the south side of the nose, as well as over the top to the east and down the northwestern flank. The Sojourner sand is in the lower Strawn of the Pennsylvanian. It was found to be more than 32 feet thick in the discovery well. Before the discovery of the Sojourner field, the area was surveyed geochemically and an anomaly was found on the basis of which a block was leased. Subsequently a reflection seismograph survey was made and the discovery well drilled. The righthand map in Figure 28.6 shows the relationship between the subsequently developed producing area and the original geochemical anomaly (Ransone, 1958).

Gas also may accumulate in lenticular sands. A few miles to the west of the Ranger oil field, in the southwestern corner of Stephens County and on the west flank of the Bend arch, is the *Calloway-Henry* gas field (Kendrick and Russell, 1935, p. 633–634). The gas trap is a northwestward plunging anticlinal nose, sealed to the east by the lensing out of the reservoir sands. Production comes from three Strawn sand lenses ranging in thickness (where productive) from 10 to 60 feet. The field was discovered by drilling the plunging anticline, which was mapped at the surface.

Limestone reservoirs also illustrate varying permeability trapping in West Central Texas. The *Noodle Creek* pool in the southwest corner of Jones County has accumulation of this type in a nonreef limestone (Imholz, 1941). The reservoir is the Noodle Creek Limestone, which is now classified as Wolfcampian (Permian) in age. The regional dip is westward off the Bend arch, but it is interrupted by downdip furrowing caused by the presence of a series of anticlinal noses. Furthermore, as in the area of the Noodle Creek pool, there is also terracing parallel to the regional strike, which may create minor closures on the plunging anticlines. Oil moving up the regional dip has been caught in part by local

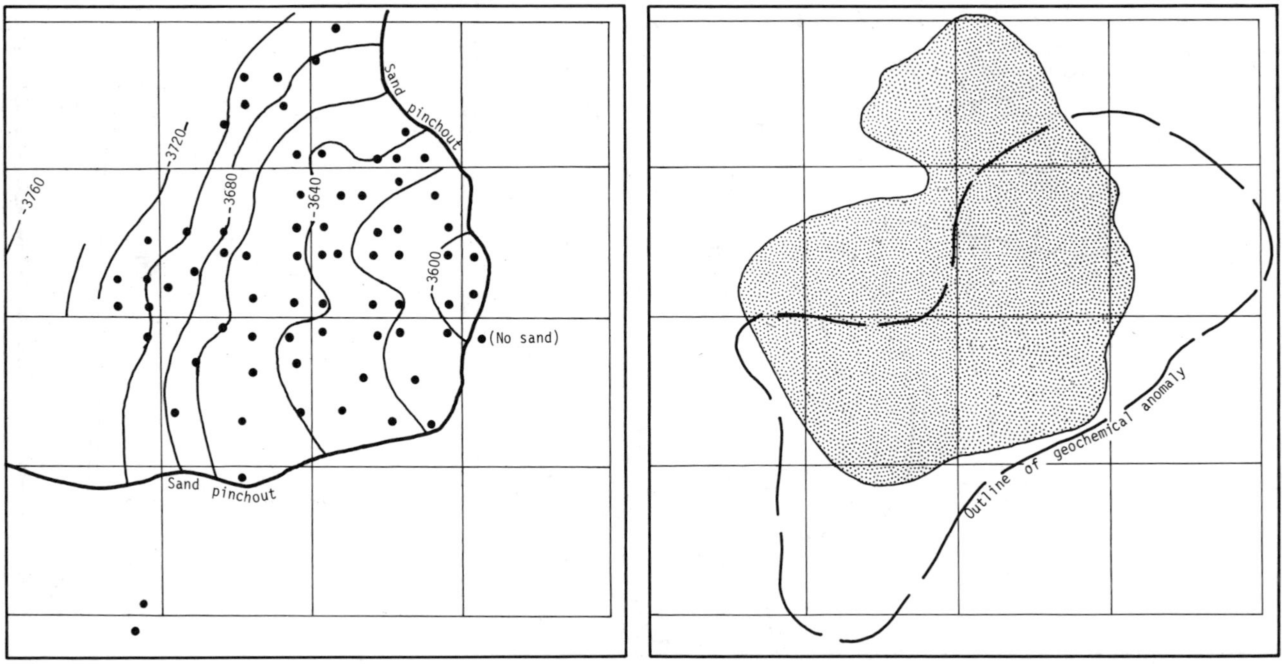

FIGURE 28.6 Sojourner field, Haskell County. Left: contour map based on top of Sojourner sand. Interval, 20 feet. Right: Sojourner producing area (stippled), showing relationship of field to earlier mapped geochemical anomaly. Squares on both maps are 1 mile (1.6 km) on a side. *Courtesy W. R. Ransome (1958, Figures 4 and 5) and the Society of Exploration Geophysicists.*

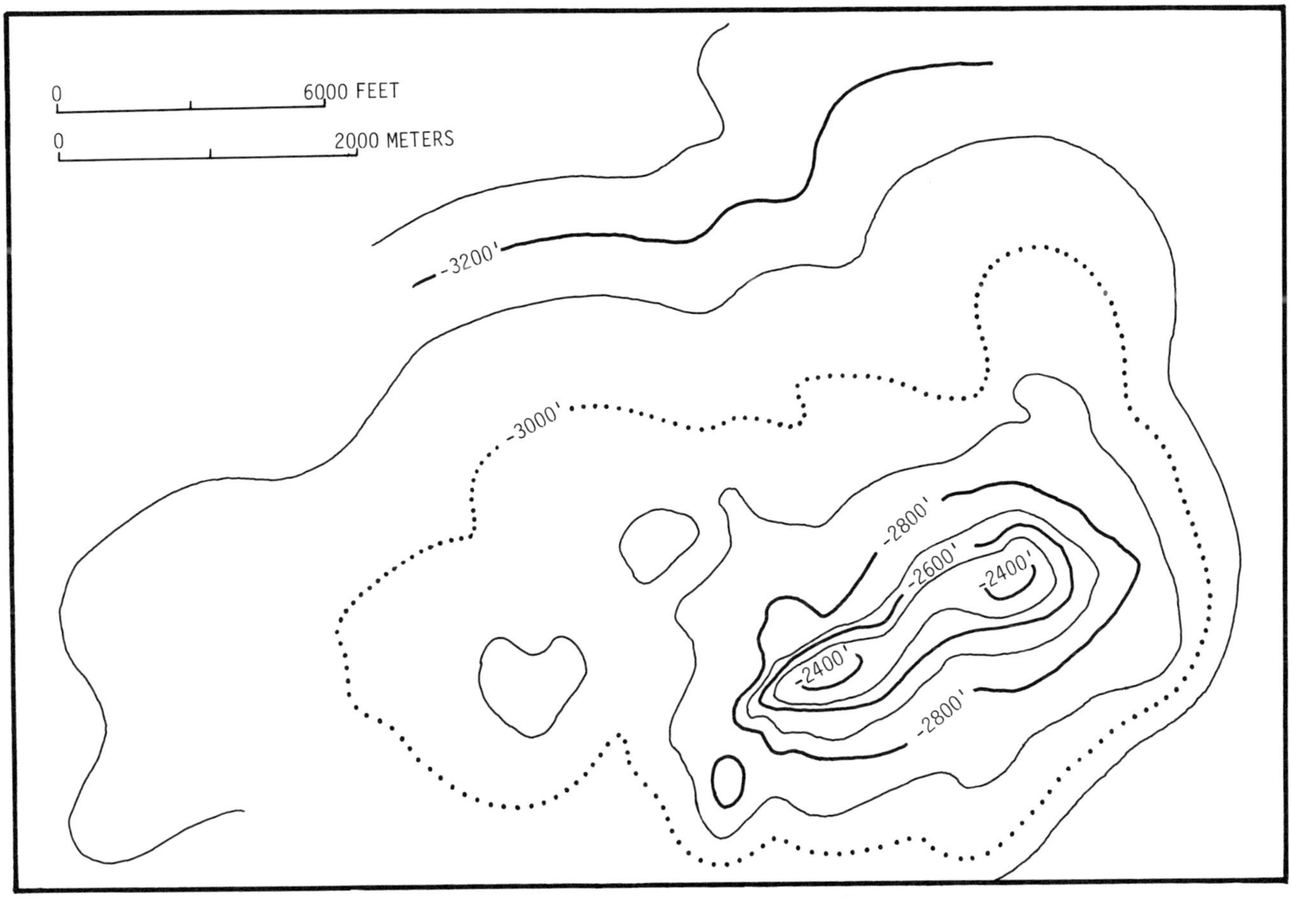

FIGURE 28.7 **Round Top pool, Fisher County, Texas. Structure contour map based on top of Canyon Reef reservoir. Interval, 100 feet. Dashed (−3,000 foot) contour encloses productive area.** ***Courtesy Willard Day and George Galbraith (1950, map following p. 67) and the Abilene Geological Society.***

closure, but mainly by the updip lensing out of porosity in the reservoir limestone.

Limestone reefs are important reservoirs on the western side of the district. Farther west, in West Texas, is Scurry County, famous for its "horseshoe" of prolific reef fields. During Pennsylvanian time this area was covered by a shelf sea in which conditions were favorable for reef development (Conselman, 1960), especially during the Canyon interval. The counties in West Central Texas with productive Pennsylvanian reefs include Stonewall, Fisher, Nolan, Haskell, Jones, and Taylor. Shackleford County, immediately to the east of Jones, has Mississippian reef production. Unfortunately not all of the reefs are productive. The East Anson reef in Jones County, the largest of them all, measuring 16 by 7 miles with more than 100 feet of relief, remained nonproductive after "almost a hundred rotary and cable tool tests" (Conselman, 1960, p. 8).

The *Aspermont Lake* field in central Stonewall County illustrates one type of productive reef (Richards, 1952). Here the reef, of Canyon age, has a so-called pinnacle that rises above the oil-water contact. Some oil is produced from the pinnacle, but most is obtained in a sand extending laterally outward from the upward part of this limestone protuberance. Another example of Canyon reef oil accumulation is the *Double Mountain* pool about 12 miles southwest in southwestern Stonewall County (Conley, 1952). Here the entire production is from the reef, which is nearly circular in plan and has a diameter of about $1\frac{1}{2}$ miles with a vertical relief approaching 500 feet. The reservoir limestone ranges in texture from fine to coarse. Some zones in the limestone contain abundant fossils and have considerable intergranular porosity; vuggy porosity and vertical fracturing are also present.

To the south of Stonewall County in northeastern Fisher County is *Round Top* (Day and Galbraith, 1950), one of the larger fields in the West Central district. The major reservoir is a reef limestone of Palo Pinto (Canyon) age that is erratic in permeability. The porosity ranges from pinpoint to vuggy; there is also some evidence of vertical fracturing. The reef axis trends northeasterly; the northwest and especially the west slopes of the reef are much more gentle than the slope to the southeast (Figure 28.7). The vertical distance between the top of the reef and the oil-water contact is 450 feet. This, combined with an area of approximately 3000 acres, explains the relatively large yield. Some production is obtained also from a higher sand (Swastika) of Cisco age, which is present in disconnected lenses.

The Mississippian reefs lie farther up on the west flank of the Bend arch in Shackleford County (Wilson, 1966). Here, in an area approximately 5 by 7 miles, is a cluster of nine fields. Most of the production comes from Mississippian bioherms. The individual reefs are sharp features, with dips on the upper surface as great as 30°. The maximum thickness is 340 feet. Wilson (1966, p. 139) describes the reefs as "great piles of crinoid stems which in some places have been reworked at the top." Compaction draping has arched the overlying strata, and some production is obtained from younger Pennsylvanian limestone and sandstone reservoirs as well as from Mississippian reefs. Eighteen new reef fields were discovered here recently in less than 2 years (Oil and Gas Jour., Apr. 22, 1968, p. 110).

Future. The drill hole density on the Bend arch in the West Central Texas district is probably great enough to preclude many new discoveries in the shallower reservoirs in this area. However, the density of probes into pre-Pennsylvanian reservoirs is inadequate both on the arch and down the flanks. The Ordovician Ellenburger, and even potential Cambrian reservoirs (Conselman, 1955), have not been tested with any degree of thoroughness as yet. Continued success can be anticipated in discovering accumulations in Pennsylvanian and Mississippian reservoirs down the flanks of the arch, especially in reefs (Conselman, 1960) and lenticular sands.

REFERENCES CONSULTED IN PREPARING CHAPTER 28

Abilene Geological Society, 1952.
Abilene Geological Society, Stratigraphic and Study Group, 1960.
Cheney, M. G., and Goss, Louis F., 1952.
Conley, J. N., 1952.
Conselman, Frank B., 1955, 1960.
Day, Willard, and Galbraith, George, 1950.
Ellison, Samuel P., Jr., 1960.
Esgen, W. K., 1929.
Frazell, W. D., 1952.
Galley, John E., 1958.
Gibson, H. A., 1946.
Imholz, H. W., 1941.
Jackson, William E., 1964.
Kendrick, Frank E., and Russell, Philip G., 1935.
Klinger, Edgar D., 1941.
Makins, Robert L., 1968.
Maxwell, Riley G., 1952.
Munn, J. K., and Riddle, B. D., 1957.
Ransone, W. R., 1958.
Reeves, Frank, 1923.
Richards, O. H., 1952.
Rister, Carl Coke, 1949.
Storm, Willis, 1929.
Vickers, R. B., and Ratliffe, Fred, 1952.
Wilkinson, Walter M., 1952.
Wilson, W. S., 1966.

Chapter

29

NORTH TEXAS

The North Texas district corresponds with District 9 of the Texas Railroad Commission. It is the smallest of those in the Texas chapters, containing fifteen counties (Figure 29.1). It is bounded on the north by the Red River and Oklahoma, on the east by Northeast Texas, on the south by both Northeast and West Central Texas, and on the west by West Texas and the Texas Panhandle.

The North Texas district ranks fifth (out of seven) in recent annual petroleum output, with between 70 and 80 million barrels annually. It has three giant fields. In first place in both annual and cumulative production is K.M.A. in Wichita County, which has already yielded nearly 150 million barrels of oil. Second and third in both categories are the Hull-Silk field in Archer County and the Walnut Bend field in Cooke County. Nineteen discoveries were made in North Texas in 1967 (Makins, 1968).

Attention was called to the oil and gas possibilities in this district as early as 1871 through the presence of oil and gas in water wells in Young County. However, North Texas did not produce commercially until 1904, when the Petrolia field in Clay County began marketing oil (Ellison, 1960, p. 19), and did not become an important source of oil until 1911, after the discovery of the Electra field in Wichita County. Seven years later S. L. Fowler drilled a wildcat on his farm on the northern outskirts of Burkburnett, also in Wichita County. He was prodded into this venture by his wife, who had noted an oil scum on the water in the farm well (Shelton, 1964, p. 16). On July 29, 1918, this well began gushing 2200 barrels a day from a sand at a depth of 1734 feet, thereby triggering one of the most frenzied town lot drilling campaigns in history. After the drilling of the second well on the Fowler farm, the Fowler Oil Company, originally capitalized at $12,000, was sold for $1,800,000 (Rister, 1949, p. 110).

Subsequently the North Texas district declined in oil

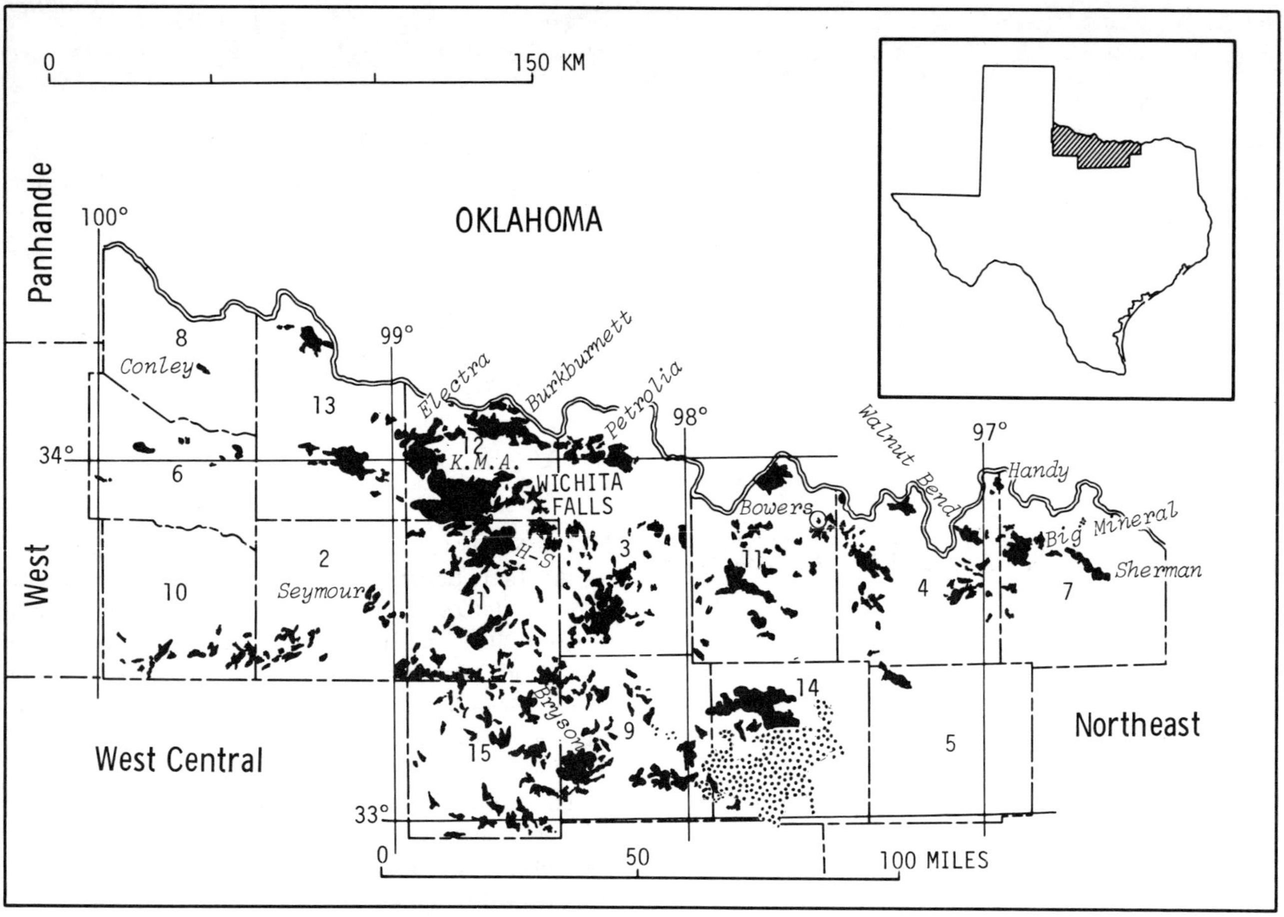

FIGURE 29.1 North Texas district. Index map with oil and gas fields. Named fields are mentioned in text. *H-S* is the *Hull-Silk* field. County Key:

1	Archer	6	Foard	11	Montague
2	Baylor	7	Grayson	12	Wichita
3	Clay	8	Hardeman	13	Wilbarger
4	Cooke	9	Jack	14	Wise
5	Denton	10	Knox	15	Young

production until the discovery of a deep Pennsylvanian reservoir in the K.M.A. field in 1937. Two years later a still deeper Ellenburger (Ordovician) reservoir was found, leading to greatly renewed activity, not only in the K.M.A. field, but throughout the district.

TECTONICS. In spite of its relatively small size, the North Texas district is fairly well creased by regional anticlines and synclines. Figure 29.2 shows by structure contours (with 5000 foot interval) the major features. The southern part of the district lies on the low end of the northward plunging Bend arch. Beyond, to the north and northwest, are the Red River uplift and the Hardeman basin. To the northeast are the Fort Worth basin and the Muenster arch. The Hardeman basin, the Red River uplift, and the Muenster arch cross into Oklahoma and are mentioned in the discussion of the petroleum geology of that state.

The few major faults are confined to the northern and especially the northeastern part of the district. The Ouachita tectonic belt of disturbed Paleozoic rocks cuts across the easternmost edge of the district beneath the overlapping Upper Cretaceous sediments.

STRATIGRAPHY. The eastern end of the North Texas district is overlapped by lowermost Upper Cretaceous sediments of the Gulf Coast embayment. At the extreme eastern border formations belonging in the Austin Group lie at the surface; these are succeeded to the westward by the older Eagle Ford and Woodbine sediments (Figure 29.2). Crossing the east-central part of the district is a tongue of Comanchean (Lower Cretaceous) formations about two counties in width. These sediments overlap unconformably Pennsylvanian rocks that crop out in the south-central part of the district and Permian rocks that extend from central North Texas westward to and beyond the district line. The Permian rocks in the western part of the district are masked here and there by erratic erosional remnants of Quaternary sand deposits (not shown on Figure 29.2).

Figure 29.3 (after Ellison, 1960, p. 23) is a section showing both the structure and the subsurface stratigraphy on both sides of the border between West Central Texas and North Texas from southwest to northeast. It shows the truncation to obliteration of the Ellenburger to the west, the thinning of the Ellenburger over the Bend arch, the horizontal shifting of the axis of the Bend arch between the Precambrian and the surface, and the thickening of the Ellenburger into the Fort Worth basin. The cross section also shows the Red River mobile belt, the core of the Muenster arch, the faulted zone in the pre-Pennsylvanian sediments east of the Muenster arch, and the Ouachita fold belt that affected the entire Paleozoic column at the eastern end of the section.

The subsurface stratigraphy in the central and eastern parts of the two districts is dominated by Lower Pennsylvanian sediments, especially shales, sandstones, and local limestones in the Atoka-Bend and Strawn. On the west flank of the Bend arch the Pennsylvanian section is overlain by a thick Permian section with Triassic and Cretaceous sediments barely lapping upon the Permian at the extreme western end. Cretaceous sediments overlap the Pennsylvanian sediments from the east end of the cross section to almost the center, which is also close to the Bend arch axis in the Pennsylvanian formations uncovered by erosion.

Carbonate rocks dominate in the thick deposits of Permian age on the west side. The Cretaceous sediments on the east consist of both sandstones and limestones.

The reservoirs beneath North Texas range in age from Permian red beds to Ordovician Ellenburger Dolomite. Shallow sands in the Wichita (Lower Permian) Formation were the source of the first oil and gas in the Petrolia field (Reutinger, 1968). Most of the district's oil has come

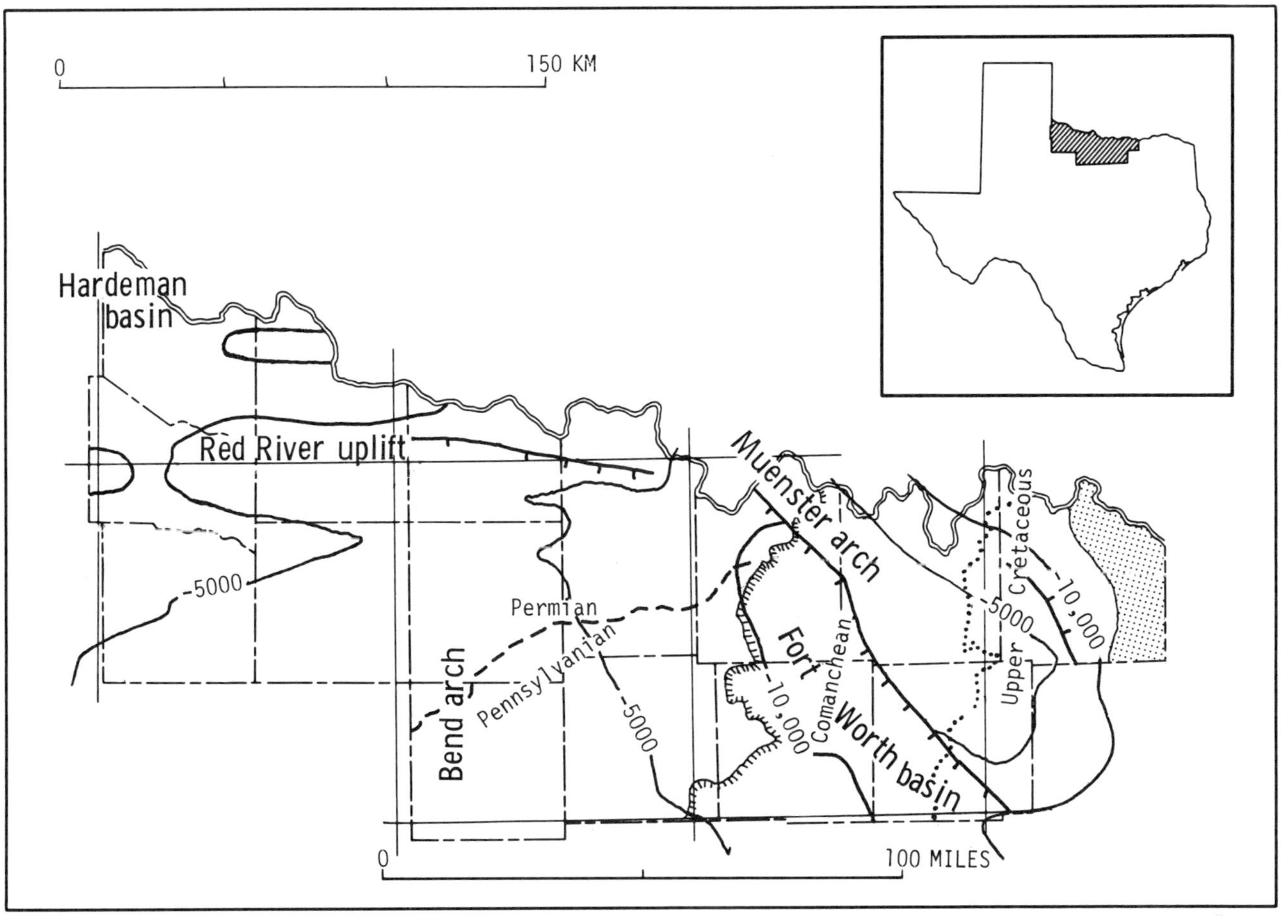

FIGURE 29.2 North Texas. Geologic and tectonic map. Outcrop areas for Permian, Pennsylvanian, Comanchean, and Upper Cretaceous sediments are shown. Isolated outliers of Quaternary in western part of district are not shown. Stippled area in east is subcrop of Ouachita facies. Structure contours on top of Ellenburger or pre-Simpson eroded surface. Interval, 5000 feet. *Courtesy H. H. Bradfield (1964, Figure 1) and the Tulsa Geological Society.*

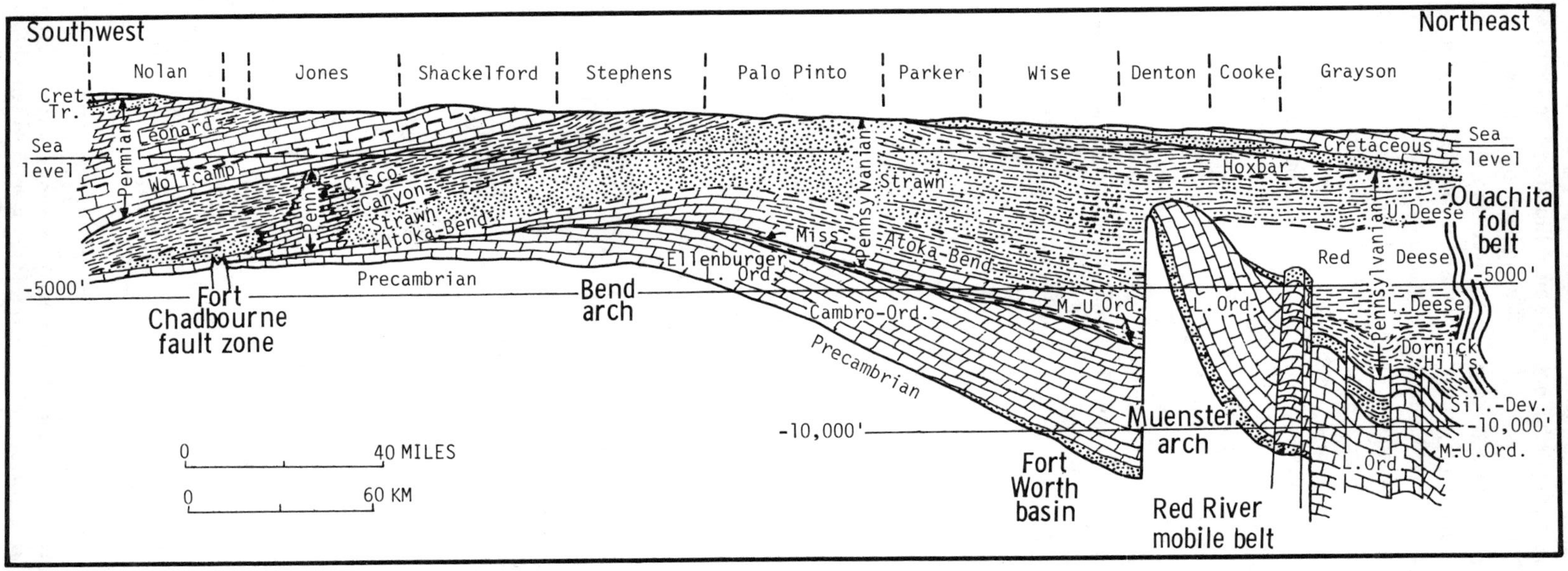

Figure 29.3 Cross section, southwest to northeast, across western West Central Texas and southern North Texas. Section crosses into North Texas at Parker-Wise county line. *Courtesy Samuel P. Ellison (1960, Figure 3, upper, p. 23).*

from Pennsylvanian sandstones. These reservoirs occur in the Cisco, Canyon, Strawn, and Bend Groups (see Figure 28.3). However, more oil has been obtained from sands in the Strawn Group of Middle Pennsylvanian age than from any other series of reservoirs. Some production has also been obtained from Pennsylvanian limestones, including especially the Caddo of the lower Strawn or Upper Bend. In the Lower Bend, lenticular sandstones and conglomerates produce gas in the Fort Worth basin (Blanchard et al., 1968).

Crinoidal bioherms of Mississippian age are small but prolific pays in the central and western parts of the district. Of greater importance is the Oil Creek sand of Ordovician age, which underlies the angular unconformity separating the Pennsylvanian from the Ordovician in the eastern part of the district, northeast of the Muenster arch.

The still older Ordovician Ellenburger Limestone is the most important deep reservoir in North Texas. Figure 29.4 (Bradfield, 1964, Figure 2) is a thickness map of the Ellenburger Group in this district. It shows a range in thickness in North Texas from less than 500 feet along the west line to over 5000 feet in the northeastern corner of the district. The minimum thickness in western North Texas is due in part to the fact that this area lies high on the Red River arch. In fact, along the arch axis in Foard County two nodes on the arch have no Ellenburger (or older Paleozoics) whatsoever, and in a considerable area pre-Simpson Paleozoic rocks are absent in one county to the west in the Texas Panhandle district.

A local beveling of the Ellenburger Group also takes place in eastern North Texas along the axis of the Muenster arch. On the southwest side of the Muenster arch the Ellenburger is thicker in the Fort Worth basin; it reaches its maximum thickness in the district to the northeast in the Marietta basin. At the extreme eastern end of the district the isopach lines abut against the western edge of the Ouachita fold belt.

The Precambrian basement rock shown in Figure 29.3 is granite except for metasediments in the Red River mobile belt, in the core of the Muenster arch, and near the western end of the cross section beneath Jones County.

Traps And Selected Fields. All three giant fields in North Texas—Hull-Silk, K.M.A., and Walnut Bend—are anticlinal accumulations, as are most of the other oil fields in the district. However, erratic permeabilities of the reservoir rocks, both sandstone and limestone, tend to control hydrocarbon distribution on the upfolds. The producing anticlines range from simple to complex; some are faulted, but most are not. Compressive forces have created most of the folds, but draping over buried hills

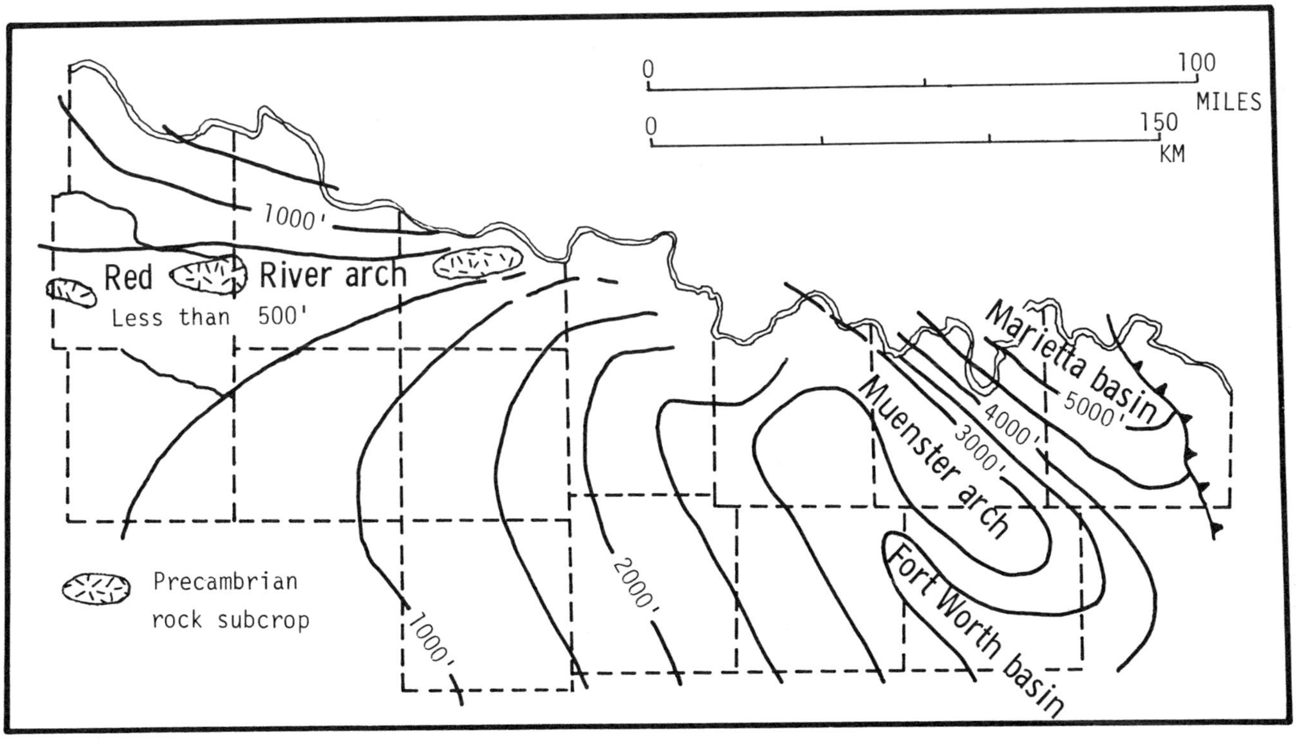

Figure 29.4 North Texas. Thickness map of subsurface Ellenburger Group (Lower Ordovician). Interval, 500 feet. Precambrian rock windows shown along Red River arch, where Cambrian and Ordovician deposits were eroded during pre-Mississippian time. *Courtesy H. H. Bradfield (1964, Figure 2, p. 117) and the Tulsa Geological Society.*

accounts for the anticlinal structure in some instances. Limestone reefs and sand lenticularity, including updip wedgouts, are responsible for trapping in a few fields.

In 1965 the largest field in North Texas in oil production was the *Walnut Bend* in northeastern Cooke County. The yield that year was over 3,700,000 barrels; Walnut Bend was also in a virtual tie with Hull-Silk for second place in cumulative production, with a total yield of nearly 66 million barrels. The Walnut Bend field gets its name from its position within a large meander of the Red River (Hilseweck, 1941, p. 776). The field is northeast of the Muenster arch. The Walnut Bend anticline, which is responsible for most of the trapping in the field, was found by surface and seismograph surveys. The discovery well, the J. M. Best No. 1, was completed in June, 1938, in a sand in the lower Strawn (Lower Pennsylvanian) at a depth of 4887 to 4900 feet. This sand, known both as the 4900 foot and the "Best," along with two overlying sand zones constitutes the "Walnut Bend regular" reservoir, which is the leading reservoir for the field. Five years after the completion of Best No. 1, the Winger sand, the deepest of the Strawn reservoirs, was found; additional discoveries since then have raised the number of reservoirs to an even dozen. They are all of Strawn age except the relatively shallow "2600 foot" sand, which lies at the base of the Hoxbar or Canyon Group above the Strawn, and the deepest reservoirs, which are Ordovician. The latter include the Oil Creek sand of the Simpson Formation and the Ellenburger, which is productive in a few wells in the northwest sector of the field. This is the only carbonate rock reservoir so far productive in Walnut Bend. The Ordovician rocks are separated from the Pennsylvanian sediments by an erosional unconformity.

Figure 29.5 is a structure contour map (Bradfield and Seale, 1957) of the Walnut Bend field, based on the Best sand. The oil-water contact in this reservoir originally was at approximately −4250 feet, so that contour outlines the Best sand pool. However, this sand, although the most consistent of the Strawn reservoirs, grades into sandy shale in some places on the anticline, and hence the yield is not completely continuous across the trap above the water level. The other reservoir sands are much more erratic; although the accumulation within them is still

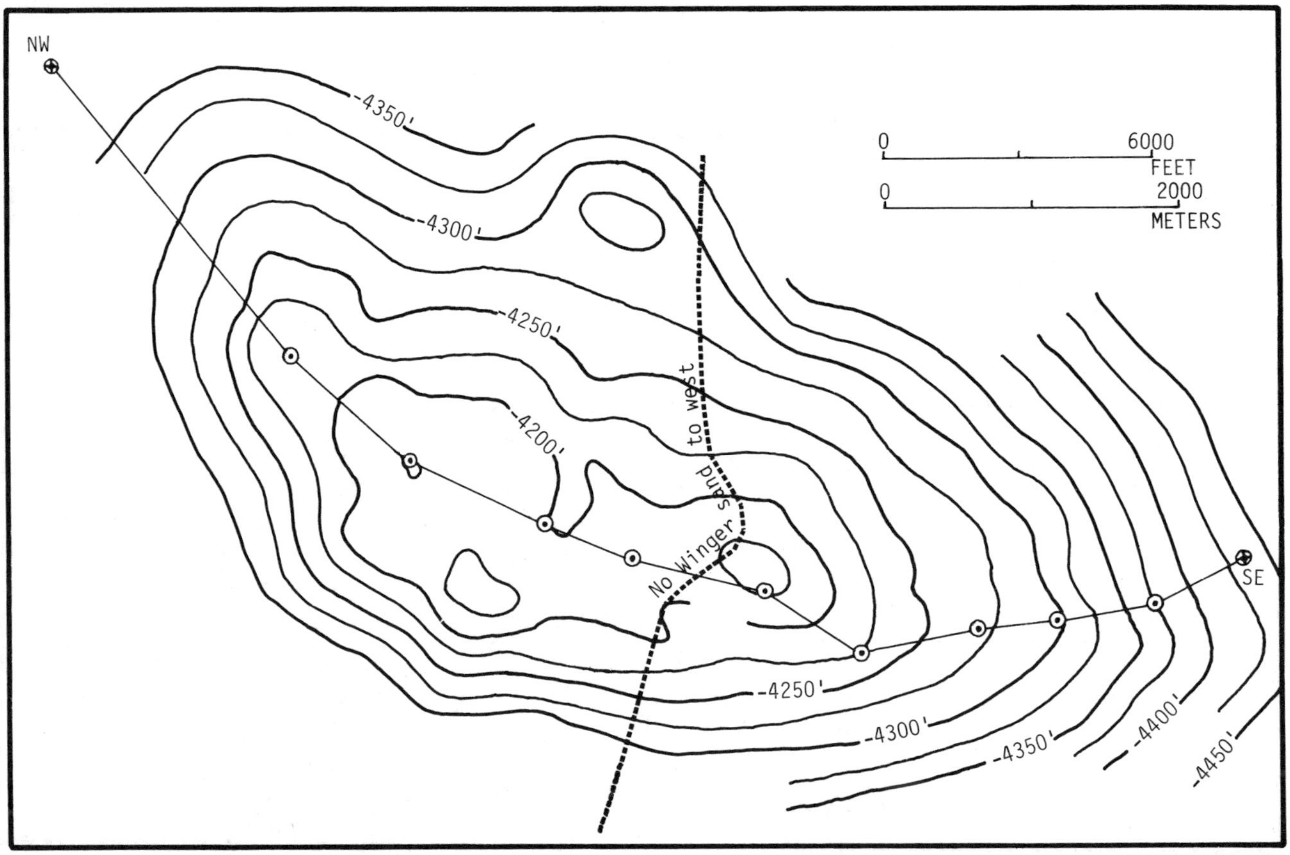

FIGURE 29.5 Walnut Bend field, Cooke County. Structure contour map based on top of Best sand. Interval, 25 feet. Oil in the Best sand is trapped above the −4250 contour. North-south dashed line shows terminus of westward transgressing Winger sea. Northwest-southeast line is line of longitudinal section shown in Figure 29.6. ***Courtesy H. H. Bradfield and R. I. Seale (1957, Figure 4, p. 111) and the Dallas Geological and Geophysical Societies.***

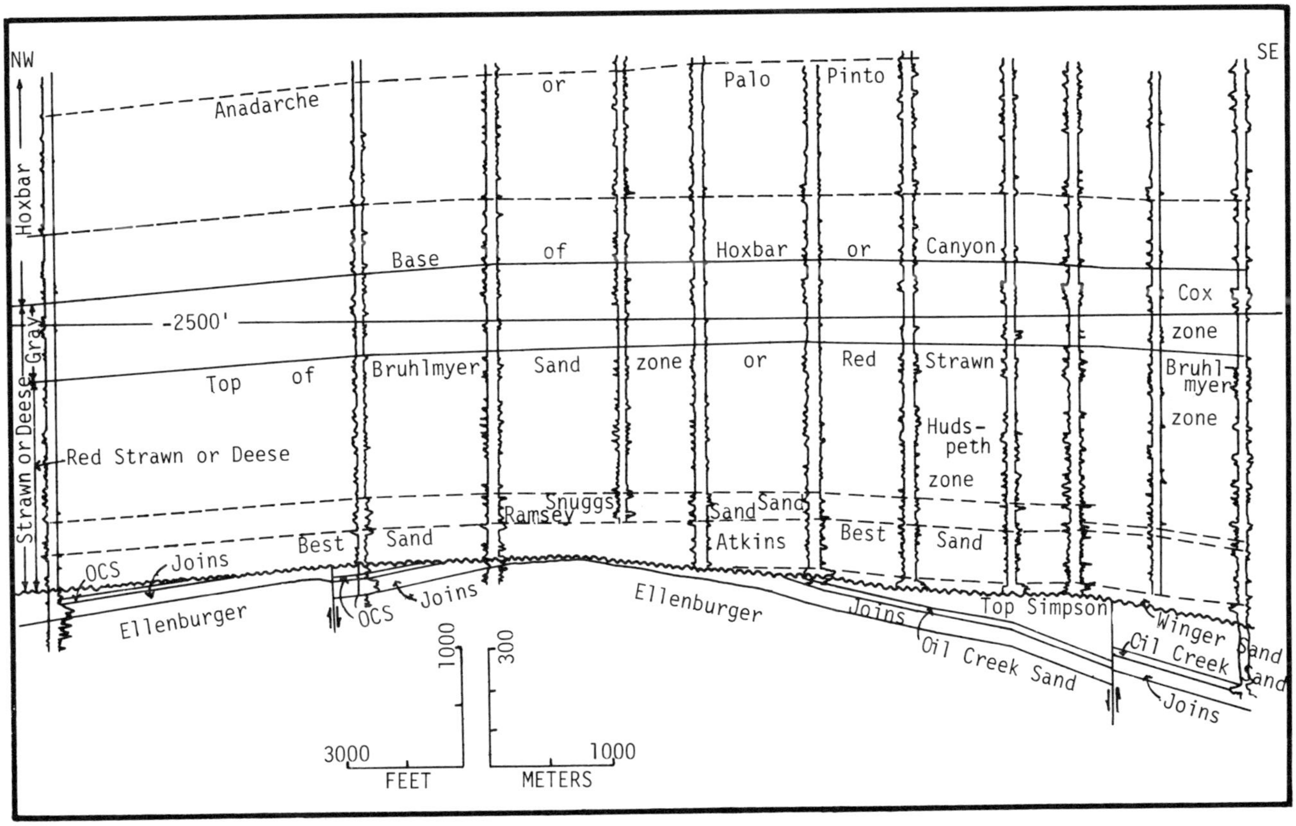

FIGURE 29.6 **Walnut Bend field; longitudinal section along the northwest-southeast line in Figure 29.5, showing Best reservoir, wedgout of Winger sand, truncation of Oil Creek reservoir beneath Pennsylvanian-Ordovician unconformity, and Ellenburger reservoir.** ***After H. H. Bradfield and R. I. Seale*** **(1957, Figure 7, p. 117).**

anticlinal, the oil distribution is dependent on local conditions of permeability. Oil can and does occur in these sands far down the flanks of the anticline where permeability and water level permit.

The Winger sand is present across the eastern half of the Walnut Bend anticline only; it was not deposited west of a nearly north-south line (shown in Figure 29.5) bifurcating the upfold. A longitudinal section (Figure 29.6) shows the Winger sand lapping on the east flank of a pre-Pennsylvanian anticline. Beneath the Winger, the Oil Creek sand reservoir of Simpson age is truncated on both flanks of the structure by the unconformity at the base of the Pennsylvanian. Obviously the Walnut Bend anticline was initiated before submergence beneath the Pennsylvanian sea. Faulting also took place during this interval. Subsequently emergence and erosion produced the truncation of Ordovician formations; this was followed by submergence and deposition of sandstones and shales of Strawn age. A renewal of folding produced the more subdued anticline in the Pennsylvanian sediments shown by the structure contour map.

Hull–Silk (Thompson, 1941) in northern Archer County is a giant field illustrating anticlinal trapping, but with variable permeability determining the distribution of oil in some of the reservoirs. In 1965 this field was second in the North Texas district in current production and third in cumulative production.

The first production in the area now covered by the Hull–Silk field came in 1916, when the Panther field was discovered. Production in this field was from shallow Pennsylvanian sands ranging in depth from 1600 to 1750 feet. The name Hull–Silk was not applied until 1938, when an exploratory well struck oil in a Strawn sand (the "Hull–Silk sand") just below 3800 feet. This exploration was motivated by the discovery in 1937 of oil in quantity in the Strawn formation in the K.M.A. field a few miles to the northwest. A short time later a well drilled on the same 40 acre lease discovered a deeper and richer sand in the Strawn at 4340 to 4415 feet. This reservoir, now referred to as the "4300 foot sand," is the most prolific in the field. It varies in thickness from 160 feet on the east and northeast sides of the pool to almost zero on the southwest. This thinning is brought about by a lithologic change as the sandstone is replaced by shale from the top downward.

Other reservoirs discovered subsequently include the "4000 foot limestone," which ranges from 40 to 100 feet in thickness and which also occurs in the Strawn, between the two previously discovered sand reservoirs. The only pre-Strawn reservoir of importance up to now is the "Caddo Limestone" from 50 to 75 feet in thickness in the underlying Bend Group (also lower Pennsylvanian).

Figure 29.7 is a structure contour map of the Hull–Silk field, based on the top of the 4300 foot sandstone (Thompson, 1941, Figure 8). The stippled pattern designates the area in which production has been obtained from this reservoir. It will be noted that this is a closed anticline with very little oil found below elevation −3350 feet. Drilled wells found the oil-water interface at −3380 feet in the southwestern part of the field and at about −3415 feet in the northeastern end. The Hull–Silk (−3800 foot) sand is missing across the top of the anticline, and oil has been trapped by the updip wedgout of the sand, mainly on the southern flank but also, to a lesser degree, on the northeast and north flanks.

In the intermediate (4000 foot) limestone, production is confined to the closed part of the structure, but only where the limestone is of adequate thickness and porosity. In the deeper Caddo Limestone the anticline is more pronounced than in the Strawn reservoirs, and production is confined to the structural closure.

The *Conley* field (Freeman, 1962, 1964) in east-central Hardeman County is currently producing about 1 million barrels a year. Trapping is in a nearly circular anticline (Figure 29.8) where oil was discovered in 1959 after a

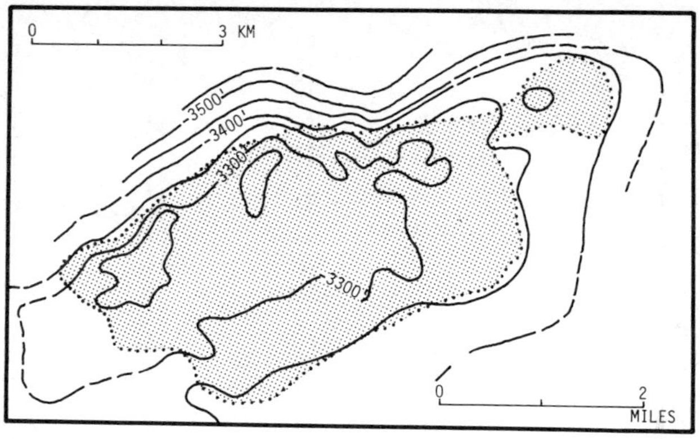

Figure 29.7 **Hull-Silk field, Archer County. Structure contour map based on top of "4300 foot" reservoir sandstone. Interval, 50 feet. Productive area in this sand is stippled.** ***Courtesy E. I. Thompson (1941, Figure 8, p. 675) and the American Association of Petroleum Geologists.***

series of seismograph surveys and two dry exploratory wells. The discovery well found production in the Chappell Limestone of Mississippian age. Subsequently oil was discovered in the deeper Ellenburger Dolomite, in Palo Pinto Limestone of Pennsylvanian age, and in two other Mississippian reservoirs—the Chester Sandstone, which lies above the Chappell Limestone, and the Osage Limestone, which lies below. The largest cumulative productions are from the Ellenburger and Mississippian reservoirs.

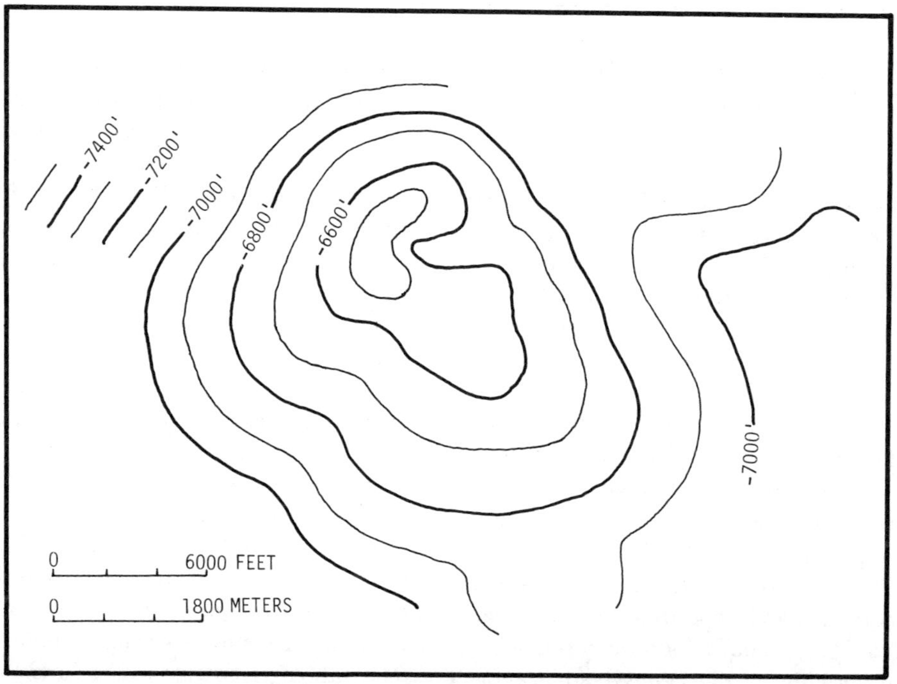

Figure 29.8 **Conley field, Hardeman County, Texas. Structure contour map based on top of Ellenburger. Interval, 100 feet. Original oil-water contact at approximately −6800 feet.** ***Courtesy John C. Freeman (1964, Figure 3, p. 130) and the Tulsa Geological Society.***

The Conley anticline is superimposed on the south flank of the Hardeman basin, which is the eastward extension of the Palo Duro basin of the Texas Panhandle. The anticline is present as a pronounced feature in the Pennsylvanian, Mississippian, and Ordovician formations, but the highest point of the fold shifts about a mile to the northwest between the Palo Pinto Limestone (Pennsylvanian) and the Ellenburger Dolomite of the Lower Ordovician.

The *K.M.A.* field in Wichita County is truly a giant with a cumulative production at the end of 1965 of more than 145 million barrels of oil. During that year it was third in production in the North Texas district, with a yield of nearly $3\frac{1}{2}$ million barrels, just slightly less than that of a second-place Hull–Silk. The K.M.A. field covers the southwestern corner of Wichita County and extends over the line into northwestern Archer County. K.M.A. became a shallow oil pool in 1920, producing from Pennsylvanian sands ranging in depth from 450 to 1800 feet. The name of the field is derived from the initials of Kemp, Munger, and Allen, the principal landowners in the area. The development of the shallow pool confirmed the fact that the production, even though obtained from lenticular sand reservoirs, was coming from a structurally high area. This knowledge led to the location and drilling in 1930 of an exploratory well that reached a total depth of 5430 feet in Ellenburger Dolomite. The hole was dry in the Ellenburger but was plugged back up to about 3650 feet, where oil showings had been noted. It was completed at that level in a limestone-sandstone section of Strawn (Pennsylvanian) age, which became known as the "K.M.A. lime," as a 125 barrel oil well. The development of this deeper pay zone was extremely slow for several years, primarily because of the concurrent discovery and flush production of super-giant East Texas field. However, in late 1937 four widely scattered wildcats testing the K.M.A. lime came in as "impressive producers" (Oil and Gas Jour. Jan. 20, 1938, p. 50); this started a boom that lasted for several years, during which hundreds of producing wells were drilled.

Subsequently, in spite of previous dry holes, production was found in the deeper Ellenburger (Ordovician) Dolomite. The Ellenburger discovery well was drilled in the southwestern part of the K.M.A. field in 1940; it flowed oil from a saturated dolomite reservoir at a depth of 4350 feet (Autry, 1940). About 300 producing wells were drilled into this reservoir during the next 4 years, thereby maintaining the K.M.A. annual output at a high level.

However, the major reservoir in cumulative production is still the K.M.A. lime. Over most of the K.M.A. field it ranges from 200 to 250 feet in thickness and contains at least six separate pay zones, of which the two most important are a sandy oölitic limestone in the upper part of the reservoir complex and a relatively pure sand in the lower part. The sandstone is the most prolific of the pays, primarily because of its 24 per cent porosity. The oölitic lime ranges from 8 to 13 per cent porosity (Tucker, 1938, p. 43). The reservoir zones within the K.M.A. lime are erratic. "Each new location is a semi-wildcat with respect to the sand zone within the lime, there being, at present, no definite method of forecasting just where in the pay zone the sand will occur—if it occurs at all" (Dally, 1938, p. 45).

The K.M.A. field lies on a spur (the Electra arch) of the Red River uplift. This spur plunges to the south, and K.M.A. lies down the plunge in the western end of the Fort Worth syncline, which separates the east-west–trending Red River uplift from the northward plunging Bend arch.

The trap at K.M.A. is an anticlinal complex containing various nodes, with intervening saddles, and covering a considerable area. Accumulation in the Strawn reservoirs is anticlinal, but the erratic behavior of the individual producing zones plays an important role in the distribution of oil within the traps. The situation is even more stringent in the Ellenburger reservoir, where trapping is likewise anticlinal but "production is highly erratic, a prolific well being offset by a poor one or a dry hole" (Bradfield, 1964, p. 120).

The discovery field for the North Texas district, *Petrolia* in northern Clay County, not only is an excellent example of anticlinal trapping but also illustrates the role that subseeps have played in the discovery of oil and gas. For some time before the advent of commercial production, farmers in the Petrolia vicinity had been troubled by the presence of oil in their water wells. Then in 1902 a well drilled for water found no water but only oil at a depth of about 150 feet. Within a few years several hundred wells were producing oil at depths of between 150 and 700 feet (Kendrick and McLaughlin, 1929, p. 543) from Permian red beds that are the surface rocks in this part of Texas. Deeper exploration led to the discovery of gas in quantity, along with oil, at depths of 1500 to 1750 feet. Helium is also present. These deeper sands constitute the major reservoir at Petrolia. They are in the lower part of the Cisco Formation, which is classified as Upper Pennsylvanian. The individual sand reservoirs are lenticular, but the lower Cisco zone in which they occur is continuous across the field.

The Petrolia anticline, a nearly circular dome with a maximum diameter of about $4\frac{1}{2}$ miles, is one of several similar oil-bearing structures that lie along the axis of the Red River arch. Each individual anticline contains Pennsylvanian and Permian sediments draped over a buried hill of Ordovician limestone that is also anticlinal and that rests on an arched surface of Precambrian rock.

The *Bowers* field (Brown, 1943) in northeastern Montague County overlies a buried Precambrian hill along the Muenster arch. Figure 29.9 is a contour map of the top of the granite basement. This hill is overlain by granite wash, which is absent over the tops of the "peaks" but approaches 200 feet in thickness on the flanks, and by middle Pennsylvanian sediments, which immediately overlie the granite wash. The lowest group, the Strawn, contains six separate oil-bearing sands in addition to the basal productive granite wash. The eighth and youngest reservoir is either latest Strawn or early Canyon in age. The sedimentary formations are draped over the Precambrian hill by differential compaction, producing a domal structure. Accumulation in the various reservoirs, all of which are discontinuous, has taken place as high in the Bowers dome as local permeability will permit. Although the granite wash varies most in thickness, it is the least erratic in permeability, and 44 per cent of the producing wells were completed in this reservoir.

Grayson County, the easternmost county in the North Texas district, is notable for its faulted anticlinal traps. The largest of these, the Sherman anticline (Bradfield, 1957d, Figure 12), is illustrated in Figure 29.10. It lies close to the center of Grayson County and extends northwest to southeast for a distance of about 12 miles. "With 2000 feet of structural closure, it ranks as one of the largest anticlinal features in this part of the country" (Bradfield, 1957d, p. 62). The higher part of the anticline is occupied by the *Sherman* oil field, which was fourth among the fields of the district in production in 1965 with a yield of nearly 3 million barrels. Oil is obtained from a series of reservoirs ranging from shallow sands in the Strawn Group down to the Simpson (Ordovician) Oil Creek sand, which yields gas and condensate from depths between 9500 and 10,500 feet. The leading oil reservoirs lie between these extremes in pre-Strawn (Lower Pennsylvanian) sands. Figure 29.10 is contoured on the Baker sand zone, which lies at a depth of about 7500 feet. This zone is as much as 180 feet thick, of which

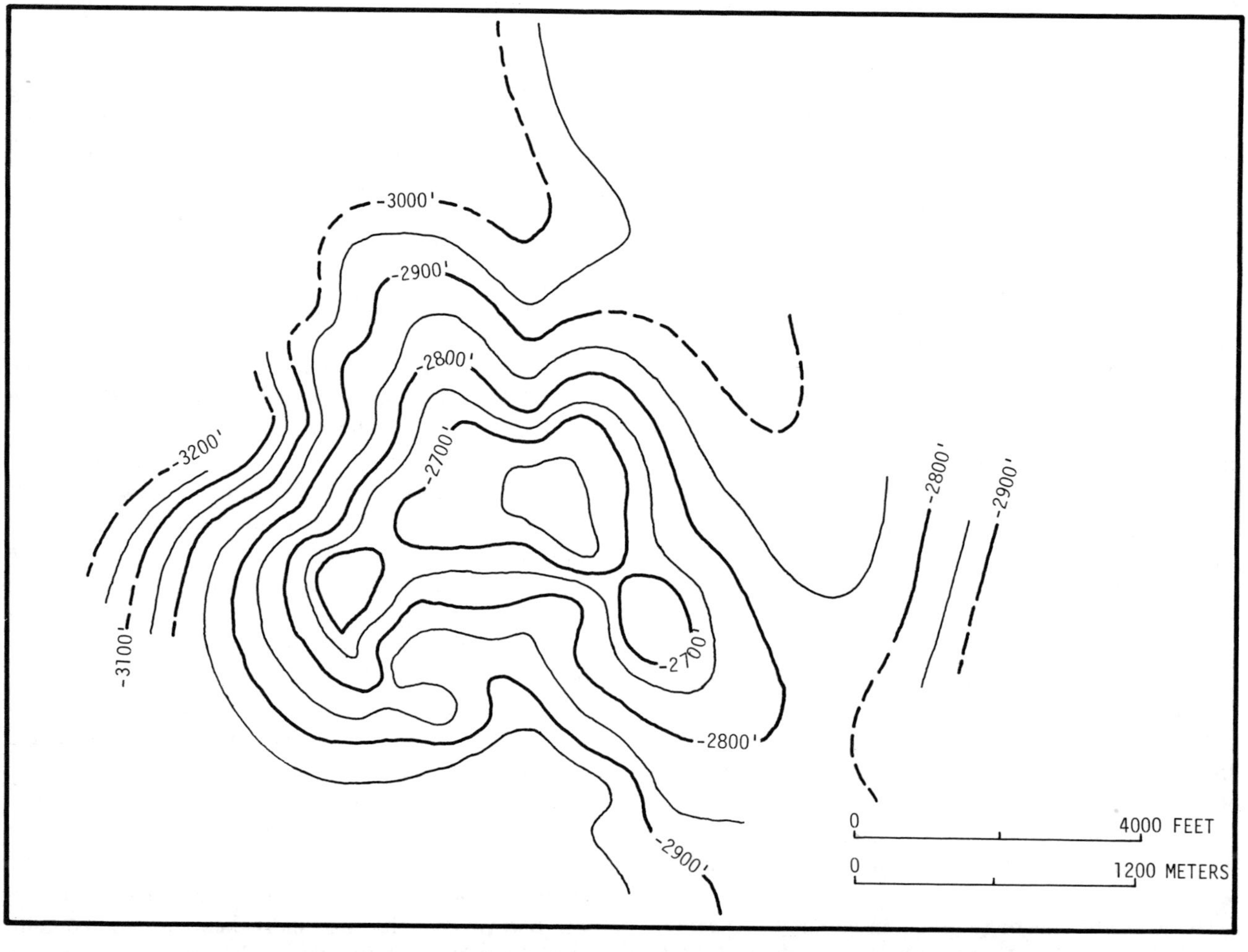

FIGURE 29.9 **Bowers field, Montague County, Texas. Structure map contoured on Precambrian granite surface. Interval, 500 feet.** ***Courtesy A. B. Brown (1943, Figure 9, p. 23) and the American Association of Petroleum Geologists.***

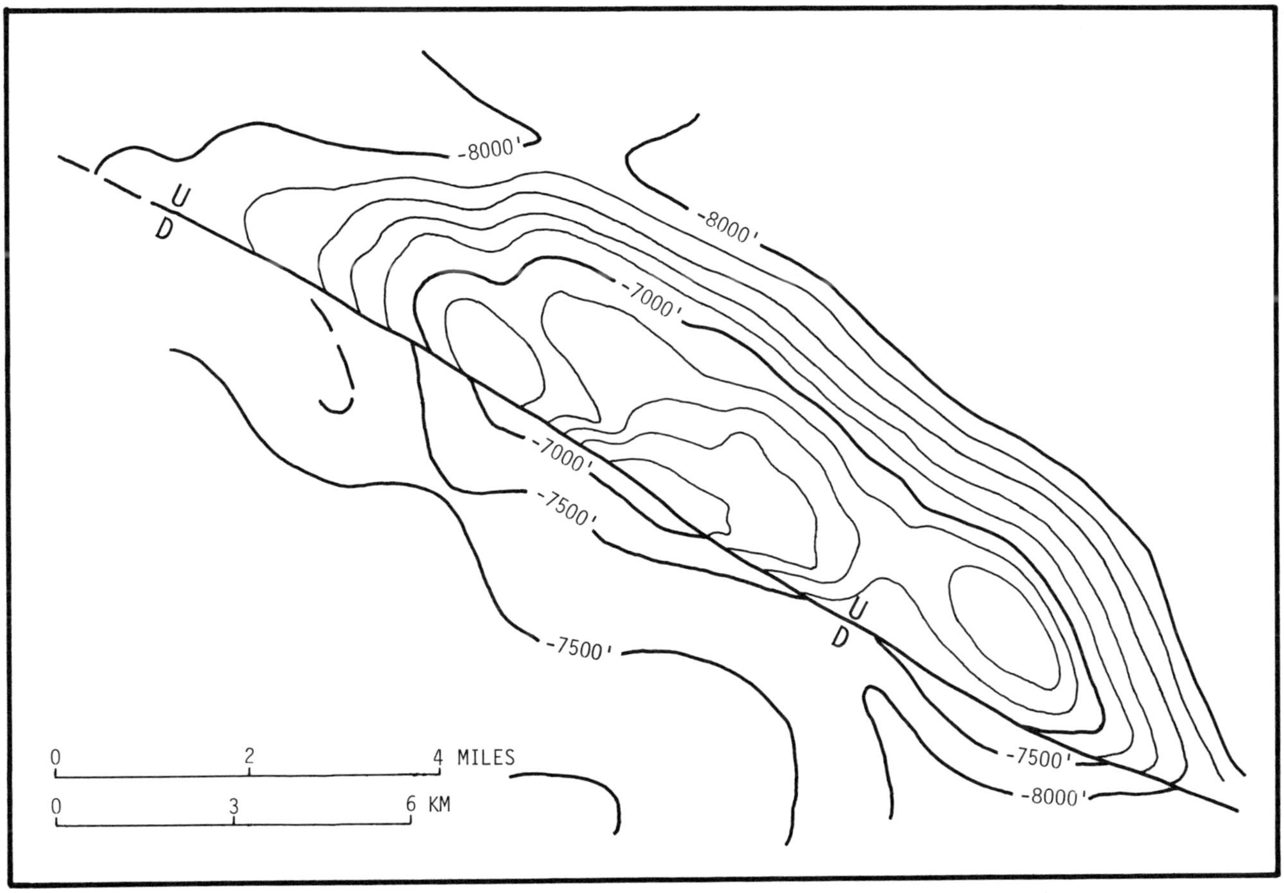

FIGURE 29.10 Sherman anticline, Grayson County, Texas, contoured on top of the Baker sand zone. Interval, 200 feet north of fault, 500 feet to south. Most of the production from this reservoir has been obtained above the −7500 foot contour. Node at eastern end dry because of sand tightness. ***Courtesy H. H. Bradfield (1957, Figure 12, p. 63) and the Dallas Geological and Geophysical Societies.***

about 120 feet is sand. However, the sand is fairly tight so that production has to be stimulated by sand fracturing. Two deeper Pennsylvanian sand reservoirs, the Davis and the Cordell, also tend to be tight because of shaliness.

The southwestern flank of the Sherman anticline is cut by an overthrust fault that has elevated the main part of the fold about 400 feet above the southwest or footwall side. This fault and the erratic permeability of most of the sandstone reservoirs limit the distribution of oil, but the trap is the Sherman anticline.

The *Big Mineral* field lies in east-central Grayson County, where it is partially inundated by the impounded waters of Lake Texoma (Garlick, 1954, p. 9). This is one of the larger fields in North Texas with a production during 1965 of more than 1,650,000 barrels. The trap is an anticline sliced by three faults that cut the east, southwest, and north flanks. The structural closure is in the neighborhood of 1000 to 1200 feet, and oil may occur throughout this vertical distance, depending on local permeability conditions. The producing reservoirs have been divided into seven zones; the upper four consist of from two to six or more individual sands (Bradfield, 1957b, p. 59). These reservoirs range in depth from 3375 to more than 10,000 feet. All except the deepest are Middle and Early Pennsylvanian in age; the deepest reservoir is the Oil Creek (Ordovician). It produces gas distillate.

Tucked into the Delaware Bend of the Red River in the extreme northwestern tip of Grayson County is the *Handy* field (Bradfield, 1957 a and c). The trapping here has been due to the impounding of oil beneath the footwall of a major overthrust fault (Figure 29.11). The formations in the thrust plate have been shifted upward and southwestward nearly 5000 feet. Earlier wells in this field lie in a row at the surface parallel to and immediately northeast of the major fault. They pass through the fault plane and intercept a reservoir sand above the oil-water interface and below the truncated terminus of the sandstone layer. The trap is closed to the northwest and

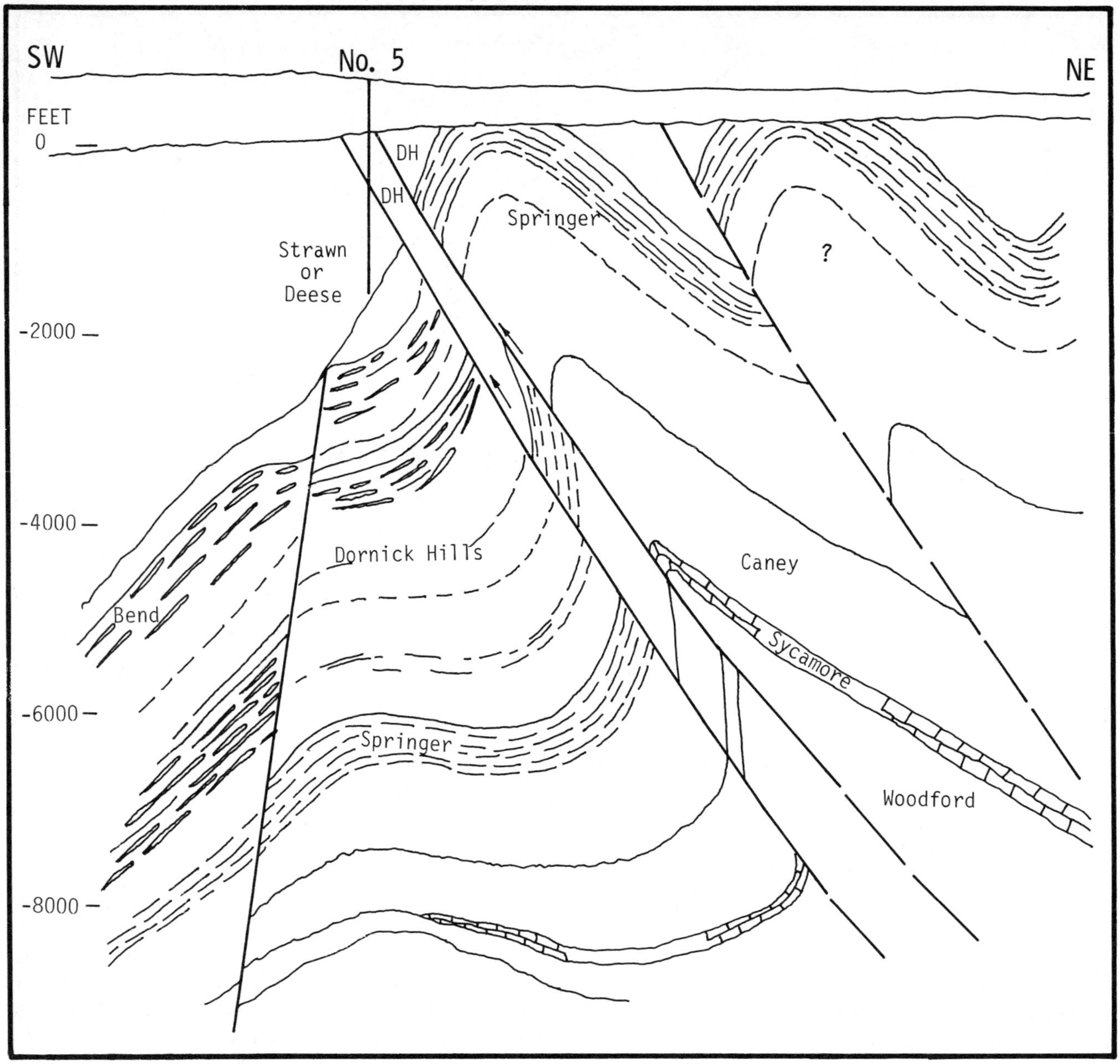

FIGURE 29.11 Handy field, Grayson County, Texas. This illustration is a highly interpretative and generalized southwest to northeast cross section through the fault trap. Well No. 5 is bottomed in, and produces from, the Handy sand, which is truncated by thrust fault. Since the preparation of this cross section numerous producing wells have been directionally drilled from southwest of No. 5 in a northeasterly direction far beneath the overthrust sheet, finding many pay zones in the lower Deese and Dornick Hills. *Courtesy H. H. Bradfield (1957c) and the Dallas Geological and Geophysical Societies.*

southeast by the arching of the structure contours into the fault plane, caused by the fault slicing across an anticlinal nose. Minor cross faulting, and lenticularity of sands, are also important in closing traps. In more recent years "production has been extended southeastward across the Red River (Lake Texoma) 4 or 5 miles where some wells have cut more than one fault" (H. H. Bradfield, letter dated 1 September, 1968).

The reservoir rock in the well shown in Figure 29.11 is a sand in the lower part of the Strawn (or Deese) Group. The major producing sand near the base of the Strawn is locally referred to as the Handy. Other nearby Strawn sands, as well as sands in the deeper Dornick Hills Formation, may also be reservoirs.

The *Seymour* group of small fields (Murphy et al., 1941; Haeberle, 1956) is in east-central Baylor County.

Within a 12 square mile rectangle (4 miles north by 3 east) are five separate oil pools illustrating three types of trap and three types of reservoir. The traps illustrated are limestone reef, anticline, and sand pinchout. The reservoir rocks (Figure 29.12) are a reef limestone of Canyon (Pennsylvanian) age, known as Seymour reef; a sand ("Seymour") below the Canyon in the Strawn Group of the Lower Pennsylvanian; and the Caddo Limestone, which lies still lower in the Strawn about 200 feet above the top of the Mississippian.

A reef field was discovered first, in 1938, at a depth of 2605 feet. The accompanying map (Figure 29.13) shows the structure contours of the Seymour area, based on the top of the Canyon reef. The discovery pool is the "Seymour reef field" in the southwestern part of the map area. There is no surface expression of this anticline. Subsequent seismograph surveys led to the discovery of the Caddo pool down the flank of the Maybelle reef and the Maybelle reef field itself. A structure contour map on top of the Caddo Limestone indicates that the small Caddo pool is the result of trapping at the top of a nearly circular anticline that does not show on the structure contour map of the younger Canyon reef. The Caddo reservoir has pinpoint porosity and probably fracture porosity as well. The Canyon reefs consist of an upper 575 feet of light-colored crystalline and fossiliferous limestone, of which only the top 15 feet is productive. Away from the reef apices this upper Canyon limestone disappears abruptly in all directions.

The Seymour sand, which stratigraphically lies between the Canyon reef and the Caddo Limestone, was first found to be productive in what became known as the Seymour Strawn field in 1951. Previously, various borings in the area drilled to test the Caddo Limestone had found good showings of oil in this sand. The East Seymour Strawn field was discovered 2 years later. In both of these sand accumulations, the trapping conditions are the same. The sandstone reservoir wedges out along a strand line that forms the southeastern edge of these two pools. Both pools are on anticlinal noses that plunge to the northwest. In both cases the oil-water interface is a contour line approximately −3420 feet in elevation. Oil moving up the regional dip in the sandstone has been trapped at the highest point on these two noses by pinchout of the reservoir sand.

The *Bryson* field (Hiestand, 1941), which lies near the center of the western boundary of Jack County, is a varying permeability trap. Jack County is at the northern end of the northward plunging Bend arch, where the regional dip is northerly into the Fort Worth syncline. The reservoirs are in the "Bryson sand zone," which is near the middle of the Strawn Series. Production depths range from 2900 to 3200 feet. Within the sand zone there may be one or more sandstone members that carry hydrocarbons and that have, in a producing well, an aggregate footage ranging from 35 to 60 feet. Where multiple sands are present, one sand may produce gas and the others oil.

Trapping is due to the combination of the updip wedgout of the sands in the Bryson zone and a structural nosing northward down the regional dip. Closure laterally is assisted also by updip convexity in the pinchout line where the sand grades into siltstone and shale. Permeability also varies downdip; many dry holes have been drilled within the area of the trap with the "dryness" due either to local impermeability in the Bryson sand zone or to the presence of an abnormally high oil-water interface.

Archer County contains numerous lenticular sand pools at depths ranging from 900 to 1700 feet (Thompson and Hubbard, 1929, p. 421). These sand reservoirs lie in the lower half of the Cisco Formation of Late Pennsylvanian age. Archer County is in an area affected by the differential movements that created both the Bend arch and the Red River uplift. Most of the producing areas lie on the higher parts of the Bend arch but show a northwesterly trend paralleling the Red River arch axis. The oil occurs in lenticular sand bodies that have been deposited on, and are confined to, the axes of these relatively low elongate anticlines.

Future. There is no indication that the days of discovery in North Texas are over. It is true that exploration well density is high across the uplift areas, where the major fields have been found to date, but most of this exploration has been relatively shallow in terms of modern drilling depths. Off the uplift areas, the drill hole density is inadequate for thorough exploration of potential Permian, Pennsylvanian, and Mississippian reservoirs. Throughout the district, both in structurally high and especially in structurally low areas, many untested possibilities exist for production in the older rocks. There must be many more beveled Simpson Sandstone prospects. The erratic permeability of the Ellenburger makes multiple tests within a single trap essential, as K.M.A. proved. To date, the pre-Ellenburger section in favorable areas has hardly been examined.

Recent successful exploration in the Hardeman basin demonstrates the potential of off-uplift stratigraphic sections. Thirty million barrels of oil has been produced in this basin from over six superimposed reservoirs ranging in age from Pennsylvanian to Cambro–Ordovician (Oil and Gas Jour., Feb. 5, 1968, p. 193). The sand wedgout and limestone reef traps, which are hard to find, are probably present in much greater abundance than the discoveries to date indicate.

Prospects are especially good in the Marietta syncline for condensate and gas accumulation in deep Ordovician Oil Creek and Ellenburger reservoirs (Bakker, 1968).

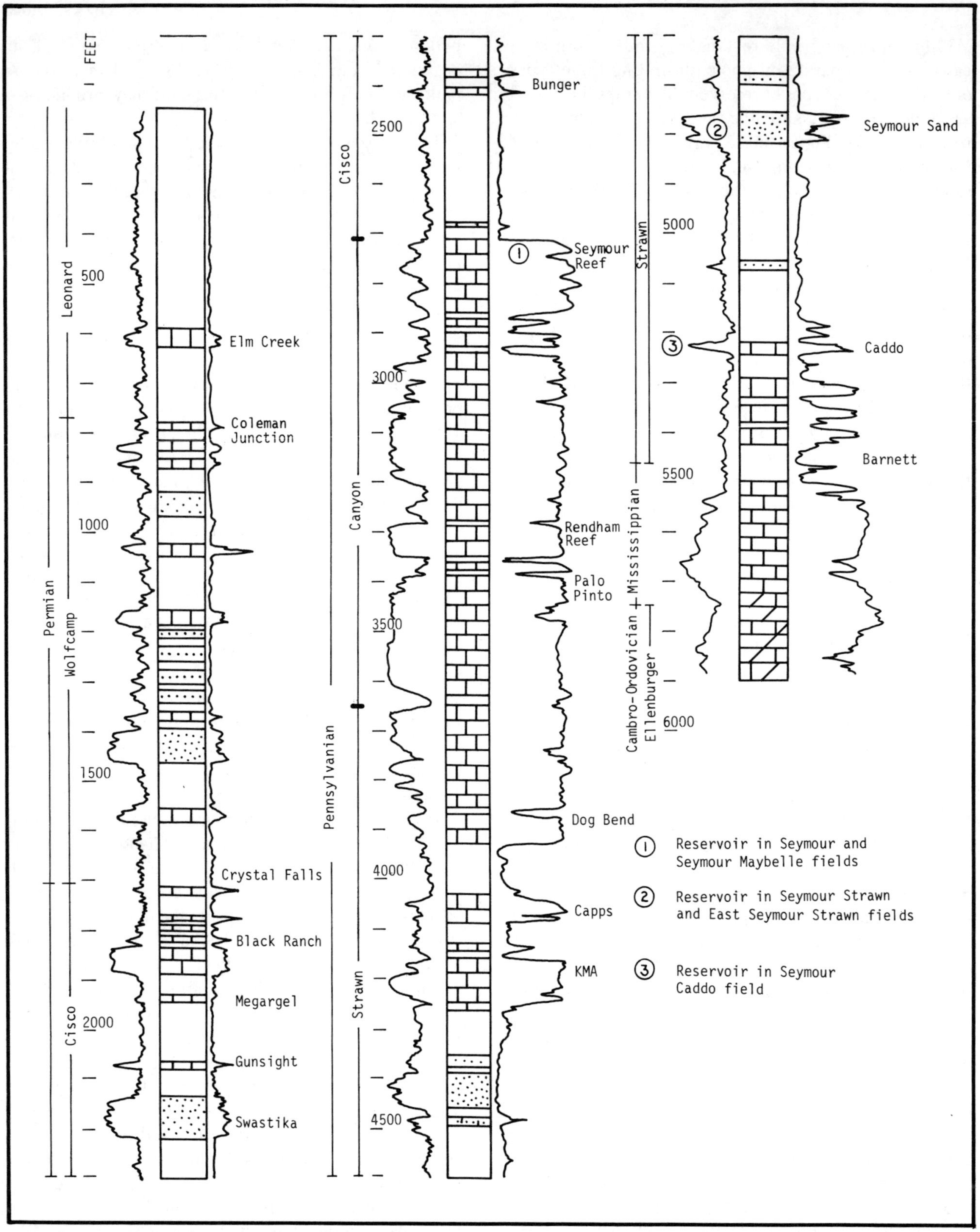

Figure 29.12 Seymour fields, Baylor County. Columnar section, showing reservoirs. *Courtesy Fred R. Haeberle (1956, Figure 2, p. 881) and the American Association of Petroleum Geologists.*

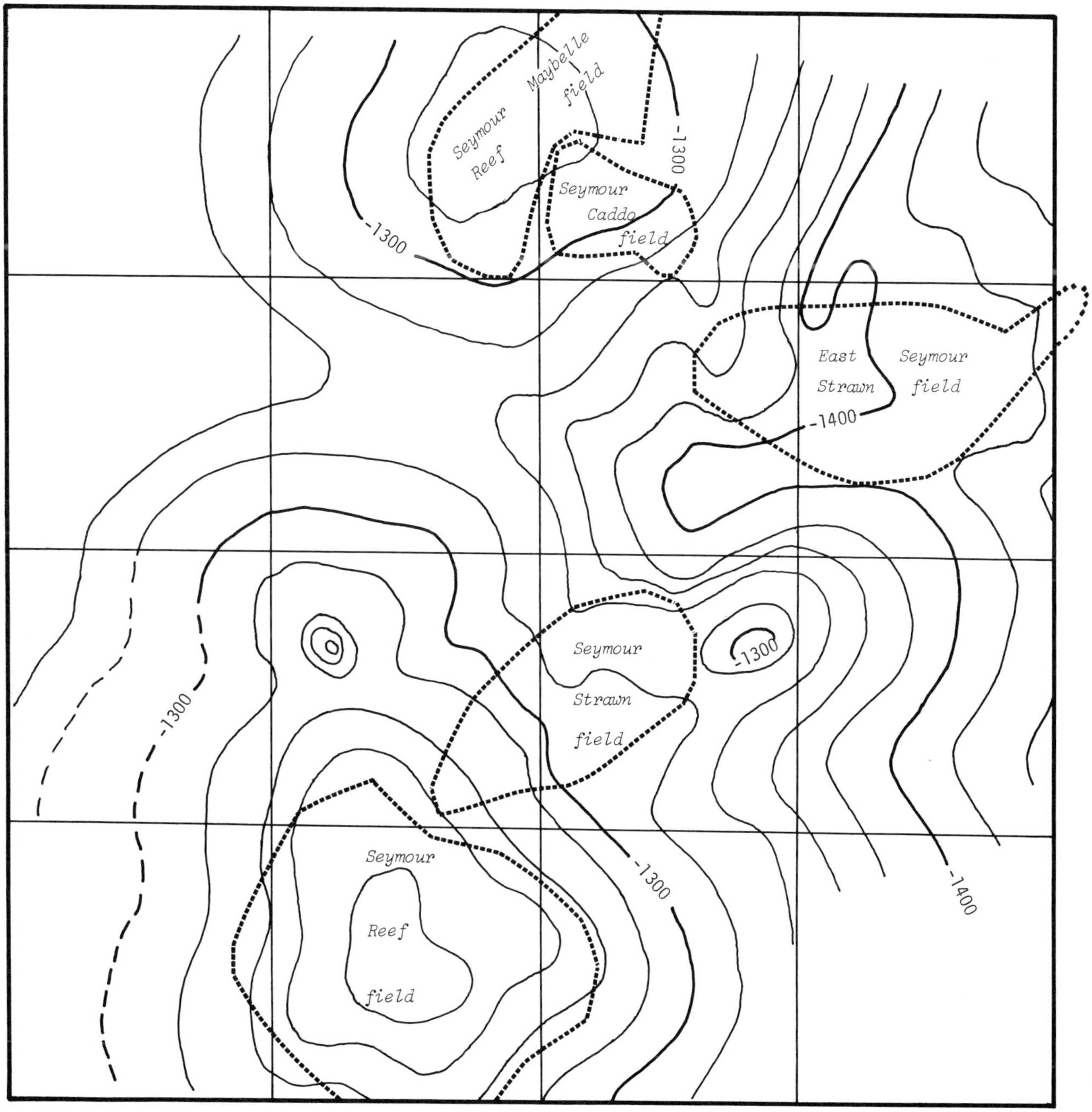

Figure 29.13 Seymour fields, Baylor County. Contour map on top of Canyon reef. Interval, 20 feet. Area 4 miles (6.44 km) square. Dashed lines enclose productive areas and field names. *Courtesy Fred R. Haeberle (1956, Figure 3, p. 883) and the American Association of Petroleum Geologists.*

REFERENCES CONSULTED IN PREPARING CHAPTER 29

Autry, Vernon E., 1940.

Bakker, D., 1968.

Blanchard, K. S., Denman, Orval, and Knight, A. S., 1968.

Bradfield, H. H., 1957a, 1957b, 1957c, 1957d, 1959, 1964.

Bradfield, H. H., and Seale, R. I., 1957.

Brown, A. B., 1943.

Dallas Geological Society and Dallas Geophysical Society, 1957.

Dally, Claude F., 1938.

Ellison, Samuel P., Jr., 1960.

Flawn, Peter T., 1956.

Freeman, John C., 1962, 1964.

Fuqua, H. B., and Thompson, B. E., 1929.

Garlick, J. M., 1954.

Haeberle, Fred R., 1956.
Hendricks, Leo, 1964.
Hiestand, T. C., 1941.
Hilseweck, William J., 1941.
Kendrick, Frank E., and McLaughlin, H. C., 1929.
Makins, Robert L., 1968.
Murphy, James K., Purcell, Paul E. M., and Barton, H. E., 1941.
North Texas Geological Society, 1962.
Reutinger, C. A., 1968.
Rister, Carl Coke, 1949.
Sellards, E. H., and Baker, C. L., 1934.
Shelton, Stanley O., 1964.
Thompson, Edwin I., 1941.
Thompson, W. C., and Hubbard, W. E., 1929.
Tucker, Mitchell, 1938.

Chapter

30

TEXAS PANHANDLE

The Panhandle district includes the northern twenty-six counties in northwesternmost Texas and constitutes District 10 of the Texas Railroad Commission. The Panhandle district is bounded on the north and east by Oklahoma, at the southeast corner by the North Texas district, on the south by the West Texas district, and on the west by New Mexico (Figure 30.1).

The distribution of oil and gas fields in this district is also shown in Figure 30.1. The Texas Panhandle is one of the largest gas-producing areas in the world. This is the major reason that Texas ranks first in natural gas yield in the United States, with 40 per cent of the national total. The Panhandle district produces about 4 per cent of the annual Texas output of oil, the least of the seven districts. However, the annual production of 35 to 40 million barrels puts this district ahead of all but eight states outside of Texas.

Tectonics. The tectonics (Eddleman, 1961, p. 65–67) of the Panhandle (Figure 30.2) is dominated by the Amarillo uplift, which crosses the central Panhandle from east to west except for the westernmost tier of counties (Figure 30.3). The uplift is cored with Precambrian crystalline rocks that in earlier geologic time were sculptured into the Amarillo Mountains. Although completely buried in Texas, this topographic feature rises above the surface in southwestern Oklahoma, where it is known as the Wichita Mountains (see Chapter 18). To the west of the western end of the Amarillo uplift across a structural saddle is the Bravo dome, or Oldham nose (Nicholson, 1960, p. 53), which crosses into the Panhandle from New Mexico. This anticlinal feature is *en échelon* with the Amarillo uplift axis rather than a continuation thereof.

To the north of the Amarillo uplift is the western end of the great Anadarko basin, which has its nadir to the east in western Oklahoma. The Anadarko basin terminates westward against the northward trending Cimarron uplift. Across this anticline, in the northwestern corner of the Panhandle, is the relatively small Dalhart basin.

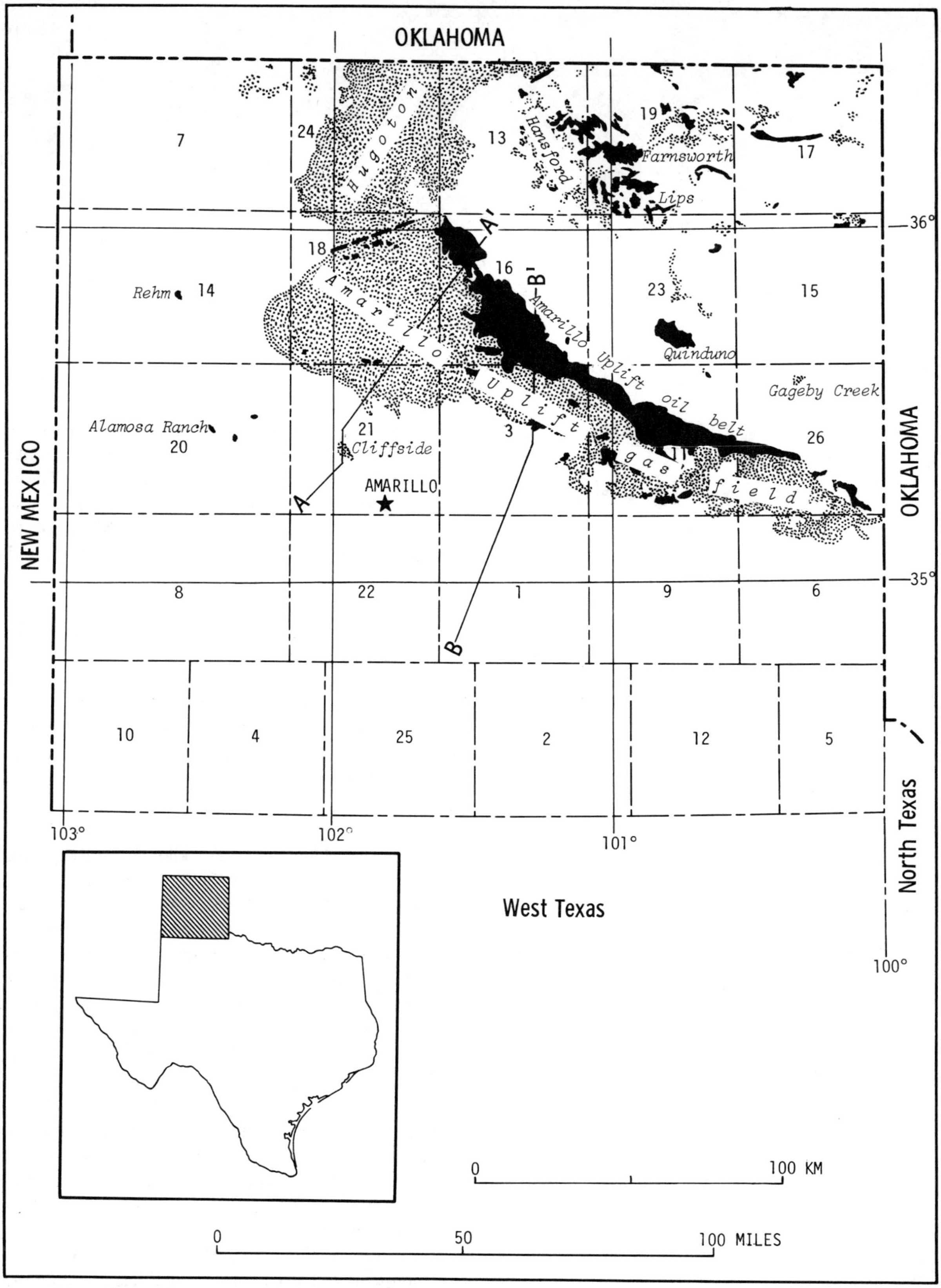

FIGURE 30.1 Texas Panhandle. Index map with oil and gas fields (Vlissides, 1964). Named fields are discussed in text. Map also shows location of cross sections *A–A′* (Figure 30.6) and *B–B′* (Figure 30.7). County Key:

1	Armstrong	5	Childress	9	Donley	13	Hansford	17	Lipscomb	21	Potter	25	Swisher
2	Briscoe	6	Collingsworth	10	Farmer	14	Hartley	18	Moore	22	Randall	26	Wheeler
3	Carson	7	Dallam	11	Gray	15	Hemphill	19	Ochiltree	23	Roberts		
4	Castro	8	Deaf Smith	12	Hall	16	Hutchinson	20	Oldham	24	Sherman		

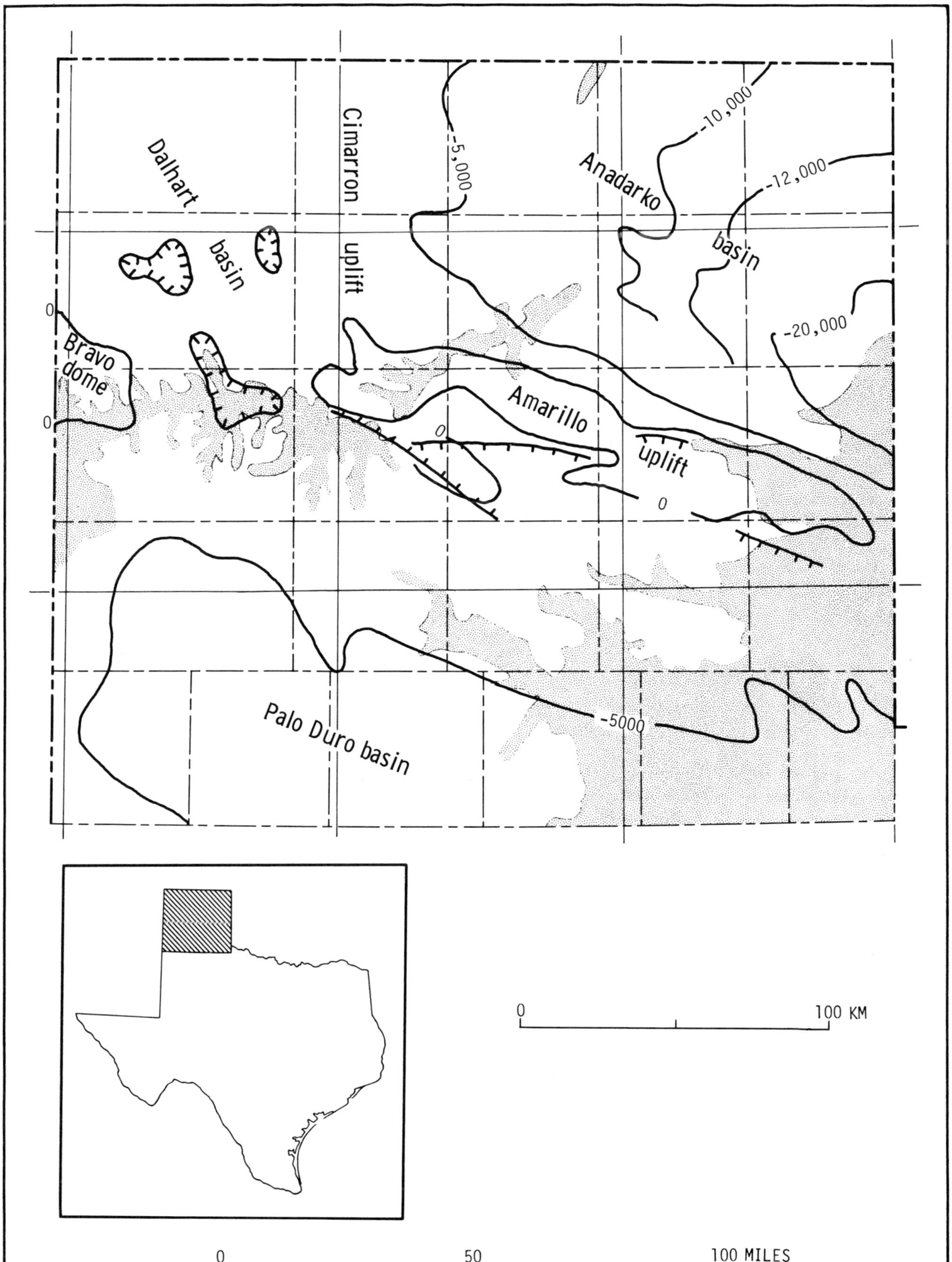

FIGURE 30.2 Texas Panhandle. Tectonic and geologic map. Shows major faults and structure contours (interval variable). Entire district covered by Tertiary veneer except where stippled (Permian and Triassic outcrops).

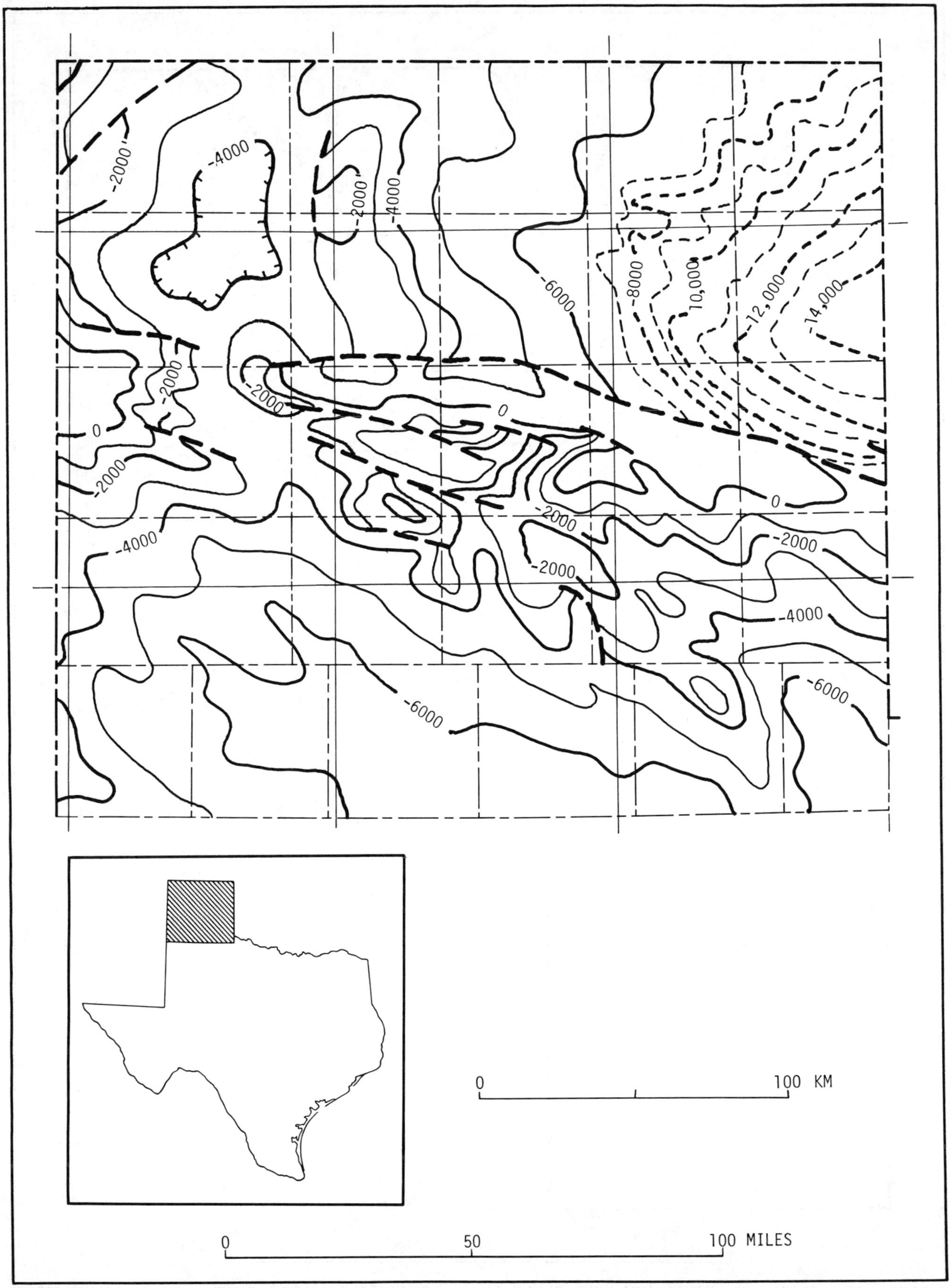

Figure 30.3 Texas Panhandle. Structure contour map on top of Precambrian surface. Interval, 1000 feet. *Courtesy John H. Nicholson (1960, Figure 43, p. 54) and the Bureau of Economic Geology.*

The southern Panhandle is downwarped into the Palo Duro basin, the northernmost subdivision of the Permian basin. The eastern extension of the Palo Duro basin, which has also been referred to as the Plainville basin (Jones, 1951), in north Texas and southwestern Oklahoma is called the Hardeman basin. Off the map to the south in West Texas is the east-west Matador arch, which separates the Palo Duro from the Midland basin.

Almost all of the oil and gas produced so far in the Texas Panhandle has been obtained from sediments either on the Amarillo uplift or in the Anadarko basin.

Stratigraphy. The geologic map (Figure 30.2) of the Texas Panhandle shows that the greater part of the district is veneered by the continental Tertiary clastic and caliche type of sediments characteristic of the high plains. The largest area where the Tertiary cover has been stripped off by erosion is the southeastern part of the district, where erosion in the drainage basin of the Red River headwaters has exposed the underlying Permian bedrock formations. Permian rocks are also exposed in the valley of the Canadian River, especially in the central part of the district.

In addition to Permian outcrops, the Dockum (Triassic) Group of variegated sands and shales reaches the surface in places along the eroded edge of the Tertiary veneer, such as in the upper valley of the Canadian River and on both sides of Palo Duro Canyon of the Prairie Dog fork of the Red River in the south-central part of the Panhandle district.

A subcrop map below the Tertiary veneer would show that most of the Panhandle district contains Permian rocks belonging to the Whitehorse Group of Late Permian age. The Triassic Dockum sediments overlie the Permian in the western part of the district. These formations are shown in the stratigraphic column (Figure 30.4). Not appearing in the column are Cretaceous and Jurassic rocks, which underlie the Tertiary in a few small areas in Dallam County, the northwesternmost county in the Panhandle district (Cunningham, 1961, p. 58).

Figure 30.4 shows the stratigraphy in the Panhandle basins (Nicholson, 1960, p. 58). It will be noted that in the largest basin (western Anadarko) the stratigraphic column continues downward from the subcropping Triassic and Upper Permian to Precambrian, with very few gaps in the record (Cunningham, 1961). The reservoirs first appear downward in the Lower Permian (Wolfcamp Series); they are principally carbonate rocks. The thick Pennsylvanian section below contains the largest number of the reservoirs so far developed in the Anadarko basin. Most numerous are sands in the Virgil and Missouri of the Upper Pennsylvanian and in the Lower Pennsylvanian Morrow. However, the intervening Des Moines and Atoka Series also contain reservoir sands.

So far, very little production has been obtained from pre-Pennsylvanian sediments. Some oil has been found in Upper Mississippian (Chester) and Lower Ordovician (Ellenburger) carbonate rock reservoirs. The Pennsylvanian subcrop map (Figure 30.5) shows the distribution of these older rocks beneath the basal Pennsylvanian unconformity. They are dragged up and truncated around the edge of the Precambrian-cored Amarillo uplift with the result that Pennsylvanian rocks can be underlain by anything from Mississippian to Precambrian. Mississippian rocks are present in the larger basins; the older rocks subcrop in relatively narrow belts around the basin peripheries. Therefore, to obtain production in older Paleozoic reservoirs within reasonable depths, it is necessary to probe the basin borders for trapped hydrocarbons, for in the basins these rocks dip downward to expensive depths.

The reservoirs in the relatively shallow accumulations over the top and flanks of the Amarillo uplift are not delineated in the stratigraphic column (Figure 30.4). They consist dominantly of arkosic granite wash of Virgil age (Late Pennsylvanian) and overlying Lower Permian carbonate rocks. These reservoirs are described subsequently in the discussion of the Amarillo uplift petroleum province. In Figure 30.4 they fall in the part of the column extending upward from the Pennsylvanian Virgil through the Permian Wolfcamp and Wichita Groups to the Red Cave at the base of the Clear Fork.

Fields and Traps. The first discovery in the Texas Panhandle was a gas well completed in 1918 in northern Potter County (Totten, 1956, 1961). This wildcat was located on a surface anticline mapped by geologist C. N. Gould. Further drilling confirmed the presence of a great gas field crossing the central Panhandle. Although the first oil well was drilled in 1921, the boom did not start until 1925, when further discoveries indicated the presence of a large and prolific oil field along the northern edge of the gas field. Meantime the Hugoton gas field, discovered in Kansas in 1922, spread southward across the Oklahoma panhandle and linked up with the western end of the Amarillo uplift gas field. These fields were the sole sources of hydrocarbons in the Texas Panhandle district from 1918 until 1949, when the Lips gas and oil field (Figure 30.1) in Roberts County in the western end of the Anadarko basin was discovered. Subsequently many additional fields have been found in the Panhandle sector of the Anadarko basin. Included among these is the *Gageby Creek* gas field in Wheeler County, in which one of the biggest gas wells in the world was drilled in 1966. This well had an estimated open flow potential of 1.74 billion cubic feet of gas daily from Hunton Dolomite (Siluro-Devonian) and Simpson Sandstone (Ordovician) reservoirs at depths of 15,000 to 16,000 feet.

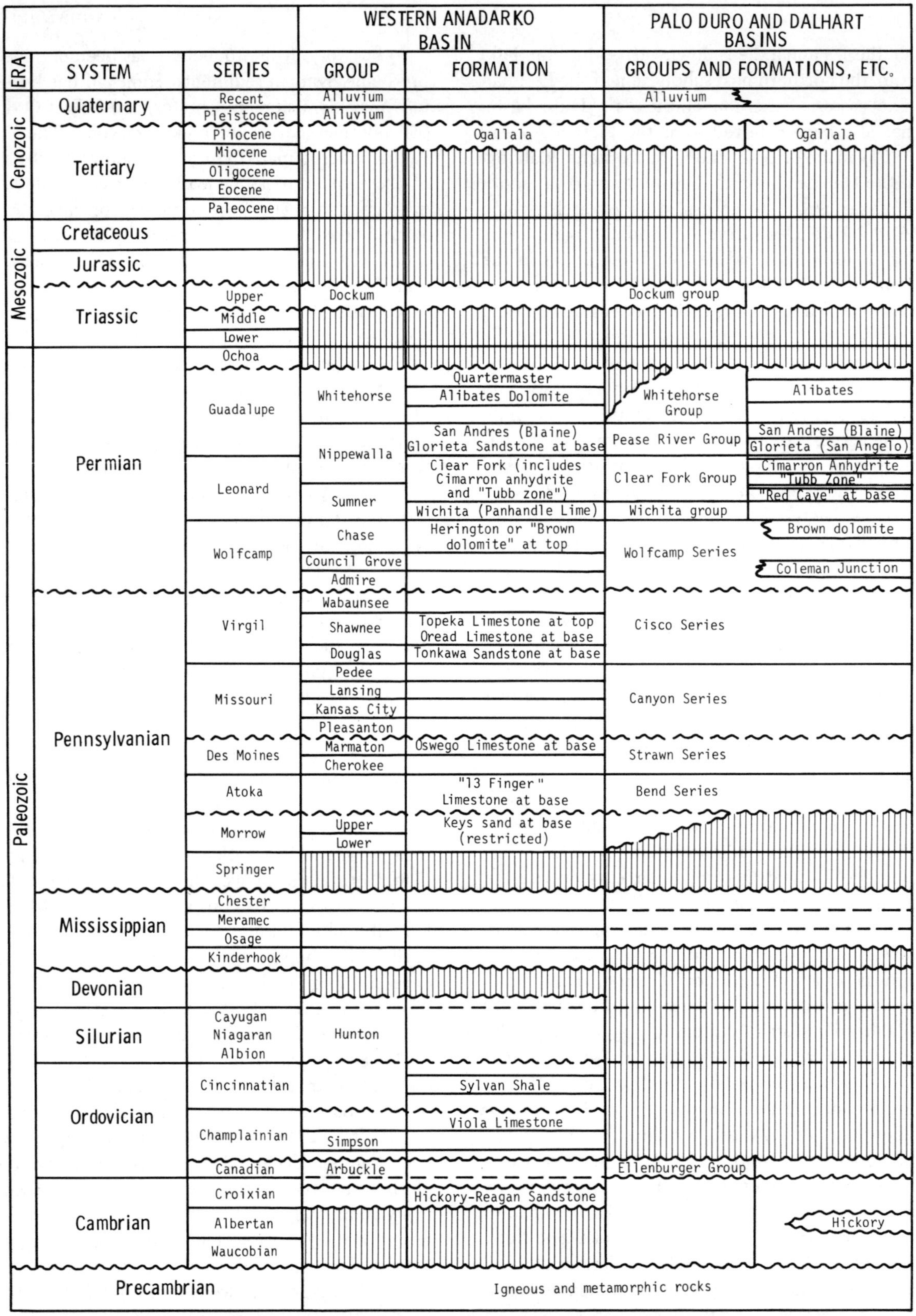

Figure 30.4 Stratigraphic column in Texas Panhandle basins. *Courtesy John H. Nicholson (1960, Figure 45, p. 58) and the Bureau of Economic Geology.* Note: many now consider the Springer to be Mississippian in age.

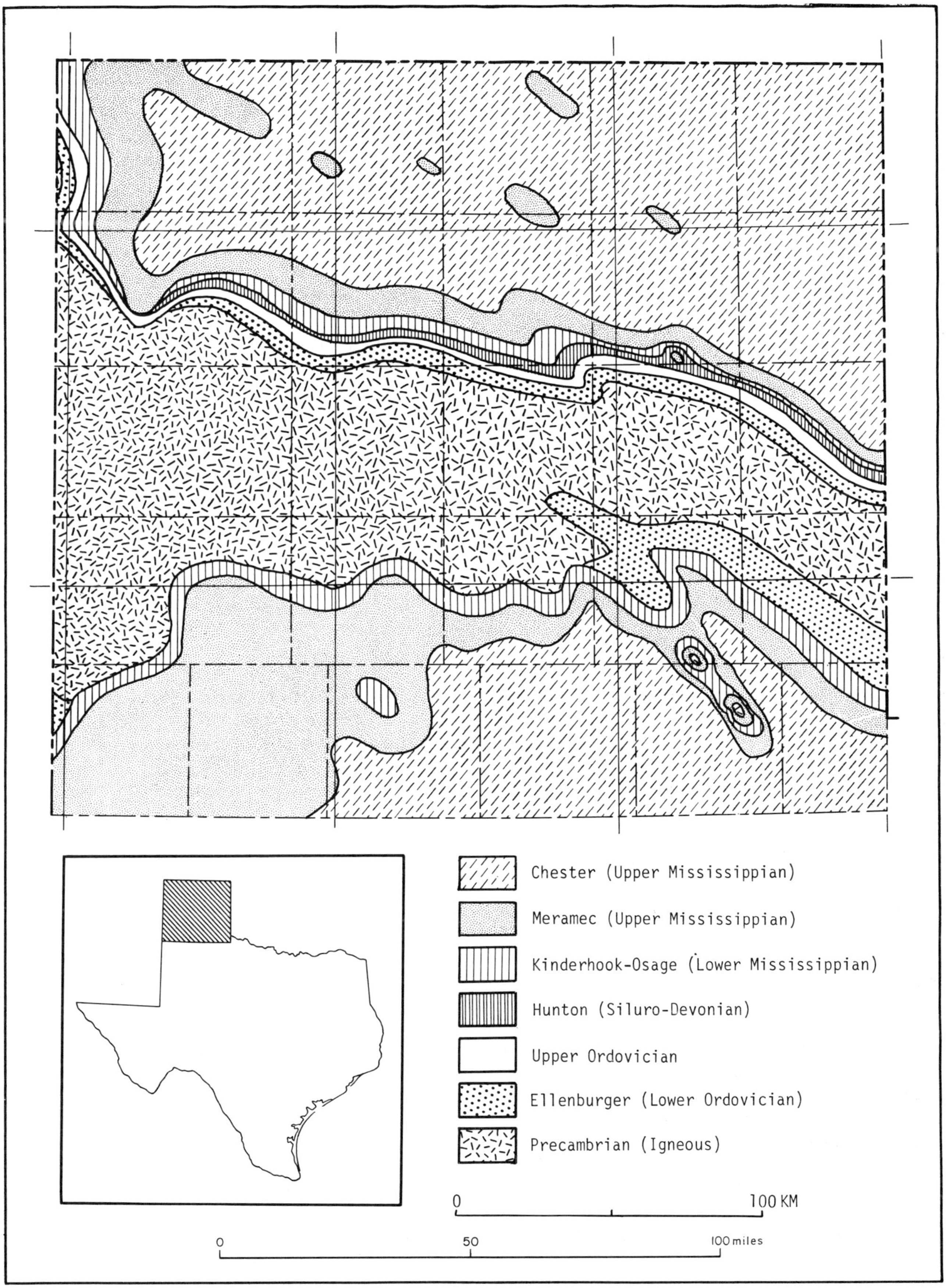

FIGURE 30.5 Texas Panhandle. Subcrop map beneath the Pennsylvanian formations. *Courtesy John H. Nicholson (1960, Figure 46, p. 59) and the Bureau of Economic Geology.*

The trap is "an anticlinal fold of a horst which was formed by the Wichita orogeny during the early Pennsylvanian" (Dietz, 1967).

The first field in the Dalhart basin, the Rehm oil field in Hartley County, was discovered in 1954. First oil production in the Palo Duro basin was obtained at Alamosa Ranch in Oldham County in 1957.

The Amarillo uplift petroleum province includes the hydrocarbon accumulations that occur over the axis and flanks of this mega-anticline. The gas-producing area has been referred to since 1918 as the "Panhandle gas field," and the oil belt on the north flank as the "Panhandle oil field." However, the many discoveries of gas and oil in recent years in the Anadarko basin and elsewhere in the Texas Panhandle district render this terminology obsolete. Rogatz (1961) applied the term "shallow oil and gas fields" to distinguish the older producing areas from the new. It is true that production depths average around 3000 feet over the uplifts, as against approximately 8000 feet in the basins. However, a depth distinction presents a problem as more and more shallow production is discovered in the basins. Already gas has been found at depths as little as 2600 feet in the Anadarko basin (Totten and Horn, 1961).

In order to avoid confusion in the following discussions the older, shallow, gas-producing area will be referred to as the "Amarillo uplift gas field," and the older oil area as the "Amarillo uplift oil belt." For the Texas part of the Hugoton gas district the present term, "Hugoton gas field," will be continued. The latter field merges production-wise with the northwestern end of the Amarillo uplift gas field, but geologically it overlies the eastern flank of the Cimarron uplift. Furthermore, trapping is due to updip diminution of permeability rather than to structural closure. Accumulation in the Hugoton field of Texas is the same as in Kansas; description will be found in Chapter 16.

The Amarillo uplift–Hugoton gas field complex has an overall length in the Texas Panhandle alone of 160 miles and a maximum width of 48 miles. The estimated ultimate recovery for this great area is 60 trillion cubic feet of gas, of which about one-third has been produced to date (Rogatz, 1961, p. 8). The *Amarillo uplift* gas field alone has produced over 70 per cent of the 25 trillion cubic foot output of the tristate Hugoton–Panhandle field (Mason, 1968, p. 1545). It is a virtually continuous producing area from one end to the other, and from one side to the other. Trapping is anticlinal; however, in this case the anticline is a giant one, the uplift itself. The only barren areas above the gas-water level occur where the drill failed to find adequate permeability in any one of several possible reservoirs, which are described on page 319.

The *Cliffside* gas field in west-central Potter occupies the Bush dome (Figure 30.6), which is offset to the south of the main Amarillo uplift (Cotner and Crum, 1935, p. 388) but nonetheless is a part thereof. This field is notable for the fact that the natural gas contains enough helium to make extraction worthwhile; after the helium has been separated at a government plant near Amarillo, the remaining gas is purchased by a gas pipeline company.

The *Amarillo uplift* oil belt on the north flank of the uplift is a giant in every respect. It is a little over 100 miles long and has a maximum width of 12 miles. By the end of 1966 this area had produced over 1,150,000,000 barrels of oil, and the remaining reserves were estimated at $\frac{1}{2}$ billion barrels. Nearly 23 million barrels of oil was produced during 1966, the forty-fifth year of productivity in the belt.

The Amarillo uplift oil belt is not a single field, as implied by the map (Figure 30.1), but is a series of closely spaced, separate oil accumulations. Most of the trapping has been anticlinal, each pool occupying a dome or other type of upfold. Some of the accumulations, however, are due to the lenticularity of the reservoir rock, especially where the reservoir is granite wash.

The stratigraphic column on the Amarillo uplift is understandably quite different from that shown for the basins (Figure 30.4). During much of the reservoir-depositing time span on the uplift, the tops of Precambrian granite mountains were above water and were contributing sediment to the surrounding submerged areas. Figure 30.5 is an areal geologic map below the unconformity at the base of the Pennsylvanian (Nicholson, 1960, p. 59). The pre-Pennsylvanian reservoirs that are productive in the Anadarko basin have been eroded off the Amarillo uplift. The reservoirs in the uplift oil and gas fields are in the Lower Permian and Upper Pennsylvanian formations, which lap upon and overlie the eroded granite core. These relationships are shown in cross sections *A–A′* (Figure 30.6) and *B–B′* (Figure 30.7); the location of the cross sections appears on the index map (Figure 30.1). Section *A–A′* (Cotner and Crum, 1935, p. 388) cuts across the western end of the uplift, including the helium-bearing Bush dome on the south. The granite core, topped by weathered granite and flanked by granite wash, can be seen in this cross section. The granite wash is the result of simultaneous erosion of the Amarillo Mountains and deposition on the nearby sea floor during Pennsylvanian time.

Cross section *B–B′* (Rogatz, 1935, p. 1095) cuts the uplift a little farther to the east. It passes through a relatively high peak on the Precambrian floor and shows dolomites as well as granite wash lapping upon the Precambrian. Figure 30.6 also shows zones A to E, used by Rogatz (1935, 1939, 1961) in his descriptions of the uplift ("shallow") oil and gas fields. These divisions correspond stratigraphically to the lower part of the

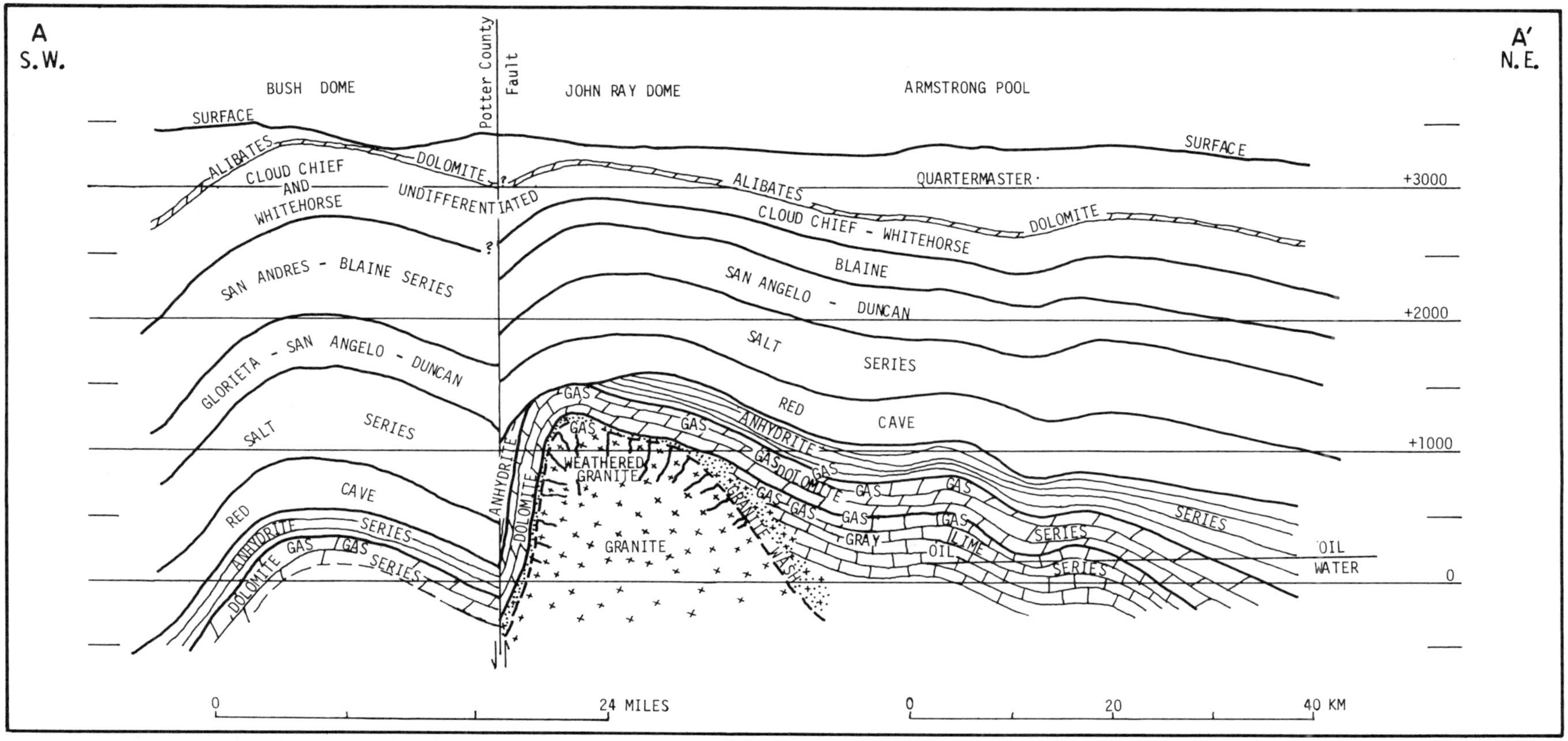

Figure 30.6 Southwest-northeast section across Amarillo uplift in Texas Panhandle. Location shown in Figure 30.1 (line A–A′). *After V. Cotner and H. E. Crum (1935, Figure 2, p. 388). Courtesy American Association of Petroleum Geologists.*

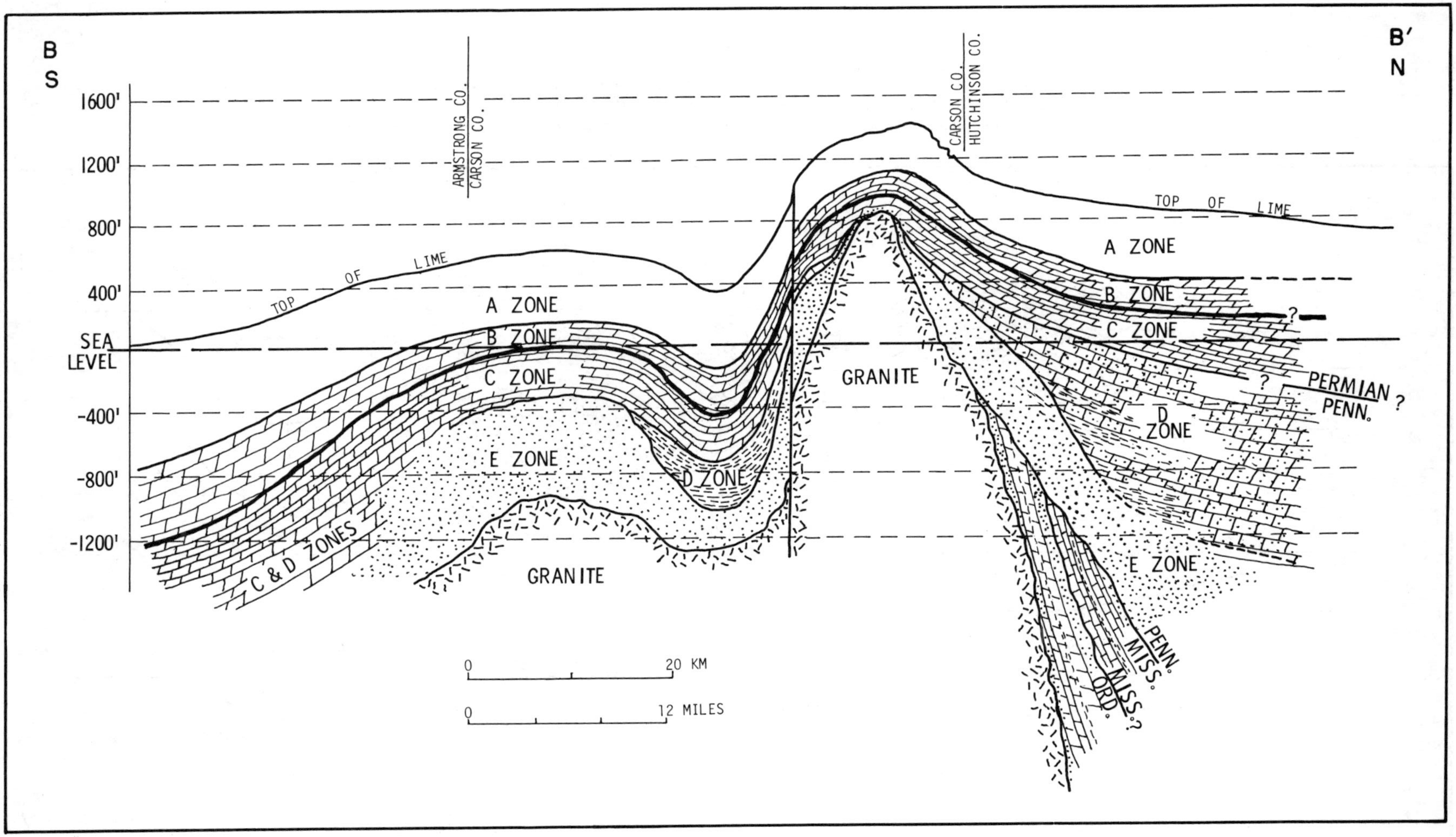

FIGURE 30.7 South-north section across Amarillo uplift to east of Figure 30.6 section. Location shown in Figure 30.1 (line B–B′). *Courtesy Henry Rogatz (1935, Figure 3, p. 1095) and the American Association of Petroleum Geologists.*

Leonard and the Wolfcamp of the Lower Permian and the Virgil Series in the Upper Pennsylvanian in the bordering basin areas.

Immediately above the highest zone delineated by Rogatz is the recently rediscovered Red Cave reservoir. The history of this development is given subsequently on pages 320–322. Red Cave production comes from sand lenses in the formation, which consists predominantly of red and green mudstone and shale. These lenses lie at different levels and vary greatly in areal extent. The most extensive sand zone is in the upper 100 feet of the Red Cave.

The highest Rogatz zone, A, is the seal rock for the underlying reservoirs. It is an evaporite section that not only covers the flanks but also crosses the granite ridge. The upper boundary is referred to as the "top of Panhandle Big Lime;" it is used as a datum for subsurface structure contour maps. Zone B, with an average thickness of 150 feet, is a dolomite-anhydrite section with inadequate porosity for a reservoir rock except in a few localities.

Zone C is the leading producer of oil and gas. It consists of a thick series of buff and white dolomite beds containing chert nodules. The upper part of this zone is a 50 to 75 foot section of oölitic dolomite with porosity as high as 15 percent. This upper C reservoir is the most extensive areally and the highest in total recovery per acre. The lower part of the zone ranks third among the reservoirs in recovery per acre; it is as much as 300 feet in thickness. Close to the uplift lower Zone C is an arkose derived from the granite. The grains are of sand bound together by dolomitic cement. Farther from the source the sand grains disappear, and the rock becomes entirely white dolomite. Lower C produces some oil and gas but is relatively unimportant as a reservoir. It has maximum thickness of 800 feet.

Zone E is the characteristic granite wash, a pink to red arkosic sand. It may be completely absent across the tops of buried peaks, and up to 4500 feet thick down the flanks. Red shales are interfingered with the wash, and calcitic cement may be present. The permeability is variable, depending on the amount of red shale or calcite cement. However, porosity is very high locally, and as a result this reservoir ranks second in the recovery per acre of oil.

Beneath the granite wash is the granite itself. Some gas and oil are produced from the crevices in the granite where accumulation has taken place in the topographically higher parts of the buried mountain range.

The discovery of the *Lips* field (Brashear, 1961) in Roberts and Ochiltree Counties in 1949 opened up the Texas sector of the Anadarko basin for development. This field was also the first to produce from pre-Pennsylvania reservoirs (other than creviced Precambrian granite) in the Texas Panhandle. The initial production was oil from fractured St. Louis (Mississippian) Limestone; subsequently the discovery well was plugged back and completed as a gasser in the Douglas sand of Late Pennsylvanian age. However, the principal reservoir of both oil and gas in the Lips field to date has been a sand in the upper Morrow (Lower Pennsylvanian) occurring at depths of 8000 to 9100 feet.

The structure at Lips is a plunging anticline that crosses the Ochiltree–Roberts county line with a southwesterly trend. The discovery well was drilled in a node on the axis of this fold, and trapping is anticlinal in the reservoirs that cross the node. Accumulation in the Morrow sand has been on the flanks of the anticline with trapping brought about by updip thinning of the reservoir.

The *Quinduno* oil and gas field (Little, 1961) in southern Roberts County on the southern flank of the Anadarko basin was discovered in 1952. The principal reservoir is a porous dolomite in the lower Albany of the upper Wolfcamp (Lower Permian). Oil is obtained in sandstones in the Upper Pennsylvanian Virgil Series. Trapping in the dolomite has been in a dome (Figure 30.8) 8 miles long and 5 miles across with a closure of at least 250 feet. The stippled area in the figure encloses wells producing oil and gas from the dolomite. The oil reservoir averages 20 feet in thickness on the west side of the field and 50 feet on the east side. It is underlain by a tilted water table with an elevation on the west side of less than 1000 feet, which dips to below 1030 feet on the east side. Above the oil is a gas cap 100 feet in average thickness; the gas-oil contact is at elevation −980 feet. Trapping in the deeper sands is also as close to the top of the anticline as local reservoir permeability will permit.

The *Hansford* gas field (Stratigraphic Committee, 1961) spreads over about 300 square miles, covering a large part of the eastern half of Hansford County. By the end of 1960, 7 years after discovery, it had produced more than 95 billion cubic feet of gas, of which about 70 per cent had come from the upper Morrow sand (Lower Pennsylvanian). The average reservoir thickness is 15 feet; both porosity and permeability are variable. The reservoir second in yield to date, with about 25 per cent of the total, is a 15 to 20 foot sand at the base of the Morrow (and the Pennsylvanian) unconformably overlying Mississippian rocks. The remaining 5 per cent of total cumulative production has come from a 15 foot sand in the middle Morrow. The trapping in all three reservoirs has been due to sandstone lensing up the regional dip. The dip is easterly into the Anadarko basin and averages about 50 feet to the mile.

The *Farnsworth* oil field (Thalman, 1961) in western Ochiltree County immediately to the east of the Hans-

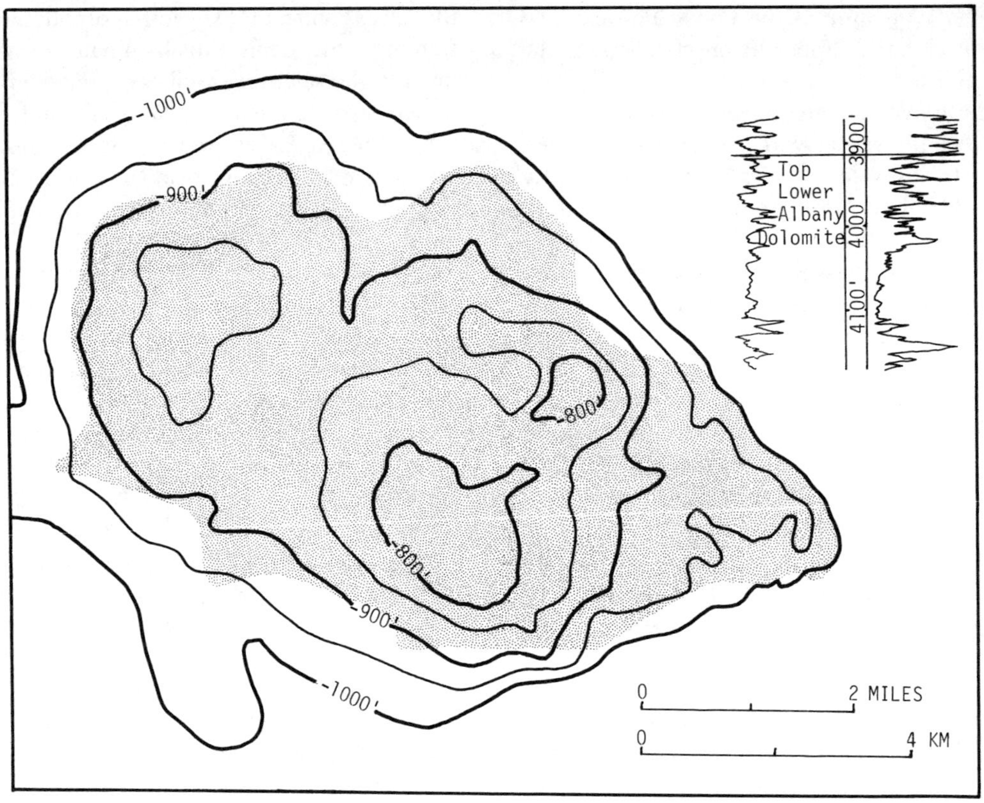

Figure 30.8 Quinduno field, Roberts County, Texas. Structure contour map based on top of lower Albany Dolomite. Interval, 50 feet. Shading covers area in which lower Albany Dolomite reservoir is productive. *Courtesy C. E. Little (1961, p. 184) and the Panhandle Geological Society.*

ford field is an excellent example of lenticular sand trapping. Figure 30.9 (upper) is a thickness map of the reservoir, a sand in the upper Morrow of the Lower Pennsylvanian. The isopach lines range from zero around the periphery of the shale-encased sand bar to more than 50 feet on the eastern summit of this bar (or dune?). The reservoir rock is a coarse-grained sand, slightly arkosic, with clay cement.

The lower map in Figure 30.9 shows structure contours over the same area. The datum is the top of the Morrow, which lies from 80 to 130 feet above the sand, depending upon the thickness of the latter. The regional dip is slightly south of east, averaging about 75 feet per mile to the northeast and southwest of the buried bar. The periphery (zero isopach) of the sand lens, which also encloses the Farnsworth field, has been superimposed (dotted line) on the structure map. Obviously the upward bulge created by the sand body has been reflected in the overlying structure by an anticlinal nose plunging down the regional dip. The summit peak at the western end of the sand accumulation has produced a node on the anticlinal axis; the higher sand hill near the eastern end is reflected in a terrace.

One of the few fields so far discovered in the Texas part of the Dalhart basin is the *Rehm* (Smith, 1961) in south-central Hartley County. This seven-well oil field, discovered in late 1954, had a cumulative production of more than 350,000 barrels by the end of 1961. The reservoir is a pink granite wash, about 40 feet thick, which was deposited during Canyon (Late Pennsylvanian) time. The trap is north-south elliptical anticline at least $2\frac{1}{2}$ miles long by $1\frac{1}{2}$ miles wide, with over 150 feet of closure. The oil occurs only in the highest part of the anticline.

Future. In spite of its dry holes, the Texas Panhandle has many prospects, downhole, uphole, intervening, and out beyond present production. Even the densely drilled Amarillo uplift still has possibilities. It is true that there is not much undrilled area for intervening discoveries, but the uphole and downhole prospects have not been exhausted.

There may be other Red Caves. The Red Cave Formation in the lower Clear Fork of the Leonard (Permian) was an unrecognized reservoir until 1959, when a gas well was knowingly completed in the formation in northern Potter County at a depth of about 2000 feet. In the early days of the Amarillo uplift gas field many wells drilled in the western sector by cable tools found gas at this level, and some were completed there and others produced dually from the Red Cave and the deeper Wolfcamp–Granite Wash reservoirs (Rogers, 1961, p. 38). However, the Red Cave production was merged with that of the deeper, more extensive, and more prolific gas reservoirs without its separate entity being recognized. Further development of the Panhandle gas field with rotary-drilled wells followed; but with this method of drilling no gas shows were noted, and the Red Cave reservoir was bypassed by many hundreds of wells

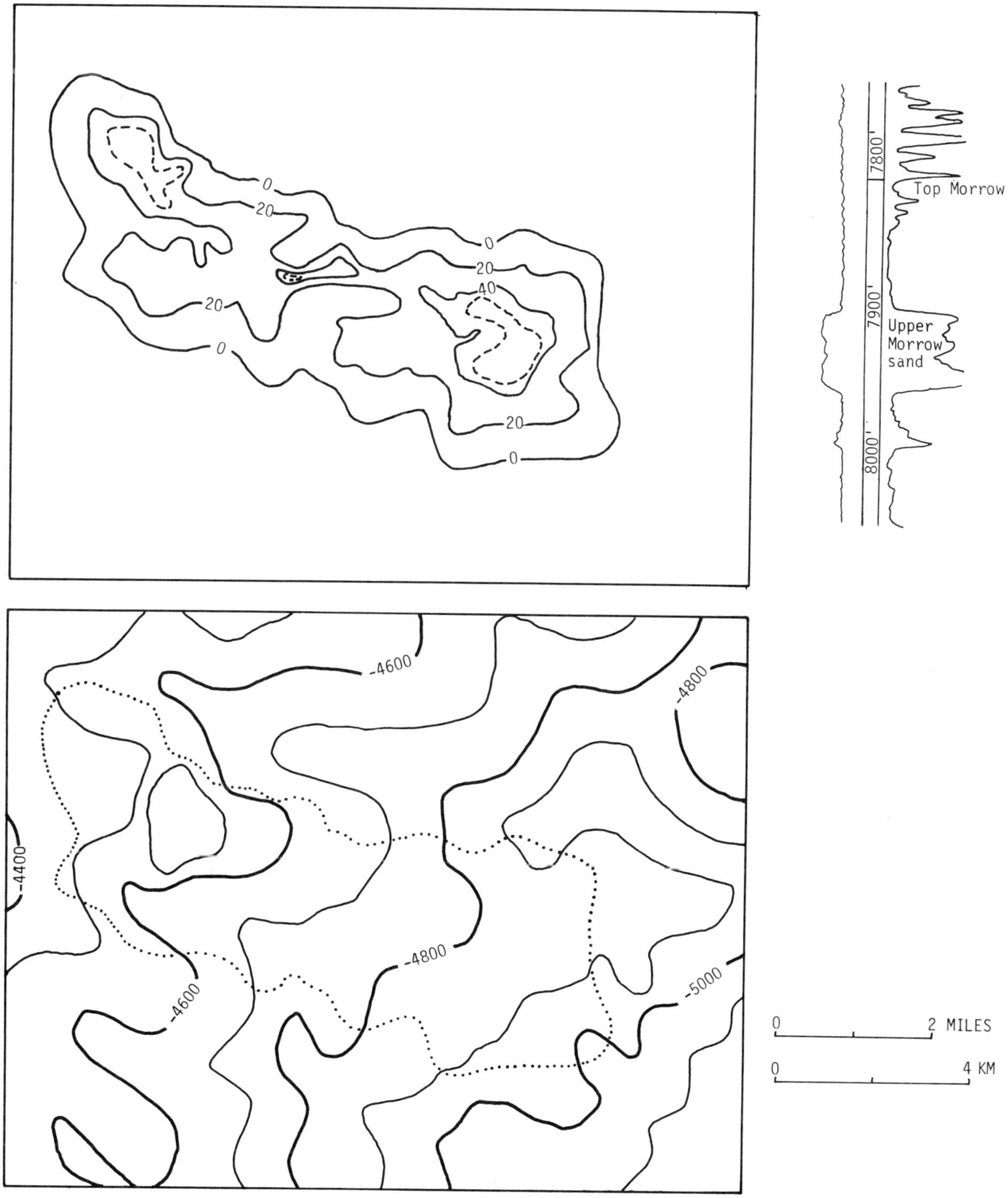

Figure 30.9 Farnsworth field, Ochiltree County, Texas. Upper: thickness map of reservoir sand, the upper Morrow; interval 10 feet. Lower: structure map of same area, contoured on top of the Morrow; interval 100 feet. *Courtesy A. L. Thalman (1961) and the Panhandle Geological Society.*

that were completed in gas sands at the lower levels.

The rediscovery, in April, 1959, of gas at the Red Cave stratigraphic level was followed by its recognition as a distinct, relatively shallow gas reservoir in the Amarillo uplift gas field. During the next 2 years the uphole discoveries exceeded the deeper discoveries in the Panhandle district, and many new shallow gas wells were drilled in the western part of this field. Depths to the gas range from 1400 to 2700 feet, and initial productions were as high as 19 million cubic feet per day (Stevens, 1961, p. 104).

Downhole the reservoirs have been probed to the granite, but has there been adequate exploration of the granite itself for crevice-stored oil or gas? Certainly some production has already been obtained from this reservoir (Cotner and Crum, 1935, p. 392; Landes et al., 1960, p. 1690).

The Hugoton field will see discoveries of oil lying deeper than the gas if the history of this field in Kansas and Oklahoma repeats itself in Texas.

The Anadarko basin, scene of many discoveries in recent years, is still promising. In addition to infilling, peripheral extensions can be anticipated. The deeper reservoirs have not been probed to anything approaching adequacy as yet. Older reservoirs yet to be explored thoroughly range in age from Mississippian to Ordovician. Some of these are upturned (or upfaulted), truncated, and overlain by Pennsylvanian sediments on the north flank of the Amarillo uplift. The discovery of the multireservoir Gageby Creek field has confirmed that these pre-Pennsylvanian rocks may be loaded. One of the six new fields discovered in the Texas Panhandle district in 1967 was Buffalo Wallow in Hemphill County in the deeper part of the Anadarko basin. The discovery well produced initially at the rate of 73 million cubic feet of gas per day from Hunton reservoirs at a depth of 19,876 feet (Summers, 1968). It is very possible that the deeper Anadarko will join the Val Verde and Delaware basins of West Texas as an important source of natural gas.

Throughout the Anadarko basin rapid facies changes, which create varying permeability traps, can be anticipated (Swanson, 1967).

It is doubtful whether the 30 year span between the development of the Amarillo uplift and the Anadarko basin oil and gas production will be repeated before a sizable yield is obtained in the Panhandle sector of the Palo Duro basin. The fact remains that, in spite of many failures to date, adequate source and reservoir rocks were deposited in this basin, as they were across the Amarillo uplift to the north and the Matador uplift to the south. This is proved by the presence of oil fields in the eastern Palo Duro basin in northeast West Texas, North Texas, and southwestern Oklahoma (Stevens, 1962). An intensive search in the Panhandle sector for traps should be successful here, as it has been to the east and in the Anadarko basin. The long-held assumption that the oil on the Amarillo uplift must have come exclusively from the Anadarko basin because all the fields were on the north flank of the uplift is no longer valid, because of the discovery of oil accumulation on the south flank as well. This oil must have migrated updip from the Palo Duro basin.

The smaller Dalhart basin likewise is endowed with source rocks, reservoir rocks, and traps.

REFERENCES CONSULTED IN PREPARING CHAPTER 30

Best, John B., Jr., 1964.
Brashear, William W., 1961.
Cotner, V., and Crum, H. E., 1935.
Cunningham, B. J., 1961.
Dietz, Gerald P. 1967.
Eddleman, M. W., 1961.
Jones, T. S., 1951.
Landes, Kenneth K., Amoruso, J. J., Charlesworth, L. J., Jr., Heany, F., and Lesperance, P. J., 1960.
Little, C. E., 1961.
Mason, John W., 1968.
Nicholson, John H., 1960.
Panhandle Geological Society, 1961.
Rogatz, H., 1935, 1939, 1961.
Rogers, R. G., 1961.
Roth, R., 1955, 1957.
Smith, G. W., 1961.
Stevens, Curtis, 1961, 1962.
Stratigraphic Committee, 1961.
Summers, Fred C., J., 1968.
Swanson, Donald C., 1967.
Thalman, A. L., 1961.
Totten, R. B., 1956, 1961.
Totten, Robert B., and Horn, Paul H., 1961, 1968.
Vlissides, Sophie D., 1964.

Chapter

31

WEST TEXAS

West Texas is bounded on the north by the Panhandle, on the northeast by North Texas, on the east by West Central Texas, on the southeast by South Texas, on the south by the Rio Grande and the Republic of Mexico, and on the northwest by New Mexico (Figure 31.1). This area contains 55 counties; it includes Districts 7c and 8 of the Texas Railroad Commission. The major structural feature of West Texas is the Permian basin, which extends westward to include the 8 counties in the southeastern corner of New Mexico.

During 1968 West Texas produced more than 530 million barrels of oil—47 per cent of the entire Texas production for the year. If West Texas were a separate state, it would rank below Louisiana and ahead of California in annual oil output. One reason, in addition to area, for the magnitude of West Texas production is its phenomenal number of giant fields. So far 23 fields in this district have already produced over 100 million barrels of oil, and another 22 fields are on their way, with proved reserves to warrant "giant" classification. The largest field of them all is Yates (Figure 31.6), which has already produced more than ½ billion barrels.

West Texas did not become oil country until January, 1921, when the Westbrook field in Mitchell County was discovered (West Texas Geol. Soc., 1951, p. 1). Dolomite production was found at depths of about 3000 feet. Exploration of this large district was accelerated in 1923, when prolific shallow production was found at the Big Lake field in Reagan County. Five years later this field had a second birthday when the first Ellenburger (Ordovician) production was discovered at depths in the 8500 to 8600 foot range. For several years the Big Lake field held the world's producing depth record. More importantly, the prolific production from this Ordovician reservoir led to successful search in other areas.

West Texas is still good hunting ground. Thirty-six significant discoveries were made in this district in 1966 (Sax, 1967, p. 1058).

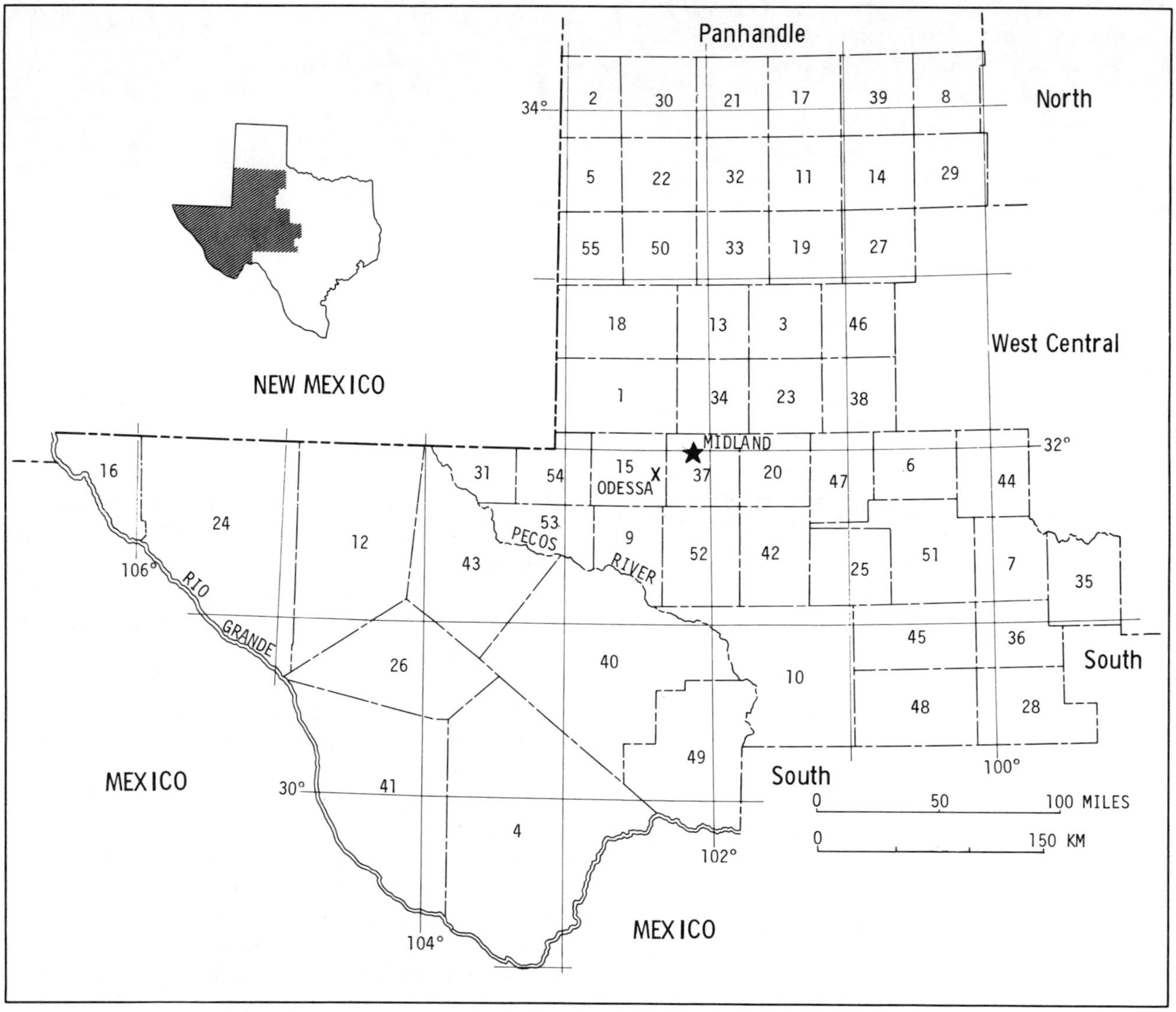

Figure 31.1 West Texas. Index map. County Key:

1	Andrews	9	Crane	17	Floyd	25	Irion	33	Lynn	41	Presidio	49	Terrell
2	Bailey	10	Crockett	18	Gaines	26	Jeff Davis	34	Martin	42	Reagan	50	Terry
3	Borden	11	Crosby	19	Garza	27	Kent	35	McCulloch	43	Reeves	51	Tom Green
4	Brewster	12	Culberson	20	Glasscock	28	Kimble	36	Menard	44	Runnels	52	Upton
5	Cochran	13	Dawson	21	Hale	29	King	37	Midland	45	Schleicher	53	Ward
6	Coke	14	Dickens	22	Hockley	30	Lamb	38	Mitchell	46	Scurry	54	Winkler
7	Concho	15	Ector	23	Howard	31	Loving	39	Motley	47	Sterling	55	Yoakum
8	Cottle	16	El Paso	24	Hudspeth	32	Lubbock	40	Pecos	48	Sutton		

Tectonics. The greater part of West Texas, as well as adjacent southeastern New Mexico, is in the Permian basin, a large sag in the earth's crust lying between the Bend arch on the east and the outliers of the Rocky Mountain geanticline to the west. The north flank of the basin terminates against the Amarillo–Ouachita uplift of the Texas Panhandle and southwestern Oklahoma. To the southwest and south the basin is limited by the Diablo platform and the Marathon thrust belt. The Permian basin, in turn, is furrowed by a number of named secondary tectonic features (Figure 31.2). They are, across the center of the district from east to west, the Eastern shelf, the Midland basin, the Central Basin platform, the Delaware basin, and the Diablo platform. The northern part of West Texas is crossed by the Matador arch, north of which is the Palo Duro basin.

Northwest of the Central Basin platform is the Northwestern shelf. To the south of the Midland basin, and east of the southern end of the Central Basin platform, is the Ozona platform, which was a part of the Central Basin platform until Late Pennsylvanian time. Farther south is the Val Verde basin, which lies to the north of the Marathon thrust belt. The Marfa basin is at the southeastern end of the Diablo platform. To the southwest of the latter platform, and largely in Mexico, is the Mesozoic geosyncline. Line *A–A′* in Figure 31.2 indicates the location of a southwest-northeast cross section shown in Figure 31.3.

All of the tectonic subdivisions of West Texas are productive except the Palo Duro basin on the northern edge, and those along the Rio Grande. Igneous rock crops out over a considerable part of southwestern West Texas (Figure 31.4).

Stratigraphy. The geologic column in West Texas ranges from Recent sediments of the latest Cenozoic down to Precambrian basement rock (Jones, 1953). The surface distribution of these rocks is shown in Figure 31.4. Nonmarine Cenozoic deposits veneer the surface over a wide belt from the southwestern Panhandle boundary southward, and thence westward along the New Mexican boundary. These sediments reach a maximum thickness of 5000 feet in the intermontane depressions of trans-Pecos Texas. Southern and southeastern West Texas are floored with Cretaceous rocks. Much of

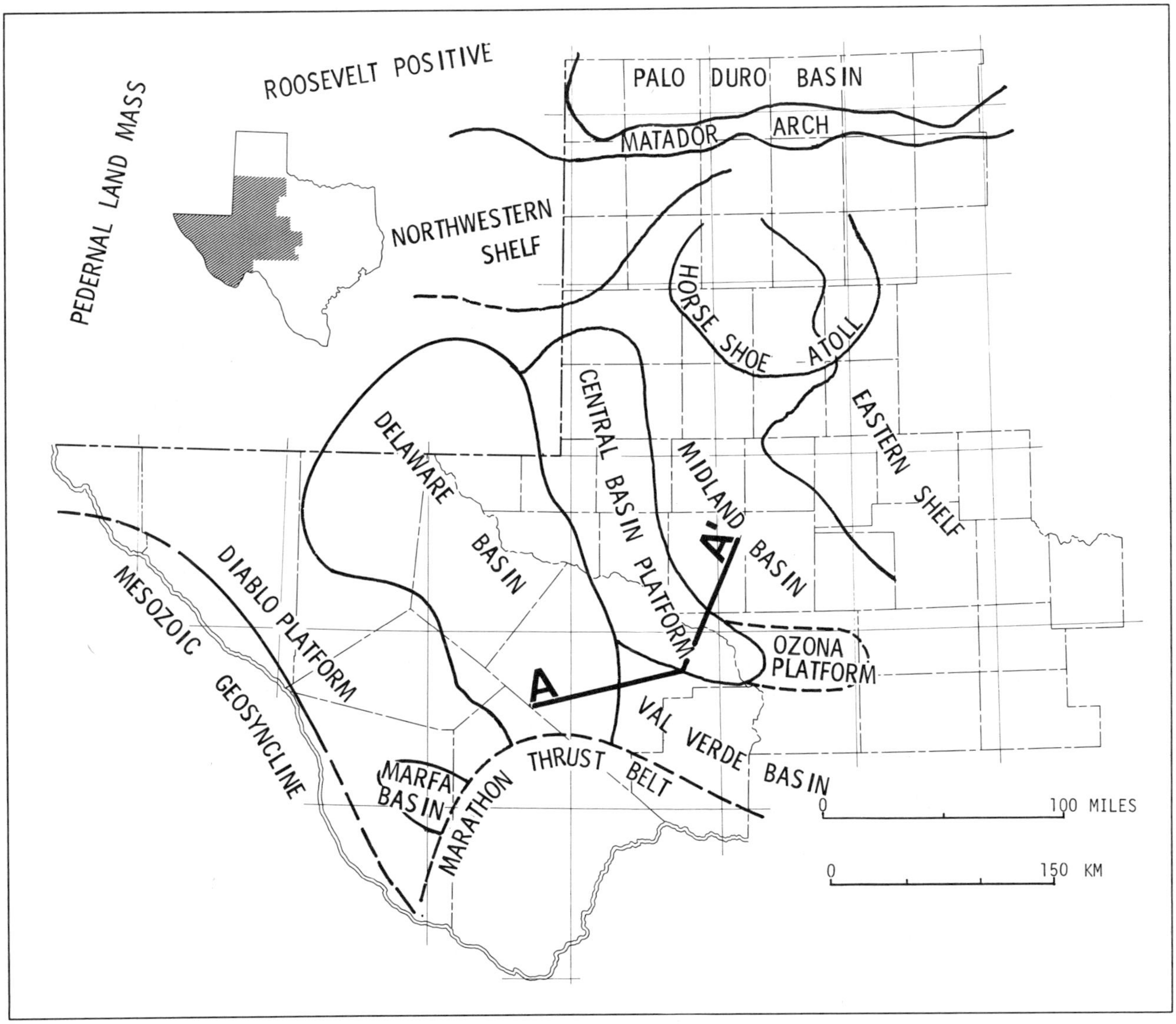

Figure 31.2 West Texas. Tectonic map, showing basins, shelves, platforms and uplifts. Also shows line of cross section (*A–A′*) shown in Figure 31.3. *Courtesy Norman A. Sax (1967, p. 1056–1057) and the American Association of Petroleum Geologists.*

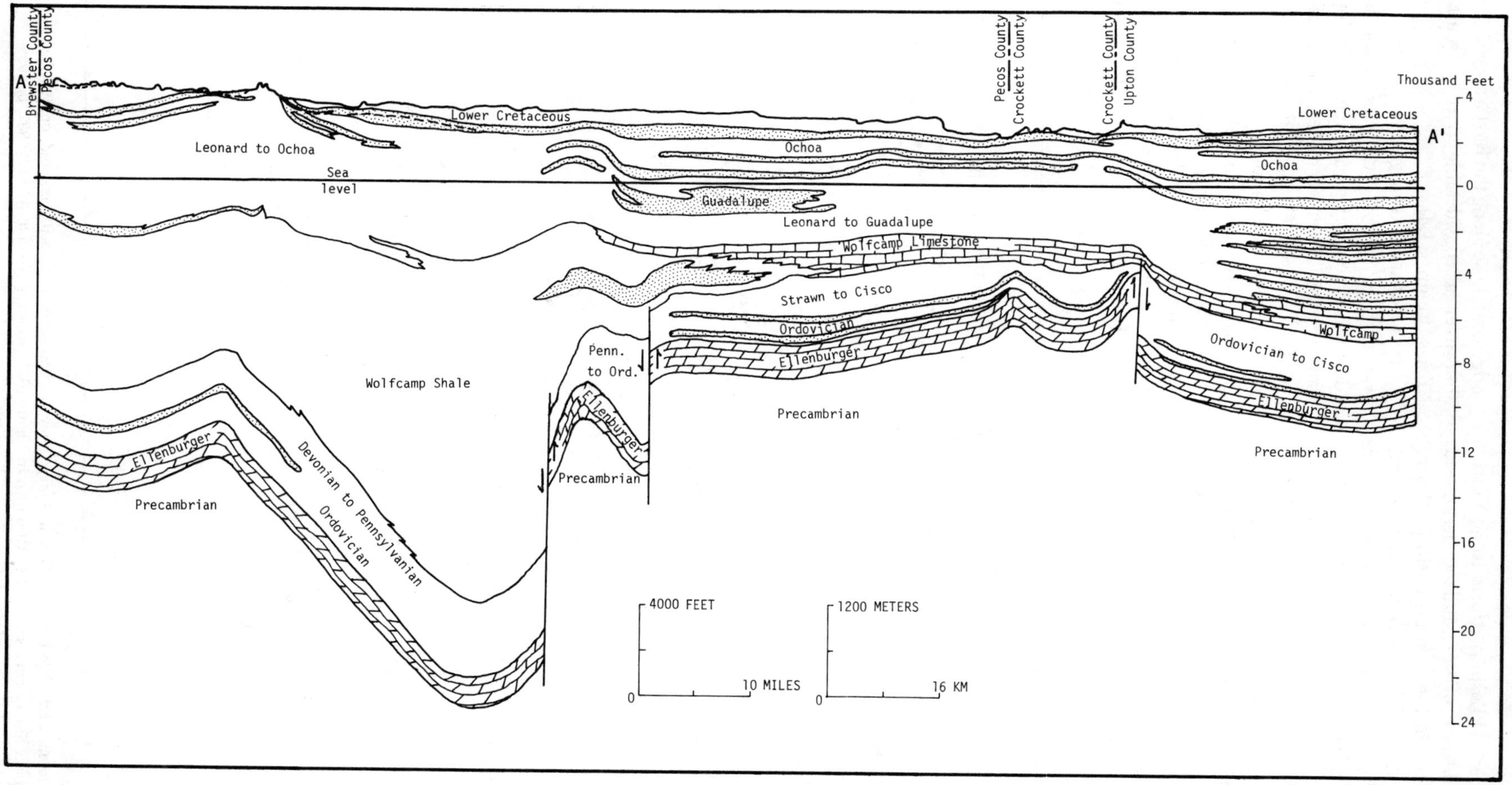

FIGURE 31.3 Cross section along line *A–A′* (Figure 31.2), showing Val Verde basin on left, Central Basin platform in middle, and Midland basin to right. *After West Texas Geological Society (1962).*

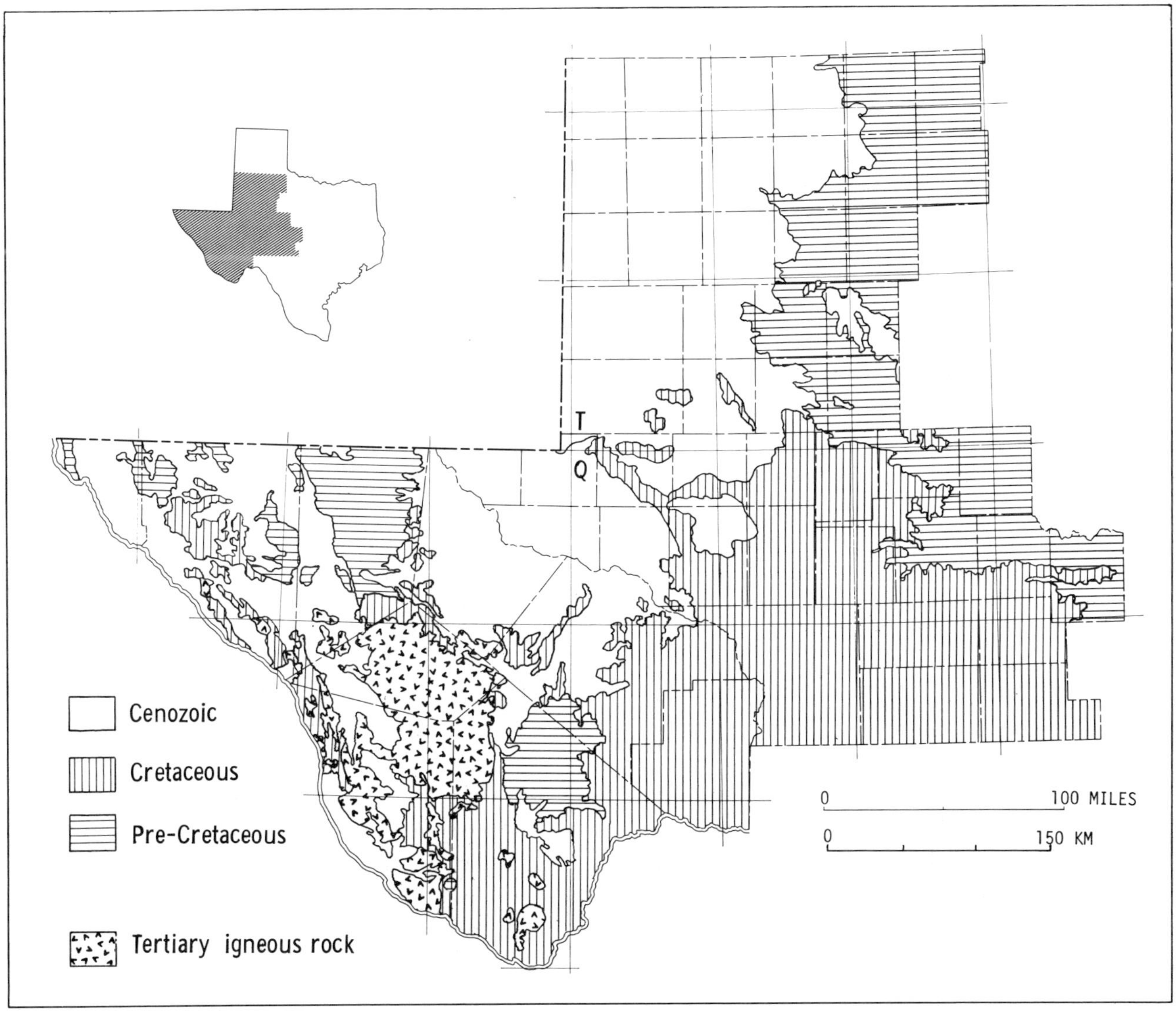

Figure 31.4 West Texas. Geologic map. Line between Q and T just off southeastern corner of New Mexico separates Cenozoic sediments into predominantly Quaternary and predominantly Tertiary.

the Midland basin in east-central West Texas is covered with Triassic rocks. Pre-Cretaceous sediments, mainly Permian, underlie much of eastern West Texas. The previously mentioned igneous rocks in southwestern West Texas are post-Cretaceous volcanics.

The primary oil and gas reservoirs so far found in West Texas are confined to the Paleozoic column (Figure 31.5). They range in age from latest Permian to Cambrian. Of the different Paleozoic periods the Permian contains not only the thickest section but also the largest number of reservoirs. It has been stated that in West Texas and southeastern New Mexico "the Upper Permian contributes the greatest percentage of the total production from depths not exceeding 5000 feet" (Wagner, 1966, p. 1195). The fact that the Permian sediments were deposited in a subsiding basin explains their great thickness.

The youngest Permian rocks are the Ochoa Series, which are confined to the deeper part of the Permian basin, in the southeastern corner of New Mexico and adjacent Texas. Here the sediments of the Ochoa Series reach a thickness of more than 4000 feet. They thin out and disappear completely, because of truncation, about 100 miles outward to the west and south and as much as 200 miles to the northeast. Sediments deposited during this time interval consist of evaporites (Hills, 1968b), clastics, and minor carbonate rocks. Relatively insignificant amounts of oil have been found in Ochoa reservoirs.

The next older Permian series, the Guadalupe, has produced more than half of the oil found in the Permian basin (Galley, 1958). One reason is the fact that the reservoirs in this part of the stratigraphic section are the shallowest and hence were among the earliest to be

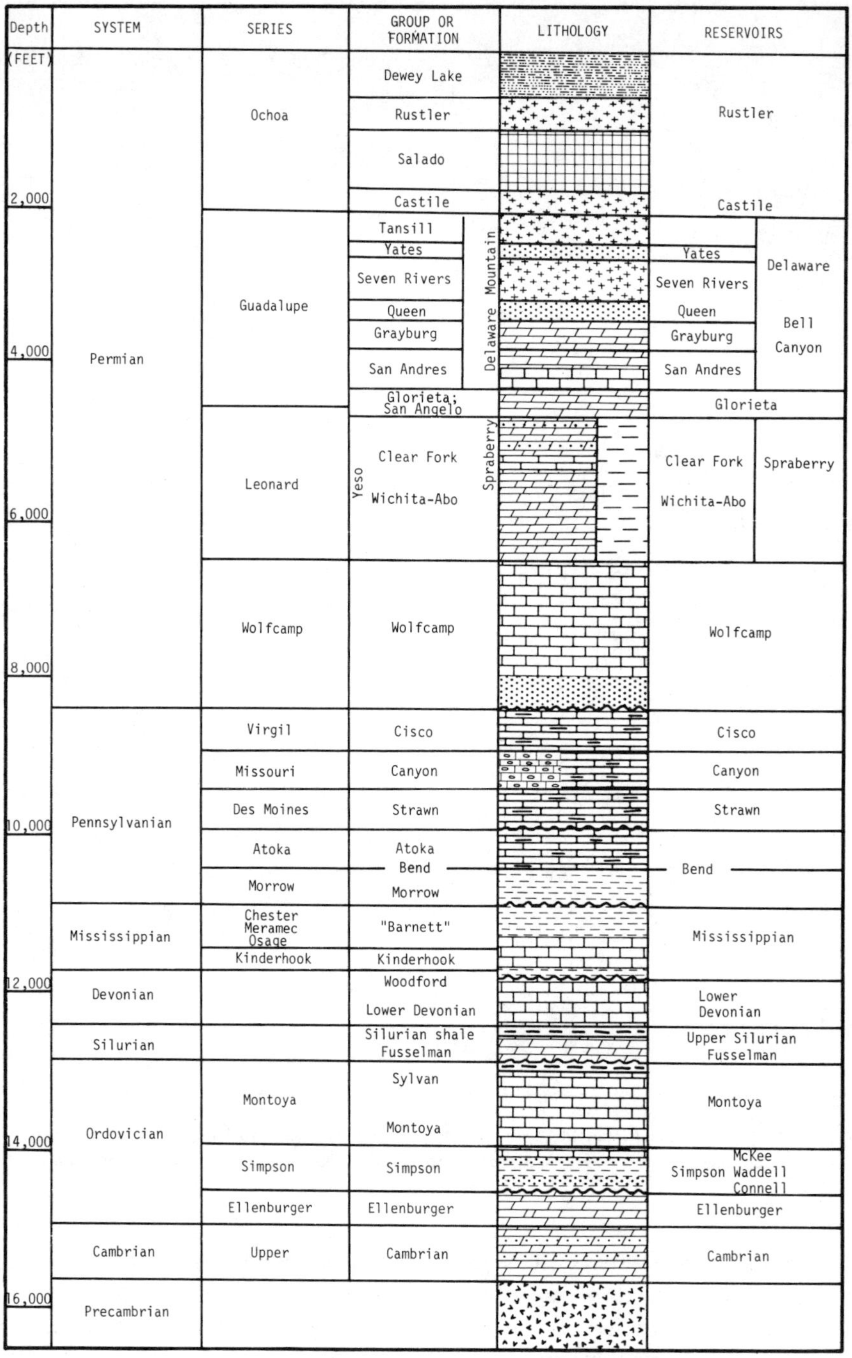

FIGURE 31.5 Generalized stratigraphic column of West Texas, showing principal reservoirs. Almost every group or formation is productive somewhere in West Texas; the Salado salt, the Woodford and Sylvan shales, and the Precambrian are exceptions. *After John E. Galley (1958, Figure 5), the American Association of Petroleum Geologists (1960, Plate 1), T. S. Jones and Harold M. Smith (1965), the West Texas Geological Society (1966, p. 6), and T. S. Jones (1968, personal communication).*

discovered and exploited. The Guadalupian sediments consist mostly of cherty carbonate rocks (especially dolomite), sandstones, and anhydrite. This series is thickest in southeastern New Mexico, where it exceeds 4000 feet. The principal Guadalupian pay zones are the Yates Sandstone and the Grayburg and San Andres Dolomites. The upper San Andres is an important reservoir in a row of fields extending north for a distance of more than 70 miles across Ector and Andrews Counties (Young, 1965).

The massive Capitan reef at El Capitan near the northwestern corner of Culberson County "was formed along the rim of the Delaware basin, growing basinward from its roots in lower Guadalupian carbonates" (Galley, 1958, p. 432). Some Upper Permian pays, especially in the Grayburg and San Andres,

> are dolomitized limestone sands sealed updip by lime muds deposited farther in back of the reef. Carbonate muds also overlie the pays because, as the reefs or shelf edges move forward, the back reef mud-sand contact moves forward also. Some of the porosity is filled with anhydrite from the later more saline waters (T. S. Jones, letter dated 4 September, 1968).

The Leonard Series underlies the Guadalupian. This section exceeds 4000 feet in the deepest part of the Delaware basin. About one-third of the total volume is shale, the remainder mainly carbonate rock. The shale may be the source rock for the oil in the Leonard reservoirs; it may also be possible that these rocks furnish at least a part of the oil found in the shale-poor Guadalupian section. Perhaps the best known of the Leonard reservoirs is the Spraberry, a siltstone in which the producible oil occurs in fractures.

The basal Permian series, the Wolfcamp, has a maximum thickness of more than 8000 feet in the deepest parts of the Delaware and Val Verde basins. The rocks are principally shales, limestones, and sandstones. The shales are thought to be the source rocks for the oil reservoirs. The pay zones are mostly limestones.

The uppermost Pennsylvanian section includes the great Scurry reef reservoir rock, which is described subsequently in the discussion of the Horseshoe atoll (p. 332). These reefs, rooted in the Des Moines Series of uppermost Pennsylvanian, jut upward into the Lower Permian section.

The upper Pennsylvanian, which consists of the Virgil, Missouri, and Des Moines Series, contains principally shale and sandstone with lesser amounts of limestone. The aggregate thickness exceeds 6000 feet. About one-third of the known Upper Pennsylvanian oil is in the Scurry reef; both sandstones and bedded limestones contain the rest. The shales, many of which are dark in color and bituminous, are the most likely source rocks.

The Lower Pennsylvanian consists of the Atoka-Morrow Series. The rocks are predominantly dark shale, shaly limestone, and sandstone. The shales are the probable source rocks, with the oil finding sanctuary in sandstone and limestone reservoirs. The best hunting for Atokan oil has been in limestones on the western flank of the Midland basin; however, some oil has also been produced from sandstones on the Eastern shelf.

The Mississippian section consists principally of the Chester, Meramec, and Osage Series. It is a thick (1600 feet plus) limestone section underlain by a thin silty gray limestone of Kinderhook age. Conformably underlying the Mississippian is the Woodford Shale, up to 750 feet of black bituminous shale that may be the source rock for hydrocarbon accumulations in pre-Mississippian rocks. The Mississippian limestones tend to be nonporous and have produced only relatively insignificant volumes of oil or gas.

The Silurian and Devonian rocks occur in a nearly circular lens that extends outward about 100 miles from the southeastern corner of New Mexico. This lenticular body of rock is an erosional remnant of cherty limestones and dolomites. It consists of a "tripolitic" chert, which probably was originally made up mainly of spicules and radiolaria, from which the tiny calcareous shell fragments were leached during exposure, leaving a fine-grained, spongelike mass. Porosity in the Devonian chert in the deep gas field of the Delaware basin is due largely to fractures (T. S. Jones, letter dated 4 September, 1968). Since the Siluro–Devonian section does not contain likely source rocks, the accumulations found within must be due to expulsion downward from above or to migration upward from beneath.

The Upper Ordovician Montoya rocks also occur in a residual lens with a distribution pattern similar to that of the Siluro–Devonian limestones. The Montoya cherty carbonates are largely nonporous, and very little in the way of oil and gas accumulations has been found within them.

Below the Montoya are the Simpson and Ellenburger Formations, important reservoirs for both oil and gas. The Simpson Formation, consisting of sandstone, green shale, and limestone, is another erosional remnant occurring in southeastern New Mexico and in western and southwestern West Texas. Three sandstone members of the formation are the reservoirs. The principal producing area is the Central Basin platform, where accumulation is mainly in anticlines.

The Simpson rocks are separated from the underlying Lower Ordovician Ellenberger in the Permian basin by an unconformity so extensive that the Ellenburger may be directly overlain by much younger rocks. The Ellenburger consists of cherty carbonate rocks, predominantly dolomite. Most of the oil so far found has been on the Central Basin platform, where the reservoirs are overlain by Simpson shale, the probable source rock (Jones and Smith, 1965). Ellenburger accumulations east of the Simpson edge may have resulted from downward expulsion of oil from younger source rocks. It has been concluded from crude oil carbon isotopic studies that "the Ellenburger Group series is a reservoir for oils of many ages, from Early Ordovician through Early Permian" (Kvenvolden and Squires, 1967, p. 1293). The Ellenburger porosity is due to the presence of vugs and fractures. Production in West Texas is prolific; many Ellenburger wells have produced more than a million barrels of oil. The oil-water interface may be as much as 1000 feet below the top of the trap.

The carbonate rocks of the Ellenburger merge downward into the clastic rocks of Cambrian age that overlie the basement in the southern part of the area. The thickness of the Cambrian clastics varies greatly, exceeding a thousand feet in the southeastern part of the district, where obviously a sedimentary trough existed. The sea in which the basal Paleozoic sediments were deposited was a transgressing one, so these deposits are not everywhere the same age.

TRAPS AND SELECTED FIELDS

Over the wide expanse of West Texas, with its hundreds of oil and gas fields (Figure 31.6), hydrocarbons have been trapped in various ways. So far, the largest amount of trapping has been in anticlines, but a huge volume of oil has accumulated in reefs and lenticular sands as well. Although many of the producing anticlines are faulted, true fault traps are scarce. Salt dome traps are absent, even though the Permian basin is the location of one of the world's great salt deposits.

The oil and gas fields of western Texas are confined largely to a belt about four counties wide extending north and south from the southeastern corner of New Mexico. The area of hydrocarbon occurrence overlaps parts of at least nine of the tectonic subdivisions shown in Figure 31.2. The fields and traps to be described have been classified by these subdivisions. The latter are discussed, so far as possible, from east to west across West Texas.

Major Fields in West Texas. The four leading fields, as of 1967, in both cumulative and annual production are given in Table 31.1.

Eastern Shelf. The Eastern shelf is a depositional terrace that extends westward from the Llano uplift, the

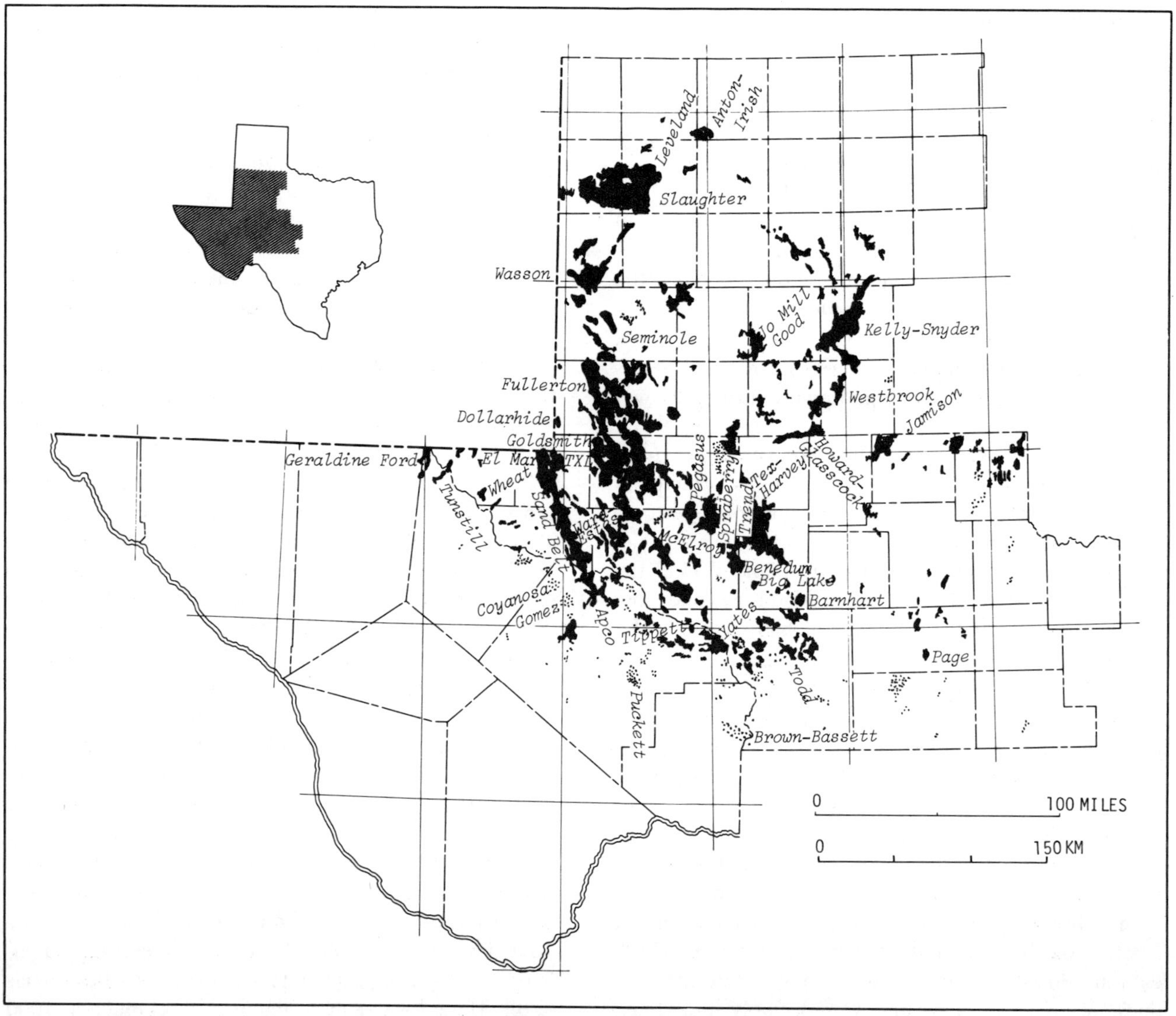

Figure 31.6 West Texas. Oil and gas fields.

Table 31.1 Four Leading Fields (Two Categories) of West Texas in 1967

Rank	Field	Production in Million Barrels	Subdistrict
	Cumulative Production		
1	Yates	512	Central Basin platform
2	Wasson	472	Northwestern shelf
3	Goldsmith	412	Central Basin platform
4	Kelly-Snyder	409	Horseshoe atoll
	Production during 1967		
1	Kelly-Snyder	37	Horseshoe atoll
2	Spraberry	28	Midland basin
3	Goldsmith	26	Central Basin platform
4	Slaughter	24	Northwestern shelf

eastern boundary of the Permian basin. The Eastern shelf, which is underlain predominantly by carbonate rocks, flexes downward gently to the west into the Midland basin, where fine clastic sediment is more abundant. Northwestward, the Eastern shelf curves around the northern end of the Midland basin and merges with the Northwestern shelf.

The *Page* oil and gas field is in south-central Schleicher County on the southern part of the Eastern shelf. The field presumably was discovered by the presence of a northwestward plunging nose in the Cretaceous rocks mapped at the surface (Ellison, 1957). The reservoirs are permeable zones in a Pennsylvanian (Strawn–Canyon) reef. The permeable zones in the upper part of the reef are filled with gas; oil-bearing reservoirs lie at lower depths. This is shown in the cross section (Figure 31.7). Two porosity zones containing gas and two con-

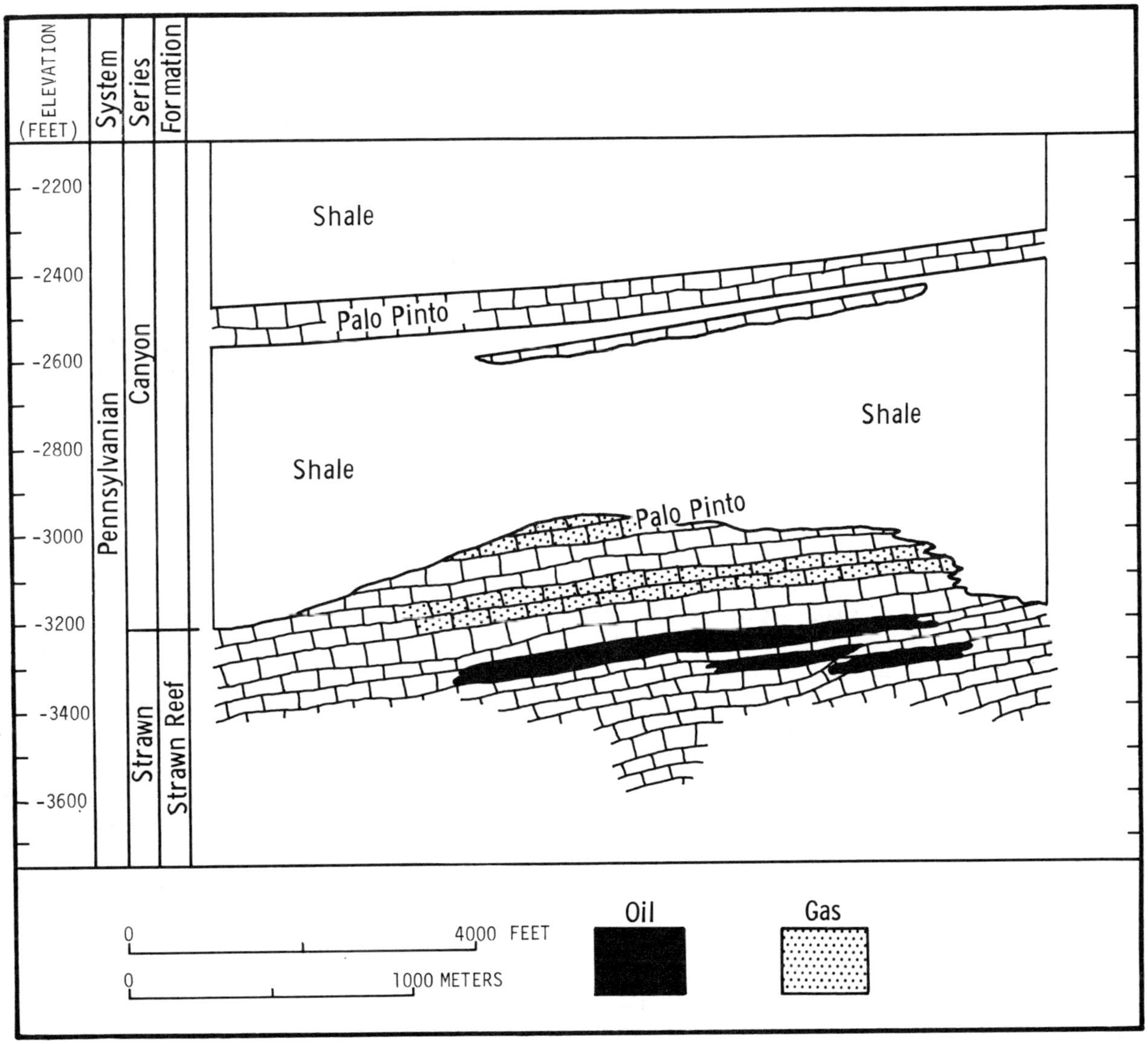

Figure 31.7 Page field, Schleicher County. Cross section showing oil and gas accumulations in a Strawn reef. *Courtesy Samuel P. Ellison, Jr. (1957, p. 250, Cross Section B-B′) and the University of Texas.*

taining oil are persistent throughout the field. The third and lowest oil zone shown in the illustration does not have field-wide continuity. Trapping in the gas zones is due to buried topography, with the porous zones sealed updip against the overlapping Canyon shale, which surrounds the reef. Oil trapping is due to updip lensing out of porosity within the reef.

In Coke County to the north of Schleicher the *Jameson* field produces from Upper Pennsylvanian reef limestone. On the west Wolfcamp sands pinch out on the flank of the reef (Conselman, 1955). Farther to the north and east, extending into western West Central Texas, is a north-south row of small fields producing from Cambrian sands. In western Runnels County, immediately to the east of Coke County, the lower Canyon Palo Pinto Limestone (Late Pennsylvanian) produces from two porous zones in the uppermost 50 feet. Trapping is in low-relief anticlines, with variable permeability playing a secondary role (Doss and Spiva, 1968).

The *Howard–Glasscock* field occupies a westward bulge of the Eastern shelf jutting into the Midland basin. Most of the field is in southeasternmost Howard County, but it overlaps to the southward into Glasscock County. Howard–Glasscock (Mooney, 1966) is one of the great fields of West Texas. Discovered in 1925, it has subsequently produced nearly $\frac{1}{4}$ billion barrels of oil; the yield in 1966 was between 7 and 8 million barrels. Five major reservoirs have contributed to this total; they are, from youngest to oldest, the Yates, Seven Rivers, Queen, Grayburg–San Andres, and the Glorieta–Clear Fork. All are Permian in age, occurring in the Guadalupe and Leonard Series at depths mostly less than 2000 feet. The trap is a 15-mile-long east-west anticline (Figure 31.8). The maximum productive width is about $2\frac{1}{2}$ miles. The crest is undulating, containing several nodes. The Grayburg–San Andres reservoir section, on which the structure contours are based, is a finely crystalline, in places sandy, dolomite with anhydrite and shale streaks. Locally it becomes oölitic and more productive. The gross reservoir thickness exceeds 800 feet, of which 80 feet is net pay. Because of lateral variations in permeability, the different reservoirs have different producing well distribution patterns, but the trapping is anticlinal throughout.

Horseshoe Atoll. One of the more notable geological features in West Texas is an extensive, completely buried reef complex that has been aptly named the Horseshoe atoll. Figure 31.2 shows the approximate location of this atoll in respect to the major tectonic subdivisions of West Texas. The highest and most productive part of the reef complex is on the east side of the Midland basin. Figure 31.9 shows the location of the oil accumulations on the atoll.

The reef rock was formed from the individual shells of marine organisms, especially crinoids and forams, that died and accumulated where they had lived. It has been suggested that the reef cohesion may have been achieved by "welding of the lime mud and constant precipitation of calcite cement" (Bergenback and Terriere, 1953, p. 1014). The reef porosity has been principally the result of (1) fracturing after lithification, and (2) leaching, which widened fractures and developed vugs. The effective porosity ranges from almost nothing to about 30 per cent, with an average of 6 per cent. Along the crest of the atoll some pores have been filled by clay, probably squeezed in from the overlying shale (Myers et al., 1956, p. 24).

The Horseshoe atoll is U-shaped with the open end to the north. It is 70 to 90 miles across, covering parts of twelve counties (Figure 31.9). The reef crest running around the U is extremely irregular, consisting of a complex of mounds and hollows; the highest points along the crest are as much as a thousand feet above the limestone platform on which the reef complex is built. The surface of the reef slopes from the crest down to the foundation limestone at angles mostly less than 6°. Subsurface geological studies have shown that the reef was a prominent topographic feature, consisting of fossiliferous limestone on the floor of the Midland basin during Late Pennsylvanian and Early Permian time. Periodic changes in sea level may have brought about differences in lithologies and porosities. During Early Permian time the atoll was buried by a thicker covering of mud and silt. This blanket served two important geologic functions: (1) it probably was the source rock for the enormous quantity of oil that found sanctuary in the reef, and (2) upon compaction this fine-grained rock became the seal that kept the hydrocarbons within the reef.

The highest parts of the reef complex are in the southeastern section, mainly in Scurry County. It is here that oil was first found in 1948, and here, during the following years, twenty-three fields were discovered and developed. This aggregation of fields, extending from southeastern Borden County across Scurry County and into Kent County, has been referred to by Stafford (1957, 1959) as the "Scurry field." The largest continuous producing area is now called the *Kelly–Snyder* field (Vest, 1966). The Kelly–Snyder is a giant field, with a production of 22 million barrels in 1966 and a cumulative yield approaching 400 million barrels. With estimated reserves in 1966 of 815 million barrels of oil, there is no doubt but that Kelly–Snyder will end up in the billion barrel production classification. It covers an area of 51,000 acres in western Scurry County.

Figure 31.10 is a structure contour map based on the top of the reef, the outstanding reservoir in the field. The productive area of this reservoir is shown by shading. The oil-water contact is at an approximate elevation of −4500 feet; the average depth to this level is about 7000

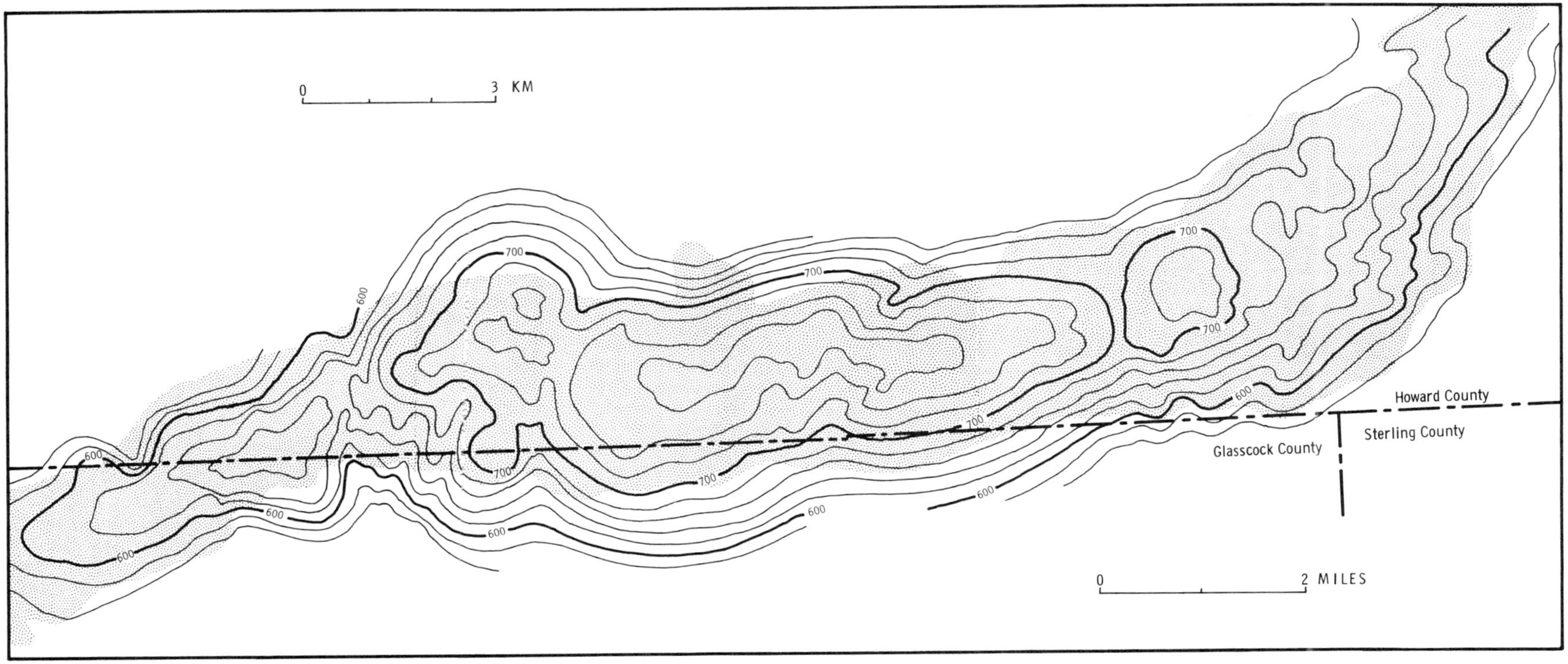

Figure 31.8 **Howard-Glasscock field, Howard and Glasscock Counties. Contour structure map based on top of Grayburg–San Andres. Interval, 20 feet. Area productive from Grayburg–San Andres is stippled.** ***Courtesy Tom D. Mooney (1966, p. 186) and the West Texas Geological Society.***

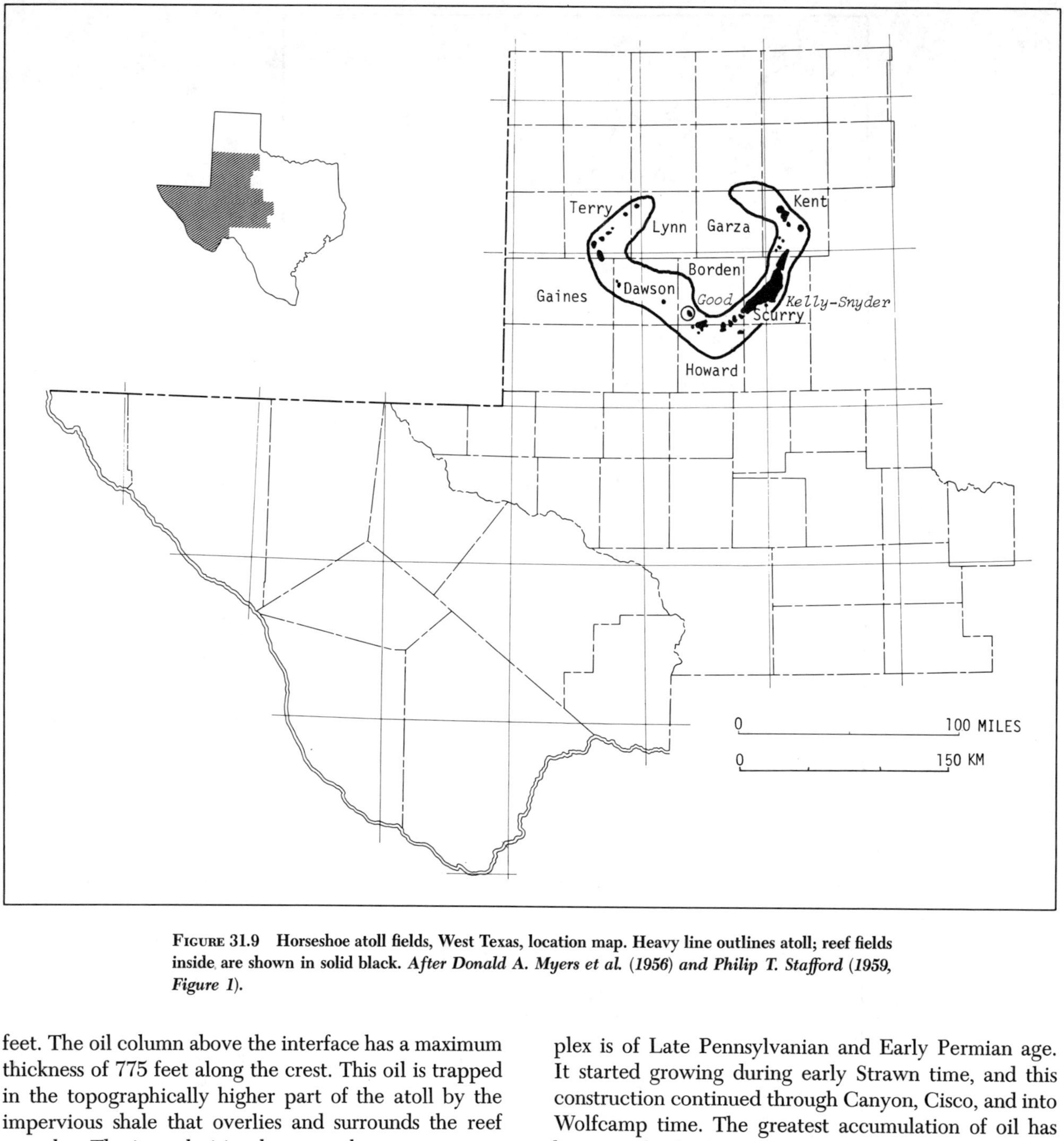

FIGURE 31.9 Horseshoe atoll fields, West Texas, location map. Heavy line outlines atoll; reef fields inside are shown in solid black. ***After Donald A. Myers et al. (1956) and Philip T. Stafford (1959, Figure 1).***

feet. The oil column above the interface has a maximum thickness of 775 feet along the crest. This oil is trapped in the topographically higher part of the atoll by the impervious shale that overlies and surrounds the reef complex. The irregularities shown on the structure contour map of the reef surface are characteristic of such surfaces.

A cross section of the northern part of the Kelly–Snyder field is shown in Figure 31.11; the location of the section is given in Figure 31.10. Although the principal reservoir is the reef, some production is obtained from erratic sands enclosed in shale, which have been arched by differential compaction of the shale.

The stratigraphy in the Scurry County sector of the Horseshoe atoll is shown in Figure 31.12. The reef complex is of Late Pennsylvanian and Early Permian age. It started growing during early Strawn time, and this construction continued through Canyon, Cisco, and into Wolfcamp time. The greatest accumulation of oil has been in the highly porous reef limestones deposited during Canyon time. This is the reservoir designated by the shading in Figure 31.10. Actually six different zones of porosity have been recognized and named (A to F). Although the Canyon limestones have an average porosity of over 10 per cent, interbeds of impervious lithographic limestone and shale are present (Stafford, 1957, p. 296). Other reservoirs have been found to be productive in some parts of the field. These include four zones in the underlying Strawn and three sands in the higher Wolfcamp. No great success has attended deeper

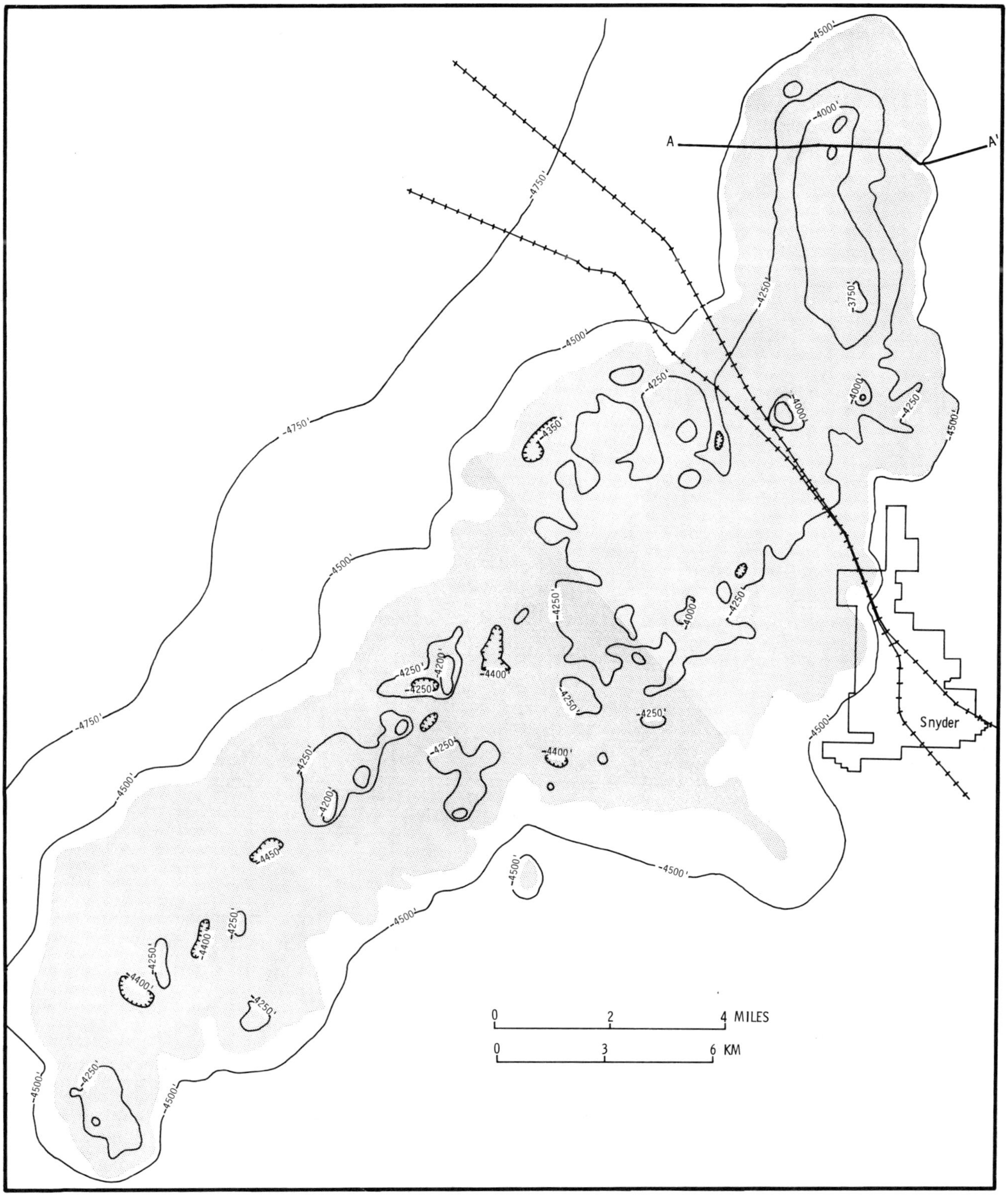

FIGURE 31.10 Kelly-Snyder (Scurry Reef) field, Scurry County. Contour map based on top of reef ("Canyon" reservoir). Interval, 250 feet, with 50 feet on locally closed contours where indicated. Gray shade denotes productive area of reef. Line *A–A′* is line of section shown in Figure 31.11. *After Philip T. Stafford (1957, p. 301). Courtesy University of Texas, Bureau of Economic Geology.*

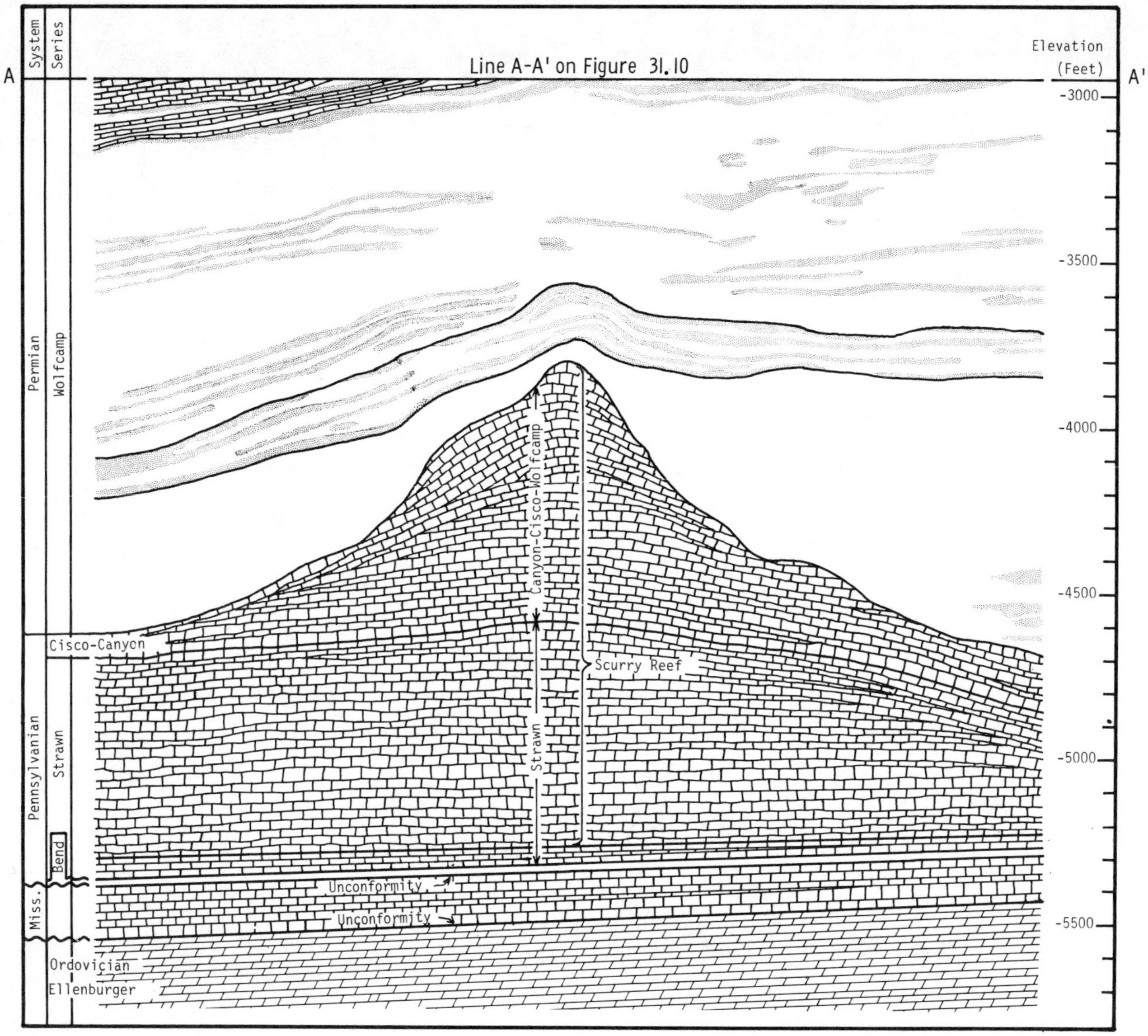

Figure 31.11 Kelly-Snyder oil field, cross section along line *A–A′* in Figure 31.10. West-east section through Scurry reef. *After Philip T. Stafford (1957; section B–B′ on tip-in following p. 300).*

drilling; the major anticlinal trap created by the reef does not extend below the reef floor.

Atoll accumulations in the western sector of the Horseshoe atoll are confined to the highest nodes along the reef complex (Burnside, 1959). Counties with production from reef traps include northwestern Howard, southwestern Borden, Dawson, northeastern Gaines, southeastern Terry, and northwestern Lynn.

The *Good* field (Schreiner, 1957) in southwestern Borden County is an example of a reef trap in the western sector. The structure contours drawn on the top of the Pennsylvanian reef show an almost perfect quaquaversal dome with a total productive area of about 2000 acres. The vertical distance from the top of the reef to the oil-water interface originally was nearly 500 feet. The reservoir thins to zero thickness, however, only a short distance beyond the periphery of the producing area, so that the trap was filled almost to capacity. The reef reservoir is a crystalline to chalky limestone with pinpoint to vuggy porosity. There is also some evidence of fractures. The age of the limestone is mainly Cisco, which is the top of the Pennsylvanian section in this area. Although there are a few local thin black shales within the reef, the source rock is presumably the much thicker, compacted black Permian shale that surrounds and overlies the reef.

Matador Arch. The Matador arch is a structural divide lying between the Northwestern shelf and the

Midland basin on the south, and the Palo Duro basin to the north. It crosses the West Texas district from east to west through the northernmost tier of counties. Trapping on the Matador arch is illustrated by the *Anton–Irish* field. The main part of this oil field is in the southwestern corner of Hale County, with extensions to the west into southeastern Lamb County and to the south into Lubbock County. The discovery well, which was located by a combination of geophysical data and subsurface geology, was completed in January, 1945. Subsequently this field has produced nearly 50 million barrels of oil.

Figure 31.13 is a structure contour map of the Anton–Irish trap, based on the top of the Lower Clear Fork (Permian) section, in which the reservoir rock occurs (Robinson, 1966). The reservoir is a finely crystalline dolomite with an overall thickness of about 400 feet; its top is about 300 feet below the top of the Lower Clear Fork. The dolomite occurs in three facies, shelf, porous shelf-edge, and basin, with production coming from the porous shelf-edge facies. This section is present along the crest of the elongate Anton–Irish anticline and also extends down a southeasterly plunging nose some 700 feet in the Lubbock County sector of the oil field. Trapping is anticlinal, with porosity controlling distribution on the structure. The Anton–Irish anticline coincides with the axis of the Matador uplift; beneath the field Wolfcampian (Lower Permian) sediments rest directly upon Precambrian basement rock.

SYSTEM	HORSESHOE ATOLL AREA	
	Rock Type	Age
Permian	Bedded carbonate and terrigenous rocks	Leonard
Permian	Bedded carbonate and terrigenous rocks	Wolfcamp
Pennsylvanian	Reef complex (mainly nonbedded limestone)	Cisco
Pennsylvanian	Reef complex (mainly nonbedded limestone)	Canyon
Pennsylvanian	Reef complex (mainly nonbedded limestone)	Strawn
Pennsylvanian	Bedded limestone	Strawn
Pennsylvanian	Limestone and shale	Bend
Mississippian	Limestone	

FIGURE 31.12 **Kelly-Snyder stratigraphic column.** ***After Philip T. Stafford (1959, Table 2, p. 6).***

MIDLAND BASIN. The Midland basin and the Delaware basin are major downwarps in the Permian basin; they are separated by the Central Basin platform. The foremost "field" in the Midland basin is the Spraberry trend, which was second in West Texas production in 1966, with 20 million barrels. The *Spraberry trend* (Elkins, 1953) contains a considerable number of oil fields near the middle of the Midland basin (Figure 31.14) that produce partly or entirely from a fine-grained sandstone in the Spraberry Formation in the Leonard Group of the Permian (Marshall, 1952). This reservoir is notable for its large area of productivity, both actual and potential, and for the reluctance with which it yields its oil. Some have referred to the Spraberry trend as "the world's largest reserve of unrecoverable oil."

The discovery well for the Spraberry trend was also the discovery well for the *Tex–Harvey* field in eastern Midland County. It was completed in February, 1949, and initially yielded 135 barrels of oil daily by pumping. Subsequent development of the Tex–Harvey field showed that the average depth to the top of the Spraberry is about 7000 feet; the formation extends downward to approximately 8000 feet, with over 300 feet of pay in two sections separated by a shale break. Typical wells came in with a potential of about 250 barrels a day. Subsequently, development proceeded to both the south and the north of Tex–Harvey, with the greatest concentration to the south in contiguous parts of Midland, Glasscock, Reagan, and Upton Counties. To the north production moved into southeastern Martin County, and still farther north into southeastern Dawson County and western Borden County. The Joe Mill field in southwestern Borden County is described subsequently.

The rapid expansion of Spraberry production was followed by a period of doubt as to whether or not exploitation of the trend would ever be profitable on account of erratic permeability and relatively low average yield per well. Subsequently, starting about 1955, a series of encouraging developments took place so that the Spraberry began to pay out, for some operators at least. New discoveries included production, in part of the trend area, from upper Leonard limestone about 500 feet above the Spraberry (Figure 31.14). More recently oil has been found in the deeper Dean Sand. Refracturing of the Spraberry has been found helpful in some areas, and waterflood projects are now under way.

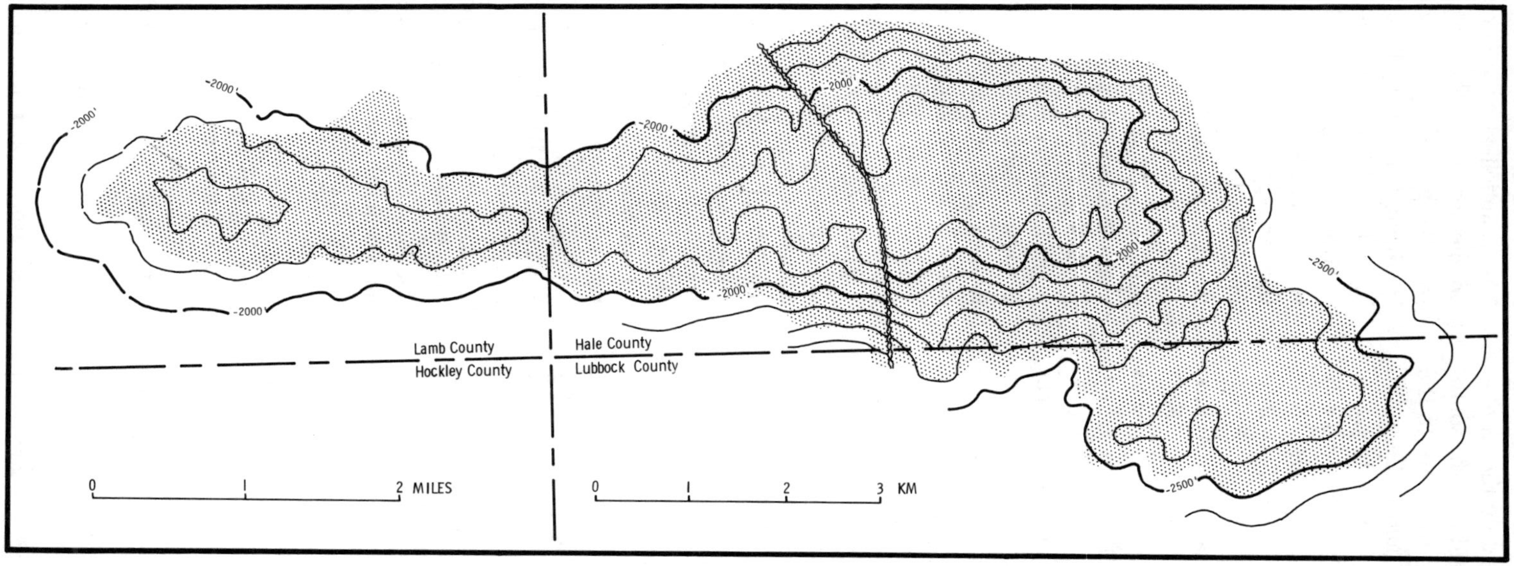

FIGURE 31.13 Anton-Irish field, Hale, Lamb, and Lubbock Counties. Structure contour map based on two beds (different datum areas separated by chain line) in lower Clear Fork Formation of Permian age. Contour interval, 100 feet. Producing area stippled. *Courtesy Joe D. Robinson (1966, tip-in following p. 22) and the West Texas Geological Society.*

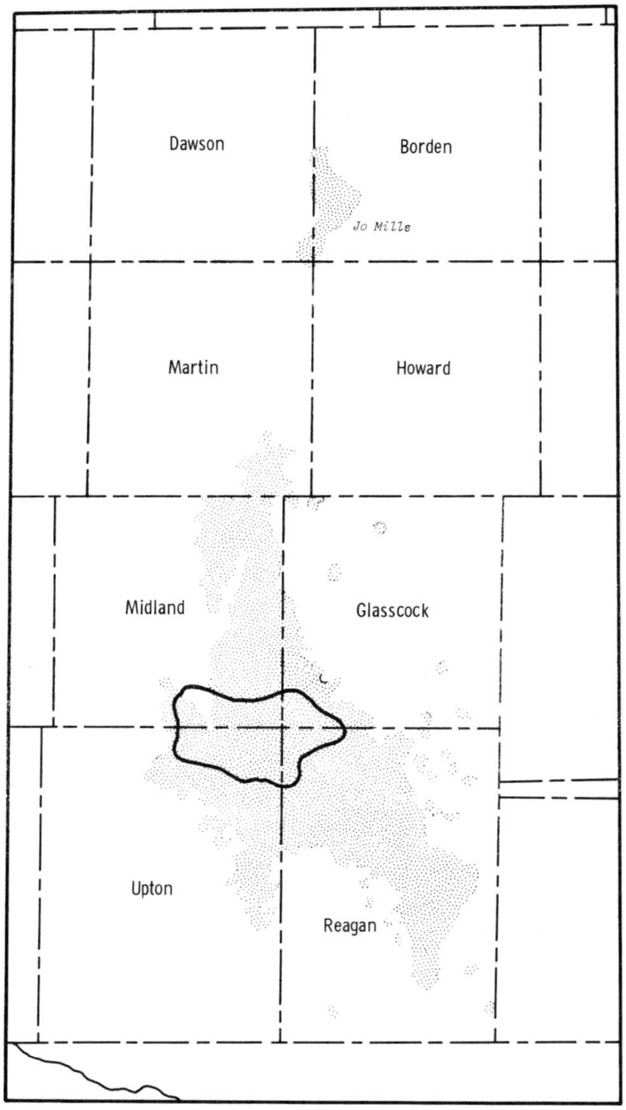

FIGURE 31.14 Spraberry trend map, Midland basin. Stippling shows areas of Spraberry production. Heavy line encloses area that is also producing from higher upper Leonard pay zone. Four square counties to north are each approximately 30 miles (48.3 km) on a side. *After Robert J. Enright (1957, p. 90) and the Oil and Gas Journal* (Aug. 15, 1966, p. 47).

Regionally the Spraberry trend occupies a monocline that dips westward at an average rate of 50 feet per mile (Bartley, 1952). Locally, as in the Tex–Harvey field, accumulation has taken place on a terrace (Warn, 1953), but this appears to be the exception rather than the rule. The Spraberry reservoir is more of a siltstone than a true sandstone. It has intergranular porosity of less than 10 per cent with permeability ranging from 0.01 to 0.4 md (Ogden and Locke, 1952). The major void space in the reservoir is in nearly vertical fractures that occur in both the massive siltstone and the shale and shaly limestone interbeds (Marshall, 1952, p. 2189). The fractures are interconnecting, as shown by communication between wells. Stratigraphically the Spraberry consists of an upper member about 300 feet thick, of fine sand or siltstone containing silty shale interbeds, succeeded downward by about 250 feet of shaly limestone and calcareous shale. Below is the lower Spraberry Sandstone, also approximately 300 feet thick (Schmitt, 1954). Trapping of oil in the Spraberry is due mainly to the updip lensing out of permeability. However, trapping also takes place along the trend, where anticlinal closures are present. Numerous fields in the Midland basin produce from pre-Spraberry reservoirs in anticlinal traps; before exhaustion and abandonment of these deeper reservoirs, the possibility of exploitation of the uphole Spraberry no doubt will be investigated.

The *Joe Mill* Spraberry field is in southwestern Borden County, some distance north of the main part of the Spraberry trend. It extends westward a short distance into Dawson County. This field, discovered in 1953, produced about 35 million barrels of oil in its first 10 years. The structure contour map (Figure 31.15) shows a westerly dip across the field of between 50 and 60 feet per mile. According to Jons (1966), Joe Mill produces from a series of lenticular sandstones that are separated from each other by black shales and tight sandstones. The productive part of the sandstone section diminishes to zero thickness on three sides of the field. Trapping is due to the updip wedgout of these reservoirs. The average pay thickness is 40 feet. The principal porosity is due to the presence of vertical fractures.

Another important field in the Midland basin is the *Pegasus* (Harbison, 1957). This field is in the southwest part of the Midland basin lying partly in southwest Midland County and partly in northwest Upton County. The field was discovered in 1949; the location for the discovery was made from subsurface data based on seismograph surveys and core drilling.

The Pegasus field has six reservoirs ranging downward in age from San Andres sand to Ellenburger. However, 77 per cent of the cumulative production has come from the Ellenburger and over 21 per cent from the Bend, leaving less than 2 per cent to be shared by the other four reservoirs. The Ellenburger reservoir is crystalline dolomite; the Bend, limestone in zones separated by shale breaks.

The trap for the main reservoirs at Pegasus is an elliptical north-south anticline, which bulges to the south and tapers to the north. The closed area on the top of the Ellenburger is about 7 miles long and has a maximum width of 4 miles. The crest of the anticline is a series of nodes. The maximum oil column in the Ellenburger has a height of over 800 feet.

The *Benedum* field (Deegan, 1949) lies along the Upton-Reagan county line in the southern part of the

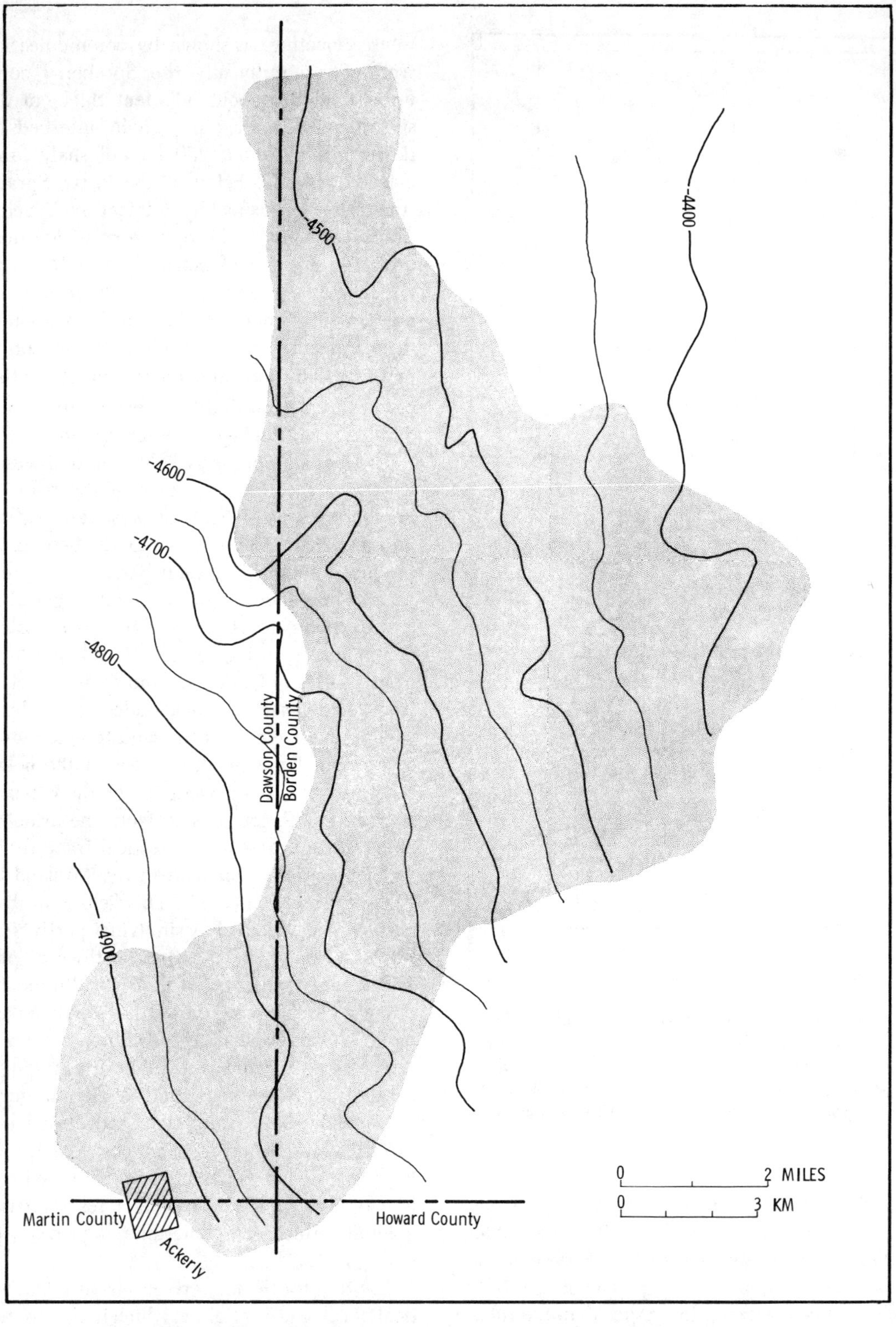

Figure 31.15 Jo Mills oil field, Borden and Dawson Counties. Structure contour map based on top of Jo Mill zone (lower Spraberry-Leonard). Interval, 50 feet. Producing area stippled. *Courtesy Richard D. Jons (1966, p. 210) and the West Texas Geological Society.*

Midland basin. It produces from Pennsylvanian (Bend), Silurian, and Ellenburger reservoirs. The trap is a pronounced dome, probably faulted on the east flank.

The *Big Lake* oil field (Hennen, 1929) in southwestern Reagan County is also in southern Midland basin. It is notable not only as a giant oil field (120 million barrels by the end of 1966) but also as the first major field in West Texas. The discovery well was completed on May 28, 1923, in the Grayburg Dolomite (Permian) at a depth of slightly over 3000 feet. This well was located from surface geology; a low anticlinal dome was recognized in the Edwards Limestone and other Lower Cretaceous rocks that crop out in this area. The closure in the surface formations is only about 30 feet, but in the Grayburg the closure is 250 feet. The Grayburg reservoir is a highly porous oölitic dolomite averaging 22 feet in thickness. It ranges in depth from 2800 to 3100 feet. Deeper reservoirs include a 15 foot sandstone of Spraberry age at a depth of 6300 feet, Fusselman Limestone low on the flank at 8950 feet, and Ellenburger (Ordovician) Dolomite at a depth of about 8500 feet.

In the southeastern corner of Reagan County is the *Barnhart* oil field (Cotton, 1966). It was discovered in 1941 after geophysical surveys. The trap is a truncated anticline (Figure 31.16) in the Ellenburger Dolomite lying at a depth of 9000 feet. Most of the oil accumulated in the uppermost 250 feet of domal closure. The reservoir permeability is largely dependent on fractures, which are developed best at the top of the structure.

Ozona Platform. The Ozona platform, which protrudes into the southern part of the Midland basin, is an older easterly extension of the southeastern corner of the Central Basin platform. Another name used for the Ozona platform is the Pecos arch.

The *Todd* field (Imbt and McCollum, 1950) in Crockett County is on the Ozona platform. It produces in part from a crinoidal reef limestone within the Strawn. The upper surface of the crinoidal limestone is domelike, but production is limited on the east side by the removal by erosion of the limestone in that direction. This reef is merely a protuberance on the west flank of a large dome in the Ordovician Ellenburger Dolomite. The Ellenburger is productive clear across the dome.

Northwestern Shelf. The Northwestern shelf connects farther east with the Eastern shelf. Although relatively small in area, the Northwestern shelf contains three gigantic fields: the Wasson (described in the following paragraphs), the Slaughter, and the Levelland. The aggregate cumulative production for these three fields was well over 900 million barrels by the end of 1966 and should reach a billion barrels before the end of the decade.

The *Wasson* oil field lies at the southeastern end of the Northwestern shelf. It is mainly in southwestern Yoakum County but extends slightly into northwestern Gaines County to the south. The western edge of the field is only about 4 miles from the New Mexico boundary. Wasson is second only to Yates in West Texas in its cumulative production. By the end of 1966 this 30 year old field had produced nearly 450 million barrels of oil. The yield during 1966 was 13 million barrels. Wasson was discovered through a combination of subsurface geology and gravity and magnetic surveys. It covers approximately 60,000 acres.

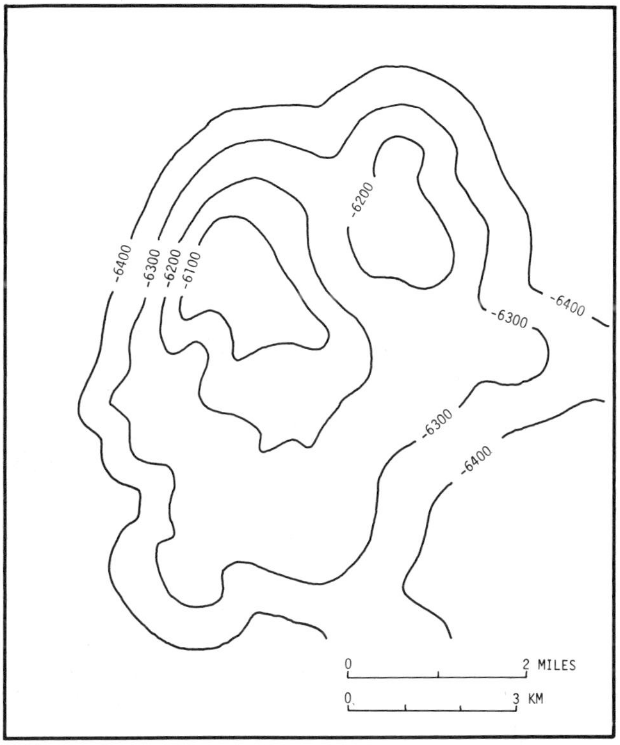

Figure 31.16 Barnhart oil field, Reagan County. Structure map based on top of eroded Ellenburger. Interval, 100 feet. Almost all of the producing wells lie inside of the −6,300 foot contour. ***Courtesy Clyde B. Cotton (1966, p. 38) and the West Texas Geological Society.***

The Wasson field is triangular in shape (Figure 31.17), "because of a tectonic axis on the southwest side of the field, and a reflection of the older Wichita–Abo reef on the southeast" (T. S. Jones, letter dated 4 September, 1968). Structure contours based on top of the principal reservoir formation, the San Andres, show an anticlinal trap with distribution controlled in part, however, by permeability. "The San Andres is backreef, and the porosity gradually disappears northwestward as the dolomite sands grade to dolomite mud" (T. S. Jones, *loc. cit.*). The top of the anticline is an undulating surface similar to the crest of the Horseshoe atoll to the east. The San Andres pay overlies earlier reef construction that took place during Permian time concurrently with the subsiding of a pre-Permian high.

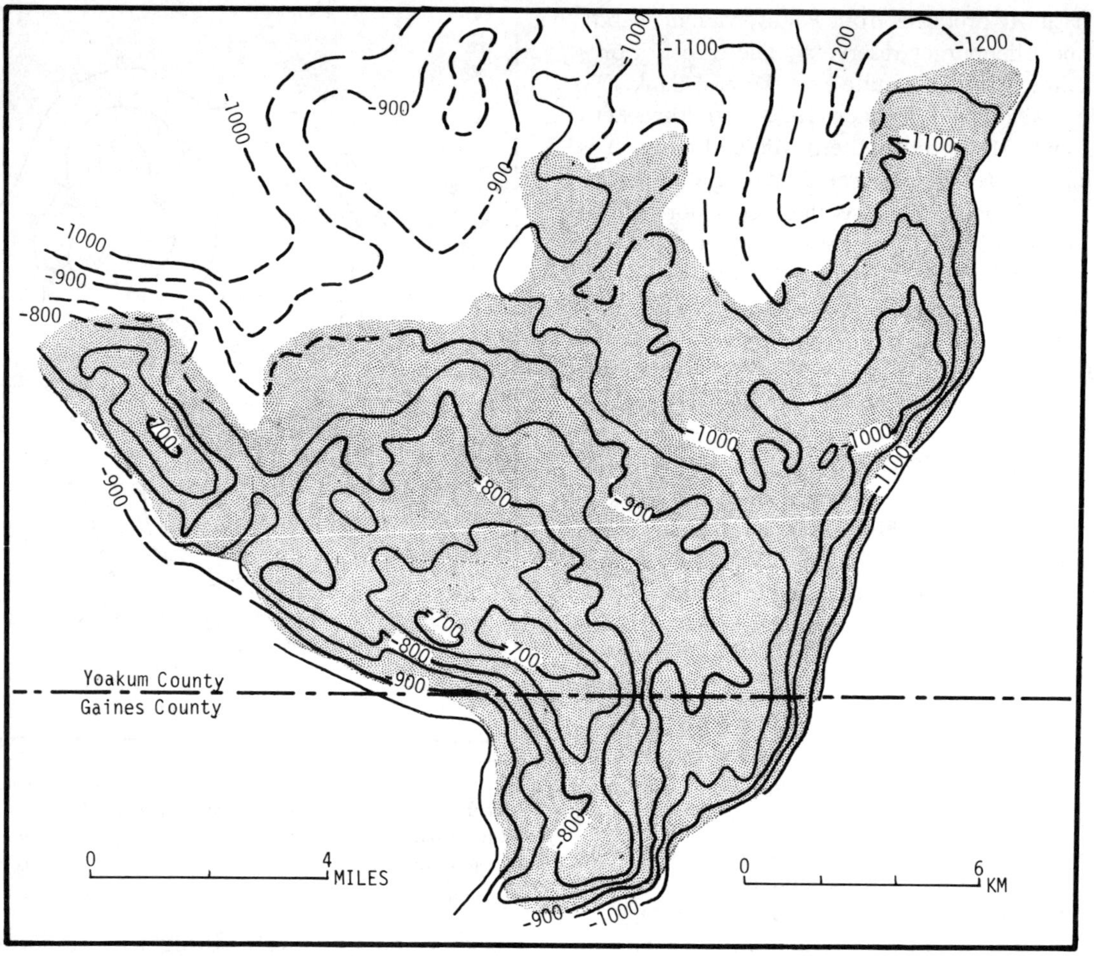

FIGURE 31.17 Wasson field, Yoakum and Gaines Counties. Structure contour map based on top of San Andres (Permian) Formation. Interval, 50 feet. Producing area shown in gray. *Courtesy W. T. Schneider (1957, p. 383) and the University of Texas.*

The principal reservoir is a finely crystalline dolomite lying from 350 to 500 feet below the top of the San Andres Formation in the Upper Permian. The overlying San Andres is also dolomite but lacks the porosity to yield oil and gas. The porosity of the reservoir rock is of three types: (1) small cavity-type voids, (2) fossil casts, and (3) intergranular pores between dolomite granules. Relatively minor production is also obtained from two somewhat deeper reservoir dolomites of the Clear Fork Group and also from the Wichita–Abo reef. A higher sand near the top of the Permian, the Yates, yields a nonflammable gas at pressures up to 2800 pounds per square inch. This gas contains 97 per cent nitrogen and is noncommercial.

A waterflood, started at Wasson in 1965, is expected to yield an additional 400 million barrels of oil.

CENTRAL BASIN PLATFORM. The Central Basin platform not only contains the greatest number and greatest density of oil and gas fields, but also leads all other West Texas subdivisions in both cumulative and current annual production. The Yates field, to be described subsequently, is the foremost field of West Texas, with more than $\frac{1}{2}$ billion barrels of oil to date. In addition numerous other giant and near-giant fields are on the Central Basin platform.

The subsurface geology explains the fecundity of the Central Basin platform. The elevation of the rocks relative to the basins on either side puts all possible reservoirs within 11,000 feet of the surface. The stratigraphic section is replete with reservoir beds, including porous and permeable carbonate rocks as well as sandstones. The post-Wolfcamp reservoirs are sealed from the surface by a thick blanket of dense dolomites and evaporites, including both anhydrite and salt. Possible source rock shales are present in the lower part of the section and on the fringes. Furthermore, these shales thicken greatly into the basins on either side, so that the platform has a large source and gathering area on both the east and the west. The tectonic history of the platform has been such as to produce both anticlines and porosity wedgouts.

The cross section (Figure 31.3) shows the higher elevation of the Central Basin platform above the Val Verde and Midland basins. Formations on the platform lie as much as 14,000 feet above contemporary deposits in the Val Verde basin, and 2000 feet above those in the Midland basin. The mechanics of the Central Basin platform folding have been discussed by Harrington (1963).

The super-giant *Yates* field is immediately to the west of the Pecos River in eastern Pecos County. Tectonically and stratigraphically it lies at the southeastern tip of the Central Basin platform. According to one account, seeps had been noted on the gigantic Yates ranch by the owner himself: "while riding on horseback across the arid mesa, he [Yates] had spotted oil seeps and had seen oil slicks on the alkaline waters of the Pecos" (Marathon Oil Co., 1966). Furthermore, erosion of the land had exposed Lower Cretaceous limestone ledges that made possible the surface mapping of the large Yates anticline. Exploration by the drill bit may have been delayed somewhat by the myth that "there is no oil west of the Pecos River." An exploratory test, located very close to the top of the anticline, by geologist Frank Clark was drilled in late 1926. It came in as a flowing well at a depth of 1004 feet for an initial production of 450 barrels a day. However, it was later deepened, raising its potential to more than 72,000 barrels a day. Three years later a possible world's record was established when well No. 30 was completed at a depth of 1070 feet with an official proration test of over 8500 barrels an hour, or a daily potential of more than 200,000 barrels. After the field had become 8 years old, the second largest well was completed with an initial production of 7000 barrels per hour (over 168,000 barrels per day).

One result of the high productivity of the Yates field was the initiation of voluntary proration, the first in Texas. By the end of 1966 the Yates field had produced more than ½ billion barrels of oil and 170 billion cubic feet of gas. Furthermore, 40 years after discovery, out of 632 wells drilled in the entire field nearly 500 were still flowing under natural pressure and only 27 had been abandoned.

Two years after the completion of the discovery well, new seepages at the surface were observed along the Pecos River. Recovery operations were started immediately, and oil was obtained by skimming the water along the river, "by trenching, by digging pits, and by drilling shallow wells" (Donoghue and Gupton, 1957, p. 427). During the calendar year 1929 over 1,700,000 barrels of oil was produced in this manner and nearly a million barrels the following year. Analyses of the oil confirmed the supposition that it had escaped from the 1000 foot deep reservoir through imperfect casing installations. Remedial work on the wells led to a diminution of seepage oil, and by 1955 the annual recovery from this source had dropped to below 3000 barrels a year. During this period of upward leakage, however, an estimated 70 million barrels of oil, of which about 25 million has so far been recovered, moved into the basal Cretaceous Toborg sand. The Yates Sandstone of Late Permian age was also a haven for this fugitive oil.

The trap at Yates is an anticline (Figure 31.18). Contours based on a marker in the principal reservoir show a high of over 1450 feet above sea level elevation (nearby Pecos River is at 2150 feet) near the southeastern end of the field with lesser nodes down the main plunge of the anticlinal axis to the northwest and also down a lesser nose to the west. The structural closure is at least 300 feet, and the structural relief is somewhat over 700 feet.

The stratigraphic section for the Yates field is shown in Figure 31.19. More than 93 per cent of the oil produced in the field has come from the Grayburg Dolomite reservoir, and this figure would be close to 100 per cent if the oil that seeped upward into younger reservoirs during the development of the field was credited to its original reservoir. The main Yates pay contains rocks of the Queen, Grayburg, and San Andres Formations. These formations all lie within the Guadalupian Series of the Upper Permian. The Queen Formation is essentially anhydrite; where this part of the Yates reservoir is productive, the oil is stored in a basal sand that immediately overlies the Grayburg. This sand is not everywhere present. Below it is a dolomite with vuggy to cavernous porosity and extremely high permeability. This section has been correlated with the dolomite pay in the Big Lake field of Reagan County. The sand reservoir above the dolomite varies from zero to 120 feet in thickness, and the producing dolomite is at least 225 feet and may be considerably greater. The dolomite reservoir is productive under nearly 2100 acres. Yates field owes its spectacular performance to a combination of megaporosity, abnormally thick reservoir section, considerable closure, and adequate oil to fill the trap.

Imediately to the north on the Pecos–Crockett county line the Tippett oil field produces principally from lowermost Wolfcamp conglomerate lenses at the base of the Permian (Hills, 1961).

The *Apco* field (Ellison, 1948), like the Yates, lies in Pecos County south of the Pecos River. It is about 50 miles to the northwest of Yates. Apco is an excellent example of trapping in an overlapped truncated dolomite wedge. The main producing zone lies within the Ellenburger Formation, which here is about 800 feet thick and consists primarily of fractured siliceous dolomite. It was weathered and leached during pre-Permian

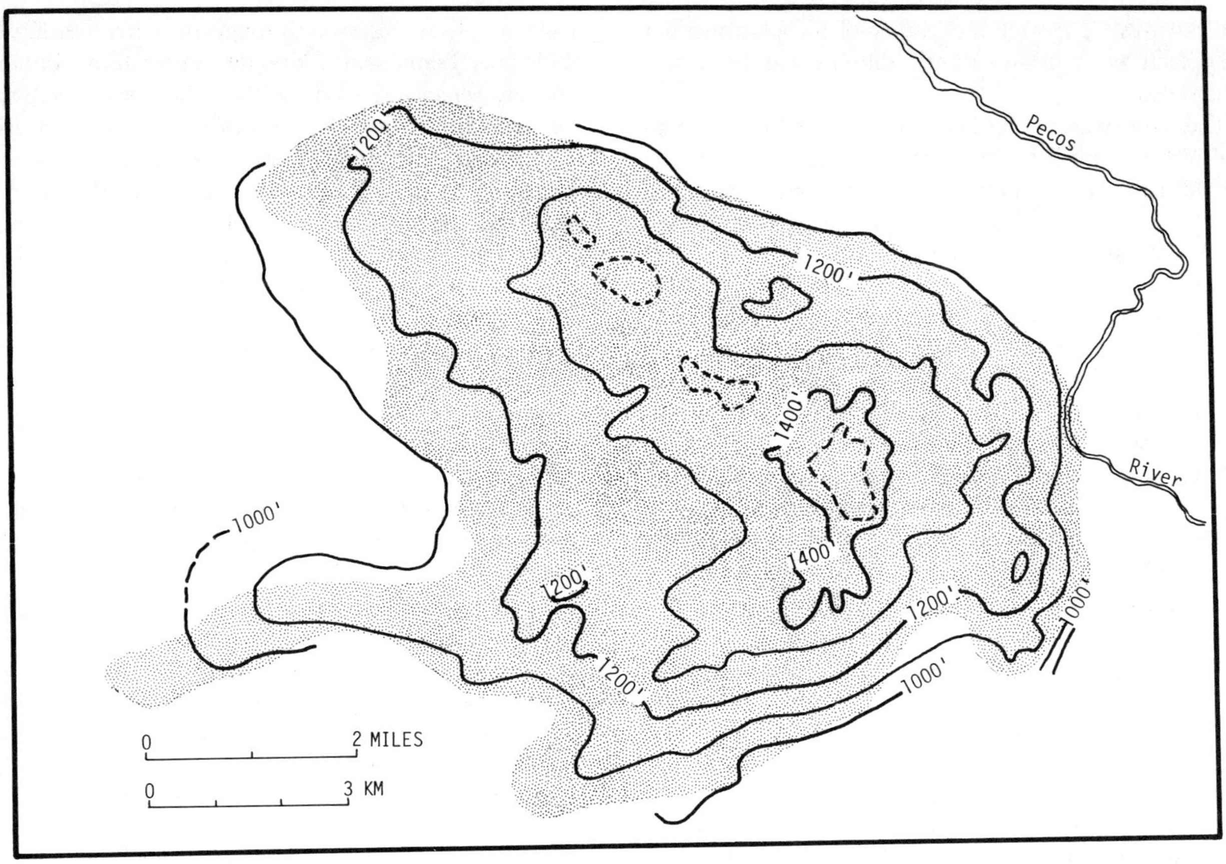

FIGURE 31.18 Yates field, Pecos County. Structure contour map based on top of Grayburg (Permian) Formation. Contour interval, 100 feet; dashed contours at 50 feet on nodes. Shading shows productive area of Yates Lime reservoir. ***Courtesy David Donaghue and W. L. Gupton, Jr. (1957, p. 431), and the University of Texas.***

emergence, which augmented the porosity. Accumulation is in a buried hogback of Ellenburger Dolomite, which laps upon the flank of a hill of Precambrian basement rock. The trap is sealed by a blanket of Permian shale overlying not only the dip slope of the hogback but also the scarp face and the Precambrian hill. Oil, probably squeezed from the overlying shale into porous zones in the tilted Ellenburger Dolomite, was impounded beneath this seal.

Across the Pecos, extending along the west side of the platform in both West Texas and southeastern New Mexico, is the *"Sand Belt,"* an area of practically continuous production. The reservoir rock along this belt is Permian sandstone, and the trapping is due to a combination of anticlinal structure and a facies change eastward from permeability to impermeability (Denham and Dougherty, 1941).

On the opposite side of the platform, extending along eastern Crane County and crossing the line into Upton County, is the *McElroy* field. This is another giant field, having contributed nearly ¼ billion barrels of oil to the West Texas production statistics. McElroy was discovered by a combination of "trend drilling in conjunction with interpretation of a limited amount of surface geological data" (McInnis and Hartman, 1957, p. 203). Since the field itself is covered with Quaternary deposits, the surface geological data must have been obtained through extrapolation. Accumulation has taken place at the top of an elongate anticline with a north-northwest trend. The productive area is about 14 miles by 3 miles at its widest point. The reservoir is the Grayburg Formation of the Guadalupe (Permian). The main productive zone is an oölitic and anhydritic fossiliferous dolomite. Porosity is both primary and secondary; the former is intergranular, and the latter has been due to the development of solution cavities.

Along the same trend, but one county to the north-northwest, is the *TXL* field (Cooper and Ferris, 1957). This giant field had produced over 200 million barrels of oil by the end of 1966. TXL is in western Ector County, about 20 miles west of Odessa, the county seat. It is in the central part of the Central Basin platform.

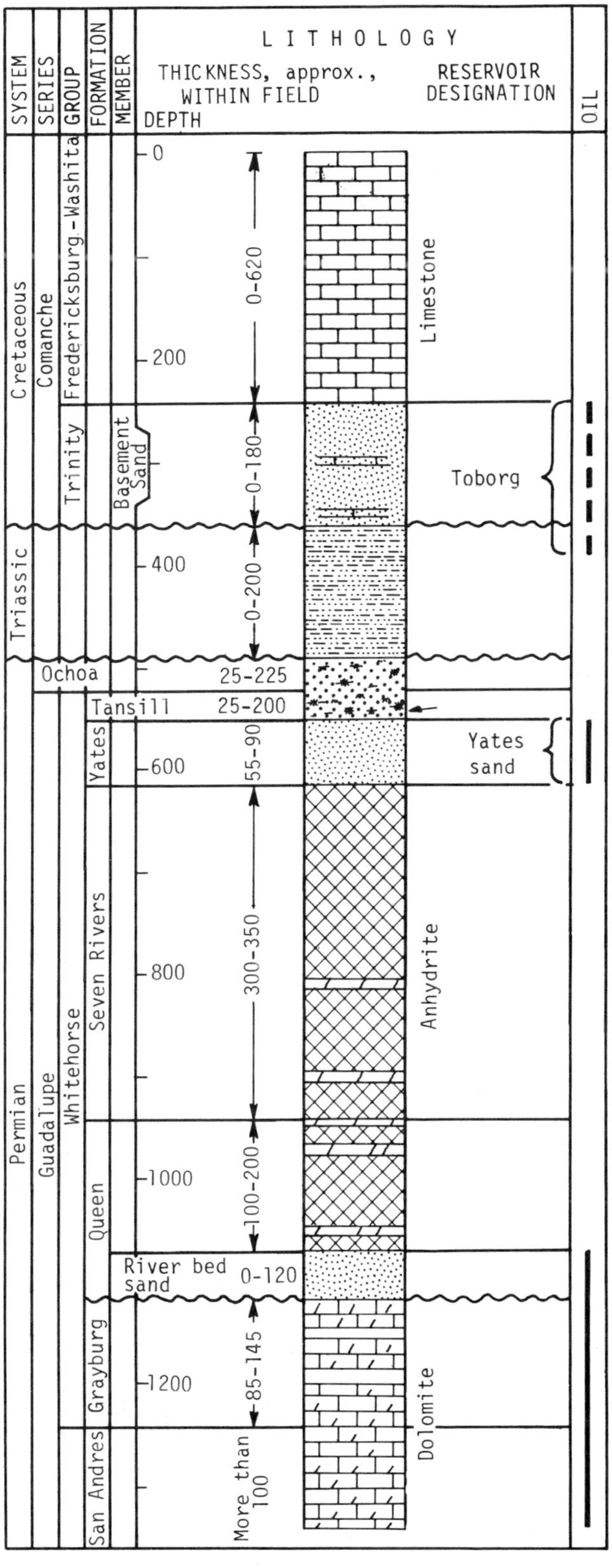

FIGURE 31.19 Yates field stratigraphic section. Crosshatched pattern is anhydrite. Vertical lines in oil column show the general location of reservoirs. *Courtesy David Donaghue and W. L. Gupton, Jr. (1957, p. 429), and the University of Texas.*

TXL, unlike McElroy, which is largely a one-reservoir field, has a history of oil and gas production from as many as eleven reservoirs. The leading sources of both oil and gas are Ellenburger and the Devonian reservoirs. The Ellenburger oil and gas come from brown to gray crystalline dolomite, with trapping due to the presence of an elongate anticline. The productive area is about 9 miles long, with a maximum width of $2\frac{1}{2}$ miles. To the south the anticline both tapers and is cut by a fault that drops the reservoir to the east and terminates productivity in that direction. The main Devonian reservoir is a highly permeable chert that has developed porosity apparently by fracturing during weathering, with the resultant crevices becoming filled with stream-carried clastics. Some additional porosity has been caused by the leaching of the chert to porous tripolite. Trapping in the Devonian reservoir has been due to updip termination of the reservoir through truncation by erosion and sealing by an overlap of tight Wolfcampian (Permian) sediments. Lateral closure is provided by the anticlinal nature of the rocks.

One of the famous fields of West Texas, noted for the cavernous nature of a part of the reservoir section, is the *Dollarhide* (Keener, 1957) in the southwestern corner of Andrews County. It is on the Central Basin platform, immediately east of the belt of fields that marks the western edge of the platform. Dollarhide was discovered by seismic data and subsurface geology; these two sources of information indicated the presence of an anticline with a structural relief of about 1200 feet. The discovery well was drilled in 1945.

The greater part of the Dollarhide output comes from pay zones in Devonian, Silurian, and Ordovician formations. The Devonian reservoir is the leading producer. It consists of crystalline limestone and dolomite with porous tripolitic chert. Second in volume of cumulative production is the Fusselman–Montoya. The Fusselman is Lower Silurian; in addition to good intergranular porosity throughout its 250 feet of section, it may have cavernous porosity locally. Open caverns 2 to 16 feet in height were encountered in fifteen wells drilled in the field. The Montoya part of the section is at the top of the Ordovician, and its porosity is continuous with that of the Fusselman.

The third reservoir, in volume of cumulative production, is the Lower Ordovician Ellenburger crystalline Dolomite. The porosity here is both vuggy and fracture. The fourth important reservoir is in the Lower Permian and is separated from the Lower Devonian reservoir over most of the field by a thin section of Woodford Shale (Upper Devonian). This reservoir includes scattered porous layers that extend through a thickness of about 1400 feet of Lower Clear Fork and Wichita–Abo. The productive zone is dolomite.

In every instance trapping is brought about by the presence of an anticline with a closure in excess of 700 feet. The main anticline is an elliptical dome with a long axis nearly north-south. A saddle, the lowest point of which coincides almost exactly with the New Mexico state line, separates this anticline from a nearly circular dome to the west, which constitutes the New Mexican Dollarhide field ("West Dollarhide"). Both the main anticline and the offspring are faulted by transverse and tangential faults. Distribution of the hydrocarbons within the trap has been modified by faulting. It is also controlled in the Devonian chert by an erosional unconformity; over the higher part of the anticline the chert is absent altogether, but where present on the flanks it has been leached of its carbonate content and has good porosity. In the Leonard, distribution is controlled partly by permeability barriers.

The Dollarhide is also a gas field, with cumulative production in excess of 200 billion cubic feet. Although all of the reservoirs produce gas along with the oil, about half of the total production has come from the Devonian.

In 1966 Andrews County was the largest oil-producing county in Texas. The western two-thirds is on the Central Basin platform, and the eastern third in the Midland basin. *Fullerton* (Moore, 1944) on the platform in northwestern Andrews County is a major oil field, having produced 186 million barrels through 1966. This field is also large in area; it is more than 12 miles long and has a maximum width of 6 miles. It trends north-northwest. Trapping is anticlinal; the principal reservoir is a porous crystalline dolomite, 350 to 400 feet thick, in the lower Clear Fork Group of Permian age. Other reservoirs are San Andres, upper Clear Fork, Wichita, Wolfcamp, Silurian, Simpson, and Ellenburger.

West Texas also has its *Seminole* field (Douglas, 1957), which is north of Fullerton in west-central Gaines county, on the northeast edge of the Central Basin platform. Here, as in Oklahoma, Seminole is a giant field with oil production well over 100 million barrels. In addition it has yielded large quantities of gas. The entire production has been obtained from three reservoirs, all of Permian age. The highest, the San Andres (Guadalupian), is anhydritic dolomite; the middle reservoir is Glorieta, which is very similar lithologically; and the lowest is Wolfcamp dolomite, which produces a little oil. The San Andres is the major source of gas. The trap is an elongate anticline with a productive area of over 13,400 acres.

Val Verde Basin (Hills, 1968a). A cross section of the Val Verde basin is shown in Figure 31.3. This basin and its sister basin to the northwest, the Delaware, reach depths greater than 25,000 feet. In fact, the deepest hole so far drilled anywhere in the world was drilled near the northwestern corner of the Val Verde basin immediately south of the Central Basin platform. It reached a total depth of 25,340 feet, still in the Ellenburger. The drill, in reaching that depth, crossed several faults "with some sections repeated as many as five times" (Peek et al., 1960, p. 782).

The Val Verde basin also contains the world's deepest production to date. The *Gomez* field at the western edge of the basin, next door to the Delaware basin, is producing gas at a maximum depth of 22,423 feet. Gomez is 9 miles northwest of Fort Stockton in Pecos County. Temperature studies of the deep holes in the Val Verde basin have led to the suggestion that the production depths are below the temperature "floor" at which liquid petroleum can exist through any length of geologic time (Landes, 1967, p. 839).

The discovery of the *Puckett* gas field (Holland, 1966) in 1952 led to intensive prospecting of the Val Verde basin. The discovery well produced at the rate of $12\frac{1}{4}$ million cubic feet of gas and 38 barrels of condensate per day from the Ordovician Ellenburger at depths of 13,100 to 14,300 feet. A second birthday took place 4 years later in 1956, when a 370 million foot gas well was completed in Devonian reservoir rock at depths between 10,350 and 10,460 feet. Some of the largest gas wells ever drilled have been completed in the Puckett field. Already this field has produced over a trillion cubic feet of gas, of which about three-fourths came from the Ellenburger.

The Puckett field is elliptical in shape, with the long axis trending northwest (Figure 31.20), and is over 8 miles long and 4 miles wide. The trap is an anticline faulted on the southwest flank. The closure is over 3000 feet on the Ellenburger. The anticline is buried beneath from 6000 to 12,000 feet of unconformable overlying Permian sediments. The buried rocks in the faulted anticline are pre-Mesozoic in age.

The Ellenburger reservoir rock is primarily dolomite some 1600 feet in thickness that is underlain by Precambrian basement rock. The Devonian reservoir consists of cherty limestone and dolomite. Ellenburger is present throughout the anticlinal area, but younger Paleozoic formations, including the Devonian reservoir, were truncated by erosion on the northeast flank of the anticline before its burial beneath Permian sediments. Because of this beveling the Devonian reservoir ranges in thickness from 130 feet to zero.

Another great gas field in the Val Verde basin is the *Brown–Bassett* (McMurtry, 1966), which was discovered in 1958. This field is in Terrell County to the southeast of the Puckett field. Most of the Brown–Bassett gas comes from Ellenburger reservoir rock. It contains 54 per cent CO_2 as against 46 per cent hydrocarbons.

Delaware Basin. The Delaware basin is larger in area and probably greater in maximum depth than the con-

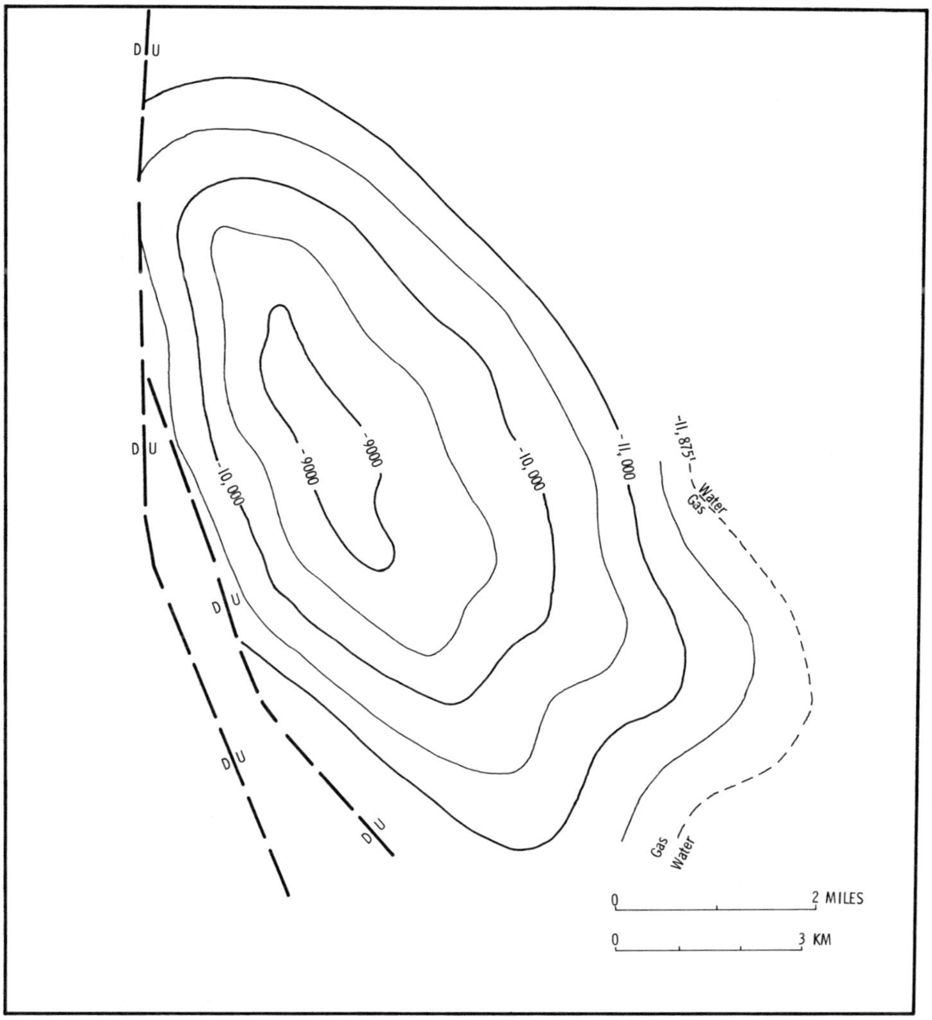

FIGURE 31.20 Puckett field, Pecos County. Structure contour map based on Ellenburger. Interval, 500 feet. Gas-water contact in Ellenburger reservoir at elevation −11,875 feet. ***Courtesy Richard R. Holland (1966, p. 290) and the West Texas Geological Society.***

tiguous Val Verde basin. So far only a start has been made in exploring the vast volume of basin sediments available here.

The geology of the Delaware basin has been discussed by Adams (1965) and Cook (1966). Holmquest (1965) describes the deep reservoirs available in both this and the Val Verde basin.

The *Coyanosa* gas and condensate field is in the southeastern sector of the Delaware basin. It is in Pecos County, about 21 miles northwest of Fort Stockton. The four reservoirs, proceeding downhole, are a Delaware sand and a conglomerate in the Wolfcamp, both of Permian age, a Devonian chert, and the Ellenburger Dolomite (Guinan, 1966). Figure 31.21 is a structure contour map of the Coyanosa area based on the top of the Ellenburger reservoir. It shows an elongate anticline covering about 9000 acres neatly packaged in a horst block. The summit area of the anticline is occupied by three nodes. The gas-water contact lies at an elevation of less than 13,170 feet, or about 1700 feet below the highest point in the trap. As in other deep Ellenburger fields, effective porosity in the dolomite is due mainly to the presence of fractures, which are most abundant along the structural crest. Producing depths are 15,000 feet and below.

To the southwest of Coyanosa, and just over the line in Reeves County, are three gas fields producing at depths in the 20,800 to 21,300 foot range. The traps are "north-south trending anticlines transected by east-west faults that are down to the south" (Langenbahn, 1968).

The *El Mar* field (Porter, 1966) is in the northeastern part of the Delaware basin. It lies in both Loving County, Texas, and Lea County, New Mexico, but the greater part of the producing area is on the Texas side

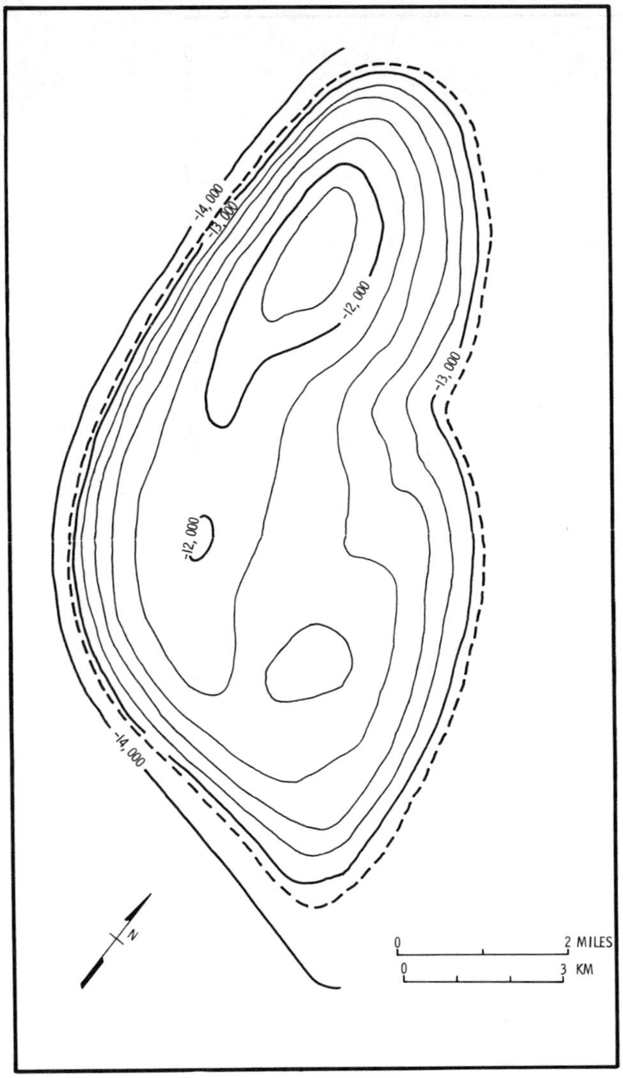

Figure 31.21 Coyanosa gas field, Pecos County. Structure contours based on top of Ellenburger. Dashed line is original gas-water interface (−13,170 feet). Contour interval, 200 feet. *Courtesy M. A. Guinan (1966, p. 104) and the West Texas Geological Society.*

of the state line. Figure 31.22 shows that the structure is a monocline with an easterly dip ranging from 75 to 100 feet to the mile. The reservoir rock is a sandstone up to 40 feet in thickness in the Bell Canyon Formation of Delaware (Permian) age. The sand is lenticular, wedging out to the northwest, south, and southeast, as shown by the bold dashed line in Figure 31.22. It is this updip loss of permeability and porosity that creates the trap. The sandstone grades into a laminated, relatively impervious siltstone. The oil column has a vertical height of approximately 200 feet; the depth to production is in the neighborhood of 4500 feet. The field was discovered by subsurface geology.

The *Ford* oil fields (Thomas, 1966b) lie on both sides of the Culberson–Reeves county line a short distance south of Eddy County, New Mexico. These fields are in the northern part of the Delaware basin. The east dip averages about 100 feet to the mile. Trapping (Figure 31.23) is due to permeability pinchouts. Reservoirs are two sands occurring within the upper Bell Canyon Formation of Delaware (Permian) age. The upper sand lies between 50 and 80 feet below the top of the Lamar Limestone. Below is 30 feet of impermeable shale, followed downward by the second sand, which is about 12 feet in thickness. Actually these sands are fairly consist-

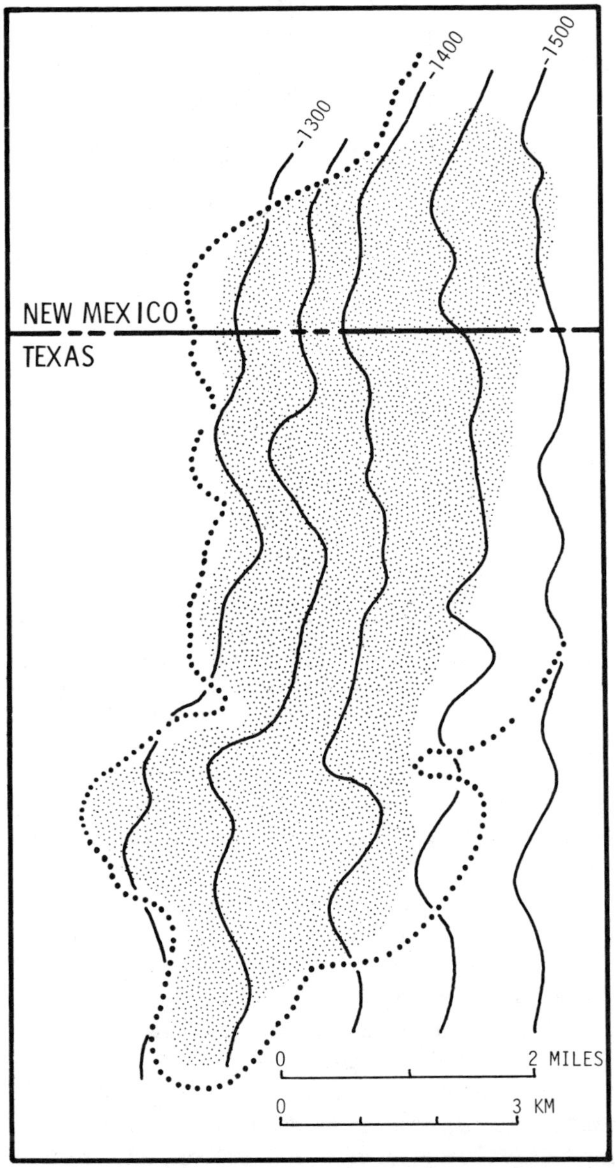

Figure 31.22 El Mar oil field, Loving County, Texas, and Lea County, New Mexico. Structure contours based on Lamar Shale. Interval, 50 feet. Stippled area is productive. Dotted line is zero isopach of reservoir sand. *Courtesy William C. Porter (1966, p. 129–130) and the West Texas Geological Society.*

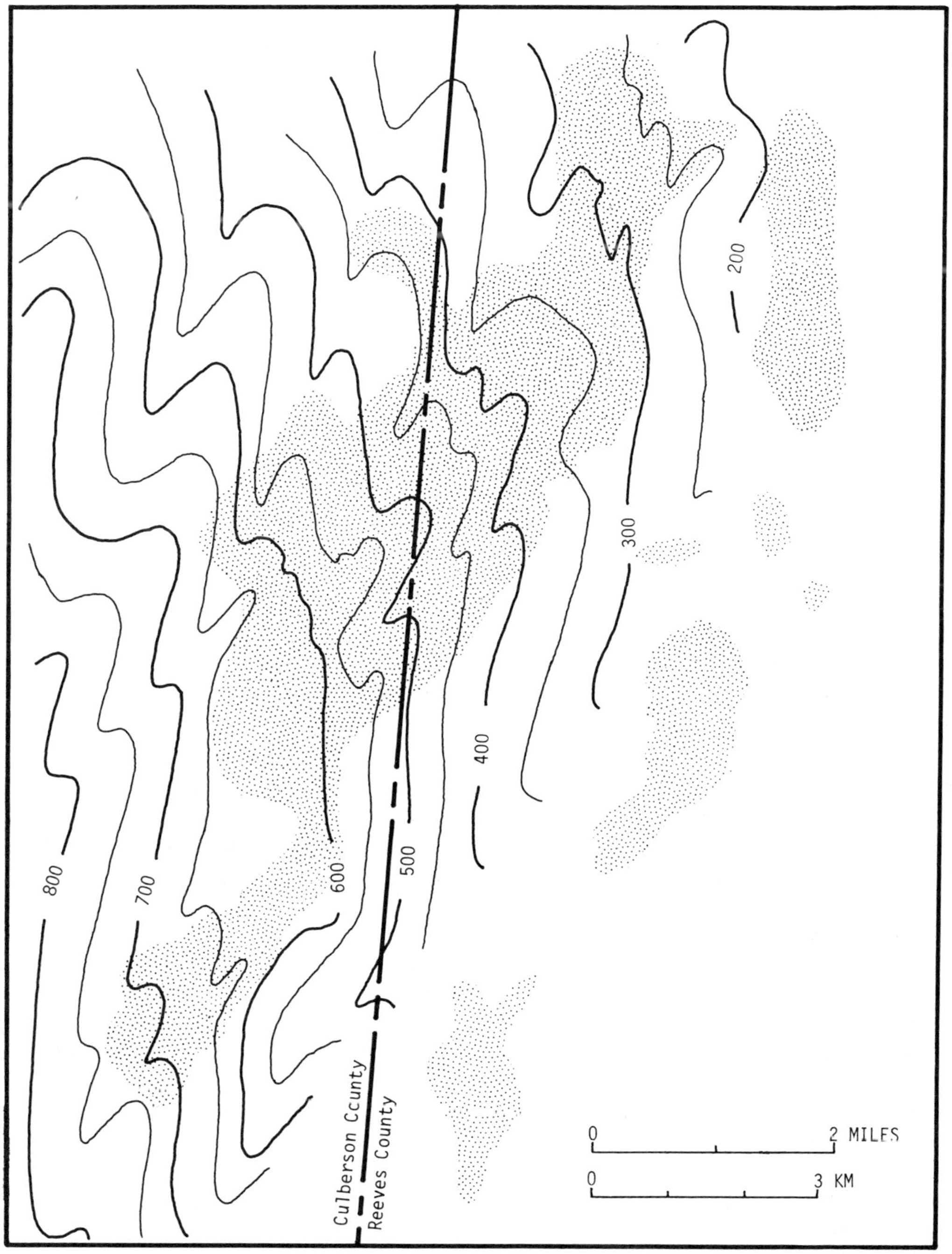

FIGURE 31.23 Ford fields, Culberson and Reeves Counties. Structure contour map based on top of Lamar Limestone. Interval, 50 feet. All contours are above sea level. Shading shows producing areas as of September, 1965. Largest accumulation is Geraldine Ford field. ***Courtesy Leonard C. Thomas (1966, p. 154) and the West Texas Geological Society.***

ent throughout the field, but they are low in permeability, because of fineness of grain, in the updip areas where nonproductive. The reservoirs lie above sea level at depths in the 2600 to 2700 foot range.

The *Tunstill* field (Thomas, 1966a) is an elongate north-south producing area bisected by the Pecos River, which is also the line between Loving and Reeves Counties. It is similar in many respects to the Ford fields, which lie a few miles to the west. The regional dip is to the east, and trapping is due to updip permeability pinchout. The principal reservoir is an upper Bell Canyon sand. It has a net pay thickness of about 14 feet. A lower and thinner sand is also productive in the northern part of the field.

The *Wheat* field (Reiter, 1957) is near the center of the Delaware basin in southwest Loving County. It produces oil from the Bell Canyon Sandstone at the top of the Guadalupe Series of the Upper Permian. Above the reservoir is about 4000 feet of evaporites belonging in the Ochoa Series of the uppermost Permian. The structure contour map of the producing area based on the formation immediately overlying the reservoir rock shows a gentle monocline dipping to the east, with a noticeable flattening in the productive area. The oil occurs where the dip is about 50 feet per mile; to the west and the east, above and below the oil accumulation, the dip is in the neighborhood of 100 feet per mile and the sandstone is water bearing. Various suggestions have been made as to the character of the trap that has produced this terrace accumulation. One is that it is a hydrodynamic trap (Hubbert, 1953); Reiter (1957) suggests either inadequate pressures to move the oil updip above the terrace or a decrease in permeability in that direction.

Future. Any district that contains a Yates field with more than ½ billion barrels of oil produced to date, and a Kelly–Snyder which is destined to exceed a billion barrels in eventual production, is obviously well endowed with source rocks, reservoirs, adequate gathering areas, and traps. This affluence is further confirmed by the large number of significant discoveries each year (36 in 1966). Most of these new finds are made in the present major producing areas, but the percentage that take place in the hitherto less productive parts of West Texas is increasing. This is especially true of the deep basins, where exploration has been slower because of higher cost involved in probing the deeper reservoirs. Continued exploration of the deep basins can be anticipated; in all probability this will result in notable increases in the domestic natural gas reserve, especially in the Val Verde and Delaware basins.

Of the many actual and potential reservoirs from Late Permian to Cambrian in age in West Texas (Wright, 1963; Reeves, 1964) none has been adequately explored throughout the district. The Abo reef (Wright, 1962) and the buried Capitan reef surrounding the Delaware basin (Wright, 1967) are examples. It was noted earlier that possible Spraberry production, bypassed on the way down to other reservoirs, will no doubt be explored and will be developed where profitable, as the output from the lower reservoirs dwindles. Furthermore, the oil fields shown in the Spraberry trend map (Figure 31.14) do not necessarily delineate the lateral extent of this reservoir. Successful probing downdip into deeper areas continues to yield additional Spraberry discoveries. Also, there is a possibility of infilling Spraberry production between southeastern Martin County and southwestern Borden County (Jo Mill field). Finally, the already large known reserve of oil in the Spraberry may be increased by successful waterflooding operations.

A most promising reservoir for production across future decades is the stratigraphically deep Ellenburger, which is a prolific source of oil or gas in many fields. It has not yet been explored extensively in the deeper basins. Likewise, the underlying Cambrian sandstone has future potential. One of the more intriguing possibilities is that of finding hydrocarbon accumulations in fractures in the Precambrian basement rock where it is topographically higher than surrounding basin sediments. The uparched basement rock beneath the entire oil rich Central Basin platform is prime prospecting ground. For example, the buried Fort Stocton mountains (Hills, 1942, p. 226) in the southwestern sector of the platform in northeastern Pecos County might be productive where weathered granite peaks are overlapped by Leonard and younger Permian sediments.

The traps in which the future oil deposits are to be found include anticlines, reefs, other porosity wedgouts in carbonate rocks, and both wedging out and lenticularity in sandstone bodies. These will be discovered by exploring both within the producing areas and around their peripheries. The lesser density of fields in some areas, such as in Menard and other counties in southeastern West Texas, is due in part at least to lesser exploration (Roberts, 1962).

Perhaps the biggest probable reserve underlies the present producing reservoirs in many of the great anticlines for which West Texas is justly famous. The outstanding potential in this regard is the Yates field. According to Donoghue and Gupton (1957, p. 428), this field has not been tested on structure to depths greater than about 300 feet below the Grayburg Formation because of a field rule prohibiting drilling more than 225 feet below the top of the "Big Lime" (top of Grayburg). A dry hole was drilled to a total of 9114 feet (374 feet into the Ellenburger), but this was on the western edge of the Yates field about a mile from the nearest producer and 400 feet below the highest closed structure contour 4 miles to the east. This lone off-structure dry hole should not condemn the trap for production at deeper levels. Although these data from Donoghue and Gupton regarding drilling go only through 1955, no deep tests in the Yates field have been reported thus far. During recent years deeper production, including some from the Ellenburger, has been found in nearby fields. It can be assumed that a second birthday (at least) at Yates is a distinct possibility.

There is also a possibility of opening up a new producing subdistrict, the Marfa basin at the southeastern end of the Diablo platform (Pearson, 1963).

REFERENCES CONSULTED IN PREPARING CHAPTER 31

Adams, John Emery, 1965.
American Association of Petroleum Geologists, 1960.
Bartley, J. H., 1952.
Bergenback, Richard E., and Terriere, Robert T., 1953.
Burnside, R. J., 1959.
Conselman, Frank B., 1955.
Cook, V. O., 1966.
Cooper, C. G., and Ferris, B. J., 1957.
Cotton, Clyde B., 1966.
Deegan, Charles J., 1949.
Denham, R. L., and Dougherty, W. E., 1941.
Donoghue, David, and Gupton, W. L., Jr., 1957.
Doss, A. K., Jr., and Spiva, Frank J., Jr., 1968.
Douglas, E. R., 1957.
Elkins, Lincoln F., 1953.
Ellison, Samuel P., Jr., 1948, 1957.
Galley, John E., 1958.
Guinan, Mark A., 1966.
Harbison, Robert R., 1957.
Harrington, John W., 1963.
Hennen, Ray V., 1929.
Herald, Frank A., ed., 1957.
Hester, Robert J., and Holland, R. R., 1960.
Hills, John M., 1942, 1961, 1968a, 1968b.
Holland, Richard R., 1966.
Holmquest, Harold J., 1965.
Hubbert, M. King, 1953.
Imbt, Robert F., and McCollum, S. V., 1950.
Jones, T. S., 1953.
Jones, Theodore S., and Smith, Harold M., 1965.
Jons, Richard D., 1966.
Keener, M. H. L., 1957.
Kvenvolden, Keith A., and Squires, Rodney M., 1967.
Landes, Kenneth K., 1967.
Langenbahn, W. E., 1968.
McInnis, A. B., and Hartman, F. H., 1957.
McMurtry, R. P., 1966.
Maher, John C. (ed.), 1960.
Marathon Oil Company, 1966.
Marshall, Joseph W., 1952.
Martin, J. Stewart, 1968.
Mooney, Tom D., 1966.
Moore, J. H., 1944.
Myers, Donald A., Stafford, Philip T., and Burnside, Robert J., 1956.
Ogden, Victor, and Locke, Jack, 1952.
Pearson, Bruce, 1963.
Peek, Charles A., et al, 1960.
Porter, William C., 1966.
Reeves, C. C., Jr., 1964.
Reiter, F. H., 1957.
Roberts, G. C., 1962.
Robinson, Joe D., 1966.
Salisbury, Gerald P., 1968.
Sax, Norman A., 1967.
Schmitt, George T., 1954.
Schneider, W. T., 1957.
Schreiner, Carl S., Jr., 1957.
Stafford, Philip T., 1957, 1959.
Texas, University of, 1957.
Thomas, Leonard C., 1966a, 1966b.
Vest, E. L., Jr., 1966.
Wagner, Ewald A., 1966.
Warn, G. Frederick, 1953.
West Texas Geological Society, 1951, 1962, 1966.
Winn, Vernard, 1968.
Wright, W. Floyd, 1962, 1963, 1967.
Young, Addison, 1965.

ROCKY MOUNTAIN

The Rocky Mountain Division of the United States consists of eight states (Figure V.1) lying between the Central and the Texas Divisions on the east and the Pacific Division on the west. To the north are Saskatchewan, Alberta, and British Columbia. The Mexican states of Chihuahua and Sonora lie to the south.

The Rocky Mountain chain extends from northern Mexico to Alaska; this continental backbone is the Great Divide, separating the Pacific drainage from the Arctic–Atlantic–Gulf of Mexico drainage. In the United States sector the mountainous zone is bordered on the west by the Western Interior basin and on the east by the Great Plains. Figure V.1 shows by crosshatching the mountain and bordering foothill areas (Austin, 1965, map in pocket). The major mountain divisions are the northern and southern Rockies. The Black Hills stand as an outlier to the east of the northern Rockies. West of the southern Rockies are the Wasatch Mountains with a northerly trend and the Uinta Mountains, which project eastward. Toward the south end of the Rocky Mountain province are the Arizona and New Mexico mountains. The state maps that follow give the names of many of the ranges.

Numerous basins, both topographic and sedimentary, intervene between individual mountain ranges. Sedimentary basins also underlie the Great Plains and are superimposed upon the Western Interior basin. The mountain peaks rise to more than 14,400 feet above sea level, and the plains and basin floors range from a minimum of about 4000 feet to over 9000 feet in elevation.

In this great area of majestic heights, an unusal variety of geology is to be found (Haun and Kent, 1965). The Rocky Mountains are sculptured from

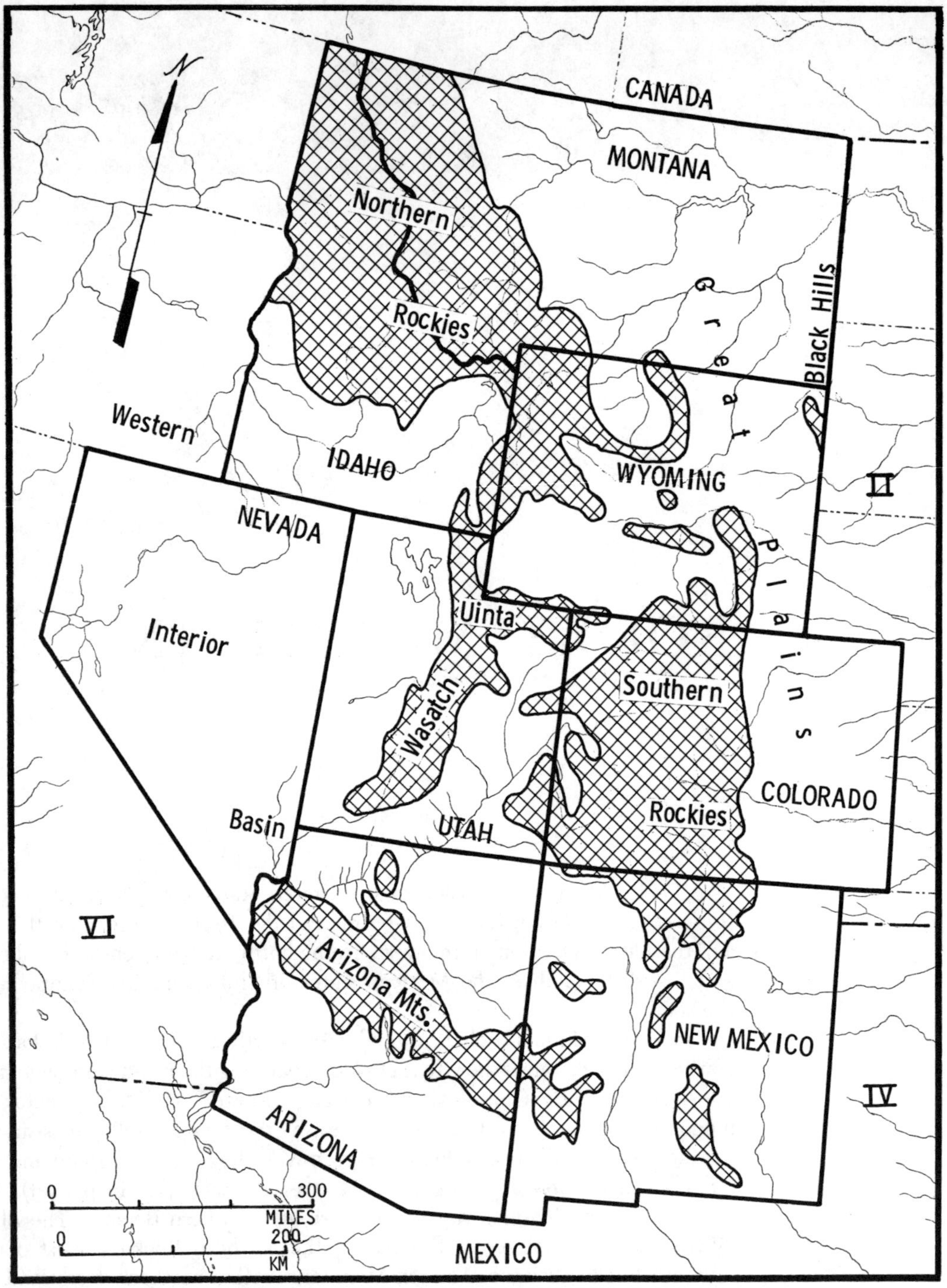

Figure V.1 Rocky Mountain division. Index map, also showing physiographic provinces. Mountainous areas are cross-hatched. ***After Morris E. Austin (1965, map in pocket).***

igneous, metamorphic, and sedimentary rocks of all types. These range in age from Precambrian to Recent, but the sedimentation that covered the widest area, and resulted in deposits of the greatest thickness, took place during the Cretaceous. The major periods of diastrophism, which produced metamorphism, overthrusting, folding, and magmatic intrusions and extrusions, occurred in the Precambrian and late Mesozoic–early Cenozoic (Laramide Revolution). Before the advent of the Paleozoic, the ancient sediments had been folded, contorted, metamorphosed, and intruded by great masses of granitic magma. Uplift, mountain sculpturing, and peneplanation followed. During the Paleozoic and Mesozoic eras, downwarpings of the crust produced the Rocky Mountain geosyncline; and multiple submergences, coupled with continental deposition during times of emergence, produced a sedimentary column with an aggregate thickness in excess of 10 miles. The Laramide Revolution arched upward the basement rocks throughout the length of the geosyncline; at the same time folding and thrust faulting, accompanied in some areas by igneous activity, took place. Before the end of the Tertiary period the mountains resulting from this revolution had been worn down, the entire province had been re-elevated, and differential erosion had begun to carve the present Rocky Mountains.

The Rocky Mountain Division stands sixth in current domestic oil production among the seven petroliferous divisions used here. It is below the Pacific Division and ahead of Eastern United States. The leading oil states in the Rocky Mountain area are Wyoming and New Mexico. Montana, Colorado, and Utah are next, with considerably less production. The annual yields from Arizona and Nevada are not significant; the eighth state, Idaho, is so far nonproductive.

This province is notable for the wide range in the geological age of its reservoir rocks; they extend from the Cambrian to the Paleocene, with every period represented.

REFERENCES CONSULTED IN PREPARING DIVISION V

Austin, Morris E., 1965.

Haun, John D., and Kent, Harry C., 1965.

Chapter

32

NEW MEXICO

In 1968 New Mexico ranked sixth in the United States in oil production, with nearly 4 per cent of the total, and fourth in gas (6 per cent of the total). Proceeding clockwise, New Mexico is bordered by Arizona, Colorado, the western tip of the Oklahoma panhandle, Texas, and Mexico (Figure 32.1). It ranges in elevation from less than 3000 feet, where the Pecos River enters Texas, to over 13,000 feet in the Sangre de Cristo Mountains in northern New Mexico. Most of the state lies between 5000 and 10,000 feet above sea level (Dane, 1965, p. 19).

Geographically and geologically the greater part of New Mexico is in the High Plains, Rocky Mountain, Colorado Plateau, and Basin and Range provinces, but 90 per cent of the oil produced in 1966 came from the southeastern part of the state, which is in the Permian basin, previously discussed under West Texas (p. 324). The gas production, however, is divided almost evenly between southeastern and northwestern New Mexico. The latter part of the state is in the Four Corners region, which also includes parts of Colorado, Utah, and Arizona.

The earliest recorded (1882) signs of hydrocarbons in New Mexico were oil springs in the northwest part of the state (Nutter, 1965, p. 41). During the first decade of the present century, oil contamination of water wells was noted in southeastern New Mexico. These signs led to exploration in both the northwest and the southeast sectors of the state that resulted eventually in commercial production.

The Ute gas field in northern San Juan county in northwestern New Mexico was discovered in 1921. The next year the Hogback field in the same county became the state's first commercial oil field. The anticlines are practically naked in this area, and the discovery of Hogback was followed by the testing of other domes, which led to the finding of the Rattlesnake field in 1924. The first large commercial discovery in southeastern

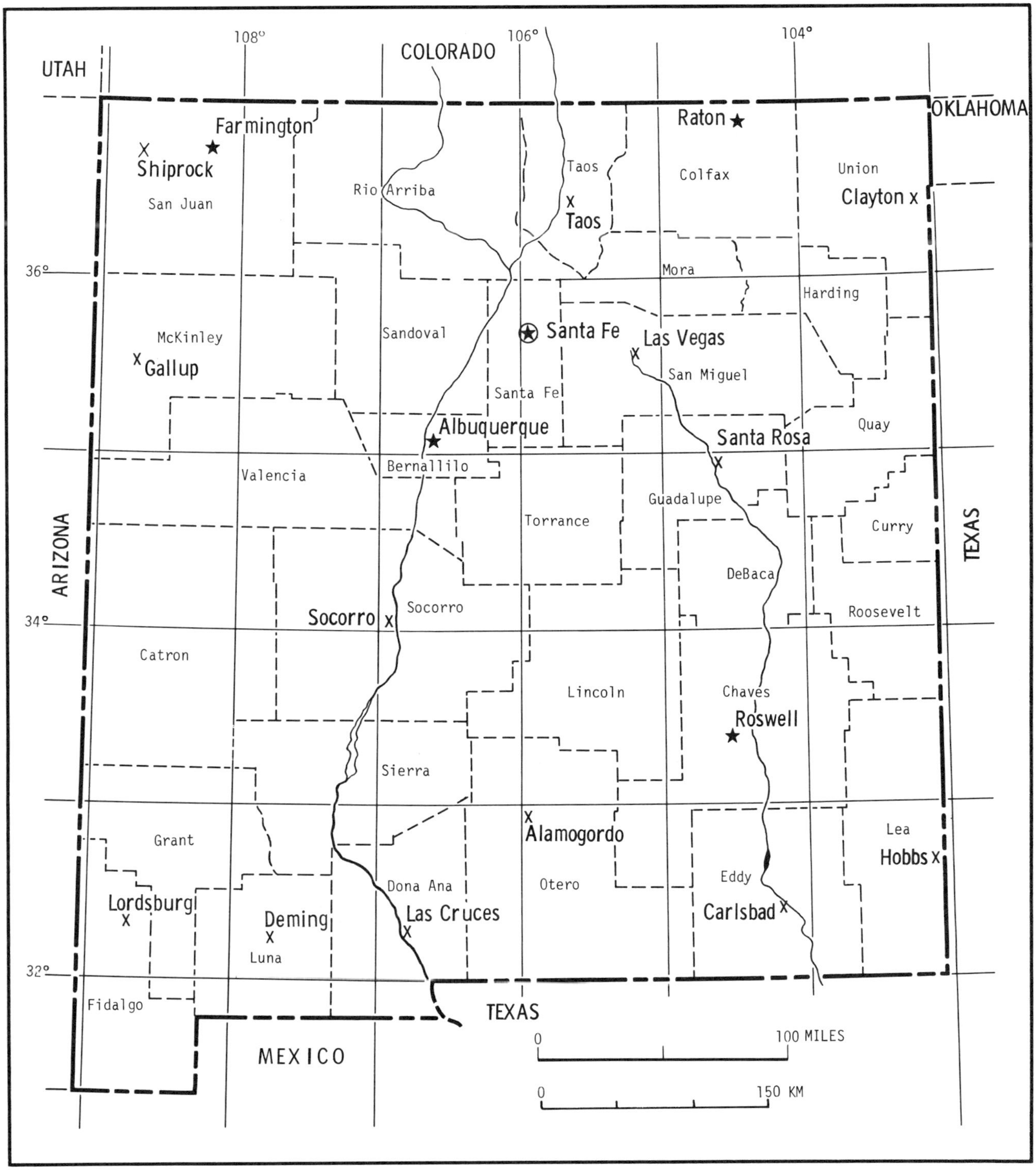

Figure 32.1 New Mexico. Index map.

New Mexico was the Artesia oil field in Eddy County, also in 1924.

Tectonics. The contour map of the Precambrian basement (Figure 32.2) shows a mountainous core extending from north to south across the center of the state, with basement rocks cropping out at various places. To the east are the Great Plains and to the west the Colorado Plateau.

The mountain core is dominated by uplifts, the most prominent being the Sangre de Cristo in northern New

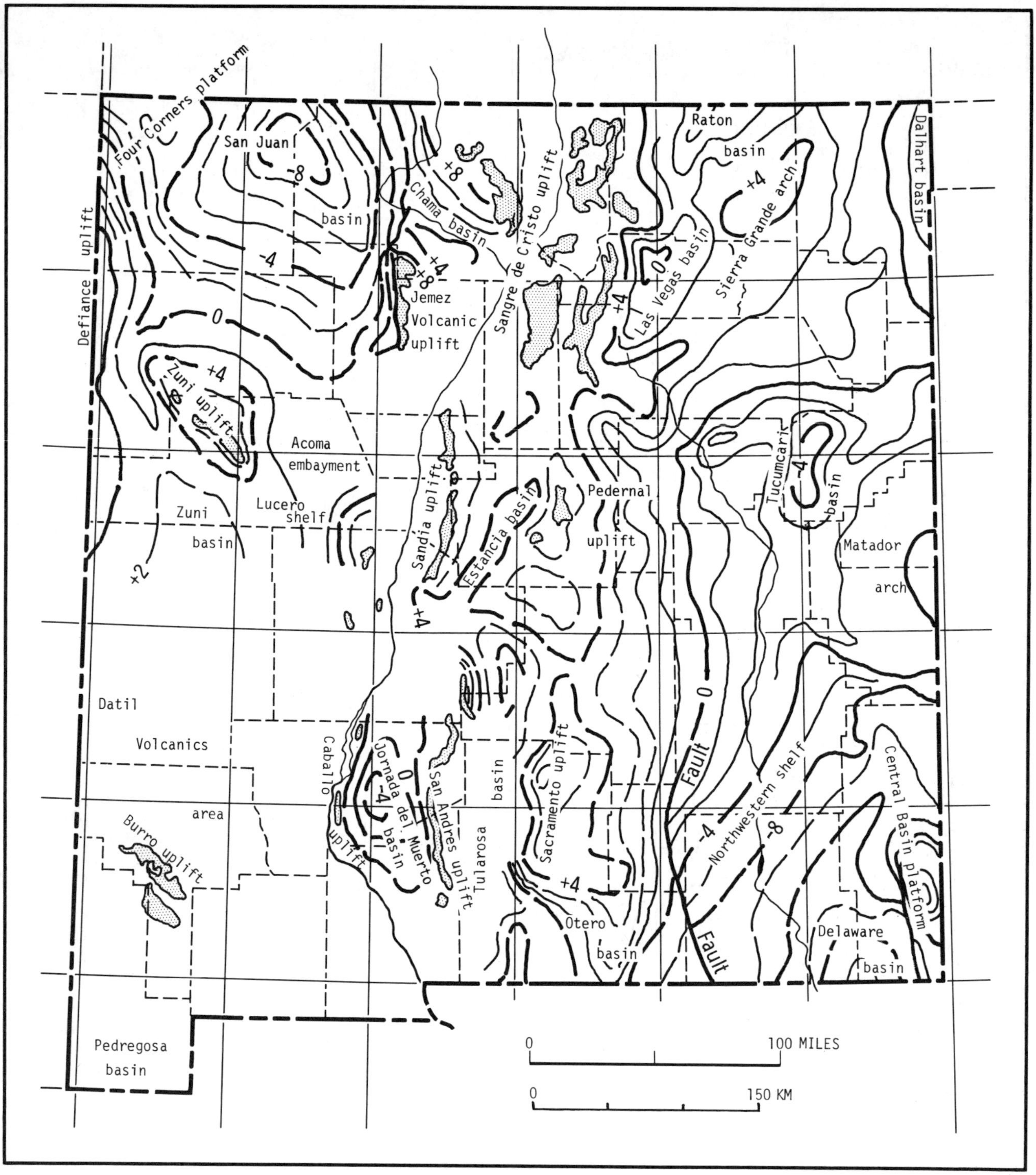

FIGURE 32.2 New Mexico. Tectonic map contoured on Precambrian basement surface. Interval variable, but mostly 2000 feet. All contour elevations are in thousands of feet. Shaded areas are basement rock outcrops. *Contours after Peter T. Flawn (1967). Names of tectonic features after George V. Cohee et al. (1962).*

Mexico and southern Colorado. Two basins, the Tularosa and the Jornada del Muerto, intervene between uplifts in the southern part of the state.

The Great Plains sector includes continuations of West Texas tectonic features along the eastern edge of New Mexico. The Delaware basin and the Central Basin platform are bordered on the northwest by the Northwestern shelf. In mideastern New Mexico is a prominent basin, the Tucumcari. A northeast-trending spur of the mountain core, the Sierra Grande arch, is bordered on the northwest by the Raton basin, most of which is in Colorado.

Western New Mexico has two uplifts, the Zuni in the northwest and the Burro in the southwest, with exposed Precambrian cores. Between is the Datil Volcanic area and the Lucero shelf (Wengerd, 1959c), where the depth to the basement ranges from 6000 feet to an unknown number. Northwestern New Mexico, east of the Zuni-Defiance uplift, is dominated by the San Juan basin. This Tertiary structural basin lies southeast of the Pennsylvanian Paradox basin. The Defiance uplift of northeastern Arizona bounds the San Juan basin on the west.

These tectonic features are the result of a series of crustal movements that took place during late Paleozoic, late Mesozoic, and Cenozoic time. The first of these movements, which produced the ancestral Rocky Mountains, took place during the Pennsylvanian and Early Permian periods (Dane, 1965, p. 32). The Sierra Grande, Pedernal, and Matador arches, as well as the Central Basin platform, were formed at this time. The Zuni uplift was also born then but was rejuvenated later, during Tertiary time.

Major diastrophism occurred during late Mesozoic–early Cenozoic time (Laramide Revolution). This consisted of folding, overthrusting, and regional uplift. The entire area was lifted above sea level; hence the youngest marine sediments are Late Cretaceous in age.

Subsequent Cenozoic time was marked by intermittent igneous activity and the late Tertiary regional uplift that instigated the sculpturing of the present face of New Mexico. Magmatic activity was both intrusive and extrusive. During the Oligocene epoch the igneous rock of Mount Taylor was intruded. On a grander scale was the extrusion of the Datil volcanics. Later in Tertiary time further magmatic intrusions and volcanic extrusions occurred and continued into the present era.

Areal Geology. A simplified geologic map of New Mexico is shown in Figure 32.3. The oldest rocks, Precambrian crystallines, are confined to the eroded cores of mountain ranges and uplifts. Paleozoic rocks, almost entirely Permian in age, dominate the pre-Tertiary outcrop picture in southern New Mexico. Pennsylvanian rocks, and to a much lesser extent older Paleozoics, fringe the Precambrian outcrop areas, especially in mountainous northern New Mexico and on the flanks of the Zuni uplift. Mesozoic sediments are confined largely to the northern half of the state, occupying most of northeastern and northwestern New Mexico.

Tertiary intrusive rocks are local in occurrence, but lavas are extensive, especially in the western and north-central parts of the state. Late Tertiary continental deposits mantle the Great Plains. Tertiary and Quaternary alluvial deposits occur also in local structural basins and along major stream courses.

The reservoir rocks in southeastern New Mexico are entirely Paleozoic; in the northwestern sector, the greater part of the hydrocarbon production has come from Cretaceous rocks, and most of the remainder from Pennsylvanian strata. The stratigraphic range in age of the reservoir rocks for the state as a whole is from uppermost Cretaceous to lowermost Ordovician.

TRAPS AND FIELDS

The two oil- and gas-producing districts in New Mexico lie at opposite (southeastern and northwestern) corners (Figure 32.4). Each consists of four counties. Much the greater part of the oil produced to date has come from southeastern New Mexico in Lea, Eddy, Chaves, and Roosevelt Counties. Lea County alone has accounted for more than 63 per cent of the total and is asserted to be the richest oil-producing county in the United States. The producing counties in the northwest corner of New Mexico are San Juan (second in the state in natural gas production), Rio Arriba, Sandoval, and McKinley. In addition, Harding County near the northeastern corner produces carbon dioxide gas in commercial quantities.

The two producing districts in New Mexico are described separately. The other two quadrants are discussed also, under "Undeveloped Quadrants" (p. 374).

Southeastern New Mexico. This part of the state is in the Permian basin, which New Mexico shares with West Texas. The geology of southeastern New Mexico is very similar to that across the state line. In fact many individual fields ignore the state line completely and produce from both sides. An example is El Mar, which was described in Chapter 31 on West Texas (Figure 31.22). The "Sand Belt" fields continue northward from Winkler County, Texas, along the eastern edge of Lea County for a distance of 40 miles (Denham and Dougherty, 1941). The West Dollarhide field in Lea County is separated from Dollarhide field in Andrews County, Texas, by a nonproductive saddle between two oil-filled anticlinal nodes.

The greater part of the production in southeastern New Mexico, as in West Texas, comes from fields on the Central Basin platform. Second in importance is the Northwestern shelf (Figure 32.2). The New Mexico sector of the Delaware basin is just beginning to come into its own as a source of hydrocarbons, especially natural gas.

The areal geology of southeastern New Mexico is

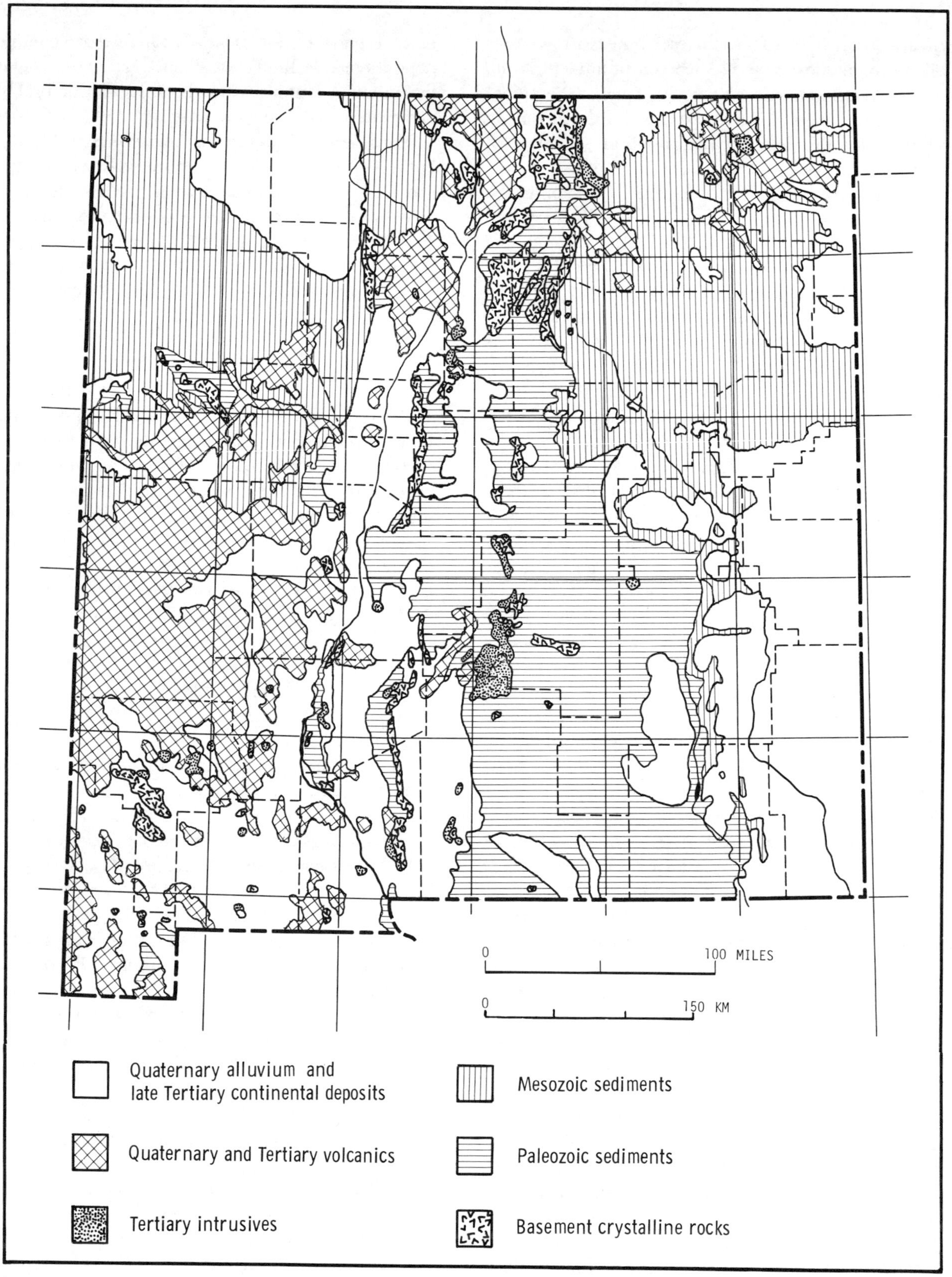

FIGURE 32.3 New Mexico. Geologic map. *After U. S. Geological Survey (1952).*

shown in Figure 32.3; the stratigraphic chart is given in Figure 32.5. Most of the important subsurface formations are described under "Stratigraphy" in Chapter 31 on West Texas and therefore will not be disscussed here. The dominant reservoir rock is carbonate rock, principally limestone. Sandstone is subordinate, largely because it is a relatively scarce rock in the subsurface of the Permian basin. Production has come largely from Permian reservoirs, over half from the Grayburg–San Andres section alone.

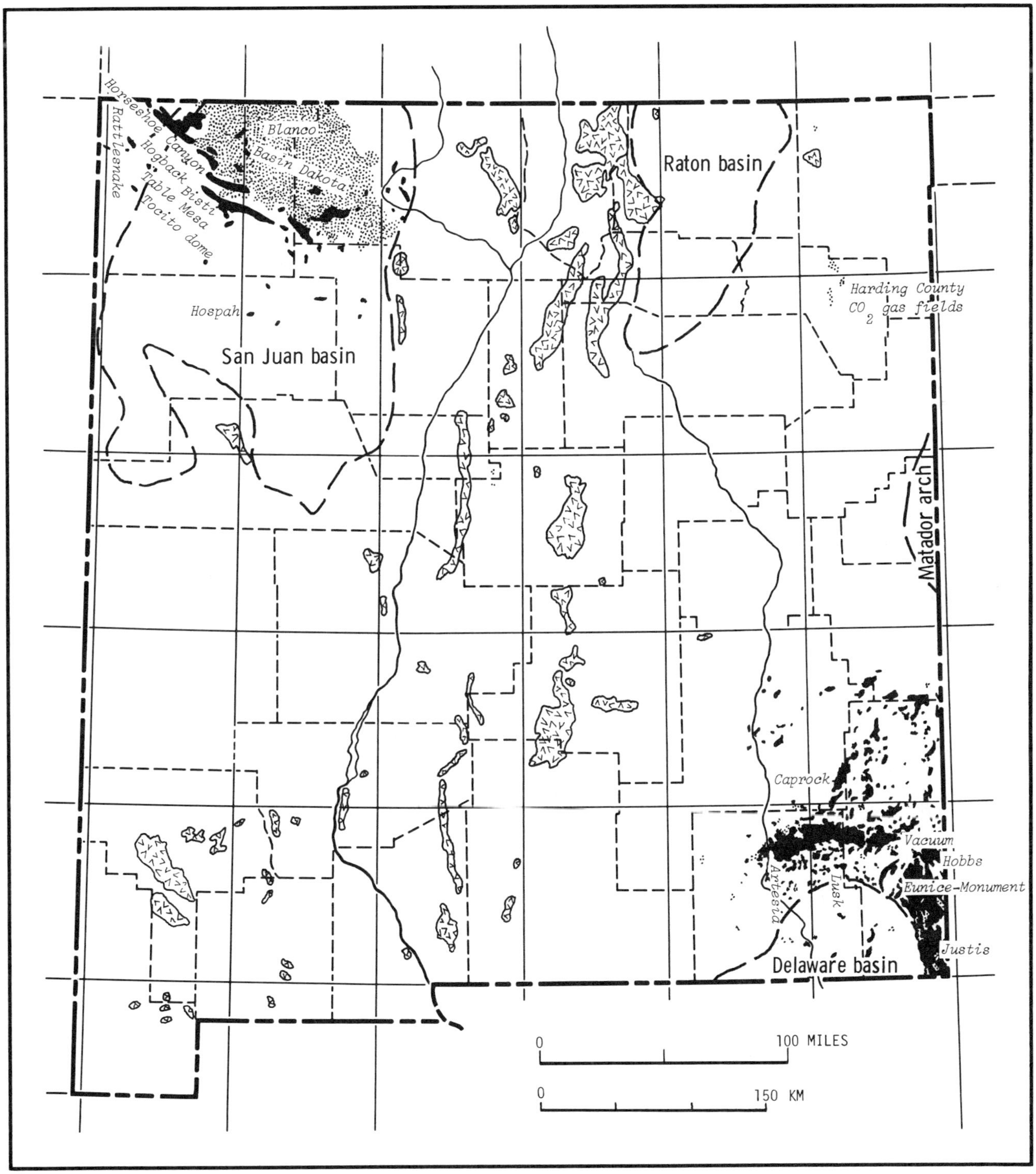

FIGURE 32.4 New Mexico. Oil and gas fields, and basement and intrusive rock outcrops.

GEOLOGIC TIME UNITS			FORMATIONS	
Cenozoic (to 70 million years)	Quaternary	Recent	Alluvium and other surficial deposits	
		Pleistocene	Gravel and associated deposits	
	Tertiary	Pliocene	Ogallala Formation	
		Miocene	Probably present (?)	
		Oligocene	Not recognized	
		Eocene	Not recognized (?)	
		Paleocene	Not recognized (?)	
Mesozoic (70–225 million years)		Cretaceous	Mesaverde Group Mancos Shale Dakota Sandstone (?)	
		Jurassic	Absent	
		Triassic	Dockum Group	
Paleozoic (225 to 600 million years)		Permian	Dewey Lake red beds Rustler Formation Salado Formation Castile Formation	
			All of the Artesia Group	Tansill Formation Yates Formation Seven Rivers Formation Queen Formation Grayburg Formation
			San Andres Limestone includes the Hondo Sandstone Member Glorieta Sandstone Yeso Formation Abo Formation Bursum Formation	Cutoff Shale Victorio Peak Limestone Bone Spring Limestone Hueco Limestone
	Carboniferous	Pennsylvanian	Includes many formations and informally named zones in the subsurface	
		Mississippian	Lake Valley Limestone and other local formations	"Mississippian Lime" of subsurface
		Devonian	Percha Shale	Woodford Shale of subsurface
		Silurian	Fusselman Dolomite	"Devonian" of subsurface
		Ordovician	Montoya Dolomite	
			Simpson Formation of subsurface	
			El Paso Formation	Ellenburger Formation of subsurface
		Ordovician and Cambrian	Bliss Sandstone	
		Precambrian	Granite and associated rocks	

FIGURE 32.5 **Southeastern New Mexico. Stratigraphic chart.** ***After C. H. Dane (1965, p. 27).***

Other leading habitats of oil accumulation in the New Mexico sector of the Permian basin are the Upper Permian Yates and Seven Rivers–Queen Formations and the Lower Permian (Leonardian) Abo reef. The Abo reef produces mainly in New Mexico; only since 1960 has production spread from west to east into Texas. This reef is now considered by some to be Leonardian in age; the Abo Formation of the surface is assigned to the underlying Wolfcampian Series (Meyer, 1966, p. 71). Fields producing from the Abo reef occur along the Vacuum–Artesia trend, which extends westward across central Lea County into northeastern Eddy County (Figure 32.4). This trend is also occupied by a series of westward trending oil-bearing anticlines; most Abo accumulations were discovered while probing beneath anticlinally trapped oil in younger reservoirs. LeMay (1960) has mapped the reef trend, showing the presence of back-reef facies to the north and forereef facies to the south. Additional facies have been suggested by Nottingham (1966). Secondary migration into the reef is described by Stenzel (1965).

The only non-Permian reservoirs with large cumulative productions of both oil and gas are limestones and dolomites in the Siluro–Devonian section (Salisbury, 1968).

So far most of the oil and gas produced in southeastern New Mexico has been trapped in anticlines. Other types of structural traps are scarce. The varying permeability (stratigraphic) traps discovered to date are sand lenses, updip wedgouts, and carbonate reefs. The leading fields in total yield are the Eunice–Monument, Hobbs, and Vacuum; these are described subsequently, along with other fields of interest.

The multipay *Justis* field lies on the Central Basin platform in the southeastern corner of New Mexico. The trap is an elongate north-south anticline, about 3 miles long by 1 mile wide. The discovery well, drilled on a node to the northwest of the main anticline in 1944, was a gas well that produced from a zone within the top 250 feet of the Glorieta Formation of Permian age (Marshall and Flotz, 1960). This gas zone, productive throughout the anticline, has produced more than 20 trillion cubic feet. At the Glorieta level the Justis anticline has about 200 feet of closure.

Deeper exploration during the years 1957 to 1959 resulted in the discovery of commercial oil reservoirs not only in the Glorieta beneath the gas pay, but also in nine deeper formations extending from Lower Permian to Lower Ordovician in age. The largest cumulative productions have come from the Ellenburger and the Fusselman (Silurian). The Ellenburger reservoir is a crystalline dolomite with both solution and vuggy porosity, but the highest permeability is in extensively fractured zones. The structural closure of the anticline at the Ellenburger level is over a thousand feet. Average depth is about 7800 feet. The Justis field is calculated to contain nearly 50 million barrels of recoverable oil.

The *Eunice–Monument* field (Babcock, 1956) is one of many examples of two or more separately discovered fields that subsequently proved to be one continuous oil and gas deposit. The total cumulative production is about 300 million barrels of oil, making this field the largest in New Mexico. It has also produced over a trillion cubic feet of sour gas. The first oil discovery was made in 1929; the combined field is still producing nearly 6 million barrels of oil a year.

The Eunice–Mounment field is on the west side of the Central Basin platform, toward the northern end in east-central Lea County. A structure map contoured on top of the Grayburg is shown in Figure 32.6. The Grayburg and immediately underlying San Andres of lower Guadalupian (Late Permian) age constitute the principal reservoir. The oil or gas pay may occur anywhere in the Grayburg and upper San Andres. In the producing area the gas-oil contact is at an approximate elevation of −150 feet and the oil-water contact lies at about −400 feet. Lesser reservoirs include the overlying Queen, Seven Rivers, and Yates, and the underlying Glorieta and upper Yeso Formations of Leonard age.

As can be seen from Figure 32.6, the trap is a north-south–trending anticline productive over a length of nearly 16 miles and a maximum width of about 6 miles. The anticline has three major nodes, but production continues across the intervening saddles. The great productivity of this field can be ascribed to the large size of the trap, the porosity and permeability of the reservoir, the thickness of the reservoir, and the large gathering area downdip to the westward into the Delaware basin.

Second only to the combined Eunice–Monument field is *Hobbs*, which has produced nearly 200 million barrels of oil. This field is likewise on an anticline on the Central Basin platform; it is at the northern tip, northeast of Eunice–Monument. Hobbs was discovered in 1928 after geophysical surveys. The principal reservoir is the "White Lime" in the San Andres (McPeters and Kelly, 1956). This is an exceptionally porous limestone that is cavernous along the crest of the trap. The overlying Grayburg Formation is also productive on the west and northwest sides of the field. The reservoir section of the Grayburg consists of sandy dolomite alternating with dolomitic sandstones. Gas is produced at Hobbs from the still higher Queen. Across the top of the anticline a reservoir in the overlying Seven Rivers produces oil at a depth of 4200 feet.

The trap at Hobbs is an elliptical anticline with a

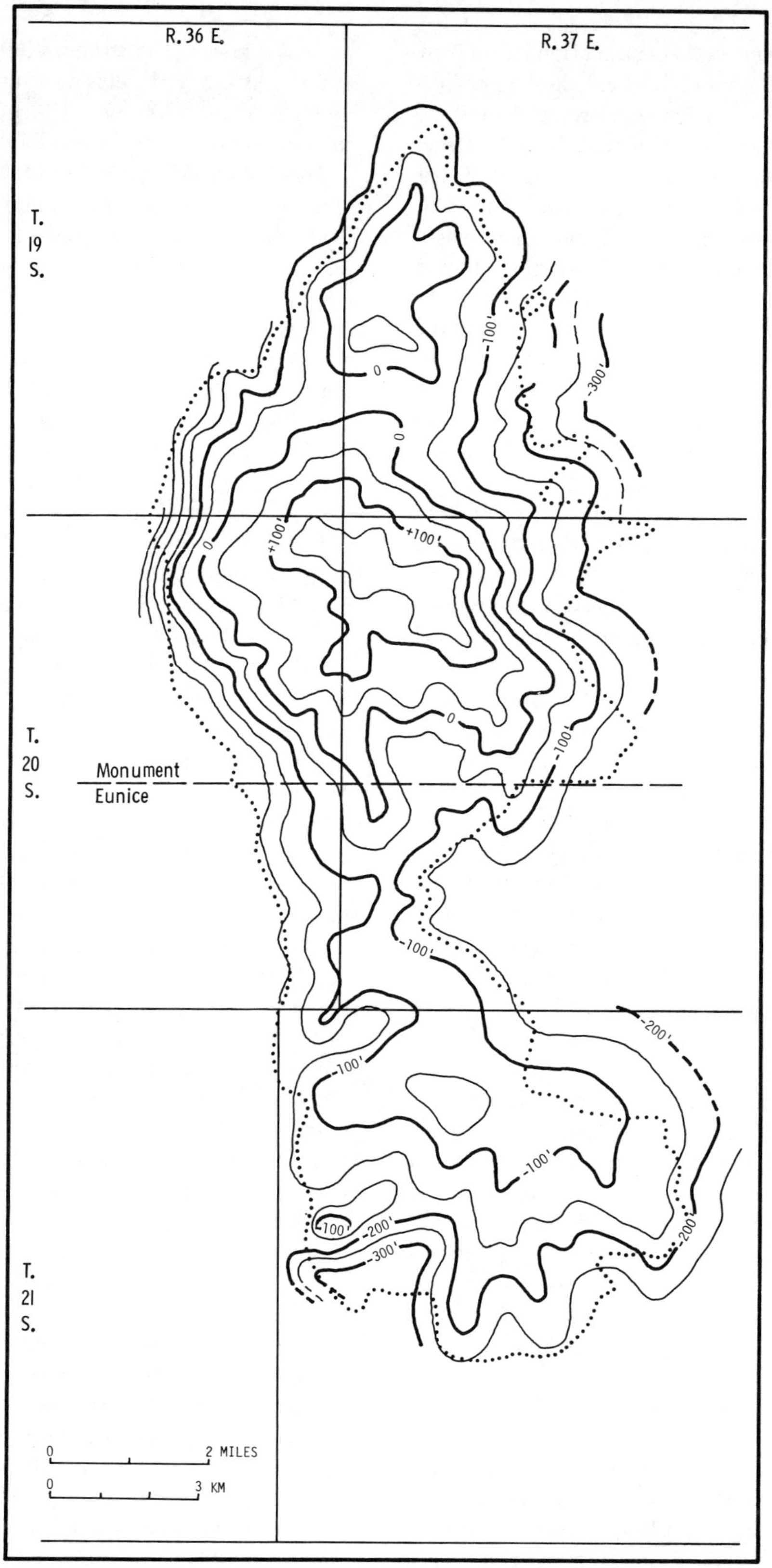

FIGURE 32.6 Eunice-Monument oil field, Lea County. Structure map contoured on top of Grayburg. Interval, 50 feet. Dotted line encloses producing area. *Courtesy Roswell Geological Society (1956, tip-in facing p. 165).*

northwest trend. Most of the oil lies above the outermost contour (−550 feet) shown on the structure map in Figure 32.7. The vertical distance of oil occurrence in the reservoir approaches 250 feet. Exceptional cavernous porosity along the crest of the anticline (DeFord and Wahlstrom, 1932) appears to be due to the fact that about one-third of the anticlinal folding occurred before the deposition of the Upper Permian evaporites; this pushed the crestal part of the anticline into a zone of active solution weathering before burial beneath evaporite seal rocks.

The *Vacuum* field in north-central Lea County is currently the most productive oil field in New Mexico, yielding nearly 14 million barrels annually under proration. It is third (after Eunice-Monument and Hobbs) in cumulative production, with more than 145 million

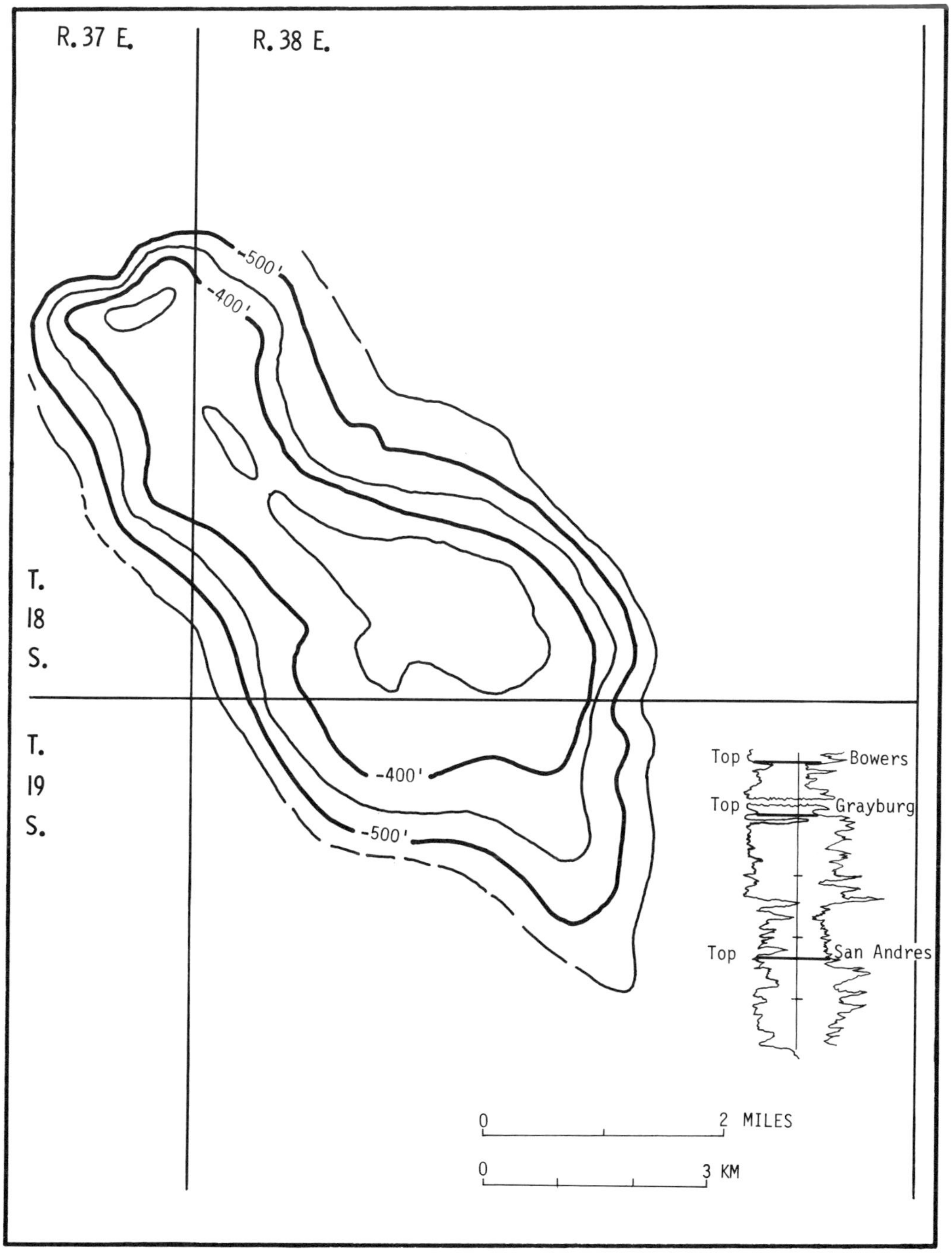

FIGURE 32.7 Hobbs field, Lea County. Structure map based on top of San Andres. Interval, 50 feet. ***Courtesy K. D. McPeters and J. M. Kelly (1956, p. 206) and the Roswell Geological Society.***

barrels. Vacuum lies northwest of Eunice–Monument. It is on the Northwestern shelf across the San Simon syncline (Stipp and Haigler, 1957) from the northern terminus of the Central Basin platform.

The Vacuum field was discovered in 1929 by a well drilled into the San Andres oölitic dolomite. This formation is still the leading reservoir, but the deeper Abo (Leonardian) is also an important source of oil. At least ten other reservoirs contain hydrocarbons over parts of the Vacuum field. The deepest of these is Devonian dolomite, which is fractured and vuggy. It lies at a depth of about 12,000 feet; a deep test in the field reached Precambrian granite at a depth of 13,790 feet.

Figure 32.8 shows structure contours drawn on the top of the San Andres reservoir. The trap is a large, irregular anticline occupying most of T. 17 S., R. 34 E., except for a closed structural depression in the north-central part of the township. The anticline also extends most of the way across the south half of T. 17 S., R. 35 E. The highest point is a node in the southeast corner of T. 17 S., R. 34 E., and the northeast corner of T. 18 S., R. 34 E. The elevation of the top of the productive section here is less than −300 feet. Oil production extends down to −600 feet, with a few wells producing from as low as −670 feet. Some wells are completed in the immediately overlying lower Grayburg (Milks, 1956); the Grayburg has much lower porosities and permeabilities, however, and its contribution to the oil in the combined reservoir is probably minor.

The deeper Abo reservoir is a reef about 9 miles long on the south flank of the anticline. The reservoir rock is a coarsely crystalline, vuggy, fractured dolomite. It is obvious from the anticlinal oil accumulation in both higher and lower (Devonian) reservoirs that structure also played an important role in trapping oil in the Abo reef. However, the reef contributed an abnormally thick (560 feet) pay zone, which compensates for the fact that the production is limited updip, south of the anticlinal axis, by pinchout of porosity. The edge of production downdip is an oil-water interface (Caffey and Dubitzky, 1967).

The *Lusk* oil and gas field (Thornton and Gaston, 1967, 1968), which straddles the line between Eddy and Lea Counties, is about 8 miles north of the rim of the Delaware basin, on the southern flank of the Northwestern shelf. It was discovered in late 1960. Subsequently, through 1966, it produced nearly 11 million barrels of oil and over 25 trillion cubic feet of natural gas. It is a northwest-trending anticline (Figure 32.9) with a long spur extending to the north, mostly on the Lea side of

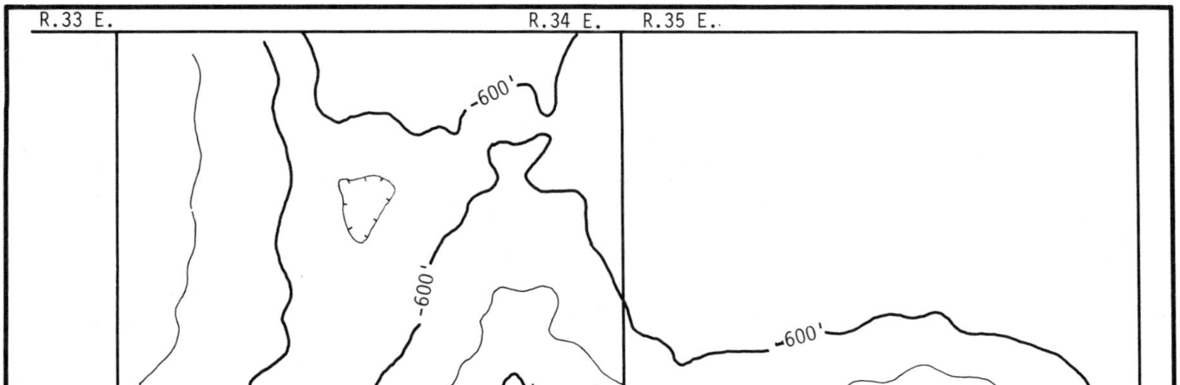

Figure 32.8 Vacuum oil field, Lea County. Structure contour map based on top of San Andres. Interval, 100 feet. *Courtesy Roy Milks (1957, tip-in before p. 354) and the Roswell Geological Society.*

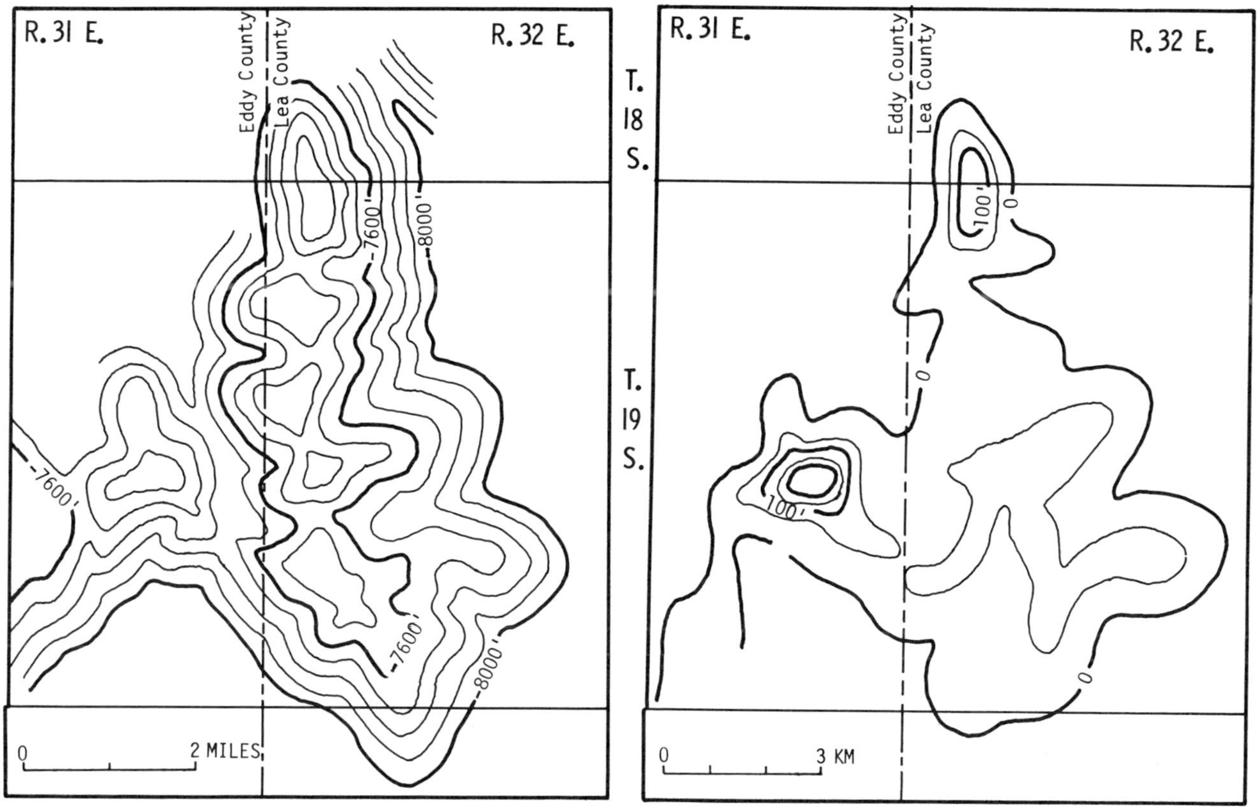

FIGURE 32.9 Lusk oil and gas field, Lea and Eddy Counties. Left: structure map based on top of Strawn limestone reservoir. Interval, 100 feet. Right: thickness map of effective Strawn porosity. Interval, 50 feet. Porosity wedgout limits field on north and west; water underlies the oil on the south and east at between −7800 and −7900 feet elevation. ***Courtesy, D. E. Thornton and H. H. Gaston (1967, p. 121–122) and the Roswell Geological Society.***

the county boundary. Production is limited on the north and west flanks of the anticline by diminution in porosity and permeability. The anticlinal trap was discovered by the presence of subsurface nosing, which led to a seismic survey.

The principal reservoir is the Strawn of Pennsylvanian age, which lies at depths in excess of 11,000 feet. It is crystalline, highly fossiliferous limestone; most of the porosity is secondary in the form of vugs, small solution cavities, and a widespread, highly effective fracture system. In a few places the fractures have been filled and are locally nonporous. Two local reefs produce sharp increases in the porosity at the western and northern ends of the anticline (Figure 32.9).

The shallowest reservoir is the Yates (Guadalupian–Permian) at depths of 2600 to 2700 feet. At this level the Lusk anticline is a southeastward plunging nose without closure. Trapping in the Yates is due to updip lensing out of porosity and permeability. The Yates reservoirs are lenticular sands and vuggy to cavernous dolomite stringers within the sands. There are three additional minor reservoirs, two in the Permian below the Yates, and one in the Morrow (Pennsylvanian) about a thousand feet below the Strawn.

The *Artesia* field in north-central Eddy County is west of Vacuum, but on the same trend. It is the first field to be discovered in southeastern New Mexico, and for a number of years was the only field in the New Mexican sector of the Permian basin (Davis, 1929). The discovery well was completed in the latter part of 1923; in the next 43 years Artesia produced over 18 million barrels of oil and is still yielding nearly 800 thousand barrels annually. The field is named after the town of Artesia, 14 miles to the northwest, which in turn owes its name to the presence of artesian water in this part of the Pecos Valley. Many of the artesian wells drilled in the Artesia area found oil and gas showings. This led to the locating and drilling of the discovery well, which produced gas and some oil from a depth of less than 2000 feet.

The reservoirs at Artesia are (1) an oölitic dolomite lying from 200 to 400 feet below the top of the San Andres Formation of early Guadalupian age (Permian), which is the major pay, and (2) sandy facies in the overlying Grayburg Dolomite (Fryman, 1956). The pores in the San Andres are the voids between oölite spheres, a perfect sandstone type of porosity. The sands in the Grayburg Dolomite are erratic and range in thickness from 1 to 25 feet.

Trapping at Artesia is due to updip lensing out of the effective porosity. The regional dip in this area of the field is easterly about 100 feet to the mile.

The *Caprock* field (Dunn, 1956), which overlaps both Chaves and Lea Counties, is a good example of sand wedgout trapping. The field is elongate with a north-northwest trend. The reservoir is a sand member of the Queen Formation (Permian–Guadalupian). Sand thickness averages 20 feet and the net pay thickness 10 feet; however, local areas may be nonproductive because of salt or anhydrite filling the pore spaces. The regional dip is easterly at 25 to 35 feet per mile. The reservoir sandstone pinches out on the west side of the field at an elevation about 125 feet above the oil-water contact downdip to the east (Stead, 1955, p. 158). The average depth to the pay is 3100 feet.

Other fields also lie along this wedgout sand trend, both to the southwest and the northeast.

In spite of the density of oil fields in Lea and northeastern Eddy Counties, the three dimensions of southeastern New Mexico have not been adequately explored. Most of the new discoveries each year fill in a part of the open space between pre-existing fields. This infilling will, of course, continue almost indefinitely. But there are also much greater open spaces northward and westward to the boundaries of the southeastern quadrant, as well as southward to the Texas line. In addition, practically nowhere in this district have the possibilities of production from deep reservoirs been exhausted.

In regard to Chaves County, lying north and west, it cannot be assumed that the fields in the southeast corner and in the Pecos Valley area have captured all of the oil in this large county. The same can be said of the fields at the southern end of Roosevelt County, even though the porosity of the foremost reservoir, the San Andres, diminishes northward (Gratton and LeMay, 1967). The San Andres trend has recently been extended westward from southern Roosevelt County into east-central Chaves County.

Farther north, in northern Roosevelt and in Curry County, the westward plunge of the Matador arch provides a locale for both structural traps and updip permeability wedgouts over a wide area, including eastern De Baca County.

Westward along and beyond the Northwestern shelf there is an inadequately explored area in contiguous parts of Eddy, Chaves, and Otero Counties. The Otero basin (Figure 32.2), floored with Paleozoic sediments, lies near the southwestern corner of the southeastern quadrant of New Mexico. Quaternary alluvium masks the bedrock over much of the area. The deepest exploration well so far drilled reached a depth of 4646 feet, bottomed in Pennsylvanian rocks (Nutter, 1965, p. 100). The Otero basin needs a much higher density of test wells, drilled to far greater stratigraphic depths. Unfortunately a part of the area lies within the White Sands Missile Range. Lincoln County to the north of Otero, although partially penetrated by Tertiary intrusives, also has hydrocarbon possibilities. The presence of oil seeps has led to some exploration, so far unsuccessful (Havenor, 1964).

The prow of the Delaware basin, which extends northward into western Lea and eastern Eddy Counties, has a greater thickness of relatively unexplored favorable sedimentary column than any other part of southeastern New Mexico. Recent developments in the Texas sector of this basin show that New Mexico can anticipate many new discoveries on its side of the line. The parade of new gas fields is already advancing northward. The deepest producer in New Mexico to date is a gas well at 17,684 feet in the Delaware basin. The reservoirs include Devonian limestones and, high up the prow, Lower Pennsylvanian lenticular sandstones. Trapping is due both to the presence of anticlines and to updip shaling of sandstone (Hills, 1968, p. 1423–1425).

Only a beginning has been made in seeking oil and gas below the prolific mid-Permian reservoirs in southeastern New Mexico. The entire column from the Wolfcamp to the basement rock is loaded with potential reservoir rocks. Deeper drilling on the Central Basin platform opened up worthwhile Siluro–Devonian and Ordovician Ellenburger production. Exploring around the rim of the Delaware basin has uncovered large gas reserves in both Pennsylvanian and Siluro–Devonian dolomites. The amount of gas that is stored in the still deeper crystalline dolomites of the Ellenburger down the flanks of the Delaware basin may well be astronomic.

Oil and gas accumulations in fractured crystalline basement rock are a distinct possibility on the Central Basin platform. According to Stipp and Haigler (1957), "Precambrian basement rocks are relatively high structurally and in places the Permian strata rest on Precambrian, or the truncated edges of older Paleozoic strata." This environment can be described as fraught with possibilities.

Northwestern New Mexico. The greater part of northwestern New Mexico is in the Colorado Plateau physiographic province (Arnold, 1965, p. 87). Although the northwestern quadrant of New Mexico is considerably larger than the San Juan basin, the basin is the dominant feature (Wengerd, 1958a; Peterson, 1959). It is from 15 to 20 thousand square miles in area and covers parts of San Juan, Rio Arriba, Sandoval, and McKinley Counties. So far, the oil and gas fields of northwestern New Mexico are confined to these counties. The two largest oil fields in total production to date are Bisti and Horseshoe (Figure 32.4), both in San Juan County. However, the largest accumulation of hydrocarbons in terms of heat units occurs in the Basin Dakota and higher gas fields, which so far cover more than 25 per cent of the total basin area.

Basement rock outcrops fringe the San Juan basin to the south and east. These are, counterclockwise, the eroded cores of the Zuni, Jemez, and Sangre de Cristo uplifts. The Zuni uplift, in turn, is fringed on the south and east by volcanics, and similar rocks cover a wide area immediately east of the Jemez basement rock exposures.

The northern flank of the San Juan basin extends into southwestern Colorado. A structural divide, the Four Corners platform (Figure 32.2), separates the San Juan from the Paradox basin of Colorado and southeastern Utah. The northeast-striking Hogback monocline, with a relatively steep dip to the southeast, marks the boundary between the Four Corners platform and the San Juan basin. This line can be seen in Figure 32.4; a half-dozen oil fields lie in the New Mexican sector of the Four Corners platform. The western flank of the San Juan basin merges to the west with the eastern flank of the Defiance uplift of northeastern Arizona.

With the exception of the monocline terminating the San Juan basin on the northwest and north, the dips into the basin are gentle, especially on the south flank, where they average about 120 feet to the mile (less than $1\frac{1}{2}°$). Reversals, resulting in anticlines, are scarce in the basin except toward the rim. In consequence, trapping of hydrocarbons out in the basin has been due in large part to stratigraphic factors.

The stratigraphic column applicable to northwestern New Mexico is given in Figure 32.10 (Dane, 1965). It is possible to compile a nearly complete chart of the formations in this area by surface observation, because of the geologic structure and the topographic relief. Between the basin centers and the eroded uplifts sediments from Recent to Precambrian are exposed.

The San Juan basin contains an aggregate thickness of sedimentary rocks of about 15,000 feet (Peterson et al., 1965). During the early Paleozoic, a sandstone was deposited in a short-lived Cambrian sea. The area was emergent during the Ordovician and Silurian periods. A marine sea inundated this part of the world during Late Devonian and Early Mississippian time; carbonate rocks were the result. An uplift followed, and during Later Mississippian and Early Pennsylvanian the emergent limestone terrane was chemically weathered, producing caverns and sinkholes. The first downwarping of the San Juan basin took place during Pennsylvanian–Permian time, as a southeastern sag of the Paradox basin. Mostly clastic sediments were deposited in the resulting sea, which transgressed over the karstic landscape. The Triassic–Jurassic interval was mainly one of emergence with wind-blown sand deposition. Major downwarping in the Cretaceous resulted in the accumulation of a thick section of sandstones and shales, starting with the Dakota Formation. It was during this period that the greatest volume of both source and reservoir rock was deposited. The cyclic sedimentation during this period has been described by Sabins (1964). Transgression by marine waters ceased with the Laramide uplift, which started in Late Cretaceous time. Subsequently a considerable volume of continental Paleocene and Eocene sediments was deposited in the San Juan basin (Baltz, 1967).

Going down the stratigraphic column in northwestern New Mexico, the first reservoir of importance is the Upper Cretaceous Pictured Cliffs Sandstone. This is the upper gas reservoir in the Blanco field on the north flank of the San Juan basin. Below the Pictured Cliffs Sandstone is the thick Lewis Shale, beneath which is the Mesaverde Group of sandstones. Mesaverde reservoirs are gas productive in the Blanco field over an area of nearly 1 million acres. The main gas-producing sandstones are the Cliff House at the top of the group and the Point Lookout at the bottom. These are not blanket sands but are discontinuous shoreline deposits (Hollenshead and Pritchard, 1961).

Beneath the Mesaverde Group is the Manocs Shale. Within the Mancos are sand bars, collectively known as the Gallup Sandstone, from which most of the oil so far produced in northwestern New Mexico has come. The Mancos Shale itself contains some oil in fractures. Under the Mancos is the widespread Dakota Series of sands, which is gas productive in the Basin Dakota gas field over a probable area of 1,400,000 acres (Arnold, 1965, p. 96).

Both oil and gas (Picard, 1960a, 1968) are being produced from carbonate rock reservoirs in the Pennsylvanian Hermosa Formation. The principal accumulations are in anticlinal traps high on the western flank of the San Juan basin. Helium has also been found in this reservoir, as well as in the deeper Devonian Ouray Limestone.

Horseshoe Canyon (Knight and Budd, 1959) in San Juan County was the largest oil field in northwestern New Mexico in 1966 with a production of more than 3,200,000 barrels. In this area the dip is eastward into the basin. Trapping is due to the lenticularity of the reservoir sands, which are in the Upper Cretaceous Gallup Formation. Each producing lens is an elongate sand bar with a northwest trend completely encased in the marine shale of the Mancos Formation.

The *Rattlesnake* field (Hinson, 1947), in the Navajo Indian reservation southwest of Shiprock in northwestern New Mexico, lies high on the northwest flank of the San Juan basin. The Rattlesnake anticline was first mapped

GEOLOGIC TIME UNITS			FORMATIONS	
Cenozoic (to 70 million years)	Quaternary	Recent	Alluvium	
		Pleistocene	Gravel, sand, clay and volcanic deposits	
	Tertiary	Pliocene	Santa Fe Group	Chuska Sandstone
		Miocene		
		Oligocene	May be present (?)	
		Eocene	Galisteo Formation (Eocene and questionably Oligocene); Baca Formation (questionably Eocene); San Jose Formation	
		Paleocene	Nacimiento Formation; Animas Formation (Upper Cretaceous and Paleocene)	
Mesozoic (70–225 million years)	Cretaceous		Ojo Alamo Sandstone McDermott Formation Kirtland Shale and Fruitland Formation Pictured Cliffs Sandstone Lewis Shale	
			Cliff House Sandstone Menefee Formation Point Lookout Formation Crevasse Canyon Formation	All of the Mesaverde Group
			Mancos Shale with Gallup Sandstone Dakota Sandstone	
	Jurassic		San Rafael Group: Morrison Formation Bluff Sandstone Summerville Formation Todilto Formation Entrada Sandstone	Zuni Sandstone
	Triassic		Glen Canyon Group Wingate Sandstone Chinle Formation	
Paleozoic (225 to 600 million years)	Permian		San Andres Limestone Glorieta Sandstone	
			Yeso Formation Abo Formation	Cutler Formation Rico Formation
	Carboniferous	Pennsylvanian	Madera Formation Sandia Formation	Hermosa Formation Paradox Member Molas Formation
		Mississippian	Arroyo Penasco Formation	Present in subsurface (includes the Leadville Limestone)
	Devonian		Present in subsurface	Ouray Limestone (includes the Elbert Formation)
	Silurian		Absent	
	Ordovician		Absent	
	Cambrian		Present in subsurface (includes the Ignacio Quartzite)	
	Precambrian		Granite, quartzite, pegmatites, etc.	

Figure 32.10 **Northwestern New Mexico. Stratigraphic chart.** ***After C. D. Dane (1965, p. 24).***

by surface observation, using a sandstone member of the outcropping Mancos Shale as a structure contour datum. The testing of this structure began in 1923, and the first well, which proved to be a small producer, was completed in February, 1924, in the Dakota Sandstone at a depth of 840 feet. A Pennsylvanian reservoir, the Hermosa Limestone, was found to be productive in 1929 at a depth of about 6600 feet. Since discovery, Rattlesnake has produced about 5 million barrels of oil, over 90 per cent of which has come from the Dakota reservoir.

Maps of the Rattlesnake field are shown in Figure 32.11. The upper figure is a subsurface structure map contoured on top of the deep Ouray Limestone of Devonian age. Contouring is based on data from four deep wells and a seismograph survey. Helium-bearing natural gas occurs in four porous zones in the upper part of the Ouray Limestone at depths in the neighborhood of 7000 feet. Porosity is of two types: solution porosity and porosity due to openings between crystals of dolomite. The Rattlesnake anticline is about 6 miles long and asymmetrical, with the steeper dip to the southwest. The apex at the Ouray level is near the southeast end of the axis.

The lower figure shows the structure contour map based on the First Dakota Sandstone. This sandstone lies more than 6100 feet above the Ouray Limestone. In the intervening distance the axis of the anticline has shifted northeast $\frac{5}{8}$ mile, owing to the asymmetry of the fold, and the apex has moved $2\frac{1}{2}$ miles to the northwest. Production has been obtained from three different sandstones within the Dakota. Accumulation has taken place in the highest parts of the anticline, but only where the sandstones have adequate porosity and permeability.

Tocito dome is a surface structural feature that was mapped by various geologists some years before New Mexico became an oil- or a gas-producing state. It was first drilled in 1943, but without success, although noncommercial gas was found in the Mississippian. Twenty years later, in 1963, the discovery well was drilled and came in for over 5 million cubic feet of wet gas in a Pennsylvanian reservoir. A year later, oil was discovered, also in the Pennsylvanian Paradox Formation, at a depth of around 6300 feet. Subsequently more than a dozen additional oil wells have been drilled on the dome; the yield from these wells put Tocito in second place in annual production in 1966.

The *Hospah* oil field (King and Wengerd, 1957) is in northeastern McKinley county. It occupies the Hospah dome, which was mapped at the surface in 1924 by recording outcrop strikes and dips. At the surface the dome is more than 3 miles in diameter with a closure in excess of 300 feet. The field was discovered in 1927 but was not brought into extensive production until a pipeline was constructed in 1940.

The producing formation, locally known as the Hospah, is a sandstone just above the upper Gallup section. It lies at a depth of 1500 to 1600 feet. The reservoir sandstone ranges from 25 to 60 feet, with an average productive thickness above the oil-water contact of 40 feet. The trapping is anticlinal, but along the eastern side of the field the reservoir sandstone declines markedly in permeability because of increase in clay content.

At the reservoir level the Hospah anticline is about $2\frac{1}{2}$ miles long in a north-south direction and $\frac{1}{2}$ mile wide. The vertical closure is over 100 feet. A northeasterly trending fault cuts the anticline on the southeast flank, dropping the northwest side about 200 feet. A lesser anticline, parallel in trend, cuts the anticline high on the northwest flank. The displacement here is about 30 feet.

The *Bisti* field (Sabins, 1963) in east-central San Juan County is the leading field in northwest New Mexico in cumulative oil yield, having produced more than 30 million barrels between discovery in 1955 and 1966. Bisti is a "shoestring" field, over 30 miles long with a maximum width of 3 miles. It trends west-northwest across the west-central part of the San Juan basin. The regional dip here is 50 to 75 feet per mile northeastward. Trapping at Bisti is in elongate sand lenses in the Gallup Sandstone of Cretaceous age. Hydrodynamic assistance has been suggested (McNeal, 1961).

Sabins has succeeded in differentiating five facies in the Gallup Sandstone, which intervenes between the upper and the lower Mancos Shale. They are, from south to north on Figure 32.12, the beach, back-bar, bar-sand, and forebar facies. These are distinguished one from the other by different petrographic characteristics, even though they may look alike megascopically and have identical electric log reaction patterns.

The oil reservoirs at Bisti, which are within the beach-sand facies, are three in number. They are shown in plan in Figure 32.12 and in section in Figure 32.13. Little or no water underlies the oil; gas in solution provides the drive. Most wells have been completed by sand fracturing to establish production (Tomkins, 1957, p. 906). Repressuring the reservoir by gas injection and LPG floods started at Bisti within a year of discovery.

The outstanding hydrocarbon accumulation in northwestern New Mexico is the large gas-producing area that covers the greater part of the San Juan basin in San Juan and Rio Arriba Counties and penetrates a short distance south of Rio Arriba County into Sandoval County. The gas area also continues northward into La Plata County, Colorado. This is one of the giant gas fields of the United

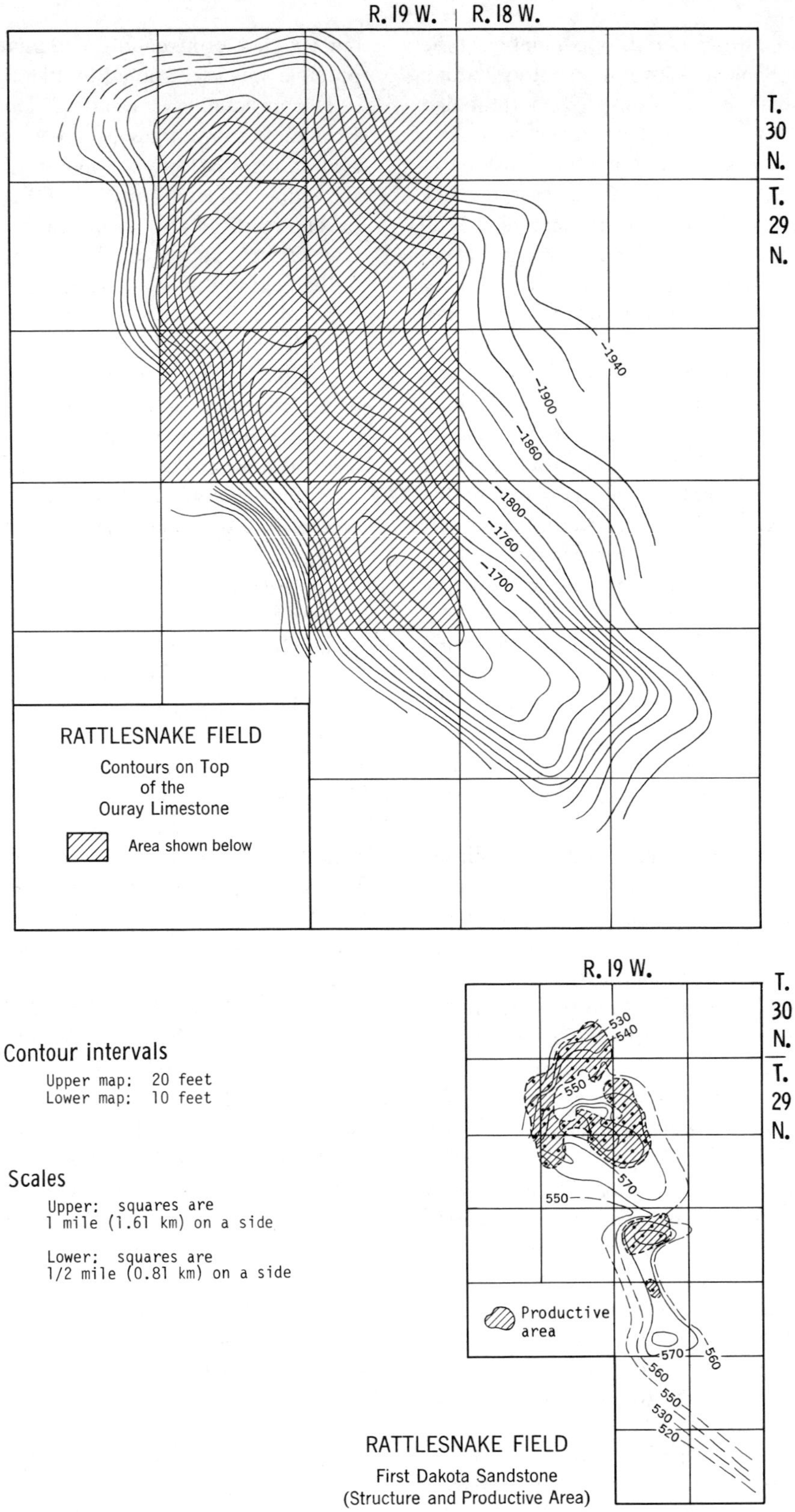

Figure 32.11 Rattlesnake field, San Juan County. Structure contour maps. ***After H. H. Hinson (1947).***

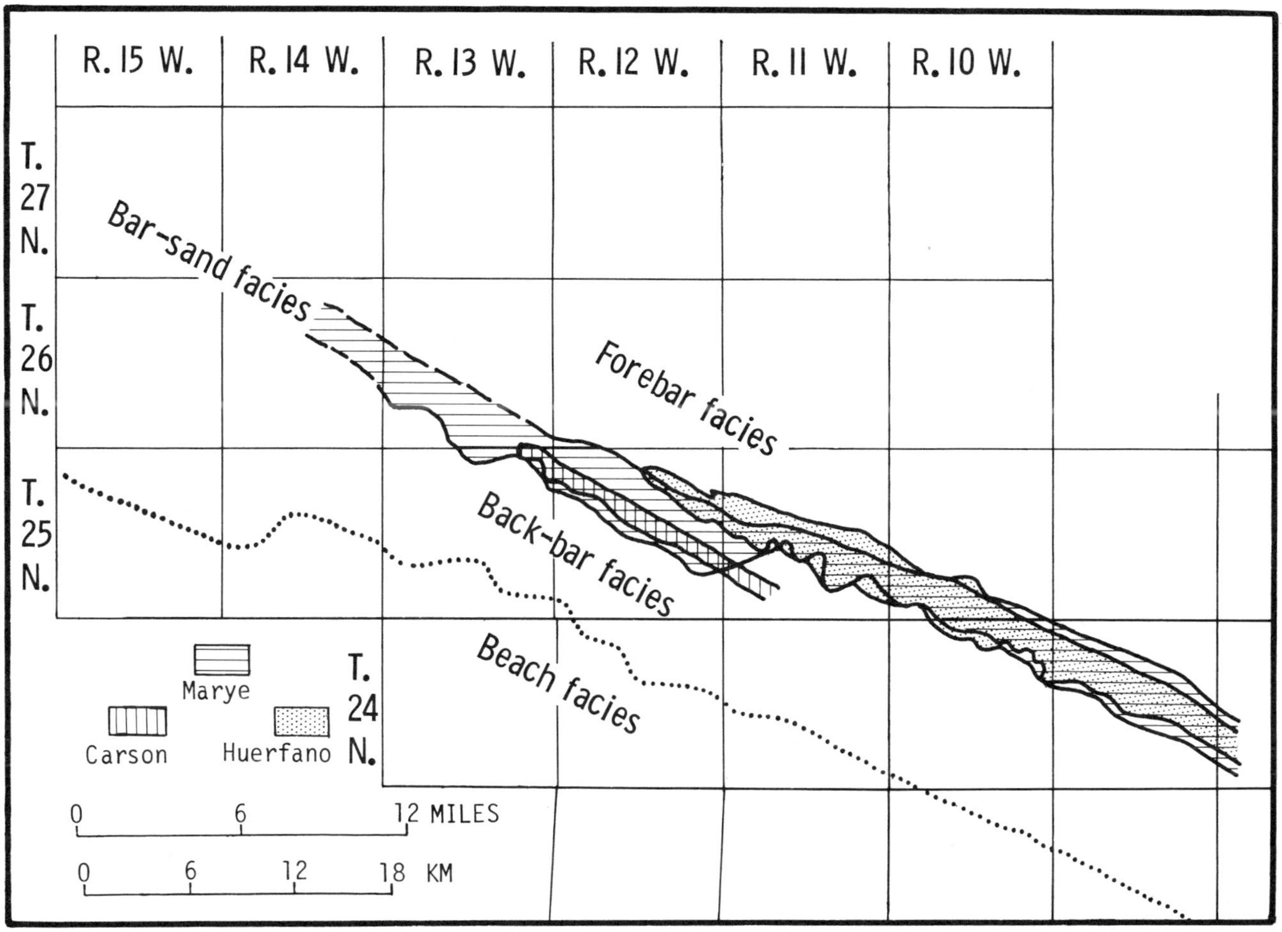

FIGURE 32.12 Bisti oil field, San Juan County. Facies map. Patterns show areal distribution of three named sand bars in the Bar sand facies; these are the oil pay zones of the Bisti field. *Courtesy Floyd F. Sabins, Jr. (1963, Figure 4, p. 197) and the American Association of Petroleum Geologists.*

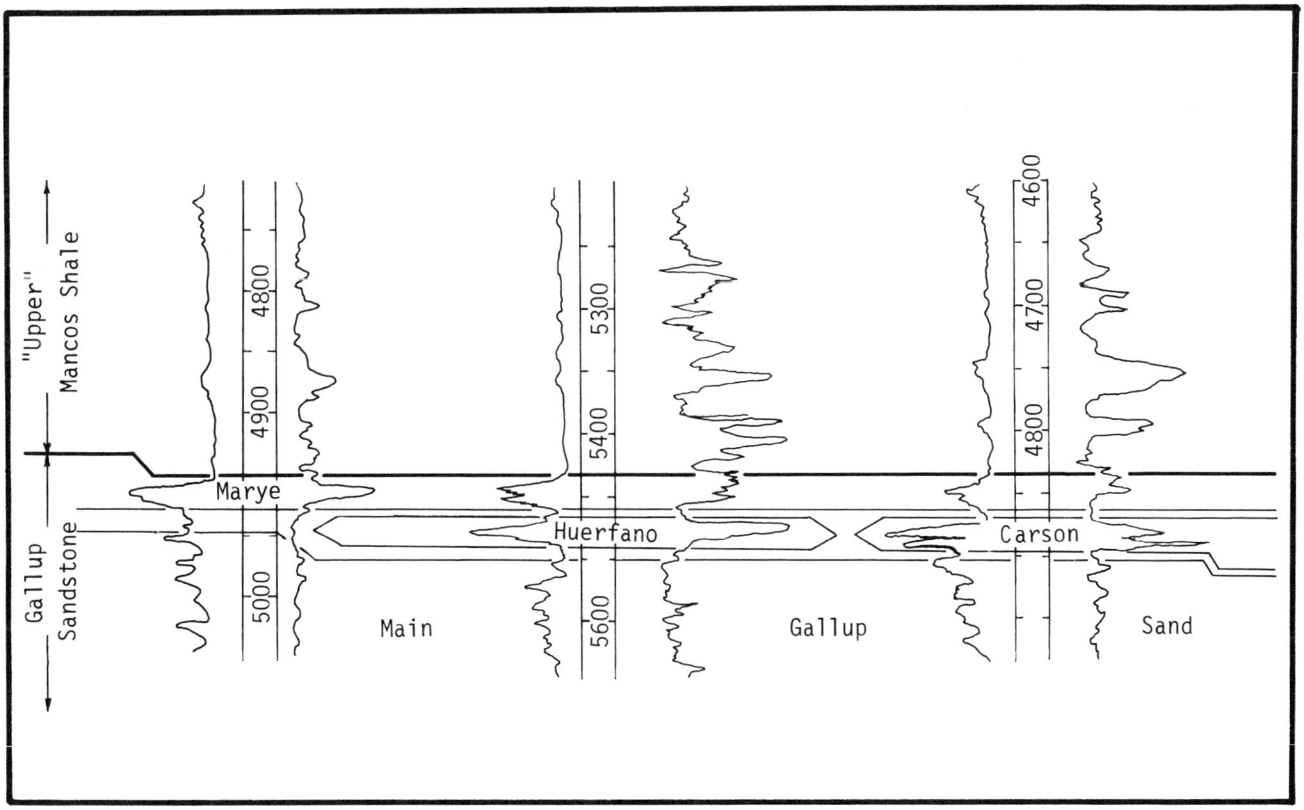

FIGURE 32.13 Bisti field. Reservoir cross section constructed from type electric logs, showing relationships between named sand bars. *Courtesy Floyd F. Sabins, Jr. (1963, Figure 5, p. 199) and the American Association of Petroleum Geologists.*

States, having a diameter of 100 miles and covering considerably more than 1 million acres.

The first gas discovery in this vast area was made in the *Blanco* field on the north flank of the basin in late 1927. Gas was found in a Mesaverde sandstone at a depth of 4600 feet. The Mesaverde has a total thickness of 800 feet in this area. Subsequently, in early 1951, more gas was discovered uphole in the Pictured Cliffs Sandstone above the Mesaverde Group at a depth of 2500 feet. This shallow reservoir has an average thickness of 80 feet. The third reservoir strike came in October, 1952, when gas and distillate were found at 7800 feet in a 200 foot Dakota section.

Earlier, in 1947, gas had been found elsewhere in the San Juan basin in the Dakota Formation, but development came slowly because of both the depth and tightness of the sands and the distance to market. In the late 1950s, the application of high-volume, high-pressure hydraulic fracturing made the exploitation of the tight Dakota reservoirs economically feasible. The deep sand gas and distillate field, which became known as *Basin Dakota*, spread rapidly to its present limits, merging with the Blanco Dakota field on the north flank of the basin in so doing.

Trapping throughout the gas area is stratigraphic. In the highest Cretaceous reservoir, the Pictured Cliffs Sandstone, production is confined to elongate northwest-trending belts of permeable sands surrounded by impermeable sands that were the result of concurrent deposition of silt and shale. The Mesaverde reservoir is about 70 miles long, with a maximum width of 40 miles. This extensive gas deposit trends westward on the south-dipping monoclinal flank of the basin. Sealing of the gas is brought about by updip wedgout of porosity.

The basin-wide Dakota reservoir is filled with gas, except locally where the clay content is too high for commercial development even when stimulated by fracturing. It is this decrease in permeability that seals in the gas and distillate. Silver (1968) believes that the gas accumulated before the "development of the present basin structure."

The discovery and exploitation of helium has been a by-product of the search for oil and gas in northwestern New Mexico. The first discovery was made in the previously mentioned Rattlesnake field, where helium was found in Ouray Limestone of Devonian age (Pierce, 1965, p. 124). The U.S. Government built a helium extraction plant at Shiprock to process this gas. Subsequently, while the *Table Mesa* field in western San Juan county was being developed in 1961, helium was found in the Hermosa Formation of Pennsylvanian age. This is the present source of helium for the Shiprock plant. Table Mesa produces petroleum from Dakota sandstones, where the accumulation is in a faulted anticline.

Helium gas has been found also in Mississippian (Picard, 1962) and Triassic rocks in northwestern New Mexico. In 1967 exploration along the southeast-trending Toadlena anticline resulted in the discovery of the unusual Dineh bi Keyah oil field in northeastern Arizona (see Chapter 37); this led to the finding later that year of oil and helium gas across the state line in New Mexico. Both helium and oil were discovered in a Devonian sandstone, and helium only was found in Precambrian quartzite.

The San Juan basin contains a large volume of sedimentary rock as yet unexplored. Potential reservoirs in the central basin area lie beneath the Dakota sandstones. In this area exploration and discovery moved down the column through the Upper Cretaceous with significant success. Yet to be thoroughly explored are older reservoirs, especially in the Pennsylvanian section (Wengerd, 1959a). Extensions to the Basin Dakota gas field southward, especially farther into Sandoval County, are possible. The entire south flank of the San Juan basin merits more intensive exploration; a large area is involved. There are no doubt wedgout traps, in addition to anticlinal accumulations, yet to be discovered around the basin perimeter.

The Chama basin (Figure 32.2) is also in the northwestern quadrant of New Mexico. It lies east of, and across a structural divide from, the San Juan basin. Shallow drilling has found gas shows; the basin has not yet been explored at depth (Chasteen, 1961).

The Estancia basin is in the southeastern corner of the northwestern quadrant. It underlies the Estancia Valley, which is a broad topographic basin. Below and to the sides of the valley fill are Paleozoic sediments. Shows of oil have been encountered during the drilling of almost all Pennsylvanian test wells. The stratigraphic section from surface to Precambrian granite floor is at least 5900 feet in thickness (Nutter, 1965, p. 99).

Halfway across the southern edge of the northwest quadrant are the Acoma embayment, Lucero shelf, and Zuni basin (Figure 32.2). These structural features lie immediately east and south of the Zuni uplift and may be considered as a southward extension of the San Juan basin (Figure 32.4). Cretaceous rocks crop out at the surface, where not masked by Tertiary and Quaternary lava flows. At least 5000 feet of marine sediments are known in this area. Tectonically and stratigraphically the setting here is similar to the oil productive western rim of the San Juan basin.

UNDEVELOPED QUADRANTS

Northeastern New Mexico. The northeastern quadrant of the state is dominated by the Sangre de

Cristo and Pedernal uplifts and the Sierra Grande and Matador arches. Between these positive features, however, are sedimentary basins. The Raton–Las Vegas basin lies between the Sangre de Cristo uplift and the Sierra Grande arch. The Tucumcari basin is midway between three positive features, Matador, Pedernal, and Sierra Grande (Figure 32.2). In the northeastern corner is the western flank of the Dalhart basin of the Texas Panhandle.

The general stratigraphy for northeastern New Mexico is shown in Figure 32.14. It differs from the columns to the east and south in the greater development of Mesozoic formations and in the absence of lower Paleozoic sediments, except for local occurrences of Devonian and Mississippian carbonate rocks.

The Tertiary Raton structural basin extends southward into New Mexico from Colorado, where it has greatest extent and reaches maximum depth (Gabelman, 1956). It is asymmetric, with the axis on the west side near the Sangre de Cristo mountain range. The Las Vegas basin to the south in New Mexico is on the same trend and has been referred to as a "sub-basin" (Baltz, 1965). The Raton downwarp has been an area of sedimentation since Early Pennsylvanian time. It contains a thick section of potentially oil-bearing marine sediments of Pennsylvanian and Cretaceous age. The thickest section explored to date by the drill was 7225 feet in the Raton basin proper. So far, the greatest encouragement has been obtained in the shallower Las Vegas sector, where about 5000 feet has been explored. Several wells logged oil shows, and one well reported 100,000 cubic feet daily of gas from Pennsylvanian rocks at between 4650 and 4800 feet depth (Nutter, 1965, p. 99). Further prospecting is indicated for this basin.

The Tucumcari basin, with a probable maximum sedimentary section of 10,000 feet, has had several deep tests, but only shows have resulted to date. The Dalhart basin might offer better hunting on its west flank in New Mexico than it has provided so far to the east in the Texas Panhandle.

The southeast flank of the Sierra Grande uplift, between Dalhart and Tucumcari basins, has possibilities for trapping in reservoir wedgouts (Wanek, 1962). Several wells drilled in this area have reported shows.

A commercially producing carbon dioxide gas area (shown in Figure 32.4) covers a part of Harding County. The reservoirs are Jurassic sandstones and Permian and Pennsylvanian arkoses, flanking the Sierra Grande arch. Tertiary and Quarternary igneous rocks, both intrusive and extrusive, are nearby and are thought to have played a role in the genesis of this gas (Pierce, 1965).

Southwestern New Mexico. This is the least likely of the four quadrants in terms of sizable oil or gas production (Wengerd, 1962). It contains mountain ranges, lava flows, igneous intrusions, extensive copper and other ore deposits, and large military reservations. Nonetheless at least three sedimentary basins and one shelf are tucked in between the uplifts and are overlapped by lava flows. These are the Tularosa, Jornado del Muerto, and Pedregosa basins, and the Lucero shelf (Figure 32.2).

The stratigraphic column for the southwestern quadrant is given in Figure 32.15. Areally, the dominant "formations" are late Tertiary lava flows mainly in the western half of the quadrant, and Quaternary alluvium in the east-central and southern parts. The only outcropping sedimentary rocks are Paleozoic in age and occur in the cores of uplifts mainly in the eastern half of the area.

All of the Paleozoic periods are represented in southwestern New Mexico, as well as the Upper Cretaceous Dakota, Mancos, and Mesaverde Formations.

> Source beds and possible petroleum reservoir rocks occur in the Paleozoic and Mesozoic sequences. The reservoir beds include porous sandstone, bioclastic calcarenite, dolomite, reef masses, porous lenses beneath truncating unformities or amid intertongued redbeds, and various limestones (Kottlowski, 1965, p. 2120).

The Tularosa basin, also referred to as the Orogrande (Nutter, 1965, p. 100), lies principally in western Otero County in the Tularosa Valley, between the San Andres uplift in the west and the Sacramento uplift to the east. Tularosa was in existence as a basin of sedimentation from Middle Mississippian to the close of Wolfcampian time (Kottlowski, 1965). As much as 7000 feet of sediment was deposited. Possible traps include anticlines, faults, and reefs (DeFord, 1951, p. 349), but exploration is impeded by the fact that much of the basin is a military reservation.

The next basin to the west, the Jornado del Muerto, was explored by two tests in 1962. Both penetrated Paleozoic sediments, one reaching a total depth of 7350 feet in igneous rock and the other being abandoned at 6500 feet in Precambrian.

The extreme southwestern tip of New Mexico is overlapped by the northern end of the Pedregosa basin of Mexico. The only deep well to be drilled in this basin reached a depth of 14,585 feet. It reported gas-cut mud at 4200 feet in a Permian limestone (Nutter, 1965, p. 100).

The Lucero shelf overlaps the northwestern and southwestern quadrants in western New Mexico. It has been described by Wengerd (1958b, 1959b). More recently Foster (1964) reported on the stratigraphy and oil possibilities of Catron County, the northern part of which is on the Lucero shelf. In this area only remnants of Cretaceous sediments remain, and the Pennsylvanian section ranges from 1400 feet to nothing where truncated

GEOLOGIC TIME UNITS			FORMATIONS
Cenozoic (to 70 million years)	Quaternary	Recent	Alluvium, other surficial deposits, and some volcanic deposits
		Pleistocene	Gravel, sand, caliche and some volcanic deposits
	Tertiary	Pliocene	Santa Fe Group: Ogallala Formation
		Miocene	Santa Fe Group
		Oligocene (?) Eocene	Galisteo Formation and other local formations
		Paleocene	Poison Canyon Formation Raton Formation
Mesozoic (70–225 million years)	Cretaceous		Vermejo Formation Trinidad Sandstone Pierre Shale Niobrara Formation Carlile Shale Greenhorn Limestone Graneros Shale Dakota Sandstone Purgatoire Formation
	Jurassic		Morrison Formation Summerville Formation Todilto Limestone Entrada Sandstone
	Triassic		Dockum Group Chinle Formation Santa Rosa Sandstone
Paleozoic (225 to 600 million years)	Permian		Artesia Formation San Andres Limestone Glorieta Sandstone Yeso Formation
	Permian and Pennsylvanian		Sangre de Cristo
	Carboniferous	Pennsylvanian	Madera Limestone Sandia Formation
		Mississippian	Tererro Formation
	Devonian		Espiritu Santo Formation
	Silurian		Absent
	Ordovician		Absent
	Cambrian		Absent
	Precambrian		Granite, diorite, other plutonic rocks, schist, quartzite, and pegmatites

FIGURE 32.14 Northeastern New Mexico. Stratigraphic chart. ***After C. D. Dane (1965, p. 25).***

<table>
<tr><th colspan="3">GEOLOGIC TIME UNITS</th><th colspan="3">FORMATIONS</th></tr>
<tr><td rowspan="8">Cenozoic
(to 70 million years)</td><td rowspan="3">Quaternary</td><td>Recent</td><td colspan="3">Alluvium, other surficial deposits and volcanic rocks</td></tr>
<tr><td rowspan="2">Pleistocene</td><td colspan="3">Gravel and volcanic rocks</td></tr>
<tr><td rowspan="3">Santa Fe Group</td><td>Gila Conglomerate</td><td rowspan="3">Datil Formation and associated volcanic rocks</td></tr>
<tr><td rowspan="5">Tertiary</td><td>Pliocene</td><td rowspan="2"></td></tr>
<tr><td>Miocene</td></tr>
<tr><td>Oligocene</td><td colspan="3">May be present (?)</td></tr>
<tr><td>Eocene</td><td colspan="3">Baca Formation of questionable Eocene age</td></tr>
<tr><td>Paleocene</td><td colspan="3">Not recognized (?)</td></tr>
<tr><td rowspan="5">Mesozoic
(70–225 million years)</td><td colspan="2" rowspan="3">Cretaceous</td><td>Mesaverde Group</td><td colspan="2">(includes some volcanic rocks)</td></tr>
<tr><td>Mancos Shale
Dakota Sandstone</td><td colspan="2">Colorado Shale
Sarten Sandstone and Beartooth Quartzite</td></tr>
<tr><td colspan="3">Bisbee Group
(includes a sequence of conglomerate, sandstone, volcanic rocks, etc., more than 10,000 feet thick in extreme southwestern New Mexico)</td></tr>
<tr><td colspan="2">Jurassic</td><td colspan="3">Absent</td></tr>
<tr><td colspan="2">Triassic</td><td colspan="3">Present in northern part of region</td></tr>
<tr><td rowspan="7">Paleozoic
(225 to 600 million years)</td><td colspan="2">Permian</td><td>Concha Formation
Scherrer Formation
Colina Limestone
Earp Formation</td><td colspan="2">San Andres Limestone
Glorieta Sandstone
Yeso Formation
Abo Formation
Bursum Formation</td></tr>
<tr><td rowspan="2">Carboniferous</td><td>Pennsylvanian</td><td>Horquilla Formation</td><td colspan="2">Magdalena Group</td></tr>
<tr><td>Mississippian</td><td>Lake Valley Limestone
Kelly Limestone</td><td colspan="2">Escabrosa Limestone</td></tr>
<tr><td colspan="2">Devonian</td><td colspan="3">Percha Shale (and other local formations)</td></tr>
<tr><td colspan="2">Silurian</td><td colspan="3">Fusselman Dolomite</td></tr>
<tr><td colspan="2">Ordovician</td><td colspan="3">Montoya Dolomite
El Paso Formation</td></tr>
<tr><td colspan="2">Ordovician and Cambrian</td><td colspan="3">Bliss Sandstone</td></tr>
<tr><td></td><td colspan="2">Precambrian</td><td colspan="3">Granite and associated rocks</td></tr>
</table>

FIGURE 32.15 Southwestern New Mexico. Stratigraphic chart. ***After C. D. Dane (1965, p. 26).***

by erosion. Permian rocks rest directly upon the Precambrian floor on the east flank of the Zuni uplift. However, there are possibilities for Permian production, providing that adequate source rocks are present. Extensive lava flows in this part of New Mexico have discouraged exploration because the lava prevents surface recognition of anticlines of pre-extrusive age. The structure of flows can be mapped. Cook (1957) has reported the tracing of a flow for 120 miles in the Colorado Plateau.

REFERENCES CONSULTED IN PREPARING CHAPTER 32

Arnold, E. C., 1965.
Babock, C. V., 1956.
Baltz, Elmer H., 1965, 1967.
Burleson, W. E., and Henkes, William C., 1967.
Caffee, E. C., and Dubitzky, R. H., 1967.
Chasteen, Kenneth, 1961.
Cone, George C., 1965.
Cook, E. F., 1957.
Dane, C. H., 1965.
Davis, Morgan J., 1929.
DeFord, Ronald K., 1951.
DeFord, Ronald K., and Wahlstrom, Edwin A., 1932.
Denham, R. L., and Dougherty, W. E., 1941.
Dunn, David A., 1956.
Flawn, Peter T., 1967.
Foster, Roy W., 1964.
Fryman, A. D., 1956.
Gabelman, John W., 1956.
Gratton, Patrick J. F., and LeMay, William J., 1967.
Havenor, Kay C., 1964.
Hills, John M., 1968.
Hinson, H. H., 1947.
Hollenshead, Charles T., and Pritchard, Roy L., 1961.
King, Vernon L., and Wengerd, Sherman A., 1957.
Knight, William V., and Budd, Harrell, 1959.
Kottlowski, Frank E., 1965.
LeMay, William J., 1960.
McNeal, Robert P., 1961.
McPeters, K. D., and Kelly, J. M., 1956.
Marshall, L. R., and Foltz, G. A., 1960.
Meyer, Richard F., 1966.
Milks, Roy, 1956.
Montgomery, R. F., 1965.
New Mexico Geological Society, 1965.
Nottingham, Marsh W., 1966.
Nutter, D. S., 1965.
Peterson, James A., 1959.
Peterson, James A., Loleit, Allan J., Spencer, Charles W., and Ullrich, Richard A., 1965.
Picard, M. Dane, 1960a, 1962, 1968.
Pierce, A. P., 1965.
Roswell Geological Society, 1956, 1960, and 1967.
Sabins, Floyd F., Jr., 1963, 1964.
Salisbury, Gerald P., 1968.
Sax, Norman A., 1967.
Silver, Caswell, 1968.
Stead, F. L., 1955.
Stenzel, William K., 1965.
Stipp, T. F., and Haigler, L. B., 1957.
Thornton, Dewey E., and Gaston, H. H., Jr., 1967, 1968.
Tomkins, J. Q., 1957.
Vlissides, Sophie D., and Bieberman, Robert A., 1961
Wanek, Alexander A., 1962.
Wengerd, Sherman A., 1958a, 1958b, 1959a, 1959b, 1959c, 1962.

Chapter

33

COLORADO

Colorado produced slightly less than 1 per cent of the country's oil in 1968, ranking eleventh. It also ranked eleventh in natural gas output. Some of Colorado's gas comes from its southwestern and southeastern corners, which are overlapped by the San Juan basin of New Mexico and the Hugoton embayment of Kansas. Oil seepages near Cañon City in Fremont County, Colorado (Figure 33.1), first excited interest in 1862, and some of this oil was exploited by shallow wells. However, it was not until 1876, when the nearby Florence pool (Figure 33.4) was developed, that Colorado began continuous oil production. No significant finds were made after Florence for 47 years. Subsequent discoveries of importance are noted under the district headings. Colorado has one giant field, Rangely in the northwestern part of the state. This field passed the 100 million barrel mark in 1952 and in 1966 was still first in the state in annual production, with a yield for that year of over 16 million barrels.

Colorado is the highest state in the Union in average elevation. It can be divided topographically into three divisions: the Great Plains province on the east, the Plateau province on the west, and the Mountain province in between (Figure 33.2). The Great Plains province occupies over three-fifths of the state. It is flat and semiarid; the underlying sedimentary rocks are likewise flat or nearly so. The plains slope easterly from 6000 to 7000 feet elevation at the mountain front to 3500 to 4000 feet at the Kansas border (Tweto, 1964).

The Rocky Mountains in west-central Colorado form the spine of the state, rising abruptly and spectacularly above the plains to the east to elevations greater than 14,000 feet. The Rocky Mountains consist of a series of ranges; when two ranges are parallel, basins ("parks") lie between. An example is North Park in Jackson County (Figure 33.4), which is both a topographic and a sedimentary basin.

Western Colorado is intermediate in elevation and topographic relief between the mountain and the plains

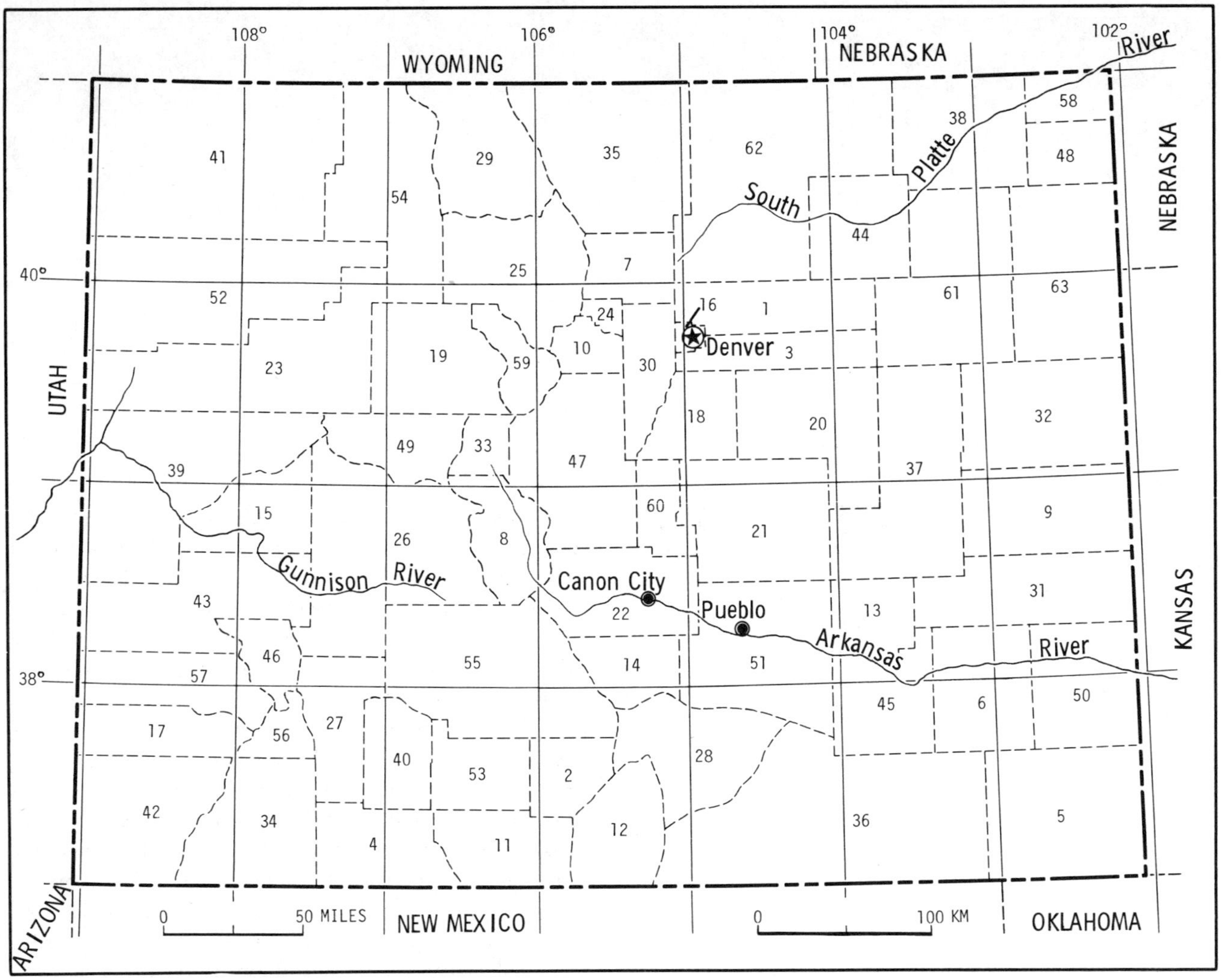

FIGURE 33.1 Colorado. Index map. County Key:

1	Adams	10	Clear Creek	19	Eagle	28	Huerfano	37	Lincoln	46	Ouray	55	Saguache
2	Alamosa	11	Conejos	20	Elbert	29	Jackson	38	Logan	47	Park	56	San Juan
3	Arapahoe	12	Costilla	21	El Paso	30	Jefferson	39	Mesa	48	Phillips	57	San Miguel
4	Archuleta	13	Crowley	22	Fremont	31	Kiowa	40	Mineral	49	Pitkin	58	Sedgwick
5	Baca	14	Custer	23	Garfield	32	Kit Carson	41	Moffat	50	Prowers	59	Summit
6	Bent	15	Delta	24	Gilpin	33	Lake	42	Montezuma	51	Pueblo	60	Teller
7	Boulder	16	Denver	25	Grand	34	La Plata	43	Montrose	52	Rio Blanco	61	Washington
8	Chaffee	17	Dolores	26	Gunnison	35	Larimer	44	Morgan	53	Rio Grande	62	Weld
9	Cheyenne	18	Douglas	27	Hinsdale	36	Las Animas	45	Otero	54	Routt	63	Yuma

provinces. This plateau area is incised by deep valleys and canyons. The sedimentary rocks are mostly flat or gently dipping, so that a mesa type of topography results from erosion in this area.

GEOLOGY. The surface as well as the subsurface geology of Colorado extends from the Recent to the Precambrian. The simplified geologic map (Figure 33.2) shows the Precambrian core of the mountain ranges cropping out extensively in central and west-central Colorado. The same pattern is used to show Tertiary intrusives, which crop out both within the Precambrian area and in surrounding sedimentary rocks. Most of the sedimentary rocks intruded are Paleozoic, Triassic, or Jurassic in age; they occur principally in submountainous uplifted areas in the plateau province.

Almost all of the Great Plains province is underlain by Cretaceous rocks, covered in part by continental Tertiary sediments that were fanned across the plains by streams emerging from the rising Rockies in the late Tertiary. Cretaceous rocks, with Tertiary rock covering

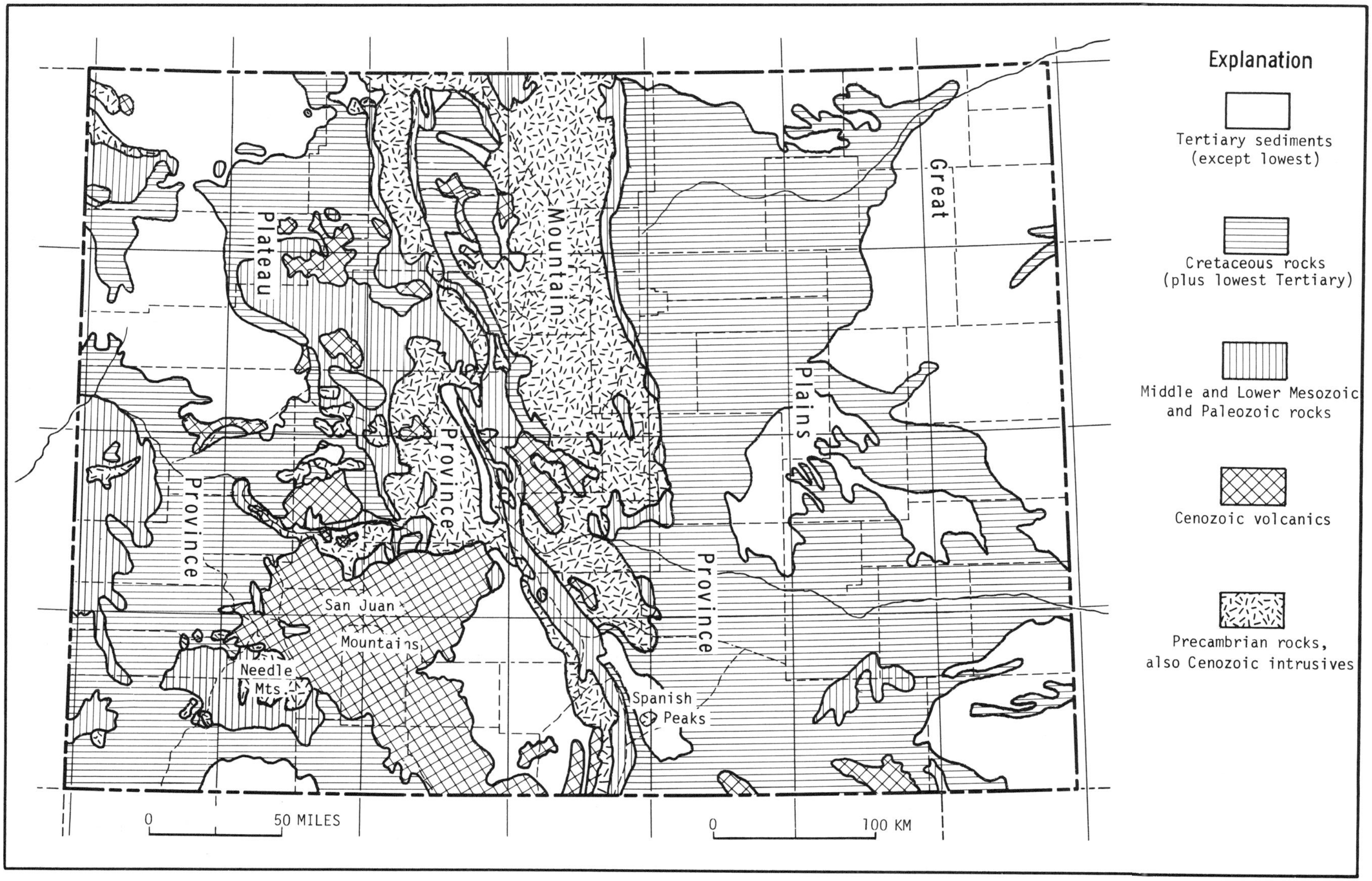

FIGURE 33.2 Colorado. Simplified geologic map. *After Ogden Tweto (1964, Figure 1).*

<table>
<tr><th colspan="3" rowspan="2">GEOLOGIC TIME UNITS</th><th colspan="2" rowspan="2">PLAINS AREA</th><th rowspan="2">MOUNTAIN AREA
(except San Juan Mts.)</th><th colspan="2">PLATEAU AREA</th></tr>
<tr><th>South Part
(including San Juan Mts.)</th><th>North Part
(including Uinta Mts.)</th></tr>
<tr><td rowspan="6">Cenozoic</td><td>Quaternary</td><td>Recent
Pleistocene</td><td colspan="2">Various alluviums
and gravels</td><td colspan="3">Various glacial deposits, alluviums, and gravels
Alamosa Formation</td></tr>
<tr><td rowspan="5">Tertiary</td><td>Pliocene</td><td colspan="2">Ogalalla Formation</td><td rowspan="5">Dry Union Formation, Santa Fe Formation
North Park Formation

White River Formation

Coalmont Formation</td><td rowspan="5">Volcanic rocks of the San Juan and West Elk Mts. (Many names)

Telluride Cg.
Blanco Basin Fm.

Wasatch Formation
Torrejon and Puerco Formations;
Ohio Creek Formation</td><td rowspan="5">Browns Park Formation

Bridger Formation
Green River Formation
Wasatch Formation

Fort Union Formation
Ohio Creek Formation</td></tr>
<tr><td>Miocene</td><td>Northern area
Arikaree Formation</td><td>Southern area
Devils Hole Formation</td></tr>
<tr><td>Oligocene</td><td>White R. Group</td><td>Farisita Conglomerate</td></tr>
<tr><td>Eocene</td><td></td><td>Huerfano Formation
Cuchara Formation</td></tr>
<tr><td>Paleocene</td><td></td><td>Poison Canyon Formation</td></tr>
<tr><td rowspan="3">Mesozoic</td><td colspan="2">Cretaceous</td><td>Denver and Dawson Formations
Arapahoe Formation
Laramie Formation
Fox Hills Sandstone

Pierre Shale
Niobrara Formation
Benton Shale { Carlile Shale, Greenhorn Limestone, Graneros Shale
Dakota Sandstone</td><td>Raton Formation

Vermejo Formation
Trinidad Sandstone</td><td>Middle Park Formation

Laramie Formation
Fox Hills Sandstone

Pierre Shale
Niobrara Formation

Benton Shale

Dakota Sandstone   Lakota Formation</td><td>Animas Formation

McDermott Member
Kirtland Shale
Fruitland Formation
Pictured Cliffs Sandstone
Lewis Shale
Mesaverde Group
Mancos Shale

Dakota Sandstone
Burro Canyon Formation</td><td>Lance Formation

Lewis Shale
Mesaverde Group
Mancos Shale

Dakota Sandstone</td></tr>
<tr><td colspan="2">Jurassic</td><td colspan="2">Morrison Formation

Ralston Creek Formation</td><td>Morrison Formation

Entrada Sandstone</td><td>Morrison Formation
Junction Creek Sandstone
Summerville Formation, Wanakah Formation
Entrada Sandstone</td><td>Morrison Formation

Curtis Formation
Entrada Sandstone
Carmel Formation</td></tr>
<tr><td colspan="2">Triassic</td><td colspan="2"></td><td>Chinle Formation</td><td>Navajo Sandstone
Kayenta Formation
Wingate Sandstone
Dolores Formation, Chinle Formation
Moenkopi Formation</td><td>Navajo Sandstone

Chinle Formation
Moenkopi Formation</td></tr>
</table>

Paleozoic	Carboniferous	Permian	Lykins Formation Lyons Sandstone Ingleside Formation	Chugwater Formation		Cutler Formation	Park City Formation
		Pennsylvanian	Fountain Formation	Maroon Formation Minturn Formation Belden Formation	Sangre de Cristo Formation Eagle Valley Evaporite Kerber Formation	Rico Formation Hermosa Formation Paradox Member Molas Formation	Weber Sandstone Morgan Formation
	Mississippian		Not exposed	Leadville Limestone		Leadville Limestone	Madison Limestone
	Devonian		Not exposed	Chaffee Formation	Dyer Dolomite Member Parting Quartzite Member	Ouray Limestone Elbert Formation	Chaffee Formation
	Silurian		Absent	Absent		Absent	Absent
	Ordovician		Not exposed	Fremont Limestone Harding Sandstone Manitou Formation		Absent	Manitou Formation
	Cambrian		Not exposed	Peerless Formation Sawatch Quartzite		Ignacio Quartzite	Sawatch Quartzite, Lodore Formation
	Precambrian		Granites and related rocks of at least three different ages. Idaho Springs Formation and various other named and unnamed metamorphic rock units.				

Figure 33.3 **Colorado. Principal stratigraphic units in major physiographic provinces of Colorado.** *After Ogden Tweto (1964, Table 1, p. 13).*

in the basins, also predominate in western Colorado. A large area in the southwestern quadrant of the state was inundated by Cenozoic lava flows, which are shown in Figure 33.2 by a separate pattern.

A chart (Tweto, 1964, p. 13) of the principal stratigraphic units in Colorado is given in Figure 33.3. This chart suffices for the subsurface as well as the surface geology except in southeastern Colorado; a separate table (Figure 33.7) covering the subsurface formations in that area is included in the discussion of fields and traps in southeastern Colorado.

Producing oil and gas reservoirs in Colorado number at least 25, ranging in age from Early Tertiary to Devonian (Bass, 1964, p. 51). The largest source of both oil and gas to date has been the Weber Sandstone of Pennsylvanian-Permian age. This statistical stature (45 per cent of the state's cumulative oil production at the end of 1962) is due to the fact that almost all of the Weber oil production has come from giant Rangely field in northwestern Colorado (Figure 33.4). Second with 32 per cent is Dakota (Cretaceous) Sandstone, the outstanding reservoir in the Denver basin. Jurassic sandstones rank third with 10 per cent of the total; the remaining 13 per cent is obtained from about 20 reservoirs, principally of Permian, Cretaceous, and Tertiary age. The main gas reservoirs are, in descending order of yield, Weber Sandstone, Dakota Sandstone, Wasatch (Tertiary) Formation, Jurassic sandstones, and Mesaverde Group (Cretaceous).

Tectonics. Figure 33.4 shows the major tectonic features, including the areas of basement rock outcrop, as well as the oil and gas fields. Dominating the scene are the Precambrian-cored mountain ranges. The major arching of the Precambrian surface took place during the Laramide Revolution in late Mesozoic and early Cenozoic time. The present mountainous topography resulted from uplift and differential erosion in late Tertiary and Quaternary time. One intermontane basin, North Park, is also an oil- and gas-producing district.

The major petroliferous districts are in the Great Plains province of eastern Colorado and the Plateau province on the west slope. Most of eastern Colorado is covered by the Denver basin, which extends northward into western Nebraska. It is terminated southward by the Apishapa uplift and on the southeast by the Sierra Grande uplift and the Las Animas arch (Bass, 1964, p. 57). Across the Apishapa uplift is the Raton basin, and across the Sierra Grande–Las Animas structural divide is the western flank of the Hugoton embayment.

Northwestern Colorado has the Sand Wash basin, which is a part of the basin and uplift complex of southwestern Wyoming. To the south, across Axial Basin anticline, is the Piceance basin, noteworthy for its large oil shale reserve. The Paradox basin of southwestern Colorado and southeastern Utah lies across the Uncompahgre uplift to the south. On the far side of the southeastern rim of the Paradox basin is the northern tip of the San Juan basin of northwestern New Mexico.

TRAPS AND FIELDS

In the discussion that follows, Colorado is divided into five producing districts: northeastern, southeastern, southwestern, northwestern, and North Park. The greater part of the state's oil and gas production comes from northwestern and northeastern Colorado, in the order named. The traps are mainly anticlinal in northwestern Colorado, and stratigraphic in the northeastern district.

Northeastern Colorado. Northeastern Colorado is arbitrarily defined as that part of the eastern Colorado plains occupied by the Denver basin (Beebe, 1968a). The reservoirs here are largely of Cretaceous age. The Denver basin (Figure 33.4) is bounded on the east by the Las Animas arch. This uplift is a distinct structural divide in southeastern Colorado, but it loses its identity as it plunges northeastward and crosses the northwestern corner of Kansas. The Denver basin is asymmetric, with the deepest part immediately in front of the mountains between Denver and Cheyenne, Wyoming. The Precambrian floor has a maximum below-sea elevation of about 7000 feet in the Denver area, which means a total thickness of the sedimentary section in this vicinity of approximately 13,000 feet.

At the southwest corner of the Denver basin is a local offshoot, the Cañon City embayment. This small structural pocket occupies a re-entrant into the Front Range between Pueblo and Cañon City. It is in this local basin that the first commercial discovery of oil was made in 1876.

Northeastern Colorado has produced somewhere between 275 and 300 million barrels of oil. The fields can be divided, geographically and geologically, into two groups—those on the steep west flank of the basin, and the fields, many in number, that lie scattered over the gentle east flank. The west flank fields will be discussed first.

One of the better known fields in Colorado, because of both antiquity and type of oil accumulation, is the *Florence–Cañon City* in the Cañon City embayment. This field was the sole source of petroleum in Colorado from 1876 until 1902. The oil is found exclusively in fractures in the Pierre Shale (Upper Cretaceous). It occurs at various stratigraphic levels, but the most prolific has been the middle zone (of three) in the Pierre. This zone is about 600 feet in thickness (Carpenter, 1961, p. 130). Many would-be development wells in this fractured shale field turned out to be dry holes because of

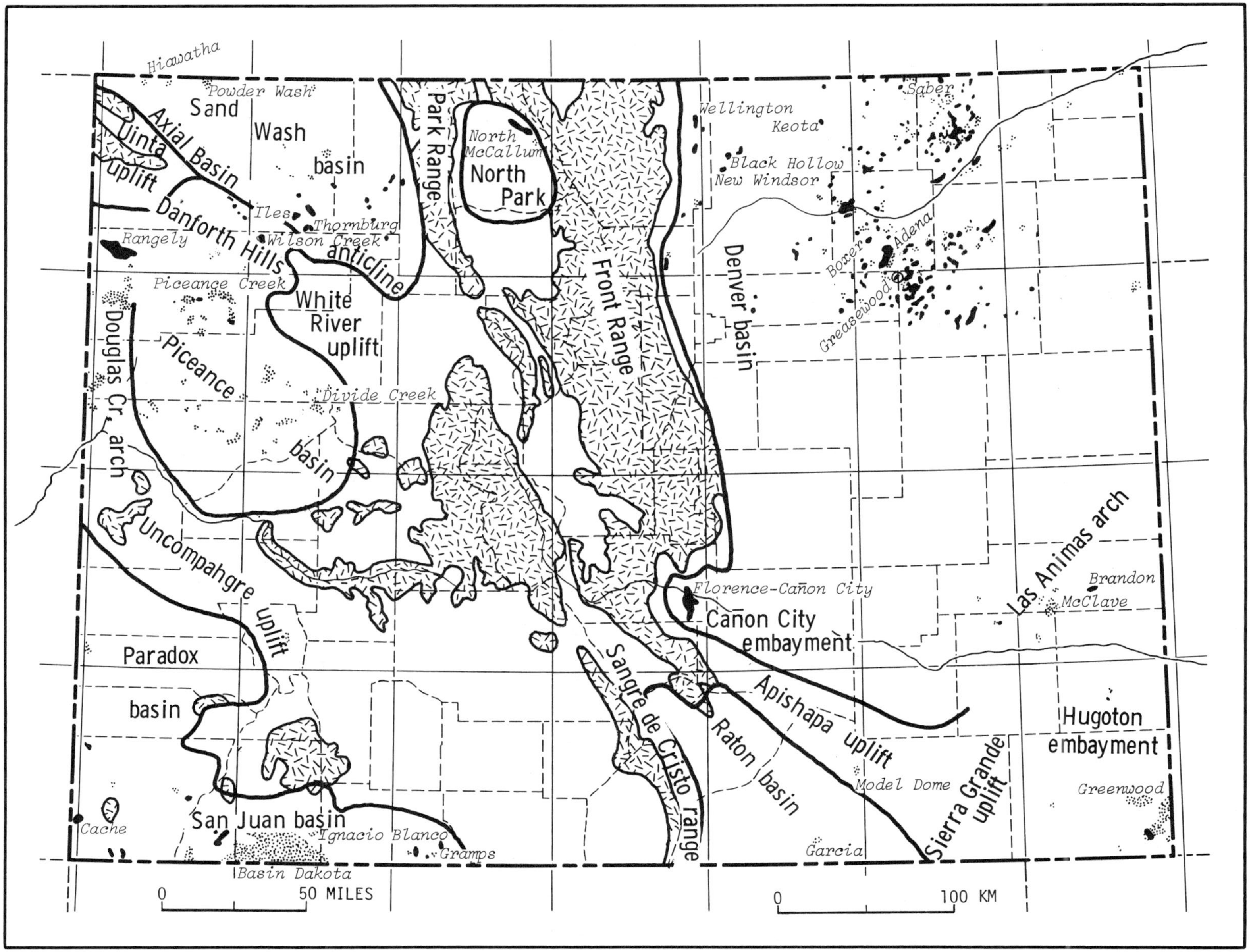

FIGURE 33.4 Colorado. Tectonic and oil and gas field map. Random dash pattern shows exposed basement rock and younger intrusives of large area. Oil fields shown in solid black; gas fields are stippled. *Tectonic designations principally from N. W. Bass (1964, Figure 10, p. 56–57).*

failure to penetrate a fracture zone (Bass, 1964, p. 66). The Florence–Cañon City field is about 15 miles long in a north-northwest direction and 3 miles wide. The regional dip, about 3°, is westerly toward the axis of the Cañon City syncline. However, the east flank of this downfold is furrowed by transverse anticlines and synclines. The Cañon City sector of the combined field lies high on the southeast flank of such an anticlinal nose. The much larger Florence field occupies a considerable area to the south and east on the opposite flank of an intervening transverse syncline (Lavington, 1954).

The other fields on the west flank are in the northern part of the Colorado sector of the Denver basin in a belt extending from 125 to 175 miles north of the Florence field. Almost all of the oil and gas in this district comes from either Dakota (Cretaceous) or Lyons (Permian) Sandstone. Trapping is anticlinal; the anticlines are exposed in the fields in Larimer and Boulder Counties, but a short distance east in Weld County the anticlines are not surface-mappable and hence have been identified by subsurface geology (Bass, 1964, p. 58).

The oldest field on the west flank of the northern Denver basin is the *Wellington* in northeastern Larimer County, which initiated Dakota oil and gas production in the Denver basin in 1923. This was also the first sandstone reservoir field in Colorado (Brainerd and Van Tuyl, 1954, p. 17). The drilling of other exposed elongate anticlines paralleling the mountain front led to similar discoveries south of the Wellington field (Figure 33.4). The first western group field to produce from a subsurface anticline was the *Black Hollow* in western Weld County in 1953. The reservoir rock here is Lyons Sandstone. Commercial oil had been discovered in the Lyons a month earlier in the Keota field on the east flank of the basin, but that field developed subsequently into a Dakota-only producer. According to Nering (1963, p. 239), oil-stained outcrops of the Lyons Sandstone along the west rim of the basin had been known to geologists for years, but apparently had not inspired testing of this sandstone in the Dakota-producing anticlines.

Since the Black Hollow discovery, four other fields have been found to be productive in the Lyons. Two of these are in older Dakota fields high on the west flank, and two are on subsurface anticlines near Black Hollow in western Weld County. In the Black Hollow and one other field the Lyons is (so far) the sole reservoir. The third field, *New Windsor*, produces from a sand in the Upper Cretaceous Pierre Shale as well as from the Lyons (Clark and Rold, 1961). In both cases trapping is anticlinal, but with distribution across the trap limited by updip pinchout of sand (Pierre) or by increase in sand tightness due to greater cementation (Lyons).

The gently dipping eastern flank of the northern Denver basin has many oil fields (Rocky Mtn. Assoc. Geologists, 1963), which produce almost exclusively from two sandstones (D and J) in the Dakota Group. Most of the oil produced in northeastern Colorado has come from this area. The largest field, Adena in Morgan County, is second to Rangely in cumulative production, with a yield of nearly 60 million barrels by the end of 1966.

The first field discovered on the east flank was *Greasewood* in eastern Weld County in 1930 (Lavington, 1941).The first well "was located on a surface anticline of doubtful closure," (Brainerd and Van Tuyl, 1954, p. 22). Oil in the Dakota reservoir sand at a depth of about 6650 feet is trapped in a westward plunging nose by a permeability pinchout updip to the east. Further exploration lagged, and it was not until a commercial discovery in the Nebraska sector of the Denver basin in 1949 that an intensive search began for west flank oil. These efforts brought in many new fields in both Nebraska and Colorado.

The Pierre Shale crops out at the surface over a large area in northeastern Colorado. Up to 300 feet of late Tertiary fluviatile sediments blankets the Pierre to the east. Figure 33.5 shows the stratigraphic column on the east flank of the Denver basin. Below the Pierre are formations of the Colorado Group and below them the sandstone-shale sequence of the Dakota. The sandstones are the reservoirs, principally the D at the top of the Dakota, and the J below the first shale. Actually these are sandstone zones consisting of siltstones, tight sandstones, and lenticular bodies of porous and permeable sandstones that are the oil and gas reservoirs. The producing thickness of the D sand averages 15 feet and that of the J sand 25 feet (MacQuown and Millikan, 1955, p. 634). The latter has produced considerably more oil, but the D sand is slightly ahead in gas yield. Drilling depths to these reservoirs vary in the main part of the basin from 4000 to 7300 feet (Bass, 1964, p. 58). The Permian Lyons Sandstone which produces on the western flank of the basin, lenses out eastward.

The discovery field, Greasewood, set the pattern for much of the oil and gas trapping on the east flank. Hydrocarbons moving eastward up the regional dip through permeable sands have been funneled into low upfolds (creases) furrowing the westward dipping formations, where they have been trapped by an eastward phasing out of permeability. Other traps include shale-encased sand lenses and channel sandstone deposits (Martin, 1965, p. 1922).

Obviously the hunt for accumulations of this type can be frustrating:

> The discovery of oil and gas bearing sand lenses is particularly difficult because their presence is not predictable from surface features and rarely in the subsurface except at the site of the pool. Hoping to find new fields, geologists project wells and dry holes into undrilled areas by theoretical sand distribution

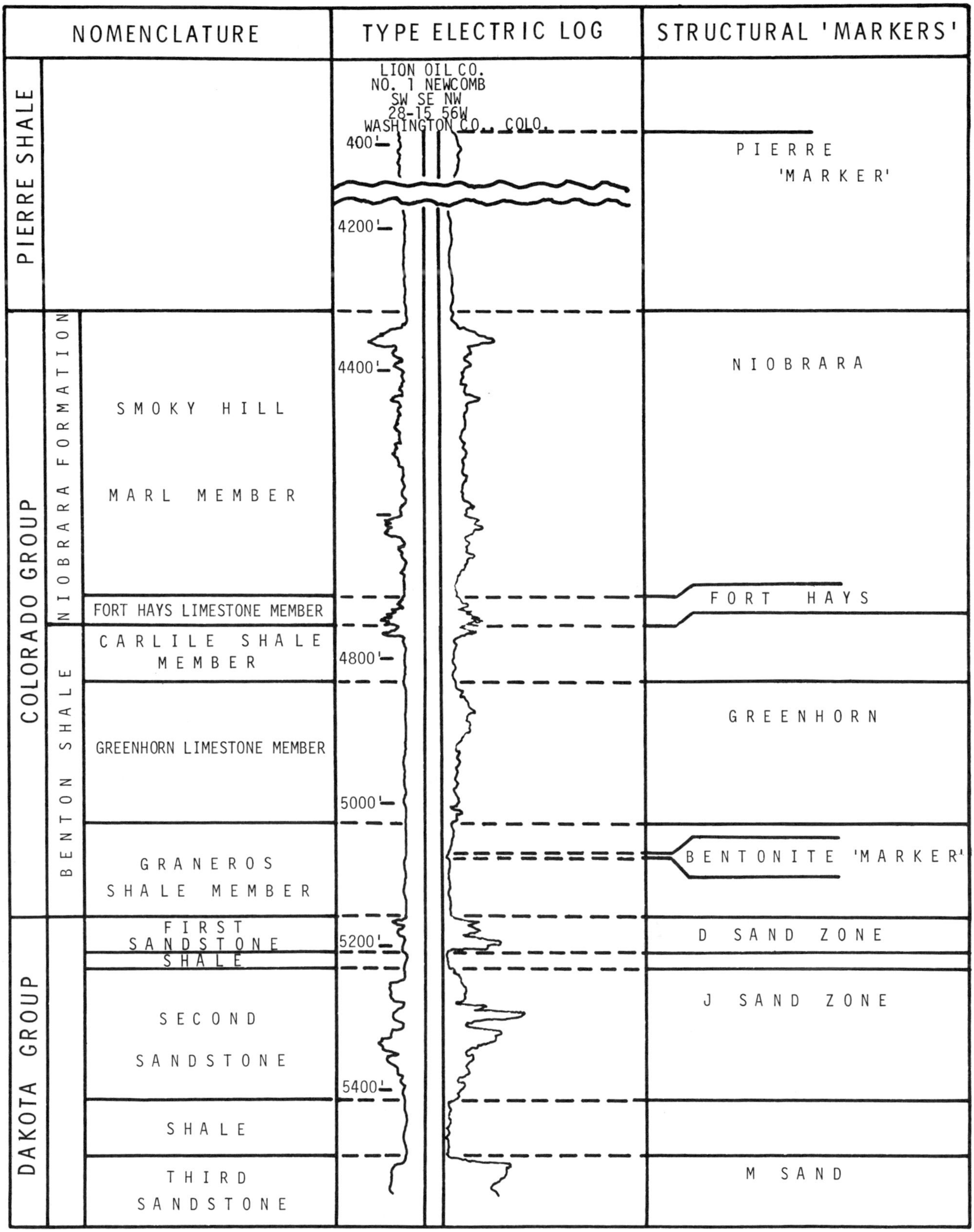

FIGURE 33.5 Northeastern Colorado. Stratigraphic units. *Courtesy W. C. MacQuown, Jr., and W. E. Millikan (1955, p. 632) and the American Association of Petroleum Geologists.*

patterns. Only one well in 10 to 20 is successful in discovering a new pool (Bass, 1964, p. 55).

Furthermore, once a pool is discovered and partially developed, its boundaries remain unclear until the producing area has been surrounded by a necklace of dry holes (Finley et al., 1955).

Most of the oil and gas fields on the eastern flank occupy a broad belt that parallels the north-northeast strike of the J sand, crossing parts of Adams, Washington, Morgan, Weld, and Logan Counties (Figure 33.4). Here the dip westerly into the basin averages 50 feet per mile. The field distribution pattern coincides at present to a remarkable degree with the area of outcrop of Pierre Shale (in Colorado, but not in Nebraska).

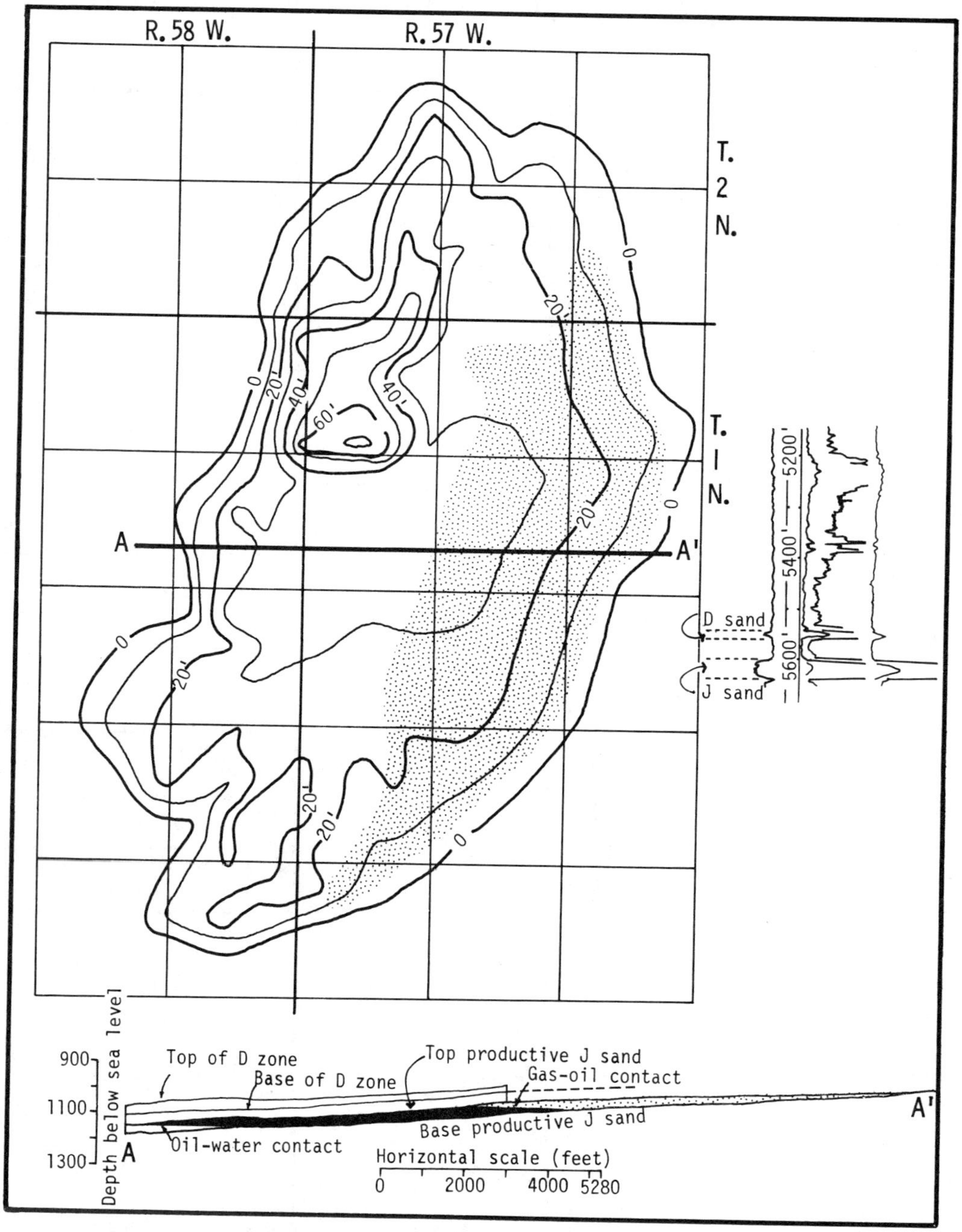

Figure 33.6 Adena field, Morgan County, Colorado. Map and cross section. Map shows thickness of oil and gas section in J reservoir sand. Interval, 20 feet. Stippled area is gas cap. Oil occupies trap to west. Sections are 1 mile (1.61 km) square. Relationships between water, oil, gas, structure, and sand pinchout to east are shown in cross section *A–A′*. *Courtesy A. T. Sayre (1961, p. 265) and the Rocky Mountain Association of Geologists.*

The outstanding field in both area and production is the *Adena* (Mygdal, 1963; Sayre, 1961). This oil and gas field in southern Morgan County was discovered by drilling a seismic anomaly, which later was shown to be unrelated to the monoclinal west dip of the reservoir sandstone. The field is 7 miles long (north-south) and $3\frac{1}{2}$ miles wide (Figure 33.6). The principal reservoir is the J sand; about 5 per cent of the total production has come from the D sand. The J sand has a maximum thickness of 72 feet. Figure 33.6 shows the thickness of the net oil- and gas-bearing pay section. A gas cap lies above the oil along the east side of the field, as can be seen in the cross section beneath the thickness map. On the west side of the field the zero thickness line is the edge of the oil-water interface at the top of the J reservoir.

Trapping at Adena is caused by the lensing out of adequate permeability up the regional dip, as elsewhere on the east flank. However, lateral sealing is not brought about by the presence of a westward plunging anticlinal nose; instead permeability wedgouts also are responsible for closing the trap at the north and south ends (McGinnis, 1958, p. 339). In other words the updip permeability barrier is, in plan, a wavy line, and Adena accumulation is in a concave (downdip) arc. Since early production from the gas cap led to rapid decline in reservoir pressure, a row of water injection wells was installed to penetrate the reservoir at the gas-oil contact. This repressuring has been successful; the dip of the reservoir of less than $\frac{1}{2}°$ no doubt assisted in making this unconventional approach workable.

Although the D sand is relatively unimportant at Adena, it is the principal reservoir in the *Boxer* oil field (Tobison, 1967) 3 miles to the northwest. This field was discovered in April, 1965, and 2 years later had become first in daily production in the Denver basin. It now covers 1600 acres; this is remarkable in that the pre-discovery dry hole average in this general area was one to 640 acres. In fact subsequent development resulted in offsetting one of these dry holes with three producers. Accumulation has taken place on a local structural terrace on the regional westerly dip. Slight reversal in dip occurs at several places along the terrace, resulting in low nodes and near-nodes. The eastern edge of the field is a permeability barrier that closes the trap just east of the node line except at the south end of the field, where the barrier curves westerly and cuts off the southernmost node.

Still another type of trap is present at the *Saber* field (Griffith, 1966), which is shared by northwestern Logan County and northeastern Weld County. Accumulation is in a D sand bar over 10 miles long, in a north-northeast direction, and 1 mile wide. It has maximum thickness of about 45 feet. In this case the reservoir sand thins to extinction down the westerly regional dip as well as updip. Griffith ascribes its origin to deposition in shallow water by longshore currents. The overlying Graneros Shale appears to be the most likely source rock for the oil that has found sanctuary in the bar.

A comparison of oil and gas maps issued in 1954 and 1964 shows a great increase in the density of fields on the east flank of the Denver basin during the decade. This expansion is still under way and can be expected to continue for a long time to come. The steady increase in subsurface data available makes it possible for geologists and geophysicists to sharpen their techniques. Combined effort will continue to produce prospects based upon sand distribution patterns and the presence of noses, terraces, and nodes. The mapping of sand wedgout lines, where adequate data are available, might lead to the finding of updip re-entrants in which other Adenas could be hiding.

More exploration of the Permian Lyons Sandstone in the western part of the basin could be profitable. Upper Cretaceous possibilities are described by Nolte (1963). Very little attention has been paid to the conceivable occurrence of oil in fractures at the top of the Niobrara, where trapping conditions exist. The overlying Pierre Shale could be an excellent source rock.

Southeastern Colorado. The part of Colorado lying east of the Rockies and not covered by the Denver basin is referred to here as southeastern Colorado. The subsurface stratigraphy (Figure 33.7) in this district is much more closely related to the midcontinent (Kansas and Oklahoma) than it is to the Rocky Mountain area. Southeastern Colorado contains two basins, the western edge of the Hugoton embayment and the Raton basin. The Las Animas arch separates the Hugoton embayment from the Denver basin, and the Apishapa uplift is the structural divide between the Raton and the Denver basins. Both the Las Animas arch and the Apishapa uplift converge to the south and merge with the Sierra Grande uplift (Figure 33.4).

The *Greenwood* sector of the great Hugoton gas field of southwestern Kansas (Figure 16.1) has expanded westward into eastern Baca County in the southeastern corner of Colorado. Fringing it to the westward are two smaller gas fields and one oil and gas field. There are five producing zones, all in the Pennsylvanian section, ranging downward from Virgilian to Morrowan (Figure 33.7). Reservoir rocks (Bass, 1964, p. 64) include oölitic limestone in the Shawnee Group (gas) and sandstone in the Morrow (both oil and gas). Trapping is due to wedgout of permeability updip to the west. The regional eastward dip in this area is approximately 30 feet to the mile (Pustmueller, 1958, p. 129).

The Las Animas arch has attracted the wildcatter for many years, but with indifferent success until recently. The *McClave* field west of the arch axis in southern

GEOLOGIC TIME UNITS	SUBSURFACE SOUTHEASTERN AND EASTERN COLORADO	MARGIN OF MOUNTAIN PROVINCE
Permian	Nippewalla Group Blaine Formation Sumner Group Chase Group Council Grove Group Admire Group	Fountain Formation
Pennsylvania	Wabaunsee Group Shawnee Group Douglas Group Lansing and Kansas City Groups Marmaton Group Cherokee Group Atoka Series Morrow Series	
Mississippian	Ste Genevieve Limestone St. Louis Limestone Salem and Warsaw Limestones Keokuk and Burlington Limestones Kinderhook Series	Leadville Limestone
Devonian	Absent in much of region	Chaffee Formation
Silurian	Absent	Absent
Ordovician	Viola Limestone Simpson Group	Fremont Limestone Harding Sandstone Manitou Formation
	Arbuckle Group	
Cambrian	Bonneterre Dolomite Lamotte Sandstone	Peerless Formation Sawatch Quartzite

FIGURE 33.7 Stratigraphic nomenclature in Paleozoic rocks, southeastern Colorado (subsurface) and along eastern edge of Rocky Mountain province (surface). ***After N. W. Bass (1964, Table 9, p. 65) and John C. Maher.***

Kiowa County (Figure 33.4) produces gas from a Morrow sand lens that pinches out laterally into shale. In the same county, but across the axis to the northeast, the *Brandon* field opened new vistas by discovering commercial oil in Osage (Mississippian) Limestone in 1965. Three years later this field had become fourth in daily production in Colorado. The reservoir is 250 feet thick, with over 70 feet of net pay. An expanded search for hydrocarbons in pre-Pennsylvanian rocks along the Las Animas arch was under way in 1968.

The Sierra Grande uplift is nonproductive, as yet at least. A part of this broad structural dome is covered with Late Cretaceous and Tertiary lava flows (Figure 33.2).

The Apishapa uplift to the northwest of the Sierra Grande arch is also nonproductive, so far as hydrocarbons are concerned. However, it does have the *Model Dome* gas field, in which the gas is nonflammable, consisting of 76 per cent nitrogen, 14 per cent carbon dioxide, 8 per cent helium, nearly 2 per cent oxygen, and only 0.1 per cent methane (Bass, 1964, p. 50). The unusually high helium percentage, coupled with the large reserve, led the U.S. Government to obtain title to the field as a standby helium reserve. The Model field is in north-central Las Animas County on the crest of Apishapa uplift. Trapping of the helium-containing gas is in the upper part of an anticline 16 miles long and from 5 to 7 miles wide (Clair and Bradish, 1956a). The reservoir is the Lyons Sandstone (Figure 33.3) at a depth of about 900 feet.

The Raton basin, which was discussed briefly in Chapter 32 on New Mexico, is a topographic and structural basin (Gabelman, 1956) lying between the Apishapa uplift and the Sangre de Cristo mountain range. Structurally it is an elongate asymmetric syncline with a north-northwest trend about 100 miles long, of which more than 65 miles is in Colorado. The keel of the syncline runs along the west side of the basin; the average easterly dip on the west flank is about 1200 feet per mile, and on the opposite flank about 200 feet. The deepest part of the trough is in the vicinity of the Spanish Peaks (Figure 33.2), which are eroded igneous stocks surrounded by radial dikes. The structural relief, as mapped on top of the Dakota Sandstone (Doeringsfeld et al., 1956), is over 6000 feet. The northern end of the basin forks to the north and northwest. The latter branch is much the larger; on the surface it is Huerfano Park in northwestern Huerfano County.

The nearest approach to commercial production so far found in the Raton basin occurs in the *Garcia* gas field, which was first explored some 75 years ago. For many years this field has supplied gas for fuel to nearby ranches, and for 10 years, between 1934 and 1944, a local natural gas gasoline plant was in operation (Clair and Bradish, 1956b, p. 77). The gas came principally from fractured Niobrara Shale. The Garcia field is in southern Las Animas County on the eastern flank of the Raton basin. The trap appears to be small nodes along the axis of a southwestward plunging anticlinal nose. Cutting the north side of the northern node is a basalt dike; lava covers the surface about 5 miles to the south.

Raton basin is one of the few bona fide sedimentary basins left in the United States without commercial production of oil and gas. The stratigraphic section contains promising source rocks and potential reservoir rocks (Harbour and Dixon, 1956). Thick sequences of marine shales and sandstones occur in both Cretaceous and Pennsylvanian rocks. Anticlines (such as the Garcia) are known on the flanks, and the geologic history is favorable to the occurrence of lenticular and wedgout traps (Lewis, 1956).

Southwestern Colorado. The actual and potential hydrocarbon-producing part of southwestern Colorado lies between the Uncompahgre uplift (Figure 33.4) on the north and the New Mexico border on the south, and the San Juan–Needle Mountain complex (Figure 33.2) on the east and Utah to the west. This area is occupied by the eastern flank of the Paradox basin and the northern flank of the San Juan basin. The two basins reach their nadirs in Utah and New Mexico, respectively, and are described in more detail in Chapters 36 and 32. They are separated one from the other, by the Four Corners platform.

The Uncompahgre uplift separates the Paradox and Piceance basins. Erosion along the crest of this great anticlinal ridge has exposed granitic basement rock. The spectacular, rugged San Juan Mountains have been carved out of pushed up volcanic rock (Figure 33.2). The Needle Mountains, which adjoin the San Juans in northern La Plata County, expose Precambrian granite and metamorphic rocks. Tertiary intrusive stocks and necks pierce the sedimentary rocks to both the northwest and the southwest of the Needle Mountains.

The northern part of the Paradox basin is creased structurally by faults and elongate salt-cored anticlines that parallel the northwest strike of the Uncompahgre uplift. The diastrophism has been much more intensive here than in the southern part of the basin.

Because of folding, faulting, unusual topographic relief, and aridity, much of the subsurface stratigraphic column in southwestern Colorado is exposed somewhere in the area. At least one Pennsylvanian formation and several thick Cambrian carbonate formations, however, do not appear as outcrops. The uplifts that expose Precambrian basement rocks are fringed by Paleozoic and Mesozoic outcrops of Cambrian, Upper Devonian, Mississippian, Pennsylvanian, Permian, Triassic, and Jurassic age. The Paradox basin is floored principally with Cretaceous deposits. The leading oil reservoir is a carbonate section in the Hermosa Formation (Pennsylvanian); the Dakota produces much gas and some oil, and the Upper Cretaceous Mesaverde Group and younger sands yield large quantities of gas. Potential reservoirs include Jurassic sandstones, Leadville (Mississippian) Limestone, and Upper Devonian carbonate rocks, including the Ouray Limestone.

In the subsurface there is an extensive development of evaporites within the Hermosa Formation. This sequence of salt, anhydrite, dolomite, and shale has been named the Paradox Member of the Hermosa (Bass, 1944).

Southwestern Colorado has more than a dozen named oil and gas fields, but most are small. The largest gas fields are the *Ignacio-Blanco* and the *Basin Dakota,* which extend northward from northwestern New Mexico (see p. 374). The maximum penetration of these fields into Colorado is about 12 miles. Production is entirely from Cretaceous sandstone reservoirs, belonging mostly in the Dakota and Mesaverde Groups. Trapping is both on the Ignacio anticline high on the north flank of the San Juan basin and southward down the flank because of lenticularity of the reservoir sands.

The largest oil field so far discovered in southwestern Colorado is the *Cache,* which between discovery in late 1964 and mid-1967 produced nearly 2 million barrels of petroleum. This field is in the Paradox basin in westernmost Montezuma County. It is an eastern outpost of the Aneth field type of accumulation that occurs a short distance to the west in southeastern Utah. Production comes "from porous strata in a group of 'stacked' biohermal carbonate mounds and local zones of leached fossiliferous dolomite" (Gray, 1967, p. 1959) at depths averaging about 5600 feet. These algal rocks belong in the Ismay zone of the upper Paradox Formation. The mound-building cycle was followed by evaporite (dolomite and anhydrite) and shale deposition, which effectively sealed the reefoid reservoirs. This completed the Ismay sedimentation.

Figure 33.8 (upper) is a structure map contoured on the top of the Ismay. The 50+ feet of closure reflects the domal character of the reservoir roof beneath the draped shale and evaporites. Below the base of the Ismay biohermal deposits the structure contours show no closure whatsoever (Gray, 1967, Figure 4, p. 1963). The regional dip in this area is about 100 feet to the mile northeast toward the keel of the Paradox basin. The lower illustration in Figure 33.8 is a cross section along line *A–A'* on the map. It shows the stacking of the biohermal mounds that constitute one of the reservoir types. The intervening sediments are principally carbonate rock lithofacies, including the "leached fossiliferous dolomite" that is the other reservoir type in the Cache field.

The second oil field in cumulative production in southwestern Colorado is *Gramps,* which lies near the rim of the San Juan basin in southwestern Archuleta County. Although this field was discovered in 1935, geological details were withheld until 1946. The reservoir rocks are Dakota and Morrison Sandstones, and the oil has accumulated against an east-west fault down to the north, which crosses a north-south anticlinal axis. Not only has the fault provided a barrier to the oil in the upthrown block, but also it has displaced the fold axis nearly a mile westward on the downthrown block. Gramps field is unique in that it is entirely privately owned and therefore free from many governmental operating regulations (Waldschmidt, 1948).

Much of southwestern Colorado has not yet been

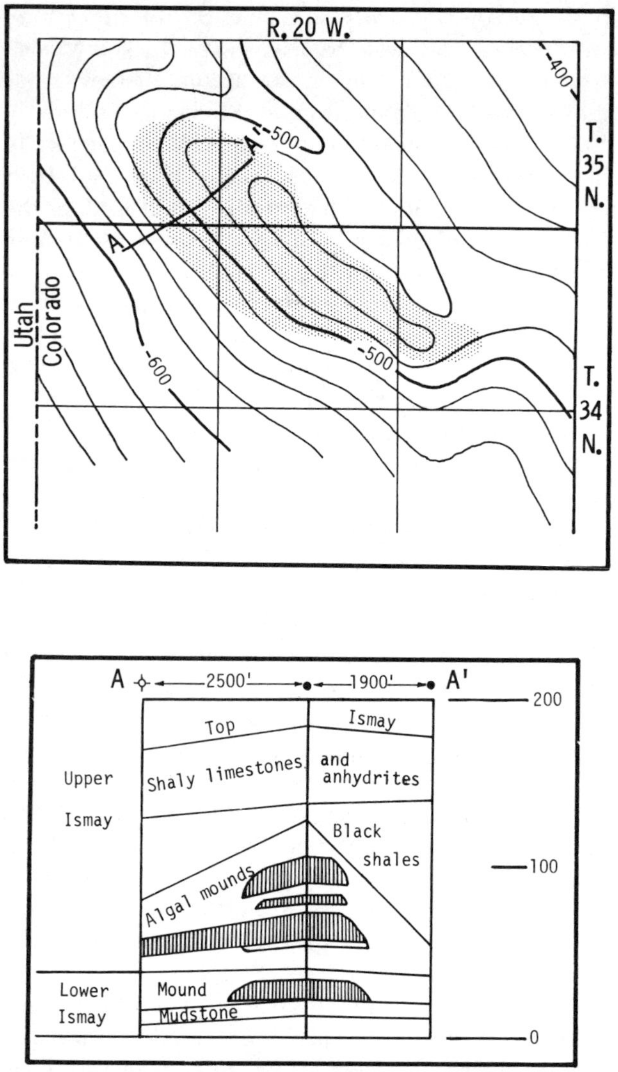

FIGURE 33.8 Cache field, Montezuma County, Colorado. Upper: structure map based on top of upper Ismay. Interval, 20 feet. Producing area is stippled. Sections on map are 1 mile (1.61 km) square. Lower: cross section along line *A–A′* on map shows superposition and draping effect of algal mounds. *Courtesy R. S. Gray (1967, Figure 6, p. 1965, and Figure 8, p. 1967) and the American Association of Petroleum Geologists.*

adequately explored. The arcuate rim of the San Juan basin northwest and west of Gramps may contain other fold and fault trap accumulations. The recent discovery of the Cache field in southern Paradox basin should encourage the search for stacked bioherms and other types of elusive traps in Pennsylvanian rocks in that area. The salt anticlines of Utah extend into and across the northern Paradox basin of southwestern Colorado. Oil is produced from relatively deep associated traps in Utah (see Chapter 36). Wedgout flank accumulations are another possibility here. Pre-Dakota, especially Jurassic, reservoirs in the San Juan basin merit intensive prospecting, as do Mississippian and Devonian reservoirs in the Paradox basin.

NORTHWESTERN COLORADO. Northwestern Colorado includes the area of the state that is west of the Rockies and north of the Uncompahgre uplift. The early explorers noted oil and gas seeps here; the first drilling was done in 1890 at the site of a gas seep. The first commercial production was found at shallow depths at Rangely in 1902 (Brainerd and Carpen, 1962, p. 23). Northwestern Colorado is the leading district in the state in total oil yield.

Structurally northwestern Colorado is dominated by two basins, the Piceance, which is entirely in Colorado, and the Sand Wash, which is shared with Wyoming (Figure 33.4). These basins are separated by the Axial Basin anticline, a structural ridge connecting the Uinta and White River uplifts. Precambrian rocks have been exposed by erosion in these arched areas, especially in the Uinta uplift. The White River uplift is covered in part by Tertiary lava flows (Figure 33.2). The Piceance basin is separated from the Uinta basin of northeastern Utah by the Douglas Creek arch, which ties the Uncompahgre and Uinta uplifts together. The geologic history of this area has been described by Quigley (1965).

The areal geology of northwestern Colorado reflects the tectonic picture. The basins are topographic as well as geologic and are floored with continental Tertiary sediments. These sediments connect the three basins, Sand Wash, Piceance, and Uinta, through low passes. Higher on the basin flanks are marine Cretaceous formations. The Precambrian cores of the uplifts are fringed by Paleozoic and pre-Cretaceous Mesozoic rocks.

Oil and gas have been produced from 12 reservoirs, ranging in age from Tertiary to Pennsylvanian-Permian. Most fields have more than one reservoir; the maximum so far is six. The principal reservoirs are shown in the column to the far right in Figure 33.3. Going down the column, the youngest productive formations are sandstones in the Green River, Wasatch, and Fort Union Formations of Eocene and Paleocene age. The oil and gas are believed to be indigenous to these Tertiary continental deposits. "Lenticular sandstone beds with interstitial porosity, fractured shale and sandstone beds, and ostracodal limestones are the predominant reservoirs" (Picard, 1956, p. 2957). Fields producing from these reservoirs are confined to the deeper parts of the Sand Wash and Piceance basins.

The Cretaceous reservoirs are in the Mesaverde, Mancos, and Dakota Formations. Of these, the Mancos is not only the largest producer but also the most interesting. It has been estimated that about 13 per cent of the oil discovered in northwest Colorado occurs in fractured shale, and the Mancos is the leading fracture reservoir (Turner, 1962, p. 30). The Rangely field has produced more than 8 million barrels of oil from Mancos fractures (McMinn and Patton, 1962, p. 106). The longevity of production from fracture reservoirs has been demonstrated at several northwestern Colorado fields, as well as in the Cañon City embayment in southeastern Colorado. The Mancos also contains sandstones that are gas reservoirs in southern Piceance basin.

The Jurassic Morrison Formation and underlying Entrada Sandstones have produced 10 per cent of Colorado's oil as well as important quantities of gas from fields, especially Wilson Creek, neighboring the Axial Basin anticline (Aaker and Jedlicka, 1968). The Morrison reservoirs are lenticular sands in the lower part of the formation (Ladd, 1957, p. 24).

The geologically oldest developed reservoir in northwestern Colorado is also the most prolific in both oil and gas. This is the Weber Sandstone of Pennsylvanian-Permian age. It is the principal reservoir at Rangely, where it is about 1150 feet thick.

A very high percentage of the oil and gas so far produced in this district has been anticlinally trapped, as can be noted in the brief review of oil and gas fields that follows. Varying permeability may, however, play an important part in controlling distribution within the structural trap. Northwestern Colorado is an arid region replete with hogbacks and dip slopes which make obvious the presence of folds. For this reason many anticlines have been tested, and a considerable percentage has been found to be oil or gas bearing.

The Sand Wash basin (Masters, 1968) lies north of the Axial Basin anticline, mostly in Moffat County, Colorado. A row of gas fields, which produce some oil, lies along, and on both sides of, the Wyoming line. The largest of these fields are Hiawatha and *Powder Wash* (Figure 33.4). The latter has produced over 72 billion cubic feet of gas and 4 million barrels of oil since development began in 1941, 10 years after discovery (Millison, 1965). The trap is a north-trending anticline containing hydrocarbons in the topmost hundred feet; the closure of the anticline at reservoir depth is about 250 feet. Between the surface and a depth of 4000 feet the apex shifts 3000 feet northeastward. More than 22 different sandstone zones in the continental Wasatch and Fort Union Formations, lying at depths between 2500 and 8000 feet, are productive. The sandstone beds within the zones are lenses, thinning to extinction in all directions.

The Axial Basin anticline has two productive nodes, Iles dome and *Thornburg* dome (Dyni, 1966). The latter lies athwart the Moffat-Rio Blanco County line where the Axial Basin anticline merges with the north end of the White River uplift. The Thornburg dome was mapped at the surface in 1924 and tested sucessfully in 1925, the discovery well showing a potential of 45 million cubic feet of gas daily from Dakota Sandstone at less than 2100 feet depth. Subsequently more gas was found at the top of the dome in the Entrada Sandstone, and oil with a gas cap was discovered in the still deeper Weber Sandstone. Accumulation in the Weber is primarily in fractures (Hurd, 1962). The anticlinal trap is elliptical, with a northwest trend. The productive closure at the Weber level is 480 feet, of which 400 feet is occupied by the gas cap.

Another spectacular structure is the Danforth Hills anticline (Grace and Nelson, 1962), which trends southeasterly across the Moffat-Rio Blanco County line (Figure 33.9). Tectonically it is a spur off the Axial Basin anticline. It forms the northeast rim of Piceance basin. This is mountainous country; the topographic relief is over 2000 feet, with elevations up to 8300 feet. The anticline is 45 miles in length; a half-dozen oil-containing nodes have been discovered so far along the axis of this upfold.

The largest field on the Danforth Hills anticline is *Wilson Creek* at the southeastern end, which was discovered in 1938. This field was second in annual yield in Colorado in 1966, and third in cumulative production. The summit of the node reaches an elevation about 1000 feet above the saddle to the northwest; the original oil-water interface in the Wilson Creek dome was a little more than 400 feet below the apex. The reservoirs are (1) sands in the lower Morrison, and (2) the underlying Entrada Sandstone, both of Jurassic age. The Morrison sands are lenticular, but generally well sorted. The Entrada Formation, about 400 feet thick, consists of massive sandstone, fine grained but with porosity of 20 per cent and permeability of 35 md in a typical pay zone (Hoffman, 1954, p. 290).

These two reservoirs are also developed elsewhere along the Danforth Hills anticline. In addition some production has been obtained in the Shinarump (Chinle-Triassic), the Weber, and more recently (late 1966) in the Dakota (Volk, 1968).

The Piceance basin and its environs are notable for the many gas fields scattered across parts of three counties, Rio Blanco, Garfield, and Mesa (Millison, 1963, p. 186). Some are out in the basin, and others lie high on the flanks of the Douglas Creek and Uncompahgre uplifts to the west and southwest. With two exceptions trapping in the basin fields is due to varying permeability of the reservoir sands. The two exceptions are the Piceance Creek (Ritzma, 1962) and the Divide Creek fields. *Piceance Creek*, discovered in 1930, was the first field in the basin. The trap is an asymmetric northwest trending anticline. Trapping in the gas fields on the west and southwest rims is also anticlinal.

The reservoir zones in the basin are at least nine in number (Millison, 1963, p. 185). Three are in the Tertiary, in the Green River, Wasatch, and Fort Union

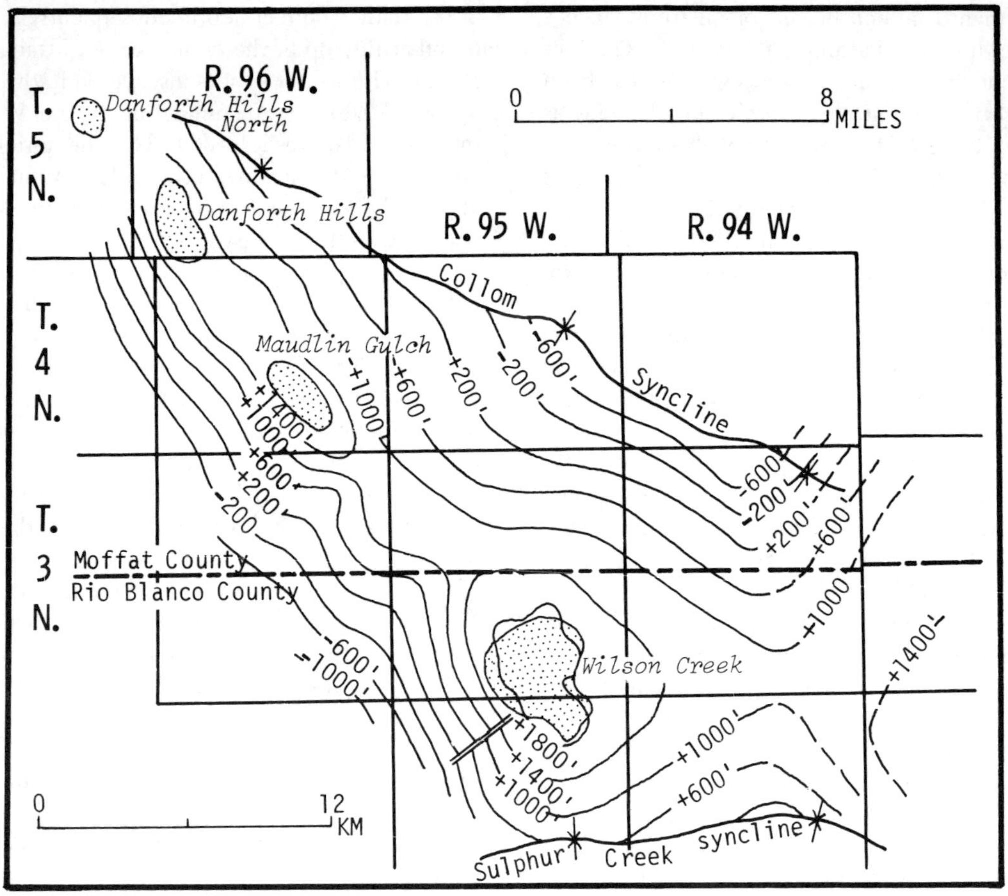

FIGURE 33.9 Danforth Hills area, northwestern Colorado. Structure map contoured on top of Dakota. Interval, 400 feet. Four oil fields occupying nodes along the axis of the Danforth Hills anticline are stippled. ***Courtesy W. P. Grace and G. Allen Nelson (1962, p. 94) and the Rocky Mountain Association of Geologists.***

Formations. The remaining six are in the Upper Cretaceous Mesaverde and Mancos Formations. The Tertiary production is concentrated in the northern part of the basin, the Mesaverde and Mancos in the southern. The western rim gas fields produce from reservoirs in the Cretaceous Mancos and Dakota, and in the Jurassic Morrison and Entrada.

Rangely, near the northwestern corner of Rio Blanco County, is the largest Colorado field in both annual and cumulative production. It passed the 400 million barrel mark in total yield in 1967; its production for 1966 was more than 16 million barrels. Shallow oil was discovered in fractured Mancos shale in 1902. "It is possible that drilling was started at Rangely because of the oil seeps on the White River where it crosses the Rangely structure" (Brainerd and Carpen, 1962, p. 23). Both oil and gas in large quantity were discovered in the deep (6335 feet) Weber Sandstone at Rangely in 1933. Because of lack of transportation facilities, remoteness from markets, and the depression then current, however, the well was shut in. Active development started in 1944; by 1949 the field had been completely developed.

The trap at Rangely is an anticline about 20 miles long in a west-northwest direction, and 8 miles wide in the central part. It lies high on the Douglas Creek arch, on the Uinta basin side. The present downdip drainage area, mostly in Utah, covers at least 1000 square miles. The anticline itself (Figure 33.10) is asymmetrical with the steeper dip on the southwest side. This dip varies from 15° to 35°; the northeasterly dip averages 5° (McMinn and Patton, 1962, p. 106). The closure as mapped on an outcropping Mesaverde bed is 1900 feet. Although several cross faults are visible at the surface, only one appears on the subsurface map contoured on the top of the Weber.

The principal reservoir, the Weber Sandstone, was originally filled with oil from elevation −1150 to elevation −330 feet. However, the Weber is a fine-grained, hard, calcareous sandstone with widely varying permeability. Increasing tightness down the southeastern flank

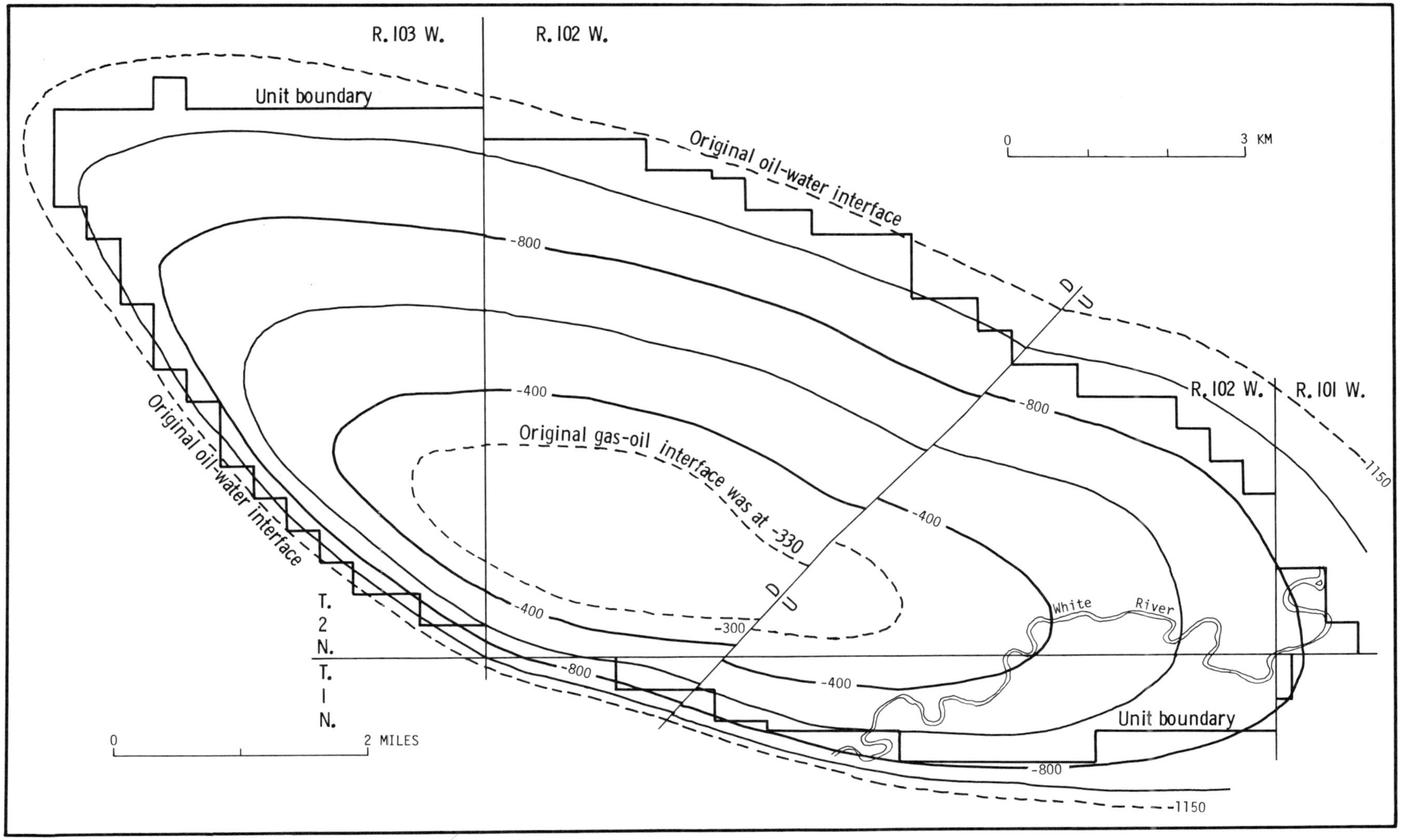

Figure 33.10 Rangely field, Rio Blanco County, Colorado. Structure contour map based on top of Weber Sandstone. Interval, 200 feet except for dashed contours for −300 and −1150 feet. Latter contour was original oil-water interface except in southeastern part of field, where oil did not extend below −800 feet because of decrease in permeability in that sector. *Courtesy P. M. McMinn and H. L. Patton (1962) and the Rocky Mountain Association of Geologists.*

of the Rangely anticline limits production in the area to about the −800 foot contour. A gas cap was found from elevation −330 to the highest point on the trap roof (at about −180 feet). Gas produced from the Weber totals about 500 billion cubic feet.

Between 2 and 3 per cent of the total Rangely production has come from fractures in the shallow Mancos Shale. This is a marine shale, dark in color and partly sandy. On the outcrop, many fractures are filled with vein calcite, which has assisted in mapping the fracture pattern at the surface.

One well, high on the anticline, was drilled to a total depth of 9360 feet, stopping in the Cambrian Sawatch Quartzite. It penetrated 2710 feet of pre-Weber sediments but found only a few slight shows.

The outlook for continuing discoveries of oil and gas in northwestern Colorado is most promising. This optimistic forecast is based on past successes and on the large area and considerable depth as yet unexplored. Whitley (1962), in discussing the Sand Wash basin, notes that three fields in this 2700 square mile area produce from traps created by the lensing out of reservoir sand lapping upon an anticlinal nose. He lists eight Tertiary and Late Cretaceous sandstone sequences in which the sands are notable for their lack of lateral continuity. Since folding is prevalent in Sand Wash basin, as elsewhere in northwestern Colorado, there must be a large number of hidden stratigraphic traps. Furthermore, the producing anticlines have not yet been adequately explored at depth; there is considerable doubt that a deep test in the Powder Wash field was high on the anticline in the deeper rocks, because of the inclination of the axial plane (Millison, 1965, p. 174–175).

Reservoir rock pinchouts are also a possibility on both flanks of the Uinta–Axial Basin–White River arch. The north side of the White River uplift has been good hunting for oil and gas in fractured shale in the Mancos Formation, including calcareous shale designated the Niobrara Member (Cummings, 1955; Coltman and Pitt, 1961). There has been little or no exploration for flank wedgouts on the Danforth Hills anticline, and the nodes themselves have not been completely tested for deeper reservoir possibilities (Grace and Nelson, 1962, p. 95).

The Piceance basin (Millison, 1968) has more stratigraphic trap fields than any other part of northwestern Colorado. In fact, as mentioned previously, all of the gas fields out in the basin except Piceance Creek and Divide Creek are of this type. In the southern half of the basin the most important reservoirs are two sandstones within the Mancos Shale Formation. According to Millison (1963, p. 184); "All wells that have been drilled to these zones in the southern part of the basin have yielded gas." If this phenomenal success record continues, Millison's prediction (p. 180) that "this basin will become one of the important gas-producing areas of the Rocky Mountains" may well come true. Gunter (1962, p. 118) believes it is probable that future drilling will result in connecting together some of the present gas fields in southern Piceance basin.

Present production and future possibilities on the north flank of the Uncompahgre arch are discussed by Krey (1962). This area is currently producing gas from Dakota and Morrison Sandstones. Trapping is mainly anticlinal, but the lenticular nature of the sands makes varying permeability trapping a definite possibility. Bond (1959, p. 132) points out the formation thinning from east to west across Piceance basin, with a good possibility for sand wedgout trapping along the east flank of the Douglas Creek arch. Across the arch to the west is the Rangely anticline, which may have trapped in quantity oil migrating up the eastern flank of the Uinta basin. If this is true, there must be other traps, stratigraphic as well as structural, to the northwest, south, and southwest of Rangely that snared some of this upbound oil and gas in the Weber. Hoffman (1957, p. 904) and others, however, have suggested that the Rangely anticline was originally an eastward dipping monocline, with oil trapped to the west by lensing permeability due to sandstone-shale interfingering. In that case there should be other Weber stratigraphic traps elsewhere in the Rangely area. Not only is the Weber reservoir inadequately explored in northwestern Colorado, but also almost no attention has been paid to the possibility of pre-Weber Paleozoic production to the east of Rangely. According to Hallgarth (1959), potential reservoir rocks occur in the Pennsylvanian below the Weber, as well as in Mississippian, Ordovician, and Cambrian rocks. There may, however, be a serious shortage of adequate source rocks in the Paleozoic section of northwestern Colorado.

North Park. The isolated North Park district (Oburn, 1968) in northern Colorado is a sedimentary basin nestled between the Front Range on the east and the Park Range on the west (Figure 33.4). It is a topographic basin as well, with the floor at an elevation of 8200 feet, more than a mile below the surrounding mountain peaks. An east-west ridge, which is also the Continental Divide and the line between Jackson and Grand Counties, separates North from barren (so far) Middle Park. North Park, which is entirely in Jackson County, contains four oil fields near its northeastern edge. The largest of these, *North McCallum,* produced some 200,000 barrels of oil and condensate (plus oil-contaminated carbon dioxide) in 1965. Accumulation in all four fields has been in asymmetric anticlines (Rocky Mtn. Assoc. Geologists, 1957, p. 109–124). The principal reservoirs are sandstones in the Dakota and Lakota Formations. The North McCallum field found oil also in the Morrison, as well as in the Entrada (Jurassic) Sandstone.

So far, discoveries have been confined to exposed anticlines around the edge of the basin, where most of the stratigraphic section has been explored. Mesozoic sediments lap upon Precambrian basement rock a short distance to the east. Out in the basin the Cretaceous rocks are buried beneath as much as 5000 feet of fresh water Tertiary sediments. The presence of an adequate section of Paleozoic rocks in this basin is doubtful, but in all probability the Mesozoic sediments are hydrocarbon bearing.

REFERENCES CONSULTED IN PREPARING CHAPTER 33

Aaker, Robert, and Jedlicka, John, 1968.
Anderman, George G., and Ackman, Edward J., 1963.
Bass, N. W., 1944, 1964.
Beebe, B. W., 1968a.
Bond, B. C., 1959.
Brainerd, A. E., and Carpen, Thad R., 1962.
Brainerd, A. E., and Van Tuyl, F. M., 1954.
Carpenter, L. C., 1961.
Clair, Joseph R., and Bradish, B. B., 1956a, 1956b.
Clark, W. W., and Rold, J. W., 1961.
Coltman, Bruce, and Pitt, William D., 1961.
Cummings, Kenneth F., 1955.
Dobbin, C. E., 1959.
Doeringsfeld, Walter W., Jr., Amuedo, C. L., and Ivey, J. G., 1956.
Dyni, John R., 1966.
Finley, E. A., Dobbin, C. E., and Richardson, E. E., 1955.
Gabelman, John W., 1956.
Grace, W. P., and Nelson, G. Allen, 1962.
Gray, R. S., 1967.
Griffith, E. G., 1966.
Gunter, Craig E., 1962.
Hallgarth, W. E., 1959.
Harbour, R. L., and Dixon, G. H., 1956.
Haun, John D., 1963.
Hoffman, Floyd H., 1954, 1957.
Hurd, Gordon, 1962.
Krey, Max, 1962.
Ladd, J. B., 1957.
Lavington, Charles S., 1941, 1954.
Lewis, Jerome A., 1956.
McGinnis, C. J., 1958.
MacKenzie, David B., 1965.
McMinn, P. M., and Patton, H. L., 1962.
MacQuown, W. C., Jr., and Millikan, W. E., 1955.
Martin, Charles A., 1965.
Masters, C. D., 1968.
Miller, Floyd, 1963.
Millison, Clark, 1963, 1965, 1968.
Monley, L. E., 1954.
Mygdal, Karl A., 1963.
Nering, Lee, 1963.
Nolte, Cliff J., 1963.
Oburn, R. C., 1968.
Picard, M. Dane, 1956.
Pustmueller, Paul S., 1958.
Quigley, M. Darwin, 1965.
Ritzma, Howard R., 1962.
Rocky Mountain Association of Geologists, 1954, 1957, 1958, 1961, 1963.
Sayre, A. T., 1961.
Tobison, N. M., 1967.
Turner, Daniel S., 1962.
Tweto, Ogden, 1964.
Volk, Richard W., 1968.
Waldschmidt, W. A., 1948.
Whitley, William W., 1962.

Chapter

34

WYOMING

Wyoming (Figure 34.1) is the leading oil-producing state in the Rocky Mountain province and is fifth in the United States. Its production in 1968 was more than 4 per cent of the domestic yield. Wyoming also accounts for about 1.1 per cent of the annual natural gas output in the United States. A total of 23 new fields was discovered in this state in 1967 (Olson et al., 1968).

The first continuous oil production in Wyoming began in 1894, but seep drilling during the preceding decade had discovered oil at several localities (Espach and Nichols 1941, p. 1–4). Lack of outlet for the oil, however, prevented development at that time. The great Salt Creek field (Figure 34.3), discovered in 1906, began large-scale production in 1917. The largest annual production in Wyoming so far was in 1963.

Twenty-one out of twenty-four counties in this state produce oil or gas. Salt Creek is still the leading oil field in Wyoming in cumulative production, with more than 450,000,000 barrels of oil. The Elk Basin field was the leading producer in 1967 and ranked second in cumulative production.

Tectonics. The broader tectonic features of Wyoming (Figure 34.2) have had a marked influence on the topography. The structural uplifts have been of such magnitude that erosion has uncovered the Precambrian crystalline cores, which have become mountain ranges. Between the uplifts are the regional synclinal belts, which contain thick series of sediments. On the west, all eras and periods are represented except the Silurian, but to the south and east, toward the Ancestral Rockies of Colorado, regional thinning eliminates the Cambrian, Ordovician, and Devonian also. These broad and thick rock deposits occupy basins, which are both structural and topographic features. It is here that the oil and gas fields of Wyoming occur (Figure 34.3) The marine Mesozoic and Paleozoic rocks are exposed, as a general rule, only around the edges of the basins; these rocks are masked toward the centers of the basins by thick

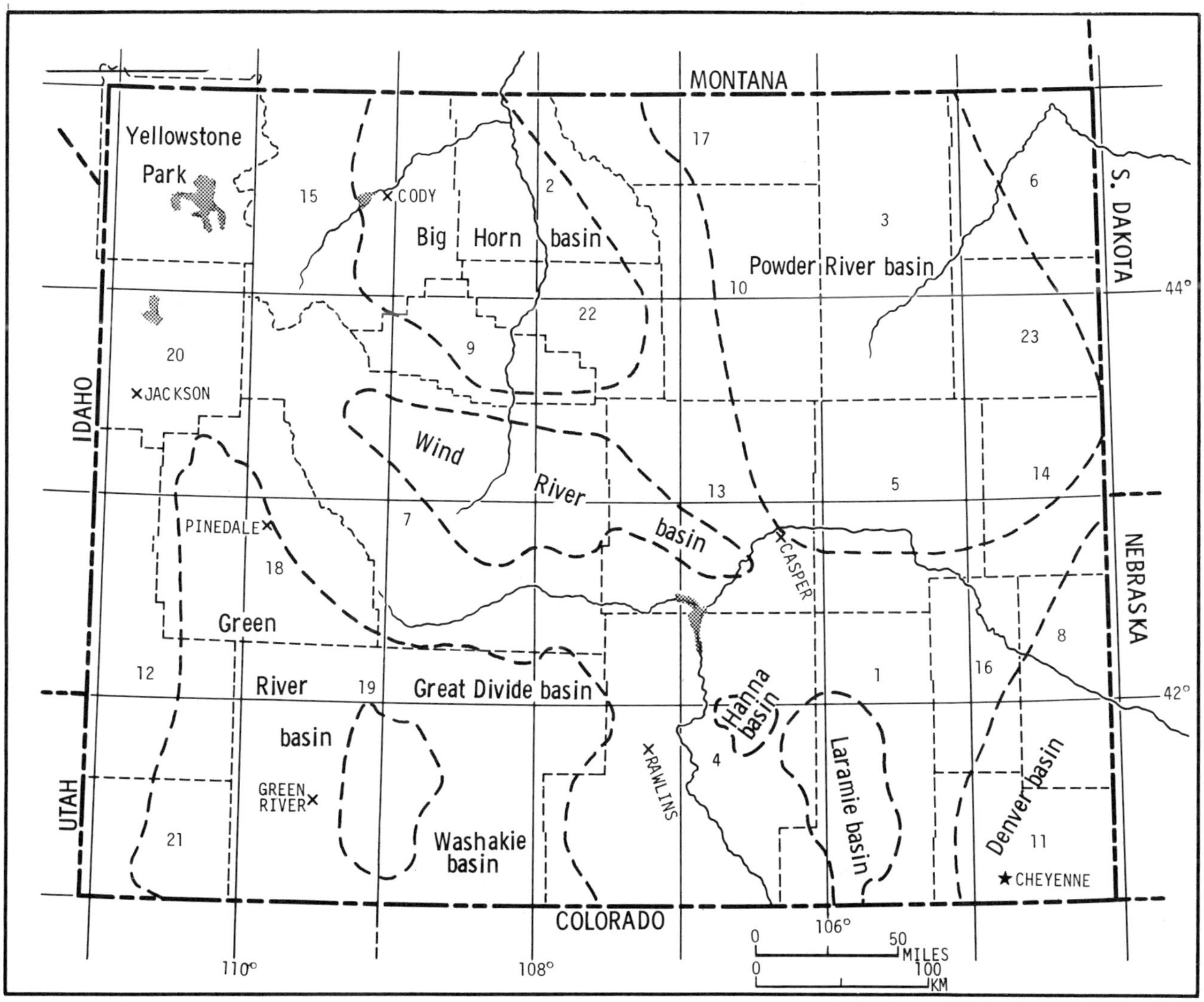

Figure 34.1 Wyoming. Index map. County Key:

1 Albany	5 Converse	9 Hot Springs	13 Natrona	17 Sheridan	21 Uinta
2 Big Horn	6 Crook	10 Johnson	14 Niobrara	18 Sublette	22 Washakie
3 Campbell	7 Fremont	11 Laramie	15 Park	19 Sweetwater	23 Weston
4 Carbon	8 Goshen	12 Lincoln	16 Platte	20 Teton	

deposits of continental Tertiary and Quaternary sediments.

The major mountain features from west to east across the northern third of Wyoming are (1) the Yellowstone Plateau and Absaroka Range, a tremendous accumulation of volcanic sediments and basaltic lava flows, and (2) the Big Horn Mountains and the Black Hills, both tectonic uplifts. South of the Yellowstone Plateau, in western Wyoming, are the great Absaroka and Darby overthrusts, which slice the mountain ranges along the Idaho border (Armstrong and Oriel, 1965). Trending east-west across northwest central Wyoming are the Washakie Range and the Owl Creek Mountains. In southwest central Wyoming are the Wind River Mountains and, to the east, the Granite Ranges of central Wyoming. South of the Wind River Mountains, in southwestern Wyoming, is the Rock Springs uplift. Southeastern Wyoming has four uplifts with northerly trends. These are, from west to east, the Sierra Madre, the Medicine Bow, the Laramie, and the Hartville. The Hartville, a structural ridge trending northeasterly, ties the Black Hills to the Front Ranges of the Rockies.

The largest basin in area is the Powder River, occupying most of the northeast quarter of the state. This basin is outlined by the Black Hills and the Hartville uplift on the east, by the northwest end of the Laramie

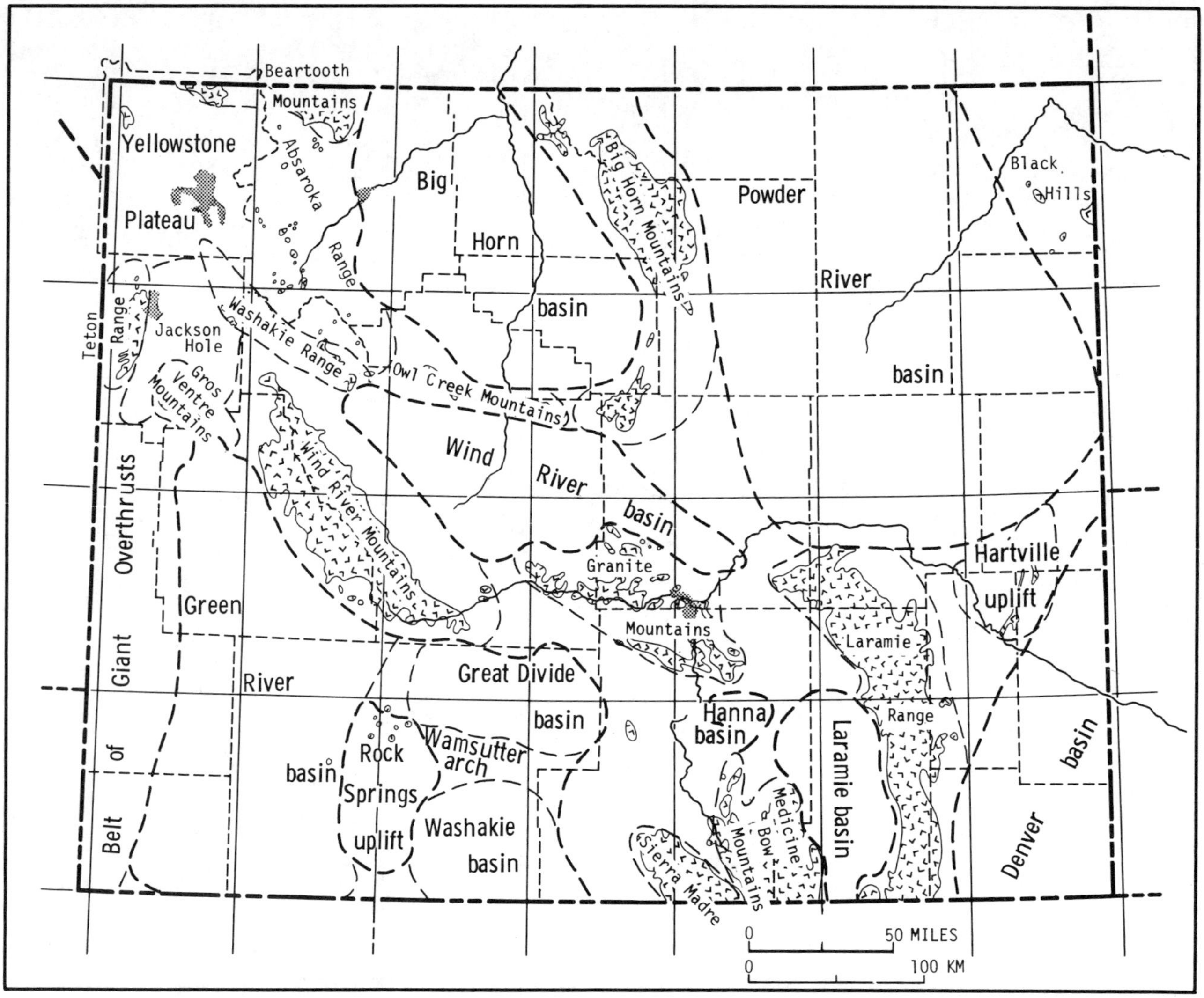

Figure 34.2 Wyoming. Tectonic map.

Mountains on the south, and by the Big Horn Mountains on the west. The Green River basin (restricted) in the opposite (southwest) corner of the state lies east of the Absaroka and Darby overthrusts, north of the Uinta uplift in northwestern Colorado, west of the Rock Springs uplift, and south of the Wind River Range.

A third large downwarp is the Big Horn basin, which is enclosed on three sides by the Yellowstone Plateau, the Owl Creek Mountains, and the Big Horn Mountains. The Wind River basin lies in central Wyoming, with the Owl Creek Mountains on the north, and the Wind River and Granite Mountains on the south. South of the latter mountains are the Great Divide and Washakie basins, separated by the Wamsutter arch. The Laramie basin lies between the Medicine Bow and the Laramie Range. A smaller basin, the Hanna, lies to the northwest. East and southeast of the Laramie Range and the Hartville uplift is the northern end of the Denver–Julesburg basin.

All the earlier discoveries in the Wyoming basins were made near the peripheries, where the pre-Tertiary sedimentary rocks are beautifully exposed and the anticlines so striking that in many places they can be easily recognized by laymen. Steep dips and large closures are the rule rather than the exception throughout the Rocky Mountain province. It was at first assumed that out toward the centers of the basins, beneath the thick cover of Tertiary continental deposits, the folds died out and no oil accumulations would be found. However, the application of geophysical instruments to Rocky Mountain exploration, plus the vastly increased depths reached by exploratory wells, has shown this assumption to be fallacious, and deep oil and gas fields have been found well toward the central parts of the basins in some instances. Furthermore, shallower Upper Cretaceous,

Paleocene, and Tertiary sediments have proved productive basinward. The Paleozoic formations, which carry lower-gravity black oil in the periphery fields as a general rule, have been found to contain high-gravity crude oil in the deeper basin accumulations.

Stratigraphy. Figure 34.4 is a table showing the geologic periods represented in Wyoming and the more important formations and hydrocarbon reservoirs (Thomas, 1958). It will be noted that the reservoirs extend from the Tertiary to the Cambrian.

The mountain range-intermontane basin patterns began to develop in Precambrian time (Blackstone, 1963; Hodgson, 1965; Prucha et al., 1965; Hoppin and Palmquist, 1965). The climax was reached during the Laramide Revolution, when, between the Middle Cretaceous and the early Tertiary, major deformation took place and the intermontane basins thus produced became filled with locally derived sediment. More than 20,000 feet of Late Cretaceous to Eocene sediment was deposited in the Green River, Wind River, and Hanna basins, to be subsequently covered by Oligocene, Miocene, and Pliocene deposits consisting largely of volcanic debris. The youngest reservoirs are nonmarine sands that were deposited in lakes during Paleocene and Eocene time in southwestern Wyoming. The most widespread reservoirs, producing in every basin, are the Upper Cretaceous sandstones. Although the Frontier sands, principal source of Salt Creek oil, have produced the most oil to date, the older Muddy and Dakota Sandstone reservoirs are also important (Wulf, 1968), and numerous discoveries of oil in younger Mesaverde sands have taken place in recent years.

Older Mesozoic reservoirs include sands in both the Jurassic (especially the Sundance) and the Triassic. Late

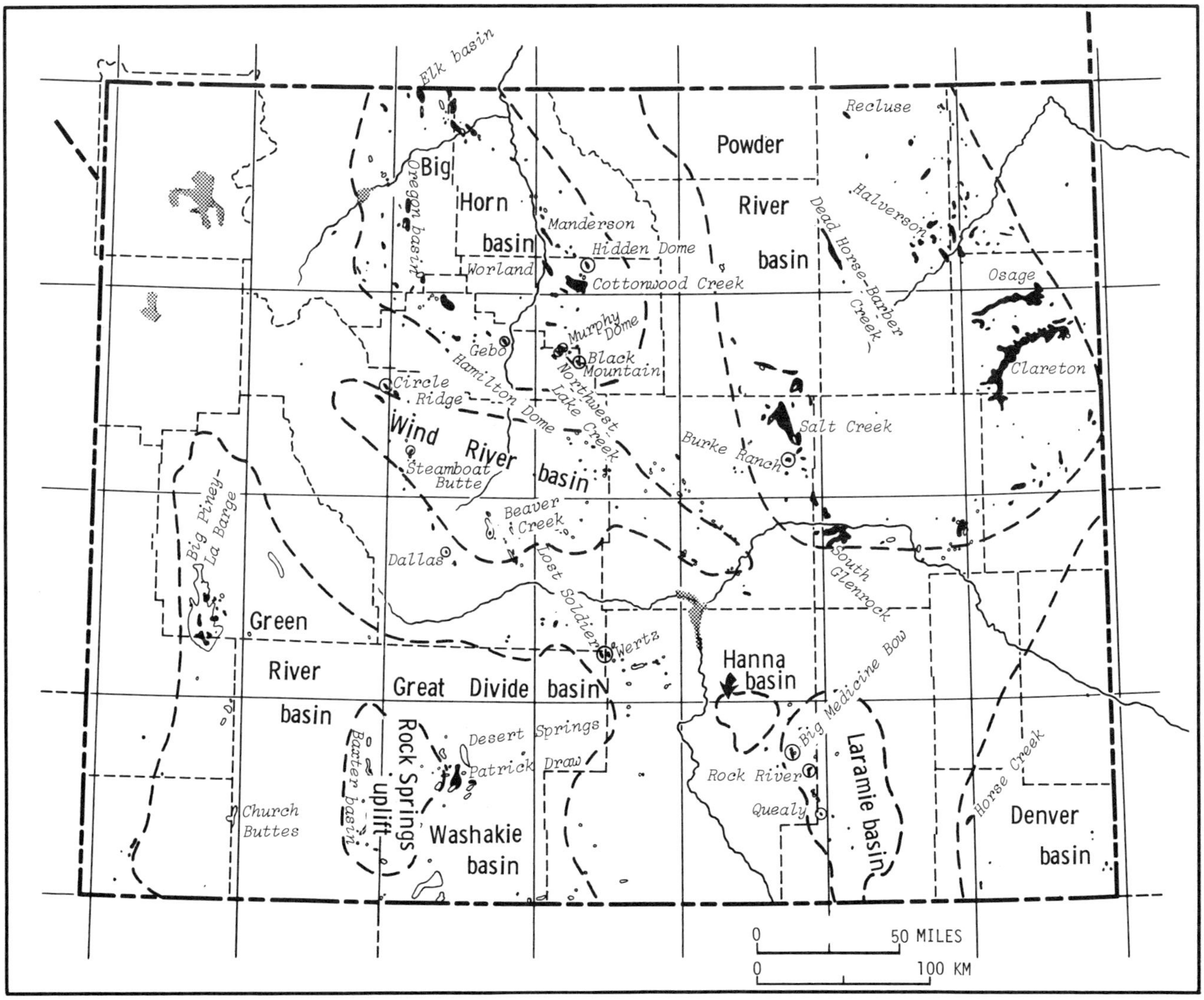

Figure 34.3 Wyoming. Oil and gas fields. Oil fields are solid black. Index map for fields mentioned in text.

ERA	PERIOD	GROUP OR FORMATION	RESERVOIRS
Cenozoic	Pliocene Miocene	North Park and/or Brown's Park	
	Oligocene	White River	
	Eocene	Bridger Green River Wind River	"Wasatch" sands
	Paleocene	Fort Union	Fort Union sands
Mesozoic	Cretaceous	Lance	
		Pierre: Lewis Mesaverde	Various sands, including Almond at top
		Pierre: Steele	Sussex Shannon Cody Shale, with reservoir sandstone near base
		Niobrara	
		Frontier	Frontier and Wall Creek sands
		Mowry	
		Muddy	Muddy sands; Newcastle Sand; Dynneson Sand
		Thermopolis	
		Inyan Kara	Fall River and "Dakota" sands Lakota sands
	Jurassic	Morrison	
		Sundance	Sundance sands
	Triassic	Nugget	Nugget sand
		Red Peak or Chugwater	Crow Mountain sand; Curtis sand
		Dinwoody	
Paleozoic	Permian	Phosphoria or Goose Egg	Phosphoria (Embar) Limestone
	Pennsylvanian	Wells or Tensleep or Casper	Tensleep sand; Minnelusa Formation
		Amsden	Darwin sand
	Mississippian	Brazer Madison	Madison Limestone
	Devonian	Darby	
	Ordovician	Big Horn	Big Horn Dolomite
	Cambrian	Gallatin	Deadwood sand
		Gros Ventre	
		Flathead	Flatsand sand

FIGURE 34.4 Wyoming. Composite stratigraphic table, emphasizing the oil and gas reservoirs. Not to scale. ***Compiled from stratigraphic nomenclature charts published in various Guidebooks of the Wyoming Geological Association.***

Paleozoic reservoirs have become of great importance. The Pennsylvanian Tensleep Sandstone is perhaps the best oil reservoir in the state, from a recovery standpoint. Its Powder River basin counterpart, the Minnelusa, a series of interbedded sandstones and dolomites, is less prolific. The Permian Phosphoria (Embar of former times) is highly productive in the northwest, where it is a porous, vuggy, dolomitic limestone. Of the older Paleozoic reservoirs, one of the most widespread is the Mississippian Madison Limestone. Sandstones of Cambrian age may also be reservoir rocks. They produce at widely spaced locations, at Lost Soldier and Wertz in central Wyoming, and at Oregon Basin in the northwestern section of the state.

The Wyoming Geological Association has published two monographs on the Late Cretaceous (1961) and Early Cretaceous (1962) rocks of Wyoming.

TRAPS AND FIELDS

Wyoming is justly famous for its large anticlinal traps, which occur as major interruptions to the regional dip on the flanks of the basins. Many smaller anticlines also have been found to be productive. Some of the anticlines are faulted, but in most of them the fold and not the fault has been responsible for trapping the oil. Much the greater part of the Wyoming production to date has come from anticlinal traps (Thomas, 1948; Dobbin, 1943). Oil and gas also found sanctuary in varying permeability traps, however, and beginning in midcentury an increasing number of such fields have been discovered (Thomas, 1961). Most of the stratigraphic accumulations found so far are either in southwestern Wyoming or on the flanks of the Powder River basin in the northeastern part of the state.

Six out of the seven oil-producing districts in Wyoming are structural basins (Figure 34.3). One of these, the Powder River basin, has produced nearly half of the total state yield of oil.

A high percentage of early discoveries in Wyoming can be credited to the presence of seeps.

Powder River Basin (Wyoming Geol. Assoc., 1949, 1958, 1961, 1962; Wyoming Geol. Assoc. and Billings Geol. Soc., 1963; Dobbin et al. 1957; Curtis et al, 1958; Wyoming Geol. Assoc. Technical Studies Committee, 1965; Tenney, 1966). The Powder River basin occupies northeastern Wyoming and southeastern Montana. Most of the fields so far discovered are in the Wyoming sector of the basin, although a recent major discovery, Bell Creek, is in Montana. The fields lie along a U-shaped belt, the axis of which is northwesterly. Outward from the U are the peripheral mountain ranges that make the basin a striking topographic as well as a structural feature. The structural basin is asymmetric, with the deepest part just inside the westerly arm of the U. The belt of fields is narrow on the western and southern sides, but wide on the eastern side, where the more gentle regional dip has encouraged the march of discovery basinward. The western side of the basin is notable for its major anticlinal traps, and the eastern side for its many varying permeability traps, most of which were discovered in recent years.

The largest field in cumulative production in the Powder River basin both in Wyoming and in the Rocky Mountain petroliferous province, is *Salt Creek*, which is on the western side. This field has produced more than 450 million barrels of oil; in 1967 the output exceeded 14 million barrels, placing Salt Creek in second place in the state. Although an oil "spring" had been known at Salt Creek before 1880 (Espach and Nichols, 1941), it was not until 1906 that the discovery well for the prolific field was drilled. The Salt Creek structure is an elliptical anticline with a steep dip on the west side (Figure 34.5) toward the Big Horn Mountains. It has approximately 1600 feet of closure, but the structural relief into the basin to the east is much greater. By far the greatest production at Salt Creek has come from the Second Wall Creek Sandstone of Frontier (Cretaceous) age (Barlow and Haun, 1966). Other reservoirs of importance, all sandstones, are the First Wall Creek of the Frontier, the deeper Cretaceous Lakota, the Third Sundance of the Jurassic, and the Pennsylvanian Tensleep (Mallory, 1949). Secondary production by gas injection became notably successful in 1939, when Salt Creek was the first large field in the United States to be unitized.

Recently numerous discoveries have been made in the Pennsylvanian Minnelusa–Tensleep Sandstone (Trotter, 1963). Several fields in Johnson County near the axis of the Powder River basin produce oil from the Minnelusa at a depth of 15,000 feet (Amsbury, 1967). Updip to the east, the Minnelusa ranges in depth from 10,000 feet to the surface outcrop. Numerous oil fields at various depths produce where sands have been trapped in buried hills beneath an erosional unconformity. The Halvorson field of eastern Campbell County is one example (Berg and Tenney, 1967).

The eastern flank of the Powder River basin is notable for its lenticular sand traps (Summerford et al., 1950). The reservoirs are Lower Cretaceous sands that may be 80 feet or more in thickness (Bolyard and McGregor, 1966). The discovery field for this type of accumulation is the *Osage* (Dobbin and Miller, 1941), where oil was found at a depth of 2000 feet in 1919. The reservoir rock is the Newcastle (Muddy) Sandstone (Lower Cretaceous), which occurs here as a lens of irregular shape, completely surrounded by shale and lying on the flank of a monocline (Figure 34.6). The oil has been trapped in

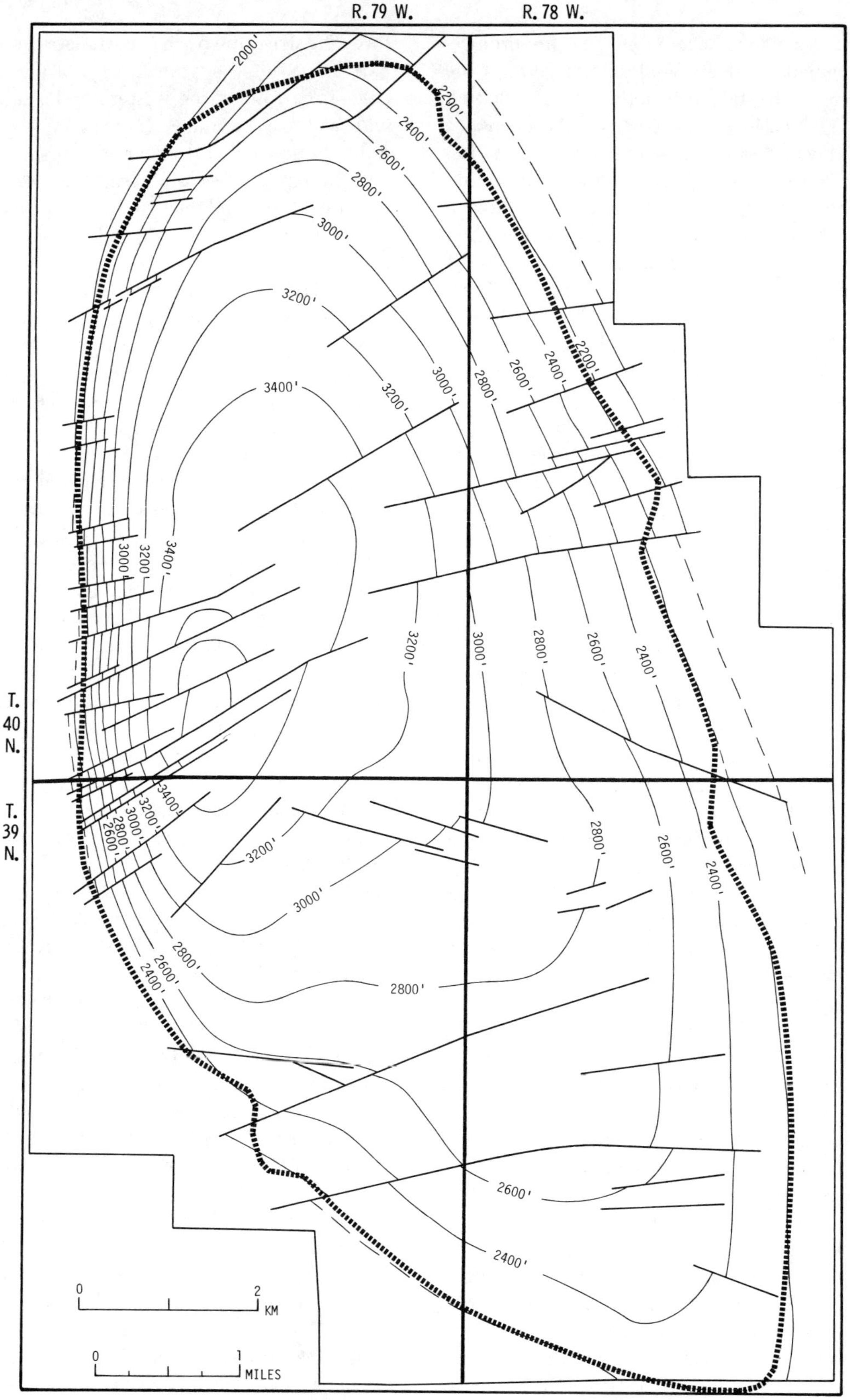

FIGURE 34.5 Salt Creek oil field, Natrona County, Wyoming. Structure contours based on Second Wall Creek sand. Heavy broken line shows original field boundary. After Elfred Beck (1929). *Courtesy author and the American Association of Petroleum Geologists.*

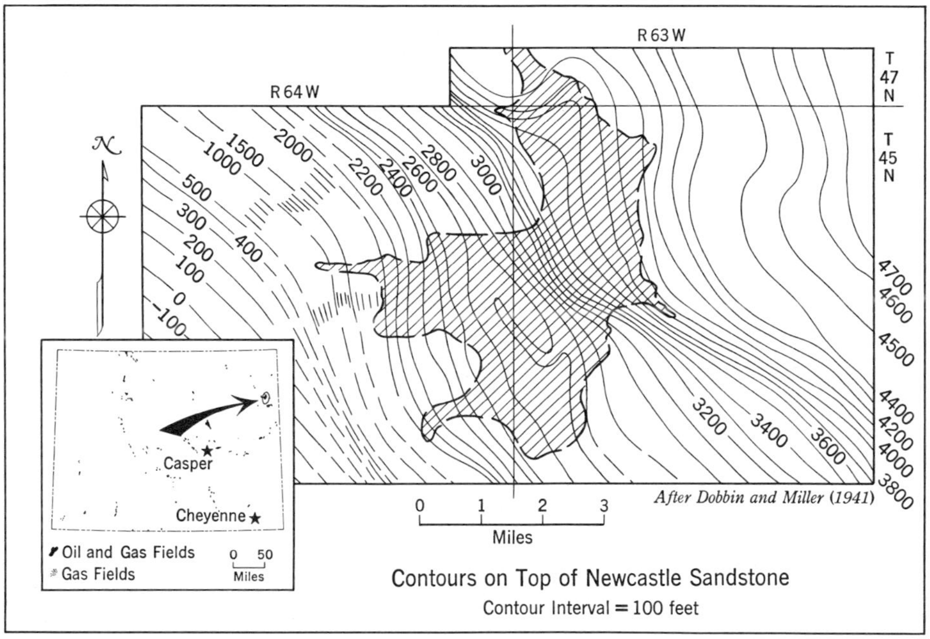

FIGURE 34.6 **Osage field, Weston County, Wyoming. Lenticular sandstone trap.** ***Courtesy American Association of Petroleum Geologists.***

this porous and permeable rock. Subsequently discovery has spread southward and westward down the regional dip of the basin. Not only many lenticular Newcastle Sandstone fields, but also other reservoirs, both younger and older, have been discovered in this area. The *Dead Horse Creek–Barber Creek* field (Sabins and Petersen 1961) produces from a 16 mile long shoestring sand in the Mesaverde Group (Upper Cretaceous).

The northern Powder River basin was the most active part of Wyoming in exploration in 1967 (Olson et al., 1968). The principal targets were stratigraphic traps in the Lower Cretaceous (Stapp, 1967). This search was stimulated by the discovery in mid-1967 of the prolific Bell Creek field Muddy sand on the northeastern flank of Powder River basin a short distance over the line in southeastern Montana. Contrary to usual procedures heretofore, exploration spread southward across the state boundary, leading to the discovery of *Recluse* (Hinton, 1968) and other Muddy reservoir fields in northern Campbell County. The slightly older Fall River ("Dakota") Formation yields oil along the eastern edge of the basin on the west flank of the Black Hills uplift (Runge, 1968). Trapping is due mainly to updip loss of adequate permeability.

The southwestern margin of the Powder River basin contains *South Glenrock* (Curry and Curry, 1954), a stratigraphic trap field discovered in 1950. The reservoirs are Lower Cretaceous sands (Dakota and Muddy). The oil is trapped by the updip shaling of the sands where they cross an anticlinal nose. Although South Glenrock is not the first stratigraphic trap pool found in Wyoming, it is the first to be discovered by deliberately exploring the flanks of an anticline for a permeability wedgout in a sand known to be tight across the top of the fold. The success here led to the search for, and the discovery of, other stratigraphic traps elsewhere in the state.

The *Burke Ranch* field in Natrona County is a similar permeability trap in the Powder River basin. Here the lensing out of permeability in the Dakota Sandstone takes place on a nose on the east limb of the Casper arch (Swirezynski, 1955).

BIG HORN BASIN. (Wyoming Geol. Assoc., 1947, 1952, 1959, 1962; Partridge, 1958; Zapp, 1956; Thomas, 1965). The Big Horn basin lies east and southeast of Yellowstone Park; its northwestern rim extends across the state boundary into Montana. The largest field, Elk Basin, also has its northern tip in Montana, but by far the greater production is obtained from the Wyoming side of the line. In 1967 the Elk Basin field led all others in Wyoming in petroleum output, with almost 20 million barrels. During the same year the Oregon Basin field was second in the state (following Salt Creek), and Hamilton Dome was fourth. Currently the Big Horn basin is the most prolific producing area in the Rocky Mountain province.

The stratigraphic column extends downward from continental Tertiary beds through a thick Mesozoic and a thinner Paleozoic section to the Precambrian. The known oil and gas reservoirs range from Cretaceous to Cambrian, but the most prolific are in the Phosphoria (Permian) and Tensleep (Pennsylvanian) rocks. The

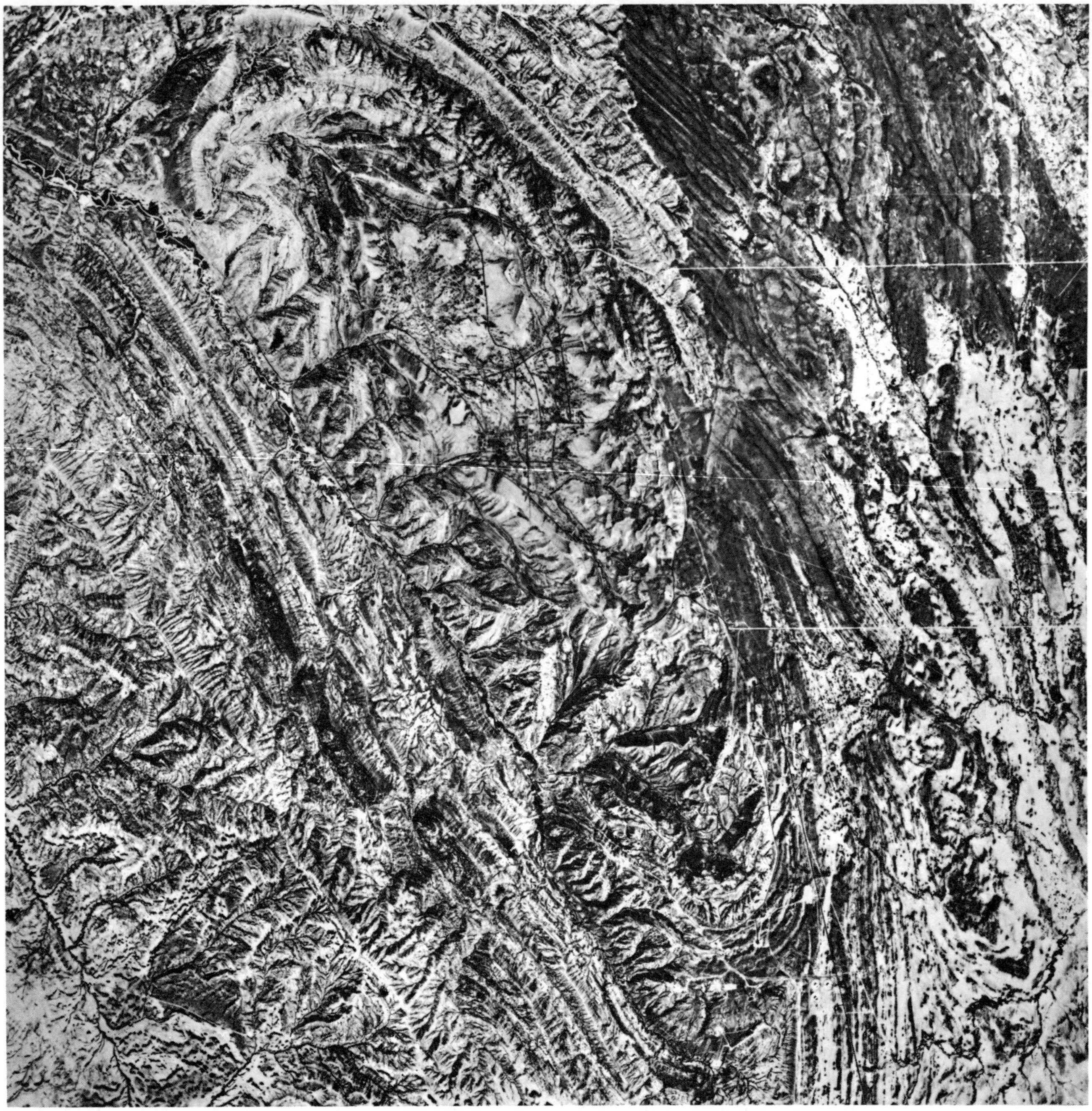

FIGURE 34.7 Elk Basin anticline, Wyoming-Montana. Aerial photograph showing differential erosion of fold. Producing oil wells are in core area, a basin carved in Upper Cretaceous (Niobrara and Carlile) shales. Surrounding hogbacks are still younger Cretaceous sandstones. Surrounding dissected plain is floored with Eocene Fort Union Formation. Area shown approximately 7 miles on a side. Photograph from Aero Service.

Phosphoria may also be the source for the oil found in adjacent reservoirs in some fields (Lawson and Smith, 1966; Stone, 1967). Other important reservoirs occur in the younger Cretaceous, especially the Wall Creek (Frontier) sands, and in the older Madison (Mississippian) Limestone.

The Big Horn basin has a structural relief of over 15,000 feet. It is elliptical in plan, with the long axis trending northwestward. Major anticlines on the flanks parallel the basin axis; it is on these that the largest oil fields so far discovered have been found. The biggest, *Elk Basin*, lies at the north end of the Big Horn basin

(McCabe, 1948). As can be seen in the air photograph (Figure 34.7), the Elk Basin anticline is a photogeologist's showpiece. The crest is a topographic basin, the result of the cropping out there of Cretaceous shales. Figure 34.8 is a subsurface structure map contoured on top of the second Wall Creek sandstone. The anticline has 5000 feet of closure, but the oil occurs only in the upper part of the structure. At least five zones have been found to be oil or gas bearing. The youngest reservoirs are the First and Second Wall Creek sands of the Frontier (Cretaceous) Formation, which lie at an average depth of 1500 feet. About 95 per cent of the Frontier oil has come from the Second Wall Creek, which is productive over an area of approximately 500 acres. The many normal faults shown on the structure contour map have affected the distribution of oil on the upthrown blocks, but the trapping has been anticlinal. These faults apparently die out in the formations below the Frontier; only one fault appears on a structure map based on the top of the Phosphoria Formation (Wyoming Geol. Assoc., Symposium Committee, 1957, maps in pocket). This map shows an elongate elliptical anticline, with two nodes, one crossing the Montana line and the other 2 miles to the southeast. The single fault mentioned apparently limits the Phosphoria–Tensleep production at the southeastern edge of the ellipse.

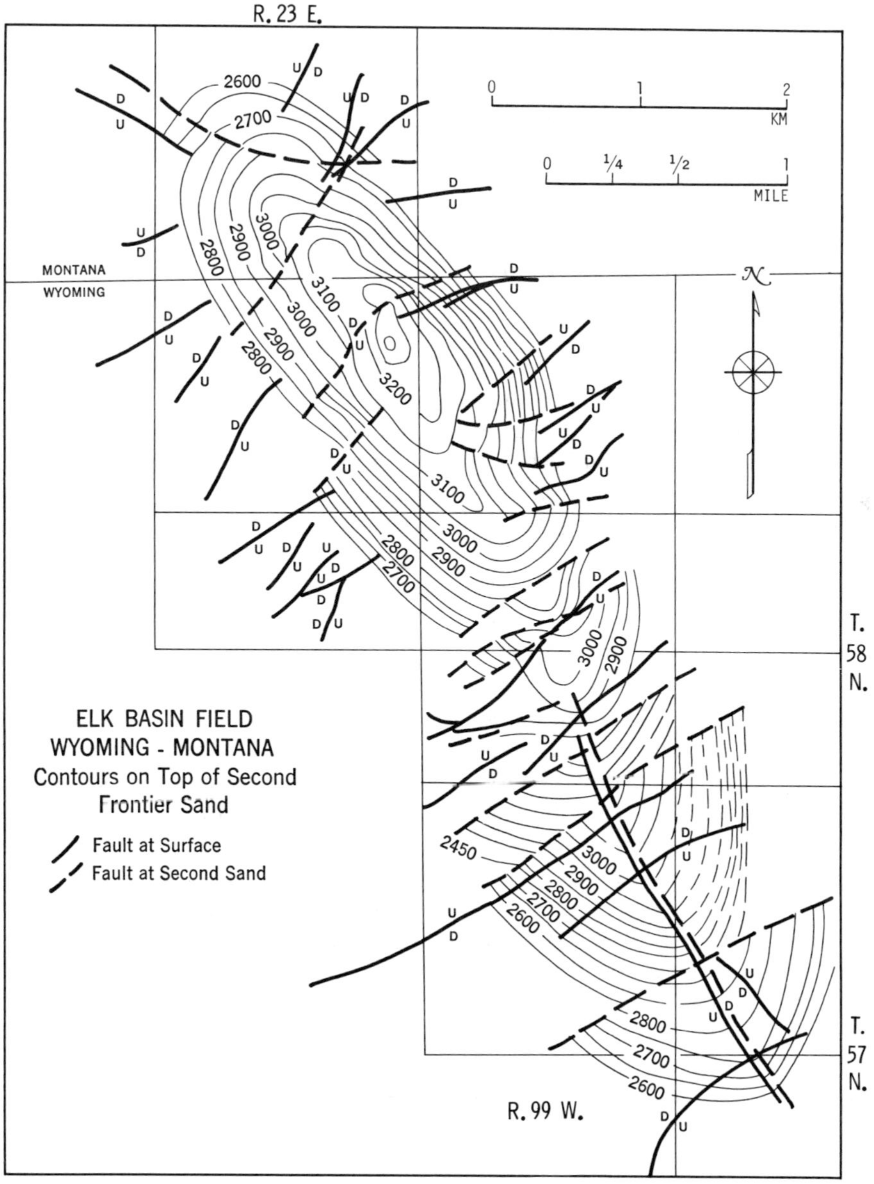

After Bartram (1929)

FIGURE 34.8

The oil in the Wall Creek sands is noticeably lighter than that in the deeper and older formations. Gas is produced from the Greybull (Cretaceous) Sandstone, reached at a depth of about 2600 feet. The Permian Phosphoria Dolomite, combined with the underlying Pennsylvanian Tensleep Sandstone, is the leading reservoir at Elk Basin. It lies at depths ranging between 3900 and 6000 feet. The producing section is about 100 feet thick, and the area is 5000 acres. About 1900 feet of the total closure is oil bearing. Oil was discovered in the Mississippian Madison Limestone between depths of 4350 and 5013 feet in 1946; by 1956 this reservoir had caught up with the Frontier sands in cumulative production. Oil is also produced from Devonian and Ordovician (Bighorn) formations.

The *Oregon Basin* field (Walton, 1947) lies on the west side of the Big Horn basin, not far west of the basin axis, which is considerably west of the geographic axis. Accumulation here is also anticlinal. The Oregon basin structure has a closure of about 1600 feet; the anticlinal axis sags in the center, dividing the fold into two domes with an intervening saddle. Black heavy oil is produced in quantity from the Permian Phosphoria Limestone, the Pennsylvanian Tensleep Sandstone, and the Mississippian Madison Limestone. High-gravity oil, as well as gas, was recently discovered at the north dome in the Cambrian Flathead Sandstone.

In 1967 the *Hamilton Dome* field (Wyoming Geol. Assoc., 1952; Kornfeld, 1954) in the southwestern part of the Big Horn basin, was the fourth field in Wyoming in annual production. The structure is an elliptical asymmetrical anticline, with the steeper dip (up to 90°) on the southwestern flank and the lesser dip (12 - 14°) on the basinward side. The northeastern flank is cut by radial step faults. The maximum closure is 3500 feet, with the oil accumulations confined to the top 1000 feet. The discovery well was completed in 1918 in a sand in the Chugwater (Triassic) Formation. Subsequently four deeper reservoirs,—the Embar Limestone in the Phosphoria (Permian) Formation, the Tensleep (Pennsylvanian) Sandstone, the Madison (Mississippian) Dolomite, and the Big Horn (Ordovician) Dolomite have been discovered. The leading producers are the Embar and the Tensleep.

One anticlinal trap, the *Worland* field in the southeastern part of the Big Horn basin, is notable for the high hydrogen sulfide content of gas that occurred in Permian Phosphoria dolomitic limestone. This gas was routed through a sulfur recovery plant that reached a peak production of approximately 300 tons of sulfur per day (Wold, 1952). Subsequently, however, the sour gas was depleted and the operation abandoned.

The Big Horn basin offers several examples of the readjustment of oil trapped in anticlines through the circulation of water entering the reservoir at the outcrop. Four fields in Hot Springs County that show a marked basinward tilting of the oil-water interfaces are *Gebo* (Mees and Bowers, 1952), *Northwest Lake* (Green and Ziener, 1953), *Black Mountain* (Horn, 1963), and *Murphy Dome* (Partridge, 1958, p. 305–306). Other instances of tilted oil-water contacts are believed by Todd (1963, p. 599) to be paleohydrodynamic tilts "maintained by secondary dolomite and solid hydrocarbon cementation."

The Big Horn basin also affords excellent examples of the loss of permeability trapping due to updip facies changes, which may or may not be related to anticlines and structural noses. The *Cottonwood Creek* field, discovered in 1953 in Washakie County on the east side of the basin, was one of the early major fields in Wyoming to be discovered by a wildcat well drilled on a stratigraphic trap prospect (Pedry, 1957). Here the migration of oil in marine dolomite of the Phosphoria (Permian) Formation up the southwest-dipping flank of the Big Horn basin was stopped by a facies change from vuggy, fractured dolomite to red shale and anhydrite (Sangbush, 1962). The field is about 8 miles long and 6 miles wide (downdip dimension). Structurally, the Cottonwood Creek producing area has a relief of more than 5000 feet.

The *Manderson* field in Big Horn County (Figure 34.9) is on a plunging anticlinal nose about 12 miles long. There is no evidence of structural closure; updip trapping again is due to a facies change in the Phosphoria Formation (Harris, 1955).

Wind River Basin. (Thompson, 1958; Wyoming Geol. Assoc., 1948, 1957, 1961; Keefer, 1965). The Wind River basin, in west-central Wyoming, ranks third in cumulative oil production in the state. Oil was first discovered in 1884 on the Dallas dome at a depth of 300 feet. The early discoveries in the Wind River basin, as elsewhere in Wyoming, were seep directed.

Tertiary rocks cover three-fourths of the basin and constitute about half of the total volume of sediments. Most of the pre-Tertiary sedimentary rock outcrops are on the southern side of the basin. The Wind River basin was emergent during much of the Paleozoic era. The Mesozoic submergences, although also intermittent in character, were of greater extent, producing a total section four times the thickness of the Paleozoic rocks.

So far, oil and gas productive reservoirs range in age from Tertiary and Paleocene Fort Union sands to the Madison Limestone in the Mississippian. The greatest oil production has come from the Tensleep Sandstone in the Pennsylvanian. Other important oil reservoirs are in the Permian Phosphoria (Embar). The Cretaceous Lakota, Dakota, and Frontier sands are important gas reservoirs.

The axis of the Wind River basin lies along the north side, paralleling the Owl Creek Mountain front. At the deepest point on the axis the sedimentary section may

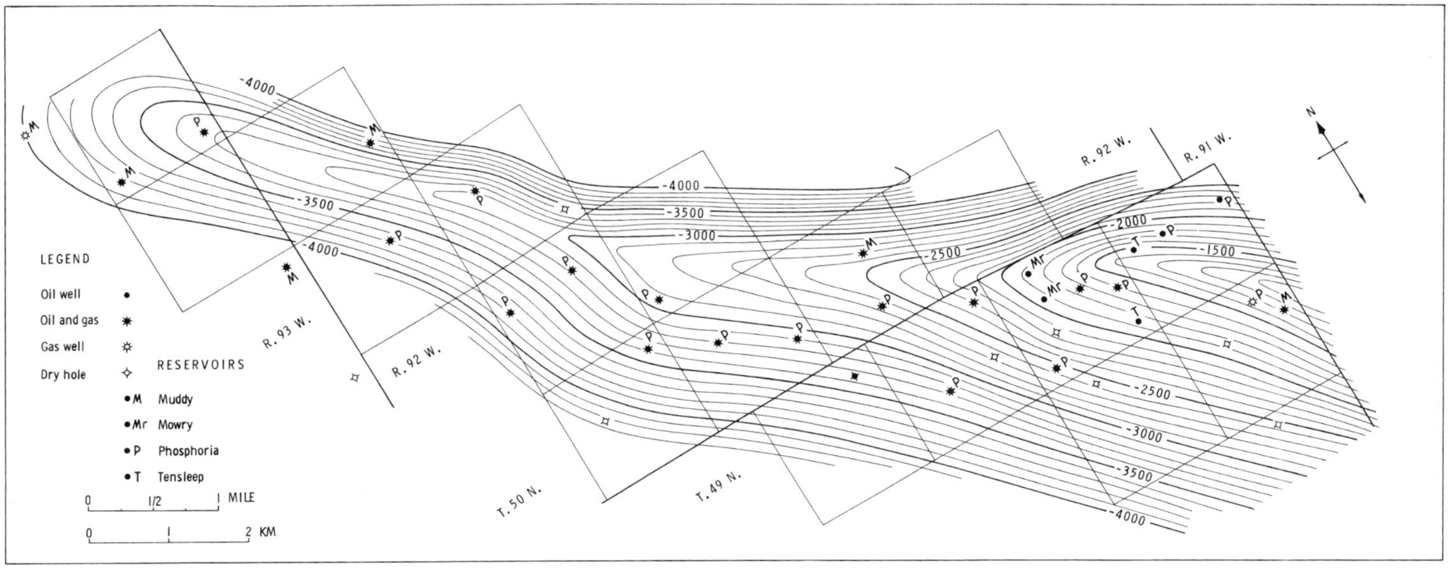

Figure 34.9 Manderson field, Big Horn County, Wyoming. Structure contour map based on top of Phosphoria. Interval, 100 feet. *Courtesy Wyoming Geological Association (1957).*

be as much as 30,000 feet in thickness. The Owl Creek Mountains have been thrust southward over the basin sediments for several miles. The wide south flank of the basin is crossed by a series of folds, most of which trend north-northwest. The typical anticline is elliptical in plan and asymmetric in cross section, with the steep dip on the west side. A reverse strike fault may parallel the anticlinal axis, so that the gentler dipping eastern flank sediments are thrust over the steeply dipping formations on the west flank.

Most of the traps so far discovered in the Wind River basin are anticlines. The *Steamboat Butte* (Murphy and Roberts, 1954; Barton, 1948) oil field is the largest in the basin in cumulative production and proven reserves and in 1964 was second in annual production. It was discovered in 1943. The trap is an elongate anticline with north-south trend and a fault on the west flank, limiting production in that direction. The estimated closure is 350 to 400 feet. The first reservoir discovered was the Nugget Sandstone of Jurassic age, but the oil reserve in the deeper Tensleep (Pennsylvanian), which was discovered about a year later, is much greater. The maximum thickness of oil-saturated Tensleep Sand is 300 feet. Other reservoirs include the Lakota (Cretaceous) and the Embar (Permian).

In 1967 *Beaver Creek* produced about 3 million barrels of oil and considerable gas. The oil reservoirs are the Mesaverde (Upper Cretaceous) Sandstone, Tensleep (Pennsylvanian) Sandstone, and Madison (Mississippian) Dolomite. The Frontier, Muddy, and Lakota Sandstones, all of Cretaceous age, produce about 30 million cubic feet of gas a day (Ross, 1957). Phosphoria (Permian) dolomitic limestone produces sour gas here and at the contiguous Riverton Dome field, to the north of Beaver Creek proper. The Beaver Creek field was discovered in 1938 by drilling on an anticline mapped on the exposed, gently folded Tertiary rocks on the assumption (subsequently proved to be correct) that the closure would increase below the Tertiary–Cretaceous unconformity (Ewing, 1955). The closure in the outcropping Tertiary sediments is a mere 100 feet; in the subsurface it is as much as 1300 feet (Wyoming Geol. Assoc., 1957, p. 63). At the Madison reservoir level (11,000 feet depth) the oil accumulation is confined largely to the west half of the domal trap because of a north-south fault that bisects the fold and drops the eastern side down (Pollock, 1960).

The *Circle Ridge* field (Beebe, 1953) is beautifully marked at the surface by differential erosion of the outcropping folded strata (Figure 34.10) The reservoirs are the Permian Phosphoria, Pennsylvanian Tensleep, and Mississippian Madison. The trap is a partly overturned faulted anticline (Figure 34.11). The fault is a thrust that causes vertical repetition of the reservoirs. Most of the producing wells were completed in the thrust block in the Phosphoria, Tensleep, and Madison reservoirs, but some are producing from the Phosphoria and Tensleep below the thrust plane. The known closure in the Phosphoria Dolomite above the fault is 1400 feet.

Southwestern Wyoming (Wyoming Geol. Assoc., 1950, 1955, 1960, 1961). Southwestern Wyoming contains one large basin to the west, the Green River, and two smaller basins, the Great Divide and the Washakie. Also present, separating the Green River basin from the two smaller basins to the east, is the Rock Springs uplift. The structural relief between the top of this uplift and the bottoms of the three basins is about 14,000 feet. The west-trending structural ridge separating the Great Divide basin on the north from the Washakie basin on the south is called the Wamsutter arch. South of the Washakie, across the state boundary in Colorado, is a third basin, the Sand Wash.

Gas fields predominate in southwestern Wyoming, especially in the Green River (Krueger, 1960, 1968) and Washakie basins, and on the Rock Springs uplift and the Wamsutter arch. Oil is produced with the gas, however, in several fields, and during 1964 five of these (Big Piney, La Barge, Patrick Draw, Lost Soldier, and Wertz) produced more than 1000 barrels a day each.

The reservoir rocks in southwestern Wyoming range in age from Tertiary to Cambrian. Paleozoic production is principally in the extreme northeastern corner of this large area; however, a recent gas completion west of Rock Springs at Church Buttes in the Pennsylvanian Morgan (Amsden) at a depth below 18,000 feet promises more widespread production from the older rocks. The leading reservoirs to the west are Eocene sands of continental origin, Upper Cretaceous sandstones (especially Frontier and Mesaverde), and the Nugget (Jurassic) Sandstone.

Until recently most of the traps successfully explored in southwestern Wyoming have been anticlines, with the closure increasing below the Tertiary–Cretaceous boundary. Most of the fields found on the Wamsutter arch during the last decade, however, are stratigraphic traps.

The largest productive area in southwestern Wyoming is *Big Piney–La Barge* (Privrasky, 1963; Shipp and Dunewald, 1962; Krueger, 1960; Wyoming Geol. Assoc., 1957). It extends about 25 miles in a north-south direction along the western edge of the Green River basin in Sublette and Lincoln Counties. The maximum width is 16 miles. The western side of the area overlaps the eastern edge of the overthrust belt, and some of the production comes from beneath the thrust planes. Paleozoic rocks have been pushed easterly over Cretaceous, and younger thrusts have shoved Cretaceous rocks over Paleocene.

The Big Piney–La Barge area contains four fields of

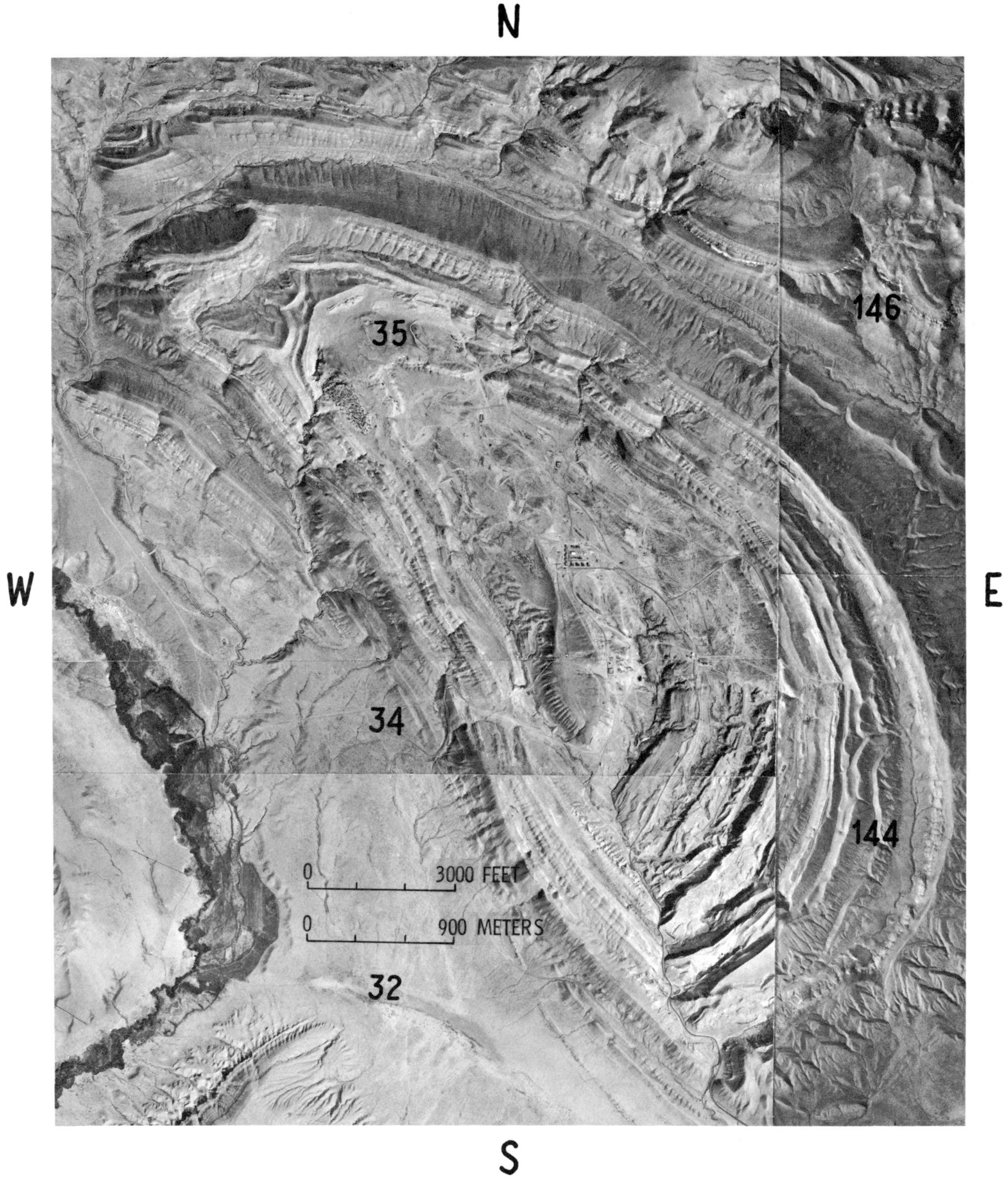

FIGURE 34.10 Air photo mosaic, with photograph numbers superimposed. Circle Ridge oil field, northern Fremont County. Southeast flank of anticline is faulted (Figure 34.11), limiting production in that direction. Beautiful example of differential erosion, with Chugwater (Triassic) sandstones making the rim-rocks and older Phosphoria carbonate rocks the central depression. *U. S. Department of Agriculture photos CCK-7N-32, 34–35 and 19N-144, 146.*

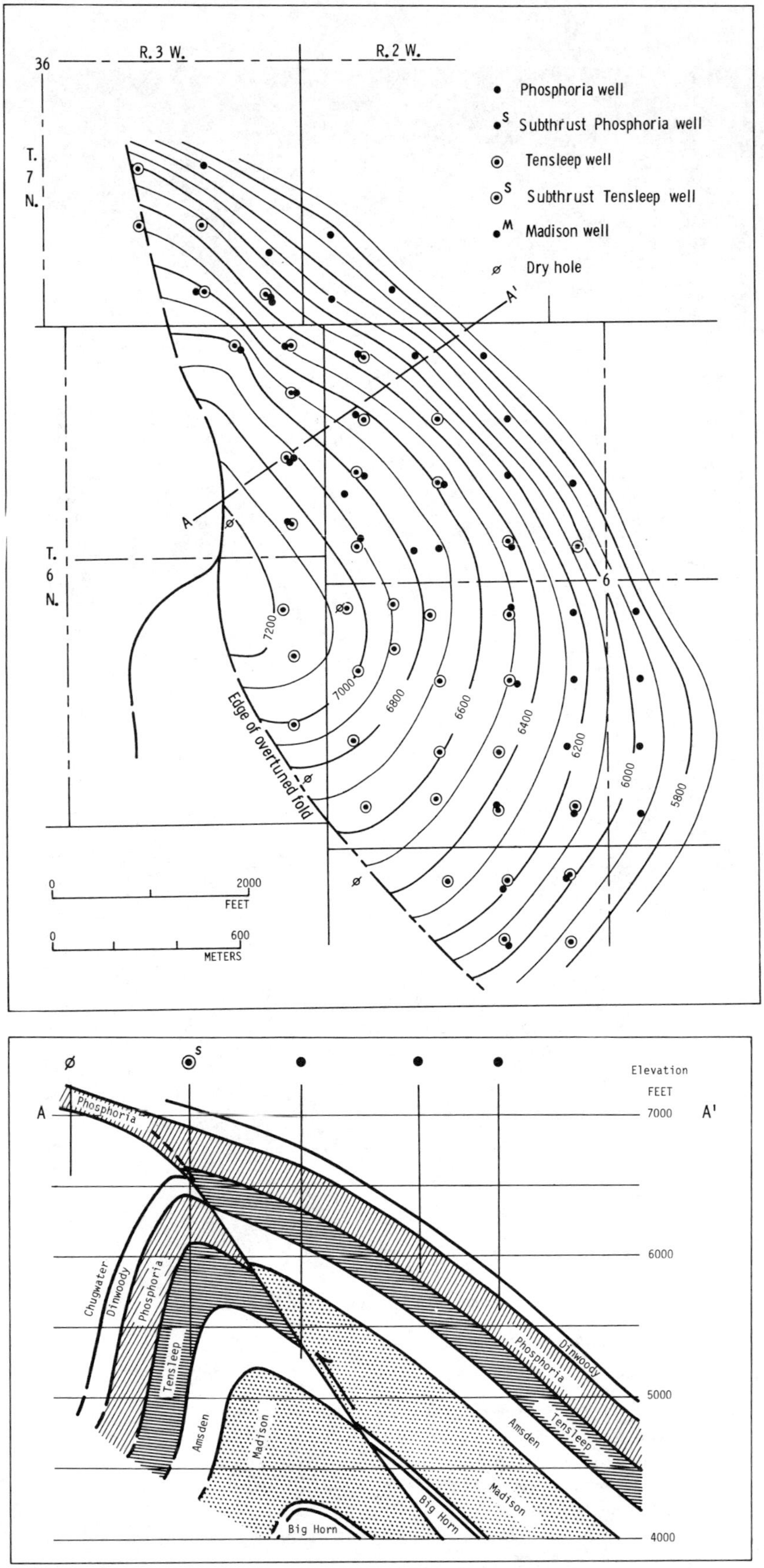

Figure 34.11 Upper: Circle Ridge field, Fremont County. Subsurface structure contour map based on top of the Permian Phosphoria Formation. Contour interval, 100 feet. Lower: Cross section along line *A–A′* in illustration above, showing thrust fault, and production on both the thrust sheet and on the subthrust block. *Both illustrations courtesy Lewis E. Beebe (1953) and the Oil and Gas Journal.*

concentrated production plus scattered gas wells between and beyond the fields named. The La Barge, in the southern part of the area, was first to be discovered, in 1924. It is primarily an oil field, as is adjacent North La Barge. The relatively small Tip Top field to the northwest produces both gas and oil. Big Piney to the northeast is primarily, but not exclusively, a gas field; it was second in Wyoming only to Salt Creek in 1963 in the number of producing wells. Much of the production comes from depths of around 900 feet.

The Big Piney–La Barge area was the first in the Rocky Mountain province to produce commercially from Tertiary reservoirs. Sands in the Wasatch (early Eocene) and in the Fort Union (Paleocene) Formations are most important. Older reservoirs include sands in the Mesaverde and Frontier Formations of Late Cretaceous age and, in Tip Top field, the Jurassic Nugget Sandstone, which is penetrated at a depth of 10,000 feet.

Near the southern end of the Big Piney–La Barge complex the close association of both oil and gas production with a Fort Union lacustrine shale suggests strongly that the shale is the source rock. It could also be the source rock for the underlying Mesaverde reservoirs (Asquith, 1966).

The trapping throughout the area is mainly anticlinal, but sand lensing plays a very important role in the distribution of the gas and oil. The oldest field, La Barge, is an anticlinal trap discovered by surface mapping 17 years after oil seeps to the west had first called attention to the area. Big Piney was discovered in 1938 by random drilling even though the anticlinal trap is mappable at the surface. The dips on the flanks, gentle at the surface, steepen with depth.

The *Baxter Basin* gas fields (Fidlar, 1962; Reese, 1968) occupy the crest of the Rock Springs uplift (Figure 34.2). Baxter Basin is a physiographic feature formed by the erosion of the Baxter Shale (Upper Cretaceous) exposed in the core of the uplift. Hogbacks of the younger Mesaverde sandstones surround the shale exposure and form the rim of the topographic basin.

Gas was first discovered on Rock Springs uplift in 1922 when a successful wildcat was drilled in the South Baxter area. The initial reservoir was the Dakota Sandstone (Lower Cretaceous) at a depth of about 2500 feet, but subsequently production has been obtained from the younger Frontier Sandstone and from the older Morrison Formation (Jurassic) and Nugget Sandstone (older Jurassic). The reservoirs in the Morrison are stray sands.

The structure map (Figure 34.12) shows a north-south row of much faulted nodes along the crest of the Rock Springs uplift. The faults are in general parallel, with a northeasterly trend. Gas accumulation is obviously anticlinal, with the deposits terminating abruptly at the faults. The lenticularity of some of the reservoirs is responsible for the somewhat erratic distribution pattern of the gas fields.

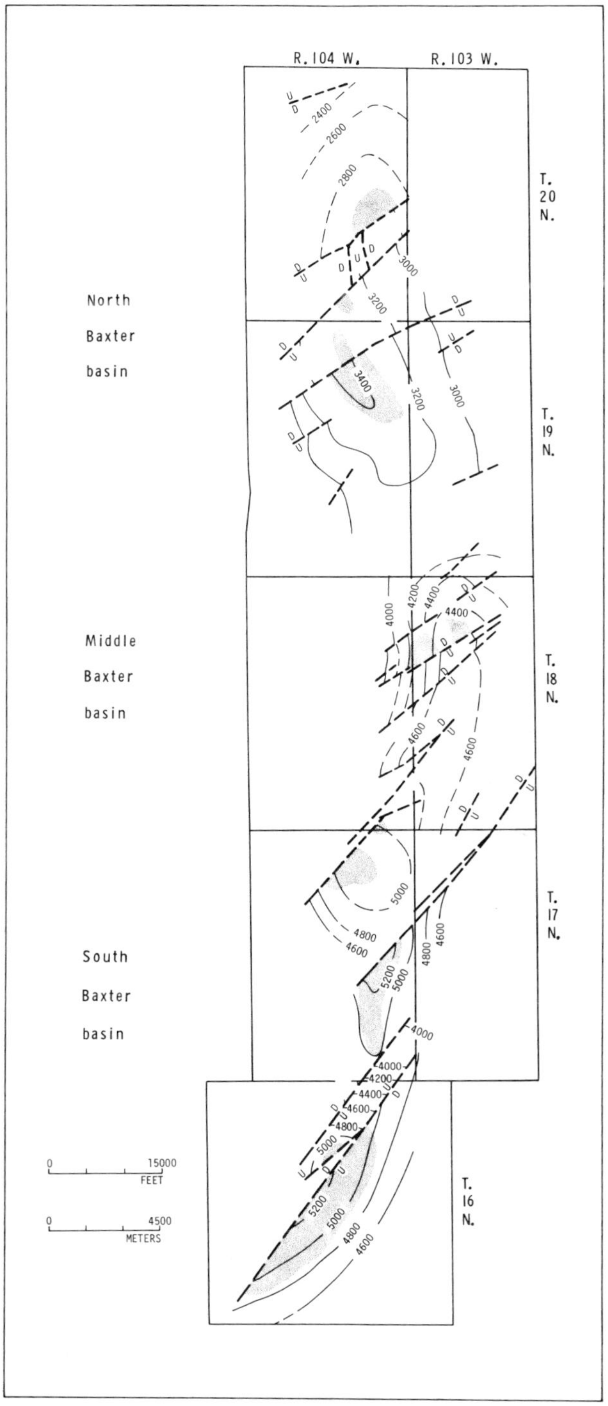

FIGURE 34.12 Baxter Basin gas fields, Sweetwater County, Wyoming. Gas-producing areas shaded. Structure contour data are top of Dakota (North Baxter) and top of Frontier (Middle and South Baxter). Interval, 200 feet. *Courtesy Wyoming Geological Association (1957).*

The oldest oil fields, and the largest in cumulative production in southwestern Wyoming, are in the northeastern part of the province. These are *Lost Soldier* (Pott and DeVore, 1952) and *Wertz*, which lie side by side on the northeastern rim of Great Divide basin in the northeastern corner of Sweetwater County, overlapping the borders of Fremont and Carbon Counties. By the close of 1964 Lost Soldier had produced over 95 million and Wertz over 50 million barrels of oil. Both are elliptical anticlines with the long axes trending northwesterly. Lost Soldier was discovered in 1916 when a 200 barrel well was completed in a Frontier (Upper Cretaceous) sand at a depth of 269 feet. Subsequently other reservoirs were found, including sands in the Dakota-Lakota (Cretaceous) and the Sundance (Jurassic). The most prolific reservoir, however, has been the Tensleep (Pennsylvanian) Sandstone, discovered in 1930 by a 2400 barrel well. The oil-filled structural closure at this level is about 2200 feet. Subsequent reservoir discoveries have been the Madison (Mississippian) Limestone in 1947 and Cambrian sandstone in 1948. The latter was an historic event, for Cambrian oil was then a rarity.

The *Patrick Draw* field (Burton, 1961) on the Wamsutter arch was discovered in 1959 and 2 years later ranked third briefly in petroleum production in Wyoming. Here, as in most of the other fields on the arch, the oil and gas occur in closed sandstone bodies deposited as offshore bars during the Late Cretaceous (Ritzma, 1963, 1968; Weimer, 1966). On the Great Divide basin flank of the arch is the *Desert Springs* (May, 1961) gas field. It was discovered by drilling on a seismic anomaly that was assumed to be a fault. Drilling showed, however, that the anomaly was due to the presence of a permeability wedgout, which created the trap (Earl, 1961).

Laramie and Hanna Basins (Wyoming Geol. Asso., 1953; Stone, 1966). The Laramie basin is hemmed in between the Medicine Bow Mountains on the west and the Laramie Mountains to the east (Blackstone, 1953). Both of these ranges trend northerly, and hence the Laramie basin does likewise. However, the Medicine Bows terminate about 45 miles north of the Colorado line, leaving room for a small basin, the Hanna, south of the next Precambrian-cored range, the Granite Mountains. The Hanna basin lies immediately to the west of the north end of the Laramie basin (Figure 34.2).

Outcropping pre-Tertiary formations in the Laramie basin consist mainly of upper Paleozoics on the east side and Mesozoic sediments on the west side. Tertiary formations unconformably overlie the older sediments in the northern part. A series of anticlinal folds extends northward on the western flank of the Laramie basin east and north of the Medicine Bows. The major oil fields occur in this belt. These include, from southwest to northeast, the Quealy, Rock River, and Big Medicine Bow (Wyoming Geol. Assoc., 1957). All three are elliptical anticlines with a north-northwest trend. Two were discovered solely by surface mapping; the third, Quealy, was discovered by both surface and seismic surveys.

The *Quealy* anticline (West, 1953) in southwestern Albany County is cut by an east-dipping thrust fault. Westward thrusting is unusual in Wyoming, which is notable for its large thrusts to the eastward. The deepest reservoir, the Casper Sandstone (Pennsylvanian), produces from both the upthrown and downthrown blocks. The shallower Lower Cretaceous sands are productive in the thrust sheet only.

Much the largest field in the basin is *Rock River* (Espach and Biggs, 1953), which lies across the Carbon County line to the northwest of Quealy. The trapping anticline is asymmetric, with dips up to 70° eastward into the basin and dips of 20 to 35° toward the Medicine Bow Range to the west. The closure at reservoir level is about 1000 feet; the field owes its more than 30 million barrel production to the fact that the trap is nearly filled. Most of the oil has come from sands (Muddy, Dakota, Lakota) in the Lower Cretaceous, with the Sundance (Jurassic) Sand supplying the rest.

The *Big Medicine Bow* field (Van Gilder, 1953) is also in eastern Carbon County about 12 miles to the northwest of Rock River. This anticlinal trap has the greatest closure, 2200 feet, in Laramie basin, but only the uppermost part of the trap is occupied. Considerable gas is produced with the oil from the Sundance Sandstone, which is the principal reservoir. The Tensleep (Pennsylvanian) Sandstone about 1400 feet deeper is the second reservoir.

Although seven nonassociated gas fields have been found in Laramie and Hanna basins, only one of them was producing in 1965 (Mitchell, 1968). This was a two-well field high on the northwest flank of Laramie basin. The trap is an elongate anticline paralleling the regional strike; the reservoir is Shannon Sandstone. Most of the other fields are depleted and abandoned. One field, lying high on the northeast flank of Hanna basin, is now used as a gas storage reservoir.

Denver Basin. The Denver basin, formerly known as the Denver-Julesburg, is an asymmetric elongate structural basin paralleling the Front Range in northeastern Colorado, southeastern Wyoming, and southwestern Nebraska. The western flank is steep, and the eastern gentle. Therefore the synclinal axis lies close to the western edge. Two antinodes of the axis produce separate basin bottoms in the vicinities of Denver and Cheyenne.

Most of the known Denver basin oil fields lie updip on the eastern flank in northeastern Colorado and southwestern Nebraska. Wyoming has a few fields across the line from Nebraska, and two fields much farther to the west on the west flank of the basin. One of these, *Horse Creek* in Laramie County, is an elliptical anticline with a northeast trend and a closure of about 2400 feet

(Peters, 1960, p. 610). The field has a length of about 4 miles and a maximum width of nearly 2 miles. The reservoir rock is the Muddy Sand of Cretaceous age. Horse Creek field was discovered by gravity and seismic surveys.

Future. At its present pace of discovery Wyoming need have no worries about its future oil production. The spate of new fields in Muddy (Cretaceous) sands and in the Minnelusa–Tensleep (Pennsylvanian) reservoir in Powder River basin, for example, indicates that additional reserves await discovery. The outlook for gas is equally encouraging. The current widespread production in southwestern Wyoming, as well as deep Paleocene gas discoveries in northeastern Fremont County in the Wind River basin, points the way to additional supplies.

The reserves still to be discovered in Wyoming are both lateral and vertical. No basin has been completely explored from rim to rim. Many untested prospects have been mapped in the Wind River alone (Van Houten and Weitz, 1956; Murphy et al., 1956; Keefer and Troyer, 1956). In the past, most exploration has involved anticlinal traps, but the list of productive and profitable stratigraphic traps is growing by geometric progression. Furthermore, the sedimentation and tectonic history of Wyoming is such as to guarantee a profusion of hidden varying permeability traps in the many basins. Searching for these traps will constitute a major activity across the years.

The third dimension, depth, is the one least explored in Wyoming. Drilling costs have discouraged exploration in the basin centers, but both traps and the most complete sedimentary sections are there. A new producing depth record for Wyoming was set in 1965, when a wildcat drilled in the deepest part of the Powder River basin came in for an initial production of 800 barrels per day from the Minnelusa at a depth range of 14,942 to 15,063 feet.

Even on the basin flanks, where depths are more moderate, there has been inadequate exploration for hydrocarbons in the older reservoirs, including especially Madison (Mississippian) Dolomite, Devonian carbonate rocks, Big Horn Dolomite and younger Ordovician carbonate rocks, and Cambrian sandstones. "Porous zones in the Ordovician formations truncated beneath Devonian and Mississippian beds may prove to be reservoirs along the flanks of large anticlinal uplifts such as the Bighorn Mountains or along smaller anticlines or faults" (Richards and Nieschmidt, 1961). Shallow gas occurrences in southwestern Wyoming (Skeeters, 1961) may overlie deeper accumulations of oil and gas. Berg (1963) suggests the possibility of both oil and gas in well sorted fluvial Paleocene sands west of Pinedale in northern Green River basin.

There are also hitherto ignored uphole possibilities. Examples include the Curtis (Triassic) Sandstone and sandstones occurring in Eagle, Judith River, and Bearpaw sections in the Montana Group of the Upper Cretaceous in the Big Horn basin (Miller et al., 1965), as well as in fractured Niobrara (also Upper Cretaceous) rocks (McAusland, 1959).

The Wind River basin contains an enormous volume of Late Cretaceous and Tertiary rocks that is virtually unexplored, in spite of the occurrence of oil-saturated sands at the outcrop. These strata exceed 20,000 feet in thickness in the deeper parts of the basin, "and significant shows or production has been found in all of these formations. Four fields are producing gas, and favorable economic conditions exist for future gas development in the basin" (Paape, 1968, p. 760).

REFERENCES CONSULTED IN PREPARING CHAPTER 34

Amsbury, David L., 1967.
Andrichuk, John M., 1958.
Armstrong, Frank C., and Oriel, Steven S., 1965.
Asquith, D. O., 1966.
Barlow, J. A., Jr., and Haun, John D., 1966.
Barton, H. E., 1948.
Bartram, John G., 1929.
Beck, Elfred, 1929.
Beebe, Lewis E., 1953.
Berg, Robert R., 1963.
Berg, Robert R., and Tenney, Charles S., 1967.
Blackstone, D. L., Jr., 1953, 1963.
Bolyard, Dudley W., and McGregor, Alexander A., 1966.
Burton, Guy, 1961.
Curry, Wm. H., Jr., and Curry, Wm. H., III, 1954.
Curtis, Bruce F., Strickland, John W., and Busby, Robert C., 1958.
Dobbin, C. E., 1943.
Dobbin, C. E., Kramer, W. B., and Horn, G. H., 1957.
Dobbin, C. E., and Miller, J. C., 1941.
Eardley, A. J., 1963.
Earl, John H., 1961.
Espach, Ralph H., and Biggs, Paul, 1953.
Espach, Ralph H., and Nichols, H. Dale, 1941.
Ewing, David J., 1955.
Fidlar, M. M., 1962.
Gras, V. B., 1968.
Green, Thomas H., and Ziener, C. W., 1953.
Harris, Loy E., 1955.
Hinton, Gene, 1968.
Hodgson, Robert A., 1965.
Hoppin, Richard A., and Palmquist, John C., 1965.
Horn, George H., 1963.
Keefer, William R., 1965.
Keefer, William R., and Troyer, Max L., 1956.

Kornfeld, Joseph A., 1954.
Krueger, Max L., 1960, 1968.
Lawson, Don E., and Smith, Jordan R., 1966.
Love, J. D., McGrew, Paul O., and Thomas, H. D., 1963.
McAusland, Edward R., 1959.
McCabe, W. S., 1948.
Mallory, R. W., 1949.
Mapel, W. J., and Pillmore, C. L., 1963.
May, B. E., 1961.
Mees, E. C., and Bowers, G. F., 1952.
Miller, D. N., Jr., Barlow, James A., Jr., and Haun, John D., 1965.
Mitchell, S. D., 1968.
Murphy, J. F., Privrasky, N. C., and Moerlein, G. A., 1956.
Murphy, John F., and Roberts, Ralph W., 1954.
Olson, Charles F., Curry, William, H., III, Norwood, E. Earl, and Slocum, R. C., 1968.
Paape, D. W., 1968.
Partridge, John F., Jr., 1958.
Pedry, John J., 1957.
Peters, Jack W., 1960.
Pollock, C. B., 1960.
Pott, Robert L., and DeVore, Stephen F., 1952.
Privrasky, Norman C., 1963.
Prucha, John James, Graham, John A., and Nickelsen, Richard P., 1965.
Reese, D. L., 1968.
Richards, P. W., and Nieschmidt, C. L., 1961.
Ritzma, Howard R., 1963, 1968.
Ross, Richard B., 1957.
Runge, John S., 1968.
Sabins, F. F., and Petersen, F. A., 1961.
Sangbush, R. L., 1962.
Shipp, B. G., and Dunnewald, J. B., 1962.
Skeeters, W. W., 1961.
Stapp, Robert W., 1967.
Stone, D. S., 1966, 1967.
Summerford, H. E., Schieck, E. E., and Hiestand, T. C., 1950.
Swirezynski, R. P., 1955.
Tenney, Charles S., 1966, 1968.
Thomas, Horace D., 1948, 1958, 1961.
Thomas, Leonard E., 1965.
Thompson, R. M., 1958.
Todd, T. W., 1963.
Trotter, John F., 1963.
Van Gilder, H. R., 1953.
Van Houten, F. B., and Weitz, J. L., 1956.
Walton, Paul T., 1947.
Weimer, Robert J., 1966.
West, W. E., Jr., 1953.
Wold, John S., 1952.
Wulf, George R., 1968.
Wyoming Geological Association, 1947, 1948, 1949, 1950, 1952, 1953, 1955, 1957, 1958, 1959, 1960, 1961, 1962, 1968.
Wyoming Geological Association, and Billings Geological Society, 1963.
Wyoming Geological Association Symposium Committee, 1957.
Wyoming Geological Association, Technical Studies Committee, 1965.
Zapp, A. D., 1956.

Chapter

35

MONTANA

The state of Montana produced nearly 50 million barrels of oil in 1968, which was about $1\frac{1}{2}$ per cent of the domestic production. It was in eleventh place among the states.

The first commercial hydrocarbon discovery in Montana was gas found in the Gas City dome (Figure 35.1) in Dawson County at the northern end of the Cedar Creek anticline (Figure 35.2) in 1913. This gas began flowing by pipeline into Glendive 10 miles to the north in 1915 (Erdmann, 1963, p. 29). During the same year gas was discovered at Baker in Fallon County. Subsequent exploration showed that this second discovery was near the middle of the 60 mile long Cedar Creek gas field (Figure 35.1). The year 1915 also saw the discovery of oil in the Elk Basin field, first on the Wyoming side of the state line (where 87 per cent of the productive acreage lies) and later, during the same year, on the Montana side.

The first exclusively Montana oil discovery was made at Cat Creek in 1920; this finding led to further search, resulting in the discovery of Kevin–Sunburst in 1922. This field reached a peak production of nearly $6\frac{1}{2}$ million barrels of oil in 1926 and established the permanence of the oil industry in Montana. Cut Bank, the state's leading oil field in cumulative production, has two discovery dates, gas in 1926 and oil in 1929.

In July, 1951, 3 months after the completion in North Dakota of the discovery oil well of the Williston basin, oil was found 125 miles to the west in the Montana sector of the basin in the Richey area, Dawson County (Shell Oil Co., 1958a). Additional discoveries during subsequent years in the Williston basin have been of such magnitude that this district now produces nearly two-thirds of Montana's oil. This has been the major cause of the state's continued increase in production during recent years while other Rocky Mountain states were declining. A contributing factor has been an expansion in secondary recovery activities, especially in waterflooding.

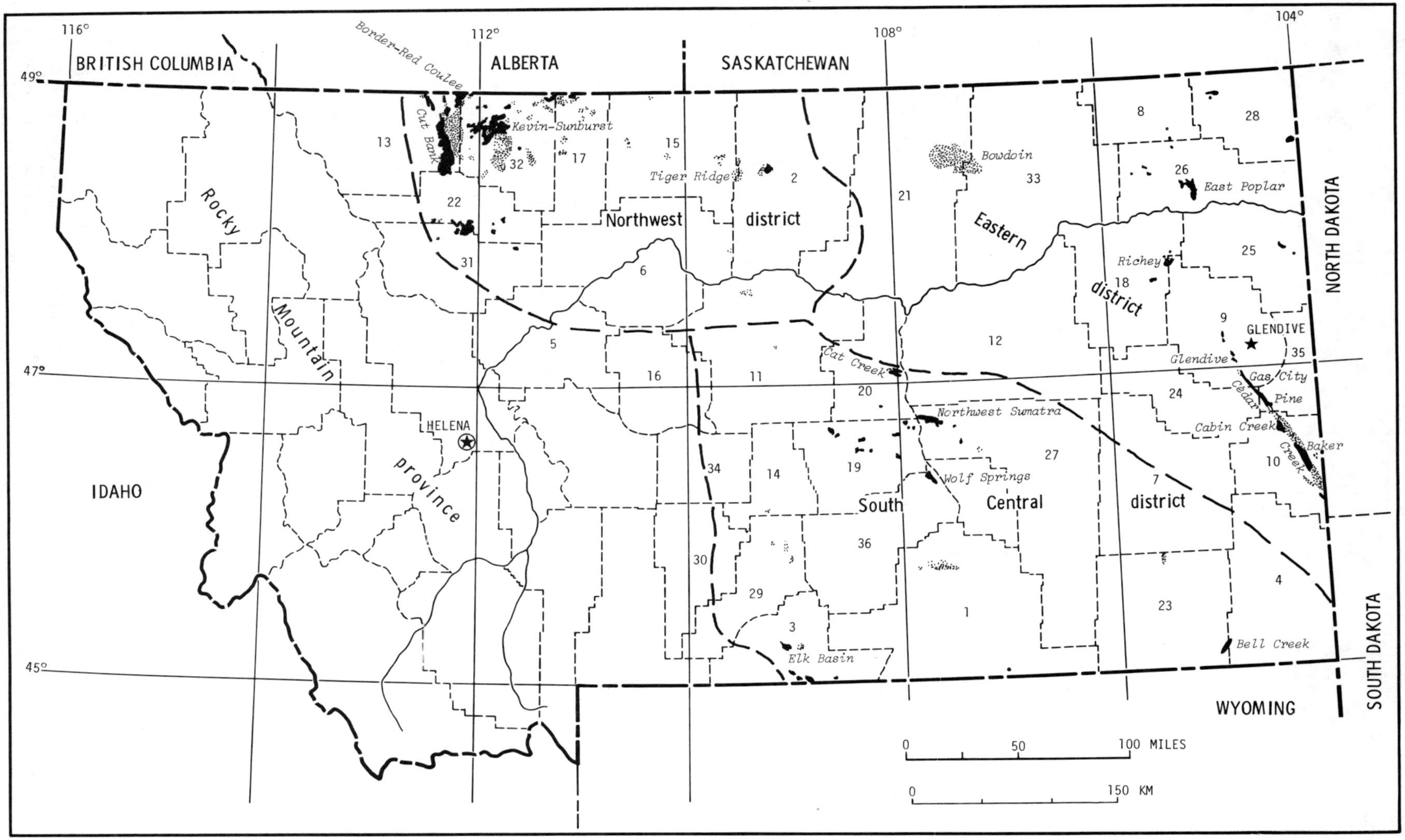

FIGURE 35.1 Montana. Index map, with oil and gas fields and producing districts. Fields discussed in text are named on map. County Key:

1	Big Horn	5	Cascade	9	Dawson	13	Glacier	17	Liberty	21	Phillips	25	Richland	29	Stillwater	33	Valley
2	Blaine	6	Choteau	10	Fallon	14	Golden Valley	18	McCone	22	Pondera	26	Roosevelt	30	Sweet Grass	34	Wheatland
3	Carbon	7	Custer	11	Fergus	15	Hill	19	Musselshell	23	Powder River	27	Rosebud	31	Teton	35	Wibaux
4	Carter	8	Daniels	12	Garfield	16	Judith Basin	20	Petroleum	24	Prairie	28	Sheridan	32	Toole	36	Yellowstone

Tectonics. The western third of Montana is in the Rocky Mountain physiographic province (Figure 35.1 and 35.2). Here the upwarped Precambrian metamorphosed sedimentary and igneous basement rocks have been stripped of their one-time Paleozoic and Mesozoic sediment cover and dissected into high mountains, deep valleys, and intermontane basins. This is not oil country.

To the east of the Rockies is the Great Plains province, covering approximately two-thirds of Montana. The transition zone, 10 to 30 miles wide, between the mountains and the plains is the Disturbed Belt (Figure 35.2), where overthrust faults dominate the tectonic picture. This belt continues northwestward for some hundreds of miles into Canada, where it is productive of oil and gas. The upturned and faulted rocks exposed along this trend range from early Paleozoic to Late Cretaceous in age.

The Plains area to the east of the Continental Divide is one of generally decreasing topographic relief and lesser diastrophism from west to east. Sharp anticlinal folds give way to nearly flat strata in the eastern part of the state. Coals drop in rank from bituminous to lignite. The Paleozoic rocks dip beneath Cretaceous east of the Disturbed Belt, and the Cretaceous formations in turn become overlapped and veneered by Cenozoic continental sediments. The dotted line in Figure 35.2 separates Tertiary (T) sediments, which are mostly Paleocene in age, from Cretaceous rocks. Windows of Cretaceous rock are exposed in the cores of major uplifts in eastern Montana, just as Paleozoic rock outcrops are surrounded by Cretaceous sediments in upwarped parts of central and southern Montana.

The east slope of Montana is structurally undulating, with both uplifts and basins (Dobbin and Erdmann, 1955); the largest of these features are shown in Figure 35.2. Dominating the picture are, proceeding clockwise around the area, the Sweetgrass arch in northwestern Montana, on which is perched the Kevin–Sunburst dome; the western rim and eastward dipping flank of the Williston basin, with the superimposed Bowdoin dome on the north, and the Cedar Creek anticline in the southeastern corner; the northwest end of the Black Hills uplift, which protrudes into the southeast corner of Montana; and the northern extremities of the Powder River and Big Horn basins, which cross into southern Montana from Wyoming. To the north of the latter basin rims, in central Montana, is an upwarped complex that includes the Big and Little Belt uplifts, involving crystalline rocks and, to the east, a sedimentary rock anticlinorium containing the Big Snowy uplift, Cat Creek anticline, and other major folds.

Stratigraphy. The major divisions of the outcropping rocks in Montana are shown in Figure 35.2. Figure 35.3 is a generalized stratigraphic column for the state; it includes both the surface rocks and the subsurface stratigraphy, especially the principal reservoir rocks. Formation thicknesses vary widely in different parts of Montana; the composite section shown gives maximum thicknesses.

Fort Union sediments of Paleocene age veneer Late Cretaceous rocks in eastern Montana, and both Cretaceous and older rocks in the southern part of the state. They consist largely of sands, shale, and lignite coal.

The Cretaceous section is almost entirely marine shale with interbeds of sandstone. It is by far the thickest sedimentary series in the state; of the two major divisions the Upper Cretaceous is the thicker, but the Lower Cretaceous has produced more oil. The youngest reservoir is a gas sand in the Judith River Formation of Montana age. Lower, at the base of the Montana Group, are the Eagle gas sand and the Shannon oil and gas reservoir of the Powder River basin. The Colorado Group includes the Frontier oil and gas sand, the Bow Island gas reservoir, which produces on the Bowdoin dome, and the upper Muddy (Dynneson) which produces at Bell Creek. Below these are the "Dakota" (First Cat Creek) sand and, more importantly, the Kootenai Group of reservoirs, including the Second and Third Cat Creek sands, the Cut Bank oil and gas sand, and the Sunburst and Greybull sands.

The Jurassic section is thinner than the Lower Cretaceous over the Sweetgrass arch but has greater thickness in the basins to the east. It contains sands in the Morrison Formation, and oil reservoirs in sands in the Sundance–Ellis Group, as well as in the Piper Limestone, which is also in the Sundance of eastern Montana. The Triassic, present only in southern and eastern Montana, is not yet productive in this state.

Pennsylvanian rocks are absent in the subsurface of western Montana, but both Tensleep Sandstone and Amsden Limestone are reservoirs to the east. Most of the Williston basin oil in eastern Montana comes from the Mississippian to Ordovician section, as in North Dakota. The Mississippian rock is thicker here than is the section of any other period except the Cretaceous.

The oil and gas reservoirs are shown in Figure 35.3. The Tyler sands and overlying Alaska Bench Formation, here placed at the top of the Mississippian, may belong in the Pennsylvanian (Krantzler, 1966). Near the base of the Upper Mississippian is the Kibbey sand reservoir. Below is the prolific Madison Group of limestone reservoirs, including the Charles, Mission Canyon, and Lodgepole. The Charles contains the Ratcliffe Reef Member (Hansen, 1966). Carbonate rock oil reservoirs also dominate the Devonian, Silurian, and Ordovician sections, including (proceeding downward) the Nisku, Dawson Bay, Winnipegosis, Interlake (Silurian), Stony Mountain, and Red River. The Red River correlates with the Big Horn Dolomite (Richards and Nieschmidt, 1961).

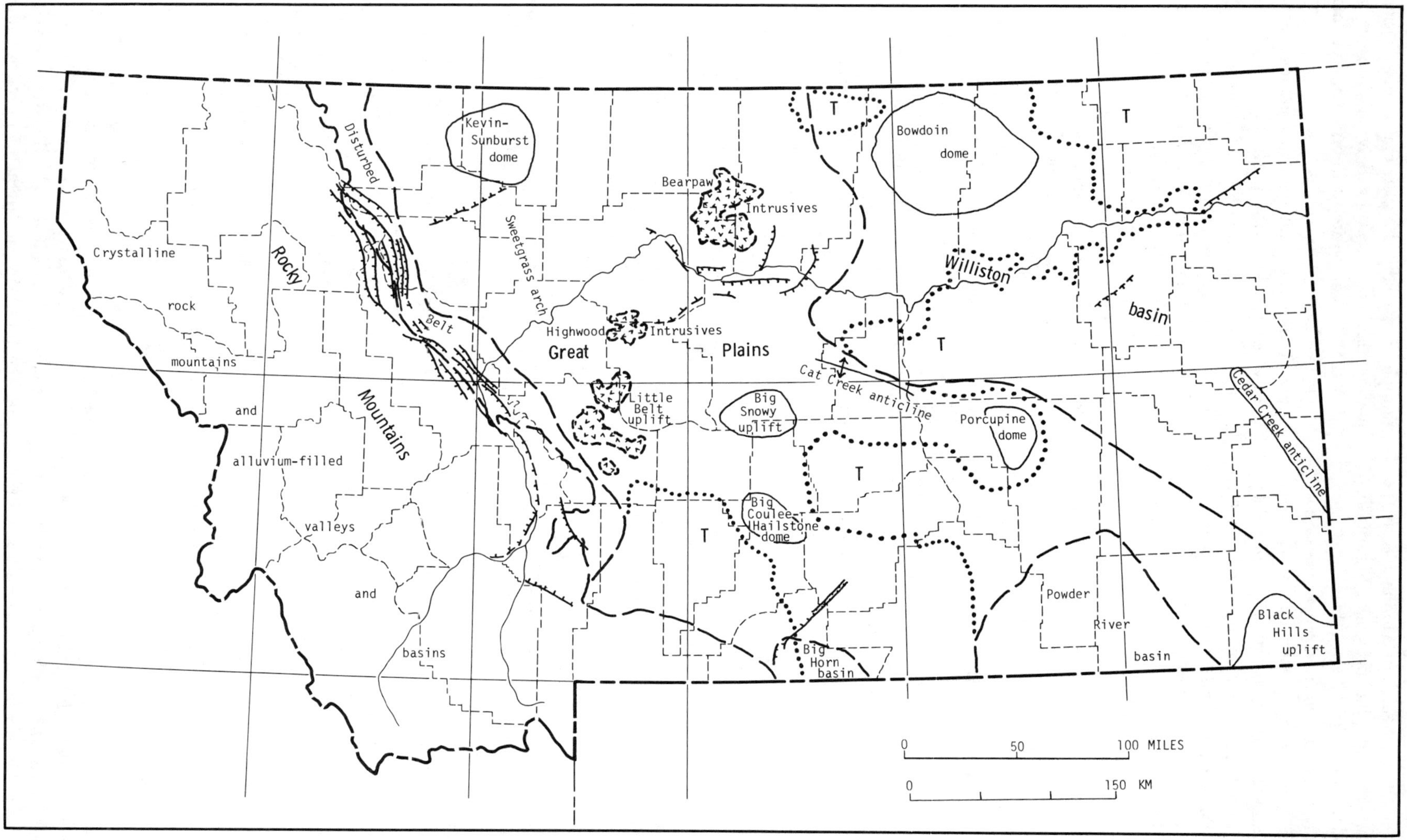

Figure 35.2 Montana. Tectonic and geologic map. T shows areas of Tertiary rock outcrops; dotted line is boundary between Tertiary and older formations.

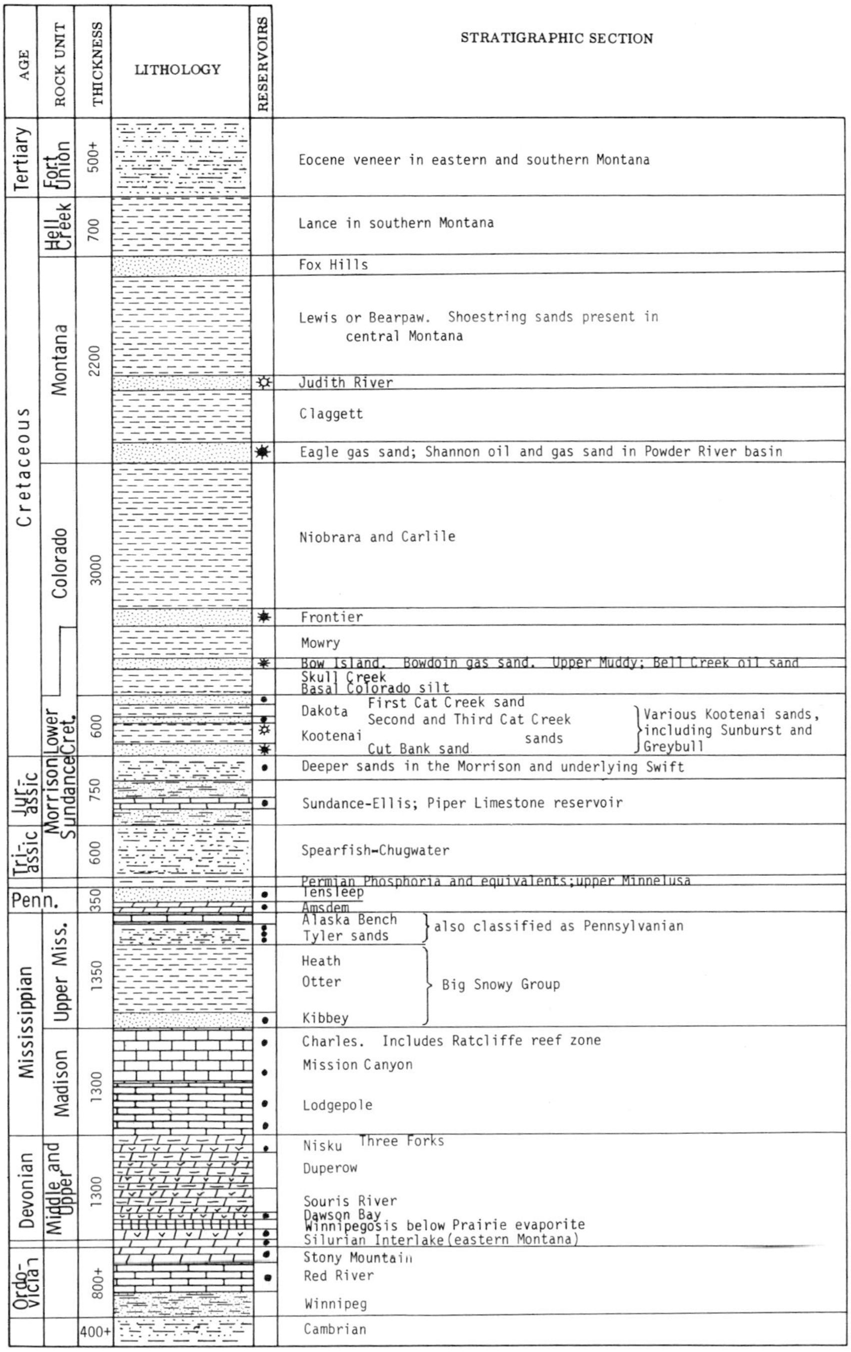

FIGURE 35.3 Montana. Generalized stratigraphic section, showing principal reservoirs. ***After Montana Oil and Gas Commission (1963), Eugene S. Perry (1953, Plate 4), Paul J. Lewis (1958), and Billings Geological Society (1958, chart between p. 232 and 233).***

To date, no pay zones have been discovered in Cambrian rocks in Montana.

FIELDS AND TRAPS. The oil and gas fields so far discovered in Montana fall naturally into three districts: northwest, eastern, and south-central. The northwest district is dominated by the Sweetgrass arch, and the eastern by the Williston basin. The south-central district is an uplifted complex against which abut the northward extensions of Wyoming basins. Some writers separate the south-central district into two parts, central and southern.

Most of the traps so far discovered in Montana are anticlinal. This is due in part to the fact that many of these upfolds are surface mappable; in fact, the larger ones can be seen on the geologic map of the United States.

However, it is also a fact that Montana's two giant fields, Cut Bank and Bell Creek, are permeability traps. The cumulative production of *Cut Bank* already exceeds 100 million barrels. This field, which is mainly in easternmost Glacier County, is still producing over 3.6 million barrels a year, an output that places it third among Montana's fields in oil yield. In addition it has produced large quantities of gas; 70 per cent of the 500 billion cubic feet of gas that has been obtained from the 24 gas fields on the Sweetgrass arch is credited to Cut Bank (Nordquist and Leskela, 1968).

The Cut Bank field lies 20 miles down the west flank of the Kevin-Sunburst dome, a node on the Sweetgrass arch. It is 31 miles long, extending to within 3 miles of the Canadian border, and has a maximum width of 10 miles. The proved acreage exceeds 100,000, of which more than half is gas field and the remainder oil and gas. The gas-only acreage is on the east (updip) side of the combined field. Cut Bank was discovered by random drilling; there is no structural indication of a trap either in the surface formations or in the subsurface. The principal producing formation is the Cut Bank sand, which lies at the base of the Kootenai (Lower Cretaceous). It is a sandstone ranging from fine grained to conglomeratic toward the base. The average pay thickness is 15 feet. Although the sand itself is continuous, it may be locally tight, a circumstance that explains the presence of about 200 dry holes within the field limits.

Other producing sandstone members of the Kootenai Formation occur above the Cut Bank sand (Cupps and Fry, 1967; Oakes, 1966). Trapping here is due to lenticularity. Recent exploration of these sands has widened the area of the Cut Bank field; one pool extends the field northwestward across the Canadian boundary (Billings Geol. Soc., 1966, p. 186–190).

The Cut Bank field also has a deeper reservoir, the Charles Formation of the upper Madison (Mississippian) Limestone. Oil and gas occur in porous zones at and near the eroded surface of the Charles Limestone.

Figure 35.4 is a structure contour map of the subsurface in the area covered by producing wells (Chickering, 1958). It is based on the top of the Ellis Shale, which immediately underlies the Cut Bank sand. The regional dip is about 100 feet per mile to the west and southwest. Various situations have created the trap, including especially pinchout of the sand body and lensing out of the permeability. It has been suggested (Cobban, 1958) that the sand was originally deposited in a shallow erosional valley cut into the crest of a gentle anticline with a north-south trend. As time went on, the velocity of the sand-carrying stream lessened, so that finer, less permeable sediment was deposited in the upper part of the

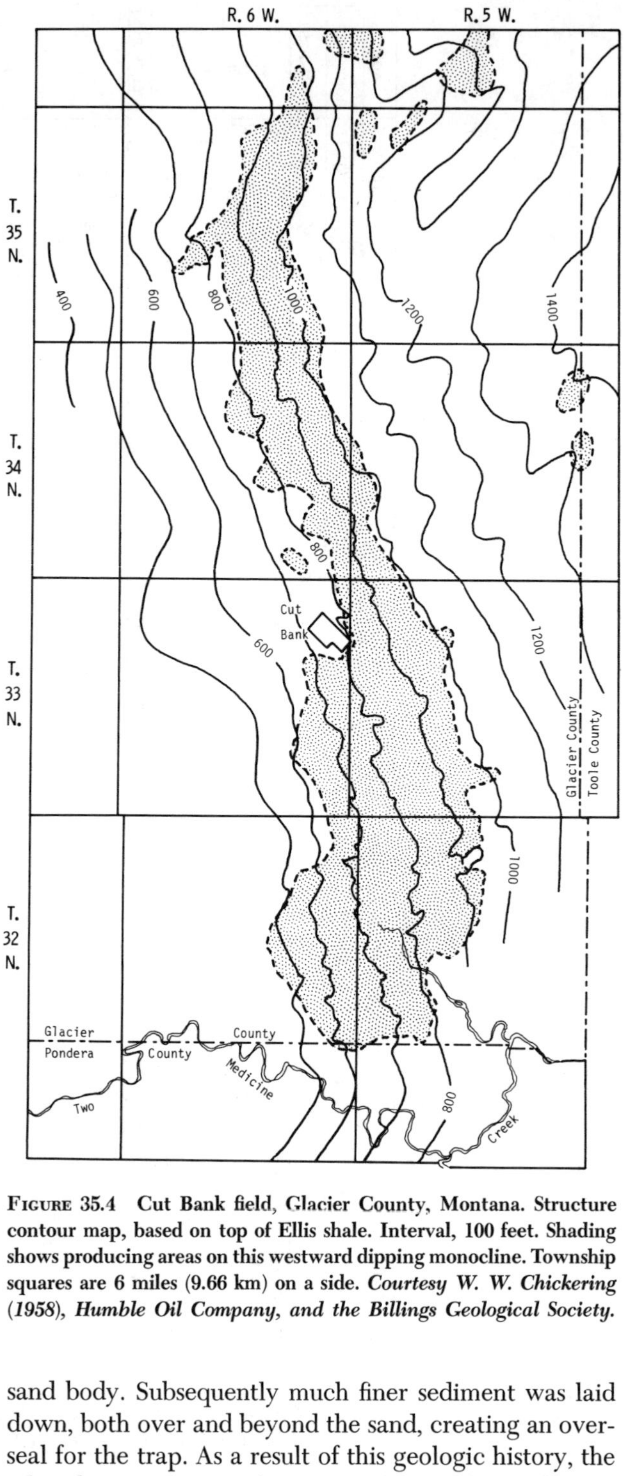

FIGURE 35.4 Cut Bank field, Glacier County, Montana. Structure contour map, based on top of Ellis shale. Interval, 100 feet. Shading shows producing areas on this westward dipping monocline. Township squares are 6 miles (9.66 km) on a side. *Courtesy W. W. Chickering (1958), Humble Oil Company, and the Billings Geological Society.*

sand body. Subsequently much finer sediment was laid down, both over and beyond the sand, creating an overseal for the trap. As a result of this geologic history, the oil and gas are trapped in permeable sand sealed in by tighter, older rocks, including the eroded face of a cuesta on the east side of the field, and by increasing impermeability upward and to the north and south. Downdip to the west, water lies beneath the oil where permeability permits.

Two other fields of the group occupying various parts of the Sweetgrass arch in northwestern Montana are discussed here. The first is an anticlinal accumulation; trapping in the second is due to updip sand wedgout. The *Kevin–Sunburst* field in Toole County in northern Montana is in reality two fields: a gas field that covers about 60,000 acres over the crest and the south flank of the Kevin–Sunburst dome, and the state's second largest oil field (in cumulative production), which occupies a considerable area on the north flank of the dome. The gas field produces from the Sunburst sand, which lies near the base of the Kootenai (Lower Cretaceous) Formation (Bartram and Erdmann, 1935, p. 269–270). The principal reservoir in the oil field is the leached top of the Madison (Mississippian) Limestone, and the irregular distribution of oil is due to erratic porosity within this reservoir. The depth to the oil pay is about 1800 feet, and the producing zone is only 5 to 10 feet in thickness.

The Kevin–Sunburst dome is a large node on the north-plunging Sweetgrass arch. Exploration on this large, readily visible structure was late in coming in spite of the fact that natural gas had been noted in water wells as early as 1912. The discovery well for the field was drilled in 1922. The trapping in both the gas and the oil fields is anticlinal with, however, local permeability controlling distribution in some places.

On the north flank of the Kevin–Sunburst dome is the *Border–Red Coulee* oil field, where trapping is due to the updip lensing out of Kootenai sands. The fact that this field straddles the international boundary between the United States and Canada had led to an unusual development history. A strip along the border about 525 feet wide has had to remain undrilled (Erdmann and Schwabrow, 1941, p. 270).

An Upper Cretaceous gas sand, the Eagle, is productive in the eastern part of the northwest district. An example is *Tiger Ridge* field (Figure 35.1), which straddles the line between Blaine and Hill Counties immediately to the north of the Bearpaw Mountain intrusive complex (Figure 35.2). This exceptionally shallow field, with the Eagle reservoir at an average depth of 1400 feet, was still being developed in 1968.

As mentioned previously, eastern Montana is occupied by the northeastward dipping flank of the Williston basin. Two major reversals occur in the regional dip in this area; one is the Bowdoin dome (Figure 35.2), and the other is the Cedar Creek anticline. In addition there are more local flexures.

The *Bowdoin gas field* (Gwynn, 1958) lies high on the Williston basin flank in eastern Phillips and western Valley Counties. It is a large field, occupying approximately 150,000 acres (more than 230 square miles). The gas has accumulated in the crest of the Bowdoin dome, which is a circular anticline with a closed area in the neighborhood of 2000 square miles. The dome is mappable at the surface; the nearly circular outcrop of the Claggett Shale in the eroded core of the dome is completely encircled by outcrops of the next younger formation, the Judith River Sandstone (both of Late Cretaceous age). The dome has a closure of at least 700 feet, but because of the great area involved dips are measurable in feet per mile rather than in degrees.

It has been suggested (Bartram and Erdmann, 1935, p. 260) that the Bowdoin dome is of deep-seated magmatic origin, because of its size and shape, alignment with known igneous features, high geothermal gradient, the presence of basaltic rock on the southeast flank of the dome, and a positive-gravity anomaly. The high geothermal gradient may explain the absence of any oil in the many test wells drilled.

The Bowdoin gas field was discovered in 1913 by a well drilled for the purpose of finding water. This well struck noncommercial gas in a sandy zone in the upper part of the Colorado Shale (Cretaceous). Subsequently commercial gas was discovered in the slightly deeper Bowdoin gas zone, which is the most persistent gas reservoir in the field. The average depth is only about 740 feet. About 250 feet deeper is the Phillips sand, which is a less extensive gas reservoir. Both the Bowdoin and the Phillips sands are in the upper part of the Colorado Shale. Both sands are fine grained, but nonetheless permeable where present. A few wells have failed to find either sand, in which case the entire section is shale or siltstone.

An unusual type of trap origin has been postulated for a group of fields in western Roosevelt County, northeastern Montana (Swenson, 1967). The traps are steep-sided flat-topped anticlines in Nisku (Upper Devonian) Dolomite. It is suggested that these structures were formed by a two-stage solution of rock salt in the deeper Prairie Formation (Middle Devonian).

The *East Poplar* field in south-central Roosevelt County is also in the Williston basin about 60 miles north of the northernmost tip of the Cedar Creek anticline. The field itself lies on the northeast flank of a southeasterly plunging regional nose known as the Poplar anticline (Powell, 1955, p. 75). The trap at East Poplar is an elliptical north-south upfold (Figure 35.5). The anticline is flat topped with two broad nodes. It is asymmetrical; the dip to the northeast is considerably greater than the dip in other directions.

The principal reservoirs are four porosity zones in the Charles Formation of the Madison Group (Mississippian). These four porous units occupy a 200 foot interval in the 5200 to 5800 foot depth range. The Charles Formation in this area consists of interbedded limestone, anhydrite, dolomite, salt, and shale. The porosity zones are in limestone beds, where the porosity is due to the presence of fractures and intercrystalline voids, plus some vugs. In addition to the Charles reservoirs at least three

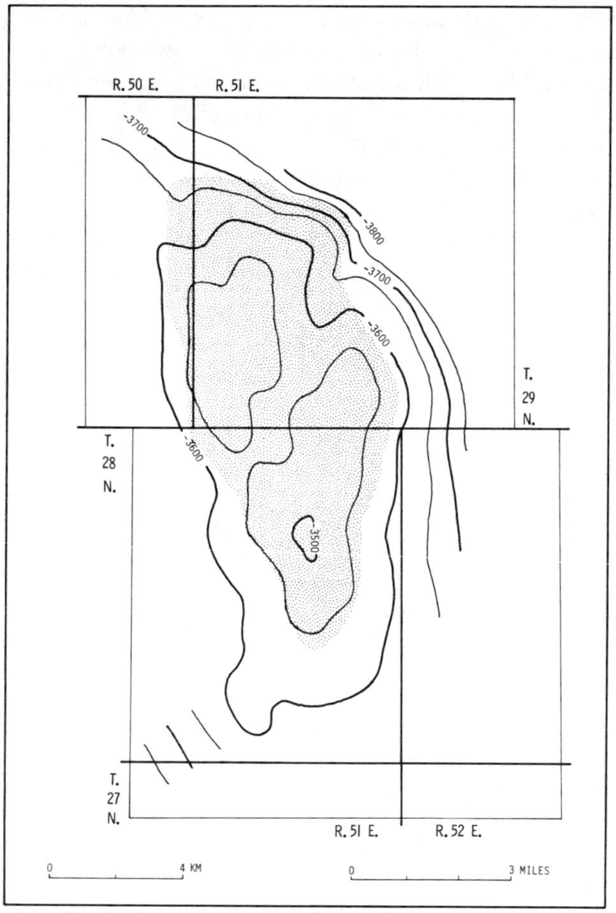

FIGURE 35.5 East Poplar field, Roosevelt County, Montana. Structure contour map based on the second zone of Madison (Mississippian) porosity. Interval, 50 feet. Productive area is shaded. ***Courtesy J. B. Powell, Jr., (1955, Figure 11, p. 86) and the American Association of Petroleum Geologists, Rocky Mountain Section.***

gas wells have been completed in the Judith River Sandstone (Upper Cretaceous) at a depth of about 850 feet.

Two interesting features of the trap at East Poplar are (1) the absence of salt across the anticline, although surrounding wells drilled through from 100 to 280 feet of salt in the upper Charles above the first porosity zone, and (2) the northward tilt of the oil-water interface of approximately 28 feet to the mile for a total relief from south to north of about 200 feet. The salt either was not deposited across the anticline or was deposited and subsequently eroded. The northward tilt of the oil-water interface has been attributed to hydrodynamic conditions (Murray, 1959).

The Cedar Creek anticline, also known as the Baker–Glendive anticline, is a remarkably straight and sharp upfold more than 100 miles long (Dobbin, 1947, p. 821–823) that creases the Williston basin flank in southeastern Montana. Its southeasterly strike carries it beyond the Montana border across the southwestern corner of North Dakota and into the northwestern corner of South Dakota. The fold is asymmetrical with the steeper side to the southwest. The anticlinal axis rises and falls along its considerable length, creating a series of nodes and antinodes. Gas was found in a node near the northwestern end in 1913 and has been produced continuously from the main *Cedar Creek* gas field to the southeast since 1915. This field is 57 miles long in Montana alone, and it extends far enough down the anticlinal flanks to occupy antinodes as well as nodes. Production, which after 50 years is still continuing at nearly a billion cubic feet annually, comes from Upper Cretaceous Judith River and Eagle sands at depths in the 2500 to 3000 foot range. Nitrogen occurs in natural gas in various Paleozoic formations, but especially in a Cambrian sandstone where an exploratory well recorded 6 million cubic feet of gas daily containing 90 per cent pure nitrogen from 10,100 feet (McCrae and Swenson, 1968). The nitrogen "probably is generated in the basement rocks" (p. 1303). Cedar Creek has been the only significant producer of non-associated gas in the western Williston basin.

Oil in quantity was first discovered on the Cedar Creek anticline in 1952, first at Glendive and then at Pine; now more than a dozen oil fields produce from nodes along the anticlinal axis (Shell Oil Co., 1958b). The pay zones are in the Mississippian-Silurian-Ordovician section at three to four times the depth of the gas-producing rocks. The permeability of the carbonate rock oil reservoirs has been increased by fracturing. It has been noted that there is a general improvement in the quality of the pay zones from south to north along the axis (Hols and Bethel, 1957). The Cedar anticline fields are now producing the greater part of Montana's oil, with Pine in first place and Cabin Creek (immediately to the southeast) in second place in 1965.

The *Pine* field (also referred to as the Pine Unit field) is described here as a typical example of oil trapping on the Cedar Creek anticline. It extends across parts of Fallon, Wibaux, and Prairie Counties. In addition to being first in annual yield, Pine ranks third in cumulative production. Figure 35.6 (Shell Oil Co., 1958c) is a structure contour map of the Pine field based on the Ordovician Red River reservoir, which is the deepest of the three principal reservoirs present. The oil occupies the higher parts of an elliptical anticline trending north-northwest which is nearly 3 miles in length. The two highest nodes along the axis of this upfold are in the middle and near the southeastern end. Both the nodes and the intervening saddle lie above the oil-water interface; hence production is continuous in that part of the anticline. A lower node at the north end contains at least one producing well. The Pine anticline, like the rest of

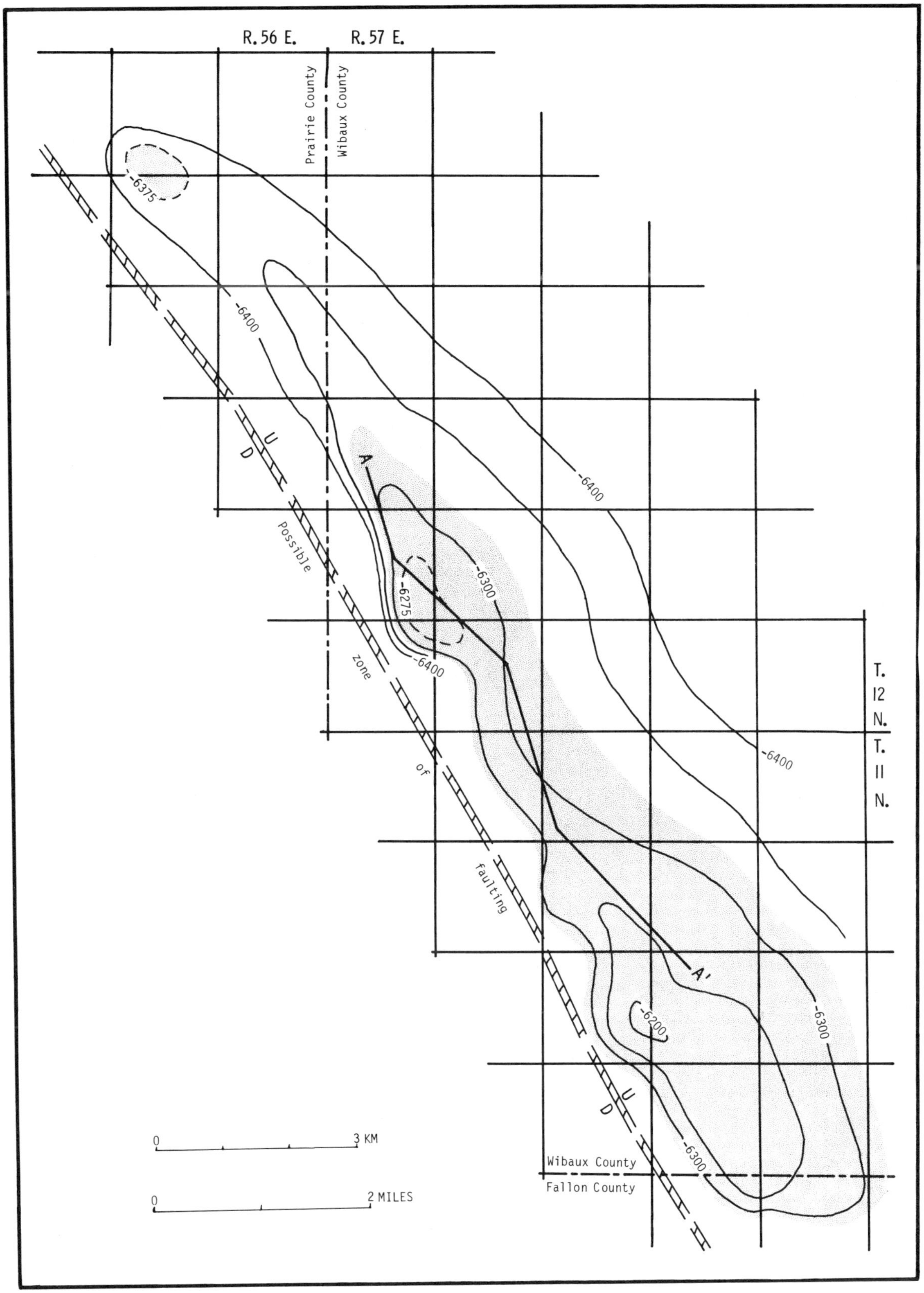

FIGURE 35.6 Pine field, mainly in Wibaux County, Montana. Structure contour map, based on Red River (Ordovician) reservoir. Interval, 50 feet (25 feet on central and north nodes). Producing areas shaded. Map also shows location of longitudinal section *A-A′* (Figure 35.7). *Courtesy Shell Oil Company (1958) and the Billings Geological Society.*

the Cedar Creek anticline, is asymmetrical with the steep slope on the southwest. The sudden dropoff on this side has led to the belief that the southwest flank may be faulted.

Figure 35.7 (Clement, 1955) is a longitudinal section along the axis of the Pine anticline. It shows the two major nodes, and the trapping of oil within the three principal reservoirs. These reservoirs are the Interlake Formation of the Upper Silurian, the Stony Mountain Formation of uppermost Ordovician, and the still deeper Ordovician Red River Formation. The upper two reservoirs are dolomite; the deepest is mainly limestone. The most productive reservoir is the Red River, with a pay zone about 225 feet in thickness. The Interlake pay has a maximum thickness of 140 feet; the intermediate Stony Mountain reservoir, of 80 feet. Trapping is strictly anticlinal, but productivity varies from well to well, depending on local differences in permeability within the three reservoirs.

A common type of trapping in the south-central Montana district is illustrated by the *Northwest Sumatra* oil field, which is on the central Montana uplift in northwestern Rosebud County (Llewellyn, 1955). It was discovered in mid-1952 after seismograph surveys. The principal reservoirs are two sands (Tyler B and Tyler C), which lie within the Tyler Shale Formation of Mississippian age. Minor production has been obtained from a higher Tyler sand (Tyler A), and from the Amsden Dolomite (Mississippian-Pennsylvanian) and the Piper Limestone (Lower Jurassic). Both the Tyler B and the Tyler C sands are coarse; the B sand has a maximum thickness of 107 feet, and the C reaches a maximum of 72 feet. These sands are productive over an area in excess of 2000 acres.

The structure map (Figure 35.8) shows that the field is largely confined to the highest part of an east-west–trending anticline with a steep northerly dip. However, the producing sands in the lower Tyler are quite lenticular, as can be seen from the cross section (Figure 35.9). It has been suggested (Foster, 1958, p. 217) that these sands were deposited in a channel cut into the Heath Shale, which underlies the Tyler. This trap appears to be another example of channel cutting and filling along the axis of an anticline.

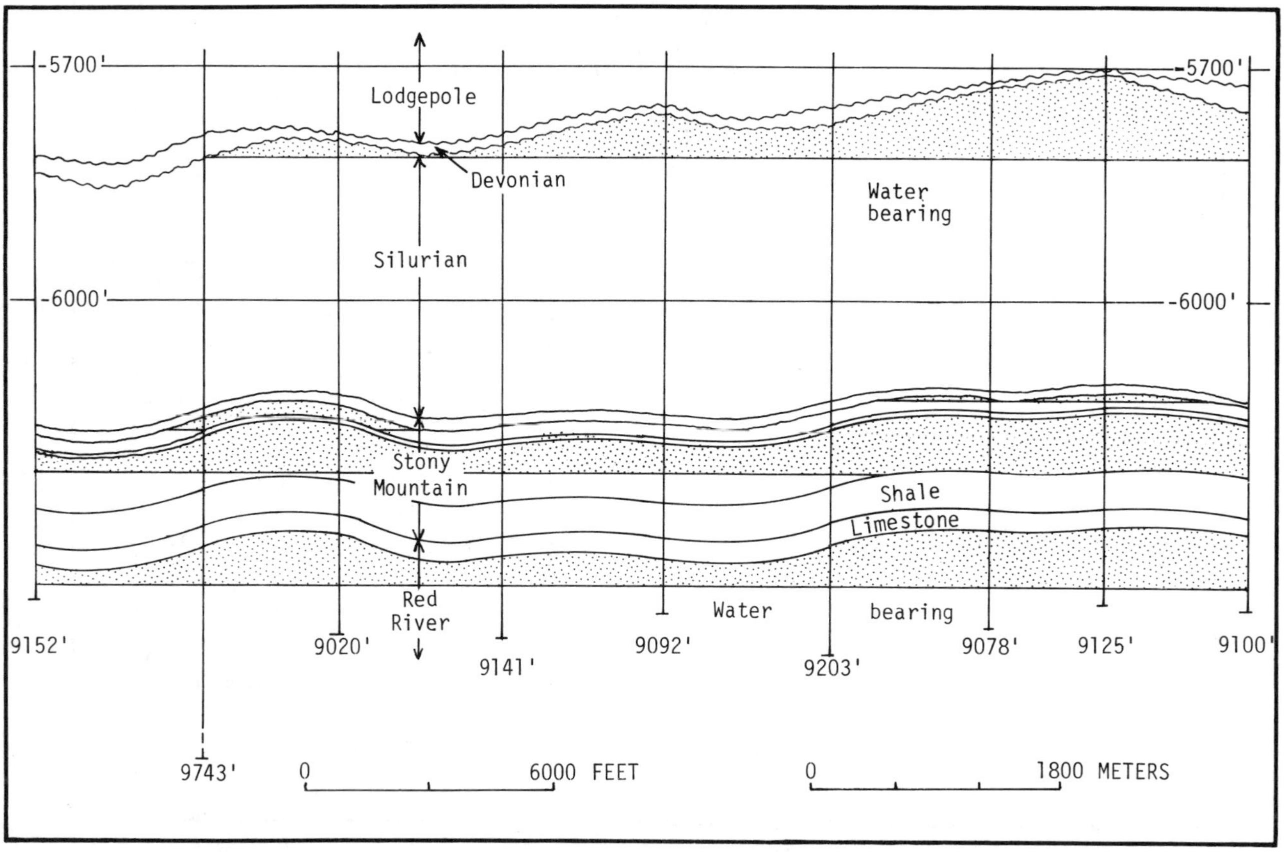

FIGURE 35.7 Pine field. Longitudinal section, along line A-A′ (Figure 35.6). Oil-bearing zones are stippled. ***Courtesy James H. Clement (1955, Figure 5, p. 170)*** **and the American Association of** ***Petroleum Geologists, Rocky Mountain Section.***

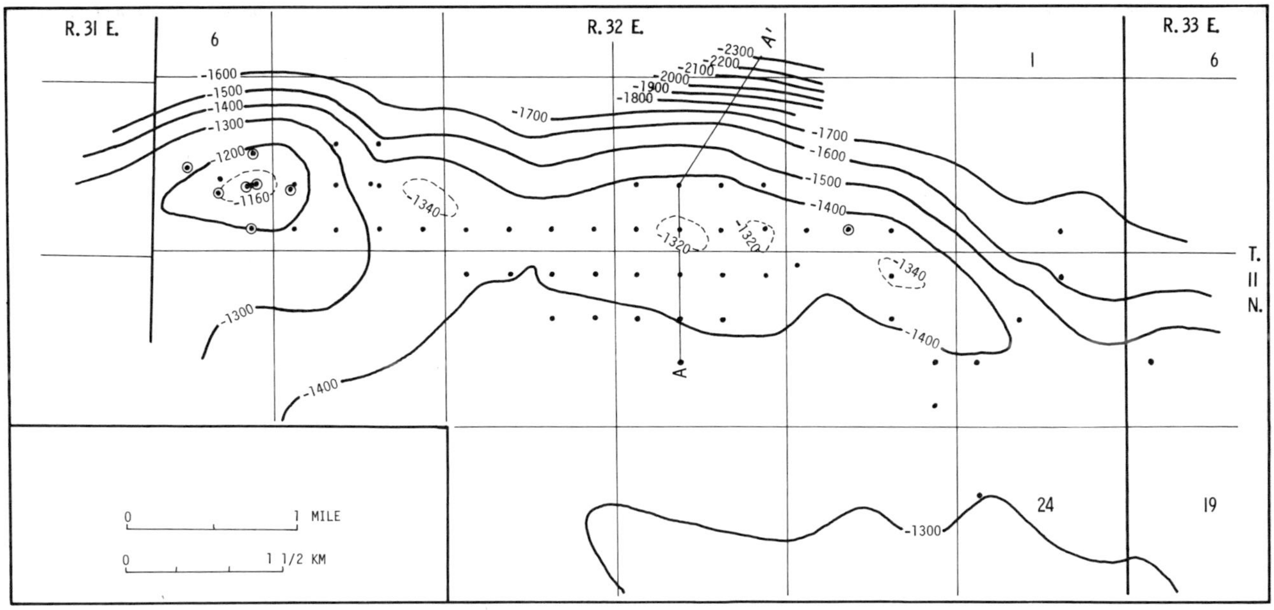

Figure 35.8 Northwest Sumatra field, Rosebud County, Montana. Structure contour map based on top of Tyler Formation. Interval, 100 feet, except on nodes, where highest 20 foot contour is shown (dashed). Solid circles are wells producing from Tyler A, B, and C reservoirs. Circled wells are Amsden producers. Location of cross section A–A′ (Figure 35.9) also shown. *Courtesy Donald I. Foster (1958, p. 216 tip-in) and the Billings Geological Society.*

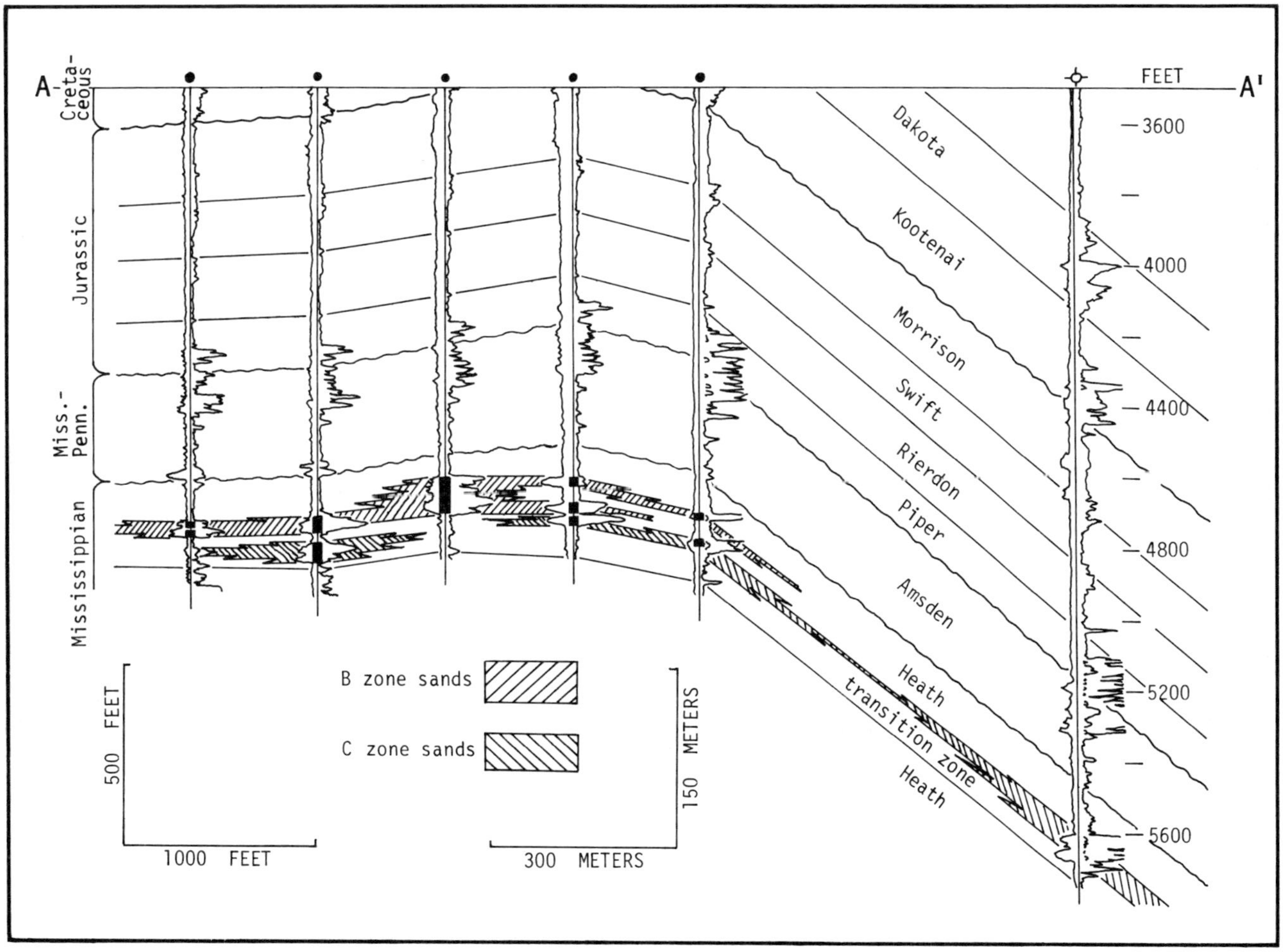

Figure 35.9 Northwest Sumatra field. Cross section, location shown by A–A′ in Figure 35.8. Solid black in borehole sections designates producing zones. *Courtesy J. Thomas Llewellyn (1955, Figure 4, p. 61) and the American Association of Petroleum Geologists, Rocky Mountain Section.*

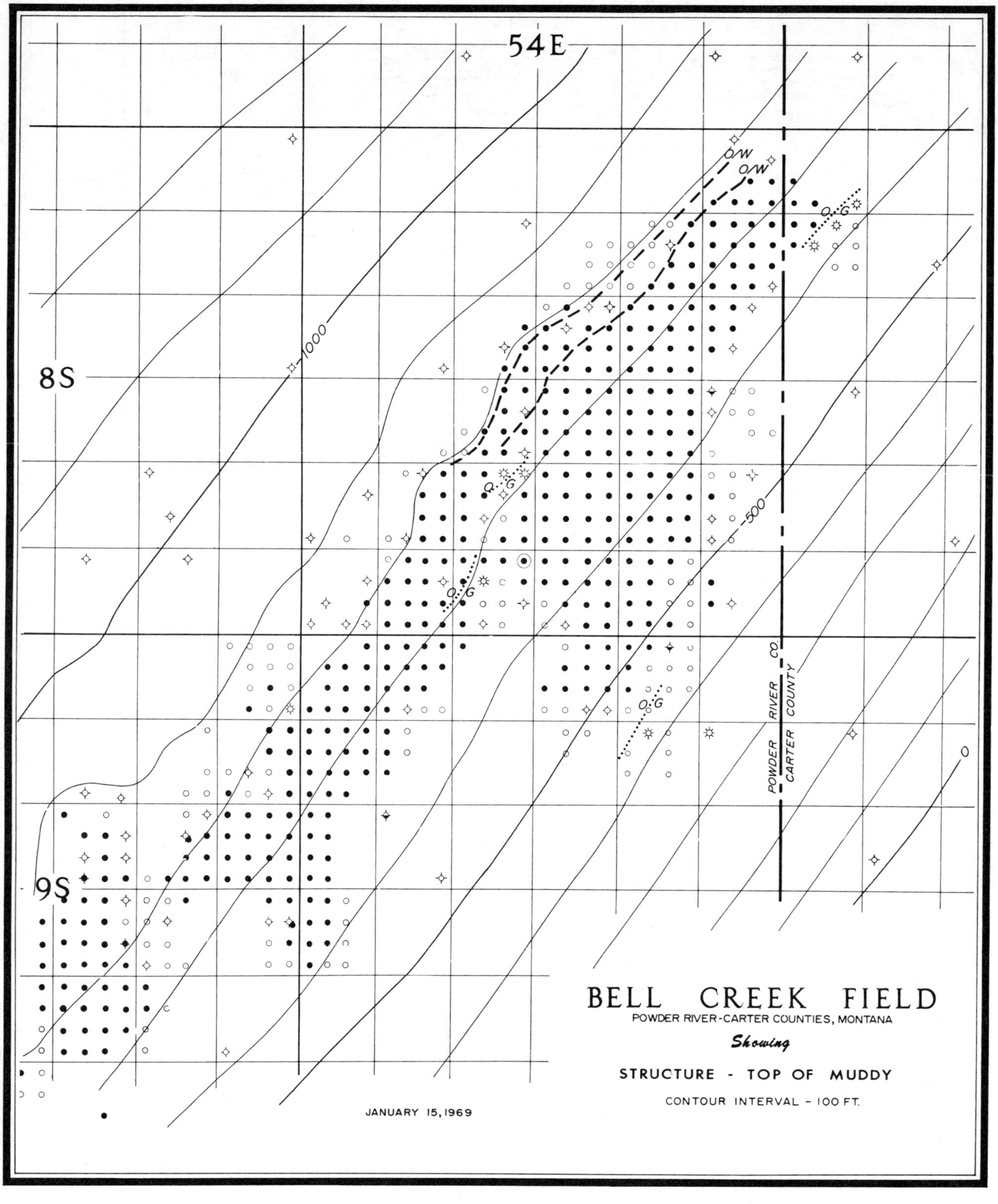

Figure 35.10 O-W = oil-water; O-G = oil-gas. *Courtesy A. A. McGregor, Charles A. Biggs, and Samuel Gary.*

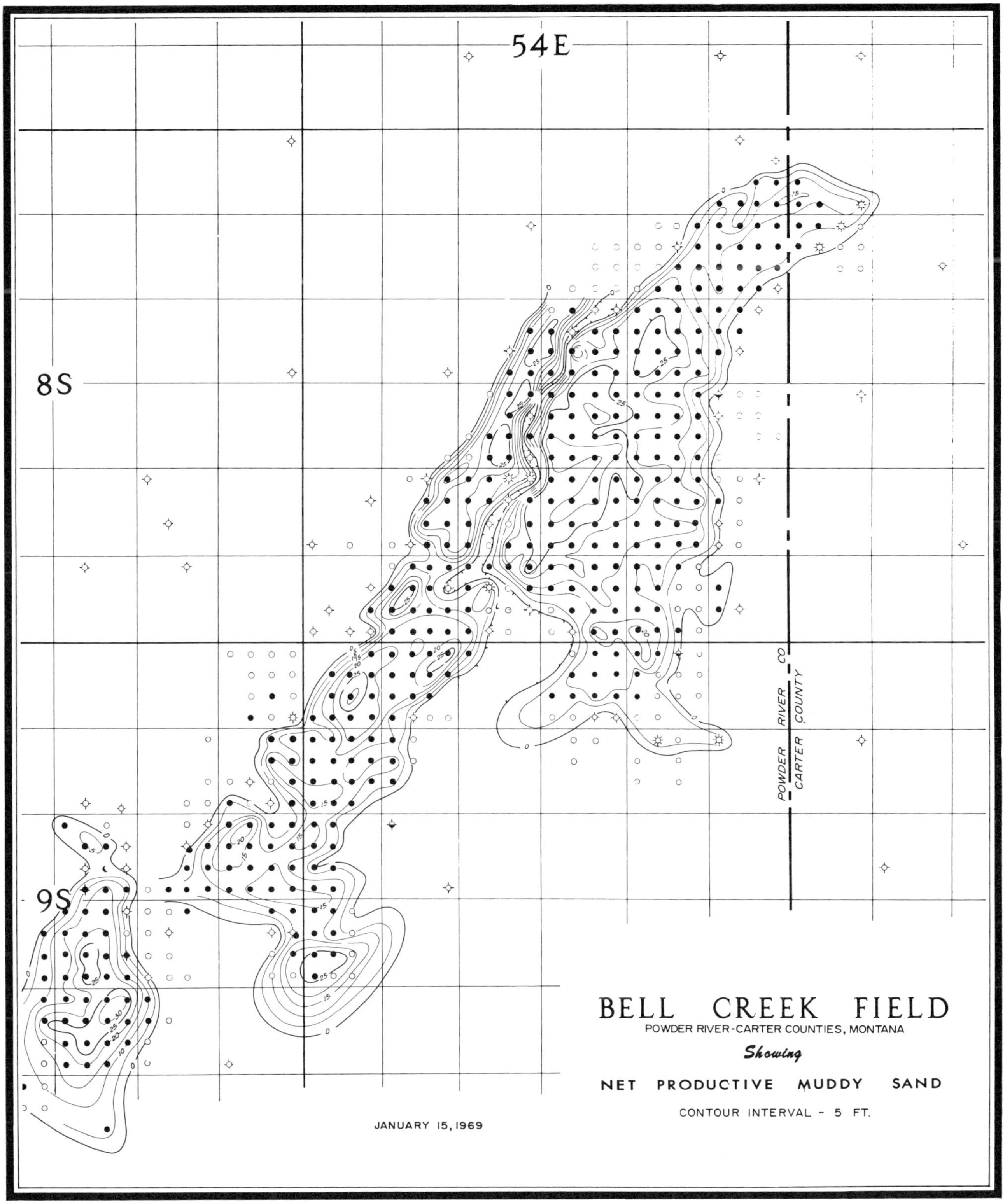

FIGURE 35.11 *Courtesy A. A. McGregor, Charles A. Biggs, and Samuel Gary.*

The older *Cat Creek* oil field (Dobbin, 1947, p. 806-807) to the northwest illustrates trapping in the higher nodes of a prominent anticline with an east-west trend. The principal reservoirs are Lower Cretaceous (Kootenai) and Upper Jurassic sands. Many transverse faults cut the anticline at the surface but appear to die out before reaching reservoir depth (Hadley, 1956).

The *Wolf Springs* oil field (Fanshawe, 1960) is close to the northern tip of Yellowstone County about 55 miles northeast of Billings. It was discovered in 1955 after both gravity and seismic surveys. The reservoir is a porous section at or near the top of the Amsden Dolomite of Pennsylvanian age. The trap is an anticlinal ridge having a northwest trend and containing three nodes. The saddle between the northernmost and the middle nodes is deep enough to cut through the oil-water interface separating the production into two areas. The middle and southern nodes are productive across the shallow intervening saddle. In both of the producing areas the oil-water interface is considerably lower on the east and northeastern flanks than on the southwestern flank. This has been ascribed both to potentiometric slope and, more recently, to the folding of a pre-existing stratigraphic trap (Fanshawe, 1960, p. 57).

In late 1968 *Bell Creek* (McGregor and Biggs, 1968; Biggs and McGregor, 1968, Gary and McGregor, 1968), one county west of the southeastern corner of Montana, was the wonder field of the Rocky Mountain province. Within 13 months of its discovery on June 6, 1967, it had more than 315 oil and gas wells and had become first in Division V in daily production, displacing the Elk Basin field. Eventual recovery is estimated to exceed 200 million barrels.

The Bell Creek field is on the Montana extension of the Powder River basin. It is near the southeastern corner of Powder River County, overflowing slightly into southwestern Carter County. The field is over 16 miles long, in a north-northeast direction, and has a maximum width of 5 miles. The pay zones are upper Muddy (Lower Cretaceous) sandstone lenses. The average thickness of the Muddy sand is 20 feet. Porosities may be as high as 33 per cent, and permeabilities as great as 13,500 md have been recorded (Berg and Davies, 1968). Reservoir depths are around 4500 feet. The productivity of the wells ranges from 100 to 1000 barrels per day.

Bell Creek lies high on the northeastern flank of Powder River basin. The regional dip is west-northwest at 100 feet per mile. The field parallels the regional strike. Trapping is brought about by updip facies change from sandstone to shale. It has been suggested that the Muddy sand reservoir at Bell Creek was deposited as a barrier bar, and that the updip permeability barrier may be lagoonal siltstone and shale (Berg and Davies, 1968). The sand itself is fine to very fine grained. At least four individual reservoirs have been delineated by different gas-oil and oil-water contact elevations. These sand bodies are both underlain and overlain by organic rich marine shale (McGregor and Biggs, 1968), providing an ideal source rock-reservoir rock relationship.

Figure 35.10, is a structure map of the Bell Creek area, showing the monoclinal dip and the absence of any reversals in it. Although there is no trapping by structure, the regional dip toward the axis of the Powder River basin supplies both a large gathering area and an adequate gradient into the stratigraphic trap. Two gas-oil contacts are shown on the east side of the field; the principal oil-water contact makes the field's western edge across T. 8 S., R. 54 E.

The opposite illustration (Figure 35.11) is an isopachous map of the gross Muddy thickness. The updip lensing out of the permeable pay zones is evident on the east side. The variable distribution of adequate thicknesses for commercial accumulation explains the pinches and the bulges in the shape of the field.

Although the Montana sector of the *Elk Basin* oil field is still an important source of oil, the production statistics are merged with the much larger output on the Wyoming side of the line. This field is described in Chapter 34.

FUTURE. Obviously, with more than 100,000 square miles of prospective hydrocarbon-producing territory, Montana still has a large area to explore. In addition, the vertical dimension has not been tested adequately. Some of the Paleozoic reservoirs in eastern Montana have not been explored in the south-central and the northwestern parts of the state. Few attempts have been made to find oil in the Cambrian rocks, in spite of their productivity in parts of Wyoming to the south.

Perhaps the most glamorous, expensive, and frustrating search has taken place in the Disturbed Belt (Langstroth, 1964). The motivation is to find deposits similar to those across the boundary in Alberta. Occurring in the Disturbed Belt on the Canadian side of the line are such prolific fields as Pincher Creek, Savanna Creek, and Turner Valley. Gas predominates in the Disturbed Belt, because of the high degree of eometamorphism of the rocks and contained hydrocarbons. So far, exploration on the Montana side has not been successful except for one gas well drilled in Teton County in 1958. This well was completed at a depth of 3800 feet with a rated initial production of 6.3 million cubic feet. The reservoir rock is Mississippian in age. Because of overthrust faulting the Mississippian section was penetrated three times during the drilling of this wildcat; it was completed as a commercial gasser in the second plate (Langstroth, 1964, p. 146). Currently, the well is shut in.

The wide top and flanks of the Sweetgrass arch continue to beckon the explorer. Anticlinal traps, lensing sands, and a Mesozoic-Paleozoic weathered angular unconformity are all present (Branch, 1966; Gribi, 1956). Profitable shallow oil in the Sunburst (Lower Cretaceous) sand was discovered in Toole County in 1963. This reservoir is a channel sandstone, highly erratic in its distribution (Runge et al., 1965).

The relatively new, but already highly productive, Lower Cretaceous sandstone oil fields in eastern Montana are indicative of what can be found on this flank of the Williston basin. The carbonate rock reservoirs in the Mississippian to Ordovician section are very much worth seeking. Certainly more reef production from the Ratcliffe Member of the Charles (Mississippian) Limestone will be found in this district (Hansen, 1966).

In south-central Montana there are both surface anticlines and subsurface wedgouts (Norwood, 1965) yet to be explored. In eastern Judith Basin County two large and several smaller anticlines on the north flank of the Little Belt Mountains have not been completely explored (Vine and Hail, 1950). The same is true of anticlines in Fergus County, the next county east (Gardner, 1959). Shoestring sands observed in the lower part of the Bearpaw Shale (Upper Cretaceous) at the outcrop in Stillwater County could be oil bearing at shallow depths to the southwest (Wulf, 1964c).

REFERENCES CONSULTED IN PREPARING CHAPTER 35

American Association of Petroleum Geologists, Rocky Mountain Section, 1955, 1959, 1960.

Anon., 1966.

Babisak, Julius, 1958.

Bartram, John G., and Erdmann, Charles E., 1935.

Beekly, Emerson K., 1956.

Berg, Robert R., and Davies, David K., 1968.

Biggs, Charles A., and McGregor, Alexander A., 1968.

Billings Geological Society (Symposium), 1958, 1966.

Blair, Arden F., 1960.

Blixt, John E., 1941.

Bolyard, Dudley W., and McGregor, Alexander A., 1966.

Branch, J. L., 1966.

Chickering, W. W., 1958.

Clement, James H., 1955.

Cobban, W. A., 1958.

Cupps, C. Q., and Fry, J., 1967.

Davis, W. E., and Hunt, R. E., 1956.

Dinkins, Donald E., 1960.

Dobbin, C. E., 1947.

Dobbin, C. E., and Erdmann, C. E., 1955.

Erdmann, Charles E., 1963.

Erdmann, Charles E., and Schwabrow, John R., 1941.

Fanshawe, John R., 1960.

Foster, Donald I., 1958.

Gardner, Louis S., 1959.

Gary, Samuel, and McGregor, A. A., 1968.

Gribi, Edward A., Jr., 1956.

Gwynn, Thomas A., 1958.

Hadley, Herbert D., 1956.

Hamke, J. R., Marchant, L. C., and Cupps, C. Q., 1966.

Hansen, Alan R., 1966.

Hols, Arnold, and Bethel, Frank T., 1957.

Howell, W. F., 1929.

Hunt, Robert E., 1956.

Hurley, G. William, 1959.

Johnson, Melvin C., 1958.

Krantzler, Irvin, 1966.

Langstroth, Ward T., 1964.

Lewis, Paul J., 1958.

Llewellyn, J. Thomas, 1955.

McCoy, Alex W., III, et al., 1951.

McCrae, R. O., and Swenson, R. E., 1968.

McGregor, Alexander A., and Biggs, Charles A., 1968.

Murray, George H., Jr., 1959.

Nordquist, J. W., and Leskela, Willard, 1968.

North Dakota Geological Society, Saskatchewan Geological Society, 1956, 1958.

Norwood, E. Earl, 1965.

Oakes, Millis H., 1966.

Olson, Charles F., Curry, William H., III, Norwood, E. Earl, and Slocum, R. C., 1968.

Perry, Eugene S., 1953.

Powell, J. B., Jr., 1955.

Richards, P. W., and Nieschmidt, C. L., 1961.

Runge, John S., Copen, James D., Tutten, William D., and Harris, Steven H., 1965.

Sahinen, U. M., Erdmann, C. E., Weissenborn, A. E., and Weiss, P. L., 1963.

Sandberg, Dorothy T., 1959.

Shell Oil Company (Billings Division Staff), 1958a, 1958b, 1958c.

Smith, G. Wendell, Summers, George E., Jr., Wallington, Dale, and Lee, Jean L., 1958.

Smith, Howard R., 1962.

Swenson, Robert E., 1967.

Towse, Donald, 1957.

Vine, James D., and Hail, William J., Jr., 1950.

Wulf, George R., 1964c.

Wyoming Geological Association, 1958.

Chapter

36

UTAH

Utah ranked fourteenth in the United States in both oil and gas production in 1967. The oil yield was over 23 million barrels, which constituted about 0.75 per cent of the country's total. More than 43 per cent of this oil came from one field, the Aneth, in southeastern Utah.

The major sedimentary basins, and almost all of the oil and gas fields, are in eastern Utah. Nonetheless the first discoveries, gas in 1891 and oil in 1907, were made in the western half of the state. The first gas was found by accident during the drilling of a water well near *Farmington Bay* (Figure 36.1) on Great Salt Lake, in Davis County north of Salt Lake City (Heylmun, 1961a). This gas was delivered to Salt Lake City by a wooden pipeline during 1895 and 1896 (Gere et al., 1964, p. 55). Dwindling reserves forced the abandonment of this short-lived field. The reservoir rock is the fresh water Pleistocene sediment with which the Salt Lake basin is floored.

The initial oil discovery, the *Virgin* field, is in a tributary valley to the Virgin River about 3 miles west of Zion National Park (established in 1919) in Washington County, southwestern Utah. Discovery was due to the presence of oil seeps. This field has produced intermittently ever since (Heylmun, 1961b; Bahr, 1963), with an average daily production estimated to be about 10 barrels a day. The reservoir is the Timpoweap Limestone Member at the base of the Triassic Moenkopi Formation. Trapping is apparently due to a lensing out of porosity updip on a broad terraced nose.

Oil seeps also led to the discovery of the second oil field in Utah, *Mexican Hat*, in 1908. The discovery was made at a depth of 225 feet in lenticular Pennsylvanian sandstones on the flank of a syncline (Kunkel, 1961). The location is in San Juan County in southeastern Utah. This field was abandoned in 1956 after producing a minuscule 18,000 barrels of oil.

No oil fields were found in Utah in the next 40 years, during which production dwindled to nothing. However,

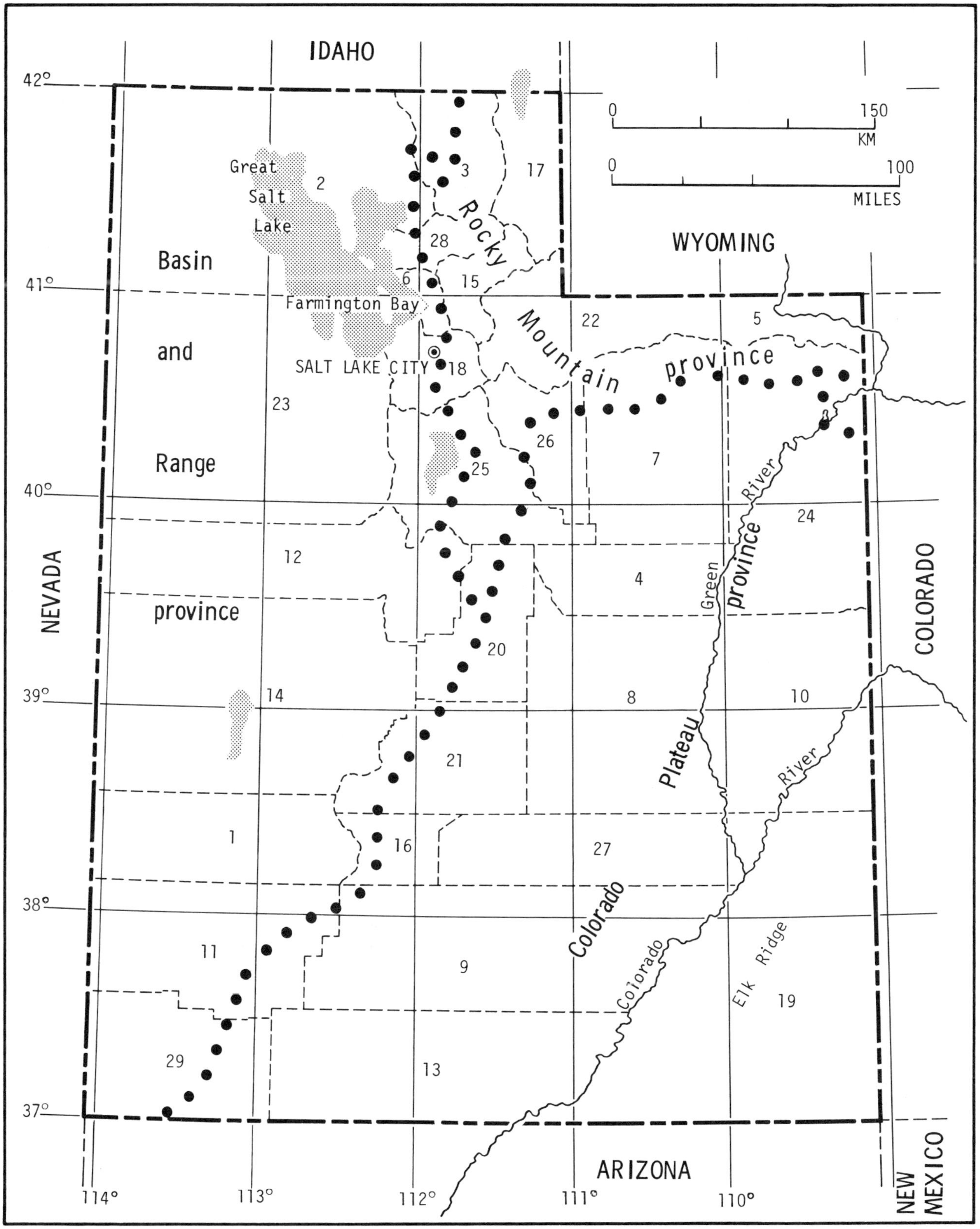

FIGURE 36.1 Utah. Index map. Dotted lines separate three major physiographic provinces. County Key:

1	Beaver	6	Davis	10	Grand	14	Millard	18	Salt Lake	22	Summit	26	Wasatch
2	Box Elder	7	Duchesne	11	Iron	15	Morgan	19	San Juan	23	Tooele	27	Wayne
3	Cache	8	Emery	12	Juab	16	Piute	20	Sanpete	24	Uintah	28	Weber
4	Carbon	9	Garfield	13	Kane	17	Rich	21	Sevier	25	Utah	29	Washington
5	Daggett												

during the period 1924 to 1927 four commercial gas fields (one of which produces carbon dioxide gas) were discovered. One of the three combustible gas fields, Clay Basin in Daggett County, is still producing (Keller, 1961). A pipeline to deliver this gas to Salt Lake City was completed in 1937 (Eardley, 1952, p. 73). Another of these gas fields, *Ashley Valley* in north-central Uintah County, became the first substantial oil field in the state. Between 1925 and 1941 Ashley Valley produced gas from Frontier (Cretaceous) and Morrison (Jurassic) sands (Peterson, 1961a). The discovery of oil in late 1948 was due to deeper drilling into the Weber (Pennsylvanian) Sandstone (Gere et al., 1964, p. 57). By the end of 1966 this field had produced nearly 16 million barrels of oil. Trapping of both gas and oil is in a faulted anticline.

Other substantial oil discoveries followed, and new ones are continuing.

Geology. Three major physiographic provinces overlap parts of Utah (Tooker and Stewart, 1964, p. 17): the Rocky Mountains, the Colorado Plateau, and the Basin and Range (Figure 36.1). The major ranges in the Rocky Mountain province are the Uintas, an east-west range, and the Wasatch Mountains with a north-south trend. The Uinta Mountains are the result of erosion of upwarped Precambrian crystalline rocks. The Wasatch Range to the west, which rises spectacularly above Great Salt Lake basin, is a dissected fault block. The great Rocky Mountain fault zone of western Wyoming continues southward across central Utah, through and beyond the Wasatch Range.

The Colorado Plateau which occupies most of eastern and southern Utah, is drained by the Colorado River and its tributaries. It is a rugged country of canyons, cliffs, mesas, laccolithic mountains, and, in the south-western part, extensive lava flows (High Plateaus Volcanic area). The Basin and Range province extends from Nevada eastward across western Utah as far as the Rocky Mountain fault zone. Flat, alluvial-floored basins, some containing large bodies of water such as Great Salt Lake, are interrupted and separated by local ranges of eroded uplifted masses of bedrock. The southern part of this province is also covered in part by widespread volcanic rocks.

The major tectonic features of Utah are shown in Figure 36.2. Currently significant production of oil and gas is confined to three sedimentary basins: Green River, Uinta, and Paradox; and two positive features: the Uncompahgre uplift and the Wasatch Plateau–Castle Valley area. In addition, minor oil production is obtained from the Kaiparowits basin, which so far has only one field, and small quantities of oil and noncombustible gases have been found on the San Rafael swell.

A simplified geologic map of Utah is shown in Figure 36.3 (Gere et al., 1964, Figure 5). A glance at this map will indicate the reason for the confinement of the known significant oil and gas deposits to eastern Utah. Most of the western half of the state is surfaced with unconsolidated Quaternary deposits, extensive lava flows, or fault block mountains. Except for the Uinta Mountains and much smaller laccolithic mountains, the eastern half is exclusively sedimentary rock, most of it deposited in subsiding sedimentary basins. The Green River and Uinta basins are floored with fresh water sediments, the Paradox with Mesozoic and late Paleozoic rocks. Cretaceous and early Tertiary formations predominate over the San Rafael swell.

The southernmost part of western Utah is also floored with Mesozoic rocks. To the north in western Utah are lavas and flat alluvium-covered topographic basins. The intervening bedrock ranges consist almost exclusively of Paleozoic sedimentary formations. Unlike the outcropping Paleozoic rocks in eastern Utah, the section to the west includes strata of Cambrian through Devonian age, as well as Carboniferous and Permian rocks.

Figure 36.4 is a stratigraphic chart for the Utah part of the Colorado Plateau. It includes both surface and subsurface formations, and especially the oil and gas reservoirs (Gere et al., 1964, Table 3). These range in age from Recent alluvium to a sandstone member in a carbonate rock formation of Late Devonian age.

The Pliocene basalt reservoir at *Rozel Point* (Figure 36.2) lies at a depth of 150 feet. The 9° oil, which occurs in vesicles in the basalt, has a sulfur content of nearly 14 per cent (Heylmun, 1961c). Rozel Point became a producing field in 1965, when a well started pumping 5 barrels a day. Seeps of heavy black oil in this vicinity (Eardley, 1963) had generated considerable interest across the years.

Lower Tertiary (Eocene) reservoirs are most important in the Uinta basin, and upper Paleozoic reservoirs in the Paradox basin. Cretaceous reservoirs are outstanding in the Green River basin, the Uncompahgre uplift, the San Rafael swell, and the Wasatch Plateau.

In terms of total production the Hermosa carbonate rock section in the Pennsylvanian is by far the most important. It is the sole reservoir at Aneth and is important elsewhere in the Paradox basin. Lenticular sandstones in the Eocene Green River Shale are second in total production. The Green River Shale, the so-called oil shale of Utah and neighboring Colorado, is actually a richly organic calcareous and dolomitic mudstone of lacustrine origin. The oil in enclosed lenticular sandstone is no doubt due to expulsion from the surrounding fine-grained bituminous rock during compaction.

Sandstones are much more abundant than carbonate rocks in this list of reservoirs. However, more than 80 per cent of Utah's cumulative production has come from carbonate rocks, especially the Hermosa. Other lime-

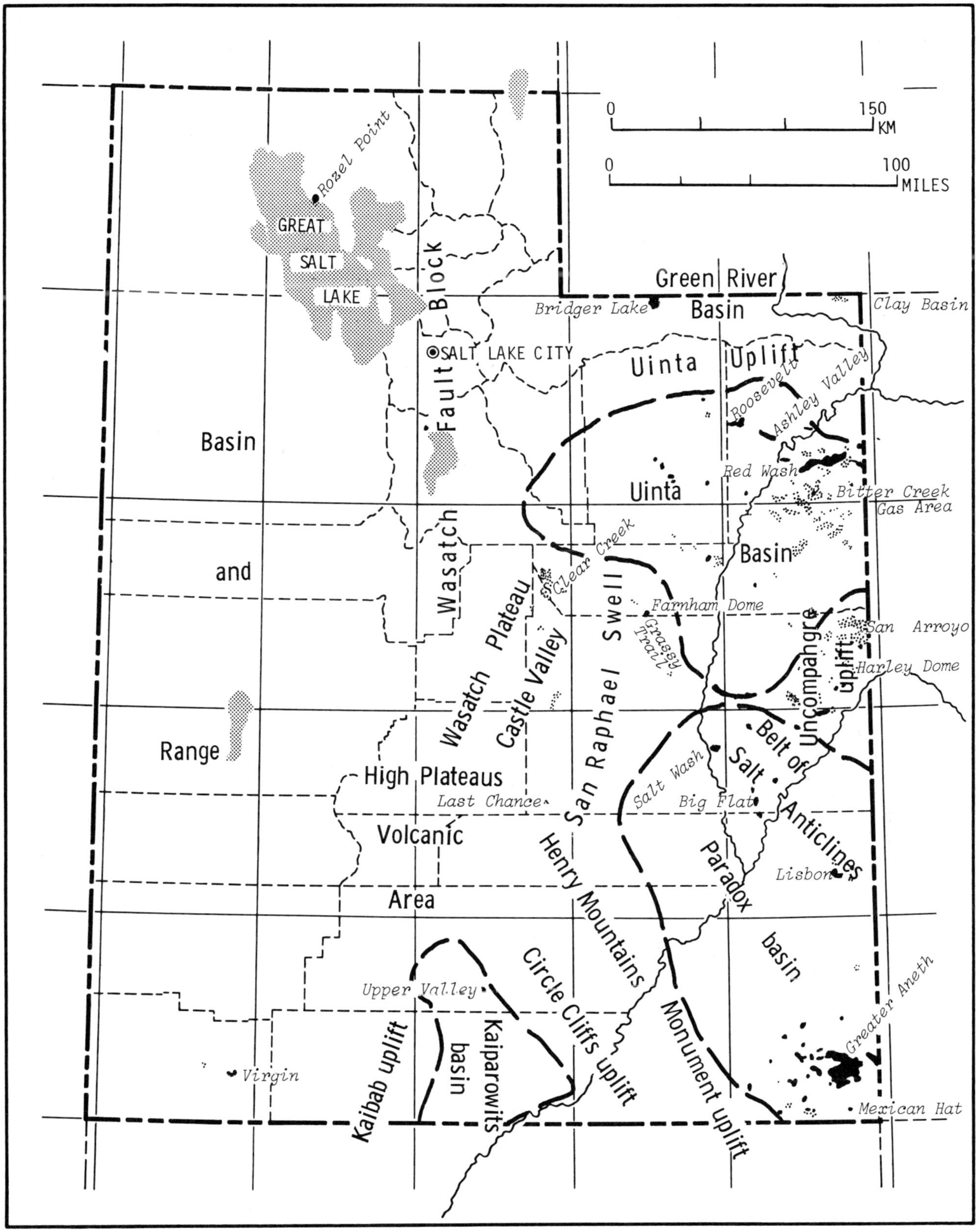

Figure 36.2 Utah. Tectonic map, with oil and gas fields.

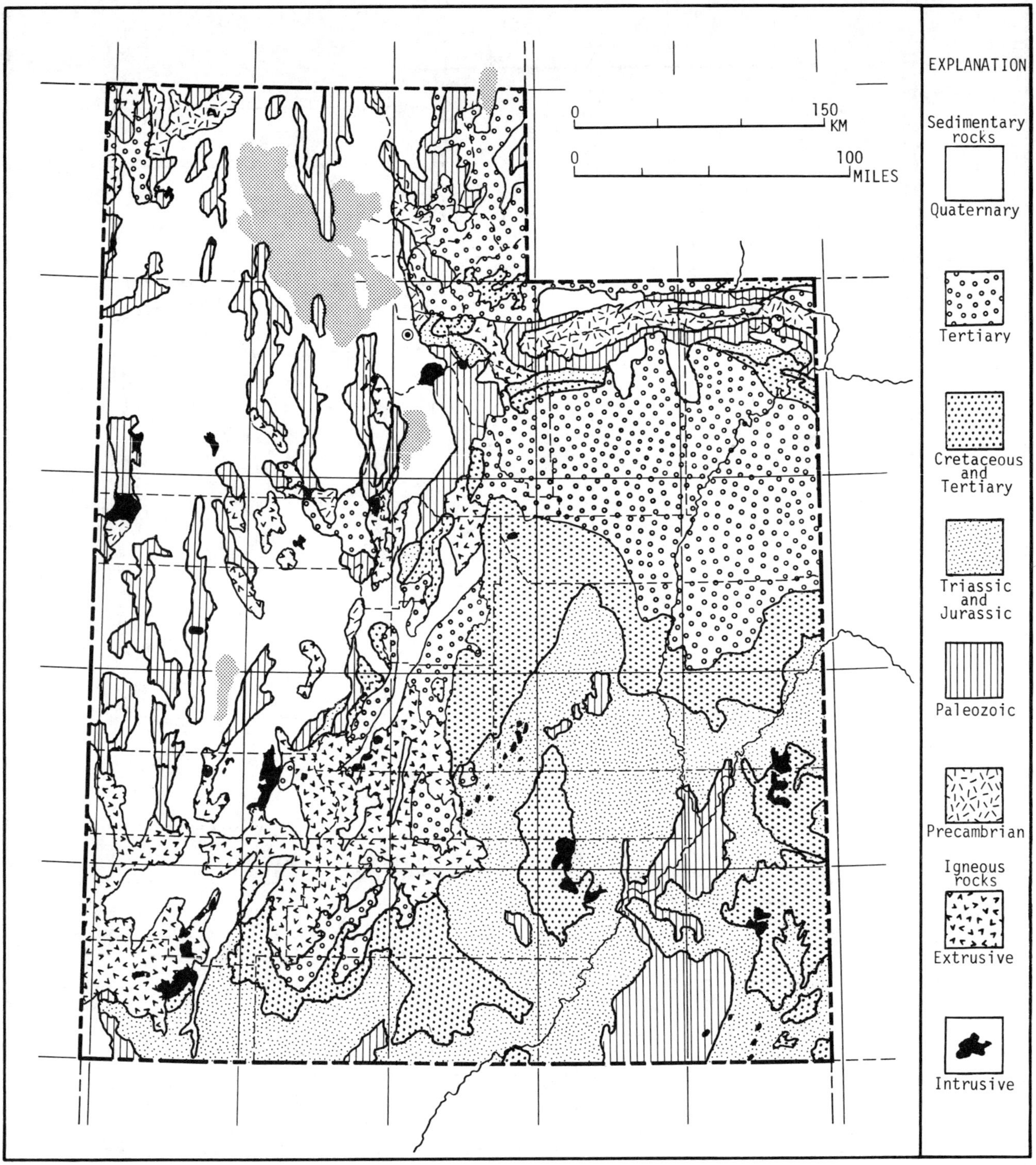

FIGURE 36.3 Utah. Geologic map. *After E. W. Tooker (1964, Figure 5).*

stone or dolomite reservoirs include the Leadville (Mississippian), Phosphoria (Permian), and Moenkopi (Triassic).

Exploration for oil and gas in Utah has led to the serendipitous discovery of three noncombustible gases—helium, carbon dioxide, and nitrogen—at various places, but especially on the San Rafael swell. The Federal Government set aside two helium reserves, one at Harley dome (Gere et al., 1964, p. 56) in Grand County and the other on the Woodside (Seeley, 1961) anticline in Emery County. The latter was subsequently sold and drilled but without success. The helium reservoir at

ERA	PERIOD	FORMATION AND LITHOLOGY	RESERVOIRS AND REMARKS
Cenozoic	Quaternary	Recent alluvium	Briefly commercial gas at Farmington, Davis County
		Pleistocene moraine	Also terrace gravel
	Pliocene	Basalt	Heavy oil at Rozel Point, Box Elder County
	Miocene	Browns Park Sandstone	
	Oligocene	Bishop Conglomerate	
	Eocene	Duchesne River Sandstone	
		Uinta Shale and Sandstone	CO_2 in Uinta basin
		Green River Shale	**Lenticular sandstones the sole reservoir at Red Wash, Uintah County**
		Wasatch Sandstone and Shale	Uinta basin
Mesozoic	Cretaceous	Mesaverde Sandstone	Especially Castlegate Member in southeastern Uinta basin
		Mancos Shale	Ferron Sandstone Member in Emery County; also fracture production
		Frontier Sandstone	Important gas reservoir in Green River basin
		Mowry Shale	
		Dakota Sandstone	Gas reservoir in Grand County; sole oil reservoir at Bridger Lake, Green River basin
		Cedar Mt. Sandstone	Minor gas reservoir in Uinta basin and on Uncompahgre uplift.
	Jurassic	Morrison Shale and Sandstone	Salt Wash Sandstone Member a gas reservoir in NE Grand County
		Summerville Mudstone	San Rafael Group
		Curtis clastics	San Rafael Group
		Entrada Sandstone	Minor gas in Grand County, also CO_2 and He. (San Rafael Group)
		Twin Cr.-Carmel Limestone	San Rafael Group
	- - - - - - - -	Navajo Sandstone	CO_2 at Farnham dome, Carbon County
	Triassic	Kayenta Sandstone	
		Wingate Sandstone	
		Chinle Sandstone	
		Ankareh red beds	
		Moenkopi carbonate rock	Oil reservoir in Emery County, and in Washington County (Virgin field). Also CO_2.
Paleozoic	Permian	Park City-Phosphoria Dolomite	Park City–Weber oil reservoir at Ashley Valley, Uintah County
		Coconino Sandstone	Gas in San Juan County, helium at Woodside, Emery County. Also CO_2 shows.
	Pennsylvanian	Hermosa: Weber Sandstone	Weber–Park City production at Ashley Valley
		Hermosa: Morgan	**Hermosa Limestone and Dolomite the sole reservoir at Aneth, San Juan County.**
		Hermosa: Round Valley	Hermosa oil reservoir also outstanding elsewhere in Paradox basin
	Mississippian	Leadville Dolomite and Limestone	Leading reservoir at Lisbon, San Juan County
	Devonian	Ouray Limestone	Small production at Lisbon
		Elbert Dolomite	McCracken Sandstone Member small producer at Lisbon
		Aneth Dolomite and Shale	
	Ordovician	Limestone in central Utah	
	Cambrian	Ophir Shale	
		Tintic Quartzite	
Precambrian		Upper	Sedimentary rock, but little metamorphosed
		Middle and Lower	Metamorphosed crystalline rocks.

FIGURE 36.4 Stratigraphic chart, Utah sector of Colorado plateau. ***Compiled from E. W. Tooker and J. H. Stewart (1964, p. 20–21).***

Harley is Entrada Sandstone; at Woodside it is Coconino. Commercial carbon dioxide gas is produced at Farnham dome (Peterson, 1961b) in Carbon County; the reservoir is Navajo Sandstone. All three of the nonflammable gases have been found in Mississippian rocks at Northwest Lisbon and elsewhere in the Four Corners region (Picard, 1962).

TRAPS AND FIELDS

Green River Basin (Gere et al., 1964, p. 56). Only a narrow strip along the southern edge of the Green River basin crosses the line from Wyoming into northeastern Utah. The basin terminates against thrust-faulted Precambrian rocks that are eroded into the Uinta Mountain range. Because of the faulting, the basin floor lies at considerable depth, even in the Utah sector of the Green River basin. At the eastern end of the Utah strip the thickness of the sedimentary column is estimated to be about 11,000 feet. Westward the thickness increases to something over 17,500 feet beneath the Bridger Lake oil field (Figure 36.2). So far, production is entirely from Cretaceous sandstones. However, Paleozoic carbonates with some sandstone and shale underlie the Cretaceous sediments in the Green River basin. At the close of 1967, there were two fields, one gas and one oil, in the Utah part of the basin.

The gas (and distillate) field, *Clay Basin* (Keller, 1961), is in Daggett County in the northeast corner of Utah. It was discovered in 1927 and began producing in 1937, when a pipeline to Salt Lake City was completed. Some gas is produced from the Frontier Sandstone of the Upper Cretaceous, but the principal reservoir is Dakota (Lower Cretaceous) Sandstone. A test to the Weber (Pennsylvanian) found noncombustible gas consisting of nitrogen and carbon dioxide. Trapping at Clay Basin is in an elliptical east-west anticline about 5 miles long with more than 300 feet of closure (Figure 36.5). Clay Basin itself is a topographic feature, created by the ero-

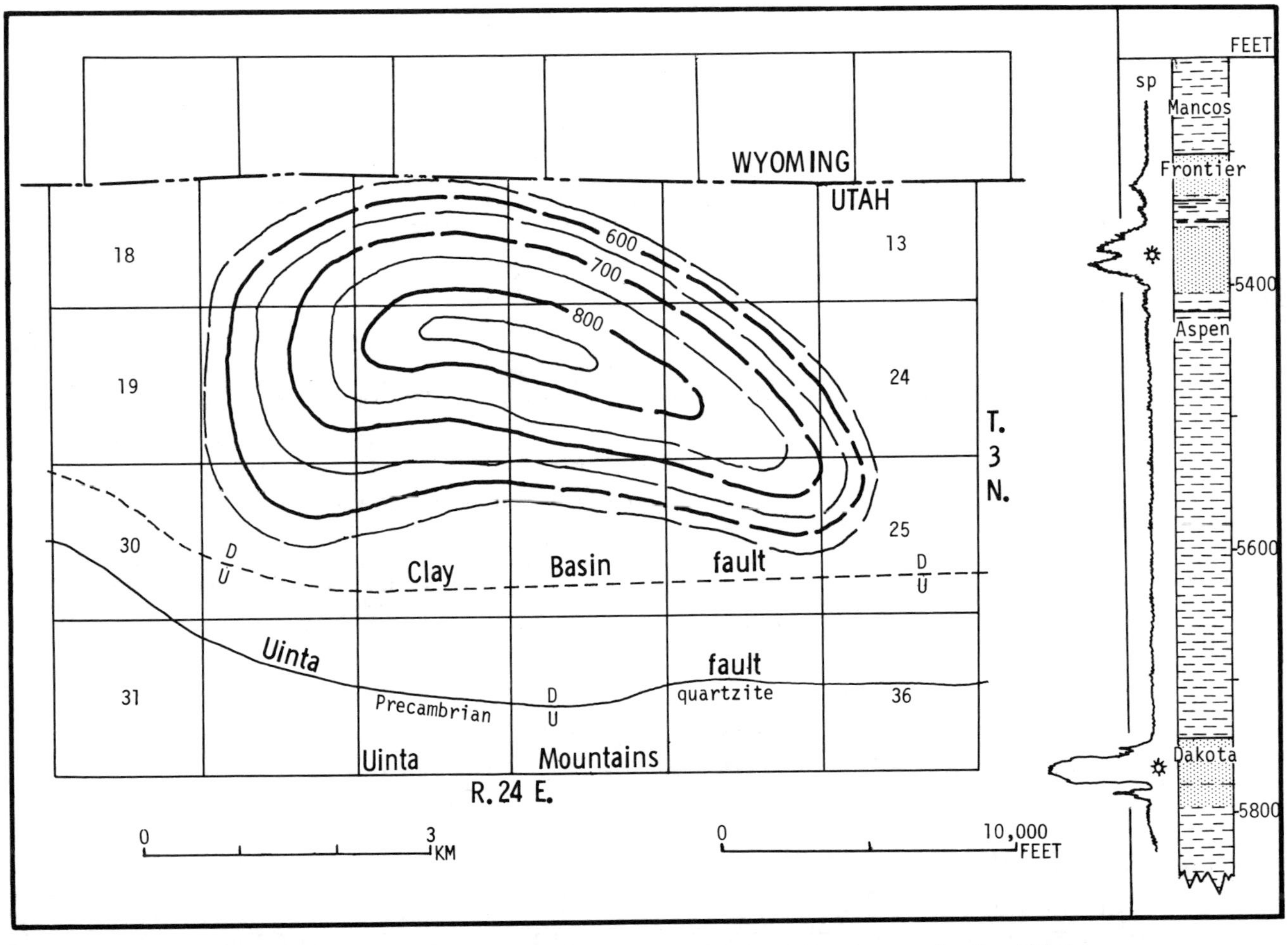

Figure 36.5 Clay Basin gas field, Daggett County, Utah. Structure contour map based on Dakota Sandstone. Interval, 50 feet. Both faults are thrusts. Position and lithology of reservoirs shown in log to right. *Courtesy E. R. Keller (1961; revised 1968) and the Intermountain Association of Petroleum Geologists.*

sion of soft Mancos (Upper Cretaceous) Shale in the core of the anticline. The basin is terminated on the south by the Uinta fault with the quartzitic Uinta Mountains beyond. Mesaverde Sandstone hogbacks rim the basin on the north.

The *Bridger Lake* oil field is in Summit County, 55 miles to the west of Clay Basin, near the western end of the Utah sector of the Green River basin. It was discovered in January, 1966, by a wildcat that flowed 2750 barrels per day from Dakota Sandstone at a depth below 15,500 feet. The field is on the north flank of the Uinta Mountains; the elevation of the discovery well is just under 9000 feet. The Uinta North Flank thrust fault is $2\frac{1}{2}$ miles south of the producing wells. Accumulation is apparently anticlinal, but the structure is complex. It has been interpreted tentatively as an anticline on the north side of a subthrust overturned and faulted syncline (Parker, 1966).

Uinta Basin (Wells, 1958; Osmond, 1965). The Uinta basin is second to the Paradox in both oil and gas production. The reserves of natural gas are estimated to exceed 1 trillion cubic feet (Folsom, 1968). This basin is in northeastern Utah, across the Uinta Mountains from the Green River basin. To the south is the Paradox basin; both basins taper toward their junction point because of the presence of the Uncompahgre uplift on the east and the San Rafael swell on the west. The Wasatch fault zone terminates the Uinta basin on the far west (Figure 36.2).

The maximum thickness of the stratigraphic section is about 26,000 feet (Gere et al., 1964, p. 56). The uppermost 11,000 feet of sediment consists of continental Tertiary deposits; the major downwarping and infilling of the Uinta basin began at the close of the Cretaceous (Wells, 1958). The basin is asymmetric with the deepest part at the northern end, close to the Uinta Mountain border. Here there is a structural relief of more than 30,000 feet in a distance of about 20 miles; the average regional dip on the north flank of the basin is therefore 1500 feet per mile. The northerly dip on the south flank ranges from 200 to 300 feet per mile.

Oil and gas production is obtained principally in varying permeability traps in Tertiary and Mesozoic sandstones (Osmond et al., 1968). The most prolific reservoir is the Tertiary (Eocene) Green River Formation. Unfortunately the oils from this formation in the Uinta basin

> are generally waxy and have a high pour point, solidifying at temperatures as high as 130° F . . . . These characteristics create many problems in the production and transportation of crude oil. Producing wells must be treated with hot oil or reamed to prevent paraffin from clogging the well bores. The crude must be dewaxed or mixed with a low pour point oil to prevent it from solidifying in the pipeline (Gere et al., 1964, p. 57).

The largest field in the Uinta basin, and the second largest in Utah, in total oil and gas yield is *Red Wash* in Uintah County. It will probably produce considerably over 100 million barrels of oil. The Red Wash field (Picard, 1957; Brace, 1961) lies immediately south of the Uinta basin axis in eastern Utah, about 10 miles west of the Colorado line. It was discovered by stratigraphic and structural mapping of outcrops, followed by a seismic survey. The surface work revealed the existence of both facies changes in the Green River Formation and an anticlinal nose. The discovery well, completed in January, 1951, confirmed the presence of a trap by producing initially 340 barrels of oil per day from lenticular sandstones in the Lower Green River Formation (Douglas Creek Member). Subsequent development found oil also in the overlying Garden Gulch Member, but Douglas Creek remains the principal reservoir. "The producing section in Greater Red Wash is composed of a great number of similar discrete sandstone bodies, each forming its own stratigraphic trap and interwoven into a vast and complicated network of individual reservoirs" (Chatfield, 1965, p. 118).

Figure 36.6 (upper) shows the structure map of the producing area, based on the top of the main producing zone. In the discovery (eastern) part of the area the oil and gas have accumulated in the upper part of a westward plunging nose. Here structure controls the distribution of hydrocarbons within each individual sandstone lens; to the west, however, production continues across a syncline. The lower illustration in Figure 36.6 is a cross section along the line shown in the upper figure. Trapping is brought about by the lensing out updip to the southeast (offshore) of sands that had been deposited along the beach of an Eocene lake.

The second field in oil production in the Uinta basin is Ashley Valley, which played an important role in the oil and gas history of Utah and therefore was described earlier. Third in oil production is the *Roosevelt* field in western Uintah County. The discovery well was completed in 1949 at a depth of 9350 feet for a flowing potential of over 1600 barrels a day. Subsequently, however, production declined, so that Roosevelt is now a minor field. The reservoir is principally a fracture system in the basal Green River dolomitic shale. Accumulation is on a broad northwest-trending nose that lies about 7 miles south of the Uinta basin synclinal axis (Rohr, 1957, p. 23). This field was the first in the basin to produce from a Tertiary reservoir. The oil has a characteristically high pour point, 90°F.

The recently developed *Bitter Creek gas area* covers parts of twelve townships in southern Uintah County, in the east-central part of Uinta basin. Principal reservoirs are lenticular sandstones in the Wasatch (Eocene) that appear to be stream channel deposits. The underlying Mesaverde Formation also contains gas reservoirs

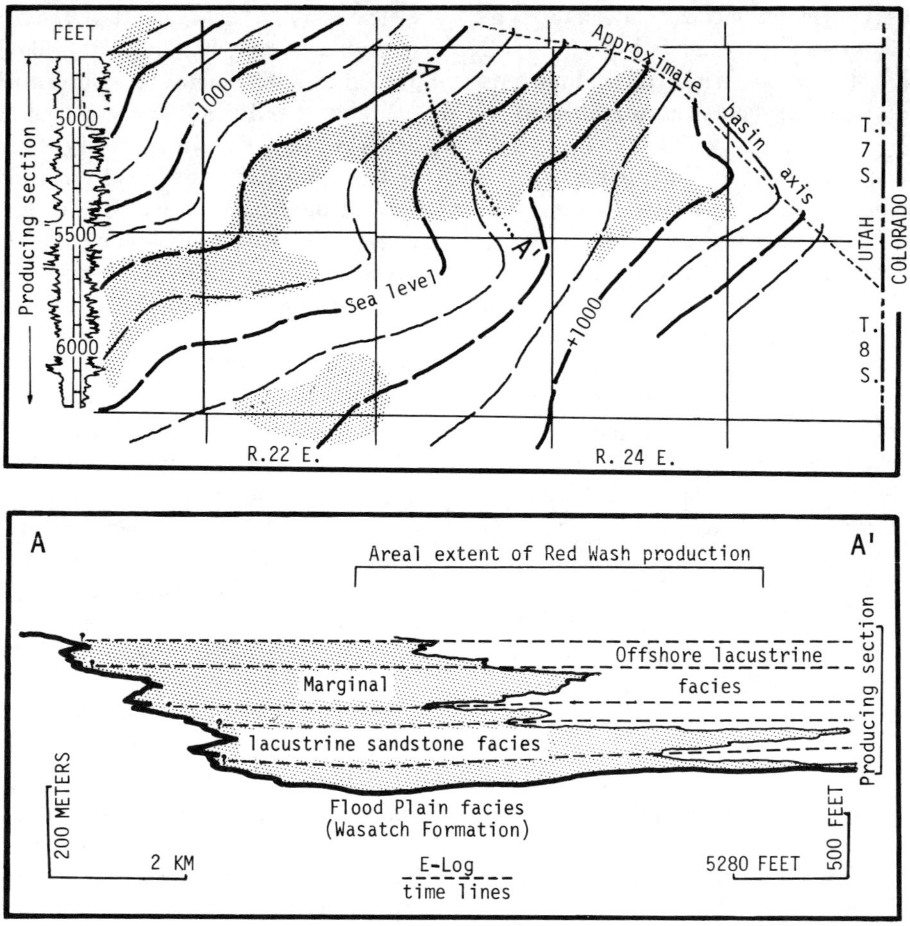

FIGURE 36.6 Greater Red Wash field, Uintah County, Utah. Upper: structure contour map based on top of main producing zone (Douglas Creek Member of lower Green River Formation—see electric log on left). Producing areas stippled. Squares are townships, approximately 6 miles (9.66 km) on a side. Lower: cross section along line A-A'. ***Courtesy John Chatfield (1965, Figures 3 and 5) and the Rocky Mountain Association of Geologists.***

in parts of the area. Throughout the Bitter Creek gas area the major factor in trapping is varying permeability.

UNCOMPAHGRE UPLIFT. The Uncompahgre uplift extends into central eastern Utah from western Colorado. In Utah it forms the structural divide separating eastern Uinta basin from eastern Paradox basin. The fields in central and northeastern Grand County shown on Figure 36.2 are on the Uncompahgre uplift, but also can be considered to be high on the flank of one basin or the other. Topographically they lie at the foot of the Book Cliffs escarpment. Many include the group of gas fields in northeastern Grand County in the Uinta basin.

The Grand County gas fields are second only to Red Wash in the cumulative production of gas in the Uinta-Uncompahgre area. Three of the fields occupy separate *en échelon* west-northwest–trending anticlines. The regional dip is northwest off the Uncompahgre uplift into the Uinta basin. The *San Arroyo* field (Williams, 1961; Horton and Gere, 1964; Owen and Whitney, 1956) is in the northeast corner of Grand County; 3 miles to the south is the *Bar-X* field (American Metal Climax, 1961), which extends across the state line into Mesa County, Colorado. Another 2 miles south is the *Westwater* anticline and gas field (Schuh, 1961). Principal reservoirs, all sandstones, are the Dakota in the Cretaceous, and the Jurassic Morrison (Salt Wash Member) and Entrada. Lesser gas reservoirs are the Castlegate Sandstone Member of the Mesaverde, fractured Mancos Shale, and Cedar Mountain Sandstone, all of Cretaceous age. Trapping in the Entrada reservoir is due entirely to anticlinal strucure; lenticularity of the other sandstone reservoirs controls the distribution of the gas. These and other producing anticlines on the Uncompahgre uplift were discovered by surface geology. In part, this is rugged country, with a topographic relief in places of 3000 feet.

PARADOX BASIN. The Paradox basin has been defined "as the area underlain by the Hermosa formation" (Gere

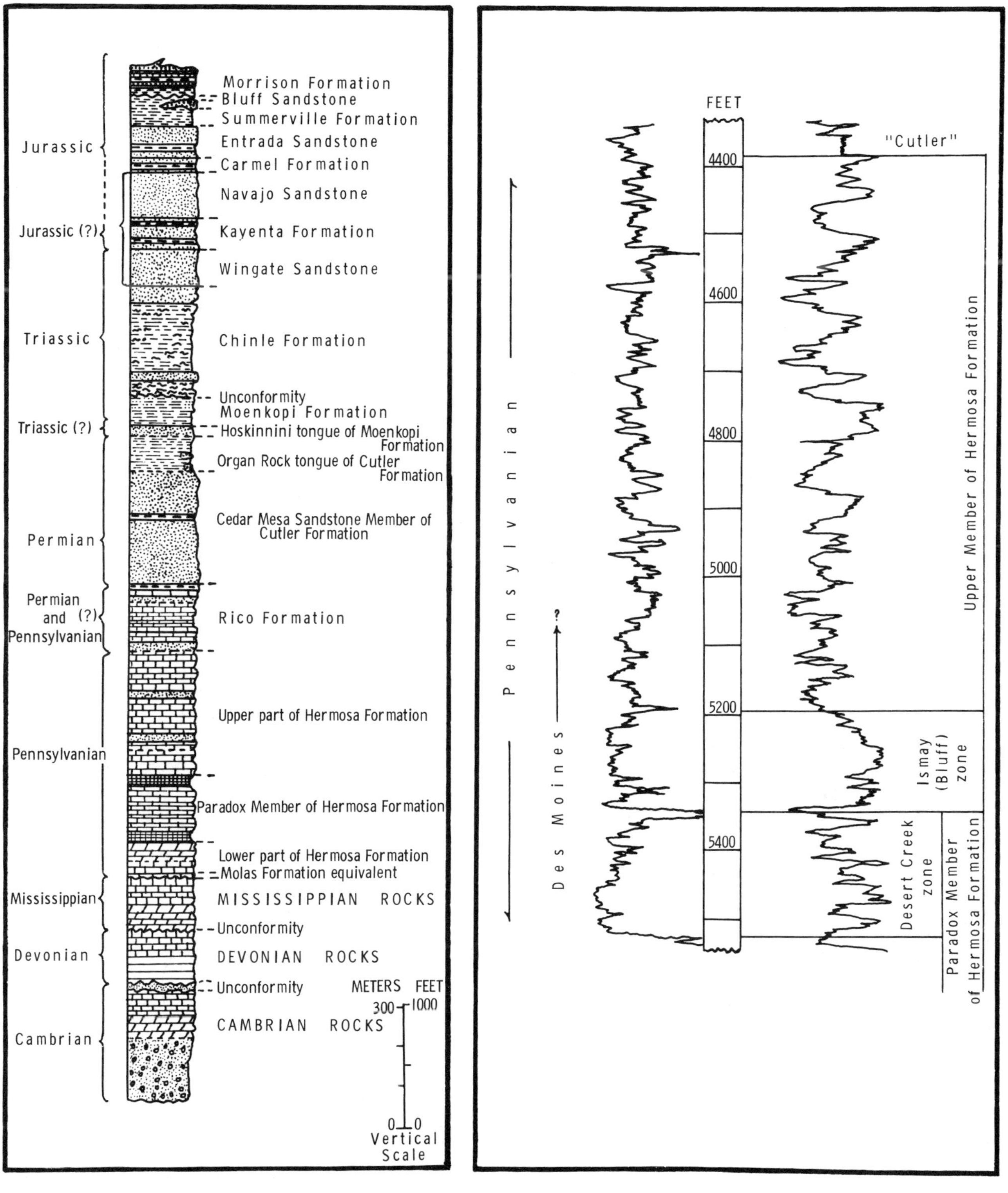

FIGURE 36.7 Paradox basin, Utah. Stratigraphy. Left: stratigraphic column from exposures along Elk Ridge, western San Juan County. *Courtesy Richard O. Lewis, Sr., (Figure 2, p. 79) and the Intermountain Association of Petroleum Geologists.* Right: subsurface section in the Greater Aneth field. *Courtesy M. Dane Picard (1959, Figure 3, p. 2461) and the American Association of Petroleum Geologists.*

et al., 1964, p. 58). The Hermosa is a Middle Pennsylvanian carbonate-evaporite rock section that includes salt and potash beds in the interior of the basin. Paradox is a sedimentary basin; the downwarping and basin filling took place in Pennsylvanian and Permian time. Pre-Pennsylvanian formations present are of Cambrian, Late Devonian, and Mississippian age. Triassic and Jurassic rocks overlie the Pennsylvanian-Permian sediments. The stratigraphic section has a total thickness of about 24,000 feet.

The Paradox basin extends eastward into southwestern Colorado. It is separated from the San Juan basin to the southeast by the Four Corners platform (Figure 32.2). Paradox is bounded on the east and northeast by the Uncompahgre uplift, and on the west and southwest by the San Rafael swell and the Monument uplift (Figure 36.2).

The stratigraphy and tectonic development of Paradox basin have been described by many, including, in more recent years, Parker and Roberts (1963), Ohlen and McIntyre (1965), and Baars (1966). Figure 36.7 (left) shows the column for Paradox basin as exposed on Elk Ridge (Figure 36.1).

The oil and gas production in the Paradox basin, the largest in Utah, is confined to two districts; the southeastern corner of the state in southeastern San Juan County, and to the north in northern San Juan and southern Grand Counties. The southern district is one of local folds, extensive reef mound development, and few faults. It also has much the greater part of the oil produced to date. The northern district has the thickest sedimentary section, most of the salt, all of the potash, and prominent salt-cored faulted anticlines. Oil and gas accumulation in the southern district will be discussed first. Field locations are given in Figure 36.2.

The discovery field for the Paradox basin, and the Four Corners area as well, was Mexican Hat (discussed on p. 432). The *Aneth* field, Utah's greatest, was discovered in 1956 and has subsequently produced over 200 million barrels of oil and has $\frac{1}{4}$ billion barrels in reserve. Subsequently to the original discovery at Aneth, a considerable number of satellitic fields with similar geology have been found, including the Cache in Colorado (see p. 391). Three close-in fields, the *McElmo Creek, Ratherford,* and *White Mesa,* merged with Aneth to become the *Greater Aneth* field (Picard, 1958). The names of the original fields have been retained as secondary recovery units. The reservoirs are two zones in the Hermosa Formation, the Ismay and the prolific Desert Creek (Figure 36.7, right). These are isolated bodies of porous carbonate rock, mainly limestone (Peterson, 1966). The carbonate reservoirs are referred to as algal mounds; they tend to be elliptical with a northwesterly trend. Black carbonaceous dolomite, commonly referred to as shale, which surrounds and overlies the mounds, is the probable source rock.

Regionally, the mounds developed on the southwest flank of the Paradox basin. The northwest trend of many of the mounds is therefore parallel to the regional strike. Figure 36.8 shows, by anticlinal axis symbols, the crests of the Desert Creek reservoir. This reservoir and the higher, scattered Ismay mounds appear to be the only sanctuaries for oil and gas in the Hermosa Formation in this area. The oil fields in Figure 36.8 are stippled. Wherever permeability permitted, the hydrocarbons accumulated in the higher parts of the reservoir. In the case of Greater Aneth, the oil and gas filled not only a large mound on the north side of the field, and the north flank of a faulted mound to the south, but also a considerable part of the intervening syncline. The impoundment of the hydrocarbons is due to both structural and stratigraphic factors. Accumulation in an Aneth type of trap is illustrated by the structure map and cross section in Figure 33.8 (Chapter 33).

The northern district of the Paradox basin is also the deepest part of the basin. It is dominated by the salt anticlines (Jones, 1959), which occur in a northwest-trending belt (Figure 36.2) about 100 miles long and 30 miles wide, nearly half of which is in Colorado. The major anticlines in this zone are elongate, ranging in length from 30 to 70 miles and in width from 2 to 5 miles (Cater and Elston, 1963). The salt cores themselves are "something like 2 miles high and 2 to 4 miles long" (Joesting and Case, 1960, p. B252). The maximum

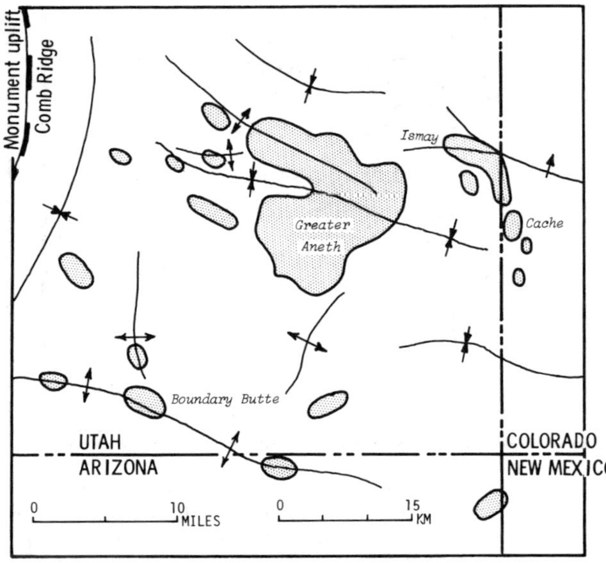

FIGURE 36.8 Aneth area, southeastern Utah and vicinity, showing major anticlinal and synclinal axes. Fields producing from Desert Creek and Ismay cycles are stippled. ***Courtesy James A. Peterson (1966, Figure 13, p. 2079) and the American Association of Petroleum Geologists.***

structural relief between the anticlinal crests and adjacent synclinal troughs is over 11,000 feet (Elston and Shoemaker, 1961). Erosion has breached the anticlines along their axes, resulting in elongate valleys in which gypsum and other rocks belonging in the Paradox Member of the Hermosa Formation (Pennsylvanian) are exposed. The salt itself has been leached; the top of the salt now lies from 400 to 1300 feet below the Paradox evaporite outcrops (Elston and Shoemaker, 1961). In one anticline (Cane Creek) commercial potash was discovered during the course of oil exploration; mine production began in 1965.

The elongate anticlines and intervening synclines are parallel with each other and with the Uncompahgre uplift, which is immediately to the northeast of the salt anticline belt. Strike faults are common. Folding, faulting, and flowage contributed to the development of the salt cores. The rocks involved in the diastrophism range in age from Tertiary to Precambrian. The thickest section is the Pennsylvanian-Permian, which is due to Pennsylvanian evaporite deposition in a sagging basin followed by inundation of clastics off the ancestral Uncompahgre uplift during the Permian. The Paleozoic below the Pennsylvanian consists of Mississipian, Devonian, and Cambrian sediments. Between these older rocks and the Hermosa Formation of the Pennsylvanian is an erosional and structural unconformity that plays an extremely important role in oil accumulation.

The anticlines started to form in Middle Pennsylvanian time, soon after the deposition of the Paradox evaporites. The arching of the Uncompahgre uplift took place at the same time. The continuing upward intrusion of the salt cores elevated the surface so that subsequent formations thin and even wedge out completely on the flanks of the major salt anticlines. Very much later, in early Tertiary time, additional folding accentuated the anticlines. This was followed, in more recent time, by leaching and collapse of the crestal parts of the folds.

Production in the northern district is relatively new, in spite of the fact that drilling of the mega-anticlines resulted in many oil shows, especially at Big Flat (Figure 36.2):

> As more exploration wells were drilled and geologic data accumulated, it became apparent that a great potential of oil and gas would be in fault blocks and folds where pre-Hermosa porous rocks are adjacent to the thick, organic-rich Hermosa Formation. Although the structural grains of the Hermosa tectonic elements parallel and influenced the development of the salt anticlines, the two are not superimposed, and the crests of the older structure are located on the flanks or between the surface expression of the salt anticlines. This knowledge, with the aid of geophysical investigations, led to the discovery of Lisbon (1960) and Salt Wash (1961) fields and further development of the Big Flat field. The accumulations in these fields formed by the migration of oil and gas from the Hermosa into porous rocks of Devonian to Pennsylvanian age (Gere et al., 1964, p. 59).

The Lisbon field is today usually referred to as Lisbon Northwest to distinguish it from the more recently discovered Lisbon Southeast field, which is on the same trend but across the state line in Colorado.

The *Lisbon Northwest* field (Parker, 1962, 1968; Pan American Petroleum Corp., 1961) in northeastern San Juan County (Figure 36.2) is by far the largest of the three fields producing in the Salt Anticline district. By the end of 1966, 6 years after discovery, it had produced nearly 16 million barrels of oil. The main reservoir is Mississippian limestone immediately beneath the Paradox Salt Member of the Hermosa Formation. The trap is illustrated in Figure 36.9. The upper drawing is a simplified longitudinal section from the Salt Wash field in southern Grand County southeasterly through the Lisbon field to the Colorado line. The lower illustration is a northeast cross section through the Northwest Lisbon field. Both show impounding of oil in porous Mississippian limestone at the top of a fault block. A gas cap (not shown) occupies the top of the trap. Sealing is both beneath and against the thick evaporites of the Paradox. The lower section, which cuts directly across the Lisbon Valley salt anticline, shows the relationship between the structure of the pre-Pennsylvanian oil-containing rocks and the younger Pennsylvanian to Cretaceous rocks. The crest of the fault-truncated anticline in which the oil has been trapped lies about 2 miles southwest of the axis of the exposed anticline.

San Rafael Swell - Castle Valley - Wasatch Plateau. The San Rafael swell is the outstanding structural and physiographic feature in south-central Utah. It is an elongate, only partly eroded mega-anticline about 80 miles long in a north-northeast direction and about 20 miles wide (Stokes, 1954, p. 6–7). So far, only two currently producing fields, both in southern Carbon County, have been found over this large area. The older is the *Farnham dome*, a carbon dioxide gas field. Discovered in 1924, it has produced more than 2 billion cubic feet of gas since the 1930s, when a dry ice plant at Wellington began processing its output. The reservoir rock is Navajo Sandstone (Jurassic); the trap, a thrust-faulted anticline (Peterson, 1961b). A small oil field, *Grassy Trail* (Gere et al., 1964, Table 3), was discovered in 1961. It produces from a limestone member of the Moenkopi (Triassic) Formation. Trapping is structural. A nonproducing field on the San Rafael swell is *Last Chance* at the southern end. Gas occurs here in lenticular sands in the Moenkopi Formation in a prominent anticline (Shannon, 1961).

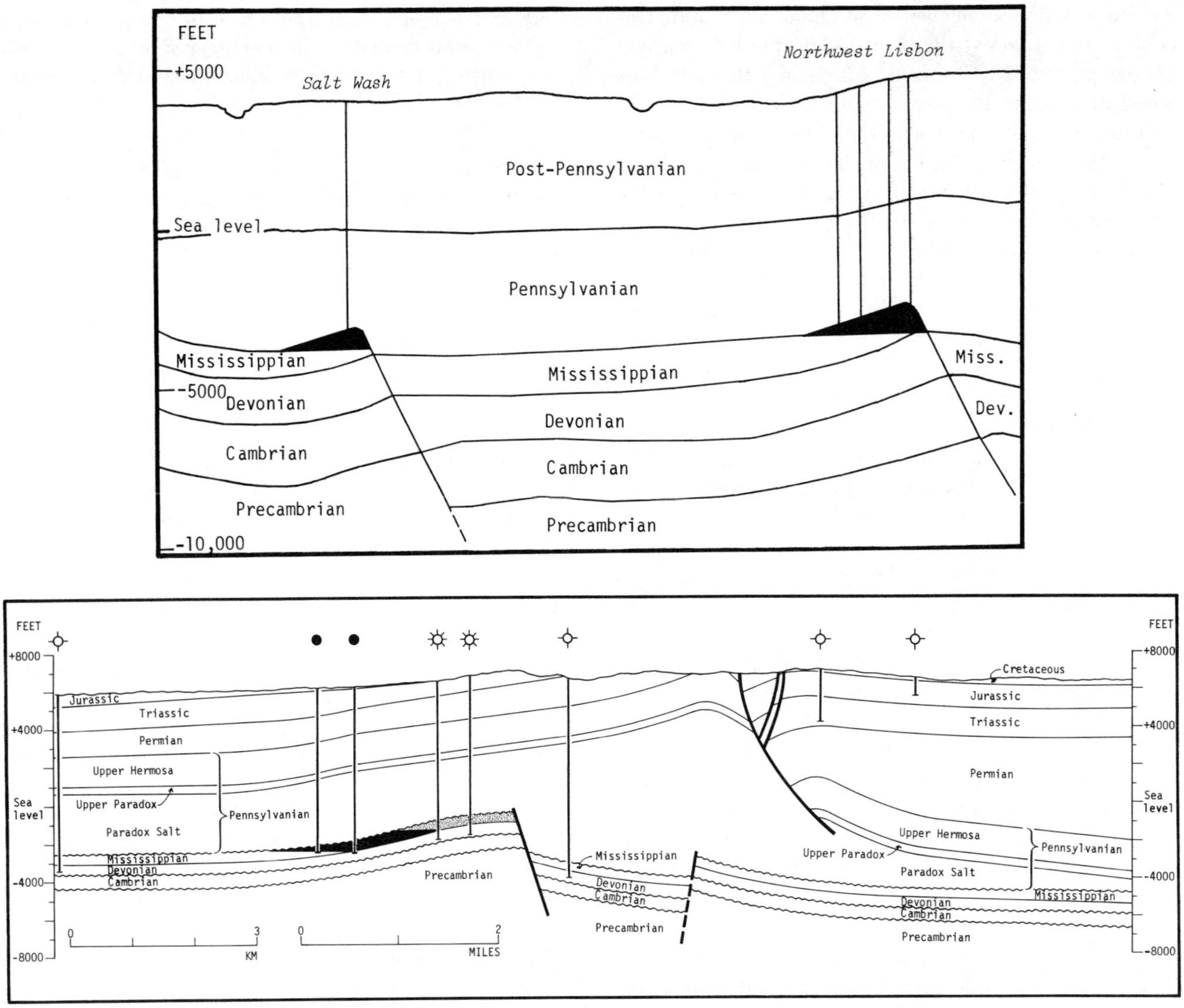

FIGURE 36.9 Salt anticline traps, northern Paradox basin, Utah. Upper: northwest-southeast (close to strike) hypothetical section through Salt Wash and Northwest Lisbon fields. Approximate distance between two fields is 60 miles (100 km). ***Courtesy Pan American Petroleum Corporation (1961, p. 1), subsidiary of Standard Oil Company (Indiana).*** **Lower: southwest-northeast cross section, Northwest Lisbon field, showing four Mississippian oil and condensate gas wells. Gas and condensate pay zone stippled.** ***Courtesy John M. Parker (1968, Figure 5, p. 1378–1379).***

To the west of San Rafael swell, in the Castle Valley and especially on the Wasatch Plateau, hunting for flammable gas has been more successful (Figure 36.2). This district contains one very large field, Clear Creek, and several smaller gas fields. Castle Valley–Wasatch Plateau is an area of extensive outcrops of Tertiary and Cretaceous rocks. The column, including underlying lower Mesozoic and Paleozoic sediments, has a total thickness of about 19,000 feet. "It is a structurally complex area of numerous north-trending horsts and grabens that generally cut older structures" (Gere et al., 1964, p. 58).

The *Clear Creek* gas field (Anon., 1961; Walton, 1955), discovered in 1951, had produced over 100 billion cubic feet by the end of 1962. The reservoirs are sandstones in the Ferron Member of the Upper Cretaceous Mancos Shale. These sandstones are low in interpore permeability; both gas storage capacity and movement are greatly enhanced by the presence of natural fractures. Accumulation is in the higher parts of the Clear Creek anticline, a major upfold in the north-central part of the Wasatch Plateau. This anticline is segmented by horst and graben faults. Trapping is in the fault slices, especially the horsts, on the anticlinal crest.

KAIPAROWITS BASIN AND SOUTHWESTERN UTAH.

Kaiparowits, a sedimentary basin in southern Utah west of the Circle Cliffs uplift, is considerably smaller than the Paradox basin to the east. The northern end of Kaiparowits basin is in west-central Garfield County; it terminates about 18 miles south of the state line in northern Coconino County, Arizona.

The basin is floored with Upper Cretaceous rocks surrounded by Jurassic and Triassic outcrops. The subsurface includes sediments representing all of the Paleozoic periods except Ordovician and Silurian. It became a producing basin in 1964 when the discovery well for the *Upper Valley* field was completed for 300 barrels a day from Kaibab (Permian) Limestone at 6700 feet. This new field is in west-central Garfield County. Subsequently two more producers have been drilled; at the end of 1965 the daily production of the Upper Valley field was "over 500 barrels" (Young, 1966, p. 1264). The trap is the large Upper Valley anticline, which is 16 miles long by 4 miles wide, and has 900 feet of closure at the surface.

The Kaiparowits basin terminates westward against the northward plunging Kaibab uplift of northern Arizona. West of the Kaibab uplift is southwestern Utah, which is in the Basin and Range province. This area is one of much folding, faulting, and igneous activity. It has one lone oil field, the hoary Virgin, which was described briefly on p. 432. The Virgin field is in east-central Washington County.

Future. None of the basins currently producing in Utah has been thoroughly explored. The discovery of deep Dakota oil in large quantity in the Green River basin at Bridger Lake 55 miles west of shallower Dakota (and Frontier) gas (Clay Basin field) indicates the probability of intervening deposits. Discoveries can be expected to the east of Clay Basin and to the west of Bridger Lake as well. They are also conceivable to the south of the surface trace of the North Flank Uinta fault, assuming this fault to be a thrust, overriding basin sediments. Unfortunately a proposed Wilderness Area might make this prospective producing belt unavailable (Ritzma, 1967a). Other reservoirs, in addition to the Frontier and Dakota, may also be oil and/or gas bearing in the Utah sector of the Green River basin.

In the Uinta basin discoveries of new fields producing from Eocene Green River and Wasatch Sandstone reservoirs take place annually, in spite of the fact that these stratigraphic traps are well hidden. Other possible reservoirs in this large basin include Mesaverde, Mancos, and Dakota of the Cretaceous, plus Entrado (Jurassic), Shinarump (Triassic), and the Pennsylvanian Weber (Gross, 1962). "Throughout most of the basin the Cretaceous and older formations have not been tested, except around the margins, and large areas of Tertiary rocks also are still untested especially in the western part of the basin" (Gere et al., 1964, p. 58).

Much of the Paradox basin likewise remains to be explored, and potential reservoirs in addition to the Hermosa should be tested. Recently the production of oil from the Coconino Sandstone at the top of the Permian was increased markedly at Boundary Butte field in the Utah sector of Four Corners by reinjecting formation water (Ritzma, 1967b). The greatest challenge as always is to find new traps. More "mound" accumulations will be sought in the southern district of Paradox basin, and more salt anticline fields in the north. The salt anticlines are similar to the Gulf Coast salt domes in that it takes considerable probing to find trapped hydrocarbons. The similarity may also extend to multiple traps around the salt structures. Another intriguing possibility is the occurrence of oil in traps produced by laccolithic-type intrusions in the Paradox basin (Heylmun, 1958a). Laccoliths are present in Utah in the LaSal, Abajo, and Henry Mountains.

Future possibilities, especially at relatively shallow depths, in western Paradox basin and on the San Rafael swell are reviewed by Heylmun (1964). Principal targets named are the Ferron Sandstone Member in the Mancos Formation (Cretaceous); the Moenkopi (Triassic) Formation, including especially the Sinbad Limestone Member; and the Kaibab Limestone and Coconino Sandstone in the Permian.

> Most drilling to date has been clustered in a few areas. There has been an average of only one well per township. . . . There is one area of over 15,000 square miles, encompassing parts of Wayne and Garfield Counties, where no wells have been drilled; and the entire region has just been scratched in regard to determining its oil and gas potential (p. 33).

Farther west, in the Castle Valley–Wasatch Plateau area, Gere et al. (1964, p. 58) and Walton (1968) advocate deeper drilling of known structural traps in order to test possible Mesozoic and Paleozoic reservoirs. The Kaiporowits basin in southern Utah covers a considerable area and without doubt has more than the one oil field so far discovered, since the stratigraphic section is adequate in both thickness and lithologies (Heylmun, 1958b). In addition to visible structural conditions favorable to oil accumulation, the lensing out of various permeable formations updip on the flanks of the basin should encourage the search for stratigraphic traps.

The Basin and Range province of western Utah (Intermountain Assoc. Petroleum Geologists, 1963; Roberts et al., 1965) covers nearly a third of the state and yet contains only two oil fields, both minuscule, and one abandoned gas field. All three (Virgin, Rozel Point, and Farmington) have been discussed previously. About 170 miles to the northwest of the Virgin field, still in the

Basin and Range province but out of the state of Utah, is the Eagle Springs field of Nye County, Nevada (Figure 38.1). In the first 10 years after its discovery Eagle Springs produced about 1 million barrels of oil; so far, this is the only substantial production in the entire Basin and Range province, which covers parts of five states. And yet there are large basins in this province, containing thick and lithologically adequate stratigraphic sections. The outstanding example is the Oquirrh, which, in northwestern Utah, contains as much as 26,000 feet of Pennsylvanian and Lower Permian sediments (Roberts et al., 1965).

Utah's dormant reserves are infinitely greater than its known petroleum and natural gas reserves, which include oil from oil shale, solid (vein) hydrocarbons, and heavy oil ("tar") in sandstones and porous limestones. The oil shale reserve is confined to the Uinta basin (Quigley and Price, 1963; Cashion, 1964; Smith and Stanfield, 1964). The formation involved is the Green River of Eocene age, the same as in the Piceance basin of northwestern Colorado. The total recoverable petroleum reserve from Green River rocks in these two basins is estimated to be in the hundreds of billions of barrels (Wells, 1958).

Gilsonite, the outstanding solid hydrocarbon, has been mined for many years in central Uintah County in the Uinta basin (Crawford and Pruit, 1963). This mineral occurs in vertical veins that cut through the sedimentary rocks. The veins trend northwesterly and occur in a zone 60 miles long and 30 miles wide (Wells, 1958, p. 359).

Heavy oil and tar deposits are also abundant in the Uinta basin (Covington, 1963, 1964), but occur in other parts of the state as well, especially the southeastern quadrant. The number of known deposits is 47 (Ritzma, 1967c). The host rock is in most instances a sandstone. The Sunnyside deposit on the southwestern rim of the Uinta basin is the largest reserve of bituminous sandstone in the United States. This and other tar sand occurrences give Utah a potential oil reserve of several billion barrels. Thermal recovery projects are planned for the richest deposits, including Sunnyside.

REFERENCES CONSULTED IN PREPARING CHAPTER 36

American Metal Climax, 1961.
Anon., 1961.
Baars, D. L., 1966.
Bahr, Carl W., 1963.
Bissell, H. J., 1967.
Brace, R. T., 1961.
Cashion, W. B., 1964.
Cater, Fred W., and Elston, D. P., 1963.
Chatfield, John, 1965.
Covington, Robert E., 1963, 1964.
Crawford, Arthur L., 1963.
Crawford, Arthur L., and Pruit, Robert G., 1963.
Eardley, Armand J., 1952, 1963.
Elston, Donald P., and Shoemaker, Eugene M., 1961.
Folsom, L. W., 1968.
Four Corners Geological Society Guidebook, 1963.
Gere, W. C., Horton, G. W., Harstead, H. H., and Russell, D. F., 1964.
Gross, Larry T., 1962.
Heylmun, Edgar B., 1958a, 1958b, 1961a, 1961b, 1961c, 1964.
Horton, G. W., and Gere, W. C., 1964.
Intermountain Association of Petroleum Geologists, 1954, 1958, 1961, 1963, 1964.
Joesting, H. R., and Case, J. E., 1960.
Jones, R. W., 1959.
Keller, E. R., 1961.
Kunkel, Robert P., 1961.
Lewis, Richard Q., Jr., 1958.
Ohlen, H. R., and McIntyre, L. B., 1965.
Osmond, John C., 1965.
Osmond, John C., Locke, Robert, Dillé, Alan C., Praetorious, Walter, and Wilkins, James G., 1968.
Owen, Arthur E., and Whitney, George W., 1956.
Pan American Petroleum Corporation, 1961.
Parker, John M., 1962, 1966, 1968.
Parker, J. Wm., and Roberts, J. W., 1963.
Peterson, James A., 1966.
Peterson, V. E., 1961a, 1961b.
Picard, M. Dane, 1957, 1958, 1959, 1962, 1968.
Quigley, M. Darwin, and Price, Jack R., 1963.
Ritzma, Howard R., 1967a, 1967b, 1967c.
Roberts, Ralph J., Crittenden, M. D., Jr., Tooker, E. W., Morris, H. T., Hose, R. K., Cheney, T. M., 1965.
Rohr, Egon H., 1957.
Schuh, E. J., 1961.
Seeley, deBenneville K., Jr., 1961.
Shannon, C. W., 1961.
Smith, John Ward, and Stanfield, Kenneth E., 1964.
Stokes, Wm. Lee, 1954.
Tooker, E. W., and Stewart, J. H., 1964.
Walton, Paul T., 1955, 1968.
Wells, Lewis F., 1958.
Williams, Floyd E., 1961.
Young, Robert T., 1966.

Chapter

37

ARIZONA

Arizona became the thirtieth oil-producing state in 1954, when the East Boundary Butte field was discovered in the northeast corner of the state. The first well produced gas and a little condensate; later some wells found oil. Two other small fields, Bita Peak (gas) and Dry Mesa (oil), were discovered subsequently nearby. The first major field, Dineh bi Keyah, was found in early 1967 by a well that initially produced 645 barrels a day, an output that subsequently rose to 1850 barrels, plus 135,000 cubic feet of gas (Pohlmann, 1967b). This field is also near the northeastern corner of Arizona (Figure 37.1). At the end of 1967 the cumulative oil production for Arizona was approximately 3½ million barrels, but by mid-1968 this state was producing at the rate of 4 million barrels per year, because of Dineh bi Keyah.

The Pinta Dome helium gas field, 100 miles to the south of the nearest oil field, was actually discovered 4 years before East Boundary Butte, but was shut in for nearly 9 years before development drilling got under way. This field lies on the southwest flank of the Defiance uplift on the east side of Black Mesa basin.

The major occurrence of sedimentary rocks is in northern, and especially northeastern, Arizona (McKee, 1951). The rest of the state is covered mainly by either lava flows, uplifted and mountain-eroded basement crystalline rocks, or horst blocks separated by deep fill graben basins. However, unmetamorphosed sedimentary formations crop out in some of the small basins in southeastern Arizona.

Northeastern Arizona is a part of the Four Corners (Picard, 1968; Peterson, 1959; Gray, 1958; DiGiambattista, 1952; Hoover, 1952) oil and gas area. Tectonically, however, it is isolated from the other three states, except for the southwestern corner of the Paradox basin, which cuts across the northeastern corner of Arizona. The western rim of the Paradox basin is the Defiance uplift; the Carrizo Mountains (Strobell, 1956), carved out of Tertiary laccolithic intrusions, are on the northern

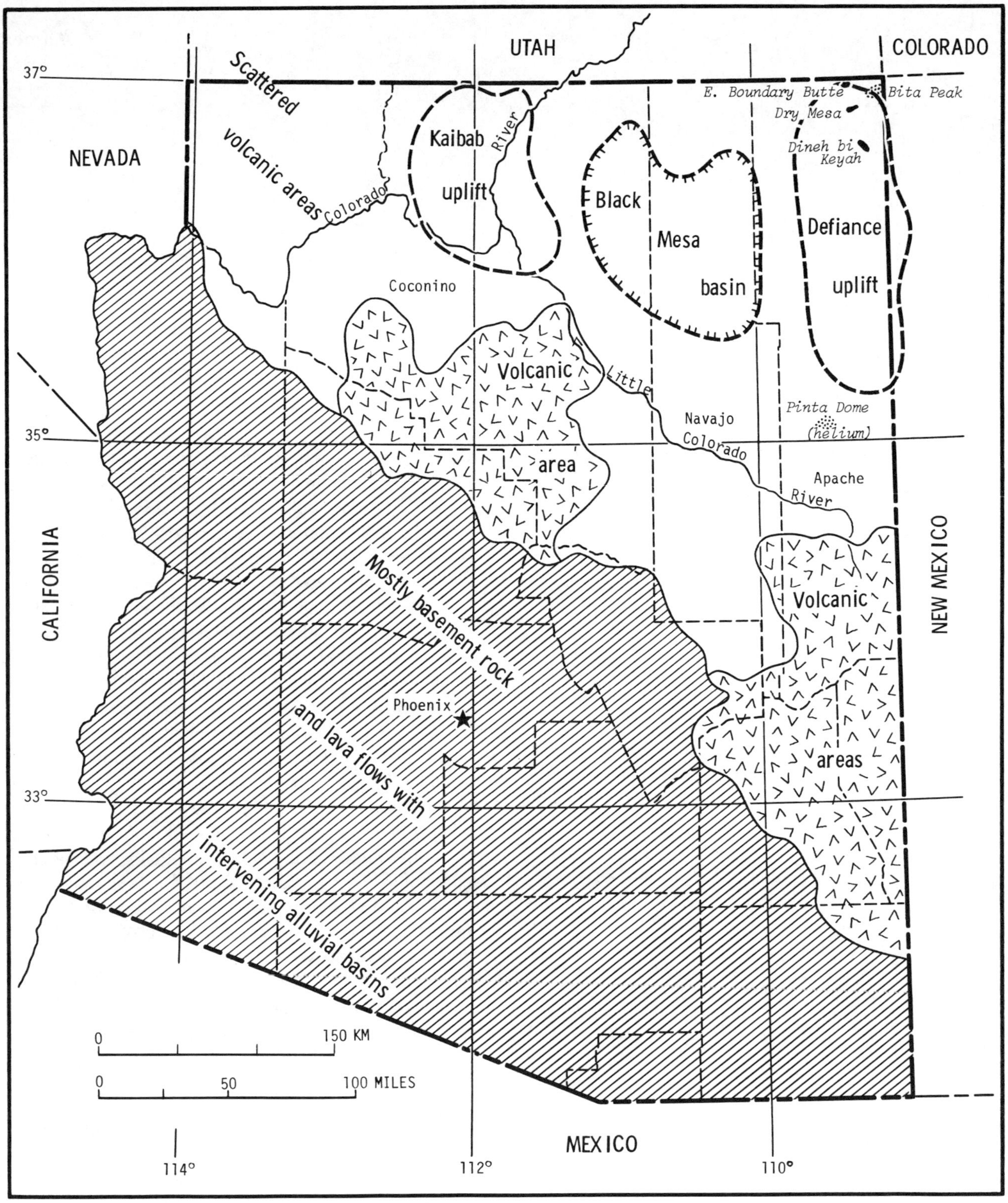

FIGURE 37.1 Arizona. Geology and index map. Oil fields solid black; gas fields stippled.

plunge of this mega-anticline. To the west and southwest of the Defiance uplift is the major tectonic feature of northeastern Arizona, the Black Mesa basin, which covers the greater part of the northeastern quadrant. It is bordered on the west by the Kaibab uplift. Much of the Black Mesa basin is Indian reservation, with a relatively small Hopi tract completely surrounded by Navajo allotment, covering many millions of acres. Unfortunately the Hopi-Navajo boundary was not surveyed; when oil and gas leasing became important, the size of the disputed no-man's-land area was found to exceed that of the undisputed Hopi territory. Exploration was therefore handicapped for more than 35 years, but in 1964 the entire area was opened for oil and gas leasing.

The basin center is occupied by Cretaceous rocks, with increasingly older Mesozoic and Paleozoic formations cropping out as the basin rim is approached. The Cretaceous rock outcrops in the central area have been eroded into buttes and mesas, which gave the basin its name (Turner, 1961). Numerous visible anticlines occur here, almost entirely within the hitherto disputed area.

The reservoir rocks in the three older fields in northeastern Arizona are in the Hermosa (Pennsylvanian) and the Mississippian. The helium reservoir at Pinta Dome is Coconino (Permian). The Pinta Dome gas has been referred to as "exotic" (Turner, 1968). The helium content varies from 1.07 to 8.29 per cent; the remainder is also nonflammable, consisting of nitrogen and carbon dioxide. All accumulations of oil and gas found so far in Arizona have been in anticlines.

Dineh bi Keyah (Navajo for "people's field") is in northeastern Apache County (Figure 37.1) about 25 miles south-southwest of the monument marking the Four Corners. It is in the Chuska Mountains, a region noted for its spectacular canyons. Within a year of discovery this field was producing 13,000 barrels daily. Primary reserves are in the range of 6 to 10 million barrels. Gas with a relatively high helium content has also been discovered in a deeper reservoir, but this resource has not yet been produced.

Geologically, Dineh bi Keyah occupies the crest of the northwest-trending Toadlena anticline, which is a fold on the northeast flank of the Defiance uplift (Figure 37.1). The reservoir rock, lying at a depth of 2800 to 4100 feet, is an unweathered igneous rock (syenite) sill intruded into a Pennsylvanian limestone and shale section. The date of the intrusion is estimated to be Oligocene. The sill is from 38 to 142 feet in thickness and is not regional in extent. Much of the effective porosity and permeability of this igneous rock is due to the presence of fractures. The oil may have migrated into the porous sill from the Pennsylvanian limestone with which it is in contact (McKenny and Masters, 1968, p. 2056). The helium-containing gas reservoir is a sandstone member of a deeper Devonian formation.

The Dineh bi Keyah discovery called attention to oil and gas possibilities along the east flank of the Defiance uplift, and future discoveries can be anticipated there, including the portion in northwestern New Mexico. The next most obvious place to prospect is the west flank, and on out into the Black Mesa basin. The objectives here can range down through the section from Cretaceous to Cambrian; only the Ordovician and Silurian periods are not represented (Brown and Louth, 1960). According to Elston (1960), Black Mesa is structural downwarp, but it is not a sedimentary basin that received more deposits than did surrounding areas. Therefore it has a thinner sedimentary section than other basins in the Four Corners area. Chances for stratigraphic traps appear best in the northern and western parts of the basin because of facies changes in various Paleozoic formations. The possibilities for structural traps are not, of course, restricted geographically.

Eventually, exploration westward into northwestern Arizona and southward into southeastern Arizona (Brown, 1952) will take place.

REFERENCES CONSULTED IN PREPARING CHAPTER 37

Brown, Silas, 1952.
Brown, Silas C., and Louth, Robert E., 1960.
DiGiambattista, C. D., 1952.
Elston, Wolfgang E., 1960.
Gray, Bert F., Jr., 1958.
Hoover, William B., 1952.
Kornfeld, Joseph A., and Travis, Maury M., 1968.
Kunkel, Robert P., Reese, Douglas, and Elias, David W., 1968.
McKee, Edwin D., 1951.
McKenny, Jere W., and Masters, John A., 1968.
Peterson, James A., 1959.
Picard, M. Dane, 1968.
Pohlmann, Henry Fred, 1967a, 1967b.
Pye, Willard, 1966, 1967.
Strobell, J. D., Jr., 1956.
Turner, Daniel S., 1961, 1968.

Chapter

38

NEVADA

With the discovery of Eagle Springs field in eastern Nevada (Figure 38.1) in June, 1954, this state became the twenty-ninth oil state (Smith, 1960). By the end of 1967 Eagle Springs had 13 producing wells with a cumulative oil production of about 1,750,000 barrels. The recoverable reserves are estimated at about 10 million barrels. As yet, no additional fields have been found.

Most of Nevada and western Utah are within the Basin and Range physiographic province. Along and just west of the Nevada-California boundary lie the Sierra Nevada, sculptured out of an upfaulted mid-Mesozoic granitic batholith. Within the Great Basin, most of the drainage is into local undrained basins where the generally arid climate causes the water to evaporate, creating mineralized lakes and playas. Separating the local basins are isolated, mostly asymmetric mountain ranges, the result of incomplete erosion of horst blocks. Most of the ranges and the intervening basins or valleys trend in a north-south direction.

The bedrock geology of Nevada includes Precambrian crystalline rocks; Paleozoic and Mesozoic sedimentary rocks, both marine and continental; Cenozoic sediments, entirely continental; and volcanic rocks in profusion. Lava flows and pyroclastic deposition took place during the Permian Period and the Mesozoic Era primarily in western Nevada, but the culmination was in mid-Tertiary, when enormous volumes of volcanic materials were extruded through most of the Basin and Range province. This was followed by the block faulting; subsequently the graben blocks have been covered by thousands of feet of post-Oligocene valley fill material from off the horst blocks and may also be veneered locally by playa evaporites (Heylmun, 1958; Horton, 1964). Some of the block faulting was contemporaneous with volcanic activity, which began in late Oligocene or early Miocene and continued into Recent time.

As a result of this geologic history, most of western Nevada is covered by ranges consisting of Tertiary vol-

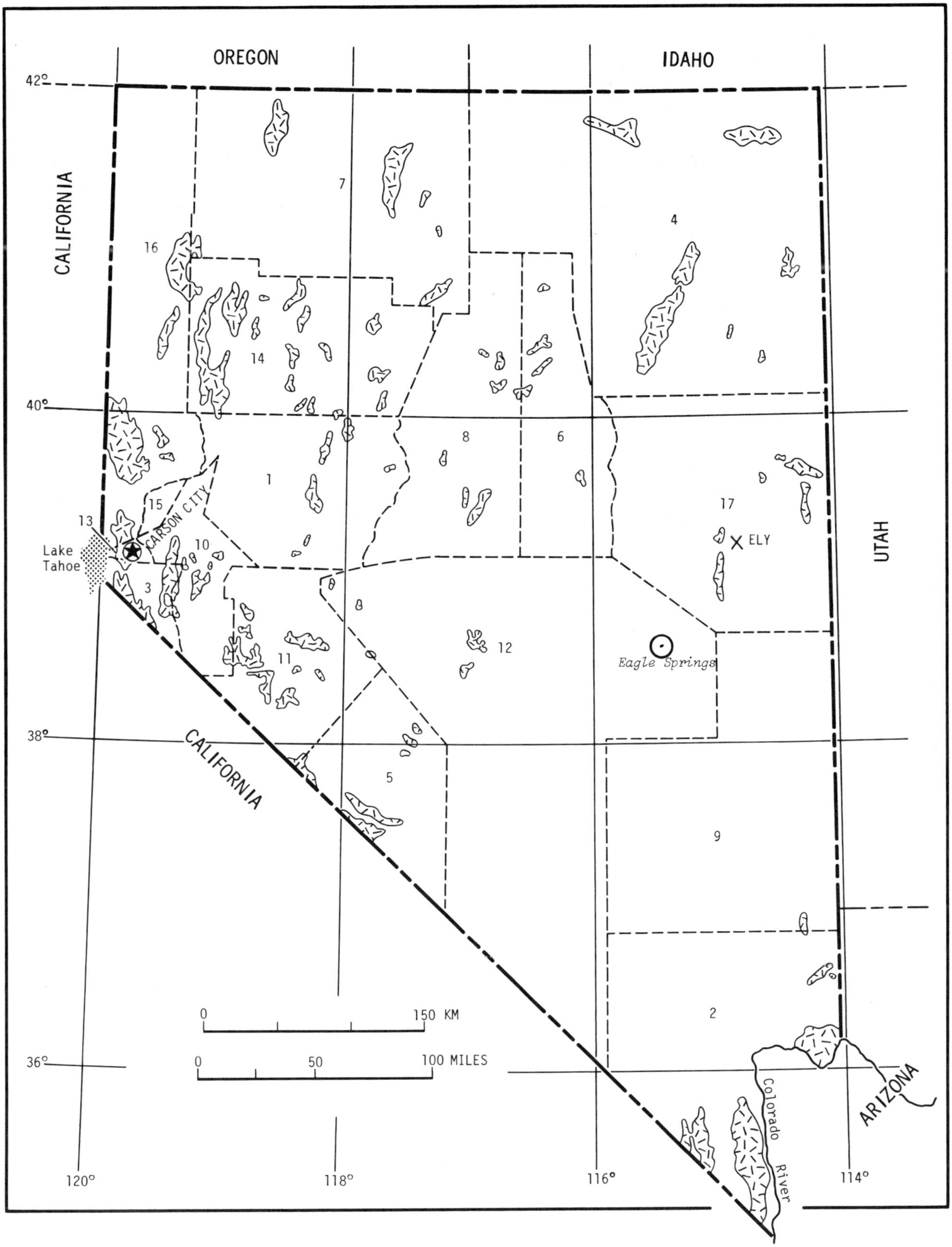

FIGURE 38.1 Nevada. Index map. Patterned areas show exposed crystalline rocks. County Key:

1 Churchill	4 Elko	7 Humboldt	10 Lyon	13 Ormsby	16 Washoe
2 Clark	5 Esmeralda	8 Lander	11 Mineral	14 Pershing	17 White Pine
3 Douglas	6 Eureka	9 Lincoln	12 Nye	15 Storey	

canic rocks overlying eugeosynclinal sediments and interbedded volcanic rocks of Paleozoic and early Mesozoic age, and intervening basins filled with Cenozoic nonmarine clastic sediment. Eastern Nevada, especially the central and northern parts, together with western Utah, offers greater potential for the generation, accumulation, and preservation of hydrocarbons. Thick sections of Paleozoic miogeosynclinal, shelf-type marine sediments, predominently carbonate rocks, as well as Tertiary fresh water limestones and pyroclastics, are present in this area.

Eagle Springs oil field is in Nye County in the central part of eastern Nevada (Figure 38.1). The discovery well produces from a Tertiary (Oligocene) volcanic welded tuff at a depth of 6450 to 6730 feet. Today the principal reservoirs are the Oligocene tuff and two lacustrine limestone members of the Eocene Sheep Pass Formation. One well produced from Pennsylvanian carbonate strata. The pay zones are discontinuous within the field boundaries. The trap appears to be a truncated, faulted, plunging fold, with sealing brought about by thick, impermeable Miocene to Recent valley fill deposited across truncated Oligocene to Upper Paleozoic strata. The trap is closed on the east by a major Basin and Range fault with a probable displacement of 10,000 to 15,000 feet (Murray and Bortz, 1967). The source of the Eagle Springs oil may be lacustrine shales and carbonate rocks of early Tertiary (Eocene) age, which underlie the Oligocene tuffs and directly overlie Mississippian to Pennsylvanian rocks (Horton, 1965; Winfrey, 1960; Picard, 1960b). Additional potential source rocks occur in the Paleozoic section (Johnson, 1959).

The best prospecting possibilities for the future appear to be in the Eagle Springs vicinity and northward into northeastern Nevada. The same source rock, reservoir, and trapping conditions that resulted in the oil accumulation at Eagle Springs no doubt exist elsewhere in eastern Nevada.

REFERENCES CONSULTED IN PREPARING CHAPTER 38

Bissell, H. J., 1967.
Heylmun, Edgar B., 1958.
Horton, Robert C., 1964, 1965.
Johnson, E. G., 1959.
Kunkel, Robert P., Reese, Douglas, and Elias, David W., 1968.
Murray, D. Keith, and Bortz, Louis C., 1967.
Osmond, J. C., 1960.
Picard, M. Dane, 1960b.
Roberts, Ralph H., Crittenden, M. D., Jr., Tooker, E. W., Morris, H. T., Hose, R. K., and Cheney, T. M., 1965.
Smith, Walter L., 1960.
Winfrey, Walter M., Jr., 1960.

Chapter

39

NONPRODUCTIVE STATE

About three-fifths of Idaho, the only nonproductive state in Division V, has igneous and metamorphic rock at the surface. The only sedimentary rock areas of any size are in (1) southeastern Idaho, which is penetrated by the northern tip of the Great Basin, and (2) southwestern Idaho, where lies the Vale–Ontario basin (Figure 39.1). The latter basin overlaps easternmost Oregon. It is filled mainly with loosely cemented Pliocene and Pleistocene stream and lake deposits. Although a few short-lived gas flows have been struck by the drill in this area (Savage, 1964), the cost of drilling the unconsolidated sediments, plus the paucity of source materials in the section so far penetrated, has discouraged further exploration.

More favorable are the early Tertiary, Mesozoic, and Paleozoic sedimentary rocks occurring in the Great Basin sector. More than 23,000 feet of sediment has been measured at the outcrop in this area. Included are age and lithologic equivalents of reservoir rocks productive of oil and gas in Wyoming, Montana, and Utah. From twenty to thirty wells have been drilled here, one to a depth of 9720 feet, but without success. One well drilled in Caribou County in southeastern Idaho recorded two shows of gas in the Twin Creek Formation (Jurassic), but the samples indicated low porosity and permeability (Newcomb, 1967). The Idaho sector of the Great Basin has been tectonically disturbed (tight folding, overthrust faulting) to such an extent that gas is a more likely prospect than oil.

REFERENCES CONSULTED IN PREPARING CHAPTER 39

Bissell, H. J., 1967.

Newcomb, J. J., 1967.

Roberts, Ralph H., Crittenden, M. D., Jr., Tooker, E. W., Morris, H. T., Hose, R. K., and Cheney, T. M., 1965.

Savage, C. H., 1964.

1 Ada
2 Adams
3 Bannock
4 Bear Lake
5 Benewah
6 Bingham
7 Blaine
8 Boise
9 Bonner
10 Bonneville
11 Boundary
12 Butte
13 Camas
14 Canyon
15 Caribou
16 Cassia
17 Clark
18 Clearwater
19 Custer
20 Elmore
21 Franklin
22 Fremont
23 Gem
24 Gooding
25 Idaho
26 Jefferson
27 Jerome
28 Kootenai
29 Latah
30 Lemhi
31 Lewis
32 Lincoln
33 Madison
34 Minidoka
35 Nez Perce
36 Oneida
37 Owyhee
38 Payette
39 Power
40 Shoshone
41 Teton
42 Twin Falls
43 Valley
44 Washington

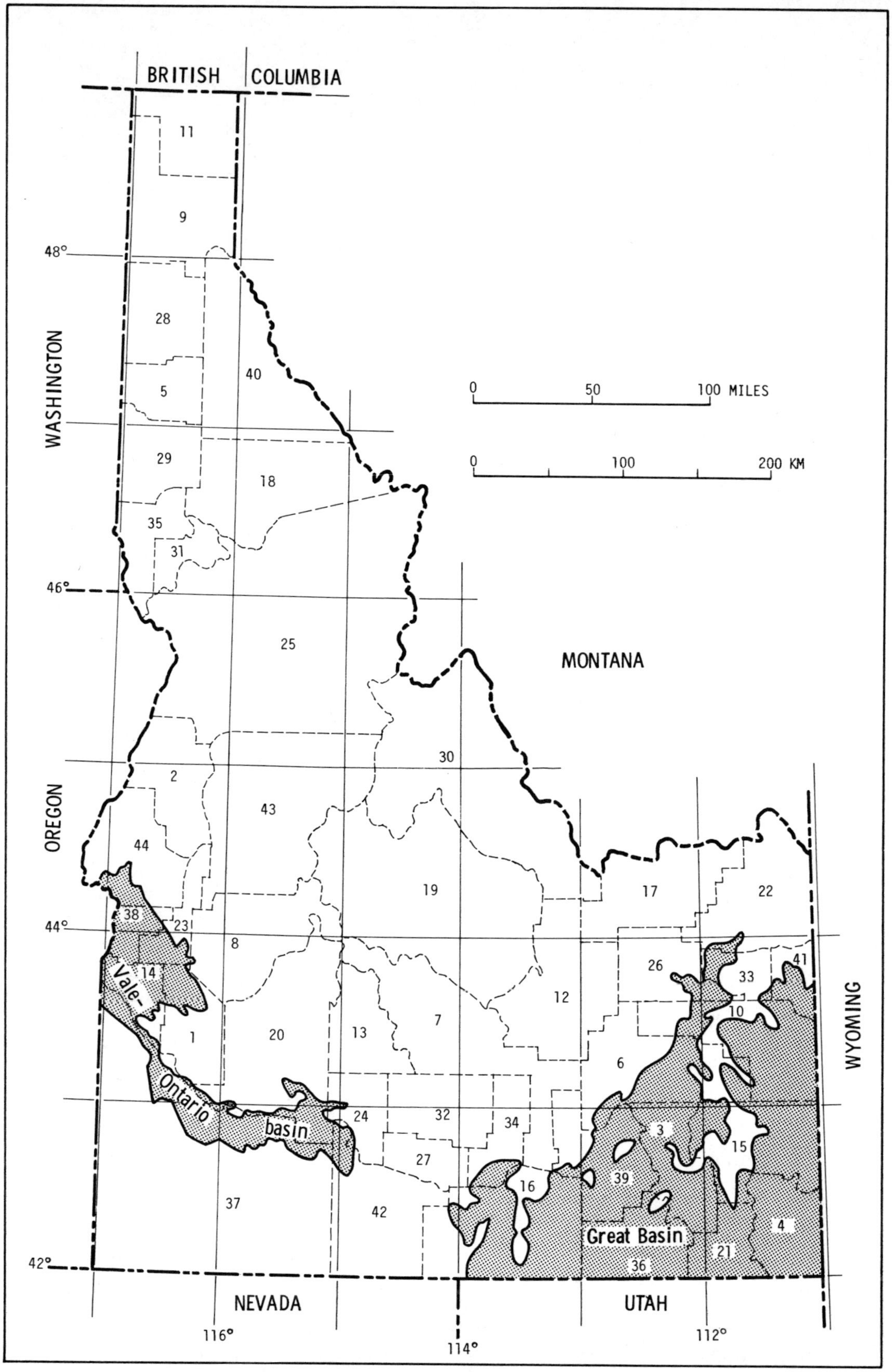

Figure 39.1 Idaho. Index map. Shaded areas show sedimentary rocks. County Key on left.

Division VI

PACIFIC

The Pacific Division (Figure VI.1) consists of three contiguous states—California, Oregon, and Washington; a fourth state, Alaska, far to the northwest and separated from the others by westernmost Canada; and Hawaii. Of these five states, California and Alaska are important producers of oil and gas. California ranked third in the United States in yield in 1968; Alaska, a spectacular newcomer, had already become eighth, up from twelfth the year before. A third state, Washington, was in the producing column briefly during the late 1950s. The nonproducing states are discussed in Chapter 42.

The Pacific Division is bounded on the east by the Rocky Mountain Division and Canada, on the south by Mexico, on the west by the Pacific Ocean, and on the north by the Arctic Ocean.

Figure VI.1, in addition to being an index map, shows the physiographic provinces cutting across the four states involved (Wahrhaftig and Gates, 1964; Fenneman, 1931). The physiographic unit along the Pacific border south of the Bering Sea is the Pacific Mountain System in Alaska, the Coast Ranges–Cascades in Washington, Oregon, and northernmost California, and the Coast Ranges–Sierra Nevadas in California. This compound mountain chain trends northwesterly from southern California to southern Alaska, where it makes a 90° left turn to the southwest and forms the backbone of the Aleutian Peninsula and island arc.

Along the Pacific coast of conterminous United States a pronounced but discontinuous tectonic trough intervenes between the Coast Ranges and the Sierra Nevada–Cascades. This is known as the Great Valley in California and as the Willamette–Puget Sound trough in Oregon and Washington.

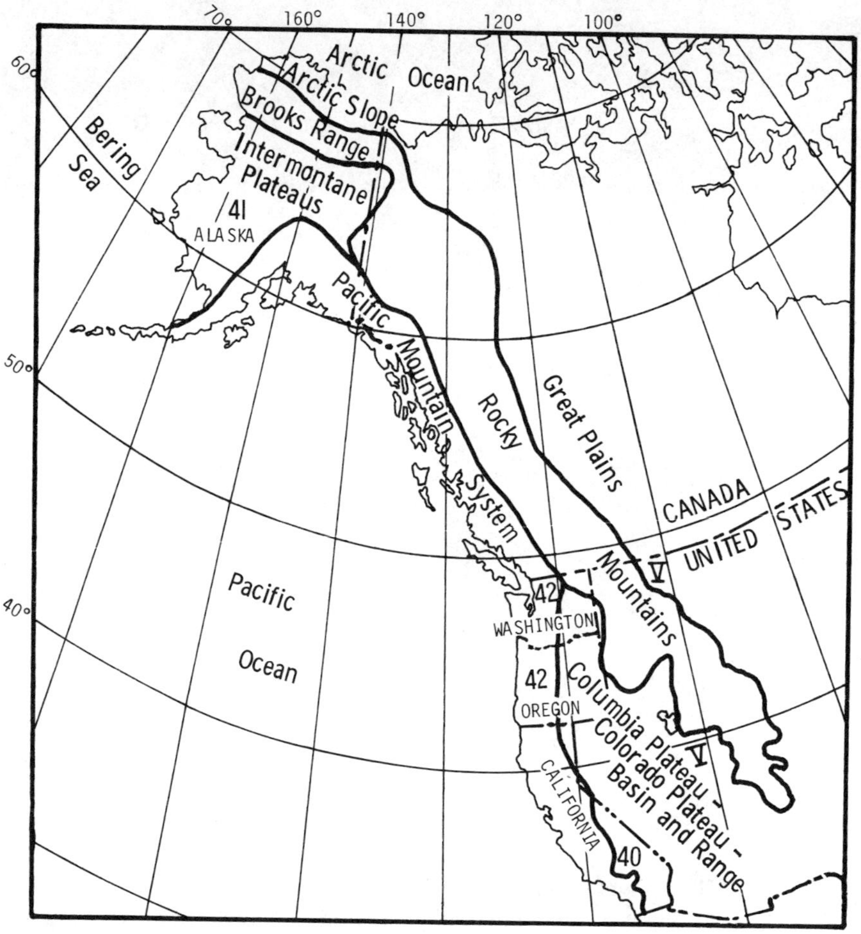

FIGURE VI.1 Pacific Division. Index and physiographic megaprovince map. *After Clyde Wahrhaftig (1954) and Nevin M. Fenneman (1931).*

To the northwest in British Columbia and southeastern Alaska this trough is mainly inundated by the Pacific Ocean, creating the Inside Passage. It reappears in southwestern Alaska, however, as the Matanuska geosyncline, which includes the petroleum-producing Cook Inlet basin.

The other major mountain province is the Rocky Mountain (Brooks Range in Alaska). This province lies east of California but cuts across eastern Oregon and Washington, as well as northern Alaska. Between the two major mountain belts in the western states is the Colorado Plateau–Basin and Range–Columbia Plateau belt. In Alaska the comparable province is the Intermontane Plateaus area. Except for strike valleys and local plateaus the two great mountain systems merge across northern Washington and western Canada.

On the eastern border of the Rocky Mountain province in conterminous United States and western Canada are the Foothills belt and the High Plains. A westerly swing of the Brooks Range sector of the Rocky Mountains in northern Alaska results in the Arctic Slope between the foothills and the Arctic Ocean; this is a northwestward extrapolation of the High Plains.

Hydrocarbons are produced in the Pacific Division within the zone occupied by the Pacific Mountain system, in both California and Alaska. In addition, the latter state is productive in the Foothills–High Plains belt, which is also a prolific source of oil and gas to the southeast in Northwest Territories, British Columbia, Alberta, Montana, Wyoming, and Colorado.

The age of the producing formations in the Pacific Division in California is Tertiary and Late Mesozoic (relatively minor). In the same province in Alaska the oil and gas so far discovered occur exclusively in Tertiary reservoirs (Gates et al., 1968). The Foothills–Arctic Slope hydrocarbons, however, are Mesozoic and Paleozoic in age, as is the production in western Canada and in west-central United States.

REFERENCES CONSULTED IN PREPARING INTRODUCTION TO DIVISION VI

Fenneman, Nevin M., 1931.
Gates, George O., Grantz, Arthur, and Patton, W. W., Jr., 1968.
Wahrhaftig, Clyde, and Gates, G. O., 1964.

Chapter

40

CALIFORNIA

California is an oil-producing state surrounded by barren Oregon on the north, almost dry Nevada on the east, nonproductive corners of Arizona and Mexico on the southeast, and the Pacific Ocean on the southwest and west (Figure 40.1). Yet this state ranked third in oil production in the United States in 1968 with 375 million barrels. It has even been first in the country in annual yield; the last time was in 1926, when its production was less than two-thirds of the 1967 output. The state reached a peak in 1953, when production exceeded 1 million barrels a day (Jennings, 1957b). During the next decade the output dropped nearly 20 per cent, and many prophesied that California was on its way out as an important source of oil. However, by 1968 California had climbed back up to over a million barrels daily, mainly because of increased offshore production, the expansion of urban fields, and greater yields from low-gravity oil fields by the use of steam injection.

California is notable for its giant fields, forty in number, and the role that they play in maintaining its rank as a leading oil state. These large fields have produced 75 per cent of the oil and contain 90 per cent of the known reserves in California.

This state also produces much gas (Hart, 1957), ranking sixth in the nation in 1968 in that commodity. Over half of this gas is produced with oil; the giant oil fields are also major sources of natural gas. The remaining gas, which is dry gas, is produced mainly in northern California, principally in the Sacramento Valley sector of the Great Valley (Figure 40.2).

Oil and gas seeps are abundant in parts of California, and the early search for buried hydrocarbon deposits was guided by the presence of these natural outcrops (Hoots and Bear, 1954). That famous animal trap, Rancho La Brea, is but one of many tarry surface accumulations that are found in California. A very active search for oil by many individuals and companies took place during the two decades from 1854 to 1874. The search extended

1 Alameda
2 Alpine
3 Amador
4 Butte
5 Calaveras
6 Colusa
7 Contra Costa
8 Del Norte
9 El Dorado
10 Fresno
11 Glenn
12 Humboldt
13 Imperial
14 Inyo
15 Kern
16 Kings
17 Lake
18 Lassen
19 Los Angeles
20 Madera
21 Marin
22 Mariposa
23 Mendocino
24 Merced
25 Modoc
26 Mono
27 Monterey
28 Napa
29 Nevada
30 Orange
31 Placer
32 Plumas
33 Riverside
34 Sacramento
35 San Benito
36 San Bernadino
37 San Diego
38 San Francisco
39 San Joaquin
40 San Luis Obispo
41 San Mateo
42 Santa Barbara
43 Santa Clara
44 Santa Cruz
45 Shasta
46 Sierra
47 Siskiyou
48 Solano
49 Sonoma
50 Stanislaus
51 Sutter
52 Tehama
53 Trinity
54 Tulare
55 Tuolumne
56 Ventura
57 Yolo
58 Yuba

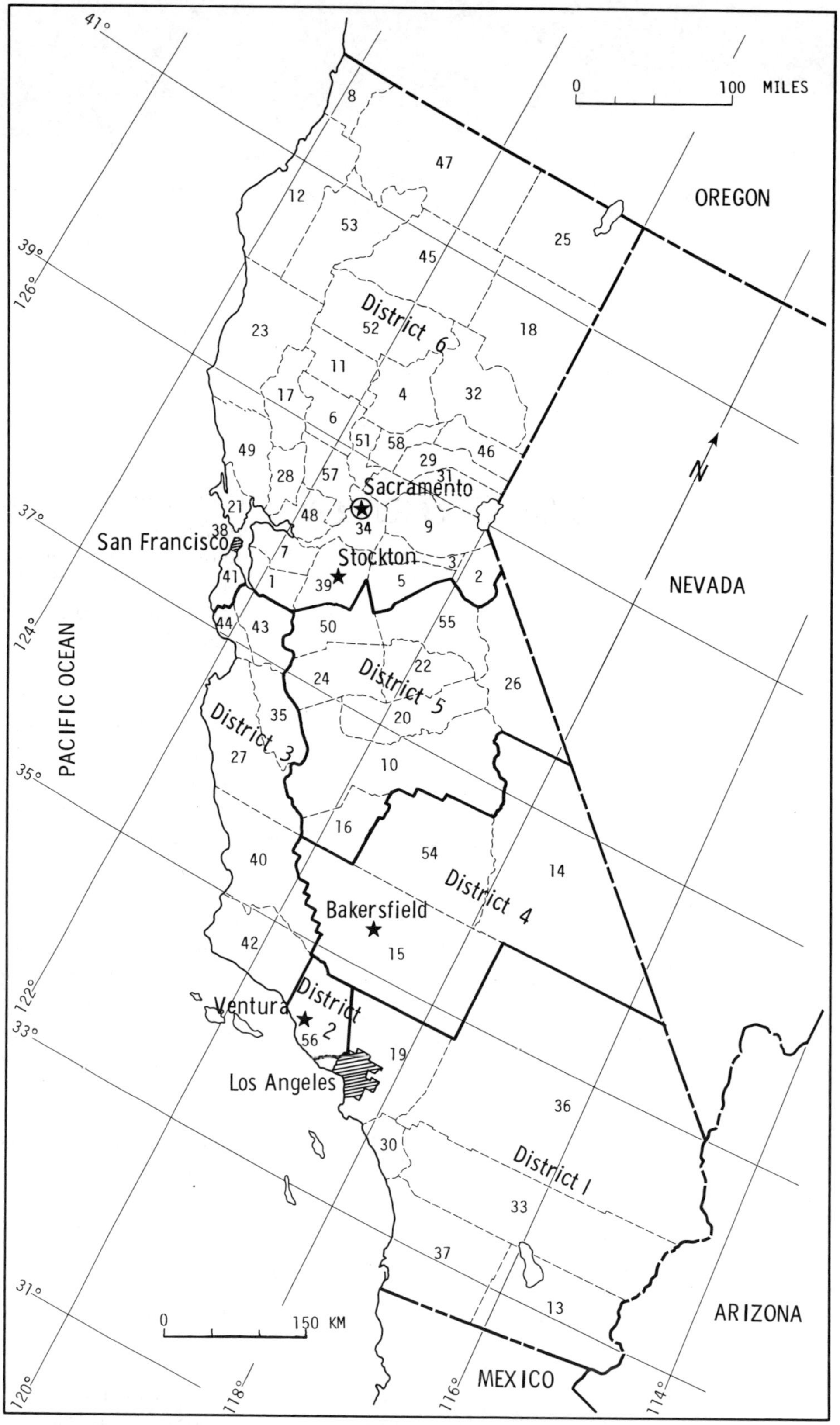

FIGURE 40.1 California. Index map. Also shows oil and gas districts administered by the California Division of Oil and Gas. County Key on left.

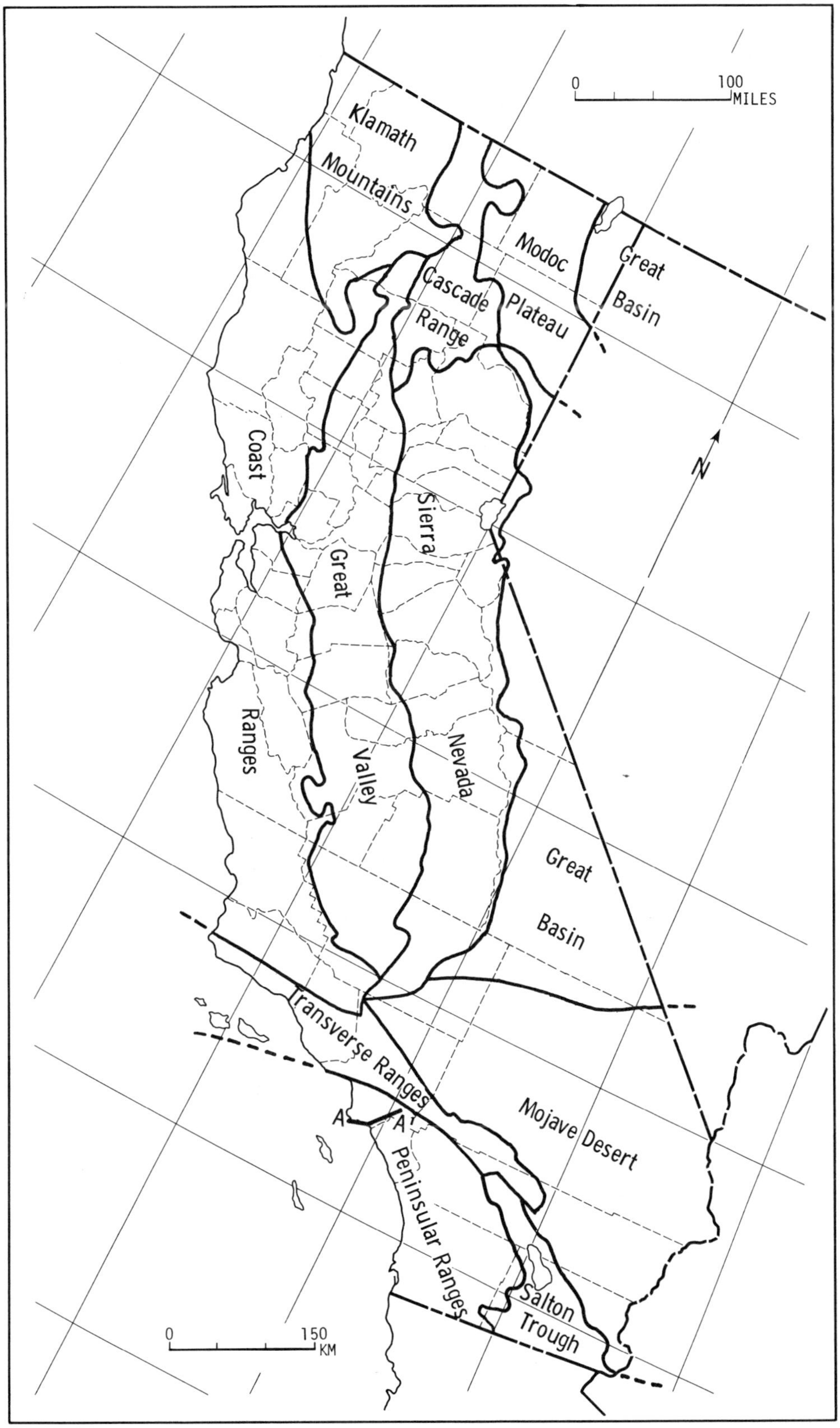

FIGURE 40.2 California. Geomorphic provinces. *After Olaf P. Jenkins (1941, Figure 37, p. 85) and J. P. Albers (1966a).*

from Los Angeles County to Humboldt County 200 miles north of San Francisco, and some small production resulted. In 1874 two successful wells were drilled in Pico Canyon near Newhall, west-central Los Angeles County. Sustained commercial production started in 1875, and fields in this vicinity have been continuously productive ever since (Anon., 1953).

Between 1875 and 1905, some thirty fields were opened up in southern California. All were situated near seeps, the presence of which was responsible for the exploration. Included among the fields discovered between 1899 and 1905 were such production giants as Coalinga, McKittrick, Midway–Sunset, Kern River, and the original Santa Maria field (now Orcutt). The traps found by seep drilling included anticlines, faults, and various types of permeability pinchouts. Concurrently the value of geology to the search for oil became recognized, and company geological staffs were organized. Surface mapping of anticlines and faults resulted in another great spurt in discovery between 1920 and 1924, when the Los Angeles basin fields of Huntington Beach, Santa Fe Springs, Long Beach, Torrance, Dominguez, and Inglewood were found. The seismograph and subsurface geology have contributed importantly to subsequent discoveries.

Oil production has moved north and west in recent years. In the late 1940s two giant oil fields, San Ardo and South Cuyama, were discovered in the Salinas Valley and Cuyama basins in the Coast Ranges (Figure 40.2).

The biggest development in recent years, however, has been the expansion of offshore drilling and production. This was a major factor in California's resurgence in output in the latter half of the 1960s.

Regional Geology. A simplified geologic map of California is given in Figure 40.3. The distribution pattern of the four map units shown resembles very closely the geomorphic province map (Figure 40.2). Basement rock, which crops out in the Coast Ranges and underlies the sedimentary basins and some of the volcanic areas, is designated by diagonal lines. It is assigned to the Franciscan–Knoxville Group of Late Jurassic age and consists of strongly deformed sediments and basic igneous rocks. Younger are the Cretaceous to Quaternary sedimentary rocks (blank pattern), where the oil and gas fields have been found. During and after the deposition of the sedimentary rocks, in Tertiary and Quaternary time, volcanic activity produced lava flows (second pattern) both locally and regionally, as in the Modoc Plateau. Granitic and metamorphic rocks (fourth pattern), ranging in age from Cretaceous to Precambrian, crop out over large areas, especially in the Sierra Nevada and adjoining physiographic provinces to the east and south. Included in this pattern are scattered local occurrences of Cenozoic nonmarine sediments and volcanic rocks in the desert areas of southeastern California.

Figure 40.4 shows the stratigraphic columns, surface and subsurface, in the major oil- and gas-producing areas. The greatest development is in the Tertiary, especially in the Pliocene and Miocene. The lower Pleistocene and Eocene are also important; so is the Cretaceous as a source of dry gas in northern California. The sediments are almost entirely clastic, consisting of shales, silts, sands, and conglomerates. They are both marine and nonmarine in origin.

It can be concluded that the Great Valley and the basins in the coastal ranges (including the Transverse and the Peninsular) were inundated during much of the Tertiary and, in places, into the Quaternary. From the organic characteristics of much of the sediment deposited in these seas, and from the enormous quantities of oil stored within these sediments, it can be concluded that the Tertiary and early Quaternary seas were teeming with aquatic life. At the same time, clastic sediment, including potential reservoir sands, was being washed in from nearby mountain areas. For these reasons, the sedimentary sections in the California oil districts contain many reservoir beds, and there is abundant oil within the reservoirs.

The reservoir rocks of California range in age from Pleistocene to basement rock (Figure 40.5). The Kern River Formation, which is productive in San Joaquin Valley, extends from the upper Pliocene into the Pleistocene. Other Pliocene reservoirs of note are the Pico, Repetto, and Etchegoin. So far, the Pliocene and the Miocene have produced most of California's oil. Important Miocene reservoirs include the Monterey, Temblor, and Vaqueros. The Oligocene Sespe Formation produces considerable oil in the Ventura–Santa Clara district. The Eocene was a very minor source of California oil until the discovery of the Gatchell zone at Coalinga, from which considerable oil has been obtained. Eocene, Paleocene, and Upper Cretaceous reservoirs are the major sources of dry gas in the Sacramento Valley. The oldest reservoir rock is pre-Cretaceous schist and other metamorphosed basement rocks.

California's reservoirs are unusual. An individual producing well may have an oil column of as much as 1500 feet, but the actual reservoir is a series of 10 to 20 foot sands, each one with its own edge water.

It has been estimated that sand or sandstone is the reservoir rock for 98 per cent of California's oil (Hoots, 1943, p. 267). The remaining 2 per cent comes from fractured chert, shale, and schist and from conglomerate composed of schist fragments. In addition, some gas is produced from siltstone and sandy shale. At least 5 per cent of the oil produced in California comes from reservoirs that are nonmarine in origin. In some places these formations grade laterally into marine facies.

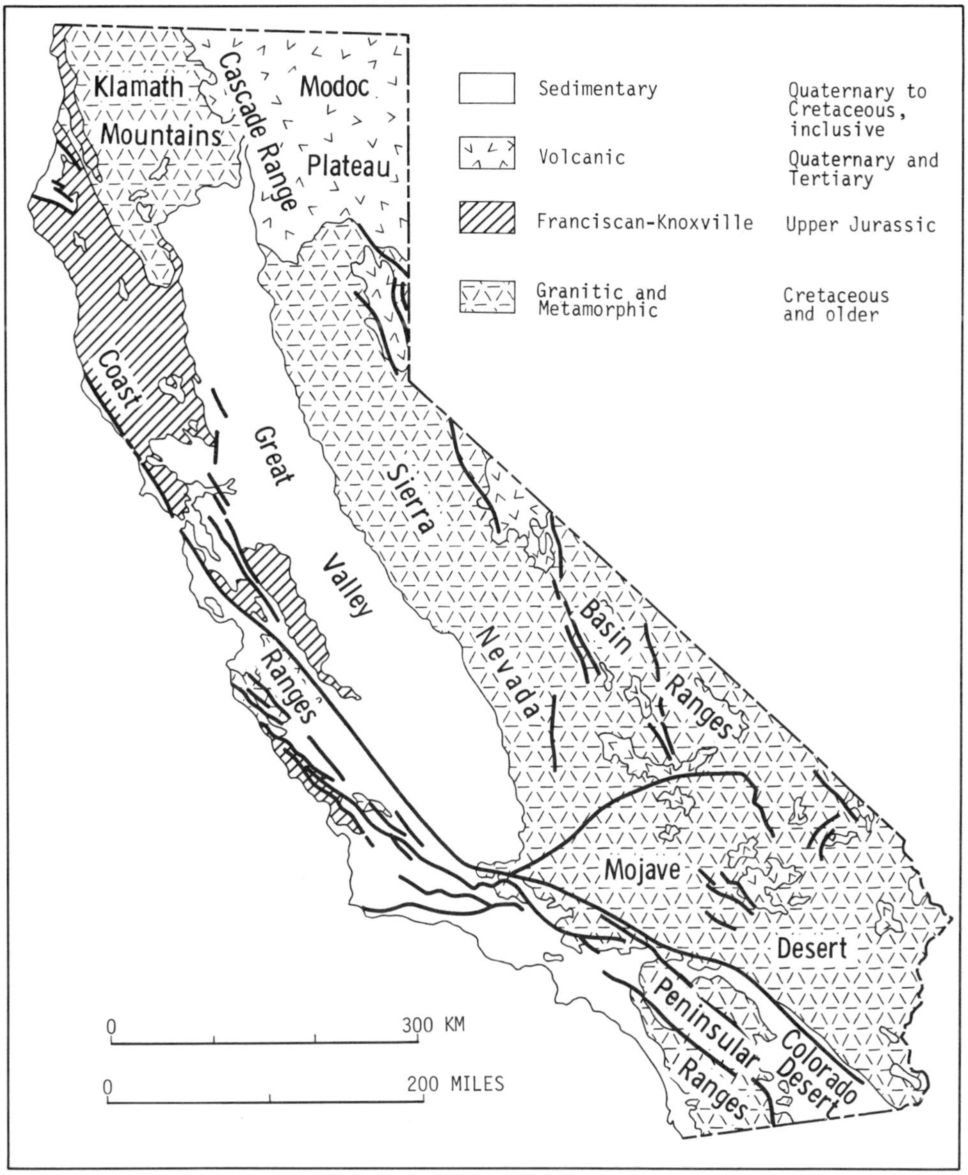

FIGURE 40.3 California. Major rock units and principal faults. Oil and gas fields are so far confined to white (sedimentary rock) areas. *After Charles F. Jennings (1957, Figure 2, p. 411). Courtesy California Division of Mines and Geology.*

Practically every type of hydrocarbon trap except those peculiar to carbonate rock reservoirs has been found in California. The anticline is the most important in terms of volume of oil (about two-thirds of the state total) so far produced. Second in importance is sealing by truncation and overlap. Faulting is common in California and has controlled the distribution of oil across anticlines in numerous fields. In addition, there are some fields in which the trap has been created by the faulting of a plunging anticline. Some trapping has been brought about by lensing of the sandstone reservoirs; in other accumulations wedgouts are the result of truncation and sealing. Although California is the best known locality for brea sealing, the actual volume of oil impounded below solid hydrocarbon barriers is relatively insignificant.

DISTRIBUTION OF OIL AND GAS FIELDS

Of the 58 counties in California, 32 produce oil or gas or both. The state is divided into 6 petroleum districts for administrative purposes (Figure 40.1). Topographically, California is divided into 11 provinces as shown in Figure 40.2 (Albers, 1966a). All of the oil and gas fields

Geologic Province			Peninsular Ranges	Transverse Ranges	California Coast Ranges
Geologic Division			SE Los Angeles Basin	Western part	
Cenozoic	Quaternary	Recent	Alluvium	Alluvium; Older alluvium	Dune sands, beach deposits alluvium, bay muds
Cenozoic	Quaternary	Pleistocene	Terrace deposits San Pedro Fm. La Habra Fm.	Santa Barbara Fm. Mud Pit Sh.	Terrace deposits, older bay muds; Orcutt Sd., Aromas Red Sds., Red Bluff Fm.; Santa Clara, Tulare, Paso Robles, Merced, San Benito, and Tehama Fms.
Cenozoic	Tertiary	Pliocene	Fernando Fm.	Pico Fm.	San Joaquin Etchegoin Jacalitos Purisima Fms.; Careago Ss. Foxen Ms.; Sisquoc Fm.
Cenozoic	Tertiary	Miocene	Puente Fm. and Monterey Sh.; Topanga Fm.; Vaqueros Fm.	Sisquoc Fm.; Monterey (or Modelo) Sh.; Volcanic rocks; Topanga Fm. Rincon Sh. Vaqueros Ss.	San Pablo, Santa Margarita, Reef Ridge, Monterey, Rincon, and Temblor Fms.; Volcanic rocks; Vaqueros Fm. San Ramon Ss.
Cenozoic	Tertiary	Oligocene	Sespe Fm.	Alegria Fm.; Sespe Fm.	Kirker, Tumey, San Ramon, and San Lorenzo Fms.; Sespe Fm.
Cenozoic	Tertiary	Eocene	Santiago Fm.	Gaviota Fm.; Coldwater Ss.; Sacate Fm. Cozy Dell Sh. Matilija Ss. Juncal Fm.	Church Creek Beds Kreyenhagen, Markley, Nortonville, Domengine, Capay, and Lodo and Meganos Fms.; Butano Ss.
Cenozoic	Tertiary	Paleocene	Silverado Fm.	"Martinez"; Jalama Fm.	Martinez, and Laguna Seca Fms.
Mesozoic	Cretaceous		Williams Fm. Ladd Fm. Trabuco Fm.; Southern California batholith; Santiago Peak Volcanics	Espada Fm.; Franciscan Fm.	Uppermost Cretaceous marine shelf deposits; Marine shelf and slope deposits equivalents to Great Valley sequence; Franciscan Fm.; Granitic rocks; Marine eugeosynclinal deposits of Franciscan Fms.; Ultramafic rocks; Knoxville Fm.
Mesozoic	Jurassic		Bedford Canyon Fm.		

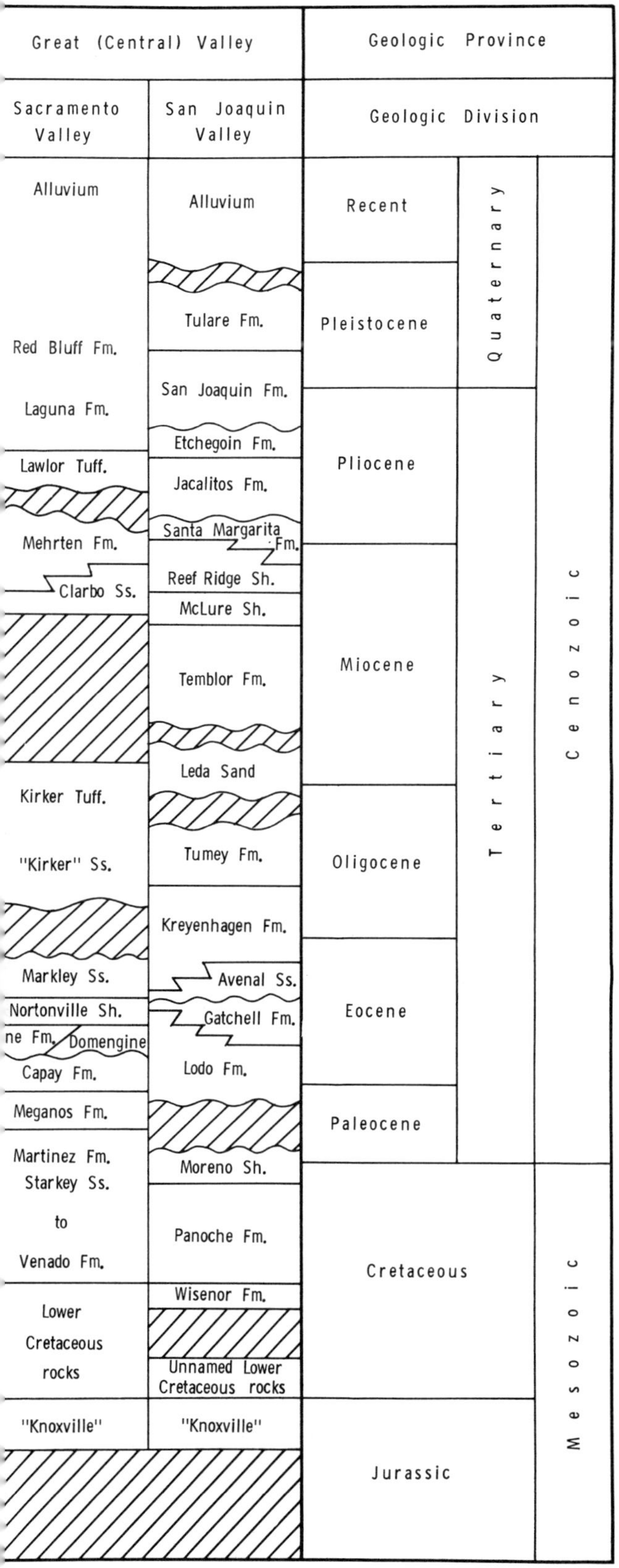

Figure 40.4 California. Stratigraphic correlation chart (on two pages), covering oil- and gas-producing areas. Diagonal lines represent missing sections, or (in Jurassic) age not identified. *After J. P. Albers (1966a, Figure 3). Courtesy California Division of Mines and Geology.*

EPOCH	LOS ANGELES	VENTURA	SAN JOAQUIN	SANTA MARIA	SALINAS-CUYAMA	SANTA CRUZ	SACRAMENTO	EEL RIVER
Pleistocene			● ✲					
Pliocene	●	●	● ☼	●	•	•	✲	✲
Miocene	●	●	● ✲	●	●	•		
Oligocene		●	●	•				
Eocene		•	● ✲			•	☼	
Paleocene		•	•				• ☼	
Late Cretaceous		•	• ☼				• ☼	
Pre-Tertiary basement rocks, age uncertain	•		•	•				

EXPLANATION

● Oil production, Large

☼ Gas production, large

• Oil production, small to moderate

✲ Gas production, small to moderate

Figure 40.5 California. Stratigraphic and geographic distribution of reservoir rocks. *Courtesy M. B. Smith and E. J. Schambeck (1966, Figure 49, p. 299), the California Division of Mines and Geology, and Earl W. Hart.*

so far discovered, however, are either in the Great Valley or in sedimentary basins occupying re-entrants into the Pacific border provinces (Figure 40.6). Most of the rest of the state is underlain by igneous rock, both lava and intrusive types; metamorphosed sedimentary rock; or metavolcanics.

The three geomorphic units along the Pacific border are, from southeast to northwest, the Peninsular Ranges, the Transverse Ranges, and the Coast Ranges provinces. The various ranges have an *en échelon* arrangement that permitted the development of elongate and wedge-shaped sedimentary basins between them during Tertiary time. Thus the juncture between the Peninsular and the Transverse Ranges provinces is marked by the prolific Los Angeles basin. The rest of the Peninsular Ranges province, to the east and the south, is occupied mainly by intrusive igneous rocks of the Southern California batholith (Peterson et al., 1966).

The eastward trending Transverse Ranges cut across the grain of the northwest trending (in southern California) Coast Ranges to the north. They consist mainly of exposed crystalline basement rock and Mesozoic sediments and volcanics. The oil-bearing, elongate Ventura basin lies mostly in between mountain ranges to the north and south.

To the northwest and north are the Coast Ranges, which separate the Great Valley from the Pacific Ocean. These ranges consist of a complex of Mesozoic and Cenozoic igneous, sedimentary, and metamorphic rocks. The igneous rocks are intrusive, both basic and acidic, and extrusive basaltic lavas. Oil and gas fields have been found in four areas in the Coast Ranges.

Production in the Great Valley province is prolific in both oil and gas in the southern part and is mostly dry gas in the northern part.

Of the remaining seven geomorphic provinces, three are probably barren of hydrocarbon deposits because of metamorphism or extensive igneous activity or both. These are (1) Klamath Mountains (both), and (2 and 3) the Cascade Range and Sierra Nevada (batholithic intrusions). A fourth province, the lava-covered Modoc Plateau, may be in part underlain by sedimentary strata. Three nonproductive provinces (Great Basin, Mojave Desert, and Salton Trough) contain Tertiary sedimentary rocks and are mentioned under "Future" (p. 488).

The oil and gas fields of California are shown in Figure 40.6. Also shown are the eight sedimentary basins in which these fields are grouped. Six of the eight basins either border the Pacific or are a short distance inland. These are, from south to north, Los Angeles, Ventura, Santa Maria, Cuyama–Salinas, Santa Cruz, and Eel River. The remaining two basins are within the Great Valley, San Joaquin to the south and Sacramento to the north.

The major oil-producing basins, with the percentage of the state production in 1964 in parentheses, are San Joaquin (41), Los Angeles (34), Ventura (15), Salinas–Cuyama (6.3), and Santa Maria (3.4) (Smith and Schambeck, 1966, p. 311). The Los Angeles basin has probably increased its percentage in more recent years because of increased production, especially offshore. The San Joaquin and Los Angeles basins are also the largest producers of wet gas, whereas the Sacramento basin is the source of 80 per cent of the dry gas. One-third or more of the virgin gas produced in California each year is dry gas.

The traps, reservoirs, and selected fields in the various basins are discussed by basins. The order of treatment is from south to north, first in the coastal provinces and then in the Great Valley.

Los Angeles Basin. Although the Los Angeles City field was discovered in 1892, it was not until the early 1920s that the Los Angeles basin assumed world importance as a source of oil. Three of the four leading fields, Long Beach, Santa Fe Springs, and Huntington Beach, were discovered during that time. Seeps played an important role in most early discoveries. The fourth and currently the largest field, Wilmington, was discovered in 1932.

Topographically, the Los Angeles basin is a plains area containing, however, various hills and knolls that reflect underlying anticlinal structures. Many of these hills, such as Signal Hill (Long Beach), are notable for their oil fields. The basin is triangular in shape, extending along the coast about 70 miles and inland for a maximum distance of 40 miles. It also extends westward beneath the waters of the Pacific. The land area is about 1450 square miles, and the maximum depth to the basement is at least 31,000 feet. The sedimentary rock volume has been estimated at about 2250 cubic miles (Smith and Schambeck, 1966, p. 312). Already more than 2 million barrels of oil has been produced for each cubic mile in the entire basin.

Geology. The Los Angeles basin (Woodford et al., 1954) is notable for the complexity of its geology, most of which is concealed beneath a cover of Quaternary alluvium (Yerkes et al., 1965). It is bounded on the north by the San Gabriel Mountains of the Transverse Ranges, and on the southeast by the Santa Ana Mountains and other segments of the Peninsular Ranges. Major faults with a northwest trend slice across the basin; the Malibu–Cucamonga fault zone cuts eastward at the northern edge of the basin at the foot of the San Gabriel Range (Smith, 1964). Prominent fold axes also trend northwesterly in the basin.

Figure 40.7 is a cross section of the Los Angeles basin cutting northeasterly from the southwest corner (location shown on Figure 40.2). It shows the major anticlines,

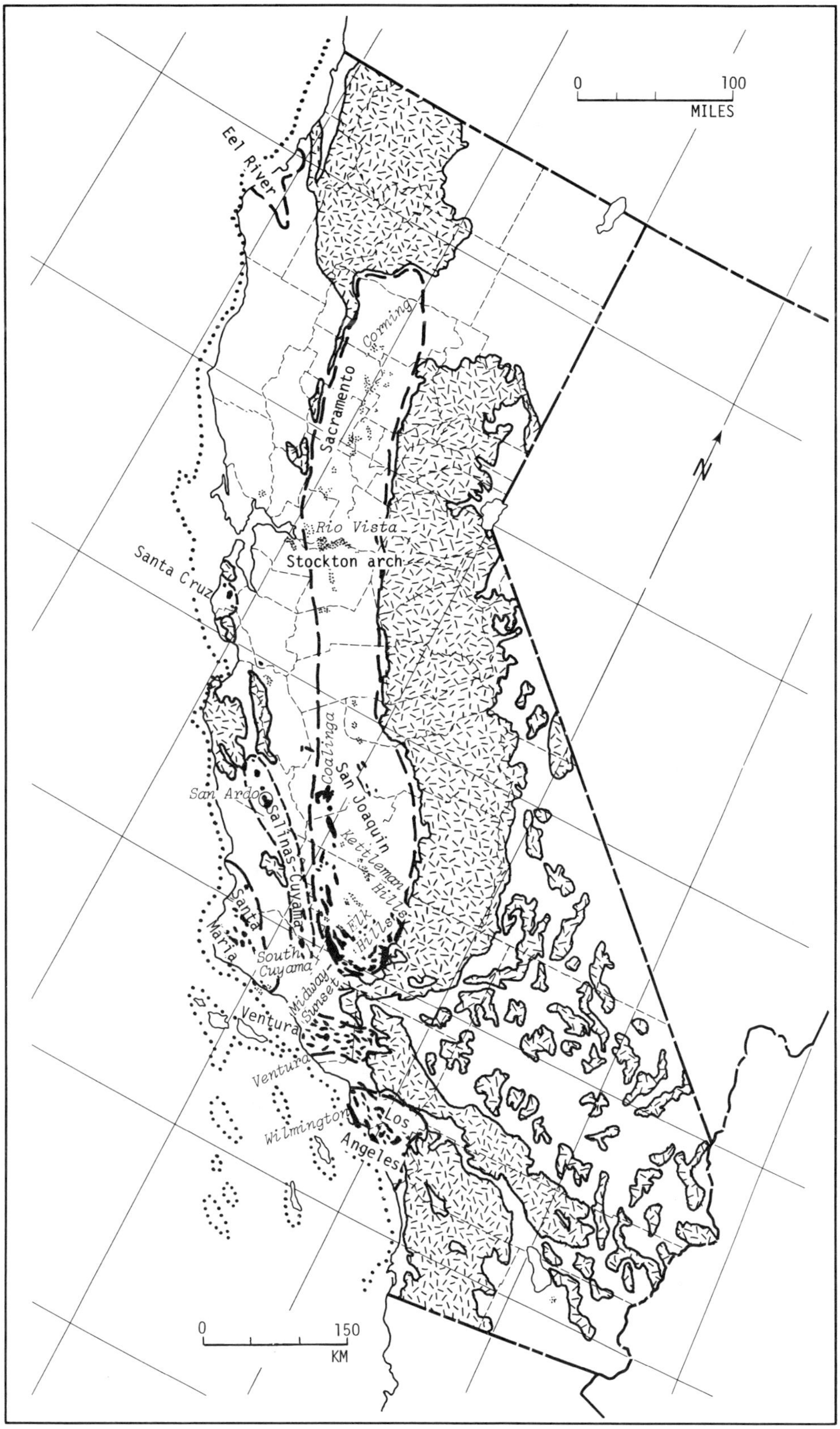

Figure 40.6 California. Oil and gas fields. Oil fields solid black; gas fields stippled. Only a few described fields are named, because of congestion. Other fields are located within counties in text. Dashed lines enclose sedimentary basins. Offshore dotted lines show approximate edge of continental shelf. Random dash pattern designates outcrops of both basement rocks and intrusive igneous rocks. *After U. S. Geological Survey (1963).*

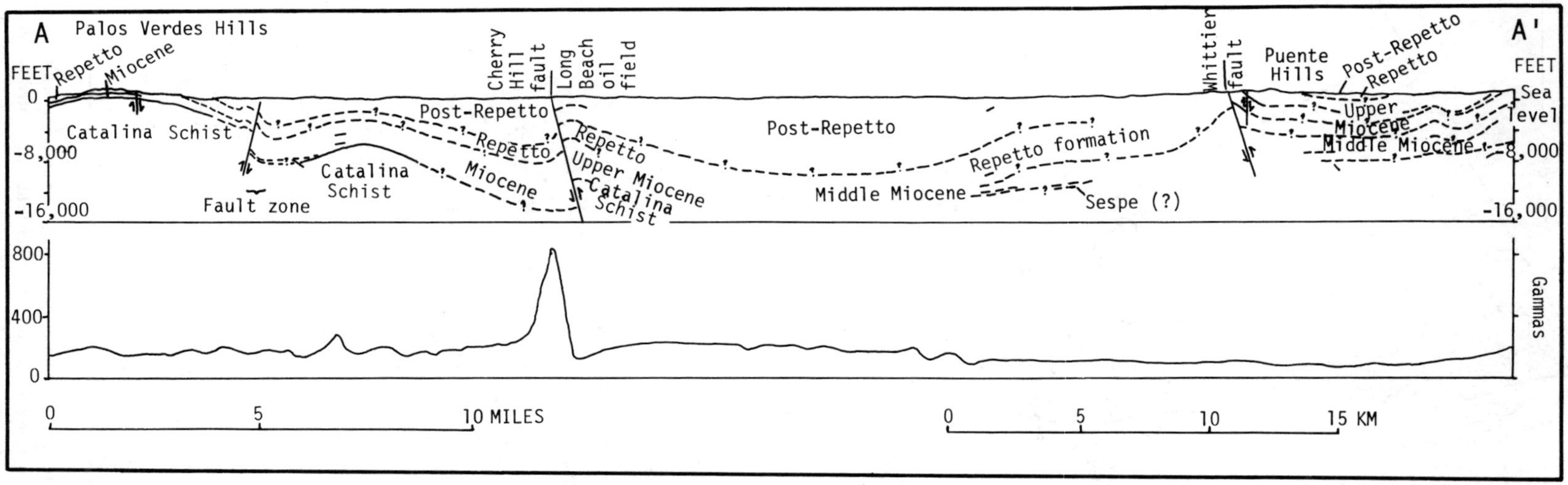

Figure 40.7 Cross sections along line *A–A'* on Figure 40.2. Upper: geologic section. Lower: aeromagnetic profile, regional gradient removed. *After Jack E. Schoellhamer and Alfred D. Woodford (1951, Sheet 1).*

synclines, and faults. The deepest part of the basin is about 10 miles east of the Los Angeles city center. The greatest deformation began in the Miocene and culminated (but did not cease) in mid-Pleistocene time.

Trapping is principally anticlinal, with faults playing an important role in some fields.

The stratigraphic columns for the western and eastern parts of Los Angeles basin are given in Figure 40.8. On the west side middle Miocene sediments rest on basement schist; in eastern Los Angeles older formations ranging in age from early Miocene to Late Cretaceous are also present. Sandstones and conglomerates in the Repetto (early Pliocene) and late Miocene formations are the reservoirs for by far the greater part of the oil.

The Los Angeles basin is perhaps the best endowed basin in terms of eventual recovery of oil and gas per cubic mile in the United States. Barbat (1958, p. 1) ascribes this lush production to the following factors:

1. Ample deposition of organic material.
2. Adequate protection from chemical and biological destruction.
3. Ample load compression, particularly in the center of the basin, to squeeze entrapped fluids from the fine-grained rocks.
4. Ample interfingering of carrier and reservoir sands of lateral persistence with the fine-grained rocks.
5. Available traps of considerable size near the margins of the basin to remove hydrocarbons from the fluids before they were expelled.
6. Relative geological youth of the area, absence of a long-standing load of superjacent rock, and a single, moderate orogeny resulting in little post-depositional alteration of the sediments.
7. Absence of large-scale uplift and major erosion, with consequent preservation of reservoir fluids and pressures.
8. Intense exploration and development.

The oil fields are fairly densely scattered over the entire basin. The greatest concentration of giant fields, however, is on the western side of the basin, where the second largest oil field in the United States, Wilmington, has as neighbors immediately to the east and north Long Beach and five other large fields strung like beads along the Newport–Inglewood fault zone.

Gas in the Los Angeles basin is discussed by Brown (1968).

Selected Fields: The *Wilmington* oil field has already produced over 1.1 billion barrels of oil, making it the foremost field in California and the second largest in the United States. Wilmington has also yielded between 850 and 900 billion cubic feet of gas (Hamilton, 1968). This field is in the Wilmington–Long Beach Harbor area of Los Angeles County, 20 miles south of Los Angeles. It extends southeastward beneath virtually all of the city of Wilmington to an offshore point southeast of Long Beach. Although accumulation in the Wilmington field is anticlinal, the anticline, unlike those in the nearby Long Beach and Torrance fields, is not obvious at the surface. The topography is only very slightly elevated, and the bedrock structure is obscured by a thick deposit of Quaternary alluvium. For this reason discovery came late.

EPOCH	WESTERN LOS ANGELES BASIN Southwest of Newport-Inglewood Fault Zone		EASTERN LOS ANGELES BASIN Northeast of Newport-Inglewood Fault Zone	
UPPER PLEISTOCENE	Terrace depsoits Palos Verdes		Terrace deposits Palos Verdes	
LOWER PLEISTOCENE	San Pedro-Timms Point-Lomita		La Habra	■
UPPER PLIOCENE	Pico	⊡ ■	La Habra Pico	■
LOWER PLIOCENE	Repetto	⊡ ■	Repetto	■
UPPER MIOCENE	Puente or Monterey	■	Puente or Monterey	■
MIDDLE MIOCENE	Monterey		Monterey Topanga	■
LOWER MIOCENE	Absent		Vaqueros	
OLIGOCENE	Absent		Sespe	
EOCENE	Absent		Santiago	
PALEOCENE	Absent		"Martinez"-Silverado	
UPPER CRETACEOUS	Absent		"Chico" (Williams-Ladd-Trabuco)	
JURASSIC (?) and older	Catalina schist (Franciscan ?)	■	Santa Monica Slate Plutonics	

■ Denotes Oil Production ⊡ Denotes Dry Gas Production

Figure 40.8 **Los Angeles basin; reservoir rocks.** ***California Division of Oil and Gas (1961, Part 2, p. 501).***

The first explorations were carried out in an attempt to extend the Torrance field to the southeast. Subsurface studies based on the information obtained during the drilling of these earlier wells, plus a seismograph survey, resulted in the discovery of the Wilmington field in January, 1932. The discovery well produced low-gravity oil, however, and did not create much excitement, although it did motivate extensive seismographic surveys. These were followed in December, 1936, by the drilling of a well that yielded 1400 barrels a day of 20° gravity oil. Rapid development ensued.

The Wilmington anticline is a large and irregular dome with the strata dipping off the axis at angles ranging from 2° to 15°. It trends southeasterly. Five major north-south normal faults divide the anticline into six fault blocks. Figure 40.9 is a section across the anticline. Most of the folding and the faulting took place before

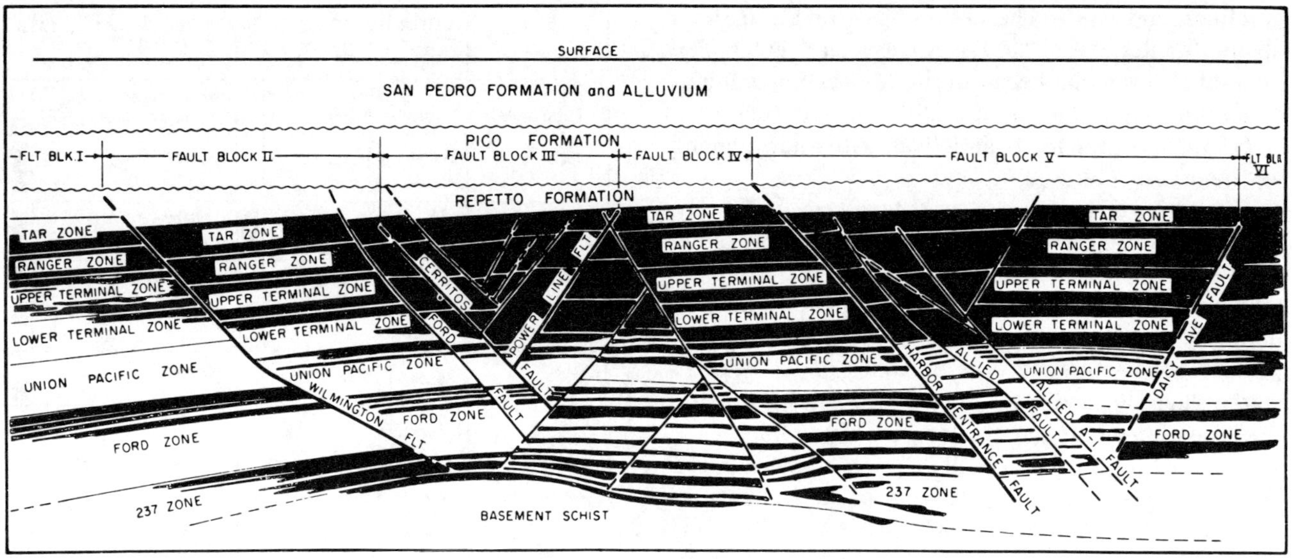

FIGURE 40.9 Wilmington field, Los Angeles County. Northwest-southeast cross section; length of section is $3\frac{1}{2}$ miles. Producing zones shown in black. *Courtesy S. V. Hamilton (1968, Figure 3, p. 167) and the American Association of Petroleum Geologists.*

the deposition of the upper Pliocene sediments. The trapping is entirely anticlinal, but in some of the faulted areas production is confined to the upblock. The oil comes from eight producing zones in the lower Pliocene Repetto Formation, in the underlying upper Miocene Puente Formation, and in the basement schist. Impermeable shale interbeds seal the oil and gas within individual sandstone layers.

Surface subsidence to a maximum of 29 feet, which accompanied the removal of oil in the Wilmington–Long Beach area, led to the injection of water at a rate of over 1 million barrels daily (in 1968) into the reservoirs. This not only has practically stopped subsidence, but also has increased daily oil production over 50,000 barrels (Smith and Schambeck, 1966, p. 313).

The development of the southeastern part of the Wilmington anticline ("East Wilmington") beneath the waters of the Pacific off Long Beach started in 1965, after agreements were reached between a consortium of five major oil companies (THUMS) and the governments of Long Beach and California. This development eventually may come close to doubling the total output of the Wilmington field. Directional drilling is done on four man-made islands in Long Beach harbor. The landscaping of these artificial islands, along with the concealment of operations inside mobile (on tracks) pseudo-apartment buildings, represents the acme of sophisticated camouflage.

Next door to the Wilmington field, to the east, is the justly famous *Long Beach* field. Through 1967 it had already produced more than ½ million barrels of oil per acre. The cumulative production of 856 million barrels places Long Beach third (after Wilmington and Midway–Sunset) in total yield in California.

The Long Beach field is about 20 miles south of Los Angeles and back of the narrow coastal plain on which most of the city of Long Beach has been built. It occupies a ridge about 5 miles long culminating in Signal Hill, which rises to a maximum of about 300 feet above the plain (Dudley, 1954). The ridge is due to the recency of the folding and is characteristic of the geotopography along the Newport–Inglewood fault zone.

The Long Beach anticline was recognized many years before the discovery well was drilled in 1921. Signal Hill and the surrounding territory had been subdivided and streets laid out, and although very few residences had been built the lots were individually owned. The discovery of oil was followed by an inefficient and chaotic town-lot drilling campaign. The anticline is 1 by 4 miles in approximate dimension and trends northwest. The dips on the southwest flank range from 40° to 45° as against 15° on the northeast flank. The closure is over 1200 feet. At least two faults cut the anticline. The longer one of these, the Long Beach fault, has a maximum displacement of 1000 feet. It is so recent in age that it makes a prominent scarp (Signal Hill) at the surface. Accumulation is anticlinal, however, rather than fault controlled.

Below the Quaternary sands and gravels exposed at the surface are from 1000 to 1500 feet of Pleistocene sediments. The underlying Pliocene is 3600 feet thick. It is divided into the upper and the lower Pliocene. The lower sands of the upper Pliocene have been found productive here and there along the top of the anticline and on the southwest flank. The most prolific reservoirs are the Alamitos zone and the underlying Brown zone of the lower Pliocene. The Alamitos zone is about 650 feet thick and consists of four sands up to 175 feet in thickness with interbedded shales. The Brown zone, which is about 1450 feet thick and extends into the upper Miocene, likewise consists of a series of thick sandstones separated by shales. Production also has been obtained from two deeper Miocene sand zones, including the DeSoto, which overlies basement schist conglomerate at depths below 10,000 feet. No oil has been found as yet in the basement rock.

Four other giant fields and several smaller ones (Allen and Joujon-Roche, 1958) lie along the Newport–Inglewood fault zone. These include *Huntington Beach* (a substantial portion of which is offshore) to the southeast of *Long Beach,* and *Dominguez* and *Inglewood* to the northwest. All are elliptical faulted anticlines, marked at the surface by hills and in a few instances by seeps. All closely parallel the trend of the fault zone except Dominguez, which is more westerly in axial direction (Graves, 1954).

Urban Los Angeles, covering over 450 square miles, contains more than a dozen oil fields. Some antedate zoning laws and building congestion; others have been found in recent years, and discovery is continuing. This upsurge in both exploration and exploitation in the "asphalt jungle" (Kilkenny and Van Amringe, 1965) has been made possible by the development of unobtrusive searching and producing techniques that led to a lessening of legal restrictions. Examples of these techniques are light-charge seismic surveys along freeway and interurban rights-of-way and camouflaged, soundproofed drilling structures.

Among the older urban fields are *Inglewood* and *Beverly Hills* at the northern end of the Newport–Inglewood fault zone. Here the trapping is similar to the described fields on the fault zone to the southeast. The newer fields produce from alluvium-concealed anticlines that are sharp and locally thrust faulted, "with steep to overturned southern flanks" (Kilkenny and Van Amringe, 1965). Sands in the upper Miocene are the principal reservoirs.

The *Santa Fe Springs* oil field, 12 miles southeast of the center of Los Angeles, is another California giant, with a cumulative production in 1968 of nearly 600 million barrels. Furthermore, all of this oil has come from an area of less than 1500 acres. Two surface indications, the presence of a slight topographic doming and of gas in water wells, led to the drilling of the discovery well in 1919. In 1923 this field produced one-sixth of the country's oil and created near panic at the market place.

The structure contour map (Figure 40.10) shows that the trap is an unusually symmetrical elliptical dome with more than 600 feet of closure. The clue to the 400,000 barrels per acre production lies in the number (nine) and the aggregate thickness (5300 feet) of the producing zones below the now abandoned shallow gas reservoir. All of these zones are in the lower Pliocene and upper Miocene. The discovery well was in the Meyer zone, and the flush production of 1923 came from the Foix, Bell, and Meyer. The lower zones were discovered subsequently, as diminishing production encouraged deeper exploration. The latest discoveries took place in 1956, when the Santa Fe zone and still deeper reservoirs were found.

The *Brea–Olinda* field (Durham and Yerkes, 1964; Scribner, 1958; English, 1926) is on the eastern side of the Los Angeles basin about 25 miles southeast of the center of Los Angeles. Tar seeping to the surface from nearly vertical strata led to the discovery of Olinda in 1897 and Brea Canyon in 1899. Development led to the merging of the two fields into one. Since discovery this field has produced over 300 millions barrels of oil in spite of accumulation in a somewhat complex trap (Figure 40.11).

The Brea–Olinda field lies immediately south of a major fault zone, the Whittier. Production comes from zones of sandstone beds in the Fernando Formation (lower Pliocene) and in the underlying Puente Formation of late Miocene age. According to Scribner (1958), these sands are lenticular because of changes in facies from sand to shale in short distances. The structure is a monocline with steep dips to the south. It has been referred to as the northern flank of the La Habra syncline (Durham and Yerkes, 1964). This monocline may also be the south flank of an anticline with its axis close to the Whittier fault zone. The oil is trapped in the steeply inclined reservoir rocks by the updip lensing out of permeability. Toward the eastern (Olinda) end of the field, closure "is provided by slivers of older strata that are thrust up along the fault plane" (Durham and Yerkes, 1964). The near verticality of the reservoir beds has led to directional drilling of production wells.

Other fields along the Whittier thrust zone include the *Yorba Linda* (Heath, 1958) to the southeast and the *Sansinena* (Woodward, 1958) to the northwest. The Whittier fault zone is over 30 miles long, trending in a west-northwest direction. Oil fields are strung out along 20 miles of that distance. The vertical displacement in the Sansinena area varies from 3500 to over 5000 feet. Trapping here is mainly in highly faulted asymmetric anticlines south of the fault, but lenticularity of the reservoir sands may be a contributing factor. The distribution of oil also is controlled to some extent by cross faulting.

Ventura Basin. The Ventura basin district lies to the west and north of the Los Angeles basin. It occupies a long re-entrant with an easterly trend lying mostly be-

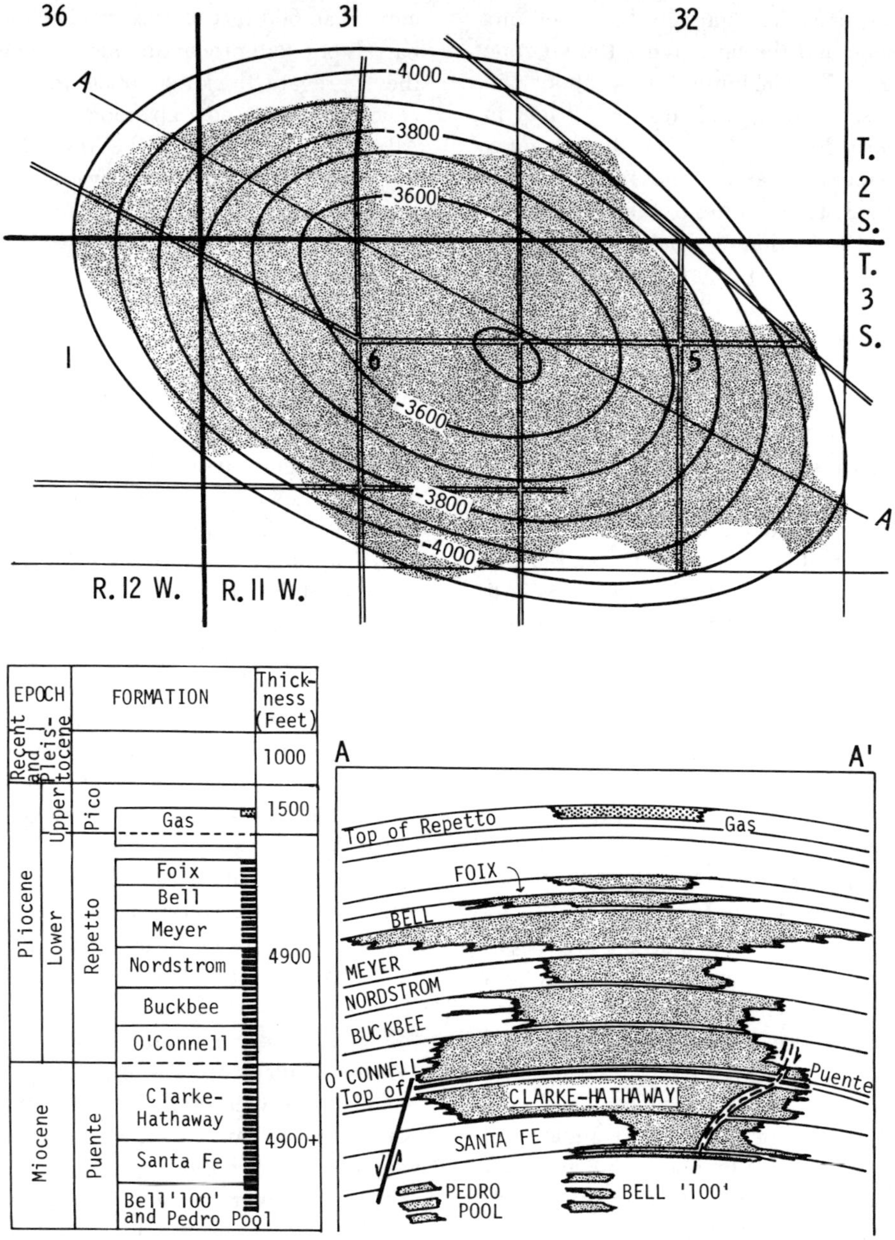

Figure 40.10 Santa Fe Springs field, Los Angeles County. Upper: structure map contoured on top of Bell zone. Interval, 100 feet. Productive area stippled. Numbered sections are 1 mile square. Lower right: cross section along line A-A' in upper drawing. Producing zones stippled. Lower left: stratigraphic section, showing age of reservoirs. *Courtesy California Division of Oil and Gas* (*1961, p. 662*).

tween two of the Transverse Ranges. Most of the emergent part of the Ventura basin is in Ventura County, but the eastern end is in Los Angeles County, 30 to 40 miles northwest of the city of Los Angeles. More than 58 per cent of this sedimentary basin, however, is offshore, beneath as much as 2000 feet of sea water (Smith and Schambeck, 1966, p. 314). The oil fields along the southern coast of Santa Barbara County are high on the north flank of the submerged Ventura basin.

Geology. Structurally the basin is a megasyncline. Compression from the north and south has created large east-trending anticlines and thrust faults. The larger valleys occupy synclines between these anticlines, and the anticlines tend to be mountainous at the surface.

The aggregate thickness of sediments deposited in the basin is about 58,000 feet. The basement rock is granite and schist. There are differences in the surface stratigraphy between the Ventura vicinity on the coast and the

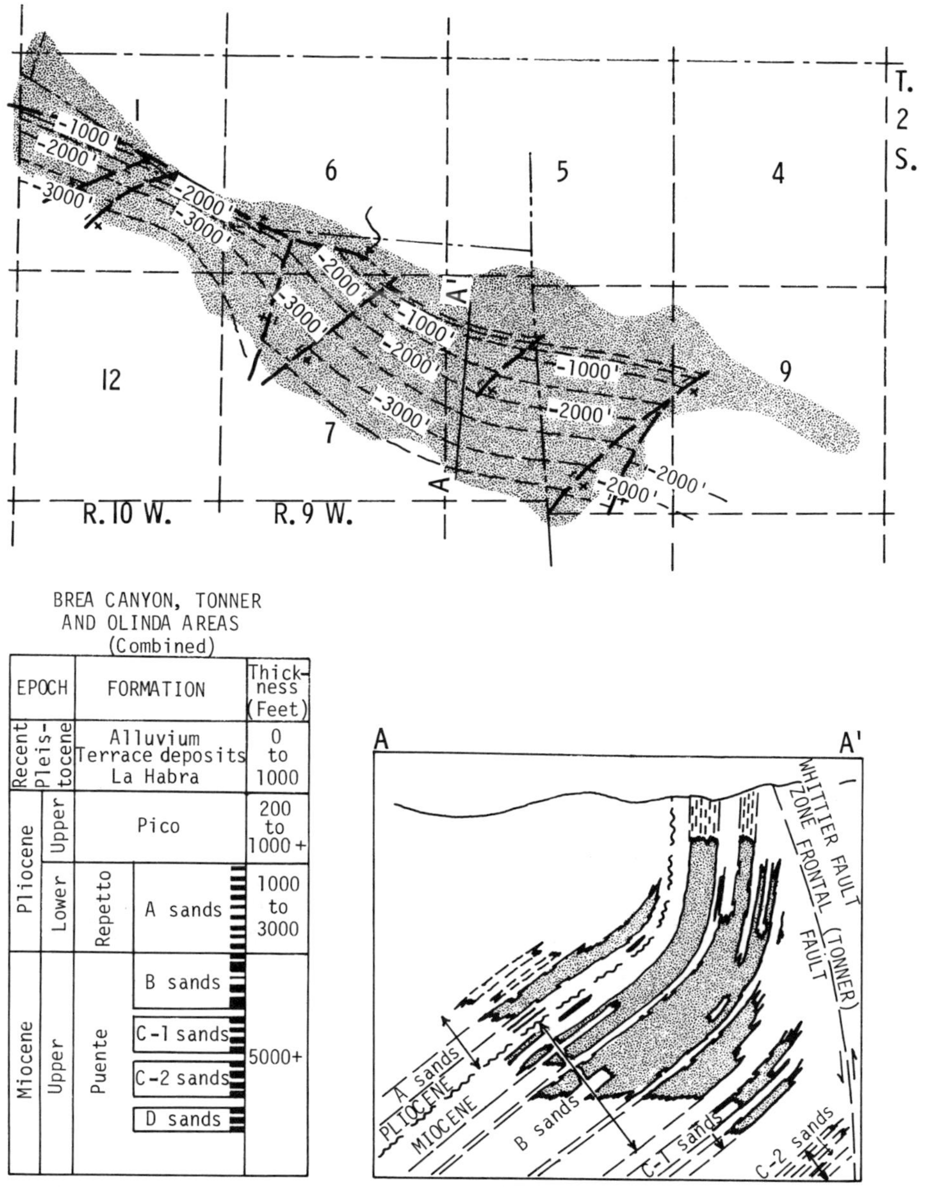

FIGURE 40.11 **Brea–Olinda field, Osage County, California. Upper: structure map contoured on top of Miocene. Beds along north side of field are vertical or nearly so. Contour interval, 500 feet. Production area stippled. Numbered sections are 1 mile square. Lower right: cross section along line A–A′. Lower left: stratigraphic section showing age of reservoirs.** ***Courtesy California Division of Oil and Gas (1961, p. 522).***

eastern part of the basin (Holman, 1958). The oil-yielding lower Pleistocene on the west grades into nonmarine sediments to the east. Likewise thick sections of Pliocene and upper Miocene sediments thin to the eastward and become more continental in type.

The most prolific oil reservoirs are Pliocene and Miocene marine sands, and nonmarine sands in the Oligocene. Lesser production comes from lower Pleistocene sands and from Eocene, Paleocene, and Upper Cretaceous reservoirs (Figure 40.12). A part of the Miocene yield is from fractures.

For the most part, trapping in the Ventura basin is either in faulted anticlines or due directly to faulting. The best examples of fault trapping in California are in the eastern end of the Ventura basin. Varying permeability sedimentary traps are also present, but are relatively minor in terms of oil so far produced. The first commercial oil well completed in southern California appears to have been a 20 barrel well drilled in 1866 in northwestern Ventura County (Carter, 1954). The district's second birthday occurred in 1876, when the Pico No. 4 was drilled near Newhall (Winterer and

EPOCH	CENTRAL VENTURA BASIN		EASTERN VENTURA BASIN Including Soledad and San Fernando Basins	
Upper Pleistocene	Terrace deposits and older alluvium		Terrace deposits	
Lower Pleistocene	Saugus San Pedro-Las Posas Santa Barbara	■	Saugus	
Upper Pliocene	Pico	■ ▩ ■	Sunshine Ranch Pico	■
Lower Pliocene	Repetto	■	Repetto	▩ ■
Upper Miocene	Santa Margarita Monterey-Modelo	■	Towsley Modelo-Monterey Castaic-Mint Canyon	■
Middle Miocene	Monterey-Modelo Conejo-Topanga	■	Modelo-Mint Canyon Topanga-Tick Canyon	
Lower Miocene	Rincon Vaqueros	■	Topanga Vaqueros-Tick Canyon	
Oligocene	Sespe	■	Vasquez-Sespe	
Eocene	Sespe-Coldwater-Cozy Dell-Matilija-Llajas- Juncal-Santa Susana	■ ■	Vasquez Llajas Santa Susana	■
Paleocene	"Martinez"	■	"Martinez"	
Upper Cretaceous	"Chico"	■	"Chico"	■
Jurassic (?) and older	Basement complex		Placerita series Granite	

■ Denotes Oil Production ▩ Denotes Dry Gas Production

FIGURE 40.12 California. Reservoir rocks, Ventura basin. ***Courtesy California Division of Oil and Gas (1961, Part 2, p. 501).***

Durham, 1962). This well, which produced 30 barrels a day, discovered the first oil field in California to become continuously productive. Now the district has more than 80 oil and gas fields and is expanding offshore.

Selected Fields. Outstanding in the Ventura basin is the *Ventura* (also "Ventura Avenue") field 2 to 3 miles north of the city on the coast after which it is named. Discovered in 1916, this field has produced more than 700 million barrels of oil and is fifth in California in cumulative production. It is 7 miles long and about a mile wide. The trap is a very elongate anticline 16 miles long that trends almost exactly east. The dips to the north and south are relatively steep, ranging from 30° to 60°. The fold is cut by several thrust faults, which have permitted some transverse migration and have affected accumulation in several of the reservoir zones. Six oil-bearing zones lie within the Pico Formation of late Pliocene age; the underlying Repetto (early Pliocene) produces a lower-gravity oil.

A second node along the Ventura anticline, to the west and about 4 miles northwest of Ventura, contains the *San Miguelito* field (Haines and Minshall, 1954). It was discovered by surface geology in 1931. This field is smaller than Ventura, about 1¾ by ½ miles, but the trapping is likewise anticlinal. An unusual feature is the presence of a thrust fault on the north side of the anticline that dips to the south beneath the fold, limiting production to some extent both by dropping the reservoirs on the north flank below the oil-water interface and through reduction of permeability by crushing reservoir beds in the fault plane.

Northwest of San Miguelito, and extending offshore into the Pacific, is the 100 million barrel *Rincon* field (Paschall, 1954), which occupies a westerly trending anticline lying *en échelon* with the western end of the Ventura fault. This anticline is also thrust faulted. The reservoirs are thick sands separated by shale interbeds in the upper Pliocene Pico Formation. Trapping is anticlinal, with some local damming along the southeastward plunging anticline by cross faults.

Still farther to the west, along the south shore of Santa Barbara County, about 12 miles west of the city of Santa Barbara, is a string of oil and gas fields. The largest of these is *Elwood*, also a 100 million barrel field. Trapping is in an elongate east-west anticline, with production coming from Miocene and Oligocene sands that lie mostly offshore. Of historical interest is the presence, about 6 miles east of the city of Santa Barbara, of the *Summerland* field. This field was discovered onshore in 1894, and by 1896 it was extended offshore by means of wells drilled from piers extending out into the ocean. This was the first offshore production in California, and perhaps in the country as well (Anon., 1959). Subsequently offshore Ventura basin has produced 150 million barrels of oil through 1967; important offshore fields include (in addition to Summerland) South Elwood and Conception.

Yet another milestone in offshore production was reached in mid-1968, when the first oil field discovered in federal waters in the Santa Barbara Channel began flowing 7000 barrels a day ashore from six wells drilled from a single platform. The platform is capable of handling 58 more directionally drilled wells. It rests in 151 feet of water 4 miles offshore, south of the Carpinteria field (Oil and Gas Jour., Aug. 5, 1968, p. 77).

Moving inland (east) from the Ventura field, the next large field is *South Mountain*, 1 mile south of Santa Paula, another production giant. Although it was discovered in 1915, large-scale production did not take place until after 1942. A second birthday came in 1955, when Pliocene production was found in the overridden block beneath the Oakridge fault (Ware and Stewart, 1958). In recent years this pool, known as the Bridge area, has accounted for more than one-half of the annual production of the South Mountain field.

Figure 40.13 (upper) is a structure contour map of South Mountain field. The anticline is a large one, "abundantly exposed in numerous rugged canyons and cliffs" (Baddley, 1954). The Oakridge thrust fault surfaces along the Santa Clara River at the northern foot of South Mountain. In the lower righthand corner of Figure 40.13

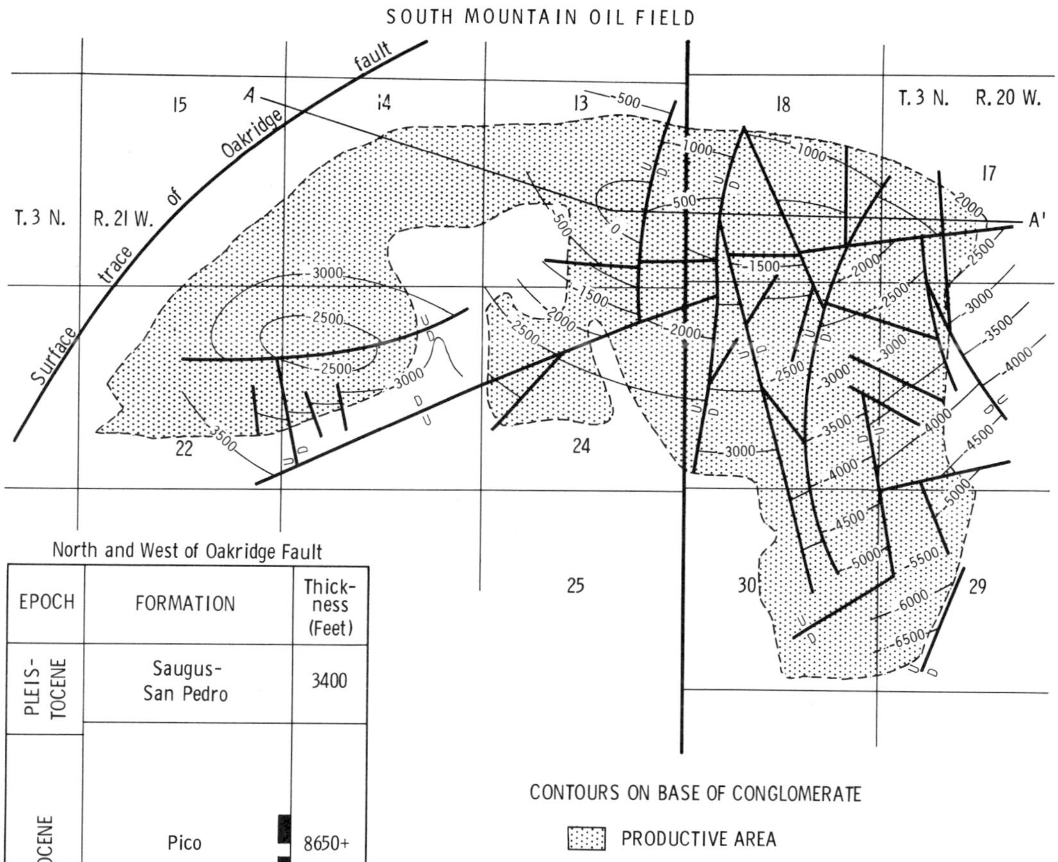

North and West of Oakridge Fault

EPOCH	FORMATION	Thickness (Feet)
PLEISTOCENE	Saugus-San Pedro	3400
PLIOCENE	Pico	8650+

South and East of Oakridge Fault

EPOCH	FORMATION	Thickness (Feet)
MIOCENE	Monterey	0 to 1200
MIOCENE	Vaqueros	0 to 1100
OLIGOCENE	Base conglomerate --- Sespe	7000
EOCENE	Llajas	2600 (drilled)

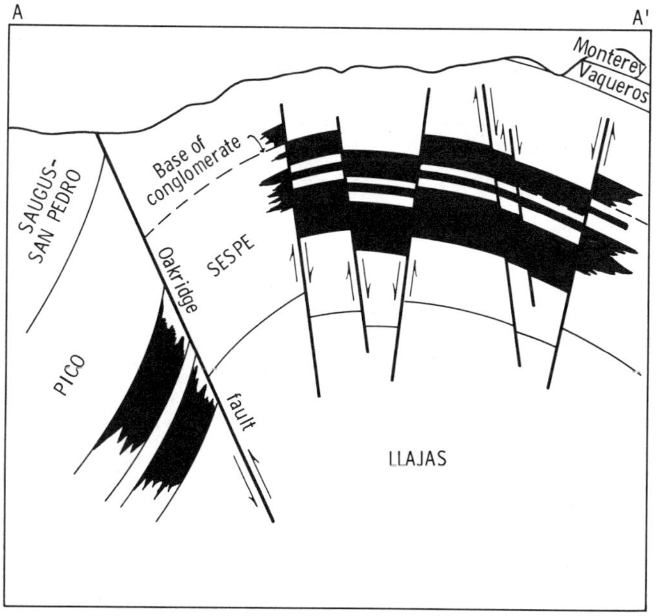

Figure 40.13 South Mountain field, Ventura County, California. Top: structure map of field. Interval, 500 feet. Numbered sections 1 mile (1.6 km) on a side. Left center: stratigraphic position of reservoirs in downthrown block. Lower left corner: reservoirs in overthrust block; shows position of conglomerate used as structure contour datum. Lower right: cross section along line A–A' on map. *Courtesy California Division of Oil and Gas (1961, p. 778).*

is a west-east cross section of the anticline. It can be seen that the thrust passes under the field so that wells drilled in the Bridge area start downward east of the fault, pass through a thick Oligocene section, and then penetrate the fault zone and enter much younger Pliocene strata. This is a displacement of some thousands of feet. The Bridge area oil is trapped in Pico sandstones sealed updip by the fault.

The oil reservoirs in the main field are confined to the Oligocene Sespe Formation, which consists of nonmarine sands and conglomerates with thin noncarbonaceous shale and clay interbeds. The oil, locally confined to the sand and gravel beds by the fine-grained sealing layers, has moved to the top of the anticline. The Sespe is devoid of possible source beds (Baddley, 1954); the discovery of oil in the Pico beds beneath the Oakridge fault tends

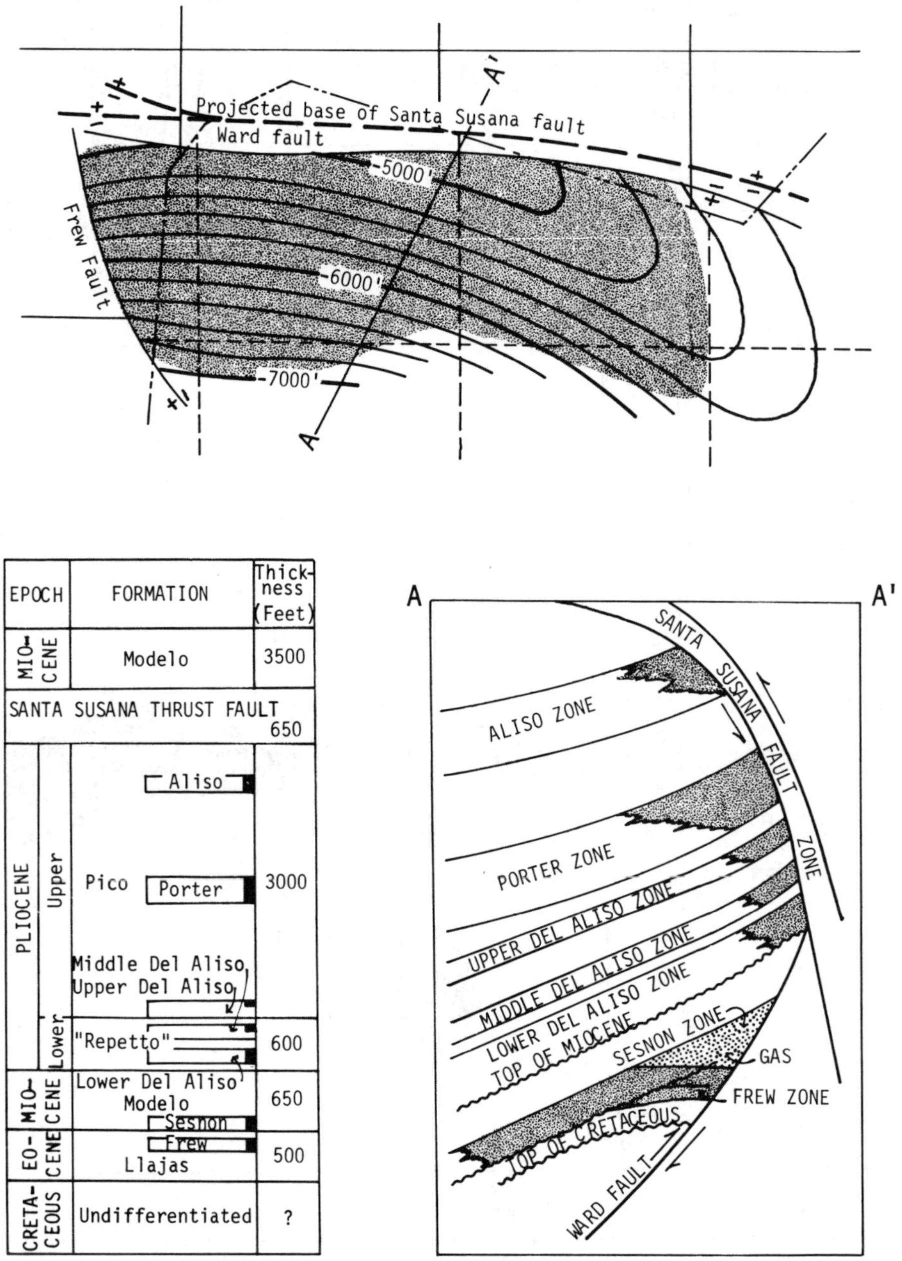

Figure 40.14 Aliso Canyon field, western Los Angeles County. Upper: structure map contoured on top of Sesnon zone. Interval, 200 feet. Productive area stippled. Vertical lines on north edge of map are 1 mile apart. Lower right: cross sections along line A–A′ on map, showing oil impoundment below fault zone. Lower left: stratigraphic section, giving age of reservoirs. *Courtesy California Division of Oil and Gas (1961, p. 506).*

to substantiate the theory that oil from the Pliocene migrated upward through the fault plane into the Sespe Formation.

The *Aliso Canyon* field (Leach, 1948) lies near the eastern end of the Ventura district in western Los Angeles County about 30 miles northwest of the center of Los Angeles and 4 miles southeast of Pico Canyon, where oil was first struck in 1874. The Aliso Canyon field was discovered in 1938. Here the district is characterized by intense folding and thrust faulting. The reservoir rocks are in the stationary block beneath the folded Santa Susana thrust block (Figure 40.14). Trapping is due to updip impoundment against three faults, the Santa Susana, the Ward, and a cross fault, the Frew. Lateral closure to the east is assisted by a nosing in the stationary block. The reservoirs are five sands in the upper and lower Pliocene, Miocene, and Eocene. The Miocene sand produces considerable gas along with oil.

Faulting has also been of major importance in trapping oil in the *Placerita* field about 5 miles northwest of Aliso Canyon. Placerita is the easternmost field in the Ventura district; crystalline basement rock is exposed in the San Gabriel Mountains only a mile to the east. Although this field was discovered in 1920, it was a minor producer until 1948, when a flowing well was completed at a depth of 580 to 717 feet. In early 1949 much lighter oil was discovered a mile to the north, and the boom was on. Two thick sand zones, one in the Pico Formation of the Upper Pliocene and the other in the Lower Pliocene Repetto, are the reservoirs. The oil is trapped below the San Gabriel fault zone to the north and the Whitney fault zone on the east; minor faults separate the pools of different gravity (Oakeshott, 1954).

Santa Maria Basin. The Santa Maria basin occupies a wedge-shaped re-entrant into the Coast Ranges immediately north of the Transverse Ranges. The area currently producing is entirely onshore in northwestern Santa Barbara County and the southwestern corner of San Luis Obispo County. Production is limited about 30 miles inland by the appearance of basement rock at the surface. To the west the basin continues offshore "for an uncertain distance" (Smith and Schambeck, 1966, p. 317). Santa Maria basin has three giant fields: Santa Maria Valley, Orcutt, and Cat Canyon West. Orcutt is the oldest (1901) and Santa Maria Valley the largest, with a cumulative production of more than 140 million barrels. The district contains 13 smaller fields. The potential of the basin offshore is as yet unknown.

Geology. Structurally the Santa Maria basin is a megasyncline creased by east-trending elongate folds. Here, as in other parts of southern California, the latest folding was so recent that most upfolds are hills and downfolds valleys. With a major exception, the Santa Maria Valley field, most of the trapping is anticlinal.

The unique feature in the district is the nature of the reservoir rocks (Figure 40.15). Some years ago it was estimated that 77 per cent of the production had come from fractured rocks and the rest from sands (Regan and Hughes, 1949). According to Smith and Schambeck (1966, p. 318), production comes from "the hard, brittle, fractured cherts and shales in the Miocene Monterey Shale, from fractured sandstones of Miocene age, from fractured sandstones of the Jurassic Knoxville Formation, and from several sands of Miocene and Pliocene ages. Also a few barrels of oil are produced from a sand that may be of Oligocene age."

Selected Fields. The *Santa Maria Valley* field, $1\frac{1}{2}$ miles south of Santa Maria, is not only the largest in the district but also the most unusual. It is a stratigraphic trap, and the first major accumulation of that type to be searched for and found in California. The field is about 7 miles long (east-west) and 2 miles wide. It lies on the floor of the Santa Maria River valley, where the bedrock is obscured by 400 to 1600 feet of alluvium and stream gravels. Structurally, it lies upon the north flank of the Santa Maria Valley syncline (Figure 40.16). Trapping updip is caused by abutment of the reservoir beds against Franciscan basement rock, and by wedgout (through truncation) of the entire Miocene reservoir section farther north where the sealing Sisquoc Formation (Pliocene) overlaps basement rock. The impounding of the oil against basement rock may be due to either uneven surfaces (Canfield, 1943) or faults (Figure 40.16, cross section).

To the south and east of Santa Maria Valley field, in central-west Santa Barbara County, are the Orcutt and the Cat Canyon group of fields. *Orcutt* produces from sands in the Sisquoc (lower Pliocene) and lower Monterey (Miocene). Between these reservoirs, in the upper Monterey is a siliceous shale and chert reservoir producing from fractures. Trapping is in an asymmetric faulted anticline (Dreyer, 1943). The Cat Canyon vicinity includes eight producing areas, all separate traps, of which *West Cat Canyon* field is by far the largest (California Div. Oil and Gas, 1961, Pt. 2, p. 808). In this field the reservoirs are sands in the lower Sisquoc shale of early Pliocene age. The trapping is apparently due to the lensing out of these sand layers on the north flank of an anticline (Manlove, 1943, p. 434). Lateral closure is due to the presence of a northward plunging nose.

Cuyama–Salinas Basin. The Cuyama–Salinas basin (Smith and Schambeck, 1966, p. 318) is inland in the Coast Ranges but parallels both the coast and the range axes. It therefore has a northwest trend. The southwestern boundary is the Naciemento fault zone, and the northeastern edge is the San Andreas fault zone. The basin has a maximum width of 28 miles and is about 160 miles long. The southern end contains the Cuyama

GENERALIZED STRATIGRAPHIC COLUMN

EPOCH	SALINAS VALLEY		SANTA MARIA BASIN		SO. SANTA BARBARA CO.	
Upper Pleistocene	Alluvium		Orcutt		Marine Terrace	
Lower Pleistocene	Paso Robles and Pancho Rico		Paso Robles		Santa Barbara	■
Upper Pliocene	(cont.)		Careaga		(cont.)	
Middle Pliocene	(cont.)		Foxen	■	Absent	
Lower Pliocene	(cont.)	■	Sisquoc	■	Sisquoc	
Upper Miocene	Santa Margarita		Sta. Margarita (?)		Monterey	
Middle Miocene	Monterey	■	Monterey	■	(cont.)	
			Point Sal	■		
Lower Miocene	Sandholdt		Sespe		Tranquillon	
					Rincon	■
	Vaqueros				Vaqueros	▩
Oligocene	Absent		(cont.)	■	Alegria-Sespe	▩
					Gaviota	
Upper Eocene	Absent		(cont.)		Sacate(Coldwater)	■
			Absent		Cozy Dell	
					Matilija	
Middle Eocene	Absent		Absent		Anita and Sierra Blanca	
Lower Eocene	Absent		Absent		(cont.)	
Paleocene	Absent		Absent		(cont.)	
Upper Cretaceous	Absent		Absent		Jalama	
Middle Cretaceous	Absent		Absent		Espada	
Lower Cretaceous	Absent		Absent		(cont.)	
Jurassic and older	Granodiorite		Knoxville		Hondo	
	Basement complex		Franciscan		Franciscan	

■ Denotes oil production ▩ Denotes dry gas production

See California Div. Oil and Gas, "Summary of Operations," V. 42, no. 1; California Div. Mines Bull. 170 (1954), Chap. 111, Plate 1; or A.A.P.G. cross sections for complete data.

FIGURE 40.15 Central coastal California. Stratigraphic column and reservoir rocks. ***Courtesy California Division of Oil and Gas*** **(*1961, p. 502*).**

Valley, and the middle and northern sections the Salinas Valley. Mountainous areas intervene, but the oil fields discovered so far are in these two valleys. The total area of the basin is about 3100 square miles, and the maximum depth over 15,000 feet. Tertiary sediments fill the basin, overlying granitic basement rock. The reservoirs currently producing are confined to the thick Miocene section (Figure 40.15).

The Cuyama Valley sector is an intermontane topographic basin about 8 miles wide by 50 miles long (Schwade, 1954). The producing belt is about 5 by 24 miles extending from northeastern Santa Barbara County into southeastern San Luis Obispo County. The oldest and second largest field, Russell Ranch, which crosses the line between Santa Barbara and San Luis Obispo Counties, was discovered in 1948. The next year the largest field, South Cuyama, which had produced about 200 million barrels of oil by the end of 1968, was found. Four much smaller fields, of which two are now abandoned, were discovered subsequently.

Selected Fields. The *South Cuyama* field, the southernmost one in the Cuyama sector, is about 35 miles due north of Santa Barbara. The field lies on the southwestern side of the valley, close to the Naciemento thrust fault (Figure 40.17). The cross section shown in the illustration is looking northwest. The oil accumulation at South Cuyama is in a dome, the southwestern flank of which has been dropped by a normal fault, the Russell (Schwade et al., 1958, p. 88). Regionally this dome is at the southwestern edge of a graben bounded by the Naciemento fault and the Morales thrust fault to the northeast. The San Andreas fault is still farther to the northeast, beyond the cross section. As can be seen in Figure 40.17, the entire stratigraphic section is com-

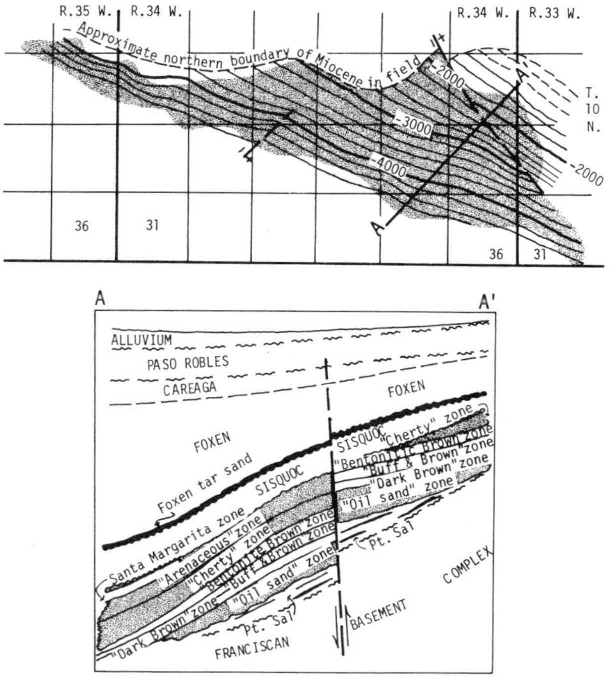

FIGURE 40.16 Santa Maria Valley field, Santa Barbara County, California. Upper: structure contour map based on top of Monterey Formation. Interval, 200 feet (north of −2000 foot contour at east end of field the interval is 100 feet). Land sections are 1 mile squares. Producing area stippled. Lower: cross section along line *A–A′* on map. Tar sand black; oil reservoirs stippled. *Courtesy California Division of Mines and Geology (1961, p. 878).*

plexly folded and faulted. Oil and gas have accumulated high on the dome, principally in two sand reservoir zones in the Vaqueros Formation (lower Miocene). The upper zone has a gas cap about 250 feet thick. The field is about 4 miles long and has a maximum width of $1\frac{3}{4}$ miles.

The southeastern end of the *Russell Ranch* field is less than 4 miles northwest of the nearest edge of South Cuyama. This field is more elongate, nearly 6 miles long and $\frac{1}{2}$ mile wide. The field abuts against the Russell fault on the southwest. The fault cuts across a pleated monocline, which originally was probably the northeastern flank of an anticline. The oil is trapped in an overlapping series of structural noses where they have been truncated updip by the fault. The principal reservoir is the same as the upper sand zone at South Cuyama; in addition there are a basal Miocene sand and an upper Miocene sand producing over a more limited area. Both fields produce 35° oil, an unusually light oil for California.

At the northwest end of the Cuyama Valley oil district the Cuyama River turns westerly to the Pacific, cutting across the grain of the rock terrane in so doing. The major faults continue to the northwest, however, and across the intervening La Panza Range is the Salinas Valley basin, which extends upriver (southward) from the river mouth for a distance of over 80 miles. The oil fields that have been found in the Salinas Valley sector are scattered along a narrow belt about 30 miles in length

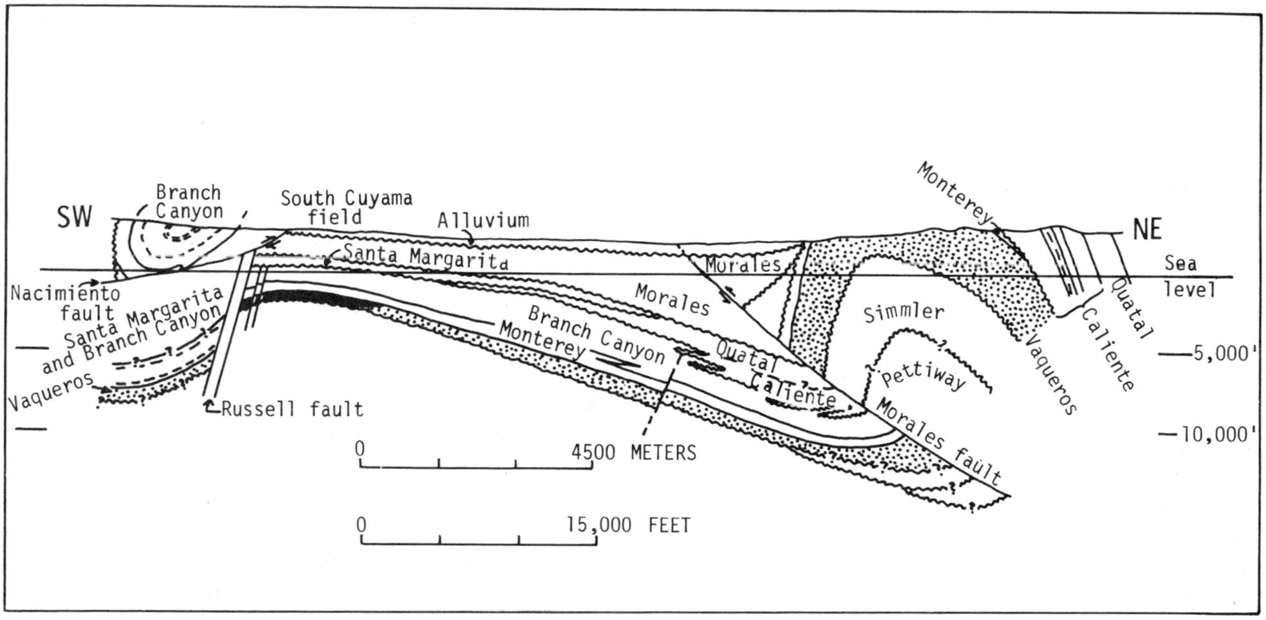

FIGURE 40.17 South Cuyama field, Santa Barbara County. Southwest-northeast structure section. Vaqueros Formation (reservoir) stippled. Oil and gas accumulation at top of anticline northeast of Russell fault shown in solid black (gas cap at crest of fold not delineated). *Courtesy Irving T. Schwade et al. (1958, Figure 8, p. 89) and the American Association of Petroleum Geologists.*

in the upper part of the valley in Monterey County. The only major field, San Ardo, is the farthest upvalley.

The *San Ardo* field, blessed by a copious quantity of shallow oil (2150 to 2300 feet) and cursed by the low gravity (11°) of the oil, allegedly reached its peak production in 1960 with 11½ million barrels. However, in 1967, as a result of added pipeline facilities, it produced over 18 million barrels to become the fifth field in yield in California for the year. The cumulative production at that time was over 200 million barrels. Discovered in 1947, San Ardo underwent slow development until 1951, when a 40 mile pipeline to the coast was completed. The high-viscosity problem has been met by heating the oil to 180° and blending it with oil of 25 per cent higher gravity before putting it in the pipeline (Miller, 1953). Four pumping stations with heating facilities along the line boost the oil toward the marine terminal. There the blending oil is removed for recycling and returned to the field through a smaller, parallel pipeline (Durham, 1966, p. B35).

The two principal reservoir sands in the San Ardo field are in the Monterey Shale of late Miocene age (Baldwin, 1950, 1953). The trap is a broad, flat anticline overlying a low buried hill of granitic basement rock (Figure 40.18). The western side of the field extends beneath a thrust fault. The producing sands are separated by 150 feet of shale. The upper sand, about 200 feet thick, lenses out into shale before reaching the fault. It had a 20 foot gas cap. The lower sand is 120 feet in thickness. Both sands have porosities in the 30 to 33 per cent range.

Many presumably closed structures have been drilled in the Salinas Valley district since the discovery of San Ardo, but results have been minimal. It has been suggested that this infertility may be due to the folding taking place after the last updip migration of hydrocarbons (Smith and Schambeck, 1966, p. 319).

Miscellaneous Coastal Districts. Northwest of the mouth of Salinas River, and about 25 miles south of San Francisco, is the *Santa Cruz* sedimentary basin covering about 260 square miles onshore (Smith and Schambeck, 1966, p. 319). This small area of present and potential oil production is bounded on the north and east by the San Andreas fault and Franciscan basement rock outcrops and on the south by Monterey Bay. The westward limit of possible future production lies at an unknown distance offshore. Two fields, La Honda and Oil Creek, are currently producing oil. Both are in southern San Mateo County. *Oil Creek* (Fothergill, 1962b) is a one-well field producing from an Eocene sandstone. The trap is the nearly vertical flank of a faulted anticline. Four other wells have been drilled, of which two penetrated the gas cap updip from the oil. Probing for this oil deposit would be like trying to find the proverbial needle standing upright in a buried haystack. The reservoir in the *La Honda* field to the northwest is also an Eocene sandstone (Fothergill, 1962a). The trap is the top of a large, elliptical anticline completely hidden from the surface beneath uncomformable westward dipping Pliocene sediments.

The *Eel River* sedimentary basin is in Humboldt County in northern California approximately 250 miles north of San Francisco and 100 miles south of the Oregon boundary. Eel River enters the Pacific about 12 miles north of Cape Mendocino. The basin on land is triangular in shape with upfaulted basement rocks limiting the district on the northeast and the south. The land area is 575 square miles (Smith and Schambeck, 1966, p. 319). Beneath the surface is as much as 12,000 feet of Quaternary, Pliocene, and upper Miocene sediments. Mid-Pleistocene diastrophism produced the present major structural features within the basin (Ogle, 1968). Dry gas is produced at two fields, Tompkins Hill and Table Bluff. The reservoirs are sandstones in the upper Pliocene. The trap at *Tompkins Hill* is an anticlinal fold, with sand lensing also playing a role. At *Table Bluff* accumulation is in sandstone lenses on an anticlinal fold.

San Joaquin Basin. The San Joaquin basin is 250 miles long and ranges in width from 50 to 60 miles. It occupies the south three-fifths of the Great Valley. To the east is the Sierra Nevada, to the west the Coast Ranges, and to the south the crosswise Tehachapi Mountains. The northern limit of the basin is the Stockton arch, about 60 miles east of San Francisco, which separates the San Joaquin and Sacramento basins. In this vicinity the San Joaquin and Sacramento Rivers join and flow into San Francisco Bay, so that the river valleys and the basins coincide except at the southern end of San Joaquin basin, where the drainage is internal (Hackel, 1966).

At the present time the northern part of the San Joaquin basin is exclusively gas producing; the southern part also contains dry gas fields but is most notable for its oil (with associated gas) fields (Rudkin, 1968). This part of the Great Valley is truly the valley of the giants, for here each of 18 fields has produced or will produce more than 100 million barrels of oil. Among them is Elk Hills, maintained as a Naval petroleum reserve, with over a billion barrels of oil in the bank after producing more than ¼ billion barrels. Here also is Midway–Sunset, for many years the nation's number one field in total production, and now fourth. The San Joaquin basin in recent years has produced about ⅓ million barrels of oil daily, or approximately 40 per cent of the total California output.

Most of the Great Valley is floored with Quaternary deposits washed in from the neighboring mountain ranges. Therefore bedrock is exposed, as a general rule, only around the periphery of the valley, and most of the earlier oil discoveries were confined to folds visible at the surface in this belt. The first discovery was Coalinga,

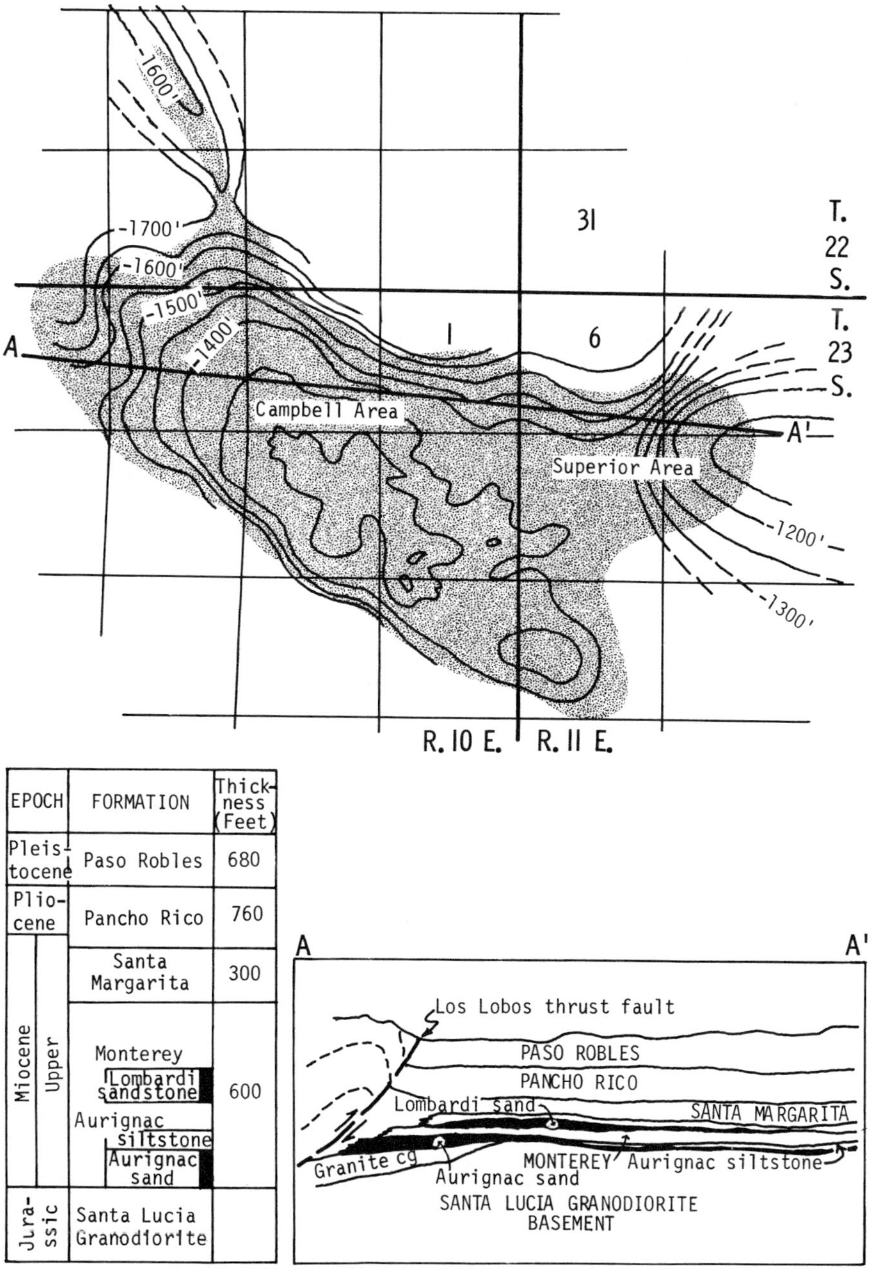

FIGURE 40.18 San Ardo field, Monterey County, California. Upper: structure contour map based on top of Lombardi sand. Interval, 50 feet, except for 1325 foot contours on anticlinal crest. Productive area stippled. Squares are 1 mile on each side. Lower right: cross section along line *A-A'* on map. Oil reservoirs shown in solid black. Lower left: stratigraphic column showing reservoirs. *Courtesy California Division of Mines and Geology (1961, p. 876).*

in 1896. McKittrick started producing in 1898, and Kern River in 1899. The first discovery in the alluviated area was a small dry gas field (Paloma) in 1934, after reflection seismograph surveys (Clark, 1940). The first significant oil discovery was the Ten Section field in 1936, also as a result of seismic mapping. This event "not only demonstrated that anticlinal folds concealed beneath the thick alluvial fill of the San Joaquin basin could be found by the reflection seismograph, but also that thick Miocene sands provide adequate reservoir rocks beneath the valley floor" (Smith and Schambeck, 1966, p. 304–305).

Geology. The Great Valley is a structural trough lying between the Sierra Nevada and the Coast Ranges. In the San Joaquin sector of this trough were deposited Cretaceous and Tertiary marine sandstones and finer clastics. Continental deposits are more abundant in the younger strata. The maximum thickness of sediments in the San Joaquin basin may be as much as 6 miles (Repenning et al., 1966). The structural axis lies near the western edge of the valley (Figure 40.19). Gentle folds and tension faults interrupt the moderate west dip of the eastern flank; to the west, as the Coast Ranges are approached, the folding is more intense and thrust faults appear.

The Stockton arch, which separates the San Joaquin basin from the Sacramento basin, unites the Diablo uplift on the west side of the Great Valley with the Sierra Nevada block. Near the southern end of the San Joaquin basin the southwestward trending Bakersfield arch crosses the basin, creating subbasins thereby (Smith and Schambeck, 1966, p. 315).

The oil production is greatest in the southern part of the Great Valley. Here there is a thick section of organic Tertiary sediments (Simonson, 1958).

Reservoir rocks in the San Joaquin basin range in age from Pleistocene through Tertiary and Upper Cretaceous to crystalline basement schist (Figure 40.20). The presence of both eastern and western sources of supply for the sediment led to separate stratigraphic columns for the two sides of the valley (Hoots et al., 1954b, Figure 2, p. 115). Ninety per cent of the oil so far produced has come from Miocene and younger reservoirs. About one-third of the total oil was obtained from nonmarine sediments (Simonson, 1958). Most of the oil comes from sand; fractures furnish the reservoir for the rest. The latter include both upper Tertiary fractured shales and basement fractured schist (Hart, 1966, p. 249).

Many varieties of traps are illustrated in the San Joaquin basin (Parker, 1954). Anticlinal accumulations are most important, with distribution across the upfold controlled in many instances by faulting or lensing permeability. Trapping in faulted monoclines occurs along the east side of the valley (Simonson, 1958). Likewise some accumulations are due to sealing by overlap across truncated reservoirs in a structural nose. Lenticular sands have trapped oil in a number of fields (Hoots et al., 1954a). Rarer types of accumulations include impoundment below a tar seal, and downdip below an outcrop due to low fluid level (Simonson, 1958).

Most of the giant accumulations of oil in the San Joaquin Valley are the result of anticlinal trapping. Some of these anticlines are elongate folds many miles in length and thousands of feet in closure. They occur on the west side, where differential erosion has carved the outcropping folded beds into ridges and valleys that show by their elliptical patterns the nature of the structure. Because of the greater resistance to erosion of some of the truncated layers in these anticlines, they tend to create hilly areas. As a result, many of the San Joaquin Valley oil wells are to be found in relatively hilly country.

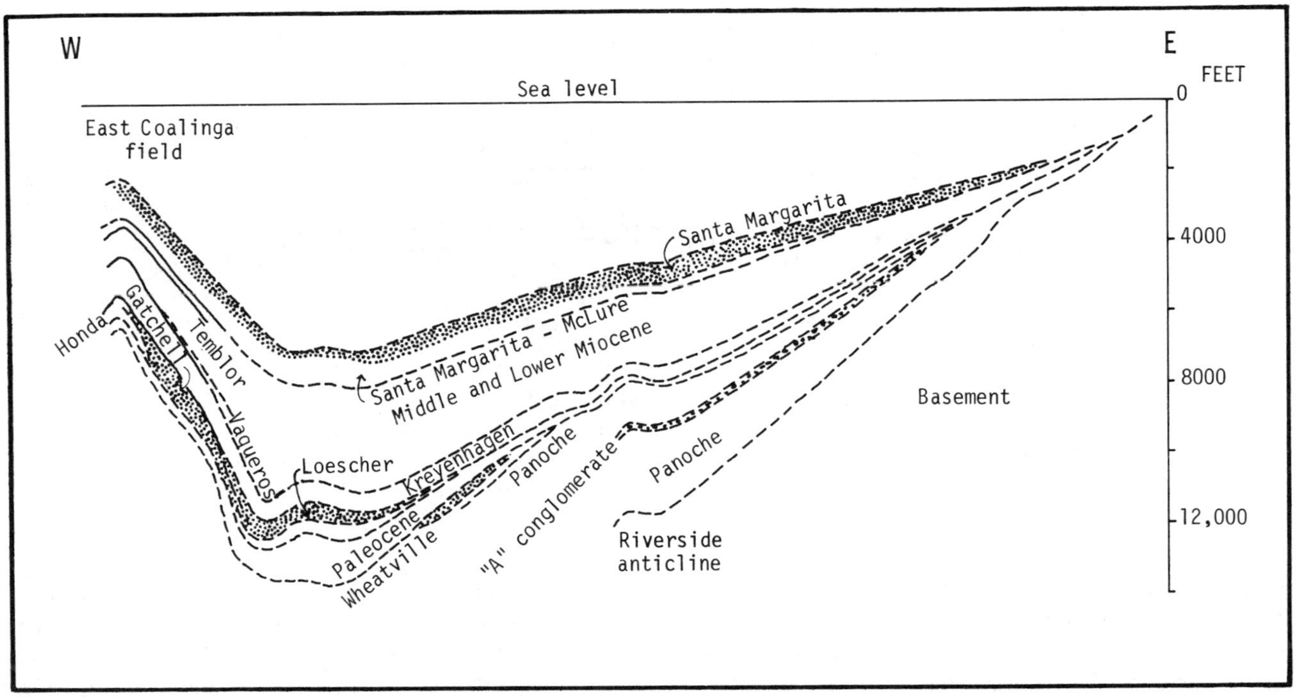

Figure 40.19 East Coalinga field. Structure section eastward across San Joaquin basin in southern Fresno County. Vertical exaggeration: 9.8×. Length of section, 60 miles. Major sandstones are stippled. *Courtesy Harold W. Hoots et al. (1954a, Contribution 8, Plate 6) and the California Division of Mines and Geology.*

EPOCH	SACRAMENTO VALLEY		SAN JOAQUIN VALLEY: WEST SIDE-CENTRAL		SAN JOAQUIN VALLEY: EAST SIDE	
Pleistocene	Red Bluff		Tulare	■	Kern River	■
Upper Pliocene	Tehama	▦	San Joaquin	■/▦	Kern River	■
Lower Pliocene	Tehama-Mehrten	■/▦	Etchegoin	■/▦	Etchegoin	■
Mio-Pliocene	Not recognized		Reef Ridge	■	Chanac	■
Upper Miocene	Mehrten		Antelope-McLure Stevens McDonald	■	Santa Margarita Fruitvale	■
Middle Miocene	Valley Springs		McDonald Devilwater Gould Button Beds	■	Round Mountain Olcese	■
Lower Miocene			Media Carneros Santos-Vaqueros Phacoides Salt Creek	■	Olcese Freeman-Jewitt Pyramid Hill Vedder Walker	■
Oligocene	Absent		Tumey-Wagonwheel Oceanic	■	Tumey Walker	■
Eocene	Markley Nortonville Domengine Ione Capay Meganos	▦	Kreyenhagen Point of Rocks Canoas Domengine-Avenal	■	Walker-Famoso	■
Paleocene	Meganos-Martinez	■/▦	Martinez	■	Absent	
Upper Cretaceous	Chico (Starkey-Winters- (Sacramento-Forbes (Guinda-Funks- (Sites-Yolo-Venado (Antelope-Salt Cr.	■/▦	Moreno Panoche	■ ▦	Absent	
Lower Cretaceous	Horsetown Paskenta		Part of Panoche and other units		Absent	
Jurassic and older	Knoxville (west side)		Basement complex (Franciscan)		Schist Granite	■

■ Denotes oil production ▦ Denotes dry gas production

FIGURE 40.20 Stratigraphic columns, Great Valley of California, including both San Joaquin and Sacramento basins. ***Courtesy California Division of Oil and Gas (1960, p. 8).***

Selected Fields. The greatest of the San Joaquin basin oil fields, both in annual production (over 29 million barrels in 1967) and in cumulative production (passing a billion barrels in 1968) is *Midway–Sunset* on the southwest end of the valley in Kern County. This field extends northwest-southeast for a distance of over 25 miles with an average width of 3 miles. Seeps and surface deposits of brea at the southeastern end of the field were exploited by Indians and settlers. During the 1890s shallow wells were drilled and produced in this area. The first of many spectacular gushers that led to the development of the Sunset field was completed in 1909. Within a few years Sunset merged with the later discovered Midway field to the northwest. Subsequently other fields have merged, and new pools have been discovered within the complex.

Structurally, the Midway–Sunset field lies along a regional homocline dipping away from an uplift in the Coast Ranges immediately to the southwest. This eastward structural slope is modified by folds, including anticlines, synclines, and noses. The stratigraphic section contains major unconformities, involving considerable truncation and overlap, with resulting wedgouts. Figure 40.21 is a generalized west-east section across the field. It shows two anticlines with an intervening syncline;

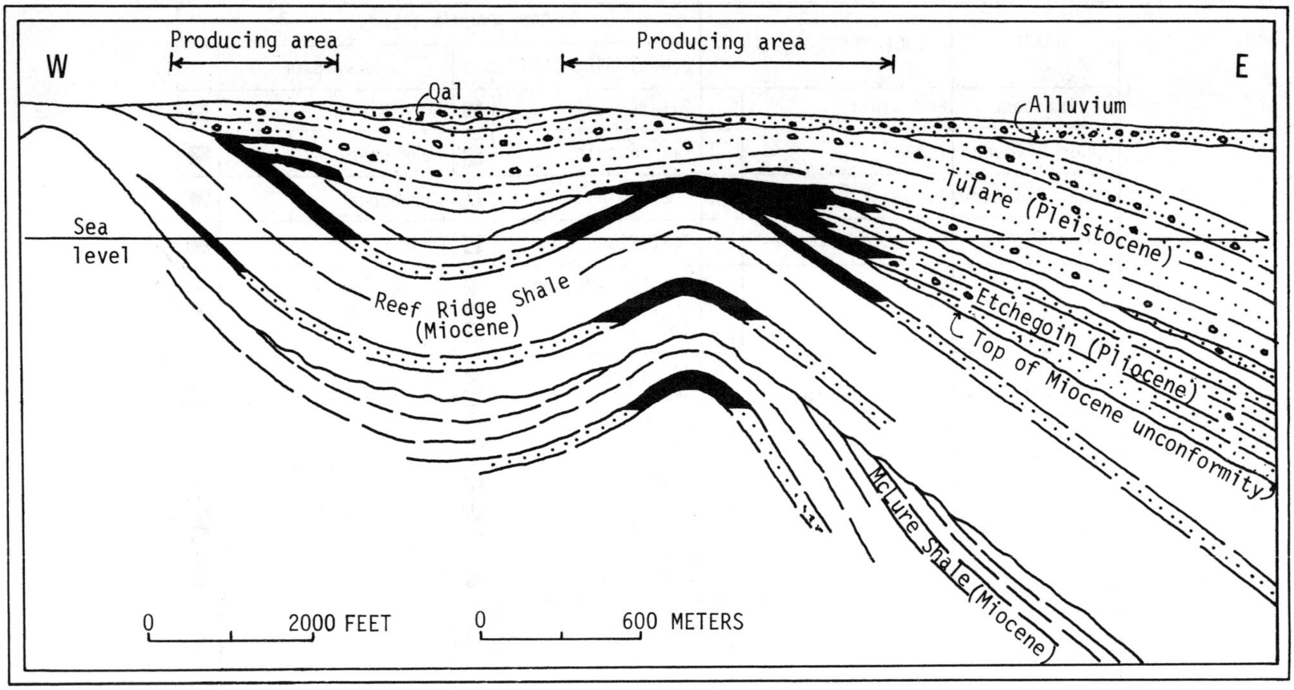

FIGURE 40.21 Midway-Sunset field, Kern County. Generalized west-east structure section. Trapped oil accumulations shown as solid black. ***Courtesy Harold W. Hoots et al. (1945b, Figure 14, p. 128) and the California Division of Mines.***

these folds are creases on the flank of the monocline. It also shows major unconformities at the Recent–Pleistocene, Quaternary–Tertiary, Pliocene–Miocene, and mid-Miocene horizons. Traps illustrated include anticlines, unconformities, and updip lensing. Along the southwest edge of the Midway field oil moving up the flank of the Midway anticline has been trapped just below the reservoir rock outcrop by the accumulation of asphalt at this point. It has been suggested that this semisolid hydrocarbon was formed by the interaction between mineralized ground waters and petroleum hydrocarbons (Pack, 1920).

The updip pinchout of the Lakeview sand (Miocene), in the Lakeview sector of the Midway–Sunset field along the gentle west flank of a syncline, trapped an enormous quantity of oil that fed the spectacular gushers mentioned previously. One of the development wells, drilled in 1910, blew out and produced an estimated 8,250,000 barrels of oil in 544 days of uncontrolled flow. The underlying water is under no great pressure, and the wild well robbed the reservoir of the larger part of its gas pressure. As a consequence, the field produces under gravity drive. The first wells to fail were those at the apex of the sand body, and presumably the last to be abandoned will be those closest to the oil-water interface (Sims and Frailing, 1950).

Another updip sand pinchout is in the Santiago pool at the southeastern end of the field. This was considered a separate field until linked up with Midway–Sunset by drilling. At Santiago a nearly vertical sandstone reservoir pinches out upward between two shales. The oil-water interface lies about 1200 feet below the point of pinchout, but because of the steep dip the field is only about 500 feet wide at its maximum point (Ledingham, 1947).

To show that old giants never die, a new pool was discovered in December, 1962, in a part of the Midway–Sunset field that had been surrounded by wells for many years. Although this pool, the 29D Monarch and 10–10 (Gallear and Kistler, 1965), is in an area with no closure in the highest Miocene strata; thickening of the deeper sediments provides local closure. Subsurface geology gave the clue to the reversal, and drilling confirmed it.

The reservoir sands at Midway–Sunset range in age from Pleistocene through Pliocene into the Miocene. Most of the production has come from a series of local reservoirs in the upper Miocene–lower Pliocene section.

A very few miles to the north of Midway–Sunset in central-west Kern County is the fabulous *Elk Hills* field, with an eventual production of more than 1¼ billion barrels. This field, discovered in 1911, is held as a Naval petroleum reserve; its annual production is purposely kept relatively small (1.6 million barrels in 1967) and is primarily for the protection of the reservoirs. The Elk Hills anticline at the surface is an elongate dome with a closure of 450 feet. It has been eroded into an isolated ridge 17 miles long and 7 miles wide at the widest point, rising out of the floor of the San Joaquin Valley. Between the surface and the reservoir formation, the dip and the closure increase greatly, and three separate nodes appear along the axis of the fold. Although trapping is entirely anticlinal, the thickness and porosity of the reservoir sands determine the volume of production from one well to the next (Pemberton, 1929).

Immediately to the northwest of both Midway–Sunset and Elk Hills is a lesser giant, *Cymric* (McMasters, 1948), which has eleven superimposed reservoirs ranging in age from Pleistocene–Pliocene (Tulare) to Eocene. The trap in the Miocene and older reservoirs is an unusually symmetrical elliptical anticline. The leading producer at Cymric is the Oceanic sand of Oligocene age. The oil here is much lighter than that in the shallow sands, running as high as 38°. Distribution of oil across the anticline is controlled to some extent by faulting, variable permeability, and truncation beneath unconformities. The anticline is not present in the post-Miocene reservoirs; trapping there has been the result of wedgout of the reservoir sands on the flank of a monocline.

Continuing up the west side of the valley, we reach the current northern limit of giant fields. These are first, extending across the Kings–Fresno county line, Kettleman North Dome, followed by the Coalinga group. *Kettleman North Dome* (Woodring et al., 1940) had produced nearly $\frac{1}{2}$ billion barrels of oil and over $3\frac{1}{2}$ trillion cubic feet of gas by the end of 1968. It is a node on a major anticline; to the southeast is Middle Dome, a minor oil field, and beyond that is South Dome, which is nonproductive. The three anticlinal nodes are slightly *en échelon* from each other. The North Dome anticline (Figure 40.22) is about 18 miles long and 5 miles wide. The dips at the surface range up to 43°. The total closure of the major anticline is in the neighborhood of 3000 feet. North Dome rises about 1600 feet above the saddle between it and Middle Dome. It is cut by many faults, but trapping is entirely anticlinal. The oil and gas come from Miocene sands ranging from 1500 to 2000 feet in thickness. The most important reservoir formation is the Temblor.

Kettleman Middle Dome (Hill, 1965) has 600 feet of closure. It produces from sands and fractured shales of the Kreyenhagen Formation (Oligocene–Eocene).

Northwest of North Dome, in southwestern Fresno County is the Coalinga group of fields. Two of these, Coalinga (Birkhauser, 1943) and Coalinga East Extension, together have a total yield of over a billion barrels. These fields are on the Coalinga anticline, which plunges southward and joins the Kettleman Hills line of folding (Chambers, 1943). Trapping in *Coalinga East Extension* is illustrated in Figure 40.23. Structurally the field is divided into two parts. The southern part (Gatchell area) is on the Coalinga anticline; north of the saddle (Amerada area) the upfold is a crease running down the eastern flank of the same anticline. The structural crest of the Gatchell is 1600 feet higher than the summit of the smaller structure to the north. The oil-water interface is low enough to permit production across the intervening saddle. The water table is tilted northward, dropping 600 feet from the southern end of the field to the northern end. Productive closures are over 1800 feet maximum in the Gatchell area, and 450 feet in the Amerada sector of which half was originally gas cap.

The sole reservoir in both areas is the Gatchell sand in the Arroyo Hondo section of the lower Eocene. It has a maximum thickness of 800 feet. This sand pinches out updip to the west, limiting production in that direction. The trapping can be classified as combination structural-stratigraphic.

The fields on the east side of the San Joaquin Valley are clustered at the southern end, in the vicinity of Bakersfield, in Kern County. Oil has been trapped in faulted monoclines in several fields in this area (Simonson, 1958). One example is the giant *Kern Front* field (Edwards, 1941a), which lies 5 miles northwest of Bakersfield. The structure is entirely monoclinal, and the oil is obtained from nonmarine sands that thin out and are faulted updip to the east. Closure to the south is due to the presence of a plunging anticline, but to the north a facies change from permeable to impermeable rock appears to close the reservoir in that direction.

The reservoir sands in the *Edison* oil field (Edwards, 1941b) about 12 miles southeast of Bakersfield illustrate not only fault seal updip to the east but also sealing by onlap and overlap of reservoirs against the basement complex. A third type of accumulation at Edison is the trapping of oil in permeable basement rock schist and felsite (Beach, 1948; May and Hewitt, 1947). The discovery well in the basement rock was completed in June, 1945, with a production of 528 barrels per day from a permeable zone 80 feet within the schist. Since then many other wells have been drilled, and a productive area of about 1500 acres of basement rock has been outlined. The successful wells penetrate as much as 1350 feet of oil-bearing crystalline rock. Without doubt this oil, which is confined to the topographically highest points of the basement rock, migrated into the fracture reservoirs from adjacent, topographically lower sedimentary layers.

Declining production in California in the early 1960s led operators to take a second look at the hitherto largely ignored heavy oil deposits in shallow sands in the major oil fields of the San Joaquin Valley. As a result much new production was obtained. Later thermal recovery ("steam soak") began to be applied on a large scale at Kern River

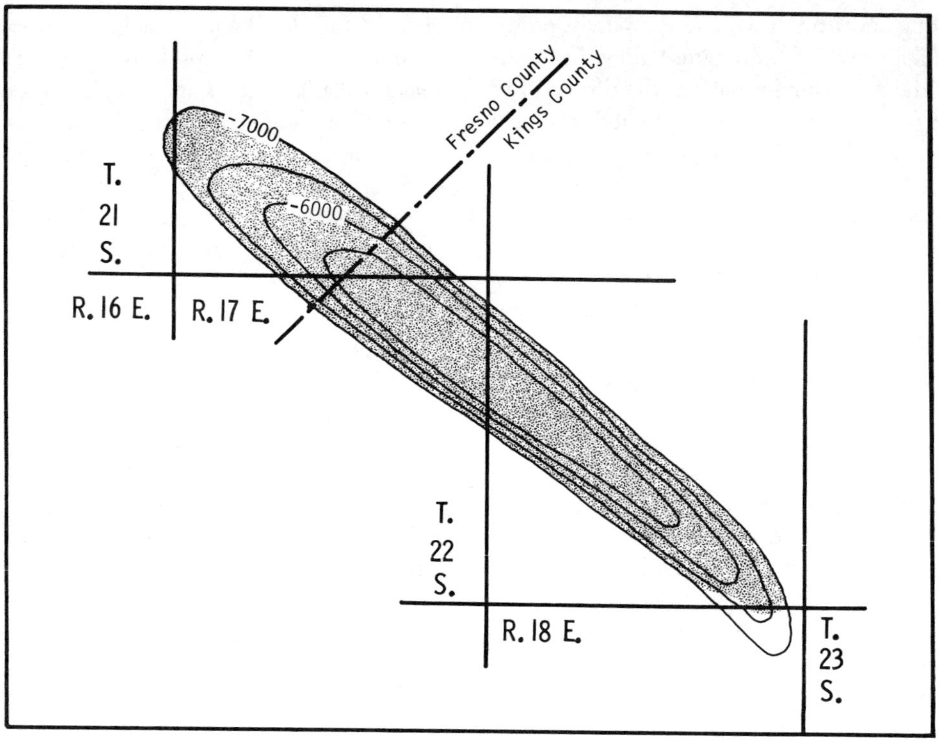

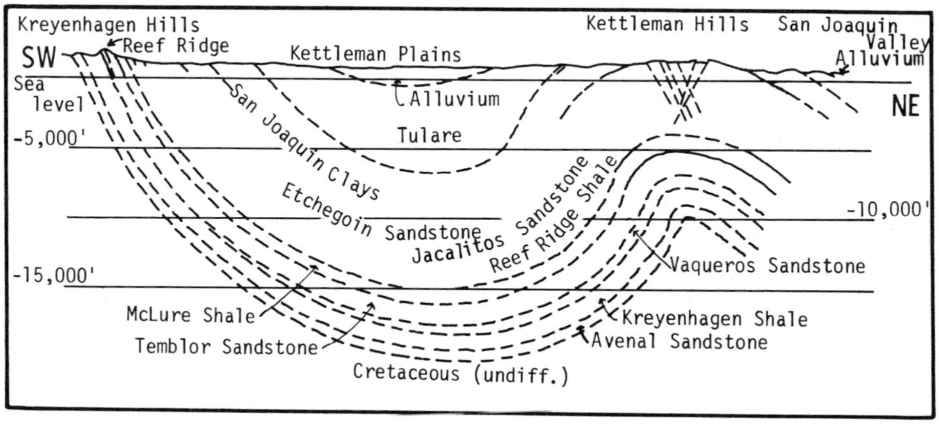

FIGURE 40.22 Kettleman Hills and North Dome field, Kings and Fresno Counties. Upper: structure map of North Dome based on top of Temblor Sandstone. Contour interval, 500 feet. Productive area stippled. Township squares are 6 miles on a side. ***Courtesy California Division of Oil and Gas (1960, p. 330).*** **Lower: southwest-northeast cross section from Kreyenhagen Hills through Kettleman Hills anticline to San Joaquin Valley.** ***Courtesy John Galloway (1943, Figure 208, p. 491) and the California Division of Mines.***

and other fields in the valley, further increasing the annual yield.

The northern end of the San Joaquin basin, immediately south of the Stockton arch, contains six or more gas fields. It is a part of the northern Great Valley dry gas province, which reaches its greater development across the arch in the Sacramento basin. The largest gas field in the San Joaquin sector is *Lathrop* about 6 miles southwest of Stockton in San Joaquin County (Teitsworth, 1968). This field, discovered in late 1961, developed within 2 years dry gas reserves second only to those of Rio Vista to the north in the Sacramento basin. The gas is trapped in an elliptical dome with an east-west trend and 600 feet of structural relief. Just beyond the field to the north is the Stockton fault. One longitudinal and two cross faults cut the dome at reservoir level, but gas production is continuous across them. There are nine distinct reservoirs in an Upper Cretaceous sand-shale zone. The Cretaceous rocks are unconformably overlain by continental Pliocene–Miocene sediments.

Sacramento Basin. The early settlers in the Sacramento Valley observed gas seeps and the presence of gas in water wells. In fact both farmhouses and several municipalities extending from Tehama County on the north to the lower end of the Sacramento River were supplied with gas and water from water wells years before the turn of the century (Carlson, 1962). Gas was first struck in water wells in Stockton in 1854, and beginning in 1889 this gas was marketed until the field was exhausted. Between 1891 and 1933, northern California was the scene of scattered small gas discoveries.

An unsuccessful attempt was made to produce gas by sinking a shaft into a seep near one of the Marysville Buttes in 1864. This same area was retested by drilling, after seismic surveys in 1933–1934, and the first commercial gas field unrelated to water wells was found as a result. Other fields were discovered in the vicinity, leading to the belief that the igneous intrusion was in

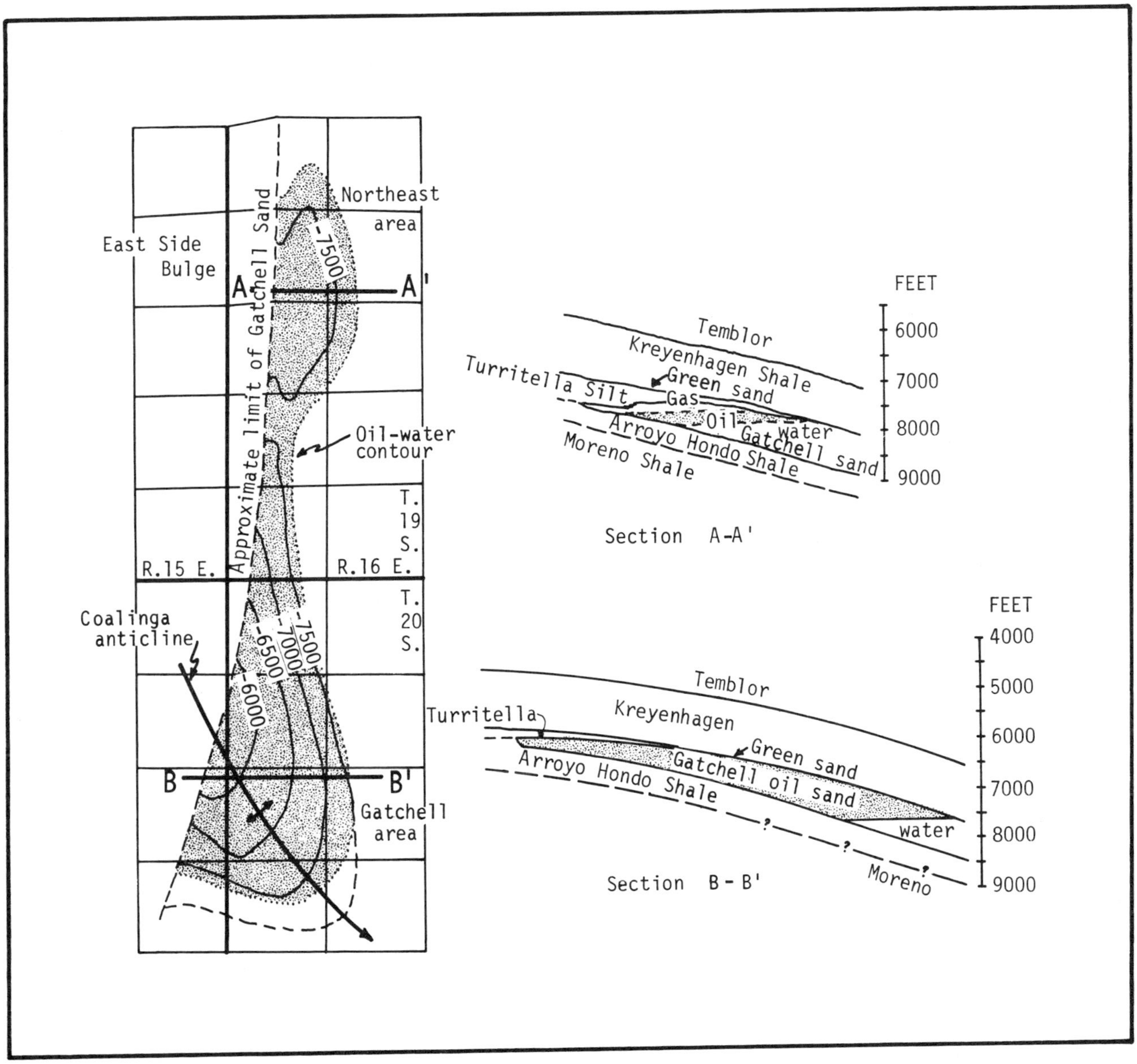

Figure 40.23 East Coalinga Extension field, Fresno County. Left: structure map contoured on top of Gatchell Sandstone. Interval, 500 feet. Productive area stippled. Squares are 1 mile on a side. *Courtesy Russell R. Simonson (1958, Figure 10, p. 110) and the American Association of Petroleum Geologists.* Right: cross sections along lines *A–A′* and *B–B′* on map, showing updip wedgout of Gatchell (reservoir) sand. Gas cap above oil in northern (*A–A′*) section. *Courtesy L. S. Chambers (1943, Figure 207, p. 489) and the California Division of Mines.*

some way responsible for the occurrence of gas. A similar density of gas fields was found to exist elsewhere than in the Marysville Buttes area, however, weakening this hypothesis somewhat. Nonetheless the upending of sedimentary rocks by the intrusion may have created some traps.

Up to the present the northernmost gas field in the Sacramento basin is Corning in Tehama County, and the southernmost fields lie on the Stockton uplift where the Sacramento and San Joaquin basins merge. It is interesting to note that this spread of gas fields coincides almost exactly with the area of water wells containing gas.

Geology. Sacramento basin (Jones, 1968), the northern sector of the Great Valley, is tectonically similar to the San Joaquin basin to the south. It is an asymmetric syncline with the axis near the western side of the valley. The basement rock floor slopes gently westward from the Sierra Nevada (Smith and Schambeck, 1966, p. 317). Stratigraphically, however, there is a major difference between the Sacramento and the San Joaquin basins. The marine Pliocene and Miocene sediments that yield both oil and gas in prolific quantities in the southern basin thin northward, become continental in type, and produce no oil and only minor quantities of gas in the Sacramento basin. On the other hand, the relatively unimportant early Tertiary and Late Cretaceous sediments in the San Joaquin basin are of major importance in the northern area, having produced about 4 trillion cubic feet of gas, although very little oil as yet.

The Sacramento basin extends about 200 miles north of the structual divide, the Stockton arch. It is up to 70 miles wide and is veneered with Recent and Pleistocene alluvial deposits. An exception near the middle of the valley is the Marysville Buttes, which rise 2000 feet above the floor, created by the erosion of an intrusive igneous plug. Between the valley floor and the basement beneath Upper Cretaceous rocks (Safonov, 1962) is a maximum thickness of sediments estimated to be at least 30,000 feet. The stratigraphic column is shown on the left side of Figure 40.20.

Selected Fields. One of the earlier gas fields found, *Rio Vista* in 1936, is also the largest. The discovery well had an initial productive capacity of over 80 million cubic feet per day, with some condensate. This field, which lies on the delta of the Sacramento River 45 miles northeast of San Francisco and 30 miles south of Sacramento, overlaps parts of three counties—Sacramento, Contra Costa, and Solano. Trapping has taken place on a broad, faulted dome (Figure 40.24). The gas is produced from six sand zones of Eocene and Paleocene age at depths ranging from 3700 to 5300 feet (Burroughs et al., 1968).

The *Willows–Beehive Bend* gas field is in southeastern Glenn County in the northern part of the gas-producing area. It is one of the larger fields in the Sacramento basin (Alkire, 1968). The gas has been trapped updip on an anticlinal nose by cross faults plus, in the lower reservoirs, sand lenticularity. All reservoirs are Upper Cretaceous in age; they occur through a vertical section about 3500 feet thick.

The trapping at Dunnigan Hills (California Div. Mines and Geol., 1962) in central Yolo County is similar, except that a third type, impoundment beneath an unconformity, is also present.

The most unusual trap, present in several fields, is the result of Paleocene stream activity, which first cut gorges up to 2500 feet deep into reservoir and seal rocks alike and then filled them with relatively impervious shale. Gas migrating updip accumulated below this barrier, where structural nosing completed the closure. This is one of the traps present in the *Brentwood* field in eastern Contra Costa County (Collins and Conner, 1966). Brentwood is close to the southwestern corner of the Sacramento basin, where the strata are upturned against the Diablo uplift and the Stockton arch. It has the distinction of being the farthest north field in California to produce oil (along with gas) commercially. About $1\frac{1}{2}$ million barrels of oil was produced between 1963, when the first well to produce oil at Brentwood was completed, and the end of 1964. A few marginal oil-producing wells also have been completed in other fields in southern Sacramento Valley. It is quite unlikely that Brentwood has the only commercial oil in this province.

FUTURE

The recent upsurge of production in California to new levels is ample evidence that this state has many years of high yield ahead. The continued following of current procedures alone will insure major production far into the future.

The major reason for California's resurgence has been the aggressive development of petroleum deposits hitherto considered unavailable or uneconomic. Examples include offshore deposits and urban fields.

Offshore Deposits. At first, offshore production came solely from onshore fields that were extended seaward. This was followed by exploration for, and discovery of, fields entirely offshore but in the seaward extension of productive onshore basins. The next phase will be the exploration of separate offshore basins (Emery, 1958).

Throughout the history of offshore development in California the industry has been handicapped by legal problems such as (1) local and state zoning and drilling restrictions; and (2) controversies as to who owned the land—the local municipality, the state of California, or the United States. Fortunately, most of these problems

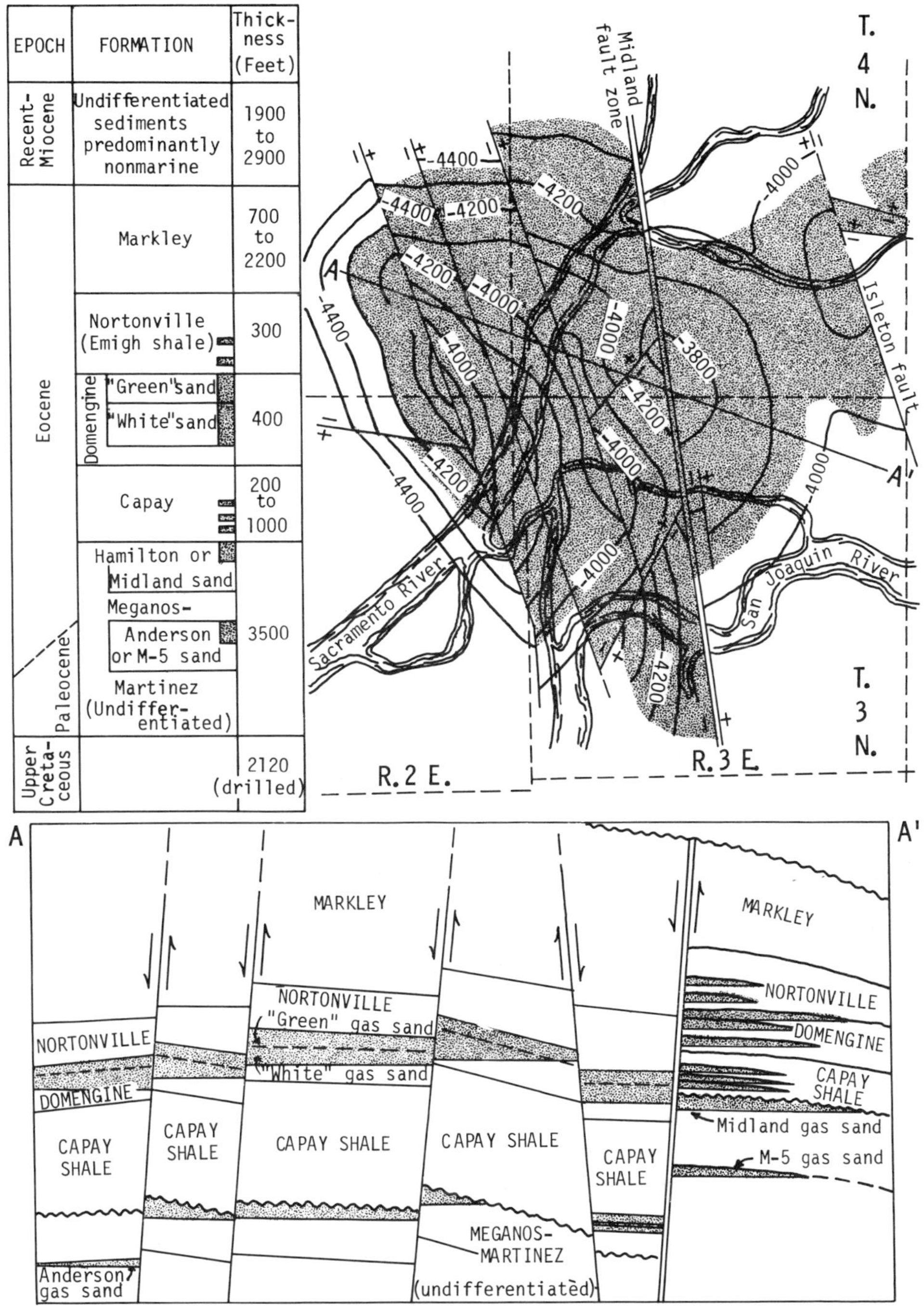

FIGURE 40.24 Rio Vista gas field, Sacramento, Solano, and Contra Costa Counties. Upper right: structure map contoured on top of Domengine gas sand. Interval, 100 feet. Productive area, covering approximately 37 square miles, is stippled. Square in southeast sector is a township, 6 miles on a side. Upper left: stratigraphic column, showing reservoir sands. Lower: cross section along line A–A′ on map. *Courtesy California Division of Oil and Gas (1960, p. 456).*

have been resolved, and a big boost to offshore search took place in February, 1968, when the U. S. government received $603 million in lease bonuses for acreage in the Santa Barbara Channel south of the city of Santa Barbara and west of Ventura (U. S. Dept. Interior, 1968, p. 59).

Between Point Conception at the southwestern corner of Santa Barbara County, where the California coast turns northward, and the Mexican boundary are 13 closed basins (in addition to Santa Barbara) on the submerged continental borderland. The ocean floor in the basin areas is covered by Pliocene and younger sediments. According to Smith and Schambeck (1966, p. 321), "It can be inferred that Miocene, and possibly older sedimentary rocks, are also present at depths in some or all of the basins. The present nearshore fields . . . produce oil from rocks of Pliocene, Miocene, and Oligocene ages." It is quite possible that the density of fields will prove to be as great offshore as onshore.

The continental shelf is much narrower north of Point Conception, except off San Francisco, where it reaches a width of 30 miles. The offshore continuations of Santa Maria, Santa Cruz, and Eel River basins have already been noted.

Ninety-five per cent of the offshore oil produced in California through 1964, nearly 800 million barrels, came from extensions of eight onshore fields, especially Wilmington and Huntington Beach (Smith and Schambeck, 1966, p. 332). The era of prolific production from entirely offshore fields lies ahead. Within 12 weeks of the leasing of federal waters in Santa Barbara Channel in December, 1966, one wildcat well had discovered a multi-pay-zone field lying beneath 190 feet of water depth, and by mid-1968 oil was flowing ashore from six wells drilled from a platform resting in 151 feet of water 4 miles offshore, south of the Carpenteria field (Oil and Gas Jour., Aug. 5, 1968, p. 77).

Urban Exploration. The city of Los Angeles and surrounding municipalities have worked out an agreement with the petroleum companies that permits the exploration and production of oil in congested areas, as was mentioned earlier in this chapter. The result has been a considerable increase in recent years in discovery and production in the "asphalt jungles" of the Los Angeles basin. There remains much prospective area to explore.

Deeper Drilling. Field extensions in the vertical have taken place innumerable times in the past, and the practice is continuing. Santa Fe Springs has had several birthdays, as deeper pay zones were found. Most fields have not yet been explored adequately to basement. Even when the basement has been reached, very little attempt has been made to develop production there, unless oil or gas was discovered by accident. Fractures in the schist conglomerate, which is overlain by oil reservoirs at Long Beach, might possibly be oil bearing.

The northern San Joaquin basin and Sacramento basin gas province merits much deeper exploration, not only for gas but also for possible "mother lode" oil.

Enlarged Peripheries. Rimward extensions within basins of the productive area are always possible except where the sedimentary section is too thin for the accumulation of commercial deposits. Two such discoveries, Livermore in northeastern Alameda County and San Emigdio in southwestern Kern County, were made in 1967 (Doell, 1968). One wonders what hydrocarbon deposits remain to be discovered north of the Corning field in southeastern Tehama County.

Infilling. No doubt the finding of new fields within the areas of older ones will continue in the future as in the past. The best hunting in this regard has been in the southern San Joaquin basin. The area between giant McKittrick and super-giant Elk Hills fields was filled in during the 1960s by several additional fields. Midway–Sunset has had a series of extensions. Deeper drilling on the flanks of the three Kettleman Hills domes might find additional stratigraphic traps (Seiden, 1964, p. 49).

Nor are infilling possibilities confined to the southern end of the Great Valley. Infilling is going on today in other areas, including urban Los Angeles. Both obscure structural highs and updip sand (or permeability wedgouts are possibilities throughout the state.

Possible New Provinces. Eastern California is overlapped by the western edge of the Great Basin, which is productive in one field in southeastern Nevada. Tertiary rocks are present also in the largely untested Mojave Desert and Salton Trough provinces. Seeps and bituminous sandstones are known in the Coast Ranges north of San Francisco.

Continued Secondary Recovery. About 23 per cent of California's oil output in 1967 was credited to secondary recovery in older fields by water and gas injection and by thermal methods. The fairly recent development of the steam stimulation technique of thermal recovery has brought about an appreciable increase in both annual production and reserve estimates in the heavy oil fields of southern San Joaquin Valley.

Bituminous Rocks. Seep oil, surface tar (brea) deposits, and sandstones impregnated with asphalt (Jennings, 1957a) have been exploited across the years, first by Indians and then by the white man. An exploitation peak was reached in 1910, when 87,000 tons of bituminous rock was quarried, and asphalt extracted by heat treatment (Adams and Beatty, 1962). Shortly thereafter production declined because of replacement at the market place by refinery asphalt, and operations ceased altogether in 1949.

California, however, still has known reserves, in place in the five largest bituminous rock deposits, of nearly 300 million barrels of oil. When the right combination of production costs and market values justifies exploitation, it can be assumed that withdrawal of this capital from the bank will be resumed.

REFERENCES CONSULTED IN PREPARING CHAPTER 40

Adams, Elmo W., and Beatty, W. B., 1962.

Albers, J. P., 1966a, 1966b.

Alkire, James, 1968.

Allen, D. B., and Joujon-Roche, J. E., 1958.

American Association of Petroleum Geologists, 1958, 1968.

American Association of Petroleum Geologists, Pacific Section, 1958.

Anon., 1953, 1959.

Baddley, E. R., 1954.

Baldwin, Thomas A., 1950, 1953.

Barbat, William F., 1958.

Beach, John H., 1948.

Beebe, B. Warren, 1968.

Birkhauser, Max, 1943.

Bowen, Oliver E. Jr., 1962.

Brown, James B., 1968.

Burroughs, Ernest, Beecroft, George W., and Barger, Ralph M., 1968.

California Division of Mines, 1940–1943, 1954, 1957.

California Division of Mines and Geology, 1962a, 1962b, 1966a, 1966b.

California Division of Oil and Gas, 1960–1961.

California Division of Oil and Gas, Annual Report, Summary of Operations, California Oil Fields.

Canfield, Charles R., 1943.

Carlson, C. R., 1962.

Carter, Frank S., 1954.

Chambers, L. S., 1943.

Clark, Robert W., 1940.

Collins, Donald F., and Conner, Donald C., 1965.

Doell, E. C., 1968.

Dreyer, F. E., 1943.

Dudley, Paul H., 1954.

Durham, David L., 1966.

Durham, D. L., and Yerkes, R. F., 1964.

Edwards, Everett C., 1941a, 1941b.

Emery, K. O., 1958.

English, Walter A., 1926.

Fothergill, Harold L., 1962a, 1962b.

Gallear, Darrell C., and Kistler, James O., 1965.

Galloway, John O., 1943.

Graves, Doyle T., 1954.

Hackel, Otto, 1966.

Haines, Richard B., and Minshall, Francis E., 1954.

Hamilton, S. V., 1968.

Hart, Earl W., 1957, 1966.

Heath, E. J., 1958.

Higgins, James W., 1958.

Hill, F. L., 1965.

Holman, William H., 1958.

Hoots, Harold W., 1943.

Hoots, Harold W., and Bear, Ted L., 1954.

Hoots, Harold W., Bear, Ted L., and Kleinpell, William D., 1954a, 1954b.

Jahns, Richard H., 1954.

Jenkins, Olaf P., 1941.

Jennings, Charles W., 1957a, 1957b.

Jones, Vern, 1968.

Kilkenny, John E., 1962, 1967.

King, P. B., 1966.

Leach, Claude E., 1948.

Ledingham, Glen W., 1947.

McMasters, John H., 1948.

Manlove, Charles F., 1943, 1968.

May, John C., and Hewitt, Robert L., 1947.

Miller, H. O., 1953.

Moody, Graham B., 1954.

Oakeshott, Gordon B., 1954, 1966.

Ogle, Burdette A., 1968.

Pack, Robert W., 1920.

Parker, Frank S., 1954.

Paschall, Robert H., 1954.

Pemberton, J. R., 1929.

Peterson, G. L., Gastil, R. G., and Allison, E. C., 1966.

Regan, Louis J., Jr., and Hughes, A. W., 1949.

Repenning, C. A., Jones, D. L., and Addicott, W. O., 1966.

Rudkin, George H., 1968.

Sacramento Petroleum Association, 1968.

Safonov, Anatole, 1962, 1968.

Schoellhamer, Jack E., and Woodford, Alfred O., 1951.

Schwade, Irving T., 1954.

Schwade, Irving T., Carlson, Stanley A., O'Flynn, James B., 1958.

Scribner, M. K., 1958.

Seiden, Hy, 1964.

Simonson, Russell R., 1958.

Sims, W. P., and Frailing, W. G., 1950.

Smith, Merritt B., 1964.

Smith, M. B., and Schambeck, F. J., 1966.

Teitsworth, R. A., 1968.

U. S. Department of the Interior, 1968.

Ware, G. C., Jr., and Stewart, R. D., 1958.

Weeks, Lewis G., 1958.

Winterer, E. L., and Durham, D. L., 1962.

Woodford, A. O., Vedder, J. G., and Yerkes, R. F., 1954.

Woodring, W. P., Stewart, Ralph, and Richards, R. W., 1940.

Woodward, A. F., 1958.

Yerkes, R. F., McCulloh, T. H., Schoellhamer, J. E., and Vedder, J. G., 1965.

Chapter

41

ALASKA

Alaska was purchased from Russia in 1867 for \$7,200,000. During 1967 this new state produced nearly 29 million barrels of oil having a value of over \$87,000,000 and natural gas worth another \$15,000,000. Gold, for which Alaska has long been justly famous, was sixth among mineral products in annual value of production, after petroleum, sand and gravel, natural gas, coal, and stone.

Petroleum production in 1968 of over 66 million barrels ranked Alaska in eighth place among the states (up from nineteenth in 3 years), ahead of Mississippi and below Kansas. The already proved oil reserves of this state virtually guarantee that Alaska will move even higher up the list of oil-producing states.

Oil seeps have been known in Alaska since 1853 (Vander Ahe, 1964, p. 19). The first discovery was a small field at Katalla (Figure 41.1) in 1902. The earliest major find was Swanson River in the Kenai Peninsula in 1957 (Grantz and Patton, 1964, p. 43). Other important discoveries followed, and within a decade the Cook Inlet basin was credited with oil reserves of about $1\frac{1}{2}$ billion barrels and gas reserves of $3\frac{1}{2}$ trillion feet (Sweet, 1968).

Regional Geology. The land area of Alaska (365 million acres) is so vast, and the geology so diverse, that only the general picture can be given here. The discussion of individual petroleum provinces will include more detail.

Figure 41.1 shows the major physiographic divisions of Alaska (Wahrhaftig and Gates, 1964). These are extensions northwestward of the physiographic provinces of western United States. Two changes take place between these areas: (1) the direction of the lineaments swings from north to northwest to west to southwest (in southwestern Alaska), and (2) the width of the three western provinces compresses from about 1000 miles (as between Denver and the coast of northern California) to 600 miles between the southern edge of the Arctic slope and the Gulf of Alaska.

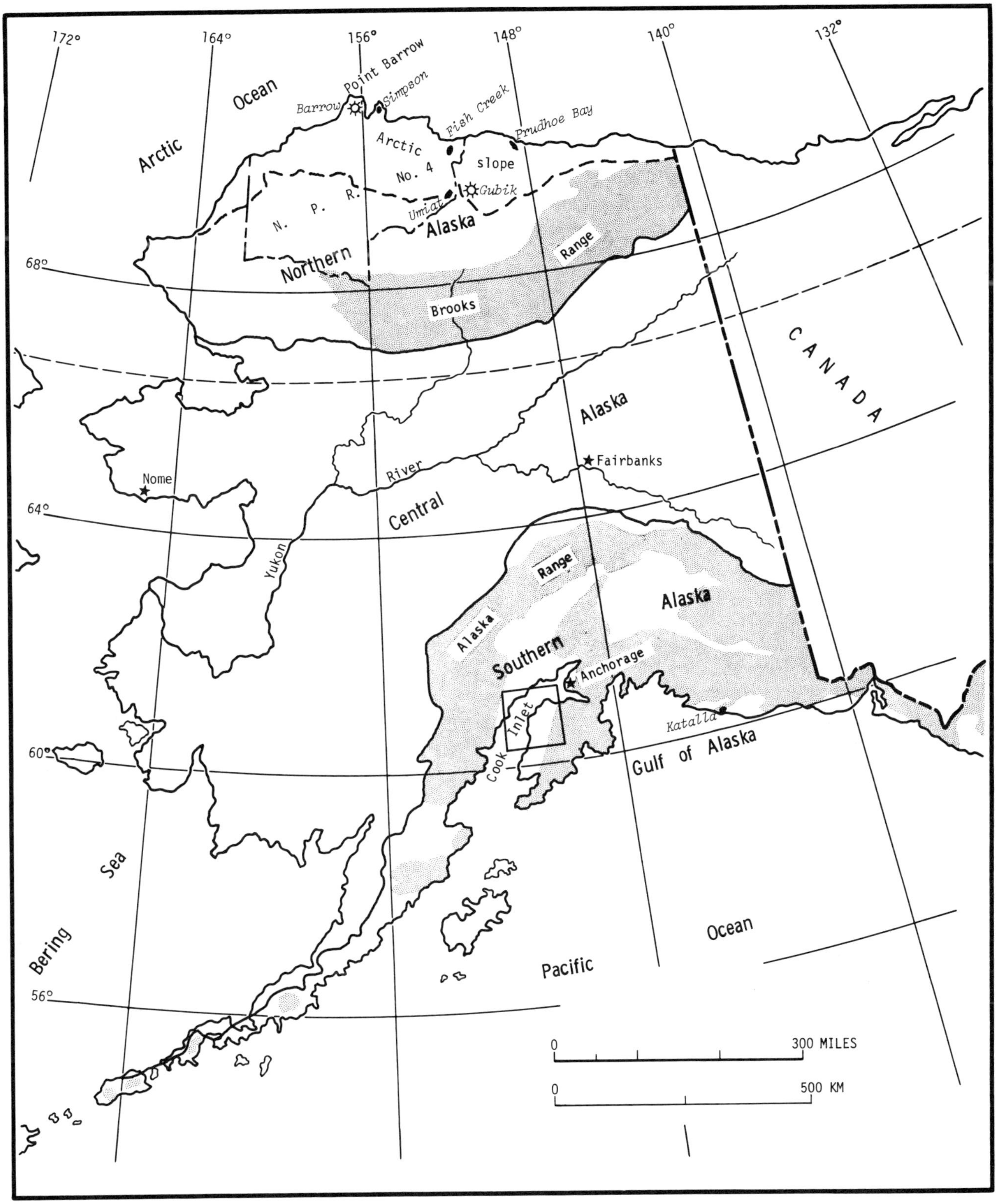

FIGURE 41.1 Alaska. Index map, showing three major divisions and locations of National Petroleum Reserve 4 (NPR 4), the Cook Inlet index map (Figure 41.3), and oil and gas fields outside of Cook Inlet. Also delineates the Arctic Coastal Plain (North Slope) and shows by shading the high mountain belts. *After U.S. Geological Survey (1964, Figures 6 and 7).*

The southernmost physiographic division in Alaska is the Pacific Mountain System. The two components in the western United States, the Coast Range–Olympic Mountain chain and the Sierra Nevada–Cascade Mountain chain, are squeezed together in southwestern Alaska. However, the Cook Inlet–Susitna and Copper River lowlands are counterparts of the Great Valley in California and the Puget–Willamette lowlands in Washington and Oregon. There are geological similarities as well. The coastal mountains in both areas are largely Mesozoic sandstones and shales, with some metamorphics, and Late Cretaceous and Tertiary basic intrusives. The Coast Range turns southwesterly north of the Gulf of Alaska and disappears beneath the Pacific beyond the island of Kodiak.

The Alaska Range lies in the interior in southern and southeastern Alaska, as do the Cascades and the Sierras in western conterminous United States. This range also swings to the southwest, north of Cook Inlet, and becomes the backbone of the Aleutian Peninsula. Beyond the tip of the peninsula only the mountain peaks protrude above water, forming the Aleutian Islands. The Alaska Range and its extensions consist mainly of deformed sediments deposited in the Alaska Range geosyncline during the Mesozoic. These sediments have been intruded by granite batholiths and pierced by lava eruptions up to and including Recent time.

The Interior Plateau physiographic province corresponds to the Intermontane province lying between the Rockies and the Sierra Nevada–Cascade chain in the western states. Much of it is occupied by the wide valley of the Yukon River drainage system, which is floored with Quaternary alluvium and eolian deposits. The uplands consist dominantly of late Mesozoic sedimentary rocks deposited in geosynclines that are separated from each other by eroded geanticlines. Some of these megastructures contain superimposed Tertiary sediments in local downwarps. The geanticlines may also be intruded by Mesozoic granitic bodies. Two major faults, roughly parallel to the arcuate mountain range complex to the south, cross the Interior Plateau province.

The northern east-west mountain system, which is principally the Brooks Range, is Alaska's Rocky Mountain physiographic province. The higher parts of this belt consist of Paleozoic clastic and carbonate rocks. The Arctic foothills zone, which occupies the northern part of the province, is analogous to the western Alberta foothills and the Montana Disturbed Belt. The rocks in this topographically more subdued area are Mesozoic clastics.

Between the foothills and the Arctic Ocean is the Arctic Slope or Coastal Plain province, which corresponds to the western sector of the Interior Plains east of the Rockies in conterminous United States (Gates and Gryc, 1963, p. 265). The Arctic slope is floored with Quaternary and Tertiary deposits.

In the following discussion, the state of Alaska is divided into three petroleum provinces: southern, central, and northern.

Southern Province (Grantz, 1964). The mountains of southern Alaska consist largely of graywacke-type sandstones of Mesozoic and Paleozoic age that have been intruded by large granite batholiths. In addition, the coast mountains have been intruded by basic igneous rocks. The lowlands between the mountain ranges and along the Gulf of Alaska shoreline are underlain by thick deposits of clastic sediments of Mesozoic and (locally) Tertiary age. It is in these lowlands that the known and anticipated petroleum deposits of southern Alaska occur. The Southern province contains, from north to south, three Mesozoic sedimentary basins: (1) the Alaska Range, (2) the Matanuska geosyncline, and (3) the Chugach geosyncline. Only the Matanuska geosyncline (Figure 41.2) has been preserved as a downwarp today. It is over 800 miles in length, extending from the tip of the Alaskan Peninsula northeasterly beyond Chitina. The Tertiary Cook Inlet basin occupies the central segment of the Matanuska geosyncline. The Yakataga basin is superimposed upon the Mesozoic Chugach Mountain geosyncline.

The known oil deposits in southern Alaska are in the Katalla area (Figure 41.1) and the Cook Inlet district (Figure 41.3). Katalla lies on the shore of the Gulf of Alaska to the east of Cordova. Oil and gas seeps were discovered there about 1896 (Saunders, 1964, p. 13), and the first successful well was drilled in 1902. A small refinery was built in 1911 to produce fuel for fishing boats. It burned down in 1933, and there has been no production at Katalla since then. In all, about 36 wells were drilled in this field; half of them produced at one time or another from a few barrels to more than 20 a day. The productive wells were all comparatively shallow, "from 550 feet to 750 feet in depth . . . the total oil produced in this field is approximately 154,000 barrels" (Saunders, 1964, p. 20). Fractured shale and sandstones of Tertiary age were the reservoirs (Miller et al., 1959, p. 1).

Oil seeps were known on the west shore of Cook Inlet as early as 1853. There have been several periods of claim staking in this district, starting in 1882. Wildcat drilling resulted in the discovery of the giant Swanson River oil and gas field on the Kenai Peninsula (Figure 41.3) late in 1957. The discovery well had a potential of over 900 barrels per day of 32° API gravity oil from a depth of about 11,000 feet. Additional oil and gas fields have been discovered subsequently (Figure 41.3).

The Cook Inlet district comprises the Cook Inlet Tertiary basin (Hill, 1963), which occupies the central

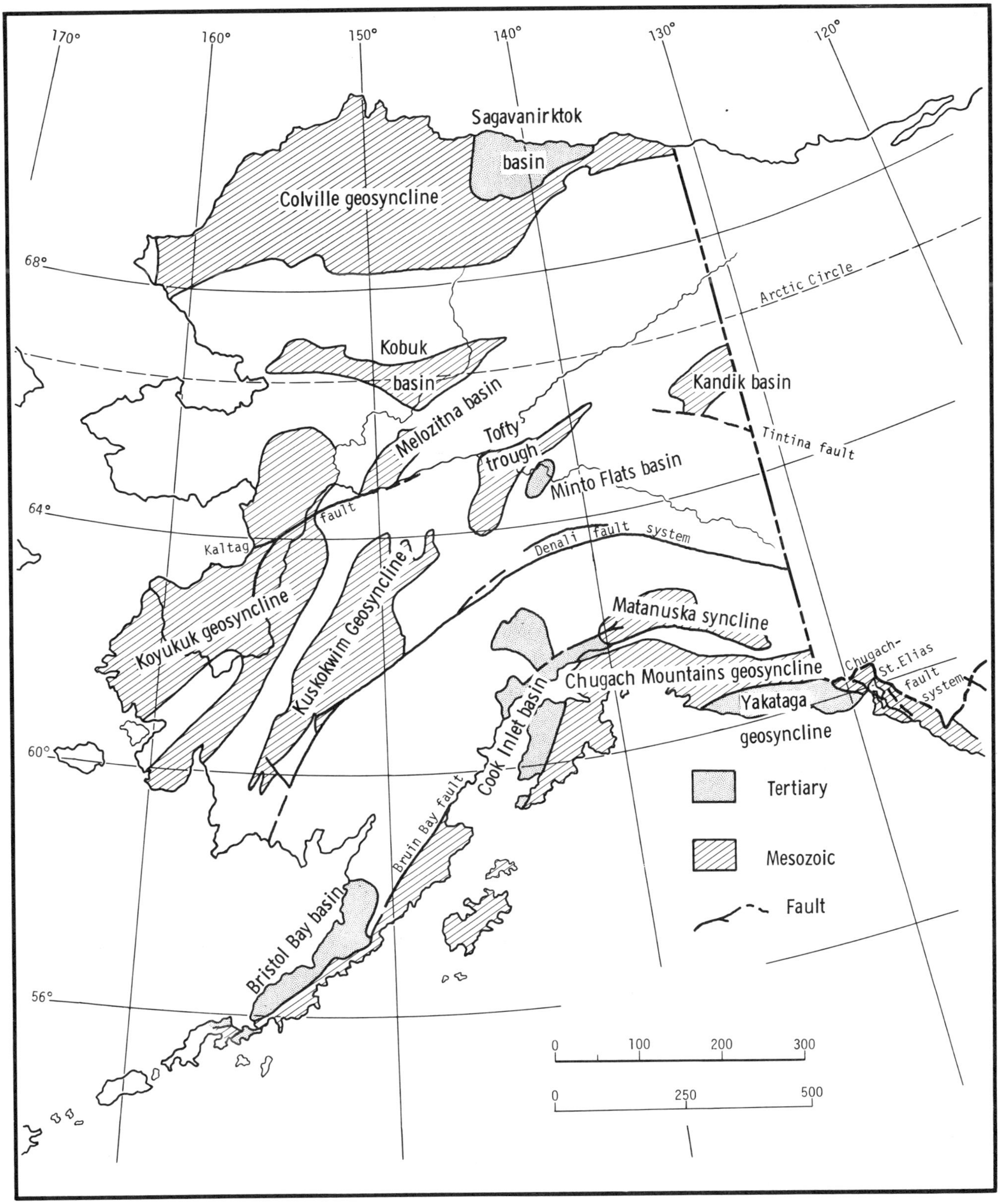

FIGURE 41.2 Alaska. Tectonic map, showing principal Mesozoic and Tertiary geosynclines, troughs, and basins, as well as major faults. *After W. W. Patton, Jr. (1964, Figure 9).*

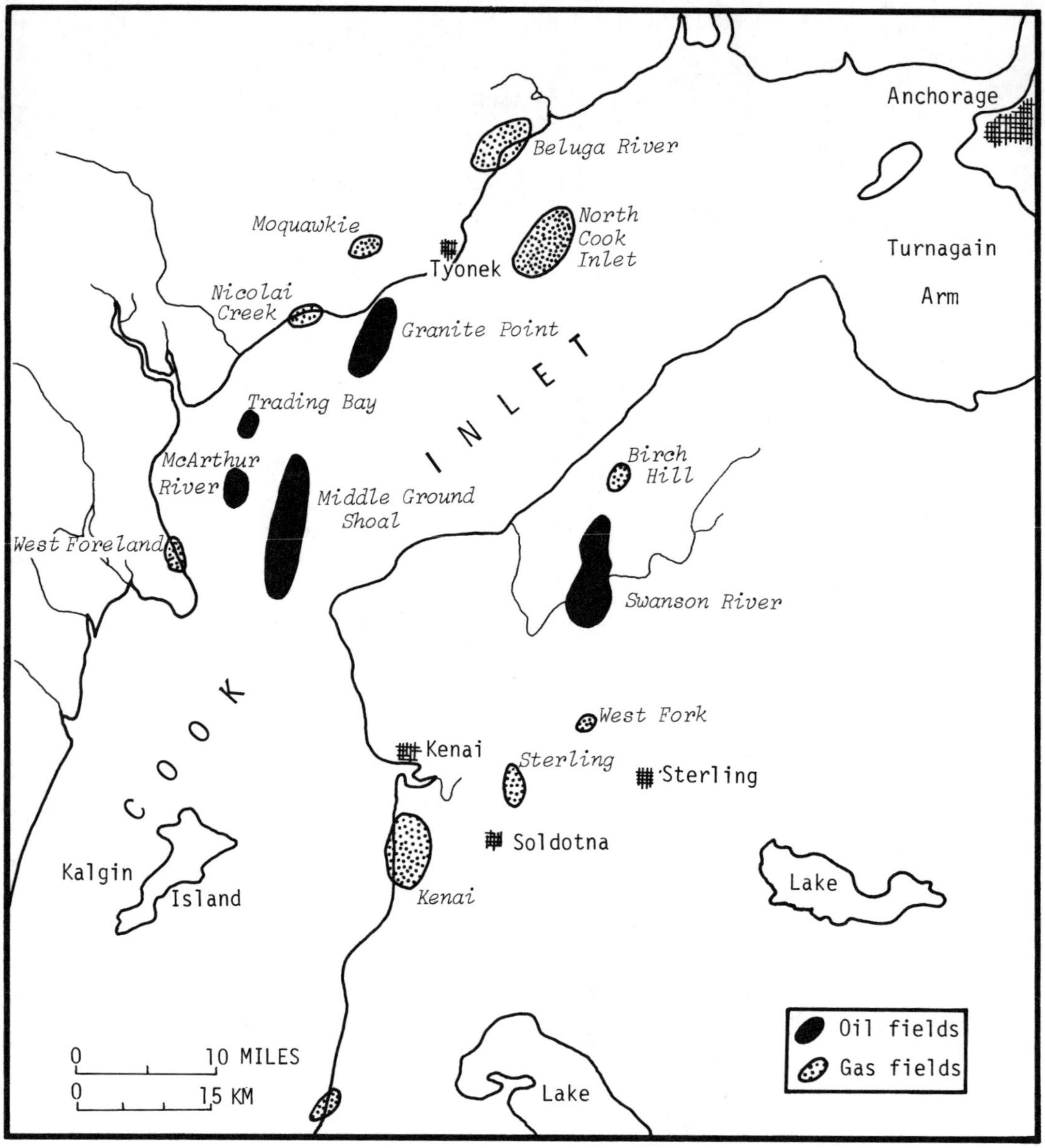

FIGURE 41.3 Cook Inlet, Alaska. Oil and gas fields. Location of map area shown on Figure 41.1. *Courtesy Ted A. Armstrong (1966, p. 82) and the Oil and Gas Journal.*

part of the Mesozoic Matanuska geosyncline. The geosyncline is also a graben, bounded on the west and north by the Bruin Bay and Castle Mountain faults, and on the east by the Chugach fault zone. Beyond the border faults are the Alaska Range and the Chugach Mountains, respectively. The closest counterpart in western United States is the Puget trough lying between the Cascade Range and the Olympic Mountains.

Bedrock in the Cook Inlet district is largely masked by Quaternary glacial deposits and the waters of the inlet. Beneath about 800 feet of this mantle rock in the Swanson River area is the extensive Tertiary Kenai Formation, which has a maximum thickness in the district of about 25,000 feet. It consists "of partly indurated sand, silt, and clay, with thin conglomeratic lenses and many thin bands of sub-bituminous and lignite coal" (Miller, 1959, p. 48). The entire formation is nonmarine in origin (Kelly, 1968).

An angular unconformity separates the Kenai Formation from the underlying Mesozoic section. Mesozoic rocks are extensively exposed in the Matanuska Valley to the northeast of Cook Inlet (Kelly, 1963, p. 281). All three periods, Cretaceous, Jurassic, and Triassic, are represented there. The age of the rocks beneath the angular unconformity at the base of the Cook Inlet Tertiary basin varies from place to place. At Swanson

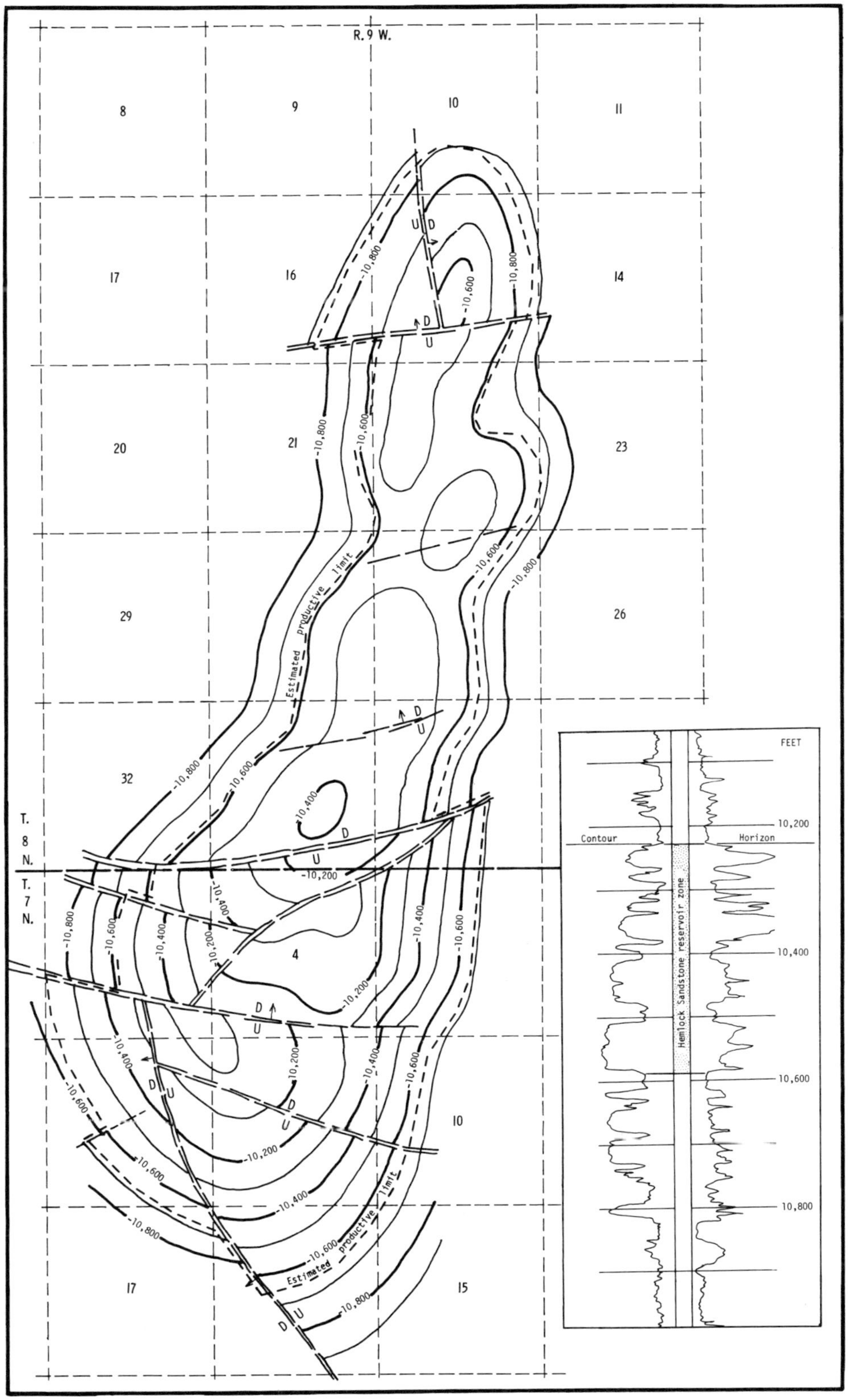

Figure 41.4 Swanson River field, Cook Inlet, Alaska. Structure contour map based on top of Hemlock zone (lower Kenai—Tertiary). Interval, 100 feet. Land sections are 1 mile (1.61 km) square. Dashed line shows approximate level of oil-water interface. At lower right is log of typical well, the Standard Oil of California, Soldotna Creek Unit 41-4, elevation 180 feet K. B. *Both map and log courtesy of Standard Oil of California, Western Operations.*

River it is Jurassic, and the rocks consist of marine clastics. The difference in dip between the Tertiary and Jurassic sedimentary strata may be as great as 20° (Hill, 1963, p. 516).

The *Swanson River* field (Figure 41.4) is an anticlinal trap. It was discovered because it is topographically high, suggesting a recent upfolding, as is the case in the Los Angeles basin. This reasoning was confirmed by a reflection seismic survey. The field as now developed has a length of about 6 miles and a width of 1 to 2 miles. More than 50 producing oil wells have been drilled, proving up over 4500 acres. Since 1962 gas has been injected into the oil reservoir to bolster the initially low fluid pressure.

The oil reservoir is the Hemlock zone in the lower Kenai. The zone itself is about 550 feet thick; the net reservoir thickness ranges from 100 to 300 feet. The rock is a sandstone, fine to coarse grained, and even conglomeratic, with interbeds of carbonaceous siltstone. The average porosity of the producing sections approaches 25 per cent; the permeability average is 100 md. The rock is undersaturated in gas. The water beneath the oil is saline (Hill, 1963, p. 513). Below the Hemlock zone is about 200 feet of basal Kenai consisting of nonmarine carbonaceous siltstone. Underneath is the eroded and truncated surface of an Upper Jurassic marine shale, considered by Hill (1963, p. 515) to be the source rock for the Swanson River oil.

The approximately 9400 feet of Kenai section that overlies the Hemlock oil zone consists of shale, siltstone, sandstone, and coal beds. It contains methane (dry) gas in four zones, which are shown on the electric log to the right of the structure contour map (Figure 41.4). At least six gas wells have been completed in upper and middle Kenai reservoirs. The methane gas appears to be indigenous in the nonmarine Kenai beds. The undersaturation of the deeper oil argues against common ancestry.

The structure map, contoured on top of the Hemlock zone, shows an elongate north-south anticline cut by transverse faults, down to the north except at the southern tip. The Hemlock oil and the higher gas have accumulated in the upper levels of the trap. The fault planes appear to be tight, as shown by variations in the oil-water level in adjacent fault slices.

So far, Swanson River is the only onshore oil field, but five additional oil fields have been discovered to the west beneath the surging tides of upper Cook Inlet. The first of these, *Middle Ground Shoal*, was discovered in 1963 by a well that flowed more than 600 barrels a day from a depth of about 9000 feet. Perhaps the most unusual success story is *McArthur River* field, immediately west of Middle Shoal field (Figure 41.3). It was discovered in November, 1965, by a wildcat that flowed oil (through a choke) at the rate of 1300 barrels per day from the Kenai Hemlock zone between 9375 and 9770 feet. Two extensions were drilled in 1966, and pipeline connections were completed the next year. By early 1968 McArthur River had become the largest field in Alaska, producing 76,000 barrels per day from 16 wells on three platforms (Oil and Gas Jour., 15 July, 1968, p. 111). The minimum estimate of ultimate yield is 300 million barrels, making this field a super-giant.

The structure at McArthur River is a prominent anticline more than 11 miles long (north-south) by 5+ miles wide. It was picked up by airborne magnetometer and confirmed by seismic surveys. The northwest flank is faulted. The productive closure is about 700 feet. The known reservoirs are 7 oil sands and 11 gas sands in the Kenai, but only the Hemlock was being produced in 1968.

At least 10 gas fields have been found in the Cook Inlet district, onshore on both sides of the inlet, and offshore as well (Figure 41.3). One of these, *Kenai*, was producing in 1966, with a pipeline to Anchorage; the others were shut in. The Kenai field, about 20 miles southwest of Swanson River, produces gas from sand zones up to 100+ feet thick in the upper Kenai at depths of around 4500 feet. A deeper sandstone in the middle Kenai at 9100 to 9700 feet was found to be a major gas reservoir in 1967 (Wilson, 1968, p. 1073). The trap is a dome containing "in excess of a 500 foot vertical gas column" (Hill, 1963, p. 515).

The future for the Cook Inlet district continues to be bright. The Tertiary Kenai and the underlying Mesozoic rocks are still to be tested in countless places not only within the perimeter of the area of discoveries to date but also up and down inlet (Detterman and Hartsock, 1967), east across the Kenai lowlands to the Chugach fault zone, and inland on the west shore of the inlet to the bounding faults in that direction. Stratigraphic traps, especially sand wedgouts, are a distinct possibility (Woncik, 1968b).

The southwestward extension of the Matanuska geosyncline is an area of folded Mesozoic rocks. A few of the upfolds have been drilled, but not always to adequate depths for thorough testing. The Kanatak district on the Alaska Peninsula west of central Kodiak has been known for its seeps for many years. Continuing search can be expected in Mesozoic rocks southwest of the Cook Inlet Tertiary district.

The Copper River district (Miller et al., 1959, p. 51–53), another possible producing area, lies in the northernmost segment of the Matanuska geosyncline, at the top of the curve (Figure 41.2). To the southwest the geosyncline continues to the tip of the Alaska Peninsula, and to the southeast it dies out in a relatively short distance. The Copper River district is a topographic basin floored with unconsolidated deposits of Quaternary age. It is completely surrounded by mountains, with the south flank of the Alaska Range to the north and the Chugach Mountains to the south. To the east and west are local ranges. These surrounding mountains consist dominantly of igneous rocks, both intrusive and extrusive, and both Paleozoic and Mesozoic in age.

In at least two localities in the south-central part of the Copper River basin flammable gas associated with saline springs seeps through the Quaternary sediments. Mesozoic sediments crop out around the periphery of the Pleistocene and Recent sediments that floor the basin, especially in the western end (Sweet, 1968). It is unlikely that any deposits of Tertiary age overlie these Mesozoic sediments (Miller et al., 1959, p. 51). A composite section of the Chitina Valley exposures (Miller et al., 1959, Pl. 4) shows a thick section of both Upper and Lower Cretaceous rocks underlain by a thinner sequence of Jurassic rocks, which in turn are underlain by a series of Upper Triassic sedimentary rocks. Both the Upper and Lower Cretaceous rocks are marine in origin and contain potential source rocks, in the form of organic shales, and potential reservoir sandstones. Age equivalents of the Jurassic formations are notable for their seeps where exposed to the southwest of Cook Inlet, and are considered to be possible reservoirs in the Copper River district also.

The major difference between the Copper River district and the productive Cook Inlet district is that in the latter area a thick Tertiary section intervenes between the Quaternary and the Mesozoic and the oil so far discovered comes from reservoirs in the lower part of the Tertiary section. It has already been noted, however, that the underlying Jurassic sediments at Cook Inlet may have been the source of oil in the Tertiary reservoirs. Therefore, even without a Tertiary cover, oil may occur in commercial volume in the Mesozoic sediments presumed to be present in the Copper River district. Seven wildcat wells have been drilled in this district in the last decade, a few to depths in excess of 8000 feet.

The Southern province contains a second Tertiary district, the Yakataga geosyncline, which lies east of Cook Inlet on the north shore of the Gulf of Alaska east of Katalla. This district, described by Miller (1957), is famous for an oil seep that may have flowed as much as a barrel of oil a day. This seep, as well as others, are in a large faulted anticline that has been explored by three dry wildcats. One, drilled at the major seep, was abandoned at 2000 feet; the other two were drilled to depths of 4800 and 10,000 feet.

To the east, across Icy Bay from the Yakataga district, is the 50 mile wide Malaspina glacier, where the oil prospects can be described only as frustrating. Seeps, and the best opportunities for traps, are in Tertiary rock hills completely surrounded by glacier ice (Plafker and Miller, 1957).

Central Province. The Central province covers the greater part of Alaska, extending from the Alaska range on the south to the Brooks range on the north and from the Canadian border to the Bering Sea. It is also referred to as Alaska's Interior, a region that is drained by the Yukon and Kuskokwim Rivers. The Central Province is physiographically varied with low-elevation flats, dissected plateaus, and highlands. The geology is correspondingly diverse. The lowlands are veneered with Quaternary deposits, and the uplands consist of Precambrian to Tertiary sediments, intrusive rocks, volcanics, and (except for the Tertiary) metamorphic rocks.

Although the largest province in area, the Central province is the least explored for oil and gas. It has a paucity of verified seeps and the least attractive geology of the three provinces. Most of it consists of eroded geanticlines, with igneous and metamorphosed sedimentary rock present in considerable volume. Such rocks are also present to an unusual degree in the sediment-filled downwarps in this province. Furthermore, the great alluvial plains that floor the major drainage systems mask the underlying bedrock over large areas.

Figure 41.2 shows eight of the geosynclines and basins present in the Central province. Six of these contain Mesozoic sediments, and two are Tertiary basins. Only two of the Mesozoic downwarps, the Kandik and the Koyukuk, hold much attraction for investigation at the present time; Bristol Bay is the only Tertiary basin of current interest. The remaining five areas of accumulation of Mesozoic and Tertiary sediments, plus smaller troughs not shown in Figure 41.2, are not discussed here for one or more of the following reasons: the presence of volcanic rocks in volume, Tertiary intrusions of granitic rock with attendant thermal metamorphism, regional metamorphism, impermeability of the sandstones, and lack of adequate source rocks for petroleum (the gas prospects may be better). Patton (1964) describes in some detail the basins not discussed here.

The 40 mile wide Mesozoic Kandik basin is the easternmost segment of the Kuskokwim geosyncline. It extends over the Canadian border and covers an estimated 3000 square miles in eastern Alaska. The known and assumed stratigraphic section includes, from the Middle Cretaceous downward, Early Cretaceous, Triassic, Permian, Pennsylvanian (?), and Late Mississippian sediments (Payne, 1959, p. 63). All of these rocks are marine except for part of the Middle Cretaceous and the Pennsylvanian (?). Permeability is in short supply in the Cretaceous section, but the Triassic and Paleozoic strata include both potential source and reservoir rocks.

The other Mesozoic downwarp discussed here is the Koyukuk geosyncline (Figure 41.2), which lies along the

Bering Sea shore both south and east of Norton Sound. It is traversed by the lower Yukon River. The Cretaceous sediments in this district are cut by the Kalteg fault and intruded in a few places by Tertiary granite. Although the total thickness of the Cretaceous section here is not known, it probably exceeds 9000 feet. The youngest sediments are largely nonmarine; coal has been mined from this formation at several places. Below are predominantly marine shale, sandstone, and conglomerate. Payne (1959, p. 66–67) compares the sandstones favorably, in terms of both age and permeability, with petroleum reservoir rocks at Umiat (Northern province) and with seepage source rocks elsewhere.

The best Tertiary possibility at the present time in the Central province is the Bristol Bay basin, which is on the north side of the Alaskan Peninsula. Much of the district is covered by Quaternary deposits. Neither the extent nor the thickness of the Tertiary deposits has been ascertained as yet, but outcrops suggest that there is an adequate thickness of interbedded marine and nonmarine rocks to merit exploration. The nonmarine rocks are coal bearing, and the carbonaceous beds in this area could be a source for methane gas. Three dry test wells have been drilled so far in the Bristol Bay district. Two of these penetrated the entire Tertiary section, verifying the presence of both possible source rocks and permeable reservoir rocks. There could be a gas province here similar to that in Cook Inlet in the nonmarine upper Tertiary, with oil accumulations in deeper rocks.

Northern Province. The Northern Province (Gryc, 1959; Patton, 1964) lies north of the 67th parallel and south of the Arctic Ocean, from the Bering Sea to the Canadian boundary. It has an area of about 125,000 square miles. This province can be subdivided into two subprovinces (Figure 41.1), the Rocky Mountain extension on the south and the coastal plain to the north. The latter is the northwestern extension of the mountain-bordering interior plains of western Canada and the United States.

The Rocky Mountain subprovince can be further subdivided into two nearly equal parts, a mountainous belt on the south and the Arctic Foothills belt on the north. Each of these segments has greater area than the coastal plain. The Brooks Range, with additional lower mountains to the west, constitutes the mountainous belt. It is a complex of Paleozoic sandstones, shales, and limestones, and metamorphic rocks of pre-Mississippian age. The sedimentary rocks are mostly marine (Gates, 1964, Fig. 8). The foothills on the north flank are built of Mesozoic sandstone and shale, both marine and nonmarine in origin. Bituminous coal is also present in considerable volume (Eberlein, 1964, p. 11). The Arctic ("North") slope is veneered by Quaternary deposits, with windows of Tertiary clastics, between the Colville River and the Canadian border.

Tectonically the Northern province contains two major elements, the geanticlinal belt on the southern side, which has been sculptured by erosion into a mountain system, and the Mesozoic Colville geosyncline to the north. The sediments that were deposited in the geosyncline are moderately to strongly folded where they crop out in the foothills belt. Mesozoic rocks are also present, flat to gently folded beneath the veneer of Quaternary deposits on the coastal plain. East of the Colville River, toward the eastern end of the Colville geosyncline, the Sagavanirktok basin (Figure 41.2) contains Tertiary sediments infolded into the Mesozoic formations. Both marine and nonmarine rocks have been observed in the Tertiary outcrops. Figure 41.5 is a composite stratigraphic section for the foothills and the coastal plain. The Paleocene–Miocene strata in the Sagavanirktok basin are shown unconformably overlying a thick Cretaceous section.

The Arctic shoreline belt contains at least ten known seeps of oil and gas. Some of these were noted as early as 1900; the seeps at Cape Simpson were described in 1909. Inland, in the foothills belt, oil-bearing sandstones and oil shales crop out at "many localities" (Miller et al., 1959, p. 2). In 1923 Naval Petroleum Reserve No. 4 was set aside by Executive order (Miller et al., 1959, p. 2). The location of this tract of public land, covering about 37,000 square miles south of Point Barrow, is shown in Figure 41.1 (NPR-4). Similar withdrawals were made in California and Wyoming to insure that the U. S. Navy, which had recently converted from coal to oil, would have sufficient fuel in the future. In 1944, during World War II, the U. S. Navy Department organized and underwrote an extensive exploration program for NPR-4. Between 1945 and 1955, when the project was terminated, a total of 82 core drill and test wells was drilled, resulting in the discovery of three noneconomic oil fields: Umiat, Cape Simpson, and Fish Creek, and two noneconomic gas fields: South Barrow and Gubik (Figure 41.1). Trapping has been in local anticlines. The reserves in these fields have been estimated at from 30 to 100 million barrels of oil and a minimum of 370 billion cubic feet of gas. A detailed history of the exploration of this petroleum reserve has been published by Reed (1958).

Both the *Umiat* oil field and the *Gubik* gas field have been described (Collins, 1958; Robinson, 1958). These two fields are about 20 miles apart (Figure 41.1). Both penetrated similar Cretaceous sections (Figure 41.6), except that the Gubik wells started higher, stratigraphically, and the Umiat wells penetrated lower formations. Both traps are elongate east-west anticlines. The Umiat anticline is 10 miles long and 3 miles wide, with 800 feet of closure. Seven out of eleven holes drilled produced some oil; in all probability the ratio would have been higher had oil-based mud been used from the start.

The principal reservoirs are sandstones in the Grandstand Formation. The oil occurs at depths of 275 to 1100 feet, and the bottom of the permanently frozen ("permafrost") ground varies from 800 to 1100 feet. It has been determined through laboratory studies that a "substantial recovery" (29 to 45 per cent) can be anticipated from reservoirs in the permafrost (Baptist, 1959). However, water-based mud cannot be used in drilling, for the water works out into the sands around the well bore and freezes, thereby effectively sealing off the oil.

The Gubik anticline, as mapped on a phantom seismic horizon in the Grandstand Formation, is shown in Figure 41.7. It is about 12 miles long, with a width of 2½ to at least 4 miles. Closure is 400 to 500 feet; the average flank dip is 3°. Only two wells have been drilled on the anticline. The first, near the top, found commercial gas in the Prince Creek Formation of the Upper Cretaceous at 890 to 1750 feet depth, and 1500 feet lower in the Ninuluk–Chandler section. The second well, on the south flank of the anticline, blew out and caught fire before it could be tested.

Another gas field, *South Barrow*, is currently producing gas for the benefit of nearby federal agencies and for sale to residents of Barrow Village 5 miles to the northwest. A fourth gas well was completed in this field in 1964, and a fifth in 1968.

During recent years the petroleum industry has drilled eleven wildcats on the north slope east of National Petroleum Reserve No. 4. Six of these were shallow Cretaceous tests; five explored older Mesozoic and Paleozoic rocks (Figure 41.5). Two of the latter, drilled 7 miles apart on an anticline 35 miles long and more than 3 miles wide, struck oil and gas in large quantity in January and June, 1968. The location is near *Prudhoe Bay* (Figure 41.1) about 150 miles southeast of Point Barrow. The principal reservoir is a Triassic sandstone that not only has high porosity and permeability but also is over 200 feet thick. The gas and underlying oil lie at depths in the 8200 to 8700 foot range. The discovery well for the Prudhoe Bay field also flowed at the rate of 1150 barrels per day from oölitic Lisburne Limestone (Mississippian) at a depth of 9500 to 9800 feet. If development wells prove to be as prolific as the discovery and stepout wells, Prudhoe Bay could become one of the great oil fields of the world. Possible reserve estimates run as high as 10 billion barrels.

The Prudhoe Bay discovery has stimulated plans for widespread exploration not only on the north slope of Alaska but in the Canadian Arctic as well. In the northern province of Alaska alone, between 60,000 and 100,000 square miles (out of a total of 125,000) is classified as potential oil country. Over much of this great area the already known Cretaceous, Triassic, and Mississippian reservoir rocks are probably present.

In the northeastern corner of Alaska possibilities occur

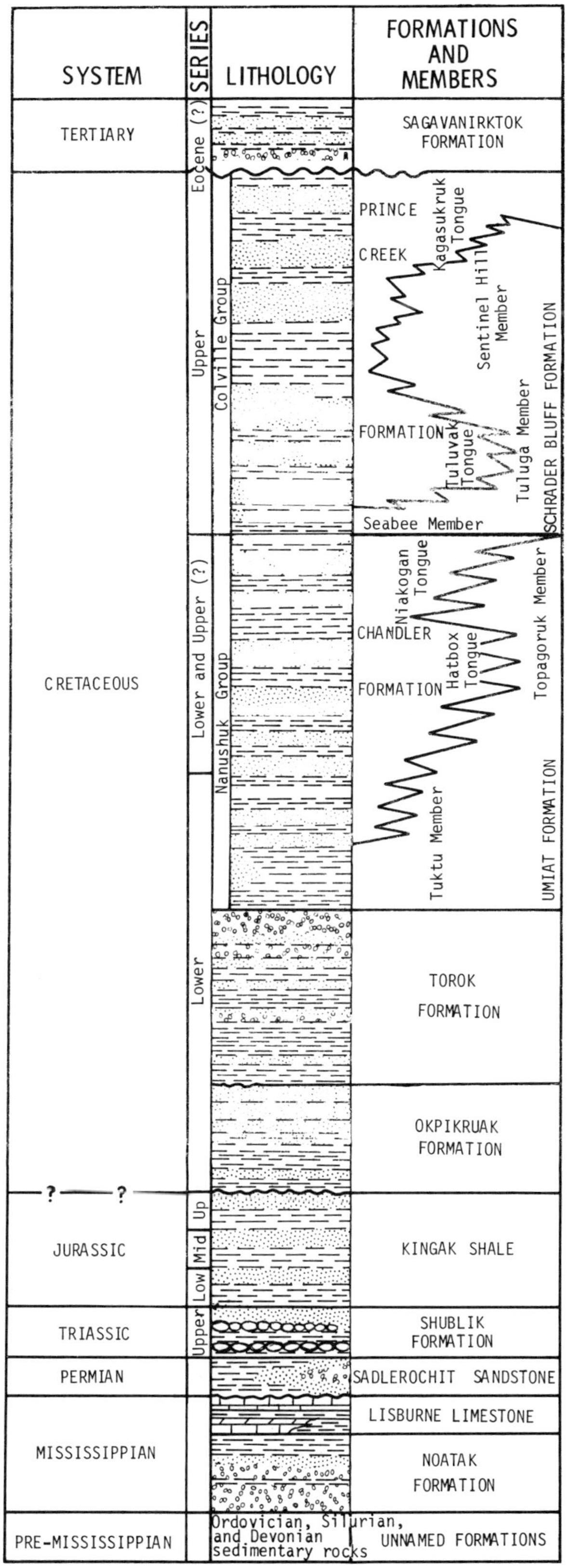

FIGURE 41.5 Composite stratigraphic section applicable to Artic (north) slope east of National Petroleum Reserve 4. ***After T. G. Payne et al. (1951, Figure 2). Courtesy R. J. Burnside (1959, p. 128, Figure 3) and the Oil and Gas Journal.***

System	Series	Group	Formation	Description
CRETACEOUS	Upper	Colville Group	Schrader Bluff Formation (Barrow Trail Member) (Rogers Creek Member)	Clay shale, siltstone, sandstone 825 feet at Gubik
			Prince Creek Formation (Tuluvak Tongue) Upper gas at Gubik	Mostly nonmarine sandstone some coal 870 feet at Gubik
			Seabee Formation	Mostly clay shale 1500 feet at Umiat; 1545 feet at Gubik
	Lower and Upper	Nanushuk Group	Ninuluk and Chandler Formations Lower gas at Gubik	Sandstone, clay, coal 260 feet (Umiat); 440 feet (Gubik)
			Grandstand Formation Oil at Umiat	Marine sandstones with clay shale interbeds 580 feet (Umiat); 760 feet (Gubik)
	Lower		Topagoruk Formation	Marine clay shale completely penetrated only at Umiat; 2800 feet thick
			Oumalik Formation	Marine clay shale. Top 400 feet only penetrated at Umiat

FIGURE 41.6 Umiat and Gubik fields, northern Alaska. Composite stratigraphic section. *After Florence R. Collins (1958) and Florence M. Robinson (1958).*

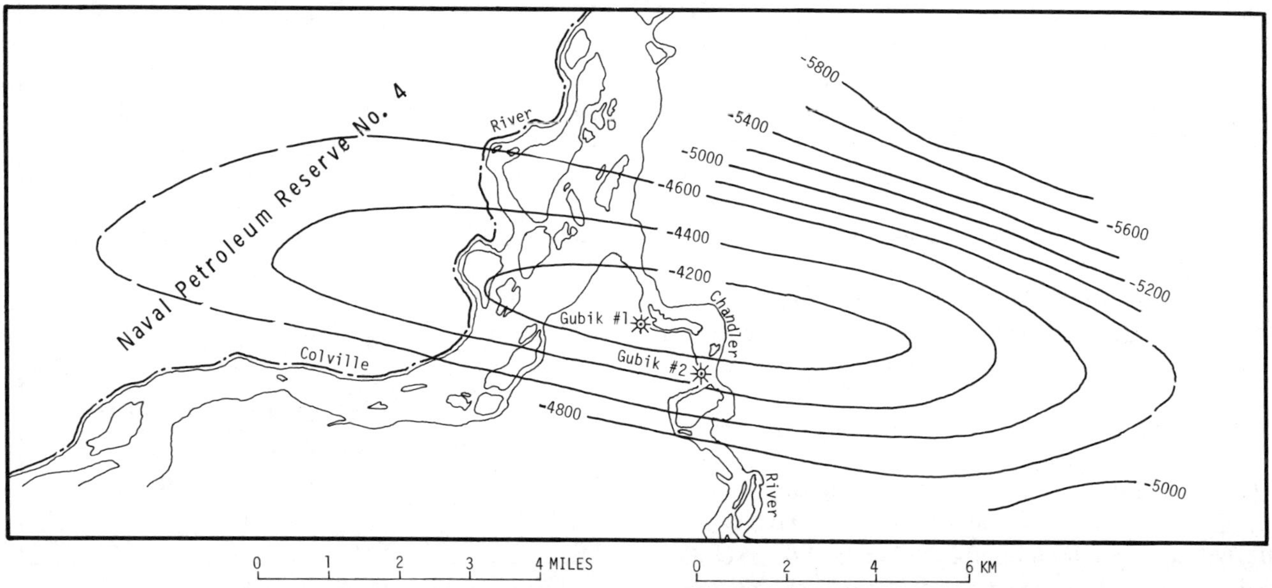

FIGURE 41.7 Gubik anticline, Arctic Alaska. Structure contour map based on phantom seismic horizon in the Grandstand (Cretaceous) Formation. Interval, 200 feet. Location of two gas wells shown. *After Florence M. Robinson (1958, Plate 13).*

for Tertiary production in the Sagavanirktok basin. Mississippian rocks may also have hydrocarbon potential in the northern part of the Brooks Range itself. Here they reach a maximum thickness of 5000 feet. The outcropping carbonate rocks have adequate permeability and "typically have a strong organic odor and locally show traces of petroleum residue" (Patton, 1964, p. 63).

Not to be ignored either is the Arctic coastal shelf, which has an area of approximately 80,000 square miles out to a depth of 600 feet, even though ice presents a prodigious challenge to development. Possible reservoirs here range downward from Tertiary to Mississippian.

REFERENCES CONSULTED IN PREPARING CHAPTER 41

Armstrong, Ted A., 1966.

Baptist, Oren C., 1959.

Brosgé, W. P., Dutro, J. T., Mangus, M. D., and Reiser, H. N., 1962.

Brosgé, William P., Whittington, Charles L., and Morris, Robert H., 1966.

Collins, Florence Rucker, 1958.

Detterman, Robert L., and Hartsock, John K., 1967.

Eberlein, G. D., 1964.

Fay, Leo F., 1966.

Gates, George O., 1964.

Gates, George O., Grantz, Arthur, and Patton, W. W., Jr., 1968.

Gates, George O., and Gryc, George, 1963.

Grantz, Arthur, 1964.

Grantz, Arthur, and Patton, W. W., Jr., 1964.

Gryc, George, 1959.

Gryc, George, et al., 1956.

Hill, M. L., 1963.

Kelly, Thomas E., 1963, 1968.

Miller, Don J., 1957, 1959.

Miller, Don J., Payne, Thomas G., and Gryc, George, 1959.

Osment, F. C., Morrow, R. M., and Craig, R. W., 1967.

Parkinson, L. J., and Helmuth, D. N., 1964.

Patton, W. W., Jr., 1964.

Patton, W. W., Jr., and Tailleur, I. L., 1964.

Payne, T. G., 1959.

Payne, T. G., et al., 1951.

Plafker, George, and Miller, John J., 1957.

Reed, John C., 1958.

Robinson, Florence M., 1958.

Saunders, R. H., 1964.

Sweet, John M., 1968.

Thompson, Raymond M., 1960.

U. S. Geological Survey, 1964.

Vander Ahe, K. L., 1964.

Wahrhaftig, Clyde, 1965.

Wahrhaftig, Clyde, and Gates, G. O., 1964.

Wilson, Thomas, Jr., 1968.

Woncik, John, 1968b.

Chapter

42

NONPRODUCTIVE STATES

There are three nonproducing states in the Pacific Division: Washington, Oregon, and Hawaii. Because the islands of Hawaii are volcanic, except for relatively minor volumes of clastic sediment and reef carbonates, this state is not discussed.

Washington. The state of Washington has the dubious distinction of being welcomed by the oil fraternity as the thirty-first petroleum state in 1957 (Wagner, 1966, p. 295) and then reverting to nonproductive status 4 years later. It has also produced gas commercially. Interest in the state was first aroused by the presence of oil and gas seeps in various parts, but especially in the Pacific border belt. By the end of 1967, 377 wells had been drilled in the search for oil and gas in the state.

Washington is divided into six physiographic provinces (Figure 42.1) that reflect three different terranes: (1) plateau lava, (2) igneous and metamorphic rock, and (3) sedimentary rock. The Columbia basin is an extensive lava plateau. The Okanogan Highlands and the Cascade Mountains are mainly igneous rocks, including eroded batholiths and volcanic cones (in the Cascades) and metamorphic rocks. The Olympic Mountains consist of metamorphic rocks and lava flows. The only extensive sedimentary terranes are in the Willapa Hills and Puget Lowlands provinces. The total area of these two provinces is about 9000 square miles, or 13 per cent of the state. However, a considerable volume of lava occurs with the sediments, especially in the Willapa Hills province, and glacial drift is abundant in the Puget Lowlands.

The sedimentary rocks are mainly Tertiary sandstones and siltstones, both continental and marine, with coal beds in the Eocene, and tuffaceous sediments ubiquitous. Wagner (1966, p. 293, 295) notes the presence of both potential source and potential reservoir rocks in the stratigraphic sections in parts of the Puget Lowlands and Willapa Hills provinces. Wolfe and McKee (1968) made similar observations in the Willapa Hills.

The now abandoned oil field, *Ocean City* in Grays

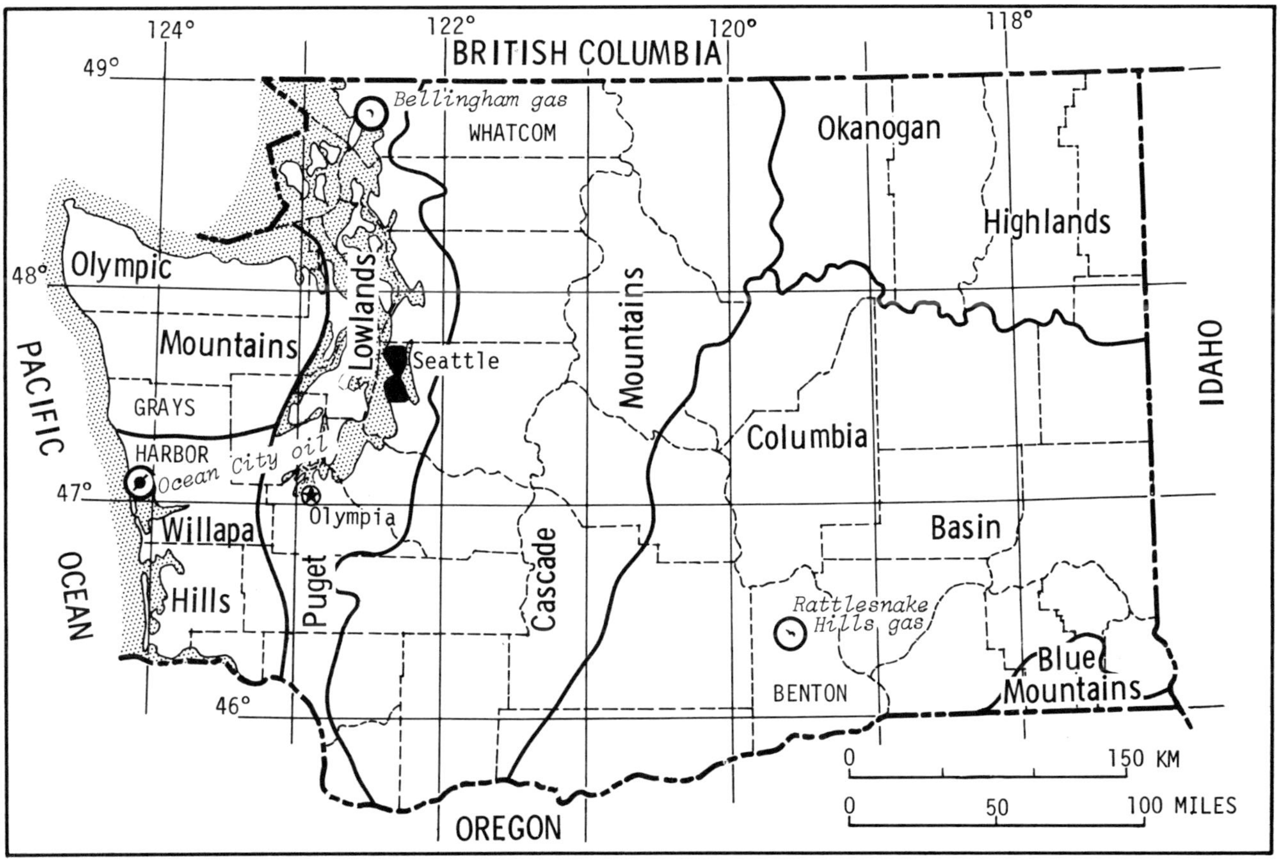

Figure 42.1 Washington. Index map, with physiographic provinces outlined.

Harbor County on the Pacific shore (Figure 42.1), was discovered in August, 1957, and its lone producer yielded about 12,500 barrels before final abandonment in 1961. Sixteen other wells drilled in this general area, including two offshore, were unsuccessful, although three produced 100 barrels or more (Wagner, 1966, p. 295). The trap is a faulted nearly circular anticline; the reservoir a fractured (but otherwise tight) sandstone of Oligocene age found at a depth between 3900 and 4000 feet (Wurden, 1959).

Washington has had two small, but commercial, gas fields (Glover, 1936). They are both very shallow, and accumulation has taken place under unusual circumstances in each case. The larger of these fields, *Rattlesnake Hills* (Hammer, 1934), is on the Columbia River lava plateau of eastern Washington. For over 20 years as many as sixteen wells produced gas from vesicular basalt. Intercalated between the flows are thin beds of lake clays that function as cap rocks. Depths to the reservoir rocks range from 700 to 1200 feet. Trapping within the lava reservoirs was due to the presence of a faulted anticline, a part of the belt of anticlinal folding that created the Rattlesnake Range of hills. This field was abandoned in 1951, after having provided a billion cubic feet of gas to seven towns in the vicinity.

The other field lies six miles north of Bellingham in Whatcom County of western Washington. At one time (1934) it contained six producing wells, most of them completed in mantle rock reservoirs lying at depths between 166 and 193 feet. Presumably the gas migrated from underlying Tertiary formations and became trapped in sand lenses in the glacial sediments.

Currently, the greatest hopes for the future of Washington as a petroleum state center on its western border, where it is already known from drilling that both potential source rocks and reservoir rocks are present. In addition to the oil and gas seeps previously mentioned, wildcat wells in this area have reported oil shows, plus of course actual production at Ocean City (Wagner, 1966, p. 294; Brown et al., 1960). Pease and Hoover (1957) call attention to oil possibilities in a part of the Willapa Hills province. Rau (1967) points out that the eight wildcats drilled in one topographic quadrangle in Grays Harbor County by no means exhaust the prospect possibilities in that area.

The western border sedimentary basins continue offshore, where, hopefully, the organic content of the source rocks increases and the induration of the reservoir

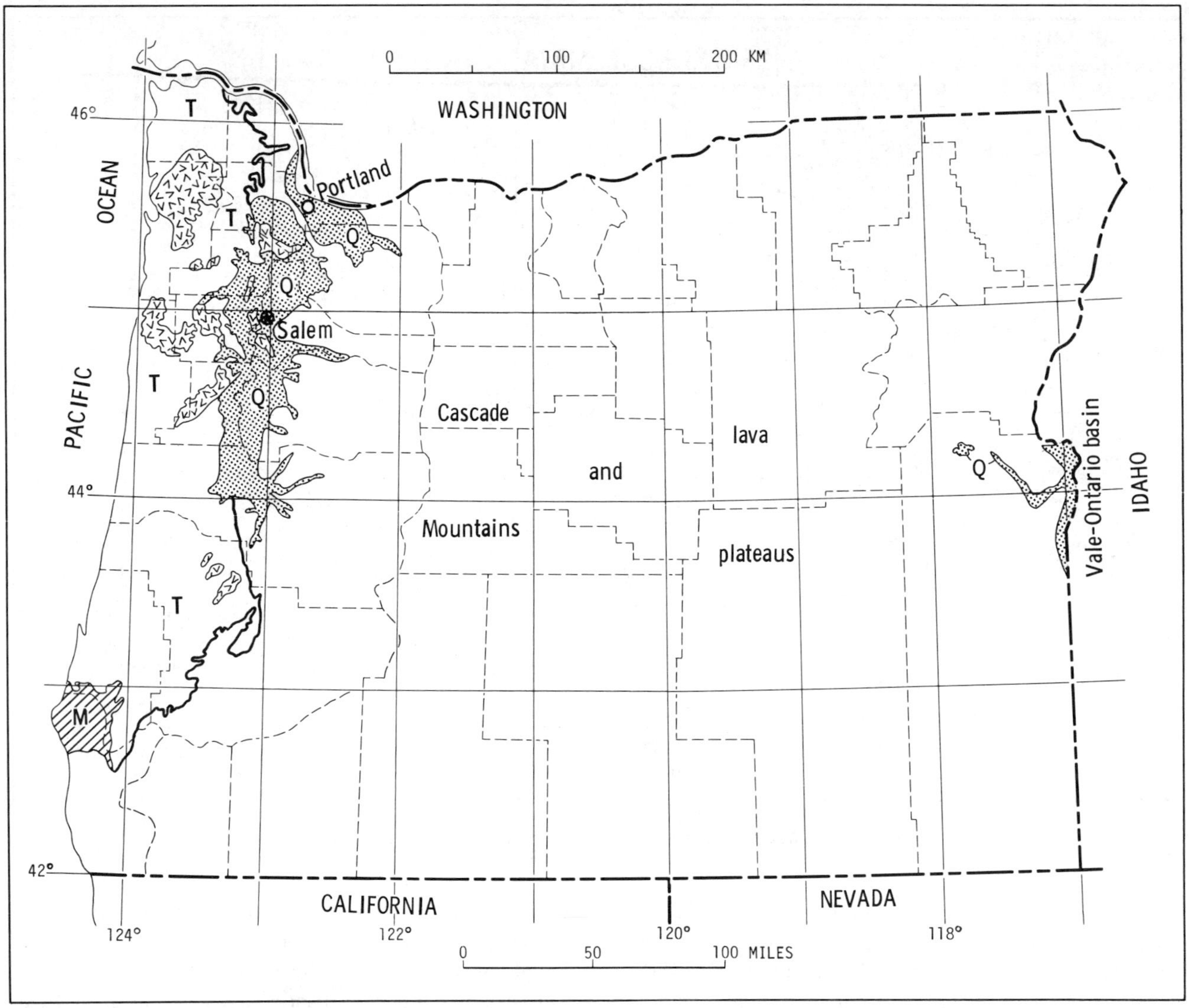

Figure 42.2 Oregon. Index and geologic map. Sedimentary rocks in M (Mesozoic), T (Tertiary), and Q (Quaternary) areas only. Large Q area on eastern edge of coastal belt is Willamette downwarp. Check pattern shows volcanic areas in coastal belt. Unmarked area east of coastal belt is almost entirely igneous.

rock decreases. However, a 4 year leasing and exploration program, from 1964 to 1967, produced two offshore dry holes at depths of approximately 10,400 and 13,200 feet, which in turn led to the quitclaiming of most of the offshore leases (Doell, 1968).

Eastward, undrilled and inadequately tested structures are known in the Puget Sound trough (Anderson, 1959). There is even the possibility that petroleum-bearing sedimentary rocks underlie the western part of the Columbia River lava plateau (Wagner, 1966, p. 290).

Oregon. Although all attempts so far to find petroleum in Oregon have been unsuccessful, the hunt continues. By the end of 70 years of intermittent exploration, 199 dry wildcats had been drilled in this state. The only positive results were small amounts of natural gas that were used locally and briefly. Most of the state is covered by volcanic rocks, both from fissure eruptions that built the great lava plateaus in eastern Oregon and from volcanos that pierced the granitic batholiths in the Cascade Mountain belt. However, sedimentary rocks do occur in the coastal belt, as well as inland in the Willamette downwarp (Figure 42.2). A small part of the Vale–Ontario basin of southwestern Idaho extends across the state line into eastern Oregon.

Western Oregon, especially the Tertiary belt, may still have oil and gas possibilities in spite of the dry hole record. The Umpqua Formation of early Eocene age contains both possible source beds and possible reservoirs, although the Tertiary sandstones in some areas are impermeable or nearly so (Baldwin, 1961). The continental shelf off the Oregon coast has a width ranging from 15 to 40 miles out to a depth of 100 fathoms (Byrne, 1962, p. 117). According to Koch (1966, p. 25), in his description of the geology of the southwestern Oregon coast, "The Miocene has favorable reservoir characteristics and should be a prime target for offshore exploration." The dry holes drilled offshore during recent years have dimmed enthusiasm, however, and most leases have been dropped.

Newton and Corcoran (1963) point out the possibilities for stratigraphic traps in the Vale–Ontario basin of easternmost Oregon.

REFERENCES CONSULTED IN PREPARING CHAPTER 42

Anderson, J. Q., 1959.

Baldwin, Ewart M., 1961.

Brown, Robert D., Jr., Gower, Howard D., and Snavely, Parke D. Jr., 1960.

Byrne, John V., 1962.

Doell, E. C., 1968.

Glover, Sheldon L., 1936.

Hammer, A. A., 1934.

Koch, John G., 1966.

Livingston, V. E., Jr., 1958.

Newton, V. C., Jr., 1963.

Newton, V. C., Jr., and Corcoran, R. E., 1963.

Pease, Maurice H., Jr., and Hoover, Linn, 1957.

Rau, W. W., 1967.

Wagner, H. C., 1966.

Wolfe, Edward W., and McKee, Edwin H., 1968.

Wurden, F. H., 1959.

REFERENCES

Aaker, Robert, and Jedlicka, John, 1968, Pre-Mancos gas fields in northwest Colorado: Am. Assoc. Petroleum Geologists, Mem. 9, v. 1, p. 856–865.

Abilene Geological Society, 1952, Geological contributions.

Abilene Geological Society, Stratigraphic and Study Group, 1960, The stratigraphic distribution of hydrocarbon production from 12 counties in the Abilene area.

Adams, Elmo W., and Beatty, W. B., 1962, Bituminous rocks in California: California Div. Mines and Geol., Mineral Inf. Service, v. 15, no. 4 (April), p. 1–9.

Adams, John, 1964, The oil boom in Morrow County, Ohio: Indep. Petroleum Monthly, April, p. 20–24.

Adams, John Emery, 1965, Stratigraphic-tectonic development of Delaware basin: Am. Assoc. Petroleum Geologists Bull., v. 49, no. 11 (November), p. 2140–2148.

Adams, William L., 1965, Apply diagenetic data to Anadarko oil hunt: Oil and Gas Jour., Apr. 19, p. 176–182.

Addison, Carl C., 1940, Buckeye oil field, Gladwin County, Michigan: Am. Assoc. Petroleum Geologists Bull., v. 24 (November), p. 1950–1982.

Adkison, W. L., 1960, Subsurface cross section of Paleozoic rocks from Barber County, Kansas, to Caddo County, Oklahoma: U.S. Geol. Survey Oil and Gas Inv. Chart OC 61.

Agnew, Allen F., and Gries, John Paul, 1960, South Dakota oil; past, present, and future *in* geologic record: Am. Assoc. Petroleum Geologists, Rocky Mtn. Sec., Geol. Record, p. 85–94.

Akkerman, R. P., 1959, Turtle Bay field, Chambers County, Texas: Gulf Coast Assoc. Geol. Socs. Trans., v. 9, p. 71–75.

Alaska Geological Society, 1964, Guide book (and reports) Anchorage, Alaska.

Albers, J. P., 1966a, Geomorphic provinces: California Div. Mines and Geol. Bull. 191, p. 35–36.

Albers, J. P., 1966b, Topography: California Div. Mines and Geology Bull. 191, p. 23–25.

Alexander, C. I., 1951, History of discovery and development of Woodbine oil fields in east Texas *in* The Woodbine and adjacent strata in the Waco area of central Texas—a symposium: East Texas Geol. Soc., Tyler, Texas.

Alkire, James, 1968, Willows-Beehive Bend gas field: Am. Assoc. Petroleum Geologists Mem. 9, v. 1, p. 639–642.

Alkire, Robert L., 1952, Oil and gas in Perry County: Ohio Geol. Survey Rept. Inv. 10.

Allan, Thomas H., and Valerius, M. M., 1929, Fairport oil field, Russell County, Kansas, *in* Structure of typical American oil fields: Am. Assoc. Petroleum Geologists, v. 1, p. 35–48.

Allen, D. B., and Joujon-Roche, J. E., 1958, West Newport oil field, *in* A guide to the geology and oil fields of the Los Angeles and Ventura regions: Am. Assoc. Petroleum Geologists, Pacific Sec., p. 143–144.

Allison, A. P., et al., 1946, Geology of Katy field, Waller, Harris, and Fort Bend Counties, Texas: Am. Assoc. Petroleum Geologists Bull., v. 30 (February), p. 157–180.

American Association of Petroleum Geologists, 1929, 1948, Structure of typical American oil fields; v. 1 and 2, 1929; v. 3, 1948.

American Association of Petroleum Geologists, 1935, Geology of natural gas.

American Association of Petroleum Geologists, 1941, Stratigraphic type oil fields.

American Association of Petroleum Geologists, 1956, 1959, Petroleum geology of southern Oklahoma, v. 1 and 2.

American Association of Petroleum Geologists, 1958, Habitat of oil (Spec. Volume); L. G. Weeks, editor, Tulsa, Oklahoma.

American Association of Petroleum Geologists, 1960, Stratigraphic cross section of Paleozoic rocks, West Texas to Northern Montana.

American Association of Petroleum Geologists, 1968, Natural gases of North America: v. 1, pt. 1, Natural gases in rocks of Cenozoic age, and pt. 2, Natural gases in rocks of Mesozoic age; v. 2, pt. 3, Natural gases in rocks of Paleozoic age, and pt. 4, Papers of general scope.

American Association of Petroleum Geologists, Pacific Section, 1958. A guide to the geology and oil fields of the Los Angeles and Ventura regions, James W. Higgins, ed., Los Angeles, California.

American Association of Petroleum Geologists, Rocky Mountain Section, 1955, 1959, 1960, Geological record.

American Metal Climax, 1961, Bar-X field *in* The oil and gas fields of Utah—a symposium: Intermountain Assoc. Petroleum Geologists, Salt Lake City, Utah.

Amoruso, John J., 1957, A structural study of the Northville oil field, Wayne, Washtenaw, and Oakland Counties, Michigan: M. S. Thesis, University of Michigan.

Amsbury, David L., 1967, Summary of oil development in the Reno area, Johnson County, Wyoming: Mtn. Geologist, v. 4, no. 1, p. 3–7.

Amsden, Thomas W., 1954, Geology of Garrett County: Maryland Dept. Geol., Mines and Water Resources Bull. 13, p. 1–95.

Amsden, Thomas W., Caplan, W. M., Hilpman, P. L., McGlasson, E. H., Rowland, T. L., and Wise, O. A., Jr., 1968, Devonian of the southern Midcontinent area, United States, *in* Internat.

Symposium on Devonian System Proc.: Alberta Soc. Petroleum Geologists, p. 913–932.

Anderman, George G., and Ackman, Edward J., 1963, Structure of the Denver Julesburg basin and surrounding areas *in* Geology of the northern Denver basin and adjacent uplifts, Rocky Mtn. Assoc. Geologists, p. 170–175.

Andersen, Harold V., 1960, Geology of Sabine Parish: Louisiana Geol. Survey Bull. 34 (February), p. 145–149, plate 17.

Anderson, J. Q., 1959, Puget Sound area has several prospective oil and gas basins: World Oil, July, p. 111–117.

Anderson, K. H., and Wells, J. S. 1964, New thinking may be key to unlocking Missouri prospects: Oil and Gas Jour., Dec. 14, p. 122–130.

Anderson, K. H., and Wells, J. S., 1968, Forest City basin of Missouri, Kansas, Nebraska, and Iowa: Am. Assoc. Petroleum Geologists Bull., v. 52, no. 2 (February), p. 264–281.

Anderson, Sidney B., 1963, Look again at North Dakota's Devonian: Oil and Gas Jour., Aug. 12, p. 178–180.

Anderson, Sidney B., 1966, A look at the petroleum potential of southwestern North Dakota: North Dakota Geol. Survey Rept. Inv. 42, Sheets 1 to 3. Report discussed by author in Oil and Gas Jour. Aug. 15, p. 134–137.

Anderson, Sidney B., 1967, Where North Dakota's best Newcastle sand trends are located: World Oil, December, p. 90–93.

Anderson, Sidney B., and Eastwood, William P., 1968, Natural gas in North Dakota: Am. Assoc. Petroleum Geologists, Mem. 9, v. 2, p. 1304–1326.

Andrews, Donald I., 1960, The Louann Salt and its relationship to Gulf Coast salt domes: Gulf Coast Assoc. Geol. Socs. Trans., v. 10, p. 215–240.

Andrews, Donald I., 1961, Indigenous Pleistocene production in offshore Louisiana: Gulf Coast Assoc. Geol. Socs. Trans., v. 11, p. 117.

Andrichuk, John M., 1958, Mississippian Madison stratigraphy and sedimentation in Wyoming and southern Montana *in* Habitat of oil: Am. Assoc. Petroleum Geologists, p. 225–267.

Anon., 1953, California's oldest well: The Standard Oiler, August, p. 8–11.

Anon., 1959, Offshore geology and oil resources: Calif. Div. Mines, Mineral Inf. Service, v. 12, no. 5 (May), p. 1–7.

Anon., 1961, Clear Creek gas field, Carbon and Emery Counties, Utah, *in* The oil and gas fields of Utah—a symposium: Intermountain Assoc. Petroleum Geologists, Salt Lake City, Utah.

Anon., 1966, The Moulton pool, Toole County, Montana and Alberta, Canada, *in* Guidebook; 17th Ann. Field Conf.: Billings Geol. Soc., p. 186–190.

Antoine, John W., and Harding, James L., 1956, Structure beneath continental shelf, northeastern Gulf of Mexico: Am. Assoc. Petroleum Geologists Bull., v. 49, no. 2 (February), p. 157–171.

Applin, Paul L., 1952, Volume of Mesozoic sediments in Florida and Georgia: Geol. Soc. America Bull., v. 63, no. 12, pt. 1 (December), p. 1159–1164.

Applin, Paul L., and Applin, Esther R., 1944, Regional subsurface stratigraphy and structure of Florida and southern Georgia: Am. Assoc. Petroleum Geologists Bull., v. 28, no. 12 (December), p. 1673–1753.

Applin, Paul L. and Applin, Esther R., 1947, Regional subsurface stratigraphy, structure, and correlation of Middle and Early Upper Cretaceous rocks in Alabama, Georgia, and north Florida: U.S. Geol. Survey Oil and Gas Inv. Prelim. Chart 26.

Armstrong, Frank C., and Oriel, Steven S., 1965, Tectonic development of Idaho–Wyoming thrust belt: Am. Assoc. Petroleum Geologists Bull., v. 49, no. 11 (November), p. 1847–1866.

Armstrong, Ted A., 1966, Alaskan oil: Oil and Gas Jour., Aug. 22, p. 77–100.

Arnold, E. C., 1965, Geology and oil and gas production in northeastern New Mexico *in* Mineral and water resources of New Mexico: U.S. Senate Rept. Comm. on Interior and Insular Affairs, 89th Cong., 1st sess. p. 87–98.

Arnold, H. H., Jr., 1939, Salem oil field, Marion County, Illinois: Am. Assoc. Petroleum Geologists Bull., v. 23 (September), p. 1352–1373.

Ashley, G. H., 1938, History of development and geologic relationships of Appalachian fields: Am. Assoc. Petroleum Geologists Bull., v. 22 (April), p. 416–430.

Asquith, D. O., 1966, Geology of Late Cretaceous Mesaverde and Paleocene Fort Union oil production, Birch Creek unit, Sublette County, Wyoming: Am. Assoc. Petroleum Geologists Bull., v. 50, no. 10 (October), p. 2176–2184.

Atwater, Gordon I., 1959, Geology and petroleum development of the continental shelf of the Gulf of Mexico: Proc. Fifth World Petrol. Cong., sec. 1, p. 409–434. Also in Gulf Coast Assoc. Geol. Socs. Trans., v. 9, p. 131–145.

Atwater, Gordon I., and Forman, McLain J., 1959, Nature of growth of southern Louisiana salt domes and its effect on petroleum accumulation: Am. Assoc. Petroleum Geologists Bull., v. 43 (November), p. 2592–2622.

Atwater, Gordon I., and Miller, E. E., 1965, The effect of decrease in porosity with depth on future development of oil and gas reserves in south Louisiana: Am. Assoc. Petroleum Geologists Bull., v. 49, no. 3 (March), p. 334.

Austin, Morris E., 1965, Land resource regions and major land resource areas of the United States: U.S. Soil Conserv. Service, Agriculture Handbook 296, December.

Autry, Vernon E., 1940, Discovery of oil in Ellenburger Formation, K.M.A. oil field, Wichita County, Texas: Am. Assoc. Petroleum Geologists Bull., v. 40 (August), p. 1494–1495.

Averitt, Paul, 1941, The Early Grove gas field, Scott and Washington Counties, Virginia: Virginia Geol. Survey Bull. 56.

Avila, John, 1967, Urbana meets basic oil-bearing criteria: Oil and Gas Jour., July 3, p. 106–110.

Avila, John, Barnes, John M., and Nuttall, Brandon, (panelists), 1963, Western Kentucky's Hanson field is a straw in the wind: Oil and Gas Jour., April 1, p. 176–179.

Baars, D. L., 1966, Pre-Pennsylvanian tectonics—key to basin evolution and petroleum occurrences in Paradox basin, Utah and Colorado: Am. Assoc. Petroleum Geologists Bull., v. 50, no. 10 (October), p. 2082–2111.

Babcock, C. V., 1956, Eunice-Monument field *in* Oil and Gas fields of southeastern New Mexico: Roswell Geol. Soc., p. 165.

Babcock, Clarence, 1965, Interview *in* World Oil, May, p. 174.

Babcock, Clarence, 1966, Florida petroleum exploration, production, and prospects, 1964: Florida Geol. Survey Inf. Circ. 49.

Babisak, Julius, 1958, Wolf Springs field *in* Montana oil and gas fields: Billings Geol. Soc., p. 231.

Baddley, E. R., 1954, Geology of the South Mountain oil field, Ventura County: California Div. Mines and Geol. Bull. 170, Map Sheet 29.

Bahr, Carl W., 1963, Virgin oil field, Washington County, Utah, *in* Guidebook to the Geology of southwestern Utah: Intermountain Assoc. Petroleum Geologists, p. 169–171.

Baker, Donald R., 1962, Organic geochemistry of Cherokee group in southeastern Kansas and northeastern Oklahoma: Am. Assoc. Petroleum Geologists Bull., v. 46 (September), p. 1621–1642.

Bakker, D., 1968, Natural gas in Texas part of Marietta syncline, Cooke and Grayson Counties, Texas: Am. Assoc. Petroleum Geologists Mem. 9, v. 2, p. 1459–1466.

Baldwin, Ewart M., 1961, Geologic map of the lower Umpqua River area, Oregon: U.S. Geol. Survey Oil and Gas Inv. Map OM-204.

Baldwin, Thomas A., 1950, San Ardo, a stratigraphic analysis of a California oil field: Am. Assoc. Petroleum Geologists Bull., v. 34, no. 10 (October), p. 1981–1989.

Baldwin, Thomas A., 1953, San Ardo field: Jour. Petroleum Technology, v. 5, no. 1, sec. 1, p. 9–10, continued, sec. 2, p. 5 (January).

Bale, Hubert E., and Williams, George O., 1951, The granite ridge of Oklahoma: Shale Shaker, v. 2 (October), p. 18–20.

Balk, Robert, 1949, Structure of Grand Saline dome, Van Zandt County, Texas: Am. Assoc. Petroleum Geologists Bull., v. 33, no. 11 (November), p. 1791–1829.

Ball, Max W., Weaver, T. J., Crider, H. D., and Ball, Douglas S., 1941, Shoestring gas fields of Michigan *in* Stratigraphic type oil fields: Am. Assoc. Petroleum Geologists, p. 237–266.

Ballard, William Wayne, 1957, Subsurface study of Morrow and Atoka series in part of Arkansas Valley of western Arkansas: Am. Assoc. Petroleum Geologists Bull., v. 41 (February), p. 263–277.

Baltz, Elmer H., 1965, Stratigraphy and history of Raton basin and notes on San Luis basin, Colorado–New Mexico: Am. Assoc. Petroleum Geologists Bull., v. 49, no. 11 (November), p. 2041–2075.

Baltz, Elmer H., 1967, Stratigraphy and regional tectonic implications of part of Upper Cretaceous and Tertiary rocks, east-central San Juan Basin, New Mexico: U.S. Geol. Survey Prof. Paper 552.

Bank of the Southwest, 1956, Production history of the Southwest, p. 5–10.

Banks, Joseph E., 1960, Petroleum in Comanche (Cretaceous) section, Bend area, Florida: Am. Assoc. Petroleum Geologists Bull., v. 44 (November), p. 1737–1748.

Baptist, Oren C., 1959, Oil production from frozen reservoir rocks, Umiat, Alaska: Petroleum Technology, v. 11, no. 11 (November), p. 85–88.

Barbat, William F., 1958, The Los Angeles basin area, California, *in* Habitat of oil: Am. Assoc. Petroleum Geologists, p. 62–77.

Barby, B. G., 1961, Panhandle gas hunt sets hot pace: Oil and Gas Jour., May 29, p. 210–217.

Barby, B. G., 1962, Reserve study of Cottage Grove, northwest Waynoka field, Woods County, Oklahoma: Tulsa Geol. Soc. Digest, v. 30, p. 82–88.

Barlow, J. A., Jr., and Haun, John D., 1966, Regional stratigraphy of Frontier Formation and relation to Salt Creek field, Wyoming: Am. Assoc. Petroleum Geologists Bull., v. 50, no. 10 (October), p. 2185–2196.

Barnes, John M., 1959, Green County sparks 500 foot drilling boom in central Kentucky: Oil and Gas Jour., July 20, p. 128–140.

Barnes, John M., 1963, Kentucky's Hopkins County has drilling spree: Oil and Gas Jour., Nov. 4, p. 180–182.

Barnes, T. R., 1953, Williston basin—new province for oil exploration: Am. Assoc. Petroleum Geologists Bull., v. 37 (February), p. 340–354.

Barnes, Virgil E., 1948, Ouachita facies in central Texas: Texas Bur. Econ. Geol. Rept. Inv. 2.

Barnes, V. E., Boucot, Arthur J., Cloud, Preston E., Jr., Miller, Richard H., and Palmer, A. R., 1966, Silurian of central Texas, a first record for the region: Science, v. 154 Nov. 25, p. 1001–1008.

Barnett, Donald C., 1941, O'Hern field, Duval and Webb Counties, Texas, *in* Stratigraphic type oil fields, Am. Assoc. Petroleum Geologists, p. 722–749.

Barr, C. R., and Bieber, A. M., 1961, Permian reefing on the Chadron arch: Oil and Gas Jour., Sept. 4, p. 212–219.

Bartlett, Charles S., Jr., 1962, Washburn anticline: Oil and Gas Jour., Aug. 20, p. 140–148.

Bartlett, Charles S., Jr., 1966, Arkansas' search for porosity broadens: Oil and Gas Jour., Jan. 24, p. 126 et seq.

Bartlett, Charles S., Jr., 1968, Washburn anticline gas fields, Arkansas: Am. Assoc. Petroleum Geologists Mem. 9, v. 2, p. 1658–1667.

Bartley, J. H., 1952, Geology of Spraberry trend of Permian basin of West Texas: Petroleum Engineer, v. 24 (July), p. B-37 et seq. Published in condensed form in Oil and Gas Jour., Apr. 21, 1952, p. 157–159.

Barton, H. E., 1948, Steamboat Butte oil field, Fremont County, Wyoming, *in* Structure of typical American oil fields: Am. Assoc. Petroleum Geologists (Spec. Volume), 3, p. 480–513.

Bartram, John G., 1929, Elk Basin oil and gas field, Park County, Wyoming, and Carbon County, Montana, *in* Structure of typical American oil fields: Am. Assoc. Petroleum Geologists, v. 2, p. 577–588.

Bartram, John G., and Erdmann, Charles E., 1935, Natural gas in Montana *in* Geology of natural gas: Am. Assoc. Petroleum Geologists, p. 245–276.

Bass, N. W., 1936, Origin of shoestring sands of Greenwood and Butler Counties, Kansas: Kansas Geol. Survey Bull. 23.

Bass, N. W., 1944, Correlation of basal Permian and older rocks in southwestern Colorado, northwestern New Mexico, northeastern Arizona, and southeastern Utah: U.S. Geol. Survey Oil and Gas Inv. Prelim. Chart 7.

Bass, N. W., 1964, Mineral fuels and associated resources—oil and gas *in* Mineral and water resources of Colorado: U.S. 88th Cong., 2d sess., Comm. Print., p. 45–67.

Bass, N. W., Goodrich, H. B., and Dillard, W. R., 1942, Subsurface geology and oil and gas resources of Osage County, Oklahoma: U.S. Geol. Survey Bull. 900.

Bates, Fred W., 1941, Geology of the Eola oil field, Avoyelles Parish, Louisiana: Am. Assoc. Petroleum Geologists Bull., v. 25 (July), p. 1363–1395.

Bates, Fred W., and Copeland, Robert R., Jr., 1952, Glenmora field, Rapides Parish, Louisiana: Am. Assoc. Petroleum Geologists Bull., v. 36 (January), p. 146–159.

Bates, Fred W., Copeland, Robert R., Jr., and Dixon, Kenneth P., 1959, Geology of Avery Island salt dome, Iberia Parish,

Louisiana: Am. Assoc. Petroleum Geologists Bull., v. 43 (May), p. 944–957.

Bates, Fred W., and Wharton, Jay B., Jr., 1943, Anse la Butte dome, St. Martin Parish, Louisiana: Am. Assoc. Petroleum Geologists Bull., v. 27 (August), p. 1123–1156.

Bauernschmitt, A. J., Jr., 1944, West Ranch oil field, Jackson County, Texas: Am. Assoc. Petroleum Geologists Bull., v. 28 (February), p. 197–216.

Bayles, Robert E., 1956, Wood County deep well; log, sample, and core description: West Virginia Geol. Survey Rept. Inv. 14, IV.

Beach, John H., 1948, Geology of Edison oil field, Kern County, California, Structure of typical American oil fields: *in* Am. Assoc. Petroleum Geologists, v. 3, p. 58–85.

Beans, Robert J., 1952, Geology of the Elk City field: World Oil, January, p. 67–72.

Beck, Elfred, 1929, Salt Creek oil field, Natrona County, Wyoming, *in* Structure of typical American oil fields: Am. Assoc. Petroleum Geologists, v. 2, pp. 589–603. Map compiled by Beck, E. W. Rumsey, and W. W. Rusk.

Beckwith, H. T., 1928, Oil and gas in Oklahoma, geology of Osage County: Oklahoma Geol. Survey Bull. 40T.

Beebe, B. W., 1957, Petroleum geology of the northwest Anadarko basin *in* Geologia del Petroleo Seccion III, p. 41–55; Congreso Geologico Internacional, Mexico.

Beebe, B. W., 1968a, Natural gas in Denver basin: Am. Assoc. Petroleum Geologists, Mem. 9, v. 1, p. 899–927.

Beebe, B. W., 1968b, Deep Edwards trend of South Texas: Am. Assoc. Petroleum Geologists, Mem. 9, v. 1, p. 961–975.

Beebe, B. W., 1968c, Natural gas in post-Paleozoic rocks of Mississippi: Am. Assoc. Petroleum Geologists, Mem. 9, v. 1, p. 1176–1226.

Beebe, Lewis E., 1953, Wyoming Circle Ridge field: Oil and Gas Jour., Sept. 14, p. 109–114.

Beekly, Emerson K., 1956, East Poplar field, Roosevelt County, Montana, *in* First internat. Williston basin symposium: North Dakota Geol. Soc. and Saskatchewan Geol. Soc., p. 61–65.

Beene, Douglas L., and Oros, Margaret L., 1967, Oil and gas developments in Kansas during 1965: Kansas Geol. Survey Bull. 185.

Beikman, Helen M., and Drakoulis, Sophie, 1958, Map of Mississippi showing oil and gas fields, unsuccessful test wells, salt domes, and pipelines: U.S. Geol. Survey Oil and Gas Inv. Map OM-200.

Bell, Alfred H., 1951, Oil possibilities of Niagaran reefs in Illinois: World Oil, January, p. 74–78.

Bell, Alfred H., Atherton, Elwood, Buschbach, T. C., and Swann, David H., 1964, Deep oil possibilities in the Illinois basin: Illinois Geol. Survey Circ. 368.

Berg, Robert R., 1963, Laramide sediments along the Wind River thrust, Wyoming, *in* Backbone of the Americas: Am. Assoc. Petroleum Geologists Mem. 2, November, p. 220–230.

Berg, Robert R., and Davies, David K., 1968, Origin of Lower Cretaceous Muddy Sandstone at Bell Creek field, Montana: Am. Assoc. Petroleum Geologists Bull., v. 52, no. 10 (October), p. 1888–1898.

Berg, Robert R., and Tenney, Charles S., 1967, Geology of Lower Permian Minnelusa oil fields, Powder River basin, Wyoming: Am. Assoc. Petroleum Geologists Bull., v. 51, no. 5 (May), p. 705–709.

Bergenback, Richard E., and Terriere, Robert T., 1953, Petrography and petrology of Scurry reef, Scurry County, Texas: Am. Assoc. Petroleum Geologists Bull., v. 37, no. 5 (May), p. 1014–1029.

Bergin, M. J., 1956, Structure and stratigraphy of the Pennsylvanian rocks exposed in the Seitz quadrangle, Breathitt, Magoffin, Morgan, and Wolfe Counties, Kentucky: U.S. Geol. Survey Oil and Gas Inv. Map OM-173.

Best, J. Boyd, 1941, Lopez oil field, Webb and Duval Counties, Texas, *in* Stratigraphic type oil fields: Am. Assoc. Petroleum Geologists Bull., p. 680–697.

Best, John B., Jr., 1964, Oil and gas prospects in the Palo Duro basin, World Oil, December, p. 97–113.

Betts, A. J., 1951, Van field, Van Zandt County, Texas (with Liddle map modified by Frank A. Herald, ed.): Univ. Texas Pub. 5116, p. 399–401.

Bevier, George M., 1925, The Damon Mound oil field, Texas: Am. Assoc. Petroleum Geologists Bull., v. 9 (May-June), p. 505–535.

Bicker, Alvin R., Jr., 1966, Claiborne County geology *in* Claiborne County geology and mineral resources: Mississippi Geol. Econ. and Topog. Survey Bull. 107, p. 9–94.

Biddick, James R., 1963, Oswego—hottest trend in the deeper part of Anadarko: World Oil, March, p. 122–126.

Biggs, Charles A., and McGregor, Alexander A., 1968, The Bell Creek oil field, Powder River and Carter Counties, Montana, *in* Wyoming Geol. Assoc. Guidebook 20th Ann. Field Conf.: p. 47–58.

Bike, Peter B., 1960, Mississippi is still a good place to look for oil: Oil and Gas Jour., Feb. 29, p. 127.

Billings Geological Society, 1958, Montana oil and gas fields—a symposium.

Billings Geological Society, 1966, Jurassic and Cretaceous stratigraphic traps, Sweetgrass arch—a symposium *in* Guidebook, 17th Ann. Field Conf.: Billings, Montana.

Billings Geological Society, North Dakota Geological Society, and Saskatchewan Geological Society, 1964, Third Internat. Williston basin symposium, Regina, Saskatchewan, Proc., Billings, Montana.

Billingsley, David L., 1964, Caddo-Pine Island field still hot after 58 years: World Oil, March, p. 86–90.

Billingsley, H. R., 1956, Sholom Alechem oil field, Stephens and Carter Counties, Oklahoma, *in* Petroleum geology of southern Oklahoma: Am. Assoc. Petroleum Geologists, v. 1, p. 294–310.

Billingsley, J. E., 1934, Occurrence of oil and gas in West Virginia, eastern Ohio, and eastern Kentucky *in* Problems of petroleum geology: Am. Assoc. Petroleum Geologists, p. 509.

Birkhauser, Max, 1943, Coalinga oil field *in* California Div. of Mines, Bull. 118, p. 484–485.

Bishop, William F., 1967, Age of pre-Smackover formations, North Louisiana and South Arkansas: Am. Assoc. Petroleum Geologists, Bull., v. 51, no. 2 (February), p. 244–245.

Bissell, H. J., 1967, Pennsylvanian and Permian basins in northwestern Utah, northeastern Nevada, and south-central Idaho: Discussion (followed by reply by Roberts et al., 1965): Am. Assoc. Petroleum Geologists Bull., v. 51, no. 5 (May), p. 791–803.

Blackstone, D. L., Jr., 1953, Tectonic map of a portion of southern Wyoming and northern Colorado *in* Wyoming Geol. Assoc. Guidebook 8th Ann. Field Conf.: Map in pocket.

Blackstone, D. L., Jr., 1963, Development of geologic structure in central Rocky Mountains *in* Backbone of Americas: Am. Assoc. Petroleum Geologists Mem. 2, p. 160–179.

Blair, Arden F., 1960, Ritchie field, Montana, *in* Geological record: Am. Assoc. Petroleum Geologists, Rocky Mtn. Sec., p. 71–77.

Blanchard, K. S., Denman, Orval, and Knight, A. S., 1968, Natural gas in Atokan (Bend) section of northern Fort Worth basin: Am. Assoc. Petroleum Geologists Mem. 9, v. 2, p. 1446–1454.

Blatchley, Raymond S., 1913, The oil fields of Crawford and Lawrence Counties: Illinois State Geol. Survey Bull. 22.

Bleakley, W. B., 1965, High temperatures licked in deep well: Oil and Gas Jour., Jan. 18, p. 70–74.

Blixt, John E., 1941, Cut Bank oil and gas field, Glacier County, Montana, *in* Stratigraphic type oil fields: Am. Assoc. Petroleum Geologists, p. 327–381.

Bolinger, John W., 1953, History of the Imogene oil field, Atascosa County, Texas: Geophysics, v. 18 (April), p. 360–370.

Bolyard, Dudley W., and McGregor, Alexander A., 1966, Stratigraphy and petroleum potential of Lower Cretaceous Inyan Kara group in northeastern Wyoming, southeastern Montana, and western South Dakota: Am. Assoc. Petroleum Geologists Bull., v. 50, no. 10 (October), p. 2221–2244.

Bond, B. C., 1959, Some remaining potentials of northwest Colorado *in* Geological record: Am. Assoc. Petroleum Geologists, Rocky Mtn. Sec., p. 131–134.

Bond, D. C., Bell, A. H., and Meents, W. F., 1968, Gas in Illinois basin: Am. Assoc. Petroleum Geologists Mem. 9, v. 2, p. 1746–1753.

Borden, Joseph L., and Brant, Ralph A., 1941, East Tuskegee pool, Creek County, Oklahoma, *in* Stratigraphic type oil fields: Am. Assoc. Petroleum Geologists, p. 436–455.

Bornhauser, Max, 1950, Oil and gas accumulation controlled by sedimentary facies in Eocene Wilcox to Cockfield formations, Louisiana Gulf Coast: Am. Assoc. Petroleum Geologists Bull., v. 34 (September), p. 1887–1896.

Bornhauser, Max, 1960, Depositional and structural history of northwest Harburg field, Newton County, Texas: Am. Assoc. Petroleum Geologists Bull., v. 44, no. 4 (April), p. 458–470.

Bowen, Oliver E., Jr., 1962, ed., *in* California Div. Mines and Geol. Bull. 181.

Bowers, Emil F., and Denman, R. H., 1953, Swan Lake viewed as discovery of major importance: Oil and Gas Jour., June 22, p. 208–220.

Boyd, Don R., and Dyer, Byron F., 1964, Frio barrier bar system of South Texas: Gulf Coast Assoc. Geol. Socs. Trans., v. 14, p. 309–322.

Boyd, W. Baxter, 1938, Jesse pool, Pontotoc and Coal Counties, Oklahoma: Am. Assoc. Petroleum Geologists Bull., v. 22 (November), p. 1560–1578.

Brace, R. T., 1961, Red Wash field *in* The oil and gas fields of Utah—a symposium: Intermountain Assoc. Petroleum Geologists, 1961.

Bradbury, James C., and Atherton, Elwood, 1965, The Precambian basement of Illinois: Illinois Geol. Survey Circ. 382.

Bradfield, H. H., 1957a, The petroleum geology of Grayson County, Texas, and the subsurface geology of Cooke County, Texas, *in* The geology and geophysics of Cooke and Grayson Counties, Texas: Dallas Geol. Soc. and Dallas Geophys. Soc., p. 15–49 and 75–100.

Bradfield, H. H., 1957b, Big Mineral field *in* The geology and geophysics of Cooke and Grayson Counties, Texas: Dallas Geol. Soc. and Dallas Geophys. Soc., p. 59–61.

Bradfield, H. H., 1957c, Handy field *in* The geology and geophysics of Cooke and Grayson Counties, Texas: Dallas Geol. Soc. and Dallas Geophys. Soc., p. 50–53.

Bradfield, H. H., 1957d, Sherman field *in* The geology and geophysics of Cooke and Grayson Counties, Texas: Dallas Geol. Soc. and Dallas Geophys. Soc., p. 62–69.

Bradfield, H. H., 1959, Petroleum geology of Grayson County, Texas, *in* v. 2 *of* Petroleum geology of southern Oklahoma: Ardmore Geol. Soc., July, p. 53–100.

Bradfield, H. H., 1964, The Ellenburger group of North Central Texas: Tulsa Geol. Soc. Digest, v. 32, p. 114–120.

Bradfield, H. H., and Seale, R. I., 1957, Walnut Bend field *in* The geology and geophysics of Cooke and Grayson Counties, Texas: Dallas Geol. Soc. and Dallas Geophys. Soc., p. 101–117.

Brainerd, A. E., and Carpen, Thad R., 1962, History of exploration and development for oil and gas in northwestern Colorado *in* Exploration for oil and gas in northwestern Colorado: Rocky Mtn. Assoc. Geologists, Denver, Colo., p. 23–28.

Brainerd, A. E., and Van Tuyl, F. M., 1954, Resumé of petroleum exploration and exploratory development in Colorado, 1862–1954, *in* Oil and gas fields of Colorado—a symposium: Rocky Mtn. Assoc. Geologists, p. 15–24.

Branan, C. B., Jr., 1968, Natural gas in Arkoma basin of Oklahoma and Arkansas: Am. Assoc. Petroleum Geologists, Mem. 9, v. 2, p. 1616–1635.

Branan, C. B., Jr., and Jordan, Louise, 1960, Southeast Oklahoma reawakens as potential gas giant: Oil and Gas Jour., Aug. 8, p. 120–122.

Branch, J. L., 1966, Drilling action on the upswing in NW Montana: Oil and Gas Jour., Nov. 7, p. 150–152.

Branson, Carl C., 1956, General geologic section of Oklahoma oil producing areas: Natl. Oil Scouts and Landmen's Assoc. Yearbook, v. 26.

Branson, Carl C., 1957, Pennsylvanian problems in eastern Oklahoma: World Oil, July, p. 87–90.

Brashear, William W., 1961, Lips field *in* Oil and gas fields of the Texas and Oklahoma Panhandles: Panhandle Geol. Soc., p. 181–188.

Braunstein, Jules, 1958, Habitat of oil in eastern gulf coast *in* Habitat of oil: Am. Assoc. Petroleum Geologists, p. 511–522.

Bristol, H. M., 1968, Structure of the base of the Mississippian Beech Creek (Barlow) Limestone in Illinois: Illinois State Geol. Survey, Illinois Petroleum 88.

Bristol, H. M., and Howard, R. H., 1966, St. Louis (Mississippian) oil production in Jasper County, Illinois: Illinois Geol. Survey, Illinois Petroleum 82.

Brooner, Frank I., 1966, The North LaWard shale diapir: South Texas Geol. Soc. Bull., v. 6, no. 19, p. 6–19.

Brosgé, W. P., Dutro, J. T., Mangus, M. D., and Reiser, H. N., 1962, Paleozoic sequence in eastern Brooks Range, Alaska: Am. Assoc. Petroleum Geologists Bull., v. 46, no. 12 (December), p. 2174–2198.

Brosgé, W. P., Whittington, Charles L., and Morris, Robert H., 1966, Geology of the Umiat–Maybe Creek region, Alaska: U.S. Geol. Survey Prof. Paper 303–H, p. 501–638.

Brown, A. B., 1943, Bowers field, Montague County, Texas: Am. Assoc. Petroleum Geologists Bull., v. 27 (January), p. 20–37.

Brown, Donald M., 1968, Forest City basin outlook: Oil and Gas Jour., pt. 1, Aug. 12, p. 108–112; pt. 2, Aug. 19, p. 110–117.

Brown, James B., 1968, Gas in Los Angeles basin, California, *in* Am. Assoc. Petroleum Geologists Mem. 9, p. 149–163.

Brown, Robert D., Jr., Gower, Howard D., and Snavely, Parke D., Jr., 1960, Geology of the Port Angeles–Lake Crescent area, Clallam County, Washington: U.S. Geol. Survey Oil and Gas Inv. Map OM-203.

Brown, Silas, 1952, Oil and gas structures of southern Arizona *in* Geological symposium of the Four Corners region: Four Corners Geol. Soc., Oct. 1, p. 61–67.

Brown, Silas C., and Louth, Robert E., 1960, Northern Arizona has good oil, gas prospects: World Oil, September, p. 98–106.

Browning, W. Fleming, 1968, Developments in southeastern states in 1967: Am. Assoc. Petroleum Geologists, Bull., v. 52, no. 6 (June), p. 1027–1032.

Bruce, Clemont H., 1949, Geology of the Utica oil field: Kentucky Geol. Survey Bull. 1.

Brucks, Ernest W., 1929, Luling oil field, Caldwell and Guadalupe Counties, Texas, *in* Structure of typical American oil fields: Am. Assoc. Petroleum Geologists, v. 1, p. 256–281.

Brundage, Harrison T., 1967, Buda limestone oil play moves toward east Texas: World Oil, September, p. 78–82.

Bryant, Leonard C., 1965, The South Copano Bay field, Aransas County, Texas: Gulf Coast Assoc. Geol. Socs. Trans., v. 15, p. 13–19.

Buehner, J. H., and Davis S. H., Jr., 1968, Albion–Pulaski–Scipio trend field *in* Oil and gas fields—a symposium: Michigan Basin Geol. Soc., p. 37–48.

Burgess, R. J., 1963, New strikes spark Cambrian play in Ohio, south Ontario: Oil and Gas Jour., Mar. 4, p. 180–183; Mar. 11, p. 128–131.

Burke, Ray A., 1958, Summary of oil occurrence in Anahuac and Frio formations of Texas and Louisiana: Am. Assoc. Petroleum Geologists Bull., v. 42, no. 12 (December), p. 2935–2950.

Burleson, W. E., and Henkes, William C., 1967, The Mineral industry of New Mexico *in* Minerals Yearbook for 1965: U.S. Bur. Mines, v. 3, p. 551–577.

Burnside, R. J., 1959, Geology of part of the Horseshoe atoll in Borden and Howard Counties, Texas, *in* Pennsylvanian and lower Permian rocks of parts of west and central Texas: U.S. Geol. Survey Prof. Paper 315-B, p. 21–35.

Burroughs, Ernest, Beecroft, George W., and Barger, Ralph M., 1968, Rio Vista gas field, Solano, Sacramento, and Contra Costa Counties, California: Am. Assoc. Petroleum Geologists Mem. 9, p. 93–101.

Burton, Guy, 1961, Patrick Draw Area, Sweetwater County, Wyoming, *in* Guidebook, 16th Ann. Field Conf.: Wyoming Geol. Assoc.

Buschbach, T. C., and Bond, D. C., 1967, Underground storage of natural gas in Illinois—1967: Illinois State Geol. Survey, Illinois Petroleum 86.

Bybee, H. H., 1948, Hitesville consolidated field, Union County, Kentucky: Am. Assoc. Petroleum Geologists Bull., v. 32 (November), p. 2063–2082.

Byrne, John V., 1962, Here's a look at offshore Oregon: Oil and Gas Jour., July 23, p. 116–119.

Caffee, E. C., and Dubitzky, R. H., 1967, Vacuum Abo reef *in* Oil and gas fields of southeastern New Mexico: Roswell Geol. Soc., p. 179.

California Division of Mines, 1940–1943, Geologic formations and economic development of the oil and gas fields of California (Olaf P. Jenkins, ed.), Bull. 118 in four parts: p. 1, Development of the Industry (April, 1940); p. 2, Geology of California and the occurrence of oil and gas (August, 1941); pts. 3 and 4, Descriptions of individual oil and gas fields and glossaries, bibliography and index (March, 1943).

California Division of Mines, 1954, Geology of southern California (Richard H. Jahns, ed.), Bull. 170.

California Division of Mines, 1957, Mineral commodities of California, geologic occurrence, economic development and utilization of the state's mineral resources, Bull. 176.

California Division of Mines and Geology, 1962a, Geologic guide to the gas and oil fields of northern California (Oliver E. Bowen, Jr., ed.), Bull. 181.

California Division of Mines and Geology, 1962b, Dunnigan Hills gas field: Bull. 181, p. 260–261.

California Division of Mines and Geology, 1966a, Geology of northern California (Edgar H. Bailey, ed.), Bull. 190.

California Division of Mines and Geology, 1966b, Bull. 191. Originally published as Mineral and water resources of California: U.S. 89th Cong., 2nd sess., Comm. Print.

California Division of Oil and Gas, 1960–1961, California oil and gas fields: p. 1, San Joaquin–Sacramento valleys and northern coastal regions (October, 1960); p. 2, Los Angeles–Ventura basins and central coastal regions (November, 1961).

California Division of Oil and Gas, Annual Report, Summary of Operations, California Oil Fields.

Calvert, Warren L., 1960, Generalized section of rocks of Ohio: Ohio Geological Survey, Oil and gas fields of Ohio (map), July.

Calvert, Warren L., 1962, Sub-Trenton rocks from Lee County, Virginia, to Fayette County, Ohio: Ohio Geological Survey Rept. Inv. 45.

Calvert, Warren L., 1964a, Morrow County, Ohio, area has about 100 active rigs: World Oil, Feb. 1, p. 79–80.

Calvert, Warren L., 1964b, Cambrian erosional remnants yield oil in central Ohio: World Oil, March, p. 78–84.

Calvert, Warren L., 1964c, Sub-Trenton rocks from Fayette County, Ohio, to Brant County, Ontario: Ohio Geol. Survey Rept. Inv. 52.

Canfield, Charles R., 1943, Santa Maria Valley oil field *in* California Div. Mines Bull. 118, p. 440–442.

Cantrell, R. B., Montgomery, J. C., and Woodard, A. E., 1959, Heterostegina reef on Nash and other piercement salt domes in northwestern Brazoria County, Texas: Gulf Coast Assoc. Geol. Socs. Trans., v. 9, p. 59–62.

Caplan, William M., 1954a, Subsurface geology and related oil and gas possibilities of northeastern Arkansas: Arkansas Resources and Development Comm., Div. Geol., Bull. 20.

Caplan, William M., 1954b, Paleozoic geology of northeastern Arkansas: Gulf Coast Assoc. Geol. Socs. Trans., v. 4, p. 201–210.

Caplan, William M., 1964, Exploration is shifting to Paleozoic prospects: World Oil, March, p. 72–76, 84.

Carlson, C. G., and Anderson, S. B., 1965, Sedimentary and tectonic history of North Dakota part of Williston basin: Am. Assoc. Petroleum Geologists Bull., v. 49, no. 11 (November), p. 1833–1846.

Carlson, Clarence G., Bakken, Wallace E., and Kume, Jack, 1960, Subsurface geology and development of petroleum in North Dakota: North Dakota Geol. Survey Bull. 34.

Carlson, C. R., 1962, Oil seeps and early petroleum development in northern California, California Div. Mines and Geol. Bull. 181, p. 23–26.

Carlton, D. P., 1929, West Columbia salt dome and oil field, Brazoria County, Texas, *in* Structure of typical American oil fields: Am. Assoc. Petroleum Geologists, v. 2, p. 451–469.

Carman, J. Ernest, and Stout, Wilber, 1934, Relationship of accumulation of oil to structure and porosity in the Lima–Indiana field *in* Problems of petroleum geology: Am. Assoc. Petroleum Geologists, p. 522.

Carpenter, G. L., 1965, Oil development and production in Indiana during 1964: Indiana Geol. Survey, Mineral Economics Ser. 11.

Carpenter, L. C., 1961, Florence–Canyon City field, Fremont County, Colorado, *in* Oil and gas field volume, Colorado-Nebraska: Rocky Mtn. Assoc. Geologists, Denver, Colorado, p. 130–131.

Carter, Frank S., 1954, Oil and gas production in California, *in* California Div. Mines, Bull. 170, p. 21–28.

Carter Oil Company, 1957, Amazing Fouke: The Link, January-February, p. 16–17.

Carver, G. E., Jr., 1948, Arcadia–Coon Creek oil pool, Oklahoma and Logan Counties, Oklahoma *in* Structure of typical American oil fields: Am. Assoc. Petroleum Geologists, v. 3, p. 319–340.

Cary, L. W., 1955, The subsurface geology of the Garber area, Garfield County, Oklahoma: Shale Shaker, v. 5 (February), p. 5–29.

Casey, S. Russel, and Cantrell, Ralph B., 1941, Davis sand lens, Hardin field, Liberty County, Texas, *in* Stratigraphic type oil fields: Am. Assoc. Petroleum Geologists, p. 564–599.

Cashion, W. B., 1964, Distribution and quality of oil shale in the Green River Formation of the Uinta basin *in* Guidebook, 13 Ann. Field Conf.: Intermountain Assoc. Geologists, p. 209–212.

Cassidy, Martin M., 1968, Excello Shale, northeastern Oklahoma: clue to locating buried reefs: Am. Assoc. Petroleum Geologists Bull., v. 52, no. 2 (February), p. 295–312.

Caswell, C. A., 1957, Arbuckle region offers good possibilities: Oil and Gas Jour., July 22, p. 158–161.

Cater, Fred W., and Elston, D. P., 1963, Structural development of salt anticlines of Colorado and Utah *in* Backbone of the Americas: Am. Assoc. Petroleum Geologists Mem. 2, November, p. 152–159.

Chambers, L. S., 1943, Coalinga East Extension area of the Coalinga oil field, *in* California Div. Mines Bull. 180, pt. 3, p. 486–490.

Champion, Beverly L., 1968, Oil and gas producing zones in Michigan *in* Oil and gas fields—a symposium: Michigan Basin Geol. Soc., p. 17–36.

Chapman, John J., 1963, Deeper oil possibilities of south Arkansas and north Louisiana: Am. Assoc. Petroleum Geologists Bull., v. 47 (November), p. 1992–1996.

Charles, Homer H., 1941, Bush City oil field, Anderson County, Kansas *in* Stratigraphic type oil fields: Am. Assoc. Petroleum Geologists, p. 43–56.

Chase, G. W., Frederickson, E. A., and Ham, W. E., 1956, Resumé of the geology of the Wichita Mountains, Oklahoma, *in* Petroleum geology of southern Oklahoma: Am. Assoc. Petroleum Geologists, v. 1, p. 36–55.

Chasteen, Kenneth, 1961, Check Chama basin, Oil and Gas Jour., Jan. 16, p. 122–125.

Chatfield, John, 1965, Petroleum geology of the Greater Red Wash Area, Uintah County, Utah: Mtn. Geologist, v. 2, no. 3, p. 115–121.

Checkley, W. G., 1968, Northville field *in* Oil and gas fields—a symposium: Michigan Basin Geol. Soc., p. 115–122.

Chen, Chih Shan, 1965, The regional lithostratigraphic analysis of Paleocene and Eocene rocks of Florida: Florida Geol. Survey Geol. Bull. 45.

Cheney, M. G., and Goss, Louis F., 1952, Tectonics of central Texas: Am. Assoc. Petroleum Geologists Bull., v. 36, no. 12 (December), p. 2237–2265.

Chenoweth, Phillip A., 1959, Is there oil and gas in the Ouachita Mountains?: Oklahoma Geol. Survey Notes (October), p. 199–209.

Chenoweth, Phillip A., 1960, Ouachita Mountains do have oil and gas potential: World Oil, Aug. 1, p. 94–100.

Cheyney, A. E., 1929, Madison Shoestring pool, Greenwood County, Kansas; *in* Structure of typical American oil fields: Am. Assoc. Petroleum Geologists, v. 2, p. 150–159.

Chickering, W. W., 1958, Cut Bank field *in* Montana oil and gas fields: Billings Geol. Soc., p. 111–112.

Chisholm, L. R., 1968, Smackover (Jurassic) reservoir of New Hope field, Franklin County, Texas, *in* Am. Assoc. Petroleum Geologists Mem. 9, v. 1, p. 1069–1073.

Cipriani, D., Jr., 1956, Southwest Randlett field, Cotton County, Oklahoma, *in* Petroleum geology of southern Oklahoma: Am. Assoc. Petroleum Geologists, v. 1, p. 311–318.

Clair, Joseph R., and Bradish, B. B., 1956a, Model Dome gas field, Las Animas County, Colorado, *in* Guidebook to the geology of the Raton basin, Colorado: Rocky Mtn. Assoc. Geologists, p. 80–81.

Clair, Joseph R., and Bradish, B. B., 1956b, Garcia gas field, Las Animas County, Colorado, *in* Guidebook to the geology of the Raton basin, Colorado: Rocky Mtn. Assoc. Geologists, p. 75–78.

Clark, C. C., 1939, Rodessa Field, Caddo Parish, Louisiana, Cass and Marion Counties, Texas, Miller County, Arkansas *in* Guidebook, 14th Ann. Field Trip: Shreveport Geol. Soc., p. 59–63.

Clark, E. W., and Rold, J. W., 1961, New Windsor, Weld County, Colorado, *in* Oil and gas field volume, Colorado-Nebraska: Rocky Mtn. Assoc. Geologists, p. 202–205.

Clark, G. C., 1961, Interior salt domes of Texas, Louisiana, and Mississippi *in* Guidebook, 1960 Field Trip: Shreveport Geol. Soc., p. 5.

Clark, Glenn C., 1926, Wilcox sand production, Tonkawa field, Oklahoma: Am. Assoc. Petroleum Geologists Bull., v. 10 (September), p. 885–891.

Clark, James A., and Halbouty, Michel T., 1952, Spindletop: Random House, New York.

Clark, R. E., 1961, Paint Creek geology of the Handyville pool, Daviess County, Kentucky: Kentucky Geol. Survey, Ser. X, Spec. Pub. 4, p. 21–31.

Clark, Robert W., 1940, Paloma oil field, Kern County, California: Am. Assoc. Petroleum Geologists Bull., v. 24, no. 4, p. 742–744.

Clark, Stuart K., Arnett, C. L., and Royds, James S., 1948, Geneseo uplift, Rice, Ellsworth, and McPherson Counties, Kansas, *in* Structure of typical American oil fields: Am. Assoc. Petroleum Geologists, v. 3, p. 225–248.

Clark, Stuart K., and Daniels, James I., 1929, Relation between structure and production in the Mervine, Ponca, Blackwell, and South Blackwell oil fields, Kay County, Oklahoma, *in* Structure of typical American oil fields: Am. Assoc. Petroleum Geologists, v. 1, p. 158–175.

Clayton, John M., 1951, New Hope field, Franklin County, Texas, Univ. Texas Pub. 5116, p. 256–261.

Clayton, John, 1965, Paleodepositional environments of the "Cherokee" sands of central Payne County, Oklahoma: Shale Shaker, v. 16, no. 3, p. 50–66.

Clegg, Kenneth E., 1965, The LaSalle anticlinal belt and adjacent structures in east-central Illinois: Illinois State Acad. Sci. Trans., v. 58, no. 2, p. 82–94.

Clement, James H., 1955, Pine field *in* Geological record: Am. Assoc. Petroleum Geologists, Rocky Mtn. Sec., p. 165–171.

Clinton, R. P., 1957, The geology of the Osage country: Shale Shaker, v. 8 (October), p. 8–21.

Cobban, W. A., 1958, quoted by Chickering (1958, p. 111) from unpublished manuscript.

Cohee, George V., 1942, Lateral variation in Chester sandstones producing oil and gas in lower Wabash River area: Am. Assoc. Petroleum Geologists Bull., v. 26 (October), p. 1606.

Cohee, George V., 1945, Stratigraphy of Lower Ordovician and Cambrian rocks in the Michigan basin: U. S. Geol. Survey Prelim. Chart 9.

Cohee, George V., 1948, Thickness and lithology of Upper Ordovician and Lower and Middle Silurian Rocks in the Michigan basin: U. S. Geol. Survey Oil and Gas Inv. Prelim. Chart 33.

Cohee, George V., 1965, Geologic history of the Michigan basin: Washington Acad. Sci. Jour., v. 55, p. 211–223.

Cohee, George V., and Landes, Kenneth K., 1958, Oil in the Michigan basin *in* Habitat of oil: Am. Assoc. Petroleum Geologists, p. 473–493.

Cohee, George V., et al, 1962, Tectonic map of the United States: U. S. Geol. Survey [and] Am. Assoc. Petroleum Geologists.

Cole, Virgil B., Merriam, Daniel F., Franks, Paul C., Hambleton, William W., and Hilpman, Paul L., 1961, Wells drilled into Precambrian rock in Kansas: Kansas Geol. Survey Bull. 150.

Cole, Virgil B., Merriam, Daniel F., and Hambleton, William W., 1965, Final report of the Kansas Geological Society Basement Rock Committee and list of Kansas wells drilled into Precambrian rocks: Kansas Geol. Survey Spec. Pub. 25.

Collins, Donald F., and Conner, Donald C., 1965, Brentwood field (abs.): Am. Assoc. Petroleum Geologists Bull., v. 49, no. 7 (July), p. 1083.

Collins, Florence R., 1958, Test wells, Umiat area, Alaska—exploration of Naval Petrol. Reserve No. 4 and adjacent areas: U. S. Geol. Survey Prof. Paper 305–B.

Coltman, Bruce, and Pitt, William D., 1961, Niobrara discoveries enhancing northwest Colorado prospects: World Oil, July.

Colton, Earl G., 1935, Natural gas in Arkansas basin of eastern Oklahoma *in* Geology of natural gas: Am. Assoc. Petroleum Geologists, p. 511–532.

Cone, George C., 1965, Generalized tectonic map of New Mexico *in* Mineral and water resources of New Mexico: U. S. Senate Rept. Comm. on Interior and Insular Affairs, 89th Cong., 1 sess., Figure 7.

Conley, J. N., 1952, Double Mountain pool, Stonewall County, Texas *in* Geological contributions: Abilene Geol. Soc., p. 17–20.

Conley, Jack N., 1962, For South Dakota oil—check the Minnelusa formation units: Oil and Gas Jour., Dec. 31, p. 188–190.

Conselman, Frank B., 1955, The geology of the Cambrian trend of West Central Texas (abst.): Geophysics, v. 20, p. 188–189.

Conselman, Frank B., 1960, Pennsylvanian reef patterns in West-Central Texas, *in* Texas Univ., Bur. Econ. Geol., Pub. 6017, p. 1–13.

Cook, Carroll E., 1938, Darrow Salt dome, Ascension Parish, Louisiana: Am. Assoc. Petroleum Geologists Bull., v. 22 (October), p. 1412–1422.

Cook, E. F., 1957, Stratigraphic and structural utility of layered volcanic rocks *in* v. 1 *of* Vulcanologia del Cenozoico: 20th Internat. Geol. Cong., Mexico, D. F., sec. 1, p. 189–207.

Cook, V. O., 1966, Regional geology of the Delaware basin: Jour. Petroleum Technology, October, p. 1260–1266.

Coon, Lester A., 1957, East Texas basin hides deep prospects: Oil and Gas Jour., Jan. 21, p. 92–96.

Cooper, C. G., and Ferris, B. J., 1957, TXL field, *in* Occurrence of oil and gas in West Texas: Univ. Texas Pub. 5716, p. 358–368.

Corpus Christi Geological Society, 1957, South Texas Domes: Annual Field Trip, Apr. 26 and 27, p. 9–30.

Corpus Christi Geological Society, 1967, Typical oil and gas fields of South Texas.

Corpus Christi Geological Society, 1968, Natural gas in post-Eocene formations of South Texas: Am. Assoc. Petroleum Geologists Mem. 9, v. 1, p. 233–263.

Cotner, V., and Crum, H. E., 1935, Geology and occurrence of natural gas in Amarillo district, Texas, *in* Geology of natural gas: Am. Assoc. Petroleum Geologists, p. 385–415.

Cotton, Clyde B., 1966, Barnhart field, Reagan County, *in* Oil and gas fields in West Texas—a symposium: West Texas Geol. Soc., p. 38–41.

Covington, Robert E., 1963, Bituminous sandstone and limestone deposits of Utah: Paper 23 *in* Utah Geol. and Mineralog. Survey Bull. 54, p. 225–248.

Covington, Robert E., 1964, Bituminous sandstones in the Uinta basin *in* Guidebook, 13th Ann. Field Conference: Intermountain Assoc. Petroleum Geologists, 1964, p. 226–242.

Craft, W. E., 1966, Channel sands are the key to Wilcox oil: Oil and Gas Jour., Apr. 11, p. 124–130.

Cram, Ira H., 1948, Cumberland oil field, Bryan and Marshall Counties, Oklahoma, *in* Structure of typical American oil fields: Am. Assoc. Petroleum Geologists, v. 3, p. 341–358.

Crawford, Arthur L., ed., 1963, Oil and gas possibilities of Utah, re-evaluated: Utah Geol. and Mineralog Survey Bull. 54 (January).

Crawford, Arthur L., and Pruit, Robert G., 1963, Gilsonite and other bituminous resources of central Uintah County, Utah: Paper 22 *in* Utah Geol. and Mineral. Survey Bull. 54, p. 215–224.

Crider, A. F., 1938, Geology of Bellevue oil field, Bossier Parish, Louisiana: Am. Assoc. Petroleum Geologists Bull., v. 22 (December), p. 1658–1681.

Criss, C. R., and McCormick, R. L., 1954, History and performance of the Coldwater oil field, Michigan: Jour. Petroleum Technology, v. 6 (February), p. 23–29.

Croneis, Carey, 1935, Natural gas in interior highlands of Arkansas *in* Geology of natural gas: Am. Assoc. Petroleum Geologists, p. 533–574.

Culp, Eugene, and Barrett, Ed, 1957, Bois d'Arc pay spurs limestone exploration: Oil and Gas Jour., June 3, p. 169–173, and June 10, p. 172–177.

Cummings, Kenneth F., 1955, Buck Peak field, Moffat County, Colorado. *in* Geological record: Am. Assoc. Petroleum Geologists, Rocky Mt. Sec.

Cunningham, B. J., 1961, Stratigraphy of Oklahoma and Texas Panhandles *in* Oil and gas fields of the Texas and Oklahoma Panhandles: Panhandle Geol. Soc., p. 45–60.

Cupps, C. Q., and Fry, J., 1967, Reservoir oil characteristics, Cut Bank field, Montana: U. S. Bur. Mines Rept. Inv. 6909.

Current, A. M., 1948, Gilbertown field, Choctaw County, Alabama, *in* Structure of typical American oil fields: Am. Assoc. Petroleum Geologists, v. 3, p. 1–4.

Currie, John B., 1956, Role of concurrent deposition and deformation of sediments in development of salt dome graben structures: Am. Assoc. Petroleum Geologists Bull., v. 40, no. 1 (January), p. 1–16.

Curry, Wm. H., Jr., and Curry, Wm. H., III, 1954, South Glenrock, a Wyoming stratigraphic oil field: Am. Assoc. Petroleum Geologists Bull., v. 38 (October), p. 2119–2156.

Curtis, Bruce F., Strickland, John W., and Busby, Robert C., 1958, Patterns of oil occurrence in the Powder River basin, *in* Habitat of oil: (Am. Assoc. Petroleum Geologists (Spec. volume), p. 268–292.

Dallas Geological Society, 1963, Geological highway map of Texas.

Dallas Geological Society, 1965, The geology of Dallas County—a symposium: Dallas, Texas, December.

Dallas Geological Society and Dallas Geophysical Society, 1957, The geology and geophysics of Cooke and Grayson Counties, Texas.

Dally, Claude F., 1938, Geology of the K.M.A. field: Oil and Gas Jour., Jan. 20, p. 47 et seq.

Dane, C. H., 1965, Topography and geology *in* Mineral and water resources of New Mexico: U.S. Senate Rept. Comm. on Interior and Insular Affairs, 89th Cong., 1st sess., p. 19–39.

Davis, Morgan J., 1929, Artesia field, Eddy County, New Mexico: Structure of typical American oil fields: Am. Assoc. Petroleum Geologists, v. 1, p. 112–123.

Davis, Ralph E., and Stephenson, Eugene A., 1929, Synclinal oil fields in southern West Virginia *in* Structure of typical American oil fields: Am. Assoc. Petroleum Geologists, v. 2, p. 571–576.

Davis, W. E., and Hunt, R. E., 1956, Geology and oil production on the northern portion of the Cedar Creek anticline, Dawson County, Montana, *in* First internat. Williston basin symposium; North Dakota Geol. Soc., and Saskatchewan Geol. Soc., p. 121–129.

Dawson, T. A., 1967, Knox oil may revive Hoosier oil hunt: Oil and Gas Jour., Dec. 18, p. 126–130.

Dawson, T. A., and Carpenter, G. L., 1963, Underground storage of natural gas in Indiana: Indiana Geol. Survey Spec. Rept. 1.

Day, Willard, and Galbraith, George, 1950, Notes on the Round Top field, Fisher County, Texas, *in* Geological contributions: Abilene Geol. Soc., p. 66–67.

Deegan, Charles J., 1945, Seeligson field outstanding on several counts, especially its multiple pay zones: Oil and Gas Jour., Aug. 11, p. 74–75, 81.

Deegan, Charles J., 1949, Benedum field is established as three-pay major reserve: Oil and Gas Jour., v. 48 (May 19), p. 183 et seq.

DeFord, Ronald K., 1951, Trans-Pecos province, West Texas and southeastern New Mexico: Am. Assoc. Petroleum Geologists Bull., v. 35, no. 2 (February), p. 347–352.

DeFord, Ronald K., and Wahlstrom, Edwin A., 1932, Hobbs field, Lea County, New Mexico: Am. Assoc. Petroleum Geologists Bull., v. 16, no. 1 (January), p. 51–90.

Denham, R. L., ed., 1962, Typical oil and gas fields of southeast Texas: Houston Geol. Soc., October.

Denham, R. L., and Dougherty, W. E., 1941, 'Sand Belt' area of Ward and Winkler Counties, Texas, and Lea County, New Mexico, *in* Stratigraphic type oil fields: Am. Assoc. Petroleum Geologists, p. 750–759.

Detterman, Robert L., and Hartsock, John K., 1967, Geology of the Iniskin-Tuxedni region, Alaska: U. S. Geol. Survey Prof. Paper 512, p. 1–78.

de Witt, Wallace, Jr., 1949, Map of the Berea sand of northern Ohio: U. S. Geol. Survey Oil and Gas Inv. Prelim. Map 99.

Dickey, Parke A., 1959, The first oil well: Jour. Petroleum Technology, v. 9 (January), p. 14–26.

Dietz, Gerald P., 1967, The geology of the Gageby Creek field, Wheeler County, Texas: Shale Shaker, v. 17, no. 7, p. 137–145.

DiGiambattista, C. D., 1952, Regional stratigraphy of Four Corners area *in* Geological symposium of the Four Corners region: Four Corners Geol. Soc., Oct. 1, p. 5–9.

Dillard, W. Reese, 1941, Olympic pool, Hughes and Okfuskee Counties, Oklahoma, *in* Stratigraphic type oil fields: Am. Assoc. Petroleum Geologists, p. 456–472.

Dille, Glen S., 1952, Oil and gas in Tulsa County: Oklahoma Geol. Survey Bull. 69, p. 104–125.

Dinkins, Donald E., 1960, The geology of the Stensvad field and its regional geologic setting with notes on the Tyler reservoir *in* Geological record: Am. Assoc. Petroleum Geologists, Rocky Mtn. Sec., p. 59–69.

Dinkins, Theo H., Jr., 1966, Subsurface stratigraphy of Claiborne County, *in* Claiborne County geology and mineral resources: Mississippi Geol. Econ. and Topog. Survey Bull. 107, p. 95–117.

Dinnean, Robert F., 1958, Geology of the Chestnut salt dome area, Natchitoches Parish, Louisiana: Louisiana Geol. Survey Geol. Summary 1, p. 8–9.

Dobbin, C. E., 1943, Structural conditions of oil and gas accumulation in Rocky Mountain region, United States, Am. Assoc. Petroleum Geologists Bull., v. 27, April, p. 440–446.

Dobbin, C. E., 1947, Exceptional oil fields in Rocky Mountain region of United States: Am. Assoc. Petroleum Geologists Bull., v. 31 (May), p. 806–807.

Dobbin, C. E., 1959, Petroleum geology of Colorado: Colorado School Mines Quart., v. 45, no. 1B (May), p. 1–28.

Dobbin, C. E., and Erdmann, C. E., 1955, Structure contour map of the Montana plains: U. S. Geol. Survey Oil and Gas Inv. Map OM–178B.

Dobbin, C. E., Kramer, W. B., and Horn, G. H., 1957, Geologic and structure map of the southwestern part of the Powder River basin, Wyoming: U. S. Geol. Survey Oil and Gas Inv. Map OM-185.

Dobbin, C. E., and Miller, J. C., 1941, Osage oil field, Weston County, Wyoming, *in* Stratigraphic type oil fields: Am. Assoc. Petroleum Geologists, p. 847–857.

Dobie, C. Walter, 1964, Erath field, Vermilion Parish, Louisiana: Lafayette Geol. Soc., p. 9–9e.

Doell, E. C., 1968, Developments in West Coast area in 1967: Am. Assoc. Petroleum Geologists Bull., v. 52, no. 6 (June), p. 1051–1058.

Doeringsfeld, Walter W., Jr., Amuedo, C. L., and Ivey, J. G., 1956, Structure contour map of Raton basin *in* Guidebook to the geology of the Raton basin, Colorado: Rocky Mtn. Assoc. Geologists, map in pocket.

Dollison, Robert S., 1965, Big Hill field, Jefferson County, Texas: Gulf Coast Assoc. Geol. Socs. Trans., v. 15, p. 25–35.

Dollison, Robert S., 1966, Developments in Upper Gulf Coast of Texas in 1965: Am. Assoc. Petroleum Geologists Bull., v. 50 (June), p. 1228–1231.

Donnan, Bryson C., 1957, Great Lakes leases offer new drilling area: Oil and Gas Jour., Oct. 14.

Donoghue, David, and Gupton, W. L., Jr., 1957, Yates field *in* Oil fields of West Texas: Univ. Texas Pub. 5716, p. 426–433.

Doss, A. K., Jr., and Spiva, Frank J., Jr., 1968, Palo Pinto limestone production, western Runnels County, Texas: Am. Assoc. Petroleum Geologists Bull., v. 52, no. 1 (January), p. 82–91.

Douglas, E. R., 1957, Seminole field *in* Occurrence of oil and gas in West Texas: Univ. Texas Pub. 5716, p. 303–306.

Doyle, William M., Jr., 1955, Tapping the Austin chalk's rich reserves: Oil and Gas Jour., Dec. 5, p. 180–182.

Dreyer, F. E., 1943, Santa Maria (Orcutt) oil field: California Div. Mines Bull. 118, p. 431.

Dudley, Paul H., 1954, Geology of the Long Beach oil field, Los Angeles County: California Div. Mines and Geol. Bull. 170, Map Sheet 34.

Dunn, David A., 1956, Caprock Queen field, Chaves and Lea Counties, *in* Oil and gas fields of southeastern New Mexico: Roswell Geol. Soc., p. 91.

Durham, Clarence O., Jr., and Smith, Chester R., 1958, Louisiana Midway-Wilcox correlation problems: Louisiana Geol. Survey Geol. Pamph. 5.

Durham, D. L., 1966, Geology of Hames Valley, Wunpost and Valleton quadrangles, Monterey County, California: U. S. Geol. Survey Bull. 1221-B, p. B-1 to B-53.

Durham, D. L., and Yerkes, R. F., 1964, Geology and oil resources of the eastern Puente Hills area, southern California: U. S. Geol. Survey Prof. Paper 420B.

Dyni, John R., 1966, Geologic map of the Thornburg oil and gas field and vicinity, Moffat and Rio Blanco Counties, Colorado: U. S. Geol. Survey Oil and gas Inv. Map OM–216.

Eardley, Armand J., 1952, Utah's oil outlook brightens, World Oil, May, p. 73–74.

Eardley, Armand J., 1963a, Oil seeps at Rozel Point: Utah Geol. and Mineral. Survey Spec. Studies No. 5.

Eardley, A. J., 1963b, Relation of uplifts to thrusts in Rocky Mountains, *in* Backbone of the Americas: Am. Assoc. Petroleum Geologists p. 209–219.

Earl, John H., 1961, Desert Springs gas field, Sweetwater County, Wyoming: Geophysics, v. 26 (December), p. 673–681.

East Texas Geological Society, 1951, The Woodbine and adjacent strata in the Waco area of central Texas—a symposium: Southern Methodist University Press, Fondren Sci. Ser., No. 4.

Eastwood, William P., 1962, How to map Wiley Madison trend in North Dakota: World Oil, Feb. 1, p. 60–67.

Eaton, R. W., 1951, Boggy Creek field, Anderson and Cherokee Counties, Texas; Univ. Texas Pub. 5116, p. 29–34.

Eaton, R. W., 1956, Resumé of subsurface geology of Northeast Texas with emphasis on salt structures: Gulf Coast Assoc. Geol. Socs. Trans., v. 6, p. 79–84.

Eberlein, G. D., 1964, Summary and Introduction *in* U. S. Senate Rept. Comm. on Interior and Insular Affairs, 58th Cong., 2nd sess.: U.S. Geol. Survey, p. 9–12.

Ebright, John R., Fettke, Charles R., and Ingham, Albert I., 1949, East Fork-Wharton gas field, Potter County, Pennsylvania: Pennsylvania Topog. and Geol. Survey Bull. M30, p. 1–43.

Ebright, John R., and Ingham, Albert I., 1951, Geology of the Leidy gas field and adjacent areas, Clinton County. Pennsylvania: Pennsylvania Topog. and Geol. Survey Bull. M34, p. 1–35.

Eddleman, M. W., 1961, Tectonics and geologic history of the Texas and Oklahoma Panhandles *in* Oil and gas fields of the Texas and Oklahoma Panhandles: Panhandle Geol. Soc., p. 61–68.

Edwards, Everett C., 1941a, Kern Front oil field, Kern County, California, *in* Stratigraphic type oil fields: Am. Assoc. Petroleum Geologists, p. 9–18.

Edwards, Everett C., 1941b, Edison oil field and vicinity, Kern County, California, *in* Stratigraphic type oil fields: Am. Assoc. Petroleum Geologists, p. 1–8.

Eisenstatt, P., 1960, Little Creek field, Lincoln and Pike Counties, Mississippi: Gulf Coast Assoc. Geol. Socs. Trans., v. 10, p. 206–213.

Elkins, Lincoln F., 1953, Reservoir performance and well spacing, Spraberry trend area field of West Texas: Jour. Petroleum Technology, v. 5, no. 7 (July), p. 177–196.

Ellison, Samuel P., Jr., 1948, Apco field, Pecos county, Texas, *in* Structure of typical American oil fields: Am. Assoc. Petroleum Geologists, v. 3, p. 399–418.

Ellison, Samuel P., Jr., 1957, Page field, Schleicher County, Texas, *in* Occurrence of oil and gas in West Texas: Univ. Texas Pub. 5716, p. 247–251.

Ellison, Samuel P., Jr., 1960, Oil and gas in North Central Texas, United States: Proc. Internat. Geol. Cong., Copenhagen, pt. XI, sec. 2, p. 19–26.

Ells, Garland D., 1959, Silurian rocks have oil and gas potential: World Oil, September, p. 88–92.

Ells, Garland D., 1962, Structures associated with the Albion-Scipio oil field trend: Michigan Geol. Survey, May, p. 1–86.

Ells, Garland D., 1967, Michigan's Silurian oil and gas pools: Michigan Geol. Survey Rept. Inv. 2.

Ells, Garland D., and Ives, Robert E., 1959, South Michigan's oil bonanza: World Oil, October, p. 136–140.

Ells, Garland D., and Layton, F., 1968, Developments in Michigan in 1967: Am. Assoc. Petroleum Geologists Bull., v. 52, no. 6 (June), p. 976–980.

Elston, Donald P., and Shoemaker, Eugene M., 1961, Preliminary structure contour map on top of salt in the Paradox member of the Hermosa Formation in the Salt Anticline region, Colorado and Utah: U. S. Geol. Survey Oil and Gas Inv., Map OM–209.

Elston, Wolfgang E., 1960, Structural development and Paleozoic stratigraphy of Black Mesa basin, northeastern Arizona, and surrounding areas: Am. Assoc. Petroleum Geologists Bull., v. 44, no. 1 (January), p. 21–36.

Emery, K. O., 1958, Southern California basins, *in* Habitat of oil: Am. Assoc. Petroleum Geologists, p. 955–967.

Emery, K. O., 1966a, Some potential mineral resources of the Atlantic continental margin *in* Geological Survey research, 1965: U. S. Geol. Survey Prof. Paper 525-C, p. C157–C160.

Emery, K. O., 1966b, Atlantic continental shelf and slope of the United States—geologic background: U. S. Geol. Survey Prof. Paper 529-A.

Emery, K. O., 1967, The Atlantic continental margin of the United States during the past 70 million years *in* Geology of the Atlantic region: Geol. Assoc. Canada Spec. Paper 4, p. 53–70.

English, R. M., and Grogan, R. M., 1948, Omaha pool and mica-peridotite intrusives, Gallatin County, Illinois, *in* Structure of typical American oil fields: Am. Assoc. Petroleum Geologists, v. 3, p. 189–212.

English, Walter A., 1926, Geology and oil resources of the Puente Hills region, southern California: U. S., Geol. Survey Bull. 768.

Enright, Robert J., 1957, Spraberry finally pays off for some: Oil and Gas Jour., Sept. 2, p. 90–92.

Erdmann, Charles E., 1963, Oil and gas in Montana, *in* Mineral and water resources of Montana: U. S. Senate Rept. Comm. on Interior and Insular Affairs, 88th Cong., 1st sess., p. 24–44.

Erdmann, Charles E., and Schwabrow, John R., 1941, Border–Red Coulee oil field, Toole County, Montana, and Alberta, Canada, *in* Stratigraphic type oil fields: Am. Assoc. Petroleum Geologists, p. 267–326.

Esgen, W. K., 1929, Relation of accumulation of petroleum to structure in Stephens County, Texas, *in* Structure of typical American oil fields: Am. Assoc. Petroleum Geologists, v. 2, p. 470–479.

Espach, Ralph H., and Biggs, Paul, 1953, The Rock River oil field, *in* Wyoming Geol. Assoc. Guidebook 8th Ann. Field Conf., p. 161–162.

Espach, Ralph H., and Nichols, H. Dale, 1941, Petroleum and natural gas fields in Wyoming, U. S. Bur. Mines Bull. 418.

Euwer, Robert M., 1964, Stay with the Osage if you want Kansas oil: Oil and Gas Jour., Mar. 2, p. 176–181.

Everett, Rizer, 1954, Oil and gas possibilities in the Warrior coal basin in Alabama: Petroleum Engineer, v. 26 (September), p. 57–69.

Ewing, David J., 1955, Beaver Creek field, Fremont County, Wyoming (abs.): Oil and Gas Jour., Apr. 11, p. 125.

Exum, F. A., and Harms, J. C., 1968, Comparison of marine bar with valley-fill stratigraphic traps, western Nebraska: Am. Assoc. Petroleum Geologists Bull., v. 52, no. 10 (October), p. 1851–1868.

Fanshawe, John R., 1960, A new interpretation—Wolf Springs field *in* Geological record: Am. Assoc. Petroleum Geol., Rocky Mtn. Sec., p. 49–57.

Farquhar, O. C., 1957, The Precambrian rocks of Kansas: Kansas Geol. Survey Bull. 127, pt. 3.

Fath, A. E., 1921, Geology of the El Dorado oil and gas field, Butler County Kansas: Kansas Geol. Survey Bull. 7.

Fay, Leo F., 1966, Developments in Alaska in 1965: Am. Assoc. Petroleum Geologists Bull., v. 50, no. 6 (June), p. 1311–1323.

Fenneman, Nevin M., 1931, Physiography of western United States: New York, McGraw-Hill.

Fenstermacher, Charles D., 1966, Large Newburg area awaits new play in West Virginia: Oil and Gas Jour., Mar. 14, p. 158–162.

Fergus, Preston, 1935, Monroe gas field, Louisiana, *in* Geology of natural gas: Am. Assoc. Petroleum Geologists, p. 741–772.

Ferris, Craig, 1961, Gravity meters uncover new reefs in Michigan: Oil and Gas Jour., Aug. 28, p. 140–142.

Fettke, Charles R., 1938, The Bradford oil field, Pennsylvania and New York: Pennsylvania Geol. Survey, 4th Ser., Bull. M-21.

Fettke, Charles R., 1941, Music Mountain oil pool, McKean County, Pennsylvania, *in* Stratigraphic type oil fields: Am. Assoc. Petroleum Geologists, p. 492–506.

Fidlar, M. M., 1962, Baxter basin gas fields, Sweetwater County, Wyoming, *in* Wyoming Geol. Assoc. Guidebook 17th Ann. Field Conf., p. 282–283.

Finley, E. A., Dobbin, C. E., and Richardson, E. E., 1955, Preliminary structure contour map of the Colorado plains: U. S. Geol. Survey Oil and Gas Inv. Map OM-176.

Finn, Fenton H., 1949, Geology and occurrence of natural gas in the Oriskany Sandstone in Pennsylvania and New York: Am. Assoc. Petroleum Geologists Bull., v. 33 (March), p. 303–335.

Fisher, Barney, 1941, La Rosa field, Refugio County, Texas: Am. Assoc. Petroleum Geologists Bull., v. 25 (February), p. 300–317.

Fisher, W. L., and McGowan, J. H., 1967, Depositional systems in the Wilcox Group of Texas and their relationship to occurrence of oil and gas: Gulf Coast Assoc. Geol. Socs. Trans., v. 17, p. 105–125.

Fisk, H. N., 1940, Geology of Avoyelles and Rapides Parishes: Louisiana Geol. Survey Bull., no. 18.

Fitts, L. E., Jr., 1951, North central Oklahoma shelf area: Shale Shaker, v. 2 (November), p. 4, 15–16.

Flawn, Peter T., 1956, Basement rocks of Texas and southeast New Mexico: Univ. Texas Pub. 5605, Mar. 1.

Flawn, Peter T., 1958, Subsurface Ouachita structural belt: World Oil, November, p. 174.

Flawn, Peter T. (Chairman Basement Rock Proj. Comm., 1967, Basement map of North America: Am. Assoc. Petroleum Geologists and U. S. Geol. Survey.

Fletcher, Corbin D., 1929, Structure of Caddo field, Caddo Parish, Louisiana, *in* Structure of typical American oil fields: Am. Assoc. Petroleum Geologists, v. 2, p. 183–195.

Folsom, Clarence B., and Anderson, Sidney B., 1955, What are prospects in Williston basin's east side?: Oil and Gas Jour., Dec. 12, p. 92–100.

Folsom, L. W., 1968, Economic aspects of Uinta basin gas development: Am. Assoc. Petroleum Geologists Mem. 9, v. 1, p. 199–208.

Foster, Donald I., 1958, Sumatra field, northwest, *in* Montana oil and gas fields: Billings Geol. Soc., p. 217–221.

Foster, Frank W., 1961, Tyler sand trend becoming better known in Dakotas: World Oil, January, p. 89–93.

Foster, Roy W., 1964, Stratigraphy and petroleum possibilities of Catron County, New Mexico: New Mexico Bur. Mines and Mineral Resources Bull. 85.

Fothergill, Harold L., 1962a, La Honda oil field, California: California Div. Mines and Geol. Bull. 181, p. 221–222.

Fothergill, Harold L., 1962b, Oil Creek oil field, California: California Div. Mines and Geol. Bull. 181, p. 223–224.

Four Corners Geological Society, 1963, Guidebook—Shelf carbonates of the Paradox basin: Durango, Colorado.

Fowler, Phillip, 1956, Faults and folds of south central Texas: Gulf Coast Assoc. Geol. Socs. Trans., v. 6, p. 37–39.

Fox, Bruce W., 1963, How many Jones reef pools hidden in Fairway region?: Oil and Gas Jour., June 10, p. 246–252.

Frascogna, Xavier M., ed., 1957, Mesozoic-Paleozoic producing areas of Mississippi and Alabama: Mississippi Geol. Soc., v. 1 (September).

Frazell, W. D., 1952, Glen Cove field, Coleman County, Texas, *in* Geological contributions: Abilene Geol. Soc., p. 33–37.

Freeman, James C., 1949, Strandline accumulation of petroleum, Jim Hogg County, Texas: Am. Assoc. Petroleum Geologists Bull., v. 33 (July), p. 1260–1270.

Freeman, John C., 1962, The Conley field, Hardeman County, Texas, *in* North Texas oil fields—a symposium: North Texas Geol. Soc., p. 26–32.

Freeman, John C., 1964, The Conley field, Hardeman County, Texas: Tulsa Geol. Soc. Digest, v. 32, p. 126–130.

Freeman, Louise Barton, 1941, Big Sinking field, Lee County, Kentucky, *in* Stratigraphic type oil fields: Am. Assoc. Petroleum Geologists, p. 166–207.

Freeman, Louise Barton, 1951, Regional aspects of Silurian-Devonian subsurface stratigraphy in Kentucky: Am. Assoc. Petroleum Geologists Bull., v. 35 (January), p. 1–61.

Fritz, Arthur J., 1968, Typical lenticular sandstone development and oil entrapment, *in* Geology and petroleum production of the Illinois basin—a symposium: Illinois and Indiana–Kentucky Geol. Socs., p. 23–28.

Frost, V. L., 1940, Ramsey oil pool, Payne County, Oklahoma: Am. Assoc. Petroleum Geologists Bull., v. 24 (November), p. 1995–2005.

Fryman, A. D., 1956, Artesia and Empire fields *in* Oil and gas fields of southeastern New Mexico: Roswell Geol. Soc., p. 55.

Fugate, George W., Jr., 1956, Geology of the area of Smith Mills north and Geneva oil pools, Henderson County, Kentucky: Kentucky Geol. Survey Bull. 18 (July), 27 p.

Fuqua, H. B., and Thompson, B. E., 1929, Relation of production to structure in central Wilbarger County, Texas, *in* Structure of typical American oil fields: Am. Assoc. Petroleum Geologists, v. 1, p. 293–303.

Furbush, Malcolm A., 1959, Hugoton field, Kansas, *in* Kansas oil and gas fields: Kansas Geol. Soc., v. 2 (Western Kansas), p. 55–64.

Gabelman, John W., 1956, Tectonic history of the Raton basin region, *in* Guidebook to the geology of the Raton basin, Colorado: Rocky Mtn. Assoc. Geologists, p. 35–39.

Gahring, Ross R., 1959, History and development of north Madill field *in* Petroleum geology of southern Oklahoma: Am. Assoc. Petroleum Geologists, v. 2, p. 274–286.

Gallear, Darrell C., and Kistler, James O., 1965, 29D Monarch and 10–10 pool, a "sleeper" in old Midway-Sunset field, Kern County, California: San Joaquin Geol. Soc. Selected Papers, v. 3, p. 19–35.

Galley, John E., 1958, Oil and geology in the Permian basin of Texas and New Mexico *in* Habitat of oil: Am. Assoc. Petroleum Geologists, p. 395–446.

Galloway, John O., 1943, Kettleman Hills oil fields: California Div. Mines. Bull. 180, pt. 3, p. 491–493.

Gardner, Frank J., 1960, Oil and Gas Jour., May 30, p. 57.

Gardner, Frank J., 1963a, South Louisiana's four barren parishes offer challenge: Oil and Gas Jour., Sept. 9, p. 155.

Gardner, Frank J., 1963b, Magnolia state gets first Smackover oiler: Oil and Gas Jour., Sept. 30, p. 115.

Gardner, Frank J., 1964, Deep drilling may hold key to dry parishes: Oil and Gas Jour., May 4, p. 115.

Gardner, Louis S., 1959, Geology of the Lewistown area, Fergus County, Montana: U. S. Geol. Survey Oil and Gas Inv. Map OM-199.

Garland, Robert F., and Hoovler, Paul B., 1965, Drillers hit north rim of Illinois basin: Oil and Gas Jour., Feb. 22, p. 166–170.

Garlick, J. M., 1954, Big Mineral field exploitation—a challenge to engineers: Jour. Petroleum Technology, v. 6, no. 12 (December), p. 9–14.

Garlough, John L., and Taylor, Garvin L., 1941, Hugoton gas field, Grant, Haskell, Morton, Stevens, and Seward Counties, Kansas, and Texas County, Oklahoma, *in* Stratigraphic type oil fields: Am. Assoc. Petroleum Geologists, p. 78–104.

Gary, Samuel, and McGregor, A. A., 1968, Exploration philosophy behind the Bell Creek oil field discovery, Powder River County, Montana: Mtn. Geology, v. 5, no. 1, p. 15–21.

Gates, George, 1964, Geologic and tectonic setting, *in* U. S. Senate Rept. Comm. on Interior and Insular Affairs, 58th Cong., 2nd sess.: U. S. Geol. Survey, p. 32–41.

Gates, George, Grantz, Arthur, and Patton, W. W., Jr., 1968, Geology and natural gas and oil resources of Alaska *in* Natural gases of North America: Am. Assoc. Petroleum Geologists Mem. 9, v. 1, p. 3–48.

Gates, George, and Gryc, George, 1963, Structure and tectonic history of Alaska, *in* Backbone of the Americas: U. S. Geol. Survey Mem. 2 (November), p. 264–277.

Gatewood, Lloyd E., 1968, Anatomy of a giant—Oklahoma City field (abs.): Am. Assoc. Petroleum Geologists Bull., v. 52, no. 3 (March), p. 528.

Gelphman, Norman Ray, 1960, West Sentinel oil field, Washita County, Oklahoma—sedimentology of the "granite wash" and structural geology: Shale Shaker, v. 10 (February), p. 2–16.

Gere, W. C., Horton, G. W., Harstead, J. H., and Russell, D. F., 1964, Oil and natural gas, *in* Mineral and water resources of Utah: U. S. 88th Cong., 2d sess., Comm. Print, p. 51–60.

Gibbon, Anthony, 1952, Smackover, Arkansas: World Oil, May, p. 260–263.

Gibson, George R., 1965, Oil and gas in southwestern region—geologic framework: Am. Assoc. Petroleum Geologists Mem. 4, p. 66–100.

Gibson, H. A., 1946, Composite columnar section of subsurface formations in the Abilene area, Texas: Abilene Geol. Soc., April 1.

Giddens, Paul, 1941, Beginnings of the petroleum industry: Pennsylvania Historical Commission.

Gilmore, R. L., 1952, Will oil be found in south-central Nebraska portion of the Salina basin?: Oil and Gas Jour., June 2, p. 91–94.

Gish, Wesley G., and Carr, Raymond M., 1929, Garber field, Garfield County, Oklahoma, *in* Structure of typical American oil fields: Am. Assoc. Petroleum Geologists, v. 1, p. 176–191.

Glover, J. E., 1955, South Texas' thriving Olmos trend: Oil and Gas Jour., Dec. 19, p. 129–134.

Glover, Sheldon L., 1936, Preliminary report on petroleum and natural gas in Washington: Washington Div. Geol. Rept. Inv. 4.

Goebel, E. D., 1966, Thermal recovery projects are increasing in Kansas, Missouri: World Oil, March, p. 78–80.

Goebel, E. D., Hilpman, P. L., Beene, D. L., and Noever, R. J., 1961, Data on secondary recovery projects in Kansas, 1960: Kansas Geol. Survey Oil and Gas Inv. 24.

Goebel, Lawrence A., 1950, Cairo field, Union County, Arkansas: Am. Assoc. Petroleum Geologists Bull., v. 34 (October), p. 1954–1980.

Goldstein, August, Jr., and Flawn, Peter T., 1958, Oil and gas possibilities of Ouachita structural belt in Texas and Oklahoma: Am. Assoc. Petroleum Geologists Bull., v. 42 (April), p. 876–881.

Gordon, Dugald, 1935, Richland gas field, Richland Parish, Louisiana, *in* Geology of natural gas: Am. Assoc. Petroleum Geologists, p. 773–786.

Gouin, Frank, 1956, Surface criteria of southern Oklahoma oil fields *in* Petroleum geology of southern Oklahoma: Am. Assoc. Petroleum Geologists, v. 1, p. 14–35.

Grace, W. P., and Nelson, G. Allen, 1962, Danforth Hills Area *in* Exploration for oil and gas in northwestern Colorado: Rocky Mtn. Assoc. Geologists, Denver, Colo., p. 92–95.

Grantz, Arthur, 1964, Southern Alaska *in* U. S. Senate Rept. Comm. on Interior and Insular Affairs, 58th Cong., 2nd sess.: U. S. Geol. Survey, p. 44–62.

Grantz, Arthur, and Patton, W. W., Jr., 1964, Petroleum and natural gas *in* U. S. Senate Rept. Comm. on Interior and Insular Affairs, 58th Cong., 2nd sess.: U. S. Geol. Survey, p. 43–44.

Gras, V. B., 1968, Church Buttes gas field, Wyoming: Am. Assoc. Petroleum Geologists Mem. 9, v. 1, p. 798–802.

Gratton, Patrick J. F., and LeMay, William J., 1967, Oil and Gas Jour., Mar. 20, p. 207–212.

Graves, Doyle T., 1954, Geology of the Dominguez oil field, Los Angeles County: California Div. Mines and Geol. Bull. 170, Map Sheet 32.

Gray, Bert F., Jr., 1958, Geology and exploration in the Four Corners: World Oil, April, p. 130–134.

Gray, R. S., 1967, Cache field—a Pennsylvanian algal reservoir in southwestern Colorado: Am. Assoc. Petroleum Geologists Bull., v. 51, no. 10 (October), p. 1959–1978.

Green, Darsie A., 1957, Trenton structure in Ohio, Indiana, and northern Illinois: Am. Assoc. Petroleum Geologists Bull., v. 41, no. 4 (April), p. 627–642.

Green, Thomas H., and Ziener, C. W., 1953, Northwest Lake Creek area, Hot Springs County, Wyoming, Oil and Gas Jour., July 13, p. 178.

Greenfield, R. E., 1958, Eastern Kentucky—a new look at an old area: Oil and Gas Jour., Aug. 11, p. 155–160.

Gribi, Edward A., Jr., 1956, Sweetgrass arch, forgotten corner of the Rockies: Oil and Gas Jour., Apr. 30, p. 174–180.

Gries, John Paul, 1964, Barker Dome oil field, Custer County, South Dakota: Mtn. Geologist, v. 1, no. 1, p. 43–46.

Griffith, E. G., 1966, Geology of Sabre bar, Logan and Weld Counties, Colorado: Am. Assoc. Petroleum Geologists, v. 50, no. 10 (October), p. 2112–2118.

Grigg, Robert P., Jr., 1957, Key characteristics of Nodosaria embayment: World Oil, July, p. 91–95.

Gross, Larry T., 1962, Mesaverde beds of the Uinta basin offer good prospects: World Oil, June, p. 87–93.

Grow, George C., 1964, Gas production near hard coal areas of Pennsylvania: World Oil, September, p. 79–81.

Gryc, George, 1959, Northern Alaska *in* Geology of possible petroleum provinces in Alaska: U. S. Geol. Survey Bull. 1094, p. 88–112.

Gryc, George, et al., 1956, Mesozoic sequence in Colville River region: Am. Assoc. Petroleum Geologists Bull., v. 40, no. 2 (February), p. 209–254.

Guinan, Mark A., 1966, Coyanosa field, Pecos County, *in* Oil and gas fields in West Texas—a symposium: West Texas Geol. Soc., p. 99–111.

Gulmon, G. W., 1967, Addendum for Mississippi: developments in southeastern states for 1966: Am. Assoc. Petroleum Geologists Bull., v. 51, no. 11 (November), p. 2311.

Gulmon, G. W., Hansen, H. E., and Simms, B. L., 1957, How pure geology works at Natchez: Oil and Gas Jour., Aug. 5, p. 72–76.

Gunter, Craig E., 1962, Oil and gas potential of Upper Cretaceous sediments, southern Piceance basin, *in* Exploration for oil and gas in northwestern Colorado: Rocky Mtn. Assoc. Geologists, Denver, Colorado, p. 114–118.

Gutstadt, Alan M., 1958, Cambrian and Ordovician stratigraphy and oil and gas possibilities in Indiana: Indiana Geol. Survey Bull. 14, p. 9.

Guy, Samuel C., 1967, Developments in southeastern states in 1966: Am. Assoc. Petroleum Geologists Bull., v. 51, no. 6 (June), p. 1102.

Gwynn, Thomas A., 1958, Bowdoin field *in* Montana oil and gas fields: Billings Geol. Soc., p. 77–78.

Hackel, Otto, 1966, Summary of the geology of the Great Valley: California Div. Mines and Geology Bull. 190, p. 217–238.

Hadley, Herbert D., 1956, Cat Creek field, *in* Montana oil and gas fields: Billings Geol. Soc., p. 99–100.

Haeberle, Fred R., 1956, Seymour Strawn and Caddo fields, Baylor County, Texas: Am. Assoc. Petroleum Geologists Bull., v. 40 (May), p. 879–889.

Hagan, Wallace W., 1960, Oil exploration in Green County, Kentucky, and adjacent areas: Am. Assoc. Petroleum Geologists Bull., v. 44, no. 7 (July), p. 1249–1250.

Hager, Dilworth S., and Burnett, Claude M., 1960, Mexia-Talco fault line in Hopkins and Delta Counties, Texas; Am. Assoc. Petroleum Geologists Bull., v. 44 (March), p. 316–356.

Haines, Richard B., and Minshall, Francis E., 1954, Geology of the San Miguelito oil field, Ventura County: California Div. Mines and Geol. Bull. 170, Map Sheet 27.

Halbouty, Michel T., 1936, Geology and geophysics showing cap rock and salt overhang of High Island dome, Galveston County, Texas, *in* Gulf Coast oil fields: Am. Assoc. Petroleum Geologists, p. 909–960.

Halbouty, Michel T., 1941, Stratigraphic reservoirs in University oil field, East Baton Rouge Parish, Louisiana, *in* Stratigraphic type oil fields: Am. Assoc. Petroleum Geologists, p. 208–236.

Halbouty, Michel T., 1962, South Liberty field, Liberty County, Texas, *in* Typical oil and gas fields of southeast Texas: Houston Geol. Soc., p. 200–205.

Halbouty, Michel T., 1966, Stratigraphic-trap possibilities in Upper Jurassic rocks, San Marcos arch, Texas: Am. Assoc. Petroleum Geologists Bull., v. 50, no. 1 (January), p. 3–24.

Halbouty, Michel T., and Barber, Thomas D., 1962, Port Acre-Port Arthur fields, Jefferson County, Texas, *in* Typical oil and gas fields of southeast Texas: Houston Geol. Soc., p. 169–173.

Halbouty, Michel T., and Simmons, Benjamin T., 1941, Hitchcock field, Galveston County, Texas, showing stratigraphic accumulation and structure, *in* Stratigraphic type oil fields: Am. Assoc. Petroleum Geologists, p. 641–660.

Hale, G. Carl, 1959, Southwest Ardmore field *in* Petroleum

geology of southern Oklahoma: Am. Assoc. Petroleum Geologists, v. 2, p. 262–273.

Hall, John F., 1952, Oriskany sand study: Ohio Geol. Survey Inv. Rept. 13, pt. II, p. 39–58.

Hallgarth, W. E., 1959, Stratigraphy of Paleozoic rocks in northwestern Colorado: U. S. Geol. Survey Oil and Gas Inv. Chart OC–59.

Ham, William E., 1958, Southern Oklahoma; golden province for oil finders: Oil and Gas Jour., Dec. 15, p. 152–156.

Ham, William E., Denison, Rodger E., and Merritt, Charles A., 1964, Basement rocks and structural evolution of southern Oklahoma: Oklahoma Geol. Survey Bull. 95.

Ham, William E., Denison, Rodger E., and Merritt, Clifford A., 1965, Basement rocks and structural evolution of southern Oklahoma—a summary: Am. Assoc. Petroleum Geologists Bull., v. 49, no. 7 (July), p. 927–934.

Ham, William E., and Wilson, James Lee, 1967, Paleozoic epeirogeny and orogeny in the central United States: Am. Jour. Sci., v. 265, (May), p. 332–407.

Hamblin, W. K., 1965, Origin of "reverse drag" on the downthrown side of normal faults: Geol. Soc. American Bull., v. 76, no. 10 (October), p. 1145–1164.

Hamilton, S. V., 1968, Wilmington field, Los Angeles County, California: Am. Assoc. Petroleum Geologists Mem. 9, p. 164–168.

Hamke, J. R., Marchant, L. C., and Cupps, C. Q., 1966, Oil fields in the Williston basin in Montana, North Dakota, and South Dakota: U. S. Bur. Mines Bull. 629.

Hammer, A. A., 1934, Rattlesnake Hills gas field, Benton County, Washington: Am. Assoc. Petroleum Geologists Bull., v. 18 (July), p. 847–859.

Hamric, Burt E., 1961, Hoffman gas area, Okmulgee and McIntosh Counties, Oklahoma: Shale Shaker, v. 11 (May), p. 2–15.

Haney, Gene, 1968, Newburg gas development, Kanawha and Jackson Counties, West Virginia, *in* Kentucky Geol. Survey, Ser. X, Spec. Pub. 15, p. 111–117.

Hansen, Alan R., 1966, Reef trends of Mississippian Ratcliffe zone, northeast Montana and northwest North Dakota: Am. Assoc. Petroleum Geologists Bull., v. 50 (October), p. 2260–2268.

Harbaugh, John W., 1957, Mississippian bioherms in northeast Oklahoma: Am. Assoc. Petroleum Geologists Bull., v. 41 (November), p. 2530–2544.

Harbison, Robert R., 1957, Pegasus field *in* Occurrence of oil and gas in west Texas: Univ. Texas Pub. 5716, p. 271–276.

Harbour, R. L., and Dixon, G. H., 1956, Geology of the Trinidad–Aguilar area, Las Animas and Huerfano Counties, Colorado: U. S. Geol. Survey Oil and Gas Inv. Map OM-174.

Hardin, Frank R., 1968, Effects of contemporaneous structural movement on sedimentation at Thornwell field, Jefferson Davis and Cameron Parishes, Louisiana: Am. Assoc. Petroleum Geologists Mem. 9, v. 1, p. 582–593.

Hardin, Frank R., and Hardin, George C., Jr., 1961, Contemporaneous normal faults of Gulf Coast and their relation to flexures: Am. Assoc. Petroleum Geologists Bull., v. 45, no. 2 (February), p. 238–248.

Hardin, George C., Jr., and Clark, W. Kenley, 1968, Gas resources of Eocene producing trend, Upper Gulf Coast of Texas: Am. Assoc. Petroleum Geologists Mem. 9, v. 1, p. 284–294.

Harding, Richard W., 1966, New York's Oriskany: Oil and Gas Jour., Mar. 21, p. 130–136.

Hargis, Richard N., 1962, Stratigraphy of the Carrizo–Wilcox of a portion of South Texas and its relationship to production, *in* Contributions to the Geology of South Texas; South Texas Geol. Soc., p. 129–145.

Harlton, Bruce H., 1960, Stratigraphy of Cement pool and adjacent area, Caddo and Grady Counties, Oklahoma: Am. Assoc. Petroleum Geologists Bull., v. 44 (February), p. 210–226.

Harlton, Bruce H., 1963, Frontal Wichita fault system of southwestern Oklahoma: Am. Assoc. Petroleum Geologists Bull., v. 47 (August), p. 1552–1580.

Harlton, Bruce H., 1966, Relation of buried Tishomingo uplift to Ardmore basin and Ouachita mountains, southeastern Oklahoma: Am. Assoc. Petroleum Geologists Bull., v. 50, no. 7 (July), p. 1365–1374.

Harms, J. C., 1966, Stratigraphic traps in a valley fill, western Nebraska: Am. Assoc. Petroleum Geologists Bull., v. 50, no. 10 (October), p. 2119–2149.

Harnsberger, W. T., 1955, Virginia's oil and gas resources: Virginia Minerals, v. 1 (January), p. 1–4.

Harrington, John W., 1963, Opinion of structural mechanics of Central basin platform area, West Texas: Am. Assoc. Petroleum Geologists Bull., v. 47, no. 12 (December), p. 2023–2038.

Harris, Lloyd A., 1968, Wapella East field, DeWitt County, Illinois, *in* Geology and petroleum production of the Illinois basin—a symposium: Illinois and Indiana–Kentucky Geol. Socs., p. 243–247.

Harris, Loy E., 1955, The Manderson field (Wyo.) *in* Am. Assoc. Petroleum Geologists Rocky Mtn. Sec., Geological record, Feb.

Harris, Steven H., Land, Cooper B., Jr., and McKeever, John H., 1966, Relation of Mission Canyon stratigraphy to oil production in north-central North Dakota: Am. Assoc. Petroleum Geologists Bull., v. 50, no. 10 (October), p. 2269–2276.

Harrison, Ray L., Jr., 1954, The story of Williston basin: Oil and Gas Jour., Apr. 19, p. 187–195.

Hart, Earl W., 1957, Natural gas: California Div. Mines Bull. 176, p. 373–384.

Hart, Earl W., 1966, Economic mineral deposits of the Great Valley: California Div. Mines and Geology Bull. 190, p. 249–252.

Hartman, F. H., 1968, Jerusalem gas field, Conway County, Arkansas: Am. Assoc. Petroleum Geologists Mem. 9, v. 2, p. 1682–1692.

Haught, Oscar L., 1953, Lithologic factors affecting prospects for early Paleozoic oil and gas in West Virginia: Am. Assoc. Petroleum Geologists, v. 37, no. 3 (March), p. 553–560.

Haught, Oscar L., 1964, Oil and gas report on Braxton and Clay Counties, West Virginia: West Virginia Geol. Survey Bull. 29.

Haught, Oscar L., 1965, Oil and gas report on Barbour and Upshur Counties, West Virginia: West Virginia Geol. Survey Bull. 31.

Haun, John D., 1963, Stratigraphy of Dakota Group and relationship to petroleum occurrence, northern Denver basin, *in* Geology of the northern Denver basin and adjacent uplifts: Rocky Mtn. Assoc. Geologists, p. 119–134.

Haun, John D., and Kent, Harry C., 1965, Geologic history of Rocky Mountain region: Am. Assoc. Petroleum Geologists Bull., v. 49, no. 11 (November), p. 1781–1800.

Hautau, Gordon H., 1954, Richfield as a major oil zone in Michigan: World Oil, December, p. 100–106.

Havenor, Kay C., 1964, Oil and gas tests in Lincoln County, New Mexico, *in* Guidebook of the Ruidoso country, New Mexico Geol. Soc., 15th Field Conf.: New Mexico Bur. Mines and Mineral Resources, p. 155–158.

Hayes, J. A., Jr., 1956, Mississippian production in the Osage: World Oil, Feb. 1, p. 85–88.

Haynes, Edward H., 1967, Developments in northwestern Colorado and western Nebraska in 1966: Am. Assoc. Petroleum Geologists Bull., v. 51, no. 6 (June), p. 1124–1128.

Headlee, A. J. W., 1960, Deep tests reflect near-peak gas yield: World Oil, June, p. 103.

Heath, E. J., 1958, Yorba Linda oil field, *in* A guide to the geology and oil fields of the Los Angeles and Ventura regions: Am. Assoc. Petroleum Geologists, Pacific Sec., p. 105–106.

Heck, E. T., 1941, Gay-Spencer-Richardson oil and gas trend, Jackson, Roane, and Calhoun Counties, West Virginia, *in* Stratigraphic type oil fields: Am. Assoc. Petroleum Geologists, p. 805–829.

Heezen, Bruce C., 1968, The Atlantic continental margin: UMR (Univ. Missouri School Mines and Metallurgy) Jour., no. 1 (April), p. 5–25.

Helmboldt, D., 1968, Porter field *in* Oil and gas fields—a symposium: Michigan Basin Geol. Soc., p. 137–142.

Hendricks, Leo, 1964, Stratigraphic summary of the Ellenburger group of North Texas: Tulsa Geol. Soc. Digest, v. 32, p. 121–125.

Hendy, W. J., 1957, Lower Cretaceous (Edwards) oil fields, Caldwell and Guadalupe Counties, Texas: Gulf Coast Assoc. Geol. Socs. Trans., v. 7, p. 23–29.

Hennen, Ray V., 1929, Big Lake oil field, Reagan County, Texas, *in* Structure of typical American oil fields: Am. Assoc. Petroleum Geologists, v. 2, p. 500–541.

Herald, Frank A., ed., 1951, Occurrence of oil and gas in northeast Texas: Texas Univ. Pub. 5116.

Herald, Frank A., ed., 1957, Occurrence of oil and gas in West Texas: Texas Univ. Pub. 5716, p. 442.

Herrick, John P., 1949, Empire oil: New York, Dodd, Mead.

Hershey, H. Garland, 1963, Interview in World Oil following Kornfeld (1963) article.

Hester, Robert J., and Holland, R. R., 1960, Structure of the Puckett field, Pecos County, Texas, *in* Natural gas in the Southwest: Southwestern Federation Geol. Socs., p. 29–37.

Heylmun, Edgar B., 1958a, Oil and gas possibilities around the laccolithic-type intrusions in the Colorado Plateau *in* Guidebook, 9th Ann. Field Conference: Intermountain Assoc. Petroleum Geologists, p. 238–240.

Heylmun, Edgar B., 1958b, Paleozoic stratigraphy and oil possibilities of Kaiparowits region, Utah: Am. Assoc. Petroleum Geologists Bull., v. 42, no. 8 (August), p. 1781–1811. Also, Kaiparowits beckons oil hunters, Oil and Gas Jour., Oct. 13, p. 180–183.

Heylmun, Edgar B., 1958c, Great Basin: giant sleeper of the west: Oil and Gas Jour., Dec. 8, p. 138–141.

Heylmun, Edgar B., 1961a, Farmington field *in* The oil and gas fields of Utah—a symposium: Intermountain Assoc. Petroleum Geologists.

Heylmun, Edgar B., 1961b, Virgin field *in* The oil and gas fields of Utah—a symposium: Intermountain Assoc. Petroleum Geologists.

Heylmun, Edgar B., 1961c, Rozel Point *in* The oil and gas fields of Utah—a symposium: Intermountain Assoc. Petroleum Geologists.

Heylmun, Edgar B., 1964, Shallow oil and gas possibilities in east and south central Utah: Utah Geol. and Mineral. Survey Spec. Studies 8.

Hiestand, T. C., 1933, Voshell field, McPherson County, Kansas: Am. Assoc. Petroleum Geologists Bull., v. 17 (February), p. 169–191.

Hiestand, T. C., 1941, Bryson oil field, Jack County, Texas, *in* Stratigraphic type oil fields: Am. Assoc. Petroleum Geologists, p. 539–547.

Higgins, James W., ed., 1958, A guide to the geology and oil fields of the Los Angeles and Ventura regions: Am. Assoc. Petroleum Geologists, Pacific Sec.

Hill, F. L., 1965, Kettleman Middle Dome oil field, *in* Summary of operations, California oil fields; California Div. Oil and Gas, v. 51, no. 1, p. 39–46.

Hill, M. L., 1963, Occurrences of petroleum in Cook Inlet area, Alaska: Proc. Sixth World Petroleum Cong., v. 1, p. 509–518.

Hills, John M., 1942, Rhythm of Permian seas–a paleogeographic study: Am. Assoc. Petroleum Geologists Bull., v. 26, no. 2 (February), p. 217–255.

Hills, John M., 1961, Tippett field, Crockett and Pecos Counties, Texas: Am. Assoc. Petroleum Geologists Bull., v. 45, no. 11 (November), p. 1859–1869.

Hills, John M., 1968a, Gas in Delaware and Val Verde basins, west Texas and southeastern New Mexico: Am. Assoc. Petroleum Geologists Mem. 9, v. 2, p. 1394–1432.

Hills, John M., 1968b, Permian basin field area, west Texas and southeastern New Mexico: Geol. Soc. America Spec. Paper 88, p. 17–27.

Hilpman, Paul, 1958, Producing zones of Kansas oil and gas fields: Kansas Geol. Survey Oil and Gas Inv. 16.

Hilseweck, William J., 1941, Walnut Bend pool, Cooke County, Texas, *in* Stratigraphic type oil fields: Am. Assoc. Petroleum Geologists, p. 776–805.

Hinson, H. H., 1947, Reservoir characteristics of Rattlesnake oil and gas field, San Juan County, New Mexico, Am. Assoc. Petroleum Geologists Bull., v. 31 (April), p. 731–771.

Hinton, Gene, 1968, Recluse oil field *in* Guidebook, 20th Ann. Field Conf.: Wyoming Geol. Assoc., p. 73–78.

Hodgson, Robert A., 1965, Genetic and geometric relations between structures in basement and overlying sedimentary rocks, with examples from Colorado Plateau and Wyoming; Am. Assoc. Petroleum Geologists Bull., y. 49, no. 7 (July), p. 935–949.

Hoffman, F. H., 1954, Wilson Creek *in* Oil and gas fields of Colorado: Rocky Mtn. Assoc. Geologists, Denver, Colorado, p. 288–290.

Hoffman, F. H., 1957, Possibilities of Weber stratigraphic traps, Rangely area, northwest Colorado: Am. Assoc. Petroleum Geologists Bull., v. 41, no. 5 (May), p. 894–905.

Holden, Richard N., 1963, Drillers take a careful look at Lower Cretaceous reef in Texas: Oil and Gas Jour., Sept. 2, p. 194–199.

Holland, Richard R., 1966, Puckett field, Pecos County, *in* Oil and gas fields in West Texas—a symposium: West Texas Geol. Soc., p. 38–41.

Holland, Wilbur C., Hough, Leo W., and Murray, Grover E., 1952,

Geology of Beauregard and Allen Parishes: Louisiana Geol. Survey Bull. 27 (May), p. 154–160 and plate 26.

Hollenshead, Charles T., and Pritchard, Roy L., 1961, Geology of producing Mesaverde sandstones, San Juan basin, *in* Geometry of sandstone bodies: Am. Assoc. Petroleum Geologists Symp., p. 98–118.

Holman, William H., 1958, Correlation of producing zones of Ventura basin oil fields, *in* A guide to the geology and oil fields of the Los Angeles and Ventura regions: Am. Assoc. Petroleum Geologists, Pacific Sec., p. 191–199.

Holmquest, Harold J., 1965, Deep pays in Delaware and Val Verde basins: Am. Assoc. Petroleum Geologists Mem. 4, p. 257–279.

Hols, Arnold, and Bethel, Frank T., 1957, Discussion of reservoir characteristics, Cedar Creek anticline fields, Montana: Petroleum Technology, v. 9 (December), p. 23–30.

Hoots, Harold W., 1943, Origin, migration, and accumulation of oil in California: California Div. Mines Bull. 118, p. 253–276.

Hoots, Harold W., and Bear, Ted L., 1954, History of oil exploration and discovery in California: California Div. Mines Bull. 170, p. 5–9.

Hoots, Harold W., Bear, Ted L., and Kleinpell, William D., 1954a, *in* California Div. Mines Bull. 170, p. 29–32.

Hoots, Harold W., Bear, Ted L., and Kleinpell, William D., 1954b, *in* California Div. Mines Bull. 170, p. 113–129.

Hoover, William B., 1952, Regional structure of the Four Corners area *in* Geological Symposium of the Four Corners region: Four Corners Geol. Soc., Oct. 1, p. 10–11.

Hoppin, Richard A. and Palmquist, John C., 1965, Basement influence on later deformation: the problem, techniques of investigation, and examples from Bighorn Mountains, Wyoming, Am. Assoc. Petroleum Geologists Bull., v. 49, no. 7 (July), p. 993–1003.

Horn, George H. (1963), Geology of the East Thermopolis area, Hot Springs and Washakie Counties, Wyoming, U. S. Geol. Survey Oil and Gas Inv. Map OM-213.

Horton, G. W., and Gere, W. C., 1964, Relation of stratigraphy and structure to gas production in the San Arroyo and East Canyon fields, Utah, *in* Guidebook, 13th Ann. Field Conf.: Intermountain Assoc. Petroleum Geologists, p. 191–195.

Horton, Robert C., 1964, Petroleum, *in* Mineral and water resources of Nevada: U. S. Senate, 88th Cong., 2nd sess., Doc. 87, p. 54–57.

Horton, Robert C., 1965, Eastern Nevada is a gamble, but production is rising: World Oil, July, p. 84–89.

Hough, Howard, and Couvillon, J. A., 1966, Deep, hot drilling marks gulf Frio play: Oil and Gas Jour., May 9, p. 142–158.

Houston Geological Society, 1962, Typical oil and gas fields of southeast Texas: Houston, Texas.

Houston Geological Society, Field Trip Committee, 1953, Am. Assoc. Petroleum Geologists Convention, Geology of the Gulf Coast: Oil and Gas Jour., June 22, p. 251.

Houston Geological Society, Study Group, 1962a, Downdip limits of production in the Oligocene of the Upper Gulf Coast of Texas: Gulf Coast Assoc. Geol. Socs. Trans., v. 12, p. 39–44.

Houston Geological Society, Study Group, 1962b, Yegua and Wilcox potential Upper Texas Gulf Coast; Gulf Coast Assoc. Geol. Socs. Trans., v. 12, p. 27–35.

Howard, Richard H., 1961, Oil and gas in the Adams–Brown-Schuyler County area, Illinois: Illinois State Geol. Survey Circ. 325.

Howard, Richard H., 1963, Wapella East oil pool, DeWitt County, Illinois—a Silurian reef: Illinois State Geol. Survey Circ. 349; also Oil and Gas Jour., May 27, p. 144–147.

Howard, Richard H., 1964, Niagaran reefs and oil accumulation in the DeWitt–McLean County area, Illinois: Illinois State Geol. Survey Circ. 377.

Howard, Richard H., 1967, Oil and gas pay maps of Illinois: Illinois Geol. Survey, Illinois Petroleum 84.

Howe, Herbert J., 1962, Subsurface geology at St. Helena, Tangipahoa, Washington and St. Tammany Parishes, Louisiana: Gulf Coast Assoc. Geol. Socs. Trans., v. 12, p. 121–155.

Howell, J. V., 1948, Geology of Benton field, Franklin County, Illinois: Am. Assoc. Petroleum Geologists Bull., v. 32 (May), p. 745–766.

Howell, W. F., 1929, Kevin–Sunburst field, Toole County, Montana, *in* Structure of typical American oil fields: Am. Assoc. Petroleum Geologists, v. 2, p. 254–268.

Hubbert, M. King, 1953, Entrapment of petroleum under hydrodynamic conditions: Am. Assoc. Petroleum Geologists Bull., v. 37 (August), p. 2016.

Huddle, J. W., Jacobsen, Eloise T., and Williamson, A. D., 1956, Oil and gas wells drilled in southwestern Virginia before 1950: U. S. Geol. Survey Bull. 1027-L.

Hudnall, James A., 1951a, East Texas field, Gregg, Rusk, Upshur, Smith, and Cherokee Counties, Texas: Univ. Texas Pub. 5116, p. 113–118.

Hudnall, James A., 1951b, Mexia field, Limestone County, Texas; Univ. Texas Pub. 5116, p. 231–235.

Huffman, George Garrett, 1951, Geology of the Ozark uplift, northeastern Oklahoma: Shale Shaker, v. 2 (November), p. 5–12.

Hull, Joseph P. D., Jr., 1962, Cretaceous Suwannee Strait, Georgia and Florida: Am. Assoc. Petroleum Geologists Bull., v. 46, no. 1 (January), p. 118–122.

Hull, Joseph P. D., Jr., 1965, Appalachian anticlines need more testing: Oil and Gas Jour., Aug. 2, p. 178–180.

Hundley, Harland B., and Shute, George C., 1963, Developments in Arkansas and North Louisiana in 1962: Am. Assoc. Petroleum Geologists Bull., v. 47 (June), p. 1097–1108.

Hunt, Robert E., 1956, Bowes field, Blaine County, Montana, *in* First international Williston basin symposium; North Dakota Geol. Soc. and Saskatchewan Geol. Soc., p. 186–191.

Hunt, V. G., and O'Connor, J. P., Jr., 1954, Developments in East Texas in 1953: Am. Assoc. Petroleum Geologists Bull., v. 38, no. 6 (June), p. 1197.

Hunter, Coleman D., 1962, Economics influences east Kentucky: Oil and Gas Jour., July 9, p. 170–174.

Hurd, Gordon, 1962, Thornburg field *in* Exploration for oil and gas in northwestern Colorado: Rocky Mtn. Assoc. Geologists, p. 37–39.

Hurley, G. William, 1959, Overthrust faulting and Paleozoic gas prospects in Montana's disturbed belt, *in* Geological record: Am. Assoc. Petroleum Geologists, Rocky Mtn. Sec., p. 91–96.

Hyatt, Don L., 1936, Preliminary report on the Fitts pool, Pontotoc County, Oklahoma: Am. Assoc. Petroleum Geologists Bull., v. 20 (July), p. 951–974.

Illinois and Indiana–Kentucky Geological Societies, 1968, Geology and petroleum production of the Illinois basin—a symposium: Evansville, Indiana.

Illinois State Geological Survey Staff, 1968, Oil and gas in Illinois, *in* Geology and petroleum production of the Illinois basin—a symposium: Illinois and Indiana-Kentucky Geol. Socs., p. 33–35.

Imbt, Robert F., and McCollum, S. V., 1950, Todd Deep field, Crockett County, Texas: Am. Assoc. Petroleum Geologists Bull., v. 34 (February), p. 239–262.

Imbt, W. C., 1941, Zenith pool, Stafford County, Kansas—an example of stratigraphic trap accumulation *in* Stratigraphic type oil fields: Am. Assoc. Petroleum Geologists, p. 139–165.

Imholz, H. W., 1941, Noodle Creek pool, Jones County, Texas, *in* Stratigraphic type oil fields: Am. Assoc. Petroleum Geologists, p. 698–721.

Independent Petroleum Association of America, Annual publication, The oil producing industry in your state: Tulsa, Oklahoma.

Ingham, Albert I., 1949, McDonald and adjacent oil fields: Pennsylvania Topog. and Geol. Survey Bull. M-29, p. 1–40.

Ingham, Albert I., 1954, The geology and development of the Leidy-South Leidy gas fields, Clinton County, Pennsylvania: Producers Monthly, v. 18, no. 10 (August), p. 22–28.

Ingham, Albert I., 1960, Oil and gas frontiers in the East: GeoTimes, v. 4, no. 6 (March), p. 14 et seq.

Ingham, A. I., Lytle, W. S., Matteson, L. S., and Sherrill, R. E., 1956, Oil and gas geology of the Sheffield quadrangle (Forest and Warren Counties), Pennsylvania: Pennsylvania Topog. and Geol. Survey Bull. M-38, p. 1–72.

Ingham, W. I., 1941, Dora oil pool, Seminole County, Oklahoma, *in* Stratigraphic type oil fields: Am. Assoc. Petroleum Geologists, p. 408–435.

Ingham, W. I., 1948, Powell's Lake oil field, Kentucky: Am. Assoc. Petroleum Geologists Bull., v. 32 (January), p. 34–51.

Intermountain Association of Petroleum Geologists, 1954, Geology of portions of the high plateaus and adjacent canyon lands, central and south-central Utah, 5th Ann. Field Conf.: Salt Lake City, Utah.

Intermountain Association of Petroleum Geologists, 1956, Geology and economic deposits of east central Utah, 7th Annual Field Conf.: Salt Lake City, Utah.

Intermountain Association of Petroleum Geologists, 1958, Guidebook to the geology of the Paradox basin, 9th Ann. Field Conf.: Salt Lake City, Utah.

Intermountain Association of Petroleum Geologists, 1961, The oil and gas fields of Utah—a symposium: Salt Lake City, Utah.

Intermountain Association of Petroleum Geologists, 1963, Guidebook to the geology of southwestern Utah: Salt Lake City, Utah.

Intermountain Association of Petroleum Geologists, 1964, Guidebook to the geology and mineral resources of the Uinta basin, 13th Ann. Field Conf.: Salt Lake City, Utah.

Ives, Robert E., 1958, Scipio—hottest thing in Michigan: Oil and Gas Jour., Sept. 1, p. 214–220.

Ives, Robert E., 1960, New Trenton extensions boost Michigan production: Oil and Gas Jour., Aug. 8, p. 122–126.

Jackson, William E., 1964, Depositional topography and cyclic deposition in West-Central Texas: Am. Assoc. Petroleum Geologists Bull., v. 48, no. 3 (March), pp. 317–328.

Jacobsen, Lynn, 1959, Sedimentation of some Springer sandstone (Mississippian-Pennsylvanian) reservoirs, southern Oklahoma: Am. Assoc. Petroleum Geologists Bull., v. 43 (November), p. 2575–2591.

Jahns, Richard H., 1954, *in* California Div. Mines Bull. 170.

Jenkins, Olaf P., 1941, Geomorphic provinces of California: California Div. Mines Bull. 118, pt. 2, p. 83–88.

Jennings, A. Ray, Payne, W. R., and Wharton, Ralph, 1968, Developments in South Texas in 1967: Am. Assoc. Petroleum Geologists Bull., v. 52, no. 6 (June), p. 1001–1005.

Jennings, Charles W., 1957a, Asphalt and bituminous rock: California Div. Mines Bull. 176, p. 59–70.

Jennings, Charles W., 1957b, Petroleum: California Div. Mines Bull. 176, p. 409–424.

Jewett, J. M., 1960, How major geologic features developed in Kansas: World Oil, January, p. 119–124, 149.

Jindra, Roy, 1967, Wilcox trend of South Texas may still have an exploration future: Oil and Gas Jour., Feb. 27, p. 189–194.

Joesting, H. R., and Case, J. E., 1960, Salt anticlines and deep-seated structures in the Paradox basin, Colorado and Utah: U. S. Geol. Survey Prof. Paper 400-B (Geol. Survey Research, 1960), p. B252–256.

Johnson, E. G., 1959, Paleozoic oil prospects look good in eastern Nevada: Oil and Gas Jour., Feb. 9, p. 148–155.

Johnson, Melvin C., 1958, Bowes field *in* Montana oil and gas fields: Billings Geol. Soc., p. 81–83.

Johnson, Ray B., and Mathy, Harold Edward, 1957, The South Texas Frio trend: Gulf Coast Assoc. Geol. Socs. Trans., v. 7, p. 207–218.

Jones, Franklin C., 1964, The Lower Wilcox trend, DeWitt County, Texas: Gulf Coast Assoc. Geol. Socs., Trans., v. 14, p. 203.

Jones, Fred B. Jr., 1968, East Durant field, Bryan County, Oklahoma: Am. Assoc. Petroleum Geologists Mem. 9, v. 2, p. 1467–1475.

Jones, Larry L., 1962, The Edwards of the Greater Person area: big potential, big problems: Oil and Gas Jour., Oct. 15, p. 266–270.

Jones, Paul H., Turcan, A. N., Jr., and Skibitzke, Herbert E., 1954, Geology and groundwater resources of southwestern Louisiana: Louisiana Geol. Survey Bull. 30, p. 47–48.

Jones, R. W., 1959, Origin of salt anticlines of Paradox basin: Am. Assoc. Petroleum Geologists Bull., v. 48, no. 8 (August), p. 1869–1895.

Jones, T. S., 1951, Plainview basin, Texas and New Mexico: Am. Assoc. Petroleum Geologists Bull., v. 35, no. 2 (February), p. 344–352.

Jones, T. S., 1953, Stratigraphy of the Permian Basin of West Texas: West Texas Geol. Soc., (February), p. 1–57.

Jones, T. S., and Smith, Harold M., 1965, Relationships of oil composition and stratigraphy in the Permian basin of West Texas and New Mexico: Am. Assoc. Petroleum Geologists Mem. 4, p. 104–224.

Jones, Vern, 1968, Natural gas in Sacramento and northern San Joaquin valleys: Am. Assoc. Petroleum Geologists, Mem. 9, v. 1, p. 597–603.

Jones, Walter B., 1952, Alabama's geology—and summary of state's oil and gas development: Oil and Gas Jour., June 16, p. 328–340.

Jones, Walter B., 1960, Warrior Basin prospects are bright, exploration needed: World Oil, June, p. 89–91.

Jones, Walter B., and McGlamery, Winnie, 1957, Major oil field is seen in Citronelle structure: World Oil, March, p. 99–102.

Jons, Richard D., 1966, Joe Mill Spraberry, southwest Borden

County, *in* Oil and gas fields in West Texas—a symposium: West Texas Geol. Soc., p. 209–212.

Jordan, Louise, 1957, Subsurface stratigraphic names of Oklahoma *in* Guidebook VI: Oklahoma Geol. Survey.

Jordan, Louise, 1962, Oklahoma's Kingfisher and Blaine Counties keep drillers interested: Oil and Gas Jour., June 18, p. 154–158.

Jordan, Louise, 1963, Statistics of Oklahoma's petroleum industry, 1962: Oklahoma Geol. Survey. *in* Oklahoma Geology Notes, v. 23, no. 6, p. 135–140.

Jordan, Louise, Pate, J. Durwood, and Williamson, Sidney R., 1959, Petroleum geology of Harper County: Oklahoma Geol. Survey Bull. 80, p. 69–92.

Kansas Geological Society, 1956, 1959, 1960, 1965, Kansas oil and gas fields: v. 1, South central Kansas, 1956; v. 2, Western Kansas, 1959; v. 3, Northeastern Kansas, 1960; v. 4, 1965.

Kansas Geological Survey, 1927, Oil and gas resources of Kansas, Anderson County: Bull. 6, pt. 7.

Kansas Geological Survey (State Geological Survey of Kansas), Oil and gas developments in Kansas: Annual publication.

Karges, H. E., 1968, Pelahatchie; the geologist's Jurassic problem child: Oil and Gas Jour., Dec. 23, p. 100–102.

Keahey, Robert, 1962, Fashing field, Atascosa-Karnes Counties, Texas: Gulf Coast Assoc. Geol. Socs. Trans., v. 12, p. 205–208.

Keahey, Robert, 1968, Fashing field, Atascosa-Karnes Counties, Texas: Am. Assoc. Petroleum Geologists Mem. 9, v. 1, p. 976–981.

Keefer, William R., 1965, Geologic history of Wind River basin, central Wyoming: Am. Assoc. Petroleum Geologists Bull., v. 49, no. 11 (November), p. 1878–1892.

Keefer, William R., and Troyer, Max L., 1956, Stratigraphy of the Upper Cretaceous and Lower Tertiary rocks of the Shotgun Butte area, Fremont County, Wyoming: U. S. Geol. Survey Oil and Gas Inv., Chart OC-56.

Keener, M. H. L., 1957, Dollarhide field *in* Occurrence of oil and gas in west Texas: Univ. Texas Pub. 5716, p. 87–96.

Keller, E. R., 1961, Clay Basin field *in* The oil and gas fields of Utah—a symposium: Intermountain Assoc. Petroleum Geologists.

Kelly, Thomas E., 1963, Geology and hydrocarbons in Cook Inlet basin, Alaska, *in* Backbone of the Americas: Am. Assoc. Petroleum Geologists Mem. 2 (November), p. 278–296.

Kelly, Thomas E., 1968, Gas accumulations in nonmarine strata, Cook Inlet basin, Alaska, *in* Natural gases of North America: Am. Assoc. Petroleum Geologists Mem. 9, v. 1, p. 49–64.

Kendrick, Frank E., and McLaughlin, H. C., 1929, Relation of petroleum accumulation to structure, Petrolia field, Clay County, Texas, *in* Structure of typical American oil fields: Am. Assoc. Petroleum Geologists, v. 2, p. 542–555.

Kendrick, Frank E., and Russell, Philip G., 1935, Natural gas in Bend Arch district, Texas, *in* Geology of natural gas: Am. Assoc. Petroleum Geologists, p. 609–649.

Kiddoo, Gordon, 1968, Peters field *in* Oil and gas fields—a symposium: Michigan Basin Geol. Soc., p. 131–136.

Kilkenny, John E., 1962, Sedimentary basins and exploration for oil and gas in California, (abs.): Am. Assoc. Petrol. Geol. Bull., v. 46, no. 2, (February), p. 110.

Kilkenny, John E., 1967, Asphalt Jungle today (abs.): Am. Assoc. Petrol. Geol. Bull. v. 51, no. 3, pt. 1 (March), p. 472.

Kilkenny, John E., and Van Amringe, John H., 1965, Exploring the "asphalt jungle" (abs.): Am. Assoc. Petroleum Geologists Bull., v. 49, no. 7 (July), p. 1085.

Kinard, John C., 1964, Recent Mission Canyon discoveries in the Roth-N.E. Landa area, Bottineau County, North Dakota, *in* Third internat. Williston basin symposium, Proc.: Billings Geol. Soc., North Dakota Geol. Soc., and Saskatchewan Geol. Soc., p. 241–245.

King, P. B., 1966, Geologic history of California: California Div. Mines and Geology Bull. 191, p. 27–34.

King, Vernon L., and Wengerd, Sherman A., 1957, Hospah oil field, McKinley County, New Mexico *in* Geology of southwestern San Juan basin, 2nd Field Conf.: Four Corners Geol. Soc., p. 155–168.

Klinger, Edgar D., 1941, Cross Cut–Blake district, Brown County, Texas, *in* Stratigraphic type oil fields: Am. Assoc. Petroleum Geologists, p. 548–563.

Klinger, Robert, and Ash, Richard, 1967, Geologic studies can project stratigraphic trap trends: World Oil, Feb. 1, p. 65–70.

Knapp, R. W., 1962, Person field: Gulf Coast Assoc. Geol. Socs. Trans., v. 12, p. 199–204.

Knebel, Robert M., and Jones, Franklin C., 1957, Edwards limestone fields of Atascosa County, Texas: Gulf Coast Assoc. Geol. Socs. Trans., v. 7, p. 35–37.

Knight, Wilbur H., 1954, Muldon field, Monroe County, Mississippi: Gulf Coast Assoc. Geol. Socs. Trans., v. 4, p. 111–119.

Knight, Wilbur H., 1955, Muldon: where Black Warrier pays off: Oil and Gas Jour., Jan. 25, p. 158–161.

Knight, William V., and Budd, Harrell, 1959, Horseshoe Canyon Gallup field; a preliminary report *in* Geological record, Am. Assoc. Petroleum Geologists, Rocky Mtn. Sec., p. 107–111.

Koch, John G., 1966, Late Mesozoic stratigraphy and tectonic history, Port Orford–Gold Beach area, southwestern Oregon coast: Am. Assoc. Petroleum Geologists Bull., v. 50, no. 1 (January), p. 25–71.

Koinm, David N., and Dickey, Parke A., 1967, Growth faulting in McAlester basin of Oklahoma: Am. Assoc. Petroleum Geologists Bull., v. 51, no. 5, (May), p. 710–718.

Kornfeld, Joseph A., 1954, New pays stimulate Hamilton dome production, Oil and Gas Jour., Jan. 11, p. 84–87.

Kornfeld, Joseph A., 1958, East Texas basin offers new type fields: World Oil, January, p. 104–110.

Kornfeld, Joseph A., 1963, Iowa's first oil well may spur three-state play: World Oil, April, p. 137–139.

Kornfeld, Joseph A., 1968, New Mississippian strikes spark southwest Kansas activity: World Oil, Feb. 1, p. 62–64.

Kornfeld, Joseph A., and Travis, Maury M., 1968, Arizona–New Mexico play expands: World Oil, January, p. 79–86.

Kornfeld, M. M., and Steinberger, C. R., 1941, Edna gas field, Jackson County, Texas: Am. Assoc. Petroleum Geologists Bull., v. 25 (January), p. 104–119.

Kottlowski, Frank E., 1965, Sedimentary basins of south-central and southwestern New Mexico: Am. Assoc. Petroleum Geologists Bull., v. 49, no. 1 (January), p. 2120–2139.

Krantzler, Irvin, 1966, Origin of oil in lower member of Tyler Formation of central Montana: Am. Assoc. Petroleum Geologists Bull., v. 50 (October), p. 2245–2259.

Kreidler, W. L., 1953, History, geology, and future possibilities of gas and oil in New York State: New York State Mus. Circ. 33 (September).

Kreidler, W. L., 1957, New York's geology: Oil and Gas Jour., Oct. 7, p. 228–233.

Kreidler, W. L., 1968, Map of oil and gas fields in southern tier of counties: N. Y. Geol. Survey.

Kreidler, W. L., Van Tyne, Arthur M., Richards, Horace G., and Grow, George C., Jr., 1968, Oil and gas developments in New York, New Jersey, and New England during 1967: Am. Assoc. Petroleum Geologists Bull., v. 52, no. 6 (June), p. 940–944.

Krey, Max, 1962, North flank Uncompahgre arch, Mesa and Garfield Counties, Colorado, *in* Exploration for oil and gas in northwestern Colorado: Rocky Mtn. Assoc. Geologists, Denver, Colorado, p. 111–113.

Krueger, M. L., 1960, Occurrence of natural gas in the western part of Green River basin *in* Wyoming Geol. Assoc. Guidebook 15th Ann. Field Conf.

Krueger, M. L., 1968, Occurrence of natural gas in Green River Basin, Wyoming: Am. Assoc. Petroleum Geologists Mem. 9, v. 1, p. 780–797.

Krusekopf, H. H., Jr. 1959, Salt domes of East Texas basin, Oil and Gas Jour., May 4, p. 143–147.

Krusekopf, H. H., Jr., 1968, Bethel field, Anderson County, Texas: Am. Assoc. Petroleum Geologists Mem. 9, v. 1, p. 1008–1012.

Kunkel, Robert P., 1961, Mexican Hat field *in* The oil and gas fields of Utah—a symposium: Intermountain Assoc. Petroleum Geologists.

Kunkel, Robert P., Reese, Douglas, and Elias, David W., 1968, Developments in Four Corners–Intermountain area in 1967: Am. Assoc. Petroleum Geologists Bull., v. 52, no. 6 (June), p. 1042–1046.

Kvenvolden, Keith A., and Squires, Rodney M., 1967, Carbon isotopic composition of crude oils from Ellenburger Group (Lower Ordovician), Permian basin, West Texas and Eastern New Mexico: Am. Assoc. Petroleum Geologists Bull., v. 51, no. 7 (July), p. 1293–1303.

Ladd, J. B., 1957, The Wilson Creek field, Rio Blanco County, Colorado: Jour. Petroleum Technology, v. 9, no. 8 (August), p. 23–27.

Lafayette Geological Society, 1963, Stratigraphic factors in southwestern Louisiana: Gulf Coast Assoc. Geol. Socs. Trans., v. 13, p. 9.

Lafayette Geological Society, 1964, Typical oil and gas fields of southwestern Louisiana: Lafayette, Louisiana, Supplement, 1967.

Lafayette and New Orleans Geological Societies, 1968, Geology of natural gas in south Louisiana: Am. Assoc. Petroleum Geologists Mem. 9, p. 376–581.

Lahee, Frederic H., 1929, Oil and gas fields of the Mexia and Tehuacana fault zones, Texas, *in* Structure of typical American oil fields: Am. Assoc. Petroleum Geologists, v. 1, p. 304–388.

Laing, William E., 1963, The Meramec trend: Oil and Gas Jour., July 29, p. 210–216.

Laird, Wilson M., 1956, The Williston basin—a backward look with a view to the future *in* First internat. Williston basin symposium: North Dakota Geol. Soc. and Saskatchewan Geol. Soc., p. 14–22.

Laird, Wilson M., and Folsom, Clarence B., Jr., 1956, North Dakota's Nesson anticline: North Dakota Geol. Survey Rept. Inv. 22.

Landes, Kenneth K., 1944, Porter oil field, Midland County, Michigan: Am. Assoc. Petroleum Geologists Bull., v. 28 (February), p. 173–196.

Landes, Kenneth K., 1946, Porosity through dolomitization: Am. Assoc. Petroleum Geologists Bull., v. 30 (March), p. 505–518.

Landes, Kenneth K., 1948, Deep River oil field, Arenac County, Michigan, *in* Structure of typical American oil fields: Am. Assoc. Petroleum Geologists, v. 3, p. 299–304.

Landes, Kenneth K., 1959, Petroleum geology, 2nd ed.: New York, John Wiley, p. 6.

Landes, Kenneth K., 1967, Eometamorphism, and oil and gas in time and space: Am. Assoc. Petroleum Geologists., v. 51, no. 6 (June), p. 828–841.

Landes, Kenneth K., Amoruso, J. J., Charlesworth, L. J., Jr., Heany, F., and Lesperance, P. J. [or et al.,], 1960, Petroleum resources in basement rocks: Am. Assoc. Petroleum Geologists Bull., v. 44, no. 10 (October), p. 1682–1691.

Landes, Kenneth K., and Jewett, John M., 1939, Oil and gas seeps in Smith County, Kansas: Kansas Geol. Survey Mineral Resources Circ. 12.

Lang, Robert C., III, 1951, A geological cross section from the Wichita Mountains to the Elk City pool: Shale Shaker, v. 1 (March), p. 6–7.

Langenbahn, W. E., 1968, Hamon, Toro, Rojo Caballos West fields, Reeves and Pecos Counties, Texas, *in* Oil and gas fields in West Texas, Supplement, West Texas Geol. Soc.

Langstroth, Ward T., 1964, Montana side (of Canadian border) lacks big production (in Disturbed Belt): Oil and Gas Jour., Apr. 13, p. 144–148.

Largent, B. C., 1968, Burbank field, Oklahoma—a giant grows (abs.): Am. Assoc. Petroleum Geologists Bull., v. 52, no. 3 (March), p. 537–538.

Larson, Wilbert S., 1962, Ackman field and environs, southwestern Nebraska: Am. Assoc. Petroleum Geologists Bull., v. 46 (November), p. 2079–2089.

Latham, Jack W., 1968, Petroleum geology of Arbuckle Group (Ordovician), Healdton field, Carter County, Oklahoma: Am. Assoc. Petroleum Geologists Bull., v. 52, no. 1 (January), p. 3–20.

Latimer, I. S., 1968, West Virginia, *in* Am. Assoc. Petroleum Geologists Bull., v. 52, no. 6 (June), p. 951–959.

Latimer, J. R., Jr., 1965, Fairway, a lesson in unitization: Oil and Gas Jour., Dec. 6, p. 114–120.

Latimer, J. R., Jr., and Oliver, Fred L., 1964, The Fairway field of east Texas; its development and efforts towards unitization: Jour. Petroleum Technology, February, p. 133–138.

Lavington, Charles, 1941, Greasewood oil field, Weld County, Colorado, *in* Stratigraphic type oil fields: Am. Assoc. Petroleum Geologists (Spec. Volume), p. 19–42.

Lavington, Charles, 1954, Structure contour map of the Florence field, Plate 2 in pocket *in* Oil and gas fields of Colorado—a symposium: Rocky Mtn. Assoc. Geologists, Denver, Colorado.

Lawry, T. F., and Mast, Richard F., 1968, Waterflood operations: Illinois State Geol. Survey, Illinois Petroleum 89, p. 67–118.

Lawson, Don E., and Smith, Jordan R., 1966, Pennsylvanian and Permian influence on Tensleep oil accumulation, Big Horn basin, Wyoming: Am. Assoc. Petroleum Geologists Bull., v. 50, no. 10 (October), p. 2197–2220.

Leach, Claude E., 1948, Geology of Aliso Canyon field, Los Angeles County, California, *in* Structure of typical American oil fields: Am. Assoc. Petroleum Geologists, v. 3.

Ledingham, Glen W., 1947, Santiago pool, Kern County, California: Am. Assoc. Petroleum Geologists Bull., v. 31, no. 11 (November), p. 2063–2067.

Lee, Lynn K., 1939, Geology of basin fields in southeastern Illinois: Am. Assoc. Petroleum Geologists Bull., v. 23 (October), p. 1493–1506.

Lee, Wallace, and Payne, Thomas G., 1944, McLouth gas and oil field, Jefferson and Leavenworth Counties, Kansas: Kansas Geol. Survey Bull. 53.

Le May, William J., 1960, Abo reef play yielding large oil fields: World Oil, August, p. 91–95.

Le Van, D. C., 1962, Wells drilled for oil and gas in Virginia prior to 1962: Virginia Div. Mineral Resources Rept. 4.

Levorsen, A. I., 1929, Greater Seminole district, Seminole and Pottawatamie Counties, Oklahoma, *in* Structure of typical American oil fields: Am. Assoc. Petroleum Geologists, v. 2, p. 315–361.

Levorsen, A. I., 1958, The mid-continent—a land of geological opportunity: World Oil, September, p. 101–104, 115.

Lewis, Jerome A., 1956, Northern Raton basin, Oil and Gas Jour., May 14, p. 128–139.

Lewis, Paul J., 1958, Generalized stratigraphic section (Outlook area) *in* Montana oil and gas fields—a symposium: Billings Geol. Soc., p. 34.

Lewis, Richard Q., Sr., 1958, Structure of the Elk Ridge–Needles area, San Juan County, Utah, *in* Guidebook 9th Ann. Field Conf.: Intermountain Assoc. Petroleum Geologists, p. 78–85.

Liddle, Ralph A., 1936, The Van oil field, Van Zandt County, Texas, Univ. Texas, Bur. Econ. Geol., Bull., 3601, p. 1–79.

Lillibridge, Harry E., 1950, Ringwood proves major Oklahoma strike: World Oil, December, p. 77 et seq.

Limes, Leonard L., and Stipe, Jack C., 1959, Occurrence of Miocene oil in south Louisiana: Gulf Coast Assoc. Geol. Socs. Trans., v. 9, p. 77.

Lindberg, George D., 1948, Deerfield oil field, Monroe County, Michigan, *in* Structure of typical American oil fields: Am. Assoc. Petroleum Geologists, v. 3, p. 305–318.

Lindberg, George D., 1963?, Future prospects in Michigan: Michigan Assoc. Petroleum Landmen, undated brochure, p. 15–18.

Little, C. E., 1961, Quinduno fields *in* Oil and gas fields of the Texas and Oklahoma Panhandles: Panhandle Geol. Soc., p. 184–196.

Livingston, V. E., Jr., 1958, Oil and gas exploration in Washington, 1900–1957: Washington Div. Mines and Geol. Inf. Circ. 29.

Llewellyn, J. Thomas, 1955, Northwest Sumatra field, Rosebud County, Montana, *in* Geological record: Am. Assoc. Petroleum Geologists, Rocky Mtn. Sec., p. 55–65.

Lochman-Balk, Christina, and Wilson, James Lee, 1967, Stratigraphy of Upper Cambrian-Lower Ordovician subsurface sequence in Williston basin: Am. Assoc. Petroleum Geologists Bull., v. 51, no. 6 (June), p. 882–917.

Lockett, J. R., 1948, Oriskany sand in Ohio: Am. Assoc. Petroleum Geologists Bull., v. 32 (November), p. 2154–2159.

Lockett, Robert, 1968, Production of gas in northern Cincinnati arch province: Am. Assoc. Petroleum Geologists Mem. 9, v. 2, p. 1716–1745.

Lofton, Coleman L., 1962, Sheridan field, Colorado County, Texas, *in* Typical oil and gas fields of southeast Texas: Houston Geol. Soc., p. 192–197.

Logsdon, Truman, and Brown, Alton R., 1967, Hunton—hottest play in Oklahoma: Shale Shaker, v. 18, no. 4 (December), p. 63–70.

Louisiana Geological Survey, 1965, Salt domes (map), Jan. 1.

Love, J. D., McGrew, Paul O., and Thomas, H. D. (1963), Relationship of latest Cretaceous and Tertiary deposition and deformation to oil and gas in Wyoming *in* Backbone of the Americas, Am. Assoc. Petroleum Geologists Mem. 2 (November), p. 196–208.

Lowenstam, H. A., 1948, Marine pool, Madison County, Illinois, Silurian reef producer *in* Structure of typical American oil fields: Am. Assoc. Petroleum Geologists, v. 3, p. 153–188.

Lowenstam, H. A., 1950, Niagaran reefs in Illinois and their relation to oil accumulation: Illinois Geol. Survey Rept. Inv. 145.

Lundy, C. L., 1958, Geology of the Fairbanks pool, Sullivan County, Indiana: M.S. Thesis, Dept. Geol., Univ. Michigan.

Lundy, C. L., 1968, Deep River field *in* Oil and gas fields—a symposium: Michigan Basin Geol. Soc., p. 61–114.

Lyons, Paul L., 1948, Geophysical case history of the Louden field, Illinois, *in* Geophysical case histories: Soc. Exploration Geophysicists, v. 1, 1948, p. 461–470.

Lytle, William S., 1958, Developments in Pennsylvania in 1957: Am. Assoc. Petroleum Geologists Bull., v. 42 (June), p. 1151.

Lytle, William S., 1963, Underground gas storage in Pennsylvania: Pennsylvania Topog. and Geol. Survey Mineral Resource Rept. M46.

Lytle, William S., 1965, Oil and gas geology of the Warren quadrangle, Pennsylvania: Penn. Topog. and Geol. Survey Mineral Resource Rept. M52.

Lytle, William S., and Heeren, Lillian A., 1964, Maps of the oil and gas fields, natural gas pipelines, and oil pipelines of Pennsylvania: Penn. Topog. and Geol. Survey Map 3.

Lytle, William S., et al, 1968, Oil and gas developments in North Carolina, Maryland, Virginia, Pennsylvania, West Virginia, and Ohio: Am. Assoc. Petroleum Geologists Bull., v. 52, no. 6 (June), p. 945–964.

McAusland, Edward R., 1959, In the Niobrara oil may yield to special methods, Oil and Gas Jour., Sept. 14, p. 158 et seq.

McBee, William, Jr., and Vaughan, Leslie G., 1956, Oil fields of the central Muenster—Waurika arch, Jefferson County, Oklahoma, and Montague County, Texas *in* Petroleum geology of southern Oklahoma: Am. Assoc. Petroleum Geologists, v. 1, p. 355–372.

McCabe, W. S., 1948, Elk Basin anticline, Park County, Wyoming, and Carbon County, Montana, Am. Assoc. Petroleum Geologists Bull., v. 32 (January), p. 52–67.

McCallum, H. D., 1933, Darst Creek oil field, Guadalupe County, Texas: Am. Assoc. Petroleum Geologists Bull., v. 17 (January), p. 16–37.

McCormick, George R., 1961, Petrology of Precambrian rocks of Ohio: Ohio Geol. Survey Rept. Inv. 41.

McCoy, Alex W., III, et al., 1951, Types of oil and gas traps in Rocky Mountain region: Am. Assoc. Petroleum Geologists Bull., v. 35 (May), p. 1000–1037.

McCracken, Earl, 1955, Northeast Missouri's oil possibilities improve: Oil and Gas Jour., January 3, p. 154–155.

McCrae, R. O., and Swenson, R. E., 1968, Geology and natural gas occurrence, western Williston basin: Am. Assoc. Petroleum Geologists Mem. 9, v. 2, p. 1288–1303.

McFarland, P. W., 1929, Laredo district, Texas, *in* Structure of typical American oil fields: Am. Assoc. Petroleum Geologists, v. 1, p. 389–408.

McGee, D. A., and Clawson, W. W., Jr., 1932, Geology and development of Oklahoma City field, Oklahoma County, Oklahoma: Am. Assoc. Petroleum Geologists Bull., v. 16 (October), p. 957–1020.

McGee, D. A., and Jenkins, H. D., 1946, West Edmond oil field, central Oklahoma: Am. Assoc. Petroleum Geologists Bull., v. 30 (November), p. 1797–1829.

McGinnis, C. J., 1958, Habitat of oil in the Denver basin *in* Habitat of oil: Am. Assoc. Petroleum Geologists (Spec. Volume), p. 328–343.

McGregor, Alexander A., and Biggs, Charles A., 1968, Bell Creek field, Montana, a rich stratigraphic trap: Am. Assoc. Petroleum Geologists Bull., v. 52, no. 10 (October), p. 1869–1887.

McGrew, Alan R., 1968, Developments in Upper Gulf Coast in 1967: Am. Assoc. Petroleum Geologists Bull., v. 52, no. 6 (June), p. 1016–1019.

McGuire, William H., 1965a, Porosity zones in the Knox Dolomite of northeastern Kentucky: Kentucky Geol. Survey, Ser. X, Spec. Pub. 10, p. 94–96.

McGuire, William H., 1965b, Relationship of oil and gas production to major gravity anomalies in western Kentucky: Kentucky Geol. Survey, Ser. X, Spec. Pub. 10, p. 65–69.

McGuire, William H., 1968, Exploration objectives in the Cambro-Ordovician of Kentucky: Kentucky Geol. Survey, Ser. X, Spec. Pub. 15, p. 64–69.

McInnis, A. B., and Hartman, F. H., 1957, McElroy field, *in* Occurrence of oil and gas in west Texas: Univ. Texas Pub. 5716, p. 203.

McKee, Edwin D., 1951, Sedimentary basins of Arizona and adjoining areas: Geol. Soc. America Bull., v. 62 (May), p. 481–506.

McKenny, Jere W., and Masters, John A., 1968, Dineh-bi-Keyah field, Apache County, Arizona: Am. Assoc. Petroleum Geologists Bull., v. 52, no. 10 (October), p. 2045–2057.

MacKenzie, David B., 1965, Depositional environments of Muddy Sandstone, western Denver basin, Colorado: Am. Assoc. Petroleum Geologists Bull., v. 49, no. 2 (February), p. 186–206.

McLellan, H. J., 1946, Raccoon Bend salt dome, Austin County, Texas: Am. Assoc. Petroleum Geologists Bull., v. 30 (August), p. 1306–1307.

McLellan, H. J., Wendtland, E. A., and Murchison, E. A., 1932, Boggy Creek salt dome, Anderson and Cherokee Counties, Texas: Am. Assoc. Petroleum Geologists Bull., v. 16 (June), p. 584–600.

McMasters, John H., 1948, Cymric oil field, Kern County, California, *in* Structure of typical American oil fields: Am. Assoc. Petroleum Geologists, v. 3, p. 38–57.

McMinn, P. M., and Patton, H. L., 1962, Rangely field, Rio Blanco County, Colorado, *in* Exploration for oil and gas in northwestern Colorado: Rocky Mtn. Assoc. Geologists, Denver, Colo., p. 104–107.

McMurtry, R. P., 1966, Brown-Bassett, *in* Oil and gas fields in West Texas: West Texas Geol. Soc., p. 83–87.

McNeal, Robert P., 1953, Subsurface geology of south-central Anardarko basin, Oklahoma: Am. Assoc. Petroleum Geologists Bull., v. 37 (December), p. 2677–2704.

McNeal, Robert P., 1961, Hydrodynamic entrapment of oil and gas in Bisti field, San Juan County, New Mexico: Am. Assoc. Petroleum Geologists Bull., v. 45, no. 3 (March), p. 315–329.

McNeil, Harold E., 1941, Wherry pool, Rice County, Kansas, *in* Stratigraphic type oil fields: Am. Assoc. Petroleum Geologists, p. 118–138.

McPeters, K. D., and Kelly, J. M., 1956, Hobbs Grayburg–San Andres field *in* Oil and gas fields of southeastern New Mexico: Roswell Geol. Soc., p. 207.

McQueen, Henry S., 1960, Deep Wilcox, Frio, and Edwards spur upper Gulf coast activity: World Oil, November, p. 119–122, 126.

MacQuown, W. C., Jr., and Millikan, W. E., 1955, Little Beaver, Badger Creek, Middlemist field area, Colorado: Am. Assoc. Petroleum Geologists Bull., v. 39, no. 5 (May), p. 630–648.

Mackey, F. L., and Bowles, Jack P. F., Jr., 1963, Oil and gas developments in Oklahoma during 1962: Am. Assoc. Petroleum Geologists Bull., v. 47, no. 6 (June), p. 1013–1034.

Magbee, Byron D., and Alkire, Robert L., 1954, Oil and gas in Morgan County: Ohio Geol. Survey Rept. Inv. 44, p. 8.

Maher, John C., ed., 1960, Stratigraphic cross section of Paleozoic rocks, West Texas to northern Montana: Comm. Stratigraphic Correlations, American Assoc. Petroleum Geologists, 18 p. plus 5 plates.

Maher, John C., 1967, Geologic framework and petroleum potential of the Atlantic Coast: Tulsa Geol. Soc. Digest, v. 35, p. 278–283.

Mais, William R., 1957, Peripheral faulting at Bayou Blue salt dome, Iberville Parish, Louisiana: Am. Assoc. Petroleum Geologists Bull., v. 41 (September), p. 1915–1951.

Makins, Robert L., 1968, Developments in North-Central Texas in 1967: Am. Assoc. Petroleum Geologists, v. 52, no. 6 (June), p. 997–1000.

Mallory, R. W., 1949, The Salt Creek oil field *in* Wyoming Geol. Assoc. Guidebook 4th Ann. Field Conf., p. 89–91.

Mallory, William W., 1949, Rocky Mountain type structure at Velma pool, Stephens County, Oklahoma: World Oil, July 1, p. 68–78.

Manlove, Charles, 1943, West Cat Canyon area of the Cat Canyon oil field, *in* California Div. Mines Bull. 118, p. 432–434.

Manlove, Charles F., 1968, Vernalis gas field, San Joaquin County, California: Am. Assoc. Petroleum Geologists Mem. 9, v. 1, p. 663–667.

Mapel, W. J., and Pillmore, C. L., 1963, Geology of the Newcastle area, Weston County, Wyoming: U. S. Geol. Survey Bull. 1141–N, p. N1–N85.

Marathon Oil Company, 1966, Yates revisited: Marathon World, winter; reprinted with permission in Ind. Petroleum Monthly, September, 1966, p. 24–26.

Marchant, L. C., 1966, Nitrogen gas in five oil fields on the Nesson anticline in North Dakota: U. S. Bur. Mines Rept. Inv. 6848.

Markham, E. O., and Lamar, L. C., 1937, South Burbank pool, Osage County, Oklahoma: Am. Assoc. Petroleum Geologists Bull., v. 21 (May), p. 560–579.

Markley, L. C., 1959, West Frederick field, Tillman County, Oklahoma, *in* Petroleum geology of southern Oklahoma: Am. Assoc. Petroleum Geologists, v. 2, p. 180–186.

Marsh, Owen T., 1967, Evidence of deep salt deposits in western Florida Panhandle: Am. Assoc. Petroleum Geologists Bull., v. 51, no. 2 (February), p. 212–222.

Marshall, Joseph W., 1952, Spraberry reservoir of West Texas: Am. Assoc. Petroleum Geologists Bull., v. 36, no. 11 (November), p. 2189–2191.

Marshall, L. R., and Foltz, G. A., 1960, Justis field *in* Oil and gas fields of southeastern New Mexico: Roswell Geol. Soc., p. 110–123.

Martin, Charles A., 1965, Denver basin: Am. Assoc. Petroleum Geologists Bull., v. 49, no. 11 (November), p. 1908–1925.

Martin, J. Stewart, 1968, Developments in West Texas and eastern New Mexico in 1967: Am. Assoc. Petroleum Geologists Bull., v. 52, no. 6 (June), p. 990–996.

Martin, James L., Hough, Leo W., Raggio, David L., and Sandberg, Adolph, 1954, Geology of Webster Parish: Louisiana Geol. Survey Bull. 29, p. 146–176.

Martyn, Phillip F., 1938, Refugio oil and gas field, Refugio County, Texas: Am. Assoc. Petroleum Geologists Bull., v. 22 (September), p. 1184–1216.

Mason, John W., 1968, Hugoton Panhandle field, Kansas, Oklahoma, and Texas: Am. Assoc. Petroleum Geologists Mem. 9, v. 2, p. 1539–1547.

Masters, C. D., 1968, Geology of natural gas occurrence in Tertiary and Late Cretaceous rocks of Sand Wash basin, Colorado and Wyoming: Am. Assoc. Petroleum Geologists Mem. 9, v. 1, p. 866–877.

Maughan, J. B., 1964, A general review of the oil and gas developments in the Province of Quebec: Canadian Mining Metall. Bull., June, p. 641–647.

Maxwell, Riley G., 1952, Katz field, Stonewall County, Texas, *in* Geological contributions: Abilene Geol. Soc., p. 47–50.

May, B. E., 1961, The Desert Springs field *in* Wyoming Geol. Assoc. Guidebook 16th Ann. Field Conf.

May, John C., and Hewitt, Robert L., 1947, The nature of the basement complex oil reservoir (abs.): Am. Assoc. Petroleum Geologists Bull., v. 31, no. 12 (December), p. 2239–2240; also California Jour. Mines and Geology, April, 1948, p. 129–158.

Meade, Wade, 1964, Stratigraphy of Lincoln Parish, Louisiana: Am. Assoc. Petroleum Geologists Bull., v. 48 (May), p. 718–721.

Meents, Wayne F., 1958, Fishhook gas pool: Illinois Geol. Survey Circ. 25A, p. 1–15.

Meents, Wayne F., 1968, Illinois glacial drift gas: Am. Assoc. Petroleum Geologists Mem. 9, v. 2, p. 1754–1758.

Meents, Wayne F., and Swann, David H., 1965, Grand Tower limestone (Devonian) of southern Illinois: Illinois Geol. Survey Circ. 389.

Mees, E. C., and Bowers, G. F., 1952, Gebo field, Hot Springs County, Wyoming: Oil and Gas Jour., Nov. 24, p. 125–126.

Melchior, Louis F., 1953, The geophysical discovery and development of the Bayou Couba dome, Louisiana: Geophysics, v. 18 (April), p. 371–382.

Mellen, F. F., 1947, Black Warrior basin, Alabama and Mississippi: Am. Assoc. Petroleum Geologists Bull., v. 31 (October), p. 1801–1816.

Mellen, F. F., 1966, Jurassic exploration warms Southeast: Oil and Gas Jour., Dec. 19, p. 96–100.

Merriam, Daniel F., 1958, Cretaceous possibilities good for northwest Kansas: Oil and Gas Jour., Mar. 31, p. 138–142.

Merriam, Daniel F., 1960, Drillers keep busy in northern Kansas: Oil and Gas Jour., Oct. 17, p. 164–170.

Merriam, Daniel F., 1963, The geologic history of Kansas: State Geol. Survey of Kansas Bull. 162.

Merriam, Daniel F., and Goebel, Edwin D., 1954, The geology of the Norton oil field, Norton County, Kansas: Kansas Geol. Survey Bull. 109, pt. 9, p. 125–152.

Merriam, Daniel F., and Goebel, Edwin D., 1956, Kansas' structural provinces offer varied types of traps: Oil and Gas Jour., Apr. 30, p. 141–154.

Merriam, Daniel F., and Goebel, Edwin D., 1959, Where's the oil in Kansas?: Oil and Gas Jour., Mar. 9, p. 212–218.

Merriam, Daniel F., and Goebel, Edwin D., 1960, Kansas' oil, gas-field discoveries continue to mount: Oil and Gas Jour., Mar. 28, p. 250–253.

Merriam, Daniel F., and Goebel, Edwin D., 1968, Occurrence of natural gas in Kansas: Am. Assoc. Petroleum Geologists Mem. 9, v. 2, p. 1548–1557.

Metzger, William J., 1965, Pennsylvania stratigraphy of the Warrior basin, Alabama: Geol. Survey Alabama Cir. 30.

Meyer, Richard F., 1966, Geology of Pennsylvanian and Wolfcampian rocks in southeast New Mexico: New Mexico Bur. Mines Mem. 17.

Meyerhoff, A. A., 1967, Future hydrocarbon provinces of Gulf of Mexico–Caribbean region: Gulf Coast Assoc. Geol. Socs. Trans., v. 17, p. 217–260.

Michaux, Frank W., Jr., and Buck, E. O., 1936, Conroe oil field, Montgomery County, Texas, *in* Gulf Coast oil fields: Am. Assoc. Petroleum Geologists, p. 789–832.

Michigan Basin Geological Society, 1968, Oil and gas fields—a symposium: Lansing, Michigan.

Michigan Geological Survey, 1964, Stratigraphic succession in Michigan: Michigan Geol. Survey Chart 1.

Michigan Geological Survey, 1968, Michigan's oil and gas fields, 1967: Michigan Geol. Survey Ann. Statistical Summ. 8.

Milhous, H. C., 1960, Tennessee, key development years just ahead: World Oil, June, p. 92–95.

Milks, Roy, 1956, Vacuum Grayburg-San Andres field *in* Oil and gas fields of southeastern New Mexico: Roswell Geol. Soc., p. 33.

Miller, D. N., Jr., Barlow, James A., Jr., and Haun, John D., 1965, Stratigraphy and petroleum potential of latest Cretaceous rocks, Big Horn Basin, Wyoming: Am. Assoc. Petroleum Geologists Bull., v. 49 (March), p. 277–285.

Miller, Don J., 1957, Geology of the southeastern part of the Robinson Mountains, Yakataga District, Alaska: U.S. Geol. Survey Oil and Gas Inv. Map OM-187.

Miller, Don J., 1959, Southern Alaska *in* U. S. Geol. Survey Bull. 1094, p. 8–53.

Miller, Don J., Payne, Thomas G., and Gryc, George, 1959, Geology of possible petroleum provinces in Alaska: U. S. Geol. Survey Bull. 1094.

Miller, Floyd, 1963, Cretaceous oil in the Denver basin *in* Geology of the northern Denver basin and adjacent uplift: Rocky Mtn. Assoc. Geologists, p. 211–214.

Miller, H. O., 1953, San Ardo field, a history of development: Jour. Petroleum Technology, January, sec. 1, p. 11–13; sec. 2, p. 5.

Miller, Lee S., 1954, Michigan hopes rise with three 1954 discoveries: Oil and Gas Jour., Oct. 4, p. 204–207.

Miller, Ralph L., 1948, Rose Hill oil field, Lee County, Virginia, *in* Structure of typical American oil fields: Am. Assoc. Petroleum Geologists, v. 3, p. 452–479.

Miller, Ralph L., 1954, Rose Hill field—unique oil occurrence?: Oil and Gas Jour., Aug. 2, p. 138–139.

Miller, Ralph L., 1956, Carbonate cementation as a sealing factor in oil accumulation, southwest Virginia: Am. Assoc. Petroleum Geologists Ann. Meeting Program, p. 38–39.

Miller, Ralph L., and Brosgé, William P., 1954, Geology and oil resources of the Jonesville district, Lee County, Virginia: U. S. Geol. Survey Bull. 990, 240 p.

Miller, Ralph L., and Fuller, J. Osborn, 1954, Geology and oil resources of the Rose Hill district—the fenster area of the Cumberland overthrust block—Lee County, Virginia: Virginia Geol. Survey Bull. 71, p. 1–383.

Miller, Robert R., 1968, Geology of Otis–Albert field, Rush and Barton Counties, Kansas: Am. Assoc. Petroleum Geologists Mem. 9, v. 2, p. 1588–1615.

Millison, Clark, 1963, Gas occurrence in Piceance basin, Colorado: Am. Assoc. Petroleum Geologists Mem. 2, p. 180–187.

Millison, Clark, 1965, Powder Wash field, Moffat County, Colorado: Mtn. Geologist, v. 2, no. 3, p. 173–179.

Millison, Clark, 1968, Gas occurrence in Upper Cretaceous and Tertiary rocks of Piceance basin, Colorado: Am. Assoc. Petroleum Geologists Mem. 9, v. 1, p. 878–898.

Minor, H. E., 1925, Goose Creek oil field, Harris County, Texas: Am. Assoc. Petroleum Geol. Bull., v. 9 (March–April), p. 286–297.

Minor, H. E. and Hanna, Marcus A., 1941, East Texas oil field, Rusk et al. Counties, Texas, *in* Stratigraphic type oil fields: Am. Assoc. Petroleum Geologists, p. 600–640.

Mississippi Geological Society, 1952, Wilcox oil fields—southern Mississippi and adjacent areas: Jackson, Mississippi.

Mississippi Geological Society, 1957, Mesozoic-Paleozoic producing areas of Mississippi and Alabama: v. 1, Jackson, Mississippi.

Mitchell, S. D., 1968, Geology and occurrence of natural gas in Laramie and Hanna basins, Wyoming: Am. Assoc. Petroleum Geologists Mem. 9, v. 1, p. 828–839.

Monahan, Rex, and Rutledge, Robert D., 1959, Mid–West's newest oil province: Oil and Gas Jour., Sept. 14, p. 167–174.

Monley, L. E., 1954, Greenwood extension *in* Oil and gas fields of Colorado—a symposium: Rocky Mountain Assoc. Geologists p. 158–159.

Montgomery, R. F., 1965, The oil and gas resources of southeastern New Mexico *in* Mineral and water resources of New Mexico: U. S. Senate Rept. Comm. Interior and Insular Affairs, 89th Cong., 1st sess., p. 74–86.

Moody, Clarence L., 1967, Gulf of Mexico distributive province: Am. Assoc. Petroleum Geologists Bull., v. 51, no. 2 (February), p. 179–199.

Moody, Graham B., 1954, Oil in California: Am. Assoc. Petroleum Geologists Bull., v. 38, no. 9 (September), p. 1888–1910.

Mooney, Tom D., 1966, Howard–Glasscock field, Howard and Glasscock Counties, *in* Oil and gas fields in West Texas—a symposium: West Texas Geol. Soc., p. 183–191.

Moore, Carl. A., 1957, Pinpointing panhandle possibilities: World Oil, September, p. 83–88.

Moore, Donald B., 1967, Subsurface geology of the Gilbertown oil field: Geol. Survey Alabama Circ. 35.

Moore, Hastings, 1951, Merigale–Paul field, Wood County, Texas: Univ. Texas Pub. 5116, p. 223–230.

Moore, J. H., 1944, Fullerton pool, Andrews County, Texas: Am. Assoc. Petroleum Geol. Bull., v. 28, no. 10 (October), p. 1541–42.

Moore, Raymond C., Frye, John C., Jewett, J. M., Lee, Wallace, and O'Connor, Howard G., 1951, The Kansas rock column: Kansas Geol. Survey Bull. 89.

Morgan, A. L., 1953, Structural analysis of Delta Farms field, Lafourche Parish, Louisiana: Am. Assoc. Petroleum Geologists Bull., v. 37 (December), p. 2649–2676.

Morgan, Donald A., 1952, Reef pools sought in Illinois: World Oil, July 1, p. 84–86.

Morgan, F. W., and Torline, M. E., 1964, Kansas' Burgess sand, elusive but profitable: Oil and Gas Jour., Feb. 3, p. 112–115.

Morgan, Henry J., Jr., 1965, Ouachita folded belt in Dallas County, *in* The geology of Dallas County—a symposium: Dallas Geol. Soc., p. 190–192.

Moritz, Carl A., 1948, Oil and gas possibilities of southwestern Tennessee: World Oil, v. 128 (May), p. 160–170.

Morrisey, Norman S., 1956, Cherokee Trend stretches out thirty miles: Oil and Gas Jour., May 7, p. 176–177.

Morrow, Ernest H., Skinner, John H., and Hinyard, Paul, 1968, Developments in Arkansas, North Louisiana, and East Texas in 1967: Am. Assoc. Petroleum Geologists, v. 52, no. 6 (June), p. 1006–1015.

Mullins, Allan T., 1968, Pennsylvanian structure of western Kentucky, *in* Kentucky Geol. Survey, Ser. X, Spec. Pub. 15. p. 81–86.

Mullins, Claude A., and Ireland, W. C., 1968, Technology gives new life to 30 year old Kansas field: Oil and Gas Jour., Mar. 11, p. 77–80.

Multer, H. Gray, 1963, Geology of the Silurian producing zones in the Moreland oil pool, Wayne County, northeastern Ohio: Ohio Geol. Survey Rept. Inv. 46.

Munn, J. K., and Riddle, B. D., 1957, The Pottsville fields, Hamilton County, Texas, *in* Joint Field Trip Guidebook: Abilene and Forth Worth Geol. Socs., p. 101–103.

Murphy, James K., Purcell, Paul E. M., and Barton, H. E., 1941, Seymour pool, Baylor County, Texas, *in* Stratigraphic type oil fields: Am. Assoc. Petroleum Geologists, p. 760–775.

Murphy, John F., Privrasky, N. C., and Moerlein, G. A., 1956, Geology of the Sheldon–Little Dome Area, Fremont County, Wyoming: U. S. Geol. Survey Oil and Gas Inv. Map OM-181.

Murphy, John F., and Roberts, Ralph W., 1954, Geology of the Steamboat Butte–Pilot Butte area, Fremont County, Wyoming: U. S. Geol. Survey Oil and Gas Inv. Ser. OM-151.

Murray, D. Keith, and Bortz, Louis C., 1967, Eagle Springs oil field, Railroad Valley, Nye County, Nevada: Am. Assoc. Petroleum Geologists Bull., v. 51, no. 10 (October), p. 2133–2145.

Murray, George H., Jr., 1959, Examples of hydro-dynamics in the Williston basin at Poplar and North Tioga fields *in* Geological record: Am. Assoc. Petroleum Geologists, Rocky Mtn. Sec., p. 55–59.

Murray, George H., Jr., 1968, Quantitative fracture study, Sanish pool, McKenzie County, North Dakota: Am. Assoc. Petroleum Geologists Bull., v. 52, no. 1 (January), p. 57–65.

Murray, Grover E., 1961, Geology of the Atlantic and Gulf coastal province of North America: New York, Harper.

Murray, Grover E., 1966, Salt structures of Gulf of Mexico basin—a review: Am. Assoc. Petroleum Geologists Bull., v. 50, no. 3 (March), p. 439–478.

Myers, Donald A., Stafford, Philip T., and Burnside, Robert J., 1956, Geology of the late Paleozoic Horseshoe atoll in West Texas: Texas Univ. Pub. 5607 (April), p. 1–113. Summary in Oil and Gas Jour., June 18, p. 248–252.

Mygdal, Karl A., 1963, Adena—largest field in Denver basin, *in* Geology of the northern Denver basin and adjacent uplifts: Rocky Mtn. Assoc. Geologists, p. 222–225.

Nanz, Robert H., Jr., 1954, Genesis of Oligocene sandstone reservoir, Seeligson field, Jim Wells and Kleberg Counties, Texas: Am. Assoc. Petroleum Geol. Bull., v. 38 (January), p. 96–117.

Neal, Eugene P., 1951, South Ceres, Oklahoma's oddest shoestring field: World Oil, December, p. 92–98.

Nelson, Robert E., 1963, Colgrade field, Winn Parish, Louisiana: Gulf Coast Assoc. Geol. Socs. Trans., v. 13, p. 3–7.

Nering, Lee, 1963, Lyons Sandstone oil production in the northern Denver Basin, Colorado, *in* Geology of the northern Denver basin and adjacent uplifts: Rocky Mtn. Assoc. Geologists, p. 239–247.

Newcomb, J. J., 1967, Elk Valley area, Caribou County, Idaho: *in* Guidebook 15th Ann. Field Conf.: Intermountain Assoc. Geologists, p. 267–270.

Newcombe, R. B., 1932, Geology of Muskegon oil field, Muskegon, Michigan: Am. Assoc. Petroleum Geologists Bull., v. 16 (February), p. 153–168.

Newcombe, R. B., 1934, Structure and accumulation in the Michigan "basin" and its relation to the Cincinnati arch *in* Problems of petroleum geology: Am. Assoc. Petroleum Geologists, p. 531–556.

New Mexico Geological Society, 1965, Guidebook of southwestern New Mexico II, 16th Field Conference, October.

New Orleans Geological Society, 1963, Stratigraphic traps in southeastern Louisiana: Gulf Coast Assoc. Geol. Socs. Trans., v. 13, p. 149.

New Orleans Geological Society, 1965, 1967, Oil and gas fields of southeastern Louisiana: v. 1 (1965) and v. 2 (1967), New Orleans, Louisiana.

Newton, V. C., Jr., 1963, Despite 60 years of dry holes, Oregon drilling continues: Oil and Gas Jour., Mar. 11, p. 131–133.

Newton, V. C., Jr., and Corcoran, R. E., 1963, Petroleum geology of the western Snake River basin: Oregon Dept. Mines and Mining Industries, Oil and Gas Inv. 1, Portland, Oregon.

Nichols, P. H., 1964, The remaining frontiers for exploration in northeastern Texas: Gulf Coast Assoc. Geol. Socs. Trans., v. 14, p. 7–22.

Nichols, P. H., Peterson, G. E., and Wuestner, C. E., 1968, Summary of subsurface geology of Northeast Texas: Am. Assoc. Petroleum Geologists Mem. 9, v. 1, p. 982–1004.

Nicholson, John H., 1960, Geology of the Texas Panhandle, *in* Aspects of the geology of Texas—a symposium: Texas Univ., Bur. Econ. Geol., Pub. 6017, p. 51–64.

Nolte, Cliff J., 1963, Potential stratigraphic accumulation of oil and gas in the Upper Cretaceous of the Denver basin, *in* Geology of the northern Denver basin and adjacent uplifts: Rocky Mtn. Assoc. Geologists, p. 156–161.

Nordquist, J. W., and Leskela, Willard, 1968, Natural gas in Sweetgrass arch area, northwestern Montana: Am. Assoc. Petroleum Geologists Mem. 9, v. 1, p. 736–759.

Norling, Donald L., Bendler, E. Perry, Buschman, William J., Jr., and Johnson, Allan W., 1968, Ohio *in* Am. Assoc. Petroleum Geologists Bull., v. 52, no. 6 (June), p. 959–964.

North Dakota Geological Society, 1959, Nesson anticline of North Dakota: North Dakota Geol. Soc., Bismark, North Dakota.

North Dakota Geological Society, 1962, Oil and gas fields of North Dakota—a symposium: North Dakota Geol. Soc., Bismarck, North Dakota.

North Dakota Geological Society and Saskatchewan Geological Society, 1956, First international Williston basin symposium.

North Dakota Geological Society and Saskatchewan Geological Society, 1958, Second international Williston basin symposium.

North Dakota Geological Survey, 1960, Oil fields in the Burke County area, North Dakota: Rept. Inv. 36.

North Texas Geological Society, 1962, North Texas oil fields—a symposium: North Texas Geol. Soc., Wichita Falls, Texas.

Norwood, E. Earl, 1965, Geological history of central and south-central Montana: Am. Assoc. Petroleum Geologists Bull., v. 49, no. 11 (November), p. 1824–1832.

Nosow, Edmund, 1959, Some Devonian-Silurian stratigraphic relationships on the west flank of the Cincinnati arch: Kentucky Geol. Survey, Ser. X, Spec. Pub. 2, p. 67–81.

Nosow, Edmund, 1960, Oil and gas developments in Kentucky in 1959: Am. Assoc. Petroleum Geologists Bull., v. 44 (June), p. 966.

Nosow, Edmund, 1961, New reservoir possibilities in Kentucky: Kentucky Geol. Survey, Ser. X, Spec. Pub. 4.

Nottingham, Marsh W., 1966. Abo reef buildup provides five stratigraphic trap zones: World Oil, May, p. 107–110.

Nutter, D. S., 1965, Oil and gas development in New Mexico *in* Mineral and water resources of New Mexico: U.S. Senate Rept. Comm. Int. and Insular Affairs, 89th Cong., 1st Sess., p. 41–74; Oil and gas in other parts of New Mexico, p. 98–100.

Oakes, Millis H., 1966, North Cut Bank field and the Moulton sandstone, *in* Guidebook, 17th Ann. Field Conference: Billings Geol. Soc., p. 191–201.

Oakeshott, Gordon B., 1954, Geology of the Placerita oil field, Los Angeles County: California Div. Mines and Geol. Bull. 170, Map Sheet 31.

Oakeshott, Gordon B., 1966, Geology of the California Coast Ranges: California Div. Mines and Geology Bull. 191, p. 36–40.

Oburn, R. C., 1968, North Park, Colorado—an oil and gas province: Am. Assoc. Petroleum Geologists Mem. 9, v. 1, p. 840–855.

Ocamb, Rayburn D., 1961, Growth faults of South Louisiana: Gulf Coast Assoc. Geol. Socs. Trans., v. 11, p. 139–175.

Ocamb, Rayburn D., and Grigg, Robert P., Jr., 1954, The Lewisburg field area, Acadia and St. Landry Parishes, Louisiana: Gulf Coast Assoc. Geol. Socs. Trans., v. 4 (November), p. 183–200.

O'Connor, J. P., Jr., 1955, Good Omen field, Smith County, Texas: Gulf Coast Assoc. Geol. Socs. Trans., v. 5 (October), p. 39–50.

Ogden, Victor, and Locke, Jack, 1952, What are the facts and figures for Spraberry Trend core analyses?: Oil and Gas Jour., Jan. 14, p. 94–97.

Ogle, Burdette A., 1968, Natural gas in Eel River basin, Humboldt County, California: Am. Assoc. Petroleum Geologists Mem. 9, p. 68–75.

Oglesby, Woodson R., 1967, A gravity profile of the south Florida shelf: Gulf Coast Assoc. Geol. Socs. Trans., v. 17, p. 278–286.

Ohio Geological Survey, 1959, Contributions to the oil and gas industry in Ohio: Inf. Circ. 23.

Ohlen, H. R., and McIntyre, L. B., 1965, Stratigraphy and tectonic features of Paradox basin, Four Corners area: Am. Assoc. Petroleum Geologists Bull., v. 49, no. 11 (November), p. 2020–2040.

Oil and Gas Journal, 1938, K.M.A. oil field reports: Jan. 20, p. 42 et seq.

Oil and Gas Journal, 1966, Where remaining reserves are stored: Jan. 31, p. 175.

Oil and Gas Journal, 1968, Forecast-review: Feb. 5, p. 138–167.

Oil and Gas Journal, 1969, Forecast-review: Jan. 27.

Oklahoma City Geological Society, 1963, Oil and gas fields of Oklahoma: Reference Rept., v. 1.

Oklahoma Geological Survey, 1928, Oil and gas in Oklahoma: Bull. 40.

Olson, Charles F., Curry, William, H., III, Norwood, E. Earl, and Slocum, R. C., 1968, Developments in Northern Rockies in 1967: Am. Assoc. Petroleum Geologists Bull., v. 52, no. 6 (June), p. 1033–1041.

O'Rourke, E. V., 1941, Lensing sands of Ohio *in* Stratigraphic type oil fields: Am. Assoc. Petroleum Geologists, p. 382–385.

Orton, Edward, 1888, The Trenton Limestone as a source of oil and gas in Ohio: Ohio Geol. Survey Rept. VI, p. 109.

Osment, F. C., Morrow, R. M., and Craig, R. W., 1967, Petroleum geology and development of the Cook Inlet basin of Alaska: 7th World Petroleum Cong. (Mexico), P.D. no. 2–B, p. 1–20.

Osmond, John C., 1960, Tectonic history of the Basin and Range province in Utah and Nevada: Mining Eng., March, p. 251–265.

Osmond, John C., 1965, Geologic history of site of Uinta basin, Utah: Am. Assoc. Petroleum Geologists Bull., v. 49, no. 11 (November), p. 1957–1973.

Osmond, John C., Locke, Robert, Dillé, Alan C., Praetorious, Walter, and Wilkins, James G., Natural gas in Uinta basin, Utah: Am. Assoc. Petroleum Geologists Mem. 9, v. 1, p. 174–198.

Overby, William K., Jr., 1962, Generalized geologic column for southern West Virginia: West Virginia Geol. Survey Chart.

Owen, Arthur E., and Whitney, George W., 1956, San Arroyo Bar-X area: Intermountain Assoc. Petroleum Geologists, p. 195–198.

Owens, Gordon L., 1967, The Precambrian surface of Ohio: Ohio Geol. Survey Rept. Inv. 64.

Owens, Harold, 1960, Florida-Bahama platform (abs.): Am. Assoc. Petroleum Geologists Bull., v. 44, no. 7 (July), p. 1254.

Oxley, Marvin L., 1959, Bois d'Arc—key to Hunton oil?: Oil and Gas Jour., May 11, p. 162–166.

Paape, D. W., 1968, Geology of Wind River basin of Wyoming and its relationship to natural gas accumulation: Am. Assoc. Petroleum Geologists Mem. 9, v. 1, p. 760–779.

Pack, Robert W., 1920, The Sunset-Midway oil field, California, pt. 1, Geology and oil resources: U. S. Geol. Survey Prof. Paper 116.

Paine, William R., 1962, Geology of Acadia and Jefferson Davis Parishes: Louisiana Geol. Survey Bull. 36.

Pampe, Carl F., 1963, Fort Trinidad's 25-sq.-mile area promises big reserves: Oil and Gas Jour. July 8, p. 162–166.

Pan-American Petroleum Corporation, 1961, Mountain geology: Horizons, v. 11, no. 6 (December), p. 1–3.

Panhandle Geological Society, 1961, Oil and gas fields of the Texas and Oklahoma Panhandles: Amarillo, Texas.

Parker, Frank S., 1954, Origin, migration, and trapping of oil in southern California: California Div. Mines Bull. 170, p. 13–19.

Parker, John M., 1962a, Pre-Pennsylvanian beds yield lion's share of oil in Lisbon area: Oil and Gas Jour., Apr. 16, p. 148–154.

Parker, John M., ed., 1962b, Oil and gas field volume, Colorado-Nebraska: Rocky Mountain Assoc. Geologists.

Parker, John M., 1966, The Bridger Lake field, Summit County, Utah: Mtn. Geologist, v. 3, no. 3, p. 139–146.

Parker, John M., 1968, Lisbon field area, San Juan County, Utah: Am. Assoc. Petroleum Geologists Mem. 9, v. 2, p. 1371–1388.

Parker, J. Wm., and Roberts, J. W., 1963, Devonian and Mississippian stratigraphy of the central part of the Colorado Plateau *in* Shelf carbonates of the Paradox basin—a symposium: Four Corners Geol. Soc. Guidebook, p. 31–60.

Parkinson, L. J., and Helmuth, D. N., 1964, The Swanson River field *in* Oil fields, earthquake, geology: Alaska Geol. Soc., p. 25–27, Anchorage.

Partridge, John F., Jr., 1958, Oil occurrence in Permian, Pennsylvanian, and Mississippian rocks, Big Horn Basin, Wyoming, *in* Habitat of oil: Am. Assoc. Petroleum Geologists (Spec. Volume), p. 293–306.

Paschall, Robert H., 1954, Geology of the Rincon oil field, Ventura County: California Div. Mines and Geol. Bull. 170, Map Sheet 26.

Patchen, Douglas G., 1968a, Oriskany Sandstone–Huntersville Chert gas production in the eastern half of West Virginia: West Virginia Geol. and Econ. Survey Circ. 9.

Patchen, Douglas G., 1968b, Newburg gas development in West Virginia: West Virginia Geol. and Econ. Survey Circ. 6, 4th ed.

Pate, Barney F., 1963, Significant North Louisiana Cotton Valley stratigraphic traps: Gulf Coast Assoc. Geol. Socs. Trans., v. 13, p. 177–183.

Pate, J. Durwood, 1959, Stratigraphic traps along north shelf of Anadarko basin, Oklahoma: Am. Assoc. Petroleum Geologists Bull., v. 43 (January), p. 39–59.

Pate, J. Durwood, 1962, Laverne gas area—four story stratigraphic trap: Tulsa Geol. Soc. Digest, v. 30, p. 65–81.

Pate, J. Durwood, 1968, Laverne gas area, Beaver and Harper Counties, Oklahoma: Am. Assoc. Petroleum Geologists Mem. 9, v. 2, p. 1509–1524.

Pate, Joe H., 1953, Subsurface geology of Carter-Knox oil field, Stephens and Grady Counties, Oklahoma: Shale Shaker, v. 4 (October), p. 4–29.

Patton, J. B., and Dawson, T. A., 1968, Paper presented before tri-state IPAA-IOPLA meeting in Evansville as reported by John C. McCaslin: Oil and Gas Jour., Sept. 9, p. 165.

Patton, W. W., Jr., 1964, Northern and interior Alaska *in* U. S. Senate Rept. Comm. on Interior and Insular Affairs, 58th Cong., 2nd sess: U. S. Geol. Survey, p. 62–77.

Patton, W. W., Jr., and Tailleur, I. L., 1964, Geology of the Killik-Itkillik region, Alaska—exploration of Naval Petroleum Reserve No. 4 and adjacent areas: U. S. Geol. Survey Prof. Paper 303-G, p. 409–500.

Payne, T. G., 1959, Central Alaska *in* Geology of possible petroleum provinces in Alaska: U. S. Geol. Survey Bull. 1094, p. 53–88.

Payne, T. G., et al., 1951, Geology of the Arctic slope of Alaska: U. S. Geol. Survey Oil and Gas Inv. Map OM-126.

Pearson, Bruce, 1963, Texas' Marfa basin may be ready to yield to oil hunters: Oil and Gas Jour., Dec. 23, p. 88–91.

Pease, Maurice H., Jr., and Hoover, Linn, 1957, Geology of the Doty-Minot Peak area, Washington: U. S. Geol. Survey Oil and Gas Inv. Map. OM-188.

Pedry, John J., 1957, Cottonwood Creek field, Washakie County, Wyoming—a carbonate stratigraphic trap: Am. Assoc. Petroleum Geologists Bull., v. 41 (May), p. 823–838.

Peek, Charles A., et al., 1960, Developments in west Texas and southeastern New Mexico in 1959: Am. Assoc. Petroleum Geologists Bull., v. 44 (June), p. 774–783.

Pemberton, J. R., 1929, Elk Hills, Kern County, California *in* Structure of typical American oil fields: Am. Assoc. Petroleum Geologists, v. 2, p. 44–61.

Pepper, James F., de Witt, Wallace, Jr., and Demarest, David F., 1954, Geology at the Bedford shale and Berea sandstone in the Appalachian basin: U. S. Geol. Survey Prof. Paper 259.

Pepper, James F., de Witt, Wallace, Jr., and Everhart, Gail M., 1955, The "Clinton" sands in Canton, Dover, Massilon, and Navarre quadrangles, Ohio: U. S. Geol. Survey Bull. 1003-A.

Perkins, Hunt, 1961, Fault closure type fields, southeast Louisiana: Gulf Coast Assoc. Geol. Socs. Trans., v. 11, p. 177–191.

Perkins, Hunt, 1968, Subsurface geology of the Sulphur Lick–Tompkinsville area, Monroe County, Kentucky: Kentucky Geol. Survey, Ser. X, Spec. Pub. 15, p. 46–55.

Perkins, Jerome Hunt, 1955, Geology of the Decide pool, Clinton County, Kentucky: Kentucky Geol. Survey Bull. 17, p. 1–33.

Perkins, Stanley L., 1964, Fairway field, Anderson and Henderson Counties, Texas, *in* Occurrence of oil and gas in northeast Texas: East Texas Geol. Soc. Pub. 5, v. 1, p. 13–25.

Perry, Eugene S., 1953, Oil and gas in Montana: Montana Bur. Mines and Geol. Mem. 35.

Perry, William J., Jr., 1964, Geology of Ray Sponaugle well, Pendleton County, West Virginia: Am. Assoc. Petroleum Geologists Bull., v. 8, no. 5 (May), p. 659–669.

Peters, Jack W., 1960, Geophysical case history of the Horse Creek field, Laramie County, Wyoming: Geophysics, v. 25 (June), p. 602–612.

Peterson, G. L., Gastil, R. G., and Allison, E. C., 1966, Geology of the Peninsular Ranges: California Div. Mines and Geology Bull. 191, p. 70–73.

Peterson, James A., 1959, Petroleum geology of the Four Corners area: 5th World Petroleum Cong. Proc., sec. 1, p. 499–523.

Peterson, James A., 1966, Stratigraphic vs. structural controls on carbonate mound hydrocarbon accumulation, Aneth area, Paradox basin: Am. Assoc. Petroleum Geologists Bull., v. 50, no. 10 (October), p. 2068–2081.

Peterson, James A., Loleit, Allan J., Spencer, Charles W., and Ullrich, Richard A., 1965, Sedimentary history and economic geology of San Juan basin: Am. Assoc. Petroleum Geologists Bull., v. 49, no. 11 (November), p. 2076–2119.

Peterson, V. E., 1961a, Ashley Valley oil field *in* The oil and gas fields of Utah—a symposium: Intermountain Assoc. Petroleum Geologists.

Peterson, V. E., 1961b, Farnham Dome field *in* The oil and gas fields of Utah—a symposium: Intermountain Assoc. of Petroleum Geologists.

Philpott, T. H., 1952, Louisiana-Arkansas region may yield additional oil: World Oil (November), p. 108–116.

Philpott, T. H., 1960, Lower Cretaceous trend of south Arkansas, north Louisiana, Mississippi, and Alabama (abs.): Am. Assoc. Petroleum Geologists Bull., v. 44 (September), p. 1603.

Picard, M. Dane, 1956, Summary of Tertiary oil and gas fields in Colorado and Utah: Am. Assoc. Petroleum Geologists Bull., v. 40, no. 12 (December), p. 2956–2960.

Picard, M. Dane, 1957, Red Wash–Walker Hollow field, stratigraphic trap, eastern Uinta basin, Utah: Am. Assoc. Petroleum Geologists Bull., v. 41, no. 5 (May), p. 923–936.

Picard, M. Dane, 1958, Subsurface structure, Aneth and adjacent areas, San Juan County, Utah, *in* Guidebook 9th Ann. Field Conf.: Intermountain Assoc. Petroleum Geologists, p. 226–230.

Picard, M. Dane, 1959, White Mesa field, environmental trap, San Juan County, Utah: Am. Assoc. Petroleum Geologists Bull., v. 43, no. 10 (October), p. 2456–2469.

Picard, M. Dane, 1960a, Geology of Pennsylvanian gas in Four Corners region: Am. Assoc. Petroleum Geologists Bull., v. 44, no. 9 (September), p. 1541–1569.

Picard, M. Dane, 1960b, On the origin of oil, Eagle Springs field, Nye County, Nevada, *in* Guidebook to geology of east central Nevada: Intermountain Assoc. Petroleum Geologists, 11th Ann. Field Conf., p. 237–244.

Picard, M. Dane, 1962, Occurrence and origin of Mississippian gas in Four Corners region: Am. Assoc. Petroleum Geologists Bull., v. 46, no. 9 (September), p. 1681–1700.

Picard, M. Dane, 1968, Outline of occurrence of Pennsylvanian gas in Four Corners region: Am. Assoc. Petroleum Geologists Mem. 9, v. 2, p. 1327–1356.

Pierce, A. P., 1965, Helium *in* Mineral and water resources of New Mexico: U. S. Senate Rept. Comm. on Interior and Insular Affairs, 89th Cong., 1st sess., p. 120–124; Carbon Dioxide, p. 125–126.

Pike, Stuart J., 1968, Black Warrior basin, northeast Mississippi and northwest Alabama: Am. Assoc. Petroleum Geologists Mem. 9, v. 2, p. 1693–1701.

Pitt, William D., 1960, Dig deeper for Ouachita reserves: Oil and Gas Jour., Nov. 14, p. 214–217.

Plafker, George, and Miller, John J., 1957, Reconnaissance geology of the Malaspina district, Alaska: U. S. Geol. Survey Oil and Gas Inv. Map OM-189.

Planalp, Roger, 1961, Oil finders making long drives on Arkoma basin's fairway: Oil and Gas Jour., Apr. 24, p. 126–128.

Planalp, Roger, 1968, Geology of Aetna gas field, Franklin and Logan Counties, Arkansas: Am. Assoc. Petroleum Geologists Mem. 9, v. 2, p. 1668–1681.

Pohlmann, Henry Fred, 1967a, The Navajo Indian Nation and Dineh bi Keyah *in* New Mexico Geol. Soc., 18th Field Conf.: New Mexico Bur. Mines and Mineral Resources, p. 63–69.

Pohlmann, Henry Fred, 1967b, Dineh bi Keyah lifts Navajo spirits: Oil and Gas Jour., Oct. 9, p. 205–211.

Pollock, C. B., 1960, Beaver Creek Madison, Wyoming's deepest water injection project: Jour. Petroleum Technology, v. 12, no. 1 (January), p. 39–41.

Poole, J. C., 1940, Saxet oil and gas field, Nueces County, Texas: Am. Assoc. Petroleum Geologists Bull., v. 24 (October), p. 1805–1835.

Porter, James L., 1968, Developments in north mid-continent in 1967: Am. Assoc. Petroleum Geologists, v. 52, no. 6 (June), p. 981–985.

Porter, William C., 1966, El Mar field, Loving County, Texas, and Lea County, New Mexico, *in* Oil and gas fields in West Texas—a symposium: West Texas Geol. Soc., p. 129–133.

Pott, Robert L., and Devore, Stephen F., 1952, Lost Soldier field, Sweetwater County, Wyoming: Oil and Gas Jour., Jan. 21, p. 80 et seq.

Powell, J. B., Jr., 1955, A case history of the East Poplar field, Roosevelt County, Montana *in* Geological record: Am. Assoc. Petroleum Geologists, Rocky Mtn. Sec., p. 75–87.

Pratt, W. E., and Johnson, D. W., 1926, Local subsidence of the Goose Creek oil field: Jour. Geology, v. 34 (October-November), p. 577–590.

Price, Paul H., 1942, Discovery of gas in Rockingham County, Virginia: Am. Assoc. Petroleum Geologists Bull., v. 26 (February), p. 275.

Price, Paul H., and Haught, Oscar L., 1964, New West Virginia area probed for gas production: Oil and Gas Jour., Aug. 29, p. 134.

Price, Paul H., Tucker, R. C., and Haught, Oscar, 1957, Oriskany looks promising in eastern West Virginia: Oil and Gas Jour., Sept. 16, p. 345.

Privrasky, Norman C., 1963, Geology of the Big Piney area, Sublette County, Wyoming: U. S. Geol. Survey Oil and Gas Inv. Map OM-205.

Prucha, John James, Graham, John A., and Nickelsen, Richard P., 1965, Basement-controlled deformation in Wyoming province of Rocky Mountains foreland: Am. Assoc. Petroleum Geologists Bull., v. 49, no. 7 (July), p. 966–992.

Puri, Harbans S., and Banks, Joseph E., 1959, Structural features of the Sunniland oil field, Collier County, Florida: Gulf Coast Assoc., Geol. Socs. Trans., v. 9, p. 121–150.

Pustmueller, Paul S., 1958, Pennsylvanian oil and gas in southwestern Colorado *in* Pennsylvanian rocks of Colorado and adjacent areas—a symposium: Rocky Mtn. Assoc. Geologists.

Putman, D. M., 1956, The West Duncan field *in* Petroleum geology of southern Oklahoma: Am. Assoc. Petroleum Geologists, v. 1, p. 319–326.

Pye, Willard D., 1955, Discovery prospects good on Williston's east flank: World Oil, March, p. 82–92.

Pye, Willard D., 1958, Habitat of oil in northern Great Plains and Rocky Mountains *in* Habitat of oil: Am. Assoc. Petroleum Geologists, p. 178–224.

Pye, Willard D., 1966, Arizona—a new frontier to explore: Interstate Oil Compact Comm. Bull., v. 8, no. 2. Abridged version in Oil and Gas Jour., May 8, 1967, p. 168–172.

Pye, Willard D., 1967, How unique Arizona syenite oil reservoir formed: World Oil, July, p. 81–83.

Quigley, M. Darwin, 1965, Geologic history of Piceance Creek–Eagle Basins: Am. Assoc. Petroleum Geologists Bull., v. 49, no. 11 (November), p. 1974–1996.

Quigley, M. Darwin, and Price, Jack R., 1963, Green River oil shale potential in Utah: Paper 21 *in* Utah Geol. and Mineralog. Survey Bull. 54, p. 207–214.

Rainwater, E. H., 1964, Regional stratigraphy of the Gulf Coast Miocene: Gulf Coast Assoc. Geol. Socs. Trans., v. 14 (October), p. 81–124.

Ransone, W. R., 1958, Case history of the Sojourner oil field, Haskell County, Texas, a geochemical confirmation: Geophysics, v. 23, no. 3 (July), p. 574–578.

Rau, H. L., and Ackley, K. A., 1939, Geology and development of Keokuk pool, Seminole and Pottawatamie Counties, Oklahoma: Am. Assoc. Petroleum Geologists Bull., v. 23 (February), p. 220–245.

Rau, W. W., 1967, Geology of Wynoochee quadrangle, Grays Harbor County, Washington: Washington Dept. Nat. Resources Bull. 56.

Read, John L., Jr., 1959, Geologic case history of Slocum dome, Anderson County, Texas: Am. Assoc. Petroleum Geologists Bull., v. 43, no. 5 (May), p. 958–973.

Reagan, M. A., Jr., and Foust, R. T., Jr., 1960, Discovery and development of the Person field, Karnes County, Texas: Gulf Coast Assoc. Geol. Socs. Trans., v. 10, p. 25–31.

Reed, E. C., 1954, Central Nebraska has possibilities: World Oil, November, p. 113–116.

Reed, E. C., Svoboda, R. F., Prichard, G. E., and Fox, Jeannette, 1958, Map of Nebraska showing areal distribution of pre-Pennsylvanian rocks, anticlines and basins, oil and gas fields, pipelines, and unsuccessful test wells: U. S. Geol. Survey Oil and Gas Inv. Map OM-198.

Reed, John C., 1958, History of the exploration of Naval Petroleum Reserve No. 4 and adjacent areas: U. S. Geol. Survey Prof. Paper 301.

Reedy, Harold J., 1958, How to make profit below 15,000 feet: Oil and Gas Jour., Apr. 14, p. 166–171.

Reedy, Harold J., and Sykes, Howard A., 1962, Carter–Knox oil field, Grady and Stephens Counties, Oklahoma: Tulsa Geol. Soc. Digest, v. 30, p. 89–107.

Reese, D. L., 1968, Gas fields of Rock Springs uplift, Sweetwater County, Wyoming: Am. Assoc. Petroleum Geologists Mem. 9, v. 1, p. 803–816.

Reeves, C. C., Jr., 1960, A closer look at Love County: Oil and Gas Jour., Jan. 4, p. 116–121, and Jan. 11, p. 124–130.

Reeves, C. C., Jr., 1961, Facies changes are key to Marietta basin's oil future: World Oil, October, p. 137–143.

Reeves, C. C., Jr., 1964, Pennsylvanian prospects of northern Midland basin: World Oil, July, p. 106–112.

Reeves, C. C., Jr., and Mount, J. Russell, 1960, Possibility of hydrocarbon accumulations along northern flank of Marietta syncline, Love County, Oklahoma: Am. Assoc. Petroleum Geologists Bull., v. 44 (January), p. 72–82.

Reeves, Frank, 1923, Geology of the Ranger oil field, Texas: U. S. Geol. Survey Bull. 736, p. 111–170.

Reeves, Frank, and Price, Paul H., 1950, Early Devonian gas in northern West Virginia and pre-Devonian oil prospects: Am. Assoc. Petroleum Geologists Bull., v. 34 (November), p. 2095–2132.

Reeves, John R., 1929, El Dorado oil field, Butler County, Kansas *in* Structure of typical American oil fields: Am. Assoc. Petroleum Geologists, v. 2, p. 160–167.

Reeves, John R., 1936, Hebron gas field, Potter County, Pennsylvania: Am. Assoc. Petroleum Geologists Bull., v. 20 (August), p. 1019–1027.

Regan, Louis J., Jr., and Hughes, A. W., 1949, Fractured reservoirs of Santa Maria district, California: Am. Assoc. Petroleum Geologists Bull., v. 33, no. 1 (January), p. 32–51.

Reger, David B., 1941, Shinnston oil pool, Harrison County, West Virginia, *in* Stratigraphic type oil fields: Am. Assoc. Petroleum Geologists, p. 830–846.

Rehn, E. E., 1956, Rim areas paying off in Illinois and Kentucky: World Oil, January, p. 48–51.

Rehn, E. E., 1968, Petroleum potential of the Rough Creek tectonic element in Kentucky *in* Kentucky Geol. Survey, Ser. X, Spec. Pub. 15, p. 12–25.

Reishus, Mark, 1968, The Newcastle Formation in the Williston basin of North Dakota: North Dakota Geol. Survey Rept. Inv. 47. Abridged version in Oil and Gas Jour., Sept. 9, p. 166–169.

Reiter, F. H., 1957, Wheat field *in* Occurrence of oil and gas in West Texas: Univ. Texas Pub. 5716, p. 397–400.

Reiter, J. O., 1959, Review of Hitchcock field, Galveston County, Texas: Gulf Coast Assoc. Geol. Socs. Trans., v. 9, p. 63–70.

Renfroe, Charles A, 1954, Novinger field *in* Kansas oil and gas fields: Kansas Geol. Soc., v. 2 (Western Kansas), p. 112–117.

Repenning, C. A., Jones, D. L., and Addicott, W. O., 1966, Geology of the Great Valley: California Div. Mines and Geology Bull. 191, p. 48–54.

Reutinger, C. A., 1968, Natural gas in Permian and Cisco beds of Red River and Electra arch: Am. Assoc. Petroleum Geologists Mem. 9, v. 2, p. 1455–1458.

Reynolds, Douglas W., and Vincent, James K., 1967, Western Kentucky's Bethel channel—the largest continuous reservoir in the Illinois basin: Kentucky Geol. Survey, Ser. X, Spec. Pub. 14, p. 19–30.

Rich, J. L., 1923, Shoestring sands of eastern Kansas: Am. Assoc. Petroleum Geologists Bull., v. 7 (March–April), p. 103–113.

Rich, J. L., 1926, Further observations on Shoestring oil pools of eastern Kansas: Am. Assoc. Petroleum Geologists Bull., v. 10 (June), p. 556–580.

Richards, O. H., 1952, Aspermount Lake field, Stonewall County, Texas, *in* Geological contributions: Abilene Geol. Soc., p. 1–3.

Richards, P. W., and Nieschmidt, C. L., 1961, The Bighorn Dolomite and correlative formations in southern Montana and northern Wyoming: U. S. Geol. Survey Oil and Gas Inv. Map OM-202.

Richardson, P. H., 1961, Minden salt dome *in* Guidebook 1960 Field Trip: Shreveport Geol. Soc., p. 17–26.

Richardson, W. E., 1965, Oswego is prolific target in Oklahoma: Oil and Gas Jour., Dec. 20, p. 96–100.

Riddell, John T., 1962, Esperson Dome field, Liberty County, Texas, *in* Typical oil and gas fields of southeast Texas: Houston Geol. Soc., p. 47–52.

Riddle, Herschel L., Slingerland, Robert E., and Groenewold, Bernard C., 1968, Developments in Louisiana Gulf Coast in 1967: Am. Assoc. Petroleum Geologists Bull., v. 52, no. 6 (June), p. 1020–1026.

Riggs, C. H., Eakin, J. L., Jr., Larman, J. Heath, and Johnston, K. H., 1958, History and potentialities of the Cushing oil field, Creek County, Oklahoma: U. S. Bur. Mines Rept. Inv. 5415.

Riggs, C. H., Wey, E., and Maude, J. V., 1954, Petroleum engineering study of Flat Rock oil field, Osage County, Oklahoma: U. S. Bur. Mines Rept. Inv. 5018.

Rister, Carl Coke, 1949, Oil! Titan of the Southwest: Norman, Oklahoma, University of Oklahoma Press.

Ritzma, Howard R., 1962, Piceance Creek gas field *in* Exploration for oil and gas in northwestern Colorado: Rocky Mtn. Assoc. Geologists, p. 96–103.

Ritzma, Howard R., 1963, Geology and occurrence of oil and gas, Wamsutter arch, Wyoming, *in* Backbone of the Americas: Am. Assoc. Petroleum Geologists Mem. 2 (November), p. 188–195.

Ritzma, Howard R., 1967a, Proposed wilderness may hide oil: Utah Geol. and Mineralog. Survey Quart. Rev., v. 1, no. 11 (February), p. 1–2.

Ritzma, Howard R., 1967b, Sinclair increases old field production and heightens interest in Four Corners: Oil and Gas Jour., May 29, p. 113.

Ritzma, Howard R., 1967c, Oil-impregnated sandstone deposits of Utah: Interstate Oil Compact Comm. Bull., v. 9, no. 2, p. 87–98.

Ritzma, Howard R., 1968, Geology and occurrence of gas, Wamsutter arch, Sweetwater County, Wyoming: Am. Assoc. Petroleum Geologists Mem. 9, v. 1, p. 817–827.

Rives, John S., 1962, Don't overlook the Hosston on the San Marcos arch: Oil and Gas Jour., Feb. 5, p. 183–185.

Roach, C. B., 1943, Subsurface study, Jennings field, Acadia Parish, Louisiana: Am. Assoc. Petroleum Geologists Bull., v. 27 (August), p. 1102–1122.

Roberts, G. C., 1962, Two Texas counties need better exploration: Oil and Gas Jour., Jan. 1, p. 122–126.

Roberts, M. C., 1960, Developments in Oklahoma in 1959: Am. Assoc. Petroleum Geologists Bull., v. 44 (June), p. 763.

Roberts, Ralph H., Crittenden, M. D., Jr., Tooker, E. W., Morris, H. T., Hose, R. K., and Cheney, T. M., 1965, Pennsylvanian and Permian basins in northwestern Utah, northeastern Nevada, and south central Idaho: Am. Assoc. Petroleum Geologists Bull., v. 49, no. 11 (November), p. 1926–1956.

Roberts, Wendell L., and Vernon, Robert O., 1961, Florida . . . more extensive drilling might discover big oil and gas producing areas: Oil and Gas Jour., Mar. 13, p. 214–219.

Robinson, Florence M., 1958, Test wells in the Gubik area, Alaska—exploration of Naval Petroleum Reserve No. 4: U. S. Geol. Survey Prof. Paper 305-C.

Robinson, Joe D., 1966, Anton–Irish field *in* Oil and gas fields in West Texas—a symposium: West Texas Geol. Soc., p. 25–28.

Rocky Mountain Association of Geologists, 1954, Oil and gas fields of Colorado—a symposium: Denver, Colorado.

Rocky Mountain Association of Geologists, 1955, The oil and gas fields of Nebraska—a symposium.

Rocky Mountain Association of Geologists, 1957, Guidebook to the Geology of North and Middle Park basins, Colorado: Denver, Colorado.

Rocky Mountain Association of Geologists, 1958, Pennsylvanian rocks of Colorado and adjacent areas—a symposium: Denver, Colorado.

Rocky Mountain Association of Geologists, 1962, Oil and gas field volume, Colorado-Nebraska: Denver, Colorado.

Rocky Mountain Association of Geologists, 1963, Guidebook to the geology of the northern Denver basin and adjacent uplifts: Denver, Colorado.

Roe, N. Charles, Shambaugh, John S., and Wood, John W., 1966, Developments in South Texas in 1965: Am. Assoc. Petroleum Geologists Bull., v. 50, no. 6 (June), p. 1208–1213.

Rogatz, Henry, 1935, Geology of Texas Panhandle oil and gas field: Am. Assoc. Petroleum Geologists Bull., v. 19, p. 1089–1109.

Rogatz, Henry, 1939, Geology of Texas Panhandle oil and gas field: Am. Assoc. Petroleum Geologists Bull., v. 23, p. 983–1053.

Rogatz, Henry, 1961, Shallow oil and gas fields of the Texas Panhandle and Hugoton *in* Oil and gas fields of the Texas and Oklahoma Panhandles: Panhandle Geol. Soc., p. 8–37.

Rogers, Leslie C., and Wilson, Howard M., 1964, Louisiana offshore boom—no end in sight: Oil and Gas Jour., June 22, p. 132–137.

Rogers, R. G., 1961, The Red Cave formation of the Texas Panhandle *in* Oil and gas fields of the Texas and Oklahoma Panhandles: Panhandle Geol. Soc., p. 38–44.

Rogers, R. Lee, 1956, North Louisiana Smackover deep tests pay off: Oil and Gas Jour., July 2, p. 150–153.

Rogers, Robert E., 1968, Carthage field, Panola County, Texas: Am. Assoc. Petroleum Geologists Mem. 9, v. 1, p. 1020–1059.

Rohr, Egon H., 1957, Reservoir study of the Roosevelt pool, Utah: Jour. Petroleum Technology, v. 9, no. 10 (October), p. 23–24.

Romberg, Frederick, and Barnes, Virgil E., 1954, A geological and geophysical study of Pilot Knob (south), Travis County, Texas: Geophysics, v. 19, no. 3 (July), p. 438–454.

Rooney, Lawrence F., 1966, Evidence of unconformity at top of Trenton limestone in Indiana and adjacent states: Am. Assoc. Petroleum Geologists Bull., v. 50, no. 3 (March), p. 533–548.

Rose, William D., 1965a, Highlights of western Kentucky oil and gas exploration in 1963 and early 1964: Kentucky Geol. Survey, Ser. X, Spec. Pub. 10, p. 41–52.

Rose, William D., 1965b, Kentucky drillers probe deep: Oil and Gas Jour., Aug. 23, p. 140–145.

Rose, William D., 1968, Gas-productive areas in western Kentucky: Am. Assoc. Petroleum Geologists Mem. 9, v. 2, p. 1759–1760.

Ross, Richard B., 1957, Beaver Creek field *in* Guidebook 12th Ann. Field Conf.: Wyoming Geol. Assoc., p. 148–149.

Roswell Geological Society, 1956, 1960, 1967, Oil and gas fields of southeastern New Mexico—a symposium with two supplements.

Roth, Ernest E., 1968, Natural gases of the Appalachian basin: Am. Assoc. Petroleum Geologists Mem. 9, v. 2, p. 1702–1715.

Roth, R., 1955, Paleogeology of Panhandle of Texas: Am. Assoc. Petroleum Geologists Bull., v. 39, p. 422–443.

Roth, R., 1957, Texas Panhandle paleogeology: pt. I, World Oil, Feb. 1, p. 82–86; pt. II, March, p. 94–98.

Rothrock, Howard E., 1949, Mayfield pool, Cuyahoga County, Ohio: Am. Assoc. Petroleum Geologists Bull., v. 33 (October), p. 1731–1746.

Rudkin, George H., 1968, Natural gas in San Joaquin Valley, California, *in* Natural gases of North America: Am. Assoc. Petroleum Geologists Mem. 9, v. 1, p. 113–134.

Rudman, A. J., Summerson, C. H., and Hintze, W. J., 1965, Geology of basement in midwestern United States: Am. Assoc. Petroleum Geologists, v. 49, no. 7 (July), p. 894–904.

Runge, John S., 1968, Exploration for "Dakota" oil traps, Black Hills area, *in* Guidebook 20th Ann. Field Conf.: Wyoming Geol. Soc., p. 79–82.

Runge, John S., Copen, James D., Tutten, William O., and Harris, Steven H., 1965, Developments in northern Rockies in 1964: Am. Assoc. Petroleum Geologists Bull., v. 49, no. 6 (June), p. 801–802.

Rutledge, Richard B., 1956, The Velma oil field, Stephens County, Oklahoma, *in* Petroleum geology of southern Oklahoma: Am. Assoc. Petroleum Geologists, v. 1, p. 260–281.

Ryniker, Charles, Shortridge, C. G., and Maxwell, R. W., 1959, Altus field, Jackson County, Oklahoma, *in* Petroleum geology of southern Oklahoma: Am. Assoc. Petroleum Geologists, v. 2, p. 165–179.

Sabins, Floyd F., Jr., 1963, Anatomy of a stratigraphic trap, Bisti field, New Mexico: Am. Assoc. Petroleum Geologists Bull., v. 47, no. 2, (February), p. 193–228.

Sabins, Floyd F., Jr., 1964, Symmetry, stratigraphy, and petrography of cyclic Cretaceous deposits in San Juan basin: Am. Assoc. Petroleum Geologists Bull., v. 48, no. 3 (March), p. 292–316.

Sabins, Floyd F., and Petersen, F. A., 1961, Geology and petrography of Dead Horse Creek and Barber Creek fields *in* Wyoming Geol. Assoc. Guidebook 16th Ann. Field Conf., p. 301–309.

Sacramento Petroleum Association, 1968, Corning and South Corning gas fields, Tehama County, California: Am. Assoc. Petroleum Geologists Mem. 9, v. 1, p. 635–638.

Safonov, Anatole, 1962, The challenge of the Sacramento Valley, California: California Div. Mines and Geol. Bull. 181, p. 77–97.

Safonov, Anatole, 1968, Stratigraphy and tectonics of Sacramento Valley: Am. Assoc. Petroleum Geologists Mem. 9, v. 1, p. 611–635.

Sahinen, U. M., Erdmann, C. E., Weissenborn, A. E., and Weiss, P. L., 1963, Geology *in* Mineral and water resources of Montana: U. S. Senate Rept. Comm. on Interior and Insular Affairs, 88th Cong., 1st sess., p. 13–21.

St. John, F. B., Jr., and Lusk, Tracy W., 1959, Developments in southeastern states in 1958: Am. Assoc. Petroleum Geologists Bull., v. 43 (June), p. 1341.

Salisbury, Gerald P., 1968, Natural gas in Devonian and Silurian rocks of Permian basin, west Texas and southeast New Mexico: Am. Assoc. Petroleum Geologists Mem. 9, v. 2, p. 1433–1445.

Sandberg, C. A., and Prichard, G. E., 1964, Petroleum and natural gas, *in* Mineral and water resources of South Dakota: U. S. Senate Rept. Comm. on Interior and Insular Affairs, 88th Cong., 2nd sess., p. 151–160.

Sandberg, Dorothy T., 1959, Structure contour map on top of the middle member of the Piper Formation of Middle Jurassic age in the Williston basin and adjacent areas in Montana, North Dakota, and South Dakota: U. S. Geol. Survey Oil and Gas Inv. Map OM-179.

Sandidge, John R., 1961, Comanchean reef trend astounds Gulf Coast observers. Just how good is it?—Here's a reappraisal: Oil and Gas Jour., July 24, p. 118–119.

Sands, J. Merville, 1927, Burbank field, Osage County, Oklahoma: Am. Assoc. Petroleum Geologists Bull., v. 11 (October), p. 1045–1054.

Sangbush, R. L., 1962, Stratigraphic trap study in Cottonwood Creek field, Big Horn basin, Wyoming: Geophysics, v. 27 (August), p. 427–444.

Sargent, Don C., 1964, Exploration will help show Pascola's potential: Oil and Gas Jour., Aug. 24, p. 128–130.

Saunders, R. H., 1964, The mineral industry in Alaska *in* U.S. Senate Rept. Comm. on Interior and Insular Affairs, 88th Cong., 2nd sess.: U. S. Geol. Survey, p. 13–18.

Savage, C. H., 1964, Petroleum and natural gas, *in* Mineral and water resources of Idaho: U. S. Senate Rept. Comm. on Interior and Insular Affairs, 88th Cong., 2nd sess., p. 145–152.

Sax, Norman A., 1967, Developments in west Texas and southeastern New Mexico in 1966: Am. Assoc. Petroleum Geologists Bull., v. 51, no. 6 (June), pt. 2, p. 1053–1061.

Sayre, A. T., 1961, Adena field, Morgan County, Colorado, *in* Oil and gas field volume, Colorado–Nebraska: Rocky Mtn. Assoc. Geologists, p. 264–265.

Scheig, Richard L., 1961, Cook field, Dewitt County, Texas: Gulf Coast Assoc. Geol. Socs. Trans., v. 11, p. 235–244.

Schmidt, R. G., and Warner, R. A., 1964, Knox is the target in northern Kentucky: Oil and Gas Jour., Apr. 20, p. 178–180.

Schmitt, George T., 1954, Genesis and depositional history of Spraberry formation, Midland basin, Texas: Am. Assoc. Petroleum Geologists Bull., v. 38, no. 9 (September), p. 1957–1958.

Schneider, Stephen J., 1959, Bay Sainte Elaine oil field, southern Louisiana: Am. Assoc. Petroleum Geologists Bull., v. 43 (October), p. 2470–2480.

Schneider, W. T., 1957, Wasson field, Yoakum and Gaines Counties, *in* Herald, Occurrence of Oil and Gas in West Texas, p. 377–384.

Schoellhamer, Jack E., and Woodford, Alfred O., 1951, The floor of the Los Angeles basin, Los Angeles, Orange, and San Bernardino Counties, California: U. S. Geol. Survey Oil and Gas Inv. Map. OM-117.

Schoon, Robert A., 1965, Selected formation tops in oil and gas tests drilled before Jan. 1, 1965: South Dakota Geol. Survey Circ. 35.

Schreiner, Carl S., Jr., 1957, Good field, Borden County, *in* Herald, Occurrence of Oil and Gas in West Texas, p. 134–136.

Schuh, E. J., 1961, Westwater field, *in* The oil and gas fields of Utah—a symposium: Intermountain Assoc. Petroleum Geologists.

Schwade, Irving T., 1954, Geology of Cuyama Valley and adjacent ranges, San Luis Obispo, Santa Barbara, Kern, and Ventura Counties: California Div. Mines and Geology Bull. 170, Map Sheet No. 1.

Schwade, Irving T., Carlson, Stanley A., O'Flynn, James B., 1958, Geologic environment of Cuyama Valley oil fields, California, *in* Habitat of oil: Am. Assoc. Petroleum Geologists (Spec. Volume), p. 78–98.

Schwalb, Howard R., 1968, Subsurface structure map on the base of New Albany Black Shale or its equivalent *in* Geology and petroleum production of the Illinois basin—a symposium: Illinois and Indiana-Kentucky Geol. Socs., map in pocket.

Scotford, David M., 1964, Cincinnati arch, mineralogical-statistical evidence of post-Ordovician origin: Am. Assoc. Petroleum Geologists Bull., v. 48, no. 4 (April), p. 427–436.

Scott, E. R., 1948, Quitman oil field, Wood County, Texas, *in* Structure of typical American oil fields: Am. Assoc. Petroleum Geologists, v. 3, p. 419–431.

Scott, K. R., Hayes, W. E., and Fietz, R. P., 1961, Geology of the Eagle Mills Formation: Gulf Coast Assoc. Geol. Socs. Trans., v. 11, p. 1–14.

Scott, Robert W., 1966, New Precambrian (?) formation in Kansas: Am. Assoc. Petroleum Geologists Bull., v. 50, no. 2 (February), p. 380–384.

Scribner, M. K., 1958, Brea Canyon area, *in* A guide to the geology and oil fields of the Los Angeles and Ventura regions: Am. Assoc. Petroleum Geologists, Pacific Sec., p. 107–108.

Scull, B. J., 1954, Oil and more gas promised in northern Arkansas: World Oil, September, p. 79–82; October, p. 117–119.

Sealy, George, 1962, Anahuac field, Chambers County, Texas, *in* Typical oil and gas fields of southeast Texas: Houston Geol. Soc., p. 1–3.

Sears, Joe M., 1951, Southwestern Oklahoma—the Hollis basin: Shale Shaker, v. 2 (September), p. 4–6.

Sears, Joe M., 1952, Hollis basin holds promise: World Oil, June, p. 90–94.

Sebring, Louie Jr., 1948, Slick-Wilcox field, Dewitt and Goliad Counties, Texas: Am. Assoc. Petroleum Geologists Bull., v. 32 (February), p. 228–251.

Seeley, deBenneville K., Jr., 1961, Woodside dome, *in* The oil and gas fields of Utah—a symposium: Intermountain Assoc. Petroleum Geologists.

Seglund, James A., 1967, South Louisiana exploration—today and tomorrow: World Oil, March, p. 83–86.

Seiden, Hy, 1964, Kettleman Hills area: San Joaquin Geol. Soc. Selected Papers, v. 2, p. 46–53.

Selk, Erwin L., 1951a, Types of oil and gas traps in southern Oklahoma: Am. Assoc. Petroleum Geologists Bull., v. 35, no. 3 (March), p. 582–606.

Sellards, E. H., 1938, Oil accumulation in igneous rocks *in* Science of petroleum: Oxford University Press, p. 261–265.

Sellards, E. H., and Baker, C. L., 1934, The geology of Texas, v. II, Structural and economic geology: Univ. Texas Bull. 3401.

Sever, Charles W., 1964–1965, The Chattahoochee anticline in Georgia: Georgia Geol. Survey Mineral Newsletter, v. 17, p. 39–43; Oil seeps along the Chattahoochee anticline in Georgia, p. 43–45.

Shannon, C. W., 1961, Last Chance gas field, Emery County, Utah, *in* The oil and gas fields of Utah—a symposium: Intermountain Assoc. Petroleum Geologists.

Sharma, G. D., 1966, Geology of Peters reef, St. Clair County, Michigan: Am. Assoc. Petroleum Geologists Bull., v. 50, no. 2 (February), p. 327–350.

Shaw, Richard F., Jr., 1955, A new look at the central Ardmore basin: World Oil, July, p. 108–112.

Shearrow, George G., 1959, Watch Ohio for deep drilling: Oil and Gas Jour., v. 57 (March), no. 13, p. 141–143.

Shearrow, George G., 1968, The story of Ohio's southeastern sleeper: Oil and Gas Jour., Apr. 1, p. 210–212.

Shearrow, George G., and Preston, Arthur F., 1965, Ohio drillers looking for "better" trap: Oil and Gas Jour., July 12, p. 142–149.

Sheets, M. M., and Cockrell, Ernest, Jr., 1962, Occurrence of oil and gas in southeast Texas *in* Typical oil and gas fields of southeast Texas: Houston Geol. Soc., p. xi–xx.

Sheldon, I. R., 1936, Driscoll pool, Duval County, Texas, *in* Gulf Coast oil fields: Am. Assoc. Petroleum Geologists, p. 620–630.

Shell Oil Company, Billings Division Staff, 1958a, Ritchie field *in* Montana oil and gas fields: Billings Geol. Soc., p. 207.

Shell Oil Company, Billings Division Staff, 1958b, Cabin Creek field *in* Montana oil and gas fields: Billings Geol. Soc., p. 95–96.

Shell Oil Company, Billings Division Staff, 1958c, Pine field *in* Montana oil and gas fields: Billings Geol. Soc., p. 189–190.

Shelton, Stanley O., 1964, Fowler's folly and the boom at Burkburnett: Indep. Petroleum Monthly, January, p. 14–17.

Sherrill, R. E., Dickey, P. A., and Matteson, L. S., 1941, Types of stratigraphic oil pools in Venango sands of northwestern Pennsylvania *in* Stratigraphic type oil fields: Am. Assoc. Petroleum Geologists, p. 507–538.

Shipp, B. G., and Dunnewald, J. B., 1962, The Big Piney–LaBarge Frontier gas field, Sublette and Lincoln Counties, Wyoming, *in* Guidebook 17th Ann. Field Conf.: Wyoming Geol. Assoc., p. 273–279.

Shreveport Geological Society, 1946, 1945 Reference report, v. 1, p. 178–181.

Shreveport Geological Society, 1947, 1945 Reference Report on certain oil and gas fields of North Louisiana, South Arkansas, Mississippi, and Alabama, v. 2.

Shreveport Geological Society, 1953, Penetration charts and reservoir data summary.

Shreveport Geological Society, 1961, Oil and gas producing zones in the Wilcox (Eocene) of central Louisiana.

Shreveport Geological Society, 1968, Stratigraphy and selected gas-field studies of North Louisiana: Am. Assoc. Petroleum Geologists Mem. 9, v. 1, p. 1099–1175.

Silver, Caswell, 1968, Principles of gas occurrence, San Juan basin: Am. Assoc. Petroleum Geologists Mem. 9, v. 1, p. 946–960.

Simmons, Kenneth A., 1967, A primer on "serpentine plugs" in South Texas: South Texas Geol. Soc. Bull., v. 7, no. 2.

Simonson, Russell R., 1958, Oil in the San Joaquin Valley, California, *in* Habitat of oil: Am. Assoc. Petroleum Geologists, p. 99–112.

Sims, W. P., and Frailing, W. G., 1950, Lakeview pool, Midway-Sunset field: Am. Inst. Mining Metall. Engineers Trans., v. 189, T.P. 2779, p. 7–18.

Singewald, Joseph T., Jr., 1954, Gas fields *in* Geology of Garrett County: Maryland Dept. Geol., Mines and Water Resources Bull. 13, p. 95–101.

Sitler, Guy F., Jr., 1963, Cambrian rocks are target no. 1 in eastern Ohio's vigorous oil search: Oil and Gas Jour., May 20, p. 198–205.

Sitler, Guy F., Jr., and Wehmeyer, Karl L., Jr., 1962, Ohio offers big potential from the bottom of the column: Oil and Gas Jour., Jan. 8, p. 111.

Six, David A., 1962, Pennsylvanian–Atoka producing sands, Red Oak–Norris gas field, Brazil anticline, Latimer and Le Flore Counties, Oklahoma: Tulsa Geol. Soc. Digest, v. 30, p. 43–55.

Six, David A., 1968, Red Oak–Norris gas field, Brazil anticline, Latimer and Le Flore Counties, Oklahoma: Am. Assoc. Petroleum Geologists Mem. 9, v. 2, p. 1644–1657.

Skarda, Everett, 1956, Oklahoma's fractured Viola limestone reservoir: Oil and Gas Jour., Feb. 27, p. 109–112.

Skeeters, W. W., 1961, Gas is target for explorers in the Green River basin, Oil and Gas Jour., Apr. 24, p. 152–158.

Skelton, Alan G., and Skelton, Martha B., 1942, A bibliography of Oklahoma oil and gas pools: Oklahoma Geol. Survey Bull. 63.

Skolfield, John L., 1968, New Harmony field, Waltersburg sand pool, White County, Illinois, and Posey County, Indiana, *in* Geology and petroleum production of the Illinois basin—a symposium: Illinois and Indiana-Kentucky Geol. Socs., p. 173–175.

Sloane, B. J., Jr., 1958, The subsurface Bodcaw sand in Louisiana: Louisiana Geol. Survey Bull. 33.

Smith, Derrell A., 1961, Geology of South Pass Block 27 field, offshore, Plaquemines Parish, Louisiana: Am. Assoc. Petroleum Geologists, v. 45 (January), p. 51–71.

Smith, Fred L., Jr., and Goodwyn, J. T., Jr., 1962, Blessing field, Matagorda County, Texas *in* Typical oil and gas fields of southeast Texas: Houston Geol. Soc., p. 18–23.

Smith, G. W., 1961, Rehm field *in* Oil and gas fields of the Texas and Oklahoma Panhandles: Panhandle Geol. Soc., p. 107–110.

Smith, G. Wendell, Summers, George E., Jr., Wallington, Dale, and Lee, Jean L., 1958, Mississippian oil reservoirs in Williston basin, *in* Habitat of oil: Am. Assoc. Petroleum Geologists, p. 149–177.

Smith, Howard R., 1962, Geology of the Melstone–Sumatra area in central Montana: U. S. Geol. Survey Oil and Gas Inv. Map O-211.

Smith, John Ward, and Stanfield, Kenneth E., 1964, Oil yields and properties of Green River oil shales in Uinta basin, Utah, *in* Guidebook 13th Ann. Field Conf.: Intermountain Assoc. Petroleum Geologists, p. 213–221.

Smith, M. B., 1964, Map showing distribution and configuration of basement rocks in California: U. S. Geol. Survey Oil and Gas Inv. Map OM-215.

Smith, M. B., and Schambeck, F. J., 1966, Petroleum and natural gas: California Div. Mines and Geology Bull. 191, p. 291–328.

Smith, R. Kenneth, and Anders, Ellis L., Jr., 1951, The geology of the Davis ranch oil pool, Wabaunsee County, Kansas: Kansas Geol. Survey Bull. 90, pt. 2.

Smith, W. K., 1951, Quitman field, Wood County, Texas: Univ. Texas Pub. 5116, p. 315–319.

Smith, Walter L., 1960, History of oil exploration in Railroad Valley, Nye County, Nevada, *in* Guidebook to geology of east central Nevada: Intermountain Assoc. Petroleum Geologists, 11th annual field Conference, p. 233–236.

Snider, L. C., 1927, A suggested explanation for the surface subsidence in the Goose Creek oil and gas field of Texas: Am. Assoc. Petroleum Geologists Bull., v. 11 (July), p. 729–745.

South Texas Geological Society, 1951, Downdip Mesozoic rocks of South Texas: Am. Assoc. Petroleum Geologists Bull., v. 35 (February), p. 357–360.

South Texas Geological Society, 1962, Contributions to the geology of South Texas: San Antonio, Texas.

Southwestern Federation of Geological Societies, 1960, Natural gas in the Southwest: Trans., v. 1, Abilene, Texas.

Spillers, James P., 1965, Distribution of hydrocarbons in south Louisiana by types of traps and trends—Frio and younger sediments: Gulf Coast Assoc. Geol. Socs. Trans., v. 15, p. 37–39. Also Am. Assoc. petroleum Geologists Bull., v. 49, no. 10 (October), p. 1749–1751.

Stafford, Philip T., 1957, Scurry field, Scurry, Kent and Borden Counties, Texas, *in* Herald: Occurrence of Oil and Gas in West Texas, p. 295–302.

Stafford, Philip T., 1959, Geology of part of the Horseshoe atoll in Scurry and Kent Counties, Texas: U. S. Geol. Survey Bull. 315-A, p. 1–18.

Stamey, R. A., Montgomery, J. C., and Easton, H. D., Jr., 1936, Greta oil field, Refugio County, Texas, *in* Gulf Coast oil fields: Am. Assoc. Petroleum Geologists, p. 648–663.

Stanonis, Frank L., 1968, Future prospective areas of western Kentucky, *in* Kentucky Geol. Survey, Ser. X, Spec. Pub. 15, p. 7–11.

Stanton, E. C., Jr., 1966, Morrow County, Ohio—case history of exploration and production (abs.): Am. Assoc. Petroleum Geologists Bull., v. 50, no. 3 (March), p. 636.

Stapp, Robert W., 1967, Relationship of Lower Cretaceous depositional environment to oil accumulation, northeastern Powder River basin, Wyoming: Am. Assoc. Petroleum Geologists Bull., v. 51, no. 10 (October), p. 2044–2055.

Stauffer, Clinton R., 1926, Minnesota's oil and gas possibilities: Am. Assoc. Petroleum Geologists, v. 10, no. 2 (February), p. 190–196.

Stead, F. L., 1955, Drickey–Caprock Queen sand trend, World Oil, June, p. 156–164.

Steig, M. H., Nichols, I. K., Gray, Shapleigh G., and Denham, R. L., 1951, Geology of Erath field, Vermilion Parish, Louisiana: Am. Assoc. Petroleum Geologists Bull., v. 35 (May), p. 943–987.

Stenzel, William K., 1965, Time of migration and accumulation of petroleum in Abo reef of southeastern New Mexico—a hypothesis, *in* Fluids in subsurface environments, Am. Assoc. Petroleum Geologists, Mem. 4, p. 243–256.

Stern, Allen R., 1955, New accumulations found beneath north flank overhang, Cote Blanche Island, St. Mary Parish, Louisiana: Gulf Coast Assoc. Geol. Socs. Trans., v. 5, p. 173–179.

Stevens, Curtis, 1961, Red Cave—Cinderella of Texas–Oklahoma Panhandles: World Oil, April, p. 104–119.

Stevens, Curtis, 1962, Major play shaping up in Palo Duro: World Oil, Feb. 1, p. 94–95.

Stipp, T. F., and Haigler, L. B., 1957, Preliminary structure contour map of a part of southeastern New Mexico, showing oil and gas development: U. S. Geol. Survey Oil and Gas Inv. Map OM-177.

Stokes, Wm. Lee, 1954, Geomorphology of south-central Utah, *in* Geology of portions of the high plateaus and adjacent canyon lands, central and south-central Utah: Intermountain Assoc. Petroleum Geologists, 1954, p. 2–14.

Stone, D. S., 1966, Geologic and economic evaluation of the Laramie–eastern Hanna Basin area, Wyoming: Mtn. Geologist, v. 3, no. 2, p. 55–73.

Stone, D. S., 1967, Theory of Paleozoic oil and gas accumulation in Big Horn basin, Wyoming: Am. Assoc. Petroleum Geologists Bull., v. 51, no. 10 (October), p. 2056–2114.

Storm, Willis, 1929, Smith–Ellis oil field, Brown County, Texas, *in* Structure of typical American oil fields: Am. Assoc. Petroleum Geologists, v. 2, p. 556–570.

Storm, Willis, Hensley, J. H., and Reiss, Bernard W., 1961, Cox and Hamon field, Duval County, Texas: Gulf Coast Assoc. Geol. Socs. Trans., v. 11, p. 251–259.

Stratigraphic Committee, 1961, Hansford fields *in* Oil and Gas fields of the Texas and Oklahoma Panhandles: Panhandle Geol. Soc., p. 87–94.

Strobell, J. D., Jr., 1956, Geology of the Carizzo Mountain area in northeastern Arizona and northwestern New Mexico: U. S. Geol. Survey Oil and Gas Inv. Map OM-160.

Summerford, H. E., Schieck, E. E., and Hiestand, T. C., 1950, Oil and gas accumulation controlled by sedimentary facies in Upper Cretaceous Newcastle sandstone, Wyoming: Am. Assoc. Petroleum Geologists Bull., v. 34 (September), p. 1850–1865.

Summers, Fred C., Jr., 1968, Exploration in Oklahoma and Panhandle of Texas in 1967: Am. Assoc. Petroleum Geologists, v. 52, no. 6 (June), p. 986–989.

Summerson, Charles H., 1962, Precambrian in Ohio and adjoining areas: Ohio Geol. Survey Rept. Inv. 44.

Sutton, Emmett, 1965, Trempealeau reservoir performance, Morrow County field, Ohio: J. Petroleum Technology, v. 17, no. 12, p. 1391–1395.

Svoboda, Richard F., 1958, Exploration possibilities in southwestern Nebraska look good: World Oil, October, p. 191–195.

Svoboda, Richard F., 1960, Interest flares in Forest City basin: Oil and Gas Jour., May 23, p. 150–153.

Svoboda, Richard F., 1962, Sleepy Hollow action sets off southwest Nebraska drilling push: Oil and Gas Jour., July 16, p. 132–137.

Swain, James F., 1950, Geology and occurrence of oil in Medina sand of Blue Rock–Salt Creek pool, Ohio: Am. Assoc. Petroleum Geologists Bull., v. 34 (September), p. 1874–1886.

Swann, David H., 1951, Waltersburg sandstone oil pools of lower Wabash area, Illinois and Indiana: Am. Assoc. Petroleum Geologists Bull., v. 35 (December), p.2561–2581.

Swann, David H., 1967, Generalized geologic column of southern Illinois *in* Oil and gas pay maps of Illinois: Illinois Geol. Survey, Illinois Petroleum 84, Figure 2.

Swann, David H., 1968, A summary of the geologic history of the Illinois basin, *in* Geology and petroleum production of the Illinois basin—a symposium: Illinois and Indiana–Kentucky Geol. Socs., p. 3–20.

Swann, David H., and Bell, Alfred H., 1958, Habitat of oil in the Illinois basin *in* Habitat of oil: Am. Assoc. Petroleum Geologists, p. 447–472.

Swann, David H., Buschbach, T. C., Atherton, Elwood, and Frye, J. C., 1966, Cambrian and Ordovician oil potential of Illinois basin (abs.): Am. Assoc. Petroleum Geologists Bull., v. 50, no. 3 (March), pt. 1, p. 636.

Swanson, Donald C., 1967, Major factors controlling Anadarko basin production: World Oil, January, p. 81–92.

Swanson, Vernon E., 1960, Oil yield and uranium in carbonaceous rocks: U. S. Geol. Survey Prof. Paper 356-A, p. 14–15.

Sweet, John M., 1968, Letter, Aug. 5.

Swenson, Robert E., 1967, Trap mechanics in Nisku Formation of northeast Montana: Am. Assoc. Petroleum Geologists Bull., v. 51, no. 10 (October), p. 1948–1958.

Swesnik, Robert M., 1948, Geology of West Edmond oil field, Oklahoma, Logan, Canadian, and Kingfisher Counties, Oklahoma, *in* Structure of typical American oil fields: Am. Assoc. Petroleum Geologists, v. 3, p. 359–398.

Swesnik, Robert M., 1950, Golden Trend of south-central Oklahoma: Am. Assoc. Petroleum Geologists Bull., v. 34 (March), p. 386–422.

Swesnik, Robert M., and Green, Thomas H., 1950, Geology of Eola area, Garvin County, Oklahoma: Am. Assoc. Petroleum Geologists Bull., v. 34 (November), p. 2176–2199.

Swirezynski, R. P., 1955, Preliminary pool report, Burke Ranch oil field, Natrona County, Wyoming (abs.): Oil and Gas Jour., Apr. 11, p. 125.

Takken, Suzanne, 1968, Subsurface geology of North Gotebo area, Kiowa and Washita Counties, Oklahoma: Am. Assoc. Petroleum Geologists Mem. 9, v. 2, p. 1492–1508.

Tanner, William F., 1956, Geology of Seminole County, Oklahoma: Oklahoma Geol. Survey Bull. 74.

Tarver, Earl A., 1966, West Virginia keeps drill busy: Oil and Gas Jour., Aug. 22, p. 152.

Taylor, Earle F., 1964, Vermont has oil, gas possibilities: World Oil, June, p. 142–146.

Teas, L. P., 1935, Natural gas of Gulf Coast salt dome area, *in* Geology of natural gas: Am. Assoc. Petroleum Geologists, p. 683–740.

Teas, L. P., and Miller, Charis R., 1936, Raccoon Bend oil field, Austin County, Texas, *in* Gulf Coast oil fields: Am. Assoc. Petroleum Geologists, p. 676–708.

Teitsworth, R. A., 1968, Lathrop gas field, San Joaquin County, California, *in* Am. Assoc. Petroleum Geologists Mem. 9, v. 1, p. 653–663.

Tenney, Charles S., 1966, Pennsylvanian and Lower Permian deposition in Wyoming and adjacent areas: Am. Assoc. Petroleum Geologists Bull., v. 50, no. 2 (February), p. 227–250.

Tenney, Charles S., 1968, Heavy oil in Minnelusa in northern Black Hills *in* Guidebook 20th Ann. Field Conf.: Wyoming Geol. Assoc., p. 95–99.

Texas, University of, 1957, Occurrence of oil and gas in West Texas: Pub. 5716, 442 p.

Thalman, A. L., 1961, Farnsworth field *in* Oil and gas fields of the Texas and Oklahoma Panhandles: Panhandle Geol. Soc., p. 152–155.

Thomas, E. P., 1950, Mississippi structures and their relation to oil accumulation: Am. Assoc. Petroleum Geologists Bull., v. 34 (July), p. 1502–1516.

Thomas, G. D., 1938, Carterville–Sarepta and Shangaloo fields, Bossier and Webster Parishes, Louisiana: Am. Assoc. Petroleum Geologists Bull., v. 22 (November), p. 1473–1503.

Thomas, George R., 1960, Geology of recent deep drilling in eastern Kentucky: Kentucky Geol. Survey, Ser. X, Spec. Pub. 3, p. 10–28.

Thomas, George R., and Nosow, Edmund, 1960, Drilling outlook remains good: World Oil, June, p. 95–100.

Thomas, Horace D., 1948, Geology and petroleum resources of Wyoming: Petroleum Eng., v. 20 (November), p. 136.

Thomas, Horace, D., 1958, Wyoming oil prospects are good: World Oil, July, p. 83–88.

Thomas, Horace D., 1961, Wyoming possibilities enhanced by variety of oil traps: World Oil, May, p. 80 et seq.

Thomas, Leonard C., 1966a, Tunstill field, Loving and Reeves Counties, *in* Oil and gas fields in West Texas—a symposium: West Texas Geol. Soc., p. 366–369.

Thomas, Leonard C., 1966b, Ford fields, Reeves and Culbertson Counties, *in* Oil and gas fields in West Texas—a symposium: West Texas Geol. Soc., p. 154–159.

Thomas, Leonard E., 1965, Sedimentation and structural development of Big Horn basin: Am. Assoc. Petroleum Geologists Bull., v. 49, no. 11 (November) p. 1867–1877.

Thomas, William A., 1968, Contemporaneous normal faults on flanks of Birmingham anticlinorium, central Alabama: Am. Assoc. Petroleum Geologists Bull., v. 52, no. 11 (November), p. 2123–2136.

Thompson, Edwin I., 1941, Hull-Silk oil field, Archer County, Texas, *in* Stratigraphic type oil fields: Am. Assoc. Petroleum Geologists, p. 661–679.

Thompson, R. M., 1958, Geology and oil and gas possibilities of the Wind River basin, Wyoming, *in* Habitat of Oil: Am. Assoc. Petroleum Geologists (Spec. Volume), p. 307–327.

Thompson, R. M., 1960, Geology and petroleum possibilities of Alaska: Internat. Geol. Cong., Copenhagen, Pt. XI, p. 27–36.

Thompson, W. C., and Hubbard, W. E., 1929, Relation of accumulation to structure in the oil fields of Archer County, Texas, *in* Structure of typical American oil fields, v. 1: Am. Assoc. Petroleum Geologists, p. 421–439.

Thornton, Dewey E., and Gaston, H. H., Jr., 1967, Lusk Strawn field *in* Roswell Geol. Soc., p. 15–20, 121–122.

Thornton, Dewey E., and Gaston, H. H., Jr., 1968, Geology and development of Lusk Strawn field, Eddy and Lea Counties, New Mexico: Am. Assoc. Petroleum Geologists Bull., v. 52, no. 1 (January), p. 66–81.

Thorsen, C. E., 1963, Age of growth faulting in southeast Louisiana: Gulf Coast Assoc. Geol. Socs. Trans., v. 13, p. 103–110.

Thralls, Hugh M., Robb, George L., and Schisler, A. J., 1962, Geology, geophysics work well together in Arkoma basin: Oil and Gas Jour., Oct. 29, p. 204–208.

Tillotson, Allen W., 1938, Olympic pool, Hughes and Okfuskee Counties, Oklahoma: Am. Assoc. Petroleum Geologists Bull., v. 22 (November), p. 1579–1587.

Tobison, N. M., 1967, Geology and development history of Boxer field, Morgan County, Colorado: Mtn. Geologist, v. 4, no. 3, p. 83–89.

Todd, T. W., 1963, Post-depositional history of Tensleep sandstone (Pennsylvanian), Big Horn Basin, Wyoming: Am. Assoc. Petroleum Geologists Bull., v. 47, no. 4 (April) p. 599–616.

Tolbert, A. M., 1958, The Frio offers lucrative hunting: Oil and Gas Jour., Jan. 6, p. 156–161.

Tomkins, J. Q., 1957, Bisti oil field, San Juan County, New Mexico: Am. Assoc. Petroleum Geologists Bull., v. 41, no. 5 (May), p. 906–922.

Tomlinson, C. W., 1935, Natural gas pools of southern Oklahoma *in* Geology of natural gas: Am. Assoc. Petroleum Geologists, p. 575–607.

Tomlinson, C. W., 1952–1953, Odd geologic structures of southern Oklahoma: World Oil, December, 1952, p. 103–110, and January, 1953, p. 90–94.

Tomlinson, C. W., and McBee, William, Jr., 1959, Pennsylvanian sediments and orogenies of Ardmore District, Oklahoma, *in* Petroleum geology of southern Oklahoma: Am. Assoc. Petroleum Geologists, v. 2, p. 3–52.

Tooker, E. W., and Stewart, J. H., 1964, Topography *in* Mineral and water resources of Utah: U. S. 88th Cong., 2d sess., Comm. Print, p. 17; Stratigraphy, p. 19–25.

Totten, Robert B., 1956, General geology and historical development, Texas and Oklahoma Panhandles: Am. Assoc. Petroleum Geologists, v. 40, p. 1945–1967.

Totten, Robert B., 1961, Review of exploration and development of oil and gas for the Texas and Oklahoma Panhandle area *in* Oil and gas fields of the Texas and Oklahoma Panhandles: Panhandle Geol. Soc., p. 69–74.

Totten, Robert B., and Horn, Paul H., 1961, Geology of gas fields of western Anadarko basin, Texas and Oklahoma Panhandles (abs.): Am. Assoc. Petroleum Geologists Bull., v. 45 (January), p. 124–125.

Totten, Robert B., and Horn, Paul H., 1968, Introduction to regional geology and typical gas fields of western Anadarko basin: Am. Assoc. Petroleum Geologists, Mem. 9, v. 2, p. 1525–1538.

Towse, Donald, 1957, Petrology of the Beaver Lodge Madison limestone reservoir: Am. Assoc. Petroleum Geologists Bull., v. 41, no. 11 (November), p. 2493–2507.

Trenchard, John, and Whisenant, J. Barney, 1936, Government wells oil field, Duval County, Texas, *in* Gulf Coast oil fields: Am. Assoc. Petroleum Geologists, p. 631–647.

Trotter, John F., 1963, The Minnelusa play of the northern Powder River, Wyoming and adjacent areas: First annual joint field conference, Wyoming Geol. Assoc. and Billings Geol. Soc.

Tucker, Mitchell, 1938, KMA now one of world's largest oil fields in proven area: Oil and Gas Jour., Jan. 20, p. 42–53, 58.

Tulsa Geological Society Digest, 1962, Symposium on natural gas in Oklahoma: v. 30, p. 41–149.

Turner, Daniel S., 1961, Black Mesa basin, gas target for tomorrow: Oil and Gas Jour., Apr. 24, p. 148–150.

Turner, Daniel S., 1962, Controls of oil and gas accumulation *in* Exploration for oil and gas in northwestern Colorado: Rocky Mtn. Assoc. Geologists, Denver, Colorado, p. 29–33.

Turner, Daniel S., 1968, Natural gas in Black Mesa basin, northeastern Arizona: Am. Assoc. Petroleum Geologists Mem. 9, v. 2, p. 1357–1370.

Tweto, Ogden, 1964, Geology *in* Mineral and water resources of Colorado: U. S. 88th Cong., 2d sess., Comm. Print., p. 11–27.

Uchupi, Elazar, and Emery, K. O., 1967, Structure of continental margin off Atlantic Coast of United States: Am. Assoc. Petroleum Geologists Bull., v. 51, no. 2 (February), p. 223–234.

Uchupi, Elazar, and Emery, K. O., 1968, Structure of continental margin off Gulf Coast of United States: Am. Assoc. Petroleum Geologists Bull., v. 52, no. 7 (July), p. 1162–1193.

Umpleby, S. S., 1956, Faulting, accumulation, fluid distribution, Ramsey pool, Payne County, Oklahoma: Am. Assoc. Petroleum Geologists Bull., v. 40 (January), p. 122–139.

United States Department of the Interior, 1968, United States petroleum through 1980: Office of Oil and Gas, Washington, D. C.

United States Geological Survey, 1932, Geologic map of the United States.

United States Geological Survey, 1963, Oil and gas fields of the United States.

United States Geological Survey, 1964, Mineral and water resources of Alaska: U. S. Senate Rept. Comm. on Interior and Insular Affairs, 88th Cong., 2nd sess., p. 1–77.

United States Geological Survey and American Association of Petroleum Geologists, 1962, Tectonic map of the United States.

Upp, J. E., Jr., 1968, Reed City field *in* Oil and gas fields—a symposium: Michigan Basin Geol. Soc., p. 149–160.

Valerius, C. N., and Von Croy, Stefan, 1951, Case history of Benton field, Bossier Parish, Louisiana: Geophysics, v. 16 (January), p. 14–28.

Van Den Berg, Jacob, 1968, Oil and gas developments: Illinois State Geol. Survey, Illinois Petroleum 89, p. 3–66.

Van Den Berg, Jacob, Swann, D. H., Carpenter, G. L., and Nosow, Edmund, 1968, Developments in east-central states in 1967: Am. Assoc. Petroleum Geologists Bull., v. 52, no. 6 (June), p. 965–975.

Vander Ahe, K. L., 1964, The petroleum industry in Alaska *in* U. S. Senate Rept. Comm. on Interior and Insular Affairs, 88th Cong., 2nd sess.: U. S. Geol. Survey, p. 19–25.

Van Gilder, H. R., 1953, The Big Medicine Bow field, Carbon County, Wyoming: Wyoming Geol. Assoc. Guidebook 8th Ann. Field Conf., p. 142–143.

Van Houten, F. B., and Weitz, J. L., 1956, Geologic map of the eastern Beaver Divide–Gas Hills area, Fremont and Natrona Counties, Wyoming: U. S. Geol. Survey Oil and Gas Inv. Map OM-180.

Varvaro, Gasper G., 1957, Geology of Evangeline and St. Landry Parishes: Louisiana Geol. Survey Bull. 31 (April), p. 184–190.

Vary, J. A., Elenbaas, J. R., and Johnson, M. A., 1968, Gas in Michigan basin: Am. Assoc. Petroleum Geologists Mem. 9, v. 2, p. 1761–1797.

Vest, E. L., Jr., 1966, Kelly–Snyder Canyon reef, western Scurry County, Texas, *in* Oil and gas fields in West Texas—a symposium: West Texas Geol. Soc., p. 213–217.

Vestal, Jack H., 1950, Petroleum geology of the Smackover formation of southern Arkansas: Div. Geol., Arkansas Resources and Devel. Comm. Inf. Circ. 14, p. 9.

Vickers, R. B., and Ratliffe, Fred, 1952, Frankirk field, Stonewall County, Texas, *in* Geological contributions: Abilene Geol. Soc., p. 28–30.

Vine, James D., and Hail, William J., Jr., 1950, Geologic map of the Hobson area, central Montana: U. S. Geol. Survey Oil and Gas Inv. Prelim. Map 108.

Vlissides, Sophie D., 1964, Map of Texas, showing oil and gas fields, pipelines, and areas of exposed basement rocks: U. S. Geol. Survey Oil and Gas Inv. Map OM-214.

Vlissides, Sophie D., and Bieberman, Robert A., 1961, Map of New Mexico, showing oil and gas fields, unsuccessful test wells, Precambrian rocks, and pipelines: U. S. Geol. Survey Oil and Gas Inv. Map OM-207.

Volk, Richard W., 1968, Developments in eastern and northwestern Colorado and western Nebraska in 1967: Am. Assoc. Petroleum Geologists, v. 52, no. 6 (June), p. 1047–1050.

Wagner, Ewald A., 1966, Developments in west Texas and southeastern New Mexico in 1965: Am. Assoc. Petroleum Geologists Bull., v. 50, no. 6 (June), p. 1195–1201.

Wagner, H. C., 1966, Petroleum and natural gas *in* Mineral and water resources in Washington: U. S. Senate Rept. Comm. on Interior and Insular Affairs, 89th Cong., 2nd sess., p. 287–296.

Wagner, Walter R., 1966, Pennsylvanians must delve into rocks of ancient age: Oil and Gas Jour., Dec. 12, p. 152–158.

Wahrhaftig, Clyde, 1965, The physiographic divisions of Alaska: U. S. Geol. Survey Prof. Paper 482.

Wahrhaftig, Clyde, and Gates, G. O., 1964, Geology *in* U. S. Senate Rept. Comm. on Interior and Insular Affairs, 88th Cong., 2nd sess.: U. S. Geol. Survey, p. 27–32.

Wakefield, W. W., 1965, Much good oil hunting land left in North Dakota's Mississippian: Oil and Gas Jour., May 31, p. 138–146.

Waldschmidt, W. A., 1948, Gramps field, Archuleta County, Colorado, *in* Structure of typical American oil fields: Am. Assoc. Petroleum Geologists, v. 3, p. 110–131.

Wallace, P. A., 1955, Chickasha: a tired gas field yields oil: Oil and Gas Jour., Jan. 17, p. 154–155.

Wallace, W. E., ed., 1962, Fault map of south Louisiana: Gulf Coast Assoc. Geol. Socs. Trans., v. 12, p. 124.

Walters, Robert F., 1953, Oil production from fractured Precambrian basement rocks in central Kansas: Am. Assoc. Petroleum Geologists Bull., v. 37, p. 300–313.

Walters, Robert F., 1958, Differential entrapment of oil and gas in Arbuckle dolomite of central Kansas: Am. Assoc. Petroleum Geologists Bull., v. 42 (September), p. 2132–2173.

Walters, Robert F., and Price, Arthur S., 1948, Kraft–Prusa oil field, Barton County, Kansas, *in* Structure of typical American oil fields: Am. Assoc. Petroleum Geologists, v. 3, p. 249–280.

Walton, Paul T., 1947, Oregon Basin oil and gas field, Park County, Wyoming: Am. Assoc. Petroleum Geologists Bull., v. 31 (August), p. 1431–1453.

Walton, Paul T., 1955, Wasatch Plateau gas fields, Utah: Am. Assoc. Petroleum Geologists Bull., v. 39, no. 4 (April), p. 385–421.

Walton, P. T., 1968, Wasatch Plateau gas fields: Am. Assoc. Petroleum Geologists, Mem. 9, v. 1, p. 928–945.

Wanek, Alexander A., 1962, Reconnaissance geologic map of parts of Harding, San Miguel, and Mora Counties, New Mexico: U. S. Geol. Survey Oil and Gas Inv. Map OM-208.

Wanenmacher, Joseph N., and Gealy, Wendell B., 1930, Surface and subsurface structure of the Tri-County oil field of southwestern Indiana: Am. Assoc. Petroleum Geologists Bull., v. 14 (April), p. 423–431.

Wang, Kia Kang, 1952, Geology of Ouachita Parish: Louisiana Geol. Survey Bull. 28.

Wanless, Harold R., 1962, Pennsylvanian rocks of Eastern Interior basin *in* Pennsylvanian System in the United States—a symposium: Am. Assoc. Petroleum Geologists, Tulsa, Oklahoma, p. 4–59.

Ware, G. C., Jr., and Stewart, R. D., 1958, Bridge area, South Mountain oil field, *in* Am. Assoc. Petroleum Geologists, Pacific Section, p. 180–181.

Warn, G. Frederick, 1953, Spraberry structural conditions: World Oil, March, p. 83–88.

Warner, C. A., 1939, Texas oil and gas since 1543: Houston, Gulf Pub. Co.

Wasson, Isabel B., 1932, Sub-Trenton formations in Ohio: Jour. Geology, v 40, no. 8, p. 673–687.

Wasson, Theron, 1948, Creole field, Gulf of Mexico, Coast of Louisiana, *in* Structure of typical American oil fields: Am. Assoc. Petroleum Geologists, v. 3, p. 281–298.

Wasson, Theron, and Wasson, Isabel B., 1927, Cabin Creek field, West Virginia: Am. Assoc. Petroleum Geologists Bull., v. 11 (July), p. 705–719.

Weaver, Kenneth N., 1966 et seq., Annual reports, Maryland Geol. Survey.

Weaver, Kenneth N., 1968, Oil and gas developments in Maryland: Am. Assoc. Petroleum Geologists Bull., v. 52, no. 6 (June), p. 945–946.

Weaver, O. D., 1964, Next major Cambrian province: Oil and Gas Jour., Aug. 10, p. 114–119.

Weaver, O. D., 1964, Tennessee could share in Appalachian basin revival: Oil and Gas Jour., Sept. 14, p. 174–179.

Weaver, O. D., 1965, North flank of Appalachian set for testing: Oil and Gas Jour., Sept. 6, p. 216–219. Also, New York offers multiple strat-trap potential in Cambro-Ordovician sediments: Oil and Gas Jour., Sept. 13, p. 166–170.

Weaver, O. D., 1966, Drillers go after shallow targets in Pennsylvania: Oil and Gas Jour., Apr. 4, p. 222–225.

Webb, Wilton H., 1968, Dale Consolidated field, Hamilton and Saline Counties, Illinois, *in* Geology and petroleum production of the Illinois basin—a symposium: Illinois and Indiana-Kentucky Geol. Socs., p. 85–89.

Weeks, Lewis G., ed., 1958, Habitat of oil: Am. Assoc. Petroleum Geologists (Spec. Volume).

Weeks, Warren B., and Alexander, Clyde W., 1942, Schuler field, Union County, Arkansas: Am. Assoc. Petroleum Geologists Bull., v. 26 (September), p. 1467–1516.

Weimer, Robert J., 1966, Time-stratigraphic analysis and petroleum accumulations, Patrick Draw field, Sweetwater County, Wyoming: Am. Assoc. Petrol. Geol. Bull., v. 50, no. 10 (October), p. 2150–2175.

Weirich, T. E., 1929, Cushing oil and gas field, Creek County, Oklahoma, *in* Structure of typical American oil fields: Am. Assoc. Petroleum Geologists, v. 2, p. 396–406.

Welch, Stewart W., 1955, Structure and stratigraphy of the outcropping Pennsylvanian rocks in the Tiptop quadrangle, Breathitt, Magoffin, and Knott Counties, Kentucky: U. S. Geol. Survey Oil and Gas Inv. Map OM-163.

Welch, Stewart W., 1958, Stratigraphy of Upper Mississippian rocks above the Tuscumbia limestone in northern Alabama and northeastern Mississippi: U. S. Geol. Survey Oil and Gas Inv. Chart OC-58.

Wellman, Dean C., 1951, Southeastern Oklahoma: Shale Shaker, v. 2 (December), p. 4–5, 14–15.

Wells, Lewis F., 1958, Petroleum occurrence in the Uinta basin *in* Habitat of oil: Am. Assoc. Petroleum Geologists (Spec. Volume).

Wendlandt, E. A., and Shelby, T. H., Jr., 1948, Talco oil field, Franklin and Titus Counties, Texas, *in* Structure of typical American oil fields: Am. Assoc. Petroleum Geologists, v. 3, p. 432–451.

Wendlandt, E. A., Shelby, T. H., Jr., and Bell, John S., 1946, Hawkins field, Wood County, Texas: Bull. Am. Assoc. Petroleum Geologists, v. 30 (November), p. 1830–1856.

Wengerd, Sherman A., 1958a, Origin and habitat of oil in San Juan basin of New Mexico and Colorado *in* Habitat of oil: Am. Assoc. Petroleum Geologists (Spec. Volume), p. 366–394.

Wengerd, Sherman A., 1958b, Lucero basin attracts wildcatters: Oil and Gas Jour., Nov. 10.

Wengerd, Sherman A., 1959a, Pennsylvanian oil possibilities of San Juan basin, Four Corners region: Am. Assoc. Petroleum Geologists Bull., v. 43, no. 9 (September), p. 2214–2227.

Wengerd, Sherman A., 1959b, Pennsylvanian geology and petroleum potential of the Lucero region, west-central New Mexico, *in* Geological record: Am. Assoc. Petroleum Geologists, Rocky Mtn. Sec., p. 25–40.

Wengerd, Sherman A., 1959c, Regional geology as related to petroleum potential of the Lucero region, west-central New Mexico: New Mexico Geol. Soc. Guidebook 10th Ann. Field Conf., p. 121–134.

Wengerd, Sherman A., 1962, Wildcat oil prospects in southwest New Mexico: World Oil, Aug. 1.

Wesson, James A., 1967, Secondary oil recovery possibilities in the basal Greenbrier dolomite zone, Sycamore-Millstone field, Sherman district, Calhoun County, West Virginia: U.S. Bur. Mines Rept. Inv. 7049 (November).

West, Thomas S., Sr., 1963, Typical stratigraphic traps—Jackson trend of south Texas: Gulf Coast Assoc. Geol. Socs. Trans., v. 13, p. 67–78.

West, W. E., Jr., 1953, The Quealy oil field, Albany County, Wyoming, Wyoming Geol. Assoc. Guidebook 8th Ann. Field Conf., 1953, p. 165–169.

West Texas Geological Society, 1951, Introduction to the petroleum geology of the Permian basin, West Texas and southeastern New Mexico: Midland, January, p. 1–51.

West Texas Geological Society, 1966, Oil and gas fields in West Texas—a symposium: Pub. 66-52, Midland, p. 1–397.

West Texas Geological Society, Stratigraphic Problems Committee (M. L. Feldman, Jr., Chairman), 1962, Southwest-northeast cross section, Marathon region to Midland basin: Pub. 62-47.

Wheeler, Robert R., 1950, Golden Trend of Oklahoma—current problems: Am. Assoc. Petroleum Geologists Bull., v. 34 (June), p. 1287–1292.

Wheeler, Robert R., 1951, Origin and oil possibilities in the Anadarko basin: Shale Shaker, v. 2 (October), p. 4–16.

Whitaker, Harvey, 1940, Hoffman field, Duval County, Texas: Am. Assoc. Petroleum Geologists Bull., v. 24 (December), p. 2126–2142.

White, B. Ross, 1966, Black Lake field, North Louisiana's main strike in past 20 years: World Oil, December, p. 81–84.

White, Stanley B., 1941, Davenport field, Lincoln County, Oklahoma, *in* Stratigraphic type oil fields: Am. Assoc. Petroleum Geologists, p. 386–407.

Whiting, Lester L., 1956, Geology and history of oil production in the Decatur-Mount Auburn-Springfield area: Illinois Geol. Survey Circ. 211, p. 1–17.

Whitley, William W., 1962, Occurrence of oil and gas in the Sand Wash basin, northwest Colorado, *in* Oil and gas fields of Colorado: Rocky Mtn. Assoc. Geologists, Denver, Colorado, p. 87–91.

Wiggins, Peter N., III, 1954, Geology of Ham Gossett oil field

Kaufman County, Texas: Am. Assoc. Petroleum Geologists Bull., v. 38 (February), p. 306–318.

Wilgus, W. L., 1950, The Elk City pool: World Oil, June, p. 123–126.

Wilkinson, Walter M., 1952, East Hamlin pool, Jones County, Texas, *in* Geological contributions: Abilene Geol. Soc., p. 21–25.

Williams, Floyd E., 1961, San Arroyo gas field, *in* The oil and gas fields of Utah—a symposium: Intermountain Assoc. Petroleum Geologists.

Williams, Neil, 1961, Deep Frio yields on "Golden Lane" along the Texas coast: Oil and Gas Jour., Apr. 24, p. 124–125.

Williams, Neil, 1962, Louisiana's big island play nets two new fields: Oil and Gas Jour., Dec. 10, p. 130–132.

Willman, H. B., and others, 1967, Geologic map of Illinois: Illinois State Geol. Survey.

Wilpolt, Ralph H., and Marden, Douglas W., 1960, Geology and oil and gas possibilities of Upper Mississippian rocks of southwestern Virginia, southern West Virginia, and eastern Kentucky: U. S. Geol. Survey Bull. 1072-K.

Wilson, Charles W., Jr., 1955, What are Tennessee's oil possibilities?: Oil and Gas Jour., Oct. 3, p. 187–190.

Wilson, Charles W., Jr., and Stearns, Richard G., 1963, Quantitative analysis of Ordovician and younger structural development of Nashville dome, Tennessee: Am. Assoc. Petroleum Geologists Bull., v. 47, no. 5 (May), p. 823–832.

Wilson, Edward N., 1968, Some aspects of drilling activities in eastern Kentucky *in* Kentucky Geol. Survey, Ser. X, Spec. Pub. 15, p. 56–63.

Wilson, Howard M., 1967, Great Katy gas reserve ready to move to market: Oil and Gas Jour., Nov. 27, p. 39–42.

Wilson, Howard M., 1967, North Louisiana's Black Lake nourishes oil starving area: Oil and Gas Jour., May 29, p. 112–113.

Wilson, R. M., 1952, Gulf Coast two-state Phoenix Lake field: Oil and Gas Jour., June 16, p. 317–326.

Wilson, Thomas, Jr., 1968, Developments in Alaska in 1967: Am. Assoc. Petroleum Geologists, v. 52, no. 6 (June), p. 1068–1080.

Wilson, W. B., 1929, Geology of Glenn pool of Oklahoma *in* Structure of typical American oil fields: Am. Assoc. Petroleum Geologists, v. 1, p. 231–242.

Wilson, W. S., 1966, Paper read before Fort Worth meeting of the Southwestern Federation of Geological Societies, Feb. 9–11: Oil and Gas Jour., Feb. 21, p. 138–139.

Winfrey, Walter M., Jr., 1960, Stratigraphy, correlation, and oil potential of the Sheep Pass formation, east-central Nevada, *in* Guidebook to geology of east central Nevada: Intermountain Assoc. Petroleum Geologists, 11th annual field conference, p. 126–133.

Wingerter, Hugh R., 1959, Greenwood gas field *in* Kansas oil and gas fields: Kansas Geol. Soc., v. 2 (Western Kansas), p. 29–36.

Wingerter, H. R., 1968, Greenwood gas field, Kansas, Colorado, and Oklahoma: Am. Assoc. Petroleum Geologists Mem. 9, v. 2, p. 1557–1566.

Winn, Vernard, 1968, Worsham gas field, Reeves County, Texas: Am. Assoc. Petroleum Geologists, Mem. 9, v. 2, p. 1389–1393.

Winter, C. Victor, 1954, Pollard field, Escambia County, Alabama: Gulf Coast Assoc. Geol. Socs. Trans., v. 4 (November), p. 121–142.

Winter, Jan A., 1961, Stratigraphy of the Lower Cretaceous (subsurface) of South Texas: Gulf Coast Assoc. Geol. Socs. Trans., v. 11, p. 15–21.

Winterer, E. L., and Durham, D. L., 1962, Geology of southeastern Ventura basin, Los Angeles County, California: U.S. Geol. Survey Prof. Paper 334-H, p. 275–366.

Wold, John S., 1952, Worland field, Big Horn and Washakie Counties, Wyoming: Oil and Gas Jour., Dec. 8, p. 112–114.

Wolfe, Edward W., and McKee, Edwin H., 1968, Geology of the Grays River quadrangle, Wahkiakum and Pacific Counties, Washington: Dept. Nat. Resources, Div. Mines and Geol., Geologic Map GM-4.

Womack, Jerry L., 1956, Aylesworth field *in* Petroleum geology of southern Oklahoma: Am. Assoc. Petroleum Geologists, v. 1, p. 373–391.

Womack, Robert, Jr., 1950, Brookhaven oil field, Lincoln County, Mississippi: Am. Assoc. Petroleum Geologists Bull., v. 34 (July), p. 1517–1529.

Woncik, John, 1961, Kingfisher County—hottest oil play in the United States: World Oil, October, p. 173–177.

Woncik, John, 1962, Kinta gas field, Haskell County, Oklahoma: Tulsa Geol. Soc. Digest, v. 30, p. 56–64.

Woncik, John, 1968a, Kinta gas field, Haskell County, Oklahoma: Am. Assoc. Petroleum Geologists, Mem. 9, v. 2, p. 1636–1643.

Woncik, John, 1968b, Cook Inlet basin has big stratigraphic traps: World Oil, July, p. 144–151.

Wood, E. Boyne, 1955, Geology of the Morganfield South oil pool, Union County, Kentucky: Kentucky Geol. Survey Bull. 14, Sec. 9, p. 1–20.

Woodford, A. O., Vedder, J. G., and Yerkes, R. F., 1954, Geology of the Los Angeles basin: California Div. Mines Bull. 170, p. 65–81.

Woodring, W. P., Stewart, Ralph, and Richards, R. W., 1940, Geology of the Kettleman Hills oil field, California: U. S. Geol. Survey Prof. Paper 195.

Woodruff, E. G., and Cooper, C. L., 1930, Geology of Rogers County: Oklahoma Geol. Survey Bull. 40, v. 3, p. 281–283.

Woods, Dalton J., 1963, Fractured chalk oil reservoirs, Sabine Parish, Louisiana: Gulf Coast Assoc. Geol. Socs. Trans., v. 13, p. 127–138.

Woodward, A. F., 1958, Sansinena oil field *in* A guide to the geology and oil fields of the Los Angeles and Ventura regions: Am. Assoc. Petroleum Geologists, Pacific Sec., p. 109–118.

Woodward, Herbert P., 1958, Emplacement of oil and gas in the Appalachian basin *in* Habitat of oil: Am. Assoc. Petroleum Geologists, p. 494–510.

Woodward, Herbert P., 1964, Central Appalachian tectonics and deep basin: Am. Assoc. Petroleum Geologists Bull., v. 48, no. 3 (March), pt. 1, p. 338–356.

World Oil, 1968, International outlook issue, Aug. 15.

Wright, A. C., 1945, Merigale field, Wood County, Texas: Am. Assoc. Petroleum Geologists Bull., v. 29 (December), p. 1779–1780.

Wright, Leo M., 1952, There's still shallow oil in northeast Oklahoma: World Oil, August, p. 114–118.

Wright, Randall, 1941, Red Fork shoestring sand pool, Pawnee, Creek, and Tulsa Counties, northeastern Oklahoma, *in* Stratigraphic type oil fields: Am. Assoc. Petroleum Geologists, p. 473–491.

Wright, W. Floyd, 1962, Abo reef: prime West Texas target: Oil and Gas Jour., July 30, p. 226–235; Aug. 6, p. 188–194.

Wright, W. Floyd, 1963, Simpson, what's the best way to tap its potential?: Oil and Gas Jour., Apr. 8, p. 138–141; Apr. 15,

p. 220–224; Siluro-Devonian rocks offer widespread prospects, Sept. 9, p. 156–159; Restudy and upgrade West Texas' Mississippian, Nov. 11, p. 224–229; Permian basin drillers look to Pennsylvanian, Dec. 16, p. 136–141.

Wright, W. Floyd, 1967, Potential West Texas giant ignored?: Oil and Gas Jour., June 19, p. 176–178.

Wroblewski, E. F., 1967, Exploration problems in the "deep" Anadarko basin: Shale Shaker, v. 17, no. 7, p. 131–136.

Wulf, George R., 1964a, Chadron arch could offer big potential to oil hunters: Oil and Gas Jour., Feb. 24, p. 148–151.

Wulf, George R., 1964b, Good Dynneson sand trend may extend northeastward: World Oil, Feb. 1, p. 53–55.

Wulf, George R., 1964c, Shoestring sandstones in Cretaceous Bearpaw shale in central Montana: Am. Assoc. Petroleum Geologists Bull., v. 48, no. 7 (July), 1196–1203.

Wulf, George R., 1968, Lower Cretaceous Muddy sandstone in the northern Rockies *in* Wyoming Geol. Assoc. Guidebook 20th Ann. Field Conference, p. 29–34.

Wulf, George R., and Gries, J. P., 1963, South Dakota, a new frontier: Oil and Gas Jour., Dec. 2, p. 192–194.

Wurden, F. H., 1959, What are the prospects in Washington State?: World Oil, July, p. 95–98.

Wyoming Geological Association, 1947, Guidebook, 2nd Ann. Field Conf.: Big Horn basin.

Wyoming, Geological Association, 1948, Guidebook, 3rd Ann. Field Conf.: Wind River basin.

Wyoming Geological Association, 1949, Guidebook, 4th Ann. Field Conf.: Powder River basin.

Wyoming, Geological Association, 1950, Guidebook, 5th Ann. Field Conf.: Southwest Wyoming.

Wyoming, Geological Association, 1952, Guidebook, 7th Ann. Field Conf.: Southern Big Horn basin.

Wyoming Geological Association, 1953, Guidebook, 8th Ann. Field Conf.: Laramie basin, Wyoming, and North Park, Colorado.

Wyoming Geological Association, 1955, Guidebook, 10th Ann. Field Conf.: Green River basin.

Wyoming Geological Association, 1957, Guidebook, 12th Ann. Field Conf.: Southwest Wind River basin.

Wyoming Geological Association, 1958, Guidebook, 13th Ann. Field Conf.: Powder River basin.

Wyoming, Geological Association, 1959, Guidebook, 14th Ann. Field Conf.: Big Horn basin.

Wyoming Geological Association, 1960, Guidebook, 15th Ann. Field Conf.: Overthrust belt of southwestern Wyoming and adjacent areas.

Wyoming Geological Association, 1961, Late Cretaceous rocks, Wyoming and adjacent areas—a symposium: Guidebook, 16th Ann. Field Conf.

Wyoming Geological Association, 1962, Early Cretaceous rocks of Wyoming and adjacent areas—a symposium: Guidebook, 17th Ann. Field Conf.

Wyoming Geological Association, 1968, Guidebook, 20th Ann. Field Conf.: Black Hills area, George R. Wulf, ed.

Wyoming Geological Association, Symposium Committee, 1957, Wyoming oil and gas fields—a symposium.

Wyoming Geological Association, Technical Studies Committee, 1965, Geologic history of Powder River basin: Am. Assoc. Petroleum Geologists Bull., v. 49, no. 11, p. 1893–1907.

Wyoming Geological Association and Billings Geological Society, 1963, First Ann. Joint Field Conf.: Northern Powder River basin.

Yerkes, R. F., McCulloh, T. H., Schoellhamer, J. E., and Vedder, J. G., 1965, Geology of the Los Angeles basin—an introduction: U.S. Geol. Survey Prof. Paper 420A, p. A1-A55.

Young, Addison, 1965, The Penwell-to-Means upper San Andres reef of West Texas *in* Fluids in subsurface environment: Am. Assoc. Petroleum Geologists Mem. 4, p. 280–293.

Young, David M., 1958, Deep drilling through Cumberland overthrust rock in southwestern Virginia: Am. Assoc. Petroleum Geologists Bull., v. 21 (November), p. 2568–2573.

Young, David M., 1968, Virginia: Am. Assoc. Petroleum Geologists Bull., v. 52, no. 6 (June), p. 946–948. Also in Virginia Minerals, v. 14, no. 2 (May), p. 13–14.

Young, Leighton F., Jr., 1966, Northeast Thompsonville field, South Texas: key to future exploration for downdip Wilcox production?: Am. Assoc. Petroleum Geologists Bull., v. 50, no. 3 (March), p. 505–517.

Young, Robert S., and Harnsberger, W. T., 1955, Geology of Bergton gas field, Rockingham County, Virginia: Am. Assoc. Petroleum Geologists Bull., v. 39 (March), p. 317–328.

Young, Robert T., 1966, Developments in Four Corners–Intermountain area in 1965: Am. Assoc. Petroleum Geologists Bull., v. 50, no. 6 (June), p. 1261–1265.

Youse, Arthur C., 1964, Gas producing zones of Greenbrier (Mississippian) limestone, southern West Virginia and eastern Kentucky: Am. Assoc. Petroleum Geologists Bull., v. 48, no. 4 (April), p. 465–486.

Zapp, A. D., 1956, Structure contour map of the Tensleep sandstone in the Big Horn Basin, Wyoming and Montana: U.S. Geol. Survey Oil and Gas Inv. Map OM-182.

INDEX

All numbers given are page numbers. Those in parenthesis refer to the page numbers of illustrations.